AF682692

PROTEIN MISFOLDING DISEASES

WILEY SERIES ON PROTEIN AND PEPTIDE SCIENCE

VLADIMIR N. UVERSKY, Series Editor

Metalloproteomics • Eugene A. Permyakov

Protein Misfolding Diseases: Current and Emerging Principles and Therapies • Edited by Marina Ramirez-Alvarado, Jeffery W. Kelly, and Christopher M. Dobson

15 OCT. 2010

DR

RAM

PROTEIN MISFOLDING DISEASES

Current and Emerging Principles and Therapies

Edited by

MARINA RAMIREZ-ALVARADO
Mayo Clinic College of Medicine

JEFFERY W. KELLY
The Scripps Research Institute

CHRISTOPHER M. DOBSON
University of Cambridge

INSTITUT
MEDIATHEQUE
PASTEUR

WILEY

A JOHN WILEY & SONS, INC., PUBLICATION

Copyright © 2010 by John Wiley & Sons, Inc. All rights reserved.

Published by John Wiley & Sons, Inc., Hoboken, New Jersey.
Published simultaneously in Canada.

No part of this publication may be reproduced, stored in a retrieval system, or transmitted in any form or by any means, electronic, mechanical, photocopying, recording, scanning, or otherwise, except as permitted under Section 107 or 108 of the 1976 United States Copyright Act, without either the prior written permission of the Publisher, or authorization through payment of the appropriate per-copy fee to the Copyright Clearance Center, Inc., 222 Rosewood Drive, Danvers, MA 01923, (978) 750-8400, fax (978) 750-4470, or on the web at www.copyright.com. Requests to the Publisher for permission should be addressed to the Permissions Department, John Willey & Sons, Inc., 111 River Street, Hoboken, NJ 07030, (201) 748-6011, fax (201) 748-6008, or online at http://www.wiley.com/go/permission.

Limit of Liability/Disclaimer of Warranty: While the publisher and author have used their best efforts in preparing this book, they make no representations or warranties with respect to the accuracy or completeness of the contents of this book and specifically disclaim any implied warranties of merchantability or fitness for a particular purpose. No warranty may be created or extended by sales representatives or written sales materials. The advice and strategies contained herein may not be suitable for your situation. You should consult with a professional where appropriate. Neither the publisher nor author shall be liable for any loss of profit or any other commercial damages, including but not limited to special, incidental, consequential, or other damages.

For general information on our other products and services or for technical support, please contact our Customer Care Department within the United States at (800) 762-2974, outside the United States at (317) 572-3993 or fax (317) 572-4002.

Wiley also publishes its books in a variety of electronic formats. Some content that appears in print may not be available in electronic formats. For more information about Wiley products, visit our web site at www.wiley.com.

Library of Congress Cataloging-in-Publication Data:

Protein misfolding diseases : current and emerging principles and therapies / [edited by] Marina Ramirez-Alvarado, Jeffery W. Kelly, Christopher M. Dobson.
p.; cm.
Includes bibliographical references and index.
ISBN 978-0-471-79928-3 (cloth)
1. Proteins—Metabolism—Disorders. 2. Protein folding. 3. Amyloidosis. I. Ramirez-Alvarado, Marina. II. Kelly, Jeffery W. III. Dobson, C. M. (Christopher M.) [DNLM: 1. Amyloidosis—etiology. 2. Protein Folding. 3. Amyloidosis—diagnosis. 4. Amyloidosis—therapy. 5. Senile Plaques. WD 205.5.A6 P967 2010]
RC632.P7P673 2010
616.3′995—dc22 2009027972

Printed in the United States of America

10 9 8 7 6 5 4 3 2 1

CONTENTS

CONTRIBUTORS xi

FOREWORD xix
R. John Ellis

PREFACE xxv

ACKNOWLEDGMENTS xxvll

INTRODUCTION TO THE WILEY SERIES ON PROTEIN AND PEPTIDE SCIENCE xxix
Vladimir N. Uversky

PART I PRINCIPLES OF PROTEIN MISFOLDING

1 **Why Proteins Misfold** 3
Silvia Campioni, Elodie Monsellier, and Fabrizio Chiti

2 **Endoplasmic Reticulum Stress and Oxidative Stress: Mechanisms and Link to Disease** 21
Jyoti D. Malhotra and Randal J. Kaufman

3 **Role of Molecular Chaperones in Protein Folding** 47
Kausik Chakraborty, Florian Georgescauld, Manajit Hayer-Hartl, and F. Ulrich Hartl

4 **Kinetic Models for Protein Misfolding and Association** 73
Evan T. Powers and Frank A. Ferrone

5 **Toxicity in Amyloid Diseases** 93
Massimo Stefani

6 **Autophagy: An Alternative Degradation Mechanism for Misfolded Proteins** 113
Maria Kon and Ana Maria Cuervo

7 **Role of Posttranslational Modifications in Amyloid Formation** 131
Andisheh Abedini, Ruchi Gupta, Peter Marek, Fanling Meng, Daniel P. Raleigh, Humeyra Taskent, and Sylvia Tracz

8 **Unraveling Molecular Mechanisms and Structures of Self-Perpetuating Prions** 145
Peter M. Tessier and Susan Lindquist

9 ***Caenorhabditis elegans* as a Model System to Study the Biology of Protein Aggregation and Toxicity** 175
Elise A. Kikis, Anat Ben-Zvi, and Richard I. Morimoto

10 **Using *Drosophila* to Reveal Insight Into Protein Misfolding Diseases** 191
Julide Bilen and Nancy M. Bonini

11 **Animal Models to Study the Biology of Amyloid-β Protein Misfolding in Alzheimer Disease** 213
Karen H. Ashe

PART II PROTEIN MISFOLDING DISEASE: GAIN-OF-FUNCTION AND LOSS-OF-FUNCTION DISEASES

12 **Alzheimer Disease: Protein Misfolding, Model Systems, and Experimental Therapeutics** 233
Donald L. Price, Alena V. Savonenko, Tong Li, Michael K. Lee, and Philip C. Wong

13 **Prion Disease Therapy: Trials and Tribulations** 259
Valerie L. Sim and Byron Caughey

14 **Misfolding and Aggregation in Huntington Disease and Other Expanded Polyglutamine Repeat Diseases** 305
Ronald Wetzel

15 **Systemic Amyloidoses** 325
Marina Ramirez-Alvarado and Joel N. Buxbaum

16 **Hemodialysis-Related Amyloidosis** 347
David P. Smith, Alison E. Ashcroft, and Sheena E. Radford

17 **Copper–Zinc Superoxide Dismutase, Its Copper Chaperone, and Familial Amyotrophic Lateral Sclerosis** 381
Duane D. Winkler, Mercedes Prudencio, Celeste Karch, David R. Borchelt, and P. John Hart

18 **Alpha-1-Antitrypsin Deficiency** 403
David A. Lomas and David H. Perlmutter

19 **Folding Biology of Cystic Fibrosis: A Consortium-Based Approach to Disease** 425
William E. Balch, Ineke Braakman, Jeff Brodsky, Raymond Frizzell, William Guggino, Gergely L. Lukacs, Christopher Penland, Harvey Pollard, William Skach, Eric Sorscher, and Philip Thomas

20 **Thiopurine S-Methyltransferase Pharmacogenomics: Protein Misfolding, Aggregation, and Degradation** 453
Fang Li and Richard M. Weinshilboum

21 **Gaucher Disease** 469
Tim Edmunds

22 **Cataract as a Protein-Aggregation Disease** 487
Yongting Wang and Jonathan A. King

23 **Islet Amyloid Polypeptide** 517
Andisheh Abedini and Daniel P. Raleigh

PART III ROLE OF ACCESSORY MOLECULES AND RISK FACTORS

24 **Role of Metals in Alzheimer Disease** 545
Blaine R. Roberts and Ashley I. Bush

25 **Why Study the Role of Heparan Sulfate in In Vivo Amyloidogenesis?** 559
Robert Kisilevsky and John Ancsin

26 **Serum Amyloid P Component** 571
Simon Kolstoe and Steve Wood

27 **Role of Oxidatively Stressed Lipids in Amyloid Formation and Toxicity** 585
Paul H. Axelsen and Hiroaki Komatsu

28 **Role of Oxidative Stress in Protein Misfolding and/or Amyloid Formation** 615
Johanna C. Scheinost, Daniel P. Witter, Grant E. Boldt, and Paul Wentworth, Jr.

29 **Aging and Aggregation-Mediated Proteotoxicity** 631
Ehud Cohen and Andrew Dillin

PART IV MEDICAL ASPECTS OF DISEASE: DIAGNOSIS AND CURRENT THERAPIES

30 **Imaging of Misfolded Proteins** 647
Harry LeVine, III

31 **Diagnosis of Systemic Amyloid Diseases** 673
Morie A. Gertz

32 **Identification of Biomarkers for Diagnosis of Amyloid Diseases: Quantitative Free Light-Chain Assays** 689
Jerry A. Katzmann

33 **Real-Time Observation of Amyloid-β Fibril Growth by Total Internal Reflection Fluorescence Microscopy** 699
Tadato Ban and Yuji Goto

34 **Current and Future Therapies for Alzheimer Disease** 711
Paramita Chakrabarty, Pritam Das, and Todd E. Golde

35 **Current Therapies for Light-Chain Amyloidosis** 775
Angela Dispenzieri and Shaji Kumar

36 **Familial and Senile Amyloidosis Caused by Transthyretin** 795
Steven R. Zeldenrust and Merrill D. Benson

37 **Identifying Targets in α-Synuclein Metabolism to Treat Parkinson Disease and Related Disorders** 817
Julianna Tomlinson, Valerie Cullen, and Michael G. Schlossmacher

38 **Emerging Molecular Targets in the Therapy of Dialysis-Related Amyloidosis** 843
Gennaro Esposito and Vittorio Bellotti

39 **Familial Amyloidosis Caused by Lysozyme** 867
Mireille Dumoulin

40 Therapeutic Prospects for Polyglutamine Disease 887
Maria Pennuto and Kenneth H. Fischbeck

PART V APPROACHES FOR NEW AND EMERGING THERAPIES

41 Chemistry and Biology of Amyloid Inhibition 905
Mark A. Findeis

42 Immunotherapy in Secondary and Light-Chain Amyloidosis 917
Jonathan Wall

43 Anti-Misfolding and Anti-Fibrillization Therapies for Protein Misfolding Disorders 933
Zane Martin and Claudio Soto

44 Therapies Aimed at Controlling Gene Expression, Including Up-Regulating a Chaperone or Down-Regulating an Amyloidogenic Protein 945
Gregor P. Lotz and Paul J. Muchowski

45 Understanding and Ameliorating the TTR Amyloidoses 967
Steven M. Johnson, R. Luke Wiseman, Natàlia Reixach, Johan F. Paulsson, Sungwook Choi, Evan T. Powers, Joel N. Buxbaum, and Jeffery W. Kelly

INDEX 1005

CONTRIBUTORS

Andisheh Abedini, Department of Surgery, College of Physicians and Surgeons, Columbia University, New York, New York

John Ancsin, Department of Pathology and Molecular Medicine, Queen's University, Kingston, Ontario, Canada; Syl and Molly Apps Research Center, Kingston General Hospital, Kingston, Ontario, Canada

Alison E. Ashcroft, Astbury Centre for Structural Molecular Biology, University of Leeds, Leeds, UK

Karen H. Ashe, Department of Neurology, University of Minnesota, Minneapolis, Minnesota

Paul H. Axelsen, Department of Pharmacology, University of Pennsylvania School of Medicine, Philadelphia, Pennsylvania

William E. Balch, Department of Cell Biology and Institute for Childhood and Neglected Diseases, The Scripps Research Institute, La Jolla, California

Tadato Ban, Institute for Protein Research, Osaka University, Osaka, Japan

Vittorio Bellotti, Dipartimento di Biochimica, Università di Pavia, Pavia, Italy

Merrill D. Benson, Department of Pathology and Laboratory Medicine, Indiana University School of Medicine, Indianapolis, Indiana

Anat Ben-Zvi, Department of Biochemistry, Molecular Biology and Cell Biology, Rice Institute for Biomedical Research, Northwestern University, Evanston, Illinois

Julide Bilen, Howard Hughes Medical Institute, Janelia Farm Research Campus, Ashburn, Virginia

Grant E. Boldt, Department of Biochemistry, The Scripps–Oxford Laboratory, University of Oxford, Oxford, UK

Nancy M. Bonini, Department of Biology, Howard Hughes Medical Institute, University of Pennsylvania, Philadelphia, Pennsylvania

David R. Borchelt, Department of Neuroscience, McKnight Brain Institute, University of Florida, Gainesville, Florida

Ineke Braakman, Department of Cellular Protein Chemistry, University of Utrecht, Utrecht, The Netherlands

Jeff Brodsky, Department of Biological Sciences, University of Pittsburgh, Pittsburgh, Pennsylvania

Ashley I. Bush, Mental Health Research Institute of Victoria, Parkville, Australia; Department of Pathology, University of Melbourne, Parkville, Victoria, Australia; Genetics and Aging Research Unit, Massachusetts General Hospital, Charlestown, Massachusetts

Joel N. Buxbaum, Departments of Molecular and Experimental Medicine and Molecular Integrative Neuroscience, and Skaggs Institute for Chemical Biology, The Scripps Research Institute, La Jolla, California

Silvia Campioni, Laboratorium für Physikalische Chemie, ETH Zürich, Switzerland

Byron Caughey, Laboratory of Persistent Viral Diseases, Rocky Mountain Laboratories, National Institute of Allergy and Infectious Diseases, National Institutes of Health, Hamilton, Montana

Paramita Chakrabarty, Department of Neuroscience, College of Medicine, Mayo Clinic Florida, Jacksonville, Florida

Kausik Chakraborty, Department of Cellular Biochemistry, Max-Planck Institute of Biochemistry, Martinsried, Germany

Fabrizio Chiti, Dipartimento di Scienze Biochimiche, Università di Firenze, Firenze, Italy

Sungwook Choi, Departments of Chemistry and Molecular and Experimental Medicine, and Skaggs Institute for Chemical Biology, The Scripps Research Institute, La Jolla, California

Ehud Cohen, The Institute for Medical Research Israel–Canada, The Hebrew University Medical School, Jerusalem, Israel

Ana Maria Cuervo, Department of Developmental and Molecular Biology, Marion Bessin Liver Research Center, Institute for Aging Research, Albert Einstein College of Medicine, Bronx, New York

Valerie Cullen, LINK Medicine, Cambridge, Massachusetts

Pritam Das, Department of Neuroscience, College of Medicine, Mayo Clinic Florida, Jacksonville, Florida

Andrew Dillin, Molecular and Cell Biology Laboratory, The Salk Institute for Biological Studies, La Jolla California

Angela Dispenzieri, Department of Medicine, Division of Hematology, Mayo Clinic, Rochester, Minnesota

Mireille Dumoulin, Centre d'Ingénierie des Protéines, Institut de Chimie, Université de Liège, Liège, Belgium

Tim Edmunds, Therapeutic Protein Research, Genzyme Corporation, Framingham, Massachusetts

R. John Ellis, Department of Biological Sciences, University of Warwick, Coventry, UK

Gennaro Esposito, Dipartimento di Scienze e Tecnologie Biomediche, Università di Udine, Udine, Italy

Frank A. Ferrone, Department of Physics, Drexel University, Philadelphia, Pennsylvania

Mark A. Findeis, Satori Pharmaceuticals Incorporated, Cambridge, Massachusetts

Kenneth H. Fischbeck, Neurogenetics Branch, National Institute of Neurological Disorders and Stroke, National Institutes of Health, Bethesda, Maryland

Raymond Frizzell, Department of Cell Biology and Physiology, School of Medicine, University of Pittsburgh, Pittsburgh, Pennsylvania

Florian Georgescauld, Department of Cellular Biochemistry, Max-Planck Institute of Biochemistry, Martinsreid, Germany

Morie A. Gertz, Department of Medicine, Division of Hematology, Mayo Clinic College of Medicine, Rochester, Minnesota

Todd E. Golde, Department of Neuroscience, College of Medicine, Mayo Clinic Florida, Jacksonville, Florida

Yuji Goto, Institute for Protein Research, Osaka University, Osaka, Japan

William Guggino, Department of Physiology, School of Medicine, Johns Hopkins University, Baltimore, Maryland

Ruchi Gupta, Department of Chemistry, State University of New York at Stony Brook, Stony Brook, New York

P. John Hart, Department of Biochemistry and the X-ray Crystallography Core Laboratory, University of Texas Health Science Center at San Antonio, San Antonio, Texas

F. Ulrich Hartl, Department of Cellular Biochemistry, Max-Planck Institute of Biochemistry, Martinsreid, Germany

Manajit Hayer-Hartl, Department of Cellular Biochemistry, Max-Planck Institute of Biochemistry, Martinsreid, Germany

Steven M. Johnson, Departments of Chemistry and Molecular and Experimental Medicine, and Skaggs Institute for Chemical Biology, The Scripps Research Institute, La Jolla, California

Celeste Karch, Department of Psychiatry, Washington University School of Medicine, St. Louis, Missouri

Jerry A. Katzmann, Department of Laboratory Medicine and Pathology, Mayo Clinic, Rochester, Minnesota

Randal J. Kaufman, Departments of Biological Chemistry and Internal Medicine, Howard Hughes Medical Institute, University of Michigan Medical School, Ann Arbor, Michigan

Jeffery W. Kelly, Departments of Chemistry and Molecular and Experimental Medicine, and Skaggs Institute for Chemical Biology, The Scripps Research Institute, La Jolla, California

Elise A. Kikis, Department of Biochemistry, Molecular Biology and Cell Biology, Rice Institute for Biomedical Research, Northwestern University, Evanston, Illinois

Jonathan A. King, Department of Biology, Massachusetts Institute of Technology, Cambridge, Massachusetts

Robert Kisilevsky, Department of Pathology and Molecular Medicine and Department of Biochemistry, Queen's University, Kingston, Ontario, Canada; Syl and Molly Apps Research Center, Kingston General Hospital, Kingston, Ontario, Canada

Simon Kolstoe, Centre for Amyloidosis and Acute Phase Proteins, University College London Medical School, London, UK

Hiroaki Komatsu, Department of Pharmacology, University of Pennsylvania School of Medicine, Philadelphia, Pennsylvania

Maria Kon, Department of Developmental and Molecular Biology, Marion Bessin Liver Research Center, Institute for Aging Research, Albert Einstein College of Medicine, Bronx, New York

Shaji Kumar, Division of Hematology, Mayo Clinic, Rochester, Minnesota

Michael K. Lee, Department of Neuropathology, Johns Hopkins University School of Medicine, Baltimore, Maryland

Harry Levine, III, Department of Molecular and Cellular Biochemistry, University of Kentucky, Lexington, Kentucky

Fang Li, Department of Molecular Pharmacology and Experimental Therapeutics, Mayo Clinic College of Medicine, Rochester, Minnesota

Tong Li, Department of Neuropathology, Johns Hopkins University School of Medicine, Baltimore, Maryland

Susan Lindquist, Whitehead Institute for Biomedical Research, Howard Hughes Medical Institute, Cambridge, Massachusetts

David A. Lomas, Department of Medicine, Cambridge Institute for Medical Research, University of Cambridge, Cambridge, UK

Gregor P. Lotz, Gladstone Institute of Neurological Disease, University of California, San Francisco, California

Gergely L. Lukacs, Department of Physiology, McGill University, Montreal, Quebec, Canada

Jyoti D. Malhotra, Department of Biological Chemistry, University of Michigan Medical School, Ann Arbor, Michigan

Peter Marek, Department of Chemistry, State University of New York at Stony Brook, Stony Brook, New York

Zane Martin, Departments of Neurology, Neuroscience and Cell Biology, and Biochemistry and Molecular Biology, George and Cynthia Mitchell Center for Neurodegenerative Diseases, University of Texas Medical Branch, Galveston, Texas

Fanling Meng, Department of Chemistry, State University of New York at Stony Brook, Stony Brook, New York

Elodie Monsellier, Dipartimento di Scienze Biochimiche, Università di Firenze, Firenze, Italy

Richard I. Morimoto, Department of Biochemistry, Molecular Biology and Cell Biology, Rice Institute for Biomedical Research, Northwestern University, Evanston, Illinois

Paul J. Muchowski, Gladstone Institute of Neurological Disease, and Departments of Biochemistry and Biophysics, and Neurology, University of California, San Francisco, California

Johan F. Paulsson, Department of Systems Biology, Harvard Medical School, Boston, Massachusetts

Christopher Penland, Cystic Fibrosis Research Laboratory, Stanford University, Stanford, California

Maria Pennuto, Department of Neuroscience, Italian Institute of Technology, Genoa, Italy

David H. Perlmutter, Departments of Pediatrics, Cell Biology and Physiology, University of Pittsburgh School of Medicine, Children's Hospital of Pittsburgh, Pittsburgh, Pennsylvania

Harvey Pollard, Department of Anatomy, Physiology and Genetics, School of Medicine, University of the Health Sciences, Bethesda, Maryland

Evan T. Powers, Department of Chemistry, The Scripps Research Institute, La Jolla, California

Donald L. Price, Department of Neuropathology, Johns Hopkins University School of Medicine, Baltimore, Maryland

Mercedes Prudencio, Department of Neuroscience, McKnight Brain Institute, University of Florida, Gainesville, Florida

Sheena E. Radford, Astbury Centre for Structural Molecular Biology, University of Leeds, Leeds, UK

Daniel P. Raleigh, Department of Chemistry, Graduate Program in Biochemistry and Structural Biology, State University of New York at Stony Brook, Stony Brook, New York

Marina Ramirez-Alvarado, Department of Biochemistry and Molecular Biology, Mayo Clinic College of Medicine, Rochester, Minnesota

Natàlia Reixach, Departments of Chemistry and Molecular and Experimental Medicine, and Skaggs Institute for Chemical Biology, The Scripps Research Institute, La Jolla, California

Blaine R. Roberts, Mental Health Research Institute of Victoria, Parkville, Victoria, Australia; Department of Pathology, University of Melbourne, Parkville, Victoria, Australia

Alena V. Savonenko, Department of Neuropathology, Johns Hopkins University School of Medicine, Baltimore, Maryland

Johanna C. Scheinost, Department of Biochemistry, The Scripps–Oxford Laboratory, University of Oxford, Oxford, UK

Michael G. Schlossmacher, Division of Neuroscience, Ottawa Health Research Institute, University of Ottawa, Ottawa, Ontario, Canada

Valerie L. Sim, Center for Prions and Protein Folding Diseases, University of Alberta, Edmonton, Alberta, Canada

William Skach, Department of Biochemistry and Molecular Biology, Oregon Health and Sciences University, Portland, Oregon

David P. Smith, Astbury Centre for Structural Molecular Biology, University of Leeds, Leeds, UK

Eric Sorscher, Department of Medicine, University of Alabama at Birmingham, Birmingham, Alabama

Claudio Soto, Departments of Neurology, Neuroscience and Cell Biology, and Biochemistry and Molecular Biology, George and Cynthia Mitchell Center for Neurodegenerative Diseases, University of Texas Medical Branch, Galveston, Texas

Massimo Stefani, Department of Biochemical Sciences and Research Centre on the Molecular Basis of Neurodegeneration, University of Florence, Florence, Italy

Humeyra Taskent, Department of Chemistry, State University of New York at Stony Brook, Stony Brook, New York

Peter M. Tessier, Department of Chemical and Biological Engineering, Center for Biotechnology and Interdisciplinary Studies, Rensselaer Polytechnic Institute, Troy, New York

Philip Thomas, Molecular Biophysics Graduate Program, University of Texas Southwestern Medical Center, Dallas, Texas

Julianna Tomlinson, Division of Neuroscience, Ottawa Health Research Institute, University of Ottawa, Ottawa, Ontario, Canada

Sylvia Tracz, Department of Chemistry, State University of New York at Stony Brook, Stony Brook, New York

Jonathan Wall, Human Immunology and Cancer Program, Preclinical and Diagnostic Molecular Imaging Laboratory, University of Tennessee Graduate School of Medicine, Knoxville, Tennessee

Yongting Wang, Department of Neuroscience, Shanghai Jiao Tong University, Shanghai, P.R. China

Richard M. Weinshilboum, Department of Molecular Pharmacology and Experimental Therapeutics, Mayo Clinic College of Medicine, Rochester, Minnesota

Paul Wentworth, Jr., Department of Biochemistry, The Scripps–Oxford Laboratory, University of Oxford, Oxford, UK; Department of Chemistry, The Scripps Research Institute, La Jolla, California

Ronald Wetzel, Department of Structural Biology and Pittsburgh Institute for Neurodegenerative Diseases, University of Pittsburgh School of Medicine, Pittsburgh, Pennsylvania

Duane D. Winkler, Department of Biochemistry and the X-ray Crystallography Core Laboratory, University of Texas Health Science Center at San Antonio, San Antonio, Texas

R. Luke Wiseman, Departments of Chemistry and Molecular and Experimental Medicine, and Skaggs Institute for Chemical Biology, The Scripps Research Institute, La Jolla, California

Daniel P. Witter, Department of Biochemistry, The Scripps–Oxford Laboratory, University of Oxford, Oxford, UK

Philip C. Wong, Department of Neuropathology, Johns Hopkins University School of Medicine, Baltimore, Maryland

Steve Wood, Centre for Amyloidosis and Acute Phase Proteins, University College London Medical School, London, UK

Steven R. Zeldenrust, Division of Hematology, Mayo Clinic College of Medicine, Rochester, Minnesota

FOREWORD

The English philosopher Cyril Joad was famous for responding to any question by saying "It all depends on what you mean by. . . ." He was often lampooned for this habit, but, of course, he was absolutely correct. Definitions are important in science because science is basically a set of ideas about how the world works, and these ideas are expressed in words. So it is not a semantic quibble to insist on defining terms—in fact, I would argue that, in the last analysis, science *is* semantics.

So what is protein misfolding, and why is it important? There is no generally agreed definition of this term in the literature. The prefix *mis* indicates that something is wrong, and what I want to suggest is that some definitions of the term *misfolding* do not reflect this. In fact, *misfolding* seems to mean different things to different people. So I want to suggest, firstly, that our definition of misfolding should be clarified, and second, that there appears to be remarkably little evidence that misfolding per se is a serious problem for the cell—the problem is misassembly, not misfolding. To explain why I make these suggestions, I need to remind you of the definitions of the terms *folding* and *assembly*, which are used commonly in relation to proteins.

Folding is defined as the collapse of an elongated primary translation product into a stable compact monomer, whereas *assembly* is the binding of monomers to one another to produce a biologically functional oligomer. The distinction between folding and assembly is not absolute but quantitative, because in both processes there are changes in the conformation of polypeptide chains, but these changes are usually much smaller during assembly than during folding. Notice that while folding is defined entirely in structural terms, the definition of assembly contains a biological criterion in addition to a chemical one. The word *functional* is used to distinguish these oligomers from nonfunctional assemblies, and to make this explicit, nonfunctional oligomers are called *misassemblies*, or more commonly, *aggregates*.

This brings me to a fundamental point about the difference between chemistry and biology. What is important in chemistry is structure, because structure determines the properties of molecules. But what is important in biology is not structure per se, but the function enabled by that structure. This is because it is function that is selected for in evolution—structure does not matter, provided that it enables functions that help the organism to survive in a competitive world. So, for example, there are at least seven different types of eye—they all form images, but they do it using different structures.

So my basic argument is that because proteins are the products of evolution by natural selection, we should include biological criteria as well as chemical criteria in our definitions. Now the term *misassembly* does include a biological criterion—*misassembly* is an older term to describe nonfunctional oligomers [1]. A more commonly used term these days is *aggregate*. A minor terminological problem with the term *aggregate* is that *functional* aggregates are being discovered in both prokaryotes and eukaryotes [2], but we can avoid this problem by reverting to the term *misassembly*. So there is no problem with the term *misassembly*, but in my view there is a problem with the term *misfolding*, in that some commonly used definitions of misfolding do not include a biological criterion.

A common view is that a misfolded conformation is one that is nonnative; that is, it has to unfold to some extent before it can reach the native conformation [3]. I propose that a better definition is that a misfolded conformation is one that cannot reach the native functional conformation on a biologically relevant time scale. I am suggesting that the prefix *mis-*, meaning "wrong," in the term *misfolding* can have a meaning only in a biological context—a misfolded structure must have a biological consequence. Moreover, such a definition aligns the meaning of the terms *misassembly* and *misfolding* in a logical symmetry. Notice also that in my definition, misfolded does not necessarily mean the same as nonnative. A misfolded conformation could be a native conformation: that is, a normal intermediate on its way to the functional conformation but too slowly for biological needs. Misassembled conformations, on the other hand, are always misfolded, by definition.

So much for theory: What about the evidence? There is evidence that the stabilization of an assembly-competent state of a polypeptide involves nonnative conformations. The assembly of the tail spike protein of phage P22 is an example [4]. It makes no sense to me to call such nonnative conformations occurring during the formation of functional oligomers misfolded. So the question is: Are there any examples of misfolded conformations as I have defined them, that is, conformations unable to reach the functional conformations rapidly enough for the needs of the organism? My interest in this question stems from my interest in molecular chaperones. Given that most denatured proteins can refold correctly in the test tube under suitable conditions, why are so many chaperones required inside the cell? Is it to prevent misfolding, or misassembly, or both? My provisional answer is that the main problem appears to be misassembly rather than misfolding per se.

A survey of the protein-refolding literature shows that misassembly is commonly observed, and chemists tackle this problem by lowering the protein concentration and/or the temperature. Inside the cell this misassembly problem is even more acute, for two reasons. First, protein synthesis in vivo takes place on polysomes, and this inevitably brings identical, partly folded polypeptides within touching distance of one another. This closeness provides the ideal circumstance for misassembly because misassembly requires identical or very similar chains [5]. Second, protein synthesis takes place in compartments highly crowded with macromolecules, and this degree of crowding enhances association reactions by one to three orders of magnitude [6].

But what is the evidence that refolding proteins can reach misfolded conformations, as I define them, yet do not misassemble, that is, reach a conformation unable to reach the functional conformation fast enough for biological purposes, yet stay monomeric? This kinetic criterion requires, of course, that we define the time scale, and I propose that one hour should cover the requirements of prokaryotic and eukaryotic cell division. A second essential criterion is that the refolding be measured under plausibly physiological conditions. This is not usually a problem for pH, temperature, and ionic strength, but it is a big problem with respect to crowding, because most protein-refolding studies continue to be done in uncrowded buffers and so underestimate the extent of the misassembly problem [7].

There is evidence that the addition of crowding agents, that is, high-molecular-mass molecules that mimic the crowding found inside cells, greatly stimulates the rate of formation of amyloid fibrils in vitro [8,9]. So crowding makes the misassembly problem worse, but it is also the case that crowding stimulates the rate of correct folding in some cases. This is because crowding favors any process that results in a reduction of excluded volume, and this includes the initial collapse of extended polypeptide chains—crowding is not all bad.

There are some potential examples of trapped misfolded conformations that remain monomeric. One that is often cited is the rubisco from the bacterium *Rhodospirillum rubrum*. When diluted out of denaturant at 25°C, this protein aggregates like crazy, even though the buffer is uncrowded, but when diluted out at 4°C it remains monomeric, as determined by FRET analysis [10]. But *R. rubrum* does not grow at 4°C, it does not make protein at that temperature, so this property of rubisco is not subject to natural selection. So this property is an example of an epiphenomenon, of interest to protein chemists but not to cell biologists.

So far I have not been able to find in the literature clear examples of misfolded conformations that stay monomeric under crowded conditions. Of course, this may just reflect my ignorance and such examples do exist and will be found, but a more interesting possibility is that a simple universal principle operates—the principle that all primary translation products know how to fold correctly inside the cell, provided that they can avoid misassembly. To demonstrate this will be a technical challenge, because ideally you would need to monitor the refolding of a single chain in the absence of any other chains but in the presence of crowding agents, effectively mimicking the refolding of a polypeptide chain released from a single ribosome. If this principle is correct, it reinforces my view that the use in the literature of descriptions of the form "chaperones fold proteins" is misleading—what many chaperones actually do is to combat aggregation.

So my current view is that chaperones exist to combat misassembly rather than misfolding, as I define it. The three lines of evidence that lead me to this view can be summarized as follows:

1. The paucity of evidence for misfolded conformations that remain monomeric under physiological crowded conditions
2. The fact that all cells make identical polypeptide chains within touching distance within crowded compartments
3. The conclusion that misassembled conformations are default states available in principle to all proteins [11]

The functioning of globular proteins often requires some degree of flexibility in the noncovalent interactions that stabilize these proteins—so-called *conformational breathing*. This allows proteins to undergo dynamic and partial changes of conformation essential to their normal functioning. So even folded globular proteins can exhibit a partly folded phenotype, thus running the risk of aggregating with identical or similar proteins. The effects of postranslational modifications and environmental stresses can also result in the partial unfolding of mature proteins and thus to their aggregation. It follows that each molecule of every protein in every cell runs the risk that at any time between its synthesis and its degradation it will bind to one or more identical molecules to form a misfolded nonfunctional aggregate.

It used to be thought that protein aggregation was merely an annoying, uninteresting complication of in vitro experiments because the emphasis in refolding research was on determining the rules of folding. It is now clear, however, that protein aggregation is a serious and universal problem for all cells because it can potentially reduce the efficiency of folding. But much more important than that from the human perspective is that some protein aggregates are toxic to cells, including neurones. Thus, protein aggregation contributes to distressing human diseases such as Alzheimer disease, Parkinson's disease, type 2 diabetes, sickle cell anemia, systemic amyloidoses, and the prion diseases [2]. The incidence of some of these diseases is rising as life span increases in the developed world, and this is why protein misfolding is important. But how important?

The number of people suffering from Alzheimer disease and related dementias in the world is currently estimated to be more than 24 million, with 4.6 million new cases being diagnosed each year [12]. In the United Kingdom, the cost of caring for the current number of 700,000 cases is predicted to double within one generation from the present level of £17 billion per year. The human cost, in terms of declining quality of life for both sufferers and carers, is impossible to quantify, but is clearly very large. It is thus regrettable that only £11 is spent on UK research into Alzheimer disease for every affected person, compared to £289 on research for every cancer patient. If treatments could be found that would reduce severe cognitive impairment in older people by just 1% per year, this would cancel out all the estimated increases in the long-term care costs due to the aging UK population.

The aim of this book is to encourage scientists to rise to this challenge by providing a comprehensive survey of many of the topics surrounding the universal biological phenomenon of protein misassembly.

Department of Biological Sciences
University of Warwick
Coventry, UK

R. JOHN ELLIS

REFERENCES

1. R. Wetzel (ed.), Protein Misassembly, Advances in Protein Chemistry, vol. 50, pp. 1–282. Academic Press, New York, 1997.
2. F. Chiti, C.M. Dobson, Protein misfolding, functional amyloid and human disease. Annu Rev Biochem 75 (2006) 333–366.
3. T. Lazardis, M. Karplus, New view of protein folding reconciled with the old through multiple unfolding simulations. Science 278 (1997) 1928–1931.
4. S. Betts, C. Haase-Pettingell, J. King, Mutational effects on inclusion body formation, Adv Protein Chem 50 (1997) 243–264.
5. J. London, C. Skrzynia, M. Goldberg, Renaturation of *Escherichia coli* tryptophanase in aqueous urea solutions. Eur J Biochem 47 (1975) 409–415.
6. R. J. Ellis, A.P. Minton, Protein aggregation in crowded environments. Biol Chem 387 (2006) 485–497.
7. R.J. Ellis, Macromolecular crowding: obvious but underappreciated. Trends Biochem Sci 26 (2000) 597–604.
8. D.M. Hatters, A.P. Minton, G.J. Howlett, Macromolecular crowding accelerates amyloid formation by human apolipoprotein C-II. J Biol Chem 277 (2002) 7824–7830.
9. V.N. Uversky, E.M. Cooper J.I. Bower, A.L. Fink, Accelerated α-synuclein fibrillation in crowded mileu. FEBS Lett 515 (2002) 99–103.
10. Z. Lin, H.S. Rye, Expansion and compression of a protein folding intermediate by GroEL, Molecular Cell 16 (2004) 23–34.
11. C.M. Dobson, Protein misfolding, evolution and disease. Trends Biochem Sci 24 (1999) 329–332.
12. http://www.alzheimers-research/org/uk/info/statistics/

PREFACE

Folding is essential to achieve the appropriate structure and function for all proteins. Despite the exquisite cellular mechanisms regulating and assisting protein folding, there are instances where accumulation of misfolded proteins occurs in different tissues, causing cell death and tissue degeneration. Protein misfolding is now viewed as an intrinsic feature of protein-folding reactions, constantly challenging the cellular environment. The goal of the book is to provide students, basic scientists, and medical professionals with an opportunity to learn about the basic principles of protein misfolding diseases as well as the current and emerging therapies being developed. Among protein misfolding diseases, the amyloid diseases are the most common and are more widely studied. There are currently 26 amyloid diseases described in the literature affecting different tissues and with diverse symptoms. Each disease is characterized by the amyloid deposition of different protein precursors. Some of these amyloid diseases are rare, whereas other conditions, such as Alzheimer disease, are very common and are considered a serious health care concern for the aging population. These diseases are generally incurable, although the various therapeutic strategies presented in this book help to slow the progression of disease.

Protein misfolding is the process in which proteins fail to adopt and maintain their folded conformation through a number of intrinsic and extrinsic mechanisms. As a result, partially folded intermediates start being populated, triggering an aggregation process that causes both loss and gain of function. In this book we offer a broad integrated overview of the process of protein misfolding, with chapters written by expert basic scientists and protein misfolding clinicians. We start by reviewing the basis of protein misfolding, the protein folding process in vivo, with an overview of the various model systems currently in use, spanning from the eukaryotic *Saccharomyces cerevisiae*, the nematode *Caenorhabditis elegans*, and the fruit fly *Drosophila melanogaster*, all the way to mammalian systems such as mouse models.

We describe in detail selected examples of protein misfolding diseases that are characterized by a gain of function, where toxic intermediates and/or amyloid fibrils cause cell damage and tissue degeneration, including diseases affecting the central nervous system, such as Alzheimer disease, spongiform encephalopathies, amyotrophic lateral sclerosis (ALS), and Huntington disease. We have included an overview of systemic amyloid diseases with a special chapter for dialysis-related amyloidosis.

We also review a wide variety of loss-of-function protein misfolding diseases, including cystic fibrosis, cataracts, type 2 diabetes, lysosomal storage diseases, and a new category of protein misfolding diseases caused by nonsynonymous single-nucleotide polymorphisms (nsSNPs) in enzymes, such as thiopurine methyltransferase, responsible for the processing of chemotherapeutic agents.

It is known that there are number of accessory molecules and risk factors that affect protein misfolding diseases, and we have explored the role of metals, glycosaminoglycans, serum amyloid P component, membranes, and oxidative stress. A special chapter is devoted to the role of aging in aggregation-mediated proteotoxicity.

We offer a complementary clinical view of protein misfolding diseases in the second part of the book, starting with an overview about diagnosis of protein misfolding diseases. Diagnosis of protein misfolding diseases is not simple. We have reviewed the various approaches followed to make an accurate diagnosis, including imaging, biomarker discovery, and a panel of clinical evaluations and tests.

We conclude the book by offering an excellent overview of the therapies currently used in a number of the protein misfolding maladies and the emerging therapeutic strategies that are now being tested for a number of protein misfolding disorders and that can easily be applied for many related diseases.

This book is the result of the hard work and efforts of many talented scientific colleagues who generously agreed to contribute to this book. I hope that their knowledge and analysis of the issues related to protein misfolding diseases may inspire your future research and medical endeavors.

Rochester, Minnesota — Marina Ramirez-Alvarado
La Jolla, California — Jeffery W. Kelly
Cambridge, United Kingdom — Christopher M. Dobson
2009

ACKNOWLEDGMENTS

This book would not have been possible without the support and guidance from my co-editors Jeff Kelly and Chris Dobson. Thank you for the opportunity to work with you and for the many lessons learned during this process. I am grateful to the Mayo Graduate School and the College of Medicine at Mayo Clinic for their institutional support for all of my research and educational work, including this book. Thanks to the National Institutes of Health and the American Heart Association for providing financial support for my research. All past and present members of the Ramirez-Alvarado team and the Amyloid Interest Group at Mayo have in different ways fueled my passion for protein misfolding diseases, and I am thankful for the opportunity to work with each of them. The preparation of this book was made easier by superb administrative support from Kristi Simmons. Finally, I want to thank Jonathan Rose at Wiley for his continuous patience and encouragement during the preparation of this book.

Marina Ramirez-Alvarado

INTRODUCTION TO THE WILEY SERIES ON PROTEIN AND PEPTIDE SCIENCE

Proteins and peptides are the major functional components of the living cell. They are involved in all aspects of the maintenance of life. Their structural and functional repertoires are endless. They may act alone or in conjunction with other proteins, peptides, nucleic acids, membranes, small molecules and ions during various stages of life. Dysfunction of proteins and peptides may result in the development of various pathological conditions and diseases. Therefore, the protein/peptide structure-function relationship is a key scientific problem lying at the junction point of modern biochemistry, biophysics, genetics, physiology, molecular and cellular biology, proteomics, and medicine.

The *Wiley Series on Protein and Peptide Science* is designed to supply a complementary perspective from current publications by focusing each volume on a specific protein- or peptide-associated question and endowing it with the broadest possible context and outlook. The volumes in this series should be considered required reading for biochemists, biophysicists, molecular biologists, geneticists, cell biologists and physiologists as well as those specialists in drug design and development, proteomics and molecular medicine with an interest in proteins and peptides. I hope that each reader will find in the volumes within this book series interesting and useful information.

First and foremost I would like to acknowledge the assistance of Anita Lekhwani, of John Wiley & Sons, Inc., throughout this project. She has guided me through countless difficulties in the preparation of this book series and her enthusiasm, input, suggestions and efforts were indispensable in bringing the *Wiley Series on Protein and Peptide Science* into existence. I would like to take this opportunity to thank everybody whose contribution in one way or another has helped and supported this project. Finally, special thank you goes to my wife, sons, and mother for their constant support, invaluable assistance, and continuous encouragement.

Vladimir N. Uversky

PART I

PRINCIPLES OF PROTEIN MISFOLDING

1

WHY PROTEINS MISFOLD

SILVIA CAMPIONI,* ELODIE MONSELLIER,* AND FABRIZIO CHITI
Dipartimento di Scienze Biochimiche, Università di Firenze, Firenze, Italy

INTRODUCTION

Protein *misfolding*, which literally means "incorrect folding," is associated with a number of pathological states in humans, collectively termed *protein misfolding diseases*. In some cases the disease arises because a specific protein is no longer functional when adopting a misfolded state or undergoes a severe trafficking impairment (*loss-of-function diseases*). In most diseases, however, the pathological state originates because misfolding occurs concomitantly with aggregation, and the underlying aggregates are detrimental (*gain-of-function diseases*). The question of why proteins misfold and self-assemble into well-organized fibrillar aggregates characterized by an extensive β sheet structure, generally referred to as *amyloid* or *amyloidlike fibrils*, is very complex and cannot be answered exhaustively even in several pages. Despite these difficulties, it is fascinating to try to address this issue, particularly by describing a few concepts that are well established, have general validity, and benefit from a general consensus in the scientific community. In this challenge we start by describing our current knowledge of why and how proteins misfold in vitro. Considerable attention will be paid to the conformational changes that represent the starting point of misfolding phenomena and to the sequence determinants that enable the resulting misfolded, yet conformational states to

* These authors contributed equally and are listed in alphabetical order.

Protein Misfolding Diseases: Current and Emerging Principles and Therapies,
Edited by Marina Ramirez-Alvarado, Jeffery W. Kelly, and Christopher M. Dobson
Copyright © 2010 John Wiley & Sons, Inc.

self-assemble into ordered fibrillar aggregates. Finally, we mention all the factors that are thought to influence protein misfolding in an in vivo context.

WHY PROTEINS MISFOLD IN VITRO

A newly synthesized polypeptide chain usually folds into a functional, globular conformational state characterized by a well-defined, persistent secondary and tertiary structure (Fig. 1). In this conformation, generally termed the *folded* or *native state*, a protein is generally stable and soluble and has a minimal propensity to undergo aberrant aggregation, particularly because most of its hydrophobic moieties and a large portion of the backbone are sequestered inside the protein (Fig. 1). Even the regions located on the protein surface that may potentially retain a residual ability to trigger an undesired intermolecular association between folded protein molecules, such as the peripheral strands of β-sheets, are protected by structural adaptations developed during evolution [1]. Therefore, the first and necessary event that generally triggers the aggregation process of a normally globular protein into structured amyloidlike fibrils is the adoption of a nonnative partially or globally unfolded state [2,3].

Aggregation from Equilibrium Partially Folded States

It is well known to experimentalists that the easiest way to induce amyloid fibril formation of a globular protein in vitro is to place the protein into environmental conditions that promote partial or substantial unfolding of the native state but still allow the formation of noncovalent bonds [4–10]. These conditions, which include extreme pH values, high temperatures, high pressures, and aqueous solutions with organic cosolvents, give rise to partially unfolded states that are significantly populated at equilibrium, in which most of the hydrophobic groups and backbone amide and carbonyl groups, normally buried in the interior of the protein, become solvent exposed. Such newly exposed moieties possess a sufficiently high conformational flexibility and are available for intermolecular interactions with other protein molecules to form the cross-β structure typical of amyloid fibrils.

Amyloid fibril formation can also be promoted, under conditions that are optimal for the conformational stability of the native state, by mutations or truncations that destabilize the folded state and thus cause the formation of analogous partially unfolded states [11–16]. Along the same lines, it has been shown that the propensity of a protein to form amyloidlike fibrils can be reduced significantly by stabilization of the native structure via specific binding to ligands or antibodies [17–21]. These observations have increased the interest in a detailed in vitro structural characterization of equilibrium misfolded states.

Among the partially folded states characterized in more detail are those populated at acidic pH at equilibrium, before aggregation occurs, by β_2-microglobulin, the protein involved in dialysis-related amyloidosis.

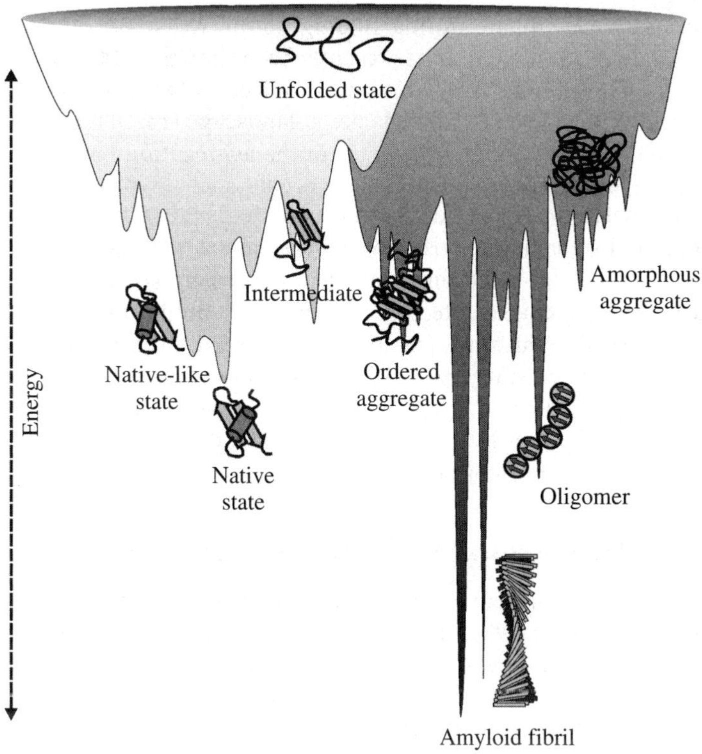

FIG. 1 Folding and aggregation: two sides of the same coin. The multitude of conformational states available to a polypeptide chain is described as a combined energy landscape for protein folding (light gray) and aggregation (dark gray). In the folding landscape the intermediate, nativelike, and fully native states are reported as different minima. The aggregation landscape shows mature amyloid fibrils as well as other ordered or amorphous aggregates. The unfolded, intermediate, and nativelike states shown in the folding landscape are able to cross the barrier and form aggregates via multiple pathways. (Adapted from [30], with permission.)

The combination of multiple techniques has allowed a deep analysis of the structural properties of partially folded states that form fibrils with different morphologies [22]. These studies have highlighted the complex, dynamic, and heterogeneous nature of these states, with different conformations interconverting into each other [23,24]. They have also shown that the amyloidogenic precursor states of this protein are hydrophobically collapsed species that contain different degrees of secondary and tertiary structure and lack the cooperativity typical of the native state [22].

Folding Intermediates as Precursors of Protein Deposition Diseases

Advances in experimental methods able to detect rare species and monitor folding events on a microsecond time scale have shown that partially folded states normally accumulate during folding before the major energy barrier, even in small proteins [25,26]. It has recently been shown that these species play a key role in determining the diversion of molecules between the folding and aggregation regions of the combined energy landscape (Fig. 1) [27–29]. Indeed, if the minima occupied by intermediates in the aggregation landscape become too deep, the likelihood of off-pathway events such as protein aggregation increases [25,30].

Recent data on the human prion protein (hPrP), whose aggregation induces fatal and infectious neurodegenerative disorders, suggest that the conformational conversion into the infectious form could be mediated by a folding intermediate [31,32]. This transient species was found to have a higher stability than the fully unfolded state and to be populated to a greater extent, relative to the wild-type protein, in many variants carrying mutations associated with inherited forms of prion diseases, including those that do not alter the global stability of the prion protein [31].

Native-State Fluctuations Can Trigger Amyloid Aggregation

The native state of a protein is not a single and rigid conformation, but rather, undergoes multiple localized unfolding reactions that give rise to an ensemble of rapidly interconverting conformations fluctuating around a free-energy minimum [30,33–35]. Some of these conformations, or ensembles of them, can also be regarded as "late intermediates" formed after the rate-limiting step of folding and remaining accessible by conformational fluctuations after the native state is fully reached [36–38]. Fluctuations that expose significant regions of the polypeptide main chain and its hydrophobic core could represent an initial step in the development of strong intermolecular interactions [39].

Two amyloidogenic variants of human lysozyme that cause autosomal-dominant hereditary amyloidosis, I56T and D67H, alter the cooperativity of the folded protein and enhance the rates of fluctuations toward partially unstructured conformations while preserving a global nativelike structure [11,39,40]. The most frequent sampling of such species, in which only the β-domain and the adjacent C-helix appear predominantly unfolded, makes the variants susceptible to aggregation. Fluctuations of the native states of monomeric transthyretin (TTR) and superoxide dismutase type I (SOD-1), associated with familial amyloid polyneuropathy and amyotrophic lateral sclerosis, respectively, cause the edge β-strands of these all-β proteins to unfold locally [41,42]. The resulting conformations have been shown to occupy the right side of the major free-energy barrier for folding and were therefore considered both as late-folding intermediates and conformational states freely accessible from the native structures through thermal fluctuations [42].

A late-folding intermediate of β_2-microglubulin, populated under native conditions and in equilibrium with the native structure, has been shown to be competent for amyloid fibril elongation [27,28,43]. The transition from the native state to this late intermediate is caused by the isomerization of a single *cis*-proline residue, but causes local unfolding of the A and D edge β-strands [27]. The acylphosphatase from *Sulfolobus solfataricus* (Sso AcP) has also been shown to aggregate at a higher rate than that of unfolding under conditions in which the native state is populated [44]. Such conditions have been shown to increase the fluctuations of the native state, as revealed by hydrogen/deuterium exchange detected with mass spectrometry, indicating that aggregation requires fluctuations within the native-state ensemble, but not necessarily unfolding [45].

Intrinsically Disordered Proteins Are at Risk for Aggregation

It has been recognized that numerous proteins lack a specific globular structure, at least when they are not associated with their specific molecular targets, and that many others contain disordered regions in addition to folded domains [46–48]. Such intrinsically disordered proteins (IDPs) have developed sequence adaptations to escape from aggregation, including a low hydrophobic residue content and a high net charge, a high content of proline residues, and a low number of aggregation-promoting regions [47,49,50]. In addition, induced folding upon binding to target molecules and interaction with membranes or macromolecules can provide further protection from aberrant aggregation [51,52]. Despite these protective mechanisms, IDPs are usually more prone than globular proteins to aggregate, as they cannot take advantage of folding. Therefore, it is not surprising to see that many proteins and peptides associated with protein-deposition diseases are indeed disordered under physiological conditions. Examples include the amyloid-beta peptide (Aβ), α-synuclein, tau, islet amyloid polypeptide (IAPP), calcitonin, and gelsolin fragments [53]. Even the monomeric precursor protein of prion diseases (PrP^C) is composed by a long N-terminal disordered region in addition to a compact C-terminal domain [54].

THE DETERMINANTS OF PROTEIN AGGREGATION FROM LARGELY OR PARTIALLY UNFOLDED STATES

We have shown that some degree of unfolding is the first prerequisite for initiating aggregation. But how do largely or partially unfolded states self-assemble into fibrillar aggregates? While it was being realised that most polypeptide chains have an inherent ability to form amyloidlike fibrils from their partially unfolded states, it was also becoming clear that this conversion occurs at different rates and with different efficiencies. Today it is well known that the tendency to form aggregates of similarly unstructured chains is indeed sequence dependent and that specific stretches of a sequence can facilitate or counteract amyloid formation [53]. Great effort has been expended to find simple rules that govern the aggregation process in order to rationalize and even predict aggregation rates and/or propensities and identify aggregation-prone segments within unfolded polypeptide chains [55–70].

Aggregation Is Promoted by High Hydrophobicity and β-Sheet Propensity and Low Net Charge

In a first attempt to analyze quantitatively the effects of mutations on aggregation, it was observed that the aggregation rates of variants of a model protein (AcP), determined under conditions in which an equilibrium denatured state is populated, depend on the changes caused by these mutations on simple physicochemical properties of the polypeptide chain: hydrophobicity, net charge, and propensity to convert from an α-helical to a β-sheet conformation [53]. On this basis, an empirical equation to predict the effect of a mutation on the aggregation rate of an unstructured peptide or protein was proposed and validated on a set of 27 single-point mutants of various unfolded proteins [53].

A conceptually similar algorithm was developed by Tartaglia and co-workers [59]. This algorithm uses charge, β-propensity, and parameters related to hydrophobicity (the polar and nonpolar water-accessible surface areas, the dipole moment, and the ρ-stacking interactions of aromatic residues) as determinants of amyloid aggregation. Their equation was able to reproduce the relative aggregation propensity of a set of variants of both disease-related and model proteins or peptides [59]. Fernandez-Escamilla and co-workers independently developed a statistical mechanics algorithm based on secondary structure propensities and estimation of a desolvatation penalty (TANGO) to predict β-sheet aggregating regions of a protein sequence as well as mutational effects [57]. TANGO evaluates the percentage of occupancy of the aggregate conformation for every residue of the chain by considering that the secondary structure of the aggregate is a β-structure and that the regions involved in aggregation are fully buried [57].

The established concept that simple physicochemical parameters of the amino acid residues of an unstructured protein influence its aggregation propensity has encouraged the search for novel and increasingly accurate algorithms. The algorithms that are available so far can predict not just the effect of a mutation on the aggregation rate of a protein, but also the absolute aggregation rate or propensity of an unstructured system [56,57,63] and, more important, the regions of the sequence that promote aggregation and form the β-core of the resulting fibrils [57,62,63,69]. All these algorithms are based on the same physicochemical parameters, or related ones, as those used in the original formula and seem to be successful in yielding predictions with reasonable accuracy.

From these studies it emerges that the sequence segments promoting aggregation of unstructured polypeptide chains have a hydrophobicity and β-sheet propensity higher than those of other regions. Mutations located in such regions can increase the aggregation propensity if they increase such factors. Interestingly, the locomotor efficiency and longevity of transgenic flies expressing mutant human Aβ(1–42) have been shown to correlate with the aggregation propensity of the expressed Aβ(1–42) variants, as deduced from one of these algorithms, indicating that these principles are largely applicable to in vivo systems [71].

Other Promising Predictive Methods

Predictive methods based on physicochemical parameters are not the only computational tools available to predict aggregation propensities and aggregation-promoting regions in a sequence. Other methods exploit empirical scales of aggregation propensities of individual amino acids based on purely experimental data [61,72], consensus sequences extracted from systematic mutagenesis of peptides [58], structural properties of protein databases [60,64,66], ensembles of structural templates derived from high-resolution amyloid fibril structures [65], or molecular dynamics simulations of protein aggregation [73,74]. Interestingly, in cases where the aggregation propensity scales of individual amino acids were determined, clear analogies exist between the results of very different approaches, ranging from empirically derived to statistical approaches [62,64,72]. This general agreement further confirms that common generic principles determine the molecular basis of amyloid aggregation.

WHY PROTEINS MISFOLD IN VIVO

Studies of protein aggregation in vitro take place in defined and extremely simple environments where important parameters such as pH, temperature, and ionic strength are controlled and modulated at order, the time scale is limited to a few weeks at most, and the only molecular species present in solution is the protein under study. However, the natural environment of a polypeptide chain—the cellular cytosol, the organelles, or the extracellular matrix—is far more complex and variable. We mention briefly the factors that are thought to contribute to protein misfolding in vivo. Several chapters of the book are dedicated to these factors. We show that although some factors causing misfolding in vitro are also able to promote aggregation in vivo, a further level of complexity lies in the biological environment in which proteins are produced and operate (Fig. 2).

Mutations Induce Aggregation

Many of the protein-deposition diseases consist of familial forms in which a mutation leads to aggregation of the mutated protein and the subsequent onset of disease. Examples include lysozyme amyloidosis and familial amyloidotic polyneuropathy [53]. Other diseases, such as amyotrophic lateral sclerosis and Creutzfeldt–Jakob disease, are mainly sporadic but include a few cases of familial forms [53]. Some of these mutations destabilize the native state of a globular protein, resulting in the formation of partially unstructured, aggregation-prone conformations (Fig. 2) [11,12,15,16,40]. Others increase the propensity to aggregate of the fully or partially unfolded state of a globular protein or of the natively unfolded state of an intrinsically disordered protein [55,75].

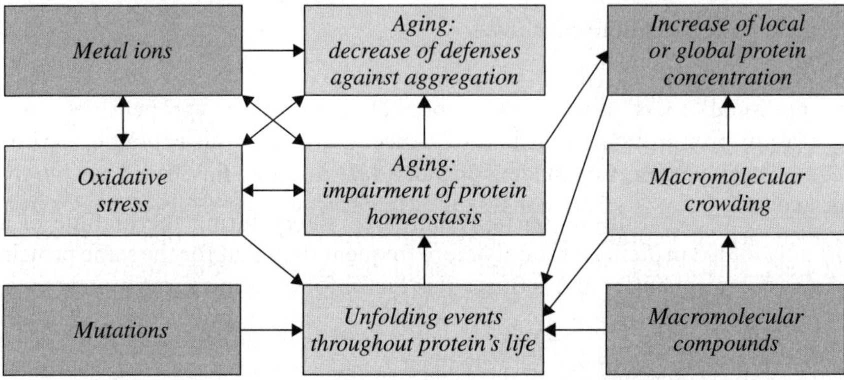

FIG. 2 Some of the factors that are thought to influence the formation of misfolded aggregates in vivo. Some act directly on the protein undergoing aggregation, and their effects can be monitored and studied in vitro (dark gray boxes). Others act at the organism level, involving a further level of complexity (light gray boxes). Arrows indicate links between the various factors.

The pathogenicity of other mutations lies in the biological context of the protein that undergoes aggregation and of its in vivo processing. For example, it has been suggested that some mutations of tau cause its dissociation from microtubules, resulting in an increase in free, intrinsically disordered tau that is susceptible to aggregation [76]. Mutations in the amyloid precursor protein APP increase the propensity for cleavage by γ-secretase, yielding a higher ratio of the more amyloidogenic Aβ(1–42) with respect to the less amyloidogenic Aβ(1–40) [77]. The D187N and D187Y mutations of gelsolin abolish the ability of domain 2 to bind calcium; the destabilized domain is more susceptible to proteolysis by furin, contributing to yield the amyloidogenic fragments of gelsolin [78].

Increase in Protein Concentration Contributes to Aggregation

As aggregation is a second- or higher-order reaction, its rate is very sensitive to the concentration of interacting chains. In vitro, increasing protein concentration decreases the lag time of aggregation and increases the elongation rate of the fibrils [79,80]. Different lines of evidence demonstrate that an increase in the in vivo concentration of a protein can also promote its aggregation. Neurodegeneration and expression levels of certain disease-related proteins, such as Aβ and α-synuclein, are clearly correlated [81]. In dialysis-related amyloidosis, the concentration of β_2-microglobulin increases up to 50-fold due to kidney failure, and this contributes to amyloid deposition [82]. Increase in the local concentration of β_2-microglobulin by collagen is thought to induce amyloid fibril formation on the surface of collagen fibrils and to explain the tissue specificity of dialysis-related amyloidosis [83]. Secondary systemic amyloidosis, also known as AA or reactive amyloidosis, is caused by increased levels of the serum amyloid A protein as a consequence of chronic inflammatory reactions [84]. In addition, increase in the local concentration of a membrane-bound protein is one mechanism by which biological membranes promote protein aggregation [85].

Unfolded States Form Obligatorily for Globular Proteins In Vivo

The occurrence of fully or partially unfolded states during the lifetime of a normally folded protein is certainly more frequent than that for the same protein in a physiological buffer and is thus a risk factor for protein aggregation in vivo. As long as translation is not terminated, a nascent polypeptide chain is susceptible to remaining unfolded in the milieu. Ongoing translation therefore offers a continuous supply of aggregation-prone species [86]. The presence of highly organized chaperone machinery coupled directly to translation witnesses to the potential deleterious effects of translation [87,88]. Secretory, mitochondrial, and membrane proteins are translocated through narrow channels and therefore need to be unfolded beforehand. Again, dedicated chaperone machinery exists to assist in these processes [89,90]. Many other circumstances can induce the transient unfolding of a protein, including the rise of stress conditions, and interaction with, or dissociation from, molecular targets (Fig. 2).

Oxidative Stress Promotes Aggregation Through a Variety of Mechanisms

Another biological factor that is potentially triggering aggregation in vivo is *oxidative stress*, defined as an imbalance between oxidant generation and antioxidant systems. Reactive oxygen species (ROS) are among the most commonly formed oxidants in biology and are generated as a by-product of normal metabolism and/or by exogenous stimuli such as ultraviolet light [91]. ROS can damage all biological polymers, including proteins. Oxidation of a protein in vitro can promote its aggregation via a variety of mechanisms, including destabilization of the native state, whose ultimate effects were described earlier in this chapter (Fig. 2) [92]. Oxidation of a protein in vivo can also promote aggregation by additional mechanisms, such as by causing cross-linking [93] or by modifying the susceptibility to proteolysis [94]. However, the exact relationship among oxidative stress, amyloid deposition, and disease onset remains unclear.

Macromolecular Crowding Favors Self-Assembly

In a typical eukaryotic cell protein, RNAs and others biopolymers occupy 30% of the volume, a situation known as *macromolecular crowding*. The direct consequence of crowding, which also exists in the extracellular space, albeit to a lessen extent, is that little space is left for additional macromolecules, which reduces their configurational entropy and therefore increases their free energy [95]. As a consequence, aggregation is theoretically favored, both kinetically and thermodynamically (Fig. 2) [95]. Macromolecular crowding has been shown to accelerate the aggregation of a wide range of proteins in vitro, leading to the suggestion that natural aggregation processes could be more favored than actually anticipated from studies performed in the test tube [95].

Interaction with Natural Compounds Promotes Fibril Formation

The excluded-volume effect is not the only mechanism by which macromolecules and other natural compounds influence protein aggregation in living organisms. Macromolecules such as glycosaminoglycans (GAGs), the serum amyloid protein (SAP), apolipoprotein E, and collagen fibers are frequently associated with amyloid deposits under pathological conditions [83,96–100]. Such species alter the kinetics of fibril formation via a direct interaction with the aggregating protein, thus sometimes being qualified as "pathological chaperones" [83,101–104]. They can also alter the susceptibility of the resulting fibrils to dissociation or proteolytic degradation [105–108]. Biological membranes also play a catalytic role in the aggregation of numerous amyloidogenic proteins (Fig. 2) [85,109].

These effects are not limited to macromolecules. Metal ions, particularly Cu^{2+} and Zn^{2+} ions, are known to accelerate fibril formation by the Aβ peptide [110], α-synuclein [111], and β_2-microglobulin [28]. In some cases, this aggregation is reversible upon chelation [110,112]. Fe^{3+} is found at abnormally high levels in Lewy bodies, suggesting that this metal ion also plays a significant role in Parkinson disease [113].

Aging Enhances the Probability of Protein Aggregation

Most protein-deposition diseases are of late onset. It is therefore obvious to draw the conclusion that aging increases the occurrence of physiological protein aggregation and is a key determinant of the pathological conditions associated with this phenomenon. Experimental evidence has been collected in animal models on the link between aging and protein aggregation [114–116]. A range of biological alterations occur, with the progression of a human being's life, that favor protein aggregation. These include a decline in the proteasome, lysosome, and chaperone activities and in the ability of these cellular machineries to respond to stress, dysregulation of metal-ion homeostasis, diminution of the antioxidant defenses, and so on. (Fig. 2) [116–118].

Although plausible, a decrease in the biological defenses against aggregation is not the only explanation that links pathology with aging. An aging person faces the progressive overwhelming of the cellular folding capacity and quality control systems by the inexorable accumulation of misfolded proteins (Fig. 2). In a *Caenorhabditis elegans* model of polyglutamine (polyQ) pathologies, polyQ aggregation increases the misfolding of otherwise soluble proteins, which in turn enhances further polyQ aggregation [119]. Therefore, it is likely that the protective mechanisms become insufficient with aging and increase the probability of further misfolding events.

CONCLUSIONS

Understanding why proteins misfold into amyloid or structurally related aggregates is of extreme importance. Although much remains to be understood, considerable progress has been achieved over the past 20 years. Some of the factors causing protein misfolding and aggregation in vitro are actually operating in vivo as well and are likely to play important roles in the pathogenesis of disease. These include destabilization of the folded states of globular proteins, increase in the intrinsic propensity of the unfolded states to aggregate, and increase in protein concentration. However, the existence of additional factors in vivo, such as oxidative stress, macromolecular crowding, interaction with biological macromolecules, and aging, adds complexity to the description of the causes of protein aggregation in vivo and of each of the pathological conditions associated with it.

REFERENCES

1. Richardson, J.S., Richardson, D.C. (2002). Natural beta-sheet proteins use negative design to avoid edge-to-edge aggregation. *Proc Natl Acad Sci USA*, *99*, 2754–2759.
2. Kelly, J.W. (1998). The alternative conformations of amyloidogenic proteins and their multi-step assembly pathways. *Curr Opin Struct Biol*, *8*, 101–106.
3. Dobson, C.M. (1999). Protein misfolding, evolution and disease. *Trend Biochem Sci*, *24*, 329–332.
4. Lai, Z., Colón, W., Kelly, J.W. (1996). The acid-mediated denaturation pathway of transthyretin yields a conformational intermediate that can self-assemble into amyloid. *Biochemistry*, *35*, 6470–6482.
5. Guijarro, J.I., Sunde, M., Jones, J.A., Campbell, I.D., Dobson, C.M. (1998). Amyloid fibril formation by an SH3 domain. *Proc Natl Acad Sci USA*, *95*, 4224–4228.
6. Chiti, F., Webster, P., Taddei, N., Clark, A., Stefani, M., Ramponi, G. (1999). Designing conditions for in vitro formation of amyloid protofilaments and fibrils. *Proc Natl Acad Sci USA*, *96*, 3590–3594.
7. McParland, V.J., Kad, N.M., Kalverda, A.P., Brown, A., Kirwin-Jones, P., Hunter, M.G., Sunde, M., Radford, S.E. (2000). Partially unfolded states of β_2-microglobulin and amyloid formation in vitro. *Biochemistry*, *39*, 8735–8746.
8. Morozova-Roche, L.A., Zurdo, J., Spencer, A., Noppe, W., Receveur, V., Archer, D.B., Joniau, M., Dobson, C.M. (2000). Amyloid fibril formation and seeding by wild-type human lysozyme and its disease-related mutational variants. *J Struct Biol*, *130*, 339–351.
9. Fandrich, M., Fletcher, M.A., Dobson, C.M. (2001). Amyloid fibrils from muscle myoglobin. *Nature*, *410*, 165–166.
10. Torrent, J., Balny, C., Lange, R. (2006). High pressure modulates amyloid formation. *Protein Pept Lett*, *13*, 271–277.

11. Booth, D.R., Sunde, M., Bellotti, V., Robinson, C.V., Hutchinson, W.L., Fraser, P.E., Hawkins, P.N., Dobson, C.M., Radford, S.E., Blake, C.C.F., Pepys, M.B. (1997). Instability, unfolding and aggregation of human lysozyme variants underlying amyloid fibrillogenesis. *Nature*, *385*, 787–793.

12. Raffen, R., Dieckman, L.J., Szpunar, M., Wunschl, C., Pokkuluri, P.R., Dave, P., Wilkins Stevens, P., Cai, X., Schiffer, M., Stevens, F.J. (1999). Physicochemical consequences of amino acid variations that contribute to fibril formation by immunoglobulin light chains. *Protein Sci*, *8*, 509–517.

13. Chiti, F., Taddei, N., Bucciantini, M., White, P., Ramponi, G., Dobson, C.M. (2000). Mutational analysis of the propensity for amyloid formation by a globular protein. *EMBO J*, *19*, 1441–1449.

14. Smith, D.P., Jones, S., Serpell, L.C., Sunde, M., Radford, S.E. (2003). A systematic investigation into the effect of protein destabilisation on beta 2-microglobulin amyloid formation. *J Mol Biol*, *330*, 943–954.

15. Stathopulos, P.B., Rumfeldt, J.A., Scholz, G.A., Irani, R.A., Frey, H.E., Hallewell, R.A., Lepock, J.R., Meiering, E.M. (2003). Cu/Zn superoxide dismutase mutants associated with amyotrophic lateral sclerosis show enhanced formation of aggregates in vitro. *Proc Natl Acad Sci USA*, *100*, 7021–7026.

16. Sekijima, Y., Wiseman, R.L., Matteson, J., Hammarström, P., Miller, S.R., Sawkar, A.R., Balch, W.E., Kelly, J.W. (2005). The biological and chemical basis for tissue-selective amyloid disease. *Cell*, *121*, 73–85.

17. Chiti, F., Taddei, N., Stefani, M., Dobson, C.M., Ramponi, G. (2001). Reduction of the amyloidogenicity of a protein by specific binding of ligands to the native conformation. *Protein Sci*, *10*, 879–886.

18. Dumoulin, M., Last, A.M., Desmyter, A., Decanniere, K., Canet, D., Larsson, G., Spencer, A., Archer, D.B., Sasse, J., Muyldermans, S., *et al.* (2003). A camelid antibody fragment inhibits the formation of amyloid fibrils by human lysozyme. *Nature*, *424*, 783–788.

19. Johnson, S.M., Wiseman, R.L., Sekijima, Y., Green, N.S., Adamski-Werner, S.L., Kelly, J.W. (2005). Native state kinetic stabilization as a strategy to ameliorate protein misfolding diseases: a focus on the transthyretin amyloidoses. *Acc Chem Res*, *38*, 911–921.

20. Ray, S.S., Nowak, R.J., Brown, R.H., Jr., Lansbury, P.T., Jr. (2005). Small-molecule-mediated stabilization of familial amyotrophic lateral sclerosis-linked superoxide dismutase mutants against unfolding and aggregation. *Proc Natl Acad Sci USA*, *102*, 3639–3644.

21. Kuwata, K., Nishida, N., Matsumoto, T., Kamatari, Y.O., Hosokawa-Muto, J., Kodama, K., Nakamura, H.K., Kimura K., Kawasaki, M., Takakura, Y., *et al.* (2007). Hot spots in prion protein for pathogenic conversion. *Proc Natl Acad Sci USA*, *104*, 11921–11926.

22. Radford, S.E., Gosal, W.S., Platt, G.W. (2005). Towards an undestranding of the structural molecular mechanism of β_2-microglobulin amyloid formation in vitro. *Biochim Biophys Acta*, *1753*, 51–63.

23. Borysik, A.J., Radford, S.E., Ashcroft, A.E. (2004). Co-populated conformational ensembles of β_2-microglobulin uncovered quantitatively by electrospray ionisation mass spectrometry. *J Biol Chem*, *279*, 27069–27077.

24. Platt, G.W., McParland, V.J., Kalverda, A.P., Homans, S.W., Radford, S.E. (2005). Dynamics in the unfolded state of β_2-microglobulin studied by NMR. *J Mol Biol, 346*, 279–294.

25. Roder, H., Maki, K., Cheng, H. (2006). Early events in protein folding explored by rapid mixing methods. *Chem Rev, 106*, 1836–1861.

26. Brockwell, D.J., Radford, S.E. (2007). Intermediates: ubiquitous species on folding energy landscapes? *Curr Opin Struct Biol, 17*, 30–37.

27. Jahn, T.R., Parker, M.J., Homans, S.W., Radford, S.E. (2006). Amyloid formation under physiological conditions proceeds via a native-like folding intermediate. *Nat Struct Biol, 13*, 195–201.

28. Eakin, C.M., Berman, A.J., Miranker, A.D. (2006). A native to amyloidogenic transition regulated by a backbone trigger. *Nat Struct Biol, 13*, 202–208.

29. Surewicz, W.K., Jones, E.M., Apetri, A.C. (2006). The emerging principles of mammalian prion propagation and transmissibility barriers: insight from studies in vitro. *Acc Chem Res, 39*, 654–662.

30. Jahn, T.R., Radford, S.E. (2008). Folding *versus* aggregation: polypeptide conformations on competing pathways. *Arch Biochem Biophys, 469*, 100–117.

31. Apetri, A.C., Surewicz, K., Surewicz, W.K. (2004). The effect of disease-associated mutations on the folding pathway of human prion protein. *J Biol Chem, 279*, 18008–18014.

32. Apetri, A.C., Maki, K., Roder, H., Surewicz, W.K. (2006). Early intermediate in human prion protein folding as evidenced by ultrarapid mixing experiments. *J Am Chem Soc, 128*, 11673–11678.

33. Tsai, C.J., Kumar, S., Ma, B., Nussinov, R. (1999). Folding funnels, binding funnels, and protein function. *Protein Sci, 8*, 1181–1190.

34. Krishna, M.M., Hoang, L., Lin, Y., Englander, S.W. (2004). Hydrogen exchange methods to study protein folding. *Methods, 34*, 51–64.

35. Bai, Y. (2006). Energy barriers, cooperativity, and hidden intermediates in the folding of small proteins. *Biochem Biophys Res Commun, 340*, 976–983.

36. Zhou, Z., Huang, Y.Z., Bai, Y.W. (2005). An on-pathway hidden intermediate and the early rate-limiting transition state of RD-apocytochrome b_{562} characterized by protein engineering. *J Mol Biol, 352*, 757–764.

37. Bollen, Y.J.M., Kamphuis, M.B., van Mierlo, C.P.M. (2006). The folding energy landscape of apoflavodoxin is rugged: hydrogen exchange reveals nonproductive misfolded intermediates. *Proc Natl Acad Sci USA, 103*, 4095–4100.

38. Teilum, K., Poulsen, F.M., Akke, M. (2006). The inverted chevron plot measured by NMR relaxation reveals a native-like unfolding intermediate in acyl-CoA binding protein. *Proc Natl Acad Sci USA, 103*, 6877–6882.

39. Canet, D., Last, A.M., Tito, P., Sunde, M., Spencer, A., Archer, D.B., Redfield, C., Robinson, C.V., Dobson, C.M. (2002). Local cooperativity in the unfolding of an amyloidogenic variant of human lysozyme. *Nat Struct Biol, 9*, 308–315.

40. Dumoulin, M., Canet, D., Last, A.M., Pardon, E., Archer, D.B., Muyldermans, S., Wyns, L., Matagne, A., Robinson, C.V., Redfield, C., Dobson, C.M. (2005). Reduced global cooperativity is a common feature underlying the amyloidogenicity of pathogenic lysozyme mutations. *J Mol Biol, 346*, 773–788.

41. Liu, K., Cho, H.S., Hoyt, D.W., Nguyen, T.N., Olds, P., Kelly, J.W., Wemmer, D.E. (2000). Deuterium-proton exchange on the native wild-type transthyretin tetramer identifies the stable core of the individual subunits and indicates mobility at the subunit interface. *J Mol Biol*, *303*, 555–565.
42. Nordlund, A., Oliveberg, M. (2006). Folding of Cu/Zn superoxide dismutase suggests structural hotspots for gain of neurotoxic function in ALS: parallels to precursors in amyloid disease. *Proc Natl Acad Sci USA*, *103*, 10218–10223.
43. Chiti, F., De Lorenzi, E., Grossi, S., Mangione, P., Giorgetti, S., Caccialanza, G., Dobson, C.M., Merlini, G., Ramponi, G., Bellotti, V. (2001). A partially structured species of beta 2-microglobulin is significantly populated under physiological conditions and involved in fibrillogenesis. *J Biol Chem*, *276*, 46714–46721.
44. Plakoutsi, G., Taddei, N., Stefani, M., Chiti, F., (2004). Aggregation of the Acylphosphatase from *Sulfolobus solfataricus*: the folded and partially unfolded states can both be precursors for amyloid formation. *J Biol Chem*, *279*, 14111–14119.
45. Plakoutsi, G., Bemporad, F., Monti, M., Pagnozzi, D., Pucci, P., Chiti, F., (2006). Exploring the mechanism of formation of native-like and precursor amyloid oligomers for the native acylphosphatase from *Sulfolobus solfataricus*. *Structure*, *14*, 993–1001.
46. Dunker, A.K., Brown, C.J., Obradovic, Z., (2002). Identification and functions of usefully disordered proteins. *Adv Protein Chem*, *62*, 25–49.
47. Uversky, V.N. (2002). Natively unfolded proteins: a point where biology waits for physics. *Protein Sci*, *11*, 739–756.
48. Fink, A.L. (2005). Natively unfolded proteins. *Curr Opin Struct Biol*, *15*, 35–41.
49. Lise, S., Jones, D.T. (2005). Sequence patterns associated with disordered regions in proteins. *Proteins*, *58*, 144–150.
50. Linding, R., Schymkowitz, J., Rousseau, F., Diella, F., Serrano, L. (2004). A comparative study of the relationship between protein structure and β-aggregation in globular and intrinsically disordered proteins. *J Mol Biol*, *342*, 345–353.
51. Dyson, H.J., Wright, P.E. (2002). Coupling of folding and binding for unstructured proteins. *Curr Opin Struct Biol*, *12*, 54–60.
52. Meszaros, B., Tompa, P., Simon, I., Dosztanyi, Z. (2007). Molecular principles of the interactions of disordered proteins. *J Mol Biol*, *372*, 549–561.
53. Chiti, F., Dobson, C.M. (2006). Protein misfolding, functional amyloid, and human disease. *Annu Rev Biochem*, *75*, 333–366.
54. Zahn, R., Liu, A., Luhrs, T., Riek, R., von Schroetter, C., Lopez Garcia, F., Billeter, M., Calzolai, L., Wider, G., Wuthrich, K. (1999). NMR solution structure of the human prion protein. *Proc Natl Acad Sci USA*, *97*, 145–150.
55. Chiti, F., Stefani, M., Taddei, N., Ramponi, G., Dobson, C.M. (2003). Rationalization of the effects of mutations on peptide and protein aggregation rates. *Nature*, *424*, 805–808.
56. DuBay K.F., Pawar, A.P., Chiti, F., Zurdo, J., Dobson, C.M., Vendruscolo, M. (2004). Prediction of the absolute aggregation rates of amyloidogenic polypeptide chains. *J Mol Biol*, *341*, 1317–1326.

57. Fernandez-Escamilla, A.M., Rousseau, F., Schymkowitz, J., Serrano L. (2004). Prediction of sequence-dependent and mutational effects on the aggregation of peptides and proteins. *Nat Biotechnol, 22*, 1302–1306.
58. Lopez de la Paz, M., Serrano, L., (2004). Sequence determinants of amyloid fibril formation. *Proc Natl Acad Sci USA, 101*, 87–92.
59. Tartaglia, G.G., Cavalli, A., Pellarin, R., Caflish, A. (2004). The role of aromaticity, exposed surface, and dipole moment in determining protein aggregation rates. *Protein Sci, 13*, 1939–1941.
60. Yoon, S., Welsh, W.J. (2004). Detecting hidden sequence propensity for amyloid fibril formation. *Protein Sci, 13*, 2149–2160.
61. de Groot, N.S., Pallares, I., Aviles, F.X., Vendrell, J., Ventura, S. (2005). Prediction of "hot spots" of aggregation in disease-linked polypeptides. *BMC Struct Biol, 5*, 18.
62. Pawar, A.P., DuBay, K.F., Zurdo, J., Chiti, F., Vendruscolo, M. (2005). Prediction of 'aggregation-prone' and 'aggregation-susceptible' regions in proteins associated with neurodegenerative diseases. *J Mol Biol, 350*, 379–392.
63. Tartaglia, G.G., Cavalli, A., Pellarin, R., Caflish, A. (2005). Prediction of aggregation rate and aggregation-prone segments in polypeptide sequences. *Protein Sci, 14*, 2723–2734.
64. Galzitskaya, O.V., Garbuzynskiy, S.O., Lobanov, M.Y. (2006). Prediction of amyloidogenic and disordered regions in protein chains. *PLoS Comput Biol, 2*, e177.
65. Thompson, M.J., Sievers, S.A., Karanicolas, J., Ivanova, M.I., Baker, D., Eisenberg, D. (2006). The 3D profile method for identifying fibril-forming segments of proteins. *Proc Natl Acad Sci USA, 103*, 4074–4078.
66. Trovato, A., Chiti, F., Maritan, A., Seno, F. (2006). Insight into the structure of amyloid fibrils from the analysis of globular proteins. *PLoS Comput Biol, 2*, e170.
67. Conchillo-Solé, O., de Groot, N.S., Aviles, F.X., Vendrell, J., Daura, X., Ventura, S. (2007). AGGRESCAN: a server for the prediction and evaluation of "hot spots" of aggregation in polypeptides. *BMC Bioinformatics, 8*, 65.
68. Monsellier, E., Ramazzotti, M., Polverino de Laureto, P., Tartaglia, G.G., Taddei, N., Fontana, A., Vendruscolo, M., Chiti, F. (2007). The distribution of residues in a polypeptide sequence is a determinant of aggregation optimized by evolution. *Biophys J, 93*, 1–10.
69. Zibaee, S., Makin, O.S., Goedert, M., Serpell, L.C. (2007). A simple algorithm locates beta-strands in the amyloid fibril core of alpha-synuclein, Abeta, and tau using the amino acid sequence alone. *Protein Sci, 16*, 906–918.
70. Zibaee, S., Jakes, R., Fraser, G., Serpell, L.C., Crowther, R.A., Goedert, M. (2007). Sequence determinants for amyloid fibrillogenesis of human alpha-synuclein. *J Mol Biol, 374*, 454–464.
71. Luheshi, L.M., Tartaglia, G.G., Brorsson, A.C., Pawar, A.P., Watson, I.E., Chiti, F., Vendruscolo, M., Lomas, D.A., Dobson, C.M., Crowther, D.C. (2007). Systematic in vivo analysis of the intrinsic determinants of amyloid beta pathogenicity. *PLoS Biol, 5*, e290.
72. de Groot, N.S., Aviles, F.X., Vendrell, J., Ventura, S. (2006). Mutagenesis of the central hydrophobic cluster in Abeta42 Alzheimer's peptide: side-chain properties correlate with aggregation propensities. *FEBS J, 273*, 658–668.

73. Khare, S.D., Wilcox, K.C., Gong, P., Dokholyan, N.V. (2005). Sequence and structural determinants of Cu, Zn superoxide dismutase aggregation. *Proteins*, *61*, 617–632.
74. Cecchini, M., Curcio, R., Pappalardo, M., Melki, R., Caflish, A. (2006). A molecular dynamic approach to the structural characterization of amyloid aggregation. *J Mol Biol*, *357*, 1306–1321.
75. Sandelin, E., Nordlund, A., Andersen, P.M., Marklund, S.S., Oliveberg, M. (2007). Amyotrophic lateral sclerosis–associated copper/zinc superoxide dismutase mutations preferentially reduce the repulsive charge of the proteins. *J Biol Chem*, *282*, 21230–21236.
76. van Swieten, J., Spillantini, M.G. (2007). Hereditary frontotemporal dementia caused by Tau gene mutations. *Brain Pathol*, *17*, 63–73.
77. De Jonghe, C., Esselens, C., Kumar-Singh, S., Craessaerts, K., Serneels, S., Checler, F., Annaert, W., Van Broeckhoven, C., De Strooper, B. (2001). Pathogenic APP mutations near the gamma-secretase cleavage site differentially affect Abeta secretion and APP C-terminal fragment stability. *Hum Mol Genet*, *10*, 1665–1671.
78. Page, L.J., Huff, M.E., Kelly, J.W., Balch, W.E. (2004). $Ca2^+$ binding protects against gelsolin amyloidosis. *Biochem Biophys Res Commun*, *322*, 1105–1110.
79. Fung, S.Y., Keyes, C., Duhamel, J., Chen, P. (2003). Concentration effect on the aggregation of a self-assembling oligopeptide. *Biophys J*, *85*, 537–548.
80. Kodaka, M. (2004). Interpretation of concentration-dependence in aggregation kinetics. *Biophys Chem*, *109*, 325–332.
81. Lansbury, P.T., Lashuel, H.A. (2006). A century-old debate on protein aggregation and neurodegeneration enters the clinic. *Nature*, *443*, 774–779.
82. Floege, J., Ehlerding, G. (1996). Beta-2-microglobulin–associated amyloidosis. *Nephron*, *72*, 9–26.
83. Relini, A., Canale, C., De Stefano, S., Rolandi, R., Giorgetti, S., Stoppini, M., Rossi, A., Fogolari, F., Corazza, A., Esposito, G., *et al.* (2006). Collagen plays an active role in the aggregation of beta2-microglobulin under physiopathological conditions of dialysis-related amyloidosis. *J Biol Chem*, *281*, 16521–16529.
84. Gillmore, J.D., Lovat, L.B., Persey, M.R., Pepys, M.B., Hawkins, P.N. (2001). Amyloid load and clinical outcome in AA amyloidosis in relation to circulating concentration of serum amyloid A protein. *Lancet*, *358*, 24–29.
85. Gorbenko, G.P., Kinnunen, P.K. (2006). The role of lipid–protein interactions in amyloid-type protein fibril formation. *Chem Phys Lipids*, *141*, 72–82.
86. Klein, J., Dhurjati, P. (1995). Protein aggregation kinetics in an *Escherichia coli* strain overexpressing a *Salmonella typhimurium* CheY mutant gene. *Appl Environ Microbiol*, *61*,1220–1225.
87. Young, J.C., Agashe, V.R., Siegers, L., Hartl, F.U. (2004). Pathways of chaperone-mediated protein folding in the cytosol. *Nat Rev Mol Cell Biol*, *5*, 781–791.
88. Bukau, B., Weissman, J., Horwich, A. (2006). Molecular chaperones and protein quality control. *Cell*, *125*, 443–451.
89. Esaki, M., Kanamori, T., Nishikawa, S., Shin, I., Schultz, P.G., Endo, T. (2003). Tom40 protein import channel binds to non-native proteins and prevents their aggregation. *Nat Struct Biol*, *10*, 988–994.

90. Prakash, S., Matouschek, A. (2004). Protein unfolding in the cell. *Trends Biochem Sci, 29*, 593–600.

91. Andreyev, A.Y., Kushnareva, Y.E., Starkov, A.A. (2005). Mitochondrial metabolism of reactive oxygen species. *Biochemistry, 70*, 200–214.

92. Granata, A., Roncone, R., Monzani, E., Casella, L. (2007). Tyrosinase-generated quinones induce covalent modification, unfolding, and aggregation of human holo-myoglobin. *Biomacromolecules, 8*, 3214–3223.

93. Zhang, H., Andrekopoulos, C., Joseph, J., Crow, J., Kalyanaraman, B. (2004). The carbonate radical anion-induced covalent aggregation of human copper, zinc superoxide dismutase, and alpha-synuclein: intermediacy of tryptophan- and tyrosine-derived oxidation products. *Free Radic Biol Med, 36*, 1355–1365.

94. Janué, A., Olive, M., Ferrer, I. (2007). Oxidative stress in desminopathies and myotilinopathies: a link between oxidative damage and abnormal protein aggregation. *Brain Pathol, 17*, 377–388.

95. Ellis, R.J., Minton, A.P. (2006). Protein aggregation in crowded environments. *Biol Chem, 387*, 485–497.

96. Snow, A.D., Mar, H., Nochlin, D., Kimata, K., Kato, M., Suzuki, S., Hassell, J., Wight, T.N. (1988). The presence of heparan sulfate proteoglycans in the neuritic plaques and congophilic angiopathy in Alzheimer's disease. *Am J Pathol, 133*, 456–463.

97. Snow, A.D., Wight, T.N., Nochlin, D., Koike, Y., Kimata, K., DeArmond, S.J., Prusiner, S.B. (1990). Immunolocalization of heparan sulfate proteoglycans to the prion protein amyloid plaques of Gerstmann–Straussler syndrome, Creutzfeldt–Jakob disease and scrapie. *Lab Invest, 63*, 601–611.

98. Young, I.D., Ailles, L., Narindrasorasak, S., Tan, R., Kisilevsky, R. (1992). Localization of the basement membrane heparan sulfate proteoglycans in islet amyloid deposits in type 2 diabetes mellitus. *Arch Pathol Lab Med, 116*, 951–954.

99. Pepys, M.B., Rademacher, T.W., Amatayakul-Chantler, S., Williams, P., Noble, G.E., Hutchinson, W.L., Hawkins, P.N., Nelson, S.R., Gallimore, J.R., Herbert, J., *et al.* (1994). Human serum amyloid P component is an invariant constituent of amyloid deposits and has a uniquely homogeneous glycostructure. *Proc Natl Acad Sci USA, 91*, 5602–5606.

100. Hashimoto, T., Wakabayashi, T., Watanabe, A., Kowa, H., Hosoda, R., Nakamura, A., Kanazawa, I., Arai, T., Takio, K., Mann, D.M., Iwatsubo, T. (2002). CLAC: a novel Alzheimer amyloid plaque component derived from a transmembrane precursor, CLAC-P/collagen type XXV. *EMBO J, 21*, 1524–1534.

101. Cotman, S.L., Halfter, W., Cole, G.J. (2000). Agrin binds to beta-amyloid (Abeta), accelerates Abeta fibril formation, and is localized to Abeta deposits in Alzheimer's disease brain. *Mol Cell Neurosci, 15*, 183–198.

102. Carter, D.B. (2005). The interaction of amyloid-beta with ApoE. *Subcell Biochem, 38*, 255–272.

103. Kakuyama, H., Soderberg, L., Horigome, K., Winblad, B., Dahlqvist, C., Naslund, J., Tjernberg, L.O. (2005). CLAC binds to aggregated Abeta and Abeta fragments, and attenuates fibril elongation. *Biochemistry, 44*, 15602–15609.

104. Liko, I., Mak, M., Klement, E., Hunyadi-Gulyas, E., Pazmany, T., Medzihradszky, K.F., Urbanyi, Z. (2007). Evidence for an extended interacting surface between beta-amyloid and serum amyloid P component. *Neurosci Lett, 412*, 51–55.

105. Gupta-Bansal, R., Frederickson, R.C., Brunden, K.R. (1995). Proteoglycan-mediated inhibition of A beta proteolysis: a potential cause of senile plaque accumulation. *J Biol Chem*, *270*, 18666–18671.

106. Tennent, G.A., Lovat, L.B., Pepys, M.B. (1995). Serum amyloid P component prevents proteolysis of the amyloid fibrils of Alzheimer disease and systemic amyloidosis. *Proc Natl Acad Sci USA*, *92*, 4299–4303.

107. Yamaguchi, I., Suda, H., Tsuzuike, N., Seto, K., Seki, M., Yamaguchi, Y., Hasegawa, K., Takahashi, N., Yamamoto, S., Gejyo, F., Naiki, H. (2003). Glycosaminoglycan and proteoglycan inhibit the depolymerization of beta2-microglobulin amyloid fibrils in vitro. *Kidney Int*, *64*, 1080–1088.

108. Soderberg, L., Dahlqvist, C., Kakuyama, H., Thyberg, J., Ito, A., Winblad, B., Naslund, J., Tjernberg, L.O. (2005). Collagenous Alzheimer amyloid plaque component assembles amyloid fibrils into protease resistant aggregates. *FEBS J*, *272*, 2231–2236.

109. Matsuzaki, K. (2007). Physicochemical interactions of amyloid beta-peptide with lipid bilayers. *Biochim Biophys Acta*, *1768*, 1935–1942.

110. Bush, A.I. (2003). The metallobiology of Alzheimer's disease. *Trends Neurosci 26*, 207–214.

111. Uversky, V.N., Li, J., Fink, A.L. (2001). Metal-triggered structural transformations, aggregation, and fibrillation of human alpha-synuclein: a possible molecular link between Parkinson's disease and heavy metal exposure. *J Biol Chem*, *27*, 44284–44296.

112. Cherny, R.A., Legg, J.T., McLean, C.A., Fairlie, D.P., Huang, X., Atwood, C.S., Beyreuther, K., Tanzi, R.E., Masters, C.L., Bush, A.I. (1999). Aqueous dissolution of Alzheimer's disease Abeta amyloid deposits by biometal depletion. *J Biol Chem*, *274*, 23223–23228.

113. Gaeta, A., Hider, R.C. (2005). The crucial role of metal ions in neurodegeneration: the basis for a promising therapeutic strategy. *Br J Pharmacol*, *146*, 1041–1059.

114. Morley, J.F., Brignull, H.R., Weyers, J.J., Morimoto, R.I. (2002). The threshold for polyglutamine-expansion protein aggregation and cellular toxicity is dynamic and influenced by aging in *Caenorhabditis elegans*. *Proc Natl Acad Sci USA*, *99*, 10417–10422.

115. Cohen, E., Bieschke, J., Perciavalle, R.M., Kelly, J.W., Dillin, A. (2006). Opposing activities protect against age-onset proteotoxicity. *Science*, *313*, 1604–1610.

116. Paz Gavilan, M., Vela, J., Castano, A., Ramos, B., del Rio, J.C., Vitorica, J., Ruano, D. (2006). Cellular environment facilitates protein accumulation in aged rat hippocampus. *Neurobiol Aging*, *27*, 973–982.

117. Keller, J.N., Gee, J., Ding, Q. (2002). The proteasome in brain aging. *Ageing Res Rev*, *1*, 279–293.

118. Soti, C., Csermely, P. (2003). Aging and molecular chaperones. *Exp Gerontol*, *38*, 1037–1040.

119. Gidalevitz, T., Ben-Zvi, A., Ho, K.H., Brignull, H.R., Morimoto, R.I. (2006). Progressive disruption of cellular protein folding in models of polyglutamine diseases. *Science*, *311*, 1471–1474.

2

ENDOPLASMIC RETICULUM STRESS AND OXIDATIVE STRESS: MECHANISMS AND LINK TO DISEASE

Jyoti D. Malhotra

Department of Biological Chemistry, University of Michigan Medical School, Ann Arbor, Michigan

Randal J. Kaufman

Departments of Biological Chemistry and Internal Medicine, Howard Hughes Medical Institute, University of Michigan Medical School, Ann Arbor, Michigan

INTRODUCTION

Protein folding is an essential process for protein function in all organisms. As a consequence, all cells have evolved sophisticated mechanisms to ensure that proper protein folding occurs and to dispose of irreversibly misfolded proteins. All proteins that transit the secretory pathway in eukaryotic cells first enter the endoplasmic reticulum (ER), where they fold and assemble into multisubunit complexes prior to transit to the Golgi compartment [1]. Quality control is a surveillance mechanism that permits only properly folded proteins to exit the ER en route to other intracellular organelles and the cell surface. Misfolded proteins are either retained within the ER lumen in a complex with molecular chaperones or are directed toward degradation through the 26S proteasome in a process called *ER-associated degradation* (ERAD) or through autophagy.

Protein Misfolding Diseases: Current and Emerging Principles and Therapies,
Edited by Marina Ramirez-Alvarado, Jeffery W. Kelly, and Christopher M. Dobson
Copyright © 2010 John Wiley & Sons, Inc.

The efficiency of protein-folding reactions depends on appropriate environmental, genetic, and metabolic conditions. Conditions that disrupt protein folding present a threat to cell viability. The ER provides a unique environment that challenges proper protein folding as nascent polypeptide chains enter the ER lumen. The high concentration of partially folded and unfolded proteins predisposes protein-folding intermediates to aggregation. Polypeptide-binding proteins, such as BiP and GRP94, act to slow protein-folding reactions and prevent aberrant interactions and aggregation. The ER lumen is an oxidizing environment, so disulfide bond formation occurs. As a consequence, cells have evolved sophisticated machinery composed of many protein disulfide isomerases (PDIs) that are required to ensure proper disulfide bond formation and to prevent formation of illegitimate disulfide bonds. The ER is also the primary Ca^{2+} storage organelle in the cell.

Both protein-folding reactions and protein chaperone functions require high levels of ER intralumenal calcium. Protein folding in the ER requires extensive amounts of energy, and depletion of energy stores prevents proper protein folding. ATP is required for chaperone function, to maintain Ca^{2+} stores and redox homeostasis, and for ERAD. Finally, proteins that enter the ER lumen are subject to numerous post-translational modifications, including N-linked glycosylation, amino acid modifications such as proline and aspartic acid hydroxylation and γ-carboxylation of glutamic acid residues, and addition of glycosylphosphatidylinositol anchors. All these processes are highly sensitive to alterations in the ER luminal environment. As a consequence, innumerable environmental insults alter protein-folding reactions in the ER through mechanisms that include depletion of ER calcium, alteration in the redox status, and energy (sugar/glucose) deprivation. In addition, gene mutations, elevated protein traffic through the ER compartment, and altered post-translational modification all contribute the accumulation of unfolded proteins in the ER lumen.

Accumulation of unfolded protein initiates activation of an adaptive signaling cascade known as the *unfolded protein response* (UPR). Appropriate adaptation to misfolded protein accumulation in the ER lumen requires regulation at all levels of gene expression, including transcription, translation, translocation into the ER lumen, and ERAD. Coordinate regulation of all these processes is required to restore proper protein folding and ER homeostasis [1–6]. Conversely, if the protein-folding defect is not resolved, chronic activation of UPR signaling occurs, which eventually induces an apoptotic (programmed cell death) response.

In this chapter we summarize the signaling pathways that mediate the UPR, mechanisms that signal cell death, the role of the UPR in mammalian physiology, and the clinical implications of the UPR in health and disease.

PROTEIN FOLDING AND QUALITY CONTROL IN THE ER

Protein folding and maturation in vivo is a highly assisted process. The ER lumen contains molecular chaperones, folding enzymes, and quality control factors that assist in folding and trafficking of newly synthesized polypeptides. Nascent polypeptide chains enter the ER lumen through a proteinaceous channel, the Sec 61 translocon complex. The nascent chains of most translocated polypeptides are subject to addition of a preassembled oligosaccharide core (*N*-acetylglucosamine$_2$–mannose$_9$–glucose$_3$) ($Glc_3Man_9GlcNac_2$) to selective asparagine (N) residues. N-Glycosylation is catalyzed by the oligosaccharyltransferase (OST), a multisubunit enzyme associated with the translocon complex. Subsequently, sequential action by the ER α-glucosidases I and II removes the two outermost glucose residues to produce a monoglucosylated core glycan. The monoglucosylated glycoprotein can then interact with two homologous ER lectins, calnexin (CNX) and calreticulin (CRT), which associate with Erp57, an oxidoreductase that catalyzes disulfide bond formation. Upon release of folding substrates from CNX and/or CRT, the innermost glucose residue is removed rapidly by glucosidase II. Two potential mechanisms by which chaperones may monitor protein folding are through exposed hydrophobic patches or through excessive surface dynamics associated with the noncompact partially folded state. The protein chaperone BiP binds to the hydrophobic patches exposed on protein-folding intermediates. Highly dynamic, non-native deglucosylated glycoproteins are recognized by the ER folding sensor UDP-glucose:glycoprotein glucosyltransferase (UGT1). UGT1 specifically re-glucosylates folding intermediates released from the CNX/CRT cycle. Reglucosylation mediates ER retention of immaturely folded glycoproteins so they enter another round of CNX/CRT-assisted folding. Polypeptides that fail to acquire their native transport-competent structure are eventually removed by ERAD. During ERAD, folding-defective proteins are retranslocated to the cytosol and are degraded by the 26S proteasome. Native, properly folded polypeptides released from CNX/CRT are transported from the ER to the Golgi compartment, in an event possibly assisted by mannose-binding lectins such as ERGIC53, VIPL, and ERGL (Fig. 1).

UPR SIGNALING

In response to ER stress, three ER-localized transmembrane signal transducers are activated to initiate adaptive responses. These transducers are two protein kinases, IRE1 (inositol-requiring kinase 1) [7,8] and PERK (double-stranded RNA-activated protein kinase-like ER kinase) [9], and the transcription factor ATF6 (activating transcription factor 6) [8,10]. These three UPR transducers are expressed constitutively in all known metazoan cells (Fig. 2). IRE1 was the first component of the UPR that was identified, initially in yeast, and is conserved in all eukaryotic cells. The essential and unique properties of IRE1 signaling in the UPR have been conserved in all eukaryotic cells, but higher eukaryotes also possess the additional sensors PERK and ATF6, which promote stress adaptation or cell death in a more complex and diverse, yet coordinated manner.

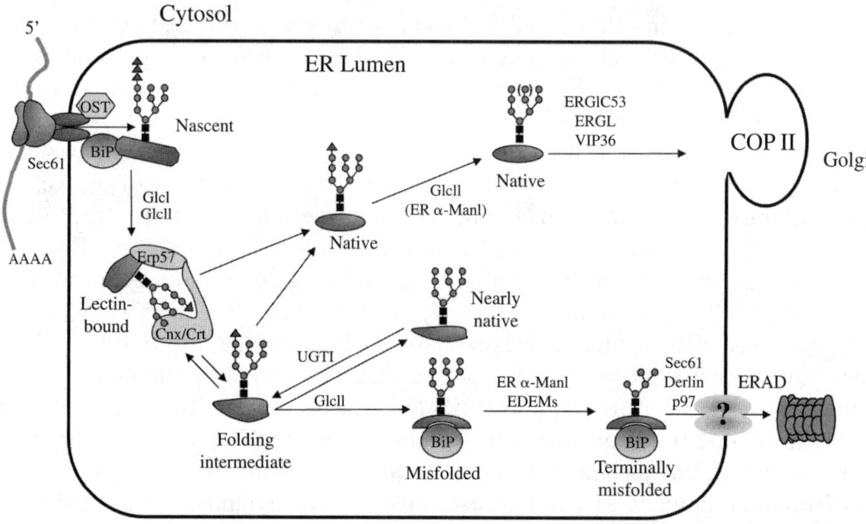

FIG. 1 Protein trafficking from the ER. Upon translocation of polypeptides through the Sec61 proteinaceous channel, asparagine residues are frequently modified by covalent addition of a preassembled oligosaccharide core (*N*-acetylglucosamine$_2$–mannose$_9$–glucose$_3$). This reaction is catalyzed by the oligosaccharyltransferase (OST), a multi-subunit complex associated with translocon. To facilitate unidirectional transport through the translocon, nascent polypeptide chains in the ER lumen interact with BiP, a molecular chaperone that binds to exposed hydrophobic residues. Subsequently, rapid deglucosylation of the two outermost glucose residues on the oligosaccharide core structures, mediated by glucosidase I and II (GlcI and GlcII), prepares glycoproteins for association with the ER lectins calnexin and calreticulin. The calnexin/calreticulin-associated oxidoreductase ERp57 facilitates protein folding by catalyzing the formation of intra-and intermolecular disulfide bonds, a rate-limiting step in the protein-folding process. Release from calnexin or calreticulin followed by glucosidase II cleavage of the innermost glucose residue prevents further interaction with calnexin and calreticulin. At this point, natively folded polypeptides transit the ER to the Golgi compartment, in a process possibly assisted by mannose-binding lectins such as ERGIC53, VIPL, and ERGL. As an essential component of protein-folding quality control, nonnative polypeptides are tagged for reassociation with calnexin and calreticulin by the UDP-glucose:glycoprotein glucosyltransferase (UGT1) to facilitate their ER retention and prevent anterograde transport. Polypeptides that are folding incompetent are targeted for degradation by retrotranslocation, possibly mediated by EDEM and Derlins, into the cytosol and delivery to the 26S proteasome. Triangles represent glucose residues, squares represent *N*-acetylglucosamine residues, and circles represent mannose residues.

PERK PHOSPHORYLATES eIF2α TO ATTENUATE mRNA TRANSLATION

The most immediate response to ER stress in metazoan cells is reversible, transient attenuation of mRNA translation, thereby preventing influx of newly synthesized polypeptides into the stressed ER lumen [11]. This translational

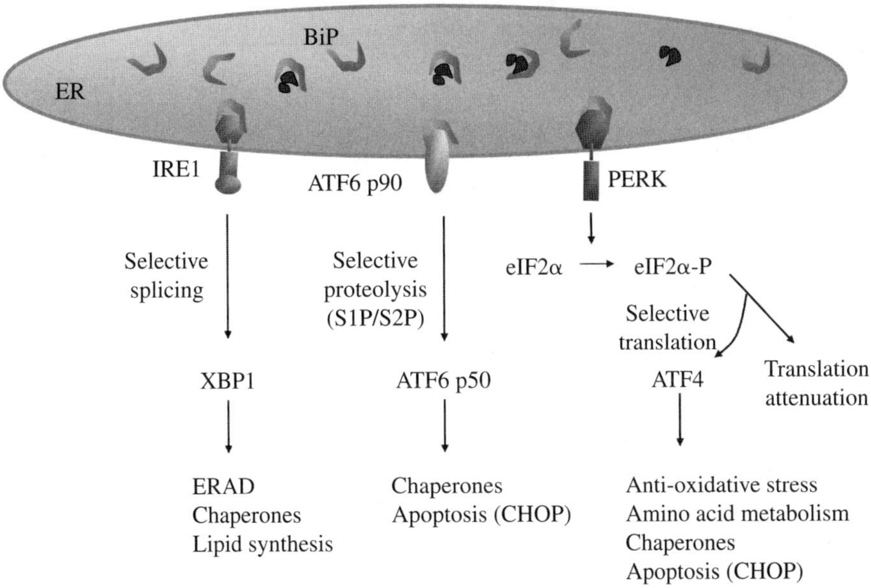

FIG. 2 Signaling the unfolded protein response. Three proximal sensors IRE1, PERK, and ATF6 act in concert to regulate the UPR through their respective signaling cascade and referred to collectively as tripartite signaling in the ER. The protein chaperone BiP is the master regulator and negatively regulates these pathways. Under nonstressed conditions, BiP binds to the lumenal domains of IRE1 and PERK to prevent their dimerization. With the accumulation of the unfolded proteins, BiP released from IRE1 permits dimerization to activate its kinase and RNase activities to initiate XBP1 mRNA splicing, thereby creating a potent transcriptional activator. Primary targets that require IRE1/XBP1 pathway for induction are genes encoding functions in ERAD. Similarly, BiP release from ATF6 permits transport to the Golgi compartment where ATF6 is cleaved by SIP and S2P proteases to yield a cytosolic fragment that migrates to the nucleus to further activate transcription of UPR-responsive genes. Finally, BiP release permits PERK dimerization and activation to phosphorylate eIF2α on Ser51, which leads to general attenuation of translational initiation. eIF2α phosphorylation induces ATF4 mRNA preferentially, and recent evidence has shown that the PERK/eIF2α/ATF4 regulatory axis induces expression of an antioxidative stress response gene pathway and promotes expression of proapoptotic transcription factor CHOP.

attenuation is signaled through PERK-mediated phosphorylation of the eukaryotic translation initiation factor 2 on the α subunit (eIF2α at Ser51). eIF2α phosphorylation inhibits the guanine nucleotide exchange factor eIF2B, which recycles the eIF2 complex to its active GTP-bound form. The formation of the ternary translation initiation complex eIF2-GTP-tRNA^{Met} is required for AUG initiation codon recognition and joining of the 60S ribosomal subunit that occurs during initiation phase of polypeptide chain synthesis. Lower levels of active ternary complex result in lower levels of translation initiation [9,12–14] (Fig. 2).

PERK is an ER-associated transmembrane serine/threonine protein kinase. Upon the accumulation of unfolded proteins in the ER lumen, PERK dimerization and trans-autophosphorylation lead to activation of its eIF2α kinase function [9,15]. In addition to translational attenuation, activation of PERK also contributes to transcriptional induction of the majority of the UPR-dependent genes [12–14,16]. Although phosphorylation of eIF2α inhibits general translation initiation, it is required for the selective translation of several mRNAs. One fundamental transcription factor for which translation is activated upon PERK-mediated phosphorylation of eIF2α is the activating transcription factor 4 (ATF4) [12–14,16]. Expression profiling identified the fact that genes encoding amino acid biosynthesis and transport functions, antioxidative stress responses, and apoptosis, such as growth arrest and DNA damage 34 (GADD34) and CAAT/Enhancer binding protein (C/EBP) homologous protein (CHOP/GADD153) [15,17], require PERK, eIF2α phosphorylation, and ATF4 [12–14,16].

Although the majority of PERK signaling is mediated through phosphorylation of eIF2α, studies suggest that the bZiP Cap 'n' Collar transcription factor nuclear respiratory factor 2 (NRF2) may also be a substrate for PERK kinase activity. NRF1 and NRF2 are transcription factors that integrate a variety of responses to oxidative stress. NRF2 is distributed in the cytoplasm through its association with the microtubule-associated protein Keap1 (Kelch-like Ech-associated protein 1). Upon ER stress, PERK phosphorylates NRF2 to promote its dissociation from Keap1, leading to the nuclear accumulation of NRF2. *Nrf2*$^{-/-}$ cells are sensitive to ER stress-induced apoptosis. NRF2 is a direct PERK substrate and effector of PERK-dependent cell survival [18]. NRF2 binds to the antioxidant response element (ARE) to activate transcription of genes encoding detoxifying enzymes such as A1 and A2 subunits of glutathione S-transferase, NAD(P)H:quinone oxidoreductase, γ-glutamylcysteine synthetase, heme oxygenase 1, and UDP-glucoronosyl transferase [19]. Possibly in a similar manner, NRF1 is localized to the ER membrane and translocates to the nucleus upon ER stress [20]. Consistent with this idea, *Perk*$^{-/-}$ cells accumulate ROS when exposed to ER stress [13].

IRE1 Initiates Unconventional Splicing of XBP1 mRNA

The first component in the UPR pathway was isolated through a genetic screen to identify mutants in UPR signaling in the budding yeast *Saccharomyces cerevisiae*. In this screen, Ire1p/Ern1p was identified as an ER transmembrane protein kinase that is required for the UPR [3,21]. Subsequently, it was discovered that Ire1p is a bifunctional protein that also has site-specific endoribonuclease (RNase) activity [3,21]. Under unstressed conditions, Ire1p protein kinase is maintained in an inactive monomeric form through interactions with the protein chaperone Kar2p/BiP. Upon accumulation of unfolded proteins in the ER lumen, Ire1p is released from Kar2p/BiP and undergoes homodimerization and trans-autophosphorylation to activate its RNase activity. The RNase activity of

Ire1p cleaves a 252-base intron from mRNA encoding the basic leucine zipper (bZIP)–containing transcription factor Hac1p. This splicing reaction alters the carboxy terminus of Hac1p to introduce a potent transcriptional activation domain. The protein encoded by spliced *HAC1* mRNA binds and activates transcription from the UPR element [UPRE, minimal motif TGACGTG(C/A)] upstream of many UPR target genes [2,22]. In *S. cerevisiae*, the UPR activates transcription of approximately 381 genes [23], more than 50% of which provide functions in the secretory pathway.

Two mammalian homologs of yeast IRE1 have been identified; IRE1α [24] and IRE1β [25]. IRE1α is expressed in most cells and tissues, with highest levels of expression in the pancreas and placenta [24]. IRE1β expression is prominent only in intestinal epithelial cells [25]. The cleavage specificities of IRE1α and IRE1β are quite similar, thereby suggesting that they do not recognize distinct substrates but, rather, confer temporal- and tissue-specific expression [26].

Transcriptional analysis of UPR gene targets, such as BiP, GRP94, and calreticulin, identified a mammalian ER stress response element [ERSE, CCAAT(N_9)CCACG] that is necessary and sufficient for UPR gene activation [27]. Yoshida et al. used a yeast one-hybrid screen to isolate factors that interact with the ERSE. This screen identified one ERSE-binding protein as the bZIP-containing transcription factor XBP1 (X-box-binding protein) [27]. Subsequently, several groups using different approaches demonstrated that *XBP1* mRNA is a substrate for the endoribonuclease activity of metazoan IRE1 [8,28–30]. Upon activation of the UPR, the IRE1 RNase activity initiates removal of a 26-nucleotide intron from *XBP1* mRNA. This splicing reaction creates a translational frameshift to produce a larger form of XBP1 that contains a novel transcriptional activation domain at its C-terminus. Spliced XBP1 is a transcriptional activator that plays a fundamental role in the activation of a wide variety of UPR target genes. Some of the genes identified that require the IRE1–XBP1 pathway are those that encode functions involved in ERAD, such as the ER degradation-enhancing mannosidase-like protein EDEM. Consistent with this observation, cells that are deficient in either IRE1 or XBP1 are defective in ERAD [31] (Fig. 2).

Analysis of gene-deleted mice has provided insight into the physiological roles of IRE1 and XBP1 in mammals. Deletion of *Ire1α* or *Xbp1* in mice creates an embryonic lethality at E11.5–E14 [29,32]. Although deletion of *Ire1β* had no developmental phenotype, $Ire1\beta^{-/-}$ mice were susceptible to experimentally induced intestinal colitis [33]. Mice with heterozygous *Xbp1* deletion appear normal but develop insulin resistance when fed a high-fat diet [34]. Thus, it was proposed that the UPR might be important in insulin signaling (see below). In addition, both IRE1 and XBP1 have critical roles in B-lymphocyte differentiation. Antigenic stimulation of mature B lymphocytes activates the UPR and signaling through IRE1-mediated *XBP1* mRNA splicing is required to drive B-lymphocyte differentiation into plasma cells [28,35–37]. These studies suggest that the IRE1/XBP1 subpathway of the UPR might be required for the differentiation of cell types that secrete high levels of protein.

This is consistent with a requirement for XBP1 in pancreatic acinar cell development [38].

ATF6-MEDIATED TRANSCRIPTIONAL ACTIVATION REQUIRES REGULATED INTRAMEMBRANE PROTEOLYSIS

The bZiP-containing activating transcription factor 6 (ATF6) was identified as another regulatory protein that, like XBP1, binds to the ERSE1 element in the promoters of UPR-responsive genes [27]. In mammals, there are two alleles of ATF6, ATF6α (90 kDa) and ATF6β (110 kDa), both synthesized in all cell types as ER transmembrane proteins. In the unstressed cells, ATF6 is localized at the ER membrane and bound to BiP. In response to ER stress, BiP dissociation permits trafficking of ATF6 to the Golgi complex, where ATF6 is cleaved sequentially by two proteases [39–41]. The serine protease site-1 protease S1P cleaves ATF6 in the luminal domain. The N-terminal portion is subsequently cleaved by the metalloprotease site-2 protease S2P [42]. The processed forms of ATF6α and ATF6β translocate to the nucleus and bind to the ATF/cAMP response element (CRE) and to the ER stress-responsive element (ERSE-1) to activate target genes [40]. ATF6α and ATF6β both require the presence of the transcription factor CBF (also called NF-Y) to bind ERSEI [39–41]. The proteases S1P and S2P were originally identified for their essential role in processing of the sterol response element–binding protein (SREBP) transcription factor, which is activated upon cholesterol deprivation [43]. Recent gene deletion studies demonstrated that ATF6α contributes significantly to UPR gene induction of protein chaperones and ERAD machinery [44,45]. As a consequence, ATF6α is required for adaptation to chronic ER stress [44]. However surprisingly, *Atf6α* deletion in the mouse did not have a significant phenotype in the absence of ER stress. Analysis of *Atf6β*-null cells did not identify any genes that require ATF6β for constitutive or UPR-induced gene expression. Although *Atf6β* deletion had no phenotype in the mouse, combined deletion of Atf6α and Atf6β produced an early embryonic lethal phenotype. Thus, ATF6α and ATF6β provide an essential complementary function early in mammalian embryogenesis.

Recently, additional bZIP-containing transcription factors that are localized to the ER and regulated by RIP have been identified. CREBH was identified as a liver-specific bZiP transcription factor of the CREB/ATF family with a transmembrane domain that directs localization to the ER [46]. Pro-inflammatory cytokines IL6, 1L1β, and TNFα increase transcription of CREBH to produce an inert protein that is localized to the ER. Upon ER stress, CREBH transits to the Golgi compartment, where it is cleaved by S1P-and S2P-processing enzymes. However, cleaved CREBH does not activate transcription of UPR genes but, rather, induces transcription of a subset of acute-phase response genes, such as C-reactive protein and murine serum amyloid P component (SAP) in hepatocytes. These studies identified CREBH as a novel ER-localized transcription factor that has an essential role in induction of innate immune response genes and for the first time, links ER stress to inflammatory responses [46].

In addition to ATF6 and CREBH, there are additional similarly related factors that are probably regulated through ER stress-induced proteolytic processing, although their physiological roles remain unknown. OASIS (old astrocyte specifically induced substance) and BBF2H7 (BBF2 human homolog on chromosome 7) are cleaved by S1P and S2P in response to ER stress in astrocytes and neurons, respectively [47,48]. Tisp40 (transcript induced in spermiogenesis 40) is cleaved by S1P and S2P to activate transcription of EDEM [49]. These tissue-specific ATF6-like molecules may contribute to the ER stress response. Finally, Luman/LZIP/CREB3 and CREB4 are also two ATF6-like molecules that are cleaved by S1P and S2P to activate UPR transcription, although their cleavage appears not to be activated by ER stress [50–52]. These transcription factors might be activated under conditions other than ER stress to activate transcription of ER chaperones.

BIP IS A MASTER REGULATOR OF UPR SENSOR ACTIVATION

In nonstressed cells, the luminal domains of IRE1, PERK, and ATF6 are bound to the protein chaperone BiP. In response to stress, unfolded proteins accumulate and bind BiP, thereby sequestering BiP and promoting BiP release from the UPR sensors. When these sensors are bound to BiP, they are maintained in an inactive state [53]. This BiP-mediated negative-regulation model for UPR activation is also supported by the observation that BiP overexpression prevented activation of the UPR upon ER stress [54]. In addition, sufficiently high levels of expression of any protein that binds BiP can activate the UPR. In contrast, the accumulation of unfolded proteins that do not bind BiP does not activate the UPR [55,56]. Analysis of the interaction between BiP and ATF6 suggested that this dissociation is not merely a consequence of competition between ATF6 and unfolded protein for binding to BiP, but rather, may involve the active ER stress-dependent release of BiP from ATF6 [57]. Recently, based on the x-ray crystal structure of the yeast Ire1p luminal domain, Credle et al. identified a deep, long MHC1-type groove that exists in an Ire1p dimer and proposed that unfolded polypeptides directly bind Ire1p to mediate its dimerization [58]. However, although x-ray crystal analysis of the human IRE1 luminal domain indicated a structure similar to that of yeast Ire1p, the MHC1-type groove was not solvent accessible [59]. In addition, the luminal domain was shown to form dimers in vitro in the absence of added polypeptide [59]. These observations bring into question the requirement for peptide binding to the MHC1-type cleft to promote dimerization. It is possible that BiP binding and peptide binding both regulate IRE1 dimerization. Future studies should resolve this issue.

ER AND OXIDATIVE PROTEIN FOLDING

Proteins that are destined for secretion depend on disulfide bonds for their maturation and function. These bonds are often crucial for the stability of the final protein structure, and the mispairing of cysteine residues can prevent proteins from attaining their native conformation and lead to misfolding. The ER is compartmentalized away from the cytosol and maintains redox conditions that enable a distinct set of folding catalysts to facilitate the formation and isomerization of disulfide bonds [60]. The process of disulfide-linked protein folding is slow, due to its dependence on a redox reaction, which requires an electron acceptor. During this folding process a protein may be oxidized to form disulfide bonds, reduced to allow isomerization of nonnative disulfide bonds or reduced to allow unfolding and subsequent degradation. All these considerations have led to the notion that disulfide bond formation is an assisted process in vivo. This was corroborated by the discovery that dsbA mutants in *Escherichia coli* display compromised disulfide bond formation [61].

There is accumulating evidence to suggest that protein folding and production of reactive oxygen species are closely linked events; however, this area of ER stress is not well explored. In eukaryotes, oxidative protein folding occurs in the ER (Fig. 3). There is a growing family of ER oxidoreductases that are supposed to be responsible for catalyzing these protein-folding reactions in mammalian cells, including PDI (protein disulfide isomerase), ERp57, ERp72, PDIR, PDIp, and P5. When disulfide bond formation occurs, cysteine residues within the active site are capable of accepting two electrons from the polypeptide chain substrate. This electron transfer results in oxidation of the substrate and reduction of the PDI active site. Despite the ability of PDI to enhance the rate of disulfide-linked folding, the mechanisms by which the ER disposes of electrons as a result of the oxidative disulfide formation reaction have remained enigmatic. Although a number of factors have been proposed to contribute to maintaining the oxidized environment of the ER, including the preferential secretion of reduced thiols and uptake of oxidized thiols, as well as a variety of different redox enzymes and small-molecule oxidants, the precise physiological relevance to oxidative folding has been absent, due to a lack of genetic evidence [62,63]. It was believed for many years that low-molecular-mass thiol glutathione is responsible for oxidizing the PDI active sites. This was contrary to observations in yeast, where depletion of glutathione did not lead to a lack of disulfide bond formation [64,65].

Extensive genetic and biochemical studies using the yeast *cerevisiae* and studies in plants and mammals have provided a detailed insight into the mechanisms underlying oxidative protein folding. The conserved ER-resident protein Ero1p plays a role in oxidative folding similar to that of the bacterial periplasmic protein DsbB. The proteins Ero1p and DsbB specifically oxidize a thioredoxin-like protein (PDI in eukaryotes, DsbA in bacteria) that serves further as an intermediary. In both prokaryotes and eukaryotes, molecular oxygen serves as the terminal electron acceptor for disulfide bond formation. Ero1p uses a flavin-dependent reaction to pass electrons directly to molecular

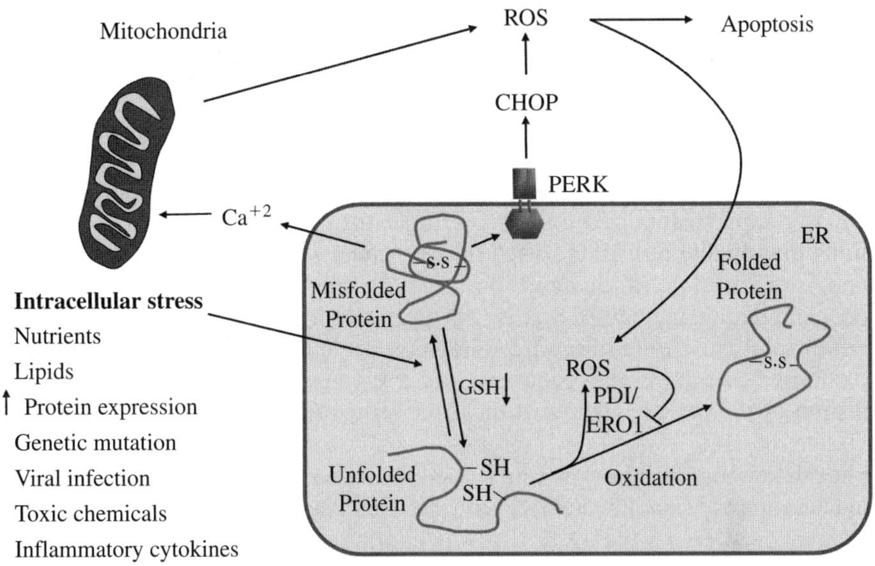

FIG. 3 Relationship between protein misfolding and oxidative stress. Protein misfolding and oxidative stress create a vicious cycle leading to ER stress and cell death. ROS are generated by exposure to multiple stresses and also as a by-product of mitochondrial respiration. Protein misfolding may cause ROS through changes in oxidative phosphorylation as a consequence of energy depletion or Ca^{2+} release from the ER. In addition, GSH can be consumed to reduce improperly paired disulfide bonds within misfolded proteins. ROS production can interfere with protein folding by inactivating PDI/ERO1 thiol–disulfide exchange reactions and/or by causing aberrant disulfide bond formation. ER stress activates CHOP expression that attenuates antioxidative stress responses. In this model, ROS can cause protein misfolding, UPR activation, and CHOP induction.

oxygen, thereby generating reactive oxygen species (ROS) that could potentially contribute to additional cellular oxidative stress. This suggests that the activity of Ero1 p must be under the tight control of the folding load in the ER [66]. There are two ERO1 isoforms in humans, hERO1-Lα and hERO1-Lβ [67,68], which differ in their tissue distribution and transcriptional regulation. Only ERO1L-β is induced by the UPR [68]; ERO1-Lα is induced during hypoxia [69]. Further studies in yeast and later in eukaryotes have identified a critical role of the flavin adenine dinucleotide (FAD) in oxidative protein folding and the discovery that purified Ero1p itself is a novel FAD-binding protein [70]. The dependency of oxidative folding on FAD defines a novel role for this versatile redox molecule in the ER lumen.

ER STRESS AND OXIDATIVE STRESS: IMPLICATIONS IN CLINICAL DISEASE

Eukaryotic cells have evolved the UPR as a homeostatic mechanism to balance the load of newly synthesized proteins with the inherent folding capacity of the ER. Increasing evidence now convincingly suggests, and it is becoming apparent, that dysfunction or dysregulation of the UPR plays an important role in various disease states, including diabetes mellitus, atherosclerosis, neoplasia, and neurodegenerative diseases.

ER Stress Contributes to Metabolic Disease

The development of type 2 diabetes is associated with a combination of insulin resistance in fat, muscle, and liver and a failure of pancreatic β-cells to compensate adequately to increase insulin production [71,72]. Insulin signaling is very sensitive to alterations in ER homeostasis and redox status. ER stress and oxidative stress, as well as inflammatory cytokines and free fatty acids, inhibit insulin signaling through activation of the protein kinase JNK. JNK phosphorylation of IRS-1 on Ser307 reduces insulin receptor–stimulated Tyr phosphorylation and insulin signaling [73,74]. Induction of ER stress may suppress insulin receptor signaling via IRE1α-dependent activation of the JNK pathway. Indeed, suppressing the JNK pathway can ameliorate insulin resistance [75], possibly by counterbalancing the deleterious effects of ER stress, oxidative stress, free fatty acids, and pro-inflammatory cytokines. The role of ER stress in insulin signaling was also suggested by the finding that ectopic expression of the molecular chaperone ORP150/GRP170 in hepatocytes improved insulin sensitivity [76]. It is possible that elevated levels of ORP150 expression improve the protein-folding capacity of the ER and reduce UPR signaling. The ability of ER stress signaling to cause insulin resistance was also suggested by recent observations showing that treatment of mice chemical chaperones that can improve protein folding in the ER and reduce ER stress and UPR signaling can increase insulin sensitivity [34]. Alternatively, there are observations which suggest that ER stress and UPR signaling through IRE1 can actually improve insulin sensitivity. Compared to control mice, when fed a high-fat diet, heterozygous *Xbp1*$^{+/-}$ mice developed insulin resistance. [77]. It is possible that reduced XBP1 signaling impairs the ER protein-folding capacity, thereby activating the UPR, which may lead to JNK activation. Therefore, ER stress signaling through IRE1-mediated XBP1 mRNA splicing may increase the ER protein-folding capacity to improve insulin signaling, whereas IRE1-mediated JNK activation could cause insulin resistance. The sum of these observations indicates that there exists a link between insulin resistance and ER stress, although the precise relationship and mechanism(s) remain to be elucidated.

A unique requirement for UPR signaling in β-cell function was first suggested by the identification of *PERK* as the gene defective in the human disease Wolcott–Rallison syndrome (WRS) [78]. Persons with WRS and *Perk*$^{-/-}$ mice develop β–cell apoptosis with early-onset insulin-dependent diabetes [9]. In addition, mice with homozygous Ser51Ala mutation at the PERK phosphorylation site in eIF2α display even greater β-cell loss that appears in utero [12]. Finally, although mice with heterozygous Ser51Ala mutation in eIF2α do not display a detectable phenotype, upon being fed a high-fat diet, they develop insulin resistance and a failure in the β-cells to produce insulin, typical of type 2 diabetes. The insulin secretion defect in high-fat-fed heterozygous Ser51Ala eIF2α mutant mice was due to an increased rate of glucose-stimulated proinsulin translation, which overwhelmed the protein-folding machinery of the ER and led to (1) a distended ER compartment, (2) prolonged association of proinsulin with the ER chaperone BiP, (3) reduced processing of proinsulin to insulin, and (4) reduced granule biogenesis [79]. Thus, regulation of translation initiation through eIF2α phosphorylation is required for ER stress signaling to prevent β-cell dysfunction when the demand for insulin is increased due to a high-fat diet and insulin resistance. These findings indicate that β-cells display a unique requirement for PERK/eIF2α-regulated translation.

Several mechanisms may explain why β-cells uniquely require the PERK/eIF2α pathway. First, β-cells may require PERK/eIF2α signaling because they are sensitive to physiological fluctuations in blood glucose. In β-cells, the generation of ATP fluctuates with blood glucose concentrations because glycolysis is controlled by glucokinase, which has a low affinity for glucose. Periodic decreases in blood glucose levels reduce the ATP/ADP ratio and would compromise protein folding in the ER so that UPR may frequently be activated. Through this mechanism, PERK/eIF2α signaling would be required in β-cells to couple protein synthesis with energy available for protein-folding reactions in the ER lumen. Alternatively, glucose stimulates insulin transcription, translation and secretion. PERK phosphorylation of eIF2α may be required for β cells to attenuate protein synthesis so that insulin production does not exceed the protein-folding capacity of the ER. Results from the high-fat-fed heterozygous Ser51Ala eIF2α mutant mice would support this hypothesis [79]. Finally, as the PERK/eIF2α pathway is known to reduce oxidative stress, it is possible that β-cells require PERK/eIF2α to minimize oxidative stress [80]. Two mechanisms have been proposed to account for the role of the PERK/eIF2α in limiting oxidative stress. First, the PERK/eIF2α pathway can prevent oxidative stress through inhibition of translation initiation when protein folding in the ER lumen is disturbed [81]. Alternatively, the PERK/eIF2α/ATF4 pathway induces expression of antioxidative stress response genes [9,81]. It is likely that both mechanisms contribute to the protective role for PERK/eIF2α in limiting ROS accumulation. There is increasing evidence which suggests that oxidative stress contributes to the β-cell failure in diabetes [82,83]. Beta cells express low levels of catalase and glutathione peroxidase, two enzymes that protect from ROS [84]. Therefore, oxidative stress would preferentially perturb β-cell function, due to their reduced capacity to neutralize ROS. Further studies are required to elucidate why the PERK/eIF2α pathway is essential for β-cell function and survival.

ER Stress Contributes to Neurodegenerative Disease

Neurodegenerative diseases such as Alzheimer disease (AD) and Parkinson disease (PD) represent a large class of conformational diseases associated with accumulation of abnormal protein aggregates in and around affected neurons.

Oxidative stress and protein misfolding play critical roles in the pathogenesis of these neurodegenerative diseases [85], which are characterized by fibrillar aggregates composed of misfolded proteins [86]. At the cellular level, neuronal death or apoptosis may be mediated by oxidative stress and/or ER stress. Up-regulation of ER stress markers has been demonstrated in postmortem brain tissues and cell culture models of many neurodegenerative disorders, including PD, AD, amyotropic lateral sclerosis (ALS), and expanded polyglutamine diseases such as Huntington disease and spinocerebellar ataxias [87]. Recent studies indicate that oligomeric forms of polypeptides predisposed to β-sheet polymerization and fibril formation may be the toxic forms that cause neuronal death. The impact of these oligomeric, potentially toxic species on ER function and the generation of ROS is presently not understood.

In vitro studies suggest these aggregates can inhibit the proteasome and ERAD. For example, in Machado–Joseph syndrome, the polyglutamine repeats present in spinocerebrocellular atrophy protein (SCA3) form cytosolic aggregates that can inhibit the proteasome. Proteasome inhibition in the cytosol can interfere with ERAD to elicit UPR activation, caspase 12 activation, and apoptosis [88,89]. Deletion of the ER stress-induced pro-apoptotic transcription factor CHOP preserved neuronal function, suggesting the importance of UPR signaling in this model.

PD is the second most common neurodegenerative disease and is characterized by a loss of dopaminergic neurons. Analyses of familial PD revealed involvement of three genes encoding α-synuclein, parkin, and ubiquitin C-terminal esterase L1 (UCH-L1). α-Synuclein is a cytoplasmic protein that forms aggregates, called *Lewy bodies*, which are characteristic of PD. Athough the link between α-synuclein and ER stress is unclear, parkin is a ubiquitin-protein ligase (E3) involved in ERAD [90]. One of the substrates of ERAD ubiquitinated by parkin is the Pael receptor, a homolog of endothelin receptor type B [91]. Interestingly, expression of parkin is induced by ER stress, and neuronal cells overexpressing parkin are resistant to ER stress [92]. UCH-L1 is an abundant protein in neurons and stabilizes a monomeric ubiquitin to ubiquitinate unfolded proteins and might be involved in ERAD [93–95]. These findings strongly suggest the involvement of ER stress in PD. In addition, there are several reports supporting the link between ER stress and PD. First, PD mimetics, such as 6-hydroxydopamine, specifically induce ER stress in neuronal cells [96]. Second, expression of ER chaperones such as PDI is up-regulated in the brain of PD patients, and PDI is accumulated in Lewy bodies [97]. The identification of PDI family member PDIp in experimental Parkinson disease and Lewy bodies suggest that oxidative protein folding in the ER may be perturbed in PD.

In humans, mutations in SIL1, which encodes an adenine nucleotide exchange factor for BiP, cause Marinesco–Sjögren syndrome, a rare disease associated with cerebellar ataxia, progressive myopathy, and cataracts [98]. In mice homozygous for a spontaneously occurring mutation in the *Sil1* transcript, cerebellar Purkinje cell degeneration and subsequent ataxia occur [99].

Analysis of *Sil1* mutant mice demonstrated that affected Purkinje cells have ubiquitinated nuclear- and ER-associated protein aggregates and display up-regulation of several ER stress markers: BiP, CHOP, and ORP150 [99]. It seems likely that the protein chaperone BiP uses ATP/ADP exchange, which is essential to preserve ER function and prevent activation of the UPR. Reduced efficiency of ATP-dependent BiP-mediated chaperone function may predispose to unfolded protein accumulation in the ER, activate the UPR, and contribute to Purkinje cell degeneration.

Oxidative stress is implicated in the pathogenesis of neurodegenerative diseases. A group of neurodegenerative diseases, including Alzheimer disease, is characterized pathologically by the deposition of intracellular aggregates containing abnormally phosphorylated forms of the microtubule-binding protein tau [100]. Using a *Drosophila* model relevant to human neurodegenerative diseases, including Alzheimer disease, it was demonstrated that oxidative stress plays a casual role in neurotoxicity and promotes tau phosphorylation. In this model, activation of the JNK pathway correlated with the degree of tau-induced neurodegeneration [101]. Although oxidative stress and ER stress have been linked to neurodegenerative diseases, at this point it is not possible to conclude that these processes are the primary cause of neuron death. However, it is possible that these stresses modify the progression and severity of these complex diseases.

Nitric oxide (NO) is a second messenger for signaling pathways that regulate a variety of physiological processes. In the brain, NO is implicated in neurotransmission, neuromodulation, and synaptic plasticity. However, excessive generation of NO and NO-derived reactive nitrogen species are implicated in the pathogenesis of neurodegenerative disorders, including AD and PD [102]. Studies now indicate that ER stress and apoptosis are critical features underlying these disorders [103]. Uehara and co-workers [104] elegantly demonstrated that NO-mediated S-nitrosylation of protein disulfide isomerase (PDI) inhibits PDI function, leads to dysregulated protein folding within the ER, elicits ER stress, and initiates neuronal cell death. A causal role for this sequence of events in neurodegenerative disease was supported by the demonstration that PDI is S-nitrosylated in the brains of patients suffering from PD or AD, but not in normal regular type brains. Thus, these findings provide additional evidence of a role for dysregulated protein S-nitrosylation (oxidative stress) in neurodegenerative disease and indicate that ER dysfunction may serve as a critical common factor that couples NO-induced cellular stress to neurodegeneration.

ER Stress Contributes to Hyperhomocysteinemia and Atherosclerosis

Elevated plasma levels of homocysteine (Hcy), a sulfur-containing amino acid, are linked to the development of ischemic heart disease; stroke and peripheral vascular disease are associated with elevated plasma levels of homocysteine (Hcy). However, it is not known whether Hcy is a primary cause of atherosclerosis and thromobosis. Hcy may mediate vascular toxicity through dysregulation of cholesterol and triglyceride biosynthesis. Hyperhomocysteinemia activates lipogenic signaling via the sterol-regulated element-binding proteins (SREBPs), leading to intracellular accumulation of cholesterol [105]. Surprisingly, ER stress appears to contribute to the activation of SREBPs by homocysteine. Livers of homocysteine-fed mice contain elevated levels of ER chaperones. In addition, overexpression of BiP prevented SREBP induction in response to homocysteine [105].

Under normal circumstances, Hcy is converted to cysteine and partly remethylated to methionine by vitamin B_{12} and folate. When normal metabolism is disturbed due to a deficiency of cystathionine-β synthase (CBS), which requires vitamin B_6 for activation, Hcy accumulates in blood and results in severe hyperhomocysteinemia. CBS condenses homocysteine and serine to form cystathionine. The harmful effects of hyperhomocysteinemia may be mediated through several processes. First, a decrease in cysteine may cause disease, due to reduced synthesis of glutathione (antioxidant). Thrombotic and cardiovascular diseases may also be encountered. Second, ROS generated during oxidation of Hcy to homocystine and disulfides may oxidize membrane lipids and proteins. Third, Hcy can react with thiols within proteins and form disulfides (thiolation) to interfere with protein folding, structure, and function. Finally, Hcy can be converted to highly reactive thiolactone, which can react with proteins forming -NH-CO- adducts, thus affecting protein structure and function. In cultured vascular endothelial cells, Hcy induces protein misfolding in the ER by interfering with disulfide bond formation [106] and activates the UPR to induce expression of several ER stress response proteins, such as BiP, GRP94, CHOP, and HERP [107–110]. Hcy can also trigger apoptosis by a signaling pathway that requires intact IRE1 [109]. These studies support the notion that Hcy can disrupt ER homoeostasis to cause UPR induction [107–109]. This is consistent with the activation of UPR markers observed in the livers of normal or *Cbs*$^{+/-}$ mice in response to hyperhomocysteinemia [111].

Atherosclerosis is caused by the abnormal deposition of cholesterol in the coronary arteries. Cholesterol accumulation in macrophages plays a critical role in the progression of atherosclerosis. Macrophages have multiple mechanisms to prevent excess cholesterol accumulation, including an increase in cholesterol esterification, induction of cellular cholesterol efflux, and the repression of lipoprotein receptor and cholesterol biosynthetic enzymes [112,113]. Upon formation of an initial atherosclerotic lesion, these mechanisms are dysregulated, thereby leading to the characteristic appearance of foam cells within the vessel intima. The macrophage-derived foam cells take up oxidized lipoprotein particles and become laden with cholesterol. The cholesterol is stored as esters within large lipid vesicles, producing a foamy appearance—hence their name. Overload of cholesterol in macrophages elicits apoptosis. Excess cholesterol must accumulate in specific pools within the cell to elicit cytotoxicity. Intracellular cholesterol is known to traffic to the plasma membrane, mitochondria, and the ER. Although the ER membrane has low levels of free cholesterol, it is particularly sensitive to cholesterol loading. Recent findings suggest that free cholesterol requires trafficking to the ER to produce its toxic effects [114]. This trafficking results in activation of UPR signaling and caspase activation, and ultimately in macrophage cell death/apoptosis. Macrophages from *Perk*$^{-/-}$ mice are hypersensitive to cholesterol-induced cell death, whereas macrophages from *Chop*$^{-/-}$ mice are highly protected.

Recent findings also suggest that defective insulin signaling and reduced AKT activity impair the ability of macrophages to deal with ER stress-induced apoptosis within atherosclerotic plaques [115]. This mechanism may contribute to the association between insulin resistance in metabolic syndrome and atherogenesis [116]. These findings suggest that the UPR plays an important role in progression of the atherogenic disease process [114]. In addition, free cholesterol loading of macrophages increases levels of cell-surface Fas ligand, activates proapoptotic Bax protein, and increases mitochondrial-dependent apoptosis. Although both the Fas death pathway and the mitochondrial cell death pathway may contribute to macrophage apoptosis, there is accumulating evidence suggesting that depletion of calcium stores in the ER and subsequent activation of the UPR is the dominant driving force in cholesterol-induced macrophage death. Finally, ER stress caused by free cholesterol loading in macrophages promotes chemokine secretion, and this may contribute to the formation of vulnerable atherosclerotic lesions. These lesions lead to an inflammatory condition with further infiltration of macrophages and lymphocytes from the blood and subsequent release of hydrolytic enzymes, cytokines, chemokines, and growth factors that can inflict more damage and eventually lead to focal necrosis [117,118].

FUTURE PERSPECTIVES

Tremendous progress has been made in clarifying the mechanisms underlying the cause of ER stress and cellular adaptive responses. Future studies will be required to understand the physiological significance of ER stress and UPR signaling in disease pathogenesis. The relationships between ER stress and apoptosis also remain to be defined. Further studies will also be required to elucidate how ER stress and UPR signaling are integrated with other stress signaling pathways, particularly those related to oxidative stress. A greater understanding of the complex interrelationship between protein misfolding and oxidative stress may lead to the development of general pharmacological agents, such as chemical chaperones to improve protein folding and/or antioxidants to reduce oxidative stress, for the treatment of human disease. A coherent mechanistic understanding of the mechanisms and pathways that signal ER stress responses should contribute to the development of more selective and specific-acting therapeutic agents targeted for diseases associated with ER pathologies.

Acknowledgments

This work was supported by Natural Institutes of Health grants DK42394, HL57346, and HL52173 (R.J. Kaufman). R.J.K. is an investigator at the Howard Hughes Medical Institute.

REFERENCES

1. Kaufman, R.J. (2002). Orchestrating the unfolded protein response in health and disease. *J Clin Invest, 110*, 1389–1398.
2. Patil, C., Walter, P. (2001). Intracellular signaling from the endoplasmic reticulum to the nucleus: the unfolded protein response in yeast and mammals. *Curr Opin Cell Biol, 13*, 349–355.
3. Mori, K., Ma, W., Gething, M.J., Sambrook, J. (1993). A transmembrane protein with a cdc2+/CDC28-related kinase activity is required for signaling from the ER to the nucleus. *Cell, 74*, 743–756.
4. Harding, H.P., Calfon, M., Urano, F., Novoa, I., Ron, D. (2002). Transcriptional and translational control in the mammalian unfolded protein response. *Annu Rev Cell Dev Biol, 18*, 575–599.
5. Schroder, M., Kaufman, R.J. (2005). The mammalian unfolded protein response. *Annu Rev Biochem, 74*, 739–789.
6. Wu, J., Kaufman, R.J. (2006). From acute ER stress to physiological roles of the unfolded protein response. *Cell Death Differ, 13*, 374–384.
7. Sidrauski, C., Walter, P. (1997). The transmembrane kinase Ire1p is a site-specific endonuclease that initiates mRNA splicing in the unfolded protein response. *Cell, 90*, 1031–1039.
8. Yoshida, H., Matsui, T., Yamamoto, A., Okada, T., Mori, K. (2001). XBP1 mRNA is induced by ATF6 and spliced by IRE1 in response to ER stress to produce a highly active transcription factor. *Cell, 107*, 881–891.
9. Harding, H.P., Zhang, Y., Bertolotti, A., Zeng, H., Ron, D. (2000). Perk is essential for translational regulation and cell survival during the unfolded protein response. *Mol Cell, 5*, 897–904.
10. Yoshida, H., Okada, T., Haze, K., Yanagi, H., Yura, T., Negishi, M., Mori, K. (2000). ATF6 activated by proteolysis binds in the presence of NF-Y (CBF) directly to the *cis*-acting element responsible for the mammalian unfolded protein response. *Mol Cell Biol, 20*, 6755–6767.
11. Kaufman, R.J. (2004). Regulation of mRNA translation by protein folding in the endoplasmic reticulum. *Trends Biochem Sci, 29*, 152–158.
12. Scheuner, D., Song, B., McEwen, E., Liu, C., Laybutt, R., Gillespie, P., Saunders, T., Bonner-Weir, S., Kaufman, R.J. (2001). Translational control is required for the unfolded protein response and in vivo glucose homeostasis. *Mol Cell, 7*, 1165–1176.
13. Harding, H.P., Zhang, Y., Zeng, H., Novoa, I., Lu, P.D., Calfon, M., Sadri, N., Yun, C., Popko, B., Paules, R., *et al.* (2003). An integrated stress response regulates amino acid metabolism and resistance to oxidative stress. *Mol Cell, 11*, 619–633.

14. Harding, H.P., Zhang, Y., Ron, D. (1999). Protein translation and folding are coupled by an endoplasmic-reticulum-resident kinase. *Nature, 397*, 271–274.

15. Harding, H.P., Novoa, I., Zhang, Y., Zeng, H., Wek, R., Schapira, M., Ron, D. (2000). Regulated translation initiation controls stress-induced gene expression in mammalian cells. *Mol Cell, 6*, 1099–1108.

16. Ron, D. (2002). Translational control in the endoplasmic reticulum stress response. *J Clin Invest, 110*, 1383–1388.

17. Ma, Y., Brewer, J.W., Diehl, J.A., Hendershot, L.M. (2002). Two distinct stress signaling pathways converge upon the CHOP promoter during the mammalian unfolded protein response. *J Mol Biol, 318*, 1351–1365.

18. Cullinan, S.B., Zhang, D., Hannink, M., Arvisais, E., Kaufman, R.J., Diehl, J.A. (2003). Nrf2 is a direct PERK substrate and effector of PERK-dependent cell survival. *Mol Cell Biol, 23*, 7198–7209.

19. Nguyen, T., Sherratt, P.J., Pickett, C.B. (2003). Regulatory mechanisms controlling gene expression mediated by the antioxidant response element. *Annu Rev Pharmacol Toxicol, 43*, 233–260.

20. Wang, W., Chan, J.Y. (2006). Nrf1 is targeted to the endoplasmic reticulum membrane by an N-terminal transmembrane domain. Inhibition of nuclear translocation and transacting function. *J Biol Chem, 281*, 19676–19687.

21. Cox, J.S., Walter, P. (1996). A novel mechanism for regulating activity of a transcription factor that controls the unfolded protein response. *Cell, 87*, 391–404.

22. Mori, K., Ogawa, N., Kawahara, T., Yanagi, H., Yura, T. (2000). mRNA splicing-mediated C-terminal replacement of transcription factor Hac1p is required for efficient activation of the unfolded protein response. *Proc Natl Acad Sci U S A, 97*, 4660–4665.

23. Travers, K.J., Patil, C.K., Wodicka, L., Lockhart, D.J., Weissman, J.S., Walter, P. (2000). Functional and genomic analyses reveal an essential coordination between the unfolded protein response and ER-associated degradation. *Cell, 101*, 249–258.

24. Tirasophon, W., Welihinda, A.A., Kaufman, R.J. (1998). A stress response pathway from the endoplasmic reticulum to the nucleus requires a novel bifunctional protein kinase/endoribonuclease (Ire1p) in mammalian cells. *Genes Dev, 12*, 1812–1824.

25. Wang, X.Z., Harding, H.P., Zhang, Y., Jolicoeur, E.M., Kuroda, M., Ron, D. (1998). Cloning of mammalian Ire1 reveals diversity in the ER stress responses. *EMBO J, 17*, 5708–5717.

26. Niwa, M., Sidrauski, C., Kaufman, R.J., Walter, P. (1999). A role for presenilin-1 in nuclear accumulation of Ire1 fragments and induction of the mammalian unfolded protein response. *Cell, 99*, 691–702.

27. Yoshida, H., Haze, K., Yanagi, H., Yura, T., Mori, K. (1998). Identification of the *cis*-acting endoplasmic reticulum stress response element responsible for transcriptional induction of mammalian glucose-regulated proteins: involvement of basic leucine zipper transcription factors. *J Biol Chem, 273*, 33741–33749.

28. Calfon, M., Zeng, H., Urano, F., Till, J.H., Hubbard, S.R., Harding, H.P., Clark, S.G., Ron, D. (2002). IRE1 couples endoplasmic reticulum load to secretory capacity by processing the XBP-1 mRNA. *Nature, 415*, 92–96.

29. Lee, K., Tirasophon, W., Shen, X., Michalak, M., Prywes, R., Okada, T., Yoshida, H., Mori, K., Kaufman, R.J. (2002). IRE1-mediated unconventional mRNA

splicing and S2P-mediated ATF6 cleavage merge to regulate XBP1 in signaling the unfolded protein response. *Genes Dev*, *16*, 452–466.

30. Shen, X., Ellis, R.E., Lee, K., Liu, C.Y., Yang, K., Solomon, A., Yoshida, H., Morimoto, R., Kurnit, D.M., Mori, K., Kaufman, R.J. (2001). Complementary signaling pathways regulate the unfolded protein response and are required for *C. elegans* development. *Cell, 107*, 893–903.
31. Yoshida, H., Matsui, T., Hosokawa, N., Kaufman, R.J., Nagata, K., Mori, K. (2003). A time-dependent phase shift in the mammalian unfolded protein response. *Dev Cell, 4*, 265–271.
32. Reimold, A.M., Etkin, A., Clauss, I., Perkins, A., Friend, D.S., Zhang, J., Horton, H.F., Scott, A., Orkin, S.H., Byrne, M.C., *et al.* (2000). An essential role in liver development for transcription factor XBP-1. *Genes Dev, 14*, 152–157.
33. Bertolotti, A., Wang, X., Novoa, I., Jungreis, R., Schlessinger, K., Cho, J.H., West, A.B., Ron, D. (2001). Increased sensitivity to dextran sodium sulfate colitis in IRE1beta-deficient mice. *J Clin Invest, 107*, 585–593.
34. Ozcan, U., Cao, Q., Yilmaz, E., Lee, A.H., Iwakoshi, N.N., Ozdelen, E., Tuncman, G., Gorgun, C., Glimcher, L.H., Hotamisligil, G.S. (2004). Endoplasmic reticulum stress links obesity, insulin action, and type 2 diabetes. *Science, 306*, 457–461.
35. Reimold, A.M., Iwakoshi, N.N., Manis, J., Vallabhajosyula, P., Szomolanyi-Tsuda, E., Gravallese, E.M., Friend, D., Grusby, M.J., Alt, F., Glimcher, L.H. (2001). Plasma cell differentiation requires the transcription factor XBP-1. *Nature, 412*, 300–307.
36. Iwakoshi, N.N., Lee, A.H., Vallabhajosyula, P., Otipoby, K.L., Rajewsky, K., Glimcher, L.H. (2003). Plasma cell differentiation and the unfolded protein response intersect at the transcription factor XBP-1. *Nat Immunol, 4*, 321–329.
37. Zhang, K., Wong, H.N., Song, B., Miller, C.N., Scheuner, D., Kaufman, R.J. (2005). The unfolded protein response sensor IRE1alpha is required at 2 distinct steps in B cell lymphopoiesis. *J Clin Invest, 115*, 268–281.
38. Lee, A.H., Chu, G.C., Iwakoshi, N.N., Glimcher, L.H. (2005). XBP-1 is required for biogenesis of cellular secretory machinery of exocrine glands. *EMBO J, 24*, 4368–4380.
39. Haze, K., Yoshida, H., Yanagi, H., Yura, T., Mori, K. (1999). Mammalian transcription factor ATF6 is synthesized as a transmembrane protein and activated by proteolysis in response to endoplasmic reticulum stress. *Mol Biol Cell, 10*, 3787–3799.
40. Yoshida, H., Okada, T., Haze, K., Yanagi, H., Yura, T., Negishi, M., Mori, K. (2001). Endoplasmic reticulum stress-induced formation of transcription factor complex ERSF including NF-Y (CBF) and activating transcription factors 6alpha and 6beta that activates the mammalian unfolded protein response. *Mol Cell Biol, 21*, 1239–1248.
41. Li, M., Baumeister, P., Roy, B., Phan, T., Foti, D., Luo, S., Lee, A.S. (2000). ATF6 as a transcription activator of the endoplasmic reticulum stress element: thapsigargin stress-induced changes and synergistic interactions with NF-Y and YY1. *Mol Cell Biol, 20*, 5096–5106.
42. Ye, J., Rawson, R.B., Komuro, R., Chen, X., Dave, U.P., Prywes, R., Brown, M.S., Goldstein, J.L. (2000). ER stress induces cleavage of membrane-bound ATF6 by the same proteases that process SREBPs. *Mol Cell, 6*, 1355–1364.

43. Goldstein, J.L., DeBose-Boyd, R.A., Brown, M.S. (2006). Protein sensors for membrane sterols. *Cell, 124*, 35–46.
44. Wu, J., Rutkowski, D.T., Dubois, M., Swathirajan, J., Saunders, T., Wang, J., Song, B., Yau, G.D., Kaufman, R.J. (2007). ATF6alpha optimizes long-term endoplasmic reticulum function to protect cells from chronic stress. *Dev Cell, 13*, 351–364.
45. Yamamoto, K., Sato, T., Matsui, T., Sato, M., Okada, T., Yoshida, H., Harada, A., Mori, K. (2007). Transcriptional induction of mammalian ER quality control proteins is mediated by single or combined action of ATF6alpha and XBP1. *Dev Cell, 13*, 365–376.
46. Zhang, K., Shen, X., Wu, J., Sakaki, K., Saunders, T., Rutkowski, D.T., Back, S.H., Kaufman, R.J. (2006). Endoplasmic reticulum stress activates cleavage of CREBH to induce a systemic inflammatory response. *Cell, 124*, 587–599.
47. Kondo, S., Murakami, T., Tatsumi, K., Ogata, M., Kanemoto, S., Otori, K., Iseki, K., Wanaka, A., Imaizumi, K. (2005). OASIS, a CREB/ATF-family member, modulates UPR signalling in astrocytes. *Nat Cell Biol, 7*, 186–194.
48. Kondo, S., Saito, A., Hino, S., Murakami, T., Ogata, M., Kanemoto, S., Nara, S., Yamashita, A., Yoshinaga, K., Hara, H., Imaizumi, K. (2007). BBF2H7, a novel transmembrane bZIP transcription factor, is a new type of endoplasmic reticulum stress transducer. *Mol Cell Biol, 27*, 1716–1729.
49. Nagamori, I., Yabuta, N., Fujii, T., Tanaka, H., Yomogida, K., Nishimune, Y., Nojima, H. (2005). Tisp40, a spermatid specific bZip transcription factor, functions by binding to the unfolded protein response element via the Rip pathway. *Genes Cells, 10*, 575–594.
50. Liang, G., Audas, T.E., Li, Y., Cockram, G.P., Dean, J.D., Martyn, A.C., Kokame, K., Lu, R. (2006). Luman/CREB3 induces transcription of the endoplasmic reticulum (ER) stress response protein Herp through an ER stress response element. *Mol Cell Biol, 26*, 7999–8010.
51. Raggo, C., Rapin, N., Stirling, J., Gobeil, P., Smith-Windsor, E., O'Hare, P., Misra, V. (2002). Luman, the cellular counterpart of herpes simplex virus VP16, is processed by regulated intramembrane proteolysis. *Mol Cell Biol, 22*, 5639–5649.
52. Stirling, J., O'Hare, P. (2006). CREB4, a transmembrane bZip transcription factor and potential new substrate for regulation and cleavage by S1P. *Mol Biol Cell, 17*, 413–426.
53. Bertolotti, A., Zhang, Y., Hendershot, L.M., Harding, H.P., Ron, D. (2000). Dynamic interaction of BiP and ER stress transducers in the unfolded-protein response. *Nat Cell Biol, 2*, 326–332.
54. Morris, J.A., Dorner, A.J., Edwards, C.A., Hendershot, L.M., Kaufman, R.J. (1997). Immunoglobulin binding protein (BiP) function is required to protect cells from endoplasmic reticulum stress but is not required for the secretion of selective proteins. *J Biol Chem, 272*, 4327–4334.
55. Ng, D.T., Watowich, S.S., Lamb, R.A. (1992). Analysis in vivo of GRP78-BiP/substrate interactions and their role in induction of the GRP78-BiP gene. *Mol Biol Cell, 3*, 143–155.
56. Graham, K.S., Le, A., Sifers, R.N. (1990). Accumulation of the insoluble PiZ variant of human alpha 1-antitrypsin within the hepatic endoplasmic reticulum

does not elevate the steady-state level of grp78/BiP. *J Biol Chem, 265*, 20463–20468.

57. Shen, J., Snapp, E.L., Lippincott-Schwartz, J., Prywes, R. (2005). Stable binding of ATF6 to BiP in the endoplasmic reticulum stress response. *Mol Cell Biol, 25*, 921–932.
58. Credle, J.J., Finer-Moore, J.S., Papa, F.R., Stroud, R.M., Walter, P. (2005). On the mechanism of sensing unfolded protein in the endoplasmic reticulum. *Proc Natl Acad Sci U S A, 102*, 18773–18784.
59. Zhou, J., Liu, C.Y., Back, S.H., Clark, R.L., Peisach, D., Xu, Z., Kaufman, R.J. (2006). The crystal structure of human IRE1 luminal domain reveals a conserved dimerization interface required for activation of the unfolded protein response. *Proc Natl Acad Sci U S A, 103*, 14343–14348.
60. Hwang, C., Sinskey, A.J., Lodish, H.F. (1992). Oxidized redox state of glutathione in the endoplasmic reticulum. *Science, 257*, 1496–1502.
61. Bardwell, J.C., McGovern, K., Beckwith, J. (1991). Identification of a protein required for disulfide bond formation in vivo. *Cell, 67*, 581–589.
62. Carelli, S., Ceriotti, A., Cabibbo, A., Fassina, G., Ruvo, M., Sitia, R. (1997). Cysteine and glutathione secretion in response to protein disulfide bond formation in the ER. *Science, 277*, 1681–1684.
63. Frand, A.R., Kaiser, C.A. (2000). Two pairs of conserved cysteines are required for the oxidative activity of Ero1p in protein disulfide bond formation in the endoplasmic reticulum. *Mol Biol Cell, 11*, 2833–2843.
64. Jessop, C.E., Chakravarthi, S., Watkins, R.H., Bulleid, N.J. (2004). Oxidative protein folding in the mammalian endoplasmic reticulum. *Biochem Soc Trans, 32*, 655–658.
65. Cuozzo, J.W., Kaiser, C.A. (1999). Competition between glutathione and protein thiols for disulphide-bond formation. *Nat Cell Biol, 1*, 130–135.
66. Tu, B.P., Weissman, J.S. (2004). Oxidative protein folding in eukaryotes: mechanisms and consequences. *J Cell Biol, 164*, 341–346.
67. Cabibbo, A., Pagani, M., Fabbri, M., Rocchi, M., Farmery, M.R., Bulleid, N.J., Sitia, R. (2000). ERO1-L, a human protein that favors disulfide bond formation in the endoplasmic reticulum. *J Biol Chem, 275*, 4827–4833.
68. Pagani, M., Fabbri, M., Benedetti, C., Fassio, A., Pilati, S., Bulleid, N.J., Cabibbo, A., Sitia, R. (2000). Endoplasmic reticulum oxidoreductin 1-lbeta (ERO1-Lbeta), a human gene induced in the course of the unfolded protein response. *J Biol Chem, 275*, 23685–23692.
69. Gess, B., Hofbauer, K.H., Wenger, R.H., Lohaus, C., Meyer, H.E., Kurtz, A. (2003). The cellular oxygen tension regulates expression of the endoplasmic oxidoreductase ERO1-Lalpha. *Eur J Biochem, 270*, 2228–2235.
70. Tu, B.P., Ho-Schleyer, S.C., Travers, K.J., Weissman, J.S. (2000). Biochemical basis of oxidative protein folding in the endoplasmic reticulum. *Science, 290*, 1571–1574.
71. Saltiel, A.R., Kahn, C.R. (2001). Insulin signalling and the regulation of glucose and lipid metabolism. *Nature, 414*, 799–806.
72. Shulman, G.I. (2000). Cellular mechanisms of insulin resistance. *J Clin Invest, 106*, 171–176.

73. Kaneto, H., Nakatani, Y., Kawamori, D., Miyatsuka, T., Matsuoka, T.A., Matsuhisa, M., Yamasaki, Y. (2006). Role of oxidative stress, endoplasmic reticulum stress, and c-Jun N-terminal kinase in pancreatic beta-cell dysfunction and insulin resistance. *Int J Biochem Cell Biol, 38*, 782–793.

74. Kaneto, H., Matsuoka, T.A., Nakatani, Y., Kawamori, D., Miyatsuka, T., Matsuhisa, M., Yamasaki, Y. (2005). Oxidative stress, ER stress, and the JNK pathway in type 2 diabetes. *J Mol Med, 83*, 429–439.

75. Kaneto, H., Nakatani, Y., Miyatsuka, T., Kawamori, D., Matsuoka, T.A., Matsuhisa, M., Kajimoto, Y., Ichijo, H., Yamasaki, Y., Hori, M. (2004). Possible novel therapy for diabetes with cell-permeable JNK-inhibitory peptide. *Nat Med, 10*, 1128–1132.

76. Nakatani, Y., Kaneto, H., Kawamori, D., Yoshiuchi, K., Hatazaki, M., Matsuoka, T.A., Ozawa, K., Ogawa, S., Hori, M., Yamasaki, Y., Matsuhisa, M. (2005). Involvement of endoplasmic reticulum stress in insulin resistance and diabetes. *J Biol Chem, 280*, 847–851.

77. Ozcan, U., Yilmaz, E., Ozcan, L., Furuhashi, M., Vaillancourt, E., Smith, R.O., Gorgun, C.Z., Hotamisligil, G.S. (2006). Chemical chaperones reduce ER stress and restore glucose homeostasis in a mouse model of type 2 diabetes. *Science, 313*, 1137–1140.

78. Delepine, M., Nicolino, M., Barrett, T., Golamaully, M., Lathrop, G.M., Julier, C. (2000). EIF2AK3, encoding translation initiation factor 2-alpha kinase 3, is mutated in patients with Wolcott–Rallison syndrome. *Nat Genet, 25*, 406–409.

79. Scheuner, D., Patel, R., Wang, F., Lee, K., Kumar, K., Wu, J., Nilsson, A., Karin, M., Kaufman, R.J. (2006). Double-stranded RNA-dependent protein kinase phosphorylation of the alpha-subunit of eukaryotic translation initiation factor 2 mediates apoptosis. *J Biol Chem, 281*, 21458–21468.

80. Marciniak, S.J., Garcia-Bonilla, L., Hu, J., Harding, H.P., Ron, D. (2006). Activation-dependent substrate recruitment by the eukaryotic translation initiation factor 2 kinase PERK. *J Cell Biol, 172*, 201–209.

81. Marciniak, S.J., Yun, C.Y., Oyadomari, S., Novoa, I., Zhang, Y., Jungreis, R., Nagata, K., Harding, H.P., Ron, D. (2004). CHOP induces death by promoting protein synthesis and oxidation in the stressed endoplasmic reticulum. *Genes Dev, 18*, 3066–3077.

82. Robertson, R.P. (2006). Oxidative stress and impaired insulin secretion in type 2 diabetes. *Curr Opin Pharmacol, 6*, 615–619.

83. Kaneto, H., Katakami, N., Kawamori, D., Miyatsuka, T., Sakamoto, K., Matsuoka, T.A., Matsuhisa, M., Yamasaki, Y. (2007). Involvement of oxidative stress in the pathogenesis of diabetes. *Antioxid Redox Signal, 9*, 355–366.

84. Sigfrid, L.A., Cunningham, J.M., Beeharry, N., Hakan Borg, L.A., Rosales Hernandez, A.L., Carlsson, C., Bone, A.J., Green, I.C. (2004). Antioxidant enzyme activity and mRNA expression in the islets of Langerhans from the BB/S rat model of type 1 diabetes and an insulin-producing cell line. *J Mol Med, 82*, 325–335.

85. Forman, M.S., Lee, V.M., Trojanowski, J.Q. (2003). 'Unfolding' pathways in neurodegenerative disease. *Trends Neurosci, 26*, 407–410.

86. Selkoe, D.J. (2003). Folding proteins in fatal ways. *Nature, 426*, 900–904.

87. Lindholm, D., Wootz, H., Korhonen, L. (2006). ER stress and neurodegenerative diseases. *Cell Death Differ, 13*, 385–392.

88. Nishitoh, H., Matsuzawa, A., Tobiume, K., Saegusa, K., Takeda, K., Inoue, K., Hori, S., Kakizuka, A., Ichijo, H. (2002). ASK1 is essential for endoplasmic reticulum stress-induced neuronal cell death triggered by expanded polyglutamine repeats. *Genes Dev, 16*, 1345–1355.

89. Bence, N.F., Sampat, R.M., Kopito, R.R. (2001). Impairment of the ubiquitin–proteasome system by protein aggregation. *Science, 292*, 1552–1555.

90. Shimura, H., Hattori, N., Kubo, S., Mizuno, Y., Asakawa, S., Minoshima, S., Shimizu, N., Iwai, K., Chiba, T., Tanaka, K., Suzuki, T. (2000). Familial Parkinson disease gene product, parkin, is a ubiquitin–protein ligase. *Nat Genet, 25*, 302–305.

91. Imai, Y., Soda, M., Inoue, H., Hattori, N., Mizuno, Y., Takahashi, R. (2001). An unfolded putative transmembrane polypeptide, which can lead to endoplasmic reticulum stress, is a substrate of parkin. *Cell, 105*, 891–902.

92. Imai, Y., Soda, M., Takahashi, R. (2000). Parkin suppresses unfolded protein stress-induced cell death through its E3 ubiquitin–protein ligase activity. *J Biol Chem, 275*, 35661–35664.

93. Osaka, H., Wang, Y.L., Takada, K., Takizawa, S., Setsuie, R., Li, H., Sato, Y., Nishikawa, K., Sun, Y.J., Sakurai, M., *et al.* (2003). Ubiquitin carboxy-terminal hydrolase L1 binds to and stabilizes monoubiquitin in neuron. *Hum Mol Genet, 12*, 1945–1958.

94. Saigoh, K., Wang, Y.L., Suh, J.G., Yamanishi, T., Sakai, Y., Kiyosawa, H., Harada, T., Ichihara, N., Wakana, S., Kikuchi, T., Wada, K. (1999). Intragenic deletion in the gene encoding ubiquitin carboxy-terminal hydrolase in gad mice. *Nat Genet, 23*, 47–51.

95. Liu, Y., Fallon, L., Lashuel, H.A., Liu, Z., Lansbury, P.T., Jr. (2002). The UCH-L1 gene encodes two opposing enzymatic activities that affect alpha-synuclein degradation and Parkinson's disease susceptibility. *Cell, 111*, 209–218.

96. Holtz, W.A., O'Malley, K.L. (2003). Parkinsonian mimetics induce aspects of unfolded protein response in death of dopaminergic neurons. *J Biol Chem, 278*, 19367–19377.

97. Conn, K.J., Gao, W., McKee, A., Lan, M.S., Ullman, M.D., Eisenhauer, P.B., Fine, R.E., Wells, J.M. (2004). Identification of the protein disulfide isomerase family member PDIp in experimental Parkinson's disease and Lewy body pathology. *Brain Res, 1022*, 164–172.

98. Anttonen, A.K., Mahjneh, I., Hamalainen, R.H., Lagier-Tourenne, C., Kopra, O., Waris, L., Anttonen, M., Joensuu, T., Kalimo, H., Paetau, A., *et al.* (2005). The gene disrupted in Marinesco–Sjögren syndrome encodes SIL1, an HSPA5 cochaperone. *Nat Genet, 37*, 1309–1311.

99. Zhao, L., Longo-Guess, C., Harris, B.S., Lee, J.W., Ackerman, S.L. (2005). Protein accumulation and neurodegeneration in the woozy mutant mouse is caused by disruption of SIL1, a cochaperone of BiP. *Nat Genet, 37*, 974–979.

100. Lee, V.M., Goedert, M., Trojanowski, J.Q. (2001). Neurodegenerative tauopathies. *Annu Rev Neurosci, 24*, 1121–1159.

101. Dias-Santagata, D., Fulga, T.A., Duttaroy, A., Feany, M.B. (2007). Oxidative stress mediates tau-induced neurodegeneration in *Drosophila. J Clin Invest*, *117*, 236–245.

102. Calabrese, V., Bates, T.E., Stella, A.M. (2000). NO synthase and NO-dependent signal pathways in brain aging and neurodegenerative disorders: the role of oxidant/antioxidant balance. *Neurochem Res, 25*, 1315–1341.

103. Paschen, W., Mengesdorf, T. (2005). Endoplasmic reticulum stress response and neurodegeneration. *Cell Calcium*, *38*, 409–415.

104. Uehara, T., Nakamura, T., Yao, D., Shi, Z. Q., Gu, Z., Ma, Y., Masliah, E., Nomura, Y., Lipton, S.A. (2006). S-Nitrosylated protein–disulphide isomerase links protein misfolding to neurodegeneration. *Nature, 441*, 513–517.

105. Werstuck, G.H., Lentz, S.R., Dayal, S., Hossain, G.S., Sood, S.K., Shi, Y.Y., Zhou, J., Maeda, N., Krisans, S.K., Malinow, M.R., Austin, R.C. (2001). Homocysteine-induced endoplasmic reticulum stress causes dysregulation of the cholesterol and triglyceride biosynthetic pathways. *J Clin Invest*, *107*, 1263–1273.

106. Lentz, S.R., Sadler, J.E. (1993). Homocysteine inhibits von Willebrand factor processing and secretion by preventing transport from the endoplasmic reticulum. *Blood, 81*, 683–689.

107. Outinen, P.A., Sood, S.K., Pfeifer, S.I., Pamidi, S., Podor, T.J., Li, J., Weitz, J.I., Austin, R.C. (1999). Homocysteine-induced endoplasmic reticulum stress and growth arrest leads to specific changes in gene expression in human vascular endothelial cells. *Blood, 94*, 959–967.

108. Huang, R.F., Huang, S.M., Lin, B.S., Wei, J.S., Liu, T.Z. (2001). Homocysteine thiolactone induces apoptotic DNA damage mediated by increased intracellular hydrogen peroxide and caspase 3 activation in HL-60 cells. *Life Sci, 68*, 2799–2811.

109. Zhang, C., Cai, Y., Adachi, M.T., Oshiro, S., Aso, T., Kaufman, R.J., Kitajima, S. (2001). Homocysteine induces programmed cell death in human vascular endothelial cells through activation of the unfolded protein response. *J Biol Chem*, *276*, 35867–35874.

110. Austin, R.C., Lentz, S.R., Werstuck, G.H. (2004). Role of hyperhomocysteinemia in endothelial dysfunction and atherothrombotic disease. *Cell Death Differ*, *11*(Suppl 1), S56–S64.

111. Hofmann, M.A., Lalla, E., Lu, Y., Gleason, M.R., Wolf, B.M., Tanji, N., Ferran, L.J., Jr., Kohl, B., Rao, V., Kisiel, W., *et al.* (2001). Hyperhomocysteinemia enhances vascular inflammation and accelerates atherosclerosis in a murine model. *J Clin Invest, 107*, 675–683.

112. Tabas, I. (2002). Consequences of cellular cholesterol accumulation: basic concepts and physiological implications. *J Clin Invest, 110*, 905–911.

113. Tabas, I. (2002). Cholesterol in health and disease. *J Clin Invest, 110*, 583–590.

114. Feng, B., Yao, P.M., Li, Y., Devlin, C.M., Zhang, D., Harding, H.P., Sweeney, M., Rong, J.X., Kuriakose, G., Fisher, E.A., *et al.* (2003). The endoplasmic reticulum is the site of cholesterol-induced cytotoxicity in macrophages. *Nat Cell Biol, 5*, 781–792.

115. Han, S., Liang, C.P., DeVries-Seimon, T., Ranalletta, M., Welch, C.L., Collins-Fletcher, K., Accili, D., Tabas, I., Tall, A.R. (2006). Macrophage insulin receptor

deficiency increases ER stress-induced apoptosis and necrotic core formation in advanced atherosclerotic lesions. *Cell Metab, 3*, 257–266.

116. Pirro, M., Mauriege, P., Tchernof, A., Cantin, B., Dagenais, G.R., Despres, J.P., Lamarche, B. (2002). Plasma free fatty acid levels and the risk of ischemic heart disease in men: prospective results from the Quebec Cardiovascular Study. *Atherosclerosis, 160*, 377–384.
117. Li, Y., Schwabe, R.F., DeVries-Seimon, T., Yao, P.M., Gerbod-Giannone, M.C., Tall, A.R., Davis, R.J., Flavell, R., Brenner, D.A., Tabas, I. (2005). Free cholesterol-loaded macrophages are an abundant source of tumor necrosis factor-alpha and interleukin-6: model of NF-kappaB- and map kinase-dependent inflammation in advanced atherosclerosis. *J Biol Chem, 280*, 21763–21772.
118. Ross, R. (1999). Atherosclerosis: an inflammatory disease. *N Engl J Med, 340*, 115–126.

3

ROLE OF MOLECULAR CHAPERONES IN PROTEIN FOLDING

KAUSIK CHAKRABORTY, FLORIAN GEORGESCAULD, MANAJIT HAYER-HARTL, AND F. ULRICH HARTL

Department of Cellular Biochemistry, Max-Planck Institute of Biochemistry, Martinsried, Germany

INTRODUCTION

To become functionally active, newly synthesized proteins generally must fold into a unique three-dimensional structure. More than 40 years ago, Anfinsen and collaborators demonstrated by pioneering in vitro experiments with ribonuclease A that all the information necessary to ensure the correct folding of this protein is contained in its amino acid sequence [1]. Since then, similar results have been obtained for a large number of proteins of different structure, indicating the existence of a *folding code*, linked with the well-established genetic code that determines the amino acid sequence. However, despite important experimental and theoretical progress, this second code is far from being elucidated and we have only partial solutions to the protein-folding problem [2] (see Chapter 1). Moreover, in the more recent past it has become clear that the Anfinsen view of protein folding is incomplete: We now know that in the cell a large fraction of newly synthesized proteins require assistance by molecular chaperones in order to reach their folded states efficiently and at a biologically relevant time scale [3–6]. This chapter focuses on the mechanisms by which different chaperone classes function in protein folding in the cytosol and on their cooperation in cellular folding pathways. We end with a brief

Protein Misfolding Diseases: Current and Emerging Principles and Therapies,
Edited by Marina Ramirez-Alvarado, Jeffery W. Kelly, and Christopher M. Dobson
Copyright © 2010 John Wiley & Sons, Inc.

discussion of the relationship between chaperone functions and diseases caused by aberrant protein folding.

Protein folding in the test tube differs significantly from folding in vivo, the main differences relating to the physical properties of the cellular solution in which folding takes place and the fact that in vivo, folding is coupled to translation. Whereas in vitro refolding experiments are typically performed in dilute solution at low protein concentrations of 10 to 50 μg/mL, folding in the cell occurs in a highly crowded environment containing 300 to 400 mg/mL of protein and other macromolecules [7,8]. This condition of macromolecular crowding or excluded volume greatly enhances the affinities between interacting macromolecules and thus increases the propensity of folding protein chains to aggregate, an off-pathway reaction that is mediated mainly by interactions between hydrophobic amino acid side chains and regions of unstructured polypeptide backbone that are exposed by unfolded proteins and normally become buried during the folding process (see Chapter 2) (Fig. 1). Because translating polypeptide chains can fold only when a complete protein domain or folding unit (ca. 100 to 300 amino acids in length) has emerged from the narrow ribosomal exit tunnel, unfolded protein chains are exposed to a potentially hostile environment for prolonged periods of time compared to refolding in vitro. Moreover, since during translation the structural information determining the native fold is not available all at once as it is during in vitro refolding, the process of translation may favor the formation of nonnative contacts within nascent chains. This may generate misfolded states that, if not resolved, would tend to aggregate upon release from the ribosome. To combat these hazards of misfolding and aggregation, cells have evolved a diverse set of molecular chaperones [6,9].

We define a *molecular chaperone* as "any protein which interacts, stabilizes or helps a non-folded protein to acquire its native conformation, but is not present in the final functional structure" [4,10]. Chaperones are involved in a multitude of cellular functions, including de novo folding, refolding of stress-denatured proteins, oligomeric assembly, stabilization of proteins during intracellular transport and membrane translocation, and assistance in proteolytic degradation. They typically recognize the exposed hydrophobic side chains of nonnative proteins, and by transiently shielding these features they suppress aggregation and promote productive folding through cycles of substrate binding and release (Fig. 1). Substrate binding by chaperone blocks aggregation but retards folding, and thus chaperone release is required to allow folding. These cycles are often regulated by ATP-dependent conformational switching of the chaperone and through regulation by various cofactors (co-chaperones). Folding is promoted as long as substrate rebinding following release is sufficiently delayed to provide time for folding (i.e., burial of hydrophobic amino acid residues) but fast enough to prevent aggregation. Numerous classes of chaperone have been described, unrelated in sequence and differing in mechanism [11,12]. Many, but not all, of these components are known as *stress proteins* or *heat-shock proteins* (Hsp's), because they are synthesized increasingly by cells under conditions of conformational stress, such as exposure to elevated temperature or oxidative

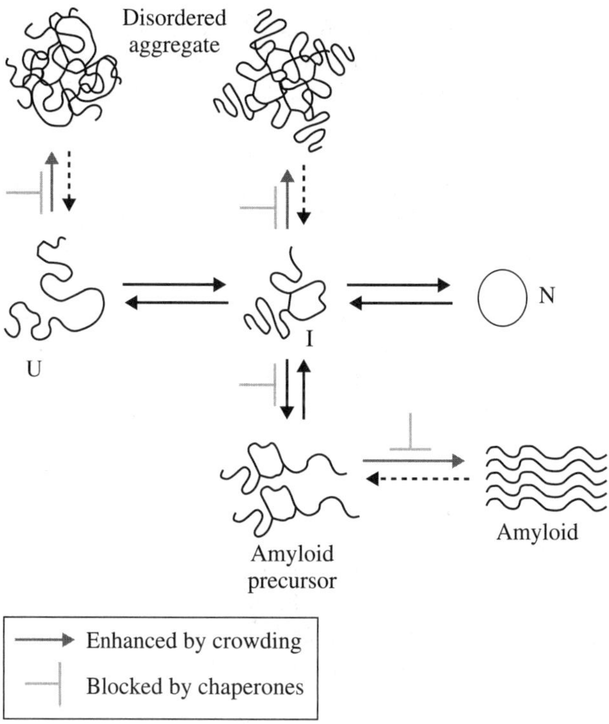

FIG. 1 Aggregation competes with productive protein folding. Aggregation of nonnative protein chains as a side reaction of productive folding in the crowded environment of the cell. Enhancement of aggregation and chain compaction by macromolecular crowding and effects of chaperones are indicated. U, unfolded protein chain released from ribosome; I, partially folded intermediate; N, native, folded protein. Crowding is also predicted to enhance the formation of amyloid fibrils. (Adapted from [6], with permission of Science AAAS.)

conditions. They are usually classified according to their molecular weight (Hsp40, Hsp60, Hsp70, Hsp90, Hsp100, and the so-called "small" Hsp's). Different chaperones often have overlapping functions such that typically only the combined deletion of components results in severe defects or lethality. Chaperone deficiency not only results in the insufficient production of proteins with essential functions, but may also lead to the accumulation of misfolded protein species that may have pronounced cell toxicity. Based on the insight that chaperone capacity decreases during aging, the latter phenomenon plays a role in the manifestation of late-onset neurodegenerative diseases, including Parkinson, Huntington and probably, Alzheimer diseases [13,14] (see Chapter 9).

PROTEIN FLUX THROUGH THE CHAPERONE NETWORK

The principal features of the chaperone pathways and networks acting in de novo protein folding and refolding have been highly conserved in evolution

[5,6,9]. In the cytosol of bacteria, archaea, and eukarya, we find chaperones that act upstream by protecting nascent polypeptides on ribosomes and by initiating folding, and components that act downstream in completing the folding process (Fig. 2). In each system the first category of components includes chaperones that bind directly to the large ribosomal subunit in close proximity to the polypeptide exit site (e.g., Trigger factor in bacteria, RAC in *Saccharomyces cerevisiae*) [11,12]. Following their interaction with these factors, nascent chains often bind chaperones of the Hsp70 system and their cofactors, which mediate co- and posttranslational folding or the transfer of proteins to downstream chaperones [5,15]. These downstream components include most prominently the cylindrical chaperonin complexes (Hsp60s), such as GroEL in bacteria and TRiC/CCT in eukarya, which provide nanocompartments for single protein molecules to fold unimpaired by aggregation. Another important chaperone system that functions downstream of the nascent chain-binding chaperones is Hsp90, which has a critical role in the folding and conformational regulation of signaling molecules [16].

The flux of newly synthesized proteins through the chaperone system is best understood for the bacterial cytosol (Fig. 2). In *Escherichia. coli*, the majority of cytosolic proteins (ca. 80%) are thought to interact with the ATP-independent, ribosome-binding Trigger factor (TF), thereby avoiding misfolding and aggregation. Small, single-domain proteins are likely to require only the interaction with TF for efficient folding, whereas longer chains (20 to 30 kDa) may interact subsequently with the ATP-regulated Hsp70, DnaK, and its Hsp40 co-chaperone, DnaJ. Based on estimates from co-immunoprecipitation experiments, about 20% by mass of newly synthesized polypeptides fold through cycles of Hsp70 binding and release [17]. The GroEL chaperonin is involved in the folding of about 10% of proteins. It receives its substrates from TF and the DnaK/DnaJ system [18]. Similar quantities of protein have been shown to interact with the eukaryotic Hsp70 and chaperonin systems [19]. In *E. coli*, GroEL and its cofactor GroES are absolutely essential [20], consistent with the identification of about 85 proteins that are predicted strictly to require chaperonin for folding [21]. In contrast, TF and DnaK can be deleted individually, but their combined deletion at 30°C and above results in lethality [17,22].

Besides their role in de novo folding, chaperones are involved in various other cellular processes by modulating the conformational states of target proteins. For example, Hsp70 proteins in the eukaryotic cytosol stabilize certain proteins destined for posttranslational uptake into the mitochondria, chloroplasts, or endoplasmic reticulum (ER) in an unfolded, translocation-competent conformation [23–25]. Some mitochondrial membrane proteins delivered by Hsp70 to import receptors on the outer mitochondrial membrane interact specifically with the C-terminus of Hsp70 [26]. Conceptually similar is the cooperation of Hsp70 with the ubiquitin–proteasome system (UPS) of protein degradation. Here, Hsp70 interacts with cofactors of the UPS to target folding-incompetent or permanently misfolded proteins for destruction [27].

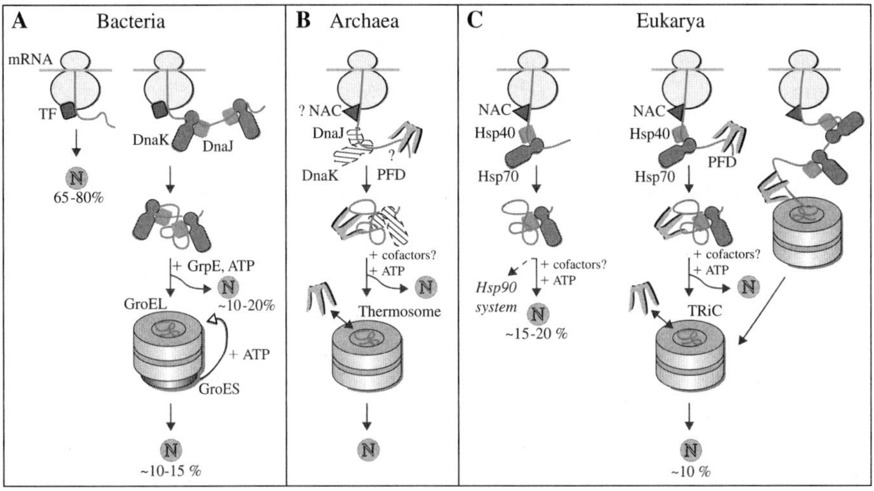

FIG. 2 Chaperone pathways of protein folding in the cytosol. Models for the chaperone-assisted folding of newly synthesized polypeptides in the cytosol. (A) Bacteria. TF, trigger factor; N, native protein. Nascent chains probably interact generally with TF, and most small proteins (65 to 80% of total) may fold rapidly upon synthesis without further assistance. Longer chains (10 to 20% of total) interact subsequently with DnaK and DnaJ and fold upon one or several cycles of ATP-dependent binding and release. About 10 to 15% of chains transit the chaperonin system (GroEL and GroES) for folding. GroEL does not bind to nascent chains and is thus likely to receive a substantial fraction of its substrates after their interaction with DnaK. (B) Archaea. PFD, prefoldin; NAC, nascent chain–associated complex. Only some archaeal species contain DnaK/DnaJ. (C) Eukarya (the example of the mammalian cytosol). Like TF, NAC probably interacts generally with nascent chains. The majority of small chains may fold upon ribosome release without further assistance. About 15 to 20% of chains reach their native states in a reaction assisted by Hsp70 and Hsp40, and a fraction of these must be transferred to Hsp90 for folding. About 10% of chains are co- or posttranslationally passed on to the chaperonin TRiC in a reaction mediated by Hsp70 and PFD. (Adapted from [6], with permission of Science AAAS.) (*See insert for color representation of figure.*)

RIBOSOME-BINDING CHAPERONES

In *E. coli*, Trigger factor (TF) is the first chaperone encountered by a nascent chain emerging from the ribosome. Nascent chains as short as 57 amino acids have been shown to interact with TF by cross-linking [28]. Immediate interaction with the nascent chain is facilitated by the positioning of TF near the opening of the ribosomal exit tunnel, where TF binds to proteins L23 and L29 of the large ribosomal subunit [29–32]. In the recent crystal structure, *E. coli* TF has an elongated shape and is divided into three domains: The N-terminal domain provides the ribosomal-binding site [30]. A peptidyl-prolyl isomerase domain, which follows the N domain in sequence, is located at the opposite end of the molecule and is connected with the N domain via a long linker segment. While this domain exhibits PPIase activity in vitro [33], in vivo it participates in nascent chain binding independent of the occurrence of proline residues [34–36]. The C-terminal domain is positioned between the N and PPIase domains and forms two protrusions or tips, which delineate a hydrophobic region involved in polypeptide binding [35,37]. The interaction of TF with the ribosome and nascent chain substrate is ATP independent but is modulated by the presence of nascent polypeptide, with translating ribosomes exhibiting substantially higher affinities for TF than do non-translating ribosomes [34].

Different models for how TF assists protein folding have been proposed. In one model, TF performs a default cycle characterized by about a 10 to 15-second residence time on the ribosome, followed by release from the L23/29 ribosomal binding site. The presence of a nascent chain predominantly increases the "on" rate for TF binding. TF undergoes a conformational opening upon ribosome binding, based on intramolecular FRET measurements, and this aperture is thought to activate TF for the association with hydrophobic segments of nascent chains as they emerge from the ribosome [34]. Provided that the hydrophobicity of exposed chain segments exceeds a certain threshold, TF may leave its ribosome docking site but remain associated with the growing nascent chain. The eventual dissociation of TF from the nascent chain would facilitate folding or polypeptide transfer to downstream chaperones such as the DnaK–DnaJ system. In this model, a primary role of TF is to shield hydrophobic chain segments, thereby preventing aggregation or misfolding and maintaining the nascent polypeptide competent for productive folding. In an alternative model, supported by the proposed topology of TF relative to the ribosome exit site, TF is thought to provide a shielded, cagelike environment in which a domain of an emerging nascent chain may fold unhindered by aggregation [30,36]. These two models are not mutually exclusive, and the specific mode of TF action may depend on the structural properties of a nascent chain substrate. Nascent chains that expose extensive hydrophobic regions may follow the first model, whereas chains without extensive hydrophobicity may interact only weakly with the binding surface of TF, allowing co-translational compaction in close proximity to the chaperone.

The mode of recognition of peptide segments by TF has been studied by cross-linking and nascent chain–binding experiments [34–36,38,39]. TF binds preferentially to segments of a nascent chain about 15 residues long that have a central hydrophobic region. The lifetime of the TF–nascent chain complex is determined by the hydrophobicity of the peptide fragment, and bioinformatic analysis predicted that about 15% of cytosolic proteins in *E. coli* have high-affinity TF-binding sites [34]. The structural properties recognized by TF resemble those of peptides that interact with Hsp70. This is consistent with the finding that TF and the Hsp70 system overlap functionally, with *E. coli* cells tolerating individual deletions of TF or DnaK, while the combined deletion of both components results in bulk protein aggregation and lethality at growth temperatures above 30°C [17,22,40–42]. The eukaryotic cytosol lacks TF but contains the structurally unrelated nascent chain–associated complex (NAC), which consists of an α (33 kDa) and a β subunit (22 kDa) [43]. Although the exact function of NAC in protein biogenesis and quality control remains to be defined, NAC associates with short nascent chains and dissociates upon chain release from the ribosome [44]. Furthermore, NAC has been reported to interact with ribosomal protein L25, the homolog of bacterial L23 [45]. In addition to NAC, yeast and other fungi have specialized Hsp70s that specifically associate with ribosomes. Ssb1 and Ssb2, two members of the Hsp70 family of chaperones in *S. cerevisiae*, interact with nascent chains and ribosomes [46]. The function of the Ssb chaperones is modulated by another Hsp70, Sszl, that forms a stable ribosome-associated complex (RAC) with zuotin [47,48], the Hsp40 partner of Ssb1 and Ssb2 [49]. A homolog of RAC also exists in mammalian cells [50,51].

THE HSP70 SYSTEM

Non-ribosome-associated members of the Hsp70 family of chaperones exist in the cytosol of eubacteria, eukarya, some archaea, and in organelles of prokaryotic origin. *S. cerevisiae* has four cytosolic versions of this protein, Ssa1 to Ssa4. Besides assisting de novo folding, members of this chaperone family have evolved to perform specialized functions in the various subcellular compartments, including a role in driving protein translocation across the mitochondrial and ER membranes for the Hsp70 homologs in the respective organelles [23,24]. The cytosol of higher eukaryotes contains both constitutively expressed as well as stress-inducible Hsp70 members. Hsp70s generally collaborate with J-domain-containing chaperones of the Hsp40 (DnaJ) family in the ATP-regulated binding and release of substrates with an exposed hydrophobic sequence. The Hsp40s form a highly diverse group of proteins with specialized functions and can recruit their Hsp70 partners to specific cellular locations and substrates [9].

Structure and Reaction Cycle

Structural aspects and the mechanism of the Hsp70 reaction cycle have been widely studied using the eubacterial homolog of Hsp70, DnaK, its Hsp40 co-chaperone DnaJ, and the nucleotide exchange factor GrpE as a paradigm. Like all Hsp70s, DnaK is divided structurally into an N-terminal nucleotide-binding domain and a C-terminal peptide-binding domain [52] (Fig. 3A). The approximate 40-kDa ATPase domain is structurally homologous to actin. The approximate 25-kDa peptide-binding domain consists of a β-sandwich subdomain and an α-helical lid segment [53]. The β-sandwich is involved primarily in interaction with approximate seven-residue extended peptide segments, and the α-helical segment is thought to latch onto the peptide-binding site (Fig. 3A). Substrate peptides contain hydrophobic residues in the central region and are typically flanked by positively charged residues; they bind to DnaK with affinities ranging from nM to μM [33,52]. Such segments are exposed by most unfolded polypeptides and occur on average every 50 residues in proteins.

The β-sandwich subdomain makes contacts with the hydrophobic side chains of bound peptide substrates and forms hydrogen bonds with the peptide backbone. As a result, bound segments adopt an extended conformation.

The substrate affinity of the Hsp70 peptide-binding domain is allosterically regulated by ATP binding and hydrolysis in the ATPase domain. In the ATP state, the α-helical latch segment of the C-terminal domain is thought to be in an open conformation, resulting in high on and off rates for peptide binding. Hydrolysis of ATP to ADP leads to lid closure and stable peptide binding (low on and off rates) (Fig. 3B). Allosteric interdomain signaling is believed to be mediated by a conformational change in the hydrophobic linker region between the ATPase and peptide-binding domains. ATP hydrolysis is strongly accelerated by the interaction of DnaJ (Hsp40) with DnaK. The N-terminal J-domain of DnaJ, the signature domain of all members of this protein family, is critical in this reaction [54,55]. In addition to stimulating the Hsp70 ATPase, DnaJ and other Hsp40s also interact directly with unfolded polypeptides through their C-terminal domain and hence can recruit DnaK to nascent chains or other protein substrates [15,33,56]. The nucleotide exchange factor (NEF), GrpE, a homodimer of 22-kDa subunits, subsequently binds to DnaK and catalyzes the exchange of ADP with ATP, which results in lid opening and substrate release, thereby completing the reaction cycle [52,57,58].

Whereas the Hsp70 chaperones generally cooperate with J-domain proteins, GrpE is restricted to the bacterial cytosol and to mitochondria and chloroplasts in eukaryotes. Three classes of proteins have been shown to catalyze Hsp70 nucleotide exchange in the eukaryotic cytosol: Bag-1 and HspBP1 (Fes1p in *S. cerevisiae*) proteins, as well as the members of the Hsp110 family [59–64]. All three groups of proteins are structurally distinct. They interact with the Hsp70 ATPase domain by different mechanisms but effect a similar structural opening of the two lobes of the domain that facilitates ADP dissociation [58,65–68]. Surprisingly, the Hsp110 proteins (Sse1/2p in *S. cerevisiae*) have a domain structure homologous to Hsp70, consisting of an N-terminal ATP-binding domain and a putative peptide-binding domain [67,68]. Recent structural analysis showed that both the N-terminal domain and the three-helical segment corresponding to the α-helical lid of canonical Hsp70s interact with the Hsp70 ATPase domain [68,69]. Hsp70 and Hsp110 chaperones may cooperate in protein folding by releasing bound substrate segments in a coordinated fashion [63,64,68].

Substrates and Mechanism of Folding

DnaK is a highly abundant protein, exceeding the concentration of ribosomes in the cytosol [6]. Although direct proteomic studies of DnaK-bound substrates are not yet available, pulse labeling of *E. coli* coupled with co-immunoprecipitation of DnaK–substrate complexes suggested that 15 to 20% of proteins by mass transit DnaK upon synthesis at 30°C [17]. DnaK preferentially binds to polypeptides larger than 20 to 30 kDa [17]. Upon deletion of TF, there is a

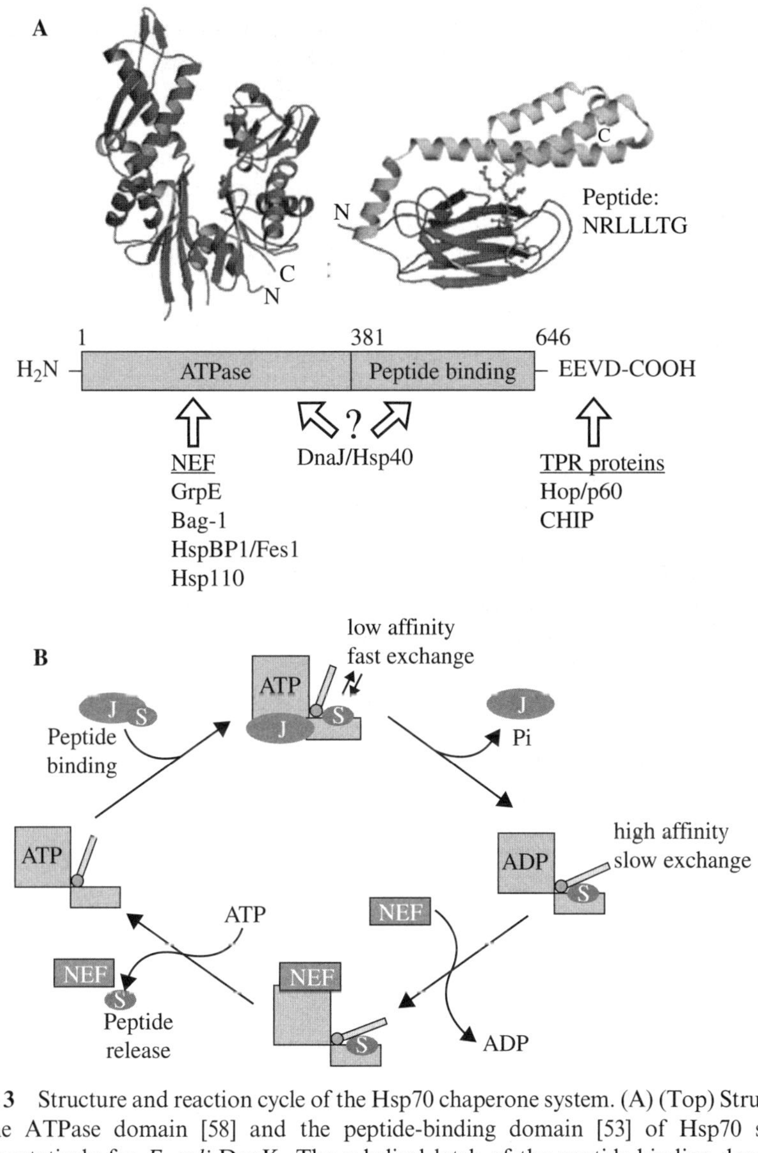

FIG. 3 Structure and reaction cycle of the Hsp70 chaperone system. (A) (Top) Structures of the ATPase domain [58] and the peptide-binding domain [53] of Hsp70 shown representatively for *E. coli* DnaK. The α-helical latch of the peptide-binding domain is shown in yellow and a ball-and-stick model of the extended peptide substrate in pink. ATP indicates the position of the nucleotide binding site. The amino acid sequence of the peptide is indicated in single-letter code. (Bottom) The interaction of prokaryotic and eukaryotic cofactors with Hsp70 is shown schematically. Residue numbers refer to human Hsp70. NEF, nucleotide exchange factors (GrpE in case of *E. coli* DnaK; Bag, HspBP1, and Hsp110 in case of eukaryotic cytosolic Hsp70). TPR, tetratricopeptide repeat domain: Hop, Hsp organizing protein; CHIP, C-terminus of Hsp70 interacting protein. Only the Hsp70 proteins of the eukaryotic cytosol have the COOH-terminal sequence EEVD that is involved in binding of TPR cofactors [142]. (B) Hsp70 reaction cycle with Hsp70 colored as in (A). J, DnaJ; NEF, nucleotide exchange factor; S, substrate peptide. (Adapted from [6], with permission of Science AAAS.) (*See insert for color representation of figure.*)

significant increase in the number and amount of proteins interacting with DnaK, consistent with the overlap in function between these two chaperones that is indicated by the lethality of their combined deletion [17,22]. TF and DnaK were shown to delay the folding of certain proteins relative to translation and to suppress the formation of nonproductive folding intermediates [41]. In the case of firefly luciferase, a eukaryotic multidomain protein, this resulted in an overall inefficient folding pathway compared to the preferred co-translational mode of folding of this protein in the eukaryotic system. However, most bacterial proteins that are larger than 60 kDa, and thus cannot be accommodated within the central cavity of the chaperonin, appear to have adapted to the TF–DnaK system. For example, the large protein β-galactosidase has been shown to use TF and DnaK effectively for folding in vivo and upon translation in bacterial lysate in vitro [41].

How exactly Hsp70 mediates the folding of a substrate protein is not yet understood, but a process of kinetic partitioning is generally believed to form the basic mechanism: Binding of Hsp70 to nonnative substrate hinders aggregation by transiently shielding exposed hydrophobic segments. Repeated cycles of binding and release then allow the protein to partition into populations that fold fast upon release, burying their hydrophobic residues, whereas molecules that fold slowly rebind to Hsp70 and continue to cycle. Binding may result in conformational remodeling, thereby perhaps removing kinetic barriers to fast folding. Proteins that are unable to partition to fast-folding species within a biologically relevant time scale are stabilized by Hsp70 in a nonaggregated state and are transferred into the specialized environment of the chaperonin folding cage for final folding [15,21].

CHAPERONINS

Chaperonins are large double-ring complexes of about 800 kDa with ATPase function that have been highly conserved in evolution and are found in all domains of life. These proteins can be subdivided into two groups that are similar in architecture but distantly related in sequence. Group I chaperonins, also known as Hsp60s, occur in bacteria (GroEL) and organelles of endosymbiotic origin. These chaperonins have seven-membered rings and cooperate functionally with Hsp10 proteins (bacterial GroES). Group II chaperonins exist in archaea and in the cytosol of eukaryotes. They consist of eight- or nine-membered rings and are independent of the assistance of Hsp10 homologs. As in the case of Hsp70, substrate binding to chaperonins is ATP regulated, but the mode of action is distinctly different. Interestingly, archaea of the genus *Methanosarcina* have acquired GroEL/GroES by lateral gene transfer and contain both group I and group II chaperonin systems in the cytosol [70].

Group I Chaperonins

GroEL and GroES have been studied most extensively [3,6,52]. GroEL consists of two heptameric rings of 57-kDa subunits that are stacked back to back [71]. Each subunit is composed of an equatorial domain, an apical domain, and an intermediate domain. The equatorial domain harbors the ATPase site and is connected to the apical domain through an intermediate hingelike domain. The apical domains form the ring opening and expose hydrophobic amino acid residues for substrate binding toward the ring center [72]. GroES is a single heptameric ring of subunits of about 10 kDa that binds to the ends of the GroEL cylinder [73,74]. Upon substrate binding, hydrophobic residues exposed by folding intermediates interact primarily with the hydrophobic surfaces of two amphiphilic helices in the apical GroEL domains [52,75,76]. GroEL-bound substrates populate a dynamic ensemble of molten-globule-like states that lack stable tertiary interactions [77–81]. Typical substrates interact with more than one apical domain, resulting in high-affinity multivalent interactions [82,83]. The GroES subunits have mobile sequence loops that contact the apical GroEL domains and mediate a conformational change that results in substrate release [52,84,85]. The basic chaperonin reaction comprises the GroES-mediated encapsulation of a single protein molecule in a cagelike structure for folding to occur unimpaired by aggregation [86–88]. GroES dissociates from GroEL in an ATP-regulated manner, allowing substrate exit (Fig. 4).

GroEL–GroES Reaction Cycle

GroEL and GroES form an allosterically highly regulated system, with GroEL undergoing dramatic conformational changes upon ATP and GroES binding [89–91]. There are two levels of allostery in GroEL which function in a coupled manner: An intraring positive allostery ensures a nearly simultaneous anticlockwise turn and upward movement of the seven apical domains of one ring upon cooperative ATP binding. This step prepares the ring for GroES binding, which is accompanied by a 120° clockwise turn of the apical domains and a further upward and outward movement. In contrast, negative allostery between the rings renders the double-ring system asymmetrical and prevents the two rings from populating the same nucleotide-bound state simultaneously. Initially, the polypeptide binds into the center of the open GroEL ring. This step may result in some degree of unfolding, with the protein occupying an ensemble of locally expanded and more compact conformations [81] (Fig. 4). Substrate binding and the binding of seven ATPs to the same GroEL ring causes the release of GroES and ADP from the opposite ring. GroES will then bind to the polypeptide and ATP-containing ring, leading to the encapsulation of the bound protein [92] (Fig. 4). Global encapsulation by GroES may be preceded by local stretching and partial compaction of substrate protein, mediated by ATP-dependent movements of the apical GroEL domains [81,93]. Importantly, upon

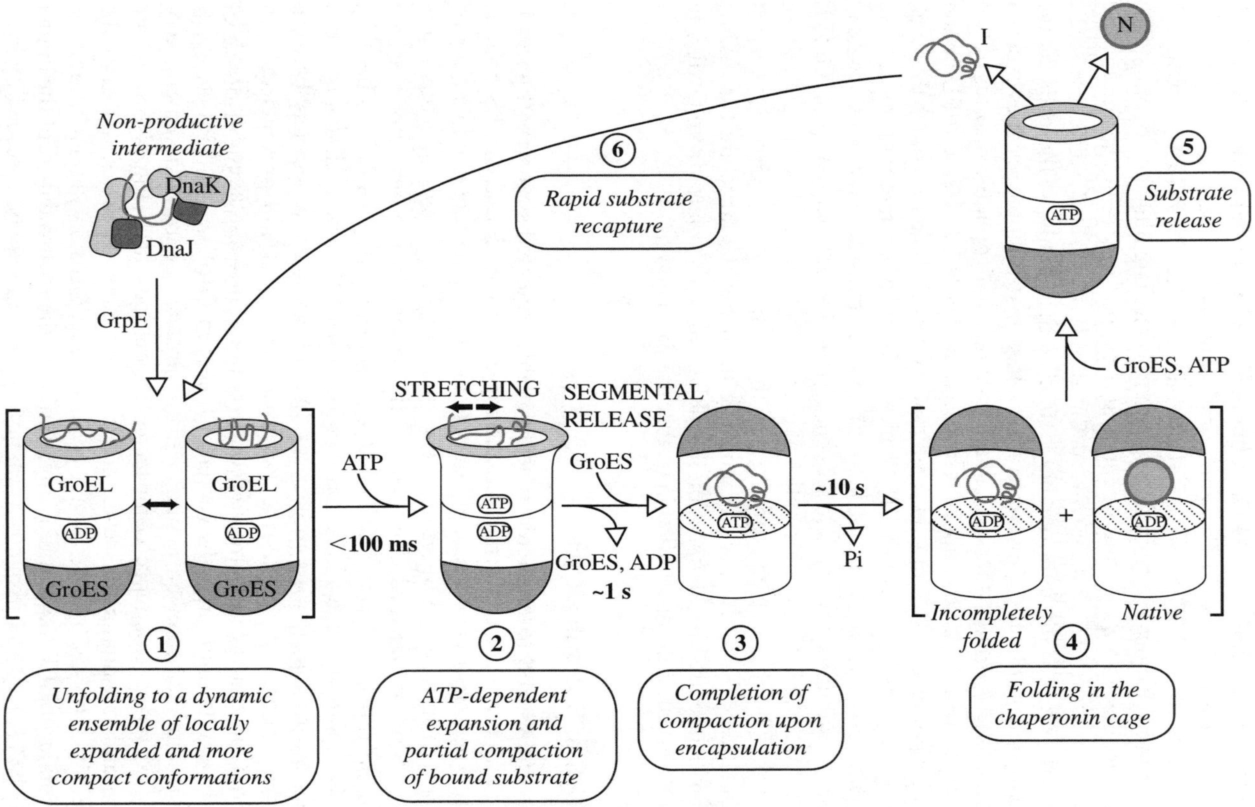

FIG. 4 Protein folding with the GroEL–GroES chaperonin system. Working model summarizing the conformational changes in a substrate protein upon transfer from DnaK–DnaJ (Hsp70 system) to GroEL and during GroEL–GroES-mediated folding. Note that binding of a second substrate molecule to the open ring of GroEL in steps 4 and 5 is omitted for simplicity. N, native state; I, folding intermediate. (From [81], with permission of Elsevier.)

GroES binding, the volume of the chaperonin cavity approximately doubles, accommodating proteins up to about 60 kDa, and the surface property of the cavity shifts from hydrophobic to hydrophilic [85]. The encapsulated protein is allowed to fold unhindered by aggregation for a period of 10 to 15 s [87,88,94], the time needed for the hydrolysis of the seven ATP molecules. At the end of the cycle, both nonnative and native proteins exit the cavity triggered by binding of unfolded protein and ATP in the opposite ring [92,95]. Nonnative intermediates still exposing hydrophobic surfaces rebind immediately to GroEL for further cycles of encapsulation until they reach the native state.

Substrates and Folding Mechanism

Approximately 15% of cytosolic proteins by mass interact with GroEL upon synthesis, as revealed by co-immunoprecipitation of GroEL–substrate complexes from pulse-labeled *E. coli* spheroplasts [18]. Proteomic studies led to the identification of about 250 different substrates which could be divided into three classes, based on their dependency on GroEL for folding [21,96]. Class I substrates are abundant cytosolic proteins of which only a minute fraction interacts with GroEL. These proteins were shown to fold spontaneously in vitro and were unaffected in folding and stability upon GroEL depletion in vivo [21]. Class II proteins are of intermediate abundance. They are aggregation-prone but can utilize either the DnaK (Hsp70) system or GroEL–GroES for folding. Finally, class III substrates are highly aggregation-prone proteins of low to intermediate abundance that are fully dependent on GroEL–GroES for folding in vivo and in vitro (obligate GroEL substrates). These approximately 85 proteins occupy about 80% of the GroEL capacity and include 13 proteins of essential function. They are unable to use the DnaK system for folding but are maintained in a folding-competent state by binding to DnaK–DnaJ. Transfer to GroEL results in efficient folding, as demonstrated with a subset of validated class III proteins. Most class III proteins have complex folds of α/β or $\alpha + \beta$ domain topology, with a distinct enrichment of the $(\beta/\alpha)_8$ TIM barrel fold [21]. Such proteins often display a pronounced tendency to populate kinetically trapped states, which appear to have particularly high affinity for GroEL [21,97].

The mechanism by which the GroEL–GroES system assists the refolding of such proteins is still a matter of debate [98]. Clearly, encapsulating a single protein molecule inside the chaperonin cage serves as an essential mechanism in preventing aggregation during folding [99,100]. This was demonstrated with mitochondrial rhodanese as a model substrate and with bacterial ribulose bisphosphate carboxylase oxygenase (Rubisco), an authentic class III protein in vitro [99]. Under nonpermissive conditions, in which spontaneous folding is blocked by aggregation, efficient folding of these proteins was observed only as long as they were enclosed in the chaperonin cavity. In a variation of these experiments, folding was measured at very low protein concentrations, where aggregation is not limiting (permissive conditions). These experiments showed that the chaperonin system accelerated the folding of the 50-kDa protein Rubisco three-to four fold, whereas for the 33-kDa rhodanese the rates of spontaneous and assisted folding were indistinguishable. Importantly, the rate acceleration of folding depended critically on encapsulation [99], and it was suggested that an effect of steric confinement inside the cage directs the protein to a faster folding track by entropically destabilizing the unfolded state and promoting the acquisition of compact native conformation. This conclusion was supported by theoretical modeling of the relationship between cage size and folding rates [101–103]. Recent experiments in which the size of the GroEL cavity was modulated by mutation demonstrated that decreasing the size of the cage can result in a 1.5- to 2-fold acceleration of folding for smaller proteins such as rhodanese [100,104], consistent with confinement theory [105]. In contrast, the folding of larger proteins was inhibited under these conditions. In addition to cage volume, the negative net charge of the GroEL cavity wall was shown to contribute to efficient folding [100,104]. Steric confinement would bring hydrophobic amino acid residues exposed by the folding protein into very close proximity to water molecules that are ordered by the negative charge clusters of the cavity wall, resulting in an enhancement of the hydrophobic effect [106]. A recent study showed that the folding pathway of the small model protein DHFR (ca.26 kDa) is unaffected by confinement, consistent with the finding that folding of this protein is not accelerated by GroEL [80]. Whether and how the folding pathway of larger proteins is altered by the chaperonin system remains to be investigated.

An alternative framework for how GroEL may accelerate folding is provided by the iterative annealing model [107,108]. In this model, the chaperonin is thought to achieve efficient folding of a substrate protein by repeated cycles of unfolding of kinetically trapped, misfolded states, followed by a partitioning between fast and slow folding routes. Indeed, compact folding intermediates are destabilized upon GroEL binding, and some conformational stretching of bound substrate has been observed upon ATP- and GroES-dependent domain movements of GroEL, potentially repositioning the protein to a higher level in the energy landscape [81,93,107,109]. However, the extent to which these steps may contribute to chaperonin-assisted folding remains to be investigated. Notably, repeated cycles of substrate binding and release are not required for accelerated folding. This was demonstrated with a single-ring mutant of GroEL [88], in which fully efficient folding occurs upon a single round of encapsulation for several model proteins tested [99,100,104]. The confinement and unfolding models are not mutually exclusive. Both mechanisms may contribute, depending on the folding properties of specific substrates.

Certain proteins larger than 60 kDa, such as the model protein mitochondrial aconitase (82 kDa), can nevertheless utilize the GroEL system for folding in a mechanism that is independent of encapsulation [110]. In this trans-folding mechanism, substrate protein and GroES interact with opposite GroEL rings, and folding is promoted by ATP- and GroES-dependent binding and release cycles. GroEL has been found to interact with several proteins greater than 60 kDa in size which may use this mechanism [21,96].

Group II Chaperonins

The group II chaperonin in the eukaryotic cytosol is called TRiC (TCP-1 ring complex) or CCT (chaperonin containing TCP-1). It is heterooligomeric and consists of eight paralogous subunits per ring, which differ mostly in their apical substrate binding domains and are all encoded by essential genes [5]. A specific subunit arrangement has been proposed based on co-immunoprecipitation experiments [111]. The general domain structure of the group II chaperonins is similar to that of GroEL, as seen from the crystal structure of the thermosome, the group II chaperonin in archaea [112]. However, all group II chaperonins deviate from GroEL in that their apical domains contain fingerlike protrusion, which act as a built-in lid and replace the function of GroES. These segments open and close in an ATP-dependent reaction cycle [113], adopting an α-helical, irislike arrangement in the closed state [113]. Class II chaperonins interact functionally with the co-chaperone prefoldin [114–117], which functions in substrate transfer to chaperonin. Prefoldin has a remarkable jellyfish-like structure consisting of six subunits with long α-helical coiled-coil domains [118]. These segments are partially unwound at their tips, exposing hydrophobic residues for the binding of nonnative protein. Prefoldin is thought to interact directly with the chaperonin [119]. Substrate transfer to TRiC is also mediated by Hsp70, consistent with the general sequence of chaperone interactions in the cytosolic folding pathway [15,120] (Fig. 2). Interestingly, Hsp70 appears to fulfill this role by interacting directly with TRiC [121].

TRiC–CCT interacts with approximately 10% of newly synthesized cytosolic proteins [19,122], including actin and tubulins as the most abundant substrates [123,124]. Efforts to refold actin and tubulin in vitro in the absence of TRiC have failed, indicating that these proteins represent obligate chaperonin substrates. TRiC substrates do not share any apparent similarity in sequence or structure, except for a certain preponderance of proteins with WD40 β-propeller domains [125,126]. Most TRiC-interacting proteins form homo- or heterooligomeric complexes [127]. Several substrates are 100 to 120 kDa in size, suggesting that TRiC may be able to assist the cotranslational refolding of individual domains of larger proteins, consistent with its capacity to interact with nascent polypeptide chains [120,128]. The apical domains of the chaperonin appear to differ in their substrate-binding specificity, as has been demonstrated for actin and for the tumor suppressor protein VHL [119,129]. For example, in the case of VHL, only subunits 1 and 7 of TRiC interact. The binding sites identified belong to a helical region of the apical domain with hydrophobic character and are analogous to those of the GroEL chaperonin [129].

As in the case of the group I chaperonin GroEL, protein folding by TRiC involves ATP-dependent substrate encapsulation in the chaperonin cavity. Consistent with a negative allostery between rings, only one ring appears to be closed at a time, as indicated by experiments with the ATP-hydrolysis transition state analog ADP-AlFx [113]. Interestingly, the binding of nonhydrolyzable

ATP was neither sufficient to close the lid nor to support actin refolding, indicating that ATP hydrolysis is required for ring closure, a deviation from the GroEL mechanism [113]. However, the binding of nonhydrolyzable ATP analog or of ADP induces a conformational change in TRiC [130]. Whether the more dramatic structural rearrangement upon ATP hydrolysis occurs in all eight subunits simultaneously or follows a sequential mechanism remains to be determined [113], but an allosteric communication between subunits has been proposed [131]. The TRiC reaction cycle is much slower than that of GroEL [132], potentially providing a substantially longer period of protein encapsulation and folding in the chaperonin cage.

MEDICAL SIGNIFICANCE OF CHAPERONES

An increasing number of diseases are recognized to be associated with aberrant protein folding and aggregation. Heritable mutations in key proteins may lead to loss of function, as in the case of the cystic fibrosis conductance regulator, a chloride membrane channel whose functional impairment results in cystic fibrosis. Such loss-of-function diseases must be distinguished from pathological states in which aberrant folding results in a toxic gain of function (see Part II). The latter group includes some of the most debilitating neurodegenerative disorders, such as Alzheimer, Parkinson, and Huntington diseases, which are all associated with the deposition of ordered, fibrillar aggregates (amyloid) within and around neuronal cells. Although the toxic principle operating in these disorders is far from being understood, a consensus is emerging that oligomeric soluble states of the respective disease protein are the primary cytotoxic species. These aggregation intermediates are thought to interact aberrantly with other proteins or membranes, altering their functional properties [13]. Interestingly, molecular chaperones of several classes, most prominently the Hsp70s, have been shown to inhibit the formation of such oligomers and to deviate the aggregation pathway from the amyloidogenic route toward the formation of amorphous aggregates [133–137]. In the case of polyglutamine-extension proteins, which cause Huntington disease and several related disorders (see Chapter lb), Hsp70 may cooperate with the chaperonin TRiC in modulating aggregation and facilitating the formation of benign oligomers of the disease protein [135–137]. Interestingly, recent progress in understanding the molecular genetic mechanisms and the signaling pathways underlying the aging process indicate that the functional capacity of molecular chaperones and other aspects of protein quality control decreases during aging [138]. This interesting connection would provide a plausible explanation for the late onset of many neurodegenerative diseases caused by aberrant protein folding [139] (see Chapter 29). It suggests that searching for ways to reestablish the normal balance between chaperone capacity and the production of misfolded proteins (e.g., by up-regulating the expression of chaperones) offers a promising therapeutic strategy [140,141].

Acknowledgments

Work in the authors' laboratory is supported by the Max-Planck-Gesellschaft, the Deutsche Forschungsgemeinschaft, the European Commission, the Körber Foundation, and the Ernst-Jung Foundation.

REFERENCES

1. Anfinsen, C.B. (1973). Principles that govern the folding of protein chains. *Science, 181*, 223–230.
2. Dobson, C.M., Karplus, M. (1999). The fundamentals of protein folding: bringing together theory and experiment. *Curr Opin Struct Biol, 9*, 92–101.
3. Hartl, F.U. (1996). Molecular chaperones in cellular protein folding. *Nature, 381*, 571–579.
4. Ellis, R.J., Hartl, F.U. (1996). Protein folding in the cell: competing models of chaperonin function. *FASEB J 10*, 20–26.
5. Frydman, J. (2001). Folding of newly translated proteins in vivo: the role of molecular chaperones. *Annu Rev Biochem, 70*, 603–647.
6. Hartl, F.U., Hayer-Hartl, M. (2002). Molecular chaperones in the cytosol: from nascent chain to folded protein. *Science, 295*, 1852–1858.
7. Ellis, R.J. (2001). Macromolecular crowding: an important but neglected aspect of the intracellular environment. *Curr Opin Struct Biology, 11*, 114–119.
8. Minton, A.P. (2006). Macromolecular crowding. *Curr Biol, 16*, R269–R271.
9. Young, J.C., Agashe, V.R., Siegers, K., Hartl, F.U. (2004). Pathways of chaperone-mediated protein folding in the cytosol. *Nat Rev Mol Cell Biol, 5*, 781–791.
10. Martin, J., Hartl, F.U. (1997). Chaperone-assisted protein folding. *Curr Opin Struct Biol, 7*, 41-52.
11. Chang, H.C., Tang, Y.C., Hayer-Hartl, M., Hartl, F.U. (2007). SnapShot: molecular chaperones, Part I. *Cell, 128*, 212.
12. Tang, Y.C., Chang, H.C., Hayer-Hartl, M., Hartl, F.U. (2007). SnapShot: mole cular chaperones, Part II. *Cell, 128*, 412.
13. Barral, J.M., Broadley, S.A., Schaffar, G., Hartl, F.U. (2004). Roles of molecular chaperones in protein misfolding diseases. *Semin Cell Dev Biol, 15*, 17–29.
14. Morley, J.F., Morimoto, R.I. (2004). Regulation of longevity in *Caenorhabditis elegans* by heat shock factor and molecular chaperones. *Mol Biol Cell, 15*, 657–664.
15. Langer, T., Lu, C., Echols, H., Flanagan, J., Hayer, M.K., Hartl, F.U. (1992). Successive action of DnaK, DnaJ and GroEL along the pathway of chaperone-mediated protein folding. *Nature, 356*, 683–689.
16. Wegele, H., Muller, L., Buchner, J. (2004). Hsp70 and Hsp90: a relay team for protein folding. *Rev Physiol, Biochem Pharmacol, 151*, 1–44.
17. Teter, S.A., Houry, W.A., Ang, D., Tradler, T., Rockabrand, D., Fischer, G., Blum, P., Georgopoulos, C., Hartl, F.U. (1999). Polypeptide flux through bacterial Hsp70: DnaK cooperates with trigger factor in chaperoning nascent chains. *Cell, 97*, 755–765.
18. Ewalt, K.L., Hendrick, J.P., Houry, W.A., Hartl, F.U. (1997). In vivo observation of polypeptide flux through the bacterial chaperonin system. *Cell, 90*, 491–500.

19. Thulasiraman, V., Yang, C.F., Frydman, J. (1999). In vivo newly translated polypeptides are sequestered in a protected folding environment. *EMBO J, 18*, 85–95.
20. Fayet, O., Ziegelhoffer, T., Georgopoulos, C. (1989). The groES and groEL heat shock gene products of *Escherichia coli* are essential for bacterial growth at all temperatures. *J Bacteriol, 171*, 1379–1385.
21. Kerner, M.J., Naylor, D.J., Ishihama, Y., Maier, T., Chang, H.C., Stines, A.P., Georgopoulos, C., Frishman, D., Hayer-Hartl, M., Mann, M., Hartl, F.U. (2005). Proteome-wide analysis of chaperonin-dependent protein folding in *Escherichia coli. Cell, 122*, 209–220.
22. Deuerling, E., Schulze-Specking, A., Tomoyasu, T., Mogk, A., Bukau, B. (1999). Trigger factor and DnaK cooperate in folding of newly synthesized proteins. *Nature, 400*, 693–696.
23. Neupert, W., Herrmann, J.M. (2007). Translocation of proteins into mitochondria. *Annu Rev Biochem, 76*, 723–749.
24. Rapoport, T.A. (2007). Protein translocation across the eukaryotic endoplasmic reticulum and bacterial plasma membranes. *Nature, 450*, 663–669.
25. Stengel, A., Soll, J., Bolter, B. (2007). Protein import into chloroplasts: new aspects of a well-known topic. *Biol Chem, 388*, 765–772.
26. Young, J.C., Hoogenraad, N.J. Hartl, F.U. (2003). Molecular chaperones Hsp90 and Hsp70 deliver preproteins to the mitochondrial import receptor Tom70. *Cell, 112*, 41–50.
27. Arndt, V., Rogon, C., Hohfeld, J. (2007). To be, or not to be: molecular chaperones in protein degradation. *Cell Mol Life Sci, 64*, 2525–2541.
28. Hesterkamp, T., Hauser, S., Lutcke, H., Bukau, B. (1996). *Escherichia coli* trigger factor is a prolyl isomerase that associates with nascent polypeptide chains. *Proc Nat Acad Sci USA, 93*, 4437–4441.
29. Kramer, G., Rauch, T., Rist, W., Vorderwulbecke, S., Patzelt, H., Schulze-Specking, A., Ban, N., Deuerling, E., Bukau, B. (2002). L23 protein functions as a chaperone docking site on the ribosome. *Nature, 419*, 171–174.
30. Ferbitz, L., Maier, T., Patzelt, H., Bukau, B., Deuerling, E., Ban, N. (2004). Trigger factor in complex with the ribosome forms a molecular cradle for nascent proteins. *Nature, 431*, 590–596.
31. Schlunzen, F., Wilson, D.N., Tian, P., Harms, J.M., McInnes, S.J., Hansen, H.A., Albrecht, R., Buerger, J., Wilbanks, S.M., Fucini, P. (2005). The binding mode of the trigger factor on the ribosome: implications for protein folding and SRP interaction. *Structure, 13*, 1685–1694.
32. Baram, D., Pyetan, E., Sittner, A., Auerbach-Nevo, T., Bashan, A., Yonath, A. (2005). Structure of trigger factor binding domain in biologically homologous complex with eubacterial ribosome reveals its chaperone action. *Proc Nat Acad Sci, USA, 102*, 12017–12022.
33. Rudiger, S., Germeroth, L., Schneider-Mergener, J., Bukau, B. (1997). Substrate specificity of the DnaK chaperone determined by screening cellulose-bound peptide libraries. *EMBO J, 16*, 1501–1507.
34. Kaiser, C.M., Chang, H.C., Agashe, V.R., Lakshmipathy, S.K., Etchells, S.A., Hayer-Hartl, M., Hartl, F.U., Barral, J.M. (2006). Real-time observation of trigger factor function on translating ribosomes. *Nature, 444*, 455–460.

35. Lakshmipathy, S.K., Tomic, S., Kaiser, C.M., Chang, H.C., Genevaux, P., Georgopoulos, C., Barral, J.M., Johnson, A.E., Hartl, F.U., Etchells, S.A. (2007). Identification of nascent chain interaction sites on Trigger factor. *J Biol Chem, 282*, 12186–12193.

36. Merz, F., Boehringer, D., Schaffitzel, C., Preissler, S., Hoffmann, A., Maier, T., Rutkowska, A., Lozza, J., Ban, N., Bukau, B., Deuerling, E. (2008). Molecular mechanism and structure of Trigger factor bound to the translating ribosome. *EMBO J, 27*, 1622–1632.

37. Merz, F., Hoffmann, A., Rutkowska, A., Zachmann-Brand, B., Bukau, B., Deuerling, E. (2006). The C-terminal domain of *Escherichia coli* Trigger factor represents the central module of its chaperone activity. *J Biol Chem, 281*, 31963–31971.

38. Hesterkamp, T., Deuerling, E., Bukau, B. (1997). The amino-terminal 118 amino acids of *Escherichia coli* trigger factor constitute a domain that is necessary and sufficient for binding to ribosomes. *J Biol Chem, 272*, 21865–21871.

39. Tomic, S., Johnson, A.E., Hartl, F.U., Etchells, S.A. (2006). Exploring the capacity of Trigger factor to function as a shield for ribosome bound polypeptide chains. *FEBS Lett, 580*, 72–76.

40. Deuerling, E., Patzelt, H., Vorderwulbecke, S., Rauch, T., Kramer, G., Schaffitzel, E., Mogk, A., Schulze-Specking, A., Langen, H., Bukau, B. (2003). Trigger factor and DnaK possess overlapping substrate pools and binding specificities. *Mol Microbiol, 47*, 1317–1328.

41. Agashe, V.R., Guha, S., Chang, H.C., Genevaux, P., Hayer-Hartl, M., Stemp, M., Georgopoulos, C., Hartl, F.U., Barral, J.M. (2004). Function of Trigger factor and DnaK in multidomain protein folding: increase in yield at the expense of folding speed. *Cell, 117*, 199–209.

42. Genevaux, P., Keppel, F., Schwager, F., Langendijk-Genevaux, P.S., Hartl, F.U., Georgopoulos, C. (2004). In vivo analysis of the overlapping functions of DnaK and Trigger factor. *EMBO Rep, 5*, 1–6.

43. Wiedmann, B., Sakai, H., Davis, T.A., Wiedmann, M. (1994). A protein complex required for signal-sequence-specific sorting and translocation. *Nature, 370*, 434–440.

44. Beatrix, B., Sakai, H., Wiedmann, M. (2000). The alpha and beta subunit of the nascent polypeptide-associated complex have distinct functions. *J Biol Chem, 275*, 37838–37845.

45. Wegrzyn, R.D., Hofmann, D., Merz, F., Nikolay, R., Rauch, T., Graf, C., Deuerling, E. (2006). A conserved motif is prerequisite for the interaction of NAC with ribosomal protein L23 and nascent chains. *J Biol Chem, 281*, 2847–2857.

46. Pfund, C., Lopez-Hoyo, N., Ziegelhoffer, T., Schilke, B.A., Lopez-Buesa, P., Walter, W.A., Wiedmann, M., Craig, E.A. (1998). The molecular chaperone Ssb from *Saccharomyces cerevisiae* is a component of the ribosome-nascent chain complex. *EMBO J, 17*, 3981–3989.

47. Gautschi, M., Lilie, H., Funfschilling, U., Mun, A., Ross, S., Lithgow, T., Rucknagel, P., Rospert, S. (2001). RAC, a stable ribosome-associated complex in yeast formed by the DnaK–DnaJ homologs Ssz1p and zuotin. *Proc Nat Acad Sci, USA, 98*, 3762–3767.

48. Gautschi, M., Mun, A., Ross, S., Rospert, S. (2002). A functional chaperone triad on the yeast ribosome. *Proc Nat Acad Sci, USA, 99*, 4209–4214.

49. Bukau, B., Deuerling, E., Pfund, C., Craig, E.A. (2000). Getting newly synthesized proteins into shape. *Cell, 101*, 119–122.
50. Hundley, H.A., Walter, W., Bairstow, S., Craig, E.A. (2005). Human Mpp11 J protein: ribosome-tethered molecular chaperones are ubiquitous. *Science, 308*, 1032–1034.
51. Otto, H., Conz, C., Maier, P., Wolfle, T., Suzuki, C.K., Jeno, P., Rucknagel, P., Stahl, J., Rospert, S. (2005). The chaperones MPP11 and Hsp70L1 form the mammalian ribosome-associated complex. *Proc Nat Acad Sci, USA, 102*, 10064–10069.
52. Bukau, B., Horwich, A.L. (1998). The Hsp70 and Hsp60 chaperone machines. *Cell, 92*, 351–366.
53. Zhu, X.T., Zhao, X., Burkholder, W.F., Gragerov, A., Ogata, C.M., Gottesman, M.E., Hendrickson, W.A. (1996). Structural analysis of substrate binding by the molecular chaperone DnaK. *Science, 272*, 1606–1614.
54. Mayer, M.P., Schroder, H., Rudiger, S., Paal, K., Laufen, T., Bukau, B. (2000). Multistep mechanism of substrate binding determines chaperone activity of Hsp70. *Nat Struct Biol, 7*, 586–593.
55. Pellecchia, M., Montgomery, D.L., Stevens, S.Y., Vander Kooi, C.W., Feng, H.P., Gierasch, L.M., Zuiderweg, E.R. (2000). Structural insights into substrate binding by the molecular chaperone DnaK. *Nat Struct Biol, 7*, 298–303.
56. Sha, B., Lee, S., Cyr, D.M. (2000). The crystal structure of the peptide-binding fragment from the yeast Hsp40 protein Sis1. *Structure, 8*, 799–807.
57. Szabo, A., Langer, T., Schroder, H., Flanagan, J., Bukau, B., Hartl, F.U. (1994). The ATP hydrolysis-dependent reaction cycle of the *Escherichia coli* Hsp70 system DnaK, DnaJ, and GrpE. *Proc Nat Acad of Sci, USA., 91*, 10345–10349.
58. Harrison, C.J., Hayer-Hartl, M., Di Liberto, M., Hartl, F., Kuriyan, J. (1997). Crystal structure of the nucleotide exchange factor GrpE bound to the ATPase domain of the molecular chaperone DnaK. *Science, 276*, 431–435.
59. Hohfeld, J., Jentsch, S. (1997). GrpE-like regulation of the hsc70 chaperone by the anti-apoptotic protein BAG-1. *EMBO J, 16*, 6209–6216.
60. Raynes, D.A., Guerriero, V., Jr. (1998). Inhibition of Hsp70 ATPase activity and protein renaturation by a novel Hsp70-binding protein. *J Biol Chem, 273*, 32883–32888.
61. Kabani, M., McLellan, C., Raynes, D.A., Guerriero, V., Brodsky, J.L. (2002). HspBP1, a homologue of the yeast Fes1 and Sls1 proteins, is an Hsc70 nucleotide exchange factor. *FEBS Lett, 531*, 339–342.
62. Steel, G.J., Fullerton, D.M., Tyson, J.R., Stirling, C.J. (2004). Coordinated activation of Hsp70 chaperones. *Science, 303*, 98–101.
63. Dragovic, Z., Broadley, S.A., Shomura, Y., Bracher, A., Hartl, F.U. (2006). Molecular chaperones of the Hsp110 family act as nucleotide exchange factors of Hsp70s. *EMBO J, 25*, 2519–2528.
64. Raviol, H., Sadlish, H., Rodriguez, F., Mayer, M.P., Bukau, B. (2006). Chaperone network in the yeast cytosol: Hsp110 is revealed as an Hsp70 nucleotide exchange factor. *EMBO J, 25*, 2510–2518.
65. Sondermann, H., Scheufler, C., Schneider, C., Hohfeld, J., Hartl, F.U., Moarefi, I. (2001). Structure of a Bag/Hsc70 complex: convergent functional evolution of Hsp70 nucleotide exchange factors. *Science, 291*, 1553–1557.

66. Shomura, Y., Dragovic, Z., Chang, H.C., Tzvetkov, N., Young, J.C., Brodsky, J.L., Guerriero, V., Hartl, F.U., Bracher, A. (2005). Regulation of Hsp70 function by HspBP1: structural analysis reveals an alternate mechanism for Hsp70 nucleotide exchange. *Mol Cell, 17*, 367–379.

67. Liu, Q., Hendrickson, W.A. (2007). Insights into Hsp70 chaperone activity from a crystal structure of the yeast Hsp110 Sse1. *Cell, 131*, 106–120.

68. Polier, S., Dragovic, Z., Hartl, F.U., Bracher, A. (2008). Structural basis for the cooperation of Hsp70 and Hsp110 chaperones in protein folding. *Cell, 133*, 1068–1079.

69. Schuermann, J.P., Jiang, J.W., Cuellar, J., Llorca, O., Wang, L.P., Gimenez, L.E., Jin, S.P., Taylor, A.B., Demeler, B., Morano, K.A., et al. (2008). Structure of the Hsp110:Hsc70 nucleotide exchange machine. *Mol Cell, 31*, 232–243.

70. Klunker, D., Haas, B., Hirtreiter, A., Figueiredo, L., Naylor, D.J., Pfeifer, G., Muller, V., Deppenmeier, U., Gottschalk, G., Hartl, F.U., Hayer-Hartl, M. (2003). Coexistence of group I and group II chaperonins in the archaeon *Methanosarcina mazei. J Biol Chem, 278*, 33256–33267.

71. Braig, K., Otwinowski, Z., Hegde, R., Boisvert, D.C., Joachimiak, A., Horwich, A.L., Sigler, P.B. (1994). The crystal structure of the bacterial chaperonin GroEL at 2.8 A. *Nature, 371*, 578–586.

72. Fenton, W.A., Kashi, Y., Furtak, K., Horwich, A.L. (1994). Residues in chaperonin GroEL required for polypeptide binding and release. *Nature, 371*, 614–619.

73. Langer, T., Pfeifer, G., Martin, J., Baumeister, W., Hartl, F.U. (1992). Chaperonin-mediated protein folding: GroES binds to one end of the GroEL cylinder, which accommodates the protein substrate within its central cavity. *EMBO J, 11*, 4757–4765.

74. Hunt, J.F., Weaver, A.J., Landry, S.J., Gierasch, L., Deisenhofer, J. (1996). The crystal structure of the GroES co-chaperonin at 2.8 angstrom resolution. *Nature, 379*, 37–45.

75. Hlodan, R., Tempst, P., Hartl, F.U. (1995). Binding of defined regions of a polypeptide to GroEL and its implications for chaperonin-mediated protein folding. *Nat Struct Biol, 2*, 587–595.

76. Sigler, P.B., Xu, Z., Rye, H.S., Burston, S.G., Fenton, W.A., Horwich, A.L. (1998). Structure and function in GroEL-mediated protein folding. *Annu Rev Biochem*, 67, 581–608.

77. Martin, J., Langer, T., Boteva, R., Schramel, A., Horwich, A.L., Hartl, F.U. (1991). Chaperonin-mediated protein folding at the surface of groEL through a "molten globule"-like intermediate. *Nature, 352*, 36–42.

78. Robinson, C.V., Gross, M., Eyles, S.J., Ewbank, J.J., Mayhew, M., Hartl, F.U., Dobson, C.M., Radford, S.E. (1994). Conformation of GroEL-bound alpha-lactalbumin probed by mass spectrometry. *Nature, 372*, 646–651.

79. Coyle, J.E., Jaeger, J., Groß, M., Robinson, C.V., Radford, S.E. (1997). Structural and mechanistic consequences of polypeptide binding by GroEL. *Fold Des, 2*, R93–R104.

80. Horst, R., Fenton, W.A., Englander, S.W., Wuthrich, K., Horwich, A.L. (2007). Folding trajectories of human dihydrofolate reductase inside the GroEL GroES chaperonin cavity and free in solution. *Proc Nat Acad Sci, USA, 104*, 20788–20792.

81. Sharma, S., Chakraborty, K., Mueller, B.K., Astola, N., Tang, Y.C., Lamb, D.C., Hayer-Hartl, M., Hartl, F.U. (2008). Monitoring protein conformation along the pathway of chaperonin-assisted folding. *Cell, 133*, 142–153.
82. Farr, G.W., Furtak, K., Rowland, M.B., Ranson, N.A., Saibil, H.R., Kirchhausen, T., Horwich, A.L. (2000). Multivalent binding of nonnative substrate proteins by the chaperonin GroEL. *Cell, 100*, 561–573.
83. Elad, N., Farr, G.W., Clare, D.K., Orlova, E.V., Horwich, A.L., Saibil, H.R. (2007). Topologies of a substrate protein bound to the chaperonin GroEL. *Mol Cell, 26*, 415–426.
84. Landry, S.J., Zeilstra, R.J., Fayet, O., Georgopoulos, C., Gierasch, L.M. (1993). Characterization of a functionally important mobile domain of GroES. *Nature, 364*, 255–258.
85. Xu, Z., Horwich, A.L., Sigler, P.B. (1997). The crystal structure of the asymmetric GroEL–GroES–(ADP)7 chaperonin complex. *Nature, 388*, 741–750.
86. Martin, J., Mayhew, M., Langer, T., Hartl, F.U. (1993). The reaction cycle of GroEL and GroES in chaperonin-assisted protein folding. *Nature, 366*, 228–233.
87. Mayhew, M., da Silva, A.C., Martin, J., Erdjument-Bromage, H., Tempst, P., Hartl, F.U. (1996). Protein folding in the central cavity of the GroEL–GroES chaperonin complex. *Nature, 379*, 420–426.
88. Weissman, J.S., Rye, H.S., Fenton, W.A., Beechem, J.M., Horwich, A.L. (1996). Characterization of the active intermediate of a GroEL–GroES-mediated protein folding reaction. *Cell, 84*, 481–490.
89. Yifrach, O., Horovitz, A. (1995). Nested cooperativity in the ATPase activity of the oligomeric chaperonin GroEL. *Biochemistry, 34*, 5303–5308.
90. Horovitz, A., Fridmann, Y., Kafri, G., Yifrach, O. (2001). Review: allostery in chaperonins. *J Struct Biol, 135*, 104–114.
91. Saibil, H.R., Horwich, A.L., Fenton, W.A. (2001). Allostery and protein substrate conformational change during GroEL/GroES-mediated protein folding. *Adv Prot Chem, 59*, 45–72.
92. Rye, H.S., Roseman, A.M., Chen, S., Furtak, K., Fenton, W.A., Saibil, H.R., Horwich, A.L. (1999). GroEL–GroES cycling: ATP and nonnative polypeptide direct alternation of folding-active rings. *Cell, 97*, 325–338.
93. Lin, Z., Madan, D., Rye, H.S. (2008). GroEL stimulates protein folding through forced unfolding. *Nat Struct Mol Biol, 15*, 303–311.
94. Ellis, R.J. (2001). Molecular chaperones: inside and outside the Anfinsen cage. *Cur Biol, 11*, R1038–R1040.
95. Hayer-Hartl, M.K., Martin, J., Hartl, F.U. (1995). Asymmetrical interaction of GroEL and GroES in the ATPase cycle of assisted protein folding. *Science, 269*, 836–841.
96. Houry, W.A., Frishman, D., Eckerskorn, C., Lottspeich, F., Hartl, F.U. (1999). Identification of in vivo substrates of the chaperonin GroEL. *Nature, 402*, 147–154.
97. Gromiha, M.M., Pujadas, G., Magyar, C., Selvaraj, S., Simon, I. (2004). Locating the stabilizing residues in (alpha/beta)8 barrel proteins based on hydrophobicity, long-range interactions, and sequence conservation. *Proteins, 55*, 316–329.

98. Horwich, A.L., Fenton, W.A., Chapman, E., Farr, G.W. (2007). Two families of chaperonin: physiology and mechanism. *Annu Rev Cell Dev Biol, 23*, 115–145.

99. Brinker, A., Pfeifer, G., Kerner, M.J., Naylor, D.J., Hartl, F.U., Hayer-Hartl, M. (2001). Dual function of protein confinement in chaperonin-assisted protein folding. *Cell, 107*, 223–233.

100. Tang, Y.C., Chang, H.C., Chakraborty, K., Hartl, F.U., Hayer-Hartl, M. (2008). Essential role of the chaperonin folding compartment in vivo. *EMBO J, 27*, 1458–1468.

101. Baumketner, A., Jewett, A., Shea, J.E. (2003). Effects of confinement in chaperonin assisted protein folding: rate enhancement by decreasing the roughness of the folding energy landscape. *J Mol Biol, 332*, 701–713.

102. Jewett, A.I., Baumketner, A., Shea, J.E. (2004). Accelerated folding in the weak hydrophobic environment of a chaperonin cavity: creation of an alternate fast folding pathway. *Proc Nat Acad Sci, USA, 101*, 13192–13197.

103. Hayer-Hartl, M., Minton, A.P. (2006). A simple semiempirical model for the effect of molecular confinement upon the rate of protein folding. *Biochemistry, 45*, 13356–13360.

104. Tang, Y.C., Chang, H.C., Roeben, A., Wischnewski, D., Wischnewski, N., Kerner, M.J., Hartl, F.U., Hayer-Hartl, M. (2006). Structural features of the GroEL–GroES nano-cage required for rapid folding of encapsulated protein. *Cell, 125*, 903–914.

105. Minton, A.P. (2001). The influence of macromolecular crowding and macromolecular confinement on biochemical reactions in physiological media. *J Biol Chem, 276*, 10577–10580.

106. England, J., Lucent, D., Pande, V. (2008). Rattling the cage: computational models of chaperonin-mediated protein folding. *Curr Opin Struct Biol, 18*, 163–169.

107. Shtilerman, M., Lorimer, G.H., Englander, S.W. (1999). Chaperonin function: folding by forced unfolding. *Science, 284*, 822–825.

108. Thirumalai, D., Lorimer, G.H. (2001). Chaperonin-mediated protein folding. *Annu Rev Biophys Biomol Struct, 30*, 245–269.

109. Lin, Z., Rye, H.S. (2004). Expansion and compression of a protein folding intermediate by GroEL. *Mol Cell, 16*, 23–34.

110. Chaudhuri, T.K., Farr, G.W., Fenton, W.A., Rospert, S., Horwich, A.L. (2001). GroEL/GroES-mediated folding of a protein too large to be encapsulated. *Cell, 107*, 235–246.

111. Liou, A.K.F., Willison, K.R. (1997). Elucidation of the subunit orientation in CCT (chaperonin containing Tcp1) from the subunit composition of CCT microcomplexes. *EMBO J, 16*, 4311–4316.

112. Ditzel, L., Lowe, J., Stock, D., Stetter, K.O., Huber, H., Huber, R., Steinbacher, S. (1998). Crystal structure of the thermosome, the archaeal chaperonin and homolog of CCT. *Cell, 93*, 125–138.

113. Meyer, A.S., Gillespie, J.R., Walther, D., Millet, I.S., Doniach, S., Frydman, J. (2003). Closing the folding chamber of the eukaryotic chaperonin requires the transition state of ATP hydrolysis. *Cell, 113*, 369–381.

114. Vainberg, I.E., Lewis, S.A., Rommelaere, H., Ampe, C., Vandekerckhove, J., Klein, H.L., Cowan, N.J. (1998). Prefoldin, a chaperone that delivers unfolded proteins to cytosolic chaperonin. *Cell, 93*, 863–873.

115. Geissler, S., Siegers, K., Schiebel, E. (1998). A novel protein complex promoting formation of functional alpha- and gamma-tubulin. *EMBO J, 17*, 952–966.

116. Siegers, K., Waldmann, T., Leroux, M.R., Grein, K., Shevchenko, A., Schiebel, E., Hartl, F.U. (1999). Compartmentation of protein folding in vivo: sequestration of non-native polypeptide by the chaperonin–GimC system. *EMBO Journal, 18*, 75–84.

117. Leroux, M.R., Fandrich, M., Klunker, D., Siegers, K., Lupas, A.N., Brown, J.R., Schiebel, E., Dobson, C.M., Hartl, F.U. (1999). MtGimC, a novel archaeal chaperone related to the eukaryotic chaperonin cofactor GimC/prefoldin. *EMBO J, 18*, 6730–6743.

118. Siegert, R., Leroux, M.R., Scheufler, C., Hartl, F.U., Moarefi, I. (2000). Structure of the molecular chaperone prefoldin: unique interaction of multiple coiled coil tentacles with unfolded proteins. *Cell, 103*, 621–632.

119. Martin-Benito, J., Boskovic, J., Gomez-Puertas, P., Carrascosa, J.L., Simons, C.T., Lewis, S.A., Bartolini, F., Cowan, N.J., Valpuesta, J.M. (2002). Structure of eukaryotic prefoldin and of its complexes with unfolded actin and the cytosolic chaperonin CCT. *EMBO J, 21*, 6377–6386.

120. Frydman, J., Nimmesgern, E., Ohtsuka, K., Hartl, F.U. (1994). Folding of nascent polypeptide chains in a high molecular mass assembly with molecular chaperones. *Nature, 370*, 111–117.

121. Cuellar, J., Martin-Benito, J., Scheres, S.H.W., Sousa, R., Moro, F., Lopez-Vinas, E., Gomez-Puertas, P., Muga, A., Carrascosa, J.L., Valpuesta, J.M. (2008). The structure of CCT-Hsc70(NBD) suggests a mechanism for Hsp70 delivery of substrates to the chaperonin. *Nat Struct Mol Biol, 15*, 858–864.

122. Gavin, A.C., Bosche, M., Krause, R., Grandi, P., Marzioch, M., Bauer, A., Schultz, J., Rick, J.M., Michon, A.M., Cruciat, C.M., et al. (2002). Functional organization of the yeast proteome by systematic analysis of protein complexes. *Nature, 415*, 141–147.

123. Gao, Y., Thomas, J.O., Chow, R.L., Lee, G.H., Cowan, N.J. (1992). A cytoplasmic chaperonin that catalyzes beta-actin folding. *Cell, 69*, 1043–1050.

124. Yaffe, M.B., Farr, G.W., Miklos, D., Horwich, A.L., Sternlicht, M.L., Sternlicht, H. (1992). TCP1 complex is a molecular chaperone in tubulin biogenesis. *Nature, 358*, 245–248.

125. Ho, Y., Gruhler, A., Heilbut, A., Bader, G.D., Moore, L., Adams, S.L., Millar, A., Taylor, P., Bennett, K., Boutilier, K., et al. (2002). Systematic identification of protein complexes in *Saccharomyces cerevisiae* by mass spectrometry. *Nature, 415*, 180–183.

126. Valpuesta, J.M., Martin-Benito, J., Gomez-Puertas, P., Carrascosa, J.L., Willison, K.R. (2002). Structure and function of a protein folding machine: the eukaryotic cytosolic chaperonin CCT. *FEBS Lett, 529*, 11–16.

127. Spiess, C., Meyer, A.S., Reissmann, S., Frydman, J. (2004). Mechanism of the eukaryotic chaperonin: protein folding in the chamber of secrets. *Trends Cell Biol, 14*, 598–604.

128. Etchells, S.A., Meyer, A.S., Yam, A.Y., Roobol, A., Miao, Y.W., Shao, Y.L., Carden, M.J., Skach, W.R., Frydman, J., Johnson, A.E. (2005). The cotranslational contacts between ribosome-bound nascent polypeptides and the subunits of the hetero-oligomeric chaperonin TRiC probed by photocross-linking. *J Biol Chem, 280*, 28118–28126.

129. Spiess, C., Miller, E.J., McClellan, A.J., Frydman, J. (2006). Identification of the TRiC/CCT substrate binding sites uncovers the function of subunit diversity in eukaryotic chaperonins. *Mol Cell, 24*, 25–37.

130. Kafri, G., Horovitz, A. (2003). Transient kinetic analysis of ATP-induced allosteric transitions in the eukaryotic chaperonin containing TCP-1. *J Mol Biol, 326*, 981–987.

131. Shimon, L., Hynes, G.M., McCormack, E.A., Willison, K.R., Horovitz, A. (2008). ATP-induced allostery in the eukaryotic chaperonin CCT is abolished by the mutation G345D in CCT4 that renders yeast temperature-sensitive for growth. *J Mol Biol, 377*, 469–477.

132. Melki, R., Cowan, N.J. (1994). Facilitated folding of actins and tubulins occurs via a nucleotide-dependent interaction between cytoplasmic chaperonin and distinctive folding intermediates. *Mol Cell Biol, 14*, 2895–2904.

133. Muchowski, P.J., Schaffar, G., Sittler, A., Wanker, E.E., Hayer-Hartl, M.K., Hartl, F.U. (2000). Hsp70 and hsp40 chaperones can inhibit self-assembly of polyglutamine proteins into amyloid-like fibrils. *Proc Nat Acad Sci, USA, 97*, 7841–7846.

134. Schaffar, G., Breuer, P., Boteva, R., Behrends, C., Tzvetkov, N., Strippel, N., Sakahira, H., Siegers, K., Hayer-Hartl, M., Hartl, F.U. (2004). Cellular toxicity of polyglutamine expansion proteins: mechanism of transcription factor deactivation. *Mol Cell, 15*, 95–105.

135. Behrends, C., Langer, C.A., Boteva, R., Bottcher, U.M., Stemp, M.J., Schaffar, G., Rao, B.V., Giese, A., Kretzschmar, H., Siegers, K., Hartl, F.U. (2006). Chaperonin TRiC promotes the assembly of polyQ expansion proteins into nontoxic oligomers. *Mol Cell, 23*, 887–897.

136. Tam, S., Geller, R., Spiess, C., Frydman, J. (2006). The chaperonin TRiC controls polyglutamine aggregation and toxicity through subunit-specific interactions. *Nat Cell Biol, 8*, 1155–1162.

137. Kitamura, A., Kubota, H., Pack, C.G., Matsumoto, G., Hirayama, S., Takahashi, Y., Kimura, H., Kinjo, M., Morimoto, R.I., Nagata, K. (2006). Cytosolic chaperonin prevents polyglutamine toxicity with altering the aggregation state. *Nat Cell Biol, 8*, 1163–1170.

138. Balch, W.E., Morimoto, R.I., Dillin, A., Kelly, J.W. (2008). Adapting proteostasis for disease intervention. *Science, 319*, 916–919.

139. Sakahira, H., Breuer, P., Hayer-Hartl, M.K., Hartl, F.U. (2002). Molecular chaperones as modulators of polyglutamine protein aggregation and toxicity. *Proc Nat Acad Sci, USA, 99*, 16412–16418.

140. Sittler, A., Lurz, R., Lueder, G., Priller, J., Hayer-Hartl, M.K., Hartl, F.U., Lehrach, H., Wanker, E.E. (2001). Geldanamycin activates a heat shock response and inhibits huntingtin aggregation in a cell culture model of Huntington's disease. *Hum Mol Gene, 10*, 1307–1315.

141. Westerheide, S.D., Morimoto, R.I. (2005). Heat shock response modulators as therapeutic tools for diseases of protein conformation. *J Biol Chem, 280*, 33097–33100.

142. Scheufler, C., Brinker, A., Bourenkov, G., Pegoraro, S., Moroder, L., Bartunik, H., Hartl, F.U., Moarefi, I. (2000). Structure of TPR domain-peptide complexes: critical elements in the assembly of the Hsp70–Hsp90 multichaperone machine. *Cell, 101*, 199–210.

4

KINETIC MODELS FOR PROTEIN MISFOLDING AND ASSOCIATION

EVAN T. POWERS
Department of Chemistry, The Scripps Research Institute, La Jolla, California

FRANK A. FERRONE
Department of Physics, Drexel University, Philadelphia, Pennsylvania

INTRODUCTION

The veritable explosion in known protein aggregation diseases has led to intense work in attempting to understand the way in which the aggregated forms appear, with the ultimate goal of deciphering the underlying physical and chemical motifs that lead to these assembly processes. An important tool is the study of the kinetics of aggregate formation. The literature on protein aggregation is already rich, and the presence of powerful personal computers and robust and user-friendly software packages permits formulation and testing of many models never before constructed. Such approaches, although ill suited at excluding models, can surely demonstrate models that do reproduce the data. However, this creates its own sense of uneasiness, for the premises on which the models are based may not be readily apparent, and often the reader is left with the sole recourse of reconstructing the simulation in his or her own computing environment if further testing is sought. The purpose of this chapter is to summarize the main principles on which analysis rests, in order to present

Protein Misfolding Diseases: Current and Emerging Principles and Therapies,
Edited by Marina Ramirez-Alvarado, Jeffery W. Kelly, and Christopher M. Dobson
Copyright © 2010 John Wiley & Sons, Inc.

not only a broad survey of what is known but, also, a more intuitive feel for how certain assumptions become manifest in the resulting kinetic observations. Biochemists have long had such intuitive senses about reactions or methods, and this has served the community well. Yet it is important that this intuition be firmly grounded. For example, a false intuition is the often-quoted error that a reaction curve that has a "delay" achieves that shape due to nucleation. Thus, it is our goal to present the conceptual framework along with the rigorous descriptions and illustrative examples that can assist in developing appropriate instincts in this growing field. First, we illustrate the principles of linear aggregation processes, then focus on specific examples. Next, we consider theory and examples for processes that are more complex, such as secondary polymerization and alternative aggregation pathways. Finally, we consider briefly issues of still more intricacy, such as stochastics and crowding.

PRINCIPLES

The rate at which polymer mass accretes depends on the number of polymers, but not their length. Let the pool of monomers able to add to polymers be denoted A and have concentration a. Monomers attach to the ends at a rate $k_{+}a$ and leave the ends at a rate k_{-}. Let the mass incorporated into the polymers be designated as Δ. Then the rate of mass addition to the polymer is

$$\frac{d\Delta}{dt} = (k_{+}a - k_{-})p \equiv Jp \tag{1}$$

where p is the concentration of polymers and J is thus the *net* rate of elongation of a single fiber (once molar units are converted). A single fiber thus grows linearly in time, and differential interference contrast (DIC) microscopy images of fiber growth confirm this nicely [1]. A simple way to count polymers is by enumerating their ends, so that p may be considered the concentration of polymer ends. Net elongation includes a depolymerization rate, and the ratio k_{-}/k_{+} is the solubility (c_s) or critical concentration. Often (e.g., Aβ aggregation), k_{-} is almost zero, making the reactions seem irreversible, but with ingenuity and care, c_s can be evaluated [2,3].

The initial species may not be the species competent to aggregate. If the initial species is denoted C and its concentration is c, then C and A may be in rapid or slow equilibrium. If rapid, then a is simply proportional to c; that is, $a = Kc$ throughout the reaction, where K is the equilibrium constant between a and c. If the equilibrium is not rapid, a is kinetically related to c and

$$\frac{dc}{dt} = -k_{ca}c + k_{ac}a \tag{2}$$

Δ, c, and a represent all the possible states of monomers, so that their sum must be constant, that is,

$$c_0 = c(t) + a(t) + \Delta(t) \tag{3}$$

This represents an approximation that no significant oligomers exist that are not counted as polymers. This is not just a matter of definition. Equation (1) assumes that the length of the polymer is irrelevant to the addition and loss rates. That would not be true for small aggregates [4], so that if they were to exist in substantial numbers, additional assumptions would be necessary. A further assumption is that the initial concentration of a is negligible, so that effectively all the initial species are in the c conformation.

Polymer Formation

In kinetic models, the mechanism is usually related to the rate of formation of the polymers themselves, dp/dt, which is often physically governed by nucleation [5]. In directed polymerizations such as in actin or tubulin, a nucleus is a *stable* entity that serves to locate the polymer formation process at a biologically significant site. In pathological assembly processes, the nucleus may be an *unstable* species, which exerts control of the reaction by being the least likely step in the process.

A free-energy barrier diagram is helpful in understanding the kinetic process that underlies nucleation. Such a diagram, shown in Figure 1a, displays the free-energy change entailed in making an aggregate, ΔG. By definition, the monomer ΔG is zero. Positive free energy describes unfavorable species, and the presence of an unfavorable region between monomers and very long polymers creates a barrier. The nucleus, which acts as the rate-limiting species, is therefore the species at which the barrier peaks since the probability of finding a given species is proportional to $\exp(-\Delta G/RT)$. At sizes a few times larger than the nucleus, the free-energy curve approaches a straight line, meaning that each added monomer is equally stabilized. The rate of polymer elongation is related to the slope of that final curve. Thus, the linearity of that portion of the free-energy curve is what is measured when elongation is shown to proceed at a constant rate. Note further that in such barrier curves, a significant number of species have a free energy above zero. All these species (e.g., smaller than 14 mers in Figure 1) are suppressed relative to monomers, and may be relatively difficult to observe the higher the barrier gets. The reasons for such a barrier are discussed elsewhere [5], but, in brief, come from the changing balance between the entropy the monomers have when free versus the energy of the bonds made in the aggregate. As aggregates become larger, more bonds are made per molecule, eventually overcoming the entropic cost that makes smaller aggregates unstable.

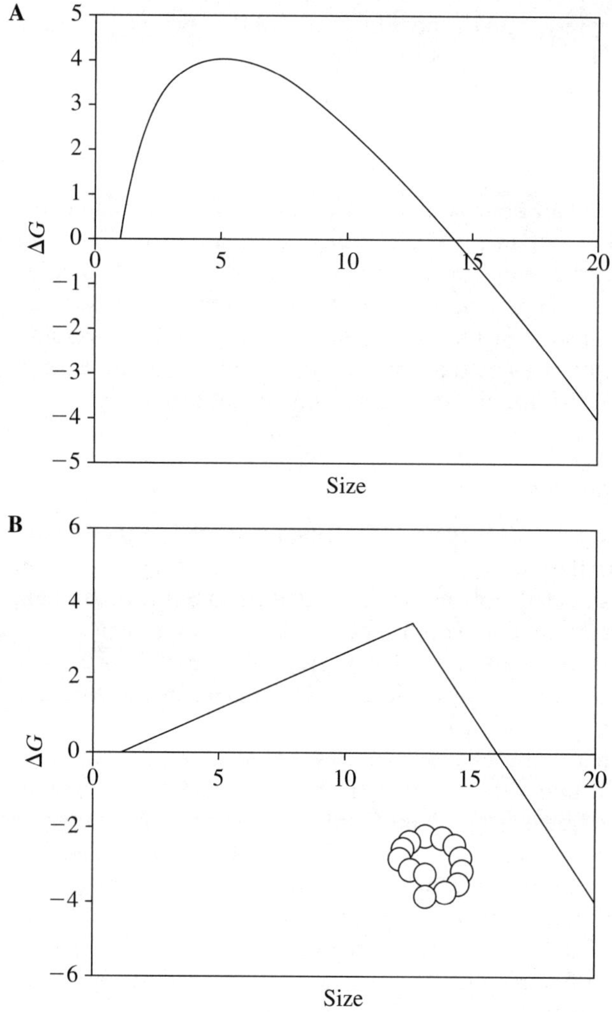

FIG. 1 Nucleation barriers. The free-energy change, ΔG, for making a given aggregate relative to the monomer as a reference state is shown for different-size aggregates. (a) Barrier for cluster formation, in which bond gain offsets losses due to entropic effects of restricting the monomers from their prior freedom of motions; (b) barrier formation for helix closure, in which an additional stabilization occurs when a linear strand can make up–down bonds once the strand is long enough. In both cases, the curve approaches a straight line for large sizes.

An often-used illustration of the barrier process based on a different principle is a ring or helix closure, in which the nucleus is the species just before the ring closure occurs, as shown in Figure 1b. This makes specific demands on the contacts [5]. In the plausible-looking model above, for example, the up–down contact made upon closure accounts for 0.6 unit of stability (seen by comparing the actual curve past the nucleus to where linear extrapolation would have placed it). It is difficult to see why, given the choice, a monomer would add at the end of the chain since it is clearly energetically favorable to add at the top and bottom. It is assumed that the linear chain that will eventually close is more stable than a "patch" hypothetically excised from the surface of the tubular polymer of similar size. The alternative is to have the nucleus be a part of the surface, an assumption used recently in viral capsid nucleation [6]. In that case, its turning point comes for the same reasons as seen in clusters: namely, that contact energy catches up with loss of free energy from the entropy of free motion of the monomers [5]. In short, a closure mechanism makes rather stringent physical demands on the system under consideration, despite appearing intuitive as well as mathematically simple.

The rate of formation of polymers, dp/dt, is given by

$$\frac{dp}{dt} \equiv f = k_+ a K_n a^n \tag{4}$$

This is the product of the addition of a polymerization-competent monomer to a nucleus, times the concentration of nuclei, taken as being in equilibrium with monomers. Note that no reverse rate is included (even though it may often be small). This is a conceptually important point. It assumes that the nucleus is the point at which the system "commits" and the reaction becomes locally downhill. Thus, the rate of polymer formation is just $k_+ K_n a^{n+1}$. This rate is labeled f: that is, the formation rate irrespective of its details, such as whether or not it is a nucleation process.

The solution of the nucleation process of equations (1) to (4) for the initial period of polymerization [7] is given by

$$\Delta = A(1 - \cos Bt) \approx \frac{1}{2} J k_+ K_n a^{n+1} t^2 \tag{5}$$

where $A = a/(n+1)$ and $B^2 = (n+1)Jk_+ K_n a^n$. The cosine solution is accurate to a slightly higher fractional extent than the t^2 solution, but it does not go to the point at which the oscillation of the cosine would appear.

The presence of a barrier accounts for the well-known effect of seeding. Introducing polymers to a supersaturated solution permits those polymers to grow at once, without the kinetic barrier that slows their initial production. Mathematically, it would add a term Jp^*t to equation (5), where p^* is the concentration of polymers that were seeded. Δ would thus begin with a term linear in t.

It might be tempting to write a composite equation that shows the rate of monomer disappearance proportional to a power of monomer concentration, such as [8,9]

$$-\frac{da}{dt} = \alpha a^n \tag{6}$$

Not only does this appear simple, it even permits a simple analytic solution in this irreversible limit. The concept behind the equation is that a monomer adds to an aggregate or nucleus of size $n-1$, and the aggregate of that size is taken as being in equilibrium with the monomers. Superficially this seems equivalent to the preceding ideas, but there are important differences. Equation (4) places the *rate* of polymer formation proportional to some power of a. Equation (6) places the concentration of polymers proportional to some power of a. However, polymer concentration does not obey such an equilibrium rule, which, for example, would imply that the concentration of polymers would diminish as a deceased. Equation (6) simply states that the *rate* at which polymers form diminishes but leaves intact those already made. Equation (6) *can* apply if $n = 1$, in which case the conceptual meaning is that a structural change is the rate-limiting step in an otherwise downhill polymerization process.

The key to nucleation is that monomers begin to "leak" across the barrier at a steady rate once equilibrium is established below the barrier (assumed rapid). As a result, the concentration of polymers p accumulates in proportion to time. With the number of polymers p thus rising in proportion to time [equation (4)], and each of the polymers itself growing at a constant rate, the mass of polymers represents one more integration [of equation (1)] with respect to time (taking J as a constant at its initial value), and thus leads to the t^2 dependence of the initial rate.

It is useful to extend this conceptual view to a downhill polymerization with an irreversible first step, as shown in Figure 2. Once the sample is prepared, every aggregated species is more stable than the next smallest species. If aggregates are defined as any species larger than monomers, monomer depletion is equivalent to polymer mass. Since the initial species aggregate, there is no distinction between C and A in this model. If we consider the rate of movement out of monomers, this leak is the same as that seen in the barrier-crossing leak in Figure 1a. Polymerization again proceeds as t^2. This will be shown rigorously shortly.

Finally, let us consider almost the same system but one in which a conformational rearrangement is needed, so the monomers competent to polymerize are less stable and plentiful than the monomers that are unconverted. Now the entire free-energy curve shifts upward, as shown by the dashed curve in Figure 2. Immediately after the system is prepared, molecules begin to leak into the rearranged state, from which they polymerize. The similarity to the process seen in classical nucleation as in Figure 1a is evident, but the nucleus size is 1. This is because it is really a conformational change rather than formation of an aggregate that is the rate-limiting step in creating the polymers.

These processes—downhill polymerization and monomeric nucleus—can be distinguished mathematically from one another and from nucleation per se. Beginning with equations (1) to (3), we consider the case where a monomeric

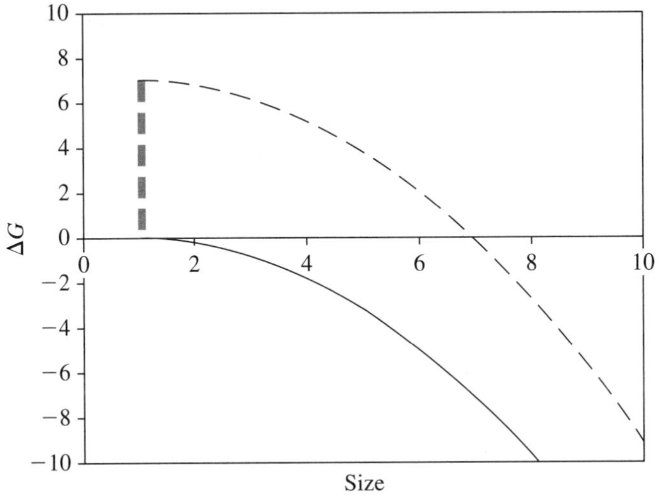

FIG. 2 Free-energy pictures for downhill polymerization and monomeric nuclei. The solid curve describes downhill polymerization in which each species is more stable than the next smaller one. If the species present initially must undergo an unfavorable conformational change, the dashed curve results. This is thus similar to the nucleation barriers of Figure 1 in that initially prepared monomers must pass though an unfavorable state before larger aggregates can be formed.

structural conversion gives rise to a species of concentration a which then polymerizes directly. There are two ways to describe the formation of polymers in lieu of equation (4). In one case we consider the formation limited by the rate at which species a encounter one another, which, lacking an established name, we shall call "double jeopardy." Thus, forming a dimer is the gateway to the existence of polymers, and the rate of polymer formation is expressed as

$$\frac{dp}{dt} = ka^2 \tag{7a}$$

Alternatively, in a "dock-and-lock" scenario [10], a species a need only encounter the original species c, in which case the rate of polymer formation becomes

$$\frac{dp}{dt} = kca \tag{7b}$$

Thereafter, polymer mass Δ forms according to equation (1). [There is a slight subtlety here. Equation (1) allows for reversible growth, but equation (7a) or (7b) involve irreversible polymer formation in dimerization. Thus, technically, the species a_2 is handled differently from the larger sizes. Given that this will be at a small concentration, the error in the simplification should be small.] k is the rate constant for dimer formation. It may be equivalent to the addition rate constant for infinite polymers, k_+, but this is not required.

Rigorous solution of these equations is not simple. Even adopting a perturbation approach, solution of the set using equation (7a) requires solution of second-order terms, because of the presence of a^2. However, insight can be had by breaking down the problem in the following way. Let the process of equation (7) be so small that we regard equations (2) and (3) as a complete system. Then

$$c(t) = c_0 - \frac{k_{ca}c_0}{k_{ca}+k_{ac}}\left(1 - e^{-(k_{ca}+k_{ac})t}\right) \tag{8}$$

and $a = c_0 - c$. Two extremes are of interest. First, suppose that $k_{ca}+k_{ac}$ is fast relative to the association process. Then c and a are in equilibrium, and

$$a = \frac{k_{ca}c_0}{k_{ca}+k_{ac}} \tag{9}$$

If we now add a slow leak to the system, so that association occurs, dp/dt will have a constant term [in either (7a) or (7b)], with a small leak term. So, to lowest order,

$$p = ka^2 t \quad \text{or} \quad p = kcat \tag{10}$$

from which it follows directly that

$$\Delta = \frac{1}{2}Jka^2t^2 \quad \text{or} \quad \Delta = \frac{1}{2}Jkc_0at^2 \tag{11}$$

where the first option is the double-jeopardy model and the second is the dock-and-lock model. This makes the rate of polymer formation look formally like a nucleation process since it has t^2 dependence. Moreover, if the dissociation rate (k_-) is small, J will be proportional to concentration, as are c and a, with the net effect that the concentration dependence is c^3. Finite k_- will make the apparent power of c less than 3. These results are seen in poly(Gln) [11], which gave rise to the monomeric nucleus concept. Note that this description does not distinguish a dock-and-lock mechanism from double jeopardy.

The other extreme is slow equilibration between C and A. Then

$$a = k_{ca}c_0t \tag{12}$$

Now

$$p = \frac{1}{3}k(k_{ca}c_0)^2t^3 \quad \text{or} \quad p = \frac{1}{2}kk_{ca}c_0^2t^2 \tag{13}$$

depending on whether double jeopardy (7a) or dock-and-lock (7b) is operative, and then

$$\Delta = \frac{1}{12}Jk(k_{ca}c_0)^2t^4 \quad \text{or} \quad \Delta = \frac{1}{6}Jkk_{ca}c_0^2t^3 \tag{14}$$

In this case, the dock-and-lock model should be distinguishable from the double-jeopardy model by its leading-term time dependence. Again the concentration dependence is about third order (depending on the back rate in J). (Since a rigorous perturbation expansion has not been carried out for this system, the results above may have additional terms in a more complete treatment.) Figure 3 illustrates the various time dependences.

Finally, we ask how the preceding should be modified if the polymerization is downhill. The simplest downhill system is one in which there is no distinction between A and C, (i.e., the monomers as prepared are competent to polymerize). Now polymers form according to equation (7a), but using c rather than a.

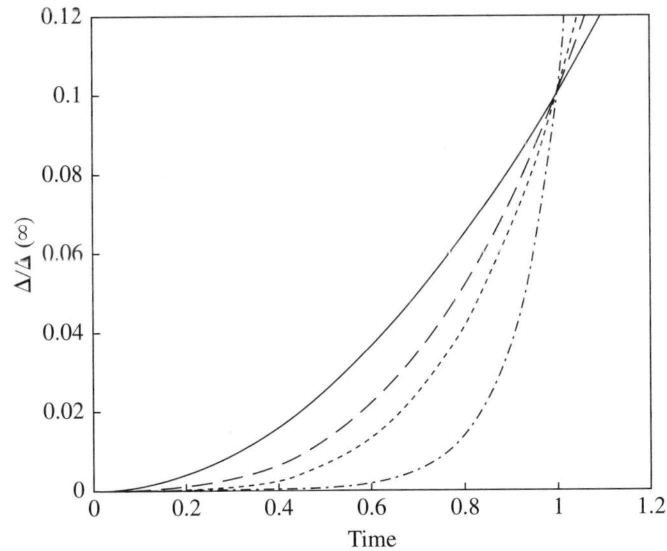

FIG. 3 Polymerization with different time dependences. All curves have been constructed to meet at the same final point (e.g., 1/10 of the reaction.) The least sharp curve (solid line) is t^2, followed by t^3 (dashed line) and t^4 (dotted line). The latter displays a much more pronounced lag phase. Most apparent as a lag is the exponential curve (dotted-dashed line), which is constructed arbitrarily here as A exp($10t$), with $A = 4.5 \times 10^{-5}$.

With no difference in a and c, (7b) is now equivalent to (7a). The resulting equations can be solved in a perturbation approach with the result

$$\Delta(t) = \frac{1}{2} c_0(1 - \cos Bt) \tag{15}$$

where $B^2 = 2kc_0J$, with k being the dimer association rate constant as in equation (7). The form of this equation is identical to that of nucleated polymerization in equation (5). It has t^2 as its leading term, as anticipated, and from inspection of the kinetic curves and their concentration dependence, is indistinguishable from the behavior of a system with monomeric nucleus. As a final modification of downhill polymerization, we consider the case of slow conformational change, which creates species that associate directly, and do so more rapidly than the conformational change. The conformational change is rate limiting, and the process is simple decay. Monomers disappear at the rate $dc/dt = -k_{ca}c$, and their disappearance corresponds exactly to formation of polymers. Thus,

$$\Delta = c_0[1 - \exp(-k_{ca}t)] \tag{16}$$

This behavior has its leading term proportional to t, with linear dependence on c_0. Such a system, rate-limited by spontaneous monomer conformational change, would not be affected by seeding.

A fairly common tool for analysis is to record the time at which the polymer mass reaches a fixed fractional extent of the reaction. Often, this is the halfway point, but can be another value, such as 1/10. Assuming (for simplicity and clarity) that there is no depolymerization rate ($k_- = 0$), the fractional extent is $\Delta/c = \beta c^2 t^m$, where m can be 2, 3, or 4, depending on the particulars of the case above, and β is simply a collection of all the rate constants. [This does not apply for the case described in equation (16), however.] To make the argument less abstract, suppose that we seek a time τ when 10% of the reaction is complete. This tenth time satisfies the relation $0.1 = \alpha c^2 \tau^{\mathrm{m}}$, from which

$$\log \tau = -\frac{2}{m} \log c + \frac{1}{m} \log \frac{0.1}{\alpha} \tag{17}$$

Therefore, a plot of log τ as a function of log c will have a slope of $-2/m$. With the exceptions of equations (5) and (16), this result applies to all the cases considered above.

EXAMPLES

Monomeric Nuclei

Poly(Gln) association was the first system that behaves as a nucleated polymerization process with monomeric nuclei [11]. Subsequently, a slow-folding mutant of a beta-clam protein was found to have a monomeric nucleus [12], as was ataxin

3 [13]. The case of poly(Gln) illustrates several important points. Nucleation was suspected initially because of a lag time that could be diminished or abolished by the introduction of seeds, which is suggestive of nucleation. Monomer depletion proceeded as t^2, satisfying an essential feature of simple nucleation but with the unexpected feature that the concentration of nuclei was directly proportional to the monomer concentration rather than concentration to a power greater than 1. Although one might take exception with the naming of such a reaction as nucleation, it maintains the advantage of showing that the mechanism can be deduced in a very robust fashion, which could have distinguished conventional nucleation ($n > 1$) from this process. Two experimental details are of paramount importance. First, a rigorous disaggregation procedure is essential here as for many systems (and differences in the assurance of disaggregation immediately before the experiment may account for apparent discrepancies reported recently [14]). Second, the monomeric species were assayed by high-performance liquid chromatography, which has the virtue of linearity in response. This issue will be reconsidered shortly for other probes, such as light scattering.

Downhill Polymerization

Transthyretin, a component of plasma, has been implicated in several amyloid diseases. A monomeric variant (M-TTR) was investigated at low pH, with a number of unexpected consequences [15]. In contrast to poly(Gln), the reaction did not show acceleration upon seeding, arguing that nucleation did not limit the rate of growth. Kinetic studies showed a linear early time course (rather than t^2) when studied by TfT fluorescence. This led to the conclusion the polymerization was a downhill process. A striking feature, however, was that light scattering showed a substantial delay, with kinetic behavior well above t or t^2. Analysis of scattering using the half-times of the reaction gave results similar to those of TfT fluorescence, supporting the downhill conclusion. The concentration dependence was further in accord with the results described in equation (15).

Light scattering of long polymers is a good probe of mass, but as Hall and Minton have found [16], small polymers do not show the same scattering efficiency, so a kinetic process that generates short fibers that elongate could well have early values suppressed. This cautionary note is worthy of constant recollection, for the best kinetic model is no better than the linearity of the probes by which it is measured. In fact, neither the linear dependence of the TfT fluorescence nor the sharper delay seen in light scattering are directly consistent with downhill polymerization, suggesting that further inspection of the nature of both probes could be fruitful.

BEYOND LINEAR POLYMERIZATION

The foregoing describes the behavior of a system that essentially has a linear reaction coordinate, so that progress can simply be mapped as a movement through a series of steps that correspond to specific size aggregates. But two other important options must be discussed. In some cases, a secondary pathway becomes accessible once the primary pathways begin to be traversed. That is, polymers may in some way give rise to additional polymers by their own breakage, branching, or secondary nucleation. In other cases, aggregates can serve as competitors for the growing fibrillar species and thus also create a landscape that is richer than the one-dimensional barriers we have viewed thus far.

Secondary Processes

In such cases, polymers may be created by processes other than the primary ones (in this case not only the usual nucleation but perhaps also, downhill aggregation or monomeric nucleation). The hallmark of a secondary process is that the rate of creation of polymers is proportional to the mass of polymers already in place. This represents a unifying assumption that covers several options. For example, fragmentation may be proportional to length, which in turn is proportional to total mass. Lateral growth of new fibers may be proportional to surface area, which for a linear fixed-diameter polymer is again proportional to total polymer mass. Thus, the secondary polymer formation rate is given by

$$\left(\frac{dp}{dt}\right)_{\text{II}} = Q\Delta \tag{18}$$

We have denoted the rate with the subscript II rather than the label p, for a very profound reason: The polymers formed either way are deemed to be equivalent, so by later inspection there is no way to tell how a polymer formed. This principle may fail if fibril conversion to thicker polymers works by a sheath-type process, in which case the large polymers can in fact be distinguished from homogeneously nucleated ones. But, for example, a broken fiber is not, in principle, a distinguishable species from one of similar size grown de novo.

When polymers simply fragment, $Q = k_f$, the fragmentation rate constant. When new polymers branch, in downhill fashion, then $Q = k'_+ a$, where k'_+ is the rate of adding a monomer to the surface. When new polymers form by a heterogeneous nucleation process (i.e., on the surface of another polymer), $Q = k'_+ K' \; a^{m+1}$, where m is the heterogeneous nucleus size and K' is the equilibrium constant for forming that nucleus on a polymer, including the possibility that not all surface sites are equally effective. It is remarkable in a way that these three processes, as different as they are, give rise to kinetic curves of the same shape.

The usual characteristic that identifies a secondary pathway is exponential growth and correspondingly dramatic lag times, as seen in the exponential curve in Figure 3. In this figure it is immediately clear how distinct the delay is here from the t^2 curves that characterize linear polymerization. It has been shown [17] that the initial reaction proceeds according to

$$\Delta = A(\cosh Bt - 1) \tag{19}$$

in which $A = f/(Q - \partial f/\partial a)$ and $B^2 = J(Q - \partial f/\partial a)$. From these relations it is immediately clear that

$$B^2A = Jf \tag{20}$$

At short times this is again t^2 but rapidly becomes exponential. If A and B can be extracted from the kinetic curves, relationship (20) permits the primary polymer formation rate f to be known irrespective of how the secondary process behaves.

This is very important. Consider a strategy of taking a certain fractional extent of the reaction and measuring the time it takes Δ to reach that fractional extent. Although it is often the case that a halfway point is taken, this may fall outside the realm over which the approximations are valid, and thus a 10 or 15% extent may be more in keeping with the solution. The fractional extent is given by $\Delta/(c_0 - c_s)$, where s designates the solubility. (Of course, if the reaction is simple and its entire time course is described, the caveat above is moot.)

If the fractional extent is denoted F, the time to reach that extent can be shown to be

$$\tau_F = B^{-1} \ln \frac{2(c_0 - c_s)F}{A} \tag{21}$$

The dominant term is $1/B$, which is much more dependent on Q than on f. Thus, if Q is the result of breakage, it is easily possible to have the time to reach 10 or 15% with $a^{1/2}$ dependence even when the nucleation rate f is strongly concentration dependent. Fortunately, the concentration dependence of f can be extracted if the quantity B^2A is examined. However, this example illustrates the importance of determining, first and foremost, the time dependence of the reaction, since the implication of the concentration dependence of τ_F is very different for a reaction that has t^2 as its leading dependence than for a reaction with exponential growth.

It is also possible that f may not be nucleation (with $n > 1$) at all. Thus, a reaction might display a sharp lag time, with a sensitivity to seeding, which conventionally would be interpreted as nucleation driven. But with dependence on the square root of initial concentration, it could have no nucleation processes at all, thanks to downhill polymerization with fragmentation. Fortunately, the protocol described above will sort rigorously through such complications.

Secondary Reaction Examples

The classic secondary reaction is that of sickle hemoglobin (HbS) double nucleation [18,19]. Observation of light-scattering intensity showed progress curves with striking initial quiescence that proved to be exponential when examined with high resolution, yet had high concentration dependence. The model developed on the basis of the kinetic inference was ultimately observed directly [1] and rationalized in structural terms [20].

Because one polymer gives rise to others, the limit of dilute polymers usually sought for light scattering is intrinsically impossible. As time progresses, scattering overshoot occurs despite a demonstrable increase in polymer mass [21]. Especially due to concern for the overshoot, as well as complex events such as monomeric diffusion into growing domains, analysis has been focused on the initial 10 to 15% of the reaction. Fortunately, it has been found, by observing diffusion into regions of polymer formation, that the linear assumption is reasonable (Aprelev, Rotter, and Ferrone, unpublished results). However, this issue is intrinsic to branching or heterogeneous nucleation, for fibers in arrays cannot act as independent scatterers. Even fragmentation is suspect on that account, for broken segments are likely to be closer to one another than is desirable for independence of scattering. This again underlies the importance of multiple probes to ensure that detection does not contaminate the observations.

A secondary nucleation process has also been implicated for fiber formation by IAPP [22,23] and insulin [24,25]. Interestingly, the work on both IAPP [22] and HbS find that primary and secondary nucleation events have more similarities than differences. Fragmentation of fibers has been observed in actin filaments [26] and, more recently, invoked in yeast prion aggregation [9].

Nonfibrillar Aggregates in Protein Fibril Formation: Off- and On-Pathway Aggregates

Nonfibrillar species are often observed during the course of protein fibril formation reactions before mature fibrils appear [27–40]. In the literature these are variously called protofibrils (especially when they are elongated), micelles, spherical aggregates/oligomers, or simply aggregates/oligomers; we refer to them collectively as aggregates. Such aggregates are generally formed rapidly because they lack rigid internal structure; thus, they do not require structured nuclei, and their formation is downhill [41]. The effect of such aggregates on fibril formation kinetics depends on their role in the mechanism. The two most likely possibilities are that the aggregates are off-pathway (incapable of converting directly to fibrils) or on-pathway (capable of converting directly to fibrils, but not necessary for fibril formation).

Off-pathway aggregates exert their influence on fibril formation kinetics through their equilibrium with monomer [41]. This equilibrium has two effects. First, it limits the concentration of monomer available for fibril nucleation and growth. This limitation arises because the concentration of monomer in equilibrium with off-pathway aggregates cannot be higher than the critical concentration for aggregate formation (in the same way that the concentration of monomer in equilibrium with fibrils cannot be higher than the critical concentration for fibril formation). Second, as monomer is depleted by fibril formation, off-pathway aggregates release monomers into solution to reestablish equilibrium. Thus, off-pathway aggregates buffer the monomer concentration, so that a decreases more slowly as fibril formation proceeds than it would in the absence of off-pathway aggregates. In fact, a can be nearly constant over much of the fibril formation time course when the total protein concentration is much higher than the critical concentration for aggregation.

The equilibrium between monomer and off-pathway aggregates manifests itself in fibril formation kinetics in several ways [41]. The limitation of the monomer concentration by off-pathway aggregation causes fibril formation to be much slower than it would be otherwise. It also leads to unusual behavior in log-log plots of the time required to reach a certain fraction completion (log τ) versus the total protein concentration (log c_0). Such plots are expected to have a negative slope, indicating that the time required for fibril formation decreases as the concentration increases. When there is off-pathway aggregation, however, this is true only if the protein concentration is below a certain value. Above this concentration, log τ actually increases as the protein concentration increases. It has been shown [41] that this "turnaround" value of the protein concentration is given approximately by

$$c_0 = \frac{c_A(n^2 + 4n + 3)}{4} \tag{22}$$

where c_A is the critical concentration for off-pathway aggregation and n is the size of the on-pathway nucleus. Finally, the constancy of a as fibril formation proceeds causes Δ to obey equation (5) over most of the fibril formation time course, rather than only the first 10 to 20% as in a simple nucleated polymerization.

To our knowledge, the general effects of on-pathway aggregates on fibril formation kinetics have not been studied in depth (although specific instances have been characterized; for example, see the work of Lomakin and co-workers [42,43]). It seems likely, however, that on- and off-pathway aggregates should have the same effects on the monomer concentration. Thus, on-pathway aggregates should both limit and buffer the monomer concentration. Unlike off-pathway aggregates, however, the monomer–aggregate equilibrium is not the only way that on-pathway aggregates affect fibril-formation kinetics. The direct conversion of on-pathway aggregates to fibrils can circumvent homogeneous nucleation, thereby accelerating fibril formation. The degree of acceleration depends on the concentration of on-pathway aggregates and on how fast they convert to fibrils. If we assume that on-pathway aggregate conversion to fibrils is much faster than nucleation and that it proceeds (by analogy with protein folding) by first-order kinetics, then

$$p = [Z]_0(1 - e^{-k_{con}t}) \tag{23}$$

where $[Z]_0$ is the initial concentration of aggregates and k_{con} is the first-order rate constant for the conversion of on-pathway aggregates to fibrils. Inserting equation (23) into (1) and solving (assuming that $k_- \sim 0$) yields

$$\begin{aligned} \Delta &= a[Z]_0 k_+ \left(\frac{e^{-k_{con}t} - 1}{k_{con}} + t \right) \\ &\approx \frac{1}{2} a[Z_0] k_+ k_{con} t^2 \end{aligned} \tag{24}$$

Equation (24) is valid for the early part of the fibril formation reaction, but recall that the on-pathway aggregates buffer the monomer concentration. Thus, equation (24) is valid over at least the first 10 to 20% of the fibril formation time course, and possibly substantially more.

OTHER SPECIAL CONSIDERATIONS

In this section we simply make mention of two increasingly important concepts and provide the reader with some appropriate references. The first issue is variously called *molecular crowding* or *solution nonideality*. This refers to the necessity of modifying a number of equations when solutions are substantially crowded (i.e., the surface-to-surface distance of molecules is significantly less than the center-to-center distance). Although present in vitro studies are in dilute solutions, there is some thought that in vivo the situation may not be so simple [44]. Fortunately, a number of treatments exist describing how one addresses nucleation and growth of polymers [45,46]. Unfortunately, the results are not especially simple although there are expressions that one can use for the calculations, as opposed to full numerical solutions.

A second issue is stochastic polymerization, the random behavior that arises when the observations all trace back to single molecular events. This was observed first in sickle hemoglobin polymerization because polymer formation from a single homogeneous nucleus could create large numbers of polymers by heterogeneous nucleation (thus rendering the originating event visible) [19,47,48]. With microscopic observation it was thus possible to observe a sample volume small enough that only one homogeneous event occurred. Thus, the hallmark of this behavior is randomness in polymerization that can be attributed to a single event in the volume observed. For example, multiple polymers in stirred solutions might all arise from an initial nucleation accompanied by substantial breakage. It is difficult to imagine a stochastic system in nonmicroscopic observation that does not employ a secondary pathway for amplification. Given the difficulties of dealing with many of the systems studied, however, it must be stressed that variation of response per se cannot distinguish microscopic stochastic variability from true sample-to-sample variation.

REFERENCES

1. Samuel, R.E., Salmon, E.D., Briehl, R.W. (1990). Nucleation and growth of fibres and gel formation in sickle cell haemoglobin. *Nature*, *345*, 833–835.
2. O'Nuallain, B., Shivaprasad, S., Kheterpal, I., Wetzel, R. (2005). Thermodynamics of Aβ (1–40) amyloid fibril elongation. *Biochemistry*, *44*, 12709–12718.
3. Sengupta, P., Garai, K., Sahoo, B., Shi, Y., Callaway, D.J., Maiti, S. (2003). The amyloid beta peptide (Abeta(1–40)) is thermodynamically soluble at physiological concentrations. *Biochemistry*, *42*, 10506–10513.
4. Hill, T.L. (1983). Length dependence of rate constants for end-to-end association and dissociation of equilibrium linear aggregates. *Biophys J*, *44*, 285–288.
5. Ferrone, F.A. (2006). Nucleation: the connections between equilibrium and kinetic behavior. *Methods Enzymol*, *412*, 285–299.
6. Zandi, R., van der Schoot, P., Reguera, D., Kegel, W., Reiss, H. (2006). Classical nucleation theory of virus capsids. *Biophys J*, *90*, 1939–1948.
7. Ferrone, F. (1999). Analysis of protein aggregation kinetics. *Methods Enzymol*, *309*, 256–274.
8. Andrews, J.M., Roberts, C.J. (2007). A Lumry–Eyring nucleated polymerization model of protein aggregation kinetics: 1. Aggregation with pre-equilibrated unfolding. *J Phys Chem B*, *111*, 7897–7913.
9. Collins, S.R., Douglass, A., Vale, R.D., Weissman, J.S. (2004). Mechanism of prion propagation: amyloid growth occurs by monomer addition. *PLoS Biol*, *2*, e321.
10. Esler, W.P., Stimson, E.R., Jennings, J.M., Vinters, H.V., Ghilardi, J.R., Lee, J.P., Mantyh, P.W., Maggio, J.E. (2000). Alzheimer's disease amyloid propagation by a template-dependent dock–lock mechanism. *Biochemistry*, *39*, 6288–6295.
11. Chen, S., Ferrone, F.A., Wetzel, R. (2002). Huntington's disease age-of-onset linked to polyglutamine aggregation nucleation. *Proc Natl Acad Sci USA*, *99*, 11884–11889.
12. Ignatova, Z., Gierasch, L.M. (2005). Aggregation of a slow-folding mutant of a beta-clam protein proceeds through a monomeric nucleus. *Biochemistry*, *44*, 7266–7274.
13. Ellisdon, A.M., Pearce, M.C., Bottomley, S.P. (2007). Mechanisms of ataxin-3 misfolding and fibril formation: kinetic analysis of a disease-associated polyglutamine protein. *J Mol Biol*, *368*, 595–605.
14. Lee, C.C., Walters, R.H., Murphy, R.M. (2007). Reconsidering the mechanism of polyglutamine peptide aggregation. *Biochemistry*. *46* 12810–12820.
15. Hurshman, A.R., White, J.T., Powers, E.T., Kelly, J.W. (2004). Transthyretin aggregation under partially denaturing conditions is a downhill polymerization. *Biochemistry*, *43*, 7365–7381.

16. Hall, D., Minton, A.P. (2005). Turbidity as a probe of tubulin polymerization kinetics: a theoretical and experimental re-examination. *Anal Biochem*, *345*, 198–213.

17. Bishop, M.F., Ferrone, F.A. (1984). Kinetics of nucleation controlled polymerization: a perturbation treatment for use with a secondary pathway. *Biophys J*, *46*, 631–644.

18. Ferrone, F.A., Hofrichter, J., Eaton, W.A. (1985). Kinetics of sickle hemoglobin polymerization: II. A double nucleation mechanism. *J Mol Biol*, *183*, 611–631.

19. Ferrone, F.A., Hofrichter, J., Eaton, W.A. (1985). Kinetics of sickle hemoglobin polymerization: I. studies using temperature-jump and laser photolysis techniques. *J Mol Biol*, *183*, 591–610.

20. Mirchev, R., Ferrone, F.A. (1997). The structural origin of heterogeneous nucleation and polymer cross linking in sickle hemoglobin. *J Mol Biol*, *265*, 475–479.

21. Cho, M.R., Ferrone, F.A. (1990). Monomer diffusion into polymer domains in sickle hemoglobin. *Biophys J*, *58*, 1067–1073.

22. Ruschak, A.M., Miranker, A.D. (2007). Fiber-dependent amyloid formation as catalysis of an existing reaction pathway. *Proc Natl Acad Sci USA*, *104*, 12341–12346.

23. Padrick, S.B., Miranker, A.D. (2002). Islet amyloid: phase partitioning and secondary nucleation are central to the mechanism of fibrillogenesis. *Biochemistry*, *41*, 4694–4703.

24. Mauro, M., Craparo, E.F., Podesta, A., Bulone, D., Carrotta, R., Martorana, V., Tiana, G., San Biagio, P.L. (2007). Kinetics of different processes in human insulin amyloid formation. *J Mol Biol*, *366*, 258–274.

25. Librizzi, F., Rischel, C. (2005). The kinetic behavior of insulin fibrillation is determined by heterogeneous nucleation pathways. *Protein Sci*, *14*, 3129–3134.

26. Wegner, A., Savko, P. (1982). Fragmentation of actin filaments. *Biochemistry*, *21*, 1909–1913.

27. Ahmad, A., Uversky, V.N., Hong, D., and Fink, A.L. (2005). Early events in the fibrillation of monomeric insulin. *J Biol Chem*, *280*, 42669–42675.

28. Baskakov, I.V., Legname, G., Baldwin, M.A., Prusiner, S.B., Cohen, F.E. (2002). Pathway complexity of prison protein assembly into amyloid. *J Biol Chem*, *277*, 21140–21148.

29. Bieschke, J., Zhang, Q.H., Powers, E.T., Lerner, R.A., Kelly, J.W. (2005). Oxidative metabolites accelerate Alzheimer's amyloidogenesis by a two-step mechanism, eliminating the requirement for nucleation. *Biochemistry*, *44*, 4977–4983.

30. Bitan, G., Lomakin, A., Teplow, D.B. (2001). Amyloid beta-protein oligomerization – prenucleation interactions revealed by photo-induced cross-linking of unmodified proteins. *J Biol Chem*, *276*, 35176–35184.

31. Conway, K.A., Lee, S.J., Rochet, J.C., Ding, T.T., Williamson, R.E., Lansbury, P.T. (2000). Acceleration of oligomerization, not fibrillization, is a shared property of both alpha-synuclein mutations linked to early-onset Parkinson's disease: Implications for pathogenesis and therapy. *Proc Natl Acad Sci USA*, *97*, 571–576.

32. Kaylor, J., Bodner, N., Edridge, S., Yamin, G., Hong, D.P., Fink, A.L. (2005). Characterization of oligomeric intermediates in alpha-synuclein fibrillation: FRET studies of Y125W/Y133F/Y136F alpha-synuclein. *J Mol Biol, 353*, 357–372.
33. Kirkitadze, M.D., Condron, M.M., Teplow, D.B. (2001). Identification and characterization of key kinetic intermediates in amyloid beta-protein fibrillogensis. *J Mol Biol, 312*, 1103–1119.
34. Plakoutsi, G., Bemporad, F., Calamai, M., Taddei, N., Dobson, C.M., Chiti, F. (2005). Evidence for a mechanism of amyloid formation involing molecular reorganisation within native-like precursor aggregates. *J Mol Biol, 351*, 910–922.
35. Rhoades, E., Agarwal, J., Gafni, A. (2000). Aggregation of an amyloidogenic fragment of human islet amyloid polypetide. *Biochim Biophys Acta, 1476*, 230–238.
36. Serio, T.R., Cashikar, A.G., Kowal, A.S., Sawicki, G.J., Moslehi, J.J., Serpell, L., Amsdorf, M.F., Lindquist, S.L. (2000). Nucleated conformational conversion and the replication of conformational information by a prion determinant. *Science, 289*, 1317–1321.
37. Sokolowski, F., Modler, A.J., Masuch, R., Zirwer, D., Baier, M., Lutsch, G., Moss, D.A., Gast, K., Naumann, D. (2003). Formation of critical oligomers is a key event during conformational transition of recombinant Syrian hamster prion protein. *J Biol Chem, 278*, 40481–40492.
38. Souillac, P.O., Uversky, V.N., Fink, A.L. (2003). Structural transformation of oligomeric intermediates in the fibrillation of the immunoglobulin light chain LEN. *Biochemistry, 42*, 8094–8104.
39. Souillac, P.O., Uversky, V.N., Millett, I.S., Khurana, R., Doniach, S., Fink, A.L. (2002). Elucidation of the molecular mechanism during the early events in immunoglobulin light chain amyloid fibrillation. Evidence for an off-pathway oligomer at acidic pH. *J Biol Chem, 277*, 12666–12679.
40. Teplow, D.B., Lazo, N.D., Bitan, G., Bernstein, S., Wyttenbach, T., Bowers, M.T., Baumketner, A., Shea, J.E., Urbanc, B., Cruz, L., Borreguero, J., Stanley, H.E. (2006). Elucidating amyloid beta-protein folding and assembly: A multidisciplinary approach. *Accounts Chem Res, 39*, 635–645.
41. Powers, E.T., Powers, D.L. (2008). Mechanisms of protein fibril formation: nucleated polymerization with competing off-pathway aggregation. *Biophys J, 94*, 379–391.
42. Lomakin, A., Chung, D.S., Benedek, G.B., Kirschner, D.A., Teplow, D.B. (1996). On the nucleation and growth of amyloid beta-protein fibrills: Detection of nuclei and quantitation of rate constants. *Proc Natl Acad Sci USA, 93*, 1125–1129.
43. Lomakin, A., Teplow, D.B., Krischner, D.A., Benedek, G.B. (1997). Kinetic theory of fibrillogenesis of amyloid beta-protein *Proc Natl Acad Sci USA, 94*, 7942–7947.
44. Minton, A.P. (2000). Implications of macromolecular crowding for protein assembly. *Curr Opin Struct Biol, 10*, 34–39.
45. Ferrone, F.A., Rotter, M.A. (2004). Crowding and the polymerization of sickle hemoglobin. *J Mol Recognit, 17*, 497–504.

46. Hall, D., Minton, A.P. (2004). Effects of inert volume-excluding macromolecules on protein fiber formation: II. Kinetic models for nucleated fiber growth. *Biophys Chem*, *107*, 299–316.
47. Hofrichter, J. (1986). Kinetics of sickle hemoglobin polymerization: III. Nucleation rates determined from stochastic fluctuations in polymerization progress curves. *J Mol Biol*, *189*, 553–571.
48. Szabo, A. (1988). Fluctuations in the polymerization of sickle hemoglobin: a simple analytical model. *J Mol Biol*, *199*, 539–542.

5

TOXICITY IN AMYLOID DISEASES

Massimo Stefani

Department of Biochemical Sciences and Research Centre on the Molecular Basis of Neurodegeneration, University of Florence, Florence, Italy

INTRODUCTION

Amyloid diseases are degenerative conditions resulting from the aggregation of one of approximately 20 misfolded peptides or proteins, each characteristic of a specific disease, into oligomers and ordered fibrillar polymers with peculiar structural features [1]. The fibrillar aggregates are found as cytoplasmic (aggresomes, Lewy or Russell bodies, fibrillary tangles) or nuclear inclusions [2,3] or, together with other molecules, as deposits of extracellular material (amyloid). The clinical importance of amyloid diseases stems from the high prevalence in the developed countries of some of them, including Alzheimer and Parkinson diseases and type 2 diabetes mellitus. Presently, over 20 different amyloid diseases, either familial, sporadic, or transmissible, are known [1], although other degenerative diseases with amyloid deposition have been described [4].

The existence of a direct link between aggregate formation and development of clinical symptoms (the amyloid hypothesis) is supported by an increasing number of biochemical and genetic studies [2,3,5,6]. Nonetheless, the causes favoring the appearance in tissue of the most highly pathogenic species, their structural features, and the molecular basis of their cytotoxicity are still under intense investigation. This chapter focuses on recent data addressing some of these issues.

Protein Misfolding Diseases: Current and Emerging Principles and Therapies,
Edited by Marina Ramirez-Alvarado, Jeffery W. Kelly, and Christopher M. Dobson
Copyright © 2010 John Wiley & Sons, Inc.

BIOLOGICAL SURFACES CAN BE KEY TRIGGERS OF AMYLOID PRECURSOR PRODUCTION, AGGREGATION, AND TOXICITY

Recent research on the molecular basis of protein aggregation has highlighted surfaces as key enhancers of aggregate nucleation, in most cases the limiting step of protein fibrillization. Actually, synthetic or biological surfaces appear able to speed up and to favor protein and peptide aggregation in various ways. Surfaces, including macromolecules, can bind unfolded monomers, increasing their local concentration, modifying their structural features, and favoring aggregate nucleation [7,8] (Fig. 1A). These effects depend on the chemical features of the monomer, its folded or unfolded state, the way it interacts with the surface and the physicochemical properties of the latter, including density of charge and hydrophobicity; for lipid membranes, density of lipid packing, curvature, rigidity, or fluidity can also be important in determining the mode of monomer interaction (Fig. 1A). In turn, the adsorption of unfolded monomers or their aggregates can modify membrane features, affecting the activities of specific membrane proteins [9]; it can also favor secondary structure in the polypeptide chain [10,11], thus modifying lipid arrangement, with possible disruption of membrane integrity [12,13]. Conversely, any interaction of folded proteins with surfaces, notably with charged or hydrophobic surfaces, can alter the protein native structure, with exposure of hydrophobic core residues and the peptide backbone favoring aggregation-prone conformational states (Fig. 1B).

Actually, the physicochemical properties of the two-dimensional environment of a surface can be very different from those of the bulk aqueous phase. The strong electrostatic field or the nonpolar environment of heavily charged or hydrophobic surfaces can modify the protein fold with exposure to the surface of regions that normally are associated with each other through electrostatic or hydrophobic interactions [8]. This view agrees with experimental data showing that surfaces can catalyze the formation of amyloid aggregates by a mechanism substantially different from that occurring in bulk solution [7]. In turn, early oligomeric assemblies can interact with phospholipid bilayers, modifying their structure and permeability and impairing the function of membrane-bound proteins [14].

The relation among membrane lipid composition, the nucleation of early peptide–protein aggregates, and their ability to bind to, and to disassemble, membranes are supported by strong evidence that protein aggregation is stimulated by anionic surfaces, particularly by anionic lipid membranes or even by anionic macromolecules [7,14–16]. Accordingly, negatively charged surfaces have been proposed as potent inductors of β-sheet structures by acting as conformational catalysts for amyloids [14,15] and anionic lipid clusters in cell membranes as sites of preferential interaction with prefibrillar aggregates [15]. The enrichment in anionic phospholipids of the neuronal membranes of people with Alzheimer disease (AD) [17] agrees with these data, which also provide clues to explain the different vulnerability of neuronal populations to the toxic insult given by protein folding variants and their early aggregates.

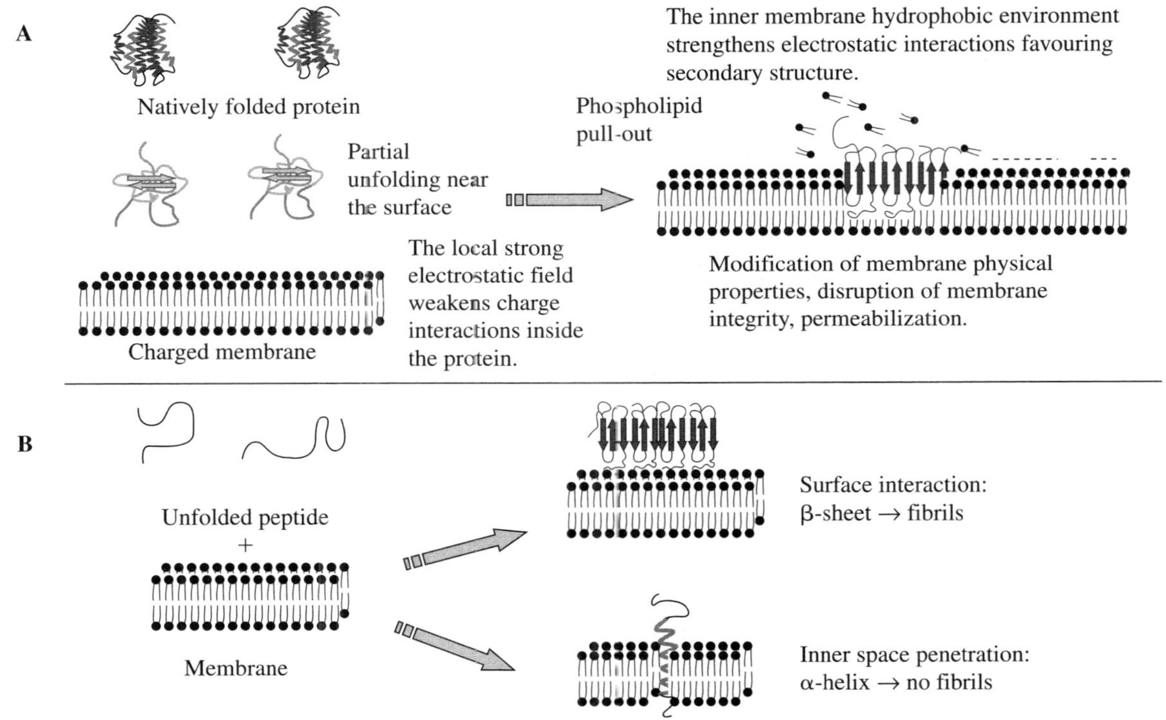

FIG. 1 (A) Natively unfolded proteins or peptides can gain different secondary structure upon interaction with a charged or hydrophobic surface; in most cases, in the presence of phospholipid surfaces, β-sheet structure, followed by oligomerization, predominates when the peptide molecules remain at the surface, whereas monomeric α-helix results when the peptide penetrates the hydrophobic interior. (B) In close proximity to a charged or hydrophobic surface, folded proteins can undergo partial unfolding following weakening of intramolecular electrostatic or hydrophobic interactions; under these conditions, nonnative intermolecular interactions are favored with possible amyloid nucleation. In the case of phospholipid bilayers or biological membranes, partial unfolding can favor penetration into the hydrophobic interior, where secondary interactions are strengthened.

Membrane Cholesterol Affects Cell Production of, and Resistance to, Aβ Peptides

The role of membrane cholesterol and gangliosides in modulating protein aggregation at the membrane level and aggregate interaction with the cell membranes has been investigated in detail. However, conflicting data on the effect of cholesterol on aggregate production and toxicity have been reported [18,19]. The relation between cholesterol and AD is paradigmatic. The positive relationship between hypercholesterolemia and risk of sporadic AD has long been known; however, a loss of cholesterol in brain leads to neurodegeneration [20], and reduced levels of cholesterol are found in brains from AD patients [21]. Possible clues can be given by lipid rafts, ganglioside- and cholesterol-enriched dynamic membrane microdomains harboring many membrane proteins, including APP and secretases [22]. The involvement of lipid rafts in APP processing could provide a rationale for the cholesterol–AD link; actually, altered cholesterol content could favor the amyloidogenic or nonamyloidogenic pathway of APP processing with increased or reduced Aβ(40/42) production [23–25]. Two alternative hypotheses on the effect of cholesterol on AD pathogenesis have been proposed.

The high-cholesterol hypothesis stems from several sets of data indicating that in peripheral and neuronal cell lines low cholesterol and increased membrane fluidity stimulate the nonamyloidogenic pathway (α/γ-secretase cleavage) following either reduced activity of BACE1 and increased α-secretase activity and the amount of APP at the cell surface, where it can undergo α-secretase cleavage [23]. The generation of amyloidogenic peptides arising from membrane processing of other proteins [26] could be affected by membrane lipid composition as well.

The alternative low-cholesterol hypothesis is supported by many data indicating that cholesterol loss in neuronal membranes enhances amyloid peptide generation [20,27]. Moreover, in the hippocampus of humans and transgenic mice, only a very moderate amount of APP and BACE1 is found in the same membrane environment, and the colocalized APP and BACE1 fraction increases markedly in AD patients and in rodents with a moderate reduction of brain cholesterol [24]. Altered lipid composition with ganglioside increase, cholesterol decrease, and lipid raft alterations has been found in temporal cortex samples from AD brains [28]. Finally, a recent study has shown that the brain cholesterol metabolite 24S-hydroxycholesterol in cultured SH-SY5Y neuroblastoma cells exerts a unique modulatory role on APP processing by reducing BACE1 activity and increasing α-secretase activity [29]. The protective effect of cell cholesterol against Aβ production and cytotoxicity is supported further by more recent investigations on the relation between seladin-1, a desmosterol reductase involved in cholesterol synthesis, lipid raft organization, cholesterol levels, and Aβ generation in neuronal cells [25]. The beneficial effect of increased cholesterol following seladin-1 overexpression has recently been confirmed in a study carried out on cultured cells [30].

Amyloid Receptors

The possible presence of amyloid aggregate receptors or specific interaction sites is another way that cell membranes can be involved in amyloid toxicity. These receptors could be specific for the shared cross-β fold rather than for any peculiar structural feature of the monomer, although in some cases they could also be monomer-specific, as in the Aβ-APP or Aβ-TNFR1 interactions proposed to be at the origin of Aβ cytotoxicity [31,32]. However, other studies argue against the presence of specific Aβ receptors [33]. Several cell surface proteins have been considered as possible receptors or preferential interaction sites for amyloids, including voltage-gated [34] or ligand-gated calcium channels such as the glutamate NMDA and AMPA receptors [35,36]; however, since 1996 the receptor for advanced glycation end products (RAGE) has been proposed as a major candidate as an amyloid receptor [37]. RAGE is increased in systemic amyloidoses, is able to interact with amyloid assemblies made from serum amyloid A, amylin, and prion-derived peptides [38], and appears to be involved in Alzheimer and Creutzfeldt–Jakob diseases as well [39,40]. By competing for ligand binding with cell-surface RAGE, its plasma-soluble form, sRAGE, might trap circulating ligands, preventing their interaction with cell surface receptors. Actually, sRAGE appears protective against cytotoxicity of transthyretin aggregates [41], and its high plasma levels are associated with a reduced risk of several diseases, including AD. Increasing plasma sRAGE is therefore considered a promising therapeutic target, potentially preventing vascular damage and neurodegeneration [42]. Finally, tissue-type plasminogen activator (tPA) has also been proposed as a multiligand site specific for the cross-β structure [43].

The presence of specific effects mediated through preferential, or even specific, interaction with membrane proteins could, at least in part, explain the variable vulnerability to amyloids of different cell types (see later). However, despite these and other data on specific interaction sites for amyloids [44], the tendency of early amyloid aggregates to interact with synthetic lipid membranes supports the idea that the interaction can be nonspecific but, possibly, modulated by the membrane lipid content (see above); it can also be able, by itself, to impair cell viability by altering membrane structure and permeability (see below).

AMYLOID AGGREGATES DISPLAY GENERIC TOXICITY

Presently, the amyloid hypothesis is supported by a large number of data on many in vivo and in vitro amyloidogenic proteins, indicating direct cytotoxicity of amyloids [45]. Increasing information supports the idea that oligomers arising in (or possibly off) the path of aggregation are more toxic than mature fibrils [46–48], although direct cytotoxicity of mature fibrils has also been reported [49–51], and the mechanisms and severity of cell impairment can vary in the presence of different aggregate conformations [52]. Moreover, in systemic amyloidoses, tissue impairment is believed to arise mainly from the diffusion barrier created by the huge amount of fibrillar material deposited, hindering secretion and gas or nutrient absorption [53].

Data that appeared in the last decade indicate that in most cases, cells experiencing toxic aggregates made from different peptides and proteins display similar early biochemical modifications [1,54]. This does not rule out the possibility of cell-specific early biochemical alterations or that different cells display differing downstream modifications as a consequence of similar early alterations. Nonetheless, many data support a scenario whereby in most cases early aggregates affect initially the functionality of cell membranes and some membrane proteins; the resulting derangement of ion homeostasis, redox status, and possibly, signaling pathways triggers a chain of events eventually culminating in cell death. Therefore, similar to the key role performed in protein–peptide aggregation (see above), cell membranes could play a central role even as targets of aggregate toxicity.

Cell Membranes Can Be Primary Targets of Aggregate Cytotoxicity

Amyloid aggregates can interact with cell components, including the plasma membrane (see above), microtubules [55], the endoplasmic reticularn (ER) [56], and mitochondria [57], where they can accumulate, causing dysfunction [58]. Increasing information highlighting similar toxic effects of early aggregates of peptides and proteins not associated with amyloid disease strongly supports the idea that amyloid cytotoxicity may be inherent in some shared structural feature of amyloids [54,59,60]. Although further studies on a wider range of peptides and proteins are required to better substantiate such a hypothesis, it appears that the cytotoxicity of protein folding variants and their early aggregates is generic [61]. It could arise from the "misfolded" nature of the monomeric or aggregated species with exposure of patches of hydrophobic residues and the polypeptide backbone normally buried into the compactly folded native state of the protein [48]. Many of these regions are likely to favor aggregation and interaction with cell membranes and other cellular components [59,61], triggering complex biochemical modifications culminating in cell death. The latter idea is supported by the intrinsic instability of misfolded monomers and prefibrillar species, which enables them to further organise into more stable ordered structures and/or to interact with cellular components [62]. It also agrees with the notion that in general, stable, harmless mature fibrils are unable to stick to cell membranes.

Prefibrillar Amyloid Aggregates Can Permeabilize the Cell Membrane

One of the most intriguing proposals put forth to explain amyloid toxicity is the channel hypothesis. This proposes that a subpopulation of protofibrils typically composed of five to seven globular elements around 2.5 nm in diameter surrounding a central hole penetrate the cell membrane, providing it with nonspecific

INSTITUT MEDIATHEQUE PASTEUR

pores that impair selective permeability and modify cell ion content [14,63–65] (Fig. 2). The channel hypothesis stems from several lines of evidence. First, one of the shared early modifications in cells exposed to toxic aggregates is membrane permeabilization with passive, nonspecific Ca^{2+} influx, suggesting the occurrence of some alteration in the structural organization of the cell membrane (see below). Second, in their path of fibrillization, many proteins and peptides either associated or notassociated with amyloid diseases typically form doughnut-shaped early oligomeric assemblies with a central pore. These assemblies permeabilize synthetic lipid membranes and have been imaged either in the latter or in the plasma membranes of exposed cells [63,66–68]. Third, protein–peptide oligomerization in the cell membrane, providing it with pores, possibly of amyloid type, is widely found in living systems, from bacteria to mammals, where it represents a way in which the cell kills foreign or competing cells or suicides [1,69].

However, although the above-mentioned doughnut-shaped prefibrillar aggregates have been imaged as in-or off-path intermediates during the fibrillization of many peptides and proteins, it is not clear whether they are, indeed, formed and are directly responsible for amyloid cytotoxicity in vivo. Often, modifications of membrane permeability have been observed in cells exposed to misfolded monomers or their prefibrillar aggregates with no porelike appearance. In general, it is believed that interaction of the aggregates or their precursors with the cell membranes can, by itself, destroy the structural arrangement of the bilayer, creating disordered areas with loss of ion-selective permeability [52,70] (Fig. 2). Furthermore, at least in some cases, the mechanisms of aggregate toxicity can be much more complex than simply relying in some nonspecific cell permeabilization.

SHARED BIOCHEMICAL MODIFICATIONS IN CELLS EXPOSED TO TOXIC AGGREGATES

The intra- or extracellular presence of toxic aggregates can impair a number of cell functions, eventually leading to cell death by apoptosis or necrosis [71,72]. However, in most cases, initial alterations of fundamental cellular processes associated with membrane perturbation appear to underlie subsequent cell impairment. As reported above, increasing information points to a central role performed by early membrane permeabilization by toxic aggregates with intracellular redox status and free Ca^{2+} alterations [65,67,70], yet other causes have also been proposed. These include transcriptional derangement in polyQ extension diseases, microtubular transport alterations in polyQ, SOD, AD and tauopathies, excitotoxicity through early deregulation of the NMDA or AMPA receptors, and the cytotoxic effect of pro-inflammatory factors secreted by microglia in AD [35,36,55,73–77].

In general, cells experiencing toxic aggregates display a sharp increase in the levels of reactive oxygen and nitrogen species (ROS, RNS) with modification of

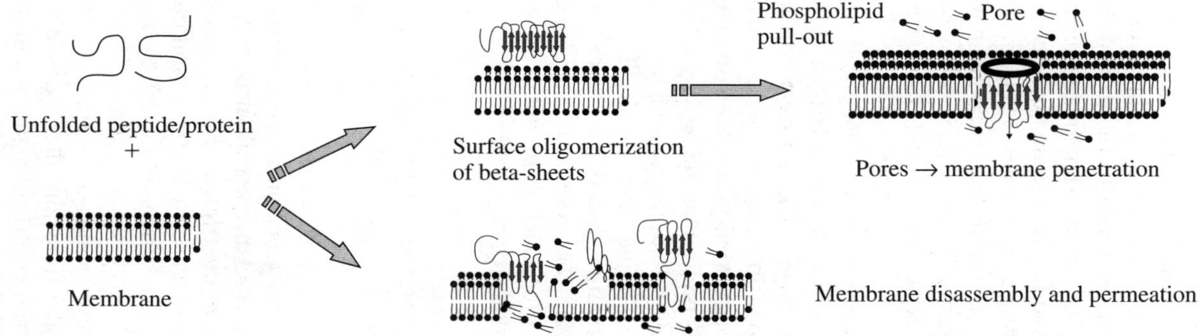

FIG. 2 Protein–peptide oligomerization on a phospholipid bilayer or a cell membrane results in membrane permeabilization. This can follow peptide assembly onto the membrane into β-sheets subsequently organizing inside the membrane into amyloid pores; permeabilization can also be the result of preformed amyloid pore interaction with the membrane, or membrane disassembly by peptide monomers or oligomers. These possibilities could be not mutually elusive. Most of the data on membrane permeabilization in cells exposed to amyloids highlight cytosolic-free Ca^{2+} increase.

the intracellular redox status. Oxidative stress results in overstimulation of excitatory glutamate receptors [77], lipid peroxidation, deregulation of NO metabolism, protein nitrosylation [76], and up-regulation of heme oxygenase-1 [78]. The key role of oxidative stress in amyloid aggregate cytotoxicity is underscored by many experimental data, including cell protection by antioxidants such as tocopherol, lipoic acid, or reduced glutathione [61,79]. In prion-infected mice, brain damage was shown to result from increased free-radical production and reduced efficacy of the antioxidant defenses in mitochondria [80]. Recent data in the nervous tissue point to direct effects of aging and oxidative stress on the activity and expression levels of the proteasome as well [81]. Proteasome inhibition can result in the accumulation of oxidized or otherwise damaged or misfolded proteins, reinforcing the detrimental effects of ROS. In this regard, a possible role of Hsp27 in preventing polyQ extension cytotoxicity by suppressing ROS production has been proposed [82].

The link between protein aggregation and ROS production needs further investigation. The generation of very reactive hydroxide radicals from hydrogen peroxide by metal ions has been proposed as a cause of oxidative stress [83]; in the case of Aβ42 or α-synuclein, specific mechanisms have also been claimed [84,85]. Finally, an up-regulation of membrane enzymes producing H_2O_2 (NADPH oxidase, cytochrome P450 reductase) has been reported in Aβ-exposed microglia and cortical neurons [86,87]. More generally, intracellular oxidative stress can be related to some structural and functional alteration of the cell membranes by the toxic aggregates (see above) with ion permeabilization, loss of regulation of membrane receptors and ion pumps [88], and/or mitochondrial function impairment. Mitochondria play a well-recognized role in oxidative stress and apoptosis; in this regard, a key factor in aggregate cytotoxicity could be the opening of permeability transition pores upon Ca^{2+} entry in neuronal mitochondria [89], with release of cytochrome *c* and other inducers of apoptosis.

It has been suggested that intracellular ROS elevation following exposure to amyloid aggregates can be an immediate consequence of the intracellular Ca^{2+} increase following membrane permeabilization through nonspecific pores or activation of glutamate-gated calcium channels (see above). Increased levels of free Ca^{2+} can stimulate the oxidative metabolism and ROS generation by activating the dehydrogenases of the citric acid cycle, providing the ATP needed by the ion pumps to clear the excess Ca^{2+}. In turn, oxidative stress can result in Ca^{2+} increase by alteration of the physicochemical (lipid peroxidation) and/or functional (decrease in ion pump activities) features of the cell membrane [90,91] with further intracellular free Ca^{2+} increase in a self-reinforcing loop [92]. Such a scenario can explain the relationship among ROS, intracellular Ca^{2+} increase, mitochondrial damage, and apoptosis described in cells exposed to toxic amyloid aggregates [60,61,91–93]. A similar chain of events could occur in old age, when cells are more susceptible to oxidative stress and their energy load is lower. Actually, many studies support a close link between Alzheimer, Parkinson, and prion diseases and calcium homeostasis deregulation. Recent data show that cells exposed to early aggregates of proteins unrelated to amyloid disease display similarly increased ROS and free Ca^{2+} levels with an apoptotic or a necrotic final outcome [59–61,72], further underscoring the generality of these modifications.

It is known that different cell types in tissue are variously affected by amyloids; for example, AD preferentially affects neurons, whereas glial cells remain intact, supporting the idea that synaptic dysfunction can be at the basis of the disease [44]. This and other findings support the necessity to investigate the biochemical features underlying the differing vulnerability of varying cell types to the same toxic prefibrillar aggregates of disease-associated or not-associated proteins. A recent report shows significant correlations between cell vulnerability, aggregate interaction with the plasma membrane, cholesterol content, total antioxidant capacity, and basal Ca^{2+}-ATPase activity in a panel of cultured cells [94]. These data support the importance of the aggregate-cell membrane interaction and the subsequent early modifications of free Ca^{2+} and redox status in triggering the chain of events culminating with cell death; they also agree with the idea that aggregate cytotoxicity and its early manifestations are generic features of amyloids.

CELLULAR CONTEXT

Intracellular Environment

Most of the cytotoxicity studies reported above have been carried out primarily by exposing cells to aggregates added to their culture media, whereas it is substantially unknown whether similar effects result from the presence of similar aggregates inside the cell. Besides favoring aggregate nucleation, any increase of the intracellular load of misfolded and aggregated proteins can saturate the ER and cytosolic quality control of protein folding, resulting in activation of the unfolded protein response or the heat-shock response, respectively [95].

Macromolecular crowding is a key, yet poorly considered factor affecting protein folding, stability, aggregation propensity, and aggregate interaction with cell surfaces [96]. The excluded volume effect in the cytosol and the ER can modify the rate of aggregation and aggregate stability, at least in the case of natively unfolded proteins and peptides. Actually, it can increase not only polypeptide folding into compact globular states but also their aggregation rate [97], possibly favoring closely associated states (mature fibrils) over asymmetric loosely folded states (early aggregates). Perhaps it can be worth noting that most proteins and peptides associated with amyloid diseases are natively unfolded or contain large unstructured regions [1]. Furthermore, the crowded intracellular milieu is very different from the cell-culture media where toxic amyloid oligomers are delivered in most toxicity experiments; here the outer surface of the plasma membrane can be reached more easily than the inner surface by the same aggregates that are present in the highly crowded intracellular space.

It must also be considered that the leaflets of the cell membranes display a characteristic asymmetry not only in protein, but also in lipid composition. For example, lipid rafts are found only in the outer leaflet of the cell membrane, whereas most anionic phospholipids are present in the inner leaflet in all cells other than cancer and apoptotic cells [98]. Accordingly, the inner leaflet displays a remarkable density of negative charge, whereas the outer leaflet is more rigid. These and other asymmetries can modulate monomer recruitment and aggregation at the cell membrane as well as aggregate interaction with the latter resulting in disassembly of the bilayer. Much less information is available on monomer recruitment on, and aggregate interaction with, the intracellular membranes and the resulting effects, as well as on the importance of the subcellular site(s) where the aggregating peptide or protein is generated. Recently, such an issue has been investigated for APP processing [25,99].

Finally, fibrils of the same monomer arising under different conditions can differ in stability and hence in resistance against breakage [100], a process possibly of importance in fibril proliferation. Actually, fibrils with a high frequency of breakage result in increased free ends, favoring not only further protein misfolding but also binding of misfolded monomers, competing with their oligomerization into toxic assemblies. Fibril stability can therefore affect the rate of progression of systemic amyloidoses and the transmission of specific prion strains [100].

Role of the Cell Functional State

The complexity of protein aggregate toxicity in vivo is also exemplified by its relation with the cell phenotype and functional state (including differentiation, resting, or proliferating state); in the case of dividing cells, the cell-cycle phase can also be of importance. It has been pointed out above that different cell types can be variously affected by the same toxic aggregates, depending on either specific biochemical features [94] or the efficiency of the machineries aimed at clearing misfolded proteins [95]. However, a given cell type can be in different functional and phenotypic states during its lifespan and it is intriguing (yet poorly investigated) to assess whether some of these states match different vulnerabilities to the aggregates.

The biochemical modifications underlying cell differentiation appear to modify cell vulnerability to early aggregates. Recent data support the idea that at least in some cases, undifferentiated cells can be less resistant to Aβ aggregates than differentiated cells [101], although different differentiation-associated phenotypes may result in variable effects [102]. Recent findings indicate that seladin-1, a desmosterol reductase involved in cholesterol synthesis, is highly expressed in undifferentiated neurogenic staminal cells found in AD-relevant brain areas, whereas its expression decreases after cell differentiation [103]; it also appears decreased in AD brains, possibly due to the impairment or loss of neurogenic staminal cells [103]. The latter possibility agrees with the reduced proliferation of neuronal precursors in animal models of AD [104]. However, extensive studies on the relation between cell differentiation and vulnerability to aggregate toxicity are still lacking, and different types of staminal cells could exhibit higher or reduced resistance to amyloid aggregates, with respect to their differentiated counterparts.

The cell-cycle hypothesis of amyloid toxicity in AD and, possibly, other neurodegenerative disorders proposes that the neurotoxicity of Aβ peptides and their aggregates arises from their ability to reactivate the cell cycle in terminally differentiated neurons [105]. The hypothesis stems from several sets of experimental evidence. The appearance of some phenotypic markers of cell cycling in vulnerable areas of AD brains has been demonstrated [106] and recent evidence indicates that postmitotic neurons can reenter the cell cycle [106]; pure cortical neurons or SH-SY5Y neuroblastoma cells challenged with Aβ peptides enter a mitotic cell cycle that does not progress beyond the S-phase and culminates in cell death [107]; neuronal death in early AD stages and in a *Drosophila* tauopathy model also results from mitotic activation of postmitotic neurons [108]; recent data on synchronized SH-SY5Y neuroblastoma cells suggest that S-phase cells display biochemical features, making them more vulnerable to amyloid toxicity [109]. These and other findings support previous data on cell-cycle-dependent apoptosis in several neurotoxic states [110] and in response to amylin aggregates [111]. The cell-cycle hypothesis provides a new perspective on amyloid toxicity, suggesting cell-cycle reactivation as a possible cause of death of differentiated cells in neuronal tissue, leading one to consider it as a potential therapeutic target in neurodegeneration.

FINAL CONSIDERATIONS

Increasing knowledge of protein folding, misfolding, and aggregation has provided a theoretical basis on which to better understand aggregate cytotoxicity and to start unraveling the molecular basis of amyloid diseases. Some key points are emerging. Biological surfaces, particularly cell membranes, play an important role in affecting protein and peptide aggregation and aggregate toxicity; early prefibrillar aggregates are primarily responsible for cell impairment, and their cytotoxicity is considered somehow generic; different exposed cells can display similar early biochemical alterations; shared biochemical and/or functional features can modulate cell resistance to aggregate toxicity.

The intrinsic potential of polypeptide chains to misfold and to polymerize into toxic aggregates and the generic toxicity of the latter suggest the existence of a negative selection against the appearance of amino acid sequences endowed with a significant tendency to aggregate; it also leads to a consideration of the molecular chaperones and folding quality control coevolution as a key driving force in protein evolution [95]. The phenotypic manifestations of mutations affecting the overall load of chaperones and the efficiency of the ubiquitin–proteasome system are likely to worsen during aging [95], when modifications of the intracellular environment can also occur. The resulting buildup of protein aggregates within the cell can further impair the ubiquitin–proteasome system, explaining the dramatic loss of neuronal function frequently underlying the progression of many neurodegenerative diseases [95,112,113].

Extension of these ideas has led us to hypothesize that a larger number of degenerative and other diseases than is presently known could result from cell viability impairment following continuous exposure to minute amounts of protein aggregates [59] resulting from specific mutations, chemical modification, or the general effects of aging. Actually, the number of diseases associated with the presence, in tissue, of protein aggregates is increasing steadily [4]. Furthermore, it is generally accepted that the rising prevalence of amyloid diseases in developed countries results from the remarkable increase in human life expectancy. Other causes can stem from unnatural practices or from medical treatments arising from evolutionary pressure. Examples are the recent outbreaks of borine spongitorin encephalapathy (BSE) (resulting from feeding bovines with prion-contaminated remains of old ones) and, in the past, cases of kuru or Creutzfeldt–Jakob disease associated with ritual cannibalism or to treatments with contaminated growth hormone, respectively [114]. Other medical treatments resulting in disease include insulin or β_2-microglobulin amyloidosis [1]. These considerations support the view that most sporadic amyloid conditions can be considered "civilization" or "postevolutionary" diseases, as they have became prevalent in developed countries due to the recent extension of our life span or to the introduction of new medical and farming practices [114,115].

Acknowledgments

The author is supported by grants from the Ènte Cassa di Rispàrmio di Firenze.

REFERENCES

1. Stefani, M., Dobson, C.M. (2003). Protein aggregation and aggregate toxicity: new insights into protein folding, misfolding diseases and biological evolution. *J Mol Med*, *81*, 678–699.
2. Kelly, J. (1998). Alternative conformation of amyloidogenic proteins and their multi-step assembly pathways. *Curr Opin Struct Biol*, *8*, 101–106.
3. Dobson, C.M. (2001). The structural basis of protein folding and its links with human disease *Philos Trans R Soc London*, *B*, *356*, 133–145.
4. Stefani, M. (2004). Protein misfolding and aggregation: new examples in medicine and biology of the dark side of the protein world. *Biochim Biophys Acta*, *1739*, 5–25.
5. van Horssen, J., Wilhelmus, M.M., Heljasvaara, R., Pihlajaniemi, T., Wesseling, P., de Waal, R.M., Verbeek, M.M. (2002). Collagen XVIII: a novel heparan sulfate proteoglycan associated with vascular amyloid depositions and senile plaques in Alzheimer's disease brains. *Brain Pathol*, *12*, 456–462.

6. Reilly, M.M. (1998). Genetically determined neuropathies *J Neurol, 245*, 6–13.
7. Zhu, M., Souillac, P.O., Ionesco-Zanetti, C., Carter, S.A., Fink, A.L. (2002). Surface-catalyzed amyloid fibril formation. *J Biol Chem, 277*, 50914–50922.
8. Sethuraman, A., Belfort, G. (2005). Protein structural perturbation and aggregation on homogeneous surfaces. *Biophys J, 88*, 1322–1333.
9. Hou, X., Richardson, S.J., Aguilar, M.I., Small, D.H. (2005). Binding of amyloidogenic transthyretin to the plasma membrane alters membrane fluidity and induces neurotoxicity. *Biochemistry, 44*, 11618–11627.
10. Bokvist, M., Lindström, F., Watts, A., Gröbner, G. (2004). Two types of Alzheimer's β-amyloid(1–40) peptide membrane interactions: aggregation preventing transmembrane anchoring versus accelerated surface fibril formation. *J Mol Biol, 335*, 1039–1049.
11. Yip, C.M., Elton, E.A., Darabie, A.A., Morrison, M.R., McLaurin, J. (2001). Cholesterol, a modulator of membrane-associated Aβ-fibrillogenesis and neurotoxicity. *J Mol Biol, 311*, 723–734.
12. Kazlauskaite, J., Senghera, N., Sylvester, I., Vénien-Bryan, C., Pinheiro, T.J.T. (2003). Structural changes of the prion protein in lipid membranes leading to aggregation and fibrillization. *Biochemistry, 42*, 3295–3304.
13. Porat, Y., Kolusheva, S., Jelinek, R., Gazit, E. (2003). The human islet amyloid polypeptide forms transient membrane-active prefibrillar assemblies. *Biochemistry, 42*, 10971–10977.
14. Stefani, M. (2007). Generic cell dysfunction in neurodegenerative disorders: role of surfaces in early protein misfolding, aggregation, and aggregate cytotoxicity. *Neuroscientist, 13*, 519–531.
15. Zhao, H., Tuominen, E.K.J., Kinnunen, P.K.J. (2004). Formation of amyloid fibers triggered by phosphatidylserine-containing membranes. *Biochemistry, 43*, 10302–10307.
16. Suk, J.Y., Zhang, F., Balch, W.E., Linhardt, R.J., Kelly, J.W. (2006). Heparin accelerates gelsolin amyloidogenesis. *Biochemistry, 45*, 2234–2242.
17. Wells, K., Farooqui, A.A., Liss, L., Horrocks, L.A. (1995). Neural membrane phospholipids in Alzheimer disease. *Neurochem Res, 20*, 1329–1333.
18. Koudinov, A.L., Koudinova, N.V. (2005). Cholesterol homeostasis failure as a unifying cause of synaptic degeneration. *J Neurol Sci, 229–230*, 233–240.
19. Sjögren, M., Mielke, M., Gustafson, D., Zandi, P., Skoog, I. (2006). Cholesterol and Alzheimer's disease: Is there a relation? *Mech Ageing Dev, 127*, 138–147.
20. Arispe, N., Doh, M. (2002). Plasma membrane cholesterol controls the cytotoxicity of Alzheimer's disease Aβ (1–40) and (1–42) peptides. *FASEB J, 16*, 1526–1536.
21. Mason, R.P., Shoemaker, W.J., Shajenko, L., Chambers, T.E., Herbette, L.G. (1992). Evidence for changes in the Alzheimer's disease brain cortical membrane structure mediated by cholesterol. *Neurobiol Aging, 13*, 413–419.
22. Allen, J.A., Halverson-Tamboli, R.A., Rasenick, M.M. (2007). Lipid raft microdomains and neurotransmitter signalling. *Nat Rev, 8*, 128–140.
23. Kojro, E., Gimpl, G., Lammich, S., März, W., Fahrenholz, F. (2001). Low cholesterol stimulates the nonamyloidogenic pathway by its effect on the α-secretase ADAM 10. *Proc Natl Acad Sci U S A, 98*, 5815–5820.

24. Abad-Rodriguez, J., Ledesma, M.D., Craessaerts, K., Perga, S., Medina, M., Delacourte, A., Dingwall, C., De Strooper, B., Dotti, C.G. (2004). Neuronal membrane cholesterol loss enhances amyloid peptide generation. *J Cell Biol, 167*, 953–960.

25. Crameri, A., Biondi, E., Kuehnle, K., Lutjohann, D., Thelen, K.M., Perga, S., Dotti, C.G., Nitsch, R.M., Ledesma, M.D., Mohajeri, M.H. (2006). The role of seladin-1/DHCR24 in cholesterol biosynthesis, APP processing and Abeta generation in vivo. *EMBO J, 25*, 432–443.

26. Townsend, M., Shankar, G.M., Mehta, T., Walsh, D.M., Selkoe, D.J. (2006). Effects of secreted oligomers on amyloid β-protein on hippocampal synaptic plasticity: a potent role for trimers. *J Physiol, 572(2)*, 477–492.

27. Ledesma, M.D., Dotti, C.G. (2005). The conflicting role of brain cholesterol in Alzheimer's disease: lessons from the brain plasminogen system. *Biochem Soc Symp, 72*, 129–138.

28. Molander-Melin, M., Blennow, K., Bogdanovic, N., Dellheden, B., Månsson, J.E., Fredman, P. (2005). Structural membrane alterations in Alzheimer brains found to be associated with regional disease development; increased density of gangliosides GM1 and GM2 and loss of cholesterol in detergent-resistant membrane domains. *J Neurochem, 92*, 171–182.

29. Famer, D., Meaney, S., Mousavi, M., Nordberg, A., Björkhem, I., Crisby, M. (2007). Regulation of alpha- and beta-secretase activity by oxysterols: cerebrosterol stimulates processing of APP via the alpha-secretase pathway. *Biochem Biophys Res Commun, 359*, 46–50.

30. Cecchi, C., Rosati, F., Pensalfini, A., Formigli, L., Nosi, D., Liguri, G., Dichiara, F., Morello, M., Danza, G., Pieraccini, G., et al. (2008). Seladin-1/DHCR24 protects neuroblastoma cells against abeta toxicity by increasing membrane cholesterol content. *J Cell Mol Med, 12*. 1990–2002.

31. Shadek, G.M., Kummer, M.P., Lu, D.C., Galvan, V., Bredesen, D.E., Koo, E.H. (2006). Aβ induces cell death by direct interaction with its cognate extracellular domain on APP (APP 597–624). *FASEB J, 20*, 1254–1266.

32. He, P., Zhong, Z., Lindholm, K., Berning, L., Lee, W., Lemere, C., Staufenbiel, M., Li, R., Shen, Y. (2004). Tumor necrosis factor death receptor signaling cascade is required for amyloid-β protein-induced neuron death. *J Neurosci, 24*, 1760–1771.

33. Cribbs, D.H., Pike, C.J., Weinstein, S.L., Velazquez, P., Cotman, C.W. (1997). All-D-enantiomers of beta-amyloid exhibit similar biological properties to all-L-beta-amyloids. *J Biol Chem, 272*, 7431–7436.

34. Hou, X., Parkington, H.C., Coleman, H.A., Mechler, A., Martin, L.L., Aguilar, M.-I., Small, D.H. (2007). Transthyretin oligomers induce calcium influx via voltage-gated calcium channels. *J Neurochem, 100*, 446–457.

35. Hsieh, H., Boehm, J., Sato, C., Iwatsubo, T., Tomita, T., Sisodia, S., Malinow, R. (2006). AMPAR removal underlies Aβ-induced synaptic depression and dendritic spine loss. *Neuron, 52*, 831–843.

36. De Felice, F.G., Velasco, P.T., Lambert, M.P., Viola, K., Fernandez, S.J., Ferreira, S.T., Klein, W.L. (2007). Aβ oligomers induce neuronal oxidative stress through an *N*-methyl-D-aspartate receptor-dependent mechanism that is blocked by the Alzheimer drug memantine. *J Biol Chem, 282*, 11590–11601.

37. Yan, S.D., Chen, X., Fu, J., Chen, M., Zhu, H., Roher, A., Slattery, T., Zhan, M., Nagashima, J., Morser, A., *et al.* (1996). RAGE and amyloid-beta peptide neurotoxicity in Alzheimer's disease. *Nature, 382*, 685–691.
38. Yan, S.D., Zhu, H., Zhu, A., Golabek, A., Du, H., Roher, A., Yu, J., Soto, C., Schmidt, A.M., Stern, D., Kindy, M. (2000). Receptor-dependent cell stress and amyloid accumulation in systemic amyloidosis. *Nat Med, 6*, 643–651.
39. Yan, S.D., Stern, D., Kane, M.D., Kuo, Y.M., Lampert, H.C, Roher, A.E. (1998). RAGE-Abeta interactions in the pathophysiology of Alzheimer's disease. *Restor Neurol Neurosci, 12*, 167–173.
40. Sasaki, N., Takeuchi, M., Chowei, H., Kikuchi, S., Hayashi, N., Nakano, H., Ikeda, S., Yamagishi, T., Kitamoto, T., Saito, T., Makita, Z. (2002). Advanced glycation end products (AGE) and their receptor (RAGE) in the brain of patients with Creutzfeldt–Jakob disease with prion plaques. *Neurosci Lett, 326*, 117–120.
41. Monteiro, F.A., Cardoso, I., Mendes Sousa, M., Saraiva, M.J. (2006). In vitro inhibition of transthyretin aggregate-induced cytotoxicity by full and peptide derived forms of the soluble receptor for advanced glycation end products (RAGE). *FEBS Lett, 580*, 3451–3456.
42. Geroldi, D., Falcone, C., Emanuele, E. (2006). Soluble receptor for advanced glycation end products: from disease marker to potential therapeutic target. *Curr Med Chem, 13*, 1971–1978.
43. Kranenburg, O., Bouma, B., Kroon-Batenburg, L.M.J., Reijerkerk, A., Wu, Y.-P., Voest, E.E., Gebbink, M.F.B.G. (2002). Tissue-type plasminogen activator is a multiligand cross-beta structure receptor. *Curr Biol, 12*, 1833–1839.
44. Lacor, P.N., Buniel, M.C., Chang, L., Fernandez, S.J., Gong, Y., Viola, K.L., Lambert, M., Velasco, P.T., Bigio, E.H., Finch, C.E., *et al.* (2004). Synaptic targeting by Alzheimer's-related amyloid β oligomers. *J Neurosci, 24*, 10191–10200.
45. Selkoe, D.J. (2003). Folding proteins in fatal ways. *Nature, 426*, 900–904.
46. Walsh, D.M., Selkoe, D.J. (2004). Oligomers on the brain: the emerging role of soluble protein aggregates in neurodegeneration. *Protein Pept Lett, 11*, 213–228.
47. Gazit, E. (2004). The role of prefibrillar assemblies in the pathogenesis of amyloid diseases. *Drugs Fut, 29*, 613.
48. Chiti, F., Dobson, C.M. (2006). Protein misfolding, functional amyloid, and human disease. *Annu Rev Biochem, 75*, 333–366.
49. Gharibyan, A.L., Zamotin, V., Yanamandra, K., Moskaleva, O.S., Margulis, B.A., Kostanyan, I.A., Morozova-Roche, L.A. (2007). Lysozyme amyloid oligomers and fibrils induce cellular death via different apoptotic/necrotic pathways *J Mol Biol, 365*, 1337–13349.
50. Novitskaya, V., Bocharova, O.V., Bronstein, I., Baskakov, I.V. (2006). Amyloid fibrils of mammalian prion protein are highly toxic to cultured cells and primary neurons. *J Biol Chem, 281*, 13828–13836.
51. Grudzielanek, S., Velkova, A., Shukla, A., Smirnovas, V., Tatarek-Nossol, M., Rehage, H., Kaporniotu, A., Winter, R. (2007). Cytotoxicity of insulin within its self-assembly and amyloidogenic pathways. *J Mol Biol, 370*, 372–384.
52. Deshpande, A., Mina, E., Glabe, C., Busciglio, J. (2006). Different conformations of amyloid β induce neurotoxicity by distinct mechanisms in human cortical neurons. *J Neurosci, 26*, 6011–6018.

53. Pepys, M.B. (1995). In *Oxford Textbook of Medicine,* 3rd ed.(Wheaterall, D.J., Ledingham, J.G., Warrel, D.A. Eds.), Oxford University Press, Oxford, UK, pp. 1512–1524.

54. Kayed, R., Head, E., Thompson, J.L., McIntire, T.M., Milton, S.C., Cotman, C.W., Glabe, C.G. (2003). Common structure of soluble amyloid oligomers implies common mechanisms of pathogenesis. *Science, 300*, 486–489.

55. King, M.E., Kan, H.-M., Baas, P.W., Erisir, A., Glabe, C.G., Bloom, S. (2006). Tau-dependent microtubule disassembly initiated by prefibrillar β-amyloid. *J Cell Biol, 175*, 541–546.

56. Ferreiro, E., Resende, R., Costa, R., Oliveira, C., Pereira, C.M.F. (2006). An endoplasmic-reticulum-specific apoptotic pathway is involved in prion and amyloid-beta peptides neurotoxicity. *Neurobiol Dis, 23*, 669–678.

57. Chen, X., Stern, D., Yan, S.D. (2006). Mitochondrial dysfunction and Alzheimer's disease. *Curr Alzheimer Res, 3*, 515–520.

58. Aleardi, A.M., Bernard, G., Augereau, O., Malgat, M., Talbot, J.C., Mazat, J.P., Letellier, T., Dachary-Prigent, J., Solaini, G.C., Rossignol, R. (2005). Gradual alteration of mitochondrial structure and function by beta-amyloids: importance of membrane viscosity changes, energy deprivation, reactive oxygen species production, and cytochrome *c* release. *J Bioenerg Biomembr, 37*, 207–225.

59. Bucciantini, M., Giannoni, E., Chiti, F., Baroni, F., Formigli, L., Zurdo, J., Taddei, N., Ramponi, G., Dobson, C.M., Stefani, M. (2002). Inherent toxicity of aggregates implies a common mechanism for protein misfolding diseases. *Nature, 416*, 507–511.

60. Sirangelo, I., Malmo, C., Iannuzzi, C., Mezzogiorno, A., Bianco, M.R., Papa, M., Irace, G. (2004). Fibrillogenesis and cytotoxic activity of the amyloid-forming apomyoglobin mutant W7FW14F. *J Biol Chem, 279*, 13183–13189.

61. Bucciantini, M., Calloni, G., Chiti, F., Formigli, L., Nosi, D., Dobson, C.M., Stefani, M. (2004). Pre-fibrillar amyloid protein aggregates share common features of cytotoxicity. *J Biol Chem, 279*, 31374–31382.

62. Kourie, J.I., Henry, C.L. (2002). Ion channel formation and membrane-linked pathologies of misfolded hydrophobic proteins: the role of dangerous unchaperoned molecules. *Clin Exp Pharmacol Physiol, 29*, 741–753.

63. Caughey, B., Lansbury, P.T. (2003). Protofibrils, pores, fibrils and neurodegeneration: separating the responsible protein aggregates from the innocent bystanders. *Annu Rev Neurosci, 26*, 267–298.

64. Kagan, B.L., Azimov, R., Azimova, R. (2004). Amyloid peptide channels. *J Membr Biol, 202*, 1–10.

65. Hirakura, Y., Kagan, B.L. (2001). Pore formation by beta-2-microglobulin: a mechanism for the pathogenesis of dialysis-associated amyloidosis. *Amyloid, 8*, 94–100.

66. Ding, T.T., Lee, S.-J., Rochet, J.-C., Lansbury, P.T. (2002). Annular α-synuclein protofibrils are produced when spherical protofibrils are incubated in solution or bound to brain-derived membranes. *Biochemistry, 41*, 10209–10217.

67. Quist, A., Duodevski, I., Lin, H., Azimova, R., Ng, D., Frangione, B., Kagan, B., Ghiso, J., Lal, R. (2005). Amyloid ion channels: a common structural link for protein misfolding diseases. *Proc Natl Acad Sci U S A, 102*, 10427–10432.

68. Relini, A., Torrassa, S., Rolandi, R., Ghiozzi, A., Rosano, C., Canale, C., Bolognesi, M., Plakoutsi, G., Bucciantini, M., Chiti, F., Stefani, M. (2004). Monitoring the process of HypF fibrillization and liposome permeabilization by protofibrils *J Mol Biol*, *338*, 943–957.
69. Lashuel, H.A., Lansbury, P.T. (2006). Are amyloid diseases caused by protein aggregates that mimic bacterial pore-forming toxins? *Q Rev Biophys*, *39*, 167–201.
70. Demuro, A., Mina, E., Kayed, R., Milton, S.C., Parker, I., Glabe, C.G. (2005). Calcium dysregulation and membrane disruption as a ubiquitous neurotoxic mechanism of soluble amyloid oligomers. *J Biol Chem*, *280*, 17294–17300.
71. Okouchi, M., Ekshyyan, O., Maracine, M., Aw, T.Y. (2007). Neuronal apoptosis in neurodegeneration. *Antioxid Redox Signal*, *9*, 1059–1096.
72. Bucciantini, M., Rigacci, S., Berti, A., Pieri, L., Cecchi, C., Nosi, D., Formigli, L., Chiti, F., Stefani, M. (2005). Patterns of cell death triggered in two different cell lines by HypF-N prefibrillar aggregates. *FASEB J*, *19*, 437–439.
73. Chevalier-Larsen, E., Holzbaur, E.L.F. (2006). Axonal transport and neurodegenerative disease. *Biochim Biophys Acta*, *1762*, 1094–1108.
74. Ross, C.A. (2002). Polyglutamine pathogenesis: emergence of unifying mechanisms for Huntington's disease and related disorders. *Neuron*, *35*, 819–822.
75. Brandt, R., Hundelt, M., Shahani, N. (2004). Tau alteration and neuronal degeneration in tauopathies: mechanisms and models. *Biochim Biophys Acta*, *1739*, 331–354.
76. Guentchev, M., Voigtlander, T., Haberler, C., Groschup, M.H., Budka, H. (2000). Evidence for oxidation stress in experimental prion disease. *Neurobiol Dis*, *7*, 270–273.
77. Shen, Y., He, P., Zhong, Z., McAllister, C., Lindholm, K. (2006). Distinct destructive signal pathways of neuronal death in Alzheimer's disease. *Trends Mol Med*, *12*, 574–579.
78. Choi, Y.G., Kim, J.L., Lee, H.P., Jin, J.K., Choi, E.K., Carp. R.I., Kim, Y.S. (2000). Induction of heme oxygenase-1 in the brain of scrapie-infected mice. *Neurosci Lett*, *11*, 173–176.
79. Zhang, L., Xing, G.Q., Barker, J.L., Chang, Y., Maric, D., Ma, W., Li, B.-S., Rubinow, D.R. (2001). α-Lipoic acid protects rat cortical neurons against cell death induced by amyloid and hydrogen peroxide through the Akt signalling pathway. *Neurosci Lett*, *312*, 125–128.
80. Lee, D.W., Sohn, H.O., Lim, H.B., Lee, Y.G., Kim, Y.S., Carp, R.J., Wisnievski, H.M. (1999). Alteration of free radical metabolism in the brain of mice infected with scrapie agent. *Free Radic Res*, *30*, 499–507.
81. Keller, J.N., Huang, F.F., Markesbery, W.R. (2002). Decreased levels of proteasome activity and proteasome expression in aging spinal cord. *Neuroscience*, *98*, 149–156.
82. Wyttenbach, A., Sauvageot, O., Carmichael, J., Diaz-Latoud, C., Arrigo, A.-P., Rubinsztein, D.C. (2002). Heat shock protein 27 prevents cellular polyglutamine toxicity and suppresses the increase of reactive oxygen species caused by huntingtin. *Hum Mol Genet*, *11*, 1137–1151.
83. Tabner, B.J., Turnbull, S., El-Agnaf, O.M.A., Allsop, D. (2002). Formation of hydrogen peroxide and hydroxyl radicals from Aβ and α-synuclein as a possible

mechanism of cell death in Alzheimer's disease and Parkinson's disease. *Free Radic Biol Med*, *32*, 1076–1083.

84. Brunelle, P., Rauk, A. (2002). The radical model of Alzheimer's disease: specific recognition of Gly29 and Gly33 by Met35 in a beta-sheet model of Abeta: an ONIOM study. *J Alzheimer's Dis*, *4*, 283–289.
85. Orth, M., Shapira, A.H.V. (2002). Mitochondrial involvement in Parkinson's disease. *Neurochem Int*, *40*, 533–541.
86. Pappolla, M.A., Omar, R.A., Chyan, Y.-J., Ghiso, J., Hsiao, K., Bozner, P., Perry, G., Smith, M.A., Cruz-Sanchez, F. (2001). Induction of NADPH cytochrome P450 reductase by the Alzheimer β-protein: amyloid as a "foreign body." *J Neurochem*, *78*, 121–128.
87. Qin, L., Liu, Y., Cooper, C., Liu, B., Wilson, B., Hong, J.-S. (2002). Microglia enhance β-amyloid peptide-induced toxicity in cortical and mesencephalic neurons by producing reactive oxygen species. *J Neurochem*, *83*, 973–983.
88. Mattson, M.P. (1999). Impairment of membrane transport and signal transduction systems by amyloidogenic proteins. *Methods Enzymol*, *309*, 733–768.
89. Moreira, P.I., Santos, M.S., Moreno, A., Rego, A.C., Oliveira, C. (2002). Effect of amyloid beta-peptide on permeability transition pore: a comparative study. *J Neurosci Res*, *15*, 257–267.
90. Butterfield, A.D., Drake, J., Pocernich, C., Castegna, A. (2001). Evidence of oxidative damage in Alzheimer's disease brain: central role for amyloid β-peptide. *Trends Mol Med*, *7*, 548–554.
91. Varadarajan, S., Yatin, S., Aksenova, M., Butterfield, D.A. (2000). Alzheimer's amyloid β-peptide-associated free radical oxidative stress and neurotoxicity. *J Struct Biol*, *130*, 184–208.
92. Squier, T.C. (2001). Oxidative stress and protein aggregation during biological aging. *Exp Gerontol*, *36*, 1539–1550.
93. Kawahara, M. (2004). Disruption of calcium homeostasis in the pathogenesis of Alzheimer's disease and other conformational diseases. *Curr Alzheimer Res*, *1*, 87–95.
94. Cecchi, C., Baglioni, S., Fiorillo, C., Pensalfini, A., Liguri, G., Nosi, D., Rigacci, S., Bucciantini, M., Stefani, M. (2005). Insights into the molecular basis of the differing susceptibility of varying cell types to the toxicity of amyloid aggregates *J Cell Sci*, *118*, 3459–3470.
95. Sherman, M.Y., Goldberg, A.L. (2001). Cellular defences against unfolded proteins: a cell biologist thinks about neurodegenerative diseases. *Neuron*, *29*, 15–32.
96. Ellis, R.J. (2001). Macromolecular crowding: an important but neglected aspect of the intracellular environment. *Curr Opin Struct Biol*, *11*, 114–119.
97. van den Berg, B., Ellis, J., Dobson, C.M. (1999). Effects of macromolecular crowding on protein folding and aggregation. *EMBO J*, *18*, 6927–6933.
98. Ran, S., Thorpe, P.E. (2002). Phosphatidylserine is a marker of tumour vasculature and a potential target for cancer imaging and therapy. *Int J Radiat Oncol Biol Phys*, *54*, 1479–1484.
99. Lee, E.B., Zhang, B., Liu, K., Geenbaum, E.A., Dams, R.W., Trojanowski, J.Q., Lee, V.M.-Y. (2005). BACE overexpression alters the subcellular processing of APP and inhibits Abeta deposition in vivo. *J Cell Biol*, *168*, 291–302.

100. Smith, J.F., Knowles, T.P.J., Dobson, C.M., MacPhee, C.E., Welland, M.E. (2006). Characterization of the nanoscale properties of individual amyloid fibrils. *Proc Natl Acad Sci U S A*, *103*, 15806–15811.

101. Datki, Z., Juhasz, A., Galfi, M., Soos, K., Papp, R., Zadori, D., Penke, B. (2003). Method for measuring neurotoxicity of aggregating polypeptides with the MTT assay on differentiated neuroblastoma cells. *Brain Res Bull*, *62*, 223–229.

102. Olesen, O.F., Dago, L., Mikkelsen, J.D. (1998). Amyloid beta neurotoxicity in the cholinergic but not in the serotonergic phenotype of RN46A cells. *Brain Res Mol Brain Res*, *57*, 266–274.

103. Benvenuti, S., Saccardi, R., Luciani, P., Urbani, S., Deledda, C., Cellai, I., Francini, F., Squecco, R., Rosati, F., Danza, G., et al. (2006). Neuronal differentiation of human mesenchymal stem cells: changes in the expression of the Alzheimer's disease–related gene seladin-1. *Exp Cell Res*, *312*, 2592–2604.

104. Haughey, N.J., Nath, A., Chan, S.L., Borchard, A.C., Rao, M.S., Mattson, M.P. (2002). Disruption of neurogenesis by amyloid beta-peptide, and perturbed neural progenitor cell homeostasis, in models of Alzheimer's disease. *J Neurochem*, *83*, 1509–1524.

105. Husseman, J.W., Nochlin, D., Vincent, I. (2000). Mitotic activation: a convergent mechanism for a cohort of neurodegenerative diseases. *Neurobiol Aging*, *21*, 815–828.

106. Nagy, Z. (2000). Cell cycle regulatory failure in neurones: causes and consequences. *Neurobiol Aging*, *21*, 761–769.

107. Frasca, G., Chiechio, S., Vancheri, C., Nicoletti, F., Copani, A., Sortino, M.A. (2004). Beta-amyloid-activated cell cycle in SH-SY5Y neuroblastoma cells: correlation with the MAP kinase pathway. *J Mol Neurosci*, *22*, 231–236.

108. Khurana, V., Lu, Y., Steinhilb, M.L., Oldham, S., Shulman, J.M., Feany, M.B. (2006). TOR-mediated cell-cycle activation causes neurodegeneration in a *Drosophila* tauopathy model. *Curr Biol*, *16*, 230–241.

109. Cecchi, C., Pensalfini, A., Stefani, M., Baglioni, S., Fiorillo, C., Cappadona, S., Caporale, R., Nosi, D., Ruggiero, M., Liguri, G. (2007). Vulnerability to amyloid toxicity depends on the cell-cycle phase in neuroblastoma replicating cells. *J Mol Med*, 86, 197–209.

110. Kruman, L.I., Wersto, R.P., Cardozo-Pelaez, F., Smilenov, L., Chan, S.L., Chrest, F.J., Emokpae, R., Jr., Gorospe, M., Mattson, M.P. (2004). Cell cycle activation linked to neuronal cell death initiated by DNA damage. *Neuron*, *41*, 549–561.

111. Ritzel, R.A., Butler, P.C. (2003). Replication increases beta-cell vulnerability to human islet amyloid polypeptide-induced apoptosis. *Diabetes*, *52*, 1701–1708.

112. Bence, N.F., Sampat, R.M., Kopito, R. (2001). Impairment of the ubiquitin–proteasome system by protein aggregation *Science*, *292*, 1552–1555.

113. Bonini, N.M. (2002). Chaperoning brain degeneration. *Proc Natl Acad Sci U S A*, *99* Suppl 4, 16407–16411.

114. Dobson, C.M. (2002). Getting out of shape. *Nature*, *418*, 729–730.

115. Csermely, P. (2001). Chaperone overload is a possible contributor to "civilization diseases." *Trends Genet*, *17*, 701–704.

6

AUTOPHAGY: AN ALTERNATIVE DEGRADATION MECHANISM FOR MISFOLDED PROTEINS

MARIA KON AND ANA MARIA CUERVO

Department of Developmental and Molecular Biology, Marion Bessin Liver Research Center, Institute for Aging Research, Albert Einstein College of Medicine, Bronx, New York

INTRODUCTION

Maintenance of cellular homeostasis is essential for proper cellular functioning. All cellular components undergo continuous synthesis and degradation, thus forcing a need for accurate quality control systems that guarantee the stability of the proteome in such a dynamic intracellular environment [1]. Chaperones and proteolytic systems both contribute to quality control inside cells (Fig. 1). The role of chaperones in quality control is reviewed extensively in other chapters of the book. Here we focus on the mechanisms of protein degradation, with particular emphasis on the lysosomal system. Both newly synthesized proteins, which do not reach their proper folded final conformation, and already folded functional proteins damaged by intra- or extracellular stressors, are targeted for degradation (Fig. 1) [1]. Degradation of intracellular components is also promoted under particular cellular conditions, such as when nutrients are scarce and the degradation of a protein's own proteome becomes the only source of amino acids for the synthesis of essential proteins [1]. Two main proteolytic systems account for most of the intracellular protein degradation: the ubiquitin–proteasome system (UPS) and lysosomes [1]. The proteasome is a barrel-shaped multisubunit structure in the cytosol that hosts at least

Protein Misfolding Diseases: Current and Emerging Principles and Therapies,
Edited by Marina Ramirez-Alvarado, Jeffery W. Kelly, and Christopher M. Dobson
Copyright © 2010 John Wiley & Sons, Inc.

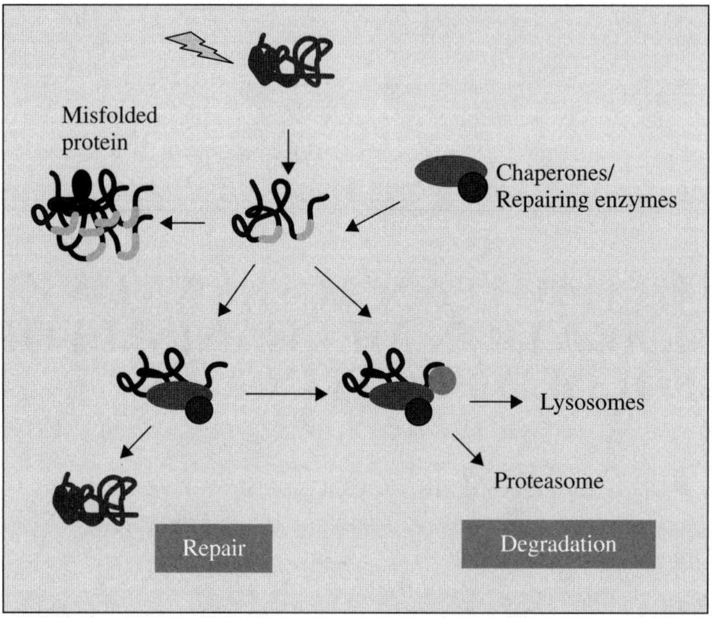

FIG. 1 Cellular quality control systems. Cells count on surveillance systems to identify prone-to-aggregate altered proteins (misfolded or damaged) and prevent their intracellular accumulation. If molecular chaperones and repairing enzymes fail to restore normal structure and function of the proteins, the altered proteins are targeted to degradation by either proteasome or lysosomes.

three different catalytic activities. Most substrate proteins are tagged for degradation by the proteasome through covalent conjugation to ubiquitin. The ubiquitin tag is then recognized by the regulatory subunits of the proteasome, and after undergoing unfolding, substrate proteins are pushed down the catalytic core of this protease for degradation [2]. The catalytic component of the other major intracellular proteolytic system resides inside lysosomes—single-membrane organelles with the highest content of digestive enzymes inside cells. These enzymes can degrade all kinds of macromolecules (proteins, lipids, nucleic acids, and carbohydrates). Lysosomal proteases, generically known as cathepsins, are maximally activated at the lysosomal acidic pH, maintained by the vacuolar ATPase [3]. In contrast to the short half-life that characterizes most of the substrate proteins for the UPS, lysosome substrates are typically long-lived proteins. Not only proteins but also a wide range of intracellular components (e.g., organelles, lipid stores, glycogen) can be degraded inside lysosomes. Although these two major proteolytic systems, the UPS and lysosomes, have traditionally been considered as completely independent units for intracellular degradation, recent work has revealed that they are tightly interconnected and that, in fact, malfunctioning of one of the systems can be compensated for, at least temporally, by up-regulation of the other [4,5]. The main focus of this chapter is on the degradation of intracellular components by the lysosomal system, a process generically known as *autophagy*, with special emphasis on the role of autophagy in the removal of misfolded proteins. However, because of the interplay between the two proteolytic systems mentioned earlier, we also comment on the consequences of the crosstalk between the UPS and lysosomes in the context of proteotoxicity.

AUTOPHAGY

Autophagy, "self-eating," is the evolutionarily conserved process of degradation of intracellular components by lysosomes [3]. The cellular relevance of this catabolic process goes beyond cellular recycling, as autophagy has been shown to participate in cellular differentiation, tissue remodeling, growth control, cellular defense against microbial invasion, antigen presentation, and adaptation to adverse environments [6–8]. Consequently, autophagic dysfunction affects essential cellular functions and contributes to pathogenesis in many different diseases [6,7]. In the following sections we review briefly the molecular mechanisms and pathophysiology of autophagy, then focus on its critical role in the removal of pathogenic proteins in the context of protein conformational disorders.

Types of Autophagy

In mammals, three autophagic pathways converge in the lysosomal compartment: microautophagy, macroautophagy, and chaperone-mediated autophagy (CMA) (Fig. 2) [3,6,8]. These pathways have unique constituents and regulators but share molecular components and substrates, thus contributing to the dynamic and flexible nature of this lysosomal system. It has traditionally been accepted that basal lysosomal degradation is accomplished by microautophagy, a constitutively active form of autophagy in which substrates get directly engulfed by invaginations at the lysosomal membrane [9]. The molecular mechanisms behind this process are for the most part unknown, making it difficult to evaluate the contribution of this pathway to the clearance of misfolded proteins.

Although basal activity of the other two autophagic pathways has been reported in different tissues, they are both maximally activated in response to stress. Macroautophagy involves the sequestration of cytosolic contents by a de novo formed double-membrane vesicle called an autophagic vacuole (AV) or autophagosome [10]. Fusion with lysosomes provides AVs with all the needed hydrolases to degrade their contents completely. A specific subset of proteins, known generically as autophagy-related proteins (Atgs) [11], have been identified to participate in the formation, maturation, and clearance of AVs. These proteins are conserved from yeast to mammals and have been divided functionally into four groups: two conjugation cascades, a nucleation complex, and a regulatory complex [12]. Of the two conjugation cascades underlying AV formation, one

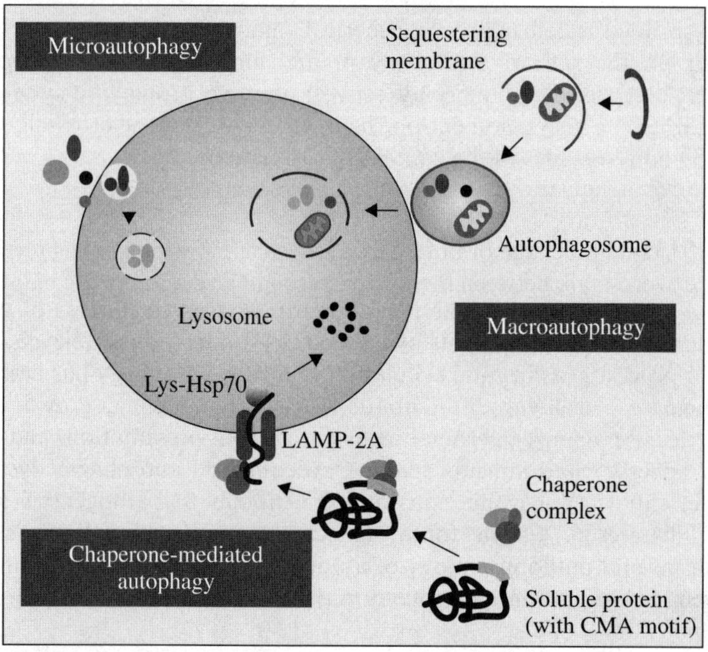

FIG. 2 Types of autophagy. Schematic model of the three different autophagic pathways described in mammals: (1) Microautophagy is a constitutive nonselective sequestration of cytosolic contents directly by invaginations at the lysosomal membrane; (2) macroautophagy is an inducible form of "in bulk" degradation of cytosolic cargo sequestered inside a double-membrane vesicle (autophagosome) which acquires the necessary hydrolases by fusion with lysosomes; (3) chaperone-mediated autophagy (CMA) is a selective and inducible type of autophagy that involves a direct translocation of unfolded cytosolic protein substrates through the lysosome membrane.

modulates the formation of a protein complex (Atg5/12/16), while the other leads to the conjugation of a protein (Atg8 or its mammalian homolog LC3) to a lipid (phosphatidylethanolamine) [10]. This LC3–lipid conjugate, termed LC3-II, localizes to AVs and is often used as a marker for this compartment.

The main component of the regulatory complex is the mammalian target of rapamycin (mTOR), a well–characterized nutrient and energy sensor that represses macroautophagy under normal nutritional conditions [13]. The complex formed by beclin-1 and class III phosphatidylinositol 3-kinase (PI3K) promotes nucleation upon dissociation from the endogenous inhibitor Bcl-2 [14]. The components that participate in lysosomal fusion of AVs to promote their clearance are less well characterized, but this process depends heavily on the cytoskeleton machinery [13].

CMA is the only form of autophagy that can selectively degrade soluble cytosolic proteins without altering other neighboring cytosolic components [15,16]. The selectivity of this autophagic pathway is conferred through the recognition of a targeting motif in the substrate proteins by a cytosolic chaperone. The chaperone–substrate protein complex is then brought to the surface of lysosomes, where it binds to the lysosome-associated membrane protein type 2A (LAMP-2A), which acts as a receptor for this pathway. Chaperones on both sides of the lysosomal membrane facilitate substrate unfolding and translocation into the lysosomal lumen [15,16]. CMA activity is directly dependent on the availability of the receptor LAMP-2A at the lysosomal membrane.

Experimental blockage of both CMA and macroautophagy has revealed the existence of crosstalk between these two stress-induced autophagic pathways [4]. However, although one type of autophagy can partially compensate for the other, they are not redundant, as cells with compromise in one of these pathways are at a disadvantage during stress conditions [4]. The mechanisms that mediate this autophagic crosstalk are currently under investigation because they could offer new means of modulating the activity of these pathways.

Physiological Functions of Autophagy

Autophagy contributes both to continuous cellular housekeeping as well as to cellular adaptation to intra– and extracellular stressors. Autophagy fulfills the need for energy and essential constituents, such as amino acids, when nutrients are scarce [8]. Although early in starvation macroautophagy is the major catabolic mechanism, if the nutritional stress persists, the need to preserve essential cellular components arises and this "in-bulk" form of intracellular degradation is replaced progressively by CMA, which allows discrimination between essential and dispensable components [15,16]. Both macroautophagy and CMA are also activated under conditions resulting in cellular damage to remove the altered components [16]. Failure to activate autophagy compromises cell viability upon exposure to different stressors, due to the buildup of damaged or toxic products inside cells [4].

Recent studies have revealed that the functions of autophagy go beyond the turnover and clearance of intracellular components. Autophagy has been shown to participate in conditions associated with cell and tissue remodeling, such as in development, and autophagic features can be observed in what is known as programmed cell death type II [7,17]. Moreover, autophagy participates in both acquired and innate immunity. Proper autophagic functioning is required not only for the elimination of different infectious organisms, but also for the presentation of both exogenous and self-antigens by major histocompatibility complex (MHC) class II molecules [18]. Interestingly, some pathogens have evolved to subvert this autophagic defense and even utilize the autophagic compartments as a "hiding" place to evade the immune system [18].

Autophagy and Disease

Recent advances in our understanding of the molecular mechanisms behind the autophagic process have facilitated the establishment of clear connections between autophagic dysfunction and different pathological processes. Autophagy has received much attention in the field of cancer biology, since the upstream regulators of macroautophagy are prominent tumor suppressors (beclin-1, p53, PTEN) and oncogenes (Bcl-2, mTOR) [19]. However, whether autophagy acts as anti- or pro-oncogenic is currently controversial. Factors such as the type and stage of cancer, and whether or not treatment has been inflicted, probably determine this dual role of autophagy in cancer [19].

The involvement of autophagy in certain muscle pathologies was first established in mice knocked out for a lysosomal membrane protein (LAMP-2), as they showed massive accumulation of AVs in liver, muscle, and heart cells [20]. Mutations in this gene were then identified in patients affected by a vacuolar myopathy known as Danon disease. In fact, several inherited myopathies, such as rimmed vacuolar myopathy, Danon disease, and X-linked myopathy with excess autophagy, manifest with an abnormal increase in AVs [21]. In addition to these genetic myopathies, autophagy has also been shown to be important in the maintenance of cardiac muscle homeostasis and as a defense against apoptotic cell death during episodes of ischemia [22].

Despite the long-term known inhibitory effect of insulin on macroautophagy, only recently have connections between autophagy and diabetes pathogenesis become evident. Impaired CMA contributes to the pathological shift to protein accumulation and cellular swelling in renal hypertrophy, as a secondary manifestation of diabetes, although the reasons for the decline in CMA remain unknown [23]. Of particular interest for the role of autophagy in the removal of misfolded proteins, the main topic of this chapter, is the fact that macroautophagy has been shown to act as a pro-survival mechanism in familial neurohypophyseal diabetes insipidus, as it functions to degrade the cytoplasmic aggregates of the mutant vasopressin protein [24].

The ability of autophagy to remove aberrant proteins constitutes the basis for the strong relationship established recently between autophagic malfunctioning and neurodegeneration, as an example of the most frequent and severe protein conformational disorders. These many ties between proteolytic pathways, aberrant protein conformations, and various neurodegenerative disorders have been reinforced strongly by the fact that mouse models with selective deficiency of macroautophagy in neurons display massive neuronal loss and rapidly develop neurodegeneration, thus confirming that autophagy is essential for neuronal cell survival [25,26].

AUTOPHAGY AND PROTEIN CONFORMATION DISORDERS

The role of autophagy in abnormal protein removal has linked this catabolic process with a series of disorders generically known as protein conformation disorders (PCDs). In PCDs, genetic or acquired alterations in a protein or group of proteins lead to cell toxicity and eventually to cell death. Identification of the pathogenic products by quality control systems and repairing—by chaperones and repairing enzymes—or removal—through the proteolytic systems—of these toxic products allows cells to maintain normal cellular function during the asymptomatic stages of the disease. However, the progressive accumulation of these misbehaving proteins, often precipitated by different types of aggravating agents, at some point overcomes the ability of recognition and clearance systems to handle these toxic products. PCDs have been described as affecting a broad range of organs and tissues, depending on the tissue specificity of the altered proteins. Those affecting preferentially the central nervous system, such as major neurodegenerative disorders, including Alzheimer, Parkinson, and Huntington diseases, are particularly prominent and have recently been submitted to intense scrutiny by groups interested in protein degradation systems. The main reason for this focused interest is the fact that the accumulation of pathogenic proteins is particularly detrimental in differentiated nondividing cells such as neurons, which lack the ability to decrease their content of toxic protein products by distribution between the daughter cells resulting from cell division. We comment here on the general role of the various different types of autophagy in handling pathogenic proteins and then illustrate the current status of our understanding of the contribution of changes in the autophagic systems to the pathogenesis of common neurodegenerative disorders.

Autophagy and Protein Misfolding

The UPS constitutes a common front of defense against misfolded proteins, as this system has the ability to discriminate selectively between abnormally and properly folded proteins. In this respect, as described above, CMA provides a similar level of selectivity to the autophagic system (Fig. 3, left). For both systems, complete disassembly and unfolding of the substrate proteins is a requirement to reach the catalytic core—in the case of the UPS—or the enzymes in the lysosomal lumen—in CMA. Consequently, once the aberrantly folded proteins organize in higher-order protein complexes (oligomers, protofibers, fibers, or aggregates), they are no longer amenable for degradation through any of these two selective pathways. Different reasons can promote an increase in the levels of the aberrant proteins and their organization in irreversible protein complexes. Among them, changes in the efficiency of the proteolytic systems that normally take care of the clearance of the abnormal proteins, when still in a soluble conformation, can contribute to their increased cytosolic levels. In fact, as detailed below, both UPS and CMA are common targets for pathogenic proteins, which can act as blockers or inhibitors of these systems.

Once irreversible protein organization occurs, and as the activity of the UPS and CMA becomes compromised, macroautophagy is the only feasible mechanism left for protein removal [27] (Fig. 3, middle). Misfolded insoluble proteins are transported along microtubules to form pericentriolar aggresomes [28]. Aggresome formation has been suggested to be protective for cells, because it sequesters the pathogenic proteins and favors their removal by

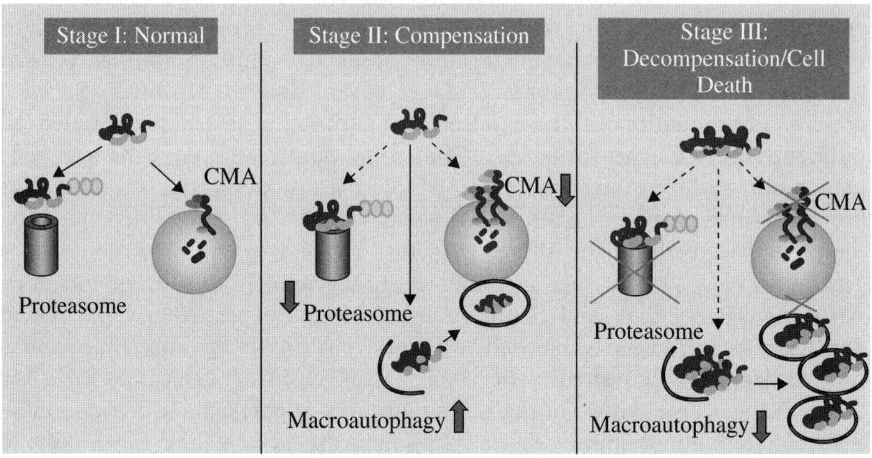

FIG. 3 Autophagy and protein conformational disorders. Soluble misfolded proteins can be removed selectively by either proteasome or lysosomes via CMA (Stage I). When the altered proteins interfere with the activity of other proteolytic systems, macroautophagy is up-regulated to compensate and facilitate aggregate protein removal (Stage II). Eventually, different factors contribute to failure of macroautophagy, with the consequent intracellular accumulation of toxic protein products that lead to cellular decompensation and, often, cell death (Stage III).

macroautophagy by confining them in a discrete cellular region [28]. In fact, pharmacological up-regulation of macroautophagy under these conditions has been shown to reduce the number of intracellular aggregates in both cultured cells and animal models of aggregopathies [29]. The integrity of the cellular cytoskeleton is essential for aggregate clearance, as it mediates both their concentration in particular cellular regions accessible to the autophagic system and efficient clearance of the aggregate sequestered inside AVs through fusion with secondary lysosomes [30,31]. One of the pending questions is whether or not clearance of protein aggregates by macroautophagy is a selective process, and if it is, what the molecular markers are that allow aggregate recognition by the macroautophagic machinery. Different disease-associated protein inclusions have been found to be enriched in p62, a novel protein able to bind both polyubiquitin stretches and LC3, an essential component of the forming autophagosome [32]. This dual interaction of p62 with the autophagic machinery and with polyubiquitinated proteins within the aggregates, along with the fact that p62 itself is degraded by autophagy, has placed p62 as a putative marker for aggregate clearance by macroautophagy [32]. However, the fact that p62 is also a critical regulator of inclusion body formation [33], and that p62 is also present in protein aggregates not amenable for macroautophagy degradation, has raised the possibility that other factors, yet to be identified, may contribute to aggregate recognition by the autophagic system. Both basal and inducible macroautophagy contribute to preventing intracellular accumulation of protein inclusions. Although cells have been shown to respond to the presence of intracellular aggregates by up-regulating macroautophagy, blockage of basal macroautophagy in different tissues, even in the absence of a pathogenic protein background, results in the accumulation of protein inclusions, supporting a role for constitutive macroautophagy in preventing protein aggregation [25,26].

Consequences of the Failure of the Autophagic Systems

The large capability of macroautophagy allows this system to handle the degradation of misfolded and pathogenic proteins even when organized in complex insoluble structures such as aggregates. This property becomes essential to guarantee cellular survival in different PCDs as impairment in other proteolytic and autophagic mechanisms progresses. Failure of macroautophagy under these conditions can thus contribute to cell toxicity and eventually to the cell death observed in many of these disorders (Fig. 3, right). Different factors can precipitate, or at least accelerate, the autophagic failure. Among them, oxidative stress and aging have been revealed as intermingling candidates that can interfere with proper autophagic activity.

The lysosomal system undergoes noticeable changes as cells age: decrease of lysosomal stability, lipofuscin accumulation, and impaired maintenance of lysosomal pH [34]. Macroautophagy and CMA activity both decrease with age in almost all tissues and organisms analyzed [34]. In fact, macroautophagy has been proposed as an important antiaging mechanism, because its blockage in long-lived *C. elegans* mutants shortens their life span [35]. The gradual decrease in the activity of the autophagic systems contributes to the slower rates of long-lived protein degradation, resulting in an increase in their cytosolic levels. This increase in the total proteome, along with the accumulation of autophagic compartments that failed to be cleared up by the lysosomal system, could generate an intracellular environment unfavorable for protein folding or for proper functioning of chaperones. As a result, cells are less able to adapt to stressful changes and handle damaged or altered proteins.

Oxidative stress can be added to the aging environment both as a cause and as a consequence of the failure of the intracellular proteolytic systems. Removal of oxidized proteins has been a well-accepted function of the proteasome for a long time, whereas lysosomes were considered primary targets of oxidizing agents (which lead to lysosomal membrane destabilization and leakage of the luminal enzymes into cytosol) [36]. However, recent studies support an active role of lysosomes in oxidized protein removal (via CMA) and in the clearance of organelles, such as mitochondria, which are frequent targets of oxidative reactions (via macroautophagy) [34,37]. As the activity of these two autophagic pathways decreases with age, it is easy to predict that cells become progressively more susceptible to oxidative damage and that the persistence of oxidizing and oxidized components inside cells will increase the chances of damage of essential autophagic molecular players, thus perpetuating this failure stage (Fig. 3, right).

All of the findings above become relevant in the pathogenesis of some late-onset neurodegenerative disorders, since aging neurons are particularly vulnerable to the progressive dysfunction of protein degradation systems. Next, we review the common and differential aspects of the changes in the autophagic system in various different neurodegenerative disorders.

Parkinson Disease

Parkinsons disease (PD), a common neurodegenerative disorder characterized by the loss of dopaminergic neurons in the substantia nigra and the norepinephrinergic neurons of the locus coeruleus areas of the brain [38], is a typical example of PCD. PD–affected neurons present cytoplasmic deposition of protein aggregates (Lewy bodies), particularly enriched in α-synuclein, a neuron–specific protein involved in vesicular trafficking. In PD, α-synuclein aggregates are considered less toxic than some of the other forms of aberrant α-synuclein, such as nonfibrillar oligomer intermediates, which are able to form porelike structures in membranes [38]. Organization of α-synuclein in aggregates thus serves a dual function of preventing this gain of toxic function and facilitating elimination of the aberrant protein via macroautophagy. In fact, even though soluble molecules of α-synuclein have been shown to be degraded by both the proteasome and the lysosomes via CMA [39,40], none of these pathways can handle the aggregated protein. Pathogenic α-synucleins, due to genetic mutations or undesired posttranslational modifications such as those induced by free dopamine, have been shown to alter CMA activity [40,41]. These abnormal α-synucleins bind to LAMP-2A at the lysosomal membrane, as any other CMA substrate, but cannot be translocated into the lysosomal lumen. The persistent binding of the pathogenic proteins to the CMA receptor blocks degradation of other substrates through this pathway [40,41]. Experimental blockage of CMA in culture cells, mimicking that observed in PD neurons, renders cells more susceptible to cellular stressors [4]. The fact that of the different posttranslational modifications of α-synuclein detected in the disease, reaction with dopamine is the one that alters the turnover of α-synuclein and impairs CMA activity could provide an explanation for the initial selective degeneration of dopaminergic neurons in this disease [41].

Numerous studies have established a clear link between malfunctioning of the UPS at various levels and PD pathogenesis. The role of UPS in PD is described in detail in other chapters of the book, but of relevance for this chapter is the fact that functional blockage not only of CMA, but also of the catalytic activities of the proteasome, have been shown to induce activation of macroautophagy [4,30]. Elucidation of the yet unknown mechanisms that mediate this crosstalk between the different proteolytic pathways should provide valuable clues on how to potentiate and/or extend this compensatory stage, in which aggregates, and possibly pro-aggregating complexes of α-synuclein, are removed efficiently by macroautophagy.

Alzheimer Disease

Alterations in autophagy have also been identified in Alzheimer disease (AD)—the most common neurodegenerative disorder—in which two pathogenic proteins [tau and amyloid precursor protein (APP)] organize into intraneuronal fibrillary tangles (aggregated hyperphosphorylated tau protein) and extracellular senile plaques [amyloid-β peptide of APP (Aβ)]. In contrast to PD, in which alterations in CMA and the UPS precede the compromise in macroautophagy, a primary defect in this last autophagic system has been identified in AD brains and in different mouse models of this pathology [42]. A dramatic increase in the number of AVs has been detected in the AD dystrophic neurons [43]. Despite this large expansion of the autophagic system, total rates of protein degradation and organelle turnover are severely compromised in these neurons, as the immature AVs accumulate in neuritic processes and synapses because they fail to be cleared up by the lysosomal system [43,44]. The series of events leading to cell death under these conditions are not elucidated completely. It has been proposed that the slow clearance of AVs in the AD neurons, which can take up to days in contrast to 10 to 20 minutes in normal neurons, transform them into an internal source for the toxic Aβ, because they contain both APP and the proteases required for this cleavage [42]. Interestingly, cytosolic Aβ has been shown to cause AV accumulation in a *C. elegans* model, thus providing a possible mechanism to perpetuate the autophagic clog [45]. Moreover, excess AVs can disrupt normal intracellular trafficking, thus altering basic cellular functions and leading to induction of apoptosis. In light of this autophagic jam, further up-regulation of macroautophagy should be avoided at this stage; instead, efforts should concentrate on promoting clearance of the AVs accumulated in the neurons affected. The primary defect in macroautophagy in AD is still unknown, but as in any other late-onset disease, the age-dependent decline in macroautophagy could make aging an important aggravating factor and provide an explanation for the late onset of AD.

Huntington Disease

Huntington disease (HD) constitutes another typical example of PCD with cytosolic formation of protein inclusions, as in PD, but in this case the modification of huntingtin, the pathogenic protein in this disorder, is always inherited genetically. Mutant huntingtin becomes prone to forming insoluble aggregate inclusions, due to the presence of a stretch of glutamines (more than 37 glutamine repeats or polyQ) in its N-terminus [46]. Along with huntingtin, inclusions in HD cells also contain proteins known to interact functionally with huntingtin, such as transcription factors, protein adaptors, and chaperones, hence explaining the transcription deregulation, altered vesicular trafficking, and loss of protein folding quality control observed in HD cells [46]. Huntingtin protein aggregates have been shown to stimulate macroautophagy [47], possibly through the sequestration of negative regulators of this pathway, such as mTOR [29], although the inhibitory effect of mutant huntingtin fragments on the UPS could also contribute to macroautophagy up-regulation [30]. This autophagic response to the cytosolic aggregates has been shown to be cytoprotective in both cell culture and animal models [29,48,49]. Different proteins, such as p62, described above, and a chaperone complex, HspB8/Bag3, have been proposed to be necessary for the recognition of polyQ aggregates by the autophagic machinery [32,50], and as for other types of aggresomes, microtubules and a microtubule-associated deacetylase (HDAC6) are essential for aggregate removal by macroautophagy [30,51].

As the disease progresses, not only the UPS but also macroautophagy seem to be the target of the toxic effect of the mutant huntingtin. Thus, this protein has been shown to sequester macroautophagy up-regulators [52] and effectors [5], resulting in compromised function of this clearance system. Whether or not the recently observed association of the wild-type protein, but not of mutant huntingtin, to the surface of AVs [53] also contributes to the failure of this system requires further investigation.

Therapeutic Implications

Except for late stages of PCD, when the ability to clear up AVs is already compromised and further induction in AV formation could have detrimental effects, there is general agreement that activation of macroautophagy in early stages of PCD could improve the clearance of aggregates and other toxic protein conformations and prevent their toxic cellular effects. Manipulations of the mTOR pathway as a major regulator of macroautophagy have been widely explored as a potential target for therapy. Indeed, common mTOR inhibitors, rapamycin and its analogs, were the first compounds shown to reduce aggregate number in different PCD models [29]. Although the effect of these compounds has recently been proposed to be through a decrease in protein synthesis (which reduces the rate of aggregate formation) rather than directly on their clearance [54], the value of macroautophagy up-regulation in aggregate clearance still stands. Compounds such as the natural disaccharide trehalose and lithium, known to increase autophagic activity in an mTOR-independent manner, have also been shown to facilitate aggregate removal [48,49]. This beneficial effect of activation of macroautophagy has promoted active searches for chemical inducers of autophagy. Particularly encouraging for the possible use of up-regulators of macroautophagy as therapeutics is the fact that in a recent small-molecule screening for inducers of autophagy, seven of the eight compounds identified were already U.S. Food and Drug Administration–approved drugs on the market for other diseases [55]. This should accelerate clinical testing as possible treatment for these otherwise untreatable diseases.

Nevertheless, therapeutic efforts should focus on stages before the compensatory activation of macroautophagy. Interventions aimed at enhancing degradation of the still-soluble altered proteins by the UPS or CMA, or at preventing the inhibitory effect of the pathogenic proteins on these systems, should be explored as therapeutic approaches for these devastating disorders.

CONCLUDING REMARKS

The contribution of the autophagic system to intracellular quality control has been known for a relatively long time, but it is only recently that the connection between malfunctioning of this system and severe PCDs has been established. One of the main reasons for the current revitalization of the study of the role of alterations in autophagy in the pathogenesis of these disorders has been the recent better molecular characterization of the autophagic pathways. Blockage of autophagy results in accumulation of altered proteins in support of a continuous role for this pathway in the clearance of these abnormal components. Furthermore, activation of autophagy in experimental models for different aggregopathies reduced intracellular aggregates and improved cellular function, thus opening up the possibility of similar interventions for the treatment of different PCDs.

As in any starting field, there are still many aspects of the relationship of autophagy to misfolded proteins that require further clarification. Thus, it is important to establish the universality of these mechanisms. In addition to the three diseases discussed in this chapter, alterations in autophagy have also been reported in other PCDs, such as Batten disease or neuronal ceroid-lipofuscinosis, a lysosomal storage disorder, in prion-mediated transmissible spongiform encephalopathies and spinocerebellar ataxia, among others [56]. Moreover, we have discussed here only PCDs resulting from cytosolic accumulation of misfolded proteins, but connections to autophagy have also been established with a subgroup of PCDs in which protein aggregation occurs in other cellular compartments. A typical example is the activation of autophagy in response to the accumulation of a mutant secretory protein in the endoplasmic reticulum (ER) of patients with α-1-antitrypsin deficiency. As the list of proteinopathies with altered autophagy increases, it will become important to distinguish those disorders with a primary defect in the autophagic system from those in which autophagic alterations are consequence of other cellular alterations. These differences can clearly determine very distinct therapeutic approaches.

REFERENCES

1. Ciechanover, A. (2005). Proteolysis: from the lysosome to ubiquitin and the proteasome. *Nat Rev Mol Cell Biol, 6*, 79–87.
2. Lecker, S.H., Goldberg, A.L., Mitch, W.E. (2006). Protein degradation by the ubiquitin–proteasome pathway in normal and disease states. *J Am Soc Nephrol, 17,* 1807–1819.

3. Dice, J. (2000). *Lysosomal Pathways of Protein Degradation*, Molecular Biology Intelligence Unit, Landes Bioscience, Austin, TX.
4. Massey, A.C., Kaushik, S., Sovak, G., Kiffin, R., Cuervo, A.M. (2006). Consequences of the selective blockage of chaperone-mediated autophagy. *Proc Nat Acad Sci U S A, 103,* 5905–5910.
5. Iwata, A., Christianson, J.C., Bucci, M., Ellerby, L.M., Nukina, N., Forno, L.S., Kopito, R.R.. (2005). Increased susceptibility of cytoplasmic over nuclear polyglutamine aggregates to autophagic degradation. *Proc Natl Acad Sci U S A, 102,* 13135–13140.
6. Cuervo, A.M. (2004). Autophagy: in sickness and in health. *Trends Cell Biol, 14,* 70–77.
7. Levine, B., Klionsky, D. (2004). Development by self-digestion: molecular mechanisms and biological functions of autophagy. *Dev Cell, 6,* 463–477.
8. Mizushima, N., Klionsky, D. (2007). Protein turnover via autophagy: implications for metabolism. *Annu Rev Nutr, 27,* 19–40.
9. Mortimore, G.E., Lardeux, B.R., Adams, C.E. (1988). Regulation of microautophagy and basal protein turnover in rat liver: effects of short-term starvation. *J Biol Chem, 263,* 2506–2512.
10. Ohsumi, Y. (2001). Molecular dissection of autophagy: two ubiquitin-like systems. *Nat Rev Mol Cell Biol, 2,* 211–216.
11. Klionsky, D.J., Cregg, J.M., Dunn, W.A., Jr., Emr, S.D., Sakai, Y., Sandoval, I.V., Sibirny, A., Subramani, S., Thumm, M., Veenhuis, M., Ohsumi, Y. (2003). A unified nomenclature for yeast autophagy-related genes. *Dev Cell, 5,* 539–545.
12. Klionsky, D.J. (2005). The molecular machinery of autophagy: unanswered questions. *J Cell Sci, 118,* 7–18.
13. Meijer, A.J., Codogno, P. (2004). Regulation and role of autophagy in mammalian cells. *Int J Biochem Cell Biol, 36,* 2445–2462.
14. Liang, X., Jackson, S., Seaman, M., Brown, K., Kempkes, B., Hibshoosh, H., Levine, B. (1999). Induction of autophagy and inhibition of tumorigenesis by beclin 1. *Nature, 402,* 672–676.
15. Dice, J. (2007). Chaperone-mediated autophagy. *Autophagy, 3,* 295–299.
16. Massey, A., Zhang, C., Cuervo, A. (2006). Chaperone-mediated autophagy in aging and disease. *Curr Top Dev Biol, 73,* 205–235.
17. Bursch, W., Ellinger, A., Gerner, C., Frohwein, U., Schulte-Hermann, R. (2000). Programmed cell death (PCD): apoptosis, autophagic PCD, or others? *Ann N Y Acad Sci,* 926, 1–12.
18. Levine, B., Deretic, V. (2007). Unveiling the roles of autophagy in innate and adaptive immunity. *Nat Rev Immunol, 7,* 767–777.
19. Hippert, M., O'Toole, P., Thorburn, A. (2006). Autophagy in cancer: good, bad, or both? *Cancer Res, 66,* 9349–9351.
20. Tanaka, Y., Guhde, G., Suter, A., Eskelinen, E.L., Hartmann, D., Lüllmann-Rauch, R., Janssen, P.M., Blanz, J., von Figura, K., Saftig, P. (2000). Accumulation of autophagic vacuoles and cardiomyopathy in Lamp-2-deficient mice. *Nature, 406,* 902–906.
21. Nishino, I. (2003). Autophagic vacuolar myopathies. *Curr Neurol Neurosci Rep, 3,* 64–69.

22. Yan, L., Vatner, D.E., Kim, S.J., Ge, H., Masurekar, M., Massover, W.H., Yang, G., Matsui, Y., Sadoshima, J., Vatner, S.F. (2005). Autophagy in chronically ischemic myocardium. *Proc Natl Acad Sci U S A, 102*, 13807–13812.

23. Sooparb, S., Price, S.R., Shaoguang, J., Franch, H.A. (2004). Suppression of chaperone-mediated autophagy in the renal cortex during acute diabetes mellitus. *Kidney Int, 65*, 2135–2144.

24. Castino, R., Davies J., Beaucourt, S., Isidoro, C., Murphy, D. (2005). Autophagy is a prosurvival mechanism in cells expressing an autosomal dominant familial neurohypophyseal diabetes insipidus mutant vasopressin transgene. *FASEB J, 19*, 1021–1023.

25. Hara, T., Nakamura, K., Matsui, M., Yamamoto, A., Nakahara, Y., Suzuki-Migishima, R., Yokoyama, M., Mishima, K., Saito, I., Okano, H., Mizushima, N. (2006). Suppression of basal autophagy in neural cells causes neurodegenerative disease in mice. *Nature, 441*, 885–889.

26. Komatsu, M., Waguri, S., Chiba, T., Murata, S., Iwata, J., Tanida, I., Ueno, T., Koike, M., Uchiyama, Y., Kominami, E., Tanaka, K. (2006). Loss of autophagy in the central nervous system causes neurodegeneration in mice. *Nature, 441*, 880–884.

27. Martinez-Vicente, M., Cuervo, A.M. (2007). Autophagy and neurodegeneration: when the cleaning crew goes on strike. *Lancet Neurol, 6*, 352–361.

28. Johnston, J.A., Ward, C.L., Kopito, R.R. (1998). Aggresomes: a cellular response to misfolded proteins. *J Cell Biol, 143*, 1883–1898.

29. Ravikumar, B., Vacher, C., Berger, Z., Davies, J.E., Luo, S., Oroz, L.G., Scaravilli, F., Easton, D.F., Duden, R., O'Kane, C.J., Rubinsztein, D.C. (2004). Inhibition of mTOR induces autophagy and reduces toxicity of polyglutamine expansions in fly and mouse models of Huntington disease. *Nat Genet, 36*, 585–595.

30. Iwata, A., Riley, B.E., Johnston, J.A., Kopito, R.R. (2005). HDAC6 and microtubules are required for autophagic degradation of aggregated huntingtin. *J Biol Chem, 280*, 40282–40292.

31. Webb, J.L., Ravikumar, B., Rubinsztein, D.C. (2004). Microtubule disruption inhibits autophagosome–lysosome fusion: implications for studying the roles of aggresomes in polyglutamine diseases. *Int J Biochem Cell Biol, 36*, 2541–2550.

32. Bjørkøy, G., Lamark, T., Brech, A., Outzen, H., Perander, M., Overvatn, A., Stenmark, H., Johansen, T. (2005). p62/SQSTM1 forms protein aggregates degraded by autophagy and has a protective effect on huntingtin-induced cell death. *J Cell Biol, 171*, 603–614.

33. Komatsu, M., Waguri, S., Koike, M., Sou, Y.S., Ueno, T., Hara, T., Mizushima, N., Iwata, J.I., Ezaki, J., Murata, S., et al. (2007). Homeostatic levels of p62 control cytoplasmic inclusion body formation in autophagy-deficient mice. *Cell, 131*, 1149–1163.

34. Cuervo, A.M., Bergamini, E., Brunk, U.T., Droge, W., French, M., Terman, A. (2005). Autophagy and aging: the importance of maintaining "clean" cells. *Autophagy, 1*, 131–140.

35. Melendez, A., Talloczy, Z., Scaman, M., Eskelinen, E.L., Hall, D.H., Levine, B. (2003). Essential role of autophagy genes in dauer development and lifespan extension in *C. elegans*. *Science, 301*, 1387–1391.

36. Terman, A. (2006). Catabolic insufficiency and aging. *Ann N Y Acad Sci, 1067*, 27–36.

37. Lemasters, J. (2005). Selective mitochondrial autophagy, or mitophagy, as a targeted defense against oxidative stress, mitochondrial dysfunction, and aging. *Rejuvenation Res, 8*, 3–5.

38. Lee, V., Trojanowski, J. (2006). Mechanisms of Parkinson's disease linked to pathological alpha-synuclein: new targets for drug discovery. *Neuron, 52*, 33–38.

39. Webb, J., Ravikumar, B., Atkins, J., Skepper, J., Rubinsztein, D. (2003). Alpha-synuclein is degraded by both autophagy and the proteasome. *J Biol Chem, 278*, 25009–25013.

40. Cuervo, A.M., Stefanis, L., Fredenburg, R., Lansbury, P.T., Sulzer, D. (2004). Impaired degradation of mutant alpha-synuclein by chaperone-mediated autophagy. *Science, 305*, 1292–1295.

41. Martinez-Vicente, M., Talloczy, Z., Kaushik, S., Massey, A.C., Mazzulli, J., Mosharov, E.V., Hodara, R., Fredenburg, R., Wu, D.C., Follenzi, A., et al. (2008). Dopamine-modified α-synuclein blocks chaperone-mediated autophagy. *J Clin Invest, 118*, 777–788.

42. Yu, W.H., Cuervo, A.M., Kumar, A., Peterhoff, C.M., Schmidt, S.D., Lee, J.H., Mohan, P.S., Mercken, M., Farmery, M.R., Tjernberg, L.O., et al. (2005). Macroautophagy: a novel beta-amyloid peptide-generating pathway activated in Alzheimer's disease. *J Cell Biol, 171*, 87–98.

43. Nixon, R.A., Wegiel, J., Kumar, A., Yu, W.H., Peterhoff, C., Cataldo, A., Cuervo, A.M. (2005). Extensive involvement of autophagy in Alzheimer disease: an immunoelectron microscopy study. *J Neuropathol Exp Neurol, 64*, 113–122.

44. Moreira, P.I., Siedlak, S.L., Wang, X., Santos, M.S., Oliveira, C.R., Tabaton, M., Nunomura, A., Szweda, L.I., Aliev, G., Smith, M.A., et al. (2007). Increased autophagic degradation of mitochondria in Alzheimer disease. *Autophagy, 3*, 614–615.

45. Florez-McClure, M., Hohsfield, L., Fonte, G., Bealor, M., Link, C. (2007). Decreased insulin-receptor signaling promotes the autophagic degradation of beta-amyloid peptide in *C. elegans*. *Autophagy, 3*, 569–580.

46. Ramaswamy, S., Shannon, K., Kordower, J. (2007). Huntington's disease: pathological mechanisms and therapeutic strategies. *Cell Transplant, 16*, 301–312.

47. Kegel, K.B., Kim, M., Sapp, E., McIntyre, C., Castaño, J.G., Aronin, N., DiFiglia, M. (2000). Huntingtin expression stimulates endosomal-lysosomal activity, endosome tubulation, and autophagy. *J Neurosci, 20*, 7268–7278.

48. Sarkar, S., Floto, R.A., Berger, Z., Imarisio, S., Cordenier, A., Pasco, M., Cook, L.J., Rubinsztein, D.C. (2005). Lithium induces autophagy by inhibiting inositol monophosphatase. *J Cell Biol, 170*, 1101–1111.

49. Sarkar, S., Davies, J., Huang, Z., Tunnacliffe, A., Rubinsztein, D. (2007). Trehalose, a novel mTOR-independent autophagy enhancer, accelerates the clearance of mutant huntingtin and {alpha}-synuclein. *J Biol Chem, 282*, 5641–5652.

50. Carra, S., Seguin, S., Lambert, H., Landry, J. (2007). HspB8 chaperone activity toward poly(Q)-containing proteins depends on its association with Bag3, a stimulator of macroautophagy. *J Biol Chem, 283*, 1437–1444.

51. Pandey, U.B., Nie, Z., Batlevi, Y., McCray, B.A., Ritson, G.P., Nedelsky, N.B., Schwartz, S.L., DiProspero, N.A., Knight, M.A., Schuldiner, O., Padmanabhan, R., Hild, M., et al. (2007). HDAC6 rescues neurodegeneration and provides an essential link between autophagy and the UPS. *Nature, 447*, 859–863.

52. Shibata, M., Lu, T., Furuya, T., Degterev, A., Mizushima, N., Yoshimori, T., MacDonald, M., Yankner, B., Yuan, J. (2006). Regulation of intracellular accumulation of mutant huntingtin by beclin 1. *J Biol Chem, 281*, 14474–14485.
53. Atwal, R. Truant, R. (2008). A stress sensitive ER membrane-association domain in huntingtin protein defines a potential role for huntingtin in the regulation of autophagy. *Autophagy, 4*, 91–93.
54. King, M., Hands, S., Hafiz, F., Mizushima, N., Tolkovsky, A., Wyttenbach, A. (2008). Rapamycin inhibits polyglutamine aggregation independently of autophagy by reducing protein synthesis. *Mol Pharmacol, 73*, 1052–1063.
55. Zhang, L., Yu, J., Pan, H., Hu, P., Hao, Y., Cai, W., Zhu, H., Yu, A.D., Xie, X., Ma, D., Yuan, J. (2007). Small molecule regulators of autophagy identified by an image-based high-throughput screen. *Proc Natl Acad Sci U S A, 104*, 19023–19028.
56. Menzies, F., Ravikumar, B., Rubinsztein, D. (2006). Protective roles for induction of autophagy in multiple proteinopathies. *Autophagy, 2*, 224–225.

7

ROLE OF POSTTRANSLATIONAL MODIFICATIONS IN AMYLOID FORMATION

ANDISHEH ABEDINI

Department of Surgery, College of Physicians and Surgeons, Columbia University, New York, New York

RUCHI GUPTA, PETER MAREK, AND FANLING MENG

Department of Chemistry, State University of New York at Stony Brook, Stony Brook, New York

DANIEL P. RALEIGH

Department of Chemistry, Graduate Program in Biochemistry and Structural Biology, State University of New York at Stony Brook, Stony Brook, New York

HUMEYRA TASKENT AND SYLVIA TRACZ

Department of Chemistry, State University of New York at Stony Brook, Stony Brook, New York

INTRODUCTION

Proteins are subject to a tremendous range of posttranslation modifications. Enzymatically catalyzed posttranslational modifications are normally tightly regulated, since they can profoundly alter the function of proteins and are essential to control cell physiology and cell signaling. A number of "incorrect" enzymatically catalyzed posttranslation modifications have been implicated in

Protein Misfolding Diseases: Current and Emerging Principles and Therapies,
Edited by Marina Ramirez-Alvarado, Jeffery W. Kelly, and Christopher M. Dobson
Copyright © 2010 John Wiley & Sons, Inc.

amyloid formation and protein aggregation diseases [1]. Less attention has been focused on the role of nonenzymatically catalyzed posttranslation modifications, sometimes termed chemical modifications or spontaneous modifications, on amyloid formation in vitro and in vivo. A bewildering variety of spontaneous posttranslation modifications can affect proteins, some of which are relatively benign whereas others have major effects (Table 1). A partial but by no means exhaustive list includes deamidation, oxidation, β-elimination, nitration, glycation, Michael addition, racemization, peptide bond cleavage, and various cross-linking reactions. The exact role of such modifications in in vivo amyloid formation is not clear. Many ex vivo samples of amyloid deposits contain proteins that have been subjected to chemical modifications; however, it is extraordinarily difficult to establish a cause-and-effect relationship. Chemical modification may have triggered amyloid formation, but conversely, since amyloid fibrils are long-lived aggregates, the presence of modifications may simply reflect the long lifetime of the deposits, whereby significant amounts of modification accumulate over time [1,2].

In principle, the effects of chemical modifications on the in vitro production of amyloid should be easier to decipher since experiments can be conducted under carefully controlled conditions starting with well-defined precursors. Unfortunately, this is not always the case, and spontaneous modifications may have occurred without the investigator's knowledge. This may be of particular relevance in studies where proteins are induced to form amyloid under relatively harsh conditions. The vast majority of in vitro studies of amyloid formation do not report on whether or not the presence and effects of chemical modifications

TABLE 1 Selected Spontaneous Posttranslational Modifications That May Play a Role in Amyloid Formation

Residue	Modification
Asn	Deamidation
Asp	Racemization
Arg	Glycation, formation of advanced glycation end products (AGEs), including generation of cross-links
Cys	β-Elimination, improper disulfide formation
Gln	Deamidation, formation of pyroglutamic acid
His	Michael addition of 4-hydroxy-2-nonenal produced from oxidative damage of lipid membranes
Lys	Glycation, formation of advanced glycation end products (AGEs), including generation of cross-links
Met	Oxidation
Ser	Racemization
Trp	Oxidation
Tyr	Oxidation, including generation of dityrosine cross-links, nitration
Asp–Pro peptide bonds	Spontaneous cleavage of the peptide bond

were considered. This is unfortunate, especially considering that modern mass spectrometry methods are well suited for detecting and characterizing such events. This could be a significant oversight because a number of studies have shown that proteins which normally do not form amyloid in vivo can be induced to do so in vitro; however, the in vitro experiments sometimes involve prolonged incubation under conditions that may lead to spontaneous posttranslational modifications. This raises the possibility that the molecules which form amyloid or trigger amyloid formation may not in fact be the same as the species present at the start of the experiment, which in turn could influence suggestions that amyloid formation is a general property of proteins and that virtually all proteins can be induced to form amyloid if appropriate conditions can be found [3,4]. More subtle consequences could arise in studies designed to develop propensity scales for aggregation and amyloid formation since the presence of a chemically modified variant protein or peptide can affect the rate of amyloid formation [5,6]. In some cases, even low levels of modified molecules may "seed" amyloid formation or aggregation by the unmodified proteins; indeed, heterogeneous seeding is well documented [1,6].

In general, the potential role of spontaneous posttranslational modification in amyloid formation has received relatively little attention. The one field in which such effects are considered routinely is, for obvious reasons, the formulation of protein- and peptide-based drugs. There is a large literature in this area, and many of the conclusions are relevant to biophysical and biochemical studies of protein aggregation and amyloid formation. There are two major issues in protein formulation: first, maintaining the protein- or polypeptide-based pharmaceutical in a soluble state for a significant time, and second, ensuring that the possibility of deleterious posttranslational modifications are minimized. Given the practical importance of these problems, there is considerable literature on the chemical stability of protein pharmaceuticals and the formulation of protein- and polypeptide-based drugs. A number of useful review articles have been published, and the general principles are directly translatable to the field of amyloid formation [7–15].

In this chapter we comment on the role of enzymatically catalyzed posttranslational modifications in amyloid formation, and discuss the possible role of spontaneous posttranslational modifications. We do not attempt to provide a comprehensive survey of the field, but rather seek to highlight general principles and provide references to a few select examples that readers may find of interest. In 2005, Nilsson provided a review of the literature up through the early twenty-first century [1]. That review, along with several others, is recommended for a detailed listing of specific case studies [1,10,12,16–18]. There is growing literature on the role of spontaneous posttranslation modifications in protein aggregation and amyloid formation, but it is still a relatively underappreciated area; however, as noted above, a large collection of relevant literature can be found in the field of biopharmaceutical stabilization and formulation. In addition, there is a large body of work in the broad area of "protein aging." The basic mechanistic chemistry of the spontaneous modification of proteins is summarized in a number of protein chemistry textbooks and reviews (see, e.g., [17,18]).

ABERRANT ENZYMATICALLY CATALYZED POSTTRANSLATIONAL MODIFICATIONS CAN PLAY A CRITICAL ROLE IN PROTEIN AGGREGATION DISEASES

Posttranslational modifications such as phosphorylation, glycosylation, proteolytic activation, lipidation, and numerous others are essential features of cell physiology and are catalyzed by normally exquisitely regulated enzymes. Sometimes, however, inappropriate posttranslational modifications do occur, and these can have disastrous consequences, including the promotion of protein aggregation. Improper enzymatically catalyzed posttranslational modifications could lead to amyloid formation in at least three ways. Inappropriate proteolytic cleavage of a protein can generate an amyloidogenic polypeptide. The classic example is the production of the Alzheimer disease Aβ (1–40) and Aβ (1–42) polypeptides, which result from inappropriate cleavage of amyloid precursor protein [19]. Medin provides a second example of the role of proteolytic cleavage in amyloid formation. The medin polypeptide is a proteolytic fragment of lactadherin and is responsible for aortic amyloid [20]. Numerous other examples were reviewed by Nilsson in 2005 [1]. A second scenario can involve the failure of correct proteolytic processing events to occur. Islet amyloid polypeptide (IAPP, also known as amylin) may offer an example of this type of behavior. IAPP is a pancreatic hormone that is cosecreted with insulin and is responsible for islet amyloid deposition in type 2 diabetes [21,22]. Incomplete proteolytic processing of proIAPP has been proposed to play a role in islet amyloid formation in vivo [23–25]. IAPP is packaged in the insulin secretory granule as a prohormone, proIAPP. Posttranslational processing of proIAPP involves proteolytic cleavage by prohormone convertases at two conserved sites. Immunochemical studies of islet amyloid have indicated the presence of the N-terminal region of proIAPP but not the C-terminal extension, arguing that a partially processed intermediate is produced [23,25]. One hypothesis is that this incorrectly processed intermediate interacts with sulfated proteoglycans of the extracellular matrix and the immobilized form acts as a seed for amyloid formation [24]. Incompletely processed variants of other prohormones, including procalcitonin and proANF, are thought to be involved in amyloid formation [26,27]. Finally, enzymatically catalyzed posttranslation modifications can lead to amyloid formation if they occur at inappropriate sites and/or at the wrong time. For example, hyperphosphorylation of the tau protein plays a role in the formation of neurofibrillary tangles in Alzheimer disease, and it has been proposed that posttranslational modification of α-synuclein may be important in Parkinson disease [28,29]. Errors in the attachment of O-linked β-*N*-acetylglucosamine have also been proposed to play a role in Alzheimer disease [30].

PROTEINS ARE SUBJECTED TO A WIDE RANGE OF NONENZYMATIC MODIFICATIONS IN VITRO AND IN VIVO

The two most common nonenzymatic modifications of proteins are deamidation of Asn and Gln, and oxidation of a number of residues. Deamidation of the amide side chains of Asn and Gln is a well-studied process, and deamidation of Asn is faster than deamidation of Gln side chains. Mass spectrometry methods for the detection of deamidated residues have continued to develop and have facilitated the detection of these modifications [31,32]. High-performance liquid chromatographic–based assays are also useful for detecting low levels of deamidation [6]. Deamidation rates are sensitive to temperature, pH, ionic strength, and choice of buffer. One interesting consequence of the effect of environmental parameters on deamidation reactions is that there appears to be a statically significant decrease in the amount of Asn and Gln repeats in hyperthermophilic organisms, probably reflecting the fact that the rate of deamidation increases with temperature [33]. A comprehensive review of the literature on Asn and Gln deamidation up to 2004, together with a useful summary of experimental data, is provided in the monogram by Robinson and Robinson [34].

Deamidation of Asn proceeds through a succinimide intermediate involving reaction with the C-terminal neighboring peptide bond (Fig. 1). Direct hydrolysis of Asn residues at the amide side chain can also occur, resulting in formation of Asp. However, the rate at neutral pH is much lower than deamidation via the succinimide intermediate [35]. The rate of the "normal reaction" (i.e., deamidation via the succinimide intermediate) is extremely sensitive to the primary sequence and is significantly faster for Asn–Gly

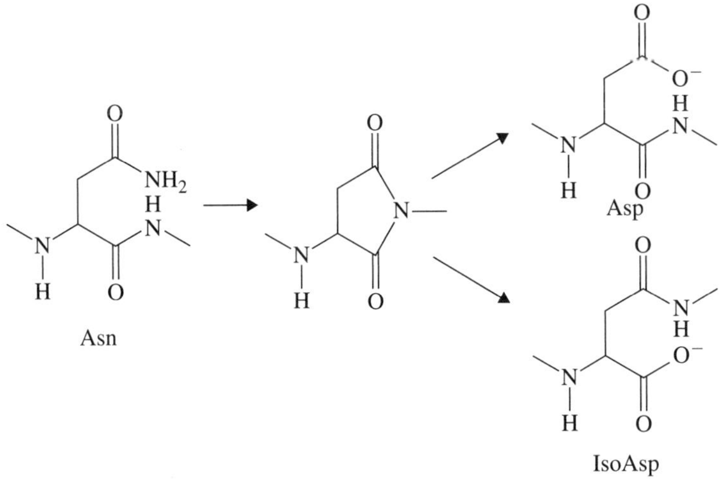

FIG. 1 Deamidation of Asn side chains proceeds via a succinimide intermediate and can lead to formation of L- and D-Asp as well as L- and D-iso-Asp.

dipeptide units than for other dipeptide linkages, owing to the steric requirements for forming the cyclic intermediate. The intrinsic sequence dependence of the rate is strongly affected by structural context. Residues that are in flexible regions of structure will deamidate more rapidly than those which are located in well-ordered elements of structure [32,34]. The succinimide ring can be open by attack by water in two ways. Hydrolysis can generate a normal peptide linkage but with the former amide side chain converted to the corresponding carboxylate. Ring opening can also lead to formation of isoaspartic acid. In an iso-Asp residue, the peptide linkage is through the former side chain; thus, this modification introduces another rotatable bond into the polypeptide backbone as well as introducing a new negative charge (Fig. 1). There is another layer of complexity since the succinimide intermediate is prone to racemization, which will lead to the production of D-Asp and/or D-iso-Asp. Thus, deamidation can lead to major changes in the properties of polypeptides, altering their net charge, potentially introducing D-amino acids and additional degrees of freedom into the polypeptide backbone.

Some proteins form amyloid starting from monomers that lack significant persistent structure (i.e., the natively unfolded proteins). In these cases, modifications such as deamidation could directly alter the amyloidogenic propensity of a polypeptide. Globular proteins that form amyloid presumably have to undergo at least partial unfolding to generate an aggregation-prone state. In this case, more indirect effects could be operative; in particular, the variety of modifications introduced by deamidation can destabilize protein structure and thus enhance the production of partially unfolded or completely unfolded species. It is important to realize that only a small fraction of modified material may be required to seed amyloid formation by the unmodified parent protein; thus, moderate levels of deamidation (or any other modification for that matter) could have a significant effect. One particularly striking example of this sort of behavior is offered by a study of small fragments of IAPP that contained proline substitutions. The mutant peptide was unable to form amyloid, but the presence of less than 5% of deamidation impurities was sufficient to seed amyloid formation and induce the normally nonamyloidogenic sequence into the fibril state [6]. This highlights the importance of considering the potential effects of spontaneous modifications, even if they are present at low levels. Given the ubiquitous nature of deamidation and the potentially serious consequences that can result, it seems unformatuate that it is not tested for routinely, especially as the assays are straightforward.

Deamidated Asn and Gln residues have been found in ex vivo samples extracted from a range of amyloids (reviewed by Nilsson [1]), but again, the question of cause and effect is difficult to address. Deamidation may trigger or promote amyloid formation, or it may be a secondary effect that occurs after the polypeptide or protein is deposited in the fibril. Indirect arguments can be made on the basis of the distribution of deamidated residues. If deamidation were a secondary effect that occurred postdeposition, a relatively equal population of deamidated products would be expected at surface-exposed sites.

Conversely, a selected subset of sites would be modified if deamidation at a specific site were responsible for triggering amyloid formation. There are reports that the distribution of modified residues in the Aβ peptide is different in neuritic than in vascular amyloid, which led to the suggestion that the modifications were preamyloid events [36,37].

One very well documented area in which deamidation appears to play an important role in vivo is in cataract formation [38–42]. The eye lens proteins, known as the crystallins, do not turn over and are thus subjected to increasing probability of chemical modification as they age. Amyloid formation and aggregation of the eye lens crystallins is potentially of considerable clinical significance, since it has been implicated in cataract formation.

Gln deamidation is rapid for N-terminal Gln residues and leads to cyclization of the side chain to produce pyroglutamic acid. This modification alters the geometry of the N-terminal residue, which could have a direct effect on the amyloidogencity of an intrinsically disordered polypeptide and could destabilize a globular protein. Deamidation of internal Gln residues is normally a slower process than deamidation of Asn, but it does occur and can have major effects on protein stability. An interesting case study of relevance to amyloid formation is provided by recent work on human γD-crystallin. Deamidation of Gln was shown to destabilize the protein in vitro and lead to more rapid unfolding, which could in turn lead to a state that is more prone to aggregate [42]. Gln can also take part in cross linking reactions with the ε-amino group of lysine side chains. Formation of these linkages can be promoted by transglutaminases.

Racemization at residues other than Asn or Gln is well known [1,17,43]. A D-amino acid residue is very destabilizing in a β-sheet comprised of L-amino acids; thus, D-amino acids might be expected to destabilize the cross β-structure of amyloid fibrils; however, amyloid fibrils also contain loops and turns. A D-Amino acid will be favorable at any site that requires a positive value of the backbone dihedral angle psi, provided that the altered side-chain geometry does not lead to any steric clashes. D-Amino acids are well known to stabilize certain types of turn conformations. Racemization of Asp and Ser in Aβ has been reported in ex vivo samples of Aβ. Again, the question of whether the modifications led to amyloid formation or occurred postdeposition is vexing. In vitro racemization has different effects on the rate of aggregation of the Aβ peptide, depending on the site modified. D-Asp Aβ is reported to aggregate faster than the unmodified peptide, and similarly, a D-Ser at position 8 is reported to increase the rate of amyloid formation in vitro, while a D-Ser modification at residue 26 is reported to slow fibril formation. The D-Ser 26 modification has also been reported to lead to chain cleavage and the generation of smaller toxic fragments [43–45].

Oxidation of protein side chains is extremely common and is well characterized. It plays a major role in aging and may well contribute to amyloid formation [1,16,17,46–51]. Arg, Cys, Gln, Glu, His, Leu, Lys, Met, Phe, Pro, Thr, Tyr, Trp, and Val are all subject to oxidation and all have been implicated in protein aging, while His, Met, Arg, Pro, and Lys are among the most common amino acids

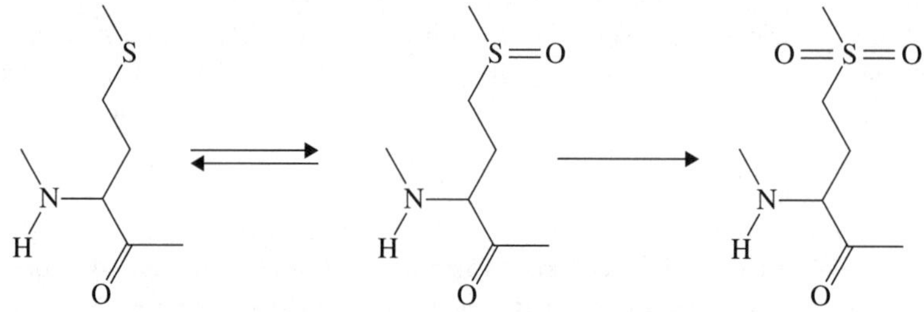

FIG. 2 Oxidation of methionine can lead to a sulfoxide and, under harsher conditions, to a sulfone.

which are susceptible to metal-catalyzed oxidation reactions [51]. A myriad of oxidative modifications of proteins have been reported. Those that have been implicated in amyloid formation include oxidation of Met side chains and oxidation of aromatic residues, particularly Tyr.

Met side-chain oxidation can lead to two products (Fig. 2). The first, which leads to the sulfoxide, is reversible and can occur under mild conditions; the second, which generates a sulfone, is normally not reversible. Oxidation to the sulfone requires harsh conditions, but formation of the sulfoxide is relatively facile and is a common nonenzymatic modification. Oxidation modifies the hydrophobicity and shape of the side chain by introducing a polar oxygen atom into the side chain. Note that, for the case of the sulfoxide formation, the new side chain is chiral. The role of Met oxidation in amyloid formation by the Aβ peptide and by α-synuclein has been examined ([50,52–54; reviewed in [1]). Oxidation of Met35 in Aβ is well documented and hinders amyloid formation. The recent solid-state nuclear magnetic resonance–based structural model of the Aβ amyloid fibril helps to rationalize these results [55]. The steric effect of the altered side chain and the introduction of the polar oxygen are expected to destabilize the fibril structure based on the packing observed for the wild-type fibril. Oxidation of Met residues in α-synuclein has also been reported to slow amyloid formation [54].

Tyrosine is subject to a number of chemical modifications, including oxidation to 3,4-dihydroxyphenylanine, nitration to 3-nitrotyrosine, chlorination to form 3,5-dichlorotyrosine, and the formation of Tyr–Tyr cross-links [16,17,50,51,56]. 3-Nitrotyrosine has been used as a stable marker for protein oxidative damage, and the potential role of nitration in amyloid formation has been considered [51–56]. Nitration of Tyr residues in α-synuclein and tau has been reported and has been proposed to affect fibril stability and/or the kinetics of fibril assembly [57–59]. In vitro studies have shown that cross-linking of α-synuclein occurs after exposure to nitrating agents and requires the presence of Tyr residues. The cross-links can influence the stability and formation of α-synuclein-derived filaments [57]. Several additional studies have examined the potential role of dityrosine cross-links on amyloid formation by α-synuclein [29,50,56–60].

Oxidative damage can affect proteins indirectly by producing species that react with susceptible side chains. One example that is relevant to amyloid formation is the generation of 4-hydroxy-2-nonenal via oxidative stress. The compound, which is produced when lipid membranes that contain omega-6 polyunsaturated fatty acyl chains are subjected to oxidative stress, has been shown to modify multiple His side chains in Aβ via Michael addition. A recent biophysical study has shown that the modified version of Aβ that is produced has an increased membrane affinity and adopts a membrane-bound conformation that is conducive to amyloid formation. This work provides a plausible mechanism for the amyloidogenic effects of oxidative lipid damage [61]. Somewhat different effects were observed with α-synuclein [62]. Incubation of α-synuclein with 4-hydroxyl-2-nonenal led to the covalent modification of the protein by formation of the Michael addition adduct, with up to six sites modified per α-synuclein molecule. The modification altered the conformational propensities of α-synuclein but inhibited fibril formation, leading instead to the formation of compact oligomers. The oligomers were toxic to primary mesencephalic cell cultures [62].

The nonenzymatically catalyzed reaction of sugars with proteins, referred to as glycation to distinguish it from glycosylation, is an important posttranslation modification in diabetes [63,64]. The reaction of reducing sugars with the amino group of lysine side chains, the N-terminal amino group, and Arg side chains is well known and was first described almost 100 years ago by Millard [65,66]. Reaction of reducing sugars with protein amino groups leads to formation of a Schiff base. The initial Schiff base adduct undergoes an Amadori rearrangement, followed by a complex set of condensation, dehydration, rearrangement, oxidation, and potential fragmentation reactions to generate a heterogeneous set of covalent adducts collectively known as advanced glycation end products (AGEs) [66]. AGEs lead to the formation of inter- and intraprotein cross-links and the cross-linked proteins can form protease-resistant intracellular aggregates. Reactive oxygen species are also produced during the formation of AGEs, which can have deleterious effects.

AGEs and the interaction of AGEs with the AGE receptor play a role in the complication of diabetes; however, their role in protein aggregation diseases and neurodegenerative disorders is less clear [50,67]. A series of papers appeared in the mid-1990s which advocated a causative role for AGEs in Alzheimer disease; however, the hypothesis was challenged in several other studies. The review article by Kikuchi and colleagues is recommended for a survey and critical analysis of that literature [67]. A number of in vitro biophysical studies have examined the effects of AGEs on the kinetics of amyloid formation, and have been briefly reviewed by Nilsson and by Kikuchi and colleagues [1,67].

CONCLUDING REMARKS

Ex vivo samples of amyloid deposits typically contain some protein that has been posttranslationally modified, but the potential role of both spontaneous and enzymatically catalyzed posttranslational modifications in amyloid formation is often underexplored. The first key issue for in vivo studies is to establish if posttranslational modifications occurred prior to fibril formation or if they occurred after amyloid formation. A deeper understanding of the role of posttranslational modifications in amyloid formation may lead to new therapeutic approaches [68].

An appreciation of the potential effects of posttranslational modifications on amyloid formation is also critical for the reliable interpretation of *in vitro* biophysical and biochemical studies. The observation of heterogeneous seeding in amyloid formation has interesting implications, since it shows that small amounts of an amyloidogenic protein can sometimes seed amyloid formation by a second protein. This leads to the possibility that a relatively low population of a posttranslationally modified protein could promote amyloid formation by rapidly assembling and then acting as a seed to induce amyloid formation by the population of unmodified molecules. Lot-to-lot variability is well known in biophysical studies of synthethic IAPP and Aβ peptides, and one possibility to consider is that it may arise, at least in part, because of variable low levels of chemically modified protein.

Acknowledgments

Our work on amyloid formation is supported by National Institutes of Health grant GM078114 to D.P.R. We thank C. Bruce Verchere and M. Nilsson for helpful discussions. A.A., S.T., and P.M. were supported in part by GAANN fellowship from the U.S. Department of Education.

REFERENCES

1. Nilsson, M.R. (2005). Role of post-translation chemical modifications in amyloid fibril formation. In *Amyloid Proteins: The Beta Sheet Conformation and Disease* (Sipe, J.D., Ed.), Wiley–VCH, Weinheim, Germany, pp. 81–109.
2. Nilsson, M.R., Dobson, C.M. (2003). Chemical modification of insulin in amyloid fibrils. *Protein Sci, 12*, 2637–2641.
3. Dobson, C.M. (1999). Protein misfolding, evolution and disease. *Trends Biochem Sci, 24*, 329–332.
4. Vendruscolo, M., Zurdo, J., Macphee, C.E., Dobson, C.M. (2003). Protein folding and misfolding: a paradigm of self-assembly and regulation in complex biological systems. *Philos Trans A, 361*, 1205–1222.
5. Chiti, F., Stefani, M., Taddei, N., Ramponi, G., Dobson, C.M. (2003). Rationalization of the effects of mutations on peptide and protein aggregation rates. *Nature, 424*, 805–808.
6. Nilsson, M.R., Driscoll, M., Raleigh, D.P. (2002). Low levels of asparagine deamidation can have dramatic effects on the aggregation of amyloidogenic peptides: implications for the study of amyloid formation. *Protein Sci, 11*, 342–349.

7. Cleland, J.L., Powell, M.F., Shire, S.J. (1993). The development of stable protein formulations: a close look at protein aggregation, deamidation, and oxidation. *Crit Rev Ther Drug Carrier Syst, 10,* 307–377.
8. Brange, J., Langkjoer, L. (1993). Insulin structure and stability. *Pharm Biotechnol, 5,* 315–350.
9. Parkins, D.A., Lashmar, U.T. (2000). The formulation of biopharmaceutical products. *Pharm Sci Technol Today, 3,* 129–137.
10. Arkawa, T., Prestreleski, S.J., Kenney, J.F., Carpenter J.F. (2001). Factors affecting short-term and long-term stabilities of proteins. *Adv Drug Deliv Rev, 46,* 307–346.
11. Chi, E.Y., Krishnan, S., Randolph, T.W., Carpenter, J.F. (2003). Physical stability of proteins in aqueous solution; mechanism and driving forces in nonnative protein aggregation. *Pharm Res, 20,* 1325–1336.
12. Wang, W. (2005). Protein aggregation and its inhibition in biopharmaceuticals. *Int J Pharm, 289,* 1–30.
13. Demeule, B., Gurny, R., Avint, T. (2006). Where disease pathogenesis meets protein formulation; renal deposition of immunoglobulin aggregates. *Eur J Pharm Biopharm, 62,* 121–130.
14. Srebalus Barnes, C.A., Lim, A. (2007). Applications of mass spectrometry for the structural characterization of recombinant protein pharmaceuticals. *Mass Spectrom Rev, 26,* 370–388.
15. Capelle, M.A.H., Gurny, R., Arvinte, T. (2007). High throughput screening of protein formulation stability: practical considerations. *Eur J Pharm Biopharm, 65,* 131–148.
16. Stadtman, E.R. (2001). Protein oxidation in ageing and age-related diseases. *Ann NY Acad Sci, 928,* 22–38.
17. Volkin, D.B., Mach, H., Middaugh, C.R. (1995). Degradative covalent reactions important to protein stability. In *Methods in Molecular Biology,* Vol. 40, *Protein Stability and Folding: Theory and Practice* (Shirley, B.A., *Ed.*), Humana Press, Totowa, NJ, pp. 35–63.
18. Reubsaet, J.L.E., Beijnen, J.H., Bult, A., van Maanen, R.J., Marchal, J.A.D., Underberg, W.J.M. (1998). Analytical techniques used to study the degradation of proteins and peptides: chemical instability. *J Pharm Biomed Anal, 17,* 955–978.
19. Haass, C., Selkoe, D.J. (1993). Cellular processing of beta-amyloid precursor protein and the genesis of amyloid beta peptide. *Cell, 75,* 1039–1042.
20. Häggqvist, B., Näslund, J., Sletten, K., Westermark, G.T., Mucchiano, G., Tjernberg, L.O., Nordstedt, C., Engström, U., Westermark, P. (1999). Medin: an integral fragment of aortic smooth muscle cell produced lactadherin forms the most common human amyloid. *Proc Natl Acad Sci U S A, 96,* 8669–8674.
21. Westermark, P., Wernstedt, C., Wilander, E., Hayden, D.W., O'Brien, T.D., Johnson, K.H. (1987). Amyloid fibrils in human insulinoma and islets of Langerhans of the diabetic cat are derived from a neuropeptide-like protein also present in normal islet cells. *Proc Natl Acad Sci U S A, 84,* 3881–3885.
22. Cooper, G.J.S., Willis, A.C., Clark, A., Turner, R.C., Sim, R.B., Reid, K.B.M. (1987). Purification and characterization of a peptide from amyloid-rich pancreases of type 2 diabetic patients. *Proc Natl Acad Sci U S A, 84,* 8628–8632.

23. Marzban, L., Rhodes, C.J., Steiner, D.F., Haataja, L., Halban, P.A., Verchere, C.B. (2006). Impaired NH2-terminal processing of human proislet amyloid polypeptide by the prohormone convertase PC2 leads to amyloid formation and cell death. *Diabetes, 55*, 2192–2201.
24. Park, K., Verchere, C.B. (2001). Identification of a heparin binding domain in the N-terminal cleavage site of pro-islet amyloid polypeptide. Implications for islet amyloid formation. *J Biol Chem, 276*, 16611–16616.
25. Paulsson, J.F., Westermark, G.T. (2005). Aberrant processing of human proislet amyloid polypeptide results in increased amyloid formation. *Diabetes, 54*, 2117–2125.
26. Pucci, A., Wharton, J., Arbustini, E., Grasso, M., Diegoli, M., Needleman, P., Viganò, M., Polak, J.M. (1991). Atrial amyloid deposits in the failing human heart display both atrial and brain natriuretic peptide like immunoreactivity. *J Pathol, 165*, 235–241.
27. Sletten, K.P., Westermark, P., Natvig, J.B. (1976). Characterization of amyloid fibril proteins from medullary carcinoma of the thyroid. *J Exp Med, 143*, 993–998.
28. Goedert, M. (1993). Tau protein and the neurofibrillary pathology of Alzheimer's disease. *Trends Neurosci, 16*, 460–465.
29. Beyer, K. (2006). Alpha-synuclein structure, posttranslational modification and alternative splicing as aggregation enhancers. *Acta Neuropathol, 112*, 237–251.
30. Dias, W.B., Hart, G.W. (2007). O-GlcNAc modification in diabetes and Alzheimer's disease. *Mol Biosyst, 3*, 766–772.
31. Cournoyer, J.J., Lin, C., O'Connor, P.B. (2006). Detecting deamidation products in proteins by electron capture dissociation. *Anal Chem, 78*, 1264–1271.
32. Zabrouskov, V., Han, X., Welker, E., Zhai, H., Lin, C., van Wijk, K.J., Scheraga, H.A., McLafferty, F.W. (2006). Stepwise deamidation of ribonuclease A at five sites determined by top down mass spectrometry. *Biochemistry, 45*, 987–992.
33. Michelitsch, M.D., Weissman, J.S. (2000). A census of glutamine/asparagine-rich regions: implications for their conserved function and the prediction of novel prions. *Proc Natl Acad Sci U S A, 97*, 11910–11915.
34. Robinson, N.E., Robinson, A.B. (2004). *Molecular Clocks: Deamidation of Asparaginyl and Glutaminyl Residues in Peptides and Proteins*, Althouse Press, Cave Junction, OR.
35. Robinson, N.E., Robinson, A.B. (2001). Deamidation of human proteins. *Proc Natl Acad Sci U S A, 98*, 12409–12413.
36. Kuo, Y.M., Emmerling, M.R., Woods, A.S., Cotter, R.J., Roher, A.E. (1997). Isolation, chemical characterization, and quantization of A beta 3-pyroglutamyl peptide from neuritic plaques and vascular amyloid deposits. *Biochem Biophys Res Commun, 237*, 188–191.
37. Shin, Y., Cho, H.S., Fukumoto, H., Shimizu, T., Shirasawa, T., Greenberg, B.D., Rebeck, G.W. (2003). Abeta species including IsoAsp23 Abeta in Iowa-type familial cerebral amyloid angiopathy. *Acta Neuropathol, 105*, 252–258.
38. Takemoto, L., Emmons, T., Granstrom, D. (1990). The sequence of two peptides from cataract lenses suggests they arise by deamidation. *Curr Eye Res, 9*, 793–797.
39. Takemoto, L., Boyle, D. (2000). Increased deamidation of asparagines during human senile cataractogenesis. *Mol Vis, 6*, 164–168.

40. Zhang, Z., Smith, D.L., Smith, J.B. (2003). Human beta-crystallins modified by backbone cleavage, deamidation and oxidation are prone to associate. *Exp Eye Res, 77*, 259–272.

41. Wilmarth, P.A. Tanner, S., Dasari, S., Nagalla, S.R., Riviere, M.A., Bafna, V., Pevzner, P.A., David, L.L. (2006). Age-related changes in human lens crystallins determined from comparative analysis of post-translational modifications in young and aged lens: does deamidation contribute to crystalline insolubility? *J Proteome Res, 5*, 2554–2566.

42. Flaugh, S.L., Mills, I.A., King, J. (2006). Glutamine deamidation destabilizes human γD-crystallin and lowers the kinetic barrier to unfolding. *J Biol Chem, 281*, 30782–30793.

43. Kubo, T., Kumagae, Y., Miller, C.A., Kaneko, I. (2003). β-Amyloid racemized at the Ser(26) residue in the brains of patients with Alzheimer disease: implications in the pathogenesis of Alzheimer disease. *J Neuropathol Exp Neurol, 62*, 248–259.

44. Shapira, R.G., Austin, G.E., Mirra, S.S. (1998). Neuritic plaque amyloid in Alzheimer's disease is highly racemized. *J Neurochem*, 50, 69–74.

45. Zagorski, M.G., Yang, J., Shao, H., Ma, K., Zang, H., Hong, A. (1999). Methodological and chemical factors affecting amyloid beta peptide amyloidogenicity. *Methods Enzymol, 309*, 189–204.

46. Harman, D. (1956). Aging: a theory based on free radical and radiation chemistry. *J Gerontol, 11*, 298–300.

47. Glazer, A.N. (1970). Specific chemical modification of proteins. *Annu Rev Biochem, 39*, 101–130.

48. Feeney, R.E., Yamasaki, R.B., Geoghegan, K.F. (1982). Chemical modification of proteins:-an overview. *Adv Chem Ser, 198*, 3–55.

49. Berlett, B.S. (1997). Protein oxidation in aging, disease, and oxidative stress. *J Biol Chem, 272*, 20313–20316.

50. Ahmed, N., Ahmed, U., Thornalley, P.J., Hager, K., Fleischer, G., Munch, G. (2005). Protein glycation, oxidation and nitration adduct residues and free adducts of cerebrospinal fluid in Alzheimer's disease and link to cognitive impairment. *J Neurochem, 92*, 255–263.

51. Soskic, V., Groebe, K., Schrattenholz, A. (2008). Nonenzymatic posttranslational protein modifications in ageing. *Exp Gernotol, 43*, 247–257.

52. Palmblad, M., Westlind-Danielsson, A., Bergquist, J. (2002). Oxidation of methionine-35 attenuates formation of amyloid beta peptide 1–40 oligomers. *J Biol Chem, 277*, 19506–19510.

53. Hou, L., Kang, I., Marchant, R.E., Zagorski, M.G. (2002). Methionine-35 oxidation reduces fibril assembly of the amyloid Abeta(1–42) peptide of Alzheimer's disease. *J Biol Chem, 277*, 40173–40176.

54. Hockenson, M.J., Uversky, V.N., Goers, J., Yamin, G., Munishkina, L.A., Fink, A.L. (2004). Role of individual methionines in the fibrillation of methionine-oxidized alpha-synuclein. *Biochemistry, 43*, 4621–4633.

55. Petkova, A.T., Ishii, Y., Balbach, J.J., Antzutkin, O.N., Leapman, R.D., Delaglio, F., Tycko, R. (2002). A structural model for Alzheimer's beta-amyloid fibrils based on experimental constraints from solid state NMR. *Proc Natl Acad Sci U S A, 99*, 16742–16747.

56. Ischiropoulous, H. (2003). Oxidative modifications of α-synuclein. *Ann N Y Acad Sci, 991*, 93–100.
57. Giasson, B.I., Duda, J.E., Murray, I.V., Chen, Q., Souza, J.M., Hurtig, H.I., Ischiropoulos, H., Trojanowski, J.Q., Lee, V.M. (2000). Oxidative damage linked to neurodegeneration by selective α-synuclein nitration in synucleinopathy lesions. *Science, 290,* 985–989.
58. Horiguchi, T., Uryu, K., Giasson, B.I., Ischiropoulos, H., Lightfoot, R., Bellmann, C., Richter-Landsberg, C., Lee, V.M., Trojanowski, JQ. (2003). Nitration of tau protein is linked to neurodegeneration in tauopathies. *Am J Pathol, 163*, 1021–1031.
59. Norris, E.H., Giasson, B.I., Ischiropoulous, H., Lee, V.M. (2003). Effects of oxidative and nitrative challenges on alpha-synuclein fibrillogenesis involve distinct mechanisms of protein modifications. *J Biol Chem, 278*, 27230–27240.
60. Krishnan, S., Chi, E.Y., Wood, S.J., Kendrick, B.S., Li, C., Garzon-Rodriguez, W., Wypych, J., Randolph, T.W., Narhi, L.O., Biere, A.L., et al. (2003). Oxidative dimer formation is the critical rate-limiting step for Parkinson's disease alpha-synuclein fibrillogenesis. *Biochemistry, 42*, 829–837.
61. Liu, L., Komatsu, H., Murray, I.V.J., Axelsen, P.H. (2008). Promotion of Amyloid β protein misfolding and fibrillogenesis by a lipid oxidation product. *J Mol Biol, 377*, 1236–1250.
62. Qin, Z., Hu, D., Han, S., Reaney, S.H., Di Monte, D.A., Fink, A.L. (2007). Effect of 4-hydroxy-2-noneal modification on alpha-synuclein aggregation. *J Biol Chem, 282*, 5862–5870.
63. Brownlee, M. (1992). Glycation products and the pathogenesis of diabetic complications. *Diabetes Care, 15*, 1835–1843.
64. Yan, S.F., Ramasamy, R., Schmidt, A.M. (2008). Mechanisms of disease: advanced glycation end-products and their receptor in inflammation and diabetes complications. *Nat Clin Pract Endocrinol Metab, 4*, 285–293.
65. Millard, M.L.-C. (1912). Action des acidesamines sur les sucres; formation de melanoidines par voie methodique. *Comp Rend, 154*, 66.
66. Cho, S.-J., Roman, G., Yeboah, F., Konishi, Y. (2007). The road to advanced glycation end products: a mechanistic perspective. *Curr Med Chem, 14*, 1653–1671.
67. Kikuchi, S., Shinpo, K., Takeuchi, M., Yamagishi, S., Makita, Z., Sasaki, N., Tashiro, K. (2003). Glycation: a sweet tempter for neuronal death. *Brain Res Rev, 41*, 306–323.
68. Rochet, J.C. (2007). Novel therapeutic strategies for the treatment of protein-misfolding diseases. *Expert Rev Mol Med, 28,* 1–34.

8

UNRAVELING MOLECULAR MECHANISMS AND STRUCTURES OF SELF-PERPETUATING PRIONS

PETER M. TESSIER

Department of Chemical and Biological Engineering, Center for Biotechnology and Interdisciplinary Studies, Rensselaer Polytechnic Institute, Troy, New York

SUSAN LINDQUIST

Whitehead Institute for Biomedical Research, Howard Hughes Medical Institute, Cambridge, Massachusetts

INTRODUCTION

The misfolding and assembly of proteins into β-sheet-rich amyloid fibers is important in both disease [1] and normal biological function [2,3]. Although many proteins form amyloid fibers in vitro, understanding the biological relevance and consequence of this process in vivo is difficult. Given that we can genetically manipulate the yeast *Saccharomyces cerevisiae* better than any other organism, analysis of protein misfolding in yeast is especially attractive. Prions are one class of naturally occurring, amyloid-forming proteins in yeast that have received much attention [3–7]. The first prion protein, PrP, was initially identified in mammals and conversion to its β-sheet-rich aggregated conformation is associated with several related infectious, neurodegenerative diseases known collectively as the spongiform encephalopathies [8]. More recently, several prions have been identified in yeast and other fungi that

Protein Misfolding Diseases: Current and Emerging Principles and Therapies,
Edited by Marina Ramirez-Alvarado, Jeffery W. Kelly, and Christopher M. Dobson
Copyright © 2010 John Wiley & Sons, Inc.

are unrelated to their mammalian counterpart in terms of primary sequence and can be beneficial biologically [3–7].

Although fungal and mammalian prions are unrelated in terms of sequence, they display many functional similarities. Most important, both prions perform functions previously reserved for nucleic acids. In the case of PrP, the protein functions as an infectious agent that causes neurodegenerative disease [8]. In the case of yeast, the prion reveals hidden genetic variation and encodes complex heritable phenotypes [9,10]. At the molecular level, both types of prions form not just one prion conformation, but a collection of structurally related yet distinct conformations (known as *prion strains*) [11–17]. Moreover, both fungal and mammalian prions display barriers to prion transmission between species [18–29]. Further, particular prion strains can overcome species barriers for both types of prions [4,8,18,22,28,30–33].

To decipher the complexities of these problems in vivo, it is necessary to analyze the biochemical properties of these proteins in vitro. Unfortunately, it has been difficult to form highly infectious mammalian prion conformers in vitro from recombinant protein (for recent progress, see [34,35]). In contrast, bona fide highly infectious prion conformers can be formed readily in vitro for fungal prions [12,13,36–38]. This has prompted many biochemical studies of these prions using diverse probes of prion assembly and prion amyloid structure, leading to many important findings and some controversy. In this chapter we review the recent biochemical analysis of fungal prions and highlight potential future areas of investigation.

KNOWN AND POTENTIAL FUNGAL PRIONS

The best studied yeast prion proteins are Sup35, Ure2, and Rnq1 [3–7]. Sup35 is a factor involved in translation termination (Fig. 1). Specifically, it acts in the recognition of stop codons during protein synthesis. Conversion of Sup35 from its soluble nonprion state, [*psi*$^-$], to its aggregated prion state, [*PSI*$^+$], causes reduced termination activity [39–41]. This results in increased read-through of stop codons and reveals complex phenotypes that, in some cases, are beneficial [9,10,42]. Sup35 contains an N-terminal domain that is rich in uncharged polar residues (29% glutamine, 16% asparagine, 16% tyrosine) and glycine (17%). This domain is natively unstructured and governs prion formation. The highly charged, middle (M) domain has a strong solubilizing activity and promotes that nonprion state [3–7]. The C-terminal folded domain encodes its translation termination activity [3–7]. Interestingly, the N-terminal domain contains 5.5 imperfect, oligopeptide repeats (PQGGYQQYN), reminiscent of the five oligopeptide repeats in PrP (PHGGGWGQ).

Ure2 is an inhibitor of Gln3, a transcription factor that represses genes involved in metabolizing poor nitrogen sources when better ones are present [3–7,43]. When Ure2 switches from it soluble nonprotein state, [*ure-o*], to its aggregated prion state, [*URE3*], the activity of Ure2 is impaired. This causes the uptake of poor nitrogen sources in the presence of good ones [44]. Ure2 contains an N-terminal domain rich in polar, uncharged residues (37% asparagines and 11% glutamine) and glycine (11%) that drives prion formation and a C-terminal domain that binds glutathione and encodes the protein's nitrogen regulation activity [3–7,43].

Rnq1 has no known function except to influence the rate at which other prion proteins, such as Sup35, can access their prion conformations [11,45–49]. This activity manifests itself when Rnq1 is in its prion state, [*RNQ*$^+$]. Rnq1, named for being rich in asparagine (N) and glutamine (Q) residues, contains a C-terminal domain that is rich in polar, uncharged residues (27% glutamine and 17% asparagines) and glycine (17%), and drives prion formation. The N-terminal domain is of unknown structure and function [3–7].

Another fungal prion protein, HET-s, exists in the filamentous fungus *Podospora anserina* and is involved in heterokaryon incompatibility [50,51]. To prevent fusion of fungal strains with different genomes, approaching *P. anserina* colonies undergo trial fusion to test for polymorphisms at a dozen loci. When the HET-s prion protein switches from its soluble nonprion state, [Het-s*], to its aggregated prion state, [Het-s], the insoluble prion protein facilitates programmed cell death for certain incompatible fusions through an unknown mechanism. Like other prions, the C-terminal prion domain of HET-s protein is rich in glycine residues (14%). However, unlike other fungal prions, this domain is relatively poor in both glutamine (3%) and asparagine (8%) residues, and rich in valine residues (11%).

Are there are other fungal prions? It seems extremely likely. Several non-Mendelian fungal phenotypes may be prion-based, including [*GR*] [52,53], [*ISP*$^+$] [54], and [*KIL-d*] [55] in *S. cerevisiae*, [*cif*] [56] in *S. pombe*, and [*C*$^+$] [57] in *P. anserina*. Many other potential prions have been identified by genomewide analysis of yeast and other organisms for proteins of sequence composition similar to that of the known yeast prions ([58] and O. King and S. Lindquist, unpublished results). The most promising candidates are currently being tested (S. Alberti, R. Halfmann, and S. Lindquist, unpublished results). Also, the unique sequence of HET-s suggests that there may be other nonglutamine/asparagine-rich fungal prions. Finally, it appears that some naturally occurring fungal prions do not require amyloid formation to perpetuate their infectious conformations (J. Brown and S. Lindquist, unpublished results).

Biochemical Commonalities of Fungal Prions

Although the three known *S. cerevisiae* prions do not share any sequence similarity with HET-s, they do exhibit many biochemical similarities that explain their biological functions [3–7]. For example, as purified proteins, their prion domains are natively unstructured [59–62] and assemble, after a lag phase, into amyloid fibers (Fig. 1) [36,63–66]. Similarly, the same proteins switch from soluble to insoluble (amyloid) conformations in vivo. Amyloids formed both in vitro and in vivo are β-sheet rich, bind dyes such as Congo Red

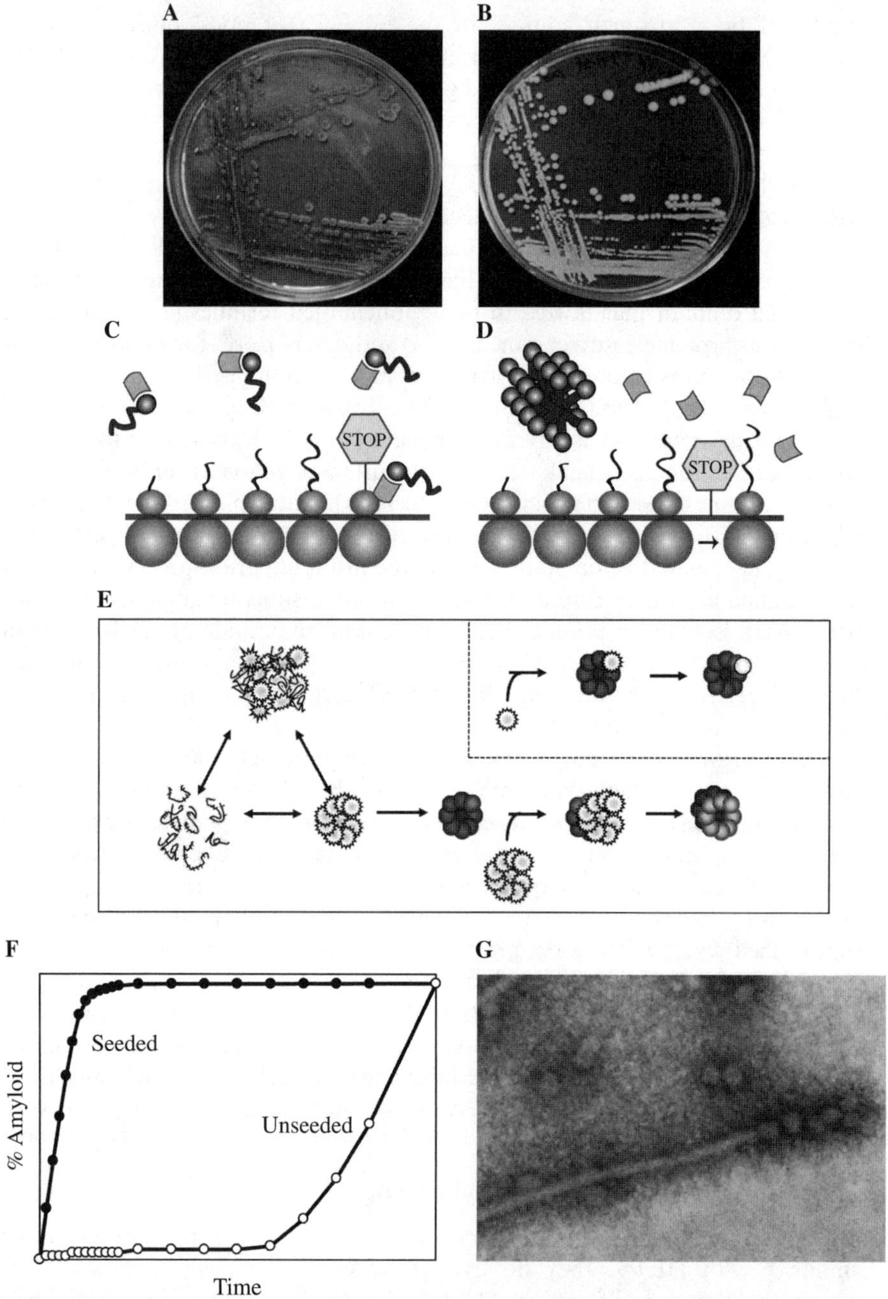

FIG. 1 Molecular basis of [PSI^+] prion propagation. Isogenic *Saccharomyces cerevisiae* in the (A) [*psi*$^-$] and (B) [PSI^+] states [3]. The protein determinant of [PSI^+], Sup35, is (C) soluble and complexed to Sup45 in the [*psi*$^-$] state and (D) insoluble and inactive in the [PSI^+] state. The inactivation of Sup35 causes read-through of stop codons and

and thioflavin T, and are resistant to dissolution in 2% sodium dodecyl sulfate at room temperature [59,64–67]. Moreover, they reduce or eliminate the lag phase and cause rapid assembly of their corresponding soluble proteins by templated polymerization (Fig. 1) [59,64–68].

Proof of Prion Hypothesis

Definitive proof that proteins, in their prion conformation, can function as infectious agents or genetic elements without the aid of nucleic acid has been the focus of intense research for both fungal [12,13,36–38] and mammalian [34,35] prions. The gold standard for verifying this hypothesis is to start with bacterially derived protein, assemble it into amyloid fibers in vitro, introduce these fibers into the host organism, and demonstrate that they induce the prion phenotype. This hypothesis has now been confirmed for all four fungal prions [12,13,36–38]. This was first accomplished for HET-s using a gene gun to transform *P. anserina* with soluble and fibrous protein, as well as amorphous aggregates; only HET-s amyloids efficiently induced conversion to the prion state [36].

The transformation of yeast with amyloids of Sup35, Ure2, and Rnq1 was accomplished using a simpler method [12,13,37,38]. After removal of the outer cell membrane, yeast cells were transformed with soluble and fibrous protein using lithium acetate and polyethylene glycol in a manner similar to a DNA transformation. In each case, the amyloid conformation was capable of inducing the prion phenotype, whereas the soluble protein was rarely able to do so. Moreover, for Sup35 this transformation was also conducted using multiple amyloid conformations; each conformation induced a unique, related prion phenotype [12,13]. This and other studies [19,69,70] for Sup35 establish that not only are amyloids bonafide infectious prions, but that it is differences in the conformations of amyloids that encode distinct prion strain phenotypes.

large phenotypic changes, some of which are beneficial. [9,10,42]. (E) The assembly of Sup35 into amyloid fibers occurs via a nucleated conformational conversion process. This involves a nucleation step where soluble Sup35 oligomeric intermediates form and mature into amyloid nuclei. (From [59], with permission of AAAS.) (F) This assembly process produces a lag phase in which the oligomers form and mature, and then a rapid assembly phase in which mature nuclei template monomeric or oligomeric Sup35 into amyloids. The lag phase is eliminated by adding preformed Sup35 amyloids to soluble Sup35. (G) Transmission electron microscopy image of amyloids of a portion of the Sup35 protein (the N-terminal and middle domains, NM). Spherical NM oligomeric intermediates are associated with the ends of the amyloids, suggesting their relevance to amyloid assembly as well as amyloid nucleation. The width of the NM fibers is approximately 10 nm. (From [59], with permission of AAAS.) (*See insert for color representation of figure.*)

PRION GENERATION AND PROPAGATION

Sup35 Nucleation and Polymerization

How are infectious Sup35 conformers generated de novo? Much research has focused on the N and M domains of Sup35, termed NM, and this work has revealed many important mechanistic details [59,63,69,71,72]. For example, the structure of NM monomers has been probed recently by single-molecule fluorescence studies at very low concentrations (subnanomolar) [73]. Since NM is devoid of cysteines, two residues in the N domain, separated by 100 residues, were mutated to cysteine. The structure of NM monomers was studied by attaching to each protein two different thiol-reactive dyes that were capable of exchanging energy in a distance-dependent manner. A principal finding was that monomeric NM rapidly collapsed into a relatively compact state after dilution from denaturant. Moreover, NM monomers assumed not only one conformation, but a suite of conformations interconverting on the nanosecond time scale. The collapsed state of NM unfolded in a very different manner than globular proteins when exposed to denaturant. NM showed noncooperative unfolding behavior with monomers expanding progressively as a function of denaturant concentration, in sharp contrast to globular proteins, which display a two-state, cooperative unfolding transition.

The mechanism of assembly of NM monomers into amyloid fibers has also received much attention. Of particular interest is the potential existence and relevance of oligomeric intermediates during amyloid nucleation. Serio et al. first identified the fact that during amyloid assembly NM forms spherical, structurally fluid oligomeric structures (Fig. 1) which are similar to those observed more recently for many other amyloid-forming proteins [59]. NM oligomers were observed initially by atomic force and transmission electron microscopy [59], and later by dynamic light scattering [74].

Several lines of evidence suggest that oligomers are on-pathway structures involved in the nucleation of NM fibers. For example, an oligomer-specific antibody recognizes an intermediate that forms at the end of the lag phase, which is consumed rapidly once fibers form [75]. Moreover, this same antibody inhibits nucleation of NM amyloid formation [76]. Both of these results have been confirmed for full-length Sup35 [76]. In addition, the lag time for unseeded assembly depends weakly on the concentration of soluble protein, suggesting a rate-limiting step involving formation of an intermediate [59]. Collectively, these results argue that NM oligomers are important on-pathway intermediates in NM amyloid assembly.

However, the importance of NM oligomers during amyloid formation is not without controversy. Collins et al. could not detect oligomers during unseeded assembly using analytical centrifugation [77]. Moreover, although they also observed the weak dependence of lag time on NM concentration for unseeded amyloid formation, they interpret this differently. Collins et al. argue that this weaker-than-expected dependence is due to fragmentation of amyloid nuclei since they observed severing of fibers after rotation [77]. In contrast, Serio et al. found that rotation did not cause fiber fragmentation [59]. The reason for the conflicting reports is unclear but may relate to small differences in residual denaturant concentration and mixing rates used in the NM assembly reactions. Further experimentation is required to resolve the precise mechanism of NM amyloid nucleation.

There is also some debate about the mechanism of elongation of preformed NM fibers. Serio et al. found that preformed fibers added to soluble NM near the end of the lag phase caused more rapid assembly than did identical experiments using soluble NM protein freshly diluted from denaturant [59]. This suggested that an intermediate competent for fiber elongation formed during the lag phase, although it did not preclude monomer addition as well. Moreover, TEM images revealed that in some cases, spherical oligomers localize near the ends of amyloids (Fig. 1) [59]. However, Collins et al. performed single-molecule fluorescence studies on NM amyloid growth and found that elongation proceeds by monomer addition rather than oligomer addition [77]. It is likely that both monomers and oligomers can add to NM fibers. Future experiments will need to ascertain which form is most relevant to in vivo prion propagation.

Extensive mutagenesis of NM has illuminated the role of specific residues in amyloid nucleation and elongation. DePace et al. conducted a random mutagenesis study of the N-terminal domain of Sup35 and identified mutations that inhibited prion propagation in vivo [71]. Interestingly, the mutations identified were predominately uncharged residues (glutamine and asparagine) switched to charged residues (arginine, lysine, and aspartate). Moreover, most of the mutations clustered near the extreme N-terminus of Sup35 (residues 8 to 33), although some have been found elsewhere [78]. Analysis of these mutants in vitro revealed nucleation and seeding defects [71], suggesting that a specific region within the N domain of Sup35 is critical for both prion nucleation and growth. More recent reports have confirmed and greatly expanded on these findings. Krishnan and Lindquist introduced single cysteine residues throughout the N domain of Sup35 and attached negatively charged (iodoacetate) and neutral (iodoacetamide) thiol-reactive labels [69]. Only cysteine mutants near the N-terminus (residues 21, 25, and 38) labeled with charged moieties showed pronounced defects in unseeded amyloid assembly [69]. Moreover, Tessier and Lindquist used a library of overlapping 20-mer peptides covering the entire sequence of NM arrayed on glass slides to study NM amyloid assembly [79]. Remarkably, soluble NM first bound to, and nucleated amyloids from, a small subset of peptides localized at the extreme N-terminus of Sup35 (residues 9 to 39), indicating that this region of Sup35 is alone sufficient to drive prion assembly of full-length NM.

Ure2 Nucleation and Polymerization

The de novo formation of Ure2 amyloids shows some similarities to Sup35. For example, the assembly of soluble Ure2 into amyloids is preceded by a lag phase during which spherical oligomers form [67,80]. These structures are also depleted upon fiber formation, suggesting that they are on-pathway intermediates [67,80]. This was confirmed by isolating these species using size-exclusion chromatography and demonstrating that they accelerate assembly of soluble Ure2 [80].

However, unlike Sup35, Ure2 natively exists as a homodimer in its soluble, nonprion state [60]. Dimerization is driven by self-association of the C-terminal domain [81]. Stabilizing this interaction by cross-linking prevents amyloid formation [60], suggesting that dissociation of dimers is a critical step in the assembly pathway. Moreover, deletions and mutations in both the N-terminal, glutamine/asparagine-rich domain and the C-terminal, α-helical domain influence prion induction [82–84]. Surprisingly, deletions or mutations in these domains can both decrease *and increase* the rate of prion induction [82–84]. Little is known about the underlying biochemical mechanism; deciphering the interplay between the N- and C-terminal domains of Ure2 during amyloid assembly represents an important area of future research.

HET-s and Rnq1 Nucleation and Polymerization

Relatively little is known about HET-s and Rnq1 assembly. For both proteins the domains that drive prion assembly are natively unstructured and located at the C-terminus [36,37,62,66,85]. Rnq1 assembles into amyloids through an oligomeric intermediate [65], while such an intermediate for HET-s has not been reported. Interestingly, the *P. anserina* protein HET-S, which only differs from HET-s by 13 amino acids, is incapable of forming prions in vivo and amyloids in vitro [50,51,62,66]. Biochemical analysis revealed that the N-terminal domain of HET-S inhibits amyloid formation of the C-terminal domain in *cis*, but not in *trans* [62]. A protein fusion containing the N-terminal domain of HET-s and the C-terminal of HET-S is capable of forming prions [62].

Protein Modulators of Fungal Prions

How does Sup35 form and propagate prion conformers in vivo? Two proteins, Rnq1 and Hsp104, are critically important [41,47,48]. The efficient formation of the prion state [PSI^+] de novo requires that Rnq1 be in its assembled prion state, [RNQ^+]. Subsequent propagation of [PSI^+], however, is independent of [RNQ^+] [47,48]. Notably, overexpression of many other glutamine/asparagine-rich proteins, including Ure2, can also facilitate [PSI^+] induction [47,48,68]. The mechanistic role of Rnq1 in de novo Sup35 assembly is controversial [47,48]. The most likely mechanism is that Rnq1 amyloids cross-seed Sup35 into self-perpetuating fibers [47]. In vitro studies provide some support for this hypothesis since Rnq1 amyloids accelerate the assembly of soluble NM into fibers, although this process is relatively inefficient [65,68].

Another possible mechanism for [PSI^+] induction is that Rnq1, when in its amyloid conformation, titrates away an inhibitor of Sup35 prion formation

[48]. No such inhibitor has been identified. However, the non-Q/N-rich protein HET-s can form bona fide prions in yeast, and the rate of prion induction in affected by the conformation of Rnq1 [86]. This dependence is puzzling and difficult to resolve with the cross-seeding mechanism given the very different amino acid sequence of HET-s. An inhibitor that binds to a common amyloid structure and that is efficiently titrated by Rnq1 amyloids may explain this conundrum [48]. Therefore, until such an inhibitor is identified or an explanation for the apparently low cross-seeding efficiency observed in vitro is given, it is not possible to discriminate definitively between these two mechanisms.

Hsp104 modulates [*PSI*$^+$] induction and propagation via a very different mechanism [41,75]. This hexameric heat-shock protein contains two ATPase domains per monomer. It is the most potent stress tolerance factor in yeast [87]. In fact, it alone can increase yeast survival 10,000-fold, due to heat treatment by disaggregating misfolded proteins that accumulate during such stressful conditions [87,88]. Deletion or overexpression of Hsp104 eliminates [*PSI*$^+$] induction and propagation [41]. To explain this unusual relationship, Shorter and Lindquist dissected the role of Hsp104 on Sup35 amyloid assembly in vitro [75,76]. Sup35 oligomer and amyloid formation were stimulated at low concentrations of Hsp104 and inhibited at high concentrations. Moreover, when Hsp104 was added to preformed Sup35 fibers at relatively high concentrations, the Sup35 fibers were fragmented and largely disassembled. These results suggest that during the unseeded assembly of Sup35, low concentrations of Hsp104 stimulate Sup35 amyloid formation both by promoting oligomer formation and fragmenting newly formed amyloid nuclei to increase the number of fiber ends. However, high concentrations of Hsp104 inhibit nucleation and seeded polymerization by disassembling nuclei and fibers into conformations that are incompetent for prion assembly.

There is some controversy about how Hsp104 regulates [*PSI*$^+$] [75,76,89,90]. Two reports had suggested that Hsp104 alone is incapable of disassembling Sup35 amyloids in vitro. However, these results are probably due to difficulties in purifying active Hsp104 from bacteria and differences in experimental conditions (e.g., 10 vs. 25°C, free vs. surface-bound Sup35) [75,76,89,90]. Nevertheless, Inoue et al. investigated if Hsp104 uniformly recognizes different NM fibers in a given assembly reaction [89]. By fluorescently labeling of both Hsp104 and NM fibers, they studied the degree of fluorescence co-localization when incubated together. Interestingly, Hsp104 bound only to a subset of the NM amyloids, suggesting that this heat-shock protein may selectively recognize specific amyloid conformations. If true, this could also explain some of the observed discrepancies (since different labs could be studying different conformations). Moreover, it represents an important area of research since the sequence and/or structural signatures of Sup35 that govern Hsp104 recognition are unknown.

Deletion of Hsp104 eliminates propagation of Ure2 prion conformers, as it does for Sup35 [91]. Curiously, unlike Sup35, overexpression of Hsp104 does not eliminate Ure2 prions [91]. Shorter and Lindquist recently investigated this paradox by investigating the effects of Hsp104 on Ure2 prion polymerization in vitro [76]. Low concentrations of Hsp104 accelerated Ure2 oligomer and amyloid formation. High concentrations of Hsp104 fragmented Ure2 amyloids. However, the Ure2 disassembly products are short fibers that are highly infectious, while the breakdown products of Sup35 amyloids are noninfectious aggregates. These results argue that Hsp104 overexpression fails to eliminate Ure2 conformers in vivo since disassembly products continue to catalyze prion polymerization. Future research will need to address why Hsp104 is incapable of inactivating Ure2 amyloids.

Less is known about the role of Hsp104 on the propagation of Rnq1 and HET-s. Deletion of Hsp104 causes loss of prion propagation for Rnq1 [92] and is required, although less stringently, for HET-s prion propagation in yeast [86]. Interestingly, HET-s in *P. anserina* can propagate without Hsp104, although some aspects of this process are compromised (e.g., lower number of amyloids per cell, lower propagation rate, lower meiotic stability) [93]. Overexpression of Hsp104 does not inhibit propagation of either Rnq1 or HET-s prions [92,93]. Importantly, Rnq1 requires Sis1, an Hsp40 chaperone, to be stably associated to propagate its prion state [45]. This interaction involves a glycine/phenyalanine-rich region in Sis1, as deletion of these residues leads to loss of the Rnq1 prion. However, the detailed mechanism of how Sis1 facilitates Rnq1 prion propagation still needs to be elucidated (for recent progress, see [94]).

PRION AMYLOID STRUCTURE

The structures of insoluble, polymeric amyloids are poorly defined since they are typically refractory to analysis by x-ray diffraction and conventional solution nuclear magnetic resonance (NMR). For years the arrangements of amino acids within amyloids has been fiercely debated [95–98]. X-ray diffraction of a wide range of amyloids frequently reveals about a 4.7-Å reflection, corresponding to spacing between β-strands perpendicular to the fiber axis, and this is widely accepted [95–98]. For some amyloids a second reflection is sometimes observed at 8 to 10 Å, corresponding to the spacing between β-sheets that stack perpendicular to the fiber axis [95–98]. The presence and relevance of this second reflection are controversial.

Peptide Amyloids

Recently, two short overlapping peptides from the extreme N-terminus of Sup35 (residues 7 to 13 and 8 to 13) have been crystallized and their structures have been studied both by x-ray diffraction [99] and solid-state NMR [100]. The crystal structures revealed that β-strands are oriented perpendicular to the long axis of the crystals (Fig. 2), as expected for amyloids. However, the key finding was that two β-sheets bond together in a self-complementing "steric zipper." Instead of opposing side-chain hydrogen bonding with each other, they

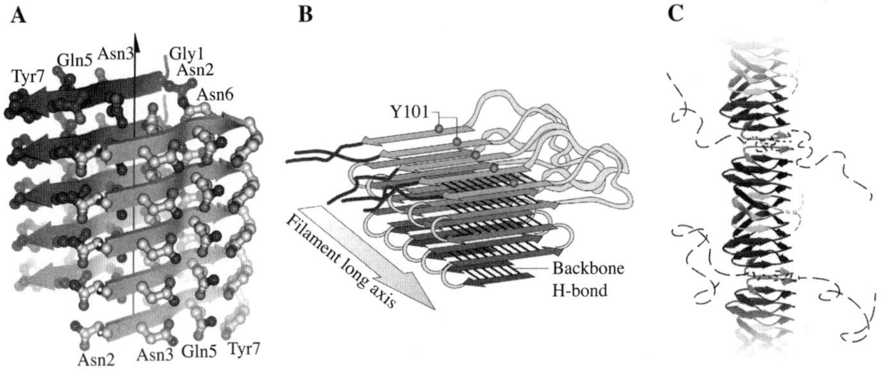

FIG. 2 Amyloid structures of Sup35 peptide and protein fragments. (A) Crystal structure of ^{7}GNNQQNY^{13}, a 7-mer peptide from the N-terminus of Sup35. The crystal structure reveals a high degree of geometric complementarity between opposing strands, which leads to exclusion of water at this interface and explains the stability of these amyloids. (B) In register, parallel β-sheet model of NM amyloid structure based on solid-state NMR results. (From [99], with permission of Macmillan Publishers Ltd.) This model proposes that most of the residues in the N domain and some residues in the M domain self-stack (such as indicated residue, Y101). (From [7], with permission of Macmillan Publishers Ltd.) (C) B-helix model of NM amyloid structure. This model proposes that two amino acid segments in the N domain are in intermolecular contact, whereas the intervening region is in intramolecular contact. (From [69].)

interdigitate with an extraordinary degree of geometric complementarity that excludes water and stabilizes the structure via van der Waals interactions. The outer faces of the two sheets are highly hydrated and may prevent lateral fiber growth. Short peptides (4 to 12 residues) from other amyloid-forming proteins have now been crystallized as well and also show steric zipper structures [101]. These results are significant since the interdigitated, dry interface observed in these structures potentially explains the remarkable stability of amyloids observed both in vitro and in vivo.

Are the crystal structures of short peptides relevant to amyloid structures for either short peptides or full-length proteins? There is evidence to suggest some relevance. For example, point mutations introduced into full-length Sup35 in this region (residues 7 to 13) inhibit prion propagation in vivo and amyloid nucleation in vitro [71]. Moreover, the x-ray diffraction patterns for crystals of the Sup35 peptide (residues 7 to 13) and amyloids assembled from full-length NM are consistent, suggesting a similar structural arrangement [99]. However, no peptide crystals by themselves have been shown to have direct biological activity (e.g., induction of [PSI^{+}] using the protein transformation method [12,13]). Collectively, these results suggest some relevance of peptide crystal structures to their amyloid counterparts. Addressing this relevance definitively will require demonstration of biological activities of peptide crystals and direct comparison of their structures to those of full-length proteins.

Sup35 Amyloids

The structural analysis of amyloids assembled from full-length proteins such as NM is extremely challenging and there is tremendous controversy over the structures proposed [69,102,103]. One prominent model is the in-register parallel β-sheet (Fig. 2) [102,103]. The crux of this model is that each residue in the amyloid core stacks on top of an identical residue from a different molecule, resulting in one molecule per 4.7 Å in the axial direction. Regions not involved in the amyloid core are expected to decorate the surface as loops or pendent chains.

Since the initial proposal of this model for NM amyloid structure [103], three principal experiments have been reported that may support its relevance. First, mass-per-unit length measurements for amyloids formed from a fragment of NM (residues 1 to 61) revealed approximately one molecule per 4.7 Å [104], consistent with the in-register parallel β-sheet model. Second, the sequence of the N domain of Sup35 was scrambled in multiple ways, and these sequences were tested for their ability to induce and propagate prions [105]. All scrambled sequences formed self-perpetuating prion conformations. However, the induction frequencies appeared much lower than observed previously for wild-type Sup35 (the wild-type control was not reported in this study). The fact that the scrambled Sup35 sequences form prions was argued to support the parallel β-sheet model since self-stacking of identical residues is assumed to be unaffected by scrambling (i.e., a residue can stack on itself regardless of the identity of neighboring residues). The low induction frequencies for the scrambled sequences suggest two alternative scenarios: Either self-stacking is influenced by neighboring residues and parallel β-sheet structures require specific sequences to form efficiently, or the parallel β-sheet model is an incomplete description of NM amyloid structure.

More recently, solid-state NMR was used to interrogate the structure of NM amyloids directly [102]. Four amino acids (phenylalanine, tyrosine, leucine, and alanine) were separately ^{13}C labeled; NM contains four phenylalanine (three in the N domain and one in the M domain), 20 tyrosine (all in N), eight leucine (one in N and seven in M), and 15 alanine (six in N and nine in M) residues. Using a recoupling method to probe ^{13}C–^{13}C separation distances selectively, the number of labeled residues within 5 Å was measured to determine which residues are in β-sheets. Since most of these residues do not neighbor identical residues, close proximity between labels must be due to intramolecular or intermolecular structure. For NM amyloids most tyrosine and leucine residues were within 5 Å of each other (21 ± 2 of 20 Tyr residues and 7 ± 1 of 8 Leu residues). A smaller fraction of phenylalanine and alanine residues were found to be in close proximity (2.5 ± 0.5 of 4 phenylalanine residues and 4 ± 2 of 15 alanine residues). Shewmaker et al. argue that the close proximity of many residues in both the N and M domains is most consistent with the in-register parallel β-sheet model [102].

One surprising aspect of this study is that most leucine residues are predicted to be in close proximity, despite the fact that seven of eight leucine residues are in the M domain [102]. The sequence of the M domain is highly charged and has been reported to be unstructured by several investigators. Based on their NMR results, Shewmaker et al. argued that the M domain contains small regions of β-sheet structure connected by loops of the most highly charged regions [102]. However, the NMR results for alanine are difficult to resolve with this proposal; only four of the 15 residues are in close proximity, despite the fact that the majority of these residues are in the M domain (nine out of 15) and several are located close to leucine residues. Further investigation of this perplexing prediction of M-domain structure is required to better understand the implications of the NMR results for the N domain.

Another prominent model of amyloid structure of NM and other proteins is the β-helix (Fig. 2) [69,98]. Crystal structures of globular β-helical proteins provide some insight into this model [98,106–108]. For example, a single rung of a β-helix typically has about 10 to 20 residues [98,106–108]. Moreover, there is a central pore inside the helix that prevents close contact of β-sheets [98,106–108]. Therefore, the β-helix model makes two predictions about NM fiber structure: (1) if the amyloid core is long enough to form more than two rungs, some residues within the core will not be in intermolecular contact and (2) the 8 to 10-Å reflection in the x-ray diffraction pattern should be absent since β-sheets are not in close contact in the direction perpendicular to the fiber axis.

Results from two studies are consistent with these predictions [69,109]. X-ray diffraction analysis of NM amyloids reveals that the reflection at 8 to 10 Å may be an artifact of drying the fibers [109]. For fibers of both N and NM, two reflections (4.7 and 8 to 10 Å) were observed for dried fibers but only one (4.7 Å) for hydrated fibers. The absence of the equatorial reflection suggests that hydrated NM amyloids are devoid of closely stacked β-sheets in the direction perpendicular to the fiber axis, consistent with the β-helix model. However, this study is controversial since the diffraction pattern is much weaker for the hydrated samples and may limit detection of the equatorial reflection [110].

The β-helix model of NM fibers is also consistent with a cysteine scanning mutagenesis study [69]. Since NM is devoid of cysteines, 37 single-cysteine mutations were introduced throughout its sequence to facilitate site-specific attachment of diverse biochemical probes. Importantly, the cysteine mutations did not influence the rate of amyloid polymerization in vitro or the fidelity of prion propagation in vivo. To access the degree of solvent accessibility of each cysteine residue, two methods were used. First, monomeric cysteine mutants were labeled with a fluorescent dye (acrylodan) sensitive to the degree of solvent exposure and then assembled into fibers. As a complementary approach, single-cysteine mutants were assembled into fibers and then labeled with fluorescent dyes (pyrene and Lucifer Yellow). For fibers assembled at 25°C, the acrylodan results revealed a contiguous, solvent-shielded amyloid core that encompasses most of the N domain (residues 21 to 121), but not the M domain. The postassembly labeling results revealed a smaller amyloid core (residues 2 to 73), but residues 77 to 121 also showed some protection. Given the length of this amyloid core, a β-helix structure would probably require more than two rungs, and therefore it is expected that not all the residues in the amyloid core would be in intermolecular contact.

Indeed, analysis of the intermolecular proximity of identical residues within NM fibers suggests that not all of the amyloid core is in self-contact [69]. Single-cysteine mutants were labeled with a fluorophore (pyrene) sensitive to interdye spacings and assembled into amyloid fibers. Two regions within the N domain (residues 20 to 40 approximately denoted as the "head" and 90 to 110 denoted as the "tail") were in close self-intermolecular contact (4 to 10 Å), while the intervening region (residues 40 to 90 approximately) and the M domain were not. Moreover, cross-linking monomeric cysteine mutants in the head and tail regions produced NM dimers that greatly accelerated amyloid formation. However, cross-linking the intervening region (residues 43 to 77) between the head and tail regions inhibited amyloid formation, again suggesting that only a subset of residues in the amyloid core form intermolecular contacts.

Krishnan and Lindquist argued that these and other results are most consistent with the β-helix model [69]; two regions are in self-intermolecular contact, while the intervening region forms intramolecular contacts and the unstructured M domain is displayed on the amyloid surface. One concern about this study is the use of large fluorescent probes and their potential influence on local amyloid structure. For defining which residues form the amyloid core, the postassembly labeling results address this concern since no fluorescent moieties were present during amyloid assembly [69]. Moreover, use of a cross-linker with a completely different structure also argues for the relevance of the head-to-head and tail-to-tail interactions and the apparent lack of self-interaction for the intervening residues [69]. Finally, NM labeled with pyrene or acylodan and then assembled into amyloids is as infectious as unlabeled amyloids, as judged by the protein transformation assay (R. Krishnan and S. Lindquist, unpublished results). Nevertheless, additional constraints on the structure of NM fibers are required to evaluate the veracity of the β-helix model.

A recent study of NM fiber structure using hydrogen/deuterium (H/D) exchange confirms some of these findings [70]. Mature NM amyloids were exposed to deuterium, dissolved in dimethyl sulfoxide and the degree of H/D exchange was probed by solution NMR. For fibers formed at 25°C, residues 4 to 70 and 110 to 128, approximately, were protected from the solvent; the importance of residues 8 to 71 in forming β-sheets was confirmed by proline scanning mutagenesis [70]. These findings agree well with the region (residues 2 to 73) predicted to be most protected (<50% accessible to fluorophore) by postassembly cysteine accessibility studies [69]. Moreover, the fiber core predicted by acrylodan fluorescence (residues 21 to 121) [69] overlaps with the H/D exchange results, although there is a discrepancy regarding whether residues 70 to 110 are shielded in the amyloid core.

Ure2 Amyloids

There is considerably less known about the structure of amyloids of Ure2 than for Sup35. Nevertheless, two controversial models of Ure2 fiber structure have been proposed [103,111–113]. The first model is the in-register parallel β-sheet (Fig. 3) [103,112,113], as proposed for Sup35 [102,103]. This model of Ure2 structure predicts that each amino acid stacks upon itself for residues 1 to 70, approximately, and the rest of the protein is displayed on the amyloid surface. The second model does not propose a specific arrangement for most residues in the N-terminal domain (Fig. 3) [111]. However, this model proposes that the Ure2 amyloid structure is not β-sheet rich, but rather, a helical arrangement of monomers whose structure is close to its soluble, native conformation. Moreover, it proposes that the N- and C-terminal domains interact in the amyloid conformation.

The parallel β-sheet model is supported by solid-state NMR data obtained for amyloids of Ure2 fragments 10 to 39 [112] and 1 to 89 [112]. For the Ure2 peptide 10 to 39, four residues (15, 18, 22, and 37) were labeled with ^{13}C or ^{15}N,

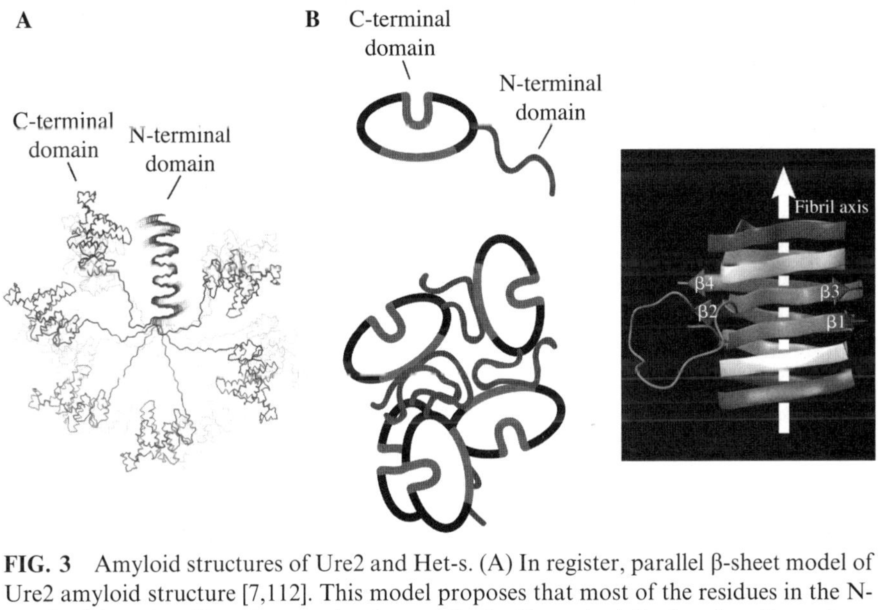

FIG. 3 Amyloid structures of Ure2 and Het-s. (A) In register, parallel β-sheet model of Ure2 amyloid structure [7,112]. This model proposes that most of the residues in the N-terminal domain self-stack on each other, while the C-terminal displayed on the surface. (From [103], with permission. Copyright © 2004 National Academy of Sciences, U.S.A.) (B) Helical model of Ure2 fiber structure. This model proposes that the Ure2 amyloid conformation is a helical assembly of Ure2 monomers in a near native conformation where the N- and C-terminal domains are in contact. (From [111], with permission). Copyright © 2005 American Society of Biochemistry and Molecular Biology.) (C) β-Solenoid model of Het-s amyloid structure. This model proposes that the amyloid conformation involves two β-strand–turn–β-strand motifs. (From [120], with permission of Macmillan Publishers Ltd.)

and the intermolecular distances between labels were measured [112]. Three of the four residues were in close proximity (= 5 Å). For the Ure2 fragment 1 to 89, a much more thorough analysis was conducted, although the assignments for specific residues were less conclusive [112]. However, Ure1 to 89 amyloids labeled with ^{13}C at seven positions (residues 9, 12, 16, 19, 43, 58, and 81) did reveal that these residues appear to be in close intermolecular proximity (= 5 Å) as well. Intriguingly, uniform ^{13}C and ^{15}N labeling of Ure(1–89) amyloids revealed that an asparagine-rich domain (^{45}NNNNNNSSSNNNN^{57}) expected to be part of the amyloid core appears to be unstructured and looped out of the fiber. Collectively, these findings support the in-register parallel β-sheet model for Ure2 amyloids where most residues are in self-intermolecular contact.

The sequence of Ure2 has been scrambled to test if randomized sequences form prions [105,114]. The induction frequencies of the scrambled sequences are similar to wild-type frequencies. As for Sup35, this is argued to support the in-register parallel β-sheet model [105]. Interestingly, scrambling the N-terminal region of Ure2 appears to affect prion induction frequency much less than for Sup35 [105]. An explanation for this is that Ure2 and Sup35 may form different types of amyloid structures. This is consistent with the fact that Sup35 contains 5.5 imperfect oligopeptide repeats (PQGGYQQYN), while Ure2 contains no such repeats.

The in-register parallel β-sheet model is also supported by three other findings. Mass-per-unit-length measurements of Ure2 fibers reveal 4.7-Å spacings between molecules [115], and x-ray diffraction of these fibers yields a reflection at 4.7 Å [116]; both of these results are consistent with the parallel β-sheet model. Moreover, amyloids of a fragment of Ure2 (residues 1 to 65) are infectious even when this fragment is fused to folded C-terminal moieties such as GFP and GST [38]. This argues that the prion conformation is formed solely by the N-terminal domain and does not require contact with the C-terminal domain of Ure2.

The x-ray diffraction results are controversial since the precise conformation of Ure2 fibers is dependent on factors such as temperature and extent of drying [117]. Bousset et al. found that Ure2 fibers formed at physiological conditions and then heated to 60°C for 1 hour yielded a reflection at 4.7 Å, whereas this reflection was absent for unheated samples [117]. Curiously, a diffuse reflection at 4.3 Å was observed for the unheated fibers, although its significance is unclear.

Several lines of evidence appear to support the helical model of Ure2 prions [103,112,113]. First, Fourier transform infrared spectroscopy (FTIR) revealed that the conformations of soluble and fibrous Ure2 are very similar [118]; both are rich in α-helical content and poor in β-sheet content. Interestingly, heat-treated fibers are rich in β-sheet content but show no seeding activity in vitro. Second, protease digestion of soluble and fibrous Ure2 shows relatively minor changes in the degree of solvent exposure [119]. Third, cysteine mutagenesis and disulfide cross-linking analysis of Ure2 fibers revealed that residues 6 (N domain) and 137 (C domain) from the same molecule are in close proximity within fibers, whereas pairs of residues (6–6 and 137–137) from different molecules are not [111]. Collectively, these results argue that the assembled conformation is similar to the native conformation: It is poor in β-sheet content and it involves an interaction between the N- and C-terminal domains of Ure2. More constraints on the Ure2 amyloid structure are needed to resolve the controversy between these two models.

HET-s Amyloids

The structure of HET-s amyloids is the best defined of the known fungal prions [120]. Analysis of solvent exposure using H/D exchange and solution NMR, and secondary structure by solid-state NMR, revealed a β-solenoid structure (Fig. 3) [120]. This structure involves two β-strand–turn–β-strand motifs (four strands total). Mutating residues in all four strands to proline disrupted prion induction and propagation, whereas mutating the turns (loops) did not. This model predicts a mass-per-unit length of one molecule per 9.4 Å, which was confirmed by scanning transmission electron microscopy [121]. Interestingly, the mass-per-unit length of Ure2 [115] and fragments of Sup35 [104] is one molecule per 4.7 Å, suggesting that the unique sequence of HET-s forms a different amyloid structure than other fungal prions.

PRION STRAINS

One of the most perplexing aspects of prions is their ability to form different structural strains [11–17]. For decades it was hypothesized that prion proteins can access not only one infectious amyloid conformation, but a suite of related, yet distinct conformations that encode different biological phenotypes [122]. Recently, this was demonstrated unequivocally by transforming yeast with NM amyloids of different structure and demonstrating that they produce distinct phenotypes (Fig. 4) [12,13]. An enabling breakthrough in this study was that different NM amyloid conformations could be formed simply by assembling fibers at different temperatures (e.g., 4 vs. 25°C) [13]. Tanaka et al. demonstrated that there are gross structural differences between the two populations of fibers by measuring differences in their stabilities (e.g., fibers formed at 4°C melt at lower temperatures than those formed at 25°C) [13]. When amyloids formed at 4°C were transformed into yeast, they produced a relatively high degree of read-through of stop codons and hence a strong [PSI^+] phenotype. Conversely, transformation of yeast with amyloids formed at 25°C encode a lower low degree of read-through and a weak [PSI^+] phenotype.

This important breakthrough of a simple protocol to form different prion strains has led to several studies of their structural differences [13,32,69,70]. Using an array of single-cysteine NM mutants to attach site-specific fluorescent labels, Krishnan and Lindquist found two important differences in the structures of NM amyloids formed at 4 and 25°C [69]. First, as shown in

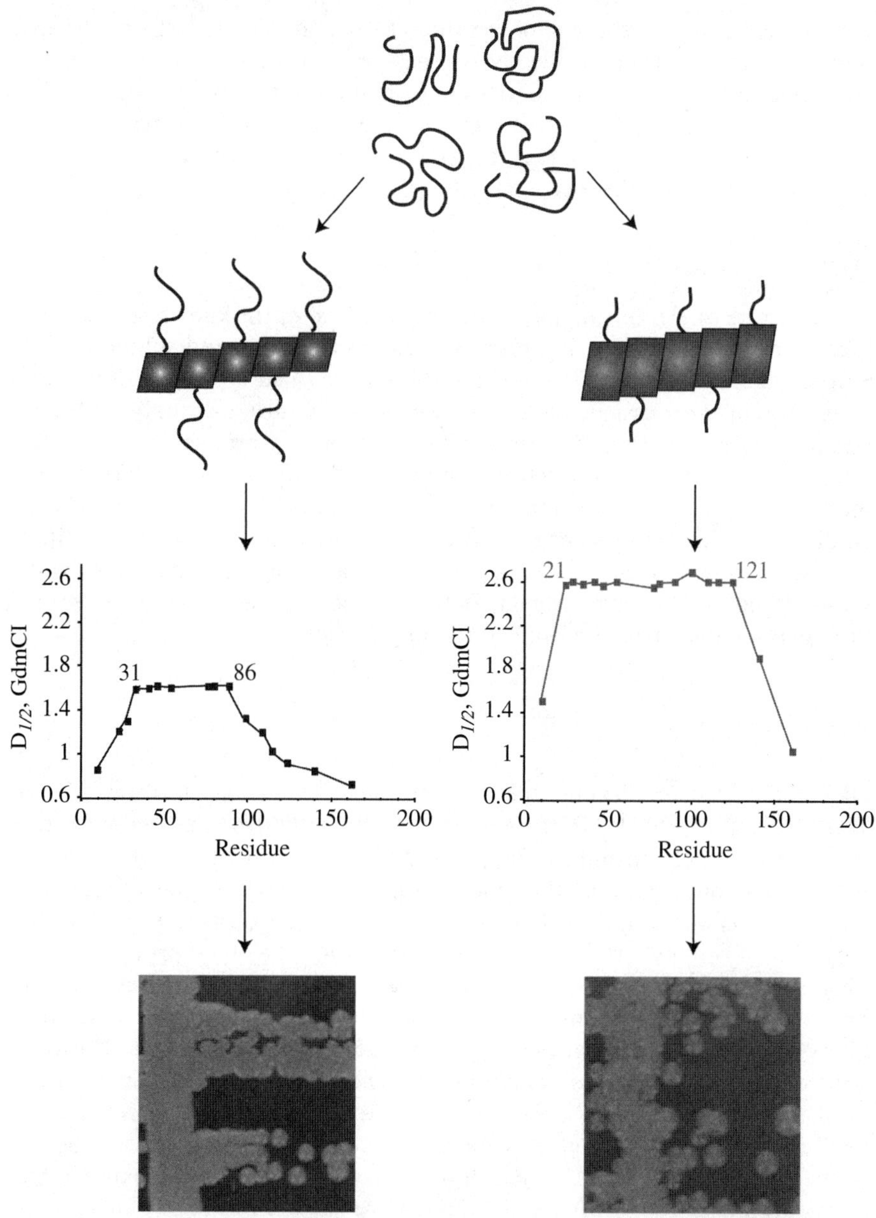

FIG. 4 Sup35 strains have different-size amyloid cores. Different Sup35 strains can be formed in vitro by assembling the protein into amyloids at different temperatures. Fibers formed at 4°C have a smaller number of residues in their core than those formed at 37°C [69,70]. To determine the size of the amyloid cores, single-cysteine mutants were introduced into the NM portion of Sup35 [69]. The single-cysteine mutants were then labeled with a fluorophore sensitive to solvent exposure (acrylodan) and assembled into fibers. By disassembling the fibers with increasing amounts of guanidine hydrochloride,

Figure 4, there were many fewer residues in the amyloid core for fibers formed at 4°C (ca. residues 31 to 86) than for those formed at 25°C (ca. residues 21 to 121). The smaller amyloid core for the 4°C fibers is consistent with their lower melting temperature and higher propensity to be fragmented in vitro relative to 25°C fibers. Second, the location of one of the self-intermolecular contacts is different. For both amyloid conformations, residues 20 to 40 (head region), approximately form an intermolecular contact. However, fibers show a second intermolecular contact (tail region) encompassing approximately residues 80 to 100 for 4°C fibers and 90 to 110 for 25°C fibers.

Some of these structural insights have been confirmed by H/D exchange experiments [70]. Residues 4 to 40 were most protected for the 4°C fibers, while residues 4 to 70 and 110 to 128 were most protected for the 25°C fibers. The importance of both of these regions in forming the amyloid core was confirmed by mutagenesis analysis; the 4°C fibers showed defects in amyloid assembly for proline mutations between residues 10 and 33, while the 25°C fibers showed similar defects for mutations between residues 8 and 71. Collectively, these results argue that the size of the fiber core is smaller for the 4°C fibers than for the 25°C fibers.

However, the H/D exchange data [70] suggest the NM amyloid core is much shorter for the 4°C fibers than previously thought [69,102]. Solid-state NMR analysis of these fibers suggests that most residues in the N domain, and some in the M domain, form β-sheets (N and M domains contain 123 and 130 residues, respectively) [102]. Fluorescence studies of similar fibers identified a smaller amyloid core (ca. residues 31 to 86) [69]. Both methods predict larger amyloid cores than proposed in the H/D exchange study (ca. residues 4 to 40) [70]. The lack of agreement highlights the difficulties in structural analysis of amyloids and the need for further research into the amino acid organization of these enigmatic conformers.

Recently, the mechanism of how different amyloid conformations impart different prion strain phenotypes in vivo has been formulated mathematically and confirmed experimentally [123]. Intuitively, it is expected that the rate of amyloid growth and amyloid division are important determinants of prion phenotypes. However, Tanaka et al. showed surprisingly that the amyloid conformation encoding the strongest [PSI^+] phenotype actually polymerized the slowest [123]. Yet the slow polymerization was more than compensated by the increased brittleness of this amyloid conformation relative to other

the midpoint of unfolding was determined using site-specific acrylodan fluorescence measurements. For fibers formed at 4°C, residues 31 to 86 formed a contiguous core of solvent shield residues, while residues 21 to 121 formed the amyloid core for fibers formed at 37°C. Transformation of unlabeled [13,69] or labeled (Krishnan and Lindquist, unpublished results) fibers formed at 4°C into yeast yielded strong [PSI^+] phenotypes; the corresponding transformations with fibers formed at 37°C produced weak [PSI^+] phenotypes. (*See insert for color representation of figure.*)

amyloid conformations encoding weaker prion phenotypes. Therefore, the ability to fragment Sup35 amyloids, governed in vivo by chaperones such as Hsp104 [41,75], is a critical determinant of prion strain phenotypes. Future work will be needed to decipher the molecular connection between the structure of amyloids and how they are fragmented.

PRION SPECIES BARRIERS

For decades it has been realized that infectious mammalian prions can be efficiently transmitted within a given species, but much less efficiently (if at all) between different species [22,26–30]. This resistance to interspecies prion infection is referred to as a *prion species barrier* or *transmission barrier* [124]. The molecular basis for species barriers is poorly understood, but the degree of sequence similarity between prion proteins appears important [19–30]. Pioneering studies in yeast have shown that fungal prions, like their mammalian counterparts, display species barriers [18–21,23–25,125]. For example, the ability of Sup35 from *Saccharomyces cerevisiae* (Sc), *Candida albicans* (Ca), and *Pichia methanolica* (Pm) to form prions and cross species barriers was studied in *S. cerevisiae* [20]. Each protein could form self-perpetuating prions when overexpressed, but none could cross-catalyze conversion of different prion proteins. The species barrier between ScSup35 and CaSup35 was confirmed in vitro; amyloid fibers of ScNM could self-template polymerization of ScNM, but not for CaNM, and vice versa [20]. This and other studies [18–21,23–25,125] established the utility of studying prion species barriers in yeast.

The biochemical basis of how prions establish species barriers has recently been probed by peptide microarray analysis (Fig. 5) [79]. A library of overlapping, 20-mer peptides covering the entire sequences of ScNM and CaNM was synthesized and arrayed on glass slides. The microarrays were incubated with soluble, fluorescently labeled ScNM and CaNM, and afterward washed stringently to remove unbound protein. Interestingly, each protein bound to only a small overlapping subset of its own peptides (ScNM residues 9 to 39 and CaNM residues 59 to 86). Despite the fact that ScNM and CaNM have very similar sequence compositions (both N domains are composed of about 50% glutamine and 50% asparagine) [20], no cross-reactivity between full-length proteins and noncognate peptides was observed (Fig. 5). The amino acid sequences bound by each protein were named *recognition elements.* Upon closer inspection, it was found that peptides within each recognition element catalyzed the nucleation of their corresponding full-length protein from soluble (monomers or oligomers) to insoluble (fibers) conformers on the surface of the arrays. Collectively, these results suggest that species barriers are governed by small, specific sequences within prion proteins that also govern their nucleation.

In some cases, mammalian prions have been transmitted between species (e.g., cattle to humans) [8,22,126]. Promiscuous prions have also been identified in yeast [18–20,25,125]. Sup35 proteins from closely related species have

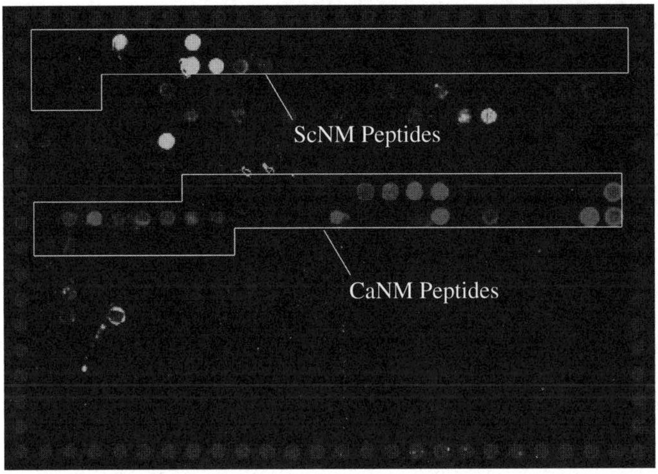

FIG. 5 Peptide microarray analysis of yeast prion species barriers. Fluorescently labeled NM domains of Sup35 from *S. cerevisiae* (ScNM) and *C. albicans* (CaNM) were incubated with overlapping peptide libraries (20 residues per peptide) derived from a sequence of each NM protein [79]. Both ScNM and CaNM bound selectively to a small subset of their own peptides but did not cross-react with peptides from the other species. (*See insert for color representation of figure.*)

recently been shown to cross transmission barriers in yeast in some cases [125]. Moreover, a protein chimera of ScNM and CaNM (Sc residues 1 to 39, Ca 40 to 140, and Sc 124 to 253 and closely related variants) can cross the transmission barrier between *S. cerevisiae* and *C. albicans* [18–20]. Peptide microarray analysis revealed the underlying mechanism. Soluble, fluorescently labeled ScNM and CaNM bound only to their own recognition sequences [79]. However, incubation of labeled Sc/Ca chimera NM with the same arrays resulted in binding to peptides in both recognition sequences. This indicates that the chimera's capacity to overcome this transmission barrier is due to it carrying prion recognition elements from both species.

Prion species barriers are also highly dependent on different strain conformations [4,8,18,22,28,30–33]. It is likely that mammalian prions were transmitted from cattle to humans through a specific, highly infectious prion conformation [8,22,28,30,31,33]. This fascinating interdependence has recently been interrogated in yeast [18,19,32]. The species barrier between ScNM and CaNM can be overcome by forming specific amyloid conformations of either protein, although the level of infectivity in this study was relatively low; this was probably due to the low sequence similarity between these proteins [32]. The Sc/Ca NM chimera can also form different amyloid conformations with unique propensities to cross species barriers [18,19]. For example, assembling soluble chimeric protein into amyloids at low (15°C) versus high (37°C) temperatures results in two different amyloid conformations, each with a unique seeding specificity; fibers formed at 15°C selectively seed ScNM, whereas fibers formed at 37°C selectively seed CaNM [18]. These species-specific chimera strains can also be assembled at room temperature by introducing specific mutations into the chimera sequence [18] or by templating chimeric protein into amyloids with small amounts of preformed fibers of either ScNM or CaNM [19].

Peptide microarrays have also been used to investigate the mechanism of how different prion strains influence species barriers [79]. Incubating the "wild-type" soluble Sc/Ca NM chimera with NM peptide arrays resulted in binding to both ScNM and CaNM recognition sequences. However, incubating these arrays with wild-type chimeric protein at different temperatures (or mutated chimeric protein at room temperature) led to unique recognition patterns. For example, incubation of the wild-type chimera at low temperature (4°C) led to selective binding to ScNM peptides, while elevated temperatures (37°C) drove selective binding of CaNM peptides. These and related results show remarkable correspondence to the species-specific seeding activities of the chimeric strains [18]; selective binding of the chimera to CaNM peptides at 37°C corresponds to conditions that promote assembly of chimeric amyloids that selectively seed CaNM, and vice versa at 4°C. These findings suggest the following mechanism: Nucleation at either recognition element initiates a chimeric amyloid conformation that retains seeding specificity for proteins carrying the same recognition sequence. It will be important to determine structurally if each chimeric strain contains a unique amyloid core containing one recognition element while the other remains unstructured or is masked from the fiber ends.

CONCLUSIONS AND PERSPECTIVES

The biochemical analysis of yeast prions has produced many astonishing findings that have shed light on their enigmatic properties. However, much remains unknown about these captivating proteins. The discovery of new prions remains one of the most important pursuits to determine how widespread are prion-based mechanisms of inheritance. The mechanistic role of oligomeric intermediates in prion assembly and propagation in vivo also needs to be resolved; this is important not only due to the relevance of these structures in prion biology but also due to the pathogenic role of similar structures in many neurodegenerative diseases. Perhaps the most daunting task in this field will be to resolve the controversy over the structures of Sup35 and Ure2 amyloids. To accomplish this, high-resolution structures should continue to be pursued while providing well-established lower-resolution constraints, such as the size of the solvent-shielded amyloid core. One challenge in defining these structures is to understand why various methods yield different structural constraints. Advances in amyloid structural analysis should enable new insights into the molecular basis of prion strains and better definition of the extent of structural differences between different prion conformers. In turn, insights into these structural differences should unlock the molecular basis of how different prion strains have unique propensities to overcome prion species barriers. In summary, heroic efforts to solve these extremely difficult problems have begun to yield fascinating discoveries in the field of prion biology. These truly impressive advances are expected to continue in the coming years and to illuminate many outstanding problems in prion biology that have broad relevance to human health.

REFERENCES

1. Selkoe, D.J. (2003). Folding proteins in fatal ways. *Nature, 426*, 900–904.
2. Fowler, D.M., Koulov, A.V., Balch, W.E., Kelly, J.W. (2007). Functional amyloid: from bacteria to humans. *Trends Biochem Sci, 32*, 217–224.
3. Shorter, J., Lindquist, S. (2005). Prions as adaptive conduits of memory and inheritance. *Nat Rev Genet, 6*, 435–450.
4. Chien, P., Weissman, J.S., DePace, A.H. (2004). Emerging principles of conformation-based prion inheritance. *Annu Rev Biochem, 73*, 617–656.
5. Uptain, S.M., Lindquist, S. (2002). Prions as protein-based genetic elements. *Annu Rev Microbiol, 56*, 703–741.
6. Tuite, M.F., Cox, B.S. (2003). Propagation of yeast prions. *Nat Rev Mol Cell Biol, 4*, 878–890.
7. Wickner, R.B., Edskes, H.K., Shewmaker, F., Nakayashiki, T. (2007). Prions of fungi: inherited structures and biological roles. *Nat Rev Microbiol, 5*, 611–618.
8. Prusiner, S.B. (1998). Prions. *Proc Natl Acad Sci USA, 95*, 13363–13383.
9. True, H.L., Berlin, I., Lindquist, S.L. (2004). Epigenetic regulation of translation reveals hidden genetic variation to produce complex traits. *Nature, 431*, 184–187.
10. True, H.L., Lindquist, S.L. (2000). A yeast prion provides a mechanism for genetic variation and phenotypic diversity. *Nature, 407*, 477–483.
11. Derkatch, I.L., Chernoff, Y.O., Kushnirov, V.V., Inge-Vechtomov, S.G., Liebman, S.W. (1996). Genesis and variability of [PSI] prion factors in *Saccharomyces cerevisiae*. *Genetics, 144*, 1375–1386.
12. King, C.Y., Diaz-Avalos, R. (2004). Protein-only transmission of three yeast prion strains. *Nature, 428*, 319–323.
13. Tanaka, M., Chien, P., Naber, N., Cooke, R., Weissman, J.S. (2004). Conformational variations in an infectious protein determine prion strain differences. *Nature, 428*, 323–328.
14. Bruce, M.E., McConnell, I., Fraser, H., Dickinson, A.G. (1991). The disease characteristics of different strains of scrapie in Sinc congenic mouse lines: implications for the nature of the agent and host control of pathogenesis. *J Gen Virol, 72*(Pt 3), 595–603.
15. Caughey, B., Raymond, G.J., Bessen, R.A. (1998). Strain-dependent differences in beta-sheet conformations of abnormal prion protein. *J Biol Chem, 273*, 32230–32235.
16. Kocisko, D.A., Come, J.H., Priola, S.A., Chesebro, B., Raymond, G.J., Lansbury, P.T., Caughey, B. (1994). Cell-free formation of protease-resistant prion protein. *Nature, 370*, 471–474.

17. Safar, J., Wille, H., Itri, V., Groth, D., Serban, H., Torchia, M., Cohen, F.E., Prusiner, S.B. (1998). Eight prion strains have PrP(Sc) molecules with different conformations. *Nat Med, 4*, 1157–1165.
18. Chien, P., DePace, A.H., Collins, S.R., Weissman, J.S. (2003). Generation of prion transmission barriers by mutational control of amyloid conformations. *Nature, 424*, 948–951.
19. Chien, P., Weissman, J.S. (2001). Conformational diversity in a yeast prion dictates its seeding specificity. *Nature, 410*, 223–227.
20. Santoso, A., Chien, P., Osherovich, L.Z., Weissman, J.S. (2000). Molecular basis of a yeast prion species barrier. *Cell, 100*, 277–288.
21. Chernoff, Y.O., Galkin, A.P., Lewitin, E., Chernova, T.A., Newnam, G.P., Belenkiy, S.M. (2000). Evolutionary conservation of prion-forming abilities of the yeast Sup35 protein. *Mol Microbiol, 35*, 865–876.
22. Collinge, J. (2001). Prion diseases of humans and animals: their causes and molecular basis. *Annu Rev Neurosci, 24*, 519–550.
23. Kushnirov, V.V., Kochneva-Pervukhova, N.V., Chechenova, M.B., Frolova, N.S., Ter-Avanesyan, M.D. (2000). Prion properties of the Sup35 protein of yeast *Pichia methanolica. EMBO J, 19*, 324–331.
24. Resende, C., Parham, S.N., Tinsley, C., Ferreira, P., Duarte, J.A., Tuite, M.F. (2002). The *Candida albicans* Sup35p protein (CaSup35p): function, prion-like behaviour and an associated polyglutamine length polymorphism. *Microbiology, 148*, 1049–1060.
25. Nakayashiki, T., Ebihara, K., Bannai, H., Nakamura, Y. (2001). Yeast [PSI+] "prions" that are crosstransmissible and susceptible beyond a species barrier through a quasi-prion state. *Mol Cell, 7*, 1121–1130.
26. Prusiner, S.B., Scott, M., Foster, D., Pan, K.-M., Groth, D., Mirenda, C., Torchia, M., Yang, S.-L., Serban, D., Carlson, G.A., et al. (1990). Transgenetic studies implicate interactions between homologous PrP isoforms in scrapie prion replication. *Cell, 63*, 673–686.
27. Scott, M., Foster, D., Mirenda, C., Serban, D., Coufal, F., Walchli, M., Torchia, M., Groth, D., Carlson, G., DeArmond, S.J., et al. (1989). Transgenic mice expressing hamster prion protein produce species-specific scrapie infectivity and amyloid plaques. *Cell, 59*, 847–857.
28. Bruce, M., Chree, A., McConnell, I., Foster, J., Pearson, G., Fraser, H. (1994). Transmission of bovine spongiform encephalopathy and scrapie to mice: strain variation and the species barrier. *Philos Trans R Soc London B, 343*, 405–411.
29. Supattapone, S., Bosque, P., Muramoto, T., Wille, H., Aagaard, C., Peretz, D., Nguyen, H.O., Heinrich, C., Torchia, M., Safar, J., et al. (1999). Prion protein of 106 residues creates an artificial transmission barrier for prion replication in transgenic mice. *Cell, 96*, 869–878.
30. Collinge, J., Palmer, M.S., Sidle, K.C., Hill, A.F., Gowland, I., Meads, J., Asante, E., Bradley, R., Doey, L.J., Lantos, P.L. (1995). Unaltered susceptibility to BSE in transgenic mice expressing human prion protein. *Nature, 378*, 779–783.
31. Hill, A.F., Desbruslais, M., Joiner, S., Sidle, K.C., Gowland, I., Collinge, J., Doey, L.J., Lantos, P. (1997). The same prion strain causes vCJD and BSE. *Nature, 389*, 448–450, 526.

32. Tanaka, M., Chien, P., Yonekura, K., Weissman, J.S. (2005). Mechanism of cross-species prion transmission: an infectious conformation compatible with two highly divergent yeast prion proteins. *Cell, 121*, 49–62.

33. Collinge, J., Sidle, K.C., Meads, J., Ironside, J., Hill, A.F. (1996). Molecular analysis of prion strain variation and the aetiology of "new variant" CJD. *Nature, 383*, 685–690.

34. Legname, G., Baskakov, I.V., Nguyen, H.O., Riesner, D., Cohen, F.E., DeArmond, S.J., Prusiner, S.B. (2004). Synthetic mammalian prions. *Science, 305*, 673–676.

35. Deleault, N.R., Harris, B.T., Rees, J.R., Supattapone, S. (2007). Formation of native prions from minimal components in vitro. *Proc Natl Acad Sci USA, 104*, 9741–9746.

36. Maddelein, M.L., Dos Reis, S., Duvezin-Caubet, S., Coulary-Salin, B., Saupe, S.J. (2002). Amyloid aggregates of the HET-s prion protein are infectious. *Proc Natl Acad Sci USA, 99*, 7402–7407.

37. Patel, B.K., Liebman, S.W. (2007). "Prion-proof" for [PIN+]: infection with in vitro-made amyloid aggregates of Rnq1p-(132–405) induces [PIN+]. *J Mol Biol, 365*, 773–782.

38. Brachmann, A., Baxa, U., Wickner, R.B. (2005). Prion generation in vitro: amyloid of Ure2p is infectious. *EMBO J, 24*, 3082–3092.

39. Patino, M.M., Liu, J.J., Glover, J.R., Lindquist, S. (1996). Support for the prion hypothesis for inheritance of a phenotypic trait in yeast. *Science, 273*, 622–626.

40. Wickner, R.B., Masison, D.C., Edskes, H.K. (1995). [PSI] and [URE3] as yeast prions. *Yeast, 11*, 1671–1685.

41. Chernoff, Y.O., Lindquist, S.L., Ono, B., Inge-Vechtomov, S.G., Liebman, S.W. (1995). Role of the chaperone protein Hsp104 in propagation of the yeast prion-like factor [psi+]. *Science, 268*, 880–884.

42. Eaglestone, S.S., Cox, B.S., Tuite, M.F. (1999). Translation termination efficiency can be regulated in *Saccharomyces cerevisiae* by environmental stress through a prion-mediated mechanism. *EMBO J, 18*, 1974–1981.

43. Lian, H.Y., Jiang, Y., Zhang, H., Jones, G.W., Perrett, S. (2006). The yeast prion protein Ure2: structure, function and folding. *Biochim Biophys Acta, 1764*, 535–545.

44. Wickner, R.B. (1994). [URE3] as an altered URE2 protein: evidence for a prion analog in *Saccharomyces cerevisiae. Science, 264*, 566–569.

45. Sondheimer, N., Lopez, N., Craig, E.A., Lindquist, S. (2001). The role of Sis1 in the maintenance of the [RNQ+] prion. *EMBO J, 20*, 2435–2442.

46. Derkatch, I.L., Bradley, M.E., Masse, S.V., Zadorsky, S.P., Polozkov, G.V., Inge-Vechtomov, S.G., Liebman, S.W. (2000). Dependence and independence of [PSI(+)] and [PIN(+)]: a two-prion system in yeast? *EMBO J, 19*, 1942–1952.

47. Derkatch, I.L., Bradley, M.E., Hong, J.Y., Liebman, S.W. (2001). Prions affect the appearance of other prions: the story of [PIN(+)]. *Cell, 106*, 171–182.

48. Osherovich, L.Z., Weissman, J.S. (2001). Multiple Gln/Asn-rich prion domains confer susceptibility to induction of the yeast [PSI(+)] prion. *Cell, 106*, 183–194.

49. Wickner, R.B., Edskes, H.K., Roberts, B.T., Pierce, M.M., Baxa, U., Ross, E. (2001). Prions beget prions: the [PIN+] mystery! *Trends Biochem Sci, 26*, 697–699.

50. Coustou, V., Deleu, C., Saupe, S., Begueret, J. (1997). The protein product of the het-s heterokaryon incompatibility gene of the fungus *Podospora anserina* behaves as a prion analog. *Proc Natl Acad Sci USA, 94*, 9773–9778.
51. Coustou, V., Deleu, C., Saupe, S.J., Begueret, J. (1999). Mutational analysis of the [Het-s] prion analog of *Podospora anserina*: a short N-terminal peptide allows prion propagation. *Genetics, 153*, 1629–1640.
52. Ball, A.J., Wong, D.K., Elliott, J.J. (1976). Glucosamine resistance in yeast: I. A preliminary genetic analysis. *Genetics, 84*, 311–317.
53. Kunz, B.A., Ball, A.J. (1977). Glucosamine resistance in yeast: II. Cytoplasmic determinants conferring resistance. *Mol Gen Genet, 153*, 169–177.
54. Volkov, K.V., Aksenova, A.Y., Soom, M.J., Osipov, K.V., Svitin, A.V., Kurischko, C., Shkundina, I.S., Ter-Avanesyan, M.D., Inge-Vechtomov, S.G., Mironova, L.N. (2002). Novel non-Mendelian determinant involved in the control of translation accuracy in *Saccharomyces cerevisiae*. *Genetics, 160*, 25–36.
55. Talloczy, Z., Menon, S., Neigeborn, L., Leibowitz, M.J. (1998). The [KIL-d] cytoplasmic genetic element of yeast results in epigenetic regulation of viral M double-stranded RNA gene expression. *Genetics, 150*, 21–30.
56. Collin, P., Beauregard, P.B., Elagoz, A., Rokeach, L.A. (2004). A non-chromosomal factor allows viability of *Schizosaccharomyces pombe* lacking the essential chaperone calnexin. *J Cell Sci, 117*, 907–918.
57. Silar, P., Haedens, V., Rossignol, M., Lalucque, H. (1999). Propagation of a novel cytoplasmic, infectious and deleterious determinant is controlled by translational accuracy in *Podospora anserina*. *Genetics, 151*, 87–95.
58. Michelitsch, M.D., Weissman, J.S. (2000). A census of glutamine/asparagine-rich regions: implications for their conserved function and the prediction of novel prions. *Proc Natl Acad Sci USA, 97*, 11910–11915.
59. Serio, T.R., Cashikar, A.G., Kowal, A.S., Sawicki, G.J., Moslehi, J.J., Serpell, L., Arnsdorf, M.F., Lindquist, S.L. (2000). Nucleated conformational conversion and the replication of conformational information by a prion determinant. *Science, 289*, 1317–1321.
60. Thual, C., Bousset, L., Komar, A.A., Walter, S., Buchner, J., Cullin, C., Melki, R. (2001). Stability, folding, dimerization, and assembly properties of the yeast prion Ure2p. *Biochemistry, 40*, 1764–1773.
61. Pierce, M.M., Baxa, U., Steven, A.C., Bax, A., Wickner, R.B. (2005). Is the prion domain of soluble Ure2p unstructured? *Biochemistry, 44*, 321–328.
62. Balguerie, A., Dos Reis, S., Ritter, C., Chaignepain, S., Coulary-Salin, B., Forge, V., Bathany, K., Lascu, I., Schmitter, J.M., Riek, R., Saupe, S.J. (2003). Domain organization and structure–function relationship of the HET-s prion protein of Podospora anserina. *EMBO J, 22*, 2071–2081.
63. Glover, J.R., Kowal, A.S., Schirmer, E.C., Patino, M.M., Liu, J.J., Lindquist, S. (1997). Self-seeded fibers formed by Sup35, the protein determinant of [PSI+], a heritable prion-like factor of *S. cerevisiae*. *Cell, 89*, 811–819.
64. Taylor, K.L., Cheng, N., Williams, R.W., Steven, A.C., Wickner, R.B. (1999). Prion domain initiation of amyloid formation in vitro from native Ure2p. *Science, 283*, 1339–1343.

65. Vitrenko, Y.A., Gracheva, E.O., Richmond, J.E., Liebman, S.W. (2007). Visualization of aggregation of the Rnq1 prion domain and cross-seeding interactions with Sup35NM. *J Biol Chem, 282*, 1779–1787.

66. Dos Reis, S., Coulary-Salin, B., Forge, V., Lascu, I., Begueret, J., Saupe, S.J. (2002). The HET-s prion protein of the filamentous fungus *Podospora anserina* aggregates in vitro into amyloid-like fibrils. *J Biol Chem, 277*, 5703–5706.

67. Thual, C., Komar, A.A., Bousset, L., Fernandez-Bellot, E., Cullin, C., Melki, R. (1999). Structural characterization of *Saccharomyces cerevisiae* prion-like protein Ure2. *J Biol Chem, 274*, 13666–13674.

68. Derkatch, I.L., Uptain, S.M., Outeiro, T.F., Krishnan, R., Lindquist, S.L., Liebman, S.W. (2004). Effects of Q/N-rich, polyQ, and non-polyQ amyloids on the de novo formation of the [PSI+] prion in yeast and aggregation of Sup35 in vitro. *Proc Natl Acad Sci USA, 101*, 12934–12939.

69. Krishnan, R., Lindquist, S.L. (2005). Structural insights into a yeast prion illuminate nucleation and strain diversity. *Nature, 435*, 765–772.

70. Toyama, B.H., Kelly, M.J., Gross, J.D., Weissman, J.S. (2007). The structural basis of yeast prion strain variants. *Nature, 449*, 233–237.

71. DePace, A.H., Santoso, A., Hillner, P., Weissman, J.S. (1998). A critical role for amino-terminal glutamine/asparagine repeats in the formation and propagation of a yeast prion. *Cell, 93*, 1241–1252.

72. King, C.Y., Tittmann, P., Gross, H., Gebert, R., Aebi, M., Wuthrich, K. (1997). Prion-inducing domain 2-114 of yeast Sup35 protein transforms in vitro into amyloid-like filaments. *Proc Natl Acad Sci USA, 94*, 6618–6622.

73. Mukhopadhyay, S., Krishnan, R., Lemke, E.A., Lindquist, S., Deniz, A.A. (2007). A natively unfolded yeast prion monomer adopts an ensemble of collapsed and rapidly fluctuating structures. *Proc Natl Acad Sci USA, 104*, 2649–2654.

74. Scheibel, T., Lindquist, S.L. (2001). The role of conformational flexibility in prion propagation and maintenance for Sup35p. *Nat Struct Biol, 8*, 958–962.

75. Shorter, J., Lindquist, S. (2004). Hsp104 catalyzes formation and elimination of self-replicating Sup35 prion conformers. *Science, 304*, 1793–1797.

76. Shorter, J., Lindquist, S. (2006). Destruction or potentiation of different prions catalyzed by similar Hsp104 remodeling activities. *Mol Cell, 23*, 425–438.

77. Collins, S.R., Douglass, A., Vale, R.D., Weissman, J.S. (2004). Mechanism of prion propagation: amyloid growth occurs by monomer addition. *PLoS Biol, 2*, e321.

78. Doel, S.M., McCready, S.J., Nierras, C.R., Cox, B.S. (1994). The dominant PNM2- mutation which eliminates the psi factor of *Saccharomyces cerevisiae* is the result of a missense mutation in the SUP35 gene. *Genetics, 137*, 659–670.

79. Tessier, P.M., Lindquist, S. (2007). Prion recognition elements govern nucleation, strain specificity and species barriers. *Nature, 447*, 556–561.

80. Catharino, S., Buchner, J., Walter, S. (2005). Characterization of oligomeric species in the fibrillization pathway of the yeast prion Ure2p. *Biol Chem, 386*, 633–641.

81. Bousset, L., Belrhali, H., Janin, J., Melki, R., Morera, S. (2001). Structure of the globular region of the prion protein Ure2 from the yeast *Saccharomyces cerevisiae*. *Structure, 9*, 39–46.

82. Fowler, D.M., Koulov, A.V., Alory-Jost, C., Marks, M.S., Balch, W.E., Kelly, J.W. (2006). Functional amyloid formation within mammalian tissue. *PLoS Biol, 4*, e6.
83. Maddelein, M.L., Wickner, R.B. (1999). Two prion-inducing regions of Ure2p are nonoverlapping. *Mol Cell Biol, 19*, 4516–4524.
84. Fernandez-Bellot, E., Guillemet, E., Cullin, C. (2000). The yeast prion [URE3] can be greatly induced by a functional mutated URE2 allele. *EMBO J, 19*, 3215–3222.
85. Sondheimer, N., Lindquist, S. (2000). Rnq1: an epigenetic modifier of protein function in yeast. *Mol Cell, 5*, 163–172.
86. Taneja, V., Maddelein, M.L., Talarek, N., Saupe, S.J., Liebman, S.W. (2007). A non-Q/N-rich prion domain of a foreign prion, [Het-s], can propagate as a prion in yeast. *Mol Cell, 27*, 67–77.
87. Sanchez, Y., Lindquist, S.L. (1990). HSP104 required for induced thermotolerance. *Science, 248*, 1112–1115.
88. Glover, J.R., Lindquist, S. (1998). Hsp104, Hsp70, and Hsp40: a novel chaperone system that rescues previously aggregated proteins. *Cell, 94*, 73–82.
89. Inoue, Y., Taguchi, H., Kishimoto, A., Yoshida, M. (2004). Hsp104 binds to yeast Sup35 prion fiber but needs other factor(s) to sever it. *J Biol Chem, 279*, 52319–52323.
90. Krzewska, J., Melki, R. (2006). Molecular chaperones and the assembly of the prion Sup35p, an in vitro study. *EMBO J, 25*, 822–833.
91. Moriyama, H., Edskes, H.K., Wickner, R.B. (2000). [URE3] prion propagation in *Saccharomyces cerevisiae*: requirement for chaperone Hsp104 and curing by over-expressed chaperone Ydj1p. *Mol Cell Biol, 20*, 8916–8922.
92. Derkatch, I.L., Bradley, M.E., Zhou, P., Chernoff, Y.O., Liebman, S.W. (1997). Genetic and environmental factors affecting the de novo appearance of the [PSI$^+$] prion in *Saccharomyces cerevisiae. Genetics, 147*, 507–519.
93. Malato, L., Dos Reis, S., Benkemoun, L., Sabate, R., Saupe, S.J. (2007). Role of Hsp104 in the propagation and inheritance of the [Het-s] prion. *Mol Biol Cell, 18*, 4803–4812.
94. Aron, R., Higurashi, T., Sahi, C., Craig, E.A. (2007). J-protein co-chaperone Sis1 required for generation of [RNQ+] seeds necessary for prion propagation. *EMBO J, 26*, 3794–3803.
95. Lansbury, P.T., Jr. (1992). In pursuit of the molecular structure of amyloid plaque: new technology provides unexpected and critical information. *Biochemistry, 31*, 6865–6870.
96. Nelson, R., Eisenberg, D. (2006). Recent atomic models of amyloid fibril structure. *Curr Opin Struct Biol, 16*, 260–265.
97. Serpell, L.C. (2000). Alzheimer's amyloid fibrils: structure and assembly. *Biochim Biophys Acta, 1502*, 16–30.
98. Wetzel, R. (2002). Ideas of order for amyloid fibril structure. *Structure, 10*, 1031–1036.
99. Nelson, R., Sawaya, M.R., Balbirnie, M., Madsen, A.O., Riekel, C., Grothe, R., Eisenberg, D. (2005). Structure of the cross-beta spine of amyloid-like fibrils. *Nature, 435*, 773–778.

100. van der Wel, P.C., Lewandowski, J.R., Griffin, R.G. (2007). Solid-state NMR study of amyloid nanocrystals and fibrils formed by the peptide GNNQQNY from yeast prion protein Sup35p. *J Am Chem Soc, 129*, 5117–5130.

101. Sawaya, M.R., Sambashivan, S., Nelson, R., Ivanova, M.I., Sievers, S.A., Apostol, M.I., Thompson, M.J., Balbirnie, M., Wiltzius, J.J., McFarlane, H.T., et al. (2007). Atomic structures of amyloid cross-beta spines reveal varied steric zippers. *Nature, 447*, 453–457.

102. Shewmaker, F., Wickner, R.B., Tycko, R. (2006). Amyloid of the prion domain of Sup35p has an in-register parallel beta-sheet structure. *Proc Natl Acad Sci USA, 103*, 19754–19759.

103. Kajava, A.V., Baxa, U., Wickner, R.B., Steven, A.C. (2004). A model for Ure2p prion filaments and other amyloids: the parallel superpleated beta-structure. *Proc Natl Acad Sci USA, 101*, 7885–7890.

104. Diaz-Avalos, R., King, C.Y., Wall, J., Simon, M., Caspar, D.L. (2005). Strain-specific morphologies of yeast prion amyloid fibrils. *Proc Natl Acad Sci USA, 102*, 10165–10170.

105. Ross, E.D., Edskes, H.K., Terry, M.J., Wickner, R.B. (2005). Primary sequence independence for prion formation. *Proc Natl Acad Sci USA, 102*, 12825–12830.

106. Jenkins, J., Pickersgill, R. (2001). The architecture of parallel beta-helices and related folds. *Prog Biophys Mol Biol, 77*, 111–175.

107. Jenkins, J., Mayans, O., Pickersgill, R. (1998). Structure and evolution of parallel beta-helix proteins. *J Struct Biol, 122*, 236–246.

108. Yoder, M.D., Jurnak, F. (1995). Protein motifs. 3. The parallel beta helix and other coiled folds. *FASEB J, 9*, 335–342.

109. Kishimoto, A., Hasegawa, K., Suzuki, H., Taguchi, H., Namba, K., Yoshida, M. (2004). Beta-helix is a likely core structure of yeast prion Sup35 amyloid fibers. *Biochem Biophys Res Commun, 315*, 739–745.

110. Makin, O.S., Sikorski, P., Serpell, L.C. (2006). Diffraction to study protein and peptide assemblies. *Curr Opin Chem Biol, 10*, 417–422.

111. Fay, N., Redeker, V., Savistchenko, J., Dubois, S., Bousset, L., Melki, R. (2005). Structure of the prion Ure2p in protein fibrils assembled in vitro. *J Biol Chem, 280*, 37149–37158.

112. Baxa, U., Wickner, R.B., Steven, A.C., Anderson, D.E., Marekov, L.N., Yau, W.M., Tycko, R. (2007). Characterization of beta-sheet structure in Ure2p1–89 yeast prion fibrils by solid-state nuclear magnetic resonance. *Biochemistry, 46*, 13149–13162

113. Chan, J.C., Oyler, N.A., Yau, W.M., Tycko, R. (2005). Parallel beta-sheets and polar zippers in amyloid fibrils formed by residues 10–39 of the yeast prion protein Ure2p. *Biochemistry, 44*, 10669–10680.

114. Ross, E.D., Baxa, U., Wickner, R.B. (2004). Scrambled prion domains form prions and amyloid. *Mol Cell Biol, 24*, 7206–7213.

115. Baxa, U., Taylor, K.L., Wall, J.S., Simon, M.N., Cheng, N., Wickner, R.B., Steven, A.C. (2003). Architecture of Ure2p prion filaments: the N-terminal domains form a central core fiber. *J Biol Chem, 278*, 43717–43727.

116. Baxa, U., Cheng, N., Winkler, D.C., Chiu, T.K., Davies, D.R., Sharma, D., Inouye, H., Kirschner, D.A., Wickner, R.B., Steven, A.C. (2005). Filaments of

the Ure2p prion protein have a cross-beta core structure. *J Struct Biol, 150*, 170–179.

117. Bousset, L., Briki, F., Doucet, J., Melki, R. (2003). The native-like conformation of Ure2p in fibrils assembled under physiologically relevant conditions switches to an amyloid-like conformation upon heat-treatment of the fibrils. *J Struct Biol, 141*, 132–142.
118. Bousset, L., Thomson, N.H., Radford, S.E., Melki, R. (2002). The yeast prion Ure2p retains its native alpha-helical conformation upon assembly into protein fibrils in vitro. *EMBO J, 21*, 2903–2911.
119. Redeker, V., Halgand, F., Le Caer, J.P., Bousset, L., Laprevote, O., Melki, R. (2007). Hydrogen/deuterium exchange mass spectrometric analysis of conformational changes accompanying the assembly of the yeast prion Ure2p into protein fibrils. *J Mol Biol, 369*, 1113–1125.
120. Ritter, C., Maddelein, M.L., Siemer, A.B., Luhrs, T., Ernst, M., Meier, B.H., Saupe, S.J., Riek, R. (2005). Correlation of structural elements and infectivity of the HET-s prion. *Nature, 435*, 844–848.
121. Sen, A., Baxa, U., Simon, M.N., Wall, J.S., Sabate, R., Saupe, S.J., Steven, A.C. (2007). Mass analysis by scanning transmission electron microscopy and electron diffraction validate predictions of stacked beta-solenoid model of HET-s prion fibrils. *J Biol Chem, 282*, 5545–5550.
122. Prusiner, S.B. (1991). Molecular biology of prion diseases. *Science, 252*, 1515–1522.
123. Tanaka, M., Collins, S.R., Toyama, B.H., Weissman, J.S. (2006). The physical basis of how prion conformations determine strain phenotypes. *Nature, 442*, 585–589.
124. Pattison, I.H., Jones, K.M. (1968). Modification of a strain of mouse-adapted scrapie by passage through rats. *Res Vet Sci, 9*, 408–410.
125. Chen, B., Newnam, G.P., Chernoff, Y.O. (2007). Prion species barrier between the closely related yeast proteins is detected despite coaggregation. *Proc Natl Acad Sci USA, 104*, 2791–2796.
126. Will, R.G., Ironside, J.W., Hornlimann, B., Zeidler, M. (1996). Creutzfeldt–Jakob disease. *Lancet, 347*, 65–66.

9

CAENORHABDITIS ELEGANS AS A MODEL SYSTEM TO STUDY THE BIOLOGY OF PROTEIN AGGREGATION AND TOXICITY

Elise A. Kikis, Anat Ben-Zvi, and Richard I. Morimoto

Department of Biochemistry, Molecular Biology and Cell Biology, Rice Institute for Biomedical Research, Northwestern University, Evanston, Illinois

INTRODUCTION

Neurodegenerative diseases of aging, including Alzheimer disease (AD), Parkinson disease (PD), amyotrophic lateral sclerosis (ALS), and polyglutamine disorders, are genetic abnormalities in which the affected mutant protein causes a gain-of-function toxicity. For each of these diseases the affected protein has a distinct function and a unique primary sequence. Studies aimed at understanding the mode of disease progression have revealed both common mechanisms involving protein misfolding and clearance as well as specific mechanisms for each disease-causing protein. Protein misfolding, as it relates to neurodegenerative diseases, generally results in the accumulation of misfolded species in either the nucleus, cytoplasm, or extracellular space that, over time, interferes with cellular function. To develop meaningful therapeutic strategies that either prevent or delay disease onset or treat disease symptoms, it will be necessary to understand the extent to which common mechanisms of protein misfolding, and the disruption of protein-folding homeostasis, can be attributed to disease pathology.

The protein folding problem, and consequently the machinery that regulates it, is nearly as ancient as life itself. Nascent polypeptide chains fold into native

Protein Misfolding Diseases: Current and Emerging Principles and Therapies,
Edited by Marina Ramirez-Alvarado, Jeffery W. Kelly, and Christopher M. Dobson
Copyright © 2010 John Wiley & Sons, Inc.

conformations, often requiring the assistance of molecular chaperones that are conserved from prokaryotes to humans [1]. Additionally, exposure to cell stress, often caused by heat shock or oxidative damage, causes an imbalance in protein homeostasis, due to the flux of damaged and misfolded proteins. The consequence at the cellular level is induction of the heat-shock response, which leads to elevated expression of molecular chaperones [2,58]. Misfolded proteins that fail to reestablish their native fold must be removed from the cell to prevent toxic consequences of the accumulation of misfolded proteins. This involves autophagy and ubiquitin-proteasome–mediated degradation of damaged species, some of which have been shown to interfere with protein clearance mechanisms [3–5].

Because of the evolutionary conservation of the protein homeostasis machinery, eukaryotic model systems including yeast, *Drosophila, and Caenorhabditis elegans* have proven invaluable in studying the mechanisms underlying protein-folding diseases. In this chapter we focus specifically on the utilization of *C. elegans* models to study protein aggregation dynamics, to identify genetic modifiers of protein aggregation/toxicity, to assess the roles of aging in the progression of aggregation/toxicity phenotypes, and to identify small molecules as potential therapeutics. Finally, we discuss how the results of such studies, in particular those of genetic screens, have been further validated in other model organisms, and how, owing to the number of genetically identified modifiers that have mammalian orthologs, their ability to modify aggregation/toxicity in mammalian models can be examined further.

C. ELEGANS MODELS FOR NEURODEGENERATIVE DISEASES OF PROTEIN FOLDING

C. elegans has become an important model system in general, owing in part to its sequenced genome and genetic tools, which together with a short life cycle, established lineage, and transparency has proven useful as a model system for human disease. More specific to its advantages as a system for the study of neurodegenerative diseases of aging, *C. elegans* has a relatively simple (but still sufficiently complex) nervous system, a genetically defined aging pathway, and live cell-imaging capabilities of fluorescent proteins for studies of disease-causing protein aggregation dynamics.

Benefiting from these characteristics, a number of *C. elegans* disease models have recently been characterized. An AD model expressing Aβ(1–42) in body-wall muscle cells was shown to form amyloid plaques that have biochemical characteristics similar to those found in the brains of AD patients [6]. In addition to amyloid plaques, AD and other neurodegenerative diseases are characterized by the formation of neurofibrillary tangles comprised of the protein tau. Expression of tau in *C. elegans* neurons resulted in motility defects indicative of tau proteotoxicity [7]. Familial forms of PD are caused by mutations in the protein α-synuclein [8,9] Expression of α-synuclein in *C. elegans* neurons was shown to cause neurodegeneration of neuronal subsets [10]. Huntington

TABLE 1 Current *C. elegans* Neurodegenerative Disease Models

Disease	Protein	Expression	Reference
Polyglutamine disorders	PolyQ-YFP	Body-wall muscle cells	17
	PolyQ-YFP	Pan neuronal	18
Huntington disease	Huntingtin (NT)	ASH sensory neurons	11
	Huntingtin (NT)	Mechanosensory neurons	12
Alzheimer disease	Aβ(1–42)	Body-wall muscle cells (constitutive)	6
	Aβ(1–42)	Body-wall muscle cells (temperature inducible)	59
Tauopathies	Tau	Pan neuronal	7
Parkinson disease	α-Synuclein	Pan neuronal	10
	α-Synuclein	Dopaminergic neurons	60
	α-Synuclein	Body-wall muscle cells	61
Amyotrophic lateral sclerosis	Mutant SOD1	Body-wall muscle cells	62
	Mutant SOD1	Pan neuronal	63

disease models have been generated via the expression, in neuronal subsets, of an N-terminal fragment of the protein huntingtin (Htt) with a long polyglutamine (polyQ) tract [11,12], which is known to be the genetic lesion leading to HD [13–15] Expression of polyQ alone fused to YFP has been expressed in *C. elegans* body-wall muscle cells [16,17] and neurons [18] as a generic model for all polyQ diseases (see Table 1 for a complete catalog of current *C. elegans* neurodegenerative disease models).

At this point it is tempting to begin drawing comparisons between the various *C. elegans* disease models with respect to aggregation/toxicity. However, in doing so, it is important to recognize that comparisons are confounded by different disease-causing proteins being expressed at different absolute amounts, under control of different promoters, and in different tissue types. That being said, some models (e.g., the neuronal expression of Htt-polyQ [11,12] and polyQ-YFP [18]), show strikingly different levels of toxicity in that polyQ-YFP began to aggregate and display proteotoxicity at Q40, whereas the Htt-polyQ model required Q128 before toxic effects of expressing the disease-causing protein were observed [12]. To verify that there is a contribution of protein context on polyQ length-dependent aggregation/toxicity, it will be necessary to perform a direct comparison by expressing polyQ-YFP and Htt-polyQ at the same amounts, under control of the same promoter.

Expansion of polyQ tracts in a number of human proteins has been shown to be the genetic determinant of neurodegenerative diseases, including HD, spinal and bulbar muscular atrophy (SBMA)/Kennedy disease, dentatorubral and pallidoluysian atrophy (DRPLA), and spinocerebellar ataxias (SCA). The expansion of a polyQ tract causes the affected protein to misfold and form aggregates, although the mechanism by which misfolded proteins leads to disease is not well understood. Recent evidence suggests that a variety of mechanisms, including the sequestration of essential proteins to aggregates [19], dramatic gene expression changes that lead to cell death [20], or a general disruption in the protein-folding environment [21], may all act together, to a greater or lesser extent, to contribute to disease. Regardless, the mechanism of disease progression seems to be largely conserved between polyQ diseases and to be polyQ-tract dependent.

POLYGLUTAMINE PROTEIN AGGREGATION DYNAMICS

To elucidate the cellular mechanism(s) underlying polyQ proteotoxicity that would be relevant to all of the CAG repeat diseases, a polyQ tract of varying length was fused to the normally inert YFP and expressed in *C. elegans* body-wall muscle cells [16,17] or neurons [18]. These analyses revealed that YFP fused to short polyQ tracts resulted in diffuse fluorescence suggestive of polyQ protein solubility. On the other hand, YFP fused to long polyQ tracts produced visible protein aggregates and caused dramatic motility defects characteristic of impaired muscle [16] or neuronal function [18]. These studies defined a threshold polyQ length for age-dependent aggregation/toxicity that correlates with the polyQ length observed in HD patients. Figure 1 shows transgenic worms expressing various lengths of polyQ fused to GFP in body-wall muscle cells (A) or neurons (B). Visualization of fluorescence shows that increasing polyQ lengths results in an increase in polyQ protein aggregate formation.

In *C. elegans* neurons, expression of polyQ lengths near the threshold for aggregation displayed a range of aggregation dynamics in different neuronal subsets. The neurons of the ventral and dorsal nerve cords were more susceptible to polyQ40 protein aggregation than were sensory interneurons, which displayed consistently diffuse soluble fluorescence of Q40-YFP. These results are intriguing and suggest that different modifiers of polyQ protein aggregation, or perhaps protein misfolding in general, act in distinct neuronal subsets. Consistent with this, the expression of the aggregation-prone protein, α-synuclein, in all *C. elegans* neurons, or specifically in dopaminergic or motor neurons, revealed differential neuronal subtype sensitivity. Namely, motor neurons expressing α-synuclein underwent significantly less dendritic loss than did dopaminergic neurons also expressing α-synuclein [10]. Therefore, the identification of cell type–specific modifiers of protein aggregation will be an important next step in understanding how the expression of certain aggregation-prone disease-causing proteins are toxic to some cells but not to others.

Finally, a distinct advantage of *C. elegans* as a model system stems from its amenability to live-cell dynamic imaging methods, due to its inherent lack of pigmentation. These methods include fluorescence recovery after photobleaching (FRAP) and fluorescence resonance energy transfer (FRET), which have been instrumental in characterizing polyQ protein aggregation dynamics. Namely, FRAP has been used to unequivocally determine whether the

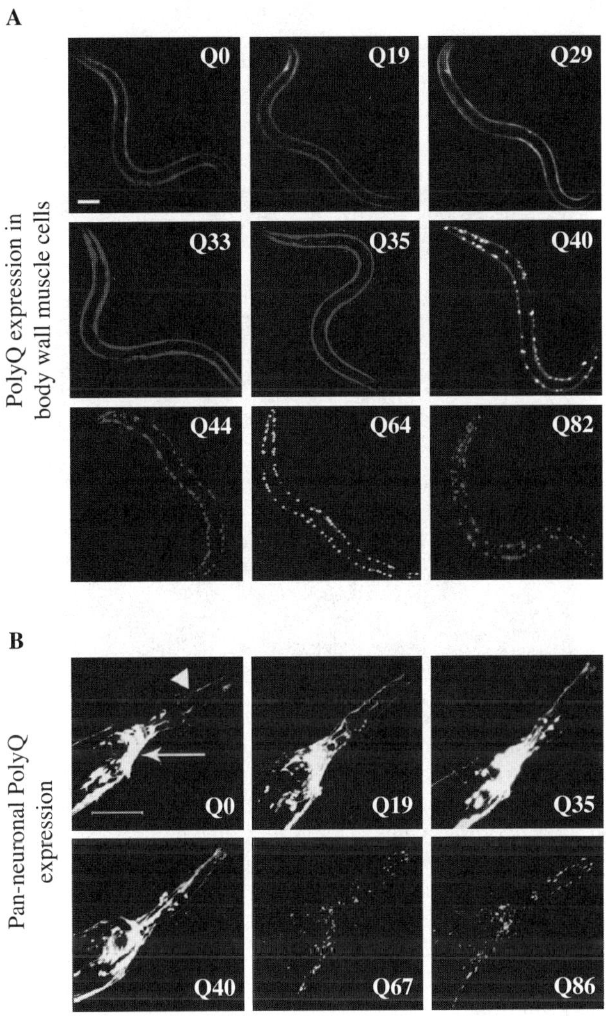

FIG. 1 Length-dependent aggregation of polyQ-YFP fusion proteins in *C. elegans* body-wall muscle cells (A) or neurons (B). Epifluoresence micrographs of 3- to 4-day-old *C. elegans* expressing different lengths of polyQ-YFP. Scale bar = 0.1 mm (A), or 50 μm (B). Arrow indicates circumpharyngeal nerve ring. Expression of a range of polyQ lengths reveals that proteins with tracts that are equal to or less than that of Q40 maintain a soluble distribution pattern, whereas those equal to or more than Q40, in body-wall muscle cells, form foci. [(A) Adapted from [16]. (B) Adapted from [18], with permission of the *Journal of Neuroscience*.]

observable shift in diffuse to punctate fluorescence with longer polyQ tract-lengths is due to a transition from a soluble to an aggregated state of polyQ proteins. Figure. 2 shows that FRAP failed to occur on visible polyQ-YFP aggregates but did occur with diffuse fluorescence, even at the threshold

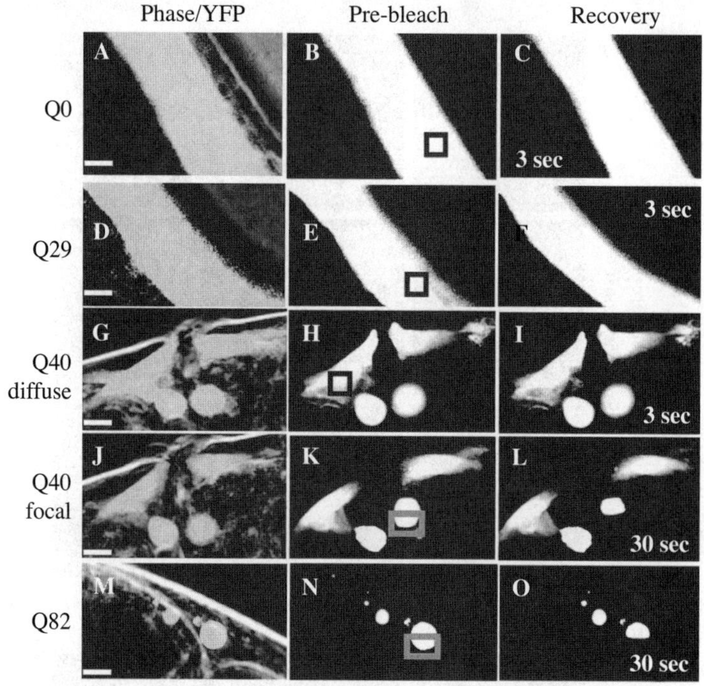

FIG. 2 Determination of polyQ-YFP solubility in living animals by using FRAP: (A, D, G, J, M) merged phase-contrast and fluorescence images; (B, E, H, K, N) fluorescence images of the same region before photobleaching (prebleach) (boxes indicate the area that was subjected to photobleaching); (C,F,I,L,O) fluorescence images of recovery at the indicated times after photobleaching. The earliest time point possible to assess recovery of the chimeric YFP signal was at 3 s. Scale bar = 3 μm. [Adapted from [16].]

for aggregation/toxicity, in body-wall muscle cells. The coexpression of polyQ-YFP with polyQ-CFP allowed for FRET analysis to determine whether polyQ aggregates in neurons display intermolecular interactions consistent with the formation of ordered species [18]. Large visible aggregates of Q86 yielded positive FRET signal indicative of close interactions (less than 100 Å apart) of stably oriented proteins [18], consistent with the hypothesis that polyQ-containing proteins form aggregation-prone β-sheet structures [22].

GENETIC SCREENS FOR MODIFIERS OF DISEASE-RELATED PHENOTYPES

The versatility of *C. elegans* as a model system to study human disease has been demonstrated in recent years with the implementation of genome-wide RNA interference (RNAi) screens aimed at the identification of genetic modifiers of disease-related phenotypes. Such screens have been facilitated by the availability of RNAi libraries, consisting of *Escherichia coli* clones containing IPTG-inducible double-stranded (ds) RNAs for the majority of *C. elegans* genes [23]. Feeding dsRNA-producing *E. coli* to *C. elegans* has proven to be a highly efficient method for targeted gene silencing, making high-throughput RNAi screens relatively straightforward [24].

One such study identified modifiers of polyQ protein aggregation [25]. The genomewide screen took advantage of a polyQ length at the threshold for aggregation (Q35), thereby allowing for a sensitized screen aimed at the identification of factors which when knocked down in the background of polyQ-YFP, led to the accumulation of visible protein aggregates. Ultimately, this screen identified 186 proteins that normally suppress age-dependent polyQ protein aggregation. The authors found that the suppressors fall into five distinct biological classes, including RNA metabolism, protein synthesis, protein folding, protein trafficking, and protein degradation. The identification of chaperones and factors involved in protein clearance were expected. However, the identification of factors involved in other biosynthetic processes led to the conclusion that protein homeostasis is more complex than previously understood and probably begins with gene expression, thus explaining the large fraction of modifiers involved in RNA and protein biosynthesis [25].

This screen also uncovered six of the eight subunits of cytosolic chaperonin CCT, whose role as a suppressor of polyQ protein aggregation was previously unknown. This finding was later validated by a number of groups using both *S. cerevisiae* and mammalian tissue culture cells expressing aggregation-prone polyQ proteins [26–28] The mere fact that the findings of the *C. elegans* RNAi screen hold true in other systems both underscores the usefulness of *C. elegans* models for the elucidation of mode of disease action, and provides support for the idea that common mechanisms underlie polyQ protein aggregation/toxicity.

An independent, RNAi screen was also performed in *C. elegans* for factors that normally suppress tau-induced motility defects [29]. Wild-type and mutant tau protein become hyperphosphorylated, aggregate, and form neurofibrilllary tangles that lead to neurodegeneration in patients suffering from AD and a number of related neurodegenerative diseases [30]. Expression of tau in *C. elegans* neurons caused motility defects that were used as the basis to identify factors, via genome-wide RNAi screening, which, when absent, enhanced the motility (unc) phenotype [29]. This analysis led to the identification of 75 suppressors of tau toxicity, falling into the following functional categories: kinases, chaperones, proteases, and phosphatases, in addition to a number of genes whose function is unknown [29]. Interestingly, the only RNAi hits in common between this and the polyQ screen described above are the Hsp70 molecular chaperone, *hsp-1*, and the heat-shock transcription factor, *hsf-1*.

The striking lack of overlap between modifiers of aggregation/toxicity for unique aggregation-prone proteins could presumably be due to different factors acting on different misfolded species, or due to different factors acting in body-wall muscle cells compared to neurons. Certainly, the identification of kinases and phosphatases in the screen for suppressors of tau toxicity provides support for the hypothesis that tau hyperphosphorylation is a prerequisite for disease. Consequently, these data also provide evidence that *C. elegans* is a valid model system for the identification of factors that are capable of acting on particular human disease–causing proteins to suppress aggregation/toxicity.

However, the extent to which the molecular mechanisms of disease are conserved between aggregation-prone proteins is unknown. To address this it will be necessary to express unrelated aggregation-prone proteins, such as tau and a polyQ-containing protein, in the same cells, for a direct comparison of modifiers of aggregation/toxicity. The expectation is that common modifiers would be those whose molecular function is in the general folding and clearance of misfolded proteins. On the other hand, modifiers that act on one or the other aggregation-prone protein would probably be more closely associated with the specific function, or mode of disease progression, of a particular protein. Importantly, a screen performed in yeast expressing mutant Htt or α-synuclein revealed an almost entirely nonoverlapping sets of genes, many with human homologs, acting as modifiers either of mutant Htt or α-synuclein toxicity [31]. The authors speculate that their modifiers probably define mechanisms or pathways that are specific for particular disease-causing proteins, such as vesicle transport playing a role in α-synuclein toxicity [31].

Interestingly, the hits obtained by an RNAi screen aimed at the identification of suppressors of osmotic stress-induced gene expression overlapped to a great extent with those identified as suppressors of polyQ protein aggregation [32]. In that study the authors identified genes, including *gpdh-1*, that are up-regulated in response to osmotic stress. They used a *gpdh-1* promoter-GFP fusion as a reporter and performed an RNAi screen to identify factors that normally function to suppress *gpdh-1* expression under conditions of no stress. Although *gpdh-1* expression does not respond to stresses other than osmotic stress [32], almost 30% of all the genes identified in this study overlap with those identified in the screen for suppressors of polyQ protein aggregation. Furthermore, 73% of the overlapping genes are predicted to fall into biological classes usually associated with protein homeostasis, including RNA processing, protein synthesis, protein folding, and degradation [32]. Ultimately, these data suggest that a core set of factors function generally in response to stress-induced protein damage, and others respond specifically to a particular stress to tailor the response to the situation at hand.

In addition to RNAi screens, forward genetic screens have also been performed to identify modulators of polyQ protein aggregation/toxicity. One such screen revealed a novel gene, *pqe-1*, which normally functions to suppress the proteotoxicity of an Htt exon 1 fragment with an expanded polyQ tract [33]. Another forward genetic screen was aimed at the identification of genes that normally function to suppress the aggregation of polyQ-YFP in body-wall muscle cells [34]. This screen uncovered mutations in *unc-30*, the transcription factor that regulates the synthesis of the neurotransmitter GABA [34]. The findings described by Garcia et al. [34] are of particular interest, because they demonstrate that the ability of an organism to manage proteotoxic stress is not a cell autonomous process as previously thought, but actually requires cell–cell communication via neuronal cholinergic signaling. Likewise, it has been shown recently that neuronal signaling is required for an organismal heat shock response [64]. Consistent with this, treating Q35-expressing *C. elegans* with small molecules, acting positively or negatively on neuronal signaling, suppressed or enhanced, respectively, polyQ protein aggregation [34]. The identification, via chemical genetic screens, of additional small molecules that alleviate disease-causing protein aggregation/toxicity will be an important step in the development of pharmaceuticals.

SMALL-MOLECULE DRUG SCREENS

Chemical genetics is a process by which small molecules having a desired effect on a given biological process are identified, often via high-throughput screening methods. Although some chemical genetic analyses have been performed using *C. elegans* as a model system [35–37], very few studies, and no comprehensive screens, have been performed with the *C. elegans* neurodegenerative disease models described here. This is in part due to anecdotal evidence that *C. elegans* does not efficiently take up small molecules from the environment. However, individual compounds have been tested in *C. elegans* models for HD, AD, and PD, and conditions under which high-throughput screens can be performed are being established.

One such study examined the effect of particular components of a *Ginkgo biloba* extract on worms expressing Aβ in body-wall muscle cells [38]. *G. biloba* extract is often given to AD patients to partially alleviate memory loss and dementia [39], and one component of the extract, ginkgolide A, was shown to substantially suppress the motility defect normally observed in *C. elegans* expressing Aβ [38]. Another study showed that treatment of Htt-expressing *C. elegans* with the antioxidant resveratrol reduced Htt toxicity in a manner dependent on the sirutiun NAD^+-dependent deacetylases [40]. Additionally, a PD model was generated by treating *C. elegans* with the neurotoxin MPP^+, which causes PD-like symptoms in vertebrates and causes motility defects in *C. elegans* [41]. The coincubation of *C. elegans* with MPP^+ and a number of established, pharmacologically active compounds normally given to PD patients, such as lisuride, apomorphine, selegiline, and nomifensine, resulted in suppression of the original toxic effects of MPP^+. Ultimately, these results demonstrate the potential viability of the MPP^+ *C. elegans* model for use in high-throughput screens to identify novel drugs for the treatment of PD.

To screen large numbers of small molecules effectively, it is essential to have a phenotype that can be scored quickly and easily. In tissue culture, screening techniques have been devised that take advantage of quantitative reporter gene expression [42], or rescue of cell death [43]. These are particularly powerful phenotypes for screening because monitoring them requires minimal human input, thereby reducing error and the time required to screen large numbers of compounds. The assays described above for *C. elegans* require visual assessment of the effect of a drug on motility, which can be laborious if large numbers of molecules are being screened.

A small-scale chemical genetics analysis was performed with a *C. elegans* HD model using a novel screening approach which circumvents the inherent problems associated with screening based on motility defects. The authors examined treated worms for lack of neuronal cell death by visualizing the loss, or lack thereof, of GFP fluorescence in ASH neurons of a sensitized line that rapidly undergoes polyQ-dependent neurodegeneration [44]. These conditions were used to validate candidate compounds identified previously in a large-scale, tissue culture–based screen [43,44], and revealed two compounds, lithium chloride and mithramycin, which suppressed HD neurotoxicity in the *C. elegans* HD model [44]. Use of this and similar assays will make it possible to screen large chemical libraries rapidly for their effect on *C. elegans* models of neurodegenerative diseases.

Because *C. elegans* is a multicellular organism, we would expect that the successful implementation of large-scale chemical genetics screens will be highly effective in identifying novel therapeutic compounds, not previously identified in cell culture models, that act either cell autonomously or cell nonautonomously. Finally, fluorescent labeling of candidate molecules will be instrumental in elucidating the mode of drug action and to determine whether the drug is acting directly or indirectly on the disease-causing proteins, and will be relatively straightforward in *C. elegans*, due to easy visualization of the fluorescent markers described above.

STRESS, PROTEIN HOMEOSTASIS, AND AGING

Many of the protein-misfolding diseases discussed in this chapter are also considered diseases of aging based on when phenotypes are observed. Expression of aggregation-prone proteins in *C. elegans* recapitulates this age-dependent onset and progression in a life span–dependent manner as demonstrated in Figure 3. More specifically, Figure 3 shows that aggregate number (A) and toxicity (B) increases as worms age, thereby suggesting that changes associated with aging may have a pivotal role in the etiology of protein-misfolding diseases. Different highly conserved genetic pathways that modulate life span were identified and studied extensively in *C. elegans*, such as the insulin-like signaling pathway (ILS), mitochondrial function, and caloric restriction [45–51] Thus, *C. elegans* offers many advantages to examine the interplay between aging pathways and age-dependent diseases.

The ILS pathway in *C. elegans* comprises the insulin-like receptor *daf-2* [46–48], a phosphoionositide-3-OH kinase, *age-1* [46,49], the forkhead transcription factor *daf-16* [47,50], and the heat shock transcription factor *hsf-1* [52,53]. Activation of *daf-2* initiates a kinase cascade that includes *age-1* and results in the phosphorylation and repression of *daf-16*. Mutation, or

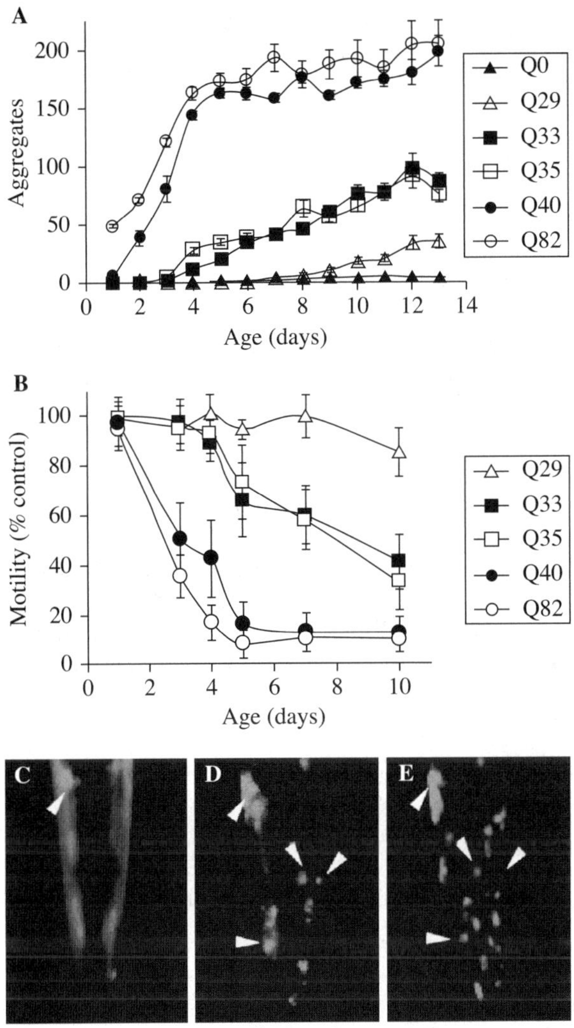

FIG. 3 Influence of aging on polyQ aggregation and toxicity (A) Accumulation of aggregates in Q82 (open circle), Q40 (filled circle), Q35 (open square), Q33 (filled square), Q29 (open triangle), and Q0 (filled triangle) during aging. Data are mean ± ⊖ SEM. Twenty-four animals of each type are represented at day 1. Cohort sizes decreased as animals died during the experiment, but each data point represents at least five animals. (B) Motility index as a function of age for the same cohorts of animals described in (A). Data are mean ± SD as a percentage of age-matched Q0 animals. (C–E) Epifluorescence micrographs of the head of an individual Q35 animal at 4 (C), 7 (D), and 10 (E) days of age, illustrating age-dependent accumulation of aggregates. Arrowheads indicate positions of the same aggregates on different days. In (E), the animal is rotated slightly relative to its position in (D). (Adapted from [16].)

down-regulation, of either *daf-2* or *age-1* results in de-repression of *daf-16* and longevity [46,47,48,49–50]. The extension of life span is also dependent on *hsf-1* activity [52,53]. Moreover, disruption of components of the ILS pathway at specific developmental time points reveals that the ILS pathway acts predominantly during adulthood to influence *C. elegans* life span [54].

To examine the role of aging pathways in misfolding diseases, the expression of genes in the ILS pathway was disrupted in animals expressing aggregation-prone polyQ or Aβ peptide [6,16]. Down-regulation of *age-1* or *daf-2* by RNAi, which extends life span, resulted in a dramatic reduction of aggregation/toxicity of polyQ proteins [16,55]. Furthermore, it has been hypothesized that the toxicity of aggregation-prone proteins may be mediated by the sequestration of chaperones [56] and the disruption of cellular folding homeostasis [21]. Consistent with this, Cohen et al. [55] demonstrated that down-regulation of *hsf-1* and *daf-16* modifies the biochemical properties of Aβ peptide and, consequently, toxicity. We therefore suggest that genetic modifiers of aging can influence the cellular environment and protein-folding homeostasis and may thereby trigger protein aggregation in old age. This remains to be further corroborated in mammalian disease models. However, the fact that celastrol, a drug that activates *hsf-1* in the absence of stress, was identified in a screen for drugs that affect HD-like phenotypes in mammalian cell lines [57] suggests that modulation of life span modifiers such as *hsf-1* may be effective for the treatment of patients.

Acknowledgments

E.A.K. was supported by an individual postdoctoral fellowship from the NIH. A.B. was supported by an EMBO long-term fellowship, the APDA, and the HDF. R.I.M. was supported by grants from the NIH (NIGMS and NIA), the HDSA Coalition for the Cure, and the Daniel F. and Ada L. Rice Foundation. We would also like to thank past and present members of the laboratory for their scientific contributions.

REFERENCES

1. Hartl, F.U., Hayer-Hartl, M. (2002). Molecular chaperones in the cytosol: from nascent chain to folded protein. *Science, 295*, 1852–1858.
2. Kim, H.J., Hwang, N.R., Lee, K.J. (2007). Heat shock responses for understanding diseases of protein denaturation. *Mol Cells, 23*, 123–131.
3. Yi, J.J., Ehlers, M.D. (2007). Emerging roles for ubiquitin and protein degradation in neuronal function. *Pharmacol Rev, 59*, 14–39.
4. Ardley, H.C., Robinson, P.A. (2004). The role of ubiquitin–protein ligases in neurodegenerative disease. *Neurodegener Dis, 1*, 71–87.
5. Nandi, D., Tahiliani, P., Kumar, A., Chandu, D. (2006). The ubiquitin–proteasome system. *J Biosci, 31*, 137–155.
6. Link, C.D. (1995). Expression of human beta-amyloid peptide in transgenic *Caenorhabditis elegans. Proc Natl Acad Sci USA, 92*, 9368–9372.

7. Kraemer, B.C., Zhang, B., Leverenz, J.B., Thomas, J.H., Trojanowski, J.Q., Schellenberg, G.D. (2003). Neurodegeneration and defective neurotransmission in a *Caenorhabditis elegans* model of tauopathy. *Proc Natl Acad Sci USA, 100*, 9980–9985.

8. Nussbaum, R.L., Polymeropoulos, M.H. (1997). Genetics of Parkinson's disease. *Hum Mol Genet, 6*, 1687–1691.

9. Chase, T.N. (1997). A gene for Parkinson disease. *Arch Neurol, 54*, 1156–1157.

10. Lakso, M., Vartiainen, S., Moilanen, A.M., Sirvio, J., Thomas, J.H., Nass, R., Blakely, R.D., Wong, G. (2003). Dopaminergic neuronal loss and motor deficits in *Caenorhabditis elegans* overexpressing human alpha-synuclein. *J Neurochem, 86*, 165–172.

11. Faber, P.W., Alter, J.R., MacDonald, M.E., Hart, A.C. (1999). Polyglutamine-mediated dysfunction and apoptotic death of a *Caenorhabditis elegans* sensory neuron. *Proc Natl Acad Sci USA, 96*, 179–184.

12. Parker, J.A., Connolly, J.B., Wellington, C., Hayden, M., Dausset, J., Neri, C. (2001). Expanded polyglutamines in *Caenorhabditis elegans* cause axonal abnormalities and severe dysfunction of PLM mechanosensory neurons without cell death. *Proc Natl Acad Sci USA, 98*, 13318–13323.

13. Andrew, S.E., Goldberg, Y.P., Kremer, B., Telenius, H., Theilmann, J., Adam, S., Starr, E., Squitieri, F., Lin, B., Kalchman, M.A., et al. (1993). The relationship between trinucleotide (CAG) repeat length and clinical features of Huntington's disease. *Nat Genet, 4*, 398–403.

14. Ashizawa, T., Wong, L.J., Richards, C.S., Caskey, C.T., Jankovic, J. (1994). CAG repeat size and clinical presentation in Huntington's disease. *Neurology, 44*, 1137–1143.

15. Myers, R.H., MacDonald, M.E., Koroshetz, W.J., Duyao, M.P., Ambrose, C.M., Taylor, S.A., Barnes, G., Srinidhi, J., Lin, C.S., Whaley, W.L., et al. (1993). De novo expansion of a $(CAG)_n$ repeat in sporadic Huntington's disease. *Nat Genet, 5*, 168–173.

16. Morley, J.F., Brignull, H.R., Weyers, J.J., Morimoto, R.I. (2002). The threshold for polyglutamine-expansion protein aggregation and cellular toxicity is dynamic and influenced by aging in *Caenorhabditis elegans. Proc Natl Acad Sci USA, 99*, 10417–10422.

17. Satyal, S.H., Schmidt, E., Kitagawa, K., Sondheimer, N., Lindquist, S., Kramer, J.M., Morimoto, R.I. (2000). Polyglutamine aggregates alter protein folding homeostasis in *Caenorhabditis elegans. Proc Natl Acad Sci USA, 97*, 5750–5755.

18. Brignull, H.R., Moore, F.E., Tang, S.J., Morimoto, R.I. (2006). Polyglutamine proteins at the pathogenic threshold display neuron-specific aggregation in a pan-neuronal *Caenorhabditis elegans* model. *J Neurosci, 26*, 7597–7606.

19. Preisinger, E., Jordan, B.M., Kazantsev, A., Housman, D. (1999). Evidence for a recruitment and sequestration mechanism in Huntington's disease. *Philos Trans R Soc London, B 354*, 1029–1034.

20. Luthi-Carter, R., Strand, A.D., Hanson, S.A., Kooperberg, C., Schilling, G., La Spada, A.R., Merry, D.E., Young, A.B., Ross, C.A., Borchelt, D.R., Olson, J.M. (2002). Polyglutamine and transcription: gene expression changes shared by DRPLA and Huntington's disease mouse models reveal context-independent effects. *Hum Mol Genet, 11*, 1927–1937.

21. Gidalevitz, T., Ben-Zvi, A., Ho, K.H., Brignull, H.R., Morimoto, R.I. (2006). Progressive disruption of cellular protein folding in models of polyglutamine diseases. *Science, 311*, 1471–1474.
22. Perutz, M.F. (1999). Glutamine repeats and neurodegenerative diseases: molecular aspects. *Trends Biochem Sci, 24*, 58–63.
23. Fraser, A.G., Kamath, R.S., Zipperlen, P., Martinez-Campos, M., Sohrmann, M., Ahringer, J. (2000). Functional genomic analysis of *C. elegans* chromosome I by systematic RNA interference. *Nature, 408*, 325–330.
24. Kamath, R.S., Martinez-Campos, M., Zipperlen, P., Fraser, A.G., Ahringer, J. (2001). Effectiveness of specific RNA-mediated interference through ingested double-stranded RNA in *Caenorhabditis elegans. Genome Biol, 2*, RESEARCH0002.
25. Nollen, E.A., Garcia, S.M., van Haaften, G., Kim, S., Chavez, A., Morimoto, R.I., Plasterk, R.H. (2004). Genome-wide RNA interference screen identifies previously undescribed regulators of polyglutamine aggregation. *Proc Natl Acad Sci USA, 101*, 6403–6408.
26. Kitamura, A., Kubota, H., Pack, C.G., Matsumoto, G., Hirayama, S., Takahashi, Y., Kimura, H., Kinjo, M., Morimoto, R.I., Nagata, K. (2006). Cytosolic chaperonin prevents polyglutamine toxicity with altering the aggregation state. *Nat Cell Biol, 8*, 1163–1170.
27. Behrends, C., Langer, C.A., Boteva, R., Bottcher, U.M., Stemp, M.J., Schaffar, G., Rao, B.V., Giese, A., Kretzschmar, H., Siegers, K., Hartl, F.U. (2006). Chaperonin TRiC promotes the assembly of polyQ expansion proteins into nontoxic oligomers. *Mol Cell, 23*, 887–897.
28. Tam, S., Geller, R., Spiess, C., Frydman, J. (2006). The chaperonin TRiC controls polyglutamine aggregation and toxicity through subunit-specific interactions. *Nat Cell Biol, 8*, 1155–1162.
29. Kraemer, B.C., Burgess, J.K., Chen, J.H., Thomas, J.H., Schellenberg, G.D. (2006). Molecular pathways that influence human tau-induced pathology in *Caenorhabditis elegans. Hum Mol Genet, 15*, 1483–1496.
30. Lee, V.M., Goedert, M., Trojanowski, J.Q. (2001). Neurodegenerative tauopathies. *Annu Rev Neurosci, 24*, 1121–1159.
31. Willingham, S., Outeiro, T.F., DeVit, M.J., Lindquist, S.L., Muchowski, P.J. (2003). Yeast genes that enhance the toxicity of a mutant huntingtin fragment or alpha-synuclein. *Science, 302*, 1769–1772.
32. Lamitina, T., Huang, C.G., Strange, K. (2006). Genome-wide RNAi screening identifies protein damage as a regulator of osmoprotective gene expression. *Proc Natl Acad Sci USA, 103*, 12173–12178.
33. Faber, P.W., Voisine, C., King, D.C., Bates, E.A., Hart, A.C. (2002). Glutamine/proline-rich PQE-1 proteins protect *Caenorhabditis elegans* neurons from huntingtin polyglutamine neurotoxicity. *Proc Natl Acad Sci USA, 99*, 17131–17136.
34. Garcia, S.M., Casanueva, M.O., Silva, M.C., Amaral, M.D., Morimoto, R.I. (2007). Neuronal signaling modulates protein homeostasis in *Caenorhabditis elegans* post-synaptic muscle cells. *Genes Dev, 21*, 3006–3016.
35. Bargmann, C.I. (1998). Neurobiology of the *Caenorhabditis elegans* genome. *Science, 282*, 2028–2033.

36. Geary, T.G., Thompson, D.P. (2001). *Caenorhabditis elegans*: how good a model for veterinary parasites? *Vet Parasitol, 101*, 371–386.

37. Rand, J.B., Duerr, J.S., Frisby, D.L. (1998). Neurogenetics of synaptic transmission in *Caenorhabditis elegans*. *Adv Pharmacol, 42*, 940–944.

38. Wu, Y., Wu, Z., Butko, P., Christen, Y., Lambert, M.P., Klein, W.L., Link, C.D., Luo, Y. (2006). Amyloid-beta-induced pathological behaviors are suppressed by *Ginkgo biloba* extract EGb 761 and ginkgolides in transgenic *Caenorhabditis elegans*. *J Neurosci, 26*, 13102–13113.

39. Christen, Y., Maixent, J.M. (2002). What is *Ginkgo biloba* extract EGb 761? An overview: from molecular biology to clinical medicine. *Cell Mol Biol (Noisy-Le-Grand), 48*, 601–611.

40. Parker, J.A., Arango, M., Abderrahmane, S., Lambert, E., Tourette, C., Catoire, H., Neri, C. (2005). Resveratrol rescues mutant polyglutamine cytotoxicity in nematode and mammalian neurons. *Nat Genet, 37*, 349–350.

41. Braungart, E., Gerlach, M., Riederer, P., Baumeister, R., Hoener, M.C. (2004). *Caenorhabditis elegans* MPP^+ model of Parkinson's disease for high-throughput drug screenings. *Neurodegener Dis, 1*, 175–183.

42. Westerheide, S.D., Kawahara, T.L., Orton, K., Morimoto, R.I. (2006). Triptolide, an inhibitor of the human heat shock response that enhances stress-induced cell death. *J Biol Chem, 281*, 9616–9622.

43. Varma, H., Voisine, C., DeMarco, C.T., Cattaneo, E., Lo, D.C., Hart, A.C., Stockwell, B.R. (2007). Selective inhibitors of death in mutant huntingtin cells. *Nat Chem Biol, 3*, 99–100.

44. Voisine, C., Varma, H., Walker, N., Bates, E.A., Stockwell, B.R., Hart, A.C. (2007). Identification of potential therapeutic drugs for Huntington's disease using *Caenorhabditis elegans*. *PLoS ONE, 2*, e504.

45. Anson, R.M., Hansford, R.G. (2004). Mitochondrial influence on aging rate in *Caenorhabditis elegans*. *Aging Cell, 3*, 29–34.

46. Dorman, J.B., Albinder, B., Shroyer, T., Kenyon, C. (1995). The *age-1* and *daf-2* genes function in a common pathway to control the lifespan of *Caenorhabditis elegans*. *Genetics, 141*, 1399–1406.

47. Kenyon, C., Chang, J., Gensch, E., Rudner, A., Tabtiang, R. (1993). A *C. elegans* mutant that lives twice as long as wild type. *Nature, 366*, 461–464.

48. Kimura, K.D., Tissenbaum, H.A., Liu, Y., Ruvkun, G. (1997). *daf-2*, an insulin receptor-like gene that regulates longevity and diapause in *Caenorhabditis elegans*. *Science, 277*, 942–946.

49. Morris, J.Z., Tissenbaum, H.A., Ruvkun, G. (1996). A phosphatidylinositol-3-OH kinase family member regulating longevity and diapause in *Caenorhabditis elegans*. *Nature, 382*, 536–539.

50. Ogg, S., Paradis, S., Gottlieb, S., Patterson, G.I., Lee, L., Tissenbaum, H.A., Ruvkun, G. (1997). The Fork head transcription factor DAF-16 transduces insulin-like metabolic and longevity signals in *C. elegans*. *Nature, 389*, 994–999.

51. Walker, G., Houthoofd, K., Vanfleteren, J.R., Gems, D. (2005). Dietary restriction in *C. elegans*: from rate-of-living effects to nutrient sensing pathways. *Mech Ageing Dev, 126*, 929–937.

52. Hsu, A.L., Murphy, C.T., Kenyon, C. (2003). Regulation of aging and age-related disease by DAF-16 and heat-shock factor. *Science, 300*, 1142–1145.

53. Morley, J.F., Morimoto, R.I. (2004). Regulation of longevity in *Caenorhabditis elegans* by heat shock factor and molecular chaperones. *Mol Biol Cell, 15*, 657–664.

54. Dillin, A., Crawford, D.K., Kenyon, C. (2002). Timing requirements for insulin/IGF-1 signaling in *C. elegans*. *Science, 298*, 830–834.

55. Cohen, E., Bieschke, J., Perciavalle, R.M., Kelly, J.W., Dillin, A. (2006). Opposing activities protect against age-onset proteotoxicity. *Science, 313*, 1604–1610.

56. Nollen, E.A., Morimoto, R.I. (2002). Chaperoning signaling pathways: molecular chaperones as stress-sensing 'heat shock' proteins. *J Cell Sci, 115*, 2809–2816.

57. Westerheide, S.D., Bosman, J.D., Mbadugha, B.N., Kawahara, T.L., Matsumoto, G., Kim, S., Gu, W., Devlin, J.P., Silverman, R.B., Morimoto, R.I. (2004). Celastrols as inducers of the heat shock response and cytoprotection. *J Biol Chem, 279*, 56053–56060.

58. Morimoto, R.I. (2008). Proteotoxic stress and inducible chaperone networks in neurodegenerative disease and aging. *Genes Dev, 22*, 1427–1438.

59. Link, C.D., Taft, A., Kapulkin, V., Duke, K., Kim, S., Fei, Q., Wood, D.E., Sahagan, B.G. (2003). Gene expression analysis in a transgenic *Caenorhabditis elegans* Alzheimer's diseae model. *Neurobiol Aging*, 24, 397–413.

60. Kuwahara, T., Koyama, A., Gengyo-Ando, K., Masuda, M., Kowa, H., Tsunoda, M., Mitani, S., Iwatsubo, T. (2006). Familial Parkinson mutant alpha-synuclein causes dopamine neuron dysfunction in transgenic *Caenorhabditis elegans*. *J Biol Chem, 281*, 334–340.

61. van Ham, T.J., Thijssen, K.L., Breitling, R., Hofstra, R.M., Plasterk, R.H., Nollen, E.A., (2008). *C. elegans* model identifies genetic modifiers of alpha-synuclein inclusion formation during aging. *PLoS Genet, 4*(3):e1000027.

62. Gidalevitz, T., Krupinski, T., Garcia, S.M., Morimoto, R.I. (2009). Destabilizing protein polymorphisms in the genetic background direct phenotypic expression of mutant SOD1 toxicity. *PLoS Genet, 5*(3):e1000399.

63. Wang J., Farr, G.W., Hall, D.H., Li, F., Furtak, K., Dreier, L., Horwich A.L. (2009). An ALS-linked mutant SOD1 produces a locomotor defect associated with aggregation and synaptic dysfunction when expressed in neurons of *Caenorhabditis elegans*. *PLoS Genet, 5*(1):e1000350.

64. Prahlad, V., Cornelius, T., Morimoto, R.I. (2008). Regulation of the cellular heat shock response in *Caenorhabditis elegans* by thermosensory neurons. *Science, 320*(5877), 811–814.

10

USING *DROSOPHILA* TO REVEAL INSIGHT INTO PROTEIN MISFOLDING DISEASES

Julide Bilen

Howard Hughes Medical Institute, Janelia Farm Research Campus, Ashburn, Virginia

Nancy M. Bonini

Department of Biology, Howard Hughes Medical Institute, University of Pennsylvania, Philadelphia, Pennsylvania

DROSOPHILA AS A MODEL SYSTEM FOR HUMAN NEURODEGENERATIVE DISEASE

Sporadic or genetically inherited human neurodegenerative diseases have devastating consequences for patients, family, and society; however, few treatments and cures are available. Identification of human genes mutated in familial disease gives us a handle to understand the biological processes that go awry in the disease situation. Pedigree analysis of families with disease can reveal two modes of genetic inheritance: dominant or recessive, indicating if the mutation leads to new function(s) or hyperactivation of normal function, or the lack of activity of gene, respectively. Based on the mode of inheritance, one can use different approaches to model the disease in animal models: targeted expression of the mutant human disease genes or knockout of the gene, followed in both cases by examining relevant effects.

Animal models, particularly *Drosophila*, have many advantages for the study of cellular and molecular pathology of human diseases. Comparison of the

Protein Misfolding Diseases: Current and Emerging Principles and Therapies,
Edited by Marina Ramirez-Alvarado, Jeffery W. Kelly, and Christopher M. Dobson
Copyright © 2010 John Wiley & Sons, Inc.

genome sequence reveals a high degree of conservation between fly and human in fundamental pathways, such that about 60% of human disease genes have obvious fly orthologs, providing the foundation to reveal basic mechanisms of human disease in the fly [1]. Almost a hundred years of intensive study on *Drosophila* provides a broad knowledge of fly anatomy, development, and signaling pathways, as well as a wide variety of genetic methods to perturb gene activity or target gene expression in tissue-specific ways [2]. The relatively fast time frame and large number of progeny animals makes it plausible to access and quantify effects rapidly and to perform large-scale unbiased forward genetic screens using either collections of mutant lines, generation of de novo mutations, or lines generated in other ways, such as with RNAi technology [3]. Here we focus on studies of protein misfolding diseases in the fly which lead to progressive neuronal dysfunction and degeneration in the brain, and highlight some of the ways that these have been used to reveal insight and the foundation for therapeutics.

FEATURES OF PROTEIN MISFOLDING DISEASES

Protein misfolding diseases include a variety of human neurodegenerative diseases, such as Alzheimer, Parkinson, and prion diseases, amyotrophic lateral sclerosis, trinucleotide expansion diseases type I (polyQ diseases such as Huntington disease and many spinocerebellar ataxias) and type II noncoding trinucleotide expansion diseases such as fragile X syndrome (for reviews, see [4–10]). These diseases lead to age-dependent progressive neurodegeneration that can be associated with movement disorders, ataxia, and cognitive dysfunction.

The diseases share common features of protein accumulation, typically into ubiquitinated aggregates. Depending on the disease, the localization of the protein accumulations can vary: extracellular plaques and neurofibrillary tangles characterize Alzheimer disease, cytoplasmic Lewy bodies typify Parkinson disease, and nuclear inclusions are found in polyQ diseases. In most cases, the disease protein per se aggregates into protein inclusions, but even in RNA-based type II trinucleotide expansion diseases, abnormal ubiquitinated protein accumulations can be seen [11], suggesting that protein homeostasis in neurons is perturbed in the disease situations.

Evidence based on crystallographic and other studies on proteins with expanded polyQ domains indicates a conformational change in protein structure associated with increased beta structure [12]. Recent studies with antibodies that recognize abnormal oligomeric protein also support conformational changes in disease proteins [13–15]. In some cases, simply an increased level of expression of the wild-type disease gene (APP, tau, α-synuclein, non-polyQ expanded SCA1) triggers neuronal toxicity in animal models [16–19] as in the disease situations (e.g., the increased gene copy number of α-synuclein associated with Parkinson disease [20–22]). This indicates that excessive normal activity of the proteins contributes to neuronal toxicity or leads to abnormal conformational changes and subsequent deleterious results. The protein aggregates are typically ubiquitinated and co-localize with select chaperones and proteasome subunits, indicating a cellular response to re-fold misfolded disease protein, or solubilize and target the protein for degradation [23–25]. FRET studies also indicate dynamic interactions of chaperones with such pathogenic protein aggregates [26].

MODELING PROTEIN MISFOLDING DISEASES IN THE FLY

Various Protein misfolding disease models have been generated in *Drosophila*, including for Huntington disease, spinocerebellar ataxias (SCA1, SCA3), noncoding repeat expansion diseases, Parkinson disease, and Alzheimer disease [16,17,27–31]. Studies using these fly models have revealed important new insights into cellular and molecular mechanisms of disease pathology.

The Critical Importance of Protein Context

At least nine human neurodegenerative diseases are due to CAG repeat expansions within the open reading frame of the respective genes, encoding an expanded-length polyQ domain within each disease protein that leads to dysfunction and degeneration in subsets of neurons in the brain. This common mechanism of mutation in these so-called polyQ diseases raises the hypothesis that an expanded polyQ domain alone can cause neuronal pathology, as reflected in the various disease situations. Indeed, ectopic expression of a CAG repeat encoding glutamine within an unrelated protein (hypoxanthine phosphotransferase) causes late-onset neurodegeneration in the mouse in a manner remarkably reminiscent of polyQ disease [32]. This suggests that expanded polyQ protein is fundamentally deleterious to neurons.

Evidence from fly models indicates that both polyQ and the non-polyQ content of the disease protein are critical for neuronal pathology. Directed expression of a pathogenic truncated spinocerebellar ataxia type 3 (SCA3) protein or the amino-terminal fragment of Huntington disease protein with a pathogenic-length CAG repeat leads to striking late-onset progressive neuronal degeneration (Fig.1A and B; [27,30]). Subsequent studies using a polyQ domain alone [33] or within an endogenous Dishevelled fly protein, which normally has a polyQ repeat [34], reveal that polyQ expansions are intrinsically toxic, but that protein content modulates degree of toxicity to neurons significantly.

Normal Function of the Host Disease Protein in Neuronal Degeneration

Expression of a C-terminal fragment of the *SCA3* gene with a pathogenic-length polyQ repeat causes striking late-onset degeneration in the fly (Fig. 1A and B; [27]). However, expression of full-length pathogenic SCA3 results in

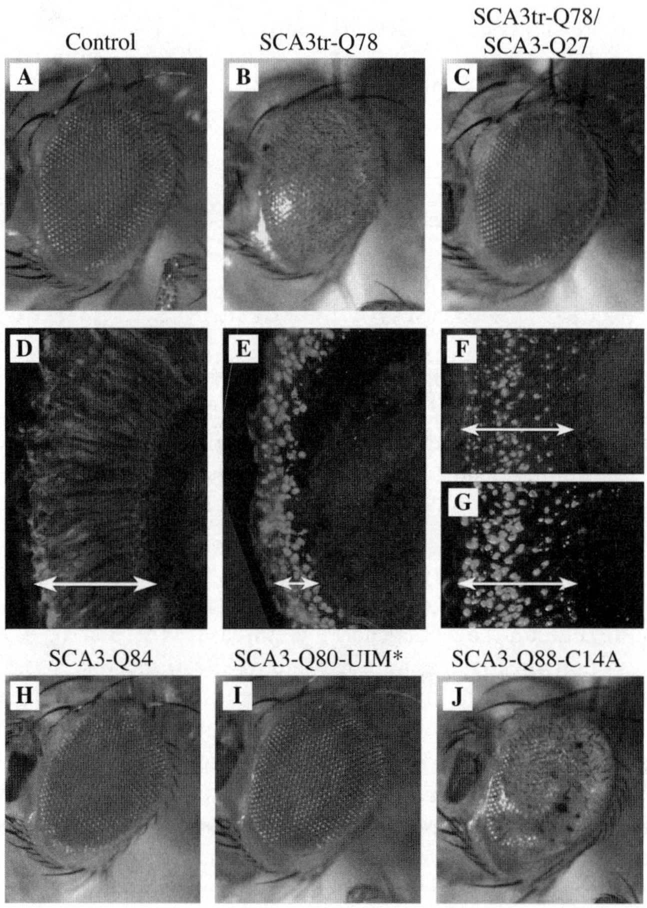

FIG. 1 Importance of protein context to disease protein toxicity. External eyes and immunostained cryosections of retinal tissue of flies expressing various combinations of pathogenic and normal ataxin-3. The white arrow in cryosections indicates the depth of the retina (as revealed by Hoechst staining and epon sectioning), which reflects the degree of retinal degeneration. (A) Normal eye and (D) eye section immunostained for a full-length ataxin-3 nonpathogenic SCA3-Q27, which is normally diffuse and highlights the normal depth of the retina in a nondegenerate situation. (B,E) The truncated pathogenic protein SCA3tr-Q78 causes severe retinal degeneration that is associated with severe collapse of the eye (white arrow) and inclusion formation by the pathogenic protein (E, anti-HA). Genotype *w; gmr-GAL4/ UAS-SCA3tr-Q78(S)*. (C,F) Coexpression of normal ataxin-3 SCA3-Q27 with the truncated pathogenic SCA3tr-Q78 shows rescue of the (C) external and (F) internal eye, with the SCA3-Q27 protein now localized to the NI formed by SCA3tr-Q78 ([F], anti-HA, SCA3tr-Q78. Normally, SCA3-Q27 is diffusely expressed within the eye. Genotype *w; gmr-GAL4 UAS-SCA3tr-Q78(S)/UAS-SCA3-Q27*.) (H–J) Eyes of flies expressing the normal pathogenic full-length ataxin-3 protein SCA3-Q84 (H), the pathogenic protein with ubiquitin interaction motifs UIMs mutated (I), and the pathogenic protein with a point mutation in the

later-onset adult-stage toxicity with greater specificity to neuronal cells than that of the truncated protein [35]. This suggests that the non-polyQ content of the protein confers some degree of tissue specificity. Strikingly, the normal ataxin-3 protein (encoded by the *SCA3* gene) suppresses neuronal toxicity mediated by the pathogenic disease protein (Fig. 1C). Studies reveal that ataxin-3 has ubiquitin protease activity and polyubiquitin binding motifs [36,38–39] in the N-terminal domain: domains that are missing from the truncated protein expressed in flies. Therefore, enhanced toxicity of the truncated protein may be the result of lack of these critical domains and associated functions of the normal protein. Indeed, point mutation of the ubiquitin-specific protease domain of ataxin-3 strikingly enhances neuronal degeneration, causing an effect that mirrors a truncation of the disease protein (Fig. 1H and J). This indicates that normal functions of ataxin-3 play a protective role in neuronal degeneration [35]. Along this line, evidence implicates N-terminal cleavage of the ataxin-3 protein in disease [40–42]; the fly experiments predict that such truncation would dramatically enhance neurotoxicity of the protein.

Epigenetic Modifications of Host Protein Modulate Neuronal Degeneration

Epigenetic protein modification of protein structure and function due to exposure to environmental changes and other signaling pathways may contribute to genetically inherited or age-associated sporadic neurodegenerative diseases. Loss of function mutations in *PARK7*, the gene encoding DJ-1, in humans causes early-onset parkinsonism [43]. Loss of function of DJ-1 in *Drosophila* increases sensitivity to environmental agents associated with oxidative stress, including paraquat and rotenone, inducing motor problems, suggesting that DJ-1 plays a protective role in response to oxidative stress [44–47]. Moreover, posttranslational modification of the conserved cysteine residues in DJ-1 may be critical for its protective role in the fly [44].

Hyperphosporylation of tau is implicated in neurodegenerative diseases, including Alzheimer disease [5]. The *Drosophila* homolog of human glycogen synthase kinase 3 (GSK-3), Shaggy, phosphorylates tau and increases tau toxicity in the fly, resulting in neurofibrillary tangles [48]. In addition, *Drosophila* PAR-1 kinase-mediated phosphorylation of tau precedes modulation by other downstream kinases, including GSK-3, and has been shown to be critically important for tau toxicity [49].

ubiquitin protease domain (J). The protease domain mutation generates a protein with dramatically enhanced toxicity, such that the toxicity of the full-length pathogenic protein now resembles that of strong expression of the N-terminally truncated protein lacking the N-terminal domain entirely [compare to (B)]. All flies in *trans* to *gmr-GAL4*. (Figures and legend adapted from [35, Figs. 1 and 5], with permission of Elsevier. Copyright © 2005.)

Phosphorylation of the host disease protein is also critical in other disease situations. Serine 776 in ataxin-1, the polyQ disease protein mutated in spinocerebellar ataxia type I (SCA1), is phosphorylated in the mouse [50]. Mutation of this serine residue to prevent phosphorylation reduces protein accumulation into inclusions, and reduces the neuronal toxicity of pathogenic ataxin-1 in Purkinje cells significantly, indicating that phosphorylation of ataxin-1 normally contributes to toxicity. Studies in a *Drosophila* model for SCA1 reveal further insights into the role of phosphorylation [51]. Co-precipitation experiments show that the regulatory protein 14-3-3, involved in binding diverse signaling proteins, binds specifically to serine 776 of ataxin-1. This interaction stabilizes the disease protein and leads to increased protein accumulation. Coexpression of ataxin-1 and 14-3-3 enhances ataxin-1-mediated toxicity in fly. In addition, interaction of ataxin-1 and 14-3-3 is regulated by Akt-1 phosphorylation, which also enhances degeneration when coexpressed with ataxin-1. Together, these results suggest that phosphorylation of disease proteins is one way to regulate stability and gradual accumulation of disease proteins, modulating protein homeostasis in neurons.

Specificity of Disease Proteins to Select Neuronal Populations

Many neurodegenerative diseases are strikingly specific to subsets of neurons in the brain, and selected *Drosophila* models for Protein misfolding diseases provide insight into cell-type specificity. The fly model of Parkinson disease using α-synuclein shows toxicity largely only to dopaminergic neurons [28], as do mutant forms of the parkinsonism protein parkin [52]. Widespread neuronal expression of wild-type and mutant tau shows degeneration in the cortex associated with vacuolization [17]. In Alzheimer disease, cholinergic neurons are particularly susceptible. Directed expression of tau to cholinergic neurons in the adult fly brain shows vacuolization and loss of cholinergic neurons, indicating specificity of disease protein toxicity [17]. A subsequent study compared the toxicity of polyglutamine disease proteins—ataxin-1 and ataxin-3—in different neuronal populations in the fly brain [53]. Directed expression of SCA3trQ78 causes degeneration in Kenyon cells and the cortex; however, no degeneration is seen with expression in the mushroom bodies or cholinergic neurons. On the other hand, expression of pathogenic SCA1 does not cause significant toxicity in Kenyon cells, but rather, cholinergic neurons are more vulnerable. Together, these results suggest that *Drosophila* models can recapitulate aspects of neuronal subpopulation specificity.

INSIGHT FROM MODIFIER SCREENS

Drosophila is a powerful genetic system in its ability to reveal cellular and molecular mechanisms of Protein misfolding diseases. Several genetic screens of disease models have been used to reveal various genes that modulate disease pathology, providing further understanding of mechanisms of protein toxicity.

Chaperones Suppress Neuronal Degeneration Mediated by Human Disease Proteins

The chaperone Hsp70 was the initial in vivo modifier shown to suppress SCA3-mediated neurodegenerative disease in the fly [54]. These studies extended from studies of HeLa cells and SCA1 patient postmortem tissue showing that Hsp70 and its co-chaperone Hsp40 co-localize to nuclear inclusions formed by pathogenic ataxin-1 protein [55]. Directed expression of human Hsp70 in the fly strikingly suppresses neuronal degeneration [54]. Subsequent studies revealed that chaperones, including Hsp70 and Hsp40, decrease ataxin-3 protein aggregates and increase protein solubility, presumably to buffer toxic conformations, refold, and/or help target the protein for degradation [35, 56–57]. These findings have been extended to the mouse, for example, to show that Hsp70 is efficacious to suppress neuronal degeneration induced by SCA1 in Purkinje cells [58].

Suppression by Hsp70 is not limited to polyQ protein-based diseases. Directed expression of Hsp70 rescues α-synuclein toxicity in *Drosophila* models of Parkinson disease, whereas reduced chaperone activity enhances toxicity [59]. Hsp70 co-localizes to the Lewy body like accumulations of α-synuclein in the fly as well as to postmortem tissue of patients with Parkinson disease and other synucleinopathies [59]. Most strikingly, Hsp70 polymorphisms are a risk factor for Parkinson disease [60].

The compounds arimoclomol, a co-inducer of heat-shock proteins, and celastrol, which induces heat-shock proteins in neurons, delay the motor neuron degeneration in SOD1 amyotrophic lateral sclerosis mouse models and increase life span [61,62]. Together, these findings emphasize broadly the critical importance of chaperone activity to human neurodegenerative disease.

MicroRNAs Modulate Neuronal Degeneration

MicroRNAs (miRNAs) are evolutionarity conserved 21- to 25-bp noncoding RNAs involved in the silencing of target mRNAs [63,64]. Precise or imprecise base complementarily between the 3′ untranslated region of the target mRNA and the miRNA results in cleavage of the mRNA or inhibition of translation, causing loss of gene function. MiRNAs have broad roles in various biological processes, including developmental timing, programmed cell death, stress, metabolism, and cancer [63,65–67].

The miRNA *bantam (ban)* was isolated as a powerful suppressor of SCA3-mediated neurodegeneration in a modifier genetic screen (Fig. 2A–C; [57]). Directed expression of *ban* rescues neuronal degeneration, whereas reduced activity enhances neuronal degeneration. Further analysis of suppression indicates that *ban* does not modulate disease protein accumulation into

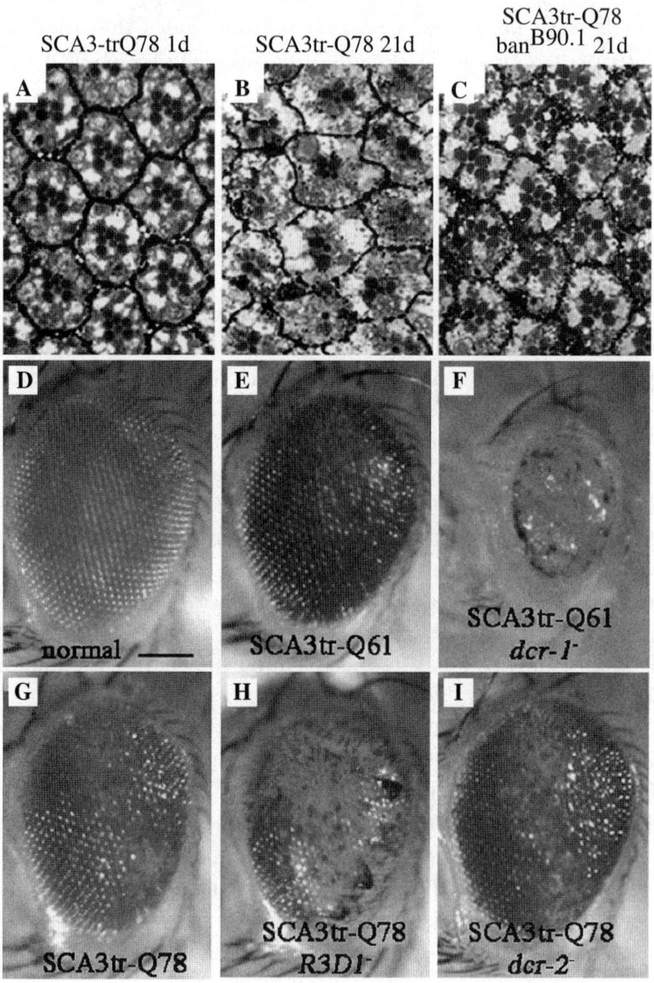

FIG. 2 Modulation of ataxin-3 pathogenicity by the miRNA *bantam* and the miRNA pathway. (A–C) Tangential sections of adult fly eyes. (A) Normally, the fly eye has a highly regular pattern of seven photoreceptor neurons in each unit eye, or ommatidium, seen here in 1-day-old flies expressing the pathogenic ataxin-3, which at this time has no effect, so that the eye is normal. (B) By 21 days, ataxin-3 has induced severe degeneration, such that the regular eye structure is quite disrupted, and each unit eye has fewer than seven photoreceptor neurons. (C) Up-regulation of the miRNA *ban* with the $ban^{B90.1}$ allele results in significantly reduced deterioration of the retinal structure by the pathogenic ataxin-3 at 21 days. Genotypes (A and B) *w; UAS-SCA3tr-Q78/+; rh1-gal4/+* and (C) *w; UAS-SCA3tr-Q78/+; rh1-gal4/ban*$^{B90.1}$. (D–I) Compromise of miRNA processing with *dcr-1* and *R3D1* mutation result in enhanced degeneration by the ataxin-3 protein. (D) A normal fly eye. (E) *SCA3tr-Q61* normally shows weak degeneration. Genotypes (D) *gmr-gal4/+*, (E) genotype: *ey-FLP; gmr-gal4 UAS-SCA3tr-Q61/+; FRT82B*. (F) *SCA3tr-Q61* degeneration in *dcr-1* is enhanced dramatically, eye genotype: *ey-FLP; gmr-gal4 UAS-SCA3tr-Q61/+; FRT82B dcr-1*^{Q1147X}.

INSTITUT MEDIATHEQUE PASTEUR

aggregates or the level of the protein; *ban* may play a role to block downstream events. This may include modulation of neuronal death pathways or increasing neuronal integrity in parallel pathways (Fig. 5B; [57,68]). *Ban* also plays a broader role in neuronal degeneration, as *ban* suppression is not specific polyQ toxicity but also suppresses tau toxicity in *Drosophila*.

miRNAs are processed into mature miRNAs by specific nuclear and cytoplasmic endonucleases [66]. To reveal a potentially broader role of miRNAs, the effect of the global loss of miRNAs was assessed for SCA3-associated neurodegeneration [57]. Strikingly, blocking miRNA processing by *dicer-1* or *R3D1/loquacious* mutation severely enhances neuronal degeneration induced by SCA3 (Fig. 2D–I) and tau. These findings extend to human cells: reduced *dicer* activity in HeLa cells strikingly enhances cell loss associated with pathogenic SCA3, while the control disease protein and *dicer* loss of function alone are normal (Fig. 3; [57]). Moreover, the small RNA fraction that includes miRNAs from HeLa cells rescues the cell loss, suggesting that select miRNAs, including potential *ban* homologs may play a role in neuronal integrity in disease situations. Along this line, subsequent studies further implicate miRNAs in neurodegeneration in the mouse, with loss of *dicer* activity in Purkinje cells causing progressive neurodegeneration and ataxia in mouse [69].

As noted above, higher-than-normal levels of expression of select disease proteins causes neuronal toxicity. In this vein, select disease proteins themselves are implicated as targets of miRNAs. For example, ataxin-1 is a target of miRNA *mir-101* [70,71]. Studies also show that atrophin, mutated in DRPLA polyQ disease, is fined-tuned by *miR-8*, with increased atrophin levels causing programmed cell death and impairing motor coordination in *Drosophila* [72].

miRNAs roles in neurodegeneration have recently been extended to Parkinson disease. *miR-133b* specifically expresses in midbrain dopaminergc neurons and is deficient in tissues from Parkinson's patients [73]. *Dicer* knockout in postmitotic midbrain dopaminergic neurons of the mouse causes progressive loss of dopaminergic neurons, suggesting multiple roles for miRNAs in Parkinson disease. Overexpression *miR-133b* increases maturation and survival of dopaminergic neurons [73,74].

(G–I) Expression of pathogenic ataxin-3 results in a mildly degenerate eye, reflected in disrupted pigmentation of the external eye. (G) SCA3tr-Q61 normally shows weak degeneration. (H) Reduction of miRNA processing results in dramatically enhanced degeneration, seen here as dramatically increased loss of external pigmentation. (I) Loss of *dcr-2*, which modulates siRNA production, has no effect on polyQ toxicity. Eye genotype: (G) *ey-FLP; gmr-gal4 UAS-SCA3tr-Q78/+; FRT82B*. (I) *FRT42D* $dcr\text{-}2^{L811fsX}$; *gmr-gal4 UAS-SCA3tr-Q78/ey-gal4 UAS-FLP*. Genotypes (H) *w; gmr-gal4 UAS-SCA3tr-Q78/+* and (H) *w*; $R3D1^{f00791}/R3D1^{f00791}$; *gmr-gal4 UAS-SCA3tr-Q78/+*. Bar is 100 μm for eyes in (D–I). [(A–C) Figures and legend adapted from [68, Fig. 1], with permission of Landes Bioscience. Copyright © 2006. (D–I), Figures and legend adapted from [57, Fig. 1], with permission of Elsevier. Copyright © 2006.]

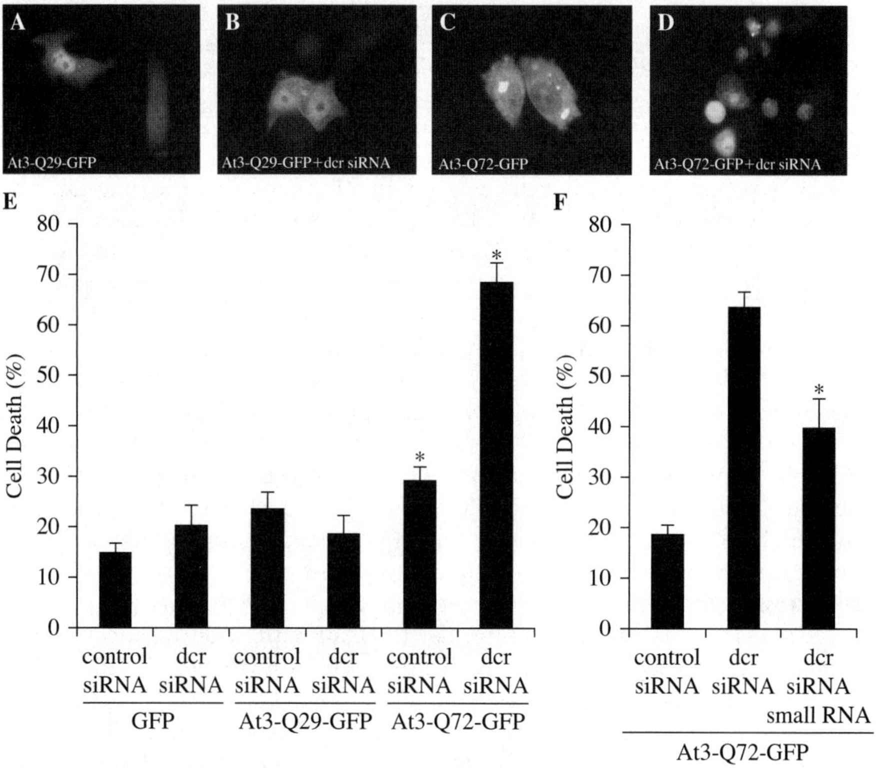

FIG. 3 Reducing miRNA processing enhances ataxin-3 toxicity in human cells. (A–D) Human HeLa cells expressing normal ataxin-3 (At3-Q29-GFP) or pathogenic ataxin-3 protein (At3-Q72-GFP) and treated with control siRNA or siRNA to *dicer*. (A,B) Cells expressing the control ataxin-3 protein have similar viability with or without *dicer* siRNA treatment. (C,D) Cells expressing pathogenic ataxin-3 protein normally (C) show little toxicity by 24 hours but (D) show dramatically enhanced death, reflected in condensed cells with altered morphology, upon *dicer* knockdown. (E) Cell death of GFP-positive cells after 24 hours detected by the uptake of propidium iodide; mean $\pm$ SEM [$n = 3$ independent experiments; $^{*}p < 0.001$ compared to GFP or At3-Q29-GFP treated with control or *dicer* siRNA; At3-Q72-GFP with *dicer* siRNA is also significantly different ($p < 0.001$) from At3-Q72-GFP with control siRNA]. (F) Increased cell death induced by pathogenic ataxin-3 protein (At3-Q72-GFP) with *dicer* knockdown is mitigated by transfection of the purified small RNA fraction back into HeLa cells ($^{*}p < 0.002$ compared to At3-Q72-GFP with *dicer* siRNA). [Figure and legend adapted from [57, Fig. 2], with permission of Elsevier. Copyright © 2006.]

Specific miRNAs may regulate various steps of neuronal degeneration. This includes neuronal specificity and maintenance as well as, for disease situations, protein homeostatic pathways, leading to changes in protein quality control. Consistent with this idea, *dicer* knockout in mouse in Purkinje cells reveals protein accumulations [69], implicating perturbation of protein homeostasis.

Globality of Protein misfolding Modifiers

Several genetic screens have been carried out for polyQ diseases and tau toxicity in *Drosophila* [16,33,75,76]. A screen against a raw polyQ protein using de novo autosomal P-insertion lines revealed that co-chaperone Hdj-1 and tetratricopeptide repeat protein Tpr2 are suppressors of toxicity [33]. This supports previous findings of the importance of chaperones in neuronal degeneration [54]. A modifier screen using an ataxin-1 fly model revealed several chaperones, components of proteasome pathways and RNA binding proteins, suggesting that chaperone and ubiquitin-dependent proteasome degradation is critical for neuronal survival in ataxin-1-mediated neurodegeneration [16]. The role(s) of RNA binding proteins in ataxin-1-induced neurodegeneration requires further investigation. These may be modulating a normal function of ataxin-1 in RNA processing [77] or may be revealing specific roles of these RNA-binding proteins in neurodegeneration.

A genome-wide screen of ataxin-3 modifiers emphasizes that genes modulating protein homeostasis are a major mechanism of protein toxicity modification [75]. Three classes of modifiers based on sequence analysis were revealed in the screen: chaperones, ubiquitin–proteasome components, and miscellaneous functions, including transcription and translational regulators. Strikingly, half of the modifiers belong to chaperones and ubiquitin pathways, raising the idea that proteins predicted to function in other biological processes may also have a role in protein quality control pathways. Directed expression of a dominant negative version of Hsp70 in the fly leads to an effect similar to polyQ toxicity [78]. To address whether the modifiers implicated in other biological processes can modulate general protein misfolding, the modifiers were tested for their ability to modulate dominant negative Hsp70. Surprisingly, all of the ataxin-3 modifiers (with the exception of the ubiquitin-specific protease Ubp64E) can also modulate this general Protein misfolding situation (Fig. 4). In some cases, rescue was better than just supplying back Hsp70. These findings provide strong evidence, along with findings in *Caenorhabditis elegans* [79], that global disruption of cellular protein folding is critical in polyQ toxicity situations.

Protein misfolding is also common to Alzheimer disease associated with tau. Phosphorylation and microtubule-binding activities have been key aspects focused on as critical to tau-associated toxicity, rather than protein misfolding. Genetic screens revealed kinases and phosphatases as the major modifiers of tau toxicity in *Drosophila* [76]. However, select ataxin-3 modifiers, including Tpr2 and polyubiquitin, suppress tau degeneration in the fly [75]; this indicates that these same pathways also modulate tau toxicity, suggesting that modulation of protein homeostasis will also be critical for Alzheimer disease situations. Indeed, vertebrate studies, as well as modifier screens of tau toxicity in *C. elegans*, reveal roles of protein misfolding in tau-associated toxic situations [80,81].

Ataxin-3 modifiers show that suppressors increase the solubility of the disease protein, whereas the enhancer decreases solubility [75]. Reduction of protein accumulations into inclusions can also correlate with reduced toxicity,

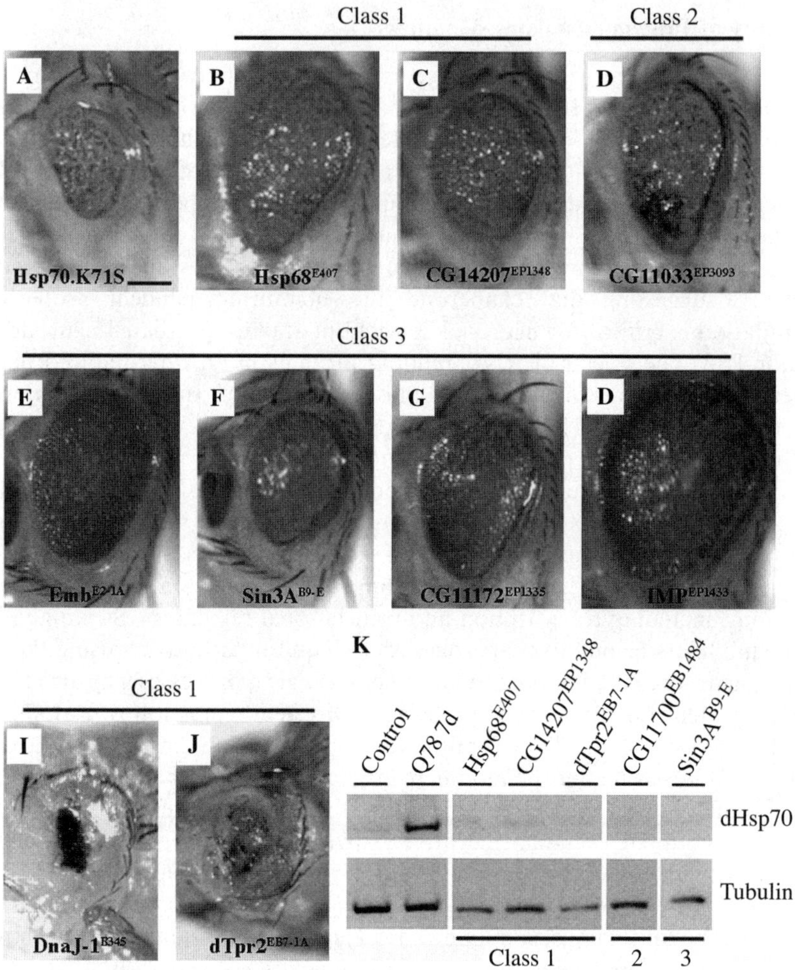

FIG. 4 Genetic modifiers of ataxin-3 toxicity modulate general protein misfolding. (A) Compromised endogenous Hsp70 via expression of a dominant negative Hsp70 transgene (*UAS-Hsp70.K71E*) leads to a severely degenerate eye situation. Genotype *w; gmr-GAL4 UAS-Hsp70.K71E/+*. (B–H) The up-regulation alleles of the chaperone modifiers (B) $Hsp68^{E407}$ and (C) $CG14207^{EP1348}$ partially rescue Hsp70.K71E. (D) An enhancer of ataxin-3 toxicity ($CG11033^{EP3093}$) also suppresses general misfolding, suggesting that ataxin-3 toxicity and misfolding do not have identical molecular mechanisms. Up-regulation modifiers (E) $emb^{E2\text{-}1A}$, (F) $Sin3A^{B9\text{-}E}$, (G) $NFAT^{EP1335}$, and (H) Imp^{EP1433} also suppress the Hsp70.K71E phenotype, suggesting a role of these modifiers in protein quality control. (I and J) Co-chaperones $DnaJ\text{-}1^{B345.2}$ and $Tpr2^{EB7\text{-}1A}$ are pupal lethal upon expression with Hsp70.K71E, causing a more severe misfolding phenotype. Bar in (A), 100 μm for (A–J). (K) Western blot indicates that the modifiers do not induce expression of Hsp70. Protein samples from the 1-day fly heads, with indicated genotypes, *gmr-GAL4* driver. Positive control, 7-day flies expressing the pathogenic truncated ataxin-3 protein-encoding transgene *UAS-SCA3trQ78*. (Figure and legend adapted from [75, Fig. 2], doi:10.1371/journal.pgen.0030177.g002.)

implicating roles of protein degradation pathways in disease pathology [75,82]. However, there are a few modifiers (i.e., miRNA ban, Tpr2, and IMP) that suppress neuronal degeneration without affecting protein accumulation, suggesting that these may act to block neuronal cell death or prevent the deleterious interactions of oligomeric or monomeric species of the disease protein [57,75].

Further analysis of ataxin-3 modifiers reveals that select modifiers are strikingly sensitive to proteasome activity [75]. This is consistent with previous evidence that subunits of the proteasome co-localize to protein inclusions in cell culture studies and patient tissue, and with the identification of components of the ubiquitin–proteasome system in genetic screens [16,55,75,80]. However, suppression by other modifiers is not affected by the proteasome, suggesting that these modifiers may use other types of protein-degradation pathways. Reducing autophagy using RNA interference lines specific for key autophagy genes *atg5* and *atg7* enhances ataxin-3-mediated neuronal degeneration [75]. Moreover, reduction of autophagy leads to increased cytoplasmic aggregates and reduced protein solubility. It is also intriguing that ataxin-3 protein itself may modulate autophagy [75], indicating yet another role for the disease protein. Further studies in *Drosophila* show that histone deacetylase 6 (HDAC6) links the ubiquitin proteasome and autophagy pathways in the regulation of toxicity of the spinobulbar muscular atrophy protein, another polyQ disease [83]. Impairment of proteasome activity is compensated for by induction of autophagy in an HDAC6-dependent manner. These findings underscore the central importance of protein quality control pathways to neuronal integrity in Protein misfolding disease situations, which, together with other pathways, like those of miRNAs, are critical to the disease situations (Fig. 5).

FROM GENES TO THE FOUNDATION FOR THERAPEUTIC COMPOUNDS

Drosophila models are useful systems to test the effects of therapeutic compounds. Previous studies show that the Huntington disease polyQ protein interacts directly with the acetyltransferase domain of CREB-binding protein (CBP) and p300/CBP associated factors, to inhibit their acetyltransferase activity [84]. Normalizing histone acetylase activity simply by feeding flies histone deacetylase (HDAC) inhibitors such as SAHA and sodium butyrate slows neuronal degeneration [84]. HDAC inhibitors increase the acetylation level of chromatin. However, directed expression of CBP may also reduce polyQ inclusions [85], suggesting that CBP and HDAC inhibitors may also affect the protein quality control pathways in neurodegenerative disease situations. The specific mTOR inhibitor rapamycin rescues polyQ toxicity in cell-culture situations and in fly and mouse models of Huntington disease through modulating autophagy [86]. Taken together, these studies indicate that modulation of multiple pathways by compounds may be effective.

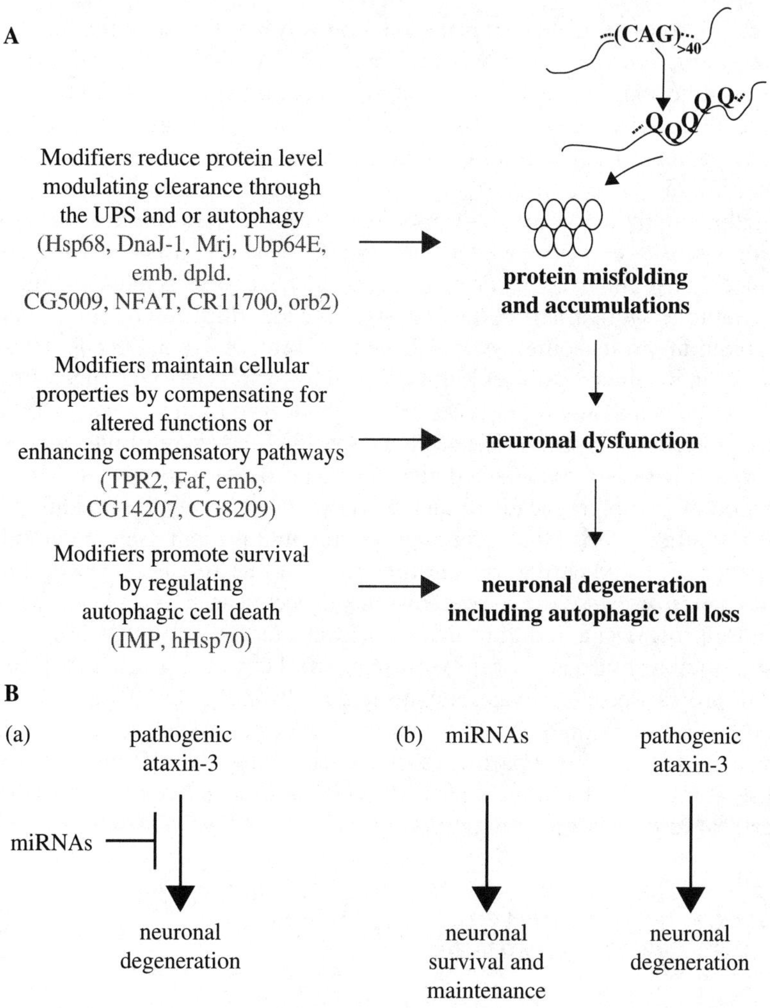

FIG. 5 Overview of biological roles of modifiers of ataxin-3-associated neurodegeneration. (A) Modifiers of pathogenic ataxin-3 toxicity may (1) reduce disease protein accumulation into inclusions in a manner sensitive to proteasome activity and/or by modulating autophagy, (2) promote cellular functionality in situations of misfolded protein, and/or (3) promote neuronal survival by regulating autophagic cell loss. (B) Possible models for how the miRNA pathway influences ataxin-3-induced degeneration. (a) miRNAs may modulate mRNA target genes that affect the pathogenicity of the ataxin-3 protein, thereby mitigating neurodegeneration. One such miRNA in *Drosophila* is *ban*, which may modulate neuronal loss. (b) miRNAs, including *ban* in *Drosophila*, may modulate mRNA target genes that influence neuronal survival or maintenance. Dysregulation of miRNA targets may promote neuronal degeneration, including enhancing degeneration induced by pathogenic ataxin-3. [(A) Figure and legend adapted from [75, Fig. 6], doi:10.1371/journal.pgen.0030177.g006. (B) Figure and legend adapted from [68, Fig. 1], with permission of Landes Bioscience. Copyright © 2006.]

As noted above, genetic studies revealed that Hsp70 suppresses α-synuclein-mediated neuronal toxicity in *Drosophila* [59]. Pharmacological enhancement of chaperone activity by geldanamycin in the fly can also rescue toxicity to dopaminergic neurons [87]. Subsequent studies showed that geldanamycin increases the stress response in the disease situation, and functions through the heat-shock factor to effect rescue [88].

Glycogen synthase kinase-3 (GSK-3) negatively regulates heat-shock factor activity, which induces expression of heat-shock proteins. Lithium, an inhibitor of GSK-3, can therefore enhance the heat-shock response by reducing negative regulation of the heat-shock factor [89]. Lithium treatment reduces nuclear inclusion formation and protects against polyQ toxicity in cell culture [89]. Subsequent studies reveal that lithium suppresses both neuronal toxicity induced by the huntingtin protein and polyalanine proteins in *Drosophila* [90]. These studies have been extended to mouse models to show that lithium treatment improves neuronal functioning and dendritic morphological loss of SCA1 mouse models [91].

SUMMARY AND THE FUTURE

We have noted a variety of Protein misfolding disease situations and touched on approaches to using these models to reveal insight into disease situations. Both gain-of-function and loss-of-function human disease situations can be modeled and studied in the fly; and a variety of approaches from genetic screens to use of compounds can be applied to reveal new insight. These studies, which can be performed more rapidly and possibly more thoroughly in the fly, can provide the foundation to test successful approaches in vertebrate models, ultimately providing the foundation for therapeutics.

Study of the fly is not limited to such Protein misfolding diseases; rather, these studies have paved the way to apply the fly in a variety of ways that parallel the human disease for the identification of critical genes (e.g., metastasis of cells in cancerlike invasive situations [92]), but also genes involved in such behaviors as sleep, aggression, and goal-directed movement [93–96]. Indeed, the power of the fly to reveal insight into problems that relate to human biology seems limited only by the creativity of scientists to apply the animal to such situations. In such a manner, the fly provides a cornerstone model of central importance to reveal biological insight into both the normal condition and disease situations.

REFERENCES

1. Rubin, G.M., Yandell, M.D., Wortman, J.R., Gabor Miklos, G.L., Nelson, C.R., Hariharan, I.K., Fortini, M.E., Li, P.W., Apweiler, R., Fleischmann, W., et al. (2000). Comparative genomics of the eukaryotes. *Science, 287*, 2204–2215.
2. Bier, E. (2005). *Drosophila*, the golden bug, emerges as a tool for human genetics. *Nat Rev Genet, 6*, 9–23.

3. Dietzl, G., Chen, D., Schnorrer, F., Su, K.C., Barinova, Y., Fellner, M., Gasser, B., Kinsey, K., Oppel, S., Scheiblauer, S., et al. (2007). A genome-wide transgenic RNAi library for conditional gene inactivation in *Drosophila. Nature, 448*, 151–156.
4. Gatchel, J.R., Zoghbi, H.Y. (2005). Diseases of unstable repeat expansion: mechanisms and common principles. *Nat Rev Genet, 6*, 743–755.
5. Lee, V.M., Goedert, M., Trojanowski, J.Q. (2001). Neurodegenerative tauopathies. *Annu Rev Neurosci, 24*, 1121–1159.
6. Aguzzi, A., Heikenwalder, M. (2006). Pathogenesis of prion diseases: current status and future outlook. *Nat Rev Microbiol, 4*, 765–775.
7. Selkoe, D.J. (2004). Cell biology of protein misfolding: the examples of Alzheimer's and Parkinson's diseases. *Nat Cell Biol, 6*, 1054–1061.
8. Taylor, J.P., Hardy, J., Fischbeck, K.H. (2002). Toxic proteins in neurodegenerative disease. *Science, 296*, 1991–1995.
9. Soong, B.W., Paulson, H.L. (2007). Spinocerebellar ataxias: an update. *Curr Opin Neurol, 20*, 438–446.
10. Ranum, L.P., Day, J.W. (2004). Pathogenic RNA repeats: an expanding role in genetic disease. *Trends Genet, 20*, 506–512.
11. Greco, C.M., Hagerman, R.J., Tassone, F., Chudley, A.E., Del Bigio, M.R., Jacquemont, S., Leehey, M., Hagerman, P.J. (2002). Neuronal intranuclear inclusions in a new cerebellar tremor/ataxia syndrome among fragile X carriers. *Brain, 125*, 1760–1771.
12. Perutz, M.F. (1999). Glutamine repeats and neurodegenerative diseases: molecular aspects. *Trends Biochem Sci, 24*, 58–63.
13. Brooks, E., Arrasate, M., Cheung, K., Finkbeiner, S.M. (2004). Using antibodies to analyze polyglutamine stretches. *Methods Mol Biol, 277*, 103–128.
14. Kayed, R., Head, E., Thompson, J.L., McIntire, T.M., Milton, S.C., Cotman, C.W., Glabe, C.G. (2003). Common structure of soluble amyloid oligomers implies common mechanism of pathogenesis. *Science, 300*, 486–489.
15. O'Nuallain, B., Wetzel, R. (2002). Conformational Abs recognizing a generic amyloid fibril epitope. *Proc Natl Acad Sci USA, 99*, 1485–1490.
16. Fernandez-Funez, P., Nino-Rosales, M.L., de Gouyon, B., She, W.C., Luchak, J.M., Martinez, P., Turiegano, E., Benito, J., Capovilla, M., Skinner, P.J., et al. (2000). Identification of genes that modify ataxin-1-induced neurodegeneration. *Nature, 408*, 101–106.
17. Wittmann, C.W., Wszolek, M.F., Shulman, J.M., Salvaterra, P.M., Lewis, J., Hutton, M., Feany, M.B. (2001). Tauopathy in *Drosophila*: neurodegeneration without neurofibrillary tangles. *Science, 293*, 711–714.
18. Dawson, T., Mandir, A., Lee, M. (2002). Animal models of PD: pieces of the same puzzle? *Neuron, 35*, 219–222.
19. Rovelet-Lecrux, A., Hannequin, D., Raux, G., Le Meur, N., Laquerriere, A., Vital, A., Dumanchin, C., Feuillette, S., Brice, A., Vercelletto, M., et al. (2006). APP locus duplication causes autosomal dominant early-onset Alzheimer disease with cerebral amyloid angiopathy. *Nat Genet, 38*, 24–26.
20. Singleton, A.B., Farrer, M., Johnson, J., Singleton, A., Hague, S., Kachergus, J., Hulihan, M., Peuralinna, T., Dutra, A., Nussbaum, R., et al. (2003). Alpha-synuclein locus triplication causes Parkinson's disease. *Science, 302*, 841.

21. Chartier-Harlin, M.C., Kachergus, J., Roumier, C., Mouroux, V., Douay, X., Lincoln, S., Levecque, C., Larvor, L., Andrieux, J., Hulihan, M., et al. (2004). Alpha-synuclein locus duplication as a cause of familial Parkinson's disease. *Lancet, 364*, 1167–1169.

22. Ibanez, P., Bonnet, A.M., Debarges, B., Lohmann, E., Tison, F., Pollak, P., Agid, Y., Durr, A., Brice, A. (2004). Causal relation between alpha-synuclein gene duplication and familial Parkinson's disease. *Lancet, 364*, 1169–1171.

23. Rubinsztein, D.C. (2006). The roles of intracellular protein-degradation pathways in neurodegeneration. *Nature, 443*, 780–786.

24. Ross, C.A., Pickart, C.M. (2004). The ubiquitin–proteasome pathway in Parkinson's disease and other neurodegenerative diseases. *Trends Cell Biol, 14*, 703–711.

25. Muchowski, P.J., Wacker, J.L. (2005). Modulation of neurodegeneration by molecular chaperones. *Nat Rev Neurosci, 6*, 11–22.

26. Satyal, S.H., Schmidt, E., Kitagawa, K., Sondheimer, N., Lindquist, S., Kramer, J.M., Morimoto, R.I. (2000). Polyglutamine aggregates alter protein folding homeostasis in *Caenorhabditis elegans. Proc Natl Acad Sci USA, 97*, 5750–5755.

27. Warrick, J.M., Paulson, H.L., Gray-Board, G.L., Bui, Q.T., Fischbeck, K.H., Pittman, R.N., Bonini, N.M. (1998). Expanded polyglutamine protein forms nuclear inclusions and causes neural degeneration in *Drosophila. Cell, 93*, 939–949.

28. Feany, M.B., Bender, W.W. (2000). A *Drosophila* model of Parkinson's disease. *Nature, 404*, 394–398.

29. Jin, P., Zarnescu, D.C., Zhang, F., Pearson, C.E., Lucchesi, J.C., Moses, K., Warren, S.T. (2003). RNA-mediated neurodegeneration caused by the fragile X premutation rCGG repeats in *Drosophila. Neuron, 39*, 739–747.

30. Jackson, G.R., Salecker, I., Dong, X., Yao, X., Arnheim, N., Faber, P.W., MacDonald, M.E., Zipursky, S.L. (1998). Polyglutamine-expanded human huntingtin transgenes induce degeneration of *Drosophila* photoreceptor neurons. *Neuron, 21*, 633–642.

31. Mutsuddi, M., Marshall, C.M., Benzow, K.A., Koob, M.D., Rebay, I. (2004). The spinocerebellar ataxia 8 noncoding RNA causes neurodegeneration and associates with staufen in *Drosophila. Curr Biol, 14*, 302–308.

32. Ordway, J.M., Tallaksen-Greene, S., Gutekunst, C.A., Bernstein, E.M., Cearley, J.A., Wiener, H.W., Dure, L.St., Lindsey, R., Hersch, S.M., Jope, R.S., et al. (1997). Ectopically expressed CAG repeats cause intranuclear inclusions and a progressive late onset neurological phenotype in the mouse. *Cell, 91*, 753–763.

33. Kazemi-Esfarjani, P., Benzer, S. (2000). Genetic suppression of polyglutamine toxicity in *Drosophila. Science, 287*, 1837–1840.

34. Marsh, J.L., Thompson, L.M. (2006). *Drosophila* in the study of neurodegenerative disease. *Neuron, 52*, 169–178.

35. Warrick, J.M., Morabito, L.M., Bilen, J., Gordesky-Gold, B., Faust, L.Z., Paulson, H.L., Bonini, N.M. (2005). Ataxin-3 suppresses polyglutamine neurodegeneration in *Drosophila* by a ubiquitin-associated mechanism. *Mol Cell, 18*, 37–48.

36. Burnett, B., Li, F., Pittman, R.N. (2003). The polyglutamine neurodegenerative protein ataxin-3 binds polyubiquitylated proteins and has ubiquitin protease activity. *Hum Mol Genet, 12*, 3195–3205.

37. Chai, Y., Berke, S.S., Cohen, R.E., Paulson, H.L. (2004). Poly-ubiquitin binding by the polyglutamine disease protein ataxin-3 links its normal function to protein surveillance pathways. *J Biol Chem, 279*, 3605–3611.
38. Donaldson, K.M., Li, W., Ching, K.A., Batalov, S., Tsai, C.C., Joazeiro, C.A. (2003). Ubiquitin-mediated sequestration of normal cellular proteins into polyglutamine aggregates. *Proc Natl Acad Sci USA, 100*, 8892–8897.
39. Scheel, H., Tomiuk, S., Hofmann, K. (2003). Elucidation of ataxin-3 and ataxin-7 function by integrative bioinformatics. *Hum Mol Genet, 12*, 2845–2852.
40. Mauri, P.L., Riva, M., Ambu, D., De Palma, A., Secundo, F., Benazzi, L., Valtorta, M., Tortora, P., Fusi, P. (2006). Ataxin-3 is subject to autolytic cleavage. *FEBS J, 273*, 4277–4286.
41. Wellington, C.L., Ellerby, L.M., Hackam, A.S., Margolis, R.L., Trifiro, M.A., Singaraja, R., McCutcheon, K., Salvesen, G.S., Propp, S.S., Bromm, M., et al. (1998). Caspase cleavage of gene products associated with triplet expansion disorders generates truncated fragments containing the polyglutamine tract. *J Biol Chem, 273*, 9158–9167.
42. Berke, S.J., Schmied, F.A., Brunt, E.R., Ellerby, L.M., Paulson, H.L. (2004). Caspase-mediated proteolysis of the polyglutamine disease protein ataxin-3. *J Neurochem, 89*, 908–918.
43. Bonifati, V., Rizzu, P., van Baren, M.J., Schaap, O., Breedveld, G.J., Krieger, E., Dekker, M.C., Squitieri, F., Ibanez, P., Joosse, M., et al. (2003). Mutations in the DJ-1 gene associated with autosomal recessive early-onset parkinsonism. *Science, 299*, 256–259.
44. Meulener, M., Whitworth, A.J., Armstrong-Gold, C.E., Rizzu, P., Heutink, P., Wes, P.D., Pallanck, L.J., Bonini, N.M. (2005). *Drosophila* DJ-1 mutants are selectively sensitive to environmental toxins associated with Parkinson's disease. *Curr Biol, 15*, 1572–1577.
45. Menzies, F.M., Yenisetti, S.C., Min, K.T. (2005). Roles of *Drosophila* DJ-1 in survival of dopaminergic neurons and oxidative stress. *Curr Biol, 15*, 1578–1582.
46. Yang, Y., Gehrke, S., Haque, M.E., Imai, Y., Kosek, J., Yang, L., Beal, M.F., Nishimura, I., Wakamatsu, K., Ito, S., et al. (2005). Inactivation of *Drosophila* DJ-1 leads to impairments of oxidative stress response and phosphatidylinositol 3-kinase/Akt signaling. *Proc Natl Acad Sci USA, 102*, 13670–13675.
47. Park, J., Kim, S.Y., Cha, G.H., Lee, S.B., Kim, S., Chung, J. (2005). *Drosophila* DJ-1 mutants show oxidative stress-sensitive locomotive dysfunction. *Gene, 361*, 133–139.
48. Jackson, G.R., Wiedau-Pazos, M., Sang, T.K., Wagle, N., Brown, C.A., Massachi, S., Geschwind, D.H. (2002). Human wild-type tau interacts with wingless pathway components and produces neurofibrillary pathology in *Drosophila*. *Neuron, 34*, 509–519.
49. Nishimura, I., Yang, Y., Lu, B. (2004). PAR-1 kinase plays an initiator role in a temporally ordered phosphorylation process that confers tau toxicity in *Drosophila*. *Cell, 116*, 671–682.
50. Emamian, E.S., Kaytor, M.D., Duvick, L.A., Zu, T., Tousey, S.K., Zoghbi, H.Y., Clark, H.B., Orr, H.T. (2003). Serine 776 of ataxin-1 is critical for polyglutamine-induced disease in SCA1 transgenic mice. *Neuron, 38*, 375–387.
51. Chen, H.K., Fernandez-Funez, P., Acevedo, S.F., Lam, Y.C., Kaytor, M.D., Fernandez, M.H., Aitken, A., Skoulakis, E.M., Orr, H.T., Botas, J., Zoghbi,

H.Y. (2003). Interaction of Akt-phosphorylated ataxin-1 with 14-3-3 mediates neurodegeneration in spinocerebellar ataxia type 1. *Cell, 113*, 457–468.

52. Sang, T.K., Chang, H.Y., Lawless, G.M., Ratnaparkhi, A., Mee, L., Ackerson, L.C., Maidment, N.T., Krantz, D.E., Jackson, G.R. (2007). A *Drosophila* model of mutant human parkin-induced toxicity demonstrates selective loss of dopaminergic neurons and dependence on cellular dopamine. *J Neurosci, 27*, 981–992.
53. Ghosh, S., Feany, M.B. (2004). Comparison of pathways controlling toxicity in the eye and brain in *Drosophila* models of human neurodegenerative diseases. *Hum Mol Genet, 13*, 2011–2018.
54. Warrick, J.M., Chan, H.Y., Gray-Board, G.L., Chai, Y., Paulson, H.L., Bonini, N.M. (1999). Suppression of polyglutamine-mediated neurodegeneration in *Drosophila* by the molecular chaperone HSP70. *Nat Genet, 23*, 425–428.
55. Cummings, C.J., Mancini, M.A., Antalffy, B., DeFranco, D.B., Orr, H.T., Zoghbi, H.Y. (1998). Chaperone suppression of aggregation and altered subcellular proteasome localization imply protein misfolding in SCA1. *Nat Genet, 19*, 148–154.
56. Chan, H.Y., Warrick, J.M., Gray-Board, G.L., Paulson, H.L., Bonini, N.M. (2000). Mechanisms of chaperone suppression of polyglutamine disease: selectivity, synergy and modulation of protein solubility in *Drosophila*. *Hum Mol Genet, 9*, 2811–2820.
57. Bilen, J., Liu, N., Burnett, B.G., Pittman, R.N., Bonini, N.M. (2006). MicroRNA pathways modulate polyglutamine-induced neurodegeneration. *Mol Cell, 24*, 157–163.
58. Cummings, C.J., Sun, Y., Opal, P., Antalffy, B., Mestril, R., Orr, H.T., Dillmann, W.H., Zoghbi, H.Y. (2001). Over-expression of inducible HSP70 chaperone suppresses neuropathology and improves motor function in SCA1 mice. *Hum Mol Genet, 10*, 1511–1518.
59. Auluck, P.K., Chan, H.Y., Trojanowski, J.Q., Lee, V.M., Bonini, N.M. (2002). Chaperone suppression of alpha-synuclein toxicity in a *Drosophila* model for Parkinson's disease. *Science, 295*, 865–868.
60. Wu, Y.R., Wang, C.K., Chen, C.M., Hsu, Y., Lin, S.J, Lin, Y.Y., Fung, H.C., Chang, K.H., Lee-Chen, G.J. (2004). Analysis of heat-shock protein 70 gene polymorphisms and the risk of Parkinson's disease. *Hum Genet, 114*, 236–241.
61. Kieran, D., Kalmar, B., Dick, J.R., Riddoch-Contreras, J., Burnstock, G., Greensmith, L. (2004). Treatment with arimoclomol, a coinducer of heat shock proteins, delays disease progression in ALS mice. *Nat Med, 10*, 402–405.
62. Kiaei, M., Kipiani, K., Petri, S., Chen, J., Calingasan, N.Y., Beal, M.F. (2005). Celastrol blocks neuronal cell death and extends life in transgenic mouse model of amyotrophic lateral sclerosis. *Neurodegener Dis, 2*, 246–254.
63. Ambros, V. (2004). The functions of animal microRNAs. *Nature, 431*, 350–355.
64. Bartel, D.P., Chen, C.Z. (2004). Micromanagers of gene expression: the potentially widespread influence of metazoan microRNAs. *Nat Rev Genet, 5*, 396–400.
65. Calin, G.A., Croce, C.M. (2006). MicroRNA signatures in human cancers. *Nat Rev Cancer, 6*, 857–866.
66. Bartel, D.P. (2004). MicroRNAs: genomics, biogenesis, mechanism, and function. *Cell, 116*, 281–297.
67. Kloosterman, W.P., Plasterk, R.H. (2006). The diverse functions of microRNAs in animal development and disease. *Dev Cell, 11*, 441–450.

68. Bilen, J., Liu, N., Bonini, N.M. (2006). A new role for microRNA pathways: modulation of degeneration induced by pathogenic human disease proteins. *Cell Cycle, 5*, 2835–2838.
69. Schaefer, A., O'Carroll, D., Tan, C.L., Hillman, D., Sugimori, M., Llinas, R., Greengard, P. (2007). Cerebellar neurodegeneration in the absence of microRNAs. *J Exp Med, 204*, 1553–1558.
70. Kiriakidou, M., Nelson, P.T., Kouranov, A., Fitziev, P., Bouyioukos, C., Mourelatos, Z., Hatzigeorgiou, A. (2004). A combined computational–experimental approach predicts human microRNA targets. *Genes Dev, 18*, 1165–1178.
71. Lewis, B.P., Shih, I.H., Jones-Rhoades, M.W., Bartel, D.P., Burge, C.B. (2003). Prediction of mammalian microRNA targets. *Cell, 115*, 787–798.
72. Karres, J.S., Hilgers, V., Carrera, I., Treisman, J., Cohen, S.M. (2007). The conserved microRNA miR-8 tunes atrophin levels to prevent neurodegeneration in *Drosophila. Cell, 131*, 136–145.
73. Kim, J., Inoue, K., Ishii, J., Vanti, W.B., Voronov, S.V., Murchison, E., Hannon, G., Abeliovich, A. (2007). A MicroRNA feedback circuit in midbrain dopamine neurons. *Science, 317*, 1220–1224.
74. Hebert, S.S., De Strooper, B. (2007). Molecular biology: miRNAs in neurodegeneration. *Science, 317*, 1179–1180.
75. Bilen, J., Bonini, N.M. (2007). Genome-wide screen for modifiers of ataxin-3 neurodegeneration in *Drosophila. PLoS Genet, 3*, 1950–1964.
76. Shulman, J.M., Feany, M.B. (2003). Genetic modifiers of tauopathy in *Drosophila. Genetics, 165*, 1233–1242.
77. Irwin, S., Vandelft, M., Pinchev, D., Howell, J.L., Graczyk, J., Orr, H.T., Truant, R. (2005). RNA association and nucleocytoplasmic shuttling by ataxin-1. *J Cell Sci, 118*, 233–242.
78. Bonini, N.M. (2002). Chaperoning brain degeneration. *Proc Natl Acad Sci USA, 99* (*Suppl 4*), 16407–16411.
79. Gidalevitz, T., Ben-Zvi, A., Ho, K.H., Brignull, H.R., Morimoto, R.I. (2006). Progressive disruption of cellular protein folding in models of polyglutamine diseases. *Science, 311*, 1471–1474.
80. Kraemer, B.C., Burgess, J.K., Chen, J.H., Thomas, J.H., Schellenberg, G.D. (2006). Molecular pathways that influence human tau-induced pathology in *Caenorhabditis elegans. Hum Mol Genet, 15*, 1483–1496.
81. Petrucelli, L., Dickson, D., Kehoe, K., Taylor, J., Snyder, H., Grover, A., De Lucia, M., McGowan, E., Lewis, J., Prihar, G., et al. (2004). CHIP and Hsp70 regulate tau ubiquitination, degradation and aggregation. *Hum Mol Genet, 13*, 703–714.
82. Bilen, J., Bonini, N.M. (2005). *Drosophila* as a model for human neurodegenerative disease. *Annu Rev Genet, 39*, 153–171.
83. Pandey, U.B., Nie, Z., Batlevi, Y., McCray, B.A., Ritson, G.P., Nedelsky, N.B., Schwartz, S.L., DiProspero, N.A., Knight, M.A., Schuldiner, O., et al. (2007). HDAC6 rescues neurodegeneration and provides an essential link between autophagy and the UPS. *Nature, 447*, 859–863.
84. Steffan, J.S., Bodai, L., Pallos, J., Poelman, M., McCampbell, A., Apostol, B.L., Kazantsev, A., Schmidt, E., Zhu, Y.Z., Greenwald, M., et al. (2001). Histone

deacetylase inhibitors arrest polyglutamine-dependent neurodegeneration in *Drosophila. Nature, 413*, 739–743.

85. Taylor, J.P., Taye, A.A., Campbell, C., Kazemi-Esfarjani, P., Fischbeck, K.H., Min, K.T. (2003). Aberrant histone acetylation, altered transcription, and retinal degeneration in a *Drosophila* model of polyglutamine disease are rescued by CREB-binding protein. *Genes Dev, 17*, 1463–1468.
86. Ravikumar, B., Vacher, C., Berger, Z., Davies, J.E., Luo, S., Oroz, L.G., Scaravilli, F., Easton, D.F., Duden, R., O'Kane, C.J., Rubinsztein, D.C. (2004). Inhibition of mTOR induces autophagy and reduces toxicity of polyglutamine expansions in fly and mouse models of Huntington disease. *Nat Genet, 36*, 585–595.
87. Auluck, P.K., Bonini, N.M. (2002). Pharmacological prevention of Parkinson disease in *Drosophila. Nat Med, 8*, 1185–1186.
88. Auluck, P.K., Meulener, M.C., Bonini, N.M. (2005). Mechanisms of suppression of {alpha}-synuclein neurotoxicity by geldanamycin in *Drosophila. J Biol Chem, 280*, 2873–2878.
89. Carmichael, J., Sugars, K.L., Bao, Y.P., Rubinsztein, D.C. (2002). Glycogen synthase kinase-3beta inhibitors prevent cellular polyglutamine toxicity caused by the Huntington's disease mutation. *J Biol Chem, 277*, 33791–33798.
90. Berger, Z., Ttofi, E.K., Michel, C.H., Pasco, M.Y., Tenant, S., Rubinsztein, D.C., O'Kane, C.J. (2005). Lithium rescues toxicity of aggregate-prone proteins in *Drosophila* by perturbing Wnt pathway. *Hum Mol Genet, 14*, 3003–3011.
91. Watase, K., Gatchel, J.R., Sun, Y., Emamian, E., Atkinson, R., Richman, R., Mizusawa, H., Orr, H.T., Shaw, C., Zoghbi, H.Y. (2007). Lithium therapy improves neurological function and hippocampal dendritic arborization in a spinocerebellar ataxia type 1 mouse model. *PLoS Med, 4*, e182.
92. Pagliarini, R.A., Xu, T. (2003). A genetic screen in *Drosophila* for metastatic behavior. *Science, 302*, 1227–1231.
93. Koh, K., Evans, J.M., Hendricks, J.C., Sehgal, A. (2006). A *Drosophila* model for age-associated changes in sleep : wake cycles. *Proc Natl Acad Sci USA, 103*, 13843–13847.
94. Suh, G.S., Wong, A.M., Hergarden, A.C., Wang, J.W., Simon, A.F., Benzer, S., Axel, R., Anderson, D.J. (2004). A single population of olfactory sensory neurons mediates an innate avoidance behaviour in *Drosophila. Nature, 431*, 854–859.
95. Chen, S., Lee, A.Y., Bowens, N.M., Huber, R., Kravitz, E.A. (2002). Fighting fruit flies: a model system for the study of aggression. *Proc Natl Acad Sci USA, 99*, 5664–5668.
96. Pick, S., Strauss, R. (2005). Goal-driven behavioral adaptations in gap-climbing *Drosophila. Curr Biol, 15*, 1473–1478.

11

ANIMAL MODELS TO STUDY THE BIOLOGY OF AMYLOID-β PROTEIN MISFOLDING IN ALZHEIMER DISEASE

Karen H. Ashe
Department of Neurology, University of Minnesota, Minneapolis, Minnesota

INTRODUCTION

In 1906, Alois Alzheimer, a psychiatrist, examined the brain of Auguste Deter, a 56-year-old demented patient for whom he cared in the years before her death, and proposed that her illness was associated with the amyloid plaques and neurofibrillary tangles that he found in her brain (Fig. 1A). Research on Alzheimer disease has focused on the structural consequences of the accumulation of plaques and tangles. Studies of plaques and tangles led to identification of the Aβ and tau proteins, the principal components of amyloid plaques and neurofibrillary tangles, respectively. How Aβ and tau proteins disrupt memory is only now becoming understood, however. Our knowledge of the manner in which these molecules disrupt brain function lags a century behind neuropathological studies of tau and Aβ, because suitable tools for examining their effects on the workings of the living brain were lacking until recently. Over the past 16 years the use of transgenic mice has permitted scientists to assay the biological activity of Aβ and tau proteins on memory and cognitive function and to understand better the adverse effects that these proteins exert on memory and cognitive function. Work in transgenic mice has also provided support for the theory that Aβ initiates Alzheimer disease and

Protein Misfolding Diseases: Current and Emerging Principles and Therapies,
Edited by Marina Ramirez-Alvarado, Jeffery W. Kelly, and Christopher M. Dobson
Copyright © 2010 John Wiley & Sons, Inc.

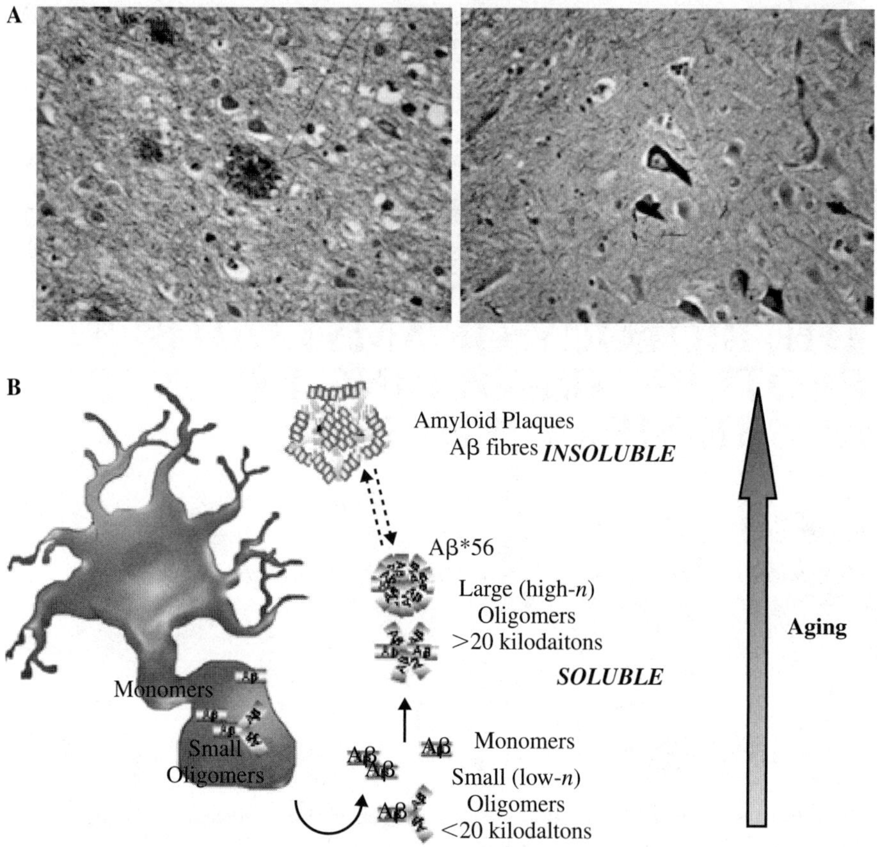

FIG. 1 Age-related changes in the brain associated with Alzheimer disease. (A) Plaques and tangles in a human brain from a patient with Alzheimer disease. Plaques are extracellular deposits of proteins (left panel); neurofibrillary tangles are fibrous collections of proteins within neurons (right panel). (From K. H., Ashe, Alzheimer's disease: transgenic mouse models, in *Encyclopedia of Neuroscience* with permission of Elsevier. Copyright © 2009). (B) The heterogeneity of Aβ proteins in the brain becomes more complex with age. The amyloid precursor protein (APP) undergoes posttranslational processing by α-, β-, and γ-secretases to generate Aβ and other APP cleavage products. Monomers and trimers have been identified inside neurons. Larger assemblies of Aβ are found extracellularly, including Aβ*56, a potentially important Aβ oligomer involved in disrupting memory formation. The monomers and oligomeric forms of Aβ are soluble in aqueous buffers. In contrast, the plaques contain Aβ fibrils that are insoluble in aqueous and detergent-containing buffers. Solubilized plaques contain monomers and low-*n* oligomers. (*See insert for color representation of figure.*)

triggers neurofibrillary tangle formation and neuronal loss as downstream events [1–5]. Because the focus of the Alzheimer research community has turned toward prevention of the disease [6], in this chapter we concentrate on recent studies leading to the identification, isolation, and characterization of a specific Aβ assembly in the brain, causing pre-dementia memory loss.

Alzheimer disease and its prodrome, amnestic mild cognitive impairment, begin when deficits in declarative memory appear. Declarative memory depends on the hippocampal formation, working in association with other limbic and cortical structures in the temporal lobe [7]. Declarative memories are conscious recollections, which elude analysis in mice for obvious reasons. Most behavioral studies in mice have therefore relied on spatial, navigational memory, which is also highly sensitive to hippocampal lesions [8]. A precise molecular explanation for why Alzheimer disease begins by targeting hippocampal-dependent memory formation remains a key unresolved conundrum.

Until a few years ago, methods for quantifying Aβ in brain tissue were limited to measurements of insoluble versus soluble Aβ monomers. To break down plaques into monomers, fibrillar Aβ is first dissociated into monomers using chaotropic agents such as guanidium hydrochloride or formic acid. However, novel protein extraction techniques now enable the measurement of Aβ species in various subcellular compartments (i.e., extracellular, intracellular, membrane, and insoluble) and the determination of their native sizes [9]. These methods have for the first time permitted the measurement and description of soluble intracellular and extracellular Aβ oligomers and monomers, and have significantly advanced our understanding of the dynamic patterns underlying the accumulation of heterogeneous Aβ proteins in the brain.

When Aβ is generated in the brain, heterogeneous forms of Aβ are produced, and this heterogeneity becomes more complex with age. A major source of heterogeneity in Aβ proteins arises from the aggregation of monomeric proteins to form higher-order structures, including soluble low-*n* oligomers, soluble high-*n* oligomers, and insoluble fibrils (Fig. 1B). In Tg2576 mice we have found intracellular monomers and trimers, whose levels remain constant with age [9]. In the extracellular space of Tg2576 mice under six months of age, low-*n* oligomers, but few high-*n* oligomers are present [9]. High-*n* oligomers appear in the extracellular space in middle-aged mice at an age (i.e., six months) coinciding with an abrupt decline in spatial reference memory [9]. One species, Aβ*56, showed the highest correlation with impaired memory and disrupted memory formation when infused into the lateral ventricles of young, healthy rats [9]. The relevance and implications of Aβ*56 may be appreciated by tracing the experiments and observations leading to its discovery.

ASSAYING THE EFFECTS ON MEMORY OF Aβ IN TRANSGENIC MICE

Amyloid plaques are used routinely for diagnosing Alzheimer disease in postmortem brain tissue [10], even though other neuropathological changes, such as neurofibrillary tangles, synaptic loss, and neuronal loss, are also usually present and sometimes correlate better with dementia [11,12]. Amyloid plaques develop in the brain with age, even in persons without dementia, but large numbers of amyloid plaques are found almost exclusively in patients with Alzheimer disease. A diagnosis of Alzheimer disease is made only if both cognitive deterioration and amyloid plaques are present [13].

The amyloid in plaques is composed of Aβ, a 39- to 43-amino acid protein released from APP following cleavage with β- and γ-secretases. When human wild-type APP was expressed in transgenic mice (JU strain) only 4% of mice older than12 months developed Aβ deposits, comprised of nonfibrillar Aβ. Yet interestingly, the mice exhibited age-related memory deficits [14,15]. Approximately 20 APP mutations have been identified in families with early-onset Alzheimer disease, all of which reside in or near the Aβ domain (http://www.molgen.ua.ac.be/ADMutations). Transgenic mice (FVB/N strain) expressing wild-type or variant human or mouse APPs develop no amyloid plaques or deposits. However, they exhibit an age-related, lethal central nervous system disorder that involves most of the corticolimbic regions of the brain, except the somatosensory-motor area, and resembles an accelerated naturally occurring senescent disorder of FVB/N mice [16]. When the same transgenes expressed in FVB/N mice were injected into embryos of a different strain background, C57B6/SJL F3, the mice tolerated and survived significantly higher levels of transgenic APP expression [16]. One such mouse gave rise to the Tg2576 line of APP transgenic mice [17].

Tg2576 mice, expressing APP with the "Swedish" mutation, which enhances overall Aβ production, develop amyloid plaques containing fibrillar β-pleated sheets of Aβ proteins, and spatial memory deficits (Fig. 2)[17]. The spatial memory deficits were measured using a version of the Morris water maze adapted for mice, in which animals were trained on successive days to locate a submerged platform. When all mice were able to locate the platform within 25 seconds, the platform was removed and the mice were released into the pool, a test known as the probe trial. The number of times the mice crossed the original platform location during the probe trial was recorded, providing an indication of how well a mouse "remembered" the location of the platform. The probe trial may distinguish the performance of mice that used a procedural strategy to locate the platform, such as swimming a certain distance from the edge of the pool, from mice that used a spatial strategy to locate the platform. Procedural and spatial learning depend on striatum and hippocampus, respectively. Animals that had learned the test using a procedural rather than a spatial strategy would be expected to perform poorly during the probe trial. The water maze data in Figure 2 showed that although the older transgenic mice learned to locate the platform, they could not remember its location, suggesting that they failed to adopt a spatial strategy to learn the task, supporting hippocampal dysfunction in the mice.

The Tg2576 mice demonstrated, for the first time, the feasibility of creating transgenic mice with robust behavioral and pathological features resembling those found in Alzheimer disease. Since performance in the water maze is sensitive to hippocampal lesions, the age-dependent deficit in Tg2576 mice

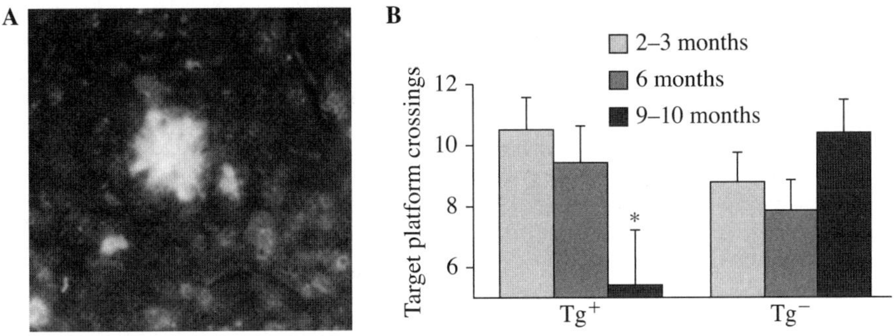

FIG. 2 Amyloid plaques and age-dependent memory loss in the Tg2576 APP transgenic mouse model of Alzheimer disease. (A) Plaques in a 354-day-old mouse stain with thioflavin S, a dye that binds β-pleated sheets. (B) With increasing age, transgene-positive, but not transgene-negative animals performing in a water maze show a dramatic decrease in platform crossings. (From [17], with permission of AAAS.)

probably represented hippocampal dysfunction, the earliest lesion in Alzheimer disease. Correlation of memory loss with deficits in long-lasting synaptic plasticity in the Schaeffer collateral pathway of the hippocampus confirmed dysfunction of this critical brain region (Fig. 3)[18]. The amyloid pathology in Tg2576 mice increased with age and could not be explained by changes in transgenic APP expression, which remained unchanged. However, these studies did not address whether the deficits in memory were caused by or merely correlated with amyloid deposition.

DIFFERENTIATING BETWEEN THE ROLES OF Aβ AND AMYLOID PLAQUES IN MEMORY LOSS

The development of memory deficits and plaques made Tg2576 mice suitable for examining the relationship between Aβ, plaques, and memory. Different forms of Aβ, biochemically distinguishable by their solubility properties in detergents, are found in varying amounts during the lifetime of Tg2576 mice [19]. Detergent [sodium dodecyl sulfate (SDS)]–soluble Aβ is present throughout life, whereas SDS-insoluble Aβ does not appear until about 6 months of age after which its levels rise rapidly [19]. Transgene-positive and transgene-negative Tg2576 mice perform equally well in the water maze until 6–11 months of age, when transgene-positive mice show a subtle but abrupt decline [20]. Memory deficits in transgene-positive mice remained small and fairly stable between 6 and 18 months of age [20]. These mice required more training than was required by transgene-negative mice to remember the platform location in the water maze, but their performance improved with continued training (Fig. 4). With transgene-positive mice older than 20 months of age, the ability to remember the platform location no longer improved with training

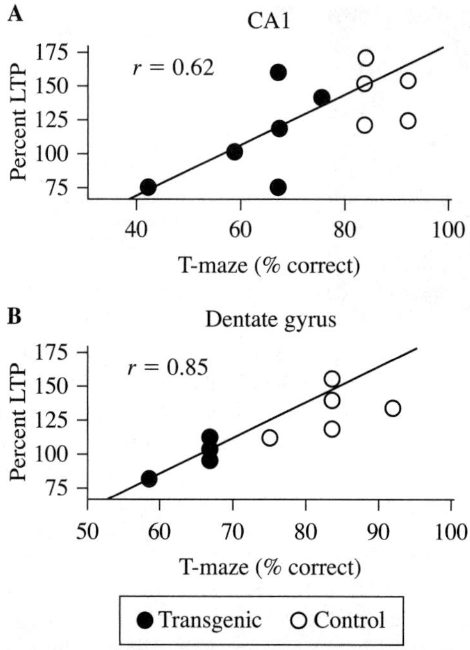

FIG. 3 Behavioral performance and long-lasting synaptic plasticity correlate. Plots of (A) the magnitude of LTP in CA1 and (B) dentate gyrus against percent correct performance in the forced-choice alternation task in the T-maze over the last two training sessions reveal a correlation between behavioral performance and neuronal plasticity. Each point represents a single animal. (From [18], with permission of Macmillan Publishers Ltd.)

(Fig. 4). The distinct patterns of memory performance at different ages suggest that more than one process is involved in disrupting memory formation.

Because spatial memory loss coincided with the appearance of SDS-insoluble Aβ, both occurring at 6 months of age, initially it appeared possible that SDS-insoluble Aβ caused the memory deficits. This idea was supported by an observation of accelerated spatial memory deficits in 4- and 5-month-old Tg2576 mice harboring a mutant presenilin-1 transgene, which induced the appearance of SDS-insoluble Aβ at younger ages [20]. However, an analysis including older mice led to rejection of the facile interpretation that memory loss and SDS-insoluble Aβ were closely connected, since no obvious correlation between SDS-insoluble Aβ and memory was apparent in a combined group of young and old mice (Fig. 5).

Interestingly, when the relationship between SDS-insoluble Aβ and memory was reanalyzed in mice grouped by age, correlations emerged in the oldest and youngest groups of mice [20]. This effect suggests that both amyloid load and SDS-insoluble Aβ are surrogate measures for one or more

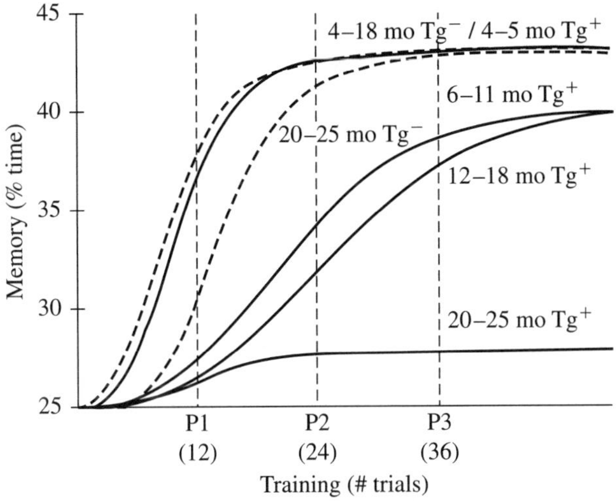

FIG. 4 Age-dependent impairment in spatial memory in transgene-positive (Tg^+) and transgene-negative (Tg^-) Tg2576 mice. The percentage of time that transgene-positive and transgene-negative mice of different ages spent in the target quadrant during three probe trials after 12 (probe 1 = P1), 24 (probe 2 = P2), and 36 (probe 3 = P3) training trials shows memory improved with training and reaches a saturating plateau in all groups. With age, there is a rightward shift of the curves. In the oldest Tg^+ mice, there is also a dramatic lowering of the saturation level of memory performance. (From [20], with permission of the *Journal of Neuroscience*. Copyright © 2002.)

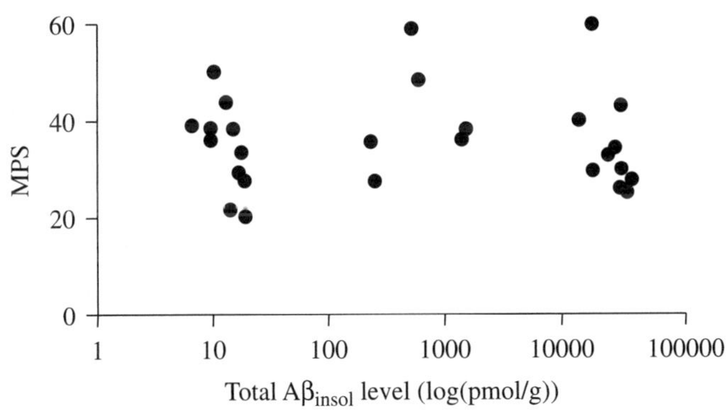

FIG. 5 Spatial memory and SDS-insoluble Aβ (Aβinsol) show no correlation. The mean probe score (MPS), which is the average of the three probe scores described in Figure 4, and SDS-insoluble Aβ levels of individual mice ranging from 5 to 22 months of age are shown. (From [20], with permission of the *Journal of Neurosciences*. Copyright © 2002.)

small Aβ assemblies that disrupt learning and memory (Fig. 6). If the formation and clearance rates of the small assemblies are equivalent and the small Aβ assemblies direct the conversion of soluble to insoluble Aβ, then, at any given age, the levels of small Aβ assemblies and the levels of

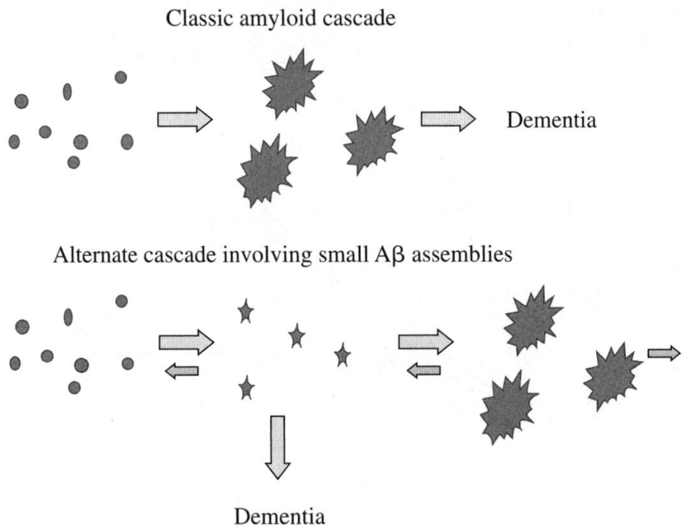

FIG. 6 Proposed model of small Aβ assemblies disrupting learning and memory. The hypothetical cascade involving small Aβ assemblies contrasts with the classic amyloid cascade hypothesis, in which cognitive dysfunction and dementia are caused by the accumulation of insoluble Aβ aggregates, resulting in neuronal destruction. In the alternate cascade, involving Aβ assemblies, memory loss is caused by a subset of soluble Aβ species. (From [20], with permission of the *Journal of Neuroscience*. Copyright © 2002.)

insoluble Aβ will correlate, but the ratio of SDS-insoluble Aβ to soluble Aβ assemblies will increase with aging [20,21].

Tg2576 mice at 16 months of age, with mature plaque deposition, show no neuronal or synaptic loss in the hippocampus [22], indicating that memory deficits in these mice might be attributable to neuronal dysfunction rather than to neuronal degeneration. Earlier studies showing deficits in long-lasting synaptic plasticity, with preservation of fast synaptic transmission, supported this idea [18]. To test the theory that memory deficits related to small Aβ assemblies occur primarily in the absence of structural damage and blockade of their effects might rapidly reverse memory deficits by neutralizing one or more critical Aβ species, BAM-10 antibodies [generated against Aβ (1–12)] were administered by intraperitoneal injection to impaired Tg2576 mice lacking mature amyloid plaques [23]. BAM-10 improved memory function to equal that of young, unimpaired mice (Fig. 7). This improvement occurred within 11 days of the first injection of BAM-10 antibodies and was not associated with a decrease in brain Aβ levels. These results support the notion that early memory deficits in Tg2576 mice are caused by non-plaque-associated Aβ assemblies. Similar conclusions were drawn from results obtained using a midregion Aβ antibody in the PDAPP transgenic mouse line [24].

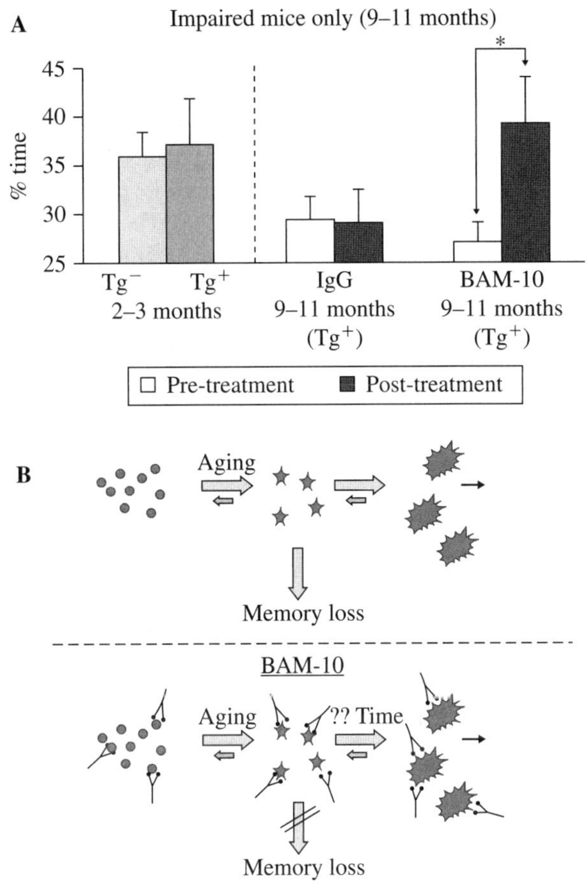

FIG. 7 Spatial memory in 9- to11-month-old Tg2576 mice before and after treatment with BAM-10 antibody. (A) In mice that were impaired at baseline (mice that spent <40% of the probe time in the target quadrant), those receiving BAM-10 ($n = 16$) showed significantly greater improvement than mice receiving nonspecific immunoglobulin ($n = 17$). (B) We hypothesized that BAM-10 penetrates the brain, where it may bind to small Aβ assemblies, neutralize their deleterious effects on cognitive function, and restore memory rapidly in Tg2576 mice. (From [23], with permission of the *Journal of Neuroscience*. Copyright © 2002.)

It should be noted that although small Aβ assemblies appear to explain the early graded loss of memory in Tg2576 mice, they cannot fully explain the later drop, occurring at about 20 months of age. The later drop might result from the accumulation of amyloid plaques, which are abundant by 20 months of age. Neurotoxicity located within the microenvironment of amyloid plaques has been proposed to underlie some aspects of dementia in AD. Once formed, amyloid plaques rapidly induce abnormal neuronal architecture and function within about a 50-μm halo surrounding the plaque core [25], including loss of dendritic spines [26,27], neuritic dystrophy, increased neurite curvature [28,29], and asynchrony of electrical activity [30]. Aβ dimers are among the leading culprits mediating these effects. Dimers of Aβ are found in close association with amyloid plaques [31], and dimers cause electrophysiological and cognitive abnormalities when applied to young, healthy rats [32]. Dimers in Tg2576 mice appear first in lipid rafts, and their subsequent levels rise in parallel with insoluble Aβ [33]. Although the levels of dimers fail to correspond to memory deficits present between 6 and 20 months of age, a potential role for dimers in Tg2576 mice older than 20 months remains an unexplored possibility. Alternatively, the later drop in Tg2576 memory may result from alteration of synaptic connections following chronic (> 14 months) exposure to other small Aβ assemblies.

ZEROING IN ON Aβ*56, A SOLUBLE Aβ ASSEMBLY IN THE BRAIN THAT IMPAIRS MEMORY

We sought to explain the first drop in memory function in Tg2576 mice. We identified Aβ*56 as a potentially important oligomeric Aβ species because its appearance in the brains of Tg2576 mice coincided with the onset of deficits in the Morris water maze, and its levels correlated with memory impairment in 6-month-old mice [9]. Most important, its levels remained stable from 6 to 14 months of age during a period of cognitive stability (Fig. 8), in contrast to a variety of other Aβ species, including insoluble Aβ, soluble monomeric Aβ, Aβ dimers, and Aβ-derived diffusible ligands (ADDLs), whose levels increase substantially during this time [19,33,34]. It should be noted that ADDLs were

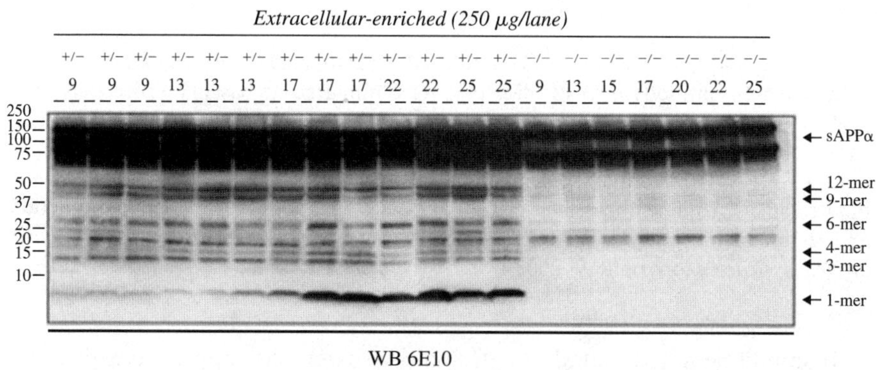

FIG. 8 Soluble extracellular-enriched Aβ species assessed by immunoblot using 6E10 in Tg2576 mice between 9 and 25 months of age. The levels of the 56-kDa species, corresponding in size approximately to a theoretical Aβ (1–42) dodecamer, remain relatively stable during this time period. In contrast, levels of soluble and insoluble Ab1-42 increase 50- and 100-fold, respectively [19]. (From [9], with permission of Macmillan Publishers Ltd.)

measured using ADDL-specific antibodies, which do not detect single specific oligomeric assemblies and are not specific to Aβ*56; anti-ADDL antibodies applied to PBS-extracted Tg2576 brain showed a 20- to 100-fold increase in signal during a time period (i.e., 13 to 17 months of age) when there was no increase in Aβ*56 levels [34]. The covariation between Aβ*56 levels and memory impairment provides necessary, albeit insufficient evidence for Aβ*56 playing a causal role in early memory loss in Tg2576 mice.

To ascertain whether Aβ*56 causes memory loss, it was applied directly to young rats [9]. Aβ*56, purified by tandem immunoaffinity purification exclusion-size chromatography (Fig. 9A), was injected into the lateral ventricles of young rats that had been pretrained in the Morris water maze. Administration of Aβ*56 2 hours before testing had no effect on the animals' ability to escape to either a hidden (Fig. 9B) or visible (Fig. 9C) platform. However, when spatial reference memory was tested 24 hours later (after a second injection of Aβ*56 given 2 hours before testing), Aβ*56-treated animals exhibited impaired spatial reference memory compared to vehicle-injected rats. Whereas the vehicle-injected animals spent significantly more time in the target quadrant that in other regions of the maze, Aβ*56-injected animals showed no such preferences (Fig. 9D and E). Importantly, Aβ*56 did not permanently impair the rats' ability to perform the task. Rats were allowed to recover for 10 days and then were tested again in the water maze over a 2-day period: on the first day, both groups again showed equal ability to escape to the visible or hidden platform; on the second day, both groups showed a preference for the target quadrant. The transient nature of the impairment induced by short-term exposure to Aβ*56 suggests that the oligomer interfered with neural activity but did not cause permanent pathological changes.

A precise molecular structure of Aβ*56 has not been determined. However, mass spectrometry of purified Aβ*56 revealed Aβ as a core component and identified no other polypeptides [9]. Aβ*56 is recognized by A11 antiserum, which specifically detects soluble amyloid assemblies distinct from fibrillar Aβ (Fig. 10)[9]. Aβ*56 remains stable in SDS–polyacrylamide gel electrophoresis containing 8 M urea [9]. However, when exposed to more than 10% hexafluoroisopropanol (HFIP), a solvent with strong hydrogen-bonding properties, Aβ*56 depolymerized and there was a parallel increase in levels of tetramers, trimers, and to a lesser extent, monomers [9]. These data suggest that Aβ*56 is held together by urea-resistant hydrogen bonds, not by covalent bonds. By atomic force microscopy, Aβ*56 appears as globular rather than fibrillar structures (Fig. 11) [35]. Cultured primary neurons from Tg2576 mice secrete monomers and trimers, but not Aβ*56 [9]. Monomers and trimers are the exclusive Aβ species found in the soluble intracellular-enriched compartment of Tg2576 mouse neurons [9]. Collectively, these data suggest that Aβ*56 forms when a quartet of Aβ trimers coalesces in the extracellular space of brains in aging animals. It is possible that this formation occurs in conjunction with other molecules or atoms, but such entities have not yet been identified.

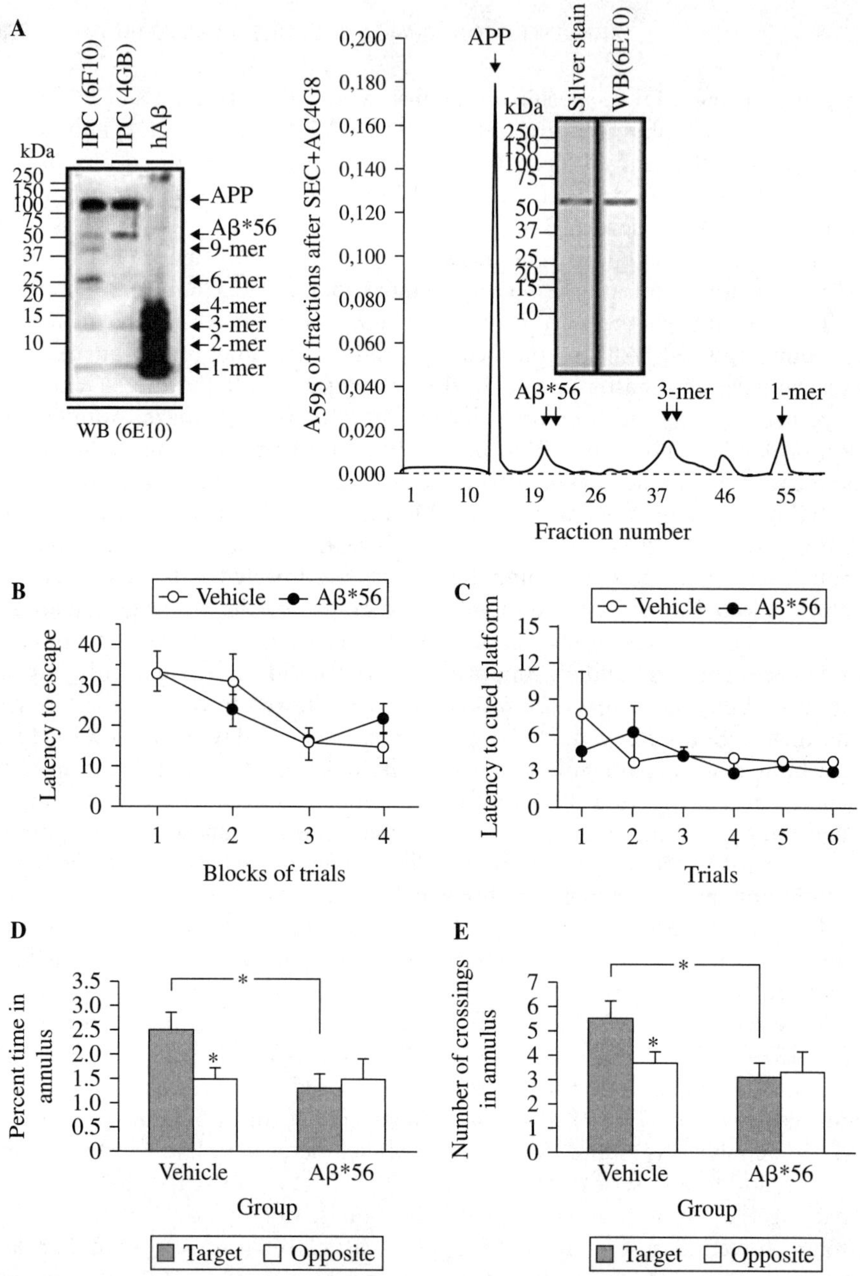

FIG. 9 Application of Aβ*56 into the lateral ventricles impairs spatial reference memory in healthy, young rats. Aβ species were immunoaffinity-purified using monoclonal antibodies 6E10 or 4G8 (A, left), then separated by size-exclusion chromatography to yield fractions containing Aβ*56 (A, right). When injected ICV into rats pretrained in the Morris water maze, Aβ*56 had no effect on the animals' ability to locate a hidden (B) or visible (C) escape platform. However, in probe trials administered 24 hours later (after a second injection of Aβ*56 or vehicle), Aβ*56-treated rats spent significantly less time in the target quadrant than did vehicle-injected rats (D,E). hAβ42, Synthetic human Aβ (1–42) peptide was used as a size marker. (From [9], with permission of Macmillan Publishers Ltd.)

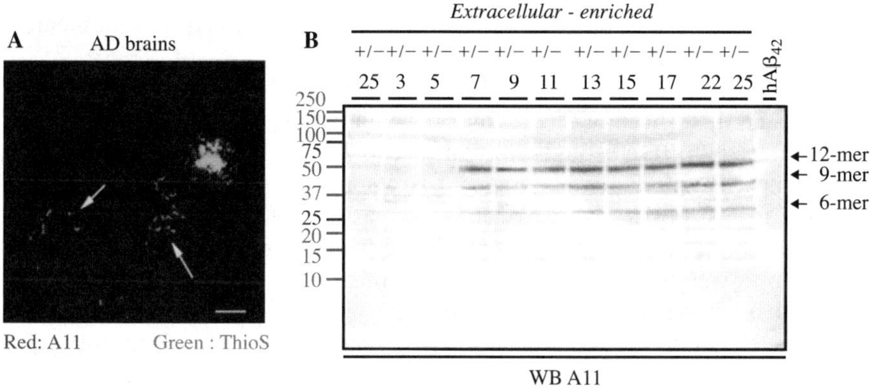

FIG. 10 The anti-oligomer antiserum A11 recognizes Aβ*56. (A) A11 staining of non-plaque-associated oligomers in brain tissue from an Alzheimer patient; (B) staining of high-N Aβ oligomers with the anti-oligomer antiserum, A11. Note that the human Aβ 42 (hAβ42) standards are not detected with the A11 antiserum (far right lane). [(A) From Kayed et al., *Science*, 2003; Apr 18; *300*(5618);486–489, with permission of AAAS. (B) from [9], with permission of Macmillan Publishers Ltd.] (*See insert for color representation of figure.*)

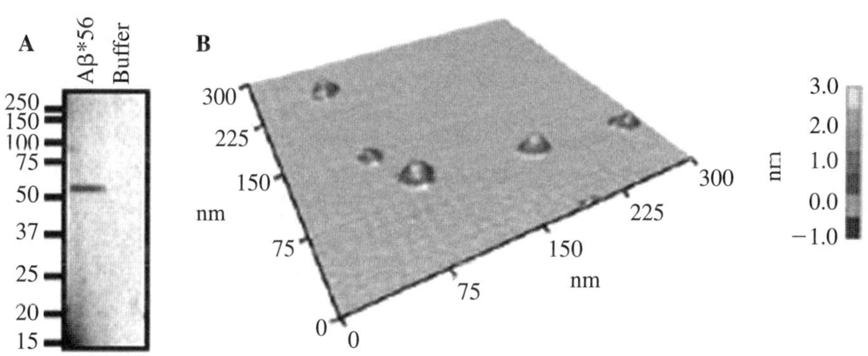

FIG. 11 Ex situ AFM height image and size distribution of Aβ*56. (A) western blot analysis (6E10 antibody) of Aβ*56 purified from detergent-soluble forebrain lysates of Tg2576 mice by tandem immunoaffinity/size-exclusion chromatography; (B) representative 1-μm^2 atomic force microscopy image of Aβ*56, demonstrating ellipsoidal shapes. (From [35]. Copyright © 2007 American Society for Biochemistry and Molecular Biology.)

Aβ*56 has been shown to correlate with memory loss not only in Tg2576 mice, but also in J20 mice expressing the Swedish and Indiana mutations, Arc48 expressing human APP with the Swedish, Indiana, and Arctic mutations, and 3xTgAD mice expressing APP with the Swedish mutation, PS1 with the M146V mutation, and tau with the P301L mutation [35–37]. The covariation of Aβ*56 with memory impairment in four distinct lines of mice in various strain backgrounds indicates that its role in at least some forms of Aβ-related memory deficits can be generalized to most APP transgenic mice.

Arc6 mice expressing human APP with the Swedish, Indiana, and Arctic at lower levels than those of Arc48 mice develop plaques more aggressively than do J20 mice [38]. Interestingly, in contrast to J20 and Arc48 mice, Arc6 mice have almost no Aβ*56, indicating that Arctic mutation-containing Aβ preferentially forms fibrils over Aβ*56 when expressed at low levels [35,38]. Consistent with the hypothesis that Aβ*56 must be present for early deficits to develop, Arc6 mice lacking Aβ*56 have normal memory function at an age (4 months) when plaque load is about 1%.

Many questions about Aβ*56 remain unanswered. The most intriguing puzzles include the factors that trigger its formation, the structural basis of its stability, and its precise neuronal targets and mechanisms of action. It will also be critical to examine whether Aβ*56 is associated with neurological status and abilities, especially in the earliest stages of Alzheimer's disease, such as amnestic mild cognitive impairment, or in asymptomatic stages of Alzheimer disease.

CONCLUSIONS

Ten years ago Alzheimer researchers argued aggressively about whether Aβ or tau played the more significant role in the pathogenesis of Alzheimer disease. Today, most Alzheimer researchers concede that both molecules are critical, but remain uncertain about *which* forms of Aβ and tau are involved at *what* stages of disease and *where* they act in the brain. Converging lines of evidence in transgenic mice and humans support the hypothesis that Aβ initiates brain dysfunction, while both Aβ and tau may mediate brain degeneration. Research performed over the past 16 years using APP transgenic mice has led to the discovery of a specific form of Aβ, Aβ*56, which may be the specific triggering event of Alzheimer disease. Much work remains to define the other forms of tau and Aβ that mediate the disease, and to develop diagnostic and therapeutic strategies aimed at the most crucial pathogenic tau and Aβ species.

REFERENCES

1. Gotz, J., Chen, F., Barmettler, R., Nitsch, R.M. (2001). Tau filament formation in transgenic mice expressing P301L tau. *J Biol Chem, 276*, 529–534.
2. Lewis, J., Dickson, D.W., Lin, W.L., Chisholm, L., Corral, A., Jones, G., Yen, S.H., Sahara, N., Skipper, L., Yager, D., et al. (2001). Enhanced neurofibrillary degeneration in transgenic mice expressing mutant tau and APP. *Science, 293*, 1487–1491.
3. Oddo, S., Caccamo, A., Shepherd, J.D., Murphy, M.P., Golde, T.E., Kayed, R., Metherate, R., Mattson, M.P., Akbari, Y., LaFerla, F.M. (2003). Triple-transgenic model of Alzheimer's disease with plaques and tangles: intracellular Abeta and synaptic dysfunction. *Neuron, 39*, 409–421.

4. Oddo, S., Billings, L., Kesslak, J.P., Cribbs, D.H., LaFerla, F.M. (2004). Abeta immunotherapy leads to clearance of early, but not late, hyperphosphorylated tau aggregates via the proteasome. *Neuron, 43*, 321–332.
5. Bolmont, T., Clavaguera, F., Meyer-Luehmann, M., Herzig, M.C., Radde, R., Staufenbiel, M., Lewis, J., Hutton, M., Tolnay, M., Jucker, M. (2007). Induction of tau pathology by intracerebral infusion of amyloid-beta-containing brain extract and by amyloid-beta deposition in APP × tau transgenic mice. *Am J Pathol, 171*, 2012–2020.
6. Khachaturian, Z.S., Petersen, R.C., Gauthier, S., Buckholtz, N., Corey-Bloom, J.P., Evans, B., Fillit, H., Foster, N., Greenberg, B., Grundman, M., et al. (2008). A roadmap for the prevention of dementia: the inaugural Leon Thal Symposium. *Alzheimer's Dement, 4*, 156–163.
7. Squire, L.R. (2004). Memory systems of the brain: a brief history and current perspective. *Neurobiol Learn Mem, 82*, 171–177.
8. de Hoz, L., Knox, J., Morris, R.G. (2003). Longitudinal axis of the hippocampus: both septal and temporal poles of the hippocampus support water maze spatial learning depending on the training protocol. *Hippocampus, 13*, 587–603.
9. Lesne, S., Koh, M.T., Kotilinek, L., Kayed, R., Glabe, C.G., Yang, A., Gallagher, M., Ashe, K.H. (2006). A specific amyloid-beta protein assembly in the brain impairs memory. *Nature, 440*, 352–357.
10. Mirra, S.S., Heyman, A., McKeel, D., Sumi, S.M., Crain, B.J., Brownlee, L.M., Vogel, F.S., Hughes, J.P., van Belle, G., Berg, L. (1991). The Consortium to Establish a Registry for Alzheimer's Disease (CERAD). Part II. Standardization of the neuropathologic assessment of Alzheimer's disease. *Neurology, 41*, 479–486.
11. Terry, R.D., Masliah, E., Salmon, D.P., Butters, N., DeTeresa, R., Hill, R., Hansen, L.A., Katzman, R. (1991). Physical basis of cognitive alterations in Alzheimer's disease: synapse loss is the major correlate of cognitive impairment. *Ann Neurol, 30*, 572–580.
12. Arriagada, P.V., Growdon, J.H., Hedley-Whyte, E.T., Hyman, B.T. (1992). Neurofibrillary tangles but not senile plaques parallel duration and severity of Alzheimer's disease. *Neurology, 42*, 631–639.
13. Khachaturian, Z.S. (1985). Diagnosis of Alzheimer's disease. *Arch Neurol, 42*, 1097–1105.
14. Higgins, L.S., Holtzman, D.M., Rabin, J., Mobley, W.C., Cordell, B. (1994). Transgenic mouse brain histopathology resembles early Alzheimer's disease. *Ann Neurol, 35*, 598–607.
15. Moran, P.M., Higgins, L.S., Cordell, B., Moser, P.C. (1995). Age-related learning deficits in transgenic mice expressing the 751-amino acid isoform of human beta-amyloid precursor protein. *Proc Natl Acad Sci U S A, 92*, 5341–5345.
16. Hsiao, K.K., Borchelt, D.R., Olson, K., Johannsdottir, R., Kitt, C., Yunis, W., Xu, S., Eckman, C., Younkin, S., Price D, et al. (1995). Age-related CNS disorder and early death in transgenic FVB/N mice overexpressing Alzheimer amyloid precursor proteins. *Neuron, 15*, 1203–1218.
17. Hsiao, K., Chapman, P., Nilsen, S., Eckman, C., Harigaya, Y., Younkin, S., Yang, F., Cole, G. (1996). Correlative memory deficits, Abeta elevation, and amyloid plaques in transgenic mice. *Science, 274*, 99–102.

18. Chapman, P.F., White, G.L., Jones, M.W., Cooper-Blacketer, D., Marshall, V.J., Irizarry, M., Younkin, L., Good, M.A., Bliss, T.V., Hyman, B.T., et al. (1999). Impaired synaptic plasticity and learning in aged amyloid precursor protein transgenic mice. *Nat Neurosci, 2*, 271–276.

19. Kawarabayashi, T., Younkin, L.H., Saido, T.C., Shoji, M., Ashe, K.H., Younkin, S.G. (2001). Age-dependent changes in brain, CSF, and plasma amyloid (beta) protein in the Tg2576 transgenic mouse model of Alzheimer's disease. *J Neurosci, 21*, 372–381.

20. Westerman, M.A., Cooper-Blacketer, D., Mariash, A., Kotilinek, L., Kawarabayashi, T., Younkin, L.H., Carlson, G.A., Younkin, S.G., Ashe, K.H. (2002). The relationship between Abeta and memory in the Tg2576 mouse model of Alzheimer's disease. *J Neurosci, 22*, 1858–1867.

21. Ashe, K.H. (2001). Learning and memory in transgenic mice modeling Alzheimer's disease. *Learn Mem, 8*, 301–308.

22. Irizarry, M.C., McNamara, M., Fedorchak, K., Hsiao, K., Hyman, B.T. (1997). APPSw transgenic mice develop age-related A beta deposits and neuropil abnormalities, but no neuronal loss in CA1. *J Neuropathol Exp Neurol, 56*, 965–973.

23. Kotilinek, L.A., Bacskai, B., Westerman, M., Kawarabayashi, T., Younkin, L., Hyman, B.T., Younkin, S., Ashe, K.H. (2002). Reversible memory loss in a mouse transgenic model of Alzheimer's disease. *J Neurosci, 22*, 6331–6335.

24. Dodart, J.C., Bales, K.R., Gannon, K.S., Greene, S.J., DeMattos, R.B., Mathis, C., DeLong, C.A., Wu, S., Wu, X., Holtzman, D.M., Paul, S.M. (2002). Immunization reverses memory deficits without reducing brain Abeta burden in Alzheimer's disease model. *Nat Neurosci, 5*, 452–457.

25. Meyer-Luehmann, M., Spires-Jones, T.L., Prada, C., Garcia-Alloza, M., de Calignon, A., Rozkalne, A., Koenigsknecht-Talboo, J., Holtzman, D.M., Bacskai, B.J., Hyman, B.T. (2008). Rapid appearance and local toxicity of amyloid-beta plaques in a mouse model of Alzheimer's disease. *Nature, 451*, 720–724.

26. Spires, T.L., Meyer-Luehmann, M., Stern, E.A., McLean, P.J., Skoch, J., Nguyen, P.T., Bacskai, B.J., Hyman, B.T. (2005). Dendritic spine abnormalities in amyloid precursor protein transgenic mice demonstrated by gene transfer and intravital multiphoton microscopy. *J Neurosci, 25*, 7278–7287.

27. Spires-Jones, T.L., Meyer-Luehmann, M., Osetek, J.D., Jones, P.B., Stern, E.A., Bacskai, B.J., Hyman, B.T. (2007). Impaired spine stability underlies plaque-related spine loss in an Alzheimer's disease mouse model. *Am J Pathol, 171*, 1304–1311.

28. Knowles, R.B., Wyart, C., Buldyrev, S.V., Cruz, L., Urbanc, B., Hasselmo, M.E., Stanley, H.E., Hyman, B.T. (1999). Plaque-induced neurite abnormalities: implications for disruption of neural networks in Alzheimer's disease. *Proc Natl Acad Sci USA, 96*, 5274–5279.

29. Le, R., Cruz, L., Urbanc, B., Knowles, R.B., Hsiao-Ashe, K., Duff, K., Irizarry, M.C., Stanley, H.E., Hyman, B.T. (2001). Plaque-induced abnormalities in neurite geometry in transgenic models of Alzheimer disease: implications for neural system disruption. *J Neuropathol Exp Neurol, 60*, 753–758.

30. Stern, E.A., Bacskai, B.J., Hickey, G.A., Attenello, F.J., Lombardo, J.A., Hyman, B.T. (2004). Cortical synaptic integration in vivo is disrupted by amyloid-beta plaques. *J Neurosci, 24*, 4535–4540.

31. Roher, A.E., Chaney, M.O., Kuo, Y.M., Webster, S.D., Stine, W.B., Haverkamp, L.J., Woods, A.S., Cotter, R.J., Tuohy, J.M., Krafft, G.A., et al. (1996). Morphology and toxicity of Abeta-(1-42) dimer derived from neuritic and vascular amyloid deposits of Alzheimer's disease. *J Biol Chem, 271*, 20631–20635.

32. Shankar, G.M., Li, S., Mehta, T.H., Garcia-Munoz, A., Shepardson, N.E., Smith, I., Brett, F.M., Farrell, M.A., Rowan, M.J., Lemere, C.A., et al. (2008). Amyloid-beta protein dimers isolated directly from Alzheimer's brains impair synaptic plasticity and memory. *Nat Med, 14*, 837–842.

33. Kawarabayashi, T., Shoji, M., Younkin, L.H., Wen-Lang, L., Dickson, D.W., Murakami, T., Matsubara, E., Abe, K., Ashe, K.H., Younkin, S.G. (2004). Dimeric amyloid beta protein rapidly accumulates in lipid rafts followed by apolipoprotein E and phosphorylated tau accumulation in the Tg2576 mouse model of Alzheimer's disease. *J Neurosci, 24*, 3801–3809.

34. Chang, L., Bakhos, L., Wang, Z., Venton, D.L., Klein, W.L. (2003). Femtomole immunodetection of synthetic and endogenous amyloid-beta oligomers and its application to Alzheimer's disease drug candidate screening. *J Mol Neurosci, 20*, 305–313.

35. Cheng, I.H., Scearce-Levie, K., Legleiter, J., Palop, J.J., Gerstein, H., Bien-Ly, N., Puolivali, J., Lesne, S., Ashe, K.H., Muchowski, P.J., Mucke, L. (2007). Accelerating amyloid-beta fibrillization reduces oligomer levels and functional deficits in Alzheimer disease mouse models. *J Biol Chem, 282*, 23818–23828.

36. Oddo, S., Vasilevko, V., Caccamo, A., Kitazawa, M., Cribbs, D.H., LaFerla, F.M. (2006). Reduction of soluble Abeta and tau, but not soluble Abeta alone, ameliorates cognitive decline in transgenic mice with plaques and tangles. *J Biol Chem, 281*, 39413–39423.

37. Billings, L.M., Green, K.N., McGaugh, J.L., LaFerla, F.M. (2007). Learning decreases A beta*56 and tau pathology and ameliorates behavioral decline in 3xTg-AD mice. *J Neurosci, 27*, 751–761.

38. Cheng, I.H., Palop, J.J., Esposito, L.A., Bien-Ly, N., Yan, F., Mucke, L. (2004). Aggressive amyloidosis in mice expressing human amyloid peptides with the Arctic mutation. *Nat Med, 10*, 1190–1192.

PART II

PROTEIN MISFOLDING DISEASE: GAIN-OF-FUNCTION AND LOSS-OF-FUNCTION DISEASES

12

ALZHEIMER DISEASE: PROTEIN MISFOLDING, MODEL SYSTEMS, AND EXPERIMENTAL THERAPEUTICS

DONALD L. PRICE, ALENA V. SAVONENKO, TONG LI, MICHAEL K. LEE, AND PHILIP C. WONG

Department of Neuropathology, Johns Hopkins University School of Medicine, Baltimore, Maryland

INTRODUCTION

This chapter focuses on Alzheimer disease (AD), one of the prototypical protein misfolding diseases of the central nervous system (CNS), on mammalian models relevant to the pathogenic mechanisms of this disorder, and on the value of these models for testing the potential of a variety of experimental therapeutic approaches [2,7,14,17,25,71,105]. AD is a major unmet medical need because of its incidence or prevalence, severity, cost, lack of mechanism-based treatments, and impacts on individuals, caregivers, and society at large [7]. The clinical syndrome (i.e., cognitive and memory disturbances progressing to dementia) results from dysfunction and death of neurons in specific brain regions and circuits important in memory and cognition. The neuropathology of AD includes [9,60]: accumulation of extracellular β–pleated Aβ (40–42) peptides which, as oligomeric assemblies and/or aggregates [71,105], are at the core of neuritic amyloid plaques, which, at some level, represent sites of synaptic disconnection [71,105]; and intracellular accumulations of conformationally altered phosphorylated tau (p-tau) assembled into paired helical filaments (PHFs) comprising neurofibrillary

Protein Misfolding Diseases: Current and Emerging Principles and Therapies,
Edited by Marina Ramirez-Alvarado, Jeffery W. Kelly, and Christopher M. Dobson
Copyright © 2010 John Wiley & Sons, Inc.

tangles (NFTs) and filament-containing neurites [2,25,59]. The mechanisms of disease are hypothesized to be related to Aβ-linked damage to synapses, alterations in the neuronal cytoskeleton, and dysfunction of nerve cells. Eventually, affected cells degenerate. This age-associated illness is influenced by genetic risk factors, with a minority of cases inherited in mendelian fashion (autosomal dominant mutations, and rarely, duplications) [29]. More commonly, putative sporadic cases are thought to be influenced by a variety of susceptibility genes (especially ApoE4), possible environmental influences, and other less well-defined factors [5,6,71,103]. Symptomatic treatments exist, but efficacious and safe mechanism-based therapies are not yet available [7].

In the autosomal dominant forms of this disorder [29], the malfolded and dysfunctional proteins and peptides (sometimes mislocalized) acquire properties that have direct or indirect impacts on the functions and viabilities of neural cells. Results of genetic investigations have led investigators to express mutant FAD (familial AD) genes in mice to model the disease [8,19,48,63,78,84,103,105] and to ablate genes in disease pathways in efforts to define the molecular participants critical to pathogenesis [46,52]. Models of this disease have provided new insights into how these altered proteins contribute to pathogenic mechanisms (gains of adverse properties, loss of functions), particularly with regard to the roles of abnormal conformations of p-tau or cleavage-generated peptides (β sheets of amyloid)[8,19,48,63,84,103,105]. Moreover, these models have been useful in identifying potential targets for therapy and novel treatments [11,48,52,56,63,78,80,82]. These models are used to assess new treatments or strategies [71,103,105].

In this Chapter we first describe the syndromes of mild cognitive impairment (MCI) and AD; diagnostic tests, including results from imaging studies; measurements of biomarkers in serum and CSF; and the neuropathology and biochemistry of the disease [7,24,28,42]. Subsequently, we focus on identified causative and risk genes. With this as a background, we detail outcomes of transgenic and gene-targeting strategies that have been used to create disease models (i.e., mice expressing mutant transgenes) and to identify potential therapeutic opportunities (targeting of genes encoding proteins implicated in disease pathways). We show how these investigations of model systems have delineated the efficacies and, on some occasions, potential toxicities of various manipulations of potential therapeutic targets. The demonstration of beneficial outcomes and the clarification of safety issues is critical for the design of new therapeutic approaches that are beginning to enter human trials. Clinical, imaging, and biomarker studies, which are proving to be of value for early diagnosis, will be critical for assessing the outcomes of these therapeutic trials. We believe that these new disease-modifying therapies will have a major impact on the lives of the elderly.

CLINICAL AND LABORATORY FEATURES OF CASES OF MCI AND AD

The index case of AD, a middle-aged woman with behavioral disturbances and dementia, was described more than 100 years ago [7]. Affecting more than

4 million people in the United States, AD is characterized by progressive impairments in memory and cognitive processes, ultimately leading to dementia [7,105]. Many elderly persons exhibit mild cognitive impairments (MCIs), characterized by memory complaints and mild impairments on formal testing, associated with intact general cognition and preserved activities of daily living [7]. Although not everyone with MCIs progress to AD, this syndrome, particularly the amnestic form of MCIs (aMCIs), is regarded as a transitional stage between normal aging and early AD or as an initial manifestation of AD [7,60]. As the illness advances, patients with AD develop progressive difficulties with memory and with a variety of cognitive functions [7]. In the late stages, affected persons become profoundly demented.

Physicians rely initially on histories, on physical, neurological, and psychiatric examinations, and on neuropsychological tests for initial diagnosis [7]. More recently, studies of biomarkers in body fluids and imaging of the brain offer great promise for early diagnosis and for assessing outcomes of antiamyloid treatments [28,42]. A recent study has demonstrated the association of low plasma A42/40 ratios with elevated risk for MCIs and AD [28]. In cases of AD, the levels of Aβ peptides in cerebrospinal fluid (CSF) are often low, and CSF levels of tau, particularly conformationally altered tau, are often elevated compared to controls [90]. Imaging studies of value include magnetic resonance imaging (MRI), which discloses progressive atrophy of specific regions of the brain, particularly the hippocampus and entorhinal cortex, and positron-emission tomography (PET) using [^{18}F] deoxyglucose (FDG) or single-photon-emission computerized tomography (SPECT), which detect decreased glucose utilization and early reductions in regional blood flow in the parietal and temporal lobes, respectively [7,42]. Studies of transgenic models of amyloidosis in the CNS suggest that efflux of Aβ from brain to plasma may serve as a measure of Aβ brain burden. Moreover, inverse relationships may exist between the amyloid load in the brain (as assessed by PET amyloid imaging) and levels (low) of Aβ in CSF. Using a novel in vivo approach, investigations have shown that the synthesis and turnover of Aβ in CSF is very rapid [24]. Biomarker and imaging studies should promote more accurate diagnosis of AD in early stages, and, presumably, will ultimately lead to more accurate assessments of the efficacies of new antiamyloid therapeutics.

Early information about the circuits damaged by disease lead to the design of symptomatic therapies for AD [7]. The demonstration of cholinergic deficits in the cortex and hippocampus and abnormalities of basal forebrain neurons led to the introduction of cholinesterase inhibitors for treatment. Evidence of involvement in glutamatergic systems in hippocampal and cortical circuits in AD, coupled with information about glutamate excitotoxicity (mediated, in part, by NMDA-R), led to trials of mementine NMDA-R antagonist [54]. The basic concept underlying the potential value of this class of drugs is activated by the pathological state that is the target of inhibition (i.e., excitotoxicity) [54]; while not affecting normal functions, the drug can serve as a neuroprotective agent in this setting. Both of these strategies are associated with modest and transient symptomatic benefits in some patients.

NEUROPATHOLOGY AND BIOCHEMISTRY

As outlined above, the clinical manifestations of AD arise from abnormalities involving brain regions and neural circuits comprised of populations of neurons that are essential for memory, learning, and cognitive performance [9,60]. Damaged neural systems include basal forebrain cholinergic neurons, circuits in amygdala and hippocampus, and predominantly glutamatergic nerve cells in the entorhinal/limbic cortices and in the neocortex [9,18,60,99]. In general, the character, distributions, and abundance of abnormalities (i.e., levels of Aβ burden, the presence of neuritic Aβ plaques and NFT, and decrements in numbers of synapses and cells) are thought to correlate with the clinical state documented in individual cases. In several cognitively characterized cohorts (consistency of controls, persons with aMCIs), cases of eAD and cases of aMCIs showed significant increases in the number of tangles in the ventral medial temporal lobe regions compared to controls [60]. Memory deficits appear to correlate most closely with the abundance of NFT in CA1 of the hippocampus and in the entorhinal cortex, suggesting that tangles, particularly in the medial temporal lobe, are more significant than amyloid deposits during the progression from normal state to MCIs to eAD [60]. It is hypothesized that the spread of NFT beyond the medial temporal lobe (i.e., to areas of neocortex) is most closely linked to the development of greater impairments in cognition, and eventually, dementia. These recent studies are consistent with the concept that aMCIs reflects a transitional state in the evolution of AD.

Cellular abnormalities within these regions include the presence within neurons of conformationally altered isoforms of tau assembled into PHF in NFT, in swollen neurites, and in neuropil threads [2,59]. Aβ-containing neuritic plaques, usually associated with both astroglial and microglial responses, are thought to represent sites of synaptic disconnection in regions receiving inputs from disease-vulnerable populations of neurons. The neuritic swellings represent degenerating axons or terminals and, possibly, dendrites [61]. Axonal varicosities are hypothesized to represent focal perturbations of axonal transport. In the target fields of damaged nerve cells, generic and transmitter-specific synaptic markers are reduced [18,91]. The mechanism of synaptic damage (whether pre- or post synaptic or both) and the molecular pathways involved in these abnormalities are very important areas of current research.

The clinical manifestations of aMCIs and AD reflect perturbations of synaptic communication within subsets of neural circuits followed by dying-back degeneration of axons and, eventually, by death of neurons. The presence of damaging Aβ peptides in terminal synaptic fields may be linked to p-tau containing PHF in neurites and cell bodies as follows: Aβ42 species, liberated at synapses, oligomerize forming extracellular Aβ assemblies or Aβ-derived diffusible ligands (ADDLs)[41,48], which affects pre-/postsynaptic targets, including glutamate receptors and other less well characterized entities, leading to synaptic dysfunction and, ultimately, disconnection of terminals from postsynaptic targets [48,61,62,81,105]. Subsequently, a retrograde signal (of uncertain nature), which originates presumably in proximity to damaged terminals, triggers signals that activate kinases (or inhibit phosphatases) in cell bodies; elevations of hyperphosphorylated tau, a microtubule-associated protein (MAP), lead to the well-established conformational changes in this protein and to the formation of PHF followed by destabilization of microtubules [2,59]. Moreover, disturbances of the cytoskeleton are presumably associated with alterations in axonal transport [47,71,105], which can, in turn, compromise the functions and viabilities of neurons. It is not known whether the P-tau-related dysfunction is a loss of function (loss of tubulin stability), a gain of an adverse property (by normal tau sequestered in PHF or by affecting transport), or the admixture of these two influences on neurons. Some of the discrepant outcomes in the literature reflect the nature of the studies (in vitro or in vivo) of model systems; the use of different experimental designs and the difficulty of identifying specific pathogenic effects linked to different forms of the toxic proteins enhance the problem of interpretation of outcomes. Moreover, the interpretation of the character, time course, and contribution of these events in AD is very difficult when relying only on postmortem human tissue. Eventually, damaged nerve cells die and extracellular "tombstone" tangles and neuritic amyloid plaques, surrounded by glial cells, represent the remains of ravages of disease.

GENETICS: FAMILIAL AD AND INFLUENCES OF RISK FACTORS

Mutant APP, PS1, and PS2

The major risk factor for putative sporadic AD is age, while inheritance of mutations or duplications of specific genes causes autosomal dominant familial AD (FAD) [29,103]. In FAD, mutation of genes encoding the amyloid precursor protein (APP) or the presenilins (PS1 and 2) influence Aβ cleavages and thus the levels and/or character (size) of Aβ peptides, which are generated by the activities of β-amyloid cleaving enzyme1 (BACE1), and, γ-secretase (a multiprotein catalytic complex comprised of PS, Nct, pen2, and Aph-1). The role of *APP* gene dosage has been documented in families with *APP* duplications and in persons with Down's syndrome (trisomy 21), who have an extra copy of *APP* [29]. Moreover, the presence of specific alleles of other genes, including ApoE4, are risk factors for putative sporadic disease [29,71,103].

The genetics of AD are complex and exhibit age-related patterns: Rare early-onset FAD mutations in *APP* and *PS* genes are transmitted as autosomal dominants; late-onset cases of AD without clear familial segregation are thought to reflect the influences of multiple risk factors [29,103]. FAD-causing mutations, occurring in three different genes located on three different chromosomes, influence a common biochemical pathway (i.e., the altered production of Aβ leading to a relative overabundance of the Aβ42 species). More than 160 mutations in these three genes (*APP, PS1, PS2*) have been reported to cause FAD. The most frequently mutated gene, *PS1*, accounts for the majority of cases with onset prior to age 50. An overview of disease-causing mutations is available at the Alzheimer Disease and Frontotemporal Dementia Mutation Database. Autosomal dominant mutations in *APP* (chromosome 21), *PS1* (chromosome 14), or *PS2* (chromosome 1) usually cause disease earlier than occurs in sporadic cases, with the majority of mutations in *APP, PS1*, and *PS2* influencing BACE1 or γ-secretase cleavages of APP to increase the levels of all Aβ species or the relative amounts of toxic Aβ42, respectively (see below). Individuals with duplications of *APP* [74] or with trisomy 21 (Down syndrome) [29] have an extra copy of *APP* and develop AD pathology relatively early in life. Cases with autosomal-dominant *APP* locus duplication often show evidence of abundant vascular and parenchymal amyloid [74].

ApoE Allele

To date, only the *Apoe4* allele of the *apolipoprotein E* gene (chromosome 19q13) has been replicated consistently in a large number of studies across many ethnic groups. While *ApoE4* is a susceptibility allele, *ApoE2*, a low-frequency allele, exhibits a weak protective effect. *ApoE*4 is neither necessary nor sufficient to cause AD, but appears to operate as a genetic risk modifier by reducing the age of onset in a gene dose-dependent manner. The biochemical consequences of the presence of *ApoE*4 in pathogenesis of AD are not yet fully understood, but this variant has been hypothesized to influence Aβ metabolism, Aβ aggregation/clearance [22,33]. A recent publication suggests that ApoE isoforms can differentially facilitate Aβ degradation by two metalloproteases, neprilysin and insulin-degrading enzyme, and thus influence clearance. ApoE4 appears to be the least effective ApoE variant. It is likely that additional late-onset AD loci remain to be identified, since APP, PS1, and 2, and ApoE account for less than 50% of the genetic variance of AD [105]. Identification of the risk genes and their function should provide new insights into disease mechanisms and potential therapeutic approaches.

Other Risk Genes

Recent research has identified gene variants encoding ubiquilin1 (*UBQLN1*) [5] and sortilin1 (*SORL1*) [73] as risk factors, that may act by influencing ubiquilin-mediated proteosomal degradation and trafficking in endosomal pathways, respectively [5,73]. The inherited variants of SORL1, documented in two clusters of the *SORL1* gene, are suggested to influence levels of expression of the protein, part of the retromer complex [89] that plays an important role in APP trafficking and pathways of recycling such that reduced expression increases entry of APP into compartments generating Aβ [73]. It is unclear how many newly recognized susceptibility loci, some of which have recently been uncovered by systemic metaanalyses [6], will prove to be significant risk factors. To date, in hundreds of independent association studies, no single gene has been demonstrated to contribute a risk approaching the same degree of consistency as *APOE*4.

APP, APLP, AND SECRETASES

Amyloid Precursor Protein

Members of the *APP* gene family (*APP, APLP1* and *2*) [95], encode type I transmembrane proteins whose functions are not fully defined [12,30,40,95,105]. APP is abundant in the nervous system, is rich in neurons, and is transported rapidly anterograde, along with secretase components, in axons to terminals [10,47,88]. At the +1 and +11 sites (see below), APP is cleaved by activities of BACE1 (β-site APP cleaving enzyme 1), producing N-terminal peptides and a secreted ectodomain (APPs). Within endocytic compartments, the γ-secretase cleavage generates the C-termini of Aβ peptides, as well as C-terminal fragments including an APP intracellular domain (AICD) [11,12,46,50,57,82,92]. There is some controversy as to the location of these biochemical events within neurons: One view holds that most of the Aβ is produced in endosomes and is released (in either pre- or postsynaptic locales) at synapses, while another school argues that Aβ peptides accumulate within neurons [45].

As described above, the *APPswe* mutation greatly enhances BACE1 cleavage at the +1 site N-terminus of Aβ, resulting in substantial elevations in levels of all Aβ peptides. The APP_{717} mutations promote γ-secretase cleavages to increase secretion of Aβ42, the most toxic Aβ peptide. While these *APP* mutations alter the processing of APP and increase the production of Aβ peptides or the amounts of the more toxic Aβ42, other *APP* mutations enhance local fibril formation and some play roles in vascular amyloidosis.

Amyloid Precursor–Like Proteins

Compared to APP, members of the amyloid precursor–like protein (APP) family, APLP1 and 2, discovered by genetic searches, exhibit both similarities and differences [95]. All have single-pass transmembrane domain and a conserved NPXY clathrin internalization signal in the conserved cytoplasmic domain. The APLPs lack the Aβ sequence of APP such that only cleavages of APP form Aβ [95]. Gene targeting studies have disclosed some redundancies in the APP family in that single knockouts are associated with mild phenotype differences, but $APP^{-/-}$, $APLP2^{-/-}$, and $APLP1^{-/-}$ do not survive, whereas $APP^{-/-}$ and $APLP2^{-/-}$ mice appear relatively normal [30]. All three family members undergo shedding of the ectodomain and cleavage by γ-secretase and release of C-terminal intracellular domain (ICD) fragments, which can serve signaling functions.

Cleavage by β- and γ-secretases releases the ectodomain of APP (APPs), liberates a cytosolic fragment termed APP intracellular domain (AICD) [12], and generates several species of Aβ peptides. In the CNS ≫ PNS, APP and the pro-amyloidogenic secretases are present in neurons and carried anterograde by fast axonal transport [10,47,88]; at terminals, Aβ peptides are generated by sequential endoproteolytic cleavages by BACE1 (at the Aβ +1 and +11 sites) to generate APP-β carboxyl-terminal fragments (APP-βCTFs) [10,11] and by the γ-secretase complex (at several sites varying from Aβ 36,38,40,42,43) to form Aβ species peptides [37,57]. The intramembranous cleavages of APP-βCTF by γ-secretase releases an APP intracellular domain (AICD) [12], which can form a complex with Fe65, a nuclear adaptor protein [12]; it is suggested that Fe65 and AICD or Fe65 alone (in a novel conformation) can gain access to the nucleus to influence gene transcription [12], a signaling mechanism analogous to that occurring in the Notch1 (NICD) pathway [55]. It has been suggested that AICD signaling may possibly play a role in learning and memory, hypothesis outlined briefly below.

BACE1 and BACE2

BACE1 is a transmembrane aspartyl protease that is directly involved in the cleavage of APP at the +11 > +1 sites of Aβ in APP [11,46,92]. In the CNS, BACE1 is demonstrable in a variety of presynaptic terminals [46]. Brain cells from *BACE1*$^{-/-}$ mice [11,46] do not produce Aβ1(40/42) and Aβ11(40/42), indicating that BACE1 is the neuronal β-secretase [11,46]. Compared to wild-type APP, APPswe is cleaved approximately 100-fold more efficiently at the +1 site, resulting in a greater increase in BACE1 cleavage products (elevating all Aβ species).

BACE2 is not an amyloidogenic enzyme in that it cleaves APP between residues 19 and 20, and 20 and 21. Although BACE2 appears in some populations of neurons in the CNS, its distribution is different from that of BACE1.

γ-Secretase

This multiprotein complex includes PS1 and [20,21,101] 2Ps; Nicastrin (Nct), a type I transmembrane glycoprotein; and Aph-1 and Pen-2, two multipass transmembrane proteins. This complex is essential for the regulated intramembranous proteolysis of a variety of transmembrane proteins, including APP and Notch [50,57,82,84]. PS1 and PS2, two highly homologous 43- to 50-kD a multipass transmembrane proteins [82,84], along with other members of this complex, are involved in regulated intramembranous cleavages of a variety of transmembrane proteins, including APP and Notch 1, which is critical for signaling necessary for cell fate decisions [50,55,82,84,101,102]. PS contains two aspartyl residues that play roles in intramembranous cleavage; substitutions of these residues (D257 in TM 6 and at D385 in TM 7) are reported to reduce secretion of Aβ and cleavage of Notch1 in vitro [101,102]. The functions of the various γ-secretase proteins and their interactions in the complex are not yet fully defined. It has been suggested that the ectodomain of Nct may be important in substrate recognition and binding of amino-terminal stubs (of APP and other transmembrane proteins) generated by a sheddases (i.e., BACE1 for APP). In one model, Aph-1 and Nct form a pre-complex that interacts with PS; subsequently, Pen-2 enters the complex, where it is critical for the "presenilinase" cleavage of PS into two fragments. PSs are endoproteolytically cleaved by a presenilinase to form an N-terminal of approximately 28 kDa fragment and a C-terminal fragment of about 18kDa, both of which are critical components of the γ-secretase complex [71,103].

As mentioned above, nearly 50% of early-onset cases of FAD are linked to over 100 different mutations in *PS1* [29,71,103]. A relatively small number of *PS2* mutations also cause autosomal dominant FAD. The majority of abnormalities in *PS* genes are missense mutations that enhance γ-secretase activities to increase the levels of Aβ42 peptides.

α-Secretase

TACE (TNFα converting enzyme) is expressed at low levels in neurons of the CNS. In other cells in other organs, APP is cleaved endoproteolytically within the Aβ sequence through alternative, nonamyloidogenic pathways. For example: α-secretase, or TACE, cleaves between Aβ residues 16 and 17 [87]. The α and BACE2 cleavages, which occur predominantly in nonneural tissues, preclude the formation of Aβ peptides and serve to protect these cells/organs from Aβ amyloidosis [104].

TRANSGENIC MODELS OF Aβ AMYLOIDOSIS AND TAUOPATHIES

Models of Aβ Amyloidosis

Investigators have taken advantage of information from genetics to create transgenic models of amyloidosis [63,78]. Mice expressing *APPswe* or APP_{717} (with or without mutant *PS1)* develop an Aβ amyloidosis in the CNS [8,48,78,79]. Mutant *APP*;*PS1* mice develop an accelerated disease secondary to increased levels of Aβ (particularly Aβ42) associated with the presence of diffuse Aβ deposits and neuritic plaques, associated with local glial responses [63] in the hippocampus and cortex. With age, levels of Aβ peptides, particularly Aβ42, increase significantly in the brain [8,78], and oligomeric species, variously termed ADDLs, $A\beta^{*}56$, and so on, appear in the CNS [41,43,48,94,96]. Depending on the nature of mouse strain, transgene construct, types of mutations, and levels of expression, some lines of mice show abundant evidence of amyloid in vessels. In forebrain regions, the density of synaptic terminals is decreased [79], and levels of transmitter markers can be modestly reduced. In some settings there are deficiencies in synaptic transmission [78], and in some lines of mice there is evidence of degeneration of subsets of neurons. *APPswe/ind* mice, whose transgene is regulated by doxycycline (Dox), have high levels of transgene expression and exhibit amyloidosis in brain [39]. Treatment with Dox decreases levels of expression (95%) accompanied by decreased Aβ production to levels of nontransgenic animals. Although the degree of amyloidosis is reduced, clearance of amyloid plaques appears to be slow, and mice with mutant *APP* expression suppressed for six months still show a significant Aβ burden.

A variety of imaging approaches have been used to examine pathology in lines of mutant mice [31,38,58,64]. Two recent studies are particularly noteworthy: first, investigators used [^{11}C]PIB to demonstrate significant retention of labeling in regions containing amyloid and to monitor responses to immunotherapy [58]; second, a two-photon imaging of labeled compounds demonstrated the rapid appearance of amyloid deposits and the subsequent recruitment of microglia and the appearance of dysmorphic neurites [64].

Behavioral studies of these lines of mice, including those generated at Johns Hopkins [78,79], disclose deficits in spatial reference memory (Morris Water Maze task) and episodic-like memory (repeated reversal and radial water maze tasks). Although *APPswe/PS1dE9* mice develop plaques at 6 months of age, all genotypes are indistinguishable, at this time, on all cognitive tests from nontransgenic animals. However, at 18months, *APPswe/PS1dE9* mice do not perform all cognitive tasks as well as mice of all other genotypes. Relationships exist between deficits in episodic-like memory tasks and total Aβ loads in the brain [78,79]. In concert, these studies of *APPswe/PS1dE9* mice suggest that some form of Aβ (ultimately associated with amyloid deposition) disrupts circuits critical for memory, with episodic-like memory being most sensitive to the toxic effects of Aβ.

The site(s) of Aβ neurotoxicity (i.e., terminal axons, presynaptic elements, and/or postsynaptic components) and the molecular interactions underlying the abnormalities remain to be defined [41,43,47,48,64,81]. Behavioral and physiological deficits have been linked to the presence of Aβ oligomers, and some of these abnormalities can be reversed by antibody-mediated reductions of levels of Aβ in the brain [43,48] (see below). Studies of TG2576 mice suggest that extracellular accumulations of 56-kDa soluble amyloid assemblies (termed Aβ*56), purified from the brains of memory-impaired mice, interfere with memory when delivered to young rats [48]. Although these transgenic lines do not reproduce the full phenotype of AD (including NFT and the death of neurons), these studies demonstrate that these mice are very useful subjects for research designed to link behavior and Aβ amyloidosis, to delineate disease mechanisms, and to test novel therapies [63,78].

As indicated above, a variety of Aβ species, ranging from monomers to oligomers, structural assemblies, and fibrillar amyloid deposits in neuritic plaques, have been suggested, at various times, to play important roles in impairing synaptic communication. The pool of insoluble Aβ (or plaques) is believed to exist in equilibrium with peptides in interstitial fluid [15]. Significantly, systemic administration of Aβ antibodies increases levels of Aβ in plasma, and the magnitude of this elevation appears to correlate with the amyloid burden in the cortex and hippocampus. Available evidence suggests that the systemic administration of antibodies facilitates movement of the peptides from the brain (the major site of production) to plasma. In one study, an Aβ peptide (naturally secreted in vitro) was injected into the ventricular system of rats and shown to inhibit LTP in the hippocampus [43]. The adverse activity of this peptide was blocked by the injection of a monoclonal Aβ antibody; active immunization was less effective in rescuing functions [43]. These observations and others are consistent with the concept that oligomeric species are toxic in the brain and are both necessary and sufficient to perturb learned behavior.

Models of Tauopathies

Tau, a low-molecular-mass microtubule-associated protein, is a key cytoskeletal protein important in protein trafficking, especially axonal transport [2,59]. Early efforts to express *tau* transgenes in mice did not lead to striking clinical phenotypes or pathology [2,63]. The paucity of tau abnormalities in various lines of mutant mice with Aβ abnormalities may be related to differences in tau isoforms expressed in this species. When prion or Thy1 promoters are used to drive tau_{P3O1L} (a mutation linked to autosomal dominant frontotemporal dementia with parkinsonism), some brain and spinal cord neurons develop tangles [2]. Aged mice expressing htau, in the absence of mouse tau, develop NFT and evidence of death of neurons [1]; this phenotype appears to be associated with reexpression of cell-cycle proteins and synthesis of DNA, which has been interpreted as consistent with the abortive efforts to reenter the cellcycle. In tau_{P301L} mice, injection of Aβ42 fibrils into specific brain regions increases the number of tangles in those neurons that project to sites of Aβ injections [27]; mice expressing $APPswe/tau_{P301L}$ exhibit enhanced tanglelike pathology in limbic system and olfactory cortex [49]. This observation is consistent with the hypothesis that the presence of Aβ in proximity to terminals is, in unknown ways, able to facilitate the formation of tangles in cell bodies of these neurons.

Limited behavioral studies in some of these models in the presence of tau pathology are associated that motor signs, a phenomenon that has been circumvented in some models. A triple transgenic mouse, created by microinjecting *APPswe* and tau_{P3O1L} into single cells derived from monozygous $PS1_{M146V}$ knock-in mice, develop age-related plaques and tangles as well as deficits in LTP, which appear to antedate overt pathology [69]. Conditional *P301 Tau* mice exhibit expression restricted to the forebrain [76]; they manifest behavioral impairments and show NFT and loss of neurons in the forebrain. After suppression of tau expression by administration of Dox, memory functions recovered and there was stabilization of numbers of neurons, but NFT continue to accumulate [76]. The authors conclude that in this model, NFT are not sufficient to cause cognitive decline or death of neurons. The various lines of mice bearing both mutant *tau* and *APP* (or *APP/PS1*) or mutant *tau* mice injected with Aβ are not ideal models of FAD because the presence of the *tau* mutations alone is associated with the development of tangles and disease. In vitro studies show that increasing levels of tau inhibit transport, particularly in anterograde direction [59]. While increasing levels of tau reduces vesicule trafficking, this manipulation does not increase generation of Aβ [26]. Finally, when endogenous tau is reduced in vivo, behavioral deficits in mutant *APP* mice are improved without altering levels of Aβ in the brains; the authors suggest that reduction in levels of tau can protect against excitotoxicity [72].

RESULTS OF TARGETING OF GENES ENCODING AMYLOIDOGENIC SECRETASES

To begin to understand the functions of some of the proteins thought to play roles in AD, investigators have targeted a variety of genes, including APP and family members, *BACE1, PS1, Nct;* and *Aph-1:* APP–DLP [30].

BACE1

BACE1$^{-/-}$ mice mate successfully and exhibit no obvious pathology [11,46,78]. *BACE1*$^{-/-}$ neurons do not cleave at the +1 and +11 sites of Aβ, and the production of Aβ peptides is abolished [11,46], observations establishing that BACE1 is the neuronal β-secretase required to generate the N-termini of Aβ. However, *BACE1*$^{-/-}$ mice show altered performance on some tests of cognition and emotion [46,78]. *BACE1* null mice manifest alterations in both hippocampal synaptic plasticity and in performance on tests of cognition and emotion [46]; the memory deficits (but not emotional alterations) in *BACE1*$^{-/-}$ mice are prevented by coexpressing *APPswe;PS1ΔE9* transgenes. This observation suggests that APP processing influences cognition/memory and that the other potential substrates of BACE1 may play roles in neural circuits related to emotion. More recently, two studies [34, 100] demonstrate that genetic deletion of *BACE1* is associated with a delay in myelination, reduced thickness of myelin sheaths, increased g-ratios, and decreased myelin markers. These abnormalities reflect alterations in the biology of neuregulin (NRG), which is known to be a signal by which axons communicate with ensheathing cells and influence myelination during development. BACE1 cleaves NRG, and processed NRG regulates myelination by phosphorylation of Akt. In BACE1$^{-/-}$ mice, NRG, cleavage products are decreased and full-length NRG is increased; levels of phosphorylated Akt are dimininished [34]. In concert, these investigations of BACE1-targeted mice suggest that BACE1 and APP/NRG processing pathways are critical for cognitive, emotional, and synaptic functions and for myelination during development of the PNS and CNS.

PS1 and PS2

PS1$^{-/-}$ embryos develop severe abnormalities of the axial skeleton, ribs, and spinal ganglia; this lethal outcome resembles a partial *Notch 1*$^{-/-}$ phenotype

[106]. *PS1*$^{-/-}$ cells secrete decreased levels of Aβ [21,50] due to the fact that PS1 (along with PS2, Nct, Aph-1, and Pen-2) is a component of the γ-secretase complex that carries out the S3 intramembranous cleavage of Notch1 [20,50,55,82]. Without γ-secretase activity, cleavage of the NEXT to NICD does not occur; NICD is not released from the plasma membrane and cannot reach the nucleus to provide a signal to initiate transcriptional processes essential for cell fate decisions. Significantly, conditional *PS1/2* targeted mice show impairments in memory and in hippocampal synaptic plasticity [77], raising important questions as to the roles of loss of PS functions in neurodegeneration and in AD [19,84]. It is important to note that *PS1*$^{-/-}$ mice whose lethal phenotype is rescued through neuronal expression of *PS1* develop skin cancer; this outcome was interpreted initially to reflect deregulation of the β-catenin pathway, but may operate through other mechanisms.

Nct

Nct$^{-/-}$ mice embryos die early and exhibit several patterning defects [50], including abnormal segmentation of somites; this phenotype closely resembles that seen in *Notch1*$^{-/-}$ and *PS* 1/2$^{-/-}$ embryos. Importantly, *Nct*$^{-/-}$ cells do not secrete Aβ peptides, whereas *NctT*$^{-/-}$ cells show reductions of about 50% [50]. The failure of *NctT*$^{-/-}$ cells to generate Aβ peptides is accompanied by accumulation of APP C-terminal fragments. Importantly, *Nct*+$^{-/-}$ mice develop tumors of the skin, a phenotype accelerated by reducing *PS1* and *P53*, both of which manipulations exacerbate the tumor phenotype [51]. The formation of these tumors appears to reflect decreased γ-secretase activities and activity of Notch1 (a tumor suppressor in the skin).

Aph-1

Aph-1a, Aph-1b, and *Aph-1c* encode four distinct Aph-1 isoforms: Aph-1aL and Aph-1aS (derived from differential splicing of *Aph-1a*), Aph-1b, and Aph-1c [57]. *Aph-1a*$^{-/-}$ embryos have patterning defects that resemble, but are not identical to, those of *Notch1, Nct*, or *PS1* null embryos [57,83]. Moreover, in *Aph-1a*$^{-/-}$-derived cells, the levels of Nct, PS fragments, and Pen-2 are decreased, and there is a concomitant reduction in levels of the high-molecular-weight γ-secretase complex and a decrease in secretion of Aβ [57]. In *Aph-1a*$^{-/-}$ cells, other mammalian Aph-1 isoforms can restore the levels of Nct, PS, and Pen-2 [57].

EXPERIMENTAL MANIPULATIONS AND POTENTIAL THERAPEUTIC STRATEGIES

Models relevant to amyloidogenesis provide an opportunity to test the influence of ablations or knockdowns of specific genes, to modulate cleavage patterns influencing generation of neurotoxic peptides; and to enhance clearance and/or degradation of Aβ [43,46,48,50,66,78]. Below we comment on selected studies that discuss several experimental strategies directed at specific therapeutic targets that hold promise for development of mechanism-based therapeutics to benefit patients with AD [78].

Reductions in β-Secretase Activity

BACE1$^{-/-}$; *APPswe;PS1ΔE9* mice do not develop Aβ deposits or age-associated abnormalities in working memory that occur in the *APPswe;PS1ΔE9* model of Aβ amyloidosis [46]. Moreover, Aβ deposits are sensitive to *BACE1* dosage and can be cleared efficiently from regions of the CNS when *BACE1* is silenced [46,86]. New approaches using conditional expression systems, RNAi silencing, or manipulations of transcription will allow investigators to examine the roles of specific proteins in the pathogenesis of diseases and to assess the degrees of reversibility of the disease processes [46,86]. The results of these approaches, along with the development of brain-penetrant inhibitors of enzyme activity in the design of new treatments, can be tested in clinical trials.

Although BACE1 is a very attractive therapeutic target [16,46], several potential problems exist with this approach. First, the BACE1 catalytic site is quite large, and it is not yet known whether it will be possible to achieve adequate brain penetration of a compound of sufficient size that it will be active in vivo. Second, BACE1 inhibitors are transported out of the brain by a p-glycoprotein; this phenomenon could be an issue in trying to maintain adequate concentrations of inhibitors in the brain. In a recent study, an inhibitor of this process was used with some success to enhance levels of a BACE1 inhibitor in the CNS [35]. Third, *BACE1* null mice manifest alterations in both hippocampal synaptic plasticity and performance on tests of cognition and emotion [46]. The memory deficits (but not emotional alterations) in *BACE1*$^{-/-}$mice are prevented by coexpressing *APPswe;PS1ΔE9* transgene; suggesting that APP processing influences cognition/memory and that the other potential substrates BACE1 may play roles in neural circuits related to cognition and emotion. Fourth, as described above, genetic deletion of *BACE1* causes hypomyelination in the developing PNS and CNS [34,100], so a phenotype is proposed to reflect alterations in the NRG–Akt pathway. Thus, although inhibition of β-secretase activity represents an exciting therapeutic opportunity, future studies will be needed to assess possible mechanism-based side effects that may occur with inhibition of BACE1[46,78,105]. Once brain-penetrant inhibitors are available, clinical trials will begin.

Reduction of γ-Secretase Activity

Both genetic and pharmaceutical lowering of γ-secretase activity decrease production of Aβ peptides in cell-free and cell-based systems and reduce levels of Aβ in mutant mice with Aβ amyloidosis, indicating that γ-secretase activity is a significant target for therapy [50,52,57,77,101,105]. However, γ-secretase activity is also essential for processing of Notch and a variety of other transmembrane proteins [55], which are critical for many properties of cells, including lineage specification and cell growth during embryonic development [50,55,57,77,82,105,106]. Significantly, one inhibitor of γ-secretase (LY–411, 575) reduces production of Aβ, but it also has profound effects on T-and B-cell development and on the appearance of intestinal mucosa (i.e., proliferation of goblet cells, increased mucin in gut lumen and crypt necrosis) [4]. Although $Nct^{+/-}$ *APPswe;PS1ΔE9* mice show reduced levels of Aβ and amyloid plaques [52], these mice also develop skin tumors, presumably due in part to reduction of γ-secretase activity and the role of Notch as a tumor suppressor in skin [51]. The mechanism whereby decrements in the activity of γ-secretase lead to squamous cell tumors is not fully understood, but appear to relate to tumor-suppressing activity of the enzyme in epithelium. In $Nct^{+/-}$ animals, Notch signaling is reduced and the epidermal growth factor receptor is activated; levels of the receptor are inversely correlated with proliferative activity in cells of the skin [51]. During trials of inhibitors, it will be necessary to be alert to potential adverse events.

Modulation of γ-Secretase Activities

Retrospective epidemiological studies suggest that significant exposure to NSAIDs reduces risk of AD, an outcome initially interpreted as related to suppression of the well-documented inflammatory process occurring in the brains of AD Patients [97]. However, in vitro studies indicate that a subset of NSAIDs modulate secretase cleavages to form shorter, less toxic Aβ species without altering processing of Notch or other transmembrane proteins [97]. Recent biochemical studies suggest that NSAID γ-secretase modulators (GSM) interact, not with γ-secretase components, but with APP, particularly with residues 28 to 36 of the Aβ domain [44]. The outcomes of these interactions are reductions of Aβ42 production as well as inhibition of Aβ aggregates. Finally, short-term treatment of mutant mice appears to have some benefit in terms of lowering levels of Aβ and the number of plaques [53]. This strategy is now being evaluated in a phase III clinical trial.

Removal of the Sources of Aβ

Investigations utilizing lesions of entorhinal cortex or perforant pathway [85] to remove APP, the source of Aβ, by lesioning cell bodies or axons/terminals involved in transport of APP to terminals significantly reduce levels of Aβ and amyloid plaques in target fields. Obviously, this strategy does not represent a therapeutic approach, but these studies do represent a proof of principle that when APP is no longer transported to target fields, Aβ can be reduced by a variety of mechanisms, including clearance.

Aβ Immunotherapy

To date, the most exciting findings regarding clearance of Aβ come from studies using active and passive Aβ immunotherapy [32,66,78,82,98]. In treatment trials in mutant mice, both Aβ immunization (with Freund's adjuvant) and passive transfer of Aβ antibodies reduce levels of Aβ and plaque burden [3,23,43,66,68,107]. The mechanisms whereby immunotherapy enhances clearance are not completely defined [71,105], and investigators have proposed at least two not mutually exclusive hypotheses: (1) a very small amount of Aβ antibody enters the brain, binds to Aβ peptides, promotes the disassembly of fibrils, and, via the Fc antibody domain, encourages activated microglia to enter the affected regions and to remove Aβ; and (2) serum antibodies serve as "a sink" for the amyloid peptides (derived from neuronal APP) drawn into the circulation, thus changing the equilibrium of Aβ in different compartments and promoting removal of Aβ from the CNS [15,23]. Whatever the mechanisms, Aβ immunotherapy in mutant mice is successful in partially clearing Aβ, in attenuating learning and behavioral deficits in several different cohorts of mutant *APP* or *APP/PS1* mice, and in partially reducing tau abnormalities in the triple transgenic mice [23,68,78].

However, several problems have been associated with Aβ immunotherapy. In the presence of congophilic angiopathy, brain hemorrhages may be associated with immunotherapy [70], perhaps because the presence of amyloid in vessels can weaken vascular walls and, potentially, immunotherapeutic removal of some intramural vascular amyloid could contribute to rupture of damaged vessels and to local bleeding. More significant is the evidence that a subset of patients receiving Aβ vaccination with certain adjuvants develop meningoencephalitis (see below). These observations indicate that although preclinical trials in mice are useful for testing efficacies, they are not necessarily predictive of adverse events in humans.

To illustrate the challenges of extrapolating outcomes in mice to trials with humans, it is useful to discuss briefly recent problems with Aβ immunotherapy. In prevention and treatment preclinical trials, both Aβ immunization (with Freund's adjuvant) and passive transfer of Aβ antibodies reduce levels of Aβ and plaque burden in mutant APP transgenic mice. Thus, immunotherapy in transgenic mice is successful in clearing Aβ and attenuating learning and behavioral deficits in at least two cohorts of mutant *APP* mice. However, patients receiving vaccinations with preaggregated Aβ and an adjuvant (followed by a booster), developed antibodies that recognize Aβ in the brain and vessels [82]. Unfortunately, although phase I vaccination trials with Aβ peptide and adjuvant were not associated with any adverse events, phase II trials detected complications (meningoencephalitis) in a subset of patients and were suspended [62,66,67]. Apparently, some changes were made in adjuvant and/or formulation during the trial. The pathology in the index case, consistent with T-cell meningitis [67], was interpreted to show some clearance of Aβ deposits, but some regions contained a relatively high density of tangles, neuropil threads, and vascular amyloid [67]. Aβ immunoreactivity was sometimes associated with microglia. T-cells were conspicuous in subarachnoid space and around some vessels [67]. In another case, there was significant reduction in amyloid deposits in the absence of clinical evidence of encephalitis [62]. Although the trial was stopped, assessment of cognitive functions in a small subset of patients (30) who received vaccination and booster immunizations, disclosed that patients who generated Aβ antibodies (as measured by a new assay) appeared to have a slower decline in several functional measures. The events occurring in this subset of patients illustrate the challenges of extrapolating outcomes in mutant mice to human trials. Investigators are attempting to make new N- and C-terminal antigens/adjuvant formulations that do not stimulate T-cell-mediated immunologic attack and are pursuing, in parallel, passive immunization approaches [66,82].

Lipoprotein Receptor Protein (LRP-IV)

An alternative clearance strategy is the systemic administration of a recombinant soluble low-density lipoprotein receptor protein (sLRP) which serves as a sink by binding Aβ peptides in the circulation [75]. This approach decreases endogenous Aβ species in the brain of control mice and in a chronic dosing paradigm, including *APPswe* mice, improved blood flow responses to stimulate performance of normal behavior. This outcome was accompanied by reduced Aβ levels in the brain and vasculature increased Aβ in plasma. Finally, in cases of AD, levels of sLRP in plasma are reduced compared to controls and there is a decrease in sLRP-bound Aβ(40/42) and an increase in free Aβ(40/42). These findings are interpreted to indicate that LRP-IV does not enter the brain and reduce Aβ via binding in the periphery.

Activation of Proteases Capable of Cleaving Aβ

Recently, investigators have attempted to influence levels of Aβ-degrading enzymes to promote amyloid degradation and clearance [36]. Increasing local levels of two metalloproteases, insulin degrading enzyme (IDE) and neprilysis (NEP), both of which cleave Aβ, reduces levels of the amyloid peptide in regions showing protease activities [36]. However, the challenge with this strategy includes difficulties in controlling the regulation of theses enzymes and the possible off-target effects of these proteases, which can also cleave non-Aβ targets (i.e., other proteins important for normal functions) [13,36,65,93].

A variety of other treatment strategies have been tested in mouse models, but space constraints limit discussion.

CONCLUSIONS

At the molecular and cellular levels AD is a protein-misfolding disease. Over the past decade, substantial progress has been made in understanding this illness.

Investigators have defined the features of MCIs and early AD, developed diagnostic and outcome measures using biomarkers and imaging, and characterized the stages of pathology that correlate clinical features of MCIs and eAD. Genetic studies have provided information regarding the roles of autosomal-dominant mutations in *APP* and *PS* genes, the dose-dependent risks of the ApoE4 alleles, and have identified other loci of risk. Parallel studies of AD and of genetically engineered models of Aβ amyloidosis (and the tauopathies) have greatly increased our understanding of pathogenic mechanisms, possible therapeutic targets, and potential mechanism-based treatments designed to benefit patients with AD. Decreasing production and assembly of misfolded protein, and the promotion of degradation and clearance of neurotoxic peptides are central to many of these strategies. This field is now on the threshold of implementing novel treatments based on an understanding of the neurobiology, neuropathology, and biochemistry of this illness. Discoveries over the next few years will lead to the design of new mechanism-based therapies that can be tested in vitro and in animal models and which will then be introduced into the clinic for the benefit of patients with this devastating illness.

Acknowledgments

The authors wish to thank the many colleagues with whom they have worked at Johns Hopkins Medical School, including Sangram Sisodia, David Borchelt, Fiona Laird, Ying Liu, Marilyn Albert, Juan Troncoso, Huaibin Cai, Lee Martin, Mohamed Farah, and Gopal Thinakaran, as well as those at other institutions, for their contributions to much of the original work cited in this chapter and for their helpful discussions. Aspects of this work were supported by grants from the U.S. Public Health Service (P50 AGO05146, R01 NS041438, P01 NS047308, R01 NS045150) as well as the Adler Foundation, the Ellison Medical Foundation, the Alzheimer's Association, and private gifts.

REFERENCES

1. Andorfer, C., Acker, C.M., Kress, Y., Hof, P.R., Duff, K., Davies, P. (2005). Cell-cycle reentry and cell death in transgenic mice expressing nonmutant human tau isoforms. *J Neurosci, 25*, 5446–5454.
2. Ballatore, C., Lee, V.M., Trojanowski J.Q. (2007). Tau-mediated neurodegeneration in Alzheimer's disease and related disorders. *Nat Rev Neurosci, 8*, 663–672.
3. Bard, F., Cannon, C., Barbour, R., Burke, R.L., Games, D., Grajeda, H., Guido, T., Hu, K., Huang, J., Johnson-Wood, K., et al. (2000). Peripherally administered antibodies against amyloid beta-peptide enter the central nervous system and reduce pathology in a mouse model of Alzheimer disease. *Nat Med, 6*, 916–919.
4. Barten, D.M., Guss, V.L., Corsa, J.A., Loo, A., Hansel, S.B., Zheng, M., Munoz, B., Srinivasan, K., Wang, B., Robertson, B.J., et al. (2005). Dynamics of {beta}-amyloid reductions in brain, cerebrospinal fluid, and plasma of {beta}-amyloid

precursor protein transgenic mice treated with a {gamma}-secretase inhibitor. *J Pharmacol Exp Ther, 312*, 635–643.

5. Bertram, L., Hiltunen, M., Parkman, M., Ingelsson, M., Lange, C., Ranasantm, K., Mullin, K., Menon, R., Sampson, A.J., Hsiao, M.Y., Elliott, et al. (2005). Family-based association between Alzheimer's disease and variants in UBQLN1. *N Engl J Med, 352*, 884–894.
6. Bertram, L., McQueen, M.B., Mullin, K., Blacker, D., Tanzi R.E. (2007). Systematic meta-analyses of Alzheimer disease genetic association studies: the AlzGene database. *Nat Genet, 39*, 17–23.
7. Blennow, K., de Leon, M.J., Zetterberg, H. (2006). Alzheimer's disease. *Lancet, 368*, 387–403.
8. Borchelt, D.R., Ratovitski, T., Van Lare, J., Lee, M.K., Gonzales, V., Jenkins, N.A., Copeland, N.G., Price, D.L., Sisodia, S.S. (1997). Accelerated amyloid deposition in the brains of transgenic mice coexpressing mutant presenilin 1 and amyloid precursor proteins. *Neuron, 19*, 939–945.
9. Braak, H., Alafuzoff, I., Arzberger, T., Kretzschmar, H., Del Tredici, K. (2006). Staging of Alzheimer disease–associated neurofibrillary pathology using paraffin sections and immunocytochemistry. *Acta Neuropathol, 112*, 389–404.
10. Buxbaum, J.D., Thinakaran, G., Koliatsos, V., O'Callahan, J., Slunt, H.H., Price, D.L., Sisodia, S.S. (1998). Alzheimer amyloid protein precursor in the rat hippocampus: transport and processing through the perforant path. *J Neurosci, 18*, 9629–9637.
11. Cai, H., Wang, Y., McCarthy, D., Wen, H., Borchelt, D.R., Price, D.L., Wong, P.C. (2001). BACE1 is the major beta-secretase for generation of Abeta peptides by neurons. *Nat Neurosci, 4*, 233–234.
12. Cao, X., Sudhof, T.C. (2001).A transcriptionally [correction of transcriptively] active complex of APP with Fe65 and histone acetyltransferase Tip60. *Science, 293*, 115–120.
13. Carson, J.A., Turner, A.J. (2002). Beta-amyloid catabolism: roles for neprilysin (NEP) and other metallopeptidases? *J Neurochem, 81*, 1–8.
14. Chiti, F., Dobson C.M. (2006). Protein misfolding, functional amyloid, and human disease. *Annu Rev Biochem, 75*, 333–366.
15. Cirrito, J.R., May, P.C., O'Dell, M.A., Taylor, J.W., Parsadanian, M., Cramer, J.W., Audia, J.E., Nissen, J.S., Bales, K.R., Paul, S.M., et al. (2003). In vivo assessment of brain interstitial fluid with microdialysis reveals plaque-associated changes in amyloid-beta metabolism and half-life. *J Neurosci, 23*, 8844–8853.
16. Citron, M. (2004). [beta]-Secretase inhibition for the treatment of Alzheimer's disease: promise and challenge. *Trends Pharmacol Sci, 25*, 92–97.
17. Cohen, F.E., Kelly J.W. (2003). Therapeutic approaches to protein-misfolding diseases. *Nature, 426*, 905–909.
18. Coyle, J.T., Price, D.L., DeLong, M.R. (1983). Alzheimer's disease: a disorder of cortical cholinergic innervation. *Science, 219*, 1184–1190.
19. De Strooper, B. (2007). Loss-of-function presenilin mutations in Alzheimer disease: talking point on the role of presenilin mutations in Alzheimer disease. *EMBO Rep, 8*, 141–146.

20. De Strooper, B., Annaert, W.G., Cupers, P., Saftig, P., Craessaerts, K., Mumm, J.S., Schroeter, E.H., Schrijvers, V., Wolfe, M.S., Ray, W.J., et al. (1999). A presenilin-1-dependent gamma-secretase-like protease mediates release of Notch intracellular domain. *Nature, 398*, 518–522.

21. De Strooper, B., Saftig, P., Craessaerts, K., Vanderstichele, H., Guhde, G., Annaert, W.G., Von Figura, K., Van Leuven, F. (1998). Deficiency of presenilin-1 inhibits the normal cleavage of amyloid precursor protein. *Nature, 391*, 387–390.

22. DeMattos, R.B., Cirrito, J.R., Parsadanian, M., May, P.C., O'Dell, M.A., Taylor, J.W., Harmony, J.A., Aronow, B.J., Bales, K.R., Paul, S.M., Holtzman, D.M. (2004). ApoE and clusterin cooperatively suppress Abeta levels and deposition: evidence that ApoE regulates extracellular Abeta metabolism in vivo. *Neuron, 41*, 193–202.

23. Dodart, J.C., Bales, K.R., Gannon, K.S., Greene, S.J., DeMattos, R.B., Mathis, C., DeLong, C.A., Wu, S., Wu, X., Holtzman, D.M., Paul, S.M. (2002). Immunization reverses memory deficits without reducing brain Abeta burden in Alzheimer's disease model. *Nat Neurosci, 5*, 452–457.

24. Fagan, A.M., Mintun, M.A., Mach, R.H., Lee, S., Dence, C.S., Shah, A.R., LaRossa, G.N., Spinner, M.L., Klunk, W.E., Mathis, C.A., et al. (2005). Inverse relation between in vivo amyloid imaging load and cerebrospinal fluid Abeta42 in humans. *Ann Neurol, 59*, 512–519.

25. Goedert, M., Spillantini, M.G. (2006). A century of Alzheimer's disease. *Science, 314*, 777–781.

26. Goldsbury, C., Mocanu, M.M., Thies, E., Kaether, C., Haass, C., Keller, P., Biernat, J., Mandelkow, E., Mandelkow, E.M. (2006). Inhibition of APP trafficking by tau protein does not increase the generation of amyloid-beta peptides. *Traffic, 7*, 873–888.

27. Gotz, J., Chen, F., Van Dorpe, J., Nitsch, R.M. (2001). Formation of neurofibrillary tangles in P3011 tau transgenic mice induced by Abeta fibrils. *Science, 293*, 1491–1495.

28. Graff-Radford, N.R., Crook, J.E., Lucas, J., Boeve, B.F., Knopman, D.S., Ivnik, R.J., Smith, G.E., Younkin, L.H., Petersen, R.C., Younkin, S.G. (2007). Association of low plasma Abeta42/Abeta40 ratios with increased imminent risk for mild cognitive impairment and Alzheimer disease. *Arch Neurol, 64*, 354–362.

29. Hardy, J. (2006). Amyloid double trouble. *Nat Genet, 38*, 11–12.

30. Heber, S., Herms, J., Gajic, V., Hainfellner, J., Aguzzi, A., Rulicke, T., von Kretzschmar, H., von Koch, C., Sisodia, S., Tremml, P., et al. (2000). Mice with combined gene knock-outs reveal essential and partially redundant functions of amyloid precursor protein family members. *J Neurosci, 20*, 7951–7963.

31. Higuchi, M., Iwata, N., Matsuba, Y., Sato, K., Sasamoto, K., Saido, T.C. (2005). (19)F and (1)H MRI detection of amyloid beta plaques in vivo. *Nat Neurosci, 8*, 527–533.

32. Hock, C., Konietzko, U., Streffer, J.R., Tracy, J., Signorell, A., Muller-Tillmanns, B., Lemke, U., Henke, K., Moritz, E., Garcia, E., et al. (2003). Antibodies against beta-amyloid slow cognitive decline in Alzheimer's disease. *Neuron, 38*, 547–554.

33. Holtzman, D.M., Bales, K.R., Tenkova, T., Fagan, A.M., Parsadanian, M., Sartorius, L.J., Mackey, B., Olney, J., McKell, D., Wozniak, D., Paul S.M.

(2000). Apolipoprotein E isoform-dependent amyloid deposition and neuritic degeneration in a mouse model of Alzheimer's disease. *Proc Natl Acad Sci USA, 97*, 2892–2897.

34. Hu, X., Hicks, C.W., He, W., Wong, P., Macklin, W.B., Trapp, B.D., Yan R. (2006). Bace1 modulates myelination in the central and peripheral nervous system. *Nat Neurosci, 9*, 1520–1525.
35. Hussain, I., Hawkins, J., Harrison, D., Hille, C., Wayne, G., Cutler, L., Buck, T., Walter, D., Demont, E., Howes, C., et al. (2007). Oral administration of a potent and selective non-peptidic BACE-1 inhibitor decreases beta-cleavage of amyloid precursor protein and amyloid-beta production in vivo. *J Neurochem, 100*, 802–809.
36. Iwata, N., Mizukami, H., Shirotani, K., Takaki, Y., Muramatsu, S., Lu, B., Gerard, N.P., Gerard, C., Ozawa, K., Saido, T.C. (2004). Presynaptic localization of neprilysin contributes to efficient clearance of amyloid-beta peptide in mouse brain. *J Neurosci, 24*, 991–998.
37. Iwatsubo, T. (2004). The gamma-secretase complex: machinery for intramembrane proteolysis. *Curr Opin Neurobiol, 14*, 379–383.
38. Jack, C.R., Jr, Garwood, M., Wengenack, T.M., Borowski, B., Curran, G.L., Lin, J., Adriany, G., Grohn, O.H., Grimm, R., Poduslo J.F. (2004). In vivo visualization of Alzheimer's amyloid plaques by magnetic resonance imaging in transgenic mice without a contrast agent. *Magn Reson Med, 52*, 1263–1271.
39. Jankowsky, J.L., Slunt, H.H., Gonzales, V., Savonenko, A.V., Wen, J.C., Jenkins, N.A., Copeland, N.G., Younkin, L.H., Lester, H.A., Younkin, S.G., Borchelt, D.R. (2005). Persistent amyloidosis following suppression of Abeta production in a transgenic model of Alzheimer disease. *PLoS Med, 2*, e355.
40. Kamenetz, F., Tomita, T., Hsieh, H., Seabrook, G., Borchelt, D., Iwatsubo, T., Sisodia, S., Malinow R. (2003). APP processing and synaptic function. *Neuron, 37*, 925–937.
41. Klein, W.L., Krafft, G.A., Finch C.E. (2001). Targeting small Aß oligomers: the solution to an Alzheimer's disease conundrum? *Trends Neurosci, 24*, 219–223.
42. Klunk, W.E., Engler, H., Nordberg, A., Wang, Y., Blomstrand, G., Holt, D.P., Bergstrom, M., Savitcheva, I., Huang, G.F., Estrada, S., et al. (2004). Imaging brain amyloid in Alzheimer's disease using the novel positron emission tomography tracer, Pittsburgh compound-B. *Ann Neurol, 55*, 1–14.
43. Klyubin, I., Walsh, D.M., Lemere, C.A., Cullen, W.K., Shankar, G.M., Betts, V., Spooner, E.T., Jiang, L., Anwyl, R., Selkoe, D.J., Rowan, M.J. (2005). Amyloid beta protein immunotherapy neutralizes Abeta oligomers that disrupt synaptic plasticity in vivo. *Nat Med, 11*, 556–561.
44. Kukar, T.L., Ladd, T.B., Bann, M.A., Fraering, P.C., Narlawar, R., Maharvi, G.M., Healy, B., Chapman, R., Welzel, A.T., Price, R.W., et al. (2008). Substrate-targeting gamma-secretase modulators. *Nature, 453*, 925–929.
45. Laferla, F.M., Green, K.N., Oddo, S. (2007). Intracellular amyloid-beta in Alzheimer's disease. *Nat Rev Neurosci, 8*, 499–509.
46. Laird, F.M., Cai, H., Savonenko, A.V., Farah, M.H., He, K., Melnikova, T., Wen, H., Chiang, H.C., Xu, G., Koliatsos, V.E., et al. (2005). BACE1, a major determinant of selective vulnerability of the brain to Aβ amyloidogenesis is essential for cognitive, emotional and synaptic functions. *J Neurosci, 25*, 11693–11709.

47. Lazarov, O., Morfini, G.A., Lee, E.B., Farah, M.H., Szodorai, A., Deboer, S.R., Koliatsos, V.E., Kins, S., Lee, V.M., Wong, P.C., et al. (2005). Axonal transport, amyloid precursor protein,kinesin-1, and the processing apparatus: revisited. *J Neurosci, 25*, 2386–2395.

48. Lesne, S., Koh, M.T., Kotilinek, L., Kayed, R., Glabe, C.G., Yang, A., Gallagher, M., Ashe, K.H. (2006). A specific amyloid-[beta] protein assembly in the brain impairs memory. *Nature, 440*, 352–357.

49. Lewis, J., Dickson, D.W., Lin, W., Chisholm, L., Corral, A., Jones, G., Yen, S., Sahara, N., Skipper, L., Yager, D., et al. (2001). Enhanced neurofibrillary degeneration in transgenic mice expressing mutat tau and APP. *Science, 293*, 1487–1491.

50. Li, T., Ma, G., Cai, H., Price, D.L., Wong, P.C. (2003). Nicastrin is required for assembly of presenilin/gamma-secretase complexes to mediate Notch signaling and for processing and trafficking of beta-amyloid precursor protein in mammals. *J Neurosci, 23*, 3272–3277.

51. Li, T., Wen, H., Brayton, C., Das, P., Smithson, L.A., Fauq, A., Fan, X., Crain, B.J., Price, D.L., Golde, T.E., et al. (2007). Epidermal growth factor receptor and notch pathways participate in the tumor suppressor function of gamma-secretase. *J Biol Chem, 282*, 32264–32273.

52. Li, T., Wen, H., Brayton, C., Laird, F.M., Fan, X., Peng, S., Placanica, L., Wu, T., Crain, B.J., Price, D.L., et al. (2007). Moderate reduction of gamma-secretase attenuates amyloid burden and limits mechanism-based liabilities. *J Neurosci, 27*, 10849–10859.

53. Lim, G.P., Yang, F., Chu, T., Chen, P., Beech, W., Teter, B., Tran, T., Ubeda, O., Ashe, K.H., Frautschy, S.A., Cole, G.M. (2000). Ibuprofen suppresses plaque pathology and inflammation in a mouse model for Alzheimer's disease. *J Neurosci, 20*, 5709–5714.

54. Lipton, S.A. (2007). Pathologically activated therapeutics for neuroprotection. *Nat Rev Neurosci, 8*, 803–808.

55. Louvi, A., Artavanis-Tsakonas, S. (2006). Notch signalling in vertebrate neural development. *Nat Rev Neurosci, 7*, 93–102.

56. Luo, Y., Bolon, B., Kahn, S., Bennett, B.D., Babu-Khan, S., Denis, P., Fan, W., Kha, H., Zhang, J., Gong, Y., et al. (2001). Mice deficient in BACE1, the Alzheimer's beta-secretase, have normal phenotype and abolished beta-amyloid generation. *Nat Neurosci, 4*, 231–232.

57. Ma, G., Li, T., Price, D.L., Wong, P.C. (2005). APH-1a is the principal mammalian APH-1 isoform present in {gamma}-secretase complexes during embryonic development. *J Neurosci, 25*, 192–198.

58. Maeda, J., Ji, B., Irie, T., Tomiyama, T., Maruyama, M., Okauchi, T., Staufenbiel, M., Iwata, N., Ono, M., Saido, T.C., et al. (2007). Longitudinal, quantitative assessment of amyloid, neuroinflammation, and anti-amyloid treatment in a living mouse model of Alzheimer's disease enabled by positron emission tomography. *J Neurosci, 27*, 10957–10968.

59. Mandkelkow, E.M, Thies, E., Mandelkow, E. (2007). Tau and axonal transport In *Alzheimer's Disease: Advances in Genetics, Molecular and Cellular Biology* Sisodia, S.S., Tonzi, R.E., Eds.), Springer, New York, pp. 237–256.

60. Markesbery, W.R., Schmitt, F.A., Kryscio, R.J., Davis, D.G., Smith, C.D., Wekstein, D.R. (2006). Neuropathologic substrate of mild cognitive impairment. *Arch Neurol, 63*, 38–46.

61. Martin, L.J., Pardo, C.A., Cork, L.C., Price, D.L. (1994). Synaptic pathology and glial responses to neuronal injury precede the formation of senile plaques and amyloid deposits in the aging cerebral cortex. *Am J Pathol, 145*, 1358–1381.

62. Masliah, E., Hansen, L., Adame, A., Crews, L., Bard, F., Lee, C., Seubert, P., Games, D., Kirby, L., Schenk D. (2005). A{beta} vaccination effects on plaque pathology in the absence of encephalitis in Alzheimer disease. *Neurology, 64*, 129–131.

63. McGowan, E., Eriksen, J., Hutton, M. (2006). A decade of modeling Alzheimer's disease in transgenic mice. *Trends Genet, 22*, 281–289.

64. Meyer-Luehmann, M., Spires-Jones, T.L., Prada, C., Garcia-Alloza, M., de Calignon, A., Rozkalne, A., Koenigsknecht-Talboo, J., Holtzman, D.M., Bacskai, B.J., Hyman B.T. (2008). Rapid appearance and local toxicity of amyloid-beta plaques in a mouse model of Alzheimer's disease. *Nature, 451*, 720–724.

65. Miller, B.C., Eckman, E.A., Sambamurti, K., Dobbs, N., Chow, K.M., Eckman, C.B., Hersh, L.B., Thiele D.L. (2003). Amyloid-beta peptide levels in brain are inversely correlated with insulysin activity levels in vivo. *Proc Natl Acad Sci U S A, 100*, 6221–6226.

66. Monsonego, A., Weiner, H.L. (2003). Immunotherapeutic approaches to Alzheimer's disease. *Science, 302*, 834–838.

67. Nicoll, J.A., Wilkinson, D., Holmes, C., Steart, P., Markham, H., Weller, R.O. (2003). Neuropathology of human Alzheimer disease after immunization with amyloid-beta peptide: a case report. *Nat Med, 9*, 448–452.

68. Oddo, S., Billings, L., Kesslak, J.P., Cribbs, D.H., LaFerla, F.M. (2004). Abeta immunotherapy leads to clearance of early, but not late, hyperphosphorylated tau aggregates via the proteasome. *Neuron, 43*, 321–332.

69. Oddo, S., Caccamo, A., Shepherd, J.D., Murphy, M.P., Golde, T.E., Kayed, R., Metherate, R., Mattson, M.P., Akbari, Y., LaFerla, F.M. (2003). Triple-transgenic model of Alzheimer's disease with plaques and tangles: intracellular Abeta and synaptic dysfunction. *Neuron, 39*, 409–421.

70. Pfeifer, M., Boncristiano, S., Bondolfi, L., Stalder, A., Deller, T., Staufenbiel, M., Mathews, P.M., Jucker M. (2002). Cerebral hemorrhage after passive anti-Abeta immunotherapy. *Science, 298*, 1379.

71. Price, D.L., Martin, L.J., Savonenko, A.V., Li, T., Laird, F.M., Wong P.C. (2008). Selected genetically engineered models relevant to human neurodegenerative disease. In *The Molecular and Genetic Basis of Neurologic and Psychiatric Disease, 4th ed.* Rosenberg R.R., Prescience S., DiMaurs, S., Barchi, R., Eds.), Lippincott Williams & Wilkins, Philadelphia, pp. 35–62.

72. Roberson, E.D., Scearce-Levie, K., Palop, J.J., Yan, F., Cheng, I.H., Wu, T., Gerstein, H., Yu, G.Q., Mucke, L. (2007). Reducing endogenous tau ameliorates amyloid beta-induced deficits in an Alzheimer's disease mouse model. *Science, 316*, 750–754.

73. Rogaeva, E., Meng, Y., Lee, J.H., Gu, Y., Kawarai, T., Zou, F., Katayama, T., Baldwin, C.T., Cheng, R., Hasegawa, H., et al. (2007). The neuronal sortilin-related

receptor SORL1 is genetically associated with Alzheimer's disease. *Nat Genet, 39*, 168.

74. Rovelet-Lecrux, A., Hannequin, D., Raux, G., Meur, N.L., Laquerriere, A., Vital, A., Dumanchin, C., Feuillette, S., Brice, A., Vercelletto, M., et al. (2006). APP locus duplication causes autosomal dominant early-onset Alzheimer disease with cerebral amyloid angiopathy. *Nat Genet, 38*, 24–26.
75. Sagare, A., Deane, R., Bell, R.D., Johnson, B., Hamm, K., Pendu, R., Marky, A., Lenting, P.J., Wu, Z., Zarcone, T., et al. (2007). Clearance of amyloid-beta by circulating lipoprotein receptors. *Nat Med, 13*, 1029–1031.
76. SantaCruz, K., Lewis, J., Spires, T., Paulson, J., Kotilinek, L., Ingelsson, M., Guimaraes, A., DeTure, M., Ramsden, M., McGowan, E., et al. (2005). Tau suppression in a neurodegenerative mouse model improves memory function. *Science, 309*, 476–481.
77. Saura, C.A., Choi, S.Y., Beglopoulos, V., Malkani, S., Zhang, D., Rao, B.S., Chattarji, S., Kelleher, R.J., III, Kandel, E.R., Duff, K., et al. (2004). Loss of presenilin function causes impairments of memory and synaptic plasticity followed by age-dependent neurodegeneration. *Neuron, 42*, 23–36.
78. Savonenko, A.V., Laird, F.M., Troncoso, J.C., Wong, P.C., Price, D.L. (2006). Role of Alzheimer's disease models in designing and testing experimental therapeutics. *Drug Discov Today, 2*, 305–312.
79. Savonenko, A., Xu, G.M., Melnikova, T., Morton, J.L., Gonzales, V., Wong, M.P.F., Price, D.L., Tang, F., Markowska, A.L., Borchelt, D.R. (2005). Episodic-like memory deficits in the APPswe/PS1dE9 mouse model of Alzheimer's disease: relationships to [beta]-amyloid deposition and neurotransmitter abnormalities. *Neurobiol Dis, 18*, 602–617.
80. Schenk, D., Barbour, R., Dunn, W., Gordon, G., Grajeda, H., Guido, T., Hu, K., Huang, J.P., Johnson-Wood, K., Khan, K., et al. (1999). Immunization with amyloid-beta attenuates Alzheimer disease-like pathology in the PDAPP mouse. *Nature, 400*, 173–177.
81. Selkoe, D.J. (2002). Alzheimer's disease is a synaptic failure. *Science, 298*, 789–791.
82. Selkoe, D.J., Schenk, D. (2003). Alzheimer's disease: molecular understanding predicts amyloid-based therapeutics. *Annu Rev Pharmacol Toxicol, 43*, 545–584.
83. Serneels, L., Dejaegere, T., Craessaerts, K., Horre, K., Jorissen, E., Tousseyn, T., Hebert, S., Coolen, M., Martens, G., Zwijsen, A., et al. (2005). Differential contribution of the three Aph1 genes to {gamma}-secretase activity in vivo. *Proc Natl Acad Sci U S A, 102*, 1719–1724.
84. Shen, J., Kelleher, R.J., 3rd. (2007). The presenilin hypothesis of Alzheimer's disease: evidence for a loss-of-function pathogenic mechanism. *Proc Natl Acad Sci U S A, 104*, 403–409.
85. Sheng, J.G., Price, D.L., Koliatsos, V.E. (2002). Disruption of corticocortical connections ameliorates amyloid burden in terminal fields in a transgenic model of Abeta amyloidosis. *J Neurosci, 22*, 9794–9799.
86. Singer, O., Marr, R.A., Rockenstein, E., Crews, L., Coufal, N.G., Gage, F.H., Verma, I.M., Masliah, E. (2005). Targeting BACE1 with siRNAs ameliorates Alzheimer disease neuropathology in a transgenic model. *Nat Neurosci, 8*, 1343–1349.

87. Sisodia, S.S., Koo, E.H., Beyreuther, K.T., Unterbeck, A., Price D.L. (1990). Evidence that β-amyloid protein in Alzheimer's disease is not derived by normal processing. *Science, 248*, 492–495.

88. Sisodia, S.S., Koo, E.H., Hoffman, P.N., Perry, G., Price D.L. (1993). Identification and transport of full-length amyloid precursor proteins in rat peripheral nervous system. *J Neurosci, 13*, 3136–3142.

89. Small, S.A., Gandy S. (2006). Sorting through the cell biology of Alzheimer's disease: intracellular pathways to pathogenesis. *Neuron, 52*, 15–31.

90. Sunderland, T., Linker, G., Mirza, N., Putnam, K.T., Friedman, D.L., Kimmel, L.H., Bergeson, J., Manetti, G.J., Zimmermann, M., Tang, B., et al. (2003). Decreased beta-amyloid1-42 and increased tau levels in cerebrospinal fluid of patients with Alzheimer disease. *JAMA, 289*, 2094–2103.

91. Sze, C., Troncoso, J.C., Kawas, C.H., Mouton, P.R., Price, D.L., Martin L.J. (1997). Loss of the presynaptic vesicle protein synaptophysin in hippocampus correlates with early cognitive decline in aged humans. *J Neuropathol Exp Neurol 56*, 933–944.

92. Vassar, R., Bennett, B.D., Babu-Khan, S., Kahn, S., Mendiaz, E.A., Denis, P., Teplow, D.B., Ross, S., Amarante, P., Loeloff, R., et al. (1999). β-Secretase cleavage of Alzheimer's amyloid precusor protein by the transmembrane aspartic protease BACE. *Science, 286*, 735–741.

93. Vekrellis, K., Ye, Z., Qiu, W.Q., Walsh, D., Hartley, D., Chesneau, V., Rosner, M.R., Selkoe, D.J. (2000). Neurons regulate extracellular levels of amyloid β-protein via proteolysis by insulin-degrading enzyme. *J Neurosci, 20*, 1657–1665.

94. Walsh, D.M., Klyubin, I., Faden, A.I., Fadeeva, J.V., Cullen, W.K., Anwyl, R., Wolfe, M.S., Rowan, M.J., Selkoe, D.J. (2002). Naturally secreted oligomers of amyloid β-protein potently inhibit hippocampal LTP in vivo. *Nature, 416*, 535–539.

95. Walsh, D.M., Minogue, A.M., Sala Frigerio, C., Fadeeva, J.V., Wasco, W., Selkoe, D.J. (2007). The APP family of proteins: similarities and differences. *Biochem Soc Trans, 35*, 416–420.

96. Wang, H.W., Pasternak, J.F., Kuo, H., Ristic, H., Lambert, M.P., Chromy, B., Viola, K.L., Klein, W.L., Stine, W.B., Krafft, G.A., Trommer, B.L. (2002). Soluble oligomers of beta amyloid (1–42) inhibit long-term potentiation but not long-term depression in rat dentate gyrus. *Brain Res, 924*, 133–140.

97. Weggen, S., Eriksen, J.L., Das, P., Sagl, S.A., Wang, R., Pietrzik, C.U., Findlay, K.A., Smith, T.E., Murphy, M.P., Bulter, T., et al. (2001). A subset of NSAIDs lower amyloidogenic Aβ42 independently of cyclooxygenase activity. *Nature, 414*, 212–216.

98. Weiner, H. L., Frenkel, D. (2006). Immunology and immunotherapy of Alzheimer's disease. *Nat Rev Immunol, 6*, 404–416.

99. Whitehouse, P.J., Price, D.L., Struble, R.G., Clark, A.W., Coyle, J.T., Delong, M.R. (1982). Alzheimer's disease and senile dementia: loss of neurons in the basal forebrain. *Science, 215*, 1237–1239.

100. Willem, M., Garratt, A.N., Novak, B., Citron, M., Kaufmann, S., Rittger, A., DeStrooper, B., Saftig, P., Birchmeier, C., Haass, C. (2006). Control of peripheral nerve myelination by the beta-secretase BACE1. *Science, 314*, 664–666.

101. Wolfe, M.S. (2006). Shutting down Alzheimer's. *Sci Am, 294*, 72.

102. Wolfe, M.S., Xia, W., Ostaszewski, B.L., Diehl, T.S., Kimberly, W.T., Selkoe, D.J. (1999). Two transmembrane aspartates in presenilin-1 required for presenilin endoproteolysis and gamma-secretase activity. *Nature, 398*, 513–517.

103. Wong, P.C., Price, D.L., Bertram, L., Tanzi, R.E. (2008). Alzheimer's disease: genetics, pathogenesis, models, and experimental therapeutics in *Molecular Biology of Aging*, (Duarente, L.P., Partridge, L., Wallace, D.C., Eds.), Cold Spring Harbor Laboratory Press, Cold Spring Harbor, NY, pp. 371–407.

104. Wong, P.C., Price, D.L., Cai, H. (2001). The brain's susceptibility to amyloid plaques. *Science, 293*, 1434–1435.

105. Wong, P.C., Price, D.L., Li, T. (2005). Neurobiology of Alzheimer's disease. In *Basic Neurochemistry*, 7th (*Eds*.), Elsevier, Burlington, MA, pp. 781–790.

106. Wong, P.C., Zheng, H., Chen, H., Becher, M.W., Sirinathsinghji, D.J., Trumbauer, M.E., Chen, H.Y., Price, D.L., Van der Ploeg, L.H., Sisodia, S.S. (1997). Presenilin 1 is required for Notch1 and DII1 expression in the paraxial mesoderm. *Nature, 387*, 288–292.

107. Zamora, E., Handisurya, A., Shafti-Keramat, S., Borchelt, D., Rudow, G., Conant, K., Cox, C., Troncoso, J.C., Kirnbauer, R. (2006). Papillomavirus-like particles are an effective platform for amyloid-beta immunization in rabbits and transgenic mice. *J Immunol*, 2662–2670.

13

PRION DISEASE THERAPY: TRIALS AND TRIBULATIONS

Valerie L. Sim and Byron Caughey

Laboratory of Persistent Viral Diseases, Rocky Mountain Laboratories, National Institute of Allergy and Infectious Diseases, National Institutes of Health, Hamilton, Montana

INTRODUCTION

Transmissible spongiform encephalopathies (TSEs) are invariably fatal neurodegenerative diseases which include Creutzfeldt–Jakob disease (CJD) in humans; bovine spongiform encephalopathy (BSE) in cattle; chronic wasting disease (CWD) in deer, elk, and moose; and scrapie in sheep. The neuropathological hallmarks of these diseases include spongiform change, neuronal loss, astrocytosis, and the accumulation of protease-resistant prion protein (PrP^{res}) aggregates. Unlike other neurodegenerative conditions, these prion diseases are transmissible, and PrP^{res} is the primary protein component of the prion agent responsible for transmission [1,2].

PrP^{res} is a corrupted and pathological form of the normal, protease-sensitive prion protein (PrP^{c}), a glycoprotein anchored to the membrane by its glycosylphosphatidyl-inositol (GPI) anchor. PrP^{c} must be expressed by a host to allow susceptibility to prion infection. The best explanation of prion pathogenicity comes from the prion hypothesis, which proposes that disease results from the repeated conformational conversion of α-helical PrP^{c} to highly β-sheet PrP^{res} through a seeded polymerization mechanism. This conversion process is facilitated by membrane interaction [3] and may occur on the cell surface or within early endosomal compartments. However, large extracellular

Protein Misfolding Diseases: Current and Emerging Principles and Therapies,
Edited by Marina Ramirez-Alvarado, Jeffery W. Kelly, and Christopher M. Dobson
Copyright © 2010 John Wiley & Sons, Inc.

plaques of PrP^{res} can develop in mice whose PrP^{c} is expressed without a GPI anchor and therefore not attached to the membrane [4]. Unfortunately, our understanding of how these events translate into cytotoxicity is not clear. Large aggregates of PrP^{res} do not necessarily lead to damage. In fact, soluble [5,6] and protease-sensitive [7,8] forms of prion infectivity have been described, and it may be that the oligomers play the most important role in toxicity [9] and conversion [6].

Although 90% of human CJD cases are classified as sporadic (sCJD), there was much anxiety over the appearance of variant CJD (vCJD) in the 1990s. Evidence supports the origin of vCJD being from exposure to BSE prions [10–12] through the food system. Of greater concern are the three cases of vCJD probably transmitted via blood transfusion that have now been reported [13]. Iatrogenic causes of CJD (iCJD) have resulted from contaminated neurosurgical instruments, hormones, corneal transplants, and dura mater grafts. There are also several familial human TSE diseases, including familial CJD (fCJD), Gerstmann–Straussler–Scheinker disease (GSS), and fatal familial insomnia (FFI). With some variability, the clinical manifestations of these diseases generally include ataxia, progressive dementia, myoclonus, and ultimately akinetic mutism. Death results from complications of being bedridden, most commonly infection.

No treatments for TSEs have been validated for use in medicine or agriculture. Part of the difficulty in designing therapeutic strategies lies in our incomplete understanding of the pathogenesis of these diseases. Nevertheless, a number of anti-TSE interventions have been pursued. One worthwhile goal is to reduce the risk of initial infection by neutralizing sources of infection, blocking infections via the most common peripheral routes, and/or blocking neuroinvasion from the periphery. Immunotherapies are being pursued with some tantalizing results, and significant progress has been made in the search for anti-TSE drugs, especially in identifying compounds with prophylactic activity. Another important but elusive goal is to be able to treat the disease after the appearance of clinical signs. This will probably involve some combination of inhibiting PrP^{res} formation, destabilizing existing PrP^{res}, blocking neurotoxic effects of the infection, and/or promoting the recovery of lost functions in the central nervous system (CNS). In this chapter we review recent progress in TSE prevention and treatment.

In Vivo or In Vitro; Searching for the Magic Bullet

In vivo tests provide the most rigorous evaluations of anti-TSE treatments. However, such experimentation tends to be quite slow and costly. The fastest animal model systems currently available are transgenic mice overexpressing either murine PrP (tga20) or hamster PrP (tg7), which require 40 to 50 days to reach symptom onset after intracerebral inoculation of rodent-adapted scrapie strains. Although these rodent models are much faster and less expensive than more relevant large-animal TSE disease models, these factors make in vivo testing impractical for high-throughput screening.

A variety of scrapie- and CWD-infected cell-culture models, and some yeast models [14], have been developed and serve as in vitro surrogates for animal studies in screening for anti-prion compounds. These models have the advantage of high throughput and low cost compared to animal testing. Extensive compound screening in cell cultures has identified a number of different classes of anti-prion compounds which have then shown efficacy in animal models. However, as expected with any in vitro drug screen, the predictive accuracy of cell-culture screens for efficacy in vivo is far from 100%. Furthermore, although some compounds appear to have broad-based anti-prion efficacy, striking variation has been observed in drug effectiveness between different prion-infected models and prion strains [15–17]. This variation emphasizes the need for testing in experimental models that approximate specific TSE disease targets as closely as possible. Unfortunately, in terms of cell culture models, no cell lines have been stably infected with two of the most important TSE agents, CJD and BSE.

Another advantage of cell cultures is the ability to probe mechanisms of prion inhibition. One can often see that anti-prion compounds that bind to PrP^c cause its clustering and internalization. The result is the sequestration of PrP^c in a state and/or subcellular location that is incompatible with conversion to PrP^{res} [18–21].

Noncellular in vitro methods have also been employed to assess a wide range of therapeutic candidates. These assays usually look for competitive binding of PrP^c and PrP^{res} or the prevention of PrP amyloid fibril formation. Recently developed techniques include surface plasmon resonance [22], fluorescence correlation spectroscopy [23], semiautomated cell-free conversion [24], and a fluorescence-polarization-based competitive binding assay [25]. Some have even turned to computer "in silico" modeling to predict binding partners [26,27].

Ultimately, compounds that look promising in vitro require testing in animals and in humans. Of the many compounds studied in rodent models, only a few have made their way into human trials or case reports. The primary outcome measures of therapy in animal studies include incubation and survival times, PrP^{res} accumulation, histopathological changes, and some take into account behavioral and clinical aspects of the disease. Key factors affecting these outcomes include the dose and route of therapeutic agent, dose and route of disease inoculation, and most important, the time at which treatment is started. Generally, the effectiveness of compounds given at the onset of clinical symptoms, or when there is significant neuropathology, is limited. However, many compounds show some effectiveness in prophylaxis or early treatment, and may have a significant role to play in decontamination of sources of infection, such as blood in the case of human variant CJD. Such drugs need not cross the blood–brain barrier, and might have value in livestock or wildlife disease management. Also, for the many prion diseases that arise following oral exposure to the infectious agent, early treatment compounds could target the peripheral replication of the agent, before neuroinvasion. Admittedly, diagnosis has lagged in this regard, with disease confirmation generally occurring well after neuroinvasion has taken place. But now there is new hope for the early preclinical TSE diagnostics, with the development of new sensitive detection assays [1,28–30], leading the way to more effective screening and testing of at-risk individuals.

There may also be hope for those with established prion disease of the CNS, including for those whose disease probably begins within the CNS, such as cases of sporadic and familial CJD. Recently several different chemotherapeutic approaches based on the direct administration of anti-prion compounds to the CNS have shown some efficacy in experimental animal models. Furthermore, Mallucci et al. have demonstrated that eliminating the expression of PrP^c in neurons in a mouse with established prion disease led to a reversal of the neuropathological spongiform changes [31]. Behavioral abnormalities also improved, in correlation with the pathological improvements [32]. Thus, a new target for therapy is the expression of PrP^c, and novel application techniques using small interfering RNAs and gene therapies may open a new era of prion disease treatment. Of course, whether any of these therapeutic approaches on their own will be adequate both to halt the disease process and to allow functional recovery remains to be seen. It is possible that combinations of these treatment arms may turn out to be the winning ticket.

IMMUNE-BASED THERAPIES

In the 1970s, Fraser and Dickinson first reported that splenectomized mice were not susceptible to scrapie [33]. Porter et al. later demonstrated a lack of humoral response in prion disease [34]. Since then, our understanding of immune system involvement in prion disease has increased. As our knowledge about the peripheral phase of prion replication and subsequent neuroinvasion grows, so do the targets for intervention and immunotherapy.

Peripheral Replication

For orally acquired prion diseases, the phase of prion replication in the periphery depends on many cells, which may be targets for early therapy. CWD [35], scrapie, BSE, and vCJD can all follow oral exposure to the infectious agent, after which the agent must replicate in the periphery and migrate to the brain. These steps depend on several key factors, at least in the models that have been studied. Within the gut, follicular dendritic cells (FDCs) are of major importance [36], and M cells may also play a role [37]. FDCs within Peyer's patches [38,39] or isolated lymphoid follicles [40] are required for host susceptibility, and infectivity and PrP^{res} accumulate at these sites. FDCs that cannot mature, in mice deficient in maturation and activation factors [including functional B cells, tumor necrosis factor-alpha (TNFα), TNF-receptor-1, lymphotoxin (LT) $\alpha+\beta$, and LT β-receptor], are associated with poor abilities to become peripherally infected [41–46]. Complement factors are also important [47], probably for facilitating antigen recognition by FDCs [48]. More specifically, the CD21/35 complement receptor has recently been proposed as the key component for targeting prions to FDCs [49]. It has been thought that FDCs must express PrP^c for prion replication to occur [41,50], but neuroinvasion was recently demonstrated in mice that express ovine PrP^c solely in neurons [51].

Neuroinvasion

Neuroinvasion can proceed via splanchnic innervation from the gut to the brainstem and spinal cord [39,52]. The autonomic nervous system, both sympathetic and parasympathetic branches, plays a key role [53,54] and has recently been implicated in BSE, where autonomic nervous system involvement precedes CNS and nonautonomic nervous system involvement [55]. Because animals and humans are asymptomatic until their nervous systems are infected, targeting peripheral replication with immunotherapies and preventing neuroinvasion is a fundamental goal of vaccine development and prophylactic or early disease treatment.

A variety of immune-based therapies have been studied, including immunostimulation, immunosuppression, and immunizations. The primary focus has been on the latter, with active, passive, and transgenic approaches to antibody administration.

A Role for Antibodies

Gabizon et al. first reported that antibodies to PrP could reduce infectivity in inoculum [56]. Later, the ability of a polyclonal serum (raised against PrP sequence 219–232) to interact with PrP^c and inhibit cell-free conversion [57] suggested that antibodies might be good treatment candidates. Enari et al. then discovered that PrP antibody 6H4 was able to cure chronically infected neuroblastoma cells [58]. It has recently been suggested that antibodies that recognize both PrP^c and PrP^{res} may be the most effective [59]. It is likely that these PrP antibodies have several mechanisms of action. In cell culture, removing PrP^c from the cell surface through cleavage of its GPI anchor is known to inhibit PrP^{res} formation [58,60]. Single-chain antibody fragments that sequester PrP^c in the endoplasmic reticulum also prevented both infection [61] and the generation of new infectivity [62]. Interfering with PrP^c–PrP^{res} interactions [58,63–65] or with PrP^c-binding receptors [63] limits conversion, and inhibiting new PrP^{res} formation has sometimes led to increased rates of PrP^c and PrP^{res} clearance [63].

Active Immunization

Prophylactic immunization has been attempted in vivo with some effect on incubation times [66–69]. (For a review of active and passive immunization, see the paper of Bade and Frey [70].) However, a major obstacle to effective active immunization is the self-tolerance to ubiquitous PrP^c [71,72]. Antibodies specific for PrP^{res} should circumvent the tolerance phenomenon, and indeed, a Tyr–Tyr–Arg peptide can induce PrP^{res}-specific antibodies in wild-type animals without toxic effect [73]. Another approach, which combined mouse PrP with immune stimulatory helper T-cell epitopes of tetanus toxin, did not improve tolerance [74], but some progress is being made with cytidyl–guanyl oligodeoxynucleotides (CpG-ODNs) which bind Toll-like receptor 9 and stimulate innate immune responses. When CpG 1826 was added to scrapie-associated fibrils from strain 139A, antibody production in wild-type mice was improved significantly [75].

Inducing antibodies to a ubiquitous protein such as PrP^c carries the risk of causing widespread complement-mediated cellular lysis. There is also the risk of inducing T-cell responses and developing side effects such as the meningoencephalitis which followed active immunization of Alzheimer patients with the Abeta vaccine AN1792 [76]. Some reassuring research from Kaiser-Schulz et al. recently found that polylactide-coglycolide microspheres, containing tandem PrP (PrP dimer) and CpG-ODN, allowed endosomal delivery of its contents to macrophages and dendritic cells, producing good humoral, $CD4^+$, and $CD8^+$ responses without any evidence of autoimmune disease [77]. Another group inserted the PrP peptide sequence 144–152 into the L1 major capsid protein of bovine papillomavirus 1, in hopes of avoiding T-cell responses. Immunization with these antigen-induced antibodies, which recognized PrP^c, inhibited PrP^{res} in cell culture and did not generate any adverse effects in vivo [78]. It will be interesting to see whether these approaches are successful in preventing prion disease in vivo.

Passive Immunization

Another approach to treatment that has been tried in rodent models is passive immunization. Immediate postexposure treatment with PrP antibodies 8B4, 8H4, and 8F9 led to a prolongation of incubation times in mice inoculated intraperitoneally with rodent-adapted scrapie [79]. In addition, strong reductions in PrP^{res} and infectivity were seen with two monoclonal IgG antibodies against recombinant α-helical PrP and recombinant β-sheet PrP [80] when intraperitoneally inoculated mice were treated early. Treatments started after clinical signs developed, or following intracerebral inoculations, did not work, possibly due to lack of antibody penetration of the blood–brain barrier.

Some antibody treatments have blocked neuroinvasion successfully by targeting non-PrP molecules. A lymphotoxin-β immunoglobulin given after intraperitoneal inoculation [36,42,81] or infection by skin scarification [82] cleared PrP^{res} from the spleen and prevented neuroinvasion, probably by interfering with FDC maturation and activation. Another candidate may be an antibody to the 37-kDa/67-kDa laminin receptor, which has a dramatic curative effect in cell culture [83], or recombinant single-chain antibodies to this receptor which were able to reduce PrP^{res} in the spleens of intraperitoneally inoculated mice [84].

Although passive immunization appears to best target the peripheral stage of prion disease, most prion diseases present after clinical signs and neuroinvasion have begun. Therefore, there is interest in introducing PrP antibodies into the CNS, where they might have further effect. A danger with this approach was realized in one study, in which hippocampal injection of high concentrations of antibodies against the PrP sequence 95–105 led to the degeneration of hippocampal and cerebellar neurons [85], probably through cross-linking of PrP^{c} by the bivalent IgG antibodies. Techniques using Fab fragments may avoid this safety concern.

Transgenic Antibodies and Interfering Peptides

Another method of introducing antibodies or other interfering peptides into rodent models is through transgenes. Mice hemizygous for the 6H4 μ-chain of IgM were completely protected from scrapie [86]. Transgenic mice containing the full-length murine PrP peptide fused to the Fc gamma tail of human IgG1 ($PrP\text{-}Fc_2$ fusion protein) expressed PrP^{c} as a soluble disulfide-linked dimer that could associate with PrP^{res} but resisted conversion. This peptide was able to delay the onset of disease in both intraperitoneally and intracerebrally inoculated mice, and prevented disease in fully transgenic animals with no wild-type PrP^{c} [87]. Although producing fully transgenic humans is clearly not a feasible therapeutic approach, these studies hold promise for gene therapy options.

Immunostimulation

Some other methods of increasing the immune response have been tried, including interferon (IFN) and adjuvant treatments. IFN and stimulators of IFN had no effect in mice [88–91], monkeys [92], or in two human cases of CJD, one sporadic and one familial [93]. Treatment with complete Freund's adjuvant alone led to an unexpected increase in incubation period following intraperitoneal or intracerebral inoculation in mice, although the mechanism of action remains unclear [94]. Treatment with CpG 1826, a phosphorothioate CpG-ODN, has increased incubation periods in mice when given early and for up to 20 days [95], but there is concern that CpG-ODNs may be toxic in long-term treatments, leading to the destruction of lymphoid follicles and thus causing immunosuppression [96]. A better use of CpG-ODNs may be as adjuvants in immunization strategies (see above) or as agents that can be mixed with fluids such as blood to decontaminate them. Interestingly, although it was assumed that CpG-ODNs exerted their antiprion effects by CpG-mediated stimulation of innate immunity [95], recent studies have shown that non-CpG-ODNs with little or no CpG-mediated immunostimulatory activity can also have profound anti-scrapie effects in vivo [97]. Furthermore, these compounds can bind directly to PrP^{c}, alter its trafficking, and inhibit PrP^{res} formation in cell cultures in the absence of immune mechanisms. These results call into question the extent to which the anti-prion effects of CpG oligonucleotides are due to stimulation of innate immunity as opposed to direct inhibition of PrP^{res} formation.

Immunosuppression

Early studies used immunosuppressive agents with limited success. Cyclophosphamide had no discernible effect [98] and prednisone could delay the disease only if given at the time of inoculation and only if the inoculation was intraperitoneal [99]. Some better results have been achieved with nonsteroidal anti-inflammatories (NSAIDs). Indomethacin, acetyl salicylic acid, and ibuprofen protected cultured neurons from the cytotoxic effects of PrP peptides and infectious preparations containing PrP, probably through inhibition of cyclo-oxygenase 1 [100]. Indomethacin has also been shown to delay disease after intracerebral inoculation [101], implying an effect beyond the simple prevention of neuroinvasion. Studies of dapsone are conflicting, with increased incubation times seen after intracerebral inoculation in rats [101], but no effect after intracerebral inoculation in mice [102].

CHEMICAL-BASED THERAPIES AND PROPHYLAXIS

The potential aims of nonimmune or drug treatment strategies for prion diseases overlap with the targets of immune therapies, including (1) decontamination of infectious sources, (2) prophylaxis against initial infections, (3) inhibition of peripheral propagation, (4) blockade of neuroinvasion, (5) inhibition of conversion to and accumulation of pathogenic PrP, (6) destabilization or enhanced clearance of pathogenic PrP, (7) blockade of direct or indirect neurotoxic effects of pathogenic PrP, and (8) compensation for damage to brain cells.

Targeting PrP Conversion

At the molecular level, one primary chemotherapeutic target is the PrP conversion reaction, and this has been the focus of the majority of screening efforts to date. Conversion inhibitors often act by binding PrP^{c} and/or PrP^{Sc} directly, thus interfering with their interactions with themselves or other influential ligands. Indirect mechanisms that block important accessory molecules or lead to the redistribution or altered expression of PrP^{c} can also affect conversion [18]. Many different chemical classes of compounds have been screened and tested in vitro (Table 1). A number of examples of these compound classes can prolong survival times of TSE-infected animals, albeit usually in a prophylaxis or early treatment manner. However, with the rising concern regarding blood transmission of vCJD, the occurrence of BSE in livestock, and the spread of CWD, in addition to the advances being made in early diagnostic testing [1, 28–30], such chemoprophylaxis compounds are becoming more relevant in the management of prion diseases.

Polyanions

Sulfated glycans and other polyanions are known inhibitors of PrP^{res} formation [103,104]. Their prophylactic effect appears to relate to their sequestration in the cells of the lymphoreticular system [105], where they can interfere with the peripheral propagation of PrP^{res}. These compounds are thought to inhibit competitive interactions between PrP molecules and the endogenous sulfated glycosaminoglycans (GAGs), which are important for PrP^{c} trafficking and PrP^{Sc} formation [21,103,104,106–108]. Of note, GAGs must be sulfated in order to exert any effect on PrP^{res} levels in cells [103], and inhibiting the sulfation of endogenous GAGs reduces PrP^{res} in culture [104]. Sulfated glycans and polyanions might also perturb PrP interactions with RNA molecules (also large polyanions), which may serve as physiological cofactors for PrP conversion [2,107,109,110]. The usefulness of this class of agents is limited by potential anticoagulant activity and poor blood–brain barrier permeability. However, the latter drawback was circumvented through intraventricular infusion in animal models [111]. Pentosan polysulfate, in particular, has been administered by this route in several human cases (see below).

Pentosan Polysulfate (PPS). One of the more widely studied compounds in this class, PPS is regularly employed to cure cell culture models of prion infection. It delays the onset of disease in mice inoculated intraperitoneally [105,112,113] and in hamsters inoculated intracerebrally or intraperitoneally [114], as long as the treatment is given around the time of inoculation. More recently, prolonged incubation times have been seen in intracerebrally inoculated tg7 mice when treatment has been administered via an intraventricular cannula and osmotic pump [111]. Although new PrP^{res} accumulation is prevented, PPS does not lead to the clearance of preexisting PrP^{res}. Effective use of this agent has now been demonstrated in a CWD-infected deer-cell model [115] as well as in a mouse model of BSE [116]. There is also now a significant amount of data from case reports of intraventricular PPS used in human prion diseases, with mixed results (see below). The ultimate role of PPS in therapy may be through using it in combination with other agents [97] (see below).

Fucoidan. This complex sulfated fucosylated polysaccharide is an edible component of seaweed which has been shown to have in vitro and in vivo

TABLE 1 Chemical-Based Therapeutic and Prophylactic Agents

Compound[a]	Activity In Vitro/In Vivo (Treated Early or Prophylaxis)[b]	Comments	Ref.
Polyanions			
Heteropolyanion-23*	+/+		103, 198, 251, 252
Dextran sulfate*	+/+	Prolongs incubation after IC inoculation if given within 2 h (hamsters)	103, 105, 114, 252
Pentosan polysulfate (PPS)**	+/+	Intraventricular infusion prolongs incubation (tg7 mice) Used in humans Inhibits PrP^{res} formation in cell culture but can stimulate cell-free conversion PPS + Fe-TSP has more than additive effects in vivo	103, 105–107, 111–114, 253
Heparan sulfate*	+/+	Inhibits PrP^{res} formation in cell culture but can stimulate cell-free conversion	103, 104, 106, 107
Heparan sulfate mimetics (e.g., HM2606, CR36)*	+/+		116, 254, 255
Fucoidan	+/+	Nontoxic Oral administration Strain dependent	117
Phosphorothioate oligonucleotides	+/+		97
RNA aptamers*	+/−		256, 257
Copaxone*	+/+		118

Sulphonated dyes and related compounds			
Congo Red*			+/+
Only modest prophylactic effects in vivo	103, 121, 122, 124–132		
		Low concentrations stimulate PrP^{res} in cell-free conversion	
		Possible teratogen and/or carcinogen	
Suramin	+/+	Only modest prophylactic effects in vivo	114, 133
Curcumin*	+/–	Strain-dependent inhibition in vitro	123
Other amyloidophilic compounds	+/+	Oral administration	136
Polycations			
Polypropyleneimine gen. 4.0, polyetheyleneimine, polyamidoamide gen. 4.0	+/?		137, 138
Phosphorus dendrimers generation 4	+/+		139
Spermine, spermidine	+/?		140
DOSPA	+/?		141
Cyclic tetrapyrroles			
PcTS, DPG_2–Fe^{3+}, TMPP–Fe^{3+}*	+/+		142–144
In-TSP	+/?		115
Fe-TAP, Fe-TSP**	+/+	Fe-TSP + PPS has more than additive effects in vivo	19
Polyene antibiotics			
Amphotericin B, MS-8209**	+/+	Human treatments ineffective	147, 149–158, 160
Mepartricin*	+/+		151
Filipin	+/?		150

(Continued)

TABLE 1. (Continued)

Compound[a]	Activity In Vitro/In Vivo (Treated Early or Prophylaxis)[b]	Comments	Ref.
Quinacrine, quinoline, acridines, phenathiazines			
Quinacrine*	+/–	Quinacrine + desipramine or simvastatin better than quinacrine alone in cells Quinacrine + rPrP-Q218K enhances inhibition in cells No human benefit seen to date; trial ongoing Increases amount of PrP^{res} in the spleen Crosses blood–brain barrier	161–163, 165–168
Chlorpromazine	+/+		162, 169
Quinine and biquinoline*	+/+		170
Mefloquine*	+/–		171
Tetracyclic antibiotics			
Tetracycline Doxycycline	+/+	Tested by mixing with inoculum only	172–174
DMSO	+/–	Conflicting treatment outcomes	175–178
Copper chelators			
D(–)Penicillamine	+/+		180
Clioquinol	?/+	Structurally similar to quinacrine	181
Copper*	#/+	Can induce conversion	182
Chrysoidine	+/?	Mechanism of action may not relate to chelating properties	183
Pyridine dicarbonitriles	+/?		26, 27

Peptide aptamers and β-sheet breakers			
Peptide aptamers	+/?		186
β-Sheet breaker peptides	+/+	Tested by mixing with inoculum only May be strain or cell model specific	187, 188
Cholesterol-depleting agents			
Lovastatin, squalestatin	+/?		189, 190
Simvastatin**	+/+	Effect may be unrelated to cholesterol lowering; brain cholesterol levels do not drop	167, 191, 192
Amiodarone, progesterone	+/?		167
Mevinolin	?/?	Causes sequestration of PrP^{c} in Golgi No infectivity experiments	193
7-Dehydrocholesterol reductase and 24-dehydrocholesterol reductase inhibitors*	+/−	No effect even if given prior to inoculation	194
Antivirals			
Adenine arabinoside**	?/+	Two out of three CJD patients had moderate temporary improvement if treated early in course Mouse studies failed to show effect	195
LRP/LR			
siRNAs, antisense RNA, and LRP antibodies	+/?		83
scFvs*	+/+	Able to reduce PrP^{res} in spleens only; no effect on survival	84

(*Continued*)

TABLE 1. (Continued)

Compound[a]	Activity In Vitro/In Vivo (Treated Early or Prophylaxis)[b]	Comments	Ref.
Intracellular mechanisms			
Tyrosine kinase inhibitors (STI571, imatinib mesylate)*	+/+		204, 205
Phospholipase inhibitors	+/?		206
P53 inhibitors (Pifitrin-)*	+/−		178
MEK ½ inhibitors (SL327)	+/?		207
Cysteine-protease inhibitors (E64d)	+/?		161
Neuroprotection			
Flupirtine maleate**	+/?	Tested in human trials; some improvement in cognition May be neuroprotective via up-regulation of bcl-2 and normalization of glutathione levels	208–210
Cannabidiol*	+/+	No apparent interaction with PrP^c or PrP^{res} May be protective via inhibition of PrP^{res}-induced microglial cell migration, or antagonism of the NMDA receptor	211
Antioxidants			
Pyrazolone derivatives	+/?	Properties other than antioxidant ones may be responsible for effect	223

#, induces conversion in vitro.

[a]*, No efficacy after late administration; **, some efficacy after late administration.

[b]+, Anti-scrapie effect demonstrated; −, anti-scrapie effect not present; ?, anti-scrapie effect not yet assessed;

anti-prion effects [117]. Incubation times in mice were prolonged if the inoculum was mixed with fucoidan prior to administration, and oral treatment begun the day after inoculation was able to delay disease. Treatment given only prior to infection did not affect disease and some strain-dependent effects were observed in cell-culture studies. Nevertheless, the safe and oral use of this compound makes it a feasible option for early treatment.

Phosphorothioate Oligonucleotides. As noted above, phosphorothioated ODNs lacking the immunostimulatory CpG motif have striking prophylactic effects in mice inoculated subcutaneously or intraperitoneally. The most effective forms have 17 or more bases [97]. The potential value of these is their reduced anticoagulant effect compared to other agents in this class.

Copaxone. This compound, a polymer of alanine, glutamate, lysine, and tyrosine, is a drug used in the treatment of multiple sclerosis. It can bind heparin and probably has anti-prion effect through competitive inhibition of PrP^{res}–GAG interactions. It can prolong the incubation period if mixed with the inoculum prior to infection or if it is given at the time of inoculation, but has no discernible effect if given later [118].

Sulfonated Dyes and Similar Compounds

Congo Red and its analogs probably have the same mechanism of action as the polyanions, in that they can compete with sulfated glycans for PrP^{c} binding [106] and thereby inhibit PrP^{res} formation. This ability may stem from the fact that these compounds can stack and mimic larger polyanions [18,119–120]. It has also been noted that Congo Red can overstabilize PrP^{res} at high concentrations, which may create an inhibitory effect [121]. A drawback of this class of agents is the potential teratogenic and/or carcinogenic activity from the benzidine moiety in Congo Red, although this concern has been addressed in some of its analogs [122,123].

Congo Red. This sulfonated amyloid stain was the first inhibitor of PrP^{res} accumulation identified [124]. It decreases PrP^{res} in cells [103,121,122,124–128] and in cell-free conversions [126,127,129,130] and decreases surface PrP^{c} in normal cells [21]. It has also been shown to prolong incubation times in hamsters after intraperitoneal or intracerebral inoculation [131,132]. Its clinical use is limited by its toxicity and its poor penetration of the blood–brain barrier.

Suramin. Polysulfonated naphthyl urea (suramin) has some structural homology to Congo Red and can decrease PrP^{res}, decrease surface PrP^{c}, and cause aggregation of recombinant PrP [133]. It has had modest effects on incubation times in hamsters, when given around the time of intraperitoneal inoculation [114,133]. Analogs of suramin, symmetrical aromatics with naphthalene or benzene sulfonic acid substitutions, have also been studied and found to decrease PrP^{res} accumulation and induce PrP aggregation, but with no effect on PrP^{c} expression [134].

Curcumin. Curcumin, the major yellow pigment in the spice turmeric, has a structure like Congo Red except that it lacks sulfonates and the benzidine moiety. The compound is nontoxic, has better blood–brain barrier permeability, and can inhibit PrP^{res} formation in vitro [123]. It is able to bind the β-sheet-rich form of PrP, as well as the α-helical intermediates, but not the native form [135]. However, cell-culture effects are strain-dependent, and no effect has been detected in hamsters inoculated intracerebrally [123].

Other Amyloidophilic Compounds

Recently, a new group of compounds were tested for their anti-prion abilities [136]. Prion-strain-dependent effects were seen in cell cultures and in intracerebrally inoculated mouse models, where oral drug administration, started early in the disease course, extended incubation times. Treated animals had less infectivity in their brains and had a preponderance of diglycosylated PrP^{res}, unlike the monoglycosylated majority in the controls.

Polycations

A group of cationic polyamines, some contained in lipid transfection media, have been shown to have some anti-prion properties. They do not affect PrP^{c} levels but may actually enhance PrP^{res} clearance [137]. Few in vivo data are available at this time.

Polypropyleneimine Generation 4.0, Polyethyleneimine, Polyamidoamide Generation 4.0. These dendritic polyamine components of SuperFect reagent were found to prevent PrP^{res} formation [137] and remove infectivity from cell culture [138]. It is proposed that they act at the level of the lysosome [138].

Phosphorus Dendrimers Generation 4. These dendritic polyamines can clear PrP^{res} and infectivity from cell cultures and can reduce the splenic content of PrP^{res} in mice inoculated intraperitoneally [139]. Being less toxic and more bioavailable than the polypropyleneimine compounds may render these polyamines more useful for future treatments.

Spermine, Spermidine. Using a screening method that employs the fluorescence of anilinonaphthalene sulfonic acid (ANS), Bera and Nandi found that spermine and spermidine are able to reduce tRNA-induced polymerization of α-PrP, highlighting these compounds as possible candidates for anti-prion activities [140].

DOSPA. This lipopolyamine can decrease PrP^{res} in cells via enhanced clearance, but also appears to block de novo formation. Its activity is dependent on membrane association [141].

Cyclic Tetrapyrroles

This diverse group of compounds tends to have highly conjugated planar aromatic ring systems that bind transition metal ions and can be circumscribed by anionic, cationic, or uncharged peripheral substituent groups. They are effective inhibitors of PrP^{res} accumulation in vitro [142] and can greatly prolong survival times in vivo [19,115,143,144]. Their inhibitory mechanism probably involves direct interactions between stacked cyclic tetrapyrroles and the flexible amino-terminal domain of PrP molecules [18,145]. Similar interactions have been observed with the natural cyclic tetrapyrrole, hemin, causing PrP^{c} to cluster and be endocytosed from the surface of cultured cells [146]. This raises the possibility that there is physiological significance to hemin–PrP^{c} interactions and that the anti-prion mechanism of cyclic tetrapyrroles involves sequestration of PrP^{c} into a nonconvertible state.

Phthalocyanin Tetrasulfonate (PcTS), Deuteroporphyrin IX 2,4-bis(ethylene glycol) Iron(III) (DPG_2-Fe^{3+}), and Meso-tetra(2-N-methylpyridyl)porphine Iron(III) (TMPP-Fe^{3+}). These compounds decreased PrP^{res} formation in cell culture and in cell-free conversion reactions [142]. They also prolong incubation periods in mice inoculated intraperitoneally with scrapie as long as treatment is initiated within several weeks of infection [143,144]. No effect was seen after intracerebral inoculation or with treatment started late [143,144], which is not surprising given that these compounds are unlikely to cross the blood–brain barrier.

Indium(III) Meso-tetra(4-sulfonatophenyl)porphine Chloride (In-TSP). This compound was recently shown to have an effect in deer cell cultures infected with CWD [115].

Tetra (4-N,N,N-trimethylanilinium)porphine (Fe-TAP), Tetra(4-sulfonatophenyl) porphine (Fe-TSP). Both of these compounds have prophylactic and early treatment effects in vivo. Fe-TAP was the more effective when given intraperitoneally to animals inoculated intraperitoneally, whereas Fe-TSP was better when administered intracerebrally starting two weeks after intracerebral inoculation [19]. Fe-TSP was even more effective in treating mice inoculated intracerebrally when it was used in combination with pentosan polysulfate [19].

Hemin. The latest player on the porphyrin stage is hemin, which has recently been shown to bind PrP^{c}, causing its aggregation and internalization [146]. It is also able to block PrP^{res} formation in cell culture (D. A. Kocisko and B. Caughey, unpublished data).

Polyene Antibiotics

Another group of agents that inhibit PrP^{res} formation and delay the onset of disease in rodents includes the antifungal drug amphotericin B and its analogs [147,148]. These polyene antibiotics probably act by perturbing the raft membrane domains with which PrP^{c} is associated [149,150].

Amphotericin B, MS-8209. Amphotericin B reduces PrP^{res} in cell culture, although it is not curative [149]. It, and its less toxic analog MS-8209, prolong incubation in hamsters treated in the preclinical phase of disease after intraperitoneal or intracerebral inoculation of 263K [147,151–156]. Effects have even been seen in the late treatment of c57bl/6 mice that have been inoculated intracerebrally [157,158]. A drawback to the use of these agents is their apparent strain specificity [152,153,159]. More important, however, amphotericin B has been used clinically in the treatment of human CJD, without success [160].

Mepartricin. This analog showed effectiveness only against intraperitoneal inoculations [151].

Filipin. This relative of amphotericin B is able to reduce PrP^{res} in cell culture [150], but no in vivo data are available at this time.

Quinacrine, Quinoline, Acridines, and Phenothiazines

Quinacrine, chlorpromazine, quinine, and related molecules are lysosomotropic factors that have inhibitory effects on PrP^{res} formation in vitro and variable effects in vivo. Quinacrine, in particular, has been used unsuccessfully to date in humans (see below).

Quinacrine. Quinacrine inhibits PrP^{res} formation in cells [161–163]. It has no effect on mice that have been inoculated intracerebrally, whether treated orally [164] or intraventricularly [111]. Interestingly, quinacrine delivered intraperitoneally to mice infected with BSE caused an increased amount of PrP^{res} in the spleen [165]. A retrospective analysis of human CJD patients treated with quinacrine, compared to untreated control cases, demonstrated no benefit [166], and a more definitive prospective controlled trial also failed to show significant effect [258] (see below). Of note, recent studies suggest that the combined use of quinacrine plus desipramine or the HMG-CoA reductase inhibitor simvastatin [167] or the recombinant mutant PrP peptide rPrP-Q218K[168], may be more effective than quinacrine alone (see below).

Chlorpromazine. Although shown to be less effective than quinacrine in treating infected cell cultures [162], chlorpromazine, then known as aminasine, was once reported to delay disease onset after intracerebral inoculation [169].

Quinine and Biquinoline. These compounds, which can bind recombinant PrP^c, were given via intraventricular cannula to mice within one month of intracerebral inoculation. Treated mice had an increased variance of incubation times, some of which were prolonged significantly if treatment was given earlier in the course. Pathological changes were reduced only in the hemisphere with the cannula [170].

Mefloquine. This antimalarial agent showed promise in cell culture but lacked efficacy in tg7 mice that were inoculated intraperitoneally [171].

Tetracyclic Compounds

Tetracycline, Doxycycline. Most studies so far on these promising compounds suggest that they may be used best as decontamination treatments, since preincubation with inocula reduces infectivity [172,173] and detectable PrP^{res} [172,174].

Dimethyl Sulfoxide

The organic solvent dimethyl sulfoxide inhibits PrP-res aggregation in cell culture [175] and reduces the infectivity titer in brain homogenate [176]. Some have reported prolonged incubation times in intracerebrally inoculated hamsters if they are treated early [177], whereas others have seen no effect [178]. A drawback of using this compound is its toxicity in long-term use [177].

Copper, Chelators

PrP has several copper-binding domains, and the role of copper in normal PrP functioning and conversion has been much debated. The resistance of PrP^{Sc} to proteinase K is enhanced in vitro with increasing copper concentrations, suggesting that a decrease in copper may be beneficial in treatment. Copper chelators have had promising effects in cell culture, although some propose that their effectiveness correlates more with superoxide dismutase ability than with chelating properties [179]. Therapies aimed at reducing copper and increasing copper have yielded results that are conflicting at present.

D(−)penicillamine. This copper chelator is able to delay disease onset in mice if given early in the course [180].

Clioquinol. This chelator also shows a modest effect in hamsters inoculated intraperitoneally or intracerebrally [181]. Of note, however, is its structural similarity to quinacrine, leaving questions as to its actual mechanism of action.

Copper. In a conflicting study, providing copper in the water of infected mice was sufficient to prolong the incubation period [182].

Chrysoidine. This chelator (although chelation may not be its sole mechanism of action) inhibits PrP^{res} in cell culture, without reducing PrP^{c} levels or destabilizing PrP^{res} [183]. It is predicted to be able to cross the blood–brain barrier and may be less toxic than quinacrine, but no in vivo data are available yet.

Pyridine Dicarbonitrile Compounds

Certain PrP mutants resist conversion, and a computational search for molecules that resembled these mutants in spatial orientation led to the discovery of several compounds with inhibitory effects in cell culture [26]. More recently, another group has identified a new pyridine dicarbonitrile with activity in cell culture [27]. No in vivo data are currently available.

Peptide Aptamers and β-Sheet Breakers

Peptides of PrP sequence 106 to 141 or 119 to 136 were found to block PrP^{res} formation in cell-free conversion systems and cell culture [184,185]. Therefore, some novel approaches to therapy involve using peptides that interfere with conversion.

Peptide Aptamers. Random 16-mer sequences were placed in a constant scaffold of thioredoxin A from *Escherichia coli.* Those that selected against the PrP sequence 23 to 231 were tested in cell culture, where they were able to bind cell surface, lysosomal, or endoplasmic reticulum (ER) PrP^{c}. Inhibition of PrP^{res} formation occurred in those that targeted the ER or lysosome [186]. These compounds have not yet been tried in vivo.

β-Sheet Breaker Peptides. These peptides represent a portion of the target protein but with prolines inserted to block incorporation into a β-sheet structure. iPrP13, one such molecule, delays incubation time if mixed with the inoculum [187]. However, there may be some variability with strain or model, as there is an effect in a Chinese hamster ovarian cell model expressing mutant PrP, but no effect on infected N2a cells [188].

Cholesterol-Depleting Agents

PrP^c is usually localized to membrane microdomains rich in sphingolipids and cholesterol, called *lipid rafts*, and conversion of membrane-associated PrP^c depends on interactions with PrP^{res} in the same membranes [3]. Agents that affect membrane cholesterol levels can therefore affect PrP^{res} formation, perhaps by redistributing the PrP^c to an inaccessible area. Lovastatin and squalestatin, which are members of the statin drug class that inhibit cholesterol synthesis, affect PrP^{res} accumulation in cell culture, an effect that is lost if cholesterol is added [189,190]. Many of the more recent studies have focused on simvastatin. An advantage of this group of agents is its documented safety in humans.

Simvastatin. This statin has the ability to cross the blood–brain barrier, although its half-life of 1.5 hours may limit its usefulness. High-dose (100 mg/kg per day) simvastatin started late in the disease course (100 days after intracerebral inoculation of mice) prolonged survival but had no effect on PrP^{res} levels [191]. It was suggested that this effect may have been more related to anti-inflammatory properties than to cholesterol lowering. Lower-dose (1 mg/kg per day) treatment begun at the time of intracerebral inoculation was able to prolong incubation and delay loss of motor coordination, but later treatment (123 days/post-inoculation) had no effect [192]. Interestingly, there was still no difference in PrP^{res} levels, and the cholesterol levels were reduced in the liver but not, in the brain. In cell-culture models, a combined treatment of simvastatin with quinacrine had synergistic inhibitory effects [167].

Amiodarone and Progesterone. Both of these compounds have cholesterol-redistributing properties and were able to clear PrP^{res} from cells [167].

Mevinolin. This HMG CoA reductase inhibitor blocks cholesterol synthesis and led to the reduction of cell surface PrP^c and accumulation of PrP^c within the Golgi [193]. Studies of this compound with PrP^{res} have not yet been reported, but the sequestration of PrP^c could have an effect on de novo PrP^{res} formation.

7-Dehydrocholesterol Reductase and 24-Dehydrocholesterol Reductase Inhibitors. These compounds were able to block PrP^{res} formation in cell culture, but had no effect in vivo, regardless of whether treatment was initiated prior to or at the time of inoculation [194]. Here there was also no drop in brain cholesterol.

Antivirals

Apart from some temporary modest improvements reported in two of three CJD patients treated intermittently with vidarabine (adenine arabinoside) [195], numerous studies of many antiviral agents, including rifampicin, cytosine arabinoside, adenosine arabinoside, isoprinosine, amantadine, methisazone, phosphonoacetic acid, virazole, methisoprinol, β-propiolactone, and thiamphenicol, have failed to show any effect in mice [147,196–199].

Targeting the Laminin Receptor Precursor Protein (LRP/LR)–PrP Interaction

The 67-kDa laminin receptor binds PrP^c as does its 37-kDa precursor [200,201]. It may also bind PrP^{res} [202] and plays a critical role in the propagation of PrP^{res} [83]. Therefore, interfering with this putative cell surface ligand for PrP may have therapeutic consequences (for a complete review of LRP/LR therapeutic approaches, see [203]).

Small Interfering RNAs (siRNAs), Antisense RNA, and LRP Antibodies. All these approaches have been used successfully to target LRP and inhibit PrP^{res} accumulation in cell culture [83]. No in vivo testing has been reported.

Recombinant Single-Chain Antibodies (scFvs). More recently, scFv's were demonstrated to affect the in vitro binding of PrP and LRP, and passive immunotransfer was able to reduce PrP^{res} in the spleens of intraperitoneally inoculated mice [84]. This treatment was given prior to inoculation, and there was no effect on incubation times or survival, so it was most likely affecting the peripheral phase of disease [84]. The researchers note that the short half-life of the antibody and its once weekly administration may have limited effectiveness.

Targeting Intracellular Enzymes and Pathways Involved in PrP^{res} Formation

A variety of enzyme and cell signaling inhibitors have been screened and tested for their anti-prion abilities in vitro, with limited in vivo testing to date. In addition to opening new possibilities for therapy, these studies have also shed light on the pathogenic mechanisms accompanying PrP^{res} accumulation.

Tyrosine Kinase Inhibitors. The inhibitor STI571, also known as imatinib mesylate, targets tyrosine kinase c-Abl and is thought to exert its anti-prion effect through lysosomal degradation pathways. It enhanced the clearance of PrP^{res} in cell culture, with complete cure after 10 days of treatment [204,205]. In vivo it decreased spleen PrP^{res} and delayed neuroinvasion if administered early and after intraperitoneal inoculation, but even intracerebral administration of the drug could not clear PrP^{res} from the brain [205].

Phospholipase Inhibitors. Both phospholipase A_2 (PLA_2) and platelet-activating factor (PAF) are required for PrP^{res} formation. In cell culture, inhibitors of PLA_2, including glucocorticoids, decreased PrP^{res} formation, PAF agonists increased PrP^{res} and PrP^c, and PAF antagonists decreased PrP^{res} and PrP^c [206].

P53 Inhibitors. Pifithrin-α, which inhibits p53, decreased PrP^{res} in cell culture, but had no major effect in infected hamsters, other than to reduce their levels of caspase-3 [178].

Mitogen-Activated Protein Kinase ½ (MEK½) Inhibitors. SL327, through inhibition of MEK½, decreased PrP^{res} in cells [207]. Given its ability to cross the blood–brain barrier, SL327 might have some in vivo effects, although no in vivo data are currently available.

Cysteine-Protease Inhibitors. L-*trans*-Epoxysuccinyl-leucylamido(4-guanidino) butane (E64d) did not affect cell-free conversion, but reduced PrP^{res} in cells, without affecting PrP^c [161].

Neuroprotection

Compounds without direct effects on PrP^c or PrP^{res} may still have therapeutic potential as neuroprotective agents and provide symptomatic treatment.

Flupirtine Maleate. This triaminopyridine analgesic is one of the few compounds that have been tested in a prospective double-blind trial of human CJD treatment. Its antiapoptotic effects were seen in cell cultures exposed to β-amyloid [208] or the prion protein fragment PrP106–126 [209]. Its mechanism of neuroprotection may be through up-regulation of the protooncogene bcl-2 and normalizing of glutathione levels [208,209]. Although there is no evidence of its PrP^{res} antagonism, given flupirtine's documented safe use in humans, a trial in 28 CJD patients was performed (see below). Significant improvements in cognitive functioning were observed, although the study was not designed to detect any differences in survival [210].

Cannabidiol. This cannabinoid molecule is a nonpsychoactive constituent of cannabis which crosses the blood–brain barrier readily, targets the brain, and has minimal toxicity. Its use as a prion therapy has recently been studied [211]. Cannabidiol did not affect cell-free conversion, the expression of PrP^c in cells or animals, or the stability of PrP^{res}. Nevertheless, over time it was able to inhibit PrP^{res} accumulation in cells, reduce PrP^{res} toxicity in primary neuronal cultures, and prolong survival in intraperitoneally inoculated mice when treatment was started in the early phase of disease. Interestingly, brain PrP^{res} levels were reduced, while spleen levels were not. It was proposed that cannabidiol may act primarily through neuroprotection, possibly secondary to its inhibition of PrP^{res}-induced microglial cell migration or its ability to antagonize the NMDA receptor.

Antioxidants

For a more complete review of the role of antioxidant defense in prion disease, see Brown's paper [212]. Some have proposed that PrP^c has superoxide dismutase (SOD) activity [213] and that its primary function is to respond to and provide protection from oxidative stress [214–217]. Cells without PrP^c are more sensitive to oxidative stress [214,218,219], and increased oxidative stress is seen in infected animals [220,221]. Brain levels of coenzyme Q_9 and Q_{10}, endogenous lipophilic antioxidants, steadily increased in BSE-infected mice expressing bovine PrP^c, beginning at the time PrP^{res} deposits began to appear (150 days post-inoculation) [222]. This suggests that the antioxidant system is not only active in prion disease, but may be overwhelmed. Antioxidants have therefore recently been investigated for possible neuroprotective or anti-prion properties.

Pyrazolone Derivatives. These compounds were derived from edaravone, a free-radical scavenger, and some have shown inhibition of PrP^{res} in cell culture, one of which had an effect at 3 nM [223]. No in vivo data are available, and it is unclear whether antioxidant properties are responsible for the effect, given that there was no correlation between inhibitory activity and the compounds' SOD activities, abilities to oxidize, or abilities to bind copper.

TARGETING PrP^c

An exciting new therapeutic prospect, with potential for treating prion diseases at later stages, is now being investigated. PrP^c must be expressed in neurons for a host to be susceptible to prion disease [224], yet turning off neuronal PrP^c expression in adult mice has no observable phenotypic consequence [225]. A reversal of neuropathological [31] and behavioral [32] abnormalities occurred in scrapie-inoculated mice when neuronal expression of PrP^c was eliminated. This resolution proceeded despite the continued presence of PrP^{res} deposits and accumulation of infectivity [31], indicating that the suppression of PrP^c expression may hold promise for prion disease therapy.

RNA Interference and Gene Therapy

One approach to lowering the amount of PrP^c in cells is to use small interfering RNA (siRNA). These short double-stranded RNAs assemble into endoribonuclease-containing complexes, which then target and cleave complementary mRNA. In cell culture, siRNA has been shown to suppress exogenous and endogenous PrP gene (Prnp) expression [226], and siRNA corresponding to the Prnp nucleotides 392 to 410 reduced PrP^{res} formation [227]. siRNA methods have also been used to target important PrP ligands such as LRP/LR, leading to inhibition of PrP^{res} accumulation in cell culture [83]. A recently developed transgenic mouse ectopically expressing LRP/LR siRNA may serve as a good model for future prion research [228].

Introducing siRNA into animal systems has been achieved using retroviruses, where an RNA polymerase III promoter and target sequence can be inserted. The retrovirus, in turn, can integrate into the genome of nondividing cells, providing a mechanism of gene therapy. A lentivirus vector, designed to express a short hairpin RNA targeted against PrP^c mRNA, was recently employed for this purpose [229]. The hairpin RNA, corresponding to Prnp nucleotides 512 to 532, reduced PrP^c expression by more than 90% and reduced PrP^{res} formation in cell culture. Intracranial injection of the vector into uninfected tga20 mice led to a reduction of PrP^c. In chimeric mice, PrP^c levels correlated with the amount of transgene expressed, and high-chimeric mice had prolonged survival times and inoculation, with reduced PrP^c and PrP^{Sc}. Although limited to a chimeric model, this study has opened the way to lentivector-based gene therapy.

An impediment to gene therapy for prion diseases is the inability of vectors to cross the blood–brain barrier easily. Stereotactic injection into the brain is possible, although it is invasive and leads to relatively localized treatment effects. A promising new way of specifically introducing siRNA into neurons has been developed by Kumar and colleagues [230]. They used the neurotropic rabies virus glycoprotein (RVG), which binds nicotinic acetylcholine receptors on neurons and which can cross the blood–brain barrier. To this they conjugated a nonamer of D-arginine (to make RVG-9R), which efficiently bound their siRNA of interest, against viral Japanese encephalitis. siRNA distribution was enhanced by RVG's ability to spread both retroaxonally and transsynaptically [231], allowing the easy transduction of neighboring neuronal cells. Targeting was specific, with no uptake in the spleen or liver, and there was no induction of an inflammatory response. They went on to treat viral Japanese encephalitis in mice successfully, with an 80% survival rate. The use of this RVG-9R with siRNA against PrP^c could be a huge step forward in prion disease treatment.

Cautionary Note on Breeding Resistant Genotypes

Much effort has been made to breed "scrapie-resistant" ARR/ARR genotypes into sheep flocks. However, a recent report cautions that we may not yet fully understand what is truly resistant to infection, as natural scrapie has now been reported in ARR/ARR sheep [232].

COMBINATION THERAPY

Given the various targets now available for prion disease treatment, including PrP conversion, PrP^{res} clearance, neuroprotection, and PrP^{c} expression, it makes sense to try combining approaches. Partial effects seen with some treatments may be enhanced by others. There are a few examples of cooperative or synergistic effects in the literature, but there is a great potential for further testing.

One apparently synergistic relationship was discovered between pentosan polysulfate and Fe(III)*meso*-tetra(4-sulfonatophenyl)porphine (Fe-TSP), which improved the survival of tg7 mice inoculated intracerebrally with scrapie [97]. In this study, effective treatments were given intracerebrally once a week, starting 2 or 4 weeks post-inoculation, and more than doubled survival times when treatment was initiated 2 weeks post-inoculation.

Two additional combination therapy reports involve the drug quinacrine. In the first, combining the conversion-resistant mutant rPrP-Q218K [168,233] with quinacrine enhanced PrP^{res} inhibition in cell culture [168]. In the second, using quinacrine with simvastatin or desipramine produced synergistic inhibition in cells and inspired the creation of a more potent cell-culture inhibitor of PrP^{res}, called quipramine, a combination of quinacrine and desipramine, created by covalently linking the acridine scaffold of one with the iminodibenzyl scaffold of the other [167]. This new molecule is thought to influence PrP by redistributing the cellular lipids via lysosomal (desipramine effect) and caveolin-1 (quinacrine effect) pathways.

HUMAN TREATMENTS

Ultimately, there is great interest in the effectiveness of treatment in the clinical setting of human prion disease. As diagnostic techniques improve, so does the possibility of initiating treatment earlier in the course of infection. It is hoped that early treatment may at least stabilize a patient at a reasonable baseline, even if a full recovery or cure is not achieved. While some human patients have received treatments earlier in the disease course than others, all have had significant disease. With a limited numbers of patients (CJD has an attack rate of 0.6 to 1.2 per million per year [234]), performing balanced controlled prospective trials is nearly impossible. Also, the variations in clinical presentation and disease duration make it difficult retrospectively to compare the small numbers of treated and untreated patients. Nevertheless, some data are available, and although no miracles have been achieved, new delivery techniques, such as intraventricular administration, have been tested and optimized and stand ready should new treatments become approved for testing.

Pentosan Polysulfate

Currently, the most human treatment data available are those for pentosan polysulphate (PPS). A recent review provides a good summary of its use in human prion diseases [235]. Of the 26 reported cases treated with intraventricular PPS, including vCJD, sCJD, iCJD, fCJD, and GSS, 11 have died. Many treatments were initiated at a later stage of disease, where only limited recovery might be possible, but cases of continuous disease progression after early treatment also occurred. Some patients had a stabilization of the neurological condition, some made small gains in swallowing reflexes, brain stem functions, speech, and had reduced myoclonus, whereas others continued to deteriorate. Side effects were not a major concern, even at doses of 200 μg/kg per day. Manageable subdural collections of fluid occurred in five, seizures developed in two (although this could have been secondary to CJD), and significant intracerebral haemorrhage occurred in one, secondary to a surgical complication. The longest-ever surviving vCJD patient, whose symptoms began in May 2001, was the first to be treated with intraventricular PPS, starting in January 2003. (For further details, see Todd et al.'s paper [236].) Despite treatment starting 19 months after onset, at which point the patient was dependent for all activities of daily living, he is still alive and stable, but fully dependent at the time of this writing (November 2007). For recent case reports of other patients on intraventricular PPS, see [237] and [238].

Clearly, the ideal treatment strategy would seek to preserve people with minimal disability, if any, rather than stabilize the disease at such a late stage, but administering PPS earlier does not necessarily improve outcomes [235]. A controlled trial has not been performed with this agent, so comparisons can only be made with case controls, but a medical research council studying PPS treatment in human patients has concluded that PPS provides no clear benefits in halting disease progression (see http://www.mrc.ac.uk/Utilities/Documentrecord/index.htm?d = MRC003453).

Quinacrine

Quinacrine has been used in several cases of CJD [239–242]. In a larger compassionate use study, quinacrine administration at a dose of 100 mg three times daily was given to 30 sCJD patients and 2 vCJD patients. No significant benefit was found compared with those in an untreated cohort [166]. Despite this unfavorable review, a prospective patient-preference trial, "PRION-1," was undertaken. Results confirmed that 300 mg of quinacrine daily had no significant effect on clinical course [258].

Flupirtine Maleate

The analgesic flupirtine maleate has been tested in a prospective double-blind trial of human CJD treatment. Twenty-six sporadic and two iatrogenic cases of CJD were treated with 300 to 400 mg/day of flupirtine. Significant improvements in ADAS-Cog scores, a measure of cognitive functioning, were observed. A survival benefit was not seen, although the study was not designed to measure this outcome [210].

Other Case Reports

Several other treatments have been tried in human CJD, with little success. Acyclovir was unsuccessful in a case of early [243] and late [244]-onset treatment. Amantadine caused transient improvements in wakefulness and mentation in some patients [245,246], but no effects on survival were seen [247–249]. Amphotericin B [160], interferon [93], and idoxuridine [250] were tried without success, and vidarabine reportedly caused repeated temporary improvements in one patient, but ultimate decline occurred, despite ongoing therapy [195].

CONCLUSIONS

As we continue to learn more about the pathogenesis of these devastating diseases, and as new techniques are developed for earlier and more sensitive detection of PrP^{res} [1,28–30], many of the compounds with prophylactic and early treatment potentials, including those immune therapies that target the peripheral phase of replication in some forms of prion disease, may come to practical use. In addition, the new methods of delivery and targeting of PrP^{c} in the central nervous system may greatly enhance our ability to have a significant impact on clinical outcomes. Combining several arms of therapy which block conversion, promote PrP^{res} clearance, and ameliorate prion disease pathogenesis might ultimately lead us to a cure.

REFERENCES

1. Castilla, J., Saa, P., Hetz, C., Soto, C. (2005). In vitro generation of infectious scrapie prions. *Cell, 121*, 195–206.
2. Deleault, N.R., Harris, B.T., Rees, J.R., Supattapone, S. (2007). From the cover: formation of native prions from minimal components in vitro. *Proc Natl Acad Sci U S A, 104*, 9741–9746.
3. Baron, G.S., Wehrly, K., Dorward, D.W., Chesebro, B., Caughey, B. (2002). Conversion of raft associated prion protein to the protease-resistant state requires insertion of PrP-res (PrP(Sc)) into contiguous membranes. *EMBO J, 21*, 1031–1040.
4. Chesebro, B., Trifilo, M., Race, R., Meade-White, K., Teng, C., LaCasse, R., Raymond, L., Favara, C., Baron, G., Priola, S., et al. (2005). Anchorless prion protein results in infectious amyloid disease without clinical scrapie. *Science, 308*, 1435–1439.
5. Berardi, V.A., Cardone, F., Valanzano, A., Lu, M., Pocchiari, M. (2006). Preparation of soluble infectious samples from scrapie-infected brain: a new tool to study the clearance of transmissible spongiform encephalopathy agents during plasma fractionation. *Transfusion, 46*, 652–658.
6. Silveira, J.R., Raymond, G.J., Hughson, A.G., Race, R.E., Sim, V.L., Hayes, S.F., Caughey, B. (2005). The most infectious prion protein particles. *Nature, 437*, 257–261.

7. Pastrana, M.A., Sajnani, G., Onisko, B., Castilla, J., Morales, R., Soto, C., Requena, J.R. (2006). Isolation and characterization of a proteinase K–sensitive PrP(Sc) fraction. *Biochemistry, 45*, 15710–15717.

8. Tzaban, S., Friedlander, G., Schonberger, O., Horonchik, L., Yedidia, Y., Shaked, G., Gabizon, R., Taraboulos, A. (2002). Protease-sensitive scrapie prion protein in aggregates of heterogeneous sizes. *Biochemistry, 41*, 12868–12875.

9. Novitskaya, V., Bocharova, O.V., Bronstein, I., Baskakov, I.V. (2006). Amyloid fibrils of mammalian prion protein are highly toxic to cultured cells and primary neurons. *J Biol Chem, 281*, 13828–13836.

10. Hill, A.F., Desbruslais, M., Joiner, S., Sidle, K.C., Gowland, I., Collinge, J., Doey, L.J., Lantos, P. (1997). The same prion strain causes vCJD and BSE. *Nature, 389*, 448–450, 526.

11. Bruce, M.E., Will, R.G., Ironside, J.W., McConnell, I., Drummond, D., Suttie, A., McCardle, L., Chree, A., Hope, J., Birkett, C., et al. (1997). Transmissions to mice indicate that "new variant" CJD is caused by the BSE agent. *Nature, 389*, 498–501.

12. Collinge, J., Sidle, K.C., Meads, J., Ironside, J., Hill, A.F. (1996). Molecular analysis of prion strain variation and the aetiology of "new variant" CJD. *Nature, 383*, 685–690.

13. Hewitt, P.E., Llewelyn, C.A., Mackenzie, J., Will, R.G. (2006). Creutzfeldt–Jakob disease and blood transfusion: results of the UK Transfusion Medicine Epidemiological Review study. *Vox Sang, 91*, 221–230.

14. Bach, S., Talarek, N., Andrieu, T., Vierfond, J.M., Mettey, Y., Galons, H., Dormont, D., Meijer, L., Cullin, C., Blondel, M. (2003). Isolation of drugs active against mammalian prions using a yeast-based screening assay. *Nat Biotechnol, 21*, 1075–1081.

15. Kocisko, D.A., Baron, G.S., Rubenstein, R., Chen, J., Kuizon, S., Caughey, B. (2003). New inhibitors of scrapie-associated prion protein formation in a library of 2000 drugs and natural products. *J Virol, 77*, 10288–10294.

16. Kocisko, D.A., Morrey, J.D., Race, R.E., Chen, J., Caughey, B. (2004). Evaluation of new cell culture inhibitors of protease-resistant prion protein against scrapie infection in mice. *J Gen Virol, 85*, 2479–2483.

17. Kocisko, D.A., Engel, A.L., Harbuck, K., Arnold, K.M., Olsen, E.A., Raymond, L.D., Vilette, D., Caughey, B. (2005). Comparison of protease-resistant prion protein inhibitors in cell cultures infected with two strains of mouse and sheep scrapie. *Neurosci Lett, 388*, 106–111.

18. Caughey, B., Caughey, W.S., Kocisko, D.A., Lee, K.S., Silveira, J.R., Morrey, J.D. (2006). Prions and transmissible spongiform encephalopathy (TSE) chemotherapeutics: a common mechanism for anti-TSE compounds? *Acc Chem Res, 39*, 646–653.

19. Kocisko, D.A., Caughey, W.S., Race, R.E., Roper, G., Caughey, B., Morrey, J.D. (2006). A porphyrin increases survival time of mice after intracerebral prion infection. *Antimicrob Agents Chemother, 50*, 759–761.

20. Kiachopoulos, S., Heske, J., Tatzelt, J., Winklhofer, K.F. (2004). Misfolding of the prion protein at the plasma membrane induces endocytosis, intracellular retention and degradation. *Traffic, 5*, 426–436.

21. Shyng, S.L., Lehmann, S., Moulder, K.L., Harris, D.A. (1995). Sulfated glycans stimulate endocytosis of the cellular isoform of the prion protein, PrPC, in cultured cells. *J Biol Chem, 270*, 30221–30229.
22. Kawatake, S., Nishimura, Y., Sakaguchi, S., Iwaki, T., Doh-ura, K. (2006). Surface plasmon resonance analysis for the screening of anti-prion compounds. *Biol Pharm Bull, 29*, 927–932.
23. Bertsch, U., Winklhofer, K.F., Hirschberger, T., Bieschke, J., Weber, P., Hartl, F.U., Tavan, P., Tatzelt, J., Kretzschmar, H.A., Giese, A. (2005). Systematic identification of antiprion drugs by high-throughput screening based on scanning for intensely fluorescent targets. *J Virol, 79*, 7785–7791.
24. Breydo, L., Bocharova, O.V., Baskakov, I.V. (2005). Semiautomated cell-free conversion of prion protein: applications for high-throughput screening of potential antiprion drugs. *Anal Biochem, 339*, 165–173.
25. Kocisko, D.A., Bertholet, N., Moore, R.A., Caughey, B., Vaillant, A. (2007). Identification of prion inhibitors by a fluorescence-polarization-based competitive binding assay. *Anal Biochem, 363*, 154–156.
26. Perrier, V., Wallace, A.C., Kaneko, K., Safar, J., Prusiner, S.B., Cohen, F.E. (2000). Mimicking dominant negative inhibition of prion replication through structure-based drug design. *Proc Natl Acad Sci U S A, 97*, 6073–6078.
27. Reddy, T.R., Mutter, R., Heal, W., Guo, K., Gillet, V.J., Pratt, S., Chen, B. (2006). Library design, synthesis, and screening: pyridine dicarbonitriles as potential prion disease therapeutics. *J Med Chem, 49*, 607–615.
28. Atarashi, R., Moore, R.A., Sim, V.L., Hughson, A.G., Dorward, D.W., Onwubiko, H.A., Priola, S.A., Caughey, B. (2007). Ultrasensitive detection of scrapie prion protein using seeded conversion of recombinant prion protein. *Nat Methods, 4*, 645–650.
29. Deleault, N.R., Geoghegan, J.C., Nishina, K., Kascsak, R., Williamson, R.A., Supattapone, S. (2005). Protease-resistant prion protein amplification reconstituted with partially purified substrates and synthetic polyanions. *J Biol Chem, 280*, 26873–26879.
30. Saa, P., Castilla, J., Soto, C. (2006). Presymptomatic detection of prions in blood. *Science, 313*, 92–94.
31. Mallucci, G., Dickinson, A., Linehan, J., Klohn, P.C., Brandner, S., Collinge, J. (2003). Depleting neuronal PrP in prion infection prevents disease and reverses spongiosis. *Science, 302*, 871–874.
32. Mallucci, G.R., White, M.D., Farmer, M., Dickinson, A., Khatun, H., Powell, A.D., Brandner, S., Jefferys, J.G., Collinge, J. (2007). Targeting cellular prion protein reverses early cognitive deficits and neurophysiological dysfunction in prion-infected mice. *Neuron, 53*, 325–335.
33. Fraser, H., Dickinson, A.G. (1970). Pathogenesis of scrapie in the mouse: the role of the spleen. *Nature, 226*, 462–463.
34. Porter, D.D., Porter, H.G., Cox, N.A. (1973). Failure to demonstrate a humoral immune response to scrapie infection in mice. *J Immunol, 111*, 1407–1410.
35. Sigurdson, C.J., Spraker, T.R., Miller, M.W., Oesch, B., Hoover, E.A. (2001). PrP(CWD) in the myenteric plexus, vagosympathetic trunk and endocrine glands of deer with chronic wasting disease. *J Gen Virol, 82*, 2327–2334.

36. Montrasio, F., Frigg, R., Glatzel, M., Klein, M.A., Mackay, F., Aguzzi, A., Weissmann, C. (2000). Impaired prion replication in spleens of mice lacking functional follicular dendritic cells. *Science, 288*, 1257–1259.
37. Heikenwalder, M., Prinz, M., Heppner, F.L., Aguzzi, A. (2004). Current concepts and controversies in prion immunopathology. *J Mol Neurosci, 23*, 3–12.
38. Beekes, M., McBride, P.A. (2000). Early accumulation of pathological PrP in the enteric nervous system and gut-associated lymphoid tissue of hamsters orally infected with scrapie. *Neurosci Lett, 278*, 181–184.
39. Mabbott, N.A., Bruce, M.E. (2001). The immunobiology of TSE diseases. *J Gen Virol, 82*, 2307–2318.
40. Glaysher, B.R., Mabbott, N.A. (2007). Role of the GALT in scrapie agent neuroinvasion from the intestine. *J Immunol, 178*, 3757–3766.
41. Brown, K.L., Stewart, K., Ritchie, D.L., Mabbott, N.A., Williams, A., Fraser, H., Morrison, W.I., Bruce, M.E. (1999). Scrapie replication in lymphoid tissues depends on prion protein-expressing follicular dendritic cells. *Nat Med, 5*, 1308–1312.
42. Mabbott, N.A., Mackay, F., Minns, F., Bruce, M.E. (2000). Temporary inactivation of follicular dendritic cells delays neuroinvasion of scrapie. *Nat Med, 6*, 719–720.
43. Klein, M.A., Frigg, R., Flechsig, E., Raeber, A.J., Kalinke, U., Bluethmann, H., Bootz, F., Suter, M., Zinkernagel, R.M., Aguzzi, A. (1997). A crucial role for B cells in neuroinvasive scrapie. *Nature, 390*, 687–690.
44. Fraser, H., Farquhar, C.F. (1987). Ionising radiation has no influence on scrapie incubation period in mice. *Vet Microbiol, 13*, 211–223.
45. Mabbott, N.A., Williams, A., Farquhar, C.F., Pasparakis, M., Kollias, G., Bruce, M.E. (2000). Tumor necrosis factor alpha-deficient, but not interleukin-6-deficient, mice resist peripheral infection with scrapie. *J Virol, 74*, 3338–3344.
46. Matsumoto, M., Mariathasan, S., Nahm, M.H., Baranyay, F., Peschon, J.J., Chaplin, D.D. (1996). Role of lymphotoxin and the type I TNF receptor in the formation of germinal centers. *Science, 271*, 1289–1291.
47. Klein, M.A., Kaeser, P.S., Schwarz, P., Weyd, H., Xenarios, I., Zinkernagel, R.M., Carroll, M.C., Verbeek, J.S., Botto, M., Walport, M.J., et al. (2001). Complement facilitates early prion pathogenesis. *Nat Med, 7*, 488–492.
48. Mabbott, N.A., Bruce, M.E., Botto, M., Walport, M.J., Pepys, M.B. (2001). Temporary depletion of complement component C3 or genetic deficiency of C1q significantly delays onset of scrapie. *Nat Med*, 7, 485–487.
49. Zabel, M.D., Heikenwalder, M., Prinz, M., Arrighi, I., Schwarz, P., Kranich, J., von, T.A., Haas, K.M., Zeller, N., Tedder, T.F., et al. (2007). Stromal complement receptor CD21/35 facilitates lymphoid prion colonization and pathogenesis. *J Immunol, 179*, 6144–6152.
50. Bruce, M.E., Brown, K.L., Mabbott, N.A., Farquhar, C.F., Jeffrey, M. (2000). Follicular dendritic cells in TSE pathogenesis. *Immunol Today, 21*, 442–446.
51. Crozet, C., Lezmi, S., Flamant, F., Samarut, J., Baron, T., Bencsik, A. (2007). Peripheral circulation of the prion infectious agent in transgenic mice expressing the ovine prion protein gene in neurons only. *J Infect Dis, 195*, 997–1006.
52. Kimberlin, R.H., Walker, C.A. (1989). Pathogenesis of scrapie in mice after intragastric infection. *Virus Res, 12*, 213–220.

53. Glatzel, M., Heppner, F.L., Albers, K.M., Aguzzi, A. (2001). Sympathetic innervation of lymphoreticular organs is rate limiting for prion neuroinvasion. *Neuron, 31*, 25–34.
54. Prinz, M., Heikenwalder, M., Junt, T., Schwarz, P., Glatzel, M., Heppner, F.L., Fu, Y.X., Lipp, M., Aguzzi, A. (2003). Positioning of follicular dendritic cells within the spleen controls prion neuroinvasion. *Nature, 425*, 957–962.
55. Hoffmann, C., Ziegler, U., Buschmann, A., Weber, A., Kupfer, L., Oelschlegel, A., Hammerschmidt, B., Groschup, M.H. (2007). Prions spread via the autonomic nervous system from the gut to the central nervous system in cattle incubating bovine spongiform encephalopathy. *J Gen Virol, 88*, 1048–1055.
56. Gabizon, R., McKinley, M.P., Groth, D., Prusiner, S.B. (1988). Immunoaffinity purification and neutralization of scrapie prion infectivity. *Proc Natl Acad Sci U S A, 85*, 6617–6621.
57. Horiuchi, M., Chabry, J., Caughey, B. (1999). Specific binding of normal prion protein to the scrapie form via a localized domain initiates its conversion to the protease-resistant state. *EMBO J, 18*, 3193–3203.
58. Enari, M., Flechsig, E., Weissmann, C. (2001). Scrapie prion protein accumulation by scrapie-infected neuroblastoma cells abrogated by exposure to a prion protein antibody. *Proc Natl Acad Sci U S A, 98*, 9295–9299.
59. Beringue, V., Vilette, D., Mallinson, G., Archer, F., Kaisar, M., Tayebi, M., Jackson, G.S., Clarke, A.R., Laude, H., Collinge, J., Hawke, S. (2004). PrPSc binding antibodies are potent inhibitors of prion replication in cell lines. *J Biol Chem, 279*, 39671–39676.
60. Caughey, B., Raymond, G.J. (1991). The scrapie-associated form of PrP is made from a cell surface precursor that is both protease- and phospholipase-sensitive. *J Biol Chem, 266*, 18217–18223.
61. Cardinale, A., Filesi, I., Vetrugno, V., Pocchiari, M., Sy, M.S., Biocca, S. (2005). Trapping prion protein in the endoplasmic reticulum impairs PrPC maturation and prevents PrPSc accumulation. *J Biol Chem, 280*, 685–694.
62. Vetrugno, V., Cardinale, A., Filesi, I., Mattei, S., Sy, M.S., Pocchiari, M., Biocca, S. (2005). KDEL-tagged anti-prion intrabodies impair PrP lysosomal degradation and inhibit scrapie infectivity. *Biochem Biophys Res Commun, 338*, 1791–1797.
63. Perrier, V., Solassol, J., Crozet, C., Frobert, Y., Mourton-Gilles, C., Grassi, J., Lehmann, S. (2004). Anti-PrP antibodies block PrPSc replication in prion-infected cell cultures by accelerating PrPC degradation. *J Neurochem, 89*, 454–463.
64. Peretz, D., Williamson, R.A., Kaneko, K., Vergara, J., Leclerc, E., Schmitt-Ulms, G., Mehlhorn, I.R., Legname, G., Wormald, M.R., Rudd, P.M., et al. (2001). Antibodies inhibit prion propagation and clear cell cultures of prion infectivity. *Nature, 412*, 739–743.
65. Horiuchi, M., Baron, G.S., Xiong, L.W., Caughey, B. (2001). Inhibition of interactions and interconversions of prion protein isoforms by peptide fragments from the C-terminal folded domain. *J Biol Chem, 276*, 15489–15497.
66. Sigurdsson, E.M., Brown, D.R., Daniels, M., Kascsak, R.J., Kascsak, R., Carp, R., Meeker, H.C., Frangione, B., Wisniewski, T. (2002). Immunization delays the onset of prion disease in mice. *Am J Pathol, 161*, 13–17.

67. Schwarz, A., Kratke, O., Burwinkel, M., Riemer, C., Schultz, J., Henklein, P., Bamme, T., Baier, M. (2003). Immunisation with a synthetic prion protein-derived peptide prolongs survival times of mice orally exposed to the scrapie agent. *Neurosci Lett, 350*, 187–189.

68. Goni, F., Knudsen, E., Schreiber, F., Scholtzova, H., Pankiewicz, J., Carp, R., Meeker, H.C., Rubenstein, R., Brown, D.R., Sy, M.S., et al. (2005). Mucosal vaccination delays or prevents prion infection via an oral route. *Neuroscience, 133*, 413–421.

69. Magri, G., Clerici, M., Dall'Ara, P., Biasin, M., Caramelli, M., Casalone, C., Giannino, M.L., Longhi, R., Piacentini, L., Della, B.S., et al. (2005). Decrease in pathology and progression of scrapie after immunisation with synthetic prion protein peptides in hamsters. *Vaccine, 23*, 2862–2868.

70. Bade, S., Frey, A. (2007). Potential of active and passive immunizations for the prevention and therapy of transmissible spongiform encephalopathies. *Expert Rev Vaccines, 6*, 153–168.

71. Souan, L., Tal, Y., Felling, Y., Cohen, I.R., Taraboulos, A., Mor, F. (2001). Modulation of proteinase-K prion protein by prion peptide immunization. *Eur J Immunol, 31*, 2338–2346.

72. Polymenidou, M., Heppner, F.L., Pellicioli, E.C., Urich, E., Miele, G., Braun, N., Wopfner, F., Schatzl, H.M., Becher, B., Aguzzi, A. (2004). Humoral immune response to native eukaryotic prion protein correlates with anti-prion protection. *Proc Natl Acad Sci U S A, 101* (*Suppl 2*), 14670–14676.

73. Paramithiotis, E., Pinard, M., Lawton, T., LaBoissiere, S., Leathers, V.L., Zou, W.Q., Estey, L.A., Lamontagne, J., Lehto, M.T., Kondejewski, L.H., et al. (2003). A prion protein epitope selective for the pathologically misfolded conformation. *Nat Med, 9*, 893–899.

74. Nitschke, C., Flechsig, E., van den, B.J., Lindner, N., Luhrs, T., Dittmer, U., Klein, M.A. (2007). Immunisation strategies against prion diseases: prime-boost immunisation with a PrP DNA vaccine containing foreign helper T-cell epitopes does not prevent mouse scrapie. *Vet Microbiol, 123*, 367–376.

75. Spinner, D.S., Kascsak, R.B., Lafauci, G., Meeker, H.C., Ye, X., Flory, M.J., Kim, J.I., Schuller-Levis, G.B., Levis, W.R., Wisniewski, T., et al. (2007). CpG oligodeoxynucleotide-enhanced humoral immune response and production of antibodies to prion protein PrPSc in mice immunized with 139Å scrapie-associated fibrils. *J Leukoc Biol, 81*, 1374–1385.

76. Orgogozo, J.M., Gilman, S., Dartigues, J.F., Laurent, B., Puel, M., Kirby, L.C., Jouanny, P., Dubois, B., Eisner, L., Flitman, S., et al. (2003). Subacute meningoencephalitis in a subset of patients with AD after Abeta42 immunization. *Neurology, 61*, 46–54.

77. Kaiser-Schulz, G., Heit, A., Quintanilla-Martinez, L., Hammerschmidt, F., Hess, S., Jennen, L., Rezaei, H., Wagner, H., Schatzl, H.M. (2007). Polylactide–coglycolide microspheres co-encapsulating recombinant tandem prion protein with CpG-oligonucleotide break self-tolerance to prion protein in wild-type mice and induce CD4 and CD8 T cell responses. *J Immunol, 179*, 2797–2807.

78. Handisurya, A., Gilch, S., Winter, D., Shafti-Keramat, S., Maurer, D., Schatzl, H.M., Kirnbauer, R. (2007). Vaccination with prion peptide-displaying

papillomavirus-like particles induces autoantibodies to normal prion protein that interfere with pathologic prion protein production in infected cells. *FEBS J, 274*, 1747–1758.

79. Sigurdsson, E.M., Sy, M.S., Li, R., Scholtzova, H., Kascsak, R.J., Kascsak, R., Carp, R., Meeker, H.C., Frangione, B., Wisniewski, T. (2003). Anti-prion antibodies for prophylaxis following prion exposure in mice. *Neurosci Lett, 336*, 185–187.
80. White, A.R., Enever, P., Tayebi, M., Mushens, R., Linehan, J., Brandner, S., Anstee, D., Collinge, J., Hawke, S. (2003). Monoclonal antibodies inhibit prion replication and delay the development of prion disease. *Nature, 422*, 80–83.
81. Mabbott, N.A., Young, J., McConnell, I., Bruce, M.E. (2003). Follicular dendritic cell dedifferentiation by treatment with an inhibitor of the lymphotoxin pathway dramatically reduces scrapie susceptibility. *J Virol, 77*, 6845–6854.
82. Mohan, J., Bruce, M.E., Mabbott, N.A. (2005). Follicular dendritic cell dedifferentiation reduces scrapie susceptibility following inoculation via the skin. *Immunology, 114*, 225–234.
83. Leucht, C., Simoneau, S., Rey, C., Vana, K., Rieger, R., Lasmezas, C.I., Weiss, S. (2003). The 37 kDa/67 kDa laminin receptor is required for PrP(Sc) propagation in scrapie-infected neuronal cells. *EMBO Rep, 4*, 290–295.
84. Zuber, C., Knackmuss, S., Rey, C., Reusch, U., Rottgen, P., Frohlich, T., Arnold, G.J., Pace, C., Mitteregger, G., Kretzschmar, H.A., et al. (2008). Single chain Fv antibodies directed against the 37 kDa/67 kDa laminin receptor as therapeutic tools in prion diseases. *Mol Immunol, 45*, 144–151.
85. Solforosi, L., Criado, J.R., McGavern, D.B., Wirz, S., Sanchez-Alavez, M., Sugama, S., DeGiorgio, L.A., Volpe, B.T., Wiseman, E., Abalos, G., et al. (2004). Cross-linking cellular prion protein triggers neuronal apoptosis in vivo. *Science, 303*, 1514–1516.
86. Heppner, F.L., Musahl, C., Arrighi, I., Klein, M.A., Rulicke, T., Oesch, B., Zinkernagel, R.M., Kalinke, U., Aguzzi, A. (2001). Prevention of scrapie pathogenesis by transgenic expression of anti-prion protein antibodies. *Science, 294*, 178–182.
87. Meier, P., Genoud, N., Prinz, M., Maissen, M., Rulicke, T., Zurbriggen, A., Raeber, A.J., Aguzzi, A. (2003). Soluble dimeric prion protein binds PrP(Sc) in vivo and antagonizes prion disease. *Cell, 113*, 49–60.
88. Gresser, I., Pattison, I.H. (1968). An attempt to modify scrapie in mice by administrtion of interferon. *J Gen Virol, 3*, 295–297.
89. Gresser, I., Maury, C., Chandler, R.L. (1983). Failure to modify scrapie in mice by administration of interferon or anti-interferon globulin. *J Gen Virol, 64(Pt 6)*, 1387–1389.
90. Field, E.J., Joyce, G., Keith, A. (1969). Failure of interferon to modify scrapie in the mouse. *J Gen Virol, 5*, 149–150.
91. Worthington, M. (1972). Interferon system in mice infected with the scrapie agent. *Infect Immun, 6*, 643–645.
92. Amyx, H., Salazar, A.M., Gajdusek, D.C., Gibbs, C.J. (1984). Chemotherapeutic trials in experimental slow virus disease. *Neurology, 34*, 149.
93. Kovanen, J., Haltia, M., Cantell, K. (1980). Failure of interferon to modify Creutzfeldt–Jakob disease. *Br Med J, 280*, 902.

94. Tal, Y., Souan, L., Cohen, I.R., Meiner, Z., Taraboulos, A., Mor, F. (2003). Complete Freund's adjuvant immunization prolongs survival in experimental prion disease in mice. *J Neurosci Res, 71*, 286–290.

95. Sethi, S., Lipford, G., Wagner, H., Kretzschmar, H. (2002). Postexposure prophylaxis against prion disease with a stimulator of innate immunity. *Lancet, 360*, 229–230.

96. Heikenwalder, M., Polymenidou, M., Junt, T., Sigurdson, C., Wagner, H., Akira, S., Zinkernagel, R., Aguzzi, A. (2004). Lymphoid follicle destruction and immunosuppression after repeated CpG oligodeoxynucleotide administration. *Nat Med, 10*, 187–192.

97. Kocisko, D.A., Vaillant, A., Lee, K.S., Arnold, K.M., Bertholet, N., Race, R.E., Olsen, E.A., Juteau, J.M., Caughey, B. (2006). Potent antiscrapie activities of degenerate phosphorothioate oligonucleotides. *Antimicrob Agents Chemother, 50*, 1034–1044.

98. Worthington, M., Clark, R. (1971). Lack of effect of immunosuppression on scrapie infection in mice. *J Gen Virol, 13*, 349–351.

99. Outram, G.W., Dickinson, A.G., Fraser, H. (1974). Reduced susceptibility to scrapie in mice after steroid administration. *Nature, 249*, 855–856.

100. Bate, C., Rutherford, S., Gravenor, M., Reid, S., Williams, A. (2002). Cyclooxygenase inhibitors protect against prion-induced neurotoxicity in vitro. *Neuroreport, 13*, 1933–1938.

101. Manuelidis, L., Fritch, W., Zaitsev, I. (1998). Dapsone to delay symptoms in Creutzfeldt–Jakob disease. *Lancet, 352*, 456.

102. Guenther, K., Deacon, R.M., Perry, V.H., Rawlins, J.N. (2001). Early behavioural changes in scrapie-affected mice and the influence of dapsone. *Eur J Neurosci, 14*, 401–409.

103. Caughey, B., Raymond, G.J. (1993). Sulfated polyanion inhibition of scrapie-associated PrP accumulation in cultured cells. *J Virol, 67*, 643–650.

104. Gabizon, R., Meiner, Z., Halimi, M., Bensasson, S.A. (1993). Heparin-like molecules bind differentially to prion proteins and change their intracellular metabolic-fate. *J Cell Physiol, 157*, 319–325.

105. Ehlers, B., Diringer, H. (1984). Dextran sulphate 500 delays and prevents mouse scrapie by impairment of agent replication in spleen. *J Gen Virol, 65*, 1325–1330.

106. Caughey, B., Brown, K., Raymond, G.J., Katzenstien, G.E., Thresher, W. (1994). Binding of the protease-sensitive form of PrP (prion protein) to sulfated glycosaminoglycan and Congo red. *J Virol, 68*, 2135–2141.

107. Wong, C., Xiong, L.-W., Horiuchi, M., Raymond, L.D., Wehrly, K., Chesebro, B., Caughey, B. (2001). Sulfated glycans and elevated temperature stimulate PrP^{Sc} dependent cell-free formation of protease-resistant prion protein. *EMBO J, 20*, 377–386.

108. Ben-Zaken, O., Tzaban, S., Tal, Y., Horonchik, L., Esko, J.D., Vlodavsky, I., Taraboulos, A. (2003). Cellular heparan sulfate participates in the metabolism of prions. *J Biol Chem, 278*, 40041–40049.

109. Deleault, N.R., Lucassen, R.W., Supattapone, S. (2003). RNA molecules stimulate prion protein conversion. *Nature, 425*, 717–720.

110. Caughey, B., Kocisko, D.A. (2003). Prion diseases: a nucleic-acid accomplice? *Nature, 425*, 673–674.

111. Doh-ura, K., Ishikawa, K., Murakami-Kubo, I., Sasaki, K., Mohri, S., Race, R., Iwaki, T. (2004). Treatment of transmissible spongiform encephalopathy by intraventricular drug infusion in animal models. *J Virol, 78*, 4999–5006.

112. Diringer, H., Ehlers, B. (1991). Chemoprophylaxis of scrapie in mice. *J Gen Virol, 72*, 457–460.

113. Farquhar, C., Dickinson, A., Bruce, M. (1999). Prophylactic potential of pentosan polysulphate in transmissible spongiform encephalopathies. *Lancet, 353*, 117.

114. Ladogana, A., Casaccia, P., Ingrosso, L., Cibati, M., Salvatore, M., Xi, Y.G., Masullo, C., Pocchiari, M. (1992). Sulphate polyanions prolong the incubation period of scrapie-infected hamsters. *J Gen Virol, 73*, 661–665.

115. Raymond, G.J., Olsen, E.A., Lee, K.S., Raymond, L.D., Bryant, P.K., III, Baron, G.S., Caughey, W.S., Kocisko, D.A., McHolland, L.E., Favara, C., et al. (2006). Inhibition of protease-resistant prion protein formation in a transformed deer cell line infected with chronic wasting disease. *J Virol, 80*, 596–604.

116. Larramendy-Gozalo, C., Barret, A., Daudigeos, E., Mathieu, E., Antonangeli, L., Riffet, C., Petit, E., Papy-Garcia, D., Barritault, D., Brown, P., Deslys, J.P. (2007). Comparison of CR36, a new heparan mimetic, and pentosan polysulfate in the treatment of prion diseases. *J Gen Virol, 88*, 1062–1067.

117. Doh-ura, K., Kuge, T., Uomoto, M., Nishizawa, K., Kawasaki, Y., Iha, M. (2007). Prophylactic effect of dietary seaweed *Fucoidan* against enteral prion infection. *Antimicrob Agents Chemother, 51*, 2274–2277.

118. Engelstein, R., Ovadia, H., Gabizon, R. (2007). Copaxone interferes with the PrP Sc-GAG interaction. *Eur J Neurol, 14*, 877–884.

119. Woody, A.M., Reisbig, R.R., Woody, R.W. (1981). Spectroscopic studies of Congo Red binding to RNA polymerase. *Biochim Biophys Acta, 655*, 82–88.

120. Priola, S.A., Caughey, B. (1994). Inhibition of scrapie-associated PrP accumulation: probing the role of glycosaminoglycans in amyloidogenesis. *Mol Neurobiol, 8*, 113–120.

121. Caspi, S., Halimi, M., Yanai, A., Sasson, S.B., Taraboulos, A., Gabizon, R. (1998). The anti-prion activity of Congo Red: putative mechanism. *J Biol Chem, 273*, 3484–3489.

122. Rudyk, H., Vasiljevic, S., Hennion, R.M., Birkett, C.R., Hope, J., Gilbert, I.H. (2000). Screening Congo red and its analogues for their ability to prevent the formation of PrP-res in scrapie-infected cells. *J Gen Virol, 81*, 1155–1164.

123. Caughey, B., Raymond, L.D., Raymond, G.J., Maxson, L., Silveira, J., Baron, G.S. (2003). Inhibition of protease-resistant prion protein accumulation in vitro by curcumin. *J Virol, 77*, 5499–5502.

124. Caughey, B., Race, R.E. (1992). Potent inhibition of scrapie-associated PrP accumulation by Congo red. *J Neurochem, 59*, 768–771.

125. Caughey, B., Ernst, D., Race, R.E. (1993). Congo red inhibition of scrapie agent replication. *J Virol, 67*, 6270–6272.

126. Demaimay, R., Harper, J., Gordon, H., Weaver, D., Chesebro, B., Caughey, B. (1998). Structural aspects of Congo Red as an inhibitor of protease-resistant prion protein formation. *J Neurochem, 71*, 2534–2541.

127. Demaimay, R., Chesebro, B., Caughey, B. (2000). Inhibition of formation of protease-resistant prion protein by trypan blue, sirius red and other Congo Red analogs. *Arch Virol Suppl, 16*, 277–283.

128. Poli, G., Ponti, W., Carcassola, G., Ceciliani, F., Colombo, L., Dall'Ara, P., Gervasoni, M., Giannino, M.L., Martino, P.A., Pollera, C., et al. (2003). In vitro evaluation of the anti-prionic activity of newly synthesized Congo Red derivatives. *Arzneimittelforschung, 53*, 875–888.

129. Kirby, L., Birkett, C.R., Rudyk, H., Gilbert, I.H., Hope, J. (2003). In vitro cell-free conversion of bacterial recombinant PrP to PrPres as a model for conversion. *J Gen Virol, 84*, 1013–1020.

130. Lucassen, R., Nishina, K., Supattapone, S. (2003). In vitro amplification of protease-resistant prion protein requires free sulfhydryl groups. *Biochemistry, 42*, 4127–4135.

131. Ingrosso, L., Ladogana, A., Pocchiari, M. (1995). Congo Red prolongs the incubation period in scrapie-infected hamsters. *J Virol, 69*, 506–508.

132. Poli, G., Martino, P.A., Villa, S., Carcassola, G., Giannino, M.L., Dall'Ara, P., Pollera, C., Iussich, S., Tranquillo, V.M., Bareggi, S., et al. (2004). Evaluation of anti-prion activity of Congo Red and its derivatives in experimentally infected hamsters. *Arzneimittelforschung, 54*, 406–415.

133. Gilch, S., Winklhofer, K.F., Groschup, M.H., Nunziante, M., Lucassen, R., Spielhaupter, C., Muranyi, W., Riesner, D., Tatzelt, J., Schatzl, H.M. (2001). Intracellular re-routing of prion protein prevents propagation of PrP(Sc) and delays onset of prion disease. *EMBO J, 20*, 3957–3966.

134. Nunziante, M., Kehler, C., Maas, E., Kassack, M.U., Groschup, M., Schatzl, H.M. (2005). Charged bipolar suramin derivatives induce aggregation of the prion protein at the cell surface and inhibit PrPSc replication. *J Cell Sci, 118*, 4959–4973.

135. Hafner-Bratkovic, I., Gaspersic, J., Smid, L.M., Bresjanac, M., Jerala, R. (2008). Curcumin binds to the alpha-helical intermediate and to the amyloid form of prion protein: a new mechanism for the inhibition of PrP(Sc) accumulation. *J Neurochem, 104*, 1553–1564. Erratum in *J Neurochem* 2008 Sep; *106*

136. Kawasaki, Y., Kawagoe, K., Chen, C.J., Teruya, K., Sakasegawa, Y., Doh-ura, K. (2007). Orally administered amyloidophilic compound is effective in prolonging the incubation periods of animals cerebrally infected with prion diseases in a prion strain-dependent manner. *J Virol, 81*, 12889–12898.

137. Supattapone, S., Nguyen, H.O., Cohen, F.E., Prusiner, S.B., Scott, M.R. (1999). Elimination of prions by branched polyamines and implications for therapeutics. *Proc Natl Acad Sci U S A, 96*, 14529–14534.

138. Supattapone, S., Wille, H., Uyechi, L., Safar, J., Tremblay, P., Szoka, F.C., Cohen, F.E., Prusiner, S.B., Scott, M.R. (2001). Branched polyamines cure prion-infected neuroblastoma cells. *J Virol, 75*, 3453–3461.

139. Solassol, J., Crozet, C., Perrier, V., Leclaire, J., Beranger, F., Caminade, A.M., Meunier, B., Dormont, D., Majoral, J.P., Lehmann, S. (2004). Cationic phosphorus-containing dendrimers reduce prion replication both in cell culture and in mice infected with scrapie. *J Gen Virol, 85*, 1791–1799.

140. Bera, A., Nandi, P.K. (2007). Biological polyamines inhibit nucleic-acid-induced polymerisation of prion protein. *Arch Virol, 152*, 655–668.

141. Winklhofer, K.F., Tatzelt, J. (2000). Cationic lipopolyamines induce degradation of PrPSc in scrapie-infected mouse neuroblastoma cells. *Biol Chem, 381*, 463–469.
142. Caughey, W.S., Raymond, L.D., Horiuchi, M., Caughey, B. (1998). Inhibition of protease-resistant prion protein formation by porphyrins and phthalocyanines. *Proc Natl Acad Sci U S A, 95*, 12117–12122.
143. Priola, S.A., Raines, A., Caughey, W.S. (2000). Porphyrin and phthalocyanine antiscrapie compounds. *Science, 287*, 1503–1506.
144. Priola, S.A., Raines, A., Caughey, W. (2003). Prophylactic and therapeutic effects of phthalocyanine tetrasulfonate in scrapie-infected mice. *J Infect Dis, 188*, 699–705.
145. Caughey, W.S., Priola, S.A., Kocisko, D.A., Raymond, L.D., Ward, A., Caughey, B. (2007). Cyclic tetrapyrrole sulfonation, metals, and oligomerization in antiprion activity. *Antimicrob Agents Chemother, 51*, 3887–3894.
146. Lee, K.S., Raymond, L.D., Schoen, B., Raymond, G.J., Kett, L., Moore, R.A., Johnson, L.M., Taubner, L., Speare, J.O., Onwubiko, H.A., et al. (2007). Hemin interactions and alterations of the subcellular localization of prion protein. *J Biol Chem, 282*, 36525–36533.
147. Pocchiari, M., Schmittinger, S., Masullo, C. (1987). Amphotericin B delays the incubation period of scrapie in intracerebrally inoculated hamsters. *J Gen Virol, 68*, 219–223.
148. Dormont, D. (2003). Approaches to prophylaxis and therapy. *Br Med Bull, 66*, 281–292.
149. Mange, A., Nishida, N., Milhavet, O., McMahon, H.E., Casanova, D., Lehmann, S. (2000). Amphotericin B inhibits the generation of the scrapie isoform of the prion protein in infected cultures. *J Virol, 74*, 3135–3140.
150. Marella, M., Lehmann, S., Grassi, J., Chabry, J. (2002). Filipin prevents pathological prion protein accumulation by reducing endocytosis and inducing cellular PrP release. *J Biol Chem, 277*, 25457–25464.
151. Pocchiari, M., Casaccia, P., Ladogana, A. (1989). Amphotericin B: a novel class of antiscrapie drugs. *J Infect Dis, 160*, 795–802.
152. Xi, Y.G., Ingrosso, L., Ladogana, A., Masullo, C., Pocchiari, M. (1992). Amphotericin B treatment dissociates in vivo replication of the scrapie agent from PrP accumulation. *Nature, 356*, 598–601.
153. McKenzie, D., Kaczkowski, J., Marsh, R., Aiken, J.M. (1994). Amphotericin B delays both scrapie agent replication and PrP-res accumulation early in infection. *J Virol, 68*, 7534–7536.
154. Adjou, K.T., Demaimay, R., Lasmezas, C., Deslys, J.-P., Seman, M., Dormont, D. (1995). MS-8209, a new amphotericin B derivative, provides enhanced efficacy in delaying hamster scrapie. *Antimicrob Agents Chemother, 39*, 2810–2812.
155. Adjou, K.T., Demaimay, R., Deslys, J.P., Lasmezas, C.I., Beringue, V., Demart, S., Lamoury, F., Seman, M., Dormont, D. (1999). MS-8209, a water-soluble amphotericin B derivative, affects both scrapie agent replication and PrPres accumulation in Syrian hamster scrapie. *J Gen Virol, 80* (*Pt 4*), 1079–1085.
156. Adjou, K.T., Privat, N., Demart, S., Deslys, J.P., Seman, M., Hauw, J.J., Dormont, D. (2000). MS-8209, an amphotericin B analogue, delays the appearance of spongiosis, astrogliosis and PrPres accumulation in the brain of scrapie-infected hamsters. *J Comp Pathol, 122*, 3–8.

157. Demaimay, R., Adjou, K., Lasmezas, C., Lazarini, F., Cherifi, K., Seman, M., Deslys, J.P., Dormont, D. (1994). Pharmacological studies of a new derivative of amphotericin B, MS-8209, in mouse and hamster scrapie. *J Gen Virol, 75*, 2499–2503.

158. Demaimay, R., Adjou, K.T., Beringue, V., Demart, S., Lasmezas, C.I., Deslys, J.-P., Seman, M., Dormont, D. (1997). Late treatment with polyene antibiotics can prolong the survival time of scrapie-infected animals. *J Virol, 71*, 9685–9689.

159. Adjou, K.T., Demaimay, R., Lasmezas, C.I., Seman, M., Deslys, J.P., Dormont, D. (1996). Differential effects of a new amphotericin B derivative, MS-8209, on mouse BSE and scrapie: implications for the mechanism of action of polyene antibiotics. *Res Virol, 147*, 213–218.

160. Masullo, C., Macchi, G., Xi, Y.G., Pocchiari, M. (1992). Failure to ameliorate Creutzfeldt–Jakob disease with amphotericin B therapy. *J Infect Dis, 165*, 784–785.

161. Doh-ura, K., Iwaki, T., Caughey, B. (2000). Lysosomotropic agents and cysteine protease inhibitors inhibit accumulation of scrapie-associated prion protein. *J Virol, 74*, 4894–4897.

162. Korth, C., May, B.C., Cohen, F.E., Prusiner, S.B. (2001). Acridine and phenothiazine derivatives as pharmacotherapeutics for prion disease. *Proc Natl Acad Sci U S A, 98*, 9836–9841.

163. Ryou, C., Legname, G., Peretz, D., Craig, J.C., Baldwin, M.A., Prusiner, S.B. (2003). Differential inhibition of prion propagation by enantiomers of quinacrine. *Lab Invest, 83*, 837–843.

164. Collins, S.J., Lewis, V., Brazier, M., Hill, A.F., Fletcher, A., Masters, C.L. (2002). Quinacrine does not prolong survival in a murine Creutzfeldt–Jakob disease model. *Ann Neurol, 52*, 503–506.

165. Barret, A., Tagliavini, F., Forloni, G., Bate, C., Salmona, M., Colombo, L., De Luigi, A., Limido, L., Suardi, S., Rossi, G., et al. (2003). Evaluation of quinacrine treatment for prion diseases. *J Virol, 77*, 8462–8469.

166. Haik, S., Brandel, J.P., Salomon, D., Sazdovitch, V., Asnerie-Laupretre, N., Laplanche, J.L., Faucheux, B.A., Soubrie, C., Boher, E., Belorgey, C., Hauw, J.J., Alperovitch, A. (2004). Compassionate use of quinacrine in Creutzfeldt–Jakob disease fails to show significant effects. *Neurology, 63*, 2413–2415.

167. Klingenstein, R., Lober, S., Kujala, P., Godsave, S., Leliveld, S.R., Gmeiner, P., Peters, P.J., Korth, C. (2006). Tricyclic antidepressants, quinacrine and a novel, synthetic chimera thereof clear prions by destabilizing detergent-resistant membrane compartments. *J Neurochem, 98*, 748–759.

168. Kishida, H., Sakasegawa, Y., Watanabe, K., Yamakawa, Y., Nishijima, M., Kuroiwa, Y., Hachiya, N.S., Kaneko, K. (2004). Non-glycosylphosphatidylinositol (GPI)-anchored recombinant prion protein with dominant-negative mutation inhibits PrPSc replication in vitro. *Amyloid, 11*, 14–20.

169. Roikhel, V.M., Fokina, G.I., Pogodina, V.V. (1984). Influence of aminasine on experimental scrapie in mice. *Acta Virol, 28*, 321–324.

170. Murakami-Kubo, I., Doh-ura, K., Ishikawa, K., Kawatake, S., Sasaki, K., Kira, J., Ohta, S., Iwaki, T. (2004). Quinoline derivatives are therapeutic candidates for transmissible spongiform encephalopathies. *J Virol, 78*, 1281–1288.

171. Kocisko, D.A., Caughey, B. (2006). Mefloquine, an antimalaria drug with antiprion activity in vitro, lacks activity in vivo. *J Virol, 80*, 1044–1046.

172. Forloni, G., Iussich, S., Awan, T., Colombo, L., Angeretti, N., Girola, L., Bertani, I., Poli, G., Caramelli, M., Grazia, B.M., et al. (2002). Tetracyclines affect prion infectivity. *Proc Natl Acad Sci U S A, 99*, 10849–10854.
173. Guo, Y.J., Han, J., Yao, H.L., Zhang, B.Y., Gao, J.M., Zhang, J., Xiao, X.L., Wang, X.F., Zhao, W.Q., Wang, D.X., Dong, X.P. (2007). Treatment of scrapie pathogen 263K with tetracycline partially abolishes protease-resistant activity in vitro and reduces infectivity in vivo. *Biomed Environ Sci, 20*, 198–202.
174. Tagliavini, F., Forloni, G., Colombo, L., Rossi, G., Girola, L., Canciani, B., Angeretti, N., Giampaolo, L., Peressini, E., Awan, T., et al. (2000). Tetracycline affects abnormal properties of synthetic PrP peptides and PrP(Sc) in vitro. *J Mol Biol, 300*, 1309–1322.
175. Tatzelt, J., Prusiner, S.B., Welch, W.J. (1996). Chemical chaperones interfere with the formation of scrapie prion protein. *EMBO J, 15*, 6363–6373.
176. Shaked, G.M., Fridlander, G., Meiner, Z., Taraboulos, A., Gabizon, R. (1999). Protease-resistant and detergent-insoluble prion protein is not necessarily associated with prion infectivity. *J Biol Chem, 274*, 17981–17986.
177. Shaked, G.M., Engelstein, R., Avraham, I., Kahana, E., Gabizon, R. (2003). Dimethyl sulfoxide delays PrP sc accumulation and disease symptoms in prion-infected hamsters. *Brain Res, 983*, 137–143.
178. Engelstein, R., Grigoriadis, N., Greig, N.H., Ovadia, H., Gabizon, R. (2005). Inhibition of P53-related apoptosis had no effect on PrP(Sc) accumulation and prion disease incubation time. *Neurobiol Dis, 18*, 282–285.
179. Fukuuchi, T., Doh-ura, K., Yoshihara, S., Ohta, S. (2006). Metal complexes with superoxide dismutase-like activity as candidates for anti-prion drug. *Bioorg Med Chem Lett, 16*, 5982–5987.
180. Sigurdsson, E.M., Brown, D.R., Alim, M.A., Scholtzova, H., Carp, R., Meeker, H.C., Prelli, F., Frangione, B., Wisniewski, T. (2003). Copper chelation delays the onset of prion disease. *J Biol Chem, 278*, 46199–46202.
181. Pollera, C., Lucchini, B., Formentin, E., Bareggi, S., Poli, G., Ponti, W. (2005). Evaluation of anti-prionic activity of clioquinol in an in vivo model (*Mesocricetus auratus*). *Vet Res Commun, 29* (*Suppl 2*), 253–255.
182. Hijazi, N., Shaked, Y., Rosenmann, H., Ben-Hur, T., Gabizon, R. (2003). Copper binding to PrPC may inhibit prion disease propagation. *Brain Res, 993*, 192–200.
183. Doh-ura, K., Tamura, K., Karube, Y., Naito, M., Tsuruo, T., Kataoka, Y. (2007). Chelating compound, chrysoidine, is more effective in both antiprion activity and brain endothelial permeability than quinacrine. *Cell Mol Neurobiol, 27*, 303–316.
184. Chabry, J., Priola, S.A., Wehrly, K., Nishio, J., Hope, J., Chesebro, B. (1999). Species-independent inhibition of abnormal prion protein (PrP) formation by a peptide containing a conserved PrP sequence. *J Virol, 73*, 6245–6250.
185. Chabry, J., Caughey, B., Chesebro, B. (1998). Specific inhibition of in vitro formation of protease-resistant prion protein by synthetic peptides. *J Biol Chem, 273*, 13203–13207.
186. Gilch, S., Kehler, C., Schatzl, H.M. (2007). Peptide aptamers expressed in the secretory pathway interfere with cellular PrPSc formation. *J Mol Biol, 371*, 362–373.
187. Soto, C., Kascsak, R.J., Saborio, G.P., Aucouturier, P., Wisniewski, T., Prelli, F., Kascsak, R., Mendez, E., Harris, D.A., Ironside, J., et al. (2000). Reversion of

INSTITUT MEDIATHEQUE PASTEUR

prion protein conformational changes by synthetic beta-sheet breaker peptides. *Lancet, 355*, 192–197.

188. Oishi, T., Hagiwara, K., Kinumi, T., Yamakawa, Y., Nishijima, M., Nakamura, K., Arimoto, H. (2003). Effects of beta-sheet breaker peptide polymers on scrapie-infected mouse neuroblastoma cells and their affinities to prion protein fragment PrP(81–145). *Org Biomol Chem, 1*, 2626–2629.
189. Taraboulos, A., Scott, M., Semenov, A., Avraham, D., Laszlo, L., Prusiner, S.B. (1995). Cholesterol depletion and modification of COOH-terminal targeting sequence of the prion protein inhibit formation of the scrapie isoform. *J Cell Biol, 129*, 121–132.
190. Bate, C., Salmona, M., Diomede, L., Williams, A. (2004). Squalestatin cures prion-infected neurones and protects against prion neurotoxicity. *J Biol Chem, 279*, 14983–14990.
191. Mok, S.W., Thelen, K.M., Riemer, C., Bamme, T., Gultner, S., Lutjohann, D., Baier, M. (2006). Simvastatin prolongs survival times in prion infections of the central nervous system. *Biochem Biophys Res Commun, 348*, 697–702.
192. Kempster, S., Bate, C., Williams, A. (2007). Simvastatin treatment prolongs the survival of scrapie-infected mice. *Neuroreport, 18*, 479–482.
193. Gilch, S., Kehler, C., Schatzl, H.M. (2006). The prion protein requires cholesterol for cell surface localization. *Mol Cell Neurosci, 31*, 346–353.
194. Hagiwara, K., Nakamura, Y., Nishijima, M., Yamakawa, Y. (2007). Prevention of prion propagation by dehydrocholesterol reductase inhibitors in cultured cells and a therapeutic trial in mice. *Biol Pharm Bull, 30*, 835–838.
195. Furlow, T.W., Jr., Whitley, R.J., Wilmes, F.J. (1982). Repeated suppression of Creutzfeldt–Jakob disease with vidarabine. *Lancet, 2*, 564–565.
196. Haig, D.A., Clarke, M.C. (1968). The effect of beta-propiolactone on the scrapie agent. *J Gen Virol, 3*, 281–283.
197. Cochran, K.W. (1971). Failure of adenine arabinoside to modify scrapie in mice. *J Gen Virol, 13*, 353–354.
198. Kimberlin, R.H., Walker, C.A. (1979). Antiviral compound effective against experimental scrapie. *Lancet, 2*, 591–592.
199. Tateishi, J. (1981). Antibiotics and antivirals do not modify experimentally-induced Creutzfeldt–Jakob disease in mice. *J Neurol Neurosurg Psychiatry, 44*, 723–724.
200. Rieger, R., Edenhofer, F., Lasmezas, C.I., Weiss, S. (1997). The human 37-kDa laminin receptor precursor interacts with the prion protein in eukaryotic cells. *Nat Med, 3*, 1383–1388.
201. Hundt, C., Peyrin, J.M., Haik, S., Gauczynski, S., Leucht, C., Rieger, R., Riley, M.L., Deslys, J.P., Dormont, D., Lasmezas, C.I., Weiss, S. (2001). Identification of interaction domains of the prion protein with its 37-kDa/67-kDa laminin receptor. *EMBO J, 20*, 5876–5886.
202. Gauczynski, S., Nikles, D., El-Gogo, S., Papy-Garcia, D., Rey, C., Alban, S., Barritault, D., Lasmezas, C.I., Weiss, S. (2006). The 37-kDa/67-kDa laminin receptor acts as a receptor for infectious prions and is inhibited by polysulfated glycanes. *J Infect Dis, 194*, 702–709.

203. Zuber, C., Ludewigs, H., Weiss, S. (2007). Therapeutic approaches targeting the prion receptor LRP/LR. *Vet Microbiol, 123*, 387–393.

204. Ertmer, A., Gilch, S., Yun, S.W., Flechsig, E., Klebl, B., Stein-Gerlach, M., Klein, M.A., Schatzl, H.M. (2004). The tyrosine kinase inhibitor STI571 induces cellular clearance of PrPSc in prion-infected cells. *J Biol Chem, 279*, 41918–41927.

205. Yun, S.W., Ertmer, A., Flechsig, E., Gilch, S., Riederer, P., Gerlach, M., Schatzl, H.M., Klein, M.A. (2007). The tyrosine kinase inhibitor imatinib mesylate delays prion neuroinvasion by inhibiting prion propagation in the periphery. *J Neurovirol, 13*, 328–337.

206. Bate, C., Reid, S., Williams, A. (2004). Phospholipase A2 inhibitors or platelet-activating factor antagonists prevent prion replication. *J Biol Chem, 279*, 36405–36411.

207. Nordstrom, E.K., Luhr, K.M., Ibanez, C., Kristensson, K. (2005). Inhibitors of the mitogen-activated protein kinase kinase 1/2 signaling pathway clear prion-infected cells from PrPSc. *J Neurosci, 25*, 8451–8456.

208. Muller, W.E., Romero, F.J., Perovic, S., Pergande, G., Pialoglou, P. (1997). Protection of flupirtine on beta-amyloid-induced apoptosis in neuronal cells in vitro: prevention of amyloid-induced glutathione depletion. *J Neurochem, 68*, 2371–2377.

209. Perovic, S., Schroder, H.C., Pergande, G., Ushijima, H., Muller, W.E. (1997). Effect of flupirtine on Bcl-2 and glutathione level in neuronal cells treated in vitro with the prion protein fragment (PrP106–126). *Exp Neurol, 147*, 518–524.

210. Otto, M., Cepek, L., Ratzka, P., Doehlinger, S., Boekhoff, I., Wiltfang, J., Irle, E., Pergande, G., Ellers-Lenz, B., Windl, O., et al. (2004). Efficacy of flupirtine on cognitive function in patients with CJD: A double-blind study. *Neurology, 62*, 714–718.

211. Dirikoc, S., Priola, S.A., Marella, M., Zsurger, N., Chabry, J. (2007). Nonpsychoactive cannabidiol prevents prion accumulation and protects neurons against prion toxicity. *J Neurosci, 27*, 9537–9544.

212. Brown, D.R. (2005). Neurodegeneration and oxidative stress: prion disease results from loss of antioxidant defence. *Folia Neuropathol, 43*, 229–243.

213. Brown, D.R., Wong, B.-S., Hafiz, F., Clive, C., Haswell, S., Jones, I.M. (1999). Normal prion protein has superoxide dismutase-like activity. *Biochem J, 344*, 1–5.

214. Brown, D.R., Schultz-Schaeffer, W.J., Schmidt, B., Kretzschmar, H.A. (1997). Prion protein deficient cells show altered response to oxidative stress due to decreased SOD-1 activity. *Exp Neurol, 146*, 104–112.

215. Brown, D.R., Schmidt, B., Kretzschmar, H.A. (1998). A prion protein fragment primes type 1 astrocytes to proliferation signals from microglia. *Neurobiol Dis, 4*, 410–422.

216. Choi, S.I., Ju, W.K., Choi, E.K., Kim, J., Lea, H.Z., Carp, R.I., Wisniewski, H.M., Kim, Y.S. (1998). Mitochondrial dysfunction induced by oxidative stress in the brains of hamsters infected with the 263 K scrapie agent. *Acta Neuropathol, 96*, 279–286.

217. Wong, B.S., Wang, H., Brown, D.R., Jones, I.M. (1999). Selective oxidation of methionine residues in prion proteins. *Biochem Biophys Res Commun, 259*, 352–355.

218. Brown, D.R., Schmidt, B., Kretzschmar, H.A. (1998). Effects of copper on survival of prion protein knockout neurons and glia. *J Neurochem, 70*, 1686–1693.

219. Brown, D.R., Nicholas, R.S., Canevari, L. (2002). Lack of prion protein expression results in a neuronal phenotype sensitive to stress. *J Neurosci Res, 67*, 211–224.

220. Wong, B.S., Brown, D.R., Pan, T., Whiteman, M., Liu, T., Bu, X., Li, R., Gambetti, P., Olesik, J., Rubenstein, R., Sy, M.S. (2001). Oxidative impairment in scrapie-infected mice is associated with brain metals perturbations and altered antioxidant activities. *J Neurochem, 79*, 689–698.

221. Kim, N.H., Park, S.J., Jin, J.K., Kwon, M.S., Choi, E.K., Carp, R.I., Kim, Y.S. (2000). Increased ferric iron content and iron-induced oxidative stress in the brains of scrapie-infected mice. *Brain Res, 884*, 98–103.

222. Martin, S.F., Buron, I., Espinosa, J.C., Castilla, J., Villalba, J.M., Torres, J.M. (2007). Coenzyme Q and protein/lipid oxidation in a BSE-infected transgenic mouse model. *Free Radic Biol Med, 42*, 1723–1729.

223. Kimata, A., Nakagawa, H., Ohyama, R., Fukuuchi, T., Ohta, S., Suzuki, T., Miyata, N. (2007). New series of antiprion compounds: pyrazolone derivatives have the potent activity of inhibiting protease-resistant prion protein accumulation. *J Med Chem, 50*, 5053–5056.

224. Brandner, S., Isenmann, S., Raeber, A., Fischer, M., Sailer, A., Kobayashi, Y., Marino, S., Weissmann, C., Aguzzi, A. (1996). Normal host prion protein necessary for scrapie-induced neurotoxicity. *Nature, 379*, 339–343.

225. Mallucci, G.R., Ratte, S., Asante, E.A., Linehan, J., Gowland, I., Jefferys, J.G., Collinge, J. (2002). Post-natal knockout of prion protein alters hippocampal CA1 properties, but does not result in neurodegeneration. *EMBO J, 21*, 202–210.

226. Tilly, G., Chapuis, J., Vilette, D., Laude, H., Vilotte, J.L. (2003). Efficient and specific down-regulation of prion protein expression by RNAi. *Biochem Biophys Res Commun, 305*, 548–551.

227. Daude, N., Marella, M., Chabry, J. (2003). Specific inhibition of pathological prion protein accumulation by small interfering RNAs. *J Cell Sci, 116*, 2775–2779.

228. Leucht, C., Vana, K., Renner-Muller, I., Dormont, D., Lasmezas, C.I., Wolf, E., Weiss, S. (2004). Knock-down of the 37-kDa/67-kDa laminin receptor in mouse brain by transgenic expression of specific antisense LRP RNA. *Transgen Res, 13*, 81–85.

229. Pfeifer, A., Eigenbrod, S., Al-Khadra, S., Hofmann, A., Mitteregger, G., Moser, M., Bertsch, U., Kretzschmar, H. (2006). Lentivector-mediated RNAi efficiently suppresses prion protein and prolongs survival of scrapie-infected mice. *J Clin Invest, 116*, 3204–3210.

230. Kumar, P., Wu, H., McBride, J.L., Jung, K.E., Kim, M.H., Davidson, B.L., Lee, S.K., Shankar, P., Manjunath, N. (2007). Transvascular delivery of small interfering RNA to the central nervous system. *Nature, 448*, 39–43.

231. Mazarakis, N.D., Azzouz, M., Rohll, J.B., Ellard, F.M., Wilkes, F.J., Olsen, A.L., Carter, E.E., Barber, R.D., Baban, D.F., Kingsman, S.M., et al. (2001). Rabies virus glycoprotein pseudotyping of lentiviral vectors enables retrograde axonal transport and access to the nervous system after peripheral delivery. *Hum Mol Genet, 10*, 2109–2121.

232. Groschup, M.H., Lacroux, C., Buschmann, A., Luhken, G., Mathey, J., Eiden, M., Lugan, S., Hoffmann, C., Espinosa, J.C., Baron, T., et al. (2007). Classic

scrapie in sheep with the ARR/ARR prion genotype in Germany and France. *Emerg Infect Dis, 13*, 1201–1207.

233. Crozet, C., Lin, Y.L., Mettling, C., Mourton-Gilles, C., Corbeau, P., Lehmann, S., Perrier, V. (2004). Inhibition of PrPSc formation by lentiviral gene transfer of PrP containing dominant negative mutations. *J Cell Sci, 117*, 5591–5597.
234. Brandel, J.P., Asnerie-Laupretre, N., Laplanche, J.L., Hauw, J.J., Alperovitch, A. (2000). Diagnosis of Creutzfeldt–Jakob disease: effect of clinical criteria on incidence estimates. *Neurology, 54*, 1095–1099.
235. Rainov, N.G., Tsuboi, Y., Krolak-Salmon, P., Vighetto, A., Doh-ura, K. (2007). Experimental treatments for human transmissible spongiform encephalopathies: Is there a role for pentosan polysulfate? *Expert Opin Biol Ther, 7*, 713–726.
236. Todd, N.V., Morrow, J., Doh-ura, K., Dealler, S., O'Hare, S., Farling, P., Duddy, M., Rainov, N.G. (2005). Cerebroventricular infusion of pentosan polysulphate in human variant Creutzfeldt–Jakob disease. *J Infect, 50*, 394–396.
237. Whittle, I.R., Knight, R.S., Will, R.G. (2006). Unsuccessful intraventricular pentosan polysulphate treatment of variant Creutzfeldt–Jakob disease. *Acta Neurochir (Wien), 148*, 677–679.
238. Parry, A., Baker, I., Stacey, R., Wimalaratna, S. (2007). Long term survival in a patient with variant Creutzfeldt–Jakob disease treated with intraventricular pentosan polysulphate. *J Neurol Neurosurg Psychiatry, 78*, 733–734.
239. Nakajima, M., Yamada, T., Kusuhara, T., Furukawa, H., Takahashi, M., Yamauchi, A., Kataoka, Y. (2004). Results of quinacrine administration to patients with Creutzfeldt–Jakob disease. *Dement Geriatr Cogn Disord, 17*, 158–163.
240. Benito-Leon, J. (2005). Compassionate use of quinacrine in Creutzfeldt–Jakob disease fails to show significant effects. *Neurology, 64*, 1824.
241. Satoh, K., Shirabe, S., Eguchi, K., Yamauchi, A., Kataoka, Y., Niwa, M., Nishida, N., Katamine, S. (2004). Toxicity of quinacrine can be reduced by co-administration of P-glycoprotein inhibitor in sporadic Creutzfeldt–Jakob disease. *Cell Mol Neurobiol, 24*, 873–875.
242. Scoazec, J.Y., Krolak-Salmon, P., Casez, O., Besson, G., Thobois, S., Kopp, N., Perret-Liaudet, A., Streichenberger, N. (2003). Quinacrine-induced cytolytic hepatitis in sporadic Creutzfeldt–Jakob disease. *Ann Neurol, 53*, 546–547.
243. Newman, P.K. (1984). Acyclovir in Creutzfeldt–Jakob disease. *Lancet, 1*, 793.
244. David, A.S., Grant, R., Ballantyne, J.P. (1984). Unsuccessful treatment of Creutzfeldt–Jakob disease with acyclovir. *Lancet, 1*, 512–513.
245. Terzano, M.G., Montanari, E., Calzetti, S., Mancia, D., Lechi, A. (1983). The effect of amantadine on arousal and EEG patterns in Creutzfeldt–Jakob disease. *Arch Neurol, 40*, 555–559.
246. Braham, J. (1971). Jakob–Creutzfeldt disease: treatment by amantadine. *Br Med J, 4*, 212–213.
247. Braham, J. (1984). Amantadine in the treatment of Creutzfeldt–Jakob disease. *Arch Neurol, 41*, 585–586.
248. Ratcliffe, J., Rittman, A., Wolf, S., Verity, M.A. (1975). Creutzfeldt–Jakob disease with focal onset unsuccessfully treated with amantadine. *Bull Los Angeles Neurol Soc, 40*, 18–20.

249. Sanders, W.L. (1979). Creutzfeldt–Jakob disease treated with amantadine. *J Neurol Neurosurg Psychiatry, 42*, 960–961.

250. Goldhammer, Y., Bubis, J.J., Sarova-Pinhas, I., Braham, J. (1972). Subacute spongiform encephalopathy and its relation to Jakob–Creutzfeldt disease: report on six cases. *J Neurol Neurosurg Psychiatry, 35*, 1–10.

251. Kimberlin, R.H., Walker, C.A. (1983). The antiviral compound HPA-23 can prevent scrapie when administered at the time of infection. *Arch Virol, 78*, 9–18.

252. Kimberlin, R.H., Walker, C.A. (1986). Suppression of scrapie infection in mice by heteropolyanion 23, dextran sulfate, and some other polyanions. *Antimicrob Agents Chemother, 30*, 409–413.

253. Birkett, C.R., Hennion, R.M., Bembridge, D.A., Clarke, M.C., Chree, A., Bruce, M.E., Bostock, C.J. (2001). Scrapie strains maintain biological phenotypes on propagation in a cell line in culture. *EMBO J, 20*, 3351–3358.

254. Adjou, K.T., Simoneau, S., Sales, N., Lamoury, F., Dormont, D., Papy-Garcia, D., Barritault, D., Deslys, J.P., Lasmezas, C.I. (2003). A novel generation of heparan sulfate mimetics for the treatment of prion diseases. *J Gen Virol, 84*, 2595–2603.

255. Schonberger, O., Horonchik, L., Gabizon, R., Papy-Garcia, D., Barritault, D., Taraboulos, A. (2003). Novel heparan mimetics potently inhibit the scrapie prion protein and its endocytosis. *Biochem Biophys Res Commun, 312*, 473–479.

256. Proske, D., Gilch, S., Wopfner, F., Schatzl, H.M., Winnacker, E.L., Famulok, M. (2002). Prion-protein-specific aptamer reduces PrPSc formation. *Chembiochem, 3*, 717–725.

257. Rhie, A., Kirby, L., Sayer, N., Wellesley, R., Disterer, P., Sylvester, I., Gill, A., Hope, J., James, W., Tahiri-Alaoui, A. (2003). Characterization of 2′-fluoro-RNA aptamers that bind preferentially to disease-associated conformations of prion protein and inhibit conversion. *J Biol Chem, 278*, 39697–39705

258. Collinge, J., Gorham, M., Hudson, F., Kennedy, A., Keogh, G., Pal, S., Rossor, M., Rudge, P., Siddique, D., Spyer, M., Thomas, D., Walker, S., Webb, T., Wroe, S., Darbyshire, J. (2009). Safety and efficacy of quinacrine in human prion disease (PRION-1 study): a patient-preference trial. *Lancet Neurol, 8*(4): 334–344.

14

MISFOLDING AND AGGREGATION IN HUNTINGTON DISEASE AND OTHER EXPANDED POLYGLUTAMINE REPEAT DISEASES

RONALD WETZEL

Department of Structural Biology and Pittsburgh Institute for Neurodegenerative Diseases, University of Pittsburgh School of Medicine, Pittsburgh, Pennsylvania

INTRODUCTION

The expanded CAG repeat diseases consist of a family of at least nine genetic neurological conditions linked to the expansion of polyglutamine (polyQ) sequences in the associated disease proteins [1]. Huntington disease (HD) is the most prevalent and best known of these diseases. Although many of the disease proteins, including the approximately 3150-amino-acid-long huntingtin, are expressed throughout the body, these conditions are primarily brain diseases. Each disease has its own unique pattern of regional brain pathology and symptoms, with the latter including variable degrees of motor deficits, including ataxia, and dementia [1]. Recent studies suggest that HD is a disease of the entire brain, with significant presymptomatic abnormalities throughout the cortex [2]. HD and related diseases appear to be primarily gain-of-function diseases [1]; at the same time, it is possible that reductions in levels of normal protein activity caused by the glutamine expansion might also contribute to expanded polyQ diseases [3].

Protein Misfolding Diseases: Current and Emerging Principles and Therapies,
Edited by Marina Ramirez-Alvarado, Jeffery W. Kelly, and Christopher M. Dobson
Copyright © 2010 John Wiley & Sons, Inc.

Although deciphering the mechanism of a gain-of-function disease is inherently difficult, it seems reasonable to hypothesize that the downstream morphological and physiological abnormalities associated with polyQ diseases are ultimately caused by some physical change in the disease protein that is intimately linked to the polyQ expansion. Since these disease proteins differ from each other in practically all aspects (i.e., size, secondary and tertiary structure, function, subcellular localization, etc.) except for their polyQ sequence, the temptation has been to seek out biophysical consequences of polyQ expansion. In this chapter we highlight efforts based on the premise that in the disease mechanism, polyQ expansion leads to immediate consequences of misfolding and/or aggregation that although centered in the polyQ sequence, can sometimes, surprisingly, reach into flanking sequences.

There are a number of arguments supporting the prevailing consensus that HD and related disorders are protein misfolding and aggregation diseases. Early on, the observation of ubiquitin-positive, polyQ-positive inclusions in neuronal nuclei in cell and animal models and in patient materials [4–6] seemed to echo other protein aggregation–related neurodegenerative diseases [7]. Further support for aggregation being an early event in the disease mechanism is in the overlap in the repeat-length dependence of disease risk and of aggregation aggressiveness in vitro [8,9] and in vivo [10]. In cell and animal models of polyQ diseases, however, a strong linkage between the appearance of large inclusions and evidence of disease-related deficits is not always observed [11], and indeed, in some cases large inclusions appear to be protective [12]. Technical limitations on our ability to detect smaller oligomers and aggregates by fluorescence microscopy, however, means that it is not possible unequivocally to rule out a role for aggregates [12]. Indeed, it is now clear that aggregation of many polyQ proteins is quite complex, such that the most toxic aggregated forms may not necessarily be the ones most visible in light microscopy. Thus, a multitude of aggregates can be observed in the test tube (see below), and more sophisticated methods are now demonstrating their presence in cell culture and in mouse models [13,14]. Because of this, as is the case in other aggregation-associated neurodegenerative diseases, attention is now shifting to earlier stages in the protein misfolding and aggregation pathway, stages populated by aggregation-prone misfolded states, oligomeric intermediates, and microaggregates [15–17].

It is also worth noting that although the foregoing logic builds a case for a critical early role for huntingtin aggregation in the disease mechanism, this can formally also be fulfilled by some hypotheses that do not explicitly require a toxic aggregate. For example, in some aggregation diseases, it has been argued that it might be the aggregation process, rather than any particular aggregate, that causes toxicity. Alternatively, one can imagine scenarios in which aggregates are a required intermediate for the formation of a nonaggregated toxic entity.

CASE FOR A TOXIC MISFOLDED MONOMER

An alternative model to those requiring aggregation as a prerequisite for toxicity is the idea that the toxic species in this family of diseases are monomeric proteins containing polyQ segments that are "misfolded." In fact, polyQ in isolation and in almost all native protein contexts appears not to be folded at all, and to undergo no apparent folding changes as repeat length increases. While some early work suggested that short, chemically synthesized polyQ peptides exist as monomers in stable β-sheet conformations [18,19], another study found a random coil conformation [20]. This discrepancy was clarified by the development of a solvent pretreatment for chemically synthesized polyQ peptides that generates peptides with aggregation tendencies similar to those of biologically derived polyQ proteins of similar repeat length; this protocol delivers soluble peptides that exhibit initial random coil circular dichroison (CD) spectra regardless of repeat length, speaking against a populated misfolded state in expanded polyQ [9,21]. Subsequent CD and nuclear magnetic resonance studies on recombinantly produced proteins confirmed that polyQ sequences regardless of repeat length exist in the monomeric state primarily in irregular structure [22–24], although a small degree of α-helix is detectible, especially in cooled solutions [25]. Diffusion times in fluorescence correlation spectroscopy suggest that polyQ in water exists in a condensed coil structure regardless of repeat length [26]. Early experiments with the anti-polyQ MAb 1C2 were interpreted to indicate that 1C2 recognizes a conformational epitope in polyQ that is enriched in expanded repeat versions of the sequence [27]. However, the enhanced binding observed is now thought to occur due to a linear lattice effect [23,24,28] on binding to short polyQ segments rather than to a populated alternative conformation. This argues that polyQ sequences in isolation do not differ conformationally with repeat length. Thus, it is possible that antibodies or other polyQ binding factors could exist which because of their extended binding surfaces are able to differentiate between short and long polyQs that, for example, might nevertheless be equally disordered in the solution ensemble [23].

Notwithstanding the above, two recently described intriguing observations should be mentioned. Nagai et al. reported that an artificial hybrid protein of thioredoxin fused to polyQ can fold into a conformation in which the polyQ accesses a highly α-helical structure. Further, on incubation, this protein rearranges into another monomeric state that is dominated by β-structure and is toxic [29]. Both of these monomer conformations appear to be stable enough to behave well in nondenaturing PAGE and size-exclusion chromatography. Although this is an interesting and, in fact, surprising result, its significance to polyQ disease pathology is not clear. The report of Nagai et al. stands in contrast to a rather extensive literature on other artificial polyQ fusion proteins, as well as with actual polyQ disease proteins and protein fragments, that teaches that the only altered states formed on incubation of the folded polyQ-containing monomeric proteins are β-sheet-rich aggregates of various types (see below). Nonetheless, the thioredoxin fusion results suggest that some proteins may contain folded domains that act as scaffolds capable of recruiting and/or stabilizing and ordering a polyQ component. These thioredoxin–polyQ fusions are not disease proteins. Whether any of the nine different polyQ disease proteins might possess such stabilizing domains remains to be seen; there are no reports, to date, that any of the disease proteins can access monomeric states containing polyQ segments in ordered secondary structure.

The other exception to the preponderance of evidence against a toxic monomer is the growing evidence that polyQ segments can influence the structure and/or stability of covalently attached folded domains. This work is reviewed in a later section on sequence context effects on aggregation. The implication is that in some proteins, the polyQ segment may alter the conformation of adjacent domains to produce an unfolded or alternatively folded monomer. Here again, however, the most widely observed outcome of this kind of indirect conformational destabilization is not a long-lived, misfolded monomer. Rather, it is either an enhancement of aggregation, or a reduction in the equilibrium amount of folded, active protein available to fulfill the protein's normal function. With respect to the latter outcome, several examples have been reported where expansion of a polyQ sequence reduces protein activity [30], and at least some expanded polyQ diseases appear to include symptoms attributable to loss of normal function [31–33].

If we knew more about the normal roles of polyQ sequences in proteins, we might better understand the nature of expanded polyQ diseases. These roles remain mysterious, however. Interestingly, homologous proteins from different organisms can exhibit quite different polyQ repeat lengths [30]. PolyQ is often found in longer elements of sequence predicted to be unstructured (R.W., unpublished observations). It may be possible that polyQ is being used simply as a spacer between folded domains and that the relative instability of the CAG repeat leads to polyQ expansions that, given their locations in flexible protein regions, are relatively tolerated unless they cross the repeat-length threshold of disease.

AGGREGATION OF SIMPLE POLYQ SEQUENCES

The only common feature of polyQ disease proteins is a polyQ sequence above a certain repeat-length threshold [1]. Therefore, the simplest aggregation-based hypothesis for molecular mechanisms of these diseases is that there is some critical generic aggregation tendency of all expanded polyQ sequences regardless of context. In fact, in vitro experiments with chemically synthesized polyQ peptides demonstrate a clear tendency to undergo spontaneous aggregation into amyloidlike aggregates, with aggregation rates increasing as repeat length increases [9]. The molecular basis for this effect has generated considerable experimental and computational work. In particular, initial aggregation rates of rigorously disaggregated peptides incubated in vitro are well modeled by equations based on a thermodynamic model of nucleation of aggregation (Fig. 1) [34]. In the analysis, polyQ molecules with repeat lengths from Q28 to Q47 engage a common aggregation mechanism in which the nucleus is a rare,

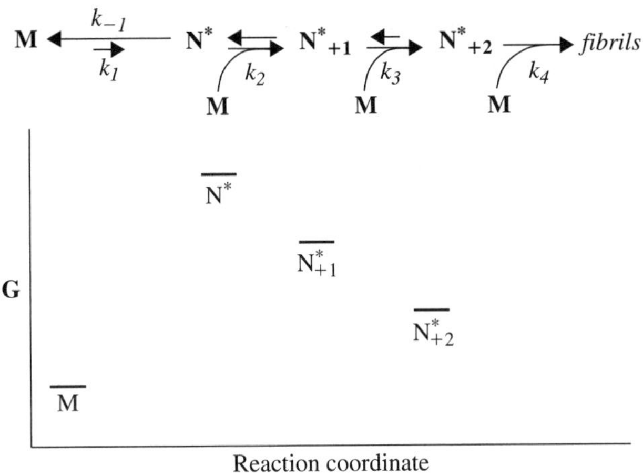

FIG. 1 Nucleation-dependent polymerization mechanism of polyGln aggregation. Formation of nucleus N^* from the monomer ensemble M is modeled as a reversible, highly unfavorable reaction. Once formed, the metastable N^* can either disintegrate back to the monomer pool or elongate, via addition of additional M molecules, to generate molecular species N^*_{+1}, N^*_{+2}, and so on, and thus decrease the free energy of the system. The number of molecules of polyGln comprising N^* (called the *critical nucleus*, or n^*) is equal to 1 for nucleation of simple polyGln aggregation [34]. (Adapted from [37], with permission. Copyright © 2006 National Academy of Sciences, U.S.A.)

aggregation-competent monomer in equilibrium with the vast sea of innocuous conformations within the monomer ensemble [34–37]. The observation that longer repeats tend to aggregate more aggressively is traced primarily to the association of longer repeat lengths with enhanced stability of the nucleus [34].

Simple polyQ aggregation kinetics data conform well to the thermodynamic nucleation model because of a couple of ususual features. First, the amyloid fibrils made in the polyQ aggregation reaction appear to be much more stable, and hence less susceptible to secondary nucleation phenomena that in other systems tend rapidly to take over even the earliest kinetic phases and confound efforts to model the primary event [38,39]. Second, in contrast to most other amyloidogenic peptides [40,41], alternatively aggregated structures (spherical oligomers and protofibrils) that do not readily conform to classical nucleation models do not appear to develop in the early phases of simple polyQ aggregation [34]. It is possible to induce simple polyQ peptides to form nonfibrillar aggregates by nonaqueous solvent treatment [42]. More commonly, however, the rigorous application of a robust disaggregation protocol [21,43] generates a well-behaved monodisperse [26] solution of disordered monomers that aggregates [9,34] according to parameters very closely replicated by recombinant polyQ peptides [24]. The nucleated growth mechanism worked out for simple polyQ peptides may contribute to some polyQ disease mechanisms and has also provided a valuable system for better understanding general aspects of the nucleated growth mechanism of spontaneous aggregation and how it is influenced by other factors [36,37].

For example, because polyQ is a homopolymer and polyQ amyloid is promiscuous in its ability to seed elongation of other repeat-length polyQs [44,45], it was possible to directly segregate and characterize the role of aggregate elongation in overall nucleation kinetics [37]. In the thermodynamic model for nucleation, whenever an unstable nucleus forms, it disintegrates rapidly, either by decaying to rejoin the ground-state monomer ensemble or by making a productive encounter with another (ground-state) polyQ to initiate the fibril elongation cascade [37] (Fig. 1). Because of this, short polyQ peptides added to the reaction are able to enhance nucleation of the aggregation of an expanded repeat polyQ peptide, specifically by increasing the frequency at which nuclei partition in the aggregation direction [37]. This has clear implications for aggregation in the cell. In fact, expression of a Q_{20} peptide in a *Drosophila* model enhances both the aggregation and toxicity of an expanded polyQ version of a fragment of the huntingtin protein [37]. The result also implies that inhibitors of the elongation step might be capable of suppressing the overall nucleation process, a prediction supported by experiment [46].

Electron micrographs of aggregates formed from simple polyQ peptides exhibit a variety of morphologies, including amyloid fibrils and other filamentous structures [47]. A mutational analysis of the ability of regularly spaced Pro–Gly pairs along a polyQ sequence to modulate aggregation kinetics led to an indication of the importance of chain reversals in aggregate structure [35]. Fiber diffraction patterns of polyQ aggregates have been interpreted as being consistent with either slabs of antiparallel β-sheet [18,48,49] or a type of β-helical structure [50]. In one analysis, there is an indication that aggregates of shorter polyQ repeat lengths form slabs that are less wide, in the β-extended chain dimension, than aggregates of longer repeat lengths [48].

ALTERED AGGREGATION OF PolyQ WITH FLANKING SEQUENCES

Complex sequences containing polyQ components can exhibit qualitatively different behavior from what is described above for simple polyQ. For example, while huntingtin exon-1, the extreme N-terminal segment of huntingtin that contains the polyQ sequence, can form amyloid fibrils in vitro [8], it can also form spherical oligomers and protofibrils similar to the aggregation "intermediates" observed in studies of other amyloidogenic proteins [15]. As discussed above, such intermediates are not seen in the aggregation of simple polyQ. Despite this apparent mechanism change, the aggregation rates for htt exon-1 sequences remain repeat-length dependent [8]. The simplest explanation for the differences in mechanism and products between chemically synthesized peptides and recombinantly produced exon1 is that flanking sequences can significantly modify both the rate and mechanism of polyQ aggregation [51].

The role of flanking sequences has been explored in detail for several expanded CAG repeat disease proteins. Huntingtin exon1 is particularly accessible to such studies, since there are only 17 amino acids on the N-terminal side of the polyQ stretch, while there is an oligoPro sequence, part of a larger Pro-rich sequence,

C-terminal to the polyQ (Table 1). Studies with chemically synthesized polyQ peptides containing an added Pro_{10} sequence C-terminal to the polyQ showed that such oligoPro sequences decrease the spontaneous aggregation rate of the peptide and, more dramatically, decrease the apparent stability of the aggregate [25]. Since this oligoPro sequence also abrogates the increase in α-helix CD that is observed when polyQ solutions are cooled (see above), the simplest explanation for all these effects is that oligoPro induces altered structure in the fluctuating polyQ monomer ensemble, decreasing the portion of molecules that are aggregation competent [25]. It would not be surprising if this alternative structure included polyproline type II conformations [52], and in fact, recent studies support this hypothesis [53]. All of these effects are lost if the oligoPro is placed on the N-terminus of the polyQ or is attached to the C-terminus through a side-chain linkage [25]. Such results are consistent with the known ability of Pro residues to influence peptide conformation only in the residues on the N-terminal side of the Pro [52]. A similar dampening effect of polyQ aggregation was observed in a yeast model for HD [54]. Expression of a FLAG-exon1-GFP fusion containing an expanded polyQ leads to aggregate formation but is not toxic (i.e., does not retard growth) in yeast. Removal of the proline-rich sequence of exon1 in the same construct, however, produces a more dispersed aggregate in the cell and exhibits growth retardation.

In this same yeast model, the presence or absence of an exogenous, negatively charged FLAG tag at the N-terminus of expanded polyQ htt can produce some surprisingly strong effects [54]. In a mammalian cell model, addition of the 17-amino acid N-terminus of htt to polyQ significantly enhances inclusion formation [55]. Such cell experiments, however, do not reveal whether the N-terminus is operating through a direct biophysical effect on the peptide or indirectly via cellular processes. In fact, recent experiments (A. K. Thakur et al., manuscript submitted) show that the htt N-terminus itself has an enormous impact on polyQ aggregation in vitro (see below).

By now there are a number of studies, in different polyQ protein systems, that reveal unusual, and strikingly similar, aggregation mechanisms. These studies, reviewed below, have two major themes: (1) that aggregation of polyQ chimeric proteins can be initiated by initial aggregation of a flanking domain rather than by the polyQ segment, and (2) that the expansion of the polyQ segment can destabilize or otherwise render such flanking domains more aggregation prone, thus providing a way for polyQ expansion to kick-start the aggregation reaction while not taking the lead in forming the aggregate.

Two early papers suggested such a domain-destabilizing effect by expansion of a covalently connected polyQ. Working with constructs in which polyQ repeats of different lengths were inserted in the interior sequence of the α-helix-rich myoglobin, Tanaka et al. observed significant, repeat-length-dependent reductions in the Cm for guanidine hydrochloride unfolding of these proteins, as monitored either by CD at 222 nm or the Soret band of the bound heme group [56]. These proteins are not perfect models for polyQ disease proteins, however, since polyQ segments in disease proteins appear to be placed not in the middle of folded domains but, rather, adjacent to them in unstructured regions (R. Wetzel, unpublished results). In the other report, Bevivino and Loll found that the Q_{78}

TABLE 1 Polyglutamine Flanking Sequences in Expanded CAG Repeat Disease Proteins[a]

Human Disease Protein	N-Terminal to Polyglutamine	C-Terminal to Polyglutamine
Huntingtin	MATLEKL MKAFESLKSF	PPPPPPPPPP PQLPQPPPQA
Atropin 1 (DRPLA)	PSTGAQSTAH PPVSTHHHHH	HHGNSGPPPP GAFPHPLEGG
Androgen receptor (SBMA)	GPRHPEAASA APPGASLLLL	ETSPRQQQQQ QGEDGSPQAH
Ataxin 1 (SCA1)	YSTLLANMGS LSQTPGHKAE	HLSRAPGLIT PGSPPPAQQN
Ataxin 2 (SCA2)	GCPRPACEPV YGPLTMSLKP	PPPAAANVRK PGGSGLLASP
Ataxin 3 (SCA3)	SGTNLTSEEL RKRREAYFEK	RDLSGQSSHP CERPATSSGA
CACNA1A (SCA6)	PRPHVSYSPV IRKAGGSGPP	AVARPGRAAT SGPRRYPGPT
Ataxin 7 (SCA7)	RGEPRRAAAA AGGAAAAAAR	PPPPQPQRQQ HPPPPPRRTR
TBP (SCA17)	LTPQPIQNTN SLSILEEQQR	AVAAAAVQQS TSQQATQGTS

Source: From [25], with permission.

[a]Only the first 20 amino acids from the polyQ are shown.

version of ataxin 3 (AT3) fused to maltose-binding protein exhibited a significantly different CD spectrum from that of the Q_{27} version of the same construct [57]. The source of the structural difference, and how any such structural differences might relate to folding stability, are not known.

A number of important studies have been conducted on AT3. This protein is relatively small compared to most CAG repeat disease proteins, consisting of a C-terminal polyQ sequence and an N-terminal folded domain called "Josephin," which is a ubiquitin dehydrolase [58] structurally homologous to the Cys family of proteases [59]. Masino et al. found no difference between the structures and folding stabilities (against thermal and urea denaturation) of the Josephin domain alone or in a Q_{18} version of AT3 [60]. Interestingly, both proteins also aggregate at the same rate, suggesting, at least in constructs not containing pathological repeat-length polyQ, that aggregation is led not by the polyQ but by the Josephin domain. No longer polyQ repeat lengths were examined by this group, however. In contrast, Chow et al. examined three polyQ repeat-length variants of AT3, including a version with a pathological repeat-length polyQ (Q_{15}, Q_{28}, and Q_{50}). These workers found that there is no overt stability difference between these forms, as indicated by acid denaturation curves [61]. At the same time, while guanidine (Gdn-HCl] melts were reversible for the two shorter polyQ versions, low concentrations of Gdn-HCl led to rapid aggregation of the Q_{50} version. The authors explain this dichotomy by suggesting that polyQ expansion may decrease the free-energy barrier for crossing the transition state for aggregation, without affecting the free-energy change for unfolding; that is, the early steps of the aggregation reaction path may not be congruent with the reversible unfolding pathway observed with chemical denaturants.

The Bottomley group went on to characterize the aggregation mechanism in more detail. They extended the suggestion by Masino et al. [60] that aggregation involving the Josephin domain plays an early role in AT3 aggregation by showing that this is even the case for expanded polyQ versions of AT3 [62]. In addition, they generated important evidence in support of a two-step mechanism for aggregation of expanded polyQ forms of AT3 (Fig. 2). The first step leads to formation of SDS-soluble worm-like aggregates that are accessible to any AT3 variant containing an intact Josephin domain. The second step, leading to formation of SDS-stable amyloidlike aggregates, is accessible only to AT3 with an expanded polyQ [62]. The Bottomley group also showed that the initial aggregation mediated by the Josephin domain occurs via nucleated growth polymerization with a critical nucleus of one [63], a nucleation mechanism similar to that of simple polyQ [34].

The two main themes of this body of work on AT3 have also been explored, and largely replicated, in work on a model system consisting of the cellular retinoic acid–binding protein I (CRABP I) fused to polyQ sequences [64–67]. CRABP I is a relatively stable protein, but a Pro39→Ala point mutant folds more slowly and tends to aggregate both in vitro and during production in a bacterial cell; in vitro aggregation kinetics are consistent with a nucleated growth polymerization pathway and a monomeric nucleus [64]. When the wild-type CRABP I protein is fused to htt exon1 containing polyQ sequences in the normal range (20 or 33), the protein aggregates no more readily than does the isolated domain lacking a polyQ tail. When the repeat length is increased

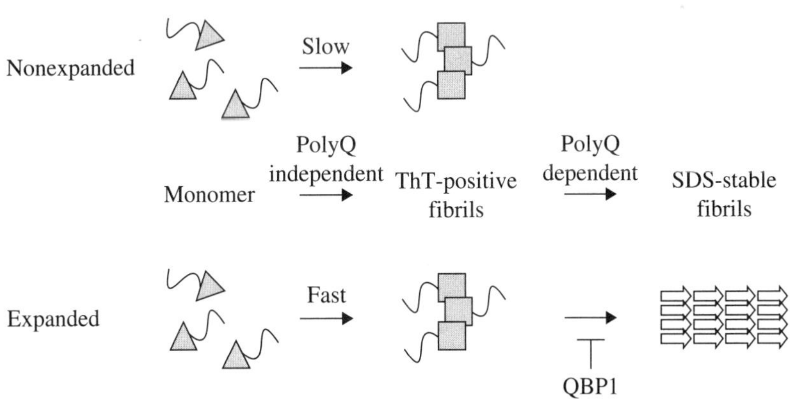

FIG. 2 Model for polyQ aggregation mediated by a flanking sequence. Shaded triangles and squares indicate two different conformations of the Josephin domain in AT3, while the curved lines indicate a fused polyQ sequence. For both normal and expanded polyQ forms, a conformational change in the Josephin domain accompanies its aggregation, whereas initially the polyQ is uninvolved. In expanded polyQ constructs, a second step is possible in which the fused polyQ sequences also undergo amyloid formation. QBP1 is a peptide-based inhibitor of polyQ aggregation [85] that blocks this second step. (From [63], with permission of Elsevier.)

(53 or 64Q), however, the sequence aggregates much more readily [65]. Interestingly, compared with either the wild-type CRABP I domain alone or the domain with a Q_{20} attached, the Q_{53} construct exhibits a urea melting curve considerably shifted to lower stability (Fig. 3) [65]. In a subsequent study it was shown that, in analogy to AT3 [62], expanded polyQ versions of the CRABP-exon1 fusions undergo a multistep aggregation pathway in which the early aggregates are SDS-labile with a core formed dominantly from CRABP-sequence elements, whereas the later aggregates are SDS-resistant, with a core dominated by polyQ segments [67]. A unique feature of the system from which many of the results above were generated is that the CRABP I domain was engineered to contain a target sequence for binding the FlAsH dye, thus allowing fluorescence visualization of folding and aggregation within the cell [66]. Interestingly, comparative studies showed that the cellular environment has a significant impact on how the aggregation proceeds [67].

The aggregation mechanism of peptides related to huntingtin exon1 shares several of the features observed in studies with AT3 and with the CRAB-P polyQ fusion proteins. In isolation, or fused to a short polyQ, the 17-residue mixed amino acid sequence of the huntingtin N-terminus (htt^{NT}) exists as a compact coil that is resistant to aggregation. In contrast, with an attached, expanded polyQ, htt^{NT} is disrupted and engages a very rapid, nonnucleated (downhill) aggregation that leads to spherical oligomers whose core appears to consist of htt^{NT} and not polyQ. In a second step that seems to proceed via the oligomeric intermediates, amyloid fibrils form that include in their cores both htt^{NT} and polyQ and that grow by monomer addition (A. K. Thakur et al.; see Note Added in Proof).

Overall, the results described here show that flanking sequences can sometimes play a critical role in controlling the aggregation of polyQ proteins.

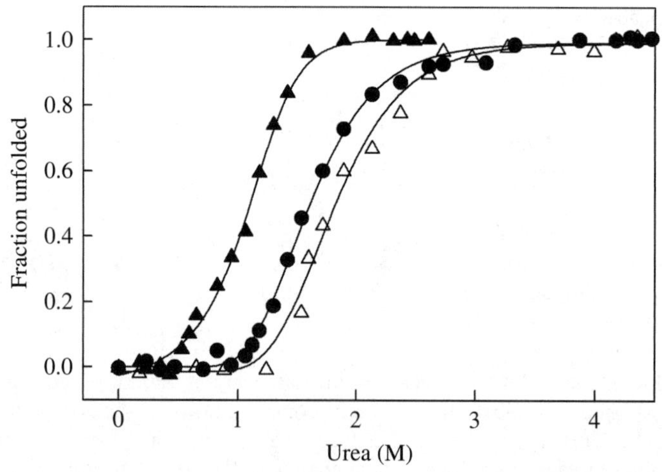

FIG. 3 Urea-induced unfolding of a CRABP domain fused to htt exon1 domains with different polyQ lengths. Filled circular, No exon1; filled triangle, exon1-Q_{20}; open triangle, exon1-Q_{53}. (Adapted from [65], with permission. Copyright © 2006 American Society for Biochemistry and Molecular Biology.)

Furthermore, the polyQ in some cases seems to engage in an intriguing reciprocal interaction with the flanking sequence—first conformationally altering and favoring the aggregation of the flanking sequence, then responding to that aggregation by itself joining in the aggregation process—with overall kinetics that are much faster than what would be exhibited by a "naked" polyQ of the same repeat length. The initial aggregates can exhibit dramatically different structures and properties compared with those of mature aggregates.

At the same time, we know that not all sequences attached to polyQ influence polyQ behavior in the manner exemplified by the Josephin domain and the huntingtin N-terminal peptide. Thus, neither an oligoproline sequence [25] nor a mutated htt N-terminal sequence (A. K. Thakur et al.; see Note Added in Proof) exhibit altered aggregation when attached to the N-terminal side of polyQ. Even for more globular flanking sequences, it may turn out that only protein domains that are marginally stable will experience the polyQ-induced conformational change that seems to enhance aggregation [65]. It therefore remains to be seen whether any of the other expanded CAG repeat disease proteins, in addition to htt and AT3, will exhibit flanking sequence modulation of polyQ behavior, or behave essentially identically to a simple polyQ sequence of the same repeat length. It should also be noted that the impact of flanking sequences on polyQ aggregation is likely to be highly susceptible to modulation by cellular factors. For example, especially because polyQ sequences tend to be located in larger regions of disorder (R. Wetzel, unpublished results), proteolysis in the cell might remove some or all of a flanking sequence, such that the now largely naked polyQ sequence would behave more like a simple polyQ. Similarly, if aggregation by the flanking sequence is triggered by some sort of domain rearrangement, it might be expected that binding of an interaction partner to the flanking sequence might lock it into an aggregation-resistant conformation. This might, for example, explain how an intracellular antibody against the N-terminal domain of htt can, under some circumstances, block aggregation and toxicity [68].

ROLE OF THE CELLULAR ENVIRONMENT

The discussion above introduces an important role for the cellular environment in potentially modulating polyQ aggregation. Direct comparison of the aggregation of CRABP-exon1 chimera showed similarities and differences between experiments conducted in *Escherichia coli* and in vitro [67]. The polyQ aggregation mechanism in one cell model [69] appears to be the same nucleated growth polymerization, with a monomeric nucleus, as seen in vitro [34]. Emerging tools for characterizing cellular aggregates, not just for size, subcellular localization, and protein constituent, but also for aggregate morphology [42] and functionality [70], should help the field move beyond simple counting of inclusions to a better understanding of the richness and subtlety of the aggregation process in the cell [13], and hence give us a better chance to identify the key toxic species.

Genetic screens have provided a tremendous insight into the range of factors influencing the aggressiveness and impact of misfolding and aggregation of expanded polyQ proteins. For example, a systematic, genome-wide RNAi screen of factors specifically affecting polyQ aggregation in *Caenorhabditis elegans* revealed five classes of factors: RNA metabolism along with protein synthesis, folding, degradation, and trafficking [71]. Thus, three types of molecular chaperone were identified, including a member of the DnaJ class, a protein family implicated in other genetic analyses of htt aggregation. In the trafficking class, factors for nuclear import and microtubule formation, both implicated in htt aggregation and/or toxicity, were identified. A variety of factors related to proteolysis, and in particular proteolytic breakdown of aggregates, were also identified, including a proteasome subunit and members of the ubiquitin family and ubiquitin E1-activating enzyme family [71].

A yeast genetic screen with a growth retardation readout uncovered 52 nonessential genes whose deletion enhances polyQ toxicity, including molecular chaperones, ubiquitin response proteins, and other proteins involved in stress responses [72]. A partial survey of protective factors has been conducted on a mammalian genome as well. Yamamoto et al. developed cell models producing an inducible, expanded polyQ form of htt exon1 associated with aggregate formation and cell death. In these cells, when htt exon1 expression is stopped, aggregates slowly disappear and cell survival is enhanced. In an siRNA screen of those genes that are up-regulated in response to aggregate formation, these workers found a number of genes whose knockdown blocks the cell's ability to recover when htt exon1 expression is stopped, including chaperones and molecules associated with protein degradation and regulation of autophagy [73].

With both flanking sequence factors and cell factors expected to modulate the aggregation of polyQ proteins, it is not surprising that different polyQ diseases can affect different populations of neurons [1] and that a wide variance of ages of onset, at least some of which is attributable to secondary genetic factors [74], is observed for patients with the same disease and the same polyQ repeat length. Mapping these complex interrelationships is important not only for better understanding disease mechanisms, but also for possible exploitation in therapeutic approaches.

TOXICITY MECHANISMS RELATED TO PROTEIN MISFOLDING

The consistent presence of protein aggregates in the pathology of major neurodegenerative diseases suggests that there may be some common disease mechanisms at work. For example, protein aggregates might inactivate, monopolize, or otherwise compromise one or more of the systems responsible for normal clearance of misfolded or aggregated proteins. Thus, polyQ aggregates have been shown to inhibit proteasome activity [75]. Even without overt inhibition, it might reasonably be expected that systems like the ubiquitin–proteasome network will fail to keep up with other important cellular functions if they are overloaded with misfolded-protein substrates. In addition to such generic mechanisms, there may also be more disease-specific toxicity mechanisms. The expanded polyQ diseases offer a particularly attractive set of mechanisms based on the promiscuity of polyQ amyloid fibril elongation [44,45]. In these "recruitment"-based mechanisms, aggregates are hypothesized to recruit other polyQ proteins into the aggregate, leading to their inactivation either by a simple sequestration and/or by a stimulation of their removal from the cell [76,77].

THERAPEUTIC POSSIBILITIES

Consistent with the discussion above, there is a growing sense that protein-misfolding diseases are associated with age-related irregularities in protein flux/homeostasis [78]. For example, in experimental models it is possible to modulate expanded polyQ effects by up-regulating the elements of the heat-shock response [79]. Although chronic, global up-regulation of Hsp proteins might be too blunt an instrument for human therapy, it might be possible to fine-tune critical elements of the response without undue toxic side effects. Another generic approach to aggregation disorders is to identify modulators of aggregation itself. In many ways, such an approach is conceptually ideal since it involves blocking something that is presumably entirely pathological and nonbeneficial. One problem that has emerged, however, is that many previously described inhibitors of amyloid formation turn out to redirect the aggregation process into another polymeric product rather than actually inhibit aggregation [80]; of course, such compounds might still be capable of delivering a therapeutic benefit [81]. In addition, some small-molecule aggregation inhibitors that have been described were subsequently shown to be part of a large class of nonspecific colloidal molecules now understood to be major nuisance compounds in high-throughput screening [82]. Despite such discouraging data, at least one specific inhibitor has also been described whose mechanism is known and which is expected to target polyQ aggregation occurring by a nucleated growth mechanism [46].

If aggregation cannot be blocked biophysically or eliminated by exploiting normal cellular mechanisms of aggregation management, there would appear to be at least three alternative therapeutic avenues: (1) restrict expression of the mutant huntingtin allele [83], (2) block the posttranslational processing that generates the most aggregation-prone forms, or (3) modulate the toxic activity of the aggregates that do form. Regarding the second possibility, identification of key protease cleavage sites in the huntingtin protein leading to formation of highly aggregation-prone htt N-terminal fragments, opens the possibility of developing protease inhibition strategies in HD [84]. If other posttranslational modifications are involved in enhancement of aggregation, blocking these might be similarly effective. The third therapeutic avenue has great potential in theory, but is severely limited in practice by our continued ignorance of the critical molecular events underlying aggregate toxicity. As we gain clarity on the currently obscure nature of these downstream mechanistic events, new opportunities for drug discovery will emerge.

Acknowledgments

I gratefully acknowledge NIH grant R01-AG019322 for support.

NOTE ADDED IN PROOF

A number of important papers have been published since this manuscript went to proof. Aggregates isolated from the brains of mouse models exhibit the great variety of htt-related aggregate morphologies possible in vivo [86]. Force microscopy experiments indicate a surprising stability to the polyQ collapsed coil [87]. The work briefly discussed in this review on the role of the htt^{NT} domain in htt fragment aggregation has now been published [88]. Another paper shows how binding of htt^{NT} to another protein (in this case the chaperone triC) can abrogate htt^{NT}-mediated aggregation [89]. A series of X-ray crystal structures of short polyQ versions of a htt N-terminal fragment show that, even though the htt^{NT} domain is disordered as a monomer in solution [88], it can exist in an α-helix conformation when packed into a trimeric helical bundle in a crystal [90]. Mass spectrometry has confirmed that, in *Drosophila*, the Thr residue of over-expressed human htt^{NT} is phosphorylated, and the initiator Met of htt^{NT} is enzymatically removed [91]. Two papers published in late 2009 address the role of phosphorylation of htt^{NT} in htt toxicity [92, 93]. In one of these papers, introduction of phosphor-mimicking Ser to Asp mutations at positions 13 and 16 within the htt^{NT} segment of full length, expanded polyQ htt in tg mice leads to abrogation of disease symptoms and brain inclusions [92]. These same mutations, in a chemically synthesized htt N-terminal fragment, lead to reduced aggregation rates and altered aggregate morphology [92].

REFERENCES

1. Bates, G.P., Benn, C. (2002). The polyglutamine diseases, In *Huntington's Disease* (Bates, G.P., Harper, P.S., Jones, L., Eds.), Oxford University Press, Oxford, UK, pp. 429–472.
2. Rosas, H.D., Hevelone, N.D., Zaleta, A.K., Greve, D.N., Salat, D.H., Fischl, B. (2005). Regional cortical thinning in preclinical Huntington disease and its relationship to cognition. *Neurology, 65*, 745–747.
3. Borrell-Pages, M., Zala, D., Humbert, S., Saudou, F. (2006). Huntington's disease: from huntingtin function and dysfunction to therapeutic strategies. *Cell Mol Life Sci, 63*, 2642–2660.
4. DiFiglia, M., Sapp, E., Chase, K.O., Davies, S.W., Bates, G.P., Vonsattel, J.P., Aronin, N. (1997). Aggregation of huntingtin in neuronal intranuclear inclusions and dystrophic neurites in brain. *Science, 277*, 1990–1993.

5. Davies, S.W., Turmaine, M., Cozens, B.A., DiFiglia, M., Sharp, A.H., Ross, C.A., Scherzinger, E., Wanker, E.E., Mangiarini, L., Bates, G.P. (1997). Formation of neuronal intranuclear inclusions underlies the neurological dysfunction in mice transgenic for the HD mutation. *Cell, 90*, 537–548.
6. Paulson, H.L., Perez, M.K., Trottier, Y., Trojanowski, J.Q., Subramony, S.H., Das, S.S., Vig, P., Mandel, J.L., Fischbeck, K.H., Pittman, R.N. (1997). Intranuclear inclusions of expanded polyglutamine protein in spinocerebellar ataxia type 3. *Neuron, 19*, 333–344.
7. Martin, J.B. (1999). Molecular basis of the neurodegenerative disorders. Erratum in *N Engl J Med* 1999 Oct 28;*341*(18); 1407. *N Engl J Med, 340*, 1970–1980.
8. Scherzinger, E., Lurz, R., Turmaine, M., Mangiarini, L., Hollenbach, B., Hasenbank, R., Bates, G.P., Davies, S.W., Lehrach, H., Wanker, E.E. (1997). Huntingtin-encoded polyglutamine expansions form amyloid-like protein aggregates in vitro and in vivo. *Cell, 90*, 549–558.
9. Chen, S., Berthelier, V., Yang, W., Wetzel, R. (2001). Polyglutamine aggregation behavior in vitro supports a recruitment mechanism of cytotoxicity. *J Mol Biol, 311*, 173–182.
10. Morley, J.F., Brignull, H.R., Weyers, J.J., Morimoto, R.I. (2002). The threshold for polyglutamine-expansion protein aggregation and cellular toxicity is dynamic and influenced by aging in *Caenorhabditis elegans*. *Proc Natl Acad Sci U S A, 99*, 10417–10422.
11. Zoghbi, H.Y., Orr, H.T. (1999). Polyglutamine diseases: protein cleavage and aggregation. *Curr Opin Neurobiol, 9*, 566–570.
12. Arrasate, M., Mitra, S., Schweitzer, E.S., Segal, M.R., Finkbeiner, S. (2004). Inclusion body formation reduces levels of mutant huntingtin and the risk of neuronal death. *Nature, 431*, 805–810.
13. Wanderer, J., Morton, A.J. (2007). Differential morphology and composition of inclusions in the R6/2 mouse and PC12 cell models of Huntington's disease. *Histochem Cell Biol, 127*, 473–484.
14. Weiss, A., Klein, C., Woodman, B., Sathasivam, K., Bibel, M., Regulier, E., Bates, G.P., Paganetti, P. (2008). Sensitive biochemical aggregate detection reveals aggregation onset before symptom development in cellular and murine models of Huntington's disease. *J Neurochem, 104*, 846–858.
15. Poirier, M.A., Li, H., Macosko, J., Cai, S., Amzel, M., Ross, C.A. (2002). Huntingtin spheroids and protofibrils as precursors in polyglutamine fibrilization. *J Biol Chem, 277*, 41032–41037.
16. Wacker, J.L., Zareie, M.H., Fong, H., Sarikaya, M., Muchowski, P.J. (2004). Hsp70 and Hsp40 attenuate formation of spherical and annular polyglutamine oligomers by partitioning monomer. *Nat Struct Mol Biol, 11*, 1215–1222.
17. Takahashi, T., Kikuchi, S., Katada, S., Nagai, Y., Nishizawa, M., Onodera, O. (2008). Soluble polyglutamine oligomers formed prior to inclusion body formation are cytotoxic. *Hum Mol Genet, 17*, 345–356.
18. Perutz, M.F., Johnson, T., Suzuki, M., Finch, J.T. (1994). Glutamine repeats as polar zippers: their possible role in inherited neurodegenerative diseases. *Proc Natl Acad Sci U S A, 91*, 5355–5358.
19. Sharma, D., Sharma, S., Pasha, S., Brahmachari, S.K. (1999). Peptide models for inherited neurodegenerative disorders: conformation and aggregation properties

of long polyglutamine peptides with and without interruptions. *FEBS Lett, 456*, 181–185.

20. Altschuler, E.L., Hud, N.V., Mazrimas, J.A., Rupp, B. (1997). Random coil conformation for extended polyglutamine stretches in aqueous soluble monomeric peptides. *J Pept Res, 50*, 73–75.
21. Chen, S., Wetzel, R. (2001). Solubilization and disaggregation of polyglutamine peptides. *Protein Sci, 10*, 887–891.
22. Masino, L., Kelly, G., Leonard, K., Trottier, Y., Pastore, A. (2002). Solution structure of polyglutamine tracts in GST–polyglutamine fusion proteins. *FEBS Lett, 513*, 267–272.
23. Bennett, M.J., Huey-Tubman, K.E., Herr, A.B., West, A.P., Ross, S.A., Bjorkman, P.J. (2002). A linear lattice model for poly-glutamine in CAG expansion diseases. *Proc Natl Acad Sci U S A, 99*, 11634–11639.
24. Klein, F.A., Pastore, A., Masino, L., Zeder-Lutz, G., Nierengarten, H., Oulad-Abdelghani, M., Altschuh, D., Mandel, J.L., Trottier, Y. (2007). Pathogenic and nonpathogenic polyglutamine tracts have similar structural properties: towards a length-dependent toxicity gradient. *J Mol Biol, 371*, 235–244.
25. Bhattacharyya, A., Thakur, A.K., Chellgren, V.M., Thiagarajan, G., Williams, A.D., Chellgren, B.W., Creamer, T.P., Wetzel, R. (2006). Oligoproline effects on polyglutamine conformation and aggregation. *J Mol Biol, 355*, 524–535.
26. Crick, S.L., Jayaraman, M., Frieden, C., Wetzel, R., Pappu, R.V. (2006). Fluorescence correlation spectroscopy shows that monomeric polyglutamine molecules form collapsed structures in aqueous solutions. *Proc Natl Acad Sci U S A, 103*, 16764–16769.
27. Trottier, Y., Lutz, Y., Stevanin, G., Imbert, G., Devys, D., Cancel, G., Saudou, F., Weber, C., David, G., Tora, L., et al. (1995). Polyglutamine expansion as a pathological epitope in Huntington's disease and four dominant cerebellar ataxias. *Nature, 378*, 403–406.
28. Li, P., Huey-Tubman, K.E., Gao, T., Li, X., West, A.P., Jr., Bennett, M.J., Bjorkman, P.J. (2007). The structure of a polyQ–anti-polyQ complex reveals binding according to a linear lattice model. *Nat Struct Mol Biol, 14*, 381–387.
29. Nagai, Y., Inui, T., Popiel, H.A., Fujikake, N., Hasegawa, K., Urade, Y., Goto, Y., Naiki, H., Toda, T. (2007). A toxic monomeric conformer of the polyglutamine protein. *Nat Struct Mol Biol, 14*, 332–340.
30. Wetzel, R. (2006). Chemical and physical properties of polyglutamine repeat sequences. In *Genetic Instabilities and Neurological Diseases*, 2nd ed. (Wells, R.D., Ashizawa, T., Eds.), Elsevier, San Diego, CA, pp. 517–534.
31. Cattaneo, E., Zuccato, C., Tartari, M. (2005). Normal huntingtin function: an alternative approach to Huntington's disease. *Nat Rev Neurosci, 6*, 919–930.
32. Chen, C., Fischbeck, K.H. (2006). Clinical features and molecular biology of Kennedy's disease. In *Genetic Instabilities and Neurological Diseases*, 2nd ed. (Wells, R.D., Ashizawa, T., Eds.) Elsevier, San Diego, CA, pp. 211–220.
33. Lim, J., Crespo-Barreto, J., Jafar-Nejad, P., Bowman, A.B., Richman, R., Hill, D.E., Orr, H.T., Zoghbi, H.Y. (2008). Opposing effects of polyglutamine expansion on native protein complexes contribute to SCA1. *Nature, 452*, 713–718.
34. Chen, S., Ferrone, F., Wetzel, R. (2002). Huntington's disease age-of-onset linked to polyglutamine aggregation nucleation. *Proc Natl Acad Sci U S A, 99*, 11884–11889.

35. Thakur, A., Wetzel, R. (2002). Mutational analysis of the structural organization of polyglutamine aggregates. *Proc Natl Acad Sci U S A, 99*, 17014–17019.

36. Bhattacharyya, A.M., Thakur, A.K., Wetzel, R. (2005). polyglutamine aggregation nucleation: thermodynamics of a highly unfavorable protein folding reaction. *Proc Natl Acad Sci U S A, 102*, 15400–15405.

37. Slepko, N., Bhattacharyya, A.M., Jackson, G.R., Steffan, J.S., Marsh, J.L., Thompson, L.M., Wetzel, R. (2006). Normal-repeat-length polyglutamine peptides accelerate aggregation nucleation and cytotoxicity of expanded polyglutamine proteins. *Proc Natl Acad Sci U S A, 103*, 14367–14372.

38. Ferrone, F. (1999). Analysis of protein aggregation kinetics. *Methods Enzymol, 309*, 256–274.

39. Collins, S.R., Douglass, A., Vale, R.D., Weissman, J.S. (2004). Mechanism of prion propagation: amyloid growth occurs by monomer addition. *PLoS Biol, 2*, e321.

40. Caughey, B., Lansbury, P.T. (2003). Protofibrils, pores, fibrils, and neurodegeneration: separating the responsible protein aggregates from the innocent bystanders. *Annu Rev Neurosci, 26*, 267–298.

41. Kirkitadze, M.D., Bitan, G., Teplow, D.B. (2002). Paradigm shifts in Alzheimer's disease and other neurodegenerative disorders: the emerging role of oligomeric assemblies. *J Neurosci Res, 69*, 567–577.

42. Kayed, R., Head, E., Thompson, J.L., McIntire, T.M., Milton, S.C., Cotman, C.W., Glabe, C.G. (2003). Common structure of soluble amyloid oligomers implies common mechanism of pathogenesis. *Science, 300*, 486–489.

43. O'Nuallain, B., Thakur, A.K., Williams, A.D., Bhattacharyya, A.M., Chen, S., Thiagarajan, G., Wetzel, R. (2006). Kinetics and thermodynamics of amyloid assembly using a high-performance liquid chromatography–based sedimentation assay. *Methods Enzymol, 413*, 34–74.

44. Perez, M.K., Paulson, H.L., Pendse, S.J., Saionz, S.J., Bonini, N.M., Pittman, R.N. (1998). Recruitment and the role of nuclear localization in polyglutamine-mediated aggregation. *J Cell Biol, 143*, 1457–1470.

45. Huang, C.C., Faber, P.W., Persichetti, F., Mittal, V., Vonsattel, J.P., MacDonald, M.E., Gusella, J.F. (1998). Amyloid formation by mutant huntingtin: threshold, progressivity and recruitment of normal polyglutamine proteins. *Somat Cell Mol Genet, 24*, 217–233.

46. Thakur, A.K., Yang, W., Wetzel, R. (2004). Inhibition of polyglutamine aggregate cytotoxicity by a structure-based elongation inhibitor. *FASEB J, 18*, 923–925.

47. Chen, S., Berthelier, V., Hamilton, J.B., O'Nuallain, B., Wetzel, R. (2002). Amyloid-like features of polyglutamine aggregates and their assembly kinetics. *Biochemistry, 41*, 7391–7399.

48. Sharma, D., Shinchuk, L.M., Inouye, H., Wetzel, R., Kirschner, D.A. (2005). Polyglutamine homopolymers having 8–45 residues form slablike beta-crystallite assemblies. *Proteins, 61*, 398–411.

49. Sikorski, P., Atkins, E. (2005). New model for crystalline polyglutamine assemblies and their connection with amyloid fibrils. *Biomacromolecules, 6*, 425–432.

50. Perutz, M.F., Finch, J.T., Berriman, J., Lesk, A. (2002). Amyloid fibers are water-filled nanotubes. *Proc Natl Acad Sci U S A, 99*, 5591–5595.

51. Nozaki, K., Onodera, O., Takano, H., Tsuji, S. (2001). Amino acid sequences flanking polyglutamine stretches influence their potential for aggregate formation. *Neuroreport, 12*, 3357–3364.

52. Hinderaker, M.P., Raines, R.T. (2003). An electronic effect on protein structure. *Protein Sci, 12*, 1188–1194.
53. Darnell, G., Orgel, J.P., Pahl, R., Meredith, S.C. (2007). Flanking polyproline sequences inhibit beta-sheet structure in polyglutamine segments by inducing PPII-like helix ptructure. *J Mol Biol, 374*, 688–704.
54. Duennwald, M.L., Jagadish, S., Muchowski, P.J., Lindquist, S. (2006). Flanking sequences profoundly alter polyglutamine toxicity in yeast. *Proc Natl Acad Sci U S A, 103*, 11045–11050.
55. Rockabrand, E., Slepko, N., Pantalone, A., Nukala, V.N., Kazantsev, A., Marsh, J.L., Sullivan, P.G., Steffan, J.S., Sensi, S.L., Thompson, L.M. (2007). The first 17 amino acids of huntingtin modulate its sub-cellular localization, aggregation and effects on calcium homeostasis. *Hum Mol Genet, 16*, 61–77.
56. Tanaka, M., Morishima, I., Akagi, T., Hashikawa, T., Nukina, N. (2001). Intra- and intermolecular beta-pleated sheet formation in glutamine-repeat inserted myoglobin as a model for polyglutamine diseases. *J Biol Chem, 276*, 45470–45475.
57. Bevivino, A.E., Loll, P.J. (2001). An expanded glutamine repeat destabilizes native ataxin-3 structure and mediates formation of parallel beta-fibrils. *Proc Natl Acad Sci U S A, 98*, 11955–11960.
58. Chai, Y., Berke, S.S., Cohen, R.E., Paulson, H.L. (2004). Poly-ubiquitin binding by the polyglutamine disease protein ataxin-3 links its normal function to protein surveillance pathways. *J Biol Chem, 279*, 3605–3611.
59. Nicastro, G., Menon, R.P., Masino, L., Knowles, P.P., McDonald, N.Q., Pastore, A. (2005). The solution structure of the Josephin domain of ataxin-3: structural determinants for molecular recognition. *Proc Natl Acad Sci U S A, 102*, 10493–10498.
60. Masino, L., Nicastro, G., Menon, R.P., Dal Piaz, F., Calder, L., Pastore, A. (2004). Characterization of the structure and the amyloidogenic properties of the Josephin domain of the polyglutamine-containing protein ataxin-3. *J Mol Biol, 344*, 1021–1035.
61. Chow, M.K., Ellisdon, A.M., Cabrita, L.D., Bottomley, S.P. (2004). Polyglutamine expansion in ataxin-3 does not affect protein stability: implications for misfolding and disease. *J Biol Chem, 279*, 47643–47651.
62. Ellisdon, A.M., Thomas, B., Bottomley, S.P. (2006). The two-stage pathway of ataxin-3 fibrillogenesis involves a polyglutamine-independent step. *J Biol Chem, 281*, 16888–16896.
63. Ellisdon, A.M., Pearce, M.C., Bottomley, S.P. (2007). Mechanisms of ataxin-3 misfolding and fibril formation: kinetic analysis of a disease-associated polyglutamine protein. *J Mol Biol, 368*, 595–605.
64. Ignatova, Z., Gierasch, L.M. (2005). Aggregation of a slow-folding mutant of a beta-clam protein proceeds through a monomeric nucleus. *Biochemistry, 44*, 7266–7274.
65. Ignatova, Z., Gierasch, L.M. (2006). Extended polyglutamine tracts cause aggregation and structural perturbation of an adjacent beta barrel protein. *J Biol Chem, 281*, 12959–12967.
66. Ignatova, Z., Gierasch, L.M. (2004). Monitoring protein stability and aggregation in vivo by real-time fluorescent labeling. *Proc Natl Acad Sci U S A, 101*, 523–528.
67. Ignatova, Z., Thakur, A.K., Wetzel, R., Gierasch, L.M. (2007). In-cell aggregation of a polyglutamine-containing chimera is a multistep process initiated by the flanking sequence. *J Biol Chem, 282*, 36736–36743.

68. Lecerf, J.M., Shirley, T.L., Zhu, Q., Kazantsev, A., Amersdorfer, P., Housman, D.E., Messer, A., Huston, J.S. (2001). Human single-chain Fv intrabodies counteract in situ huntingtin aggregation in cellular models of Huntington's disease. *Proc Natl Acad Sci U S A, 98*, 4764–4769.

69. Colby, D.W., Cassady, J.P., Lin, G.C., Ingram, V.M., Wittrup, K.D. (2006). Stochastic kinetics of intracellular huntingtin aggregate formation. *Nat Chem Biol, 2*, 319–323.

70. Osmand, A.P., Berthelier, V., Wetzel, R. (2006). Imaging polyglutamine deposits in brain tissue. *Methods Enzymol, 412*, 106–122.

71. Nollen, E.A., Garcia, S.M., van Haaften, G., Kim, S., Chavez, A., Morimoto, R.I., Plasterk, R.H. (2004). Genome-wide RNA interference screen identifies previously undescribed regulators of polyglutamine aggregation. *Proc Natl Acad Sci U S A, 101*, 6403–6408.

72. Willingham, S., Outeiro, T.F., DeVit, M.J., Lindquist, S.L., Muchowski, P.J. (2003). Yeast genes that enhance the toxicity of a mutant huntingtin fragment or alpha-synuclein. *Science, 302*, 1769–1772.

73. Yamamoto, A., Cremona, M.L., Rothman, J.E. (2006). Autophagy-mediated clearance of huntingtin aggregates triggered by the insulin-signaling pathway. *J Cell Biol, 172*, 719–731.

74. Wexler, N.S., Lorimer, J., Porter, J., Gomez, F., Moskowitz, C., Shackell, E., Marder, K., Penchaszadeh, G., Roberts, S.A., Gayan, J., et al. (2004). Venezuelan kindreds reveal that genetic and environmental factors modulate Huntington's disease age of onset. *Proc Natl Acad Sci U S A, 101*, 3498–3503.

75. Bence, N.F., Sampat, R.M., Kopito, R.R. (2001). Impairment of the ubiquitin–proteasome system by protein aggregation. *Science, 292*, 1552–1555.

76. Nucifora, F.C., Jr., Sasaki, M., Peters, M.F., Huang, H., Cooper, J.K., Yamada, M., Takahashi, H., Tsuji, S., Troncoso, J., Dawson, V.L., et al. (2001). Interference by huntingtin and atrophin-1 with cbp-mediated transcription leading to cellular toxicity. *Science, 291*, 2423–2428.

77. Jiang, H., Nucifora, F.C., Jr., Ross, C.A., DeFranco, D.B. (2003). Cell death triggered by polyglutamine-expanded huntingtin in a neuronal cell line is associated with degradation of CREB-binding protein. *Hum Mol Genet, 12*, 1–12.

78. Balch, W.E., Morimoto, R.I., Dillin, A., Kelly, J.W. (2008). Adapting proteostasis for disease intervention. *Science, 319*, 916–919.

79. Westerheide, S.D., Morimoto, R.I. (2005). Heat shock response modulators as therapeutic tools for diseases of protein conformation. *J Biol Chem, 280*, 33097–33100.

80. Kodali, R., Wetzel, R. (2007). Polymorphism in the intermediates and products of amyloid assembly. *Curr Opin Struct Biol, 17*, 48–57.

81. Bodner, R.A., Outeiro, T.F., Altmann, S., Maxwell, M.M., Cho, S.H., Hyman, B.T., McLean, P.J., Young, A.B., Housman, D.E., Kazantsev, A.G. (2006). Pharmacological promotion of inclusion formation: a therapeutic approach for Huntington's and Parkinson's diseases. *Proc Natl Acad Sci U S A, 103*, 4246–4251.

82. Feng, B.Y., Toyama, B.H., Wille, H., Colby, D.W., Collins, S.R., May, B.C., Prusiner, S.B., Weissman, J., Shoichet, B.K. (2008). Small-molecule aggregates inhibit amyloid polymerization. *Nat Chem Biol, 4*, 197–199.

83. McBride, J.L., Boudreau, R.L., Harper, S.Q., Staber, P.D., Monteys, A.M., Martins, I., Gilmore, B.L., Burstein, H., Peluso, R.W., Polisky, B., et al. (2008). Artificial miRNAs mitigate shRNA-mediated toxicity in the brain: implications for the therapeutic development of RNAi. *Proc Natl Acad Sci U S A, 105*, 5868–5873.

84. Ratovitski, T., Gucek, M., Jiang, H.B., Chighladze, E., Waldron, E., D'Ambola, J., Hou, Z.P., Liang, Y.D., Poirier, M.A., Hirschhorn, R.R., Graham, R., Hayden, M.R., Cole, R.N., Ross, C.A. (2009). Mutant huntingtin N-terminal fragments of specific size mediate aggregation and toxicity in neuronal cells. *J Biol Chem, 284*, 10855–10867.

85. Nagai, Y., Tucker, T., Ren, H., Kenan, D.J., Henderson, B.S., Keene, J.D., Strittmatter, W.J., Burke, J.R. (2000). Inhibition of polyglutamine protein aggregation and cell death by novel peptides identified by phage display screening. *J Biol Chem, 275*, 10437–10442.

86. Sathasivam, K., Lane, A., Legleiter, J., Warley, A., Woodman, B., Finkbeiner, S., Paganetti, P., Muchowski, P. J., Wilson, S., Bates, G. P. (2010). Identical oligomeric and fibrillar structures captured from the brains of R6/2 and knock-in mouse models of Huntington's disease, *Hum Mol Genet, 19,* 65–78.

87. Dougan, L., Li, J.Y., Badilla, C.L., Berne, B.J., Fernandez, J.M. (2009). Single homopolypeptide chains collapse into mechanically rigid conformations, *Proc Nat Acad Sci U S A 106,* 12605–12610.

88. Thakur, A.K., Jayaraman, M., Mishra, R., Thakur, M., Chellgren, V.M., Byeon, I.J., Anjum, D.H., Kodali, R., Creamer, T.P., Conway, J.F., Gronenborn, A.M., Wetzel, R. (2009). Polyglutamine disruption of the huntingtin exon 1 N terminus triggers a complex aggregation mechanism, *Nat Struct Mol Biol, 16,* 380–389.

89. Tam, S., Spiess, C., Auyeung, W., Joachimiak, L., Chen, B., Poirier, M.A., Frydman, J. (2009). The chaperonin TRiC blocks a huntingtin sequence element that promotes the conformational switch to aggregation, *Nat Struct Mol Biol, 16*, 1279–1285.

90. Kim, M.W., Chelliah, Y., Kim, S.W., Otwinowski, Z., Bezprozvanny, I. (2009). Secondary structure of Huntingtin amino-terminal region, *Structure, 17,* 1205–1212.

91. Aiken, C.T., Steffan, J.S., Guerrero, C.M., Khashwji, H., Lukacsovich, T., Simmons, D., Purcell, J.M., Menhaji, K., Zhu, Y.Z., Green, K., Laferla, F., Huang, L., Thompson, L.M., Marsh, J.L. (2009). Phosphorylation of threonine 3: implications for Huntingtin aggregation and neurotoxicity, *J Biol Chem 284,* 29427–29436.

92. Gu, X., Greiner, E.R., Mishra, R., Kodali, R., Osmand, A., Finkbeiner, S., Steffan, J.S., Thompson, L.M., Wetzel, R., Yang, X.W. (2009). Serines 13 and 16 are critical determinants of full-length human mutant Huntingtin induced disease pathogenesis in HD mice, *Neuron, 64,* 828–840.

93. Thompson, L.M., Aiken, C.T., Kaltenbach, L.S., Agrawal, N., Illes, K., Khoshnan, A., Martinez-Vincente, M., Arrasate, M., O'Rourke, J.G., Khashwji, H., Lukacsovich, T., Zhu, Y.Z., Lau, A.L., Massey, A., Hayden, M.R., Zeitlin, S.O., Finkbeiner, S., Green, K.N., Laferla F.M., Bates, G., Huang, L., Patterson, P.H., Lo, D.C., Cuervo, A. M., Marsh, J.L., Steffan, J.S. (2009). IKK phosphorylates Huntingtin and targets it for degradation by the proteasome and lysosome, *J Cell Biol, 187,* 1083–1099.

15

SYSTEMIC AMYLOIDOSES

MARINA RAMIREZ-ALVARADO

Department of Biochemistry and Molecular Biology, Mayo Clinic College of Medicine, Rochester, Minnesota

JOEL N. BUXBAUM

Departments of Molecular and Experimental Medicine and Molecular Integrative Neuroscience, and Skaggs Institute for Chemical Biology, The Scripps Research Institute, La Jolla, California

INTRODUCTION

The amyloidoses are disorders of protein conformation in which the misfolded precursors aggregate to form insoluble structures that displace normal tissue, causing organ dysfunction. Recent studies with cultured cells suggest that some aggregates produce tissue damage by a cytotoxic mechanism as well. The deposits are almost exclusively extracellular, bind the dye Congo Red with apple green birefringence under polarized light, and usually contain a group of nonfibrillar accessory molecules, including serum amyloid P component (SAP), apolipoprotein E, and the heparan sulfate proteoglycan perlecan. They were initially defined pathologically as systemic or localized. The systemic disorders are distinguished from the primarily localized forms by their major sites of deposition being at a distance from the site of synthesis [1].

Historically, the amyloidoses represent the first examples of disorders of protein conformation, although at the time of their original description the notion of diseases related to protein misfolding was far in the future. The demonstration that the apparent amorphous Congophilic tissue deposits had a

Protein Misfolding Diseases: Current and Emerging Principles and Therapies,
Edited by Marina Ramirez-Alvarado, Jeffery W. Kelly, and Christopher M. Dobson
Copyright © 2010 John Wiley & Sons, Inc.

well-defined physical structure, as defined by electron microscopy, was an important observation [2]. Two major technical advances conditioned our current conception of this class of diseases. First, the development of methods allowing quantitative extraction of the tissue fibrils with subsequent chemical analysis allowed the definitive demonstration that there were many amyloid precursors. Some were mutant forms of normal proteins, while others were composed of wild-type molecules [3–5].

The application of recombinant technology to produce large quantities of relatively pure proteins allowed the detailed examination of the biophysical properties of the precursors and the processes of aggregation and fibril formation. The process of in vitro fibrillogenesis has been analyzed for many of the molecules that form fibrils in vivo, providing a large body of information that has led to insights into protein structure, the development of clinical diagnostic tools, and in some cases compounds with therapeutic promise for one or more of the disorders.

Nonetheless, major questions remain, most of which require a combination of in vitro and in vivo studies to provide insight. For example, what determines the apparent tissue specificity of most misfolding diseases? Why are so many of the disorders neurodegenerative? Why are some systemic rather than local? How do misfolded proteins, or proteins with the capacity to misfold, exit the cell? What biological processes affect the tendency of a given amyloidogenic protein to deposit? Do tissue deposits have an off-rate? What are the roles of the nonfibrillar proteins that are consistently found co-deposited with fibrils? What is (are) the mechanism(s) of tissue damage? Is it only displacement of normal cellular or tissue structures, or something more actively destructive?

SYSTEMIC VS. NEURODEGENERATIVE AMYLOIDOSIS: PROTEIN SECRETION AND SITE OF DEPOSITION

The pathologically defined local neurodegenerative amyloidoses may be extracellular or intracellular [6]. Alzheimer disease and the prionoses are clearly extracellular and display the characteristic tinctorial properties of amyloid. It is currently unclear whether the initial steps in aggregation in Alzheimer disease occur intra- or extracellularly. Some laboratories have demonstrated intracellular aggregates, but it is possible that the aggregation occurs initially in the perineuronal space as the Aβ peptides are generated and aggregate rapidly and are then taken up by neuronal endocytosis. In either case, tissue culture studies suggest that it is the oligomeric aggregates rather than the fibrils per se that are neurotoxic.

In the prionoses, aggregation appears to initiate while prion protein (PrP) is tethered to the cell membrane via its phosphatidyl-inositol glycan (gpi) anchor, topographically constraining the biological effects of the aggregates. When animals are made transgenic for PrP lacking the gpi-anchoring sequence (i.e., anchorless PrP), the protein aggregates deposit systemically rather than producing spongiform change in the central nervous system (CNS) [7]. When the anchorless proteins are injected into conventional PrP mice, the new animals show the characteristic pattern of CNS aggregation and pathology, indicating that infectivity does not depend on the anchor, although neurotoxicity does.

When deposition occurs intracellularly, except for the paired helical filaments (phosphorylated tau) seen in Alzheimer's disease, the deposits are generally non-Congophilic. They are not associated with serum amyloid P component (SAP), ApoE, or perlecan. The intracellular disorders are largely neurodegenerative, although there are examples of intracellular cardiac and skeletal myopathies related to mutant and wild-type fibril-forming proteins, called *desminopathies* [8]. We believe that the major difference between the intracellular neurodegenerative and systemic amyloidoses rests with the cell type producing the precursor. In the systemic disorders such as light-chain amyloidosis (AL) and the transthyretin (TTR) familial amyloidotic polyneuropathies (FAP), the immunoglobulin light chains and TTR are synthesized and secreted by cells (plasma cells for AL and hepatocytes for FAP) which have evolved to secrete large quantities of proteins. These proteins have an organismal rather than a cell maintenance function. Since the cells synthesize and secrete large amounts of proteins, even if a portion of the protein is misfolded and retained within the secretory pathway to refold by the action of chaperones or be degraded by endoplasmic reticulum–associated degradation (ERAD), enough protein is secreted to serve the needs of the organism. Detailed studies of the unfolded protein response (UPR) in plasma cells, the β-cells of the pancreatic islets and hepatocytes have shown that each secretory cell has its own variation on the UPR theme, as well as having substantial capacity to produce chaperones and the elements of the proteasome system.

In contrast, neurons produce proteins primarily for their own functional needs, or short-lived relatively small peptides packaged in vesicles for synaptic transmission, and may not be as well equipped to deal with even a small misfolded fraction of their total protein output. The greater sensitivity of neurons to heat stress may reflect this aspect of neuronal physiology. Even a small amount of misfolded amyloid-prone precursor may overwhelm the neuron's compensatory mechanisms with intracellular aggregation and primarily local deposition. Two neurodegenerative disorders, Huntington and Parkinson diseases, both display microscopically visible intracellular aggregates (inclusion bodies) which generally do not possess the in vivo properties of amyloid fibrils described above (i.e., Congo Red binding with birefringence, ultrastructural fibrils, and the associated accessory proteins in the pathologic lesions).

Inclusion bodies provide a clear morphological marker of aggregation and whether the disease is the result of a pathological gain of function (increased aggregation potential), loss of function, or both. Nonetheless, when the apparent precursors are produced as recombinant proteins and incubated under amyloid-forming conditions in vitro, they form aggregates and fibrils with amyloid properties. Hence, the proteins can be considered amyloid precursors, but the disorders do not fit the classical pathological definition of an amyloidosis. If the processes leading to the formation of aggregates are eventually determined to be similar or identical in both sets of diseases, it will be appropriate to refer to all of these disorders as amyloidoses with differing morphological features rather than requiring that all the amyloidoses have identical histopathologies. If the processes are identical or represent variations on a common theme, one might speculate on why they differ. It is well known that in vitro amyloid fibril formation requires a specific critical concentration of precursor to progress from oligomer to fibril. It is possible that in the intracellular neurodegenerative disorders the concentration of prefibrillar oligomers is sufficiently cytotoxic to kill the cell prior to the development of mature fibrils.

SOURCES OF PROTEIN AND SECRETION

Secretory Cells and Amyloidosis

Not all differentiated cell types have the same secretory capacity. Van Anken and Braakman have reported that the mammalian endoplasmic reticulum UPR is very diverse, allowing activation of different subsets of downstream effectors, in particular during the development of a particular tissue. The IreI pathway seems to be very important during the development of secretory tissues [9]. The Ire1a and XBP-1 transcription factors are essential for liver development. The most extensively studied example of secretory cell development is the maturation of quiescent B lymphocytes into mature plasma cells. The volume of ER cisternae expands at least threefold to accommodate the bulk biosynthesis of immunoglobulin molecules. XBP-1$^{-/-}$B cells can secrete antibody only minimally. Transfection of the XBP-1 gene alone can trigger B-cell differentiation; however, the characteristic feature of induction of the UPR in these cells involves a specific splice in the XBP-1 transcript generating a molecule that can enhance transcription of downstream targets.

Studies of the secretion and folding of various clinically relevant mutants of misfolding-prone protein (i.e., TTR) by cultured cells of different lineages have shown that it is a combination of the energetics of folding and the proteostatic and secretory capacities of the cell that determines the efficiency of secretion. Only the most destabilized TTR variants are subjected to ER-associated degradation (ERAD), and then only in certain tissues [10]. Interestingly, the most destabilized mutants, measured by their low thermodynamic stability, are the mutants that are commonly associated with a clinical phenotype reflecting local leptomeningeal deposition and less pronounced systemic tissue involvement, suggesting that hepatocytes may be better able to handle the misfolding-prone cargo.

Folding, Secretion, and Degradation: A Mathematical Model

Wiseman and co-workers have proposed a model in which they have described the interdependency of global protein folding energetics and the adaptable biology of cellular folding and membrane trafficking [11]. Their model suggests that export of any particular protein is significantly less stringent than has been viewed conventionally. Its folding energetics are a function of its structure, while its trafficking, export, and/or degradation in a given cell depend on the availability of sites in the trafficking, export, and degradation apparatus and the energy of binding of the protein with those sites compared with the thermodynamics and kinetics of folding. The idea is based on the previous report by the Kelly and Balch laboratories [10], in which folding stability scores were calculated for each TTR variant, correlating with protein export. The model suggests that fast folders will be exported out of the cell, while slow folders will be degraded. This is largely independent of the protein stability and can be adjusted to the ratio between export and translocation machineries found in a particular tissue [11].

Applying these theoretical models to the TTR and AL protein secretion, both proteins are made predominantly by cells considered professional secretory cells (hepatocytes for TTR and plasma cells for free light chains in AL). It is entirely possible that misfolded- prone TTR and free light-chain AL variants are secreted because secretory cells are naturally biased toward having a more dominant export machinery and TTR and free light chains fold fast enough to be able to be secreted, regardless of their thermodynamic stability. How can we explain the TTR deposition found in the choroid plexus and leptomeningeal cells and peripheral nerves? There are at least two plausible explanations: The high levels of thyroxine, a natural TTR ligand, act as a chemical chaperone increasing the stability of TTR in the choroid plexus, allowing the molecule to be secreted. The other possibility is that TTR is not only made in the choroid plexus, but is synthesized in neurons as well. If this is the case, FAPs would actually be examples of localized disease. Several studies have shown staining of neurons with anti-TTR antibodies [12,13].

As mentioned above, Trifilo and co-workers explored the extraneural manifestations of scrapie-infected transgenic mice expressing the prion protein lacking the gpi membrane anchor. The "anchorless" protein was found in the brain, blood, and heart in the abnormal prion protease resistant form (PrPres). More important, the hearts of these transgenic mice contained PrPres-positive amyloid deposits that led to myocardial stiffness and cardiac disease [7]. This provocative work invites the reader to assess the real reasons behind the differences between localized amyloidosis that occurs in the central nervous system and the classic systemic amyloidosis that causes remote organ damage.

CAUSES OF AMYLOID DEPOSITION

Overproduction of the Protein: AL (Clone) and AA (Inflammatory Response)

In AL, an abnormal expansion of a monoclonal plasma cell population results in high levels of free light chains in the circulation. The concentration of free monoclonal light chains in the plasma and urine has been used for the last seven years to follow the progression of the disease. It has been shown that reduction of the absolute levels of free light chain (reflecting diminution in the precursor producing clone size) is considered a good prognostic marker after treatment [14]. However, it is now clear that a more detailed understanding of the physicochemical characteristics of the free light chains is required to explain subpopulations of patients where 50% reduction of free light chain does not translate to an organ response after treatment (Nelson Leung, unpublished observations).

The systemic amyloidosis initially defined, related to overproduction of a specific protein was amyloidosis A (AA), which occurs secondary to chronic inflammation. It is characterized by the tissue deposition of AA fibrils, a truncated form of the acute-phase reactant serum amyloid A (SAA), a protein apparently involved in macrophage lipid handling during the course of inflammation. AA is a serious complication of certain chronic inflammatory diseases, such as tuberculosis, leprosy, rheumatoid arthritis, and some parasitic diseases and is more common in parts of the world where chronic infection is prevalent. The cause of the disorder is the chronic inflammation that causes chronic elevation of SAA [15]. Some genetically defined isoforms of the protein are more amyloidogenic, and it is carriers of those isoforms that are predisposed to form tissue deposition–prone aggregates, with levels of SAA that do not deposit in allele-negative persons. Again, amyloidogenesis appears to be a function of qualitative (propensity of the protein to misfold) and quantitative (inflammation-induced increased production) factors and the homeostatic processes that are required to cope with them: in this case, the quantitatively insufficient proteolytic degradation of the nonfibrillogenic precursor to yield a fibrillogenic intermediate.

Under Excreted Proteins: β_2-Microglobulin

There is only one example in which it is clear that an under excreted protein is responsible for systemic amyloidoses: β_2-microglobulin (β_2m) in chronic renal failure. Dialysis-related amyloidosis (DRA) is characterized by deposition of β_2m, the light chain of the major histocompatibility complex class I molecule in the soft tissues surrounding joints. This MHC polypeptide is shed continually from the cell surface and removed from circulation by the kidney. Patients with end-stage renal disease do not have the capacity to remove β_2m, leading to a 60-fold increase in the serum concentration. This increase in the concentration in addition to some conformational changes that have been triggered by low pH and the presence of some cofactors (e.g., metals, glycosaminoglycans, serum amyloid P component, trifluorinated alcohols, sodium dodecyl sulfate) promote β_2m amyloid formation [16]. Chapter 16 offers a more comprehensive review of DRA.

In AL, the accumulation of free light chains occurs initially because of overproduction of the amyloidogenic precursor by the clonal population of plasma cells in the bone marrow. With time, however, kidney function is frequently compromised by the light-chain amyloid deposits. Since the kidney is the major site of light-chain degradation and excretion, loss of functional renal parenchyma reduces light-chain catabolism, increasing light-chain half-life and serum concentration, adding to the total body light-chain burden, thus amplifying the tendency to tissue deposition [17].

Mutations: AL, TTR, Lysozyme, ApoAI, ApoAII, Fib, Gel, and Cys

AL is distinguished from the systemic amyloidoses that reflect destabilizing mutations in protein structure in being an acquired disorder. The structural changes that lead to amyloidogenicity are due to somatic hypermutation that occurs in clonal B-cell populations during immunodifferentiation and is retained in the expanded clone of plasma cells responsible for the disease. The destabilizing nature of mutations seems to be the general characteristic of most hereditary mutations in systemic amyloidoses. In addition, it is generally accepted that somatic mutations present in amyloidogenic light-chain proteins cause loss of stability, allowing the protein to sample partially folded states that promote misfolding, with some interesting exceptions. The effect of somatic mutation on the thermodynamic stability for AL proteins has been studied extensively [18–22]. These studies have shown that not all of the somatic mutations found in AL proteins are destabilizing and that, in fact, some mutations may be protective of the integrity of the protein structure [19,22]. Raffen and co-workers determined that amyloidogenic proteins SMA and REC were less stable than the multiple myeloma control LEN and only specific amyloid-related destabilizing mutations from SMA and REC introduced in the multiple myeloma protein LEN caused amyloid fibrils in vitro [20]. Hurle and co-workers analyzed the effect of single amyloidogenic mutations (from a database search of amyloidogenic protein sequences) incorporated in the multiple myeloma protein REI. Their results suggest that different mutations have different effects on the protein, either causing large destabilization or having neutral or even protective stabilizing effects [18]. Additional mutational studies of different AL proteins are needed to identify more precisely protein regions or the nature of mutations that are destabilizing and amyloidogenic.

The single mutations found in TTR are probably the best characterized in terms of their stability, dissociation constants, and effects in amyloid formation. These mutations are listed on several Web sites and their thermodynamic stability and tetramer dissociation constants have been reported [10,23]: http://www.bumc.bu.edu/Dept/Content.aspx?DepartmentID = 354& PageID = 8850; http://www.ibmc.up.pt/mjsaraiva/ttrmut.html; www.ibmc.up.pt/mjsaraiva/ttmut.html; http://www.iupui.edu/~amyloid/ttrmutaions.htm. For a more comprehensive review of TTR, consult Chapter 45.

In the case of lysozyme-associated systemic amyloidoses, five variants have been reported, four of which are associated with systemic amyloidoses affecting the kidney, liver, and spleen. The fifth variant is nonamyloidogenic and is common within the British population [24]. Two amyloidogenic variants (I56T and D67H) have been characterized extensively using biophysical methods. These variants are less stable than wild-type lysozyme while maintaining the enzymatic activity and three-dimensional structure of the wild-type protein [25].

Apolipoprotein A-I (ApoAI) is the major protein component of high-density lipoproteins (HDL). ApoAI has several functions: It serves as a scaffold of HDL particles, as a cofactor for the enzyme lecithin:cholesterol acyltransferase (LCAT), which is essential for the reverse cholesterol transport, and it interacts with membranes as a requirement for cholesterol efflux, hepatic uptake, and interactions with other lipoprotein particles [26]. Apolipoprotein A-I hereditary amyloidosis is characterized by progressive deposition of amyloid fibrils consisting of N-terminal polypeptide fragments of the protein. The amyloid deposits predominantly affect the kidney, heart, and the liver, causing either progressive nephropathy or cardiomyopathy. In general, mutant ApoAI appears to be associated with reduced total HDL levels and changes in HDL subclasses that favor and lead to extravascular deposition of mutant protein. Eleven amyloidogenic ApoAI variants have been identified, most of them appearing in only one kindred. G26R has been found in four unrelated families of different ethnic backgrounds. A deletion–insertion mutant (Δ60–71 InsValThr) from a large Spanish kindred presents severe hepatopathy [27]. W50R has been found in a person with liver and renal involvement. The mutation R173P is present in persons with cutaneous, laryngeal, and cardiac amyloid with low circulating HDL levels [28]. Other laryngeal-related mutations are proline substitutions at residues 90 and 175 and histidine 178. Deletion of lysine at residue 107 has been associated with increased susceptibility to amyloid deposition in the aortic intima and ischemic heart disease [29]. Some ApoAI mutants have been associated with an imbalance between wild-type and mutant species in total circulation: G26R, L60R, and L174S, suggesting abnormal metabolism of the mutant compared to the wild-type protein [30]. To date, no thermodynamic or amyloid formation kinetics experiments have been conducted for human ApoAI mutants.

Other hereditary amyloidoses are generally rare and affect small numbers of kindreds discovered in various parts of the world. The L68Q variant of human cystatin C (hCC) is the causative agent of hereditary cystatin C amyloid angiopathy, where repeated hemorrhage, dementia, paralysis, and death are associated with cystatin deposits in cerebral blood vessels. It is believed that domain swapping causes the rearrangements in hCC that initiate amyloid formation [31].

Human gelsolin is a member of a large family of actin-modulating proteins. The amyloid deposits associated with familial amyloidosis of Finnish type (FAF) are found in the cornea, facial nerve, peripheral nerves, and skin. These deposits are composed of a proteolytic fragment or fragments of gelsolin mutant D187N or D187Y. Gelsolin amyloidogenesis in FAF requires furin proteolysis of the mutants to initiate formation of the amyloidogenic 71- or 53-residue peptides, followed by the proteolytic activity of MT1-matrix metalloprotease [32,33].

The Aα chain of fibrinogen has mutants that cause renal amyloidosis. Four amyloidogenic mutations have been described in eight unrelated kindreds. The most common mutation is E526V [29]. A different amyloidogenic variant consists of a deletion frameshift mutation of 49 residues whose N-terminal 23 amino acids were identical to residues 499 to 521 of normal fibrinogen Aα chain. The remainder of the peptide (26 residues) represented a completely new sequence for mammalian proteins. DNA sequencing documented that the new sequence was the result of a single nucleotide deletion at position 4897 of the fibrinogen Aα-chain gene, which gives a frameshift at codon 522 and premature termination at codon 548 [34]. Another deletion frame shift mutation involves a single nucleotide deletion at the third base of codon 524 of the fibrinogen Aα-chain genes (4904delG) that resulted in a frameshift and premature termination of the protein at codon 548 [35]. The remaining mutation involves a leucine for arginine substitution at position 554 [29]. To date, no thermodynamic or amyloid formation kinetics experiments have been conducted for fibrinogen Aα-chain mutants.

Truncations: AA, AL, ApoAI

Another cause of amyloidogenicity involves truncations in the precursor protein. These truncations are well characterized in AA and AL. In the case of AA, both human and mouse AA fibrils purified from amyloidotic tissues are heterogeneous in size (5.5 to 8 kDa) and represent the N-terminal two-thirds of the 12-kDa A-SAA molecule [36]. The complete SAA is rarely found in the amyloid deposits. In vitro, under partly denaturing conditions (low pH), both the full-length human recombinant A-SAA1 and the N-terminus 12-amino acid hydrophobic fragment corresponding to the HDL-binding portion are able to form fibrils that are structurally similar to the native AA amyloid deposits.

AL amyloid fibrils are composed of the N-terminal immunoglobulin light chain variable domain (VL) in most cases. The last β-strand in the VL, β strand G, possesses a proline residue that promotes a "kink," exposing the end of this β-strand away from the rest of the domain. This strand is then followed by a hinge region that connects with the constant domain (CL), in which there are small hydrophobic residues (L, V, A), positive residues (K, R), as well as more proline residues. This region is very susceptible to proteolytic cleavage. Protein truncation before amyloid formation provides an attractive means by which some AL proteins are destabilized, resulting in amyloid formation, although it is possible that the full-length light chain is involved in the fibril formation and the VL forms the core of the fibril. Subsequent proteolysis of the decorating CL could leave the VL as the only portion of the light chain in the fibril.

ApoAI-induced amyloid fibrils contain N-terminal fragments of the protein in addition to the point mutations described earlier in this section [26]. It is possible that a similar mechanism to the one proposed for AL is occurring in ApoAI.

Extension of Polypeptide: ApoAII and Bri

A peptide can also become amyloidogenic by extending beyond its normal C-terminus. Two examples of this type of event are described below.

Apolipoprotein AII (ApoAII) is the second most abundant apolipoprotein in plasma high-density lipoprotein (HDL), after ApoAI. ApoAII amyloidosis has been found in patients with hereditary renal amyloidosis. In each case, amyloid deposition has been associated with a peptide extension at the carboxy terminus of apoAII, as a result of the mutation in the normal stop codon (STOP78R, STOP78G, or STOP78S) [37–39].

A mouse strain with a point mutation in ApoAII (R1.P1-*Apoa2*^{c}) has been found to develop late-onset systemic amyloidosis related to the deposition of this protein. Interestingly, it is one of the two forms of murine amyloidosis that can be accelerated by either systemic or oral administration of sonicated preformed fibrils [40]. Data also suggest that natural seeding may occur when cagemates ingest fecal material from affected animals [41].

Familial British dementia (FBD) and familial Danish dementia (FDD) are considered cerebral amyloid angiopathies (CAA) characterized by ocular hemorrhages, cataracts, hearing loss, and cerebral ataxia followed by dementia. In a British family, a mutation of the stop codon has been replaced by an arginine, extending the reading frame of peptide BRI to yield a furin-processed 34-residue peptide called A-Bri, 11 residues longer than the wild type. In a Danish family, a 10-base duplication insertion before the stop codon also yields a 34-residue peptide (A-dan). The amyloid deposits are found in blood vessels; therefore, we have considered it a systemic amyloidosis.

Mutational Diversity of AL: Many Ways to Get There

As mentioned previously, AL is a disease with a large degree of mutational variability. The Ramirez-Alvarado laboratory has mapped the different regions of mutational diversity for over 20 VL domains from AL proteins [42]. They have classified mutational regions critical for the overall stability of the VL domain into four groups:

1. *Mutations distributed throughout the structure in the top and/or bottom of the immunoglobulin Greek key β-barrel, affecting the overall stability of the domain.*
2. *Mutations located in the N-terminus and/or C-terminus* β*-strands.* Loss of interactions between these two strands could enhance conformational flexibility in this region of the protein. Interestingly, O'Nuallain and co-workers have described the cryptic epitope recognized by an AL fibril-specific antibody to be part of the N-terminus strand, suggesting that this region may play a role in the initial conformational changes observed in AL amyloid formation [43].
3. *Mutations located in the* β*-hairpin formed by strands D and E.* It was reported previously that peptides with a sequence corresponding to this

region of the protein inhibit amyloid formation in vitro [21]. Increased flexibility of this 'hinge' region may cause a conformational change that exposes hydrophobic regions of the domain that could trigger amyloid formation.

4. *Mutations present in the dimer interface where heavy and light chains dimerize (strands C, C′, F,* and *G*). Mutations in the external face of the antiparallel β-sheet corresponding to the heavy chain/light chain interface may impair light-chain dimerization with the heavy chain. We have recently reported that mutations in the dimer interface cause the protein to adopt an altered dimer interface that makes the protein less stable and more amyloidogenic [44].

Role of Native Multimerization in Amyloidogenesis

Qin and co-workers have recently reported that the amyloidogenic light-chain VL SMA has a significant propensity to self-associate to form dimers with a dimerization constant of 40 μM, reflecting the strong structural homology between light- and heavy-chain variable domains, which form a tight dimer. SMA in its dimeric form is more thermodynamically stable than is its monomeric form. Moreover, the rate of amyloid formation for SMA is inversely dependent on protein concentration, in contrast to most amyloid systems. This suggests that dimerization inhibits amyloid formation [45]

The mutations in the dimer interface and their effect on light-chain dimer stability are consistent with the observation that tetramer dissociation to a misfolded monomer is typically rate limiting for TTR misassembly and leads to the formation of amyloid aggregates. All of the TTR mutations associated with familial disease studied so far decrease either the thermodynamic or kinetic stability of the tetramer, increasing the rate of dissociation and thus of amyloidogenesis. Screening and structure-based design have identified six distinct classes of small molecules that bind to the TTR tetramer and stabilize the native state, preventing amyloid formation and abrogating programmed cell death in a cell culture system[46–48].

In conclusion, multimeric native subunits or ligand-binding events act as a form of co-chaperoning for potentially amyloidogenic proteins. Mutations, truncations, and extensions of amyloidogenic polypeptide appear to lower the critical concentration of protein required to initiate misfolding, because wild-type, full-length proteins are capable of forming amyloid fibrils with a later age of onset (in the case of TTR).

SYSTEMIC DEPOSITION: TISSUE TARGETING

AL, ALys, AApoI, AApoII, ATTR, and AFib are all associated with renal amyloidosis, perhaps reflecting a physicochemical environment that is particularly conducive to protein aggregation. Amyloid is found in all compartments of the kidney, with the glomerulus being a primary site of fibril deposition [49,50]. The kidneys filter waste from blood and excrete it together with water as urine. When glomeruli are damaged, more protein may be present in the filtrate. During the passage of the glomerular filtrate down the tubules, the protein concentration of the filtrate is increased. In a similar fashion, the concentration of urea, a product of protein catabolism, is also increased (0.4 to 1.5 M) in the inner renal medulla, especially during antidiuresis [50]. Urea is a protein denaturant that causes unfolding by interacting with the peptide bond, and therefore exposure to urea is expected to increase the population of misfolding-prone partially folded states. As a result of this effect, it is possible that a light-chain variant with an intrinsic thermodynamic stability that is just sufficient to avoid amyloid fibril deposition in other areas of the body might form fibrils in the kidney [50]. Kim and co-workers report that in vitro, urea accelerates amyloid formation and decreases thermodynamic stability, while the renal osmolytes betaine and sorbitol have the opposite effect. It is worth noting that the concentrations of betaine and sorbitol in the kidney are about 10 times lower than the concentration of urea (10 to 50 mmol/kg for betaine and sorbitol; 200 to 1200 mmol/kg for urea) [51], and the studies by Kim and co-workers were done using 0.5 M betaine and sorbitol and up to 1.5 M urea. It is possible that the denaturing effect of urea in the kidney is dominant over the stabilizing effects of the osmolytes. Similar rigorous studies should be performed with other proteins that cause renal amyloidosis. More studies dissecting the effects of osmolytes on protein stability and tissue targeting are necessary. AL could be the best model for this study. It would be very interesting to see if there are any differences between light chains that deposit primarily in the kidney versus the light chains that deposit in the heart, a phenomenon that has been suggested by some analyses of light-chain VL region types [52–54]. It is possible that light chains associated with cardiac amyloidosis may be less thermodynamically stable than light chains involved in renal amyloidosis. Another way to pose this hypothesis is that the light chains that misfold and deposit in the heart are actually less sensitive to the chemical environment than are light chains that are deposited in kidneys.

Functional differences between light chains that cause different forms of renal deposition disease have been demonstrated by Keeling et al. [55]. They have shown that AL and light chain deposition disease (LCDD) light chains induce divergent phenotypical transformations of human mesangial cells grown in tissue culture. This report suggests that the human mesangial cells respond differently to light chains from different light-chain diseases and supports the notion discussed above regarding the susceptibility of the various organ environments. Future studies in which sequence determinants of the susceptibility in different organ environments involving both human mesangial cells and mammalian cardiomyocytes would be important to establish the role of specific somatic mutations in the behavior of the light chains and the effects in cell culture. However, it must be noted that there are several reported instances in which deposits of different types are seen in different tissues of the same patient (i.e., amyloid in vessels and the amorphous deposits of light-chain deposition disease in the kidney). In at least one of these persons, the amino acid sequence of the protein from both sites was identical [56], reinforcing the notion that tissue location is important in determining some aspects of aggregation and deposition. Moreover, these studies could be extended to other proteins involved in renal amyloidosis, such as lysozyme, ApoAI, ApoAII, and fibrinogen α chain. It is possible that we will observe differences in the response from these different tissue culture systems that will support the notion that some specific cell types may be better equipped than others to deal with misfolded states.

Another important aspect of tissue targeting is the observation that deposition of almost all the amyloid occurs initially in blood vessels, usually between the endothelium and the smooth muscle cell layer. The propensity for blood vessel deposition is not understood, other than that the precursors transit the vascular system continuously. Further studies are necessary to elucidate the role of endothelial cells in amyloidogenesis.

In the case of AA, chronic inflammation increases SAA gene transcription through the effects of IL-6, IL-1, and TNFα on the hepatocyte acting on HNF3α and HNF-6. Lipid carrying SAA is internalized by macrophages [57] and presumably hepatic Kupffer cells, the lipid is released, and the protein is cleaved. It is not clear whether the normal pathway is for complete cleavage to small peptides and excretion, and whether generation of the amyloidogenic AA fragment represents a relative failure of that process dealing with the amyloidogenic isoforms of SAA [58,59]. There is also evidence that SAA may be synthesized in other cells, notably cells involved in the inflammatory response. Tissue culture studies have shown that mixed cultures of hepatocytes and macrophages can generate AA fibrils.

Mechanisms of Protein Internalization

Teng and co-workers have reported that LCs that cause AL or LCDD interact with small invaginations of the plasma membrane in human renal mesangial cells with ultrastructural features of caveolae [60]. Both AL-LC and LCDD-LCs compete for the same receptor, co-localize in the plasma membrane of human renal mesangial cells, and cross-link with filamin and talin, both components of caveolin-1. Receptor-mediated internalization is suggested by these authors, although the receptor was not identified. Monis et al. performed internalization experiments which suggested that the amyloidogenic light chains are not internalized using a membrane receptor but by an active mechanism such as constitutive pinocytosis into an endosomal–lysosomal pathway mediated by microtubules [61]. Interestingly, there was a significant difference between the internalization of truncated, monomeric, and dimeric light chains. Truncated and monomeric light chains are easily internalized, whereas dimeric light chains are rarely internalized under the conditions used to conduct the experiments. It is worth mentioning that the nomenclature of monomeric and dimeric is derived from mass spectrometry data, where they are assessing the presence of covalently bound light chains. No association experiments were reported for these proteins, although the concentrations used for the experiments (1.39 μM) are close to the concentrations reported by Qin et al. for a monomer SMA VL (5 μM) [45]. Monis et al. suggest that different cell types, such as rat cardiac fibroblasts and mesangial cells, may have different modes of internalization for AL proteins [61].

Given the different modes of internalization reported for human mesangial cells and rat cardiac fibroblasts, more experiments using different amyloidogenic proteins are needed. Moreover, it would be important to couple these comparison studies with accurate determination of the oligomerization state of the starting protein used in the experiments because the oligomerization state of the protein may determine the mode of internalization. This appears quite clear from tissue culture studies of the toxicity induced by TTR or Aβ peptides [47,62].

Extracellular Chaperones: Clusterin

Clusterin or apolipoprotein J is a 449-amino acid secreted heterodimeric disulfide-linked glycoprotein that acts like a small heat-shock protein chaperone and is involved in cell death regulation in response to oxidative stress injury. The mature heterodimeric clusterin is secreted and is found in all human body fluids. Clusterin cellular uptake and degradation is mediated by a member of the low-density lipoprotein receptor gene family, the endocytic receptor gp330/megalin [63]. Clusterin interacts with an impressive array of unrelated ligands, including membrane receptors, bacterial proteins, heparin, other apolipoproteins, IgGs, and amyloid precursor proteins [64], as well as partially folded stressed proteins [65,66]. It is possible that clusterin may be assisting the cellular internalization of amyloid precursor proteins via megalin or other membrane receptors that it binds. Cell-culture studies investigating the role of clusterin in the internalization of amyloid precursor proteins in systemic amyloidosis, using different types of cellular systems (e.g., renal mesangial cells, cardiomyocytes, hepatocytes) would increase our understanding of the factors contributing to the internalization of amyloidogenic proteins in target tissues.

MECHANISMS OF ORGAN DAMAGE

Are the Toxic Events that Cause Organ Damage Common to All Systemic Amyloidosis Precursors?

The effects of amyloidogenic protein precursors, oligomeric species, and amyloid fibrils have been studied for Aβ and ATTR [47,48,62]. Reixach and co-workers have conducted a cell culture–based analysis of the effect of purified TTR species on the viability of the cells in culture [47]. Their findings indicate that neither

TTR fibrils nor soluble aggregates larger than 100 kDa were toxic to the cells. The initial event triggering the formation of cytotoxic species appears to be dissociation of the native tetramer into monomeric species, with the toxic form being either the misfolded monomer or dimer.

More recently, a *Drosophila* model for ATTR has been developed [67]. The authors used flies with different copy numbers of TTR transgenes and showed that flies expressing high protein levels had a larger proportion of high-molecular-mass aggregates that are assumed to be less toxic. The temperature-dependent phenotypes observed in this invertebrate model also correlated with lower protein expression. The authors hypothesize that there is an optimal concentration, specific to each mutated variant of TTR, that determines the rate of toxic aggregate formation and consequently its effect on the phenotype, a notion consistent with the earlier studies in cultured cells.

For AL, all published reports of the effect of AL proteins on cell culture have been conducted without defining the oligomeric state of the proteins [55,61,68,69]. Even though the proteins were assumed to be in a "soluble" state, no information is provided about the oligomerization state of these proteins, the storage conditions before the experiments were conducted, or the removal of preformed aggregates from stock solutions using filtration or ultracentrifugation. Recent reports using physiological levels of amyloidogenic light chains from patients with amyloid cardiomyopathy incubated in the presence of cardiomyocytes suggest that the presence of the "soluble" light chains altered the cellular redox state in isolated cardiomyocytes, with an increase in intracellular reactive oxygen species and up-regulation of the redox-sensitive heme oxygenase-1. The oxidant stress imposed by the light chains resulted further in the impairment of cardiomyocyte contractility and relaxation associated with alterations in intracellular calcium handling [68].

Keeling et al. have compared the effects of amyloidogenic light chains and light chains from light-chain deposition disease patients on cultured human kidney mesangial cells and amyloid deposition surrounding the cells [55]. The cells affected by amyloidogenic light chains show suppression of smooth muscle actin due to the loss of myofilaments and overexpression of CD68, which results from the acquisition of lysosomes, resulting in a macrophage-like phenotypic transformation. They suggest that endocytosis and catabolism of amyloidogenic light chains in the acquired lysosomes, as has been proposed by Shirahama and Cohen, is a key step in the formation of amyloid in the kidney [70,71]. Studies of the catabolism of human light chain in kidney show that the light chains are degraded predominantly in the lysosomal compartments and that a polymerization step seems to be part of the process [72]. A follow-up study showed that amyloid fibrils are formed by Bence Jones proteins that have been incubated in the presence of normal kidney lysosomal enzymes at low pH [73].

Cohen et al., using spleen cultures with casein-induced amyloidosis, showed that the amyloidogenic protein (SAA) is taken up by the cells before the formation of extracellular amyloid [74]. Takahashi et al. reported that in the casein-amyloid-enhancing factor-induced amyloidosis model, the amyloid fibrils were found extracellularly and in some cytoplasmic invaginations of macrophages and concluded that some amyloid fibrils are polymerized in the cytoplasm of the macrophages by the proteolytic cleavage of previously pinocytized SAA protein [75]. These observations have been supported by more recent tissue-culture studies [58]. In summary, these earlier reports show that while amyloidogenic proteins are internalized into cells, there is no support for acquisition of lysosomes in any of these systems.

Trinkaus-Randall et al. investigated the cellular response of primary cardiac fibroblasts to internalized amyloidogenic light chains. The presence of amyloidogenic light chains induced sulfation of the secreted glycosaminoglycans, which was associated with the translocation of heparan sulfate into the nucleus [69]. How this is related to either deposition or the cytotoxic activity of amyloidogenic light chains is not clear. A systematic study such as the one conducted for ATTR is needed to clarify if indeed there is a common mechanism of organ damage by small oligomeric species in AL and ATTR.

It is clear that we have learned much about the systemic amyloidoses by using reductionist in vitro and tissue culture systems. Nonetheless, by their varied nature, the systemic amyloidoses are organismal disorders and require more detailed analyses in that context to achieve full understanding. It is possible that analyses of the responses to therapy will give insights that can be extended by more reductionist experimental systems. For example, the fact that domino recipients of livers from persons with TTR mutations develop peripheral TTR deposits with the same mutant proteins as the donors has suggested that while the liver does not appear to be affected by TTR deposits, it may not be as capable of dealing with misfolded TTR precursors as it is earlier in life. More and better animal models would be helpful for both understanding of pathogenesis and testing of new therapies.

Acknowledgments

M.R.-A. has been supported by National Institutes of Health grant GM071514, American Heart Association grant AHA 06-30077N, and the Mayo Foundation. J.N.B. has received support from the National Institutes of Health (AG019259, AG030027), the W.M. Keck Foundation, and FoldRx Pharmaceutials during the preparation of this review.

REFERENCES

1. Westermark, P., Benson, M.D., Buxbaum, J.N., Cohen, A.S., Frangione, B., Ikeda, S., Masters, C.L., Merlini, G., Saraiva, M.J., Sipe, J.D. (2007). A primer of amyloid nomenclature. *Amyloid, 14*, 179–183.
2. Cohen, A.S., Calkins, E. (1959). Electron microscopic observations on a fibrous component in amyloid of diverse origins. *Nature, 183*, 1202–1203.
3. Benditt, E.P., Lagunoff, D., Eriksen, N., Iseri, O.A. (1962). Amyloid extraction and preliminary characterization of some proteins. *Arch Pathol, 74*, 323–330.

4. Glenner, G.G., Bladen, H.A. (1966). Purification and reconstitution of the periodic fibril and unit structure of human amyloid. *Science, 154*, 271–272.
5. Pras, M., Schubert, M., Zucker-Franklin, D., Rimon, A., Franklin, E.C. (1968). The characterization of soluble amyloid prepared in water. *J Clin Invest, 47*, 924–933.
6. Buxbaum, J.N. (2003). Diseases of protein conformation: What do in vitro experiments tell us about in vivo diseases? *Trends Biochem Sci, 28*, 585–592.
7. Trifilo, M.J., Yajima, T., Gu, Y., Dalton, N., Peterson, K.L., Race, R.E., Meade-White, K., Portis, J.L., Masliah, E., Knowlton, K.U., et al. (2006). Prion-induced amyloid heart disease with high blood infectivity in transgenic mice. *Science, 313*, 94–97.
8. Maloyan, A., Gulick, J., Glabe, C.G., Kayed, R., Robbins, J. (2007). Exercise reverses preamyloid oligomer and prolongs survival in alphaB-crystallin-based desmin-related cardiomyopathy. *Proc Natl Acad Sci U S A, 104*, 5995–6000.
9. van Anken, E., Braakman, I. (2005). Endoplasmic reticulum stress and the making of a professional secretory cell. *Crit Rev Biochem Mol Biol, 40*, 269–283.
10. Sekijima, Y., Wiseman, R.L., Matteson, J., Hammarstrom, P., Miller, S.R., Sawkar, A.R., Balch, W.E., Kelly, J.W. (2005). The biological and chemical basis for tissue-selective amyloid disease. *Cell, 121*, 73–85.
11. Wiseman, R.L., Powers, E.T., Buxbaum, J.N., Kelly, J.W., Balch, W.E. (2007). An adaptable standard for protein export from the endoplasmic reticulum. *Cell, 131*, 809–821.
12. Schwarzman, A.L., Tsiper, M., Wente, H., Wang, A., Vitek, M.P., Vasiliev, V., Goldgaber, D. (2004). Amyloidogenic and anti-amyloidogenic properties of recombinant transthyretin variants. *Amyloid, 11*, 1–9.
13. Stein, T.D., Johnson, J.A. (2002). Lack of neurodegeneration in transgenic mice overexpressing mutant amyloid precursor protein is associated with increased levels of transthyretin and the activation of cell survival pathways. *J Neurosci, 22*, 7380–7388.
14. Dispenzieri, A., Lacy, M.Q., Katzmann, J.A., Rajkumar, S.V., Abraham, R.S., Hayman, S.R., Kumar, S.K., Clark, R., Kyle, R.A., Litzow, M.R., et al. (2006). Absolute values of immunoglobulin free light chains are prognostic in patients with primary systemic amyloidosis undergoing peripheral blood stem cell transplantation. *Blood, 107*, 3378–3383.
15. Ali-Khan, Z. (2007). *Serum amyloid A and AA amyloidosis.* In *Protein Misfolding, Aggregation, and Conformational Diseases*, Part B, *Molecular Mechanisms of Conformational Diseases* (Uversky, V.N., Fink, A.L., Eds.), Springer, New York.
16. Morten, I.J. (2007). *β_2-microglobulin and dialysis related amyloidosis*, In *Protein Misfolding, Aggregation, and Conformational Diseases* (Uversky, V.N., Fink, A.L., Eds.), Springer, New York.
17. Solomon, A., Waldmann, T.A., Fahey, J.L., McFarlane, A.S. (1964). Metabolism of Bence Jones proteins. *J Clin Invest, 43*, 103–117.
18. Hurle, M.R., Helms, L.R., Li, L., Chan, W., Wetzel, R. (1994). A role for destabilizing amino acid replacements in light-chain amyloidosis. *Proc Natl Acad Sci U S A, 91*, 5446–5450.
19. Wall, J., Schell, M., Murphy, C., Hrncic, R., Stevens, F.J., Solomon, A. (1999). Thermodynamic instability of human lambda 6 light chains: correlation with fibrillogenicity. *Biochemistry, 38*, 14101–14108.

20. Raffen, R., Dieckman, L.J., Szpunar, M., Wunschl, C., Pokkuluri, P.R., Dave, P., Wilkins Stevens, P., Cai, X., Schiffer, M., Stevens, F.J. (1999). Physicochemical consequences of amino acid variations that contribute to fibril formation by immunoglobulin light chains. *Protein Sci, 8*, 509–517.
21. Davis, P.D., Raffen, R., Dul, L.J., Vogen, M.S., Williamson, K.E., Stevens, J.F., Argon, Y. (2000). Inhibition of amyloid fiber assembly by both BiP and its target peptide. *Immunity, 13*, 433–442.
22. Wall, J., Gupta, V., Wilkerson, M., Schell, M., Loris, R., Adams, P., Solomon, A., Stevens, F.J., Dealwis, C. (2004). Structural basis of light chain amyloidogenicity: comparison of thermodynamic properties, fibrillogenic potential and tertiary structural features of four Vl6 proteins. *J Mol Recog, 17*, 323–331.
23. Buxbaum, J.N. (2004). The systemic amyloidoses. *Curr Opin Rheumatol, 16*, 67–75.
24. Merlini, G., Bellotti, V. (2005). Lysozyme: a paradigmatic molecule for the investigation of protein structure, function and misfolding. *Clin Chim Acta, 357*, 168–172.
25. Booth, D.R., Sunde, M., Bellotti, V., Robinson, C.V., Hutchinson, W.L., Fraser, P.E., Hawkins, P.N., Dobson, C.M., Radford, S.E., Blake, C.C., Pepys, M.B. (1997). Instability, unfolding and aggregation of human lysozyme variants underlying amyloid fibrillogenesis. *Nature, 385*, 787–793.
26. Genschel, J., Haas, R., Propsting, M.J., Schmidt, H.H. (1998). Apolipoprotein A-I induced amyloidosis. *FEBS Lett, 430*, 145–149.
27. Booth, D.R., Tan, S.Y., Booth, S.E., Tennent, G.A., Hutchinson, W.L., Hsuan, J.J., Totty, N.F., Truong, O., Soutar, A.K., Hawkins, P.N., et al. (1996). Hereditary hepatic and systemic amyloidosis caused by a new deletion/insertion mutation in the apolipoprotein AI gene. *J Clin Invest, 97*, 2714–2721.
28. Buxbaum, J.N., Tagoe, C.E. (2000). The genetics of the amyloidoses. *Annu Rev Med, 51*, 543–569.
29. Hawkins, P.N. (2003). Hereditary systemic amyloidosis with renal involvement. *J Nephrol, 16*, 443–448.
30. Merlini, G., Westermark, P. (2004). The systemic amyloidoses: clearer understanding of the molecular mechanisms offers hope for more effective therapies. *J Intern Med, 255*, 159–178.
31. Staniforth, R.A., Giannini, S., Higgins, L.D., Conroy, M.J., Hounslow, A.M., Jerala, R., Craven, C.J., Waltho, J.P. (2001). Three-dimensional domain swapping in the folded and molten-globule states of cystatins, an amyloid-forming structural superfamily. *EMBO J, 20*, 4774–4781.
32. Huff, M.E., Balch, W.E., Kelly, J.W. (2003). Pathological and functional amyloid formation orchestrated by the secretory pathway. *Curr Opin Struct Biol, 13*, 674–682.
33. Page, L.J., Suk, J.Y., Huff, M.E., Lim, H.J., Venable, J., Yates, J., Kelly, J.W., Balch, W.E. (2005). Metalloendoprotease cleavage triggers gelsolin amyloidogenesis. *EMBO J, 24*, 4124–4132.
34. Hamidi Asl, L., Liepnieks, J.J., Uemichi, T., Rebibou, J.M., Justrabo, E., Droz, D., Mousson, C., Chalopin, J.M., Benson, M.D., Delpech, M., Grateau, G. (1997). Renal amyloidosis with a frame shift mutation in fibrinogen alpha-chain gene producing a novel amyloid protein. *Blood, 90*, 4799–4805.

35. Uemichi, T., Liepnieks, J.J., Yamada, T., Gertz, M.A., Bang, N., Benson, M.D. (1996). A frame shift mutation in the fibrinogen A alpha chain gene in a kindred with renal amyloidosis. *Blood, 87*, 4197–4203.

36. Rocken, C., Shakespeare, A. (2002). Pathology, diagnosis and pathogenesis of AA amyloidosis. *Virchows Arch, 440*, 111–122.

37. Benson, M.D., Liepnieks, J.J., Yazaki, M., Yamashita, T., Hamidi Asl, K., Guenther, B., Kluve-Beckerman, B. (2001). A new human hereditary amyloidosis: the result of a stop-codon mutation in the apolipoprotein AII gene. *Genomics, 72*, 272–277.

38. Yazaki, M., Liepnieks, J.J., Yamashita, T., Guenther, B., Skinner, M., Benson, M.D. (2001). Renal amyloidosis caused by a novel stop-codon mutation in the apolipoprotein A-II gene. *Kidney Int, 60*, 1658–1665.

39. Yazaki, M., Liepnieks, J.J., Barats, M.S., Cohen, A.H., Benson, M.D. (2003). Hereditary systemic amyloidosis associated with a new apolipoprotein AII stop codon mutation Stop78Arg. *Kidney Int, 64*, 11–16.

40. Fu, X., Korenaga, T., Fu, L., Xing, Y., Guo, Z., Matsushita, T., Hosokawa, M., Naiki, H., Baba, S., Kawata, Y., et al. (2004). Induction of AApoAII amyloidosis by various heterogeneous amyloid fibrils. *FEBS Lett, 563*, 179–184.

41. Xing, Y., Nakamura, A., Chiba, T., Kogishi, K., Matsushita, T., Li, F., Guo, Z., Hosokawa, M., Mori, M., Higuchi, K. (2001). Transmission of mouse senile amyloidosis. *Lab Invest, 81*, 493–499.

42. Abraham, R.S., Geyer, S.M., Ramirez-Alvarado, M., Price-Troska, T.L., Gertz, M.A., Fonseca, R. (2004). Analysis of somatic hypermutation and antigen selection in the clonal B cell in immunoglobulin light chain amyloidosis (AL). *J Clin Immunal, 24*, 340–353.

43. O'Nuallain, B., Allen, A., Kennel, S.J., Weiss, D.T., Solomon, A., Wall, J.S. (2007). Localization of a conformational epitope common to non-native and fibrillar immunoglobulin light chains. *Biochemistry, 46*, 1240–1247.

44. Baden, E.M., Owen, B.A., Peterson, F.C., Volkman, B.F., Ramirez-Alvarado, M., Thompson, J. R. (2008). Altered dimer interface decreases stability in an amyloidogenic protein *J Biol Chem*, 283, 15853–15860.

45. Qin, Z., Hu, D., Zhu, M., Fink, A.L. (2007). Structural characterization of the partially folded intermediates of an immunoglobulin light chain leading to amyloid fibrillation and amorphous aggregation. *Biochemistry, 46*, 3521–3531.

46. Cohen, F.E., Kelly, J.W. (2003). Therapeutic approaches to protein-misfolding diseases. *Nature*, *426*, 905–909.

47. Reixach, N., Deechongkit, S., Jiang, X., Kelly, J.W., Buxbaum, J.N. (2004). Tissue damage in the amyloidoses: transthyretin monomers and nonnative oligomers are the major cytotoxic species in tissue culture. *Proc Natl Acad Sci U S A, 101*, 2817–2822.

48. Reixach, N., Adamski-Werner, S.L., Kelly, J.W., Koziol, J., Buxbaum, J.N. (2006). Cell based screening of inhibitors of transthyretin aggregation. *Biochem Biophys Res Commun, 348*, 889–897.

49. Kyle, R.A. (1994). Monoclonal proteins and renal disease. *Annu Rev Med, 45*, 71–77.

50. Kim, Y.S., Cape, S.P., Chi, E., Raffen, R., Wilkins-Stevens, P., Stevens, F.J., Manning, M.C., Randolph, T.W., Solomon, A., Carpenter, J.F. (2001).

Counteracting effects of renal solutes on amyloid fibril formation by immunoglobulin light chains. *J Biol Chem, 276*, 1626–1633.

51. Garcia-Perez, A., Burg, M.B. (1991). Renal medullary organic osmolytes. *Physiol Rev, 71*, 1081–1115.
52. Comenzo, R.L., Wally, J., Kica, G., Murray, J., Ericsson, T., Skinner, M., Zhang, Y. (1999). Clonal immunoglobulin light chain variable region germline gene use in AL amyloidosis: association with dominant amyloid-related organ involvement and survival after stem cell transplantation. *Br J Haematol, 106*, 744–751.
53. Comenzo, R.L., Zhang, Y., Martinez, C., Osman, K., Herrera, G.A. (2001). The tropism of organ involvement in primary systemic amyloidosis: contributions of Ig V(L) germ line gene use and clonal plasma cell burden. *Blood, 98*, 714–720.
54. Abraham, R.S., Geyer, S.M., Price-Troska, T.L., Allmer, C., Kyle, R.A., Gertz, M.A., Fonseca, R. (2003). Immunoglobulin light chain variable (V) region genes influence clinical presentation and outcome in light chain–associated amyloidosis (AL). *Blood, 101*, 3801–3808.
55. Keeling, J., Teng, J., Herrera, G.A. (2004). AL-amyloidosis and light-chain deposition disease light chains induce divergent phenotypic transformations of human mesangial cells. *Lab Invest, 84*, 1322–1338.
56. Kaplan, B., Vidal, R., Kumar, A., Ghiso, J., Frangione, B., Gallo, G. (1997). Amino-terminal identity of co-existent amyloid and non-amyloid immunoglobulin kappa light chain deposits: a human disease to study alterations of protein conformation. *Clin Exp Immunol, 110*, 472–478.
57. Kluve-Beckerman, B., Manaloor, J., Liepnieks, J.J. (2001). Binding, trafficking and accumulation of serum amyloid A in peritoneal macrophages. *Scand J Immunol, 53*, 393–400.
58. Kluve-Beckerman, B., Manaloor, J.J., Liepnieks, J.J. (2002). A pulse-chase study tracking the conversion of macrophage-endocytosed serum amyloid A into extracellular amyloid. *Arthritis Rheum, 46*, 1905–1913.
59. Higuchi, K., Kitagawa, K., Naiki, H., Hanada, K., Hosokawa, M., Takeda, T. (1991). Polymorphism of apolipoprotein A-II (apoA-II) among inbred strains of mice: relationship between the molecular type of apoA-II and mouse senile amyloidosis. *Biochem J, 279*(Pt 2), 427–433.
60. Teng, J., Russell, W.J., Gu, X., Cardelli, J., Jones, M.L., Herrera, G.A. (2004). Different types of glomerulopathic light chains interact with mesangial cells using a common receptor but exhibit different intracellular trafficking patterns. *Lab Invest, 84*, 440–451.
61. Monis, G.F., Schultz, C., Ren, R., Eberhard, J., Costello, C., Connors, L., Skinner, M., Trinkaus-Randall, V. (2006). Role of endocytic inhibitory drugs on internalization of amyloidogenic light chains by cardiac fibroblasts. *Am J Pathol, 169*, 1939–1952.
62. Hartley, D.M., Walsh, D.M., Ye, C.P., Diehl, T., Vasquez, S., Vassilev, P.M., Teplow, D.B., Selkoe, D.J. (1999). Protofibrillar intermediates of amyloid beta-protein induce acute electrophysiological changes and progressive neurotoxicity in cortical neurons. *J Neurosci, 19*, 8876–8884.
63. Trougakos, I.P., Gonos, E.S. (2006). Regulation of clusterin/apolipoprotein J, a functional homologue to the small heat shock proteins, by oxidative stress in ageing and age-related diseases. *Free Radic Res, 40*, 1324–1334.

64. Matsubara, E., Soto, C., Governale, S., Frangione, B., Ghiso, J. (1996). Apolipoprotein J and Alzheimer's amyloid beta solubility. *Biochem J, 316*(Pt 2), 671–679.

65. Poon, S., Easterbrook-Smith, S.B., Rybchyn, M.S., Carver, J.A., Wilson, M.R. (2000). Clusterin is an ATP-independent chaperone with very broad substrate specificity that stabilizes stressed proteins in a folding-competent state. *Biochemistry, 39*, 15953–15960.

66. Lakins, J.N., Poon, S., Easterbrook-Smith, S.B., Carver, J.A., Tenniswood, M.P., Wilson, M.R. (2002). Evidence that clusterin has discrete chaperone and ligand binding sites. *Biochemistry, 41*, 282–291.

67. Pokrzywa, M., Dacklin, I., Hultmark, D., Lundgren, E. (2007). Misfolded transthyretin causes behavioral changes in a *Drosophila* model for transthyretin-associated amyloidosis. *Eur J Neurosci, 26*, 913–924.

68. Brenner, D.A., Jain, M., Pimentel, D.R., Wang, B., Connors, L.H., Skinner, M., Apstein, C.S., Liao, R. (2004). Human amyloidogenic light chains directly impair cardiomyocyte function through an increase in cellular oxidant stress. *Circ Res, 94*, 1008–1010.

69. Trinkaus-Randall, V., Walsh, M.T., Steeves, S., Monis, G., Connors, L.H., Skinner, M. (2005). Cellular response of cardiac fibroblasts to amyloidogenic light chains. *Am J Pathol, 166*, 197–208.

70. Shirahama, T., Cohen, A.S. (1975). Intralysosomal formation of amyloid fibrils *Am J Pathol, 81*, 101–116.

71. Shirahama, T., Cohen, A.S. (1973). An analysis of the close relationship of lysosomes to early deposits of amyloid: ultrastructural evidence in experimental mouse amyloidosis. *Am J Pathol*, 73, 97–114.

72. Tan, M., Epstein, W. (1972). Polymer formation during the degradation of human light chain and Bence-Jones proteins by an extract of the lysosomal fraction of normal human kidney. *Immunochemistry, 9*, 9–16.

73. Epstein, W.V., Tan, M., Wood, I.S. (1974). Formation of "amyloid" fibrils in vitro by action of human kidney lysosomal enzymes on Bence Jones proteins. *J Lab Clin Med, 84*, 107–110.

74. Cohen, A.S., Gross, E., Shirahama, T. (1965). The light and electron microscopic autoradiographic demonstration of local amyloid formation in spleen explants. *Am J Pathol, 47*, 1079–1111.

75. Takahashi, M., Yokota, T., Kawano, H., Gondo, T., Ishihara, T., Uchino, F. (1989). Ultrastructural evidence for intracellular formation of amyloid fibrils in macrophages. *Virchows Arch A, 415*, 411–419.

16

HEMODIALYSIS-RELATED AMYLOIDOSIS

David P. Smith, Alison E. Ashcroft, and Sheena E. Radford
Astbury Centre for Structural Molecular Biology, University of Leeds, Leeds, UK

INTRODUCTION

Hemodialysis-related amyloidosis (HDRA) is a serious complication of long-term hemodialysis in which beta 2-microglobulin (β_2m) has been identified as the major component of the amyloid fibrils that typically deposit in the joints of these patients [1]. The disease occurs in patients undergoing all types of renal replacement therapy, as well as uremic predialysis patients, suggesting that dialysis itself is not the primary cause of the disease [2]. Eventually, the disease leads to carpal tunnel syndrome, osteoarthropathy of peripheral joints, severe morbidity, and in rare cases, even mortality [2]. In this chapter we discuss the pathology and etiology of HDRA, the structural biology of the key causative agent, β_2m, and its amyloidogenic characteristics, and possible points for therapeutic intervention.

β_2M STRUCTURE AND FUNCTION IN VIVO

Human β_2m is a small, single-domain, nonpolymorphic, nonglycosylated protein of 99 amino acids [3]. It is the light chain of the class I human leucocyte antigen (HLA), a key cell surface protein that binds peptide in the endoplasmic reticulum lumen for presentation on the cell surface to cytotoxic T cells. X-ray crystallography and nuclear magnetic resonance (NMR) studies of β_2m both

Protein Misfolding Diseases: Current and Emerging Principles and Therapies,
Edited by Marina Ramirez-Alvarado, Jeffery W. Kelly, and Christopher M. Dobson
Copyright © 2010 John Wiley & Sons, Inc.

bound to the HLA heavy chain and free in solution [3–5] reveal a β-sandwich fold typical of the Ig family (Fig. 1). The protein consists of two antiparallel β-sheets of three and four strands; (A, B, D, E) and (C, F, G), respectively, held together by short loops. The native fold is stabilized by a single buried disulfide bond between Cys25 and Cys80 in strands B and F. In addition, the protein contains a short (two-residue) C' β-strand that is located in the loop connecting strands C and D. The fourth β-strand (D), which lies at one edge of the β-sandwich, is divided into two short two-residue β-strands (D1 and D2) by a three-residue β-bulge in HLA-bound β_2m, as well as in the monomeric protein in solution, although conformational changes occur in this region in some crystal forms of the β_2m monomer [3–7].

In vivo, β_2m is shed continuously from the surface of cells displaying class I HLA molecules. The synthesis of β_2m ranges from 2 to 4 mg/kg per day, and the protein has a half-life of 2.5 hours, with plasma concentrations varying between 2.4 and 3.7 mg/L in healthy persons [8]. Circulating β_2m exists as a monomer and is distributed throughout the extracellular space. Approximately 95% of this free β_2m is eliminated via glomerular filtration, tubular reabsorption, and subsequent intracellular proteolysis [2,9]. The remaining approximately 5% of

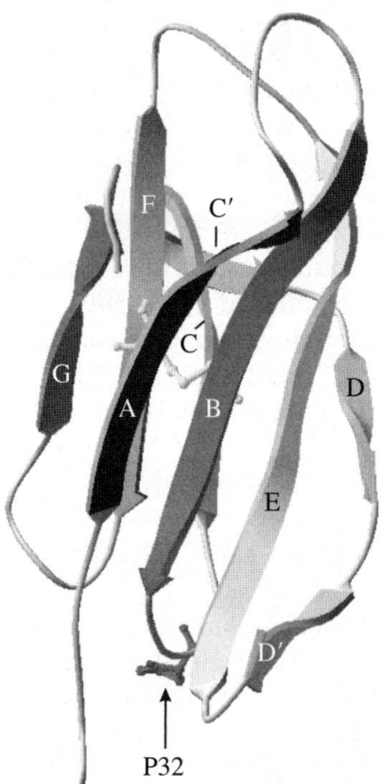

FIG. 1 Ribbon diagram of the crystal structure of β_2m taken from the class 1 HLA complex (1DUZ) [3]. Secondary structural elements are labeled A through G. The side chain of the *cis*-proline (P32) is shown in ball-and-stick representation.

β_2m is removed by a currently unknown mechanism. During renal failure the positive effects of glomerular filtration are effectively removed, increasing the half-life of β_2m in the serum to 3.6 to 5.9 days (Fig. 2) [8]. This results in an increase in β_2m concentration in the serum by up to 25- to 60-fold, depending on the degree of residual renal function [2,10], leading to the deposition of the full-length wild-type protein into amyloid plaques. The retention of β_2m in these plaques that typically, but not exclusively, accumulate in the joints, has been

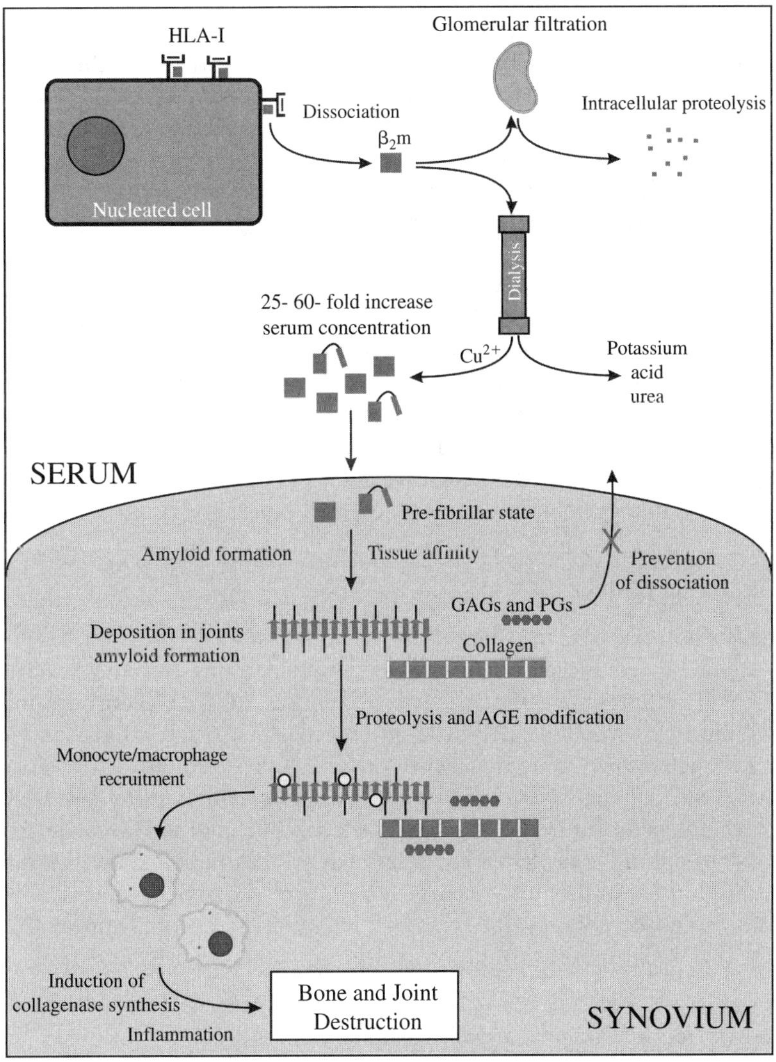

FIG. 2 Events involved in the development of HDRA. Note that the precise details of the structure of the precursor(s) are unknown, as are the details of the amyloid fibrils that form.

shown to be the main pathogenic process underlying β_2m amyloid formation [2,11]. However, there is no apparent link between plasma concentration and the extent and severity of HDRA [12], suggesting that other systemic and/or local factors could be involved in the initiation of the disease process.

CONSTITUENTS OF β_2M AMYLOID DEPOSITS

Amyloid material extracted from long-term HDRA patients is comprised of long nonbranching fibrils with a characteristic morphology, often associated with collagen fibrils (Fig. 2) [13,14]. The major constituent of these amyloid deposits is full-length wild-type β_2m [14,15]. However, both fragmentation and chemical modifications have been observed in fibrils purified from plaques of long-term hemodialysis patients. Linke et al. (1986) first observed limited proteolysis in amyloid derived from the kidney stones of uraemic patients [15]. A range of truncated species of β_2m were observed displaying cleavages at positions 6, 10, or 19. This observation was confirmed in a range of further studies, with the 7–99 fragment (ΔN6) accounting for approximately 20 to 30% of the total β_2m in ex vivo fibrils [15,16]. The loss of the C-terminal Met99 or the C-terminal 13 residues has also been reported [14]. β_2m has also been shown to be cleaved C-terminally to Lys58 by activated complement component C1s [17]. However, this cleavage is not found in ex vivo amyloid material [18], although Lys58 cleavage does increase the ability of β_2m to form amyloid in vitro [19]. In healthy persons this cleavage event is thought to induce cytotoxic activity in lymphocytes, as a response to HLA binding [17].

Chemical Modification

Acidic isoforms of β_2m in which Asn17 and/or Asn43 are deamidated to Asp have been reported as constituents of β_2m amyloid, along with the oxidation of Met99 [14]. Analysis of the in vitro properties of β_2m incorporating the deamidation of Asn17 revealed little difference in its structural and amyloidogenic characteristics compared with the wild-type protein [20]. Advanced glycation end products (AGEs) have also been detected within β_2m amyloid. These include pentosidine, N^{ε}-(carboxymethyl)lysine, and imidazolone [21] and arise due to oxidative or "carbonyl stress" in HDRA patients. AGE modification is not limited to β_2m and occurs ubiquitously in dialysis patients. Amyloid deposits are observable several years before clinical symptoms appear [22] and suggest that AGE modification occurs after amyloid deposition [23].

Serum Amyloid P Component and Apolipoprotein E

Serum amyloid P component (SAP) is an ever-present constituent of all types of amyloid [24]. SAP is a member of the pentraxin family and is composed of five identical 25-kDa subunits in a pentameric ring. It binds noncovalently to amyloid in a calcium-dependent manner and renders the fibrils resistant to proteolytic degradation [24]. SAP has been found associated with β_2m amyloid deposits in HDRA patients [14,25]. In addition, β_2m fibrils can bind to, and are stabilized by, apolipoprotein E (apoE), a cholesterol transport protein [26,27]. α2-Macroglobulin has also been identified by biochemical/immunohistochemical means in β_2m amyloid deposits [28]. However, the surrounding tissue also stained positive for these antiproteases, and hence it is unclear whether they play a role in the persistence of amyloid fibrils [2].

Proteoglycans and Collagen

The tissue-specific deposition, along with a known intransigence of pure β_2m to form fibrils in vitro at neutral pH in the absence of preformed fibrillar seeds [1,29], suggests a role for factors within cartilage in enhancing amyloid formation from this protein. β_2m Amyloid fibrils are closely associated with extracellular matrix components, including glycosaminoglycans (GAGs), proteoglycans (PGs), and collagen (Fig. 2) [30–32]. In vitro assays indicate that GAGs and PGs bind and stabilize β_2m fibrils, suggesting that these factors may promote fibril deposition by acting as a scaffold for deposition [26,33], by stabilizing fibrils once formed [34–36], or by enhancing polymerization [36]. Heparin-stabilized seeds have also been shown to promote fibril formation from monomeric β_2m at neutral pH via the stabilization of amyloid material formed from recombinant β_2m at pH 2.5 [34,37]. Nevertheless, heparin is known to have little effect on the structure or stability of monomeric β_2m [34]. Stabilization of the fibril seed with GAGs and PGs, along with partial unfolding of the monomer by sodium dodecyl sulfate (SDS), also allows fibril formation to proceed at neutral pH [38]. The addition of the GAGs chrondroitin-4 or 6-sulfate to in vitro fibril growth assays results in the spontaneous generation of amyloidlike fibrils in the absence of seeds from ΔN6 [36]. Fibril formation is not observed over the same time course in the presence of hyaluronic acid, a nonsulfated GAG that is also abundant in cartilaginous joints [36]. β_2m also shows an affinity for collagen type I with a K_d value of 0.41 mM and type II with a K_d value of 2.3 mM [39], as well as AGE-modified collagen [32,40], and once bound to collagen may convert into amyloid fibrils facilitated by other components, such as GAGs and PGs. Amyloid deposits do not occur in the shafts of long bones, indicating that the synovial joint is a necessary prerequisite for deposition, possibly due to the abundance of type II collagen fibrils therein [41]. Type I collagen has also been shown to facilitate β_2m fibril formation in vitro, and the positively charged regions along the collagen fiber may play a direct role in fibrillogenesis [13]. It appears, therefore, that factors such as sulfated GAGs, PGs, and collagen are crucially involved in amyloid formation in HDRA. These molecules may act by stabilizing preformed amyloid fibrils, preventing their dissociation by allowing fibril extension to occur [25,34,40,42,43] and possibly even promoting the initial self-assembly steps of fibril formation and deposition.

Diagnosis and Clinical Manifestations

Although HDRA is a systemic disorder, β_2m's high affinity for collagen leads to predominantly osteoarticular deposition [44]. Deposits can be identified histologically after only a few months of dialysis treatment and often precede clinical manifestations by several years [22]. Carpal tunnel syndrome is linked inherently to the clinical signs of β_2m amyloidosis. Symptoms are caused by the entrapment of the median nerve at the wrist due to the deposition of β_2m amyloid in the carpal tunnel, and treatment usually requires surgical release [2]. The surface inflammation of both bones and joints results in stiffness of large and medium-size joints and arises from the accumulation of deposits in articular cartilage, synovial membrane, villi, tendons, and subchondral bone [2]. β_2m at concentrations of 0.3 to 30 μg/mL is known to induce collagenase synthesis in human fibroblasts; it can also act as a growth factor on calvarial cell cultures and is known to release metallic metalloproteinase-1 from synovial fibroblasts [45]. Several of these properties are related to the induction of interleukin-6 release by β_2m, resulting in inflammation and pathological phase of β_2m amyloidosis (Fig. 2).

Role of Macrophages in β_2m Amyloid Deposition and Pathology

The infiltration of macrophages from peripheral blood into areas rich in collagen is a histological characteristic of HDRA and has been implicated in fibril deposition, osteoarthropathy of the joints, and inflammation [23,46,47]. Infiltration occurs in the advanced stages of amyloidosis and may play a key role in turning clinically silent deposits into symptomatic entities (Fig. 2). AGE modification of β_2m can partially account for macrophage infiltration by the induction of inflammatory cytokines [32,48]. The binding to AGE receptors induces pro-inflammatory cytokines such as interleukin-1β and tumor necrosis factor-α to be produced [49]. AGE-modified β_2m can also induce the production of all types of transforming growth factors (TGFβ1–3) [50]. It is becoming clear that the infiltration of macrophages [51] and subsequent inflammation are a direct result of AGE modification, which occurs secondary to fibril deposition [23].

Macrophages also express a number of genes implicated in amyloid production [52], such as serum amyloid A_3 and amyloid-enhancing factor [23]. β_2m Amyloid has been observed directly within macrophage lysosomes [53], raising the possibility that uptake of the protein into lysosomes may be involved in the development of β_2m amyloidosis. Recent live-cell imaging experiments have demonstrated that macrophages internalize monomeric β_2m, whereupon it is sorted to lysosomes. At lysosomal pH, β_2m self-associates in vitro to form amyloidlike fibrils with an array of morphologies [54,55].

However, cleavage of the monomeric protein by lysosomal proteases isolated from macrophages results in rapid degradation of the monomeric protein, preventing amyloid formation [55]. Incubation of macrophages with preformed fibrils revealed that macrophages readily internalize amyloid fibrils formed extracellularly, but in marked contrast with the monomeric protein, the fibrils are not degraded within macrophage lysosomes [55]. It appears therefore that the internalization of β_2m by macrophages may represent a secondary event helping to explain the known proteolytic cleavage of β_2m in amyloid deposits. The role of macrophages in osteoarthropathy of the joints and inflammation in the symptomatic stages of HDRA is now becoming clear, and the prevention of these autoimmune responses could prove beneficial in the treatment of patients suffering from HDRA.

MECHANISM OF FIBRIL FORMATION OF β_2M

The formation of amyloidlike fibrils from β_2m in vitro has been studied extensively by a number of groups following its discovery as the primary causative agent of HDRA in 1985 [1,15,56]. Experiments have been performed under a wide range of conditions to try to recapitulate amyloid formation in vitro; however, it has proved very difficult to form amyloid from pure β_2m at neutral pH (Table 1). Given the intrinsic difficulty of forming β_2m fibrils in vitro, alternative methods of studying β_2m fibril growth, by extending fibrils purified from ex vivo extracts were developed [57]. The rate of fibril extension was found to be maximal at pH 2.5 and dependent on both the temperature and concentrations of seed and monomer [57]. More recently, the seed-dependent extension of β_2m at pH 2.5 has been tracked by the use of the histological dye, thioflavinT, using total internal reflection fluorescence microscopy [58]. These data showed that extension is mostly unidirectional. The seeds must therefore be polar, with growth rates of 47.3 ± 15.0 nm/min being reported [58]. As with other amyloid-forming systems, population of one or more partially unfolded state(s) has been shown to be critically important for the formation of amyloid fibrils. For β_2m the population of the assembly-competent state can be achieved via acidification of full-length protein, resulting in the de novo formation of amyloid fibrils (i.e., in the absence of seeds), by the addition of metal ions, co-solvents, or detergents, or by mutating the protein sequence (Table 1).

Formation of amyloidlike fibrils from β_2m at neutral pH can be achieved via the addition of amyloid material from HDRA patients [59] or fibrils formed in vitro stabilized by the addition of GAGs or PGs [34,35,37]. The fibrils produced retain the morphology of their parent seeds, reaching an endpoint after a few hours or weeks, depending on the conditions, indicating that under neutral pH conditions, β_2m populates a state capable of fibril elongation. This state is rarely populated, however, and below the critical concentration needed for fibril nucleation [37]. Ultrasonication of β_2m at neutral pH in the presence of 0.5 mM sodium dodecyl sulfate (SDS) results in a marked increase of

TABLE 1 Summary of Conditions Used to Generate Amyloidlike Fibrils from β_2m Under Neutral pH Conditions In Vitro

Fibril Seed	Method	Conditions	Ref.
None	Ultrasonication induced seeds	37°C; without agitation; 50 mM Na phosphate (pH 7.0) containing 0.5 mM SDS and 100 mM NaCl	60
pH 2.5 fibrils	SDS-stabilized seeds	37°C; without agitation; 50 mM Na buffer (pH 7.5), 100 mM NaCl, and various concentrations of SDS (0–10 mM)	38
pH 2.5 fibrils or ex vivo fibrils	Trifluoroethanol <20% (v/v)	37°C; without agitation; 0–90 μg/mL heparin-stabilized or unmodified seeds, 50 mM phosphate buffer (pH 7.5), 100 mM NaCl, 0–30 % TFA	63
None	Heat-induced extension	Initial aggregation of β_2m at 310 rpm in an ITC cell at 37°C over 24 h at pH ~7.3–7.6 in 0–100 mM NaCl, followed by several cycles of heating–cooling over the range 10–100°C at a pressure of 25 psi	64
None	Cu^{2+} induced	37°C; 25 mM phosphate (pH 7.4), 150 mM KCl, 1 M urea, and 200 mM $CuCl_2$	62
None	Collagen induced	40°C; ammonium acetate 12–30 mM, pH 6.4, in the presence of type I collagen	13
pH 2.5 fibrils	Mutation P32G	37°C; with agitation at 250 rpm, 25 mM Na phosphate buffer (pH 7.0), heparin-stabilized seeds	37
None	Mutation I7A, V9A, V93A, R97A	37°C; with agitation at 200 rpm, 25 mM Na phosphate, 25 mM Na acetate buffer at pH 7.0	67
pH 2.5 fibrils	Extension of stabilized seeds	37°C; with agitation at 200 rpm, 25 mM Na phosphate, 25 mM Na acetate Stabilization of pH 2.5 fibrils by the addition of heparin, SAP, apoE, uremic serum, or synovial fluid	34
None	Truncation ΔN6	37°C; with agitation at 200 rpm, 25 mM Na phosphate, 25 mM Na acetate (low yield)	34

TABLE 1 ***(Continued)***

Fibril Seed	Method	Conditions	Ref.
None	Truncation ΔN6	37°C; with agitation at 200 rpm, 137 mM NaCl, 2.7 mM KCl, 10.1 mM Na phosphate, 1.8 mM K phosphate; pH 7.4, heparin, (GAGs) chrondroitin-4 or 6-sulfate at 5 or 0.5 mg/ml	36
Ex vivo fibrils	Truncation ΔN6	37°C; 50 mM phosphate buffer pH 7.0, 100 mM NaCl	66

thioflavinT fluorescence after 1 day, indicating that fibrils or oligomers have formed. These products of sonication caused the acceleration of fibril formation at pH 7.0, forming fibrils with a diameter of about 7 nm [60]. Several cycles of self-seeding under neutral pH conditions using SDS-stabilized seeds produced at pH 2.5 resulted in rapid and extensive fibril growth, suggesting that the maturation of fibrils is an important part of the molecular mechanism of amyloidosis [61]. Shifting the equilibrium toward precursor states resulting in fibril formation has also been achieved by the addition of Cu^{2+} ions [62], trifluoroethanol [63], SDS [38], heat-induced unfolding [64] or mutagenesis [37,65,66]. Specifically, destabilization of the native protein by mutations in the N- or C-terminal strands [67], or the mutation of Pro32 to Ala or Gly, results in an increased propensity for fibril formation [37,65]. The N-terminally truncated ΔN6 variant also has spectral properties similar to those of the amyloidogenic partially unfolded state of wild-type β_2m formed at lower pH [16,66]. ΔN6 is able to rapidly extend preformed fibrils at pH 7.0 and displays increased rates of fibrillogenesis at acidic pH [16,66]. In each of these cases, assembly is thought to proceed by increasing the concentration of one or more key precursor states, resulting in enhanced self-association into fibrillar arrays.

Detailed studies using atomic force microscopy (AFM) and electron microscopy (EM) have defined and characterized a multitude of distinct classes of β_2m fibrils formed in vitro (Fig. 3) [68,69]. The majority conform to the definition of amyloid in that they display Congo Red birefringence, bind thioflavinT, are long unbranched polymers, give rise to a cross-β-ray fiber diffraction pattern, and bind antibodies that recognize generic amyloid epitopes (Fig. 3B, E, and G) [54,70]. The fibril morphology formed is critically dependent on the conditions in which the fibrils are created. However, by careful choice of the assembly environment such as pH, buffer salts, and agitation conditions, β_2m fibrils can be generated in a morphologically homogeneous manner [54]. AFM single partial image analysis and a detailed survey of solution conditions revealed a complex mechanism of assembly involving at least two parallel routes [54]. Incubation of β_2m under acidic conditions (pH 5.0 and below) and high ionic strength (ca. 200 to 400 mM) gives rise to the spontaneous nonnucleated assembly of fibrils with a wormlike

(WL) morphology [29,70,71] (Fig. 3A and C). These fibrils are relatively short (ca. 100 to 600 nm) when formed in high-ionic-strength solutions, yet under low-ionic-strength conditions at the same pH, much shorter rodlike (RL) fibrils are produced. Both the WL and RL fibrils are flexible and appear to be constructed by the juxtaposition of spherical partials 30 nm in length [69]. By contrast, at lower pH (<3.0) and low ionic strength (<50 mM) β_2m assembles via a nucleation-dependent mechanism [54,72], forming long, straight (LS) fibrils with an axial repeat periodicity of 50 to 100 nm, depending on the morphology [69] (Fig. 3D and F) and are more reminiscent of ex vivo material. Similar to other amyloid systems, various intermediate aggregates can be formed in the lag phase of assembly depending on the conditions employed, and include amorphous species, rings, and toroids [69]. A novel fibril form has been reported via a pressure-induced structure reorganization of β_2m amyloid fibrils, before the pressure-induced unfolding occurs [73]. The changes in volume and structure indicate that the fibrils formed under ambient pressure are less tightly packed, with a large number of cavities. An amyloid structure without optimal packing enables various isoforms to form, suggesting the structural basis of multiple forms of amyloid fibrils [69,73]. Marked acceleration of fibril growth was also observed on repeated self-seeding above 300 MPa, revealing the coexistence of alternative fibril types with a similar structure but with an increased growthrate under high pressure [74]. Together these observations indicate that the assembly of β_2m amyloid fibrils occurs on a complex energy landscape involving multiple competing pathways leading to fibrils of different morphologies, depending on the assembly conditions and the initial conformation of the monomer and resulting oligomers.

β_2M FIBRIL STRUCTURE

One of the major unsolved issues in the field of amyloidosis is the atomic-level structure of an amyloid fibril formed from a full-length intact protein. Progress has been made toward determining the structure of β_2m fibrils formed in vitro using a combination of proteolysis [75,76], hydrogen exchange [77,78], and atomic force microscopy [68]. Both WL and LS amyloid fibrils formed *de novo* from β_2m display a cross-β secondary structure [70] and bind SAP [34] (Fig. 3B and E). Antibodies that bind to a common epitope of amyloid fibrils (named WO1 [79]) or to oligomers formed early during the assembly of many different proteins into amyloid fibrils (antioligomer antibodies [80]) have recently been isolated. Consistent with amyloid fibrils formed from other proteins and peptides, the antiamyloid antibody WO1 binds to both WL and LS fibrils of β_2m, while very little binding is found to monomeric β_2m at pH 7 or β_2m unfolded transiently at pH 2.5 [54] (Fig. 3G). However, whilst the antioligomer antibody also binds WL fibrils, only weak binding is found in LS fibrils, suggesting that the epitope recognized by this antibody may be masked in the higher-order assembly. Small structured RNA molecules known as *aptamers*

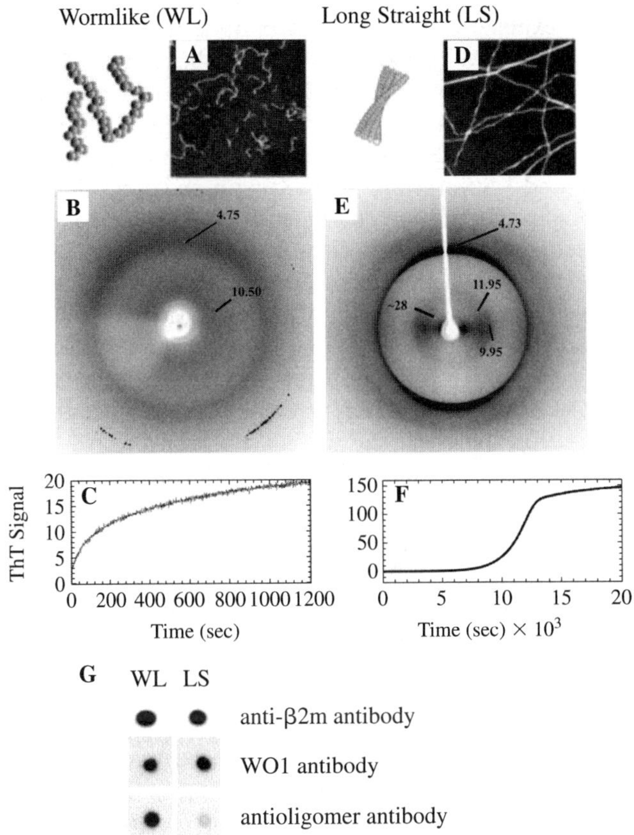

FIG. 3 Characterization of different types of amyloidlike fibrils formed from β_2m in vitro. Wormlike (WL) fibrils formed in vitro at pH 3.6 at an ionic strength of 0.4 M (left-hand panels) and long, straight (LS) fibrils formed at pH 2.5 at an ionic strength of 50 mM (right-hand panels). (A,D) Atomic force microscopy images and cartoon representation of fibrils formed under each condition. The samples were dried onto mica. (B,E) X-ray fiber diffraction image: The major reflections on the meridian and equator are labeled in angstroms. (C,F) Growth kinetics at 37°C monitored by the fluorescence of thioflavin T; (C) at pH 3.6, 0.4 M ionic strength (no agitation) and (F) pH 2.5, 50 mM ionic strength (agitation at 1400 rpm). (G) Dot-blot analysis of the WL and LS fibrils probed using an anti-β_2m antibody (positive control), antibody WO1 [79], and the generic antioligomer antibody A11 [80]. (Adapted from [54,70].)

have also been produced and have been demonstrated to bind to both the acid unfolded state and WL and LS fibrils of β_2m with nanomolar affinity [81]. Specific aptamers directed against the WL and LS fibrils do not bind the acid unfolded state and hence recognize fibril-specific epitopes [81]. The distinct kinetics and avidity of the different fibril forms for the different aptamers implies either that the anti-WL and anti-LS aptamers bind distinct epitopes, or that the solvent exposure of a common epitope differs substantially in the different fibril types. Alternatively, the LS fibrils may display a higher frequency of the epitope per unit length, implying a higher degree of order.

Depolymerization of fibrils formed via the addition of organic solvents such as dimethyl sulfoxide (DMSO) has allowed experiments to be performed in which both WL and LS fibrils are digested with proteases or exposed to D_2O and the patterns of protection are determined. Limited proteolysis data indicate that about 90% of the polypeptide chain is involved in the LS fibril structure (residues 10 to 99), while the protected core involves only about 30 residues in WL fibrils [75]. Residues close to the Cys25–Cys80 disulfide bond are more susceptible to proteolysis in the WL fibrils then in the LS fibrils (Fig. 4, open triangles) [75]. Further proteolytic digests of LS fibrils formed at pH 4.0 by extension of fibril seeds from HDRA patients reveal that both the N- and C-terminal regions are exposed to the solvent with a protected core comprising residues 20 to 87 [76] (Fig. 4, filled triangles). Cleavage sites found at Lys6 and Lys19 in vivo have also been observed in LS and WL fibrils formed at pH 4.0, suggesting a similarity of the fibrils formed in the two environments [75,76]. Fourier transform infrared spectroscopy shows that fibrils formed in vitro from natively folded or unfolded β_2m adopt an identical β-sheet architecture [82]. The same β-strand signature is observed whether fibril formation in vitro occurs spontaneously or from seeded reactions [82]. Comparison of these spectra with those of amyloid fibrils extracted from patients with DRA revealed an identical amide I absorbance maximum, suggestive of a characteristic and conserved amyloid fold between fibrils arising from the acid unfolded β_2m at pH 2.5 or native β_2m at pH 7.0 [82]. The conformation of both the LS and WL β_2m fibrils has been investigated at a residue-specific level by hydrogen exchange [77,78]. Protection factors indicate that the β-strands involved in the core of the native protein, as well as their connecting loops, comprise the structured core of the LS fibril [77,78], in accord with the proteolysis data [75,76] (Fig. 4). By contrast with the LS fibrils, a different hydrogen exchange (HX) protection pattern was obtained for the WL fibrils, with two clusters centered on Leu40 and Phe62 [78]. Residues in the vicinity of the intramolecular disulfide bond in WL fibrils undergo rapid HX exchange consistent with their apparent exposure to solvent in this fibrillar form (Fig. 4) [78].

The difference in subunit organization between the LS and WL fibrils is demonstrated by the redox potential of the disulfide bond, experiments suggesting that this bond is more exposed to solvent in the WL fibrils than in the LS fibrils, such that it can be reduced by dithiothreitol [83,84]. Reduction of the disulfide bond results in a loss in the ability of the monomeric protein to form LS fibrils at low pH, the reduced protein instead forming thinner nodular fibrils under all conditions studied [83–85]. The presence of reductants also prevents amyloid formation under neutral-pH conditions, suggesting that disulfide is exposed to solvent in amyloidogenic intermediates [86]. Further, more recent hydrogen-exchange experiments on the LS fibrils indicated that residues in the central region of the polypeptide chain (strands B to F) are strongly protected from HX exchange, while the residues at the N- and C-termini are less protected [78]. The exchange kinetics reveal large deviations from a single exponential for many residues, suggesting that the same residue exists in different environments from molecule to molecule, possibly even within a single fibril. The introduction of tryptophan residues at various positions throughout β_2m indicated that the central region of the polypeptide chain is buried in the LS fibrils (as monitored by the change in the λ_{max} of Trp fluorescence), while Trp95 located in the C-terminal region is more exposed (Table 2, Fig. 4, hexagons) [87]. Interestingly, although Trp60 is buried in the LS fibrils, its fluorescence intensity is increased significantly compared with the native protein, suggesting that this residue is distant to the disulfide bond in the fibril structure by contrast with its location in the native state [87]. Further mutagenesis studies focused on the disruptive effects of substitution with proline residues indicated that the introduction of this residue in either the N- or C-terminal strands (A and G) has little effect on fibril extension rates, yet disruption of strands B, C, E, or F by introduction of a proline has a significant effect on fibril formation. Interestingly, the introduction of proline in the vicinity of strand D has a range of effects from the suppression of fibril formation (H51P), to retardation of extension (L54P) or no observable effect (V49P), highlighting this area of the protein as holding importance in fibril formation [88].

It is clear that further experiments are required to generate a detailed model of both the LS and WL the fibrils; techniques such as cryo-EM, solid-state nuclear magnetic resonance (NMR), and electron paramagnetic resonance are sure to play a major role in providing higher-resolution detail of the intact fibril [89].

IDENTIFYING REGIONS INVOLVED IN AGGREGATION

Several groups have undertaken studies utilizing differing peptide fragments corresponding to different regions of the native protein in order to identify potentially amyloidogenic regions in the sequence of β_2m. Peptides corresponding to the native β-strand B/C [90,91], E [92], F/G (the latter only in the presence of high salt concentrations $>$ 1.5 M [93]), and various short peptides [94,95] (Fig. 4) have all been shown to form amyloidlike fibrils in vitro. In addition residues 83 to 89 have been identified as imparting amyloidogenic properties on β_2m using experiments utilizing differences in the amyloidogenic properties of mouse and human β_2m [94]. The peptides mentioned above show little correlation in secondary structure propensity, peptide length, p*I* value or hydrophobicity, yet they are all able to aggregate into amyloid. For peptide E there is a high content of aromatic sidechains, which seems to correlate closely with the ability of different sequences to aggregate into amyloid fibrils [96]. These data suggest that the ability of a sequence to undergo π-π electron stacking may be critically important in assembly [97]. The E-strand region has

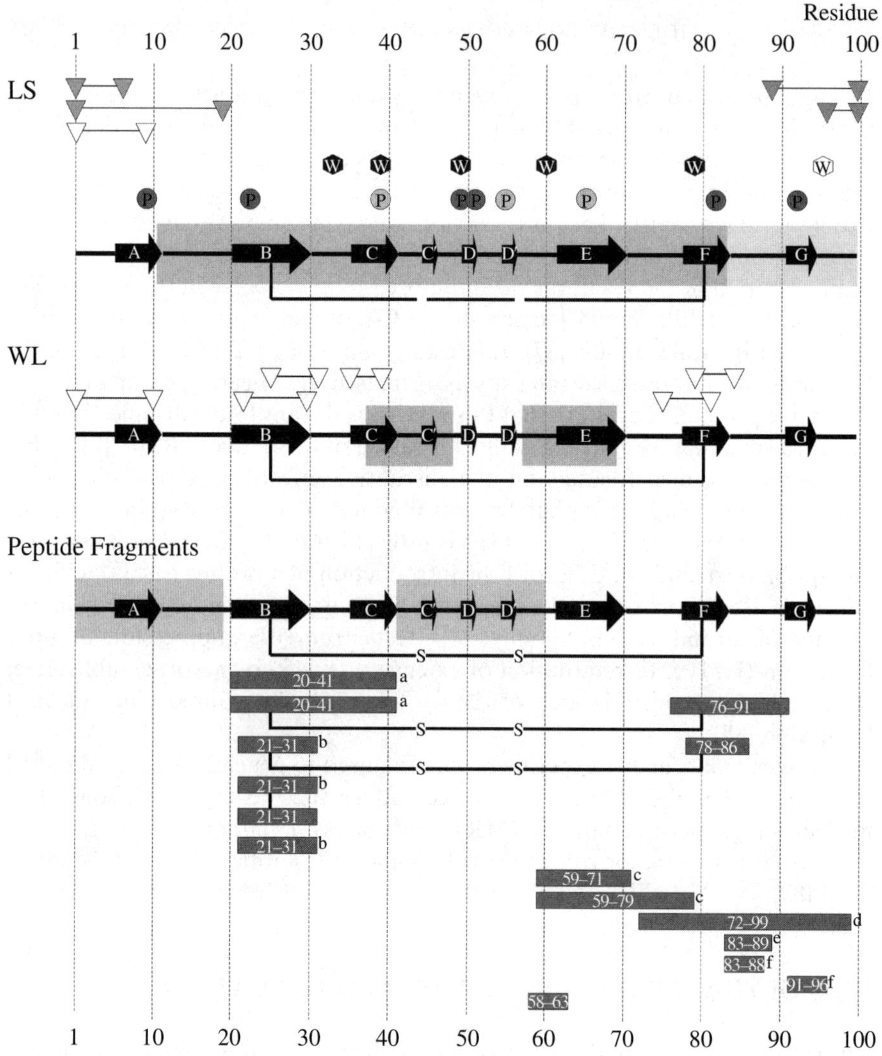

FIG. 4 Schematic diagram outlining current knowledge about the structure of β_2m amyloid fibrils gained to date from biochemical and biophysical analyses. *Top panel:* Hydrogen exchange (HX) and limited proteolysis results are shown for long, straight (LS) fibrils grown at pH 2.5 de novo; approximate regions of high protection from HX (yellow), regions of partial protection (pale yellow) [77,78]. Open triangles, limited proteolysis cut sites observed at pH 2.5 [75,77]; gray filled triangles, limited proteolysis cut sites observed using fibrils formed by extending seeds of ex vivo fibrils at pH 4.0 [76]. Hexagons; Trp residues buried in the core of the fibrils (black), Trp residues exposed to solvent (white) [87]. Proline mutations displaying significant (red), moderate (yellow), and no effect (green) on fibril extension [88]. *Middle panel:* Wormlike (WL) fibrils produced under high-salt conditions at pH 2.5 [75,78]. Regions of HX protection are shown in yellow [78], limited proteolysis cut sites are shown as open triangles [75,77]. *Bottom panel:* Peptide fragments deriving from β_2m known to form fibrils in vitro. Regions in gray are not known to form fibrils in isolation. [From (a) [90], (b) [91], (c) [92], (d) [93], (e) [94], (f) [95].] (*See insert for color representation of figure.*)

TABLE 2 Summary of β_2m Variants Created and Their Effects on the Amyloid Potential of the Protein

	Mutation	Effect	Ref.
In vivo modifications	N17D	Found in in vivo deposits; no effect on de novo fibril formation	20
	ΔK58	Found in in vivo deposits; increases the ability of the protein to form amyloid in vitro	18
Proline *cis*/–*trans* variants	P32G	Increased population of I_T state; increased rate of seeded fibril formation	37
	P32A	Increased population of I_T state; formation of a crystallographic dimer	65
	P32V	Increased population of I_T state; loss of slow refolding phase during protein refolding from acid-unfolded state	104
	P32G/I7A	Population of a partially folded state at pH 7.0 Rapid de novo fibril formation at pH 7.0	106
Alanine scan pH 7.0	R3A, F30A, V37A, L54A, Y66A, F70A, V82A, H84A	Decreased stability of the native monomer, no effect on fibril formation at pH 7.0 over 5 days	70
	I7A, V9A, V93A, R97A	Decreased stability of the native monomer, low yield of de novo fibril formation at pH 7.0	67, 70
Mutation scan of acid unfolded state	I7A, F30A, L40F	Similar fibril formation kinetics to wild-type pH 2.5	96, 99
	L40R, Y66A, Y66E, Y66S, Y67A	Disruption of the hydrophobically collapsed acid-unfolded state; reduced rates of seed elongation, reduced fibril formation rates	96, 99
	F62A, L65R, F70A	Disruption of the hydrophobically collapsed acid-unfolded state; abolition of seed elongation, severely reduced fibril formation rates	96, 99
	F62A/Y63A/Y67A (triple mutant)	Abolition of the hydrophobically collapsed acid-unfolded state; abolition of seed elongation, prevention of fibril formation rates	96, 99
Proline scan	L23P, H51P, V82P	Significant retardation of fibril extension at pH 2.5; low yields of fibrils	88
	L39P, L54P L65P	Retardation of fibril extension at pH 2.5; yield of fibrils comparable to wild-type	88

(Continued)

TABLE 2 (*Continued*)

	Mutation	Effect	Ref.
	V9P, V49P, V93P	No change to extension rates at pH 2.5; yield of fibrils comparable to wild-type	88
Histidine variants	H31F	Increased stability of the native monomer; decreased affinity for Cu^{2+} in the native state	115
	H31Y	Increased stability of monomer; decreased affinity for Cu^{2+} in the native state; formation of a crystallographic dimer; precipitation and unfolding observed < 4–6 weeks	7, 129
	H31S	Increased stability of monomer	129
	H13F, H51F, H84A	Decreased stability of the native monomer	115
	H84F	Unstructured protein	115
Tryptophan scan	S33W[a]	W33 buried in core of LS fibril	87
Tryptophan variants	W60G	Stabilizes native fold, inhibits amyloid formation	103
	W95G	Destabilizes the native fold, produces nonfibrillar aggregates	103
Double mutant background (W60F, W95F)[a]	L39W[a]	W39 fluorescence highly quenching in LS fibril; involved in core of LS fibril; lag phased observed with seeded growth	87
	V49W[a]	W49 fluorescence highly quenching in LS fibril; involved in core of LS fibril; lag phased observed with seeded growth	87
	W60F	W95 exposed in LS fibril	87
	Y78W[a]	W78 fluorescence highly quenching in LS fibril; involved in core of LS fibril	87
	W95F	W60 fluorescence significantly intense compared with wild-type; involved in core of LS fibril	87
Truncations	N3	Precipitation and unfolding observed < 4–6 weeks	129
	N6	Found in vivo deposits; increased rate of fibril extension at pH 7.0; structural properties similar to partially folded state formed at low pH	16, 34, 66
	N9/N10	Can be mimicked by limited proteolysis of LS fibrils	75
	N19	Found in vivo deposits; no in vitro studies published	130

TABLE 2 (*Continued*)

	Mutation	Effect	Ref.
Reduction	Reduction of the disulphide bond	Retains nativelike structural characteristics; LS fibril formation prevented; short WL fibril formation observed	83–85
Chimer	Human/mouse	Replacement of residues 83–89 (loop F–G) from human β_2m into nonamyloidogenic mouse β_2m imparts amyloidogenicity	94

[a]Variant protein containing two mutations W60F and W95F in which additional mutations were added. The W60F and W95F mutations substitute the two natural Trp residues.

also been found to be the most stable in the precursor state formed at pH 3.6 [98] and is also involved in nonnative residual structure in the acid-denatured state at pH 2.5 [96]. Aggregation predictions show that the residues around strand E (62–70) are the only region inherently prone to aggregation [99–101]. Mutations located outside the E-strand region have little effect on the structure in the acid denatured state or on fibril formation [99]. By contrast, mutations located within the E-strand region perturb the residual structure present in the acid-unfolded state and can have dramatic effects on seeded and de novo fibril growth, with one variant (L65R) completely preventing fibril formation (Table 1) [99]. Further, a triple mutant (F62A,Y63A,Y67A) containing substitutions of three aromatic residues in the E-strand region results in a loss of structure in the acid-denatured state and reduction of the ability to form fibrils at pH 2.5 [96,99]. This suggests a key role for this region in the early assembly and elongation of fibrils, possibly by the formation of a partially structured intermediate in which this region forms at least one interacting surface. Proline mutations (L23P (strand B), H51P (strand D), and V82P (strand F)) significantly retard fibril extension (Fig. 4) [88] and may act to perturb the precursor state [88]. Further investigation should be focused on determining the importance of each region in the context of the full-length protein and their importance in aggregation in order to identify residues involved in fibril nucleation and elongation, as well as in stabilizing the fibril structure itself.

CLUES AS TO THE IDENTITY OF THE AMYLOID PRECURSOR STATE

One of the most intriguing properties of β_2m, but not one of the most immediately obvious, is the fact that full-length wild-type protein does not readily undergo self-assembly under pH neutral conditions even at concentrations many times higher than that found in the serum of a dialysis patient

[34,37]. Hence, one or more factors must be altered by the uremic state, leading to the population of an amyloidogenic precursor and resulting in fibril formation. Acidification of the protein to around pH 3.6 results in the population of a partially unfolded state as judged by one-dimensional ^{1}H-NMR and circular dichroiem [29]. This partially unfolded state has been investigated indirectly by site-specific ^{1}H–^{15}N NMR and shown to be a noncooperatively stabilized ensemble that displays significant destabilization of both the N- and C-terminal strands, yet retains a stable, compact core corresponding to the native strands B through F [98]. Together with the observation that β_2m can be induced to form amyloidlike fibrils at neutral pH by the removal of the N-terminal six amino acids, these results suggest that strands A and G play an important role in protecting monomeric native β_2m from fibril formation [66,67]. Mutation of residues in strand A or G has yielded small amounts of amyloidlike material de novo at pH 7.0, whereas mutations within the core of the protein have little effect, consistent with the view that amyloidogenicity is not simply related to the global free energy of the protein [67,70]. Strand D has also been implicated in amyloid formation. A number of crystal and NMR structures have been solved for β_2m, and a selection is summarized in Figure 5. Crystallization of wild-type β_2m at pH 5.7 indicated the loss of the β-bulge in strand D, leading to the formation of a single seven-residue β-strand at one edge of the protein [4,6] (Fig. 5C). This β-bulge is retained in the HLA bound form of β_2m and in the crystal of the mutant H31Y (Fig. 5A and D). Hence, the D-strand can effectively adopt two alternative conformations, at least in a crystallized form. The presence of short edge strands has been proposed as an effective method of preventing edge-to-edge association of β-sheet proteins [102]. Conformational changes such as those seen in the D strand, even if only rarely populated, may increase the propensity of β_2m to self-associate. In addition to the foregoing results, and consistent with previous observations [5], the ^{1}H–^{15}N HSQC spectrum of β_2m at pH 5.7 shows line-broadening effects for residues 55 and 56 (Fig. 5B). These data suggest that these residues, which lie in strand D2 in the crystal structure of the HLA-bound form of β_2m, and in the C-terminal region of strand D in the crystal structure of monomeric β_2m, are in intermediate exchange between different conformations in solution, one of which presumably includes the continuous β-strand involving residues 51 to 57 observed by crystallography of the wild-type protein [4]. Similar to the wild-type protein, the crystal structure of the variant protein, W60G, also displays loss of the β-bulge in strand D, but by contrast with the wild-type protein does not form amyloid fibrils when seeded in the presence of TFE [103]. Hence, the absence of the β-bulge is not the only factor driving the initial stages of aggregation, and other factors, such as conformational flexibly of the protein, may play a role [103].

The folding mechanism of β_2m at pH 7.0 and 37°C has been studied in detail and a folding intermediate, trapped via a nonnative *trans*-prolyl conformation of Pro32 (Fig. 1), was shown to be populated to $3.7 \pm 1.4\%$ in the native state ensemble under physiological conditions [37]. By mutation of *cis*-proline 32 to *trans*-glycine (P32G), the population of this intermediate state was increased to 26.2$\pm$6.6%. In an elegant study in which the rate of amyloid formation (in seeded reactions) was correlated with the concentration of this intermediate, the I_{trans} (I_T) intermediate was identified as a direct precursor of amyloidlike fibrils [37]. Structural analysis using NMR shows that this species is highly nativelike, but contains perturbation of the edge strands A and D and the N-terminal region which would normally protect this β-sandwich protein from self-association (Fig. 5B), while highly nativelike structure was retained in strands B, C, E, F, and G [37]. In a separate parallel study, the role of Pro32 in amyloid formation was also investigated by mutation of Pro32 to Ala, as well as by the binding of a single Cu^{2+} to the wild-type protein [65]. Most notably, the isomerization of Pro32 from the *apo cis* configuration to a *trans* conformation was observed to occur upon Cu^{2+} binding [65]. Mutation of Pro32 to Ala (P32A) was introduced to mimic the *trans* form of the protein and resulted in a variant with increased affinity for metal which also underwent rapid oligomerization compared with the wild-type protein [65]. Interestingly, crystals of P32A revealed the atomic structure of the *trans* conformer, resulting in the formation of a crystallographic dimer (Fig. 5E). Elimination of the β-bulge in strand D resulted in a single continuous strand identical to that observed previously in the wild-type protein [4]. Dimer formation is mediated by an antiparallel interaction of two complete D strands related by a twofold axis. The variant protein H31Y, in which a critical Cu^{2+} binding ligand is removed, also displays an interesting crystal lattice packing, showing the occurrence of antiparallel pairing of two short D2 strands [7] (Fig. 5D). *Cis–trans* proline isomerization, therefore, may well control the partitioning of molecules between the folding and aggregation landscapes. The role of prolyl isomerization in the slow refolding of β_2m from the acid-unfolded state has also been confirmed by the refolding of the P32V mutant and a double-jump experiment with wild-type β_2m, both demonstrating the disappearance of the Pro-limited slow refolding phase [37,104]. It appears that isomerization of Pro32 results in the perturbation of the native A-strand and the β-budge located in D-strand, and these conformational changes may well be involved the early stages of the self-assembly process. These regions therefore represent a tempting target for the development of reagents able to prevent fibril formation.

The population of various conformational states of β_2m involved in amyloid assembly has been directly monitored via electrospray ionization–mass spectrometry (ESI-MS) between pH 6.0 and 2.0. Linear deconvolution of the charge-state distributions was utilized to show the extent to which each conformational ensemble is populated throughout this pH range [105]. Thus, at neutral pH the native protein gives rise to a narrow charge-state distribution centered on the 7+ ions. At pH 3.6, conditions under which WL fibrils are produced [29,70], the conformational ensemble broadens as the protein partially unfolds and is dominated by an ESI-MS charge-state distribution centered on the 9+ ions. By contrast, under more acidic conditions (pH 2.6), under which conditions LS fibrils are formed, the charge-state distribution is further expanded and dominated by the 10+ and 11+ ions (Fig. 6A and D). Further to this work, ion mobility spectrometry (IMS) coupled to ESI-MS has been used to separate and analyze co-populated protein conformers differing in their cross-sectional area and charge state, indicating that the partially folded state highly populated at pH 3.6 has a cross-sectional area similar to that of the native state, whereas the acid-unfolded state (populated at pH < 4) is more highly expanded (Fig. 6B, and E) [106]. Furthermore, a double-amino-acid variant (I7A/P32G) which forms fibrils de novo in vitro at neutral pH [82] was shown to have spectral properties similar to those of the acid-unfolded state of the wild-type protein at pH < 4 [106]. Hydrogen-exchange experiments carried out with wild-type

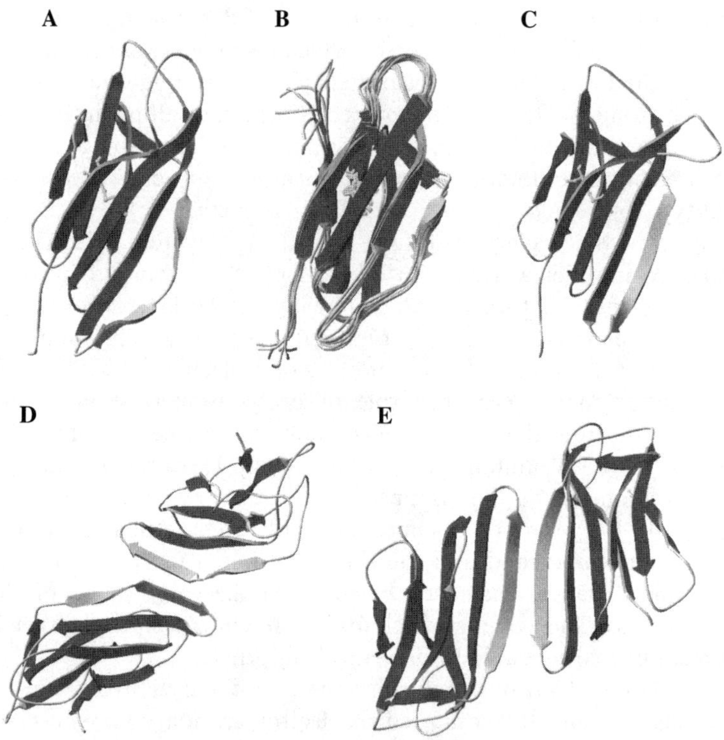

FIG. 5 Selection of different crystal structures of β_2m. Strand D is shown in orange in all cases. (A) β_2m in complex with the heavy chain of the HLA–1DUZ [3]. (B) NMR solution structure of β_2m–1JNJ [5]. (C) Crystal structure of β_2m at pH 5.6 (note the straight D strand)–1LDS [4]. Note that the crystal structure of β_2m at pH 7.0 (2YXF) is essentially the same as at pH 5.6 [6]. (D) Crystal structure of H31Y. The crystal lattice packing shows the occurrence of an antiparallel pairing of the short D2 strand–1PY4 [7]. (E) Crystallographic dimer of P32A related by a twofold axis yielding an eight-strand β-sheet comprised of strands ABED-DEBA–2F8O [65]. (*See insert for color representation of figure.*)

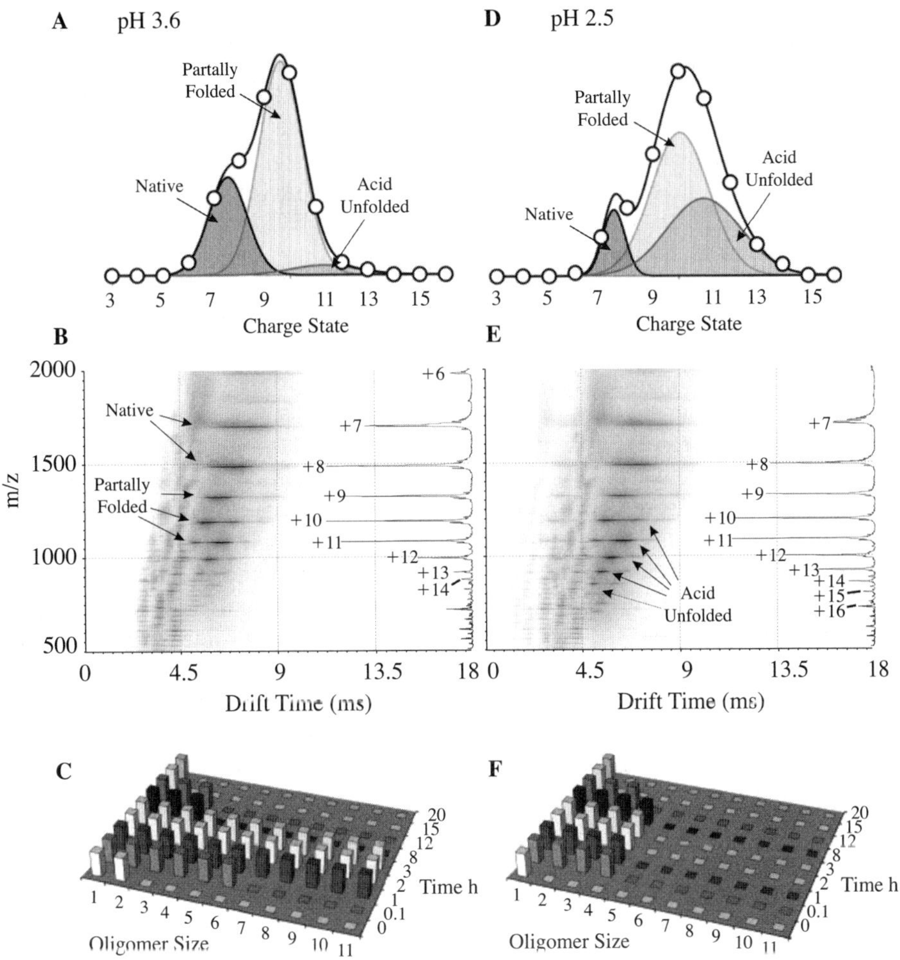

FIG. 6 ESI-MS data collected using β_2m at pH 3.6 (left-hand panels) and pH 2.5 (right-hand panels). (A,D) Co-populated conformational ensembles of β_2m uncovered quantitatively by ESI–MS [105]. (B,E) ESI–IMS–MS driftscope plots showing drift time (x axis) versus m/z (y-axis) for wild-type β_2m under each condition. Insets at the right-hand side of each plot: the summed, full-scan m/z spectra of wild-type β_2m for each data acquisition, showing the charge-state ions detected [106]. (C,F) Oligomer distributions observed during β_2m fibril assembly measured by nanoESI–MS under each condition over a range of m/z 3200 to 5500 [108]. (*See insert for color representation of figure.*)

β_2m and the more amyloidogenic cleaved variant ΔK58-β_2m exhibit both EX2 [where the lifetime of the exchange-competent state is shorter than the intrinsic chemical rate of exchange (k_{int}), indicative of local fluctuations] and EX1 kinetics (where the lifetime of the unprotected state is sufficiently long compared with k_{int} to allow complete exchange of all solvent exposed amide hydrogens, indicating cooperative unfolding of at least part of the protein) [107]. This type of exchange behavior allows the rate of transient unfolding to be obtained and correlated with amyloid formation. Accordingly, increases in the rate of EX1 kinetics brought about by cleavage of Lys58 were shown to lead to a higher aggregation propensity.

The population of oligomeric species during fibril formation under acidic conditions has also been observed directly via ESI-MS [108]. Using this technique, it was shown that WL fibrils assemble at pH 3.6 by a mechanism consistent with monomer addition, with species ranging from monomer to 13-mer being identified directly and uniquely as transient assembly intermediates (Fig. 6C). By contrast, only monomers to tetramers are observed during nucleated growth at pH 2.5, leading to the formation of LS fibrils (Fig. 6F). This suggests either that nucleation-dependent assembly occurs via an unstable (high-energy) oligomer greater than tetramer in size, or that conformational rearrangement of one or more of the species detected constitutes nucleation. Consistent with this detailed analysis of the protein concentration dependence of fibril formation in β_2m assembly is suggestive of a nucleus around the size of a hexamer [109]. Oligomers arising from the interaction of β_2m with Cu^{2+} have also been observed directly via ESI-MS and the formation of dimers and tetramers detected [110]. The partially folded and acid-unfolded states of β_2m have been highly characterized and appear to share many similarities with the rarely populated I_T species seen at pH 7.0 and controlled by the isomerization of Pro32. Hence, any perturbation of the energy landscape such that the precursor state is more highly populated will result in de novo fibril formation if the concentration of the precursor is high enough to undergo nucleation or, alternatively, to allow fibril extension to occur in the presence of fibril seeds.

ROLE OF Cu^{2+}

Transition metal cations have been shown to initiate and modulate aggregation reactions of a number of proteins through a variety of mechanisms [111,112]. β_2m has been suggested to undergo an opportunistic interaction with Cu^{2+} as part of the pathological process of HDRA [112]. The concentration of serum Cu^{2+} (15 to 25 μM) is tightly regulated such that little, if any, free Cu^{2+} is available to facilitate amyloidosis in vivo [113]. However, patients are exposed to transition metal ions during the dialysis treatment, and over the course of a single year a HDRA patient may be exposed to 15,000 to 30,000 L of water. The maximum level of Cu^{2+} allowed in the dialysate is 1.6 μM [114], which is within a factor of 2 of the measured affinity of Cu^{2+} for β_2m [115]. The addition of Cu^{2+} within these medically accepted limits in the presence of physiological concentrations of urea in uremic patients could result in significant destabilization of native β_2m and the formation of amyloid fibrils at pH 7.0 and 37°C [62]. Further, Cu^{2+} has been shown to bind preferentially to nonnative states of

β_2m, mediated by binding to His13, His51, and His84 [116,117], whereas His31 is known to bind Cu^{2+} in the native state [115,118]. Metal binding has been shown to increase the pico- to nanosecond time scale fluctuations of the β-strand D in which His51 lies, which is propagated to the core of the molecule, thus promoting global and slow fluctuations [118]. Molecular dynamics simulations have also suggested that Cu^{2+} binding perturbs the secondary structure and disrupts the native hydrophobic contacts in the neighboring segments of β_2m, including the β-strand D2 and part of the β-strands E, B, and C, resulting in greater exposure to the solvent of the D–E and B–C loops [119]. As discussed above, Cu^{2+} also plays a role in the isomerization of Pro32 from the *apo cis* to the *holo trans* state by the binding to H31 [65] and facilitates the formation of oligomeric species purported to be intermediates of amyloid formation [110,117]. Using an array of biochemical and biophysical methods, Miranker and colleagues have shown that Cu^{2+} mediates the formation of a monomeric, activated state followed by the formation of a discrete dimeric intermediate. This dimeric intermediate assembles into tetra- and hexameric forms, displaying little additional oligomerization on the time scales of their own formation (<1 hour). Fibril formation subsequently ensues, progressing from these intermediate states on much longer time scales (>1 week). These precursor species require Cu^{2+} to form, but once generated, do not require this metal cation for their stability [110,120]. Cu^{2+} binding may well contribute to the overall destabilization of β_2m by increasing the equilibrium population of intermediates, leading to amyloid formation. Hence, the elimination of this metal from the dialysis procedure would appear to be prudent as a method of reducing the incidence of amyloid deposition in dialysis patients.

IMPLICATIONS FOR THERAPY

In the absence of small-molecule therapies to prevent HDRA, current medical treatment focuses on relieving the symptoms. The administration of adrenocorticosteroids (e.g., prednisone) is highly effective at relieving pain associated with amyloid arthropathy in dialysis patients [121]. Nonsteroidal anti-inflammatory drugs (NSAIDs) also reduce the pain and swelling of HDRA-affected joints. In severe cases surgical treatment may be required to relieve pain. For example, in carpal tunnel syndrome, surgical decompression of the carpal tunnel can be performed to relieve pain and prevent irreversible neuromuscular impairment which may result from long-term medial nerve compression [121]. Due to the fact that retention is a prerequisite for amyloid formation, removal of circulating β_2m during hemodialysis is an obvious avenue for therapeutic intervention. However, no clear correlation between the type of dialysis procedure used and the onset of HDRA has been observed [122]. For dialysis patients, removal of β_2m by hemoperfusion treatment using porous Lixelle cellulose beads onto which hydrophobic hexadecyl alkyl chains are covalently bound results in the elimination of 1 mg of β_2m per milliliter of beads [123]. This treatment reduces the circulating levels of β_2m and inflammatory cytokines, thus improving the symptoms of patients with HDRA [123]. Single standard high-flux hemodialysis sessions have also been shown to reduce β_2m plasma levels by 50%, possibly by the adsorption of β_2m to the synthetic dialyzer membrane [124] in addition to diffusive clearance [125]. The development of more biocompatible high-flux hemodialyzer membranes continues to be explored as one way towards reducing the incidence of HDRA or even preventing the onset of β_2m amyloid-associated symptoms [126].

Developing effective therapies against amyloid disease remains an immense challenge. Ligands able to bind and stabilize the native state of β_2m provide one potential strategy. Novel RNA aptamers which utilize the fact that the native β_2m does not share epitopes with either LS or WL fibrils or their precursors could be useful therapeutically [81], and the first aptamer-based drugs are beginning to appear in the clinic [127]. Alternative strategies include inhibition of amyloid fibrillogenesis by novel GAG inhibitors, fibril disassembly by β-sheet breaker peptides, or enhancement of clearance of existing amyloid deposits by either immunotherapy or small-molecule inhibitors of SAP [128]. Great progress has been made toward our understanding of HDRA and the structural biology of β_2m amyloid formation in the last decade. Although many questions remain to be answered, the field as a whole is now in a position to utilize this information to generate effective therapies and therapeutics against HDRA, as well as against other protein conformational disorders.

Acknowledgments

We acknowledge financial support from the Wellcome Trust and the BBSRC. We are also grateful to many colleagues for stimulating discussions over many years and for sharing ideas with us during the course of composing this manuscript.

REFERENCES

1. Gejyo, F., Yamada, T., Odani, S., Nakagawa, Y., Arakawa, M., Kunitomo, T., Kataoka, H., Suzuki, M., Hirasawa, Y., Shirahama, T., et al. (1985). A new form of amyloid protein associated with chronic-hemodialysis was identified as beta 2-microglobulin. *Biochem Biophys Res Commun, 129*, 701–706.
2. Floege, J. Ketteler, M. (2001). Beta 2-microglobulin-derived amyloidosis: an update. *Kidney Int, 59*, 164–171.
3. Khan, A.R., Baker, B.M., Ghosh, P., Biddison, W.E., Wiley, D.C. (2000). The structure and stability of an HLA-A*0201/octameric tax peptide complex with an empty conserved peptide-N-terminal binding site. *J Immunol, 164*, 6398–6405.
4. Trinh, C.H., Smith, D.P., Kalverda, A.P., Phillips, S.E., Radford, S.E. (2002). Crystal structure of monomeric human beta 2-microglobulin reveals clues to its amyloidogenic properties. *Proc Natl Acad Sci U S A, 99*, 9771–9776.

5. Verdone, G., Corazza, A., Viglino, P., Pettirossi, F., Giorgetti, S., Mangione, P., Andreola, A., Stoppini, M., Bellotti, V., Esposito, G. (2002). The solution structure of human beta 2-microglobulin reveals the prodromes of its amyloid transition. *Protein Sci, 11*, 487–499.

6. Iwata, K., Matsuura, T., Sakurai, K., Nakagawa, A., Goto, Y. (2007). High-resolution crystal structure of beta 2-microglobulin formed at pH 7.0. *J Biochem, 142*, 413–419.

7. Rosano, C., Zuccotti, S., Mangione, P., Giorgetti, S., Bellotti, V., Pettirossi, F., Corazza, A., Viglino, P., Esposito, G., Bolognesi, M. (2004). Beta 2-microglobulin H31Y variant 3D structure highlights the protein natural propensity towards intermolecular aggregation. *J Mol Biol, 335*, 1051–1064.

8. Floege, J., Bartsch, A., Schulze, M., Shaldon, S., Koch, K.M., Smeby, L.C. (1991). Clearance and synthesis rates of beta 2-microglobulin in patients undergoing hemodialysis and in normal subjects. *J Lab Clin Med, 118*, 153–165.

9. Karlsson, F.A., Groth, T., Sege, K., Wibell, L., Peterson, P.A. (1980). Turnover in humans of beta 2-microglobulin: the constant chain of HLA-antigens. *Eur J Clin Invest, 10*, 293–300.

10. Vincent, C., Chanard, J., Caudwell, V., Lavaud, S., Wong, T., Revillard, J.P. (1992). Kinetics of I^{125} beta 2-microglobulin turnover in dialyzed patients. *Kidney Int, 42*, 1434–1443.

11. Drueke, T.B. (2000). Beta 2-microglobulin and amyloidosis. *Nephrol Dial Transplant, 15* (Suppl 1), 17–24.

12. Gejyo, F., Homma, N., Suzuki, Y., Arakawa, M. (1986). Serum levels of beta 2-microglobulin as a new form of amyloid protein in patients undergoing long-term hemodialysis. *N Engl J Med, 314*, 585–586.

13. Relini, A., Canale, C., De Stefano, S., Rolandi, R., Giorgetti, S., Stoppini, M., Rossi, A., Fogolari, F., Corazza, A., Esposito, G., *et al.* (2006). Collagen plays an active role in the aggregation of beta 2-microglobulin under physiopathological conditions of dialysis-related amyloidosis. *J Biol Chem, 281*, 16521–16529.

14. Stoppini, M., Mangione, P., Monti, M., Giorgetti, S., Marchese, L., Arcidiaco, P., Verga, L., Segagni, S., Pucci, P., Merlini, G., Bellotti, V. (2005). Proteomics of beta 2-microglobulin amyloid fibrils. *Biochim Biophys Acta, 1753*, 23–33.

15. Linke, R.P., Bommer, J., Ritz, E., Waldherr, R., Eulitz, M. (1986). Amyloid kidney stones of uremic patients consist of beta 2-microglobulin fragments. *Biochem Biophys Res Commun, 136*, 665–671.

16. Bellotti, V., Stoppini, M., Mangione, P., Sunde, M., Robinson, C., Asti, L., Brancaccio, D., Ferri, G. (1998). Beta 2-microglobulin can be refolded into a native state from ex-vivo amyloid fibrils. *Eur J Biochem, 258*, 61–67.

17. Nissen, M.H., Roepstorff, P., Thim, L., Dunbar, B., Fothergill, J.E. (1990). Limited proteolysis of beta 2-microglobulin at Lys-58 by complement component C1s. *Eur J Biochem, 189*, 423–429.

18. Giorgetti, S., Stoppini, M., Tennent, G.A., Relini, A., Marchese, L., Raimondi, S., Monti, M., Marini, S., Østergaard, O., Heegaard, N.H., *et al.* (2007). Lysine 58-cleaved beta2-microglobulin is not detectable by 2D electrophoresis in ex-vivo amyloid fibrils of two patients affected by dialysis-related amyloidosis. *Protein Sci, 16*, 343–349.

19. Heegaard, N.H., Jørgensen, T.J., Rozlosnik, N., Corlin, D.B., Pedersen, J.S., Tempesta, A.G., Roepstorff, P., Bauer, R., Nissen, M.H. (2005). Unfolding, aggregation, and seeded amyloid formation of lysine-58-cleaved beta 2-microglobulin. *Biochemistry, 44*, 4397–4407.
20. Kad, N.M., Thomson, N., Smith, D.P., Smith, D.A., Radford, S.E. (2001). Beta 2-microglobulin and its deamidated variant, N17D form amyloid fibrils with a range of morphologies in vitro. *J Mol Biol, 313*, 559–571.
21. Niwa, T., Katsuzaki, T., Miyazaki, S., Momoi, T., Akiba, T., Miyazaki, T., Nokura, K., Hayase, F., Tatemichi, N., Takei, Y. (1997). Amyloid beta 2-microglobulin is modified with imidazolone, a novel advanced glycation end product, in dialysis-related amyloidosis. *Kidney Int, 51*, 187–194.
22. Jadoul, M., Garbar, C., Noël, H., Sennesael, J., Vanholder, R., Bernaert, P., Rorive, G., Hanique, G., van Ypersele de Strihou, C. (1997). Histological prevalence of beta 2-microglobulin amyloidosis in hemodialysis: a prospective post-mortem study. *Kidney Int, 51*, 1928–1932.
23. Hou, F.F. Owan, W.F. (2002). Beta 2-microglobulin amyloidosis: role of monocytes / macrophages. *Curr Opin Nephrology and Hypertension, 11*, 417–421.
24. Pepys, M.B. (2006). Amyloidosis. *Annu Rev Med, 57*, 223–241.
25. Ono, K., Uchino, F. (1994). Formation of amyloid-like substance from beta 2-microglobulin in vitro. *Nephron, 66*, 404–407.
26. Yamamoto, S., Hasegawa, K., Yamaguchi, I., Goto, Y., Gejyo, F., Naiki, H. (2005). Kinetic analysis of the polymerization and depolymerization of beta 2-microglobulin-related amyloid fibrils in vitro. *Biochim Biophys Acta, 1753*, 34–43.
27. Mahley, R.W. (1988). Apolipoprotein E: cholesterol transport protein with expanding role in cell biology. *Science, 240*, 622–630.
28. Motomiya, Y., Ando, Y., Haraoka, K., Sun, X., Iwamoto, H., Uchimura, T., Maruyama, I. (2003). Circulating level of alpha2-macroglobulin-beta 2-microglobulin complex in hemodialysis patients. *Kidney Int, 64*, 2244–2252.
29. McParland, V.J., Kad, N.M., Kalverda, A.P., Brown, A., Kirwin-Jones, P., Hunter, M.G., Sunde, M., Radford, S.E. (2000). Partially unfolded states of beta 2-microglobulin and amyloid formation in vitro. *Biochemistry, 39*, 8735–8746.
30. Athanasou, N.A., Puddle, B., Sallie, B. (1995). Highly sulphated glycosaminoglycans in articular cartilage and other tissues containing beta 2-microglobulin dialysis amyloid deposits. *Nephrol Dial Transplant, 10*, 1672–1678.
31. Inoue, S., Kuroiwa, M., Ohashi, K., Hara, M., Kisilevsky, R. (1997). Ultrastructural organization of hemodialysis-associated beta 2-microglobulin amyloid fibrils. *Kidney Int, 52*, 1543–1549.
32. Hou, F.F., Chertow, G.M., Kay, J., Boyce, J., Lazarus, J.M., Braatz, J.A., Owen, W.F., Jr. (1997). Interaction between beta 2-microglobulin and advanced glycation end products in the development of dialysis related amyloidosis. *Kidney Int, 51*, 1514–1519.
33. Bardin, T., Kuntz, D. (1987). The arthropathy of chronic haemodialysis. *Clin Exp Rheumatol, 5*, 379–386.
34. Myers, S.L., Jones, S., Jahn, T.R., Morten, I.J., Tennent, G.A., Hewitt, E.W., Radford, S.E. (2006). A systematic study of the effect of physiological factors on

beta 2-microglobulin amyloid formation at neutral pH. *Biochemistry, 45*, 2311–2321.

35. Yamaguchi, I., Suda, H., Tsuzuike, N., Seto, K., Seki, M., Yamaguchi, Y., Hasegawa, K., Takahashi, N., Yamamoto, S., Gejyo, F., Naiki, H. (2003). Glycosaminoglycan and proteoglycan inhibit the depolymerization of beta 2-microglobulin amyloid fibrils in vitro. *Kidney Int, 64*, 1080–1088.
36. Borysik, A.J., Morten, I.J., Radford, S.E., Hewitt, E.W. (2007). Specific glycosaminoglycans promote unseeded amyloid formation from beta 2-microglobulin under physiological conditions. *Kidney Int, 72*, 174–181.
37. Jahn, T.R., Parker, M.J., Homans, S.W., Radford, S.E. (2006). Amyloid formation under physiological conditions proceeds via a native-like folding intermediate. *Nat Struct Mol Biol, 13*, 195–201.
38. Yamamoto, S., Hasegawa, K., Yamaguchi, I., Tsutsumi, S., Kardos, J., Goto, Y., Gejyo, F., Naiki, H. (2004). Low concentrations of sodium dodecyl sulfate induce the extension of beta 2-microglobulin-related amyloid fibrils at a neutral pH. *Biochemistry, 43*, 11075–11082.
39. Giorgetti, S., Rossi, A., Mangione, P., Raimondi, S., Marini, S., Stoppini, M., Corazza, A., Viglino, P., Esposito, G., Cetta, G., *et al.* (2005). Beta 2-microglobulin isoforms display an heterogeneous affinity for type I collagen. *Protein Sci, 14*, 696–702.
40. Homma, N., Gejyo, F., Isemura, M., Arakawa, M. (1989). Collagen-binding affinity of beta 2-microglobulin, a preprotein of hemodialysis-associated amyloidosis. *Nephron, 53*, 37–40.
41. Coggi, G., Dell'Orto, P., Braidotti, P., Coggi, A., Viale, G. (1989). Coexpression of intermediate filaments in normal and neoplastic human tissues: a reappraisal. *Ultrastruct Pathol, 13*, 501–514.
42. Yamaguchi, I., Hasegawa, K., Takahashi, N., Gejyo, F., Naiki, H. (2001). Apolipoprotein E inhibits the depolymerization of beta 2-microglobulin-related amyloid fibrils at a neutral pH. *Biochemistry, 40*, 8499–8507.
43. Naiki, H., Yamamoto, S., Hasegawa, K., Yamaguchi, I., Goto, Y., Gejyo, F. (2005). Molecular interactions in the formation and deposition of beta 2-microglobulin-related amyloid fibrils. *Amyloid, 12*, 15–25.
44. Linke, R.P., Schäeffer, J., Gielow, P., Lindner, P., Lottspeich, F., Plückthun, A., Weiss, E.H. (2000). Production of recombinant human beta 2-microglobulin for scintigraphic diagnosis of amyloidosis in uremia and hemodialysis. *Eur J Biochem, 267*, 627–633.
45. Bellotti, V., Gallieni, M., Giorgetti, S., Brancaccio, D. (2001). Dynamic of beta 2-microglobulin fibril formation and reabsorption: the role of proteolysis. *Semin Dial, 14*, 117–122.
46. Ayers, D.C., Athanasou, N.A., Woods, C.G., Duthie, R.B. (1993). Dialysis arthropathy of the hip. *Clin Orthop Relat Res*, 216–224.
47. Ohashi, K., Hara, M., Kawai, R., Ogura, Y., Honda, K., Nihei, H., Mimura, N. (1992). Cervical disks are most susceptible to beta 2-microglobulin amyloid deposition in the vertebral column. *Kidney Int, 41*, 1646–1652.
48. Hou, F.F., Miyata, T., Boyce, J., Yuan, Q., Chertow, G.M., Kay, J., Schmidt, A.M., Owen, W.F. (2001). Beta 2-microglobulin modified with advanced glycation end products delays monocyte apoptosis. *Kidney Int, 59*, 990–1002.

49. Miyata, T., Hori, O., Zhang, J., Yan, S.D., Ferran, L., Iida, Y., Schmidt, A.M. (1996). The receptor for advanced glycation end products (RAGE) is a central mediator of the interaction of AGE-beta 2-microglobulin with human mononuclear phagocytes via an oxidant-sensitive pathway. Implications for the pathogenesis of dialysis-related amyloidosis. *J Clin Invest, 98*, 1088–1094.

50. Matsuo, K., Ikizler, T.A., Hoover, R.L., Nakamoto, M., Yasunaga, C., Pupim, L.B., Hakim, R.M. (2000). Transforming growth factor-beta is involved in the pathogenesis of dialysis-related amyloidosis. *Kidney Int, 57*, 697–708.

51. Miyata, T., Inagi, R., Iida, Y., Sato, M., Yamada, N., Oda, O., Maeda, K., Seo, H. (1994). Involvement of beta 2-microglobulin modified with advanced glycation end products in the pathogenesis of hemodialysis-associated amyloidosis. Induction of human monocyte chemotaxis and macrophage secretion of tumor necrosis factor-alpha and interleukin-1. *J Clin Invest, 93*, 521–528.

52. Ramadori, G., Rieder, H., Sipe, J., Shirahama, T., Meyer zum Büschenfelde, K.H. (1989). Murine tissue macrophages synthesize and secrete amyloid proteins different to amyloid-A (AA). *Eur J Clin Invest, 19*, 316–322.

53. Van Ypersele, C., Drucke, T.B.(1996). *Dialysis Amyloid*, Oxford University Press, New York, pp. 34–68.

54. Gosal, W.S., Morten, I.J., Hewitt, E.W., Smith, D.A., Thomson, N.H., Radford, S.E. (2005). Competing pathways determine fibril morphology in the self-assembly of beta 2-microglobulin into amyloid. *J Mol Biol, 351*, 850–864.

55. Morten, I.J., Gosal, W.S., Radford, S.E., Hewitt, E.W. (2007). Investigation into the role of macrophages in the formation and degradation of beta 2-microglobulin amyloid fibrils. *J Biol Chem, 282*, 29691–29700.

56. Gorevic, P.D., Casey, T.T., Stone, W.J., DiRaimondo, C.R., Prelli, F.C., Frangione, B. (1985). Beta 2-microglobulin is an amyloidogenic protein in man. *J Clin Invest, 76*, 2425–2429.

57. Naiki, H., Hashimoto, N., Suzuki, S., Kimura, H., Nakakuki, K., Gejyo, F. (1997). Establishment of a kinetic model of dialysis-related amyloid fibril extension in vitro. *Amyloid:Int J Exp Clin Invest, 4*, 223–232.

58. Ban, T., Hamada, D., Hasegawa, K., Naiki, H., Goto, Y. (2003). Direct observation of amyloid fibril growth monitored by thioflavin T fluorescence. *J Biol Chem, 278*, 16462–16465.

59. Chiti, F., De Lorenzi, E., Grossi, S., Mangione, P., Giorgetti, S., Caccialanza, G., Dobson, C.M., Merlini, G., Ramponi, G., Bellotti, V. (2001). A partially structured species of beta 2-microglobulin is significantly populated under physiological conditions and involved in fibrillogenesis. *J Biol Chem, 276*, 46714–46721.

60. Ohhashi, Y., Kihara, M., Naiki, H., Goto, Y. (2005). Ultrasonication-induced amyloid fibril formation of beta 2-microglobulin. *J Biol Chem, 280*, 32843–32848.

61. Kihara, M., Chatani, E., Sakai, M., Hasegawa, K., Naiki, H., Goto, Y. (2005). Seeding-dependent maturation of beta 2-microglobulin amyloid fibrils at neutral pH. *J Biol Chem, 280*, 12012–12018.

62. Morgan, C.J., Gelfand, M., Atreya, C., Miranker, A.D. (2001). Kidney dialysis-associated amyloidosis: a molecular role for copper in fiber formation. *J Mol Biol, 309*, 339–345.

63. Yamamoto, S., Yamaguchi, I., Hasegawa, K., Tsutsumi, S., Goto, Y., Gejyo, F., Naiki, H. (2004). Glycosaminoglycans enhance the trifluoroethanol-induced extension of beta 2-microglobulin-related amyloid fibrils at a neutral pH. *J Am Soc Nephrol, 15*, 126–133.

64. Sasahara, K., Yagi, H., Naiki, H., Goto, Y. (2007). Heat-induced conversion of beta 2-microglobulin and hen egg-white lysozyme into amyloid fibrils. *J Mol Biol, 372*, 981–991.

65. Eakin, C.M., Berman, A.J., Miranker, A.D. (2006). A native to amyloidogenic transition regulated by a backbone trigger. *Nat Struct Mol Biol, 13*, 202–208.

66. Esposito, G., Michelutti, R., Verdone, G., Viglino, P., Hernández, H., Robinson, C.V., Amoresano, A., Dal Piaz, F., Monti, M., Pucci, P., *et al.* (2000). Removal of the N-terminal hexapeptide from human beta 2-microglobulin facilitates protein aggregation and fibril formation. *Protein Sci, 9*, 831–845.

67. Jones, S., Smith, D.P., Radford, S.E. (2003). Role of the N and C-terminal strands of beta 2-microglobulin in amyloid formation at neutral pH. *J Mol Biol, 330*, 935–941.

68. Gosal, W.S., Myers, S.L., Radford, S.E., Thomson, N.H. (2006). Amyloid under the atomic force microscope. *Protein Pept Lett, 13*, 261–270.

69. Kad, N.M., Myers, S.L., Smith, D.P., Smith, D.A., Radford, S.E., Thomson, N.H. (2003). Hierarchical assembly of beta 2-microglobulin amyloid in vitro revealed by atomic force microscopy. *J Mol Biol, 330*, 785–797.

70. Smith, D.P., Jones, S., Serpell, L.C., Sunde, M., Radford, S.E. (2003). A systematic investigation into the effect of protein destabilisation on beta 2-microglobulin amyloid formation. *J Mol Biol, 330*, 943–954.

71. Fabian, H., Gast, K., Laue, M., Misselwitz, R., Uchanska-Ziegler, B., Ziegler, A., Naumann, D. (2008). Early stages of misfolding and association of beta 2-microglobulin: insights from infrared spectroscopy and dynamic light scattering. *Biochemistry, 47*, 6895–6906.

72. Naiki, H., Hashimoto, N., Suzuki, S., Kimura, H., Nakakuki, K., Gejyo, F. (1997). Establishment of a kinetic model of dialysis-related amyloid fibril extension in vitro. *Amyloid Int J Exp Clin Invest, 4*, 223–232.

73. Chatani, E., Kato, M., Kawai, T., Naiki, H., Goto, Y. (2005). Main-chain dominated amyloid structures demonstrated by the effect of high pressure. *J Mol Biol, 352*, 941–951.

74. Chatani, E., Naiki, H., Goto, Y. (2006). Seeding-dependent propagation and maturation of beta 2-microglobulin amyloid fibrils under high pressure. *J Mol Biol, 359*, 1086–1096.

75. Myers, S.L., Thomson, N.H., Radford, S.E., Ashcroft, A.E. (2006). Investigating the structural properties of amyloid-like fibrils formed in vitro from beta 2-microglobulin using limited proteolysis and electrospray ionisation mass spectrometry. *Rapid Commun Mass Spectrom, 20*, 1628–1636.

76. Monti, M., Principe, S., Giorgetti, S., Mangione, P., Merlini, G., Clark, A., Bellotti, V., Amoresano, A., Pucci, P. (2002). Topological investigation of amyloid fibrils obtained from beta 2-microglobulin. *Protein Sci, 11*, 2362–2369.

77. Hoshino, M., Katou, H., Hagihara, Y., Hasegawa, K., Naiki, H., Goto, Y. (2002). Mapping the core of the beta(2)-microglobulin amyloid fibril by H/D exchange. *Nat Struct Biol, 9*, 332–336.

78. Yamaguchi, K., Katou, H., Hoshino, M., Hasegawa, K., Naiki, H., Goto, Y. (2004). Core and heterogeneity of beta 2-microglobulin amyloid fibrils as revealed by H/D exchange. *J Mol Biol, 338*, 559–571.

79. O'Nuallain, B., Wetzel, R. (2002). Conformational Abs recognizing a generic amyloid fibril epitope. *Proc Natl Acad Sci U S A, 99*, 1485–1490.

80. Kayed, R., Head, E., Thompson, J.L., McIntire, T.M., Milton, S.C., Cotman, C.W., Glabe, C.G. (2003). Common structure of soluble amyloid oligomers implies common mechanism of pathogenesis. *Science, 300*, 486–489.

81. Bunka, D.H., Mantle, B.J., Morten, I.J., Tennent, G.A., Radford, S.E., Stockley, P.G. (2007). Production and characterization of RNA aptamers specific for amyloid fibril epitopes. *J Biol Chem, 282*, 34500–34509.

82. Jahn, T.R., Tennent, G.A., Radford, S.E. (2008). A common beta-sheet architecture underlies in vitro and in vivo beta 2-microglobulin amyloid fibrils. *J Biol Chem, 283*, 17279–17286.

83. Gozu, M., Lee, Y.H., Ohhashi, Y., Hoshino, M., Naiki, H., Goto, Y. (2003). Conformational dynamics of beta 2-microglobulin analyzed by reduction and reoxidation of the disulfide bond. *J Biochem, 133*, 731–736.

84. Smith, D.P., Radford, S.E. (2001). Role of the single disulphide bond of beta 2-microglobulin in amyloidosis in vitro. *Protein Sci, 10*, 1775–1784.

85. Ohhashi, Y., Hagihara, Y., Kozhukh, G., Hoshino, M., Hasegawa, K., Yamaguchi, I., Naiki, H., Goto, Y. (2002). The intrachain disulfide bond of beta 2-microglobulin is not essential for the immunoglobulin fold at neutral pH, but is essential for amyloid fibril formation at acidic pH. *J Biochem, 131*, 45–52.

86. Yamamoto, K., Yagi, H., Ozawa, D., Sasahara, K., Naiki, H., Goto, Y. (2008). Thiol compounds inhibit the formation of amyloid fibrils by beta 2-microglobulin at neutral pH. *J Mol Biol, 376*, 258–268.

87. Kihara, M., Chatani, E., Iwata, K., Yamamoto, K., Matsuura, T., Nakagawa, A., Naiki, H., Goto, Y. (2006). Conformation of amyloid fibrils of beta 2-microglobulin probed by tryptophan mutagenesis. *J Biol Chem, 281*, 31061–31069.

88. Chiba, T., Hagihara, Y., Higurashi, T., Hasegawa, K., Naiki, H., Goto, Y. (2003). Amyloid fibril formation in the context of full-length protein: effects of proline mutations on the amyloid fibril formation of beta 2-microglobulin. *J Biol Chem, 278*, 47016–47024.

89. Fandrich, M. (2007). On the structural definition of amyloid fibrils and other polypeptide aggregates. *Cell Mol Life Sci, 64*, 2066–2078.

90. Kozhukh, G.V., Hagihara, Y., Kawakami, T., Hasegawa, K., Naiki, H., Goto, Y. (2002). Investigation of a peptide responsible for amyloid fibril formation of beta 2-microglobulin by *Achromobacter* protease I. *J Biol Chem, 277*, 1310–1315.

91. Hasegawa, K., Ohhashi, Y., Yamaguchi, I., Takahashi, N., Tsutsumi, S., Goto, Y., Gejyo, F., Naiki, H. (2003). Amyloidogenic synthetic peptides of beta 2-microglobulin: a role of the disulfide bond. *Biochem Biophys Res Commun, 304*, 101–106.

92. Jones, S., Manning, J., Kad, N.M., Radford, S.E. (2003). Amyloid-forming peptides from beta 2-microglobulin:insights into the mechanism of fibril formation in vitro. *J Mol Biol, 325*, 249–257.

93. Ivanova, M.I., Gingery, M., Whitson, L.J., Eisenberg, D. (2003). Role of the C-terminal 28 residues of beta 2-microglobulin in amyloid fibril formation. *Biochemistry, 42*, 13536–13540.

94. Ivanova, M.I., Sawaya, M.R., Gingery, M., Attinger, A., Eisenberg, D. (2004). An amyloid-forming segment of beta 2-microglobulin suggests a molecular model for the fibril. *Proc Natl Acad Sci U S A, 101*, 10584–10589.

95. Ivanova, M.I., Thompson, M.J., Eisenberg, D. (2006). A systematic screen of beta 2-microglobulin and insulin for amyloid-like segments. *Proc Natl Acad Sci U S A, 103*, 4079–4082.

96. Platt, G.W., McParland, V.J., Kalverda, A.P., Homans, S.W., Radford, S.E. (2005). Dynamics in the unfolded state of beta 2-microglobulin studied by NMR. *J Mol Biol, 346*, 279–294.

97. Gazit, E. (2002). A possible role for pi-stacking in the self-assembly of amyloid fibrils. *FASEB J, 16*, 77–83.

98. McParland, V.J., Kalverda, A.P., Homans, S.W., Radford, S.E. (2002). Structural properties of an amyloid precursor of beta 2-microglobulin. *Nat Struct Biol, 9*, 326–331.

99. Platt, G.W., Routledge, K.E., Homans, S.W., Radford, S.E. (2008). Fibril growth kinetics reveal a region of beta 2-microglobulin important for nucleation and elongation of aggregation. *J Mol Biol, 378*, 251–263.

100. Lopez de la Paz, M., Serrano, L. (2004). Sequence determinants of amyloid fibril formation. *Proc Natl Acad Sci U S A, 101*, 87–92.

101. Conchillo-Solé, O., de Groot, N.S., Avilés, F.X., Vendrell, J., Daura, X., Ventura, S. (2007). AGGRESCAN: a server for the prediction and evaluation of "hot spots" of aggregation in polypeptides. *BMC Bioinformatics, 8*, 65.

102. Richardson, J.S.,. Richardson, D.C. (2002). Natural beta-sheet proteins use negative design to avoid edge-to-edge aggregation. *Proc Natl Acad Sci U S A, 99*, 2754–2759.

103. Esposito, G., Ricagno, S., Corazza, A., Rennella, E., Gümral, D., Mimmi, M.C., Betto, E., Pucillo, C.E., Fogolari, F., Viglino, P., *et al.* (2008). The controlling roles of Trp60 and Trp95 in beta 2-microglobulin function, folding and amyloid aggregation properties. *J Mol Biol, 378*, 885–895.

104. Kameda, A., Hoshino, M., Higurashi, T., Takahashi, S., Naiki, H., Goto, Y. (2005). Nuclear magnetic resonance characterization of the refolding intermediate of beta 2-microglobulin trapped by non-native prolyl peptide bond. *J Mol Biol, 348*, 383–397.

105. Borysik, A.J., Radford, S.E., Ashcroft, A.E. (2004). Co-populated conformational ensembles of beta 2-microglobulin uncovered quantitatively by electrospray ionization mass spectrometry. *J Biol Chem, 279*, 27069–27077.

106. Smith, D.P., Giles, K., Bateman, R.H., Radford, S.E., Ashcroft, A.E. (2007). Monitoring copopulated conformational states during protein folding events using electrospray ionization–ion mobility spectrometry–mass spectrometry. *J Am Soc Mass Spectrom, 18*, 2180–2190.

107. Jørgensen, T.J.D., Chenga, L., Heegaard, N.H.H. (2007). Mass spectrometric characterization of conformational preludes to beta 2-microglobulin aggregation. *Int J Mass Spectrom, 268*, 207–216.

108. Smith, A.M., Jahn, T.R., Ashcroft, A.E., Radford, S.E. (2006). Direct observation of oligomeric species formed in the early stages of amyloid fibril formation using electrospray ionisation mass spectrometry. *J Mol Biol, 364*, 9–19.

109. Xue, W.F., Homans, S.W., Radford, S.E. (2008). Systematic analysis of nucleation-dependent polymerization reveals new insights into the mechanism of amyloid self-assembly. *Proc Natl Acad Sci U S A, 105*, 8926–8931.

110. Calabrese, M.F., Miranker, A.D. (2007). Formation of a stable oligomer of beta 2-microglobulin requires only transient encounter with Cu(II). *J Mol Biol, 367*, 1–7.

111. Adlard, P.A., Bush, A.I. (2006). Metals and Alzheimer's disease. *J Alzheimer's Dis, 10*, 145–163.

112. Eakin, C.M., Miranker, A.D. (2005). From chance to frequent encounters: origins of beta 2-microglobulin fibrillogenesis. *Biochim Biophys Acta, 1753*, 92–99.

113. Thomson, N.M., Stevens, B.J., Humphery, T.J., Atkins, R.C. (1983). Comparison of trace elements in peritoneal dialysis, hemodialysis, and uremia. *Kidney Int, 23*, 9–14.

114. Vorbeck-Meister, I., Sommer, R., Vorbeck, F., Hörl, W.H. (1999). Quality of water used for haemodialysis: bacteriological and chemical parameters. *Nephrol Dial Transplant, 14*, 666–675.

115. Eakin, C.M., Knight, J.D., Morgan, C.J., Gelfand, M.A., Miranker, A.D. (2002). Formation of a copper specific binding site in non-native states of beta 2-microglobulin. *Biochemistry, 41*, 10646–10656.

116. De Lorenzi, E., Grossi, S., Massolini, G., Giorgetti, S., Mangione, P., Andreola, A., Chiti, F., Bellotti, V., Caccialanza, G. (2002). Capillary electrophoresis investigation of a partially unfolded conformation of beta 2-microglobulin. *Electrophoresis, 23*, 918–925.

117. Eakin, C.M., Attenello, F.J., Morgan, C.J., Miranker, A.D. (2004). Oligomeric assembly of native-like precursors precedes amyloid formation by beta 2-microglobulin. *Biochemistry, 43*, 7808–7815.

118. Villanueva, J., Hoshino, M., Katou, H., Kardos, J., Hasegawa, K., Naiki, H., Goto, Y. (2004). Increase in the conformational flexibility of beta 2-microglobulin upon copper binding: a possible role for copper in dialysis-related amyloidosis. *Protein Sci, 13*, 797–809.

119. Deng, N.J., Yan, L., Singh, D., Cieplak, P. (2006). Molecular basis for the Cu^{2+} binding-induced destabilization of beta 2-microglobulin revealed by molecular dynamics simulation. *Biophys J, 90*, 3865–3879.

120. Antwi, K., Mahar, M., Srikanth, R., Olbris, M.R., Tyson, J.F., Vachet, R.W. (2008). Cu(II) organizes beta 2-microglobulin oligomers but is released upon amyloid formation. *Protein Sci, 17*, 748–759.

121. Yamamoto, S., Gejyo, F. (2005). Historical background and clinical treatment of dialysis-related amyloidosis. *Biochim Biophys Acta, 1753*, 4–10.

122. Jaradat, M.I., Moe, S.M. (2001). Effect of hemodialysis membranes on beta 2-microglobulin amyloidosis. *Semin Dial, 14*, 107–112.

123. Kutsuki, H. (2005). Beta 2-microglobulin-selective direct hemoperfusion column for the treatment of dialysis-related amyloidosis. *Biochim Biophys Acta, 1753*, 141–145.

124. Lornoy, W., Because, I., Billiouw, J.M., Sierens, L., Van Malderen, P., D'Haenens, P. (2000). On-line haemodiafiltration: remarkable removal of beta 2-microglobulin. Long-term clinical observations. *Nephrol Dial Transplant, 15*(Suppl 1), 49–54.

125. Gejyo, F., Kawaguchi, Y., Hara, S., Nakazawa, R., Azuma, N., Ogawa, H., Koda, Y., Suzuki, M., Kaneda, H., Kishimoto, H., *et al.* (2004). Arresting dialysis-related amyloidosis: a prospective multicenter controlled trial of direct hemoperfusion with a beta 2-microglobulin adsorption column. *Artif Organs, 28*, 371–380.

126. Winchester, J.F., Salsberg, J.A., Levin, N.W. (2003). Beta 2-microglobulin in ESRD: an in-depth review. *Adv Ren Replace Ther, 10*, 279–309.

127. Bunka, D.H., Stockley, P.G. (2006). Aptamers come of age—at last. *Nat Rev Microbiol, 4*, 588–596.

128. Gillmore, J.D., Hawkins, P.N. (2006). Drug insight: emerging therapies for amyloidosis. *Nat Clin Pract Nephrol, 2*, 263–270.

129. Corazza, A., Pettirossi, F., Viglino, P., Verdone, G., Garcia, J., Dumy, P., Giorgetti, S., Mangione, P., Raimondi, S., Stoppini, M., *et al.* (2004). Properties of some variants of human beta 2-microglobulin and amyloidogenesis. *J Biol Chem, 279*, 9176–9189.

130. Linke, R.P., Hampl, H., Lobeck, H., Ritz, E., Bommer, J., Waldherr, R., Eulitz, M. (1989). Lysine-specific cleavage of beta 2-microglobulin in amyloid deposits associated with hemodialysis. *Kidney Int, 36*, 675–681.

17

COPPER–ZINC SUPEROXIDE DISMUTASE, ITS COPPER CHAPERONE, AND FAMILIAL AMYOTROPHIC LATERAL SCLEROSIS

Duane D. Winkler

Department of Biochemistry and the X-ray Crystallography Core Laboratory, University of Texas Health Science Center at San Antonio, San Antonio, Texas

Mercedes Prudencio, Celeste Karch, and David R. Borchelt

Department of Neuroscience, McKnight Brain Institute, University of Florida, Gainesville, Florida

P. John Hart

Department of Biochemistry and the X-ray Crystallography Core Laboratory, University of Texas Health Science Center at San Antonio, San Antonio, Texas

STRUCTURAL PROPERTIES OF COPPER–ZINC SUPEROXIDE DISMUTASE

Human copper–zinc superoxide dismutase (SOD1) is a 32-kDa homodimeric enzyme that catalyzes the disproportionation of superoxide radical to hydrogen peroxide and molecular oxygen ($2O_2^- + 2H^+ \rightarrow O_2 + H_2O_2$) [1,2]. Each subunit folds as an eight-stranded Greek key β-barrel, binds one copper and one zinc

Protein Misfolding Diseases: Current and Emerging Principles and Therapies,
Edited by Marina Ramirez-Alvarado, Jeffery W. Kelly, and Christopher M. Dobson
Copyright © 2010 John Wiley & Sons, Inc.

ion, and contains one instrasubunit disulfide bond [3]. Figure 1A shows the mature wild-type holoenzyme (PDB code 2C9V [4]). Two lengthy loop elements project from the β-barrel that are important in zinc binding (the *zinc loop*, loop IV, residues 50 to 83) and formation of the active site (the *electrostatic loop*, loop VI, residues 121 to 142). In the mature enzyme, the *disulfide loop*, a substructure of loop IV (residues 50 to 62), is covalently linked to the β-barrel through a disulfide bond between C57 and C146 (for a review, see [5]).

The pathogenic SOD1 mutations are divided into two groups based on their positions in the structure (Fig. 1A). β-Barrel mutants are isolated with metal content nearly identical to that found for the wild-type SOD1, while metal-binding mutants tend to be deficient in copper and/or zinc [6,7]. Three-dimensional structures are known for β-barrel mutants A4V [8], G37R [9,10], H43R [11], G93A [12], and I113T [8], and these metal-replete structures reveal only slight perturbations relative to the wild-type enzyme. Structures of the metal-binding mutants H46R [13,14], H46R/H48Q double mutant [15], D125H [16], and S134N [13,17] have also been determined, and most of these are metal-ion deficient, which in turn results in conformational disorder of the electrostatic and zinc loop elements. The biophysical properties of the two classes of pathogenic SOD1 mutants are quite different in their metal-free, disulfide-reduced (nascent) forms. Newly translated β-barrel mutants tend to be substantially destabilized relative to the wild-type enzyme, while nascent metal-binding mutants tend to retain stability similar to wild-type enzyme [18]. The stability of nascent SOD1 proteins overall is dramatically enhanced via posttranslational modification to the mature holoenzyme through the action of its copper chaperone (CCS) [19] (see below).

The mature human SOD1 holoenzyme is a remarkably stable dimer, retaining enzymatic activity at elevated temperature and in high concentrations of denaturing agents [20,21]. While the stabilizing effects of metal-ion binding have long been known, more recent studies have illuminated the role of the intrasubunit disulfide bond in dimer stability. The dimer interface is formed mainly by reciprocal interactions of the disulfide loop and β-strand 8 across the molecular twofold axis (Fig. 1A) [22]. Analytical ultracentrifugation analyses have revealed that reduction of the disulfide bond in the metal-free protein results in monomerization [22]. As can be inferred from Figure 1A, a reduced disulfide bond will result in enhanced mobility of the disulfide loop, weakening the interactions across the dimer interface [23].

GENETICS AND MODELS OF SOD1-LINKED fALS

In humans, the *sod1* gene is comprised of five exons located on chromosome 21 [24]. In all mammals, SOD1 is expressed ubiquitously in all tissues, and within cells it is localized primarily to the cytosol; lesser amounts are found in the nucleus, peroxisomes, and mitochondria [25]. The amino acid sequence of SOD1 is highly conserved; 112 of 153 residues are conserved in mammals with

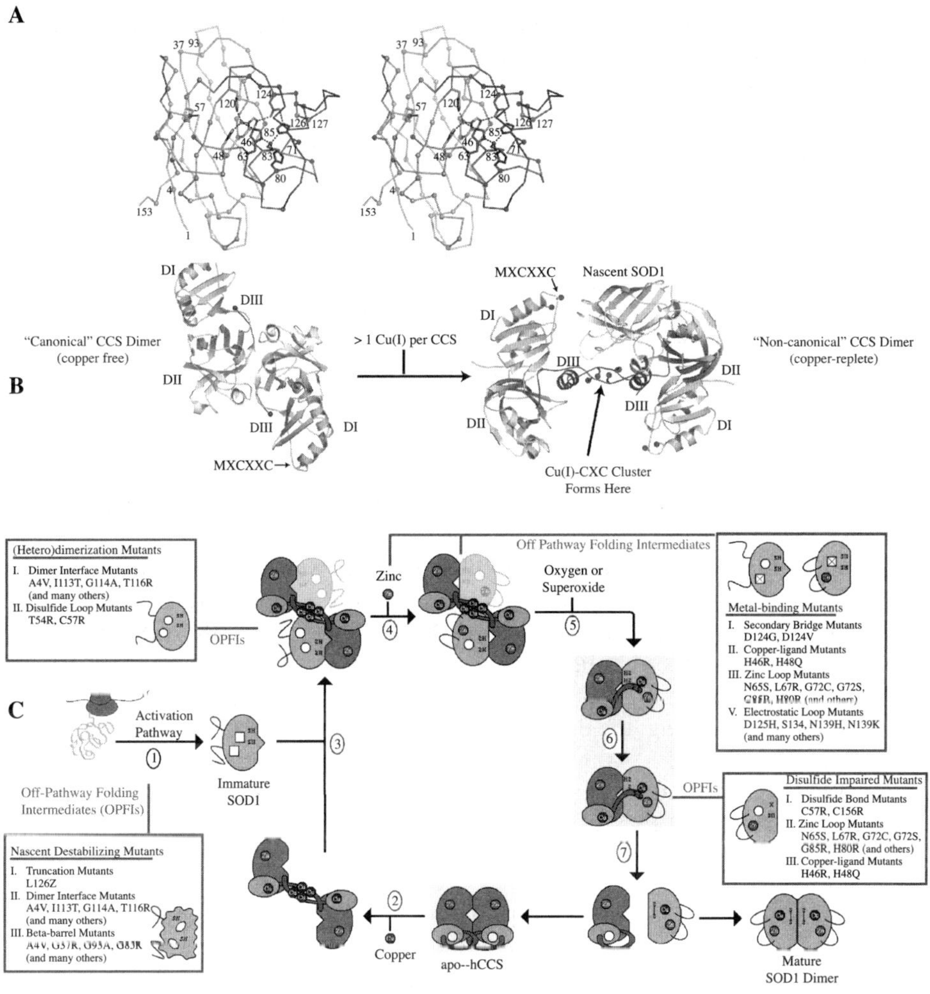

FIG. 1 (A) Stereo view of human SOD1 (PDB code 1AZV 9]). The β-barrel is shown in gray, the disulfide loop (residues 50 to 61) is shown in purple, the zinc loop (residues 62 to 83) is shown in blue, and the electrostatic loop (residues 121 to 142) is shown in red. The disulfide bond between C57 of the disulfide loop and C146 of β-strand 8 is shown as yellow sticks. Cu and Zn ligands are shown as gray and blue sticks and Cu and Zn ions are represented as cyan and magenta spheres, respectively. The α-carbon positions of β-barrel and metal-binding mutants are shown as gray and green spheres, respectively. The α-carbon positions of pathogenic SOD1 mutants for which there are mouse models (see the text) are shown as orange spheres. (B) The possible role of CCS in pathogenic SOD1 toxicity in fALS. A CCS canonical dimer (PDB code 1QUP 72])-to-noncanonical dimer (PDB code 1JK9 75]) transition upon the binding of Cu(I) 74]. (C) Alternate model of CCS action with selected off-pathway pathogenic SOD1 folding intermediates (see the text). (*See insert for color representation of figure.*)

70 absolutely conserved across eukaryotic phyla [26]. Missense mutations that cause familial amyotropic lateral sclerosis (fALS) have been documented at 68 positions in the SOD1 protein (Table 1 and Fig. 1A). Sixty-one of these mutations occur at residues conserved in mammals, with 49 mutations occurring at positions that are extremely conserved [26].

With the discovery of mutations in SOD1 as a cause of ALS, it became possible to create transgenic animal models of the disease. Transgenic mice that overexpress human SOD1-encoding mutations linked to familial ALS develop muscle loss and paralysis characteristic of the human disease. Almost all of these models are built using a 12-kb fragment of human genomic DNA, containing all regulatory elements, as the transgene vector [27]. Mutant human SOD1 genes that have been introduced into mice include A4V [28], G37R [29], H46R [30], G85R [31], G93A [32], L126Z [28,33], L126del(stop 131) [34], and Gins127TGGG [35]. Two human mutants have been expressed in transgenic rats (H46R and G93A [36,37]) and one version of mutant mouse SOD1 (G86R), using a vector derived from mouse genomic DNA, has been expressed in mice [38]. In some cases, transgenic mice have been used to express forms of mutant SOD1 that combine disease-linked mutations with experimental mutations in order to examine mechanisms of SOD1 toxicity. Mice producing mutant forms of SOD1 with two (H46R/H48Q) or four (H46R/H48Q/H63G/H120G) of the copper-binding residues mutated develop fALS-like paralysis [39,40] suggesting that the binding of copper in the active site is unlikely to be required in producing toxic proteins. The mouse model most widely used expresses the familial (f)ALS variant SOD1-G93A at very high levels and has a disease onset, marked by hindlimb weakness, at 3 to 4 months of age, with death occurring by 4 to 5 months of age [32].

AGGREGATION OF MUTANT SOD1 IN fALS

Defining protein aggregation is essential in understanding the role of protein misfolding in disease pathogenesis. In tissues from humans or from animal models, protein aggregates are usually defined by the formation of discernable macromolecular structures, often termed inclusion bodies. In human disease the availability of autopsy cases from SOD1-linked cases has been limited, but there are a number of case reports in the literature. The most consistently reported pathologic structures are hyaline or Lewy body–like inclusions that are immunoreactive to SOD1 antibodies; mutants examined include A4V [41,42], H46R [43], I113T [44], G72C [45], delL126 (stop 131) [46,47], and L126S [48]. In some cases, inclusion pathology has been noted but data on SOD1 immunoreactivity is lacking; D101N [49] and G37R [50], or inclusion pathology, is either absent or the inclusions that are present fail to react with antibodies to SOD1 [51–55]. In the fALS mouse models, pathologic inclusions are not necessarily prominent pathologic features [29,39,56,57], but have been observed as the major pathology of mice that express human SOD1-G85R [31].

Biochemically, aggregates isolated from tissues and cells are defined by several criteria (for a review, see [58]). In general, aggregates are derived from assemblies of protein that attain relatively high molecular mass (examples include filamentous aggregates as well as smaller oligomeric structures). In many cases, pathologic protein aggregates resist dissociation in detergent, and larger aggregates can be separated from tissue homogenates by ultracentrifugation or size-exclusion chromatography. Forms of mutant SOD1 that are insoluble in nonionic detergent have been detected in multiple mouse models, including mice that express the following variants: A4V [28], G37R [39], G85R [39], mouse G86R [59], G93A [39], L126Z [33], and Gins127TGGG [35]. Similar aggregates were found in spinal cord tissues of a fALS patient harboring the A4V mutation [39]. Importantly, mutant SOD1 aggregates with similar features are formed in cultured cells that express high levels of mutant protein [39,59,60]. Thus, aggregation of the mutant protein seems to be a consequence of fALS mutation.

Recent studies have investigated the role of disulfide bonding in SOD1 aggregation, and data reported to date have been interpreted as evidence that disulfide bond formation between mutant SOD1 proteins could either initiate oligomerization or stabilize aggregate structures [26,60–65]. Collectively, these studies have focused attention on cysteines 6 and 111 of SOD1 as playing important roles in modulating mutant SOD1 aggregation through aberrant intermolecular disulfide bonds. However, there have been reports of fALS mutations at all four cysteines in SOD1 (Table 1); and recent studies demonstrated that SOD1 mutants encoding disease-linked mutations at these cysteine residues (e.g., C6G, C6F, C111Y, C146R) rapidly formed aggregates when expressed in cell culture [59,60]. Moreover, Karch and Borchelt demonstrated that experimental mutants that lack all four cysteine residues (C6F/C57S/C111Y/C146R), or which encode only a single cysteine at positions 6 or 111 (C6/C57S/C111Y/C146R or C6F/C57S/C111/C146R) rapidly aggregated when expressed in cultured cells [59]. Collectively, these data suggest that extensive disulfide cross-linking is not required and may not play a significant role in the aggregation of mutant SOD1.

Little is known of the structure of mutant SOD1 aggregates that form in vivo. In vitro, metal-depleted mutant forms of SOD1 can acquire conformations of amyloidlike filaments and *amyloid pores* [13]. Amyloid pores have been observed in other neurodegenerative diseases in which protein aggregation is a characteristic feature [66]. In most but not all fALS mouse models, there is evidence of the accumulation of amyloidlike material; thioflavin-S^+ structures [33,39]. At present, we lack a sufficient understanding of the role of specific SOD1 aggregate structures in disease pathogenesis to predict whether disruption of mutant SOD1 aggregation would be beneficial or detrimental. A recent study of forms of SOD1 engineered to produce stable dimeric enzyme suggested that toxicity is not tied to aggregation [67]. However, data from cell culture models suggests that formation of large SOD1 aggregates could be a primary mechanism of toxicity [68].

TABLE 1

Mutations Exon 1	Principal Ref.
A4→S, T, or V	88–90
C6→F or G	92, 93
V7→E	95
L8→Q or V	96, 97
G10→V	98
G12→R	100
V14→G or M	102, 103
G16→A or S	97, 104
N19→S	97
F20→C	97
E21→G or K	96, 106
Q22→L	97

Mutations Exon 2	Principal Ref.
G37→R	91
L38→R or V	91, 94
G41→D or S	91
H43→R	91
F45→C	99
H46→R	101
V47→F	97
H48→Q or R	97, 105
E49→K	94
T54→R	97
C57→R	107

Mutations Exon 3	Principal Ref.
S59→I	97
N65→S	109
L67→R	94
G72→C or S	45, 110
D76→V or Y	102, 114

Mutations Exon 4	Principal Ref.
H80→R	108
L84→F or V	110, 111
G85→R	91
N86→D, K, or S	107, 112, 113
V87→A	97
A89→T or V	97, 115
D90→A or V	116, 117
G93→A, C, D, R, S, or V	91, 96, 118–120
A95→T	99
D96→N	121
V97→M	97
E100→G or K	91, 96
D101→G, H, N or Y	126–129
I104→F	130
S105→L	97
L106→V	91
G108→V	133
D109→Y	134
c111→Y	135
I112→M or T	109, 119
I113→F or T	91, 97
G114→A	97
T116→R	107
R115→G	137
V118→L	97, 138

Mutations Exon 5	Principal Ref.
D124→G or V	97, 118
D125→H	105
L126→S or stop	48, 96
S134→N	122
N139→H or K	123, 124
A140→G	125
G141→E or stop	97, 127
L144→F or S	131, 132
A145→G or T	97, 132
C146→R	96
G147→R	97
V148→G or I	130, 131
I149→T	124
I151→S or T	97, 136

TABLE 1 (*Continued*)

Deletions and Insertions Exon 4	Principal Ref.	Deletions and Insertions Exon 5	Principal Ref.
T88delTAD*	97	L126del (stop 131)	124
S105delSL*	97	E133del*	118
V118Lins (stop 122)	139	G127ins (stop 133)	102
		E132ins (stop 133)	133

COPPER CHAPERONE FOR SOD1 AND SOD1 MATURATION

The human copper chaperone for SOD1 (CCS) is a three-domain polypeptide that confers two critical stabilizing posttranslational modifications on newly synthesized SOD1: insertion of the catalytic copper ion [19] and oxidation of the SOD1 intrasubunit disulfide bond [69]. The presence of the latter is quite rare for cytosolic proteins given the strong reducing environment of the cytosol. At the protein level, the ratio of SOD1 to CCS in the cytosol is between 15- and 30-fold [70], and CCS must cycle through the nascent SOD1 pool to activate these molecules [71]. CCS domain I (residues 1 to 84) contains the copper-binding motif MXCXXC and is postulated to acquire copper ion from the membrane copper transporter CTR1 [71]. Domain II (residues 85 to 233) is remarkably similar to human SOD1 and retains amino acid residues found at the SOD1 dimer interface [72]. Domain II is thus probably responsible for the specificity of CCS/SOD1 interaction via the formation of a heterodimer. Domain III (residues 234 to 273) contains the copper-binding motif CXC, which is proposed to insert copper ion directly into nascent SOD1 [73]. The "heterodimerization" model of CCS activation of SOD1 has endured for the last 10 years, although the spatial-temporal mechanistic details of the activation process have remained elusive. Human CCS itself dimerizes though its SOD1-like domain II, which contains a zinc-binding site and disulfide bond analogous to those found in SOD1. Unanswered mechanistic questions include the following: How does CCS reorganize itself to utilize domain II as a nascent SOD1 recognition module? How is copper transferred from CCS domain I to CCS domain III prior to delivery to SOD1? What amino acid residues of both proteins participate in copper delivery and disulfide bond oxidation? Perhaps most important, how might fALS mutations in SOD1 interfere with CCS action, and what are the properties of the resulting immature SOD1 proteins?

ALTERNATE MODEL FOR CCS ACTION

Blackburn and colleagues titrated purified human CCS with increasing concentrations of Cu(I) followed by EXAFS and gel filtration studies. They noted that upon addition of a single equivalent of Cu(I) per CCS molecule, the canonical CCS domain II-mediated dimer dissociated into monomers, and that addition of additional equivalents resulted in the formation of a Cu_4S_6 copper cluster mediated by two CXC motifs of CCS domain III (Fig. 1B and C). This canonical-to-noncanonical CCS dimer transition, if it occurs in vivo, would seem to act as a functional copper-sensing switch to make CCS domain II available to nascent SOD1 binding only when sufficient copper is available [74]. A domain III–mediated noncanonical CCS dimer was also observed in the crystal structure of a yeast SOD1/yeast CCS complex [75]. These observations suggest the alternate model of CCS action shown in Figure 1B and C. However, the alternate model still does not reveal mechanistic details of posttranslational modification of SOD1 nor of how the binding of a single copper ion results in allosteric CCS dimer dissociation. The latter mechanism is particularly intriguing given that CCS contains its own intrasubunit disulfide bond and bound zinc, which, as described above, are both factors known to stabilize SOD1 dimers [22].

IMMATURE PATHOGENIC SOD1 AND TOXICITY

All pathogenic SOD1 mutations may result in off-pathway folding intermediates by hindering the action of CCS at various points in the SOD1 maturation cycle (Fig. 1B and 1C). Recent studies strongly suggest that the resulting immature SOD1 molecules eventually end up in the aggregates observed in humans and in transgenic mouse models [33,35,63]. For example, the nascent L126 truncation variant is so destabilized relative to the nascent wild-type enzyme that CCS probably encounters only a tiny fraction of these molecules before they are degraded or enter the insoluble fraction. CCS would fail to stabilize the nascent L126 molecules it does encounter because the latter molecule is completely lacking a β-strand necessary for wild-type SOD1 heterodimerization with CCS domain II. While nascent β-barrel mutants such as A4V, G37R, and G93A (among many others) are not as radically destabilized as L126Z, they are still destabilized significantly relative to the nascent wild-type enzyme [18]. We speculate that CCS is unable to cycle through the entire pool of these nascent SOD1 pathogenic mutants before they are degraded or enter the insoluble fraction. Metal-binding mutants such as H46R, H48Q, H46R/H48Q, H80R, and D124V (among many others) are not destabilized relative to the nascent wild-type enzyme, but CCS can never stabilize these molecules via posttranslational modification because these mutations directly prevent metal binding. CCS can also never stabilize the pathogenic SOD1 mutants C57R and C146R, which can never make the intrasubunit disulfide bond.

The notion that immature pathogenic SOD1 molecules may be the noxious species in SOD1-linked fALS is supported by recent studies from the laboratories of Jeffery Elliot and Valeria Culotta. Overexpression of CCS was observed to greatly accelerate disease in a G93A SOD1 mouse model in the absence of visible proteinaceous inclusions [76]. Surprisingly, CCS overexpression failed to enhance oxidation of the G93A SOD1 disulfide bond, and in fact, elevated the population of disulfide-reduced G93A SOD1 in the soluble fraction of brain and spinal cord of these animals [77]. In this model, there appears to be augmentation of the mitochondrial pathology that is inherent in the G93A mice but is not found in several other models (e.g., H46R/H48Q, G85R, L126Z, and G127insTGGG). These data suggest that CCS may be interacting with nascent G93A (which are at a stoichiometric ratio of approximately 1:1 in these animals) preventing its aggregation, and that the elevated levels of soluble disulfide-reduced G93A SOD1 observed augment the mitochondrial pathology, resulting in significantly earlier onset of paralytic symptoms. It remains unclear why CCS overexpression in these animals results in elevated levels of disulfide-reduced G93A SOD1 and additional studies aimed at understanding this phenomenon are needed.

THERAPEUTICS

There are no drugs that specifically target SOD1 that are used to treat fALS patients. The only therapy available for these patients, like all other ALS cases, is riluzole [78], which targets the glutamate neurotransmitter system. However, recent studies have explored the potential to use molecular therapies that specifically target the expression of mutant SOD1 in fALS [79–81]. Others have used approaches to modify the expression of mutant SOD1 selectively in transgenic mice; providing evidence that lowering levels of mutant SOD1 in microglia [82] or astrocytes [83] slows the later stages of disease progression. Lowering mutant SOD1 in muscle does not affect disease onset, progression, or survival [84]. The major advantage of these molecular therapies is that they target directly the cause of disease. One group has addressed whether stabilization of SOD1 structure may be a therapeutic target for treatment of fALS [85]. Small molecules that stabilize native dimeric structure of SOD1-A4V were found to slow aggregation of the protein in vitro [86,87]. Small molecules designed to block SOD1 oligomerization have yet to be identified; and to date, there is no information as to whether inhibition of SOD1 aggregation is effective in lowering toxicity in an animal model.

CONCLUSIONS

More than 100 mutations in SOD1 have been identified as causing fALS. The precise role of mutant SOD1 aggregation in disease pathogenesis remains uncertain, although there is little dispute that aggregates of mutant SOD1 are found in all of the transgenic mouse models that have been generated thus far, except in one model, where CCS is also overexpressed. Ultimately, the role of mutant SOD1 aggregation in disease may not be established until therapeutic compounds that specifically target mutant SOD1 aggregation are tested in clinical applications. We suggest a potential mechanism of toxicity involving reduced ability of the mutant proteins to interact properly with CCS, which mediates critical posttranslational modification of the SOD1 as it folds into a stable dimeric enzyme. The resulting immature pathogenic SOD1 proteins are essentially off-pathway folding intermediates that exert their toxic effects through their aggregated or soluble forms, or both.

Acknowledgments

This work has been supported over the years by the National Institutes of Health/NINDS (P.J.H./D.R.B.), the ALS Association (P.J.H./D.R.B.), the Robert A. Welch Foundation (P.J.H.), the Packard Foundation (D.R.B.), and the Judith and Jean Pape Adams Charitable Foundation (P.J.H.). Support for the X-ray Crystallography Core Laboratory by the Executive Research Council at the University of Texas Health Science Center is gratefully acknowledged.

REFERENCES

1. Fridovich, I. (1989). Superoxide dismutases: an adaptation to a paramagnetic gas. *J Biol Chem, 264*, 7761–7764.
2. Valentine, J.S., Pantoliano, M.W. (1981). *Copper proteins.* In *Metal Ions in Biology* (Spiro, T.G., Ed.), Wiley, New York, pp. 291–358.
3. Tainer, J.A., Getzoff, E.D., Beem, K.M., Richardson, J.S., Richardson, D.C. (1982). Determination and analysis of the 2 Å-structure of copper, zinc superoxide dismutase. *J Mol Biol, 160*, 181–217.
4. Strange, R.W., Antonyuk, S.V., Hough, M.A., Doucette, P.A., Valentine, J.S., Hasnain, S.S. (2006). Variable metallation of human superoxide dismutase: atomic resolution crystal structures of Cu–Zn, Zn–Zn and as-isolated wild-type enzymes. *J Mol Biol, 356*, 1152–1162.
5. Hart, P.J. (2006). Pathogenic superoxide dismutase structure, folding, aggregation and turnover. *Curr Opin Chem Biol, 10* 131–138.
6. Hayward, L.J., Rodriguez, J.A., Kim, J.W., Tiwari, A., Goto, J.J., Cabelli, D.E., Valentine, J.S., Brown, R.H., Jr. (2002). Decreased metallation and activity in subsets of mutant superoxide dismutases associated with familial amyotrophic lateral sclerosis. *J Biol Chem, 277*, 15923–15931.
7. Valentine, J.S., Hart, P.J. (2003). Misfolded CuZnSOD and amyotrophic lateral sclerosis. *Proc Natl Acad Sci U S A, 100*, 3617–3622.
8. Hough, M.A., Grossmann, J.G., Antonyuk, S.V., Strange, R.W., Doucette, P.A., Rodriguez, J.A., Whitson, L.J., Hart, P.J., Hayward, L.J., Valentine, J.S., Hasnain, S.S. (2004). Dimer destabilization in superoxide dismutase may result in disease-causing properties: structures of motor neuron disease mutants. *Proc Natl Acad Sci U S A, 101*, 5976–5981.

9. Hart, P.J., Liu, H., Pellegrini, M., Nersissian, A.M., Gralla, E.B., Valentine, J.S., Eisenberg, D. (1998). Subunit asymmetry in the three-dimensional structure of a human CuZnSOD mutant found in familial amyotrophic lateral sclerosis. *Protein Sci, 7*, 545–555.
10. Banci, L., Bertini, I., D'Amelio, N., Libralesso, E., Turano, P., Valentine, J.S. (2007). Metalation of the amyotrophic lateral sclerosis mutant glycine 37 to arginine superoxide dismutase (SOD1) apoprotein restores its structural and dynamical properties in solution to those of metalated wild-type SOD1. *Biochemistry, 46*, 9953–9962.
11. DiDonato, M., Craig, L., Huff, M.E., Thayer, M.M., Cardoso, R.M., Kassmann, C.J., Lo, T.P., Bruns, C.K., Powers, E.T., Kelly, J.W., et al. (2003). ALS mutants of human superoxide dismutase form fibrous aggregates via framework destabilization. *J Mol Biol, 332*, 601–615.
12. Shipp, E.L., Cantini, F., Bertini, I., Valentine, J.S., Banci, L. (2003). Dynamic properties of the G93A mutant of copper–zinc superoxide dismutase as detected by NMR spectroscopy: implications for the pathology of familial amyotrophic lateral sclerosis. *Biochemistry, 42*, 1890–1899.
13. Elam, J.S., Taylor, A.B., Strange, R., Antonyuk, S., Doucette, P.A., Rodriguez, J.A., Hasnain, S.S., Hayward, L.J., Valentine, J.S., Yeates, T.O., Hart, P.J. (2003). Amyloid-like filaments and water-filled nanotubes formed by SOD1 mutant proteins linked to familial ALS. *Nat Struct Biol, 10*, 461–467.
14. Antonyuk, S., Elam, J.S., Hough, M.A., Strange, R.W., Doucette, P.A., Rodriguez, J.A., Hayward, L.J., Valentine, J.S., Hart, P.J., Hasnain, S.S. (2005). Structural consequences of the familial amyotrophic lateral sclerosis SOD1 mutant His46Arg. *Protein Sci, 14*, 1201–1213.
15. Wang, J., Caruano-Yzermans, A., Rodriguez, A., Scheurmann, J.P., Slunt, H.H., Cao, X., Gitlin, J., Hart, P.J., Borchelt, D.R. (2007). Disease-associated mutations at copper ligand histidine residues of superoxide dismutase 1 diminish the binding of copper and compromise dimer stability. *J Biol Chem, 282*, 345–352.
16. Elam, J.S., Malek, K., Rodriguez, J.A., Doucette, P.A., Taylor, A.B., Hayward, L.J., Cabelli, D.E., Valentine, J.S., Hart, P.J. (2003). An alternative mechanism of bicarbonate-mediated peroxidation by copper–zinc superoxide dismutase: rates enhanced via proposed enzyme-associated peroxycarbonate intermediate. *J Biol Chem, 278*, 21032–21039.
17. Banci, L., Bertini, I., D'Amelio, N., Gaggelli, E., Libralesso, E., Matecko, I., Turano, P., Valentine, J.S. (2005). Fully metallated S134N Cu,Zn-superoxide dismutase displays abnormal mobility and intermolecular contacts in solution. *J Biol Chem, 280*, 35815–35821.
18. Rodriguez, J.A., Shaw, B.F., Durazo, A., Sohn, S.H., Doucette, P.A., Nersissian, A.M., Faull, K.F., Eggers, D.K., Tiwari, A., Hayward, L.J., Valentine, J.S. (2005). Destabilization of apoprotein is insufficient to explain Cu,Zn-superoxide dismutase-linked ALS pathogenesis. *Proc Natl Acad Sci U S A, 102*, 10516–10521.
19. Culotta, V.C., Klomp, L.W., Strain, J., Casareno, R.L., Krems, B., Gitlin, J.D. (1997). The copper chaperone for superoxide dismutase. *J Biol Chem, 272*, 23469–23472.
20. Forman, H.J., Fridovich, I. (1973). On the stability of bovine superoxide dismutase: the effects of metals. *J Biol Chem, 248*, 2645–2649.

21. Hartz, J.W., Deutsch, H.F. (1972). Subunit structure of human superoxide dismutase. *J Biol Chem, 247*, 7043–7050.

22. Doucette, P.A., Whitson, L.J., Cao, X., Schirf, V., Demeler, B., Valentine, J.S., Hansen, J.C., Hart, P.J. (2004). Dissociation of human copper–zinc superoxide dismutase dimers using chaotrope and reductant: insights into the molecular basis for dimer stability. *J Biol Chem, 279*, 54558–54566.

23. Banci, L., Bertini, I., Cantini, F., D'Amelio, N., Gaggelli, E. (2006). Human SOD1 before harboring the catalytic metal: solution structure of copper-depleted, disulfide-reduced form. *J Biol Chem, 281*, 2333–2337.

24. Levanon, D., Lieman-Hurwitz, J., Dafni, N., Wigderson, M., Sherman, L., Bernstein, Y., Laver-Rudich, Z., Danciger, E., Stein, O., Groner, Y. (1985). Architecture and anatomy of the chromosomal locus in human chromosome 21 encoding the Cu/Zn superoxide dismutase. *EMBO J, 4*, 77–84.

25. Sturtz, L.A., Diekert, K., Jensen, L.T., Lill, R., Culotta, V.C. (2001). A fraction of yeast Cu,Zn-superoxide dismutase and its metallochaperone, CCS, localize to the intermembrane space of mitochondria: a physiological role for SOD1 in guarding against mitochondrial oxidative damage. *J Biol Chem, 276*, 38084–38089.

26. Wang, J., Xu, G., Borchelt, D.R. (2006). Mapping superoxide dismutase 1 domains of non-native interaction: roles of intra- and intermolecular disulfide bonding in aggregation. *J Neurochem, 96*, 1277–1288.

27. Epstein, C.J., Avraham, K.B., Lovett, M., Smith, S., Elroy-Stein, O., Rotman, G., Bry, C., Groner, Y. (1987). Transgenic mice with increased Cu/Zn-superoxide dismutase activity: animal model of dosage effects in Down syndrome. *Proc Natl Acad Sci U S A, 84*, 8044–8048.

28. Deng, H.X., Shi, Y., Furukawa, Y., Zhai, H., Fu, R., Liu, E., Gorrie, G.H., Khan, M.S., Hung, W.Y., Bigio, E.H., et.al. (2006). Conversion to the amyotrophic lateral sclerosis phenotype is associated with intermolecular linked insoluble aggregates of SOD1 in mitochondria. *Proc Natl Acad Sci U S A, 103*, 7142–7147.

29. Wong, P.C., Pardo, C.A., Borchelt, D.R., Lee, M.K., Copeland, N.G., Jenkins, N.A., Sisodia, S.S., Cleveland, D.W., Price, D.L. (1995). An adverse property of a familial ALS-linked SOD1 mutation causes motor neuron disease characterized by vacuolar degeneration of mitochondria. *Neuron, 14*, 1105–1116.

30. Sasaki, S., Nagai, M., Aoki, M., Komori, T., Itoyama, Y., Iwata, M. (2007). Motor neuron disease in transgenic mice with an H46R mutant SOD1 gene. *J Neuropathol Exp Neurol, 66*, 517–524.

31. Bruijn, L.I., Becher, M.W., Lee, M.K., Anderson, K.L., Jenkins, N.A., Copeland, N.G., Sisodia, S.S., Rothstein, J.D., Borchelt, D.R., Price, D.L., Cleveland, D.W. (1997). ALS-linked SOD1 mutant G85R mediates damage to astrocytes and promotes rapidly progressive disease with SOD1-containing inclusions. *Neuron, 18*, 327–338.

32. Gurney, M.E., Pu, H., Chiu, A.Y., Dal Canto, M.C., Polchow, C.Y., Alexander, D.D., Caliendo, J., Hentati, A., Kwon, Y.W., Deng, H.X., et al. (1994). Motor neuron degeneration in mice that express a human Cu,Zn superoxide dismutase mutation. *Science, 264*, 1772–1775.

33. Wang, J., Xu, G., Li, H., Gonzales, V., Fromholt, D., Karch, C., Copeland, N.G., Jenkins, N.A., Borchelt, D.R. (2005). Somatodendritic accumulation of misfolded SOD1-L126Z in motor neurons mediates degeneration: alphaB-crystallin modulates aggregation. *Hum Mol Genet, 14*, 2335–2347.

34. Watanabe, Y., Yasui, K., Nakano, T., Doi, K., Fukada, Y., Kitayama, M., Ishimoto, M., Kurihara, S., Kawashima, M., Fukuda, H., et al. (2005). Mouse motor neuron disease caused by truncated SOD1 with or without C-terminal modification. *Brain Res Mol Brain Res, 135*, 12–20.

35. Jonsson, P.A., Ernhill, K., Andersen, P.M., Bergemalm, D., Brännström, T., Gredal, O., Nilsson, P., Marklund, S.L. (2004). Minute quantities of misfolded mutant superoxide dismutase-1 cause amyotrophic lateral sclerosis. *Brain, 127*, 73–88.

36. Howland, D.S., Liu, J., She, Y., Goad, B., Maragakis, N.J., Kim, B., Erickson, J., Kulik, J., DeVito, L., Psaltis, G., et al. (2002). Focal loss of the glutamate transporter EAAT2 in a transgenic rat model of SOD1 mutant-mediated amyotrophic lateral sclerosis (ALS). *Proc Natl Acad Sci U S A, 99*, 1604–1609.

37. Nagai, M., Aoki, M., Miyoshi, I., Kato, M., Pasinelli, P., Kasai, N., Brown, R.H., Jr., Itoyama, Y. (2001). Rats expressing human cytosolic copper–zinc superoxide dismutase transgenes with amyotrophic lateral sclerosis: associated mutations develop motor neuron disease. *J Neurosci, 21*, 9246–9254.

38. Ripps, M.E., Huntley, G.W., Hof, P.R., Morrison, J.H., Gordon, J.W. (1995). Transgenic mice expressing an altered murine superoxide dismutase gene provide an animal model of amyotrophic lateral sclerosis. *Proc Natl Acad Sci U S A, 92*, 689–693.

39. Wang, J., Slunt, H., Gonzales, V., Fromholt, D., Coonfield, M., Copeland, N.G., Jenkins, N.A., Borchelt, D.R. (2003). Copper-binding-site-null SOD1 causes ALS in transgenic mice: aggregates of non-native SOD1 delineate a common feature. *Hum Mol Genet, 12*, 2753–2764.

40. Wang, J., Xu, G., Gonzales, V., Coonfield, M., Fromholt, D., Copeland, N.G., Jenkins, N.A., Borchelt, D.R. (2002). Fibrillar inclusions and motor neuron degeneration in transgenic mice expressing superoxide dismutase 1 with a disrupted copper-binding site. *Neurobiol Dis, 10*, 128–138.

41. Shibata, N., Hirano, A., Kobayashi, M., Siddique, T., Deng, H.X., Hung, W.Y., Kato, T., Asayama, K. (1996). Intense superoxide dismutase-1 immunoreactivity in intracytoplasmic hyaline inclusions of familial amyotrophic lateral sclerosis with posterior column involvement. *J Neuropathol Exp Neurol, 55*, 481–490.

42. Kato, S., Nakashima, K., Horiuchi, S., Nagai, R., Cleveland, D.W., Liu, J., Hirano, A., Takikawa, M., Kato, M., Nakano, I., et al. (2001). Formation of advanced glycation end-product-modified superoxide dismutase-1 (SOD1) is one of the mechanisms responsible for inclusions common to familial amyotrophic lateral sclerosis patients with SOD1 gene mutation, and transgenic mice expressing human SOD1 gene mutation. *Neuropathology, 21*, 67–81.

43. Ohi, T., Nabeshima, K., Kato, S., Yazawa, S., Takechi, S. (2004). Familial amyotrophic lateral sclerosis with His46Arg mutation in Cu/Zn superoxide dismutase presenting characteristic clinical features and Lewy body–like hyaline inclusions. *J Neurol Sci, 225*, 19–25.

44. Kokubo, Y., Kuzuhara, S., Narita, Y., Kikugawa, K., Nakano, R., Inuzuka, T., Tsuji, S., Watanabe, M., Miyazaki, T., Murayama, S., Ihara, Y. (1999). Accumulation of neurofilaments and SOD1-immunoreactive products in a patient with familial amyotrophic lateral sclerosis with I113T SOD1 mutation. *Arch Neurol, 56*, 1506–1508.

45. Stewart, H.G., Mackenzie, I.R., Eisen, A., Brännström, T., Marklund, S.L., Andersen, P.M. (2006). Clinicopathological phenotype of ALS with a novel G72C SOD1 gene mutation mimicking a myopathy. *Muscle Nerve, 33*, 701–706.

46. Kato, S., Shimoda, M., Watanabe, Y., Nakashima, K., Takahashi, K., Ohama, E. (1996). Familial amyotrophic lateral sclerosis with a two base pair deletion in superoxide dismutase 1 gene: multisystem degeneration with intracytoplasmic hyaline inclusioins in astrocytes. *J Neuropathol Exp Neurol, 55*, 1089–1101.

47. Kato, S., Sumi-Akamaru, H., Fujimura, H., Sakoda, S., Kato, M., Hirano, A., Takikawa, M., Ohama, E. (2001). Copper chaperone for superoxide dismutase co-aggregates with superoxide dismutase 1 (SOD1) in neuronal Lewy body-like hyaline inclusions: an immunohistochemical study on familial amyotrophic lateral sclerosis with SOD1 gene mutation. *Acta Neuropathol, 102*, 233–238.

48. Takehisa, Y., Ujike, H., Ishizu, H., Terada, S., Haraguchi, T., Tanaka, Y., Nishinaka, T., Nobukuni, K., Ihara, Y., Namba, R., et al. (2001). Familial amyotrophic lateral sclerosis with a novel Leu126Ser mutation in the copper/zinc superoxide dismutase gene showing mild clinical features and Lewy body–like hyaline inclusions. *Arch Neurol, 58*, 736–740.

49. Cervenakova, L., Protas, I.I., Hirano, A., Votiakov, V.I., Nedzved, M.K., Kolomiets, N.D., Taller, I., Park, K.Y., Sambuughin, N., Gajdusek, D.C., et al. (2000). Progressive muscular atrophy variant of familial amyotrophic lateral sclerosis (PMA/ALS). *J Neurol Sci, 177*, 124–130.

50. Inoue, K., Fujimura, H., Ogawa, Y., Satoh, T., Shimada, K., Sakoda, S. (2002). Familial amyotrophic lateral sclerosis with a point mutation (G37R) of the superoxide dismutase 1 gene: a clinicopathological study. *Amyotroph Lateral Scler Other Motor Neuron Disord, 3*, 244–247.

51. Shibata, N., Asayama, K., Hirano, A., Kobayashi, M. (1996). Immunohistochemical study on superoxide dismutases in spinal cords from autopsied patients with amyotrophic lateral sclerosis. *Dev Neurosci, 18*, 492–498.

52. Ince, P.G., Shaw, P.J., Slade, J.Y., Jones, C., Hudgson, P. (1996). Familial amyotrophic lateral sclerosis with a mutation in exon 4 or the Cu/Zn superoxide dismutase gene: pathological and immunocytochemical changes. *Acta Neuropathol, 92*, 395–403.

53. Shaw, C.E., Enayat, Z.E., Powell, J.F., Anderson, V.E., Radunovic, A., al-Sarraj, S., Leigh, P.N. (1997). Familial amyotrophic lateral sclerosis: molecular pathology of a patient with a SOD1 mutation. *Neurology, 49*, 1612–1616.

54. Kadekawa, J., Fujimura, H., Ogawa, Y., Hattori, N., Kaido, M., Nishimura, T., Yoshikawa, H., Shirahata, N., Sakoda, S., Yanagihara, T. (1997). A clinicopathological study of a patient with familial amyotrophic lateral sclerosis associated with a two base pair deletion in the copper/zinc superoxide dismutase (SOD1) gene. *Acta Neuropathol (Berl), 94*, 617–622.

55. Katayama, S., Watanabe, C., Noda, K., Ohishi, H., Yamamura, Y., Nishisaka, T., Inai, K., Asayama, K., Murayama, S., Nakamura, S. (1999). Numerous conglomerate inclusions in slowly progressive familial amyotrophic lateral sclerosis with posterior column involvement. *J Neurol Sci, 171*, 72–77.

56. Gurney, M.E. (1994). Transgenic-mouse model of amyotrophic lateral sclerosis. *N Engl J Med, 331*, 1721–1722.

57. Wang, J., Xu, G., Slunt, H.H., Gonzales, V., Coonfield, M., Fromholt, D., Copeland, N.G., Jenkins, N.A., Borchelt, D.R. (2005). Coincident thresholds of mutant protein for paralytic disease and protein aggregation caused by restrictively expressed superoxide dismutase cDNA. *Neurobiol Dis, 20*, 943–952.

58. Murphy, R.M. (2002). Peptide aggregation in neurodegenerative disease. *Annu Rev Biomed Eng, 4*, 155–174.

59. Karch, C.M., Borchelt, D.R. (2008). A limited role for disulfide cross-linking in the aggregation of mutant SOD1 linked to familial amyotrophic lateral sclerosis. *J Biol Chem, 283*, 13528–13537.

60. Cozzolino, M., Amori, I., Pesaresi, M.G., Ferri, A., Nencini, M., Carrì, M.T. (2008). Cysteine 111 affects aggregation and cytotoxicity of mutant Cu,Zn-superoxide dismutase associated with familial amyotrophic lateral sclerosis. *J Biol Chem, 283*, 866–874.

61. Banci, L., Bertini, I., Durazo, A., Girotto, S., Gralla, E.B., Martinelli, M., Valentine, J.S., Vieru, M., Whitelegge, J.P. (2007). Metal-free superoxide dismutase forms soluble oligomers under physiological conditions: a possible general mechanism for familial ALS. *Proc Natl Acad Sci U S A, 104*, 11263–11267.

62. Furukawa, Y., Fu, R., Deng, H.X., Siddique, T., O'Halloran, T.V. (2006). Disulfide cross-linked protein represents a significant fraction of ALS-associated Cu, Zn-superoxide dismutase aggregates in spinal cords of model mice. *Proc Natl Acad Sci U S A, 103*, 7148–7153.

63. Jonsson, P.A., Graffmo, K.S., Andersen, P.M., Brännström, T., Lindberg, M., Oliveberg, M., Marklund, S.L. (2006). Disulphide-reduced superoxide dismutase-1 in CNS of transgenic amyotrophic lateral sclerosis models. *Brain, 129*, 451–464.

64. Niwa, J., Yamada, S., Ishigaki, S., Sone, J., Takahashi, M., Katsuno, M., Tanaka, F., Doyu, M., Sobue, G. (2007). Disulfide bond mediates aggregation, toxicity, and ubiquitylation of familial amyotrophic lateral sclerosis-linked mutant SOD1. *J Biol Chem, 282*, 28087–28095.

65. Banci, L., Bertini, I., Boca, M., Girotto, S., Martinelli, M., Valentine, J.S., Vieru, M. (2008). SOD1 and amyotrophic lateral sclerosis: mutations and oligomerization. *PLoS ONE, 3*, e1677.

66. Lashuel, H.A., Hartley, D., Petre, B.M., Walz, T., Lansbury, P.T., Jr. (2002). Neurodegenerative disease: amyloid pores from pathogenic mutations. *Nature, 418*, 291.

67. Witan, H., Kern, A., Koziollek-Drechsler, I., Wade, R., Behl, C., Clement, A.M. (2008). Heterodimer formation of wild-type and amyotrophic lateral sclerosis-causing mutant Cu/Zn-superoxide dismutase induces toxicity independent of protein aggregation. *Hum Mol Genet, 17*, 1373–1385.

68. Matsumoto, G., Stojanovic, A., Holmberg, C.I., Kim, S., Morimoto, R.I. (2005). Structural properties and neuronal toxicity of amyotrophic lateral sclerosis-associated Cu/Zn superoxide dismutase 1 aggregates. *J Cell Biol, 171*, 75–85.

69. Brown, N.M., Torres, A.S., Doan, P.E., O'Halloran, T.V. (2004). Oxygen and the copper chaperone CCS regulate posttranslational activation of Cu,Zn superoxide dismutase. *Proc Natl Acad Sci U S A, 101*, 5518–5523.

70. Rothstein, J.D., Dykes-Hoberg, M., Corson, L.B., Becker, M., Cleveland, D.W., Price, D.L., Culotta, V.C., Wong, P.C. (1999). The copper chaperone CCS is

abundant in neurons and astrocytes in human and rodent brain. *J Neurochem, 72*, 422–429.

71. Furukawa, Y., O'Halloran, T.V. (2006). Posttranslational modifications in Cu,Zn-superoxide dismutase and mutations associated with amyotrophic lateral sclerosis. *Antioxid Redox Signal, 8*, 847–867.
72. Lamb, A.L., Wernimont, A.K., Pufahl, R.A., Culotta, V.C., O'Halloran, T.V., Rosenzweig, A.C. (1999). Crystal structure of the copper chaperone for superoxide dismutase. *Nat Struct Biol, 6*, 724–729.
73. Schmidt, P.J., Rae, T.D., Pufahl, R.A., Hamma, T., Strain, J., O'Halloran, T.V., Culotta, V.C. (1999). Multiple protein domains contribute to the action of the copper chaperone for superoxide dismutase. *J Biol Chem, 274*, 23719–23725.
74. Stasser, J.P., Siluvai, G.S., Barry, A.N., Blackburn, N.J. (2007). A multinuclear copper(I) cluster forms the dimerization interface in copper-loaded human copper chaperone for superoxide dismutase. *Biochemistry, 46*, 11845–11856.
75. Lamb, A.L., Torres, A.S., O'Halloran, T.V., Rosenzweig, A.C. (2001). Heterodimeric structure of superoxide dismutase in complex with its metallochaperone. *Nat Struct Biol, 8*, 751–755.
76. Son, M., Puttaparthi, K., Kawamata, H., Rajendran, B., Boyer, P.J., Manfredi, G., Elliott, J.L. (2007). Overexpression of CCS in G93A-SOD1 mice leads to accelerated neurological deficits with severe mitochondrial pathology. *Proc Natl Acad Sci U S A, 104*, 6072–6077.
77. Proescher, J.B., Son, M., Elliott, J.L., Culotta, V.C. (2008). Biological effects of CCS in the absence of SOD1 enzyme activation: implications for disease in a mouse model of ALS. *Hum Mol Genet, 17*, 1728–1737.
78. Bensimon, G., Lacomblez, L., Meininger, V. (1994). A controlled trial of riluzole in amyotrophic lateral sclerosis. ALS/Riluzole Study Group. *N Engl J Med, 330*, 585–591.
79. Smith, R.A., Miller, T.M., Yamanaka, K., Monia, B.P., Condon, T.P., Hung, G., Lobsiger, C.S., Ward, C.M., McAlonis-Downes, M., Wei, H., et al. (2006). Antisense oligonucleotide therapy for neurodegenerative disease. *J Clin Invest, 116*, 2290–2296.
80. Ralph, G.S., Radcliffe, P.A., Day, D.M., Carthy, J.M., Leroux, M.A., Lee, D.C., Wong, L.F., Bilsland, L.G., Greensmith, L., Kingsman, S.M., et al. (2005). Silencing mutant SOD1 using RNAi protects against neurodegeneration and extends survival in an ALS model. *Nat Med, 11*, 429–433.
81. Raoul, C., Abbas-Terki, T., Bensadoun, J.C., Guillot, S., Haase, G., Szulc, J., Henderson, C.E., Aebischer, P. (2005). Lentiviral-mediated silencing of SOD1 through RNA interference retards disease onset and progression in a mouse model of ALS. *Nat Med, 11*, 423–428.
82. Boillée, S., Yamanaka, K., Lobsiger, C.S., Copeland, N.G., Jenkins, N.A., Kassiotis, G., Kollias, G., Cleveland, D.W. (2006). Onset and progression in inherited ALS determined by motor neurons and microglia. *Science, 312*, 1389–1392.
83. Lobsiger, C.S., Boillee, S., Cleveland, D.W. (2007). Inaugural Article: Toxicity from different SOD1 mutants dysregulates the complement system and the neuronal regenerative response in ALS motor neurons. *Proc Natl Acad Sci U S A, 104*, 7319–7326.

84. Miller, T.M., Smith, R.A., Cleveland, D.W. (2006). Amyotrophic lateral sclerosis and gene therapy. *Nat Clin Pract Neurol, 2*, 462–463.

85. Ray, S.S., Lansbury, P.T., Jr. (2004). A possible therapeutic target for Lou Gehrig's disease. *Proc Natl Acad Sci U S A, 101*, 5701–5702.

86. Ray, S.S., Nowak, R.J., Brown, R.H., Jr, Lansbury, P.T., Jr. (2005). Small-molecule-mediated stabilization of familial amyotrophic lateral sclerosis-linked superoxide dismutase mutants against unfolding and aggregation. *Proc Natl Acad Sci U S A, 102*, 3639–3644.

87. Ray, S.S., Nowak, R.J., Strokovich, K., Brown, R.H., Jr., Walz, T., Lansbury, P.T., Jr. (2004). An intersubunit disulfide bond prevents in vitro aggregation of a superoxide dismutase-1 mutant linked to familial amytrophic lateral sclerosis. *Biochemistry, 43*, 4899–4905.

88. Nakanishi, T., Kishikawa, M., Miyazaki, A., Shimizu, A., Ogawa, Y., Sakoda, S., Ohi, T., Shoji, H. (1998). Simple and defined method to detect the SOD-1 mutants from patients with familial amyotrophic lateral sclerosis by mass spectrometry. *J Neurosci Methods, 81*, 41–44.

89. Nakano, R., Sato, S., Inuzuka, T., Sakimura, K., Mishina, M., Takahashi, H., Ikuta, F., Honma, Y., Fujii, J., Taniguchi, N., et al. (1994). A novel mutation in Cu/Zn superoxide dismutase gene in Japanese familial amyotrophic lateral sclerosis. *Biochem Biophys Res Commun, 200*, 695–703.

90. Rosen, D.R., Bowling, A.C., Patterson, D., Usdin, T.B., Sapp, P., Mezey, E., McKenna-Yasek, D., O'Regan, J., Rahmani, Z., Ferrante, R.J., et al. (1994). A frequent ala 4 to val superoxide dismutase-1 mutation is associated with a rapidly progressive familial amyotrophic lateral sclerosis. *Hum Mol Genet, 3*, 981–987.

91. Rosen, D.R., Siddique, T., Patterson, D., Figlewicz, D.A., Sapp, P., Hentati, A., Donaldson, D., Goto, J., O'Regan, J.P., Deng, H.X., et al. (1993). Mutations in Cu/Zn superoxide dismutase gene are associated with familial amyotrophic lateral sclerosis. *Nature, 362*, 59–62.

92. Morita, M., Aoki, M., Abe, K., Hasegawa, T., Sakuma, R., Onodera, Y., Ichikawa, N., Nishizawa, M., Itoyama, Y. (1996). A novel two-base mutation in the Cu/Zn superoxide dismutase gene associated with familial amyotrophic lateral sclerosis in Japan. *Neurosci Lett, 205*, 79–82.

93. Kohno, S., Takahashi, Y., Miyajima, H., Serizawa, M., Mizoguchi, K. (1999). A novel mutation (Cys6Gly) in the Cu/Zn superoxide dismutase gene associated with rapidly progressive familial amyotrophic lateral sclerosis. *Neurosci Lett, 276*, 135–137.

94. Boukaftane, Y., Khoris, J., Moulard, B., Salachas, F., Meininger, V., Malafosse, A., Camu, W., Rouleau, G.A. (1998). Identification of six novel SOD1 gene mutations in familial amyotrophic lateral sclerosis. *Can J Neurol Sci, 25*, 192–196.

95. Hirano, M., Fujii, J., Nagai, Y., Sonobe, M., Okamoto, K., Araki, H., Taniguchi, N., Ueno, S. (1994). A new variant Cu/Zn superoxide dismutase (Val7 → Glu) deduced from lymphocyte mRNA sequences from Japanese patients with familial amyotrophic lateral sclerosis. *Biochem Biophys Res Commun, 204*, 572–577.

96. Siddique, T., Deng, H.X. (1996). Genetics of amyotrophic lateral sclerosis. *Hum Mol Genet, 5* (Spec No), 1465–1470.

97. Andersen, P.M., Sims, K.B., Xin, W.W., Kiely, R., O'Neill, G., Ravits, J., Pioro, E., Harati, Y., Brower, R.D., Levine, J.S., et al. (2003). Sixteen novel mutations in the Cu/Zn superoxide dismutase gene in amyotrophic lateral sclerosis: a decade of discoveries, defects and disputes. *Amyotroph Lateral Scler Other Motor Neuron Disord, 4*, 62–73.
98. Kim, N.H., Kim, H.J., Kim, M., Lee, K.W. (2003). A novel SOD1 gene mutation in a Korean family with amyotrophic lateral sclerosis. *J Neurol Sci, 206*, 65–69.
99. Gellera, C., Castellotti, B., Riggio, M.C., Silani, V., Morandi, L., Testa, D., Casali, C., Taroni, F., Di Donato, S., Zeviani, M., Mariotti, C. (2001). Superoxide dismutase gene mutations in Italian patients with familial and sporadic amyotrophic lateral sclerosis: identification of three novel missense mutations. *Neuromuscul Disord, 11*, 404–410.
100. Penco, S., Schenone, A., Bordo, D., Bolognesi, M., Abbruzzese, M., Bugiani, O., Ajmar, F., Garrè, C. (1999). A SOD1 gene mutation in a patient with slowly progressing familial ALS. *Neurology, 53*, 404–406.
101. Aoki, M., Ogasawara, M., Matsubara, Y., Narisawa, K., Nakamura, S., Itoyama, Y., Abe, K. (1993). Mild ALS in Japan associated with novel SOD mutation. *Nat Genet, 5*, 323–324.
102. Andersen, P.M., Nilsson, P., Keränen, M.L., Forsgren, L., Hägglund, J., Karlsborg, M., Ronnevi, L.O., Gredal, O., Marklund, S.L. (1997). Phenotypic heterogeneity in motor neuron disease patients with CuZn-superoxide dismutase mutations in Scandinavia. *Brain, 120* (Pt 10), 1723–1737.
103. Deng, H.X., Tainer, J.A., Mitsumoto, H., Ohnishi, A., He, X., Hung, W.Y., Zhao, Y., Juneja, T., Hentati, A., Siddique, T. (1995). Two novel SOD1 mutations in patients with familial amyotrophic lateral sclerosis. *Hum Mol Genet, 4*, 1113–1116.
104. Kawamata, J., Shimohama, S., Takano, S., Harada, K., Ueda, K., Kimura, J. (1997). Novel G16S (GGC-AGC) mutation in the SOD-1 gene in a patient with apparently sporadic young-onset amyotrophic lateral sclerosis. *Hum Mutat, 9*, 356–358.
105. Enayat, Z.E., Orrell, R.W., Claus, A., Ludolph, A., Bachus, R., Brockmüller, J., Ray-Chaudhuri, K., Radunovic, A., Shaw, C., Wilkinson, J., et al. (1995). Two novel mutations in the gene for copper-zinc superoxide dismutase in UK families with amyotrophic lateral sclerosis. *Hum Mol Genet, 4*, 1239–1240.
106. Jones, C.T., Swingler, R.J., Brock, D.J. (1994). Identification of a novel SOD1 mutation in an apparently sporadic amyotrophic lateral sclerosis patient and the detection of Ile113Thr in three others. *Hum Mol Genet, 3*, 649–650.
107. Andersen, P.M. (2006). Amyotrophic lateral sclerosis associated with mutations in the CuZn superoxide dismutase gene. *Curr Neurol Neurosci Rep, 6*, 37–46.
108. Alexander, M.D., Traynor, B.J., Miller, N., Corr, B., Frost, E., McQuaid, S., Brett, F.M., Green, A., Hardiman, O. (2002). "True" sporadic ALS associated with a novel SOD-1 mutation. *Ann Neurol, 52*, 680–683.
109. García-Redondo, A., Bustos, F., Juan, Y. Seva, B., Del Hoyo, P., Jiménez, S., Campos, Y., Martín, M.A., Rubio, J.C., Cañadillas, F., et al. (2002). Molecular analysis of the superoxide dismutase 1 gene in Spanish patients with sporadic or familial amyotrophic lateral sclerosis. *Muscle Nerve, 26*, 274–278.

110. Shaw, C.E., Enayat, Z.E., Chioza, B.A., Al-Chalabi, A., Radunovic, A., Powell, J.F., Leigh, P.N. (1998). Mutations in all five exons of SOD-1 may cause ALS. *Ann Neurol, 43*, 390–394.

111. Aoki, M., Abe, K., Houi, K., Ogasawara, M., Matsubara, Y., Kobayashi, T., Mochio, S., Narisawa, K., Itoyama, Y. (1995). Variance of age at onset in a Japanese family with amyotrophic lateral sclerosis associated with a novel Cu/Zn superoxide dismutase mutation. *Ann Neurol, 37*, 676–679.

112. Beck, M., Sendtner, M., Toyka, K.V. (2007). Novel SOD1 N86K mutation is associated with a severe phenotype in familial ALS. *Muscle Nerve, 36*, 111–114.

113. Hayward, C., Brock, D.J., Minns, R.A., Swingler, R.J. (1998). Homozygosity for Asn86Ser mutation in the CuZn-superoxide dismutase gene produces a severe clinical phenotype in a juvenile onset case of familial amyotrophic lateral sclerosis. *J Med Genet, 35*, 174.

114. Segovia-Silvestre, T., Andreu, A.L., Vives-Bauza, C., Garcia-Arumi, E., Cervera, C., Gamez, J. (2002). A novel exon 3 mutation (D76V) in the SOD1 gene associated with slowly progressive ALS. *Amyotroph Lateral Scler Other Motor Neuron Disord, 3*, 69–74.

115. Rezania, K., Yan, J., Dellefave, L., Deng, H.X., Siddique, N., Pascuzzi, R.T., Siddique, T., Roos, R.P. (2003). A rare Cu/Zn superoxide dismutase mutation causing familial amyotrophic lateral sclerosis with variable age of onset, incomplete penetrance and a sensory neuropathy. *Amyotroph Lateral Scler Other Motor Neuron Disord, 4*, 162–166.

116. Andersen, P.M., Nilsson, P., Ala-Hurula, V., Keränen, M.L., Tarvainen, I., Haltia, T., Nilsson, L., Binzer, M., Forsgren, L., Marklund, S.L. (1995). Amyotrophic lateral sclerosis associated with homozygosity for an Asp90Ala mutation in CuZn-superoxide dismutase. *Nat Genet, 10*, 61–66.

117. Chou, C.M., Huang, C.J., Shih, C.M., Chen, Y.P., Liu, T.P., Chen, C.T. (2005). Identification of three mutations in the Cu,Zn-superoxide dismutase (Cu,Zn-SOD) gene with familial amyotrophic lateral sclerosis: transduction of human Cu,Zn-SOD into PC12 cells by HIV-1 TAT protein basic domain. *Ann N Y Acad Sci, 1042*, 303–313.

118. Hosler, B.A., Nicholson, G.A., Sapp, P.C., Chin, W., Orrell, R.W., de Belleroche, J.S., Esteban, J., Hayward, L.J., Mckenna-Yasek, D., Yeung, L., et al. (1996). Three novel mutations and two variants in the gene for Cu/Zn superoxide dismutase in familial amyotrophic lateral sclerosis. *Neuromuscul Disord, 6*, 361–366.

119. Esteban, J., Rosen, D.R., Bowling, A.C., Sapp, P., McKenna-Yasek, D., O'Regan, J.P., Beal, M.F., Horvitz, H.R., Brown, R.H., Jr. (1994). Identification of two novel mutations and a new polymorphism in the gene for Cu/Zn superoxide dismutase in patients with amyotrophic lateral sclerosis. *Hum Mol Genet, 3*, 997–998.

120. Elshafey, A., Lanyon, W.G., Connor, J.M. (1994). Identification of a new missense point mutation in exon 4 of the Cu/Zn superoxide dismutase (SOD-1) gene in a family with amyotrophic lateral sclerosis. *Hum Mol Genet, 3*, 363–364.

121. Hand, C.K., Mayeux-Portas, V., Khoris, J., Briolotti, V., Clavelou, P., Camu, W., Rouleau, G.A. (2001). Compound heterozygous D90A and D96N SOD1 mutations in a recessive amyotrophic lateral sclerosis family. *Ann Neurol, 49*, 267–271.

122. Watanabe, M., Aoki, M., Abe, K., Shoji, M., Iizuka, T., Ikeda, Y., Hirai, S., Kurokawa, K., Kato, T., Sasaki, H., Itoyama, Y. (1997). A novel missense point mutation (S134N) of the Cu/Zn superoxide dismutase gene in a patient with familial motor neuron disease. *Hum Mutat, 9*, 69–71.

123. Nogales-Gadea, G., Garcia-Arumi, E., Andreu, A.L., Cervera, C., Gamez, J. (2004). A novel exon 5 mutation (N139H) in the SOD1 gene in a Spanish family associated with incomplete penetrance. *J Neurol Sci, 219*, 1–6.

124. Pramatarova, A., Figlewicz, D.A., Krizus, A., Han, F.Y., Ceballos-Picot, I., Nicole, A., Dib, M., Meininger, V., Brown, R.H., Rouleau, G.A. (1995). Identification of new mutations in the Cu/Zn superoxide dismutase gene of patients with familial amyotrophic lateral sclerosis. *Am J Hum Genet, 56*, 592–596.

125. Naini, A., Musumeci, O., Hayes, L., Pallotti, F., Del Bene, M., Mitsumoto, H. (2002). Identification of a novel mutation in Cu/Zn superoxide dismutase gene associated with familial amyotrophic lateral sclerosis. *J Neurol Sci, 198*, 17–19.

126. Yulug, I.G., Katsanis, N., de Belleroche, J., Collinge, J., Fisher, E.M. (1995). An improved protocol for the analysis of SOD1 gene mutations, and a new mutation in exon 4. *Hum Mol Genet, 4*, 1101–1104.

127. Sato, T., Yamamoto, Y., Nakanishi, T., Fukada, K., Sugai, F., Zhou, Z., Okuno, T., Nagano, S., Hirata, S., Shimizu, A., Sakoda, S. (2004). Identification of two novel mutations in the Cu/Zn superoxide dismutase gene with familial amyotrophic lateral sclerosis: mass spectrometric and genomic analyses. *J Neurol Sci, 218*, 79–83.

128. Jones, C.T., Shaw, P.J., Chari, G., Brock, D.J. (1994). Identification of a novel exon 4 SOD1 mutation in a sporadic amyotrophic lateral sclerosis patient. *Mol Cell Probes, 8*, 329–330.

129. Tan, C.F., Piao, Y.S., Hayashi, S., Obata, H., Umeda, Y., Sato, M., Fukushima, T., Nakano, R., Tsuji, S., Takahashi, H. (2004). Familial amyotrophic lateral sclerosis with bulbar onset and a novel Asp101Tyr Cu/Zn superoxide dismutase gene mutation. *Acta Neuropathol, 108*, 332–336.

130. Ikeda, M., Abe, K., Aoki, M., Sahara, M., Watanabe, M., Shoji, M., St. George-Hyslop, P.H., Hirai, S., Itoyama, Y. (1995). Variable clinical symptoms in familial amyotrophic lateral sclerosis with a novel point mutation in the Cu/Zn superoxide dismutase gene. *Neurology, 45*, 2038–2042.

131. Deng, H.X., Hentati, A., Tainer, J.A., Iqbal, Z., Cayabyab, A., Hung, W.Y., Getzoff, E.D., Hu, P., Herzfeldt, B., Roos, R.P., et al. (1993). Amyotrophic lateral sclerosis and structural defects in Cu,Zn superoxide dismutase. *Science, 261*, 1047–1051.

132. Sapp, P.C., Rosen, D.R., Hosler, B.A., Esteban, J., McKenna-Yasek, D., O'Regan, J.P., Horvitz, H.R., Brown, R.H., Jr. (1995). Identification of three novel mutations in the gene for Cu/Zn superoxide dismutase in patients with familial amyotrophic lateral sclerosis. *Neuromuscul Disord, 5*, 353–357.

133. Orrell, R.W., Habgood, J.J., Gardiner, I., King, A.W., Bowe, F.A., Hallewell, R.A., Marklund, S.L., Greenwood, J., Lane, R.J., deBelleroche, J. (1997). Clinical and functional investigation of 10 missense mutations and a novel frameshift insertion mutation of the gene for copper–zinc superoxide dismutase in UK families with amyotrophic lateral sclerosis. *Neurology, 48*, 746–751.

134. Naini, A., Mehrazin, M., Lu, J., Gordon, P., Mitsumoto, H. (2007). Identification of a novel D109Y mutation in Cu/Zn superoxide dismutase (sod1) gene associated with amyotrophic lateral sclerosis. *J Neurol Sci, 254*, 17–21.

135. Shibata, N., Kawaguchi, M., Uchida, K., Kakita, A., Takahashi, H., Nakano, R., Fujimura, H., Sakoda, S., Ihara, Y., Nobukuni, K., et al. (2007). Protein-bound crotonaldehyde accumulates in the spinal cord of superoxide dismutase-1 mutation-associated familial amyotrophic lateral sclerosis and its transgenic mouse model. *Neuropathology, 27*, 49–61.

136. Kostrzewa, M., Damian, M.S., Muller, U. (1996). Superoxide dismutase 1: identification of a novel mutation in a case of familial amyotrophic lateral sclerosis. *Hum Genet, 98*, 48–50.

137. Kostrzewa, M., Burck-Lehmann, U., Muller, U. (1994). Autosomal dominant amyotrophic lateral sclerosis: a novel mutation in the Cu/Zn superoxide dismutase-1 gene. *Hum Mol Genet, 3*, 2261–2262.

138. Shimizu, T., Kawata, A., Kato, S., Hayashi, M., Takamoto, K., Hayashi, H., Hirai, S., Yamaguchi, S., Komori, T., Oda, M. (2000). Autonomic failure in ALS with a novel SOD1 gene mutation. *Neurology, 54*, 1534–1537.

139. Jackson, M., Al-Chalabi, A., Enayat, Z.E., Chioza, B., Leigh, P.N., Morrison, K.E. (1997). Copper/zinc superoxide dismutase 1 and sporadic amyotrophic lateral sclerosis: analysis of 155 cases and identification of a novel insertion mutation. *Ann Neurol, 42*, 803–807.

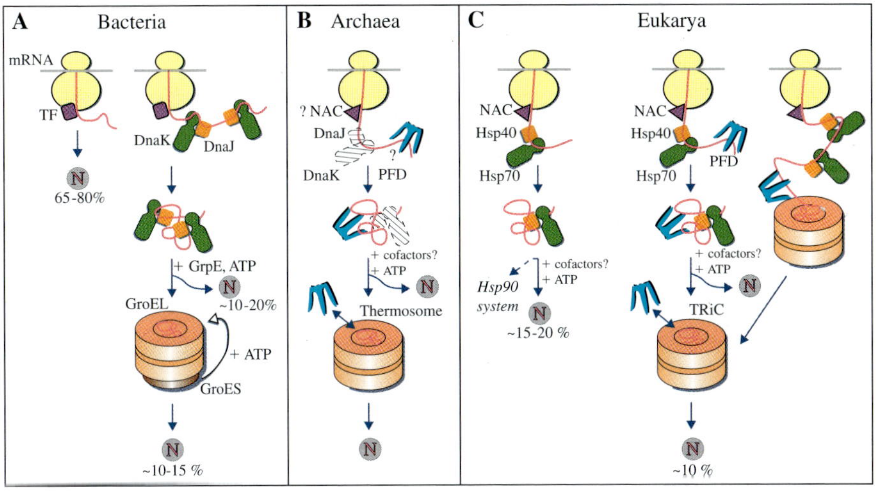

Chapter 3, Figure 2 Chaperone pathways of protein folding in the cytosol. Models for the chaperone-assisted folding of newly synthesized polypeptides in the cytosol. (A) Bacteria. TF, trigger factor; N, native protein. Nascent chains probably interact generally with TF, and most small proteins (65 to 80% of total) may fold rapidly upon synthesis without further assistance. Longer chains (10 to 20% of total) interact subsequently with DnaK and DnaJ and fold upon one or several cycles of ATP-dependent binding and release. About 10 to 15% of chains transit the chaperonin system (GroEL and GroES) for folding. GroEL does not bind to nascent chains and is thus likely to receive a substantial fraction of its substrates after their interaction with DnaK. (B) Archaea. PFD, prefoldin; NAC, nascent chain–associated complex. Only some archaeal species contain DnaK/DnaJ. (C) Eukarya (the example of the mammalian cytosol). Like TF, NAC probably interacts generally with nascent chains. The majority of small chains may fold upon ribosome release without further assistance. About 15 to 20% of chains reach their native states in a reaction assisted by Hsp70 and Hsp40, and a fraction of these must be transferred to Hsp90 for folding. About 10% of chains are co- or posttranslationally passed on to the chaperonin TRiC in a reaction mediated by Hsp70 and PFD.

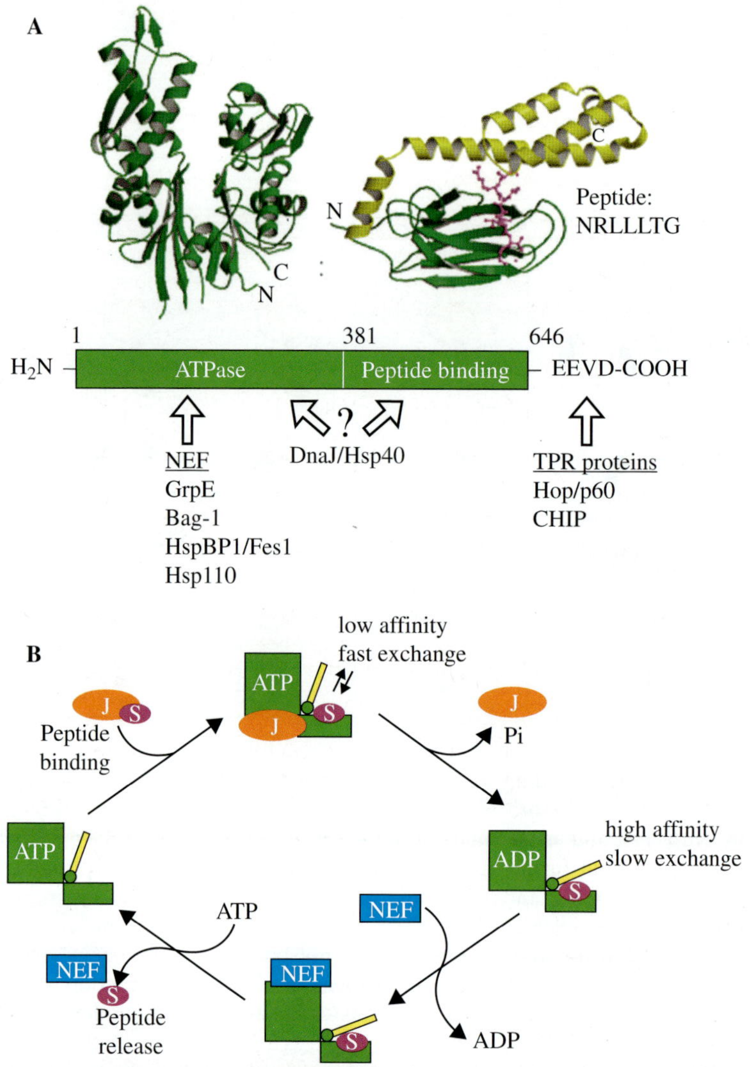

Chapter 3, Figure 3 Structure and reaction cycle of the Hsp70 chaperone system. (A) (Top) Structures of the ATPase domain [58] and the peptide-binding domain [53] of Hsp70 shown representatively for *E. coli* DnaK. The α-helical latch of the peptide-binding domain is shown in yellow and a ball-and-stick model of the extended peptide substrate in pink. ATP indicates the position of the nucleotide binding site. The amino acid sequence of the peptide is indicated in single-letter code. (Bottom) The interaction of prokaryotic and eukaryotic cofactors with Hsp70 is shown schematically. Residue numbers refer to human Hsp70. NEF, nucleotide exchange factors (GrpE in case of *E. coli* DnaK; Bag, HspBP1, and Hsp110 in case of eukaryotic cytosolic Hsp70). TPR, tetratricopeptide repeat domain: Hop, Hsp organizing protein; CHIP, C-terminus of Hsp70 interacting protein. Only the Hsp70 proteins of the eukaryotic cytosol have the COOH-terminal sequence EEVD that is involved in binding of TPR cofactors [142]. (B) Hsp70 reaction cycle with Hsp70 colored as in (A). J, DnaJ; NEF, nucleotide exchange factor; S, substrate peptide.

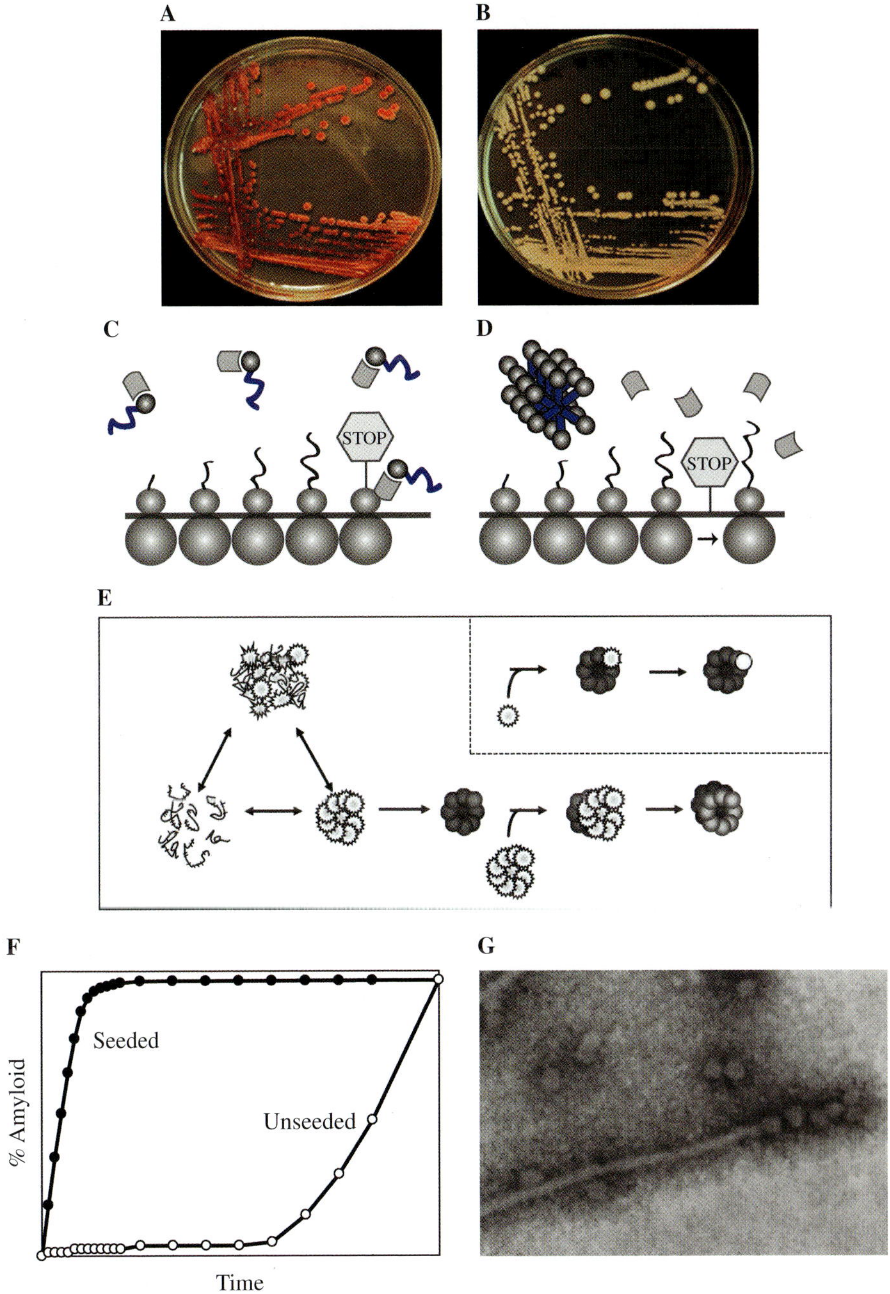

Chapter 8, Figure 1 Molecular basis of $[PSI^{+}]$ prion propagation. *See pages 148–149 for full caption.*

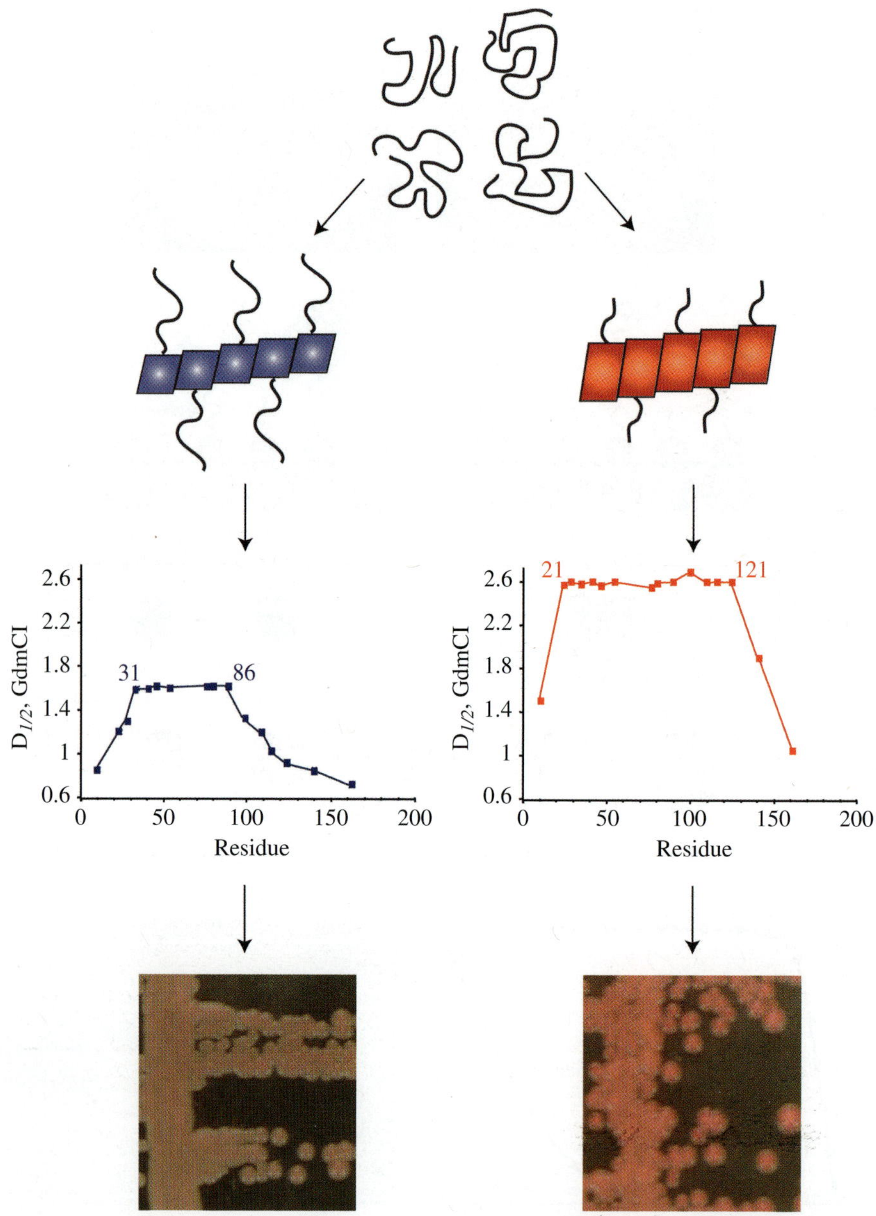

Chapter 8, Figure 4 Sup35 strains have different-size amyloid cores. *See pages 162–163 for full caption.*

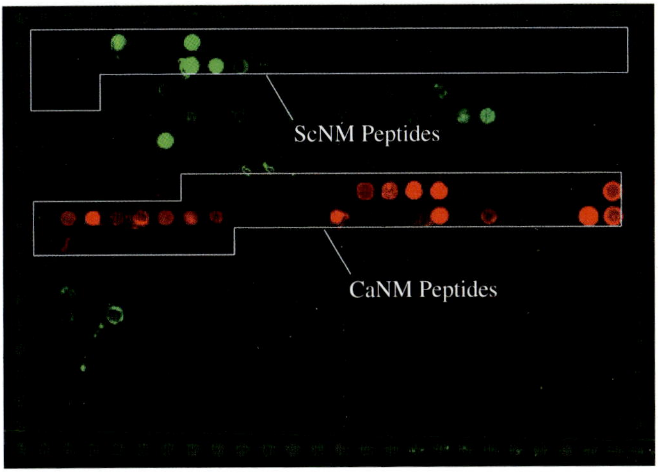

Chapter 8, Figure 5 Peptide microarray analysis of yeast prion species barriers. *See page 165 for full caption.*

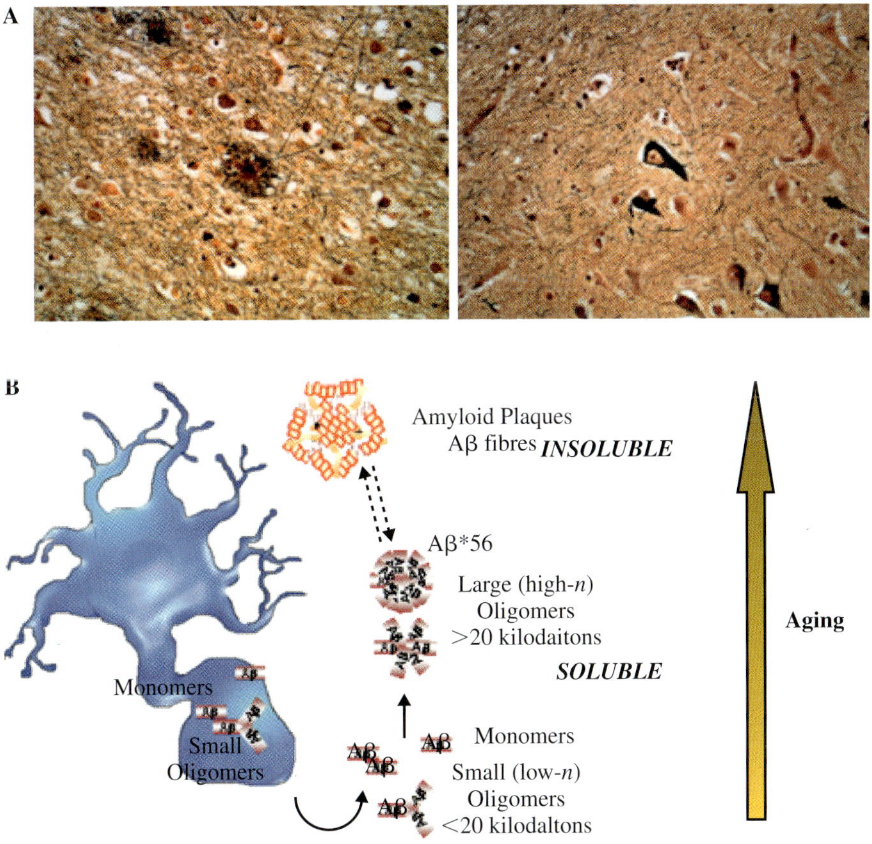

Chapter 11, Figure 1 Age-related changes in the brain associated with Alzheimer disease. *See page 214 for full caption.*

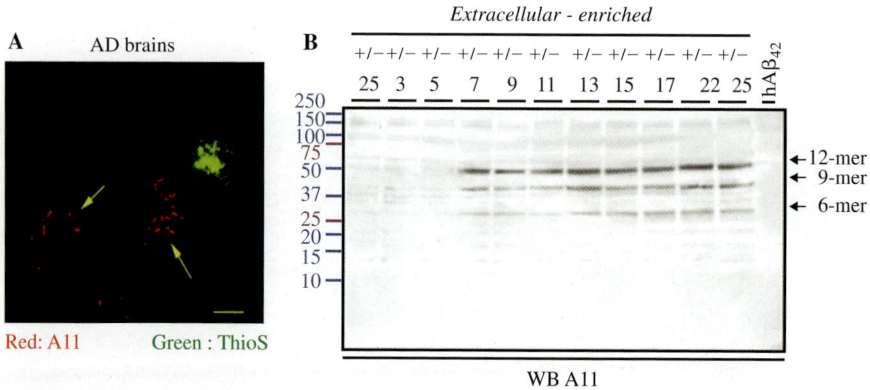

Chapter 11, Figure 10 The anti-oligomer antiserum A11 recognizes Aβ*56. (A) A11 staining of non-plaque-associated oligomers in brain tissue from an Alzheimer patient; (B) staining of high-N Aβ oligomers with the anti-oligomer antiserum, A11. Note that the human Aβ 42 (hAβ42) standards are not detected with the A11 antiserum (far right lane).

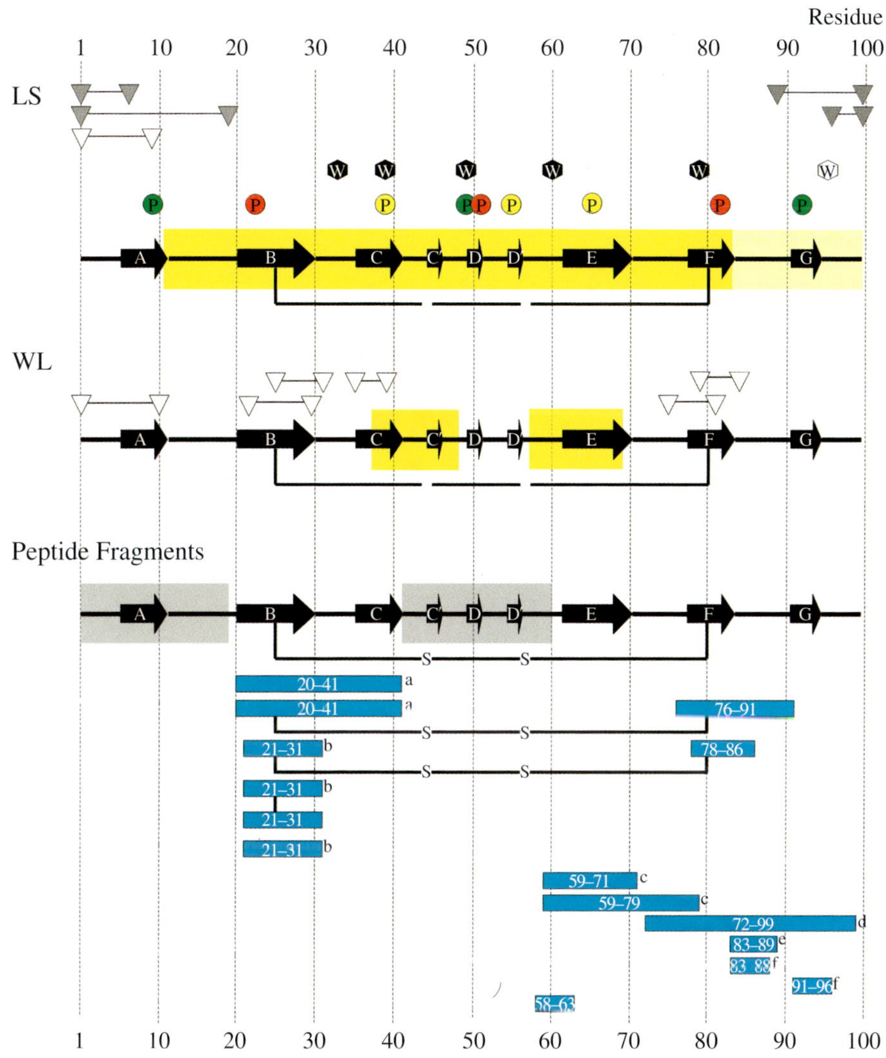

Chapter 16, Figure 4 Schematic diagram outlining current knowledge about the structure of β_2m amyloid fibrils gained to date from biochemical and biophysical analyses. *Top panel:* Hydrogen exchange (HX) and limited proteolysis results are shown for long, straight (LS) fibrils grown at pH 2.5 de novo; approximate regions of high protection from HX (yellow), regions of partial protection (pale yellow) [77,78]. Open triangles, limited proteolysis cut sites observed at pH 2.5 [75,77]; gray filled triangles, limited proteolysis cut sites observed using fibrils formed by extending seeds of ex vivo fibrils at pH 4.0 [76]. Hexagons; Trp residues buried in the core of the fibrils (black), Trp residues exposed to solvent (white) [87]. Proline mutations displaying significant (red), moderate (yellow), and no effect (green) on fibril extension [88]. *Middle panel:* Wormlike (WL) fibrils produced under high-salt conditions at pH 2.5 [75,78]. Regions of HX protection are shown in yellow [78], limited proteolysis cut sites are shown as open triangles [75,77]. *Bottom panel:* Peptide fragments deriving from β_2m known to form fibrils in vitro. Regions in gray are not known to form fibrils in isolation.

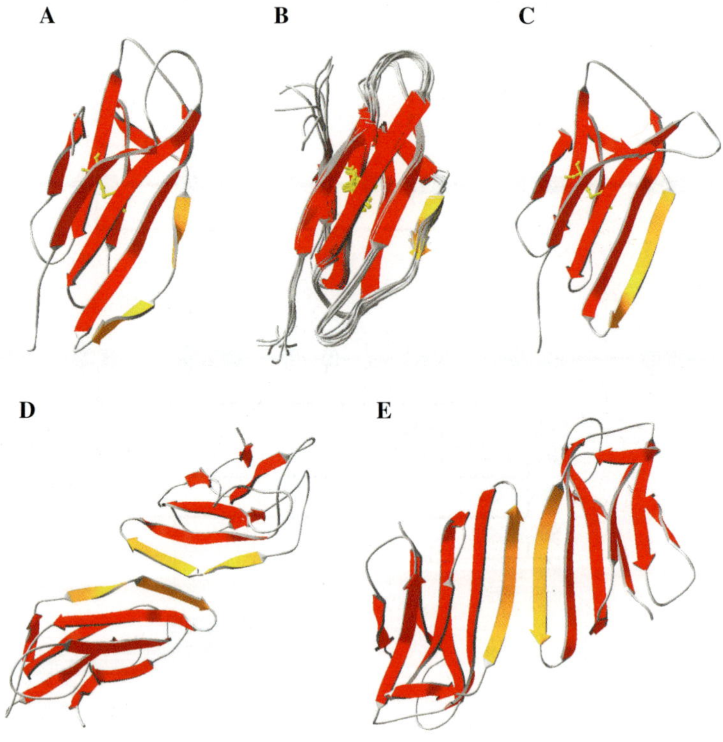

Chapter 16, Figure 5 Selection of different crystal structures of β_2m. Strand D is shown in orange in all cases. (A) β_2m in complex with the heavy chain of the HLA–1DUZ [3]. (B) NMR solution structure of β_2m–1JNJ [5]. (C) Crystal structure of β_2m at pH 5.6 (note the straight D strand)–1LDS [4]. Note that the crystal structure of β_2m at pH 7.0 (2YXF) is essentially the same as at pH 5.6 [6]. (D) Crystal structure of H31Y. The crystal lattice packing shows the occurrence of an antiparallel pairing of the short D2 strand–1PY4 [7]. (E) Crystallographic dimer of P32A related by a twofold axis yielding an eight-strand β-sheet comprised of strands ABED-DEBA–2F8O [65].

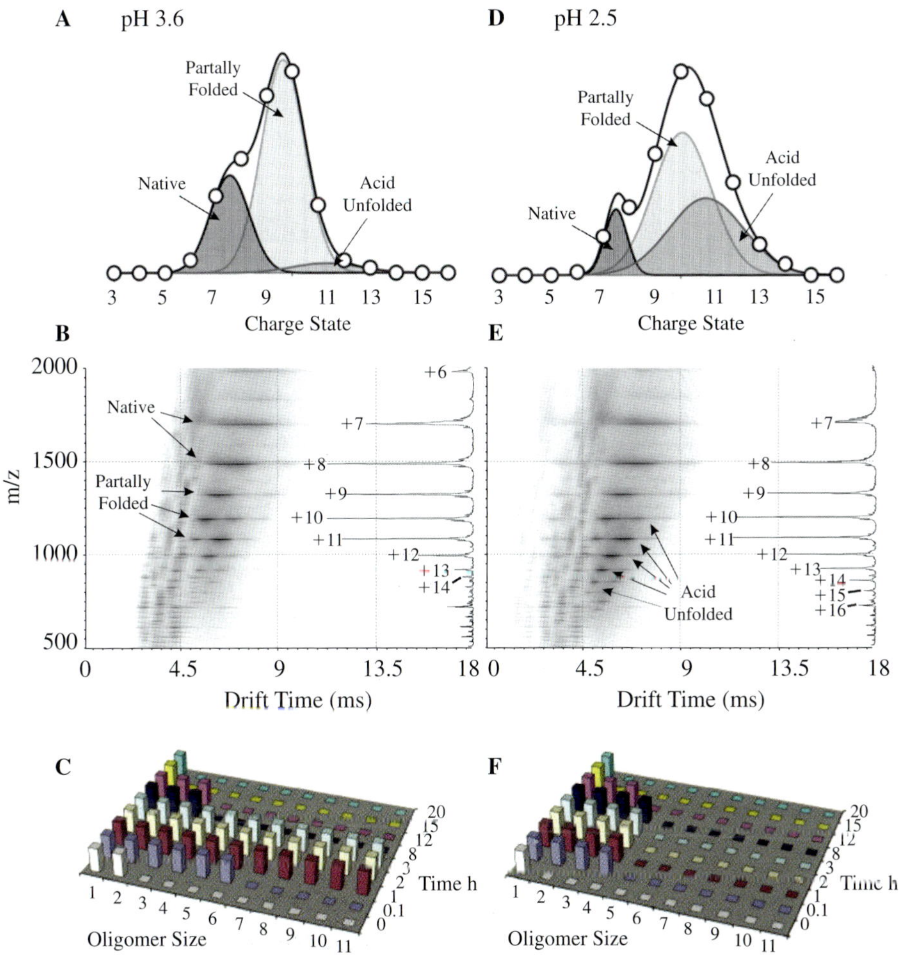

Chapter 16, Figure 6 ESI-MS data collected using β_2m at pH 3.6 (left-hand panels) and pH 2.5 (right-hand panels). (A,D) Co-populated conformational ensembles of β_2m uncovered quantitatively by ESI–MS [105]. (B,E) ESI–IMS–MS driftscope plots showing drift time (*x* axis) versus *m*/*z* (*y*-axis) for wild-type β_2m under each condition. Insets at the right-hand side of each plot: the summed, full-scan *m*/*z* spectra of wild-type β_2m for each data acquisition, showing the charge-state ions detected [106]. (C,F) Oligomer distributions observed during β_2m fibril assembly measured by nanoESI–MS under each condition over a range of *m*/*z* 3200 to 5500 [108].

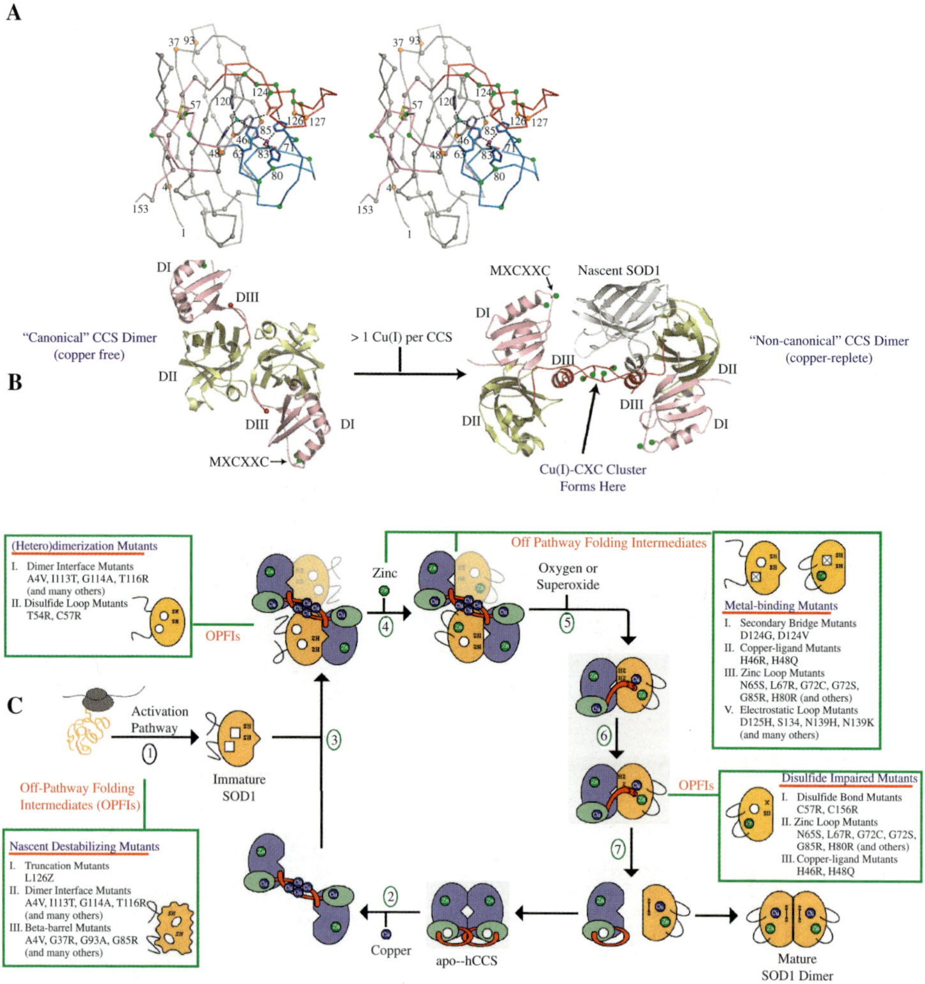

Chapter 17, Figure 1 (A) Stereo view of human SOD1 (PDB code 1AZV 9]). The β-barrel is shown in gray, the disulfide loop (residues 50 to 61) is shown in purple, the zinc loop (residues 62 to 83) is shown in blue, and the electrostatic loop (residues 121 to 142) is shown in red. The disulfide bond between C57 of the disulfide loop and C146 of β-strand 8 is shown as yellow sticks. Cu and Zn ligands are shown as gray and blue sticks and Cu and Zn ions are represented as cyan and magenta spheres, respectively. The α-carbon positions of β-barrel and metal-binding mutants are shown as gray and green spheres, respectively. The α-carbon positions of pathogenic SOD1 mutants for which there are mouse models (see the text) are shown as orange spheres. (B) The possible role of CCS in pathogenic SOD1 toxicity in fALS. A CCS canonical dimer (PDB code 1QUP 72])-to-noncanonical dimer (PDB code 1JK9 75]) transition upon the binding of Cu(I) 74]. (C) Alternate model of CCS action with selected off-pathway pathogenic SOD1 folding intermediates (see the text).

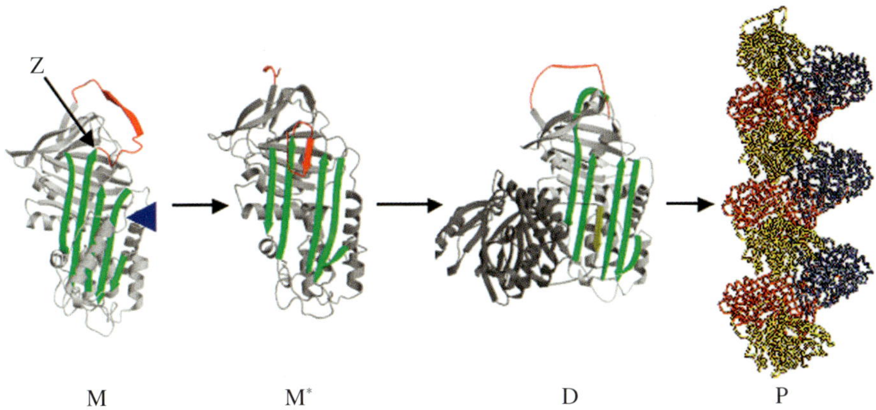

Chapter 18, Figure 1 The Z mutation distorts the relationship between the reactive center loop and β-sheet A. *See page 406 for full caption.*

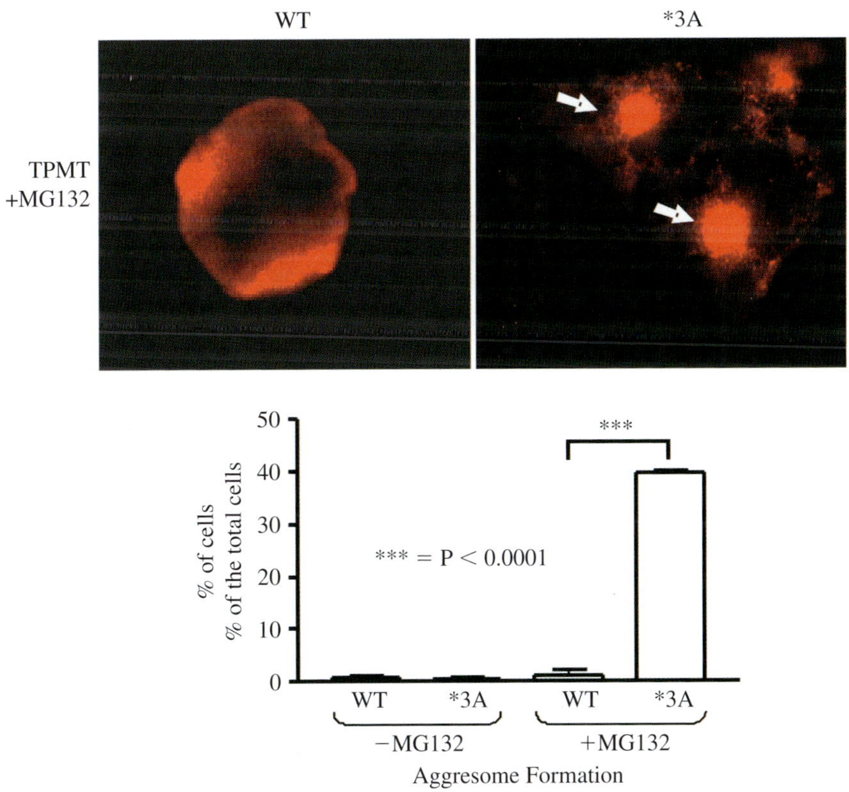

Chapter 20, Figure 5 TPMT aggresome formation. *See page 462 for full caption.*

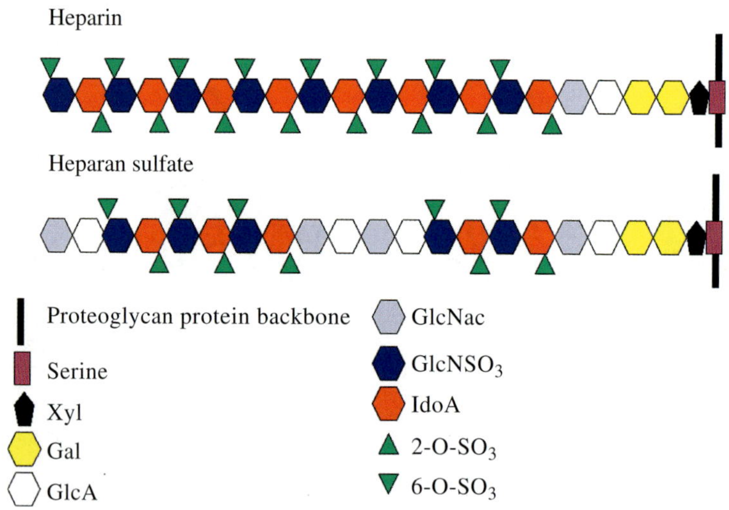

Chapter 25, Figure 1 Comparison of the linear structures of heparin and heparan sulfate. Note that beyond the tetrasaccharide linkage region heparin has a consistent pattern of repeating disaccharides. On occasion a 3-O-SO$_3$ is present in the GlcNSO$_3$ (not shown). By contrast, heparan sulfate has short stretches of heparinlike structure seperated by regions that are poorly sulfated within which is GlcA rather than sulfated IdoA.

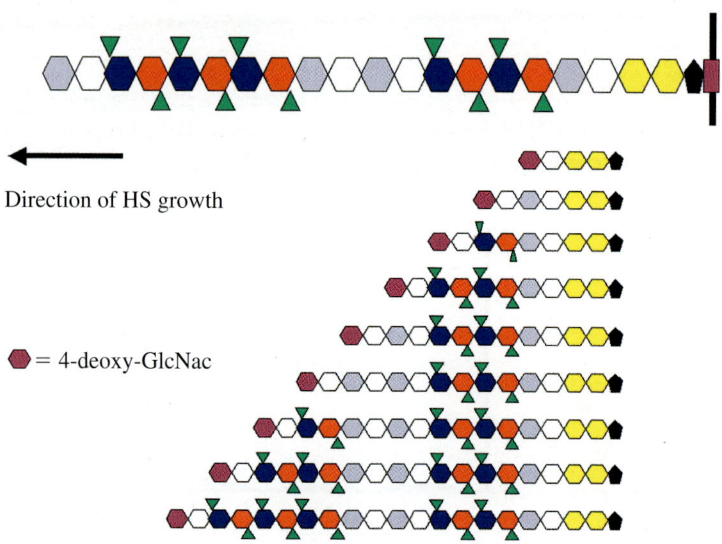

Chapter 25, Figure 2 Nested series of HS polysaccharides that would be generated by 4-deoxy-GlcNac if there is a specific HS linear structural pattern (e.g., top of figure) that exists at a specific proteoglycan serine locus. *See page 565 for full caption.*

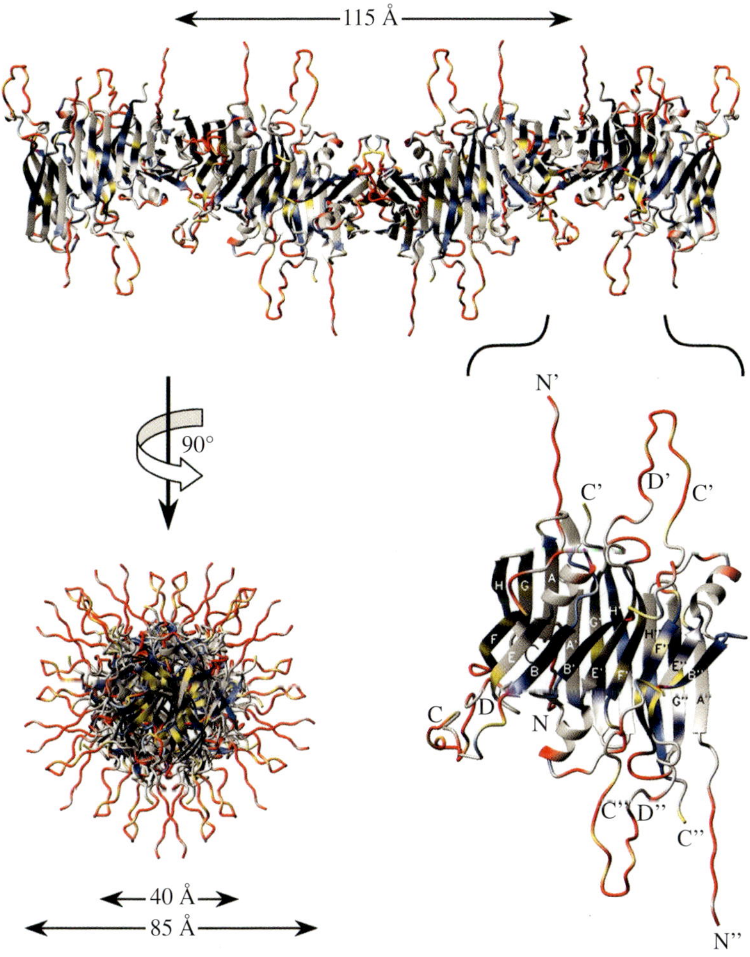

Chapter 26, Figure 1 Proposed structure of an TTR fiber based on deuterium exchange protection NMR and electron spin resonance data. Destabilization of strands C and D along with small modifications of the native fold expose edge strands A and B, leading to fiber formation and propagation. Loops between strands along with the N and C termini are exposed on the surface of the fiber, providing potential ligands for SAP.

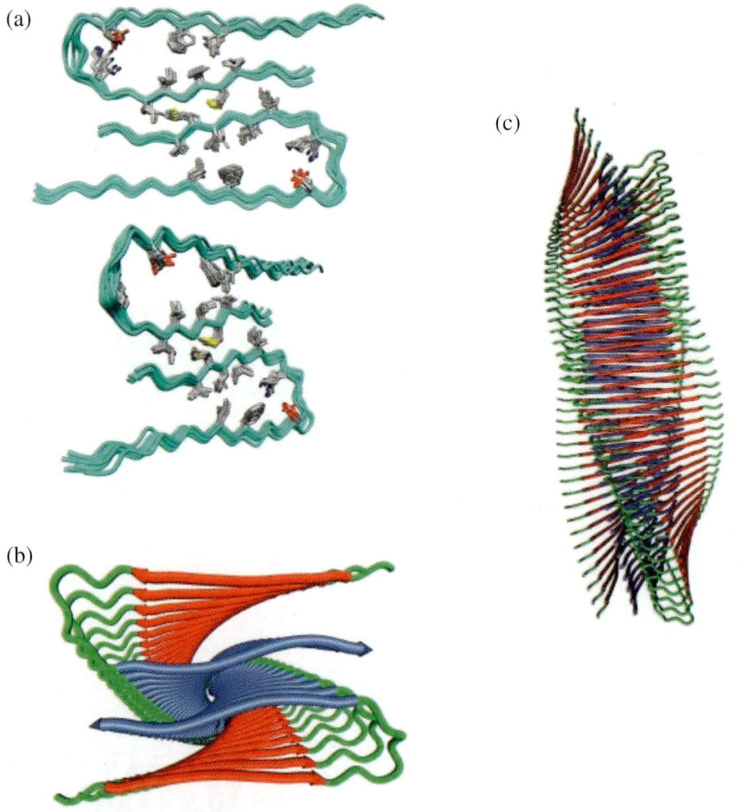

Chapter 26, Figure 2 Proposed structure of Aβ(1–40) amyloid. (a) Aβ dimers interacting via hydrophobic residues 30 to 40. Dimers can be layered (b) with (c) illustrating how stacking can lead to fibril morphology.

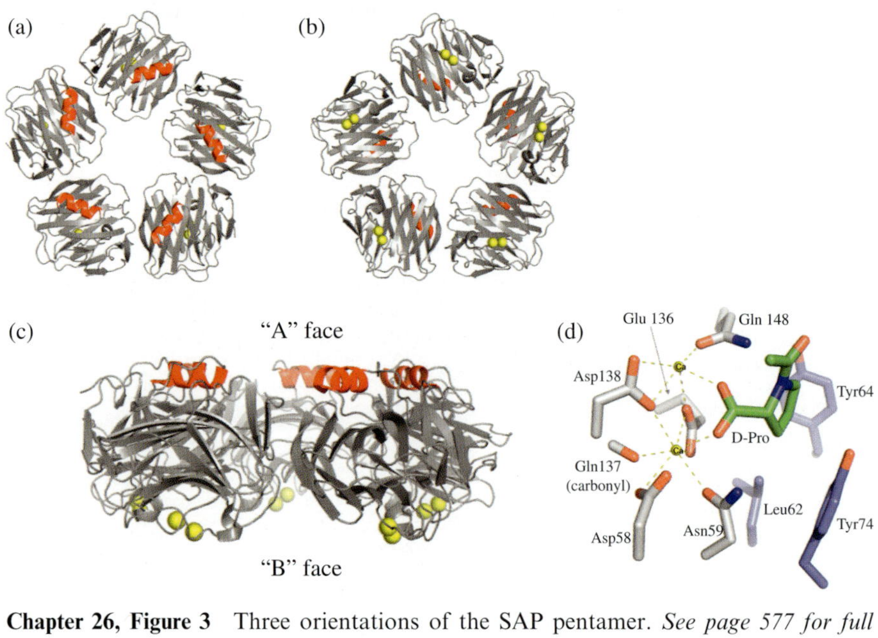

Chapter 26, Figure 3 Three orientations of the SAP pentamer. *See page 577 for full caption.*

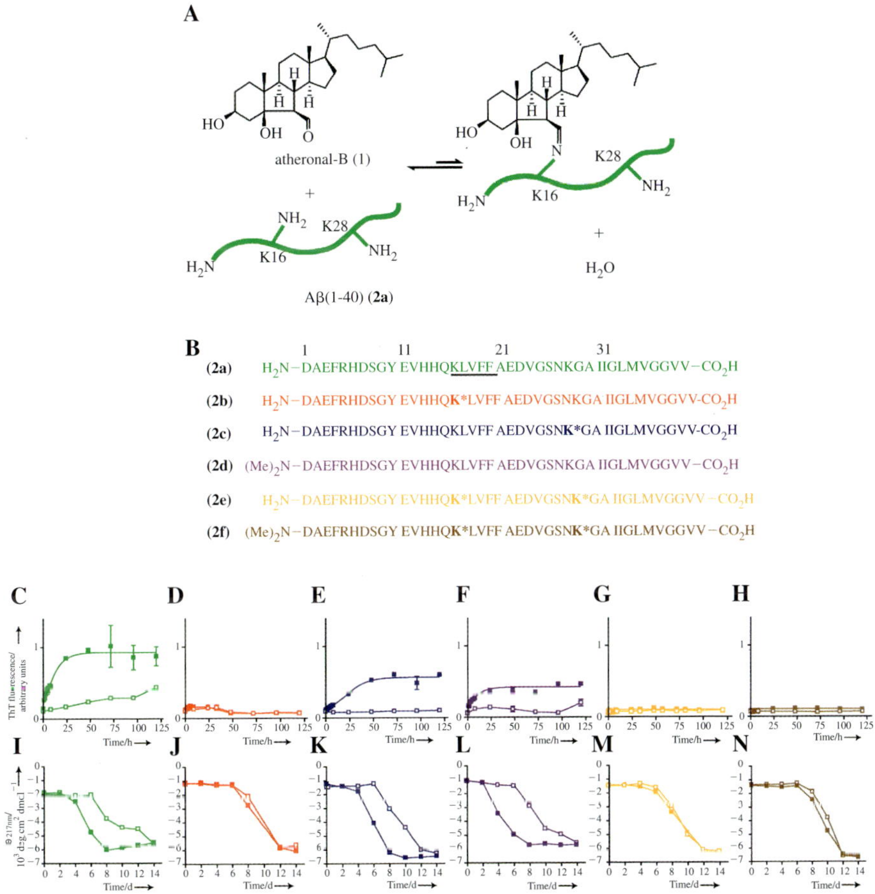

Chapter 28, Figure 4 (A) Schiff base formation between atheronal-B and primary amines on Aβ; (B) sequences of mono-, bis-, and tris-dimethylated peptide analogs (**2b** to **2f**) to native Aβ(1–40) (2a); (C–N) kinetics of atheronal-B-induced aggregation of amyloid-β peptides **2a** to **2f**. (C–H) ThT analyses, ex: 440 nm and em: 485 nm reported as mean ± SD; (I–N) far-UV CD analyses (reported as an average of three scans) of mean residue ellipticity (-) at 217 nm, of peptides **2a** (green, wild-type); **2b** (red, **K*16**); **2c** (blue, **K*28**); **2d** (purple, Me_2N-D1); **2e** (orange, **K*16**, **K*28**); **2f** (brown, Me_2N-D1, **K*16**, **K*28**). In each case, the peptide (100 μM) in phosphate-buffered saline, pH 7.4, is incubated quiescently in the presence (filled squares) or absence (open squares) of aldehdye 1 (100 μM) at 37°C.

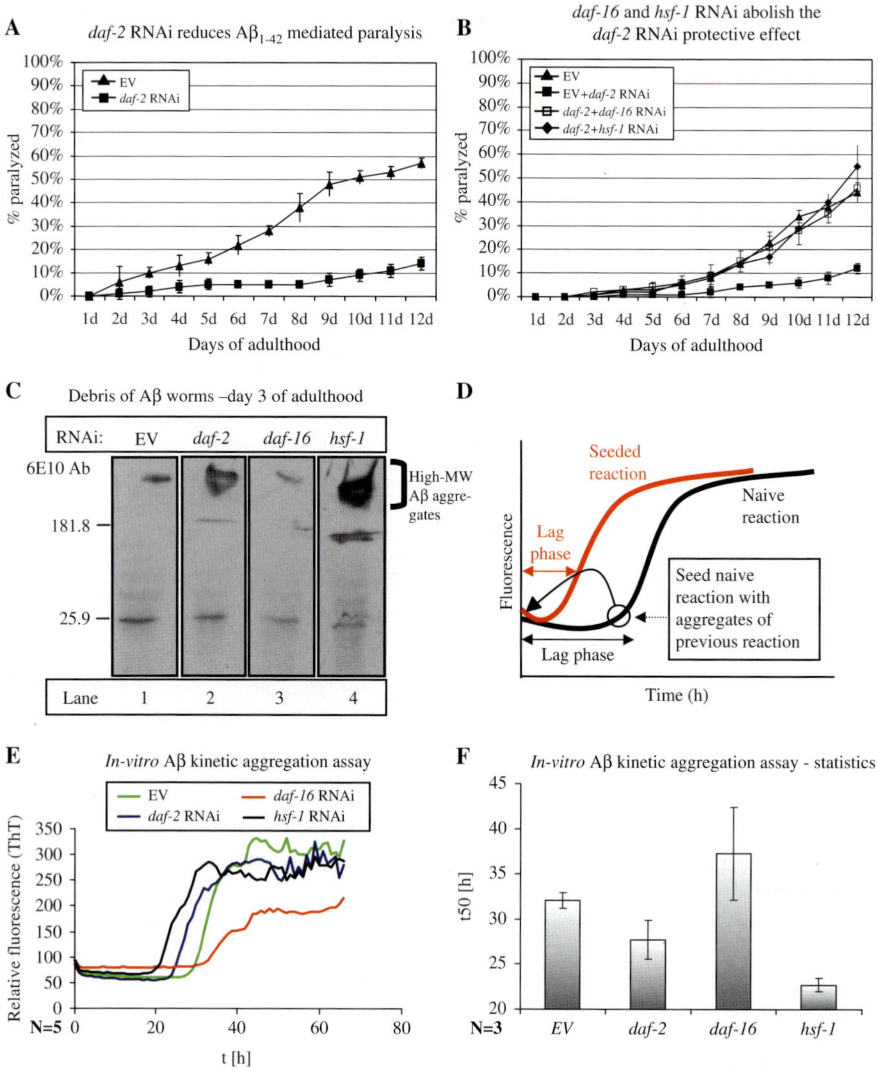

Chapter 29, Figure 1 Lack of correlation between high-molecular-mass Aβ aggregates and toxicity. (A) *daf-2* RNAi protects Aβ worms from the paralysis phenotype associated with Aβ expression. (B) The *daf-2* RNAi-mediated protective effect is *daf-16* and *hsf-1* dependent. *daf-2* RNAi protected worms from paralysis when diluted with EV bacteria but not when mixed with either *daf-16* or *hsf-1* RNAi bacteria. (C) Aβ worms were grown on RNAi bacteria as indicated. At day 3 of adulthood the worms were homogenized, spun, and debris was separated from the soluble fraction. Aβ contents in worm debris were analyzed using Western blot and 6E10 antibody. (From [10], with permission of AAAS.) (D) In vitro kinetic aggregation assay. The typical lag phase that is associated with in vitro aggregation of Aβ can be shortened by seeding of the reaction with previously aggregated Aβ. This technique has been exploited to measure Aβ aggregate content in worm samples. (E) Aβ seed contents of worms grown on RNAi bacteria (as indicated) were evaluated using kinetic aggregation assay. (From [10], with permission of AAAS.) (F) Quantification of three independent in vitro kinetic aggregation assays [as in (E)].

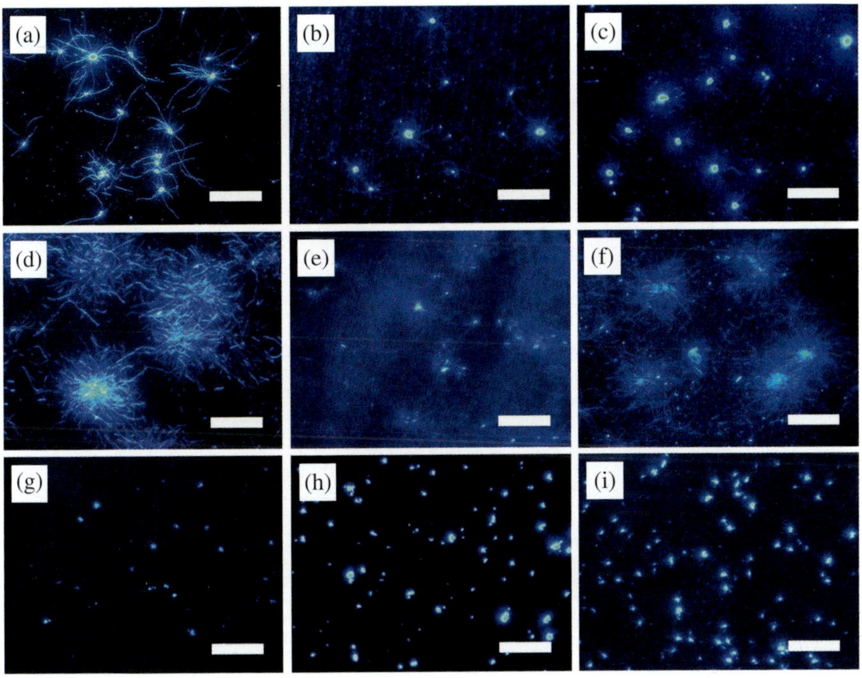

Chapter 33, Figure 4 Surface-dependent growth of Aβ(1–40) amyloid fibrils. *See page 704 for full caption.*

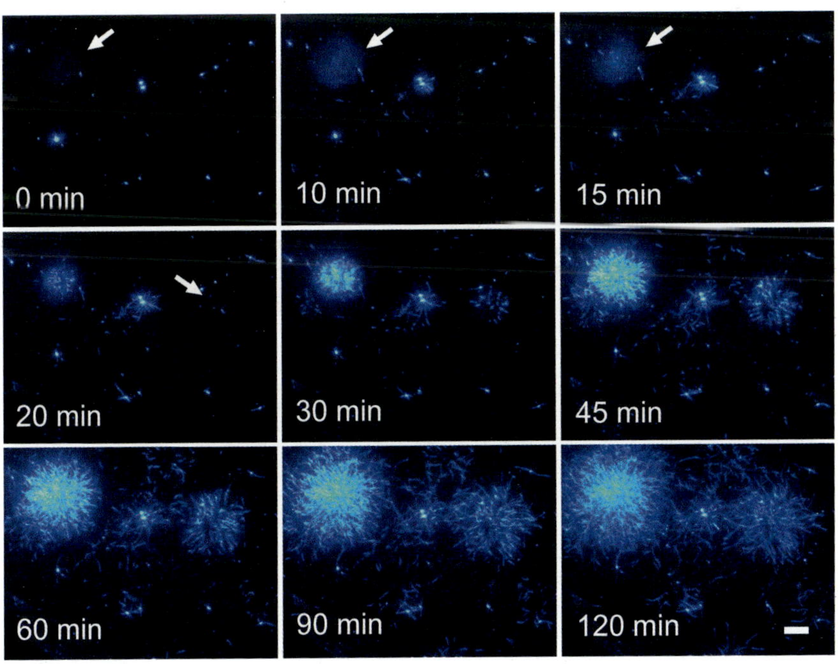

Chapter 33, Figure 5 Real-time observations of the formation of Aβ(1–40) spherulite. *See page 705 for full caption.*

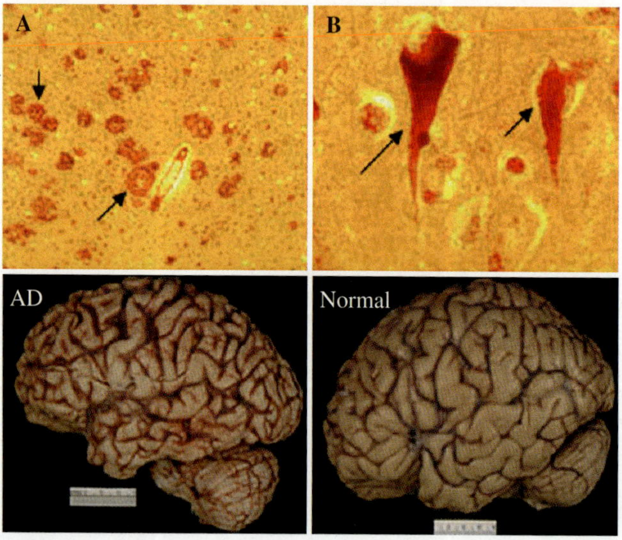

Chapter 34, Figure 1 Typical neuropathological lesions in Alzheimer disease brain. *See page 713 for full caption.*

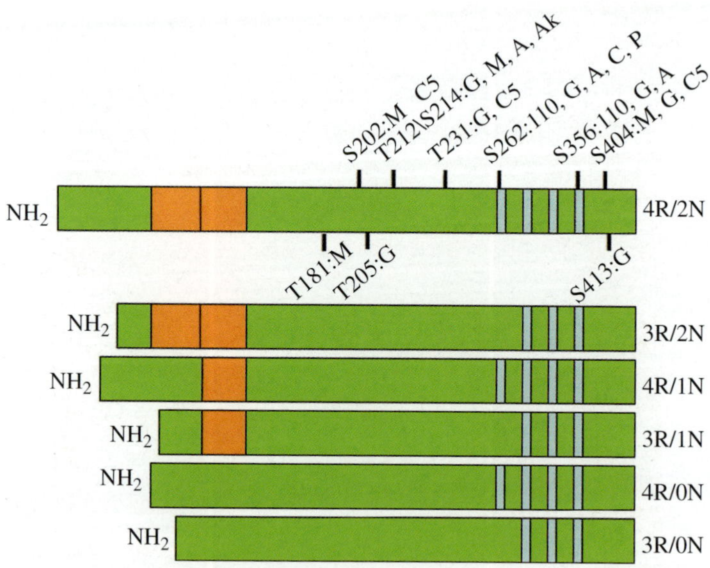

Chapter 34, Figure 4 The domain organization and some phosphorylation sites of tau isoforms expressed in adult humans. *See page 730 for full caption.*

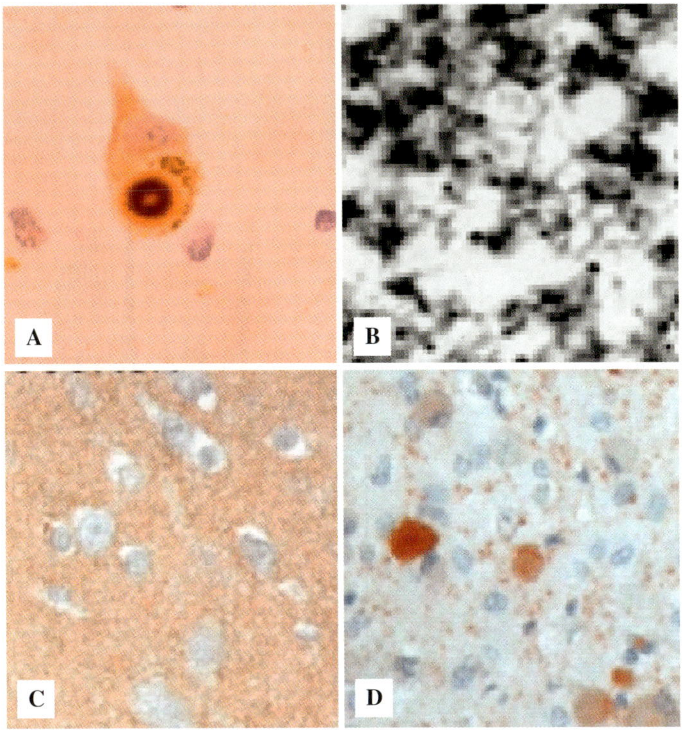

Chapter 37, Figure 1 Neuropathological findings of two human brain disorders associated with synucleinopathy. *See page 819 for full caption.*

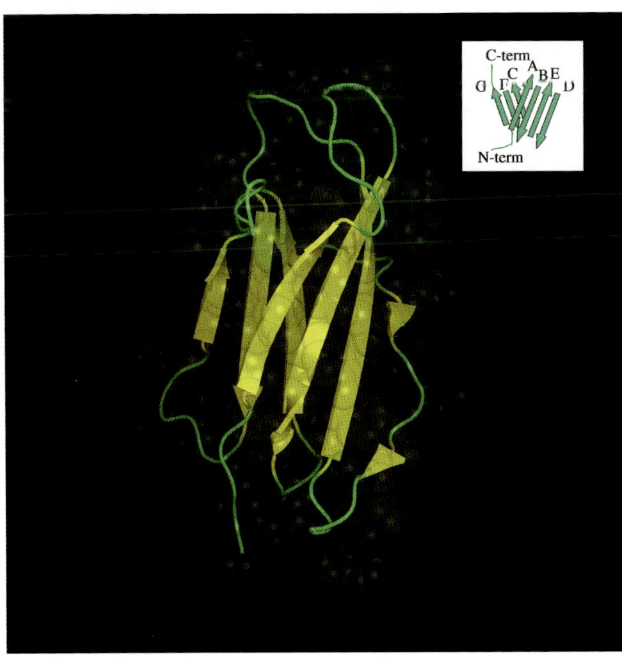

Chapter 38, Figure 1 Wild-type β_2m structure in MHC-I [20]. The strand segments are colored in yellow, the loop and bulge segments in green. The van der Waals surface of the molecule is shown in transparency. The strand naming scheme according to this representation is reported.

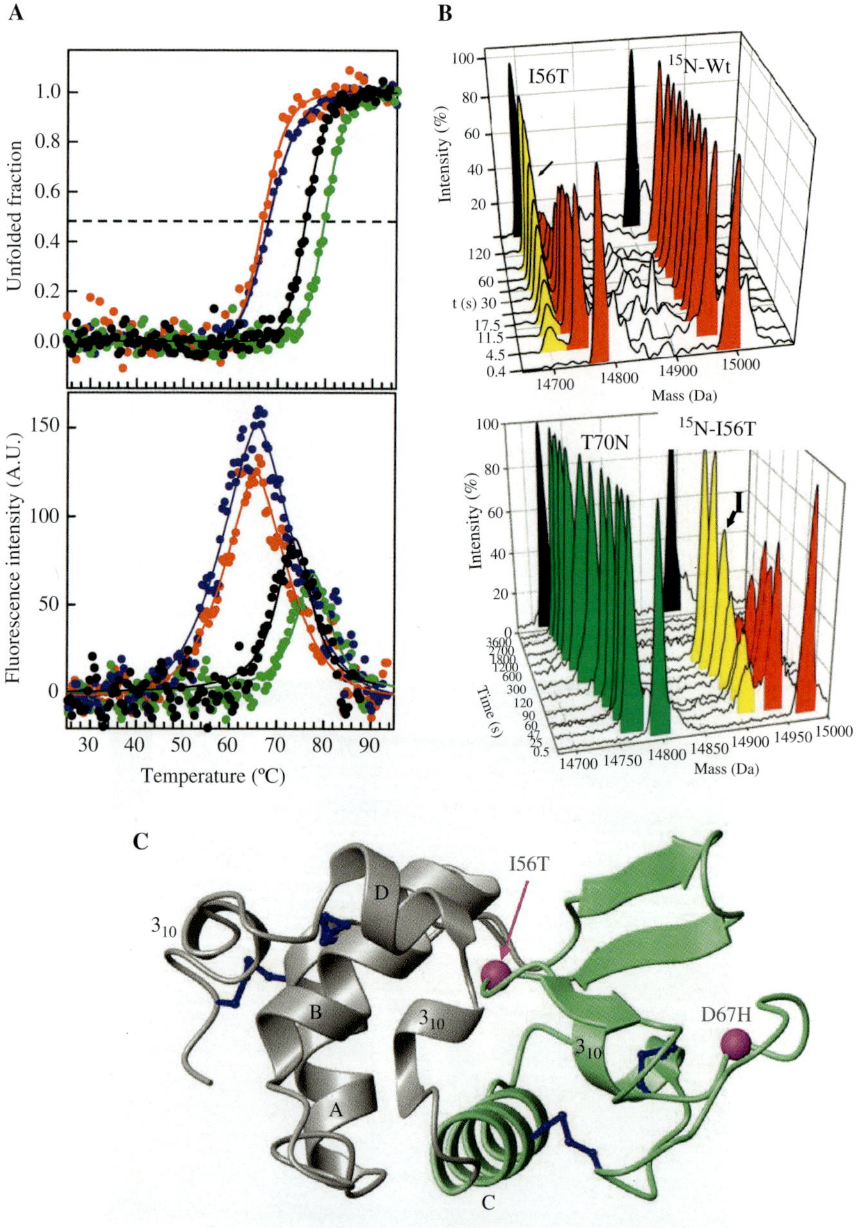

Chapter 39, Figure 2 Effects of the mutations on the stability and global cooperativity of lysozyme. *See pages 874–875 for full caption.*

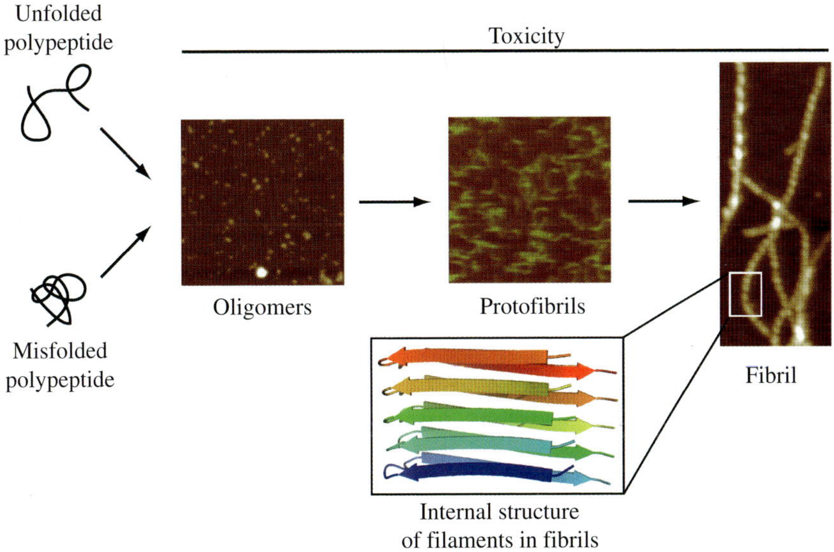

Chapter 45, Figure 1 Amyloid fibril formation arises from the concentration-dependent self-assembly of natively unfolded polypeptides (top) or partially denatured proteins (bottom) and ultimately proceeds through multiple intermediates to form the final cross-β-sheet amyloid fibril.

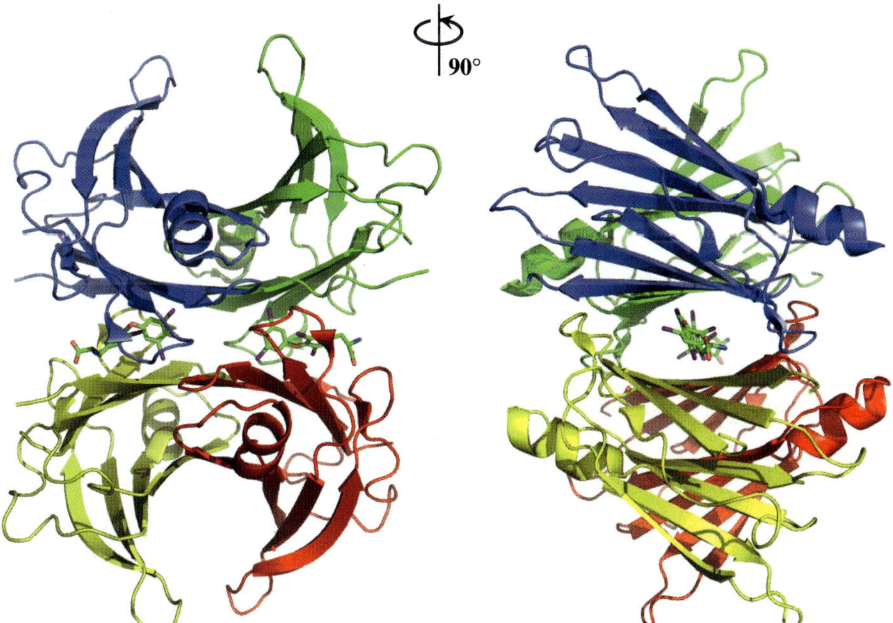

Chapter 45, Figure 2 Structure of the TTR tetramer, with each monomer depicted in a different color, with thyroxine bound along the crystallographic two fold axis in each of two symmetry related thyroxine-binding sites.

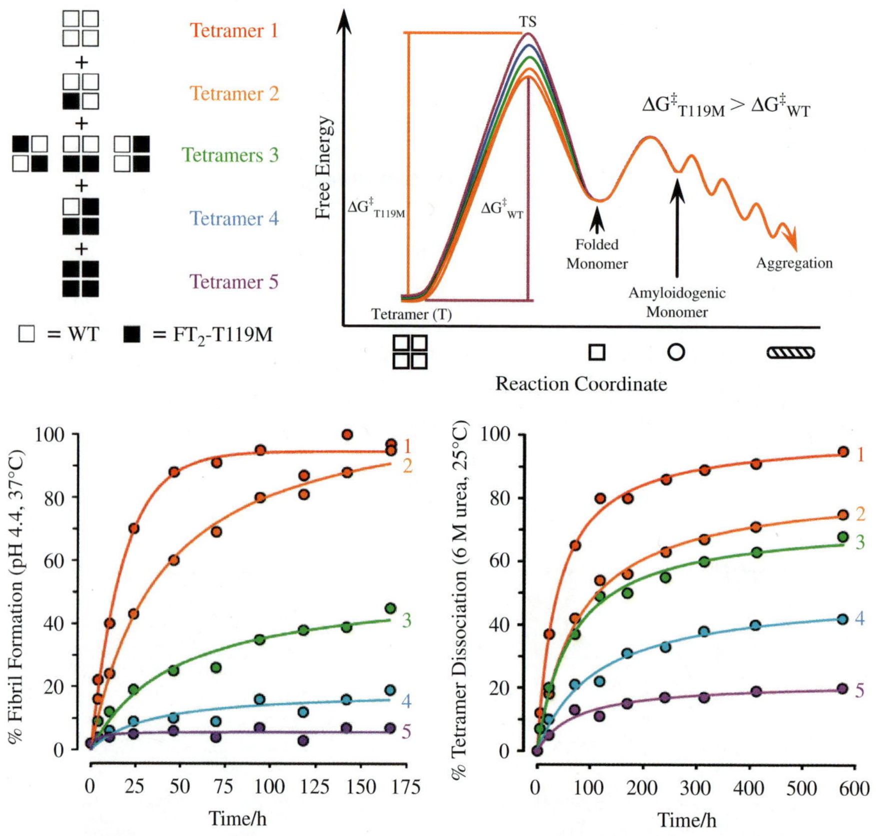

Chapter 45, Figure 5 Interallelic trans-suppression ameliorates TTR amyloid disease by making the TTR dissociation barrier increase proportional to the number of T119M TTR suppressor subunits in the tetramer otherwise composed of subunits that can engage in amyloidogenesis.

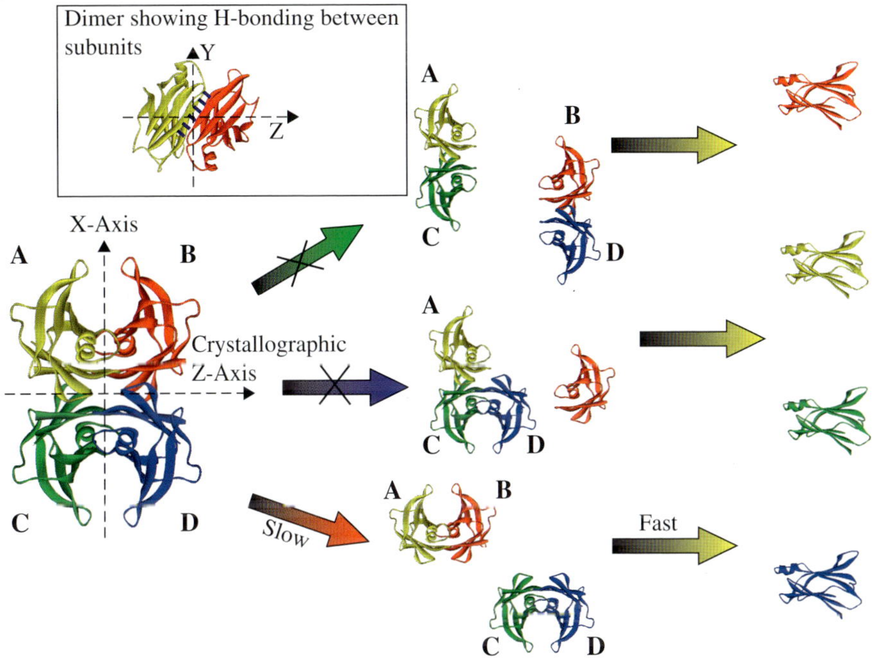

Chapter 45, Figure 6 (A) The TTR tetramer could dissociate through numerous mechanisms, many of which are shown. The operational mechanism is shown at the bottom, wherein the dimers dissociate from the tetramer about the interface incorporating the crystallographic Z-axis.

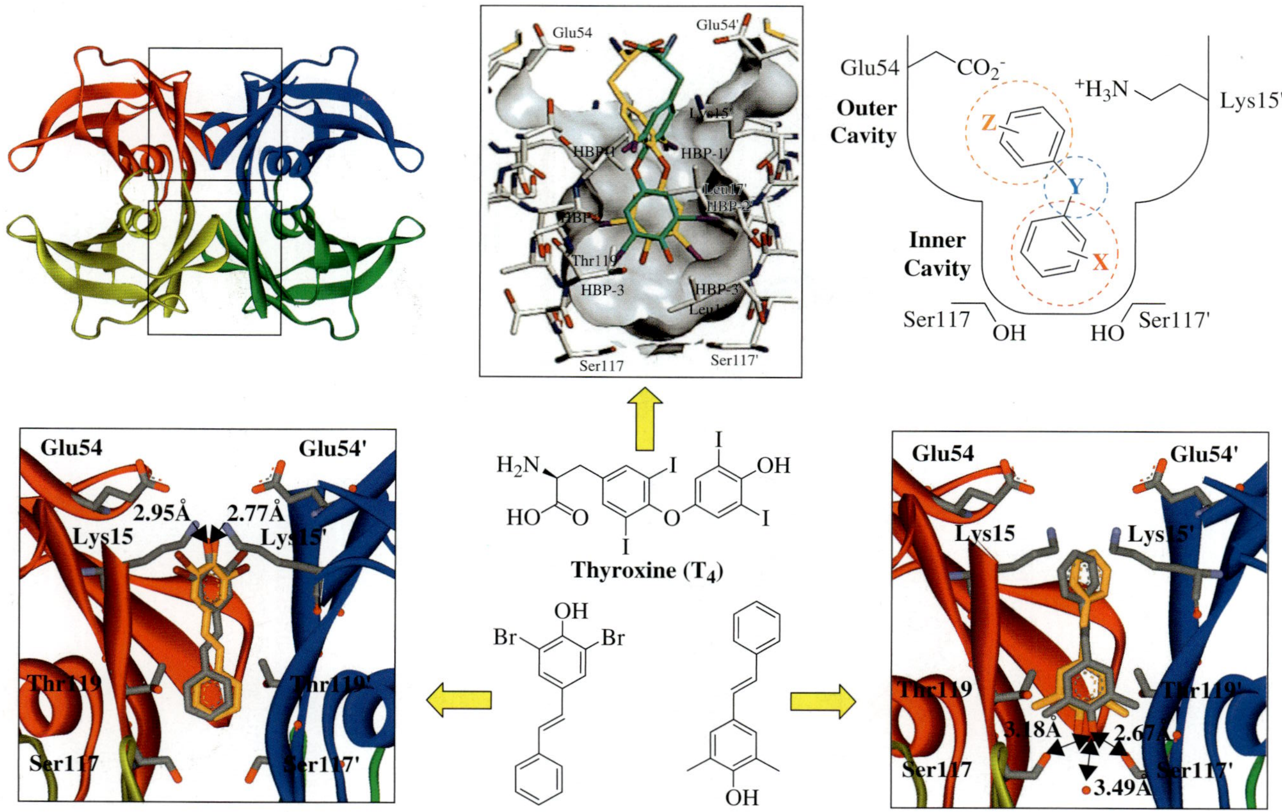

Chapter 45, Figure 6 (B) Dissociation through this mechanism is prevented through kinetic stabilization of the tetramer by binding of at least one ligand to one of the two thyroxine-binding sites.

INSTITUT MEDIATHEQUE PASTEUR

18

ALPHA-1-ANTITRYPSIN DEFICIENCY

David A. Lomas
Department of Medicine, Cambridge Institute for Medical Research, University of Cambridge, Cambridge, UK

David H. Perlmutter
Departments of Pediatrics, Cell Biology and Physiology, University of Pittsburgh School of Medicine, Children's Hospital of Pittsburgh, Pittsburgh, Pennsylvania

INTRODUCTION

The classical form of α_1-antitrypsin deficiency is unique among the protein misfolding diseases in that it causes target organ injury by both loss-of-function and gain-of-toxic function mechanisms. Lack of its antiprotease activity is associated with premature development of pulmonary emphysema by loss of function [1]. Accumulation of the mutant protein in liver cells is associated with chronic liver disease and predisposition to hepatocellular carcinoma by gain of toxic function [2].

α_1-Antitrypsin is a 55-kD secretory glycoprotein that inhibits neutrophil proteases, including elastase, cathepsin G, and proteinase 3. Plasma α_1-antitrypsin is derived predominantly from the liver and increases three- to fivefold during the host response to tissue injury and inflammation. It is the archetype of a family of structurally related serine protease inhibitors termed serpins [3]. There are more than 100 variant alleles at the human α_1-antitrypsin locus, most of which are rare in

Protein Misfolding Diseases: Current and Emerging Principles and Therapies,
Edited by Marina Ramirez-Alvarado, Jeffery W. Kelly, and Christopher M. Dobson
Copyright © 2010 John Wiley & Sons, Inc.

frequency. The Z variant (Glu342Lys) causes the classical form of the deficiency by rendering the protein prone to abnormal folding and aggregation [4]. The mutant protein accumulates in the endoplasmic reticulum (ER) of liver cells, resulting in an 85 to 90% reduction in serum concentrations of α_1-antitrypsin [2]. The deficiency is autosomal codominant and affects 1 : 1600 to 1 : 2000 live births.

LUNG DISEASE

Emphysema was noted in some of the first persons reported to have an absence of the α-1 band on serum protein electrophoresis [5]. It was confirmed by family studies [6] and is now the only genetic factor that is widely accepted to predispose to emphysema. The respiratory disease associated with α_1-antitrypsin deficiency usually presents with increasing dyspnea after the third decade, with cor pulmonale and polycythemia occurring late in the course of the disease. Chest radiographs typically show bibasal emphysema, with paucity and pruning of the basal pulmonary vessels. Lung function tests are typical for emphysema, with a reduced ratio of forced expiratory volume in 1 second to forced vital capacity (FEV1/FVC), gas trapping (raised ratio of residual volume to total lung capacity), and low gas transfer factor. The onset of respiratory disease can be delayed to the sixth decade in never-smokers who are homozygous for the Z allele, and these persons often have a normal life span [7]. High-resolution computed tomography scans with 1- to 2-mm collimation are the most accurate method of assessing the distribution of panlobular emphysema and for monitoring progress of the pulmonary disease [8,9] although this currently has little clinical value outside clinical trials.

LIVER DISEASE

A nationwide screening study of every newborn in Sweden carried out in the early 1970s by Sveger identified 127 newborns homozygous for Z α_1-antitrypsin [10]. These infants have been followed since then for liver and lung disease [11]. This unique study, lacking the inherent bias in ascertainment that characterizes all other studies of this disease, has shown that only 8% of the population develops clinically significant liver disease. This means that there is a subpopulation of homozygotes that are predisposed to liver disease and that the large majority is protected from liver disease.

In those with clinically significant liver disease, it most commonly presents in infancy with elevated transaminases and conjugated hyperbilirubinemia and progresses to chronic liver failure with signs of portal hypertension. However, liver disease can first present later in childhood, adolescence, or adult life. It can also present clinically in the form of hepatocellular carcinoma. Autopsy studies have provided evidence that this deficiency specifically predisposes to hepatocellular carcinoma [12].

The liver disease is relatively nonspecific, characterized by inflammation, mild necrosis, and steatosis as well as fibrotic change. The only specific histologic finding is the PAS$^+$/diastase-resistant globules in hepatocytes. Interestingly, only some of the hepatocytes have globules, and the number of globule-devoid hepatocytes increases with age.

One of the important advances in understanding the pathobiology of this liver disease was the demonstration of hepatic inflammation, adenomas, and carcinomas in mice transgenic for the mutant human Z α_1-antitrypsin gene [13,14]. This result provided powerful evidence that liver disease in α_1-antitrypsin deficiency involved a gain-of-toxic function mechanism because the mice retained endogenous antielastase activity.

STRUCTURAL PATHOBIOLOGY OF α_1-ANTITRYPSIN DEFICIENCY

Polymerization of Z α_1-Antitrypsin and Liver Disease

The structure of α_1-antitrypsin is composed of three β-sheets (A to C) and an exposed mobile reactive loop that presents a peptide sequence as a pseudosubstrate for the target proteinase [15–18]. After docking, the enzyme cleaves the P1-P1′ peptide bond of α_1-antitrypsin and is inactivated by a dramatic conformational transition that swings it 70Å from the upper to the lower pole of the protein in association with insertion of the reactive loop as an extra strand (s4A) in β-sheet A [19]. This remarkable conformational transition is central to the inhibitory activity of α_1-antitrypsin. However, as with most sophisticated mechanisms, the mobile domains are vulnerable to dysfunction.

The Z mutation distorts the relationship between the reactive center loop and β-sheet A (Fig. 1). Consequent perturbation in the structure allows the formation of an unstable intermediate that we have called M* [20–22]. M* molecules then link sequentially to form polymers in which the reactive center loop of one α_1-antitrypsin molecule inserts into the β-sheet A of another [4,20,23–26]. Spectroscopic analysis has demonstrated that oligomers of α_1-antitrypsin form during an initial lag phase and that these then condense to form a heterogeneous mixture of longer polymers [20,26]. It is these polymers that accumulate within the endoplasmic reticulum of hepatocytes to form the PAS$^+$ inclusions that are the hallmark of Z α_1-antitrypsin liver disease [4,27–29].

Although many α_1-antitrypsin deficiency variants have been described, only two other mutants of α_1-antitrypsin have been associated similarly with profound plasma deficiency and hepatic inclusions: α_1-antitrypsin Siiyama (Ser53Phe) [30,31] and Mmalton [32] (deletion of phenylalanine at position 52, also known as Mnichinan [33] and Mcagliari [34]). The Siiyama mutation is the commonest cause of severe α_1-antitrypsin deficiency in Japan, and the Mmalton variant is the commonest cause of severe α_1-antitrypsin deficiency in Sardinia. Both of these mutants are in the shutter domain underlying the

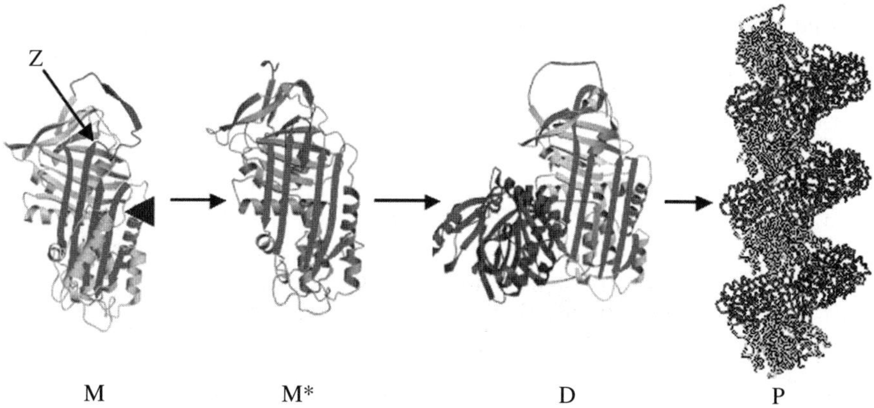

FIG. 1 The Z mutation of α_1-antitrypsin (Glu342Lys at P_{17}; arrowed) perturbs the structure of β-sheet A (green) and the mobile reactive center loop (red) to form the intermediate M*. The patent β-sheet A can then accept the loop of another molecule (as strand 4) to form a dimer (D) which then extends into polymers (P) [20–22]. It is these polymers that accumulate within hepatocytes to cause liver disease [4]. The position of the lateral hydrophobic pocket that is the target of rational drug design is shown with a blue arrowhead. Note the change in conformation in this region of the molecule as it forms M* and then dimers and polymers. (*See insert for color representation of figure.*)

bifurcation of strands 3 and 5 of β-sheet A (Fig. 1). The mutations disrupt a hydrogen-bond network that is based on residue 334His and bridge strands 3 and 5 of the A sheet [35], causing it to part to allow the formation of folding intermediates [36] and loop-sheet polymers in vivo [37,38].

Polymerization also underlies the mild plasma deficiency of other variants that perturb the shutter domain: S (Glu264Val) and I (Arg39Cys) α_1-antitrypsin [39,40]. These point mutations cause less disruption to β-sheet A than does the Z variant. Thus, the rates of polymer formation are much slower than that of Z α_1-antitrypsin [20], and this results in less retention of protein within hepatocytes, milder plasma deficiency, and the lack of a clinical phenotype. However, if a mild, slowly polymerizing I or S variant of α_1-antitrypsin is inherited with a rapidly polymerizing Z variant, the two can interact to form heteropolymers within hepatocytes, leading to inclusions and, finally, cirrhosis [40]. Thus, the severity of retention of mutants of α_1-antitrypsin within hepatocytes can be explained by the rate of polymer formation. Those mutants that cause the most rapid polymerization cause the most retention of α_1-antitrypsin within the liver. This correlates, in turn, with the greatest risk of liver damage and cirrhosis, and the most severe plasma deficiency.

Polymerization of Z α_1-Antitrypsin and Emphysema

Emphysema associated with plasma deficiency of α_1-antitrypsin is widely believed to be due to the reduction in plasma levels of α_1-antitrypsin to 10 to

15% of normal. This in turn markedly reduces the α_1-antitrypsin that is available to protect the lungs against proteolytic attack by the enzyme neutrophil elastase [41]. The situation is exacerbated as the Z mutation reduces the association rate between α_1-antitrypsin and neutrophil elastase approximately fivefold [42–45]. Thus, the α_1-antitrypsin available within the lung is not as effective as the normal M protein. This combination of α_1-antitrypsin deficiency, reduction in the efficacy of the α_1-antitrypsin molecule, and cigarette smoke can have a devastating effect on lung function [46,47], probably by allowing the unopposed action of proteolytic enzymes. The inhibitory activity of Z α_1-antitrypsin can be further reduced as the Z mutation favors the spontaneous formation of α_1-antitrypsin loop-sheet polymers within the lung [48]. This conformational transition inactivates α_1-antitrypsin as a proteinase inhibitor, thereby reducing further the already depleted levels of α_1-antitrypsin that are available to protect the alveoli (see [49] for a review). Moreover, the conversion of α_1-antitrypsin from a monomer to a polymer converts it to a chemoattractant for human neutrophils [50,51]. The magnitude of the effect is similar to that of the chemoattractant C5a and present over a range of physiological concentrations (EC_{50} 4.5 ± 2 μg/mL). Polymers also induced neutrophil shape change and stimulated myeloperoxidase release and neutrophil adhesion [50]. More recently, a monoclonal antibody has been used to demonstrate polymers in emphysematous tissue associated with Z α_1-antitrypsin deficiency but not in emphysema in people with normal levels of α_1-antitrypsin. Neutrophils co-localized with polymers in the alveoli. The pro-inflammatory properties of polymers were further confirmed by the demonstration that they caused a neutrophil influx when instilled into the lungs of mice [52]. Therefore, the chemoattractant properties of polymers may explain the excess number of neutrophils in bronchoalveolar lavage [53] and in tissue sections of lung parenchyma from people with Z α_1-antitrypsin deficiency. Any pro-inflammatory effect of polymers is likely to be exacerbated by inflammatory cytokines, cleaved or complexed α_1-antitrypsin [54], elastin degradation products [55], and cigarette smoke, which themselves cause neutrophil recruitment. Thus, emphysema associated with Z α_1-antitrypsin deficiency is likely to result from a combination of loss of function of α_1-antitrypsin (deficiency of circulating proteinase inhibitor, reduced inhibitory activity, and intraalveolar polymerization) and toxic gain of function as a result of the chemotactic properties of intraalveolar polymers.

CELLULAR PATHOBIOLOGY OF α_1-ANTITRYPSIN DEFICIENCY

Early studies have shown that the amino acid substitution of lysine for glutamate at residue 342 is sufficient for retention of the mutant Z α_1-antitrypsin molecule in the endoplasmic reticulum (ER) [1]. The mechanism for this retention is still not entirely known. We do know that Z α_1-antitrypsin polymerizes and aggregates in the ER [4], but it is not clear that polymerization is responsible for retention. The most powerful evidence that polymerization is responsible comes from studies in which secretion of Z α_1-antitrypsin is partially reconstituted by introduction of a second mutation that suppresses the polymerogenic properties [56,57]. However, these studies do not exclude the possibility that retention is caused by a distinct abnormality in folding that is also corrected partially by the second experimentally introduced mutation. Several observations suggest that polymerization is a result of retention rather than its cause. First, naturally occurring variants of α_1-antitrypsin that do not have polymerogenic properties are retained in the ER for as long as, or even longer than, Z α_1-antitrypsin [58]. Second, only 18% of the intracellular pool of Z α_1-antitrypsin at steady state is in the form of insoluble high-molecular-mass polymers in model cell lines characterized by marked ER retention [58,59]. The remainder of the pool is found in heterogeneous, soluble complexes with multiple chaperones that are also retained in the ER [59]. In either case polymerization and aggregation of Z α_1-antitrypsin appear to be critical determinants of how cells respond to this mutant protein from the perspective of adaptation and pathogenesis of liver injury/carcinoma.

Because liver disease involves a gain-of-toxic function mechanism, yet there is wide variability in hepatic phenotype, investigations have focused on the possibility that mechanisms by which cells protect themselves from mislocalized mutant proteins constitute potential disease modifiers. Moreover, we now know from a large body of basic cell biology research that cells have elaborate mechanisms for degrading mutant proteins that are retained in the ER and elaborate mechanisms for activating protective signaling pathways. One study has provided evidence for this concept by showing that there is a lag in ER degradation of Z α_1-antitrypsin after gene transfer into cell lines derived from homozygotes with established liver disease compared to cell lines from homozygotes with no apparent liver disease [60].

Early studies indicated that the proteasome played a role in degradation of mutant Z α_1-antitrypsin [61,62]. Later it was discovered that autophagy was a critical pathway for Z α_1-antitrypsin disposal [63–65]. Powerful evidence came from two completely different types of studies. In autophagy-deficient mammalian cell lines created by targeted gene disruption of ATG5, mutant Z α_1-antitrypsin was degraded more slowly than in the corresponding wild-type cell lines, resulting in massive intracellular accumulation of the mutant protein [66]. The mutant protein was also detected in autophagosomes of wild-type cell lines. Using a completely different approach, screening for yeast mutants in degradation of Z α_1-antitrypsin, Kruse et al. discovered that autophagy gene ATG6 was essential [67]. Degradation of Z α_1-antitrypsin could be reconstituted by expression of ATG6. In yeast autophagy was particularly critical for degradation of the polymerized form of Z α_1-antitrypsin that accumulated at high levels of expression [67]. Kruse et al. also found that the same pathway is essential for degradation of an aggregated mutant fibrinogen subunit that causes a rare liver disease associated with hypofibrinogenemia [68]. Together, these studies have established that autophagy participates in degradation of mutant Z α_1-antitrypsin. Further, they suggest that autophagy is responsible for piecemeal disposal of the aggregated forms of Z α_1-antitrypsin, as well as other aggregated proteins that accumulate in the ER, while the proteasomal pathway disposes of soluble forms of Z α_1-antitrypsin.

Recent experiments have also shown that accumulation of Z α_1-antitrypsin in the ER is sufficient to activate the autophagic response. For this, a mouse model of α_1-antitrypsin deficiency with inducible hepatocyte-specific expression of Z α_1-antitrypsin called the Z mouse was mated to the GFP-LC3 mouse [66]. LC3 is an autophagosomal membrane-specific protein, so the GFP-LC3 mouse is capable of making green fluorescent autophagosomes. Although the GFP-LC3 mouse develops green fluorescent autophagosomes in the liver only when it is subjected to starvation, a known stimulus for hepatic autophagy, the Z $\times$ GFP-LC3 mouse does not require starvation for the development of hepatic green fluorescent autophagosomes. Induction of Z α_1-antitrypsin gene expression in hepatocytes is sufficient in the Z $\times$ GFP-LC3 mouse.

Hepatic genomic analysis in the Z mouse has provided a clue for how autophagy is induced when Z α_1-antitrypsin accumulates in the ER [69]. The genomic analysis indicates that regulator of G-signaling RGS16, an antagonist of G protein Gαi3 [70], is markedly up-regulated (probably induced) when Z α_1-antitrypsin accumulates in the ER. Targeted disruption of Gαi3 is now known to reverse the inhibitory effect of insulin signaling on hepatic autophagy [71]. Thus, up-regulation of RGS16, and probably other members of the RGS family, is a potential mechanism for de-repressing hepatic autophagy in α_1-antitrypsin deficiency (Fig. 2).

Hidvegi et al. have used model cell lines and transgenic mice with inducible expression of Z α_1-antitrypsin to identify signaling pathways, other than autophagy, that are activated when this mutant protein accumulates in the ER and constitute potential modifiers of the hepatic phenotype. This type of model system was deemed ideal for identifying the initial cellular responses and being able to separate them from subsequent secondary and tertiary adaptive responses that evolve chronically. Two types of approaches were used. First, known cellular response pathways were examined specifically [72]. The results indicated that accumulation of mutant Z α_1-antitrypsin activated the NFκB pathway, ER caspases, BAP31, and mitochondrial caspases but not the unfolded protein response. Activation of NFκB has potentially important implications for target organ injury in the deficiency. For one thing, through downstream targets, NFκB activation could mediate inflammatory cell infiltration, particularly neutrophilic, into the liver and respiratory epithelium. For another, activation of NFκB has been shown to play a key role in inflammation-associated carcinogenesis [73,74] so it could be involved in the pathogenesis of hepatocellular carcinoma in α_1-antitrypsin deficiency. Cleavage and activation of BAP31, an integral membrane protein of the ER that has been

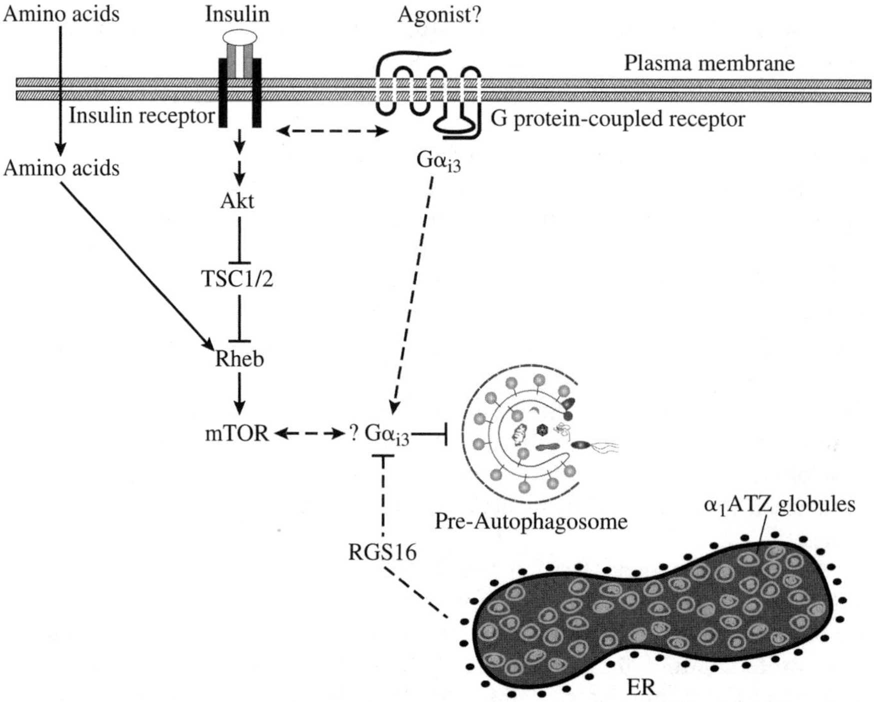

FIG. 2 Theoretical role of RGS16 in activating the hepatic autophagic response in α_1-antitrypsin deficiency. Amino acids and insulin/IGF-1 signaling inhibit autophagy through mTOR. The G protein Gαi3 plays a role in inhibition of autophagy by mTOR. Induction of RGS16 when mutant Z α_1-antitrypsin accumulates in the ER antagonizes Gαi3, thereby de-repressing autophagy. (From D. H. Perlmutter 2007, *Cell Death and Differentiation*, *16*, 39–45.)

associated with other ER retention states [75], may be particularly important because it appears to mediate proapoptotic signals from the ER to mitochondria [76]. Thus, it may provide a mechanistic basis for the mitochondrial dysfunction that has been noted in cell line and transgenic mouse models of α_1-antitrypsin deficiency as well as in the liver of deficient patients [65].

It is still not entirely clear why accumulation of Z α_1-antitrypsin in the ER does not activate the unfolded protein response. However, this result has been corroborated by a plethora of strategies and by studies done in multiple laboratories [72]. It probably has something to do with the intrinsic properties of Z α_1-antitrypsin, perhaps its polymerogenic properties, because ER accumulation of two naturally occurring nonpolymerogenic truncated variants of α_1-antitrypsin, Saar and SaarZ, does activate the UPR [72]. The most obvious putative explanation for this difference, failure of Z α_1-antitrypsin to titrate BiP or other chaperones away from the proximal transducers of the unfolded protein response, does not appear to apply. An identical profile of ER chaperones, including BiP, has been found to interact with Z α_1-antitrypsin and the nonpolymerogenic AT variants in cross-linking and co-immunoprecipitation experiments [59].

As a second approach for identification of signaling pathways activated by ER accumulation of mutant Z α_1-antitrypsin, Hidvegi et al. utilized genomic analysis of liver from the Z mouse, with its inducible, hepatocyte-specific expression of Z α_1-antitrypsin [69]. In contrast to examination of known cellular response pathways, this method was envisioned to detect changes in gene expression that could implicate signaling pathways not previously linked to α_1-antitrypsin deficiency or ER stress states in general. Because multiple control groups could be applied to this type of analysis—Z mouse with and without induction of the target gene: Z mouse with induction of the target gene for different time intervals; M mouse, with inducible, hepatocyte-specific expression of the wild-type human AT gene, with and without induction of the target gene; nontransgenic littermate at the same age—the results have been particularly informative. Importantly, the results validated a number of previous observations about accumulation of Z α_1-antitrypsin in the ER and about the effects of α_1-antitrypsin deficiency on the liver. First, it included a number of changes in gene expression that are indicative of a TGFβ effect and are consistent with the known hepatic fibrotic effect of AT deficiency. Second, there were numerous changes in gene expression that reflect a significant change in sterol and lipid metabolism and would be reflective of the mild-to-moderate hepatic steatosis that occurs in α_1-antitrypsin deficiency. Network analysis predicts that many of these changes can be explained by decreases in SREBF1 and 2. Third, there were changes in expression of NFκB target genes consistent with neutrophilic and Th2 lymphocytic inflammatory responses. Fourth, the gene expression profile was consistent with marked effects on cell proliferation/cell death and carcinogenic tendency, as expected for α_1-antitrypsin deficiency (see below). Most interesting among these were increases in expression of AP2 and SP1 and decreases in EGR1, STAT1, AP5, and Id3. Furthermore, network analysis predicts activation of ERK2, CDK inhibitor 2a, p53, fos, and myc activity together with increases in caspase 3 and APAF1 activity. Fifth, there were quite marked effects on expression of members of the cytochrome P450 family, probably reflecting the changes in ER composition imposed by the increased protein load. Sixth, the gene expression profile did not reflect an unfolded protein response.

This hepatic genomic analysis also identified changes in gene expression that would not have been predicted previously and that provide novel implications for the cellular response to ER accumulation of mutant Z α_1-antitrypsin and the pathogenesis of liver inflammation/carcinogenesis in α_1-antitrypsin deficiency. First, we noted up-regulation of RGS16 and RGS5, probably playing an important role in activation of autophagy. More detailed studies of hepatic RGS16 up-regulation indicated that it is a good marker of the distinct ER stress that mutant Z α_1-antitrypsin imposes in several ways [69]. It does not occur when nonpolymerogenic α_1-antitrypsin mutants accumulate in the ER, and it does not occur when the unfolded protein response is evoked by tunicamycin or calcium ionophore. RGS16 expression was elevated significantly in the liver of patients with α_1-antitrypsin deficiency as well. Second, the genomic analysis indicated changes in gene expression that are likely to affect the ubiquitin-dependent proteasomal pathway in a way that would not have previously predicted by increases in expression of two deubiqutinating enzymes and decreases in expression of one proteasomal regulatory subunit. Third, there were decreases in expression of a number of lysosomal enzymes. Fourth, there were striking changes in expression of a cluster of transcription factors that play a role in liver-specific gene expression and are regulated by circadian rhythm and by fasting. Because RGS16 expression is also up-regulated by starvation [77], the hepatic gene expression profile found here raises the hypothesis that intracellular accumulation of aggregated protein give the cell the false impression that it is being starved and elicits a cellular response program that reflects the response to starvation. Fifth, network analysis of the profile suggests the involvement of several other signaling pathways including the insulin-like growth factor 1, MAP kinase, stress-activated protein kinase, and hypoxic response signaling pathways.

Accumulation of mutant Z α_1-antitrypsin in the ER has marked effects on proliferation and death of hepatocytes in vivo. Using BrdU labeling, Rudnick et al. showed that there was a 5- to 10-fold increase in hepatocyte proliferation in the resting liver of the PiZ mouse model of α_1-antitrypsin deficiency [78]. Interestingly, these studies revealed that it is the globule-devoid hepatocytes that are proliferating. The regenerative response to partial hepatectomy was tested as well. Although the PiZ mice were more likely to die when subjected to partial hepatectomy, those that survived showed that both the globule-containing and globule-devoid hepatocytes proliferated under these conditions. These results, taken together with histological staining for markers of cell death, suggest that the globule-containing hepatocytes are relatively impaired in proliferation and death, while the globule-devoid hepatocytes have a selective proliferative advantage.

These studies have led to a hypothetical paradigm for the pathogenesis of hepatocellular carcinoma in α_1-antitrypsin deficiency [79]. Central to this paradigm is the globule-containing hepatocyte. It is relatively impaired in proliferative and death properties ("sick but not dead") and responsible for putative "injury/regeneration" signals. The globule-devoid cells have a selective proliferative advantage in that they are the only hepatocytes that can effectively receive and transduce these putative regenerative signals. The cancer-prone state is therefore engendered by having cells that are unable to die at the appropriate time and cells that are chronically dividing in an inflamed milieu. The relative block in proliferation and death that characterizes the globule-containing hepatocytes is a particularly important part of this paradigm. Because the block is relative, these cells eventually die but they can be replenished, so that some of them are always present and, presumably, elaborating trans-acting regenerative signals. Studies by An et al. have shown that the globule-devoid hepatocytes have a lesser amount of insoluble aggregated Z α_1-antitrypsin [29]. Presumably they are progenitors or relatively young cells for which there has not been as much time for accumulation of insoluble Z α_1-antitrypsin. Because of this, it would presumably only take one or a few of these, ultimately to lead to many more.

The mechanism by which globule-containing hepatocytes eventually die is not completely understood, but there is some evidence that mitochondrial dysfunction is the final common pathway. Mitochondrial injury has been demonstrated functionally and by detailed ultrastructural studies in cell line and mouse models as well as in the liver of affected patients [65]. Activation of mitochondrial caspases has also been described. Moreover, antagonism of mitochondrial dysfunction by administration of cyclosporine A has been associated with reduction in mitochondrial ultrastructural change, caspase-3 activation, and liver disease morbidity in the PiZ mouse model [65].

CURRENT AND POTENTIAL THERAPEUTIC STRATEGIES

The only treatment available for the cirrhosis associated with α_1-antitrypsin deficiency is supportive therapy. Liver transplantation provides definitive treatment for patients with end-stage α_1-antitrypsin deficiency–related cirrhosis, and children who receive a transplant have an excellent clinical outcome [80]. There is good evidence that many Z α_1-antitrypsin homozygotes would develop only mild lung disease if they abstain from smoking [7,81]. Patients with α_1-antitrypsin deficiency should receive advice on smoking cessation, and individuals with airflow obstruction should be assessed with lung function tests followed by trials of bronchodilators and inhaled corticosteroids [82]. Many patients benefit from pulmonary rehabilitation [83] and, where appropriate, assessment for long-term oxygen therapy [82]. The most severely affected patients should be considered for lung transplantation, which can prolong survival, improve functional capacity, and enhance quality of life [84]. However, rejection remains an obstacle to better medium-term results and, currently, heart–lung transplantation has a five-year survival of approximately 50%. The role of lung volume reduction surgery (LVRS) in patients with α_1-antitrypsin is unclear, as the benefit is far less than in patients with smoking-related upper lobe emphysema with normal levels of α_1-antitrypsin [85].

The genetic deficiency in the antielastase screen may be rectified biochemically by intravenous infusions of α_1-antitrypsin [41]. There is registry data to suggest that this therapy may slow the rate of decline in lung function in patients with an FEV_1 value of 35 to 49% predicted [86], but this has yet to be proven in randomized controlled trials. The only controlled trial that has assessed α_1-antitrypsin replacement therapy showed that infusions of α_1-antitrypsin may slow the progression of emphysema as assessed by HRCT scans but has no effect on decline in FEV_1 [87]. A larger study is required to confirm these findings [88] and α_1-antitrypsin replacement therapy (Prolastin, Bayer) is currently not available in some parts of Europe.

Other treatments at earlier stages of development include gene therapy and the administration of retinoic acid and chemical chaperones. Vectors carrying the α_1-antitrypsin gene have been targeted to liver [89], lung [90] and muscle [91,92] in animals. There is good expression of α_1-antitrypsin but additional data are required to assess whether this can be achieved in humans. In particular, it is important to determine the length of time of protein expression and whether the levels of α_1-antitrypsin in the epithelial lining fluid of the lung are sufficient to prevent ongoing proteolytic damage. Similarly, although the effects of retinoic acid on alveolar regeneration in the rat look promising [93], they have yet to be demonstrated in patients with emphysema.

Novel Strategies to Prevent Conformational Transitions and Disease

There is now strong evidence that polymers of α_1-antitrypsin form by an aberrant linkage between the reactive center loop of one molecule and β-sheet A of another [4,15,24,94–97]. This has allowed the development of new strategies to attenuate polymerization and so treat the associated disease. Three strategies have been pursued to date: (1) chemical chaperones to stabilize the unstable mutant serpin, (2) the use of reactive loop peptides that compete for binding to β-sheet A, and (3) filling a surface cavity to block the conformational transition underlying polymer formation.

1. *Chemical chaperones to stabilize serpins and block polymerization.* Chemical chaperones can stabilize intermediates on the folding pathway. Osmolytes such as betaine, trimethyamine oxide, sarcosine, glycerol, erythritol, and trehalose all stabilize α_1-antitrypsin against polymer formation [98,99]. However, the chaperone trimethyamine oxide had no effect on the secretion of Z α_1-antitrypsin in cell culture [100], as it favored the conversion of unfolded Z α_1-antitrypsin to polymers [101]. In contrast, glycerol increased the secretion of Z α_1-antitrypsin from cell lines [100] probably as it binds to, and stabilizes, β-sheet A [35,99]. 4-phenylbutyrate also increased the secretion of Z α_1-antitrypsin from cell lines and transgenic mice [100] but, unfortunately, had no effect on the secretion of α_1-antitrypsin in patients with α_1-antitrypsin deficiency [102].
2. *Peptides with homology to the reactive center loop can compete for binding to β-sheet A and block polymerization.* The polymerization of Z α_1-antitrypsin can be blocked by annealing of reactive loop peptides to β-sheet A [4,103]. Such peptides were 11 to 13 residues in length but were promiscuous and could bind to other members of the serpin superfamily [103,104]. More recently, the recognition that the Z mutation forces α_1-antitrypsin into a conformation similar to M* (see Fig. 1) has allowed the design of a 6-mer peptide that specifically anneals to the lower part of β-sheet A and blocks polymerization [22,105,106]. This peptide was specific to Z α_1-antitrypsin

and did not anneal significantly to other serpins (such as antithrombin, α_1-antichymotrypsin, and PAI-1) with a similar tertiary structure. Indeed, smaller peptides have been developed that will also anneal to a patent β-sheet A and block the polymerization of Z α_1-antitrypsin in vitro [106,107]. The aim now is to convert these peptides into small-molecule inhibitors that can be used to block aberrant polymerization in vivo.

3. *Filling a surface hydrophobic pocket to block polymerization.* A third strategy comes from the identification of a hydrophobic pocket in α_1-antitrypsin that is bounded by strand 2A and helices D and E [18,108]. The cavity is patent in the native protein but is filled as β-sheet A accepts an exogenous reactive loop peptide during polymerization [18]. The introduction of either Thr114Phe or Gly117Phe on strand 2 of β-sheet A within this cavity raised the melting temperature of α_1-antitrypsin significantly and retarded polymer formation [109]. The importance of these observations was underscored by the finding that the Thr114Phe mutation reduced polymer formation and increased the secretion of Z α_1-antitrypsin from a *Xenopus* oocyte expression system. This cavity was used as a target for rational structure-based drug design to block polymer formation [110]. Virtual ligand screening was performed on 1.2 million small molecules, and six compounds were identified that reduced polymer formation in vitro. Modeling the effects of ligand binding on the cavity and rescreening the library identified an additional 10 compounds that blocked polymerization completely. The best antagonists were effective at ratios of compound to Z α_1-antitrypsin of 2.5 : 1 and reduced the intracellular accumulation of Z α_1-antitrypsin by 70% in a cell model of disease [110]. These data demonstrate the importance of this cavity as a site for rational drug design to ameliorate polymerization and treat the associated conformational disease.

REFERENCES

1. Lomas, D.A. (2006). Parker B Francis Lectureship: Antitrypsin deficiency, the serpinopathies, and chronic obstructive pulmonary disease. *Proc Am Thorac Soc, 3*, 499–501.
2. Perlmutter, D.H. (2002). Liver injury in α_1-antitrypsin deficiency: an aggregated protein induces mitochondrial injury. *J Clin Invest, 110*, 1579–1583.
3. Carrell, R., Lomas, D., Stein, P., Whisstock, J. (1997). Dysfunctional variants and the structural biology of the serpins. *Adv Exp Med Biol, 425*, 207–222.
4. Lomas, D.A., Evans, D.L., Finch, J.T., Carrell, R.W. (1992). The mechanism of Z α_1-antitrypsin accumulation in the liver. *Nature, 357*, 605–607.
5. Laurell, C.-B., Eriksson, S. (1963). The electrophoretic α_1-globulin pattern of serum in α_1-antitrypsin deficiency. *Scand J Clin Lab Invest, 15*, 132–140.
6. Eriksson, S. (1965). Studies in α_1-antitrypsin deficiency *Acta Med Scand Suppl, 432*, 1–85.

7. Seersholm, N., Kok-Jensen, A. (1998). Clinical features and prognosis of life time non-smokers with severe α_1-antitrypsin deficiency. *Thorax, 53*, 265–268.

8. Dirksen, A., Holstein-Rathlou, N.-H., Madsen, F., Skovgaard, L.T., Ulrik, C.S., Heckscher, T., Kok-Jensen, A. (1998). Long-range correlations of serial FEV_1 measurements in emphysematous patients and normal subjects. *J Appl Physiol, 85*, 259–265.

9. Dirksen, A., Friis, M., Olesen, K.P., Skovgaard, L.T., Sørensen, K. (1997). Progress of emphysema in severe α_1-antitrypsin deficiency as assessed by annual CT. *Acta Radiol, 38*, 826–832.

10. Sveger, T. (1976). Liver disease in alpha$_1$-antitrypsin deficiency detected by screening of 200,000 infants. *N Engl J Med, 294*, 1316–1321.

11. Sveger, T. (1988). The natural history of liver disease in α_1-antitrypsin deficient children. *Acta Paediatr Scand, 77*, 847–851.

12. Eriksson, S., Carlson, J., Velez, R. (1986). Risk of cirrhosis and primary liver cancer in alpha$_1$-antitrypsin deficiency. *N Engl J Med, 314*, 736–739.

13. Carlson, J.A., Barton Rogers, B., Sifers, R.N., Finegold, M.J., Clift, S.M., De Mayo, F.J., Bullock, D.W., Woo, S.L.C. (1989). Accumulation of PiZ α_1-antitrypsin causes liver damage in transgenic mice. *J Clin Invest, 83*, 1183–1190.

14. Dycaico, M.J., Grant, S.G.N., Felts, K., Nichols, W.S., Geller, S.A., Hager, J.H., Pollard, A.J., Kohler, S.W., Short, H.P., Jirik, F.R., et al. (1988). Neonatal hepatitis induced by α_1-antitrypsin: a transgenic mouse model. *Science, 242*, 1409–1412.

15. Elliott, P.R., Lomas, D.A., Carrell, R.W., Abrahams, J.-P. (1996). Inhibitory conformation of the reactive loop of α_1-antitrypsin. *Nat Struct Biol, 3*, 676–681.

16. Ryu, S.-E., Choi, H.-J., Kwon, K.-S., Lee, K.N., Yu, M.-H. (1996). The native strains in the hydrophobic core and flexible reactive loop of a serine protease inhibitor: crystal structure of an uncleaved α_1-antitrypsin at 2.7Å. *Structure, 4*, 1181–1192.

17. Elliott, P.R., Abrahams, J.-P., Lomas, D.A. (1998). Wildtype α_1-antitrypsin is in the canonical inhibitory conformation. *J Mol Biol, 275*, 419–425.

18. Elliott, P.R., Pei, X.Y., Dafforn, T.R., Lomas, D.A. (2000). Topography of a 2.0Å structure of α1-antitrypsin reveals targets for rational drug design to prevent conformational disease. *Protein Sci, 9*, 1274–1281.

19. Huntington, J.A., Read, R.J., Carrell, R.W. (2000). Structure of a serpin–protease complex shows inhibition by deformation. *Nature, 407*, 923–926.

20. Dafforn, T.R., Mahadeva, R., Elliott, P.R., Sivasothy, P., Lomas, D.A. (1999). A kinetic mechanism for the polymerisation of α_1-antitrypsin. *J Biol Chem, 274*, 9548–9555.

21. Gooptu, B., Hazes, B., Chang, W.-S.W., Dafforn, T.R., Carrell, R.W., Read, R., Lomas, D.A. (2000). Inactive conformation of the serpin α_1-antichymotrypsin indicates two stage insertion of the reactive loop; implications for inhibitory function and conformational disease. *Proc Natl Acad Sci U S A, 97*, 67–72.

22. Mahadeva, R., Dafforn, T.R., Carrell, R.W., Lomas, D.A. (2002). Six-mer peptide selectively anneals to a pathogenic serpin conformation and blocks polymerisation: implications for the prevention of Z α_1-antitrypsin related cirrhosis. *J Biol Chem, 277*, 6771–6774.

23. Janciauskiene, S., Dominaitiene, R., Sternby, N.H., Piitulainen, E., Eriksson, S. (2002). Detection of circulating and endothelial cell polymers of Z and wildtype alpha-1-antitrypsin by a monoclonal antibody. *J Biol Chem, 277*, 26540–26546.

24. Sivasothy, P., Dafforn, T.R., Gettins, P.G.W., Lomas, D.A. (2000). Pathogenic α_1-antitrypsin polymers are formed by reactive loop-β-sheet A linkage. *J Biol Chem, 275*, 33663–33668.

25. James, E.L., Bottomley, S.P. (1998). The mechanism of α_1-antitrypsin polymerization probed by fluorescence spectroscopy. *Arch Biochem Biophys, 356*, 296–300.

26. Purkayastha, P., Klemke, J.W., Lavender, S., Oyola, R., Cooperman, B.S., Gai, F. (2005). α_1-Antitrypsin polymerisation: a fluorescence correlation spectroscopic study *Biochemistry, 44*, 2642–2649.

27. Wu, Y., Swulius, M.T., Moremen, K.W., Sifers, R.N. (2003). Elucidation of the molecular logic by which misfolded α1-antitrypsin is preferentially selected for degradation. *Proc Natl Acad Sci U S A, 100*, 8229–8234.

28. Janciauskiene, S., Eriksson, S., Callea, F., Mallya, M., Zhou, A., Seyama, K., Hata, S., Lomas, D.A. (2004). Differential detection of PAS-positive inclusions formed by the Z, Siiyama and Mmalton variants of α_1-antitrypsin. *Hepatology, 40*, 1203–1210.

29. An, J.K., Blomenkamp, K., Lindblad, D., Teckman, J.H. (2005). Quantitative isolation of alpha-l-antitrypsin mutant Z protein polymers from human and mouse livers and the effect of heat. *Hepatology, 41*, 160–167.

30. Seyama, K., Nukiwa, T., Takabe, K., Takahashi, H., Miyake, K., Kira, S. (1991). S_{iiyama} (serine 53 (TCC) to phenylalanine 53 (TTC)): a new α1-antitrypsin-deficient variant with mutation on a predicted conserved residue of the serpin backbone. *J Biol Chem, 266*, 12627–12632.

31. Seyama, K., Nukiwa, T., Souma, S., Shimizu, K., Kira, S. (1995). α_1-Antitrypsin--deficient variant Siiyama (Ser^{53}[TCC] to Phe^{53}[TTC]) is prevalent in Japan: status of α_1-antitrysin deficiency in Japan. *Am J Respir Crit Care Med, 152*, 2119–2126.

32. Roberts, E.A., Cox, D.W., Medline, A., Wanless, I.R. (1984). Occurrence of alpha-1-antitrypsin deficiency in 155 patients with alcoholic liver disease. *Am J Clin Pathol, 82*, 424–427.

33. Matsunaga, E., Shiokawa, S., Nakamura, H., Maruyama, T., Tsuda, K., Fukumaki, Y. (1990). Molecular analysis of the gene of the α_1-antitrypsin deficiency variant, Mnichinan. *Am J Hum Genet, 46*, 602–612.

34. Sergi, C., Consalez, G.C., Fabbretti, G., Brisigotti, M., Faa, G., Costa, V., Romeo, G., Callea, F. (1994). Immunohistochemical and genetic characterization of the M Cagliari α-1-antitrypsin molecule (M-like α-1-antitrypsin deficiency). *Lab Invest, 70*, 130–133.

35. Zhou, A., Stein, P.E., Huntington, J.A., Carrell, R.W. (2003). Serpin polymerisation is prevented by a hydrogen bond network that is centered on His334 and stabilized by glycereol. *J Biol Chem, 278*, 15116–15122.

36. Kang, H.A., Lee, K.N., Yu, M.-H. (1997). Folding and stability of the Z and Siiyama genetic variants of human α_1-antitrypsin. *J Biol Chem, 272*, 510–516.

37. Lomas, D.A., Finch, J.T., Seyama, K., Nukiwa, T., Carrell, R.W. (1993). α_1-Antitrypsin S_{iiyama} (Ser^{53} → Phe); further evidence for intracellular loop-sheet polymerisation. *J Biol Chem, 268*, 15333–15335.

38. Lomas, D.A., Elliott, P.R., Sidhar, S.K., Foreman, R.C., Finch, J.T., Cox, D.W., J.C., W., Carrell, R.W. (1995). Alpha$_1$-antitrypsin Mmalton (^{52}Phe deleted) forms loop-sheet polymers in vivo: evidence for the C sheet mechanism of polymerisation. *J Biol Chem, 270*, 16864–16870.

39. Elliott, P.R., Stein, P.E., Bilton, D., Carrell, R.W., Lomas, D.A. (1996). Structural explanation for the dysfunction of S α_1-antitrypsin. *Nat Struct Biol, 3*, 910–911.

40. Mahadeva, R., Chang, W.-S.W., Dafforn, T., Oakley, D.J., Foreman, R.C., Calvin, J., Wight, D., Lomas, D.A. (1999). Heteropolymerisation of S, I and Z α_1-antitrypsin and liver cirrhosis. *J Clin Invest, 103*, 999–1006.

41. Wewers, M.D., Casolaro, M.A., Sellers, S.E., Swayze, S.C., Mc.Phaul, K.M., Wittes, J.T., Crystal, R.G. (1987). Replacement therapy for alpha$_1$-antitrypsin deficiency associated with emphysema. *N Engl J Med, 316*, 1055–1062.

42. Ogushi, F., Fells, G.A., Hubbard, R.C., Straus, S.D., Crystal, R.G. (1987). Z-type α_1-antitrypsin is less competent than M1-type α_1-antitrypsin as an inhibitor of neutrophil elastase. *J Clin Invest, 80*, 1366–1374.

43. Guzdek, A., Potempa, J., Dubin, A., Travis, J. (1990). Comparative properties of human α-1-proteinase inhibitor glycosylation variants. *FEBS Lett, 272*, 125–127.

44. Lomas, D.A., Evans, D.L., Stone, S.R., Chang, W.-S.W., Carrell, R.W. (1993). Effect of the Z mutation on the physical and inhibitory properties of α_1-antitrypsin. *Biochemistry, 32*, 500–508.

45. Llewellyn-Jones, C.G., Lomas, D.A., Carrell, R.W., Stockley, R.A. (1994). The effect of the Z mutation on the ability of α_1-antitrypsin to prevent neutrophil mediated tissue damage. *Biochim Biophys Acta, 1227*, 155–160.

46. Larsson, C. (1978). Natural history and life expectancy in severe alpha$_1$-antitrypsin deficiency, PiZ. *Acta Med Scand, 204*, 345–351.

47. Janus, E.D., Phillips, N.T., Carrell, R.W. (1985). Smoking, lung function, and α_1-antitrypsin deficiency. *Lancet, 1*, 152–154.

48. Elliott, P.R., Bilton, D., Lomas, D.A. (1998). Lung polymers in Z α_1-antitrypsin related emphysema. *Am J Respir Cell Mol Biol, 18*, 670–674.

49. Gooptu, B., Lomas, D.A. (2008). Polymers and inflammation: disease mechanisms of the serpinopathies. *J Exp Med, 205*, 1529–1534.

50. Parmar, J.S., Mahadeva, R., Reed, B.J., Farahi, N., Cadwallader, K., Bilton, D., Chilvers, E.R., Lomas, D.A. (2002). Polymers of α_1-antitrypsin are chemotactic for human neutrophils: a new paradigm for the pathogenesis of emphysema. *Am J Respir Cell Mol Biol, 26*, 723–730.

51. Mulgrew, A.T., Taggart, C.C., Lawless, M.W., Greene, C.M., Brantly, M.L., O'Neill, S.J., McElvaney, N.G. (2004). Z α_1-antitrypsin polymerizes in the lung and acts as a neutrophil chemoattractant. *Chest, 125*, 1952–1957.

52. Mahadeva, R., Atkinson, C., Li, J., Stewart, S., Janciauskiene, S., Kelley, D.G., Parmar, J., Pitman, R., Shapiro, S.D., Lomas, D.A. (2005). Polymers of Z α_1-antitrypsin co-localise with neutrophils in emphysematous alveoli and are chemotactic in vivo. *Am J Pathol, 166*, 377–386.

53. Morrison, H.M., Kramps, J.A., Burnett, D., Stockley, R.A. (1987). Lung lavage fluid from patients with α_1-proteinase inhibitor deficiency or chronic obstructive bronchitis: anti-elastase function and cell profile. *Clin Sci Lond, 72*, 373–381.

54. Banda, M.J., Rice, A.G., Griffin, G.L., Senior, R.M. (1988). The inhibitory complex of human α_1-proteinase inhibitor and human leukocyte elastase is a neutrophil chemoattractant. *J Exp Med, 167*, 1608–1615.
55. Senior, R.M., Griffin, G.L., Mecham, R.P. (1980). Chemotactic activity of elastin derived peptides. *J Clin Invest, 66*, 859–862.
56. Kim, J., Lee, K.N., Yi, G.-S., Yu, M.-H. (1995). A thermostable mutation located at the hydrophobic core of α_1-antitrypsin suppresses the folding defect of the Z-type variant. *J Biol Chem, 270*, 8597–8601.
57. Sidhar, S.K., Lomas, D.A., Carrell, R.W., Foreman, R.C. (1995). Mutations which impede loop/sheet polymerisation enhance the secretion of human α_1-antitrypsin deficiency variants. *J Biol Chem, 270*, 8393–8396.
58. Lin, L., Schmidt, B., Teckman, J., Perlmutter, D.H. (2001). A naturally occurring non-polymerogenic mutant of α_1-antitypsin charcaterized by prolonged retention in the endoplasmic reticulum. *J Biol Chem, 270*, 8393–8396.
59. Schmidt, B.Z., Perlmutter, D.H. (2005). GRP78, GRP94 and GRP170 interact with α_1-antitrypsin mutants that are retained in the endoplasmic reticulum. *Am J Physiol Gastrointest Liver Physiol, 289*, G444–G455.
60. Wu, Y., Whitman, I., Molmenti, E., Moore, K., Hippenmeyer, P., Perlmutter, D. H. (1994). A lag in intracellular degradation of mutant α_1-antitrypsin correlates with liver disease phenotype in homozygous PiZZ α_1-antitrypsin deficiency. *Proc Natl Acad Sci U S A, 91*, 9014–9018.
61. Werner, E.D., Brodsky, J.L., McCracken, A.A. (1996). Proteasome-dependent endoplasmic reticulum–associated protein degradation: an unconventional route to a familiar fate. *Proc Natl Acad Sci U S A, 93*, 13797–13801.
62. Qu, D., Teckman, J.H., Omura, S., Perlmutter, D.H. (1996). Degradation of a mutant secretory protein, α_1-antitrypsin Z, in the endoplasmic reticulum requires proteosome activity. *J Biol Chem, 271*, 22791–22795.
63. Teckman, J.H., Perlmutter, D.H. (2000). Retention of mutant α_1-antitrypsin Z in endoplasmic reticulum is associated with an autophagic response. *Am J Physiol Gastrointest Liver Physiol, 279*, G961–G974.
64. Teckman, J.H., An, J.K., Loethen, S., Perlmutter, D.H. (2002). Fasting in α_1-antitrypsin deficient liver: constitutive activation of autophagy. *Am J Physiol Gastrointest Liver Physiol, 283*, G1156–G1165.
65. Teckman, J.H., An, J.K., Blomenkamp, K., Schmidt, B., Perlmutter, D. (2004). Mitochondrial autophagy and injury in the liver in α_1-antitrypsin deficiency. *Am J Physiol Gastrointest Liver Physiol, 286*, G851–G862.
66. Kamimoto, T., Shoji, S., Hidvegi, T., Mizushima, N., Umebayashi, K., Perlmutter, D.H., Yoshimori, T. (2006). Intracellular inclusions containing mutant alpha1-antitrypsin Z are propagated in the absence of autophagic activity. *J Biol Chem, 281*, 4467–4476.
67. Kruse, K.B., Brodsky, J.L., McCracken, A.A. (2006). Characterization of an ERAD gene as VPS30/ATG6 reveals two alternative and functionally distinct protein quality control pathways: one for soluble Z variant of human alpha-1 proteinase inhibitor (A1PiZ) and another for aggregates of A1PiZ. *Mol Biol Cell, 17*, 203–212.

68. Kruse, K.B., Dear, A., Kaltenbrun, E.R., Crum, B.E., George, P.M., Brennan, S.O., McCracken, A.A. (2006). Mutant fibrinogen cleared from the endoplasmic reticulum via endoplasmic reticulum-associated protein degradation and autophagy. *Am J Pathol, 168*, 1300–1308.

69. Hidvegi, T., Mirnics, K., Hale, P., Ewing, M., Beckett, C., Perlmutter, D.H. (2007). Regulator of G signaling 16 is a marker for the distinct ER stress state associated with aggregated mutant alpha 1antitrypsin Z in the classical form of α_1-antitrypsin deficiency. *J Biol Chem, 282*, 27769–27780.

70. Canhe, C., Sheng-Cai, L. (1998). The core domain of RGS16 retains G-protein binding and GAP activity in vitro, but is not functional in vivo. *FEBS Lett, 422*, 359–362.

71. Gohla, A., Klement, K., Piekorz, R.P., Pexa, K., vom Dahl, S., Spicher, K., Dreval, V., Haussinger, D., Birnbaumer, L., Nurnberg, B. (2007). An obligatory requirement for the heterotrimeric G protein G_{i3} in the antiautophagic action of insulin in the liver. *Proc Natl Acad Sci U S A, 104*, 3003–3008.

72. Hidvegi, T., Schmidt, B.Z., Hale, P., Perlmutter, D.H. (2005). Accumulation of mutant alpha1-antitrypsin Z in the endoplasmic reticulum activates caspases-4 and -12, NFkappaB, and BAP31 but not the unfolded protein response. *J Biol Chem, 280*, 39002–39015.

73. Pikarsky, E., Porat, R.M., Stein, I., Abramovitch, R., Amit, S., Kasem, S., Gutkovich-Pyest, E., Urieli-Shoval, S., Galun, E., Ben-Neriah, Y. (2004). NFκB functions as a tumor promoter in inflammation-associated cancer. *Nature, 431*, 461–466.

74. Maeda, S., Kamata, H., Luo, J.-L., Leffert, H., Karin, M. (2005). IKKβ couples hepatocyte death to cytokine-driven compensatory proliferation that promotes chemical hepatocarcinogenesis. *Cell, 121*, 977–990.

75. Schamel, W.W., Kuppig, S., Becker, B., Gimborn, K., Hauri, H.P., Reth, M.A. (2003). A high-molecular-weight complex of membrane proteins BAP29/BAP31 is involved in the retention of membrane-bound IgD in the endoplasmic reticulum. *Proc Natl Acad Sci U S A, 100*, 9861–9866.

76. Breckenridge, D.G., Stojanovic, M., Marcellus, R.C., Shore, G.C. (2003). Caspase cleavage product of BAP31 induces mitochondrial fission through endoplasmic reticulum calcium signals, enhancing cytochrome *c* release to the cytosol. *J Cell Biol, 160*, 1115–1127.

77. Huang, J., Pashkov, V., Kurrasch, D.M., Yu, K., Gold, S.J., Wilkie, T.M. (2006). Feeding and fasting controls liver expression of a regulator of G protein signaling (Rgs16) in periportal hepatocytes. *Comp Hepatol, 5*, 8–18.

78. Rudnick, D.A., Liao, Y., An, J.K., Muglia, L.J., Perlmutter, D.H., Teckman, J.H. (2004). Analyses of hepatocellular proliferation in a mouse model of α_1-antitrypsin deficiency. *Hepatology, 39*, 1048–1055.

79. Rudnick, D.A., Perlmutter, D.H. (2005). Alpha-1-antitrypsin deficiency: a new paradigm for hepatocellular carcinoma in genetic liver disease. *Hepatology, 42*, 514–521.

80. Prachalias, A.A., Kalife, M., Francavilla, R., Muiesan, P., Dhawan, A., Baker, A., Hadzic, D., Mieli-Vergani, G., Rela, M., Heaton, N.D. (2000). Liver transplantation for alpha-1-antitrypsin deficiency. *Transpl Int, 13*, 207–210.

81. Piitulainen, E., Eriksson, S. (1999). Decline in FEV1 related to smoking status in individuals with severe alpha1-antitrypsin deficiency. *Eur Respir J, 13*, 247–251.

82. British Thoracic Society. (1997). BTS guidelines for the management of chronic obstructive pulmonary disease. *Thorax, 52*(Suppl 5).

83. Lacasse, Y., Wong, E., Guyatt, G.H., King, D., Cook, D.J., Goldstein, R.S. (1996). Meta-analysis of respiratory rehabilitation in chronic obstructive pulmonary disease. *Lancet, 348*, 1115–1119.

84. Trulock E.P., 3rd. (1998). Lung transplantation for COPD. *Chest, 113* Suppl 4, 269S–276S.

85. Stoller, J.K., Gildea, T.R., Ries, A.L., Meli, Y.M., Karafa, M.T., National Emphysema Treatment Trial Research Group. (2007). Lung volume reduction surgery in patients with emphysema and alpha-1-antitrypsin deficiency. *Ann Thorac Surg, 83*, 241–251.

86. The alpha-1-Antitrypsin Deficiency Registry Study Group. (1998). Survival and FEV_1 decline in individuals with severe deficiency of α_1-antitrypsin. *Am J Respir Crit Care Med, 158*, 49–59.

87. Dirksen, A., Dijkman, J.H., Madsen, F., Stoel, B., Hutchison, D.C.S., Ulrik, C.S., Skovgaard, L.T., Kok-Jensen, A., Rudolphus, A., Seersholm, N., et al. (1999). A randomised clinical trial of α_1-antitrypsin augmentation therapy. *Am J Respir Crit Care Med, 160*, 1468–1472.

88. Schluchter, M.D., Stoller, J.K., Barker, A.F., Buist, A.S., Crystal, R.G., Donohue, J.F., Fallat, R.J., Turino, G.M., Vreim, C.E., Wu, M.C. (2000). Feasibility of a clinical trial of augmentation therapy for alpha-1-antitrypsin deficiency: the alpha-1-antitrypsin deficiency study group. *Am J Respir Crit Care Med, 161*, 796–801.

89. Song, S., Embury, J., Laipis, P.J., Berns, K.I., Crawford, J.M., Flotte, T.R. (2001). Stable therapeutic serum levels of human alpha-1 antitrypsin (AAT) after portal vein injection of recombinant adeno-associated virus (rAAV) vectors. *Gene Ther, 8*, 1299–1306.

90. Rosenfeld, M.A., Siegfried, W., Yoshimura, K., Yoneyama, K., Fukayama, M., Stier, L.E., Pääkkö, P.K., Gilardi, P., Stratford-Perricaudet, L.D., Perricaudet, M., et al. (1991). Adenovirus-mediated transfer of a recombinant α_1-antitrypsin gene to the lung epithelium in vivo. *Science, 252*, 431–434.

91. Song, S., Morgan, M., Ellis, T., Poirier, A., Chesnut, K., Wang, J., Brantly, M., Muzyczka, N., Byrne, B. J., Atkinson, M., Flotte, T.R. (1998). Sustained secretion of human alpha-1-antitrypsin from murine muscle transduced with adeno-associated virus vectors. *Proc Natl Acad Sci U S A, 95*, 14384–14388.

92. Bou-Gharios, G., Wells, D.J., Lu, Q.L., Morgan, J.E., Partridge, T. (1999). Differential expression and secretion of alpha 1 anti-trypsin between direct DNA injection and implantation of transfected myoblast. *Gene Ther, 6*, 1021–1029.

93. Massaro, G.D., Massaro, D. (1997). Retinoic acid treatment abrogates elastase-induced pulmonary emphysema in rats. *Nat Med, 3*, 675–677.

94. Schulze, A.J., Baumann, U., Knof, S., Jaeger, E., Huber, R., Laurell, C.-B. (1990). Structural transition of α_1-antitrypsin by a peptide sequentially similar to β-strand s4A. *Eur J Biochem, 194*, 51–56.

95. Mast, A.E., Enghild, J.J., Salvesen, G. (1992). Conformation of the reactive site loop of α_1-proteinase inhibitor probed by limited proteolysis. *Biochemistry, 31*, 2720–2728.

96. Huntington, J.A., Pannu, N.S., Hazes, B., Read, R., Lomas, D.A., Carrell, R.W. (1999). A 2.6Å structure of a serpin polymer and implications for conformational disease. *J Mol Biol, 293*, 449–455.

97. Dunstone, M.A., Dai, W., Whisstock, J.C., Rossjohn, J., Pike, R.N., Feil, S.C., Le Bonniec, B.F., Parker, M.W., Bottomley, S.P. (2000). Cleaved antitrypsin polymers at atomic resolution. *Protein Sci, 9*, 417–420.

98. Chow, M.K.M., Devlin, G.L., Bottomley, S.P. (2001). Osmolytes as modulators of conformational changes in the serpins. *Biol Chem, 382*, 1593–1599.

99. Sharp, L.K., Mallya, M., Kinghorn, K.J., Wang, Z., Crowther, D.C., Huntington, J.A., Belorgey, D., Lomas, D.A. (2006). Sugar and alcohol molecules provide a therapeutic strategy for the serpinopathies that cause dementia and cirrhosis. *FEBS J, 273*, 2540–2552.

100. Burrows, J.A.J., Willis, L.K., Perlmutter, D.H. (2000). Chemical chaperones mediate increased secretion of mutant α_1-antitrypsin (α_1-AT) Z: a potential pharmacologcial strategy for prevention of liver injury and emphysema. *Proc Natl Acad Sci U S A, 97*, 1796–1801.

101. Devlin, G.L., Parfrey, H., Tew, D.J., Lomas, D.A., Bottomley, S.P. (2001). Prevention of polymerization of M and Z α_1-antitrypsin (α_1-AT) with trimethylamine *N*-oxide: implications for the treatment of α_1-AT deficiency. *Am J Respir Cell Mol Biol, 24*, 727–732.

102. Teckman, J.H. (2004). Lack of effect of oral 4-phenylbutyrate on serum alpha-1-antitrypsin in patients with alpha-1-antitrypsin deficiency: a preliminary study. *J Pediatr Gastroenterol Nutr, 39*, 34–37.

103. Skinner, R., Chang, W.-S.W., Jin, L., Pei, X., Huntington, J.A., Abrahams, J.-P., Carrell, R.W., Lomas, D.A. (1998). Implications for function and therapy of a 2.9Å structure of binary-complexed antithrombin. *J Mol Biol, 283*, 9–14.

104. Fitton, H.L., Pike, R.N., Carrell, R.W., Chang, W.-S.W. (1997). Mechanisms of antithrombin polymerisation and heparin activation probed by insertion of synthetic reactive loop peptides. *Biol Chem, 378*, 1059–1063.

105. Parfrey, H., Dafforn, T.R., Belorgey, D., Lomas, D.A., Mahadeva, R. (2004). Inhibiting polymerisation: new therapeutic strategies for Z α_1-antitrypsin related emphysema. *Am J Respir Cell Mol Biol, 31*, 133–139.

106. Zhou, A., Stein, P.E., Huntington, J.A., Sivasothy, P., Lomas, D.A., Carrell, R.W. (2004). How small peptides block and reverse serpin polymerization. *J Mol Biol, 342*, 931–941.

107. Chang, Y.P., Mahadeva, R., Chang, W.S., Shukla, A., Dafforn, T.R., Chu, Y.H. (2006). Identification of a 4-mer peptide inhibitor that effectively blocks the polymerization of pathogenic Z α_1-antitrypsin. *Am J Respir Cell Mol Biol, 35*, 540–548.

108. Lee, C., Maeng, J.-S., Kocher, J.-P., Lee, B., Yu, M.-H. (2001). Cavities of α_1-antitrypsin that play structural and functional roles. *Protein Sci, 10*, 1446–1453.

109. Parfrey, H., Mahadeva, R., Ravenhill, N., Zhou, A., Dafforn, T.R., Foreman, R.C., Lomas, D.A. (2003). Targeting a surface cavity of α_1-antitrypsin to prevent conformational disease. *J Biol Chem, 278*, 33060–33066.

110. Mallya, M., Phillips, R.L., Saldanha, S.A., Gooptu, B., Leigh Brown, S.C., Termine, D.J., Shirvani, A.M., Wu, Y., Sifers, R.N., Abagyan, R., Lomas, D.A. (2007) Small molecules block the polymerisation of Z α_1-antitrypsin and increase the clearance of intracellular aggregates. *J Med Chem, 50*, 5357–5363.

19

FOLDING BIOLOGY OF CYSTIC FIBROSIS: A CONSORTIUM-BASED APPROACH TO DISEASE

WILLIAM E. BALCH

Department of Cell Biology and Institute for Childhood and Neglected Diseases, The Scripps Research Institute, La Jolla, California

INEKE BRAAKMAN

Department of Cellular Protein Chemistry, University of Utrecht, Utrecht, The Netherlands

JEFF BRODSKY

Department of Biological Sciences, University of Pittsburgh, Pittsburgh, Pennsylvania

RAYMOND FRIZZELL

Department of Cell Biology and Physiology, School of Medicine, University of Pittsburgh, Pittsburgh, Pennsylvania

WILLIAM GUGGINO

Department of Physiology, School of Medicine, Johns Hopkins University, Baltimore, Maryland

GERGELY L. LUKACS

Department of Physiology, McGill University, Montreal, Quebec, Canada

CHRISTOPHER PENLAND

Cystic Fibrosis Foundation, Bethesda, Maryland

Protein Misfolding Diseases: Current and Emerging Principles and Therapies,
Edited by Marina Ramirez-Alvarado, Jeffery W. Kelly, and Christopher M. Dobson
Copyright © 2010 John Wiley & Sons, Inc.

HARVEY POLLARD

Department of Anatomy, Physiology and Genetics, School of Medicine, University of the Health Sciences, Bethesda, Maryland

WILLIAM SKACH

Department of Biochemistry and Molecular Biology, Oregon Health and Sciences University, Portland, Oregon

ERIC SORSCHER

Department of Medicine, University of Alabama at Birmingham, Birmingham, Alabama

PHILIP THOMAS

Molecular Biophysics Graduate Program, University of Texas Southwestern Medical Center, Dallas, Texas

INTRODUCTION

Cystic fibrosis (CF) is a complex disease that has a variety of confounding features that lead to the symptoms found in the clinic. It is largely a protein misfolding disease of the cystic fibrosis transmembrane conductance regulator (CFTR) [1,169,170], but numerous components in the cell contribute to the manifestation of misfolding phenotype, including translational events, chaperone systems, and degradative pathways as well as trafficking and regulatory protein interactions that dictate its function(s) as a channel and master regulator of sodium, chloride, and bicarbonate transport at the apical cell surface [2]. Such a complex array of interactions (the CFTR interactome [3]) [4–6] emphasize the need for a concerted, consortium-based approach to encourage interaction of investigators at all levels of the disease etiology to accelerate the pace of tool, assay, and target discovery that would lead to more effective treatments. Below is a brief overview of the folding and biology of CFTR by members of the *CF*olding *C*onsortium (CFC), a group effort founded by the Cystic Fibrosis Foundation that selectively addresses contemporary issues in our understanding of the protein (mis)folding problem in this disease [171].

CYSTIC FIBROSIS AND THE CYSTIC FIBROSIS CONDUCTANCE REGULATOR: GENETICS AND CLINICAL MANIFESTATIONS DEFINING THE DEPTH OF THE PROBLEM

CFTR is an epithelial ion channel expressed in exocrine glands that conducts chloride and bicarbonate across the plasma membrane and regulates transepithelial transport of sodium [2,7]. Absence of functional CFTR in CF engenders gross dysfunction of multiple organs, including the lungs, pancreas, liver, intestine, vas deferens, and sweat glands. CF pulmonary ramifications involve deranged maintenance of the airway surface liquid, diminished mucociliary clearance, and death due to respiratory failure. Median life expectancy of individuals with cystic fibrosis is presently approximately 37 years (27 years in ΔF/ΔF homozygous patients) [8].

Based on its primary amino acid and domain structure, CFTR is a member of the ATP-binding cassette (ABC) gene family. Like other ABC proteins, CFTR includes two membrane spanning regions [transmembrane (TM) domains 1 and 2], and two ATP-binding cassettes [also known as nucleotide-binding domains (NBDs)]. The regulatory domain (RD) of CFTR is not typically observed in other ABC proteins and contains a number of protein kinase A phosphorylation sites that contribute to anion conductance activation [9–15].

CFTR is synthesized as a single 150-kDa polypeptide chain. As with other ABC proteins, when CFTR NBDs and TMs are expressed as "half-molecules," the domains find each other, reassociating into functional ion channels at the cell surface [9,10]. Domain interactions are critical for CFTR channel activity. For example, both TM1 and TM2 are postulated to contribute α-helices to the transmembrane chloride conductive pore [11,12] and to interact with the NBDs [16]. Cyclic AMP (cAMP)–dependent phosphorylation by protein kinase A (PKA) is thought to change the orientation of the R-region to allow access to the pore [13,14]. Mutant CFTR lacking an R-region no longer requires phosphorylation for activity [15]. NBDs 1 and 2 dimerize and jointly hydrolyze ATP to serve as a switch that permits ion-channel gating.

The biochemical and structural basis underlying CFTR domain interactions, role in channel activity, and interaction with other proteins comprising the CFTR interactome are an area of intense interest in a number of laboratories. Such an understanding is essential for discovering small molecules (biologics and chemicals) capable of restoring CFTR function in disease. The CFTR gene comprises approximately 250,000 base pairs on chromosome 7 that encode the 1480 amino acids of the intact protein. Over 1500 disease-associated mutations have been identified and characterized based on their underlying pathogenic mechanism. The most common ΔF508 mutation is found in approximately 70% of defective *CFTR* alleles and is defined as a class II defect, leading to an ion channel with substantial residual function, but which is degraded by the proteasome early in biogenesis and before it can arrive at the cell surface [2]. Premature truncation alleles in CFTR are noted by the convention "X" (e.g., the G542X defect) and constitute mutations in class I. Some premature truncation mutants arise from frameshift mutations (e.g., 621 + 1G→T) and 1717-1G→A) and are not denoted as "X" mutations. Certain of these have been observed with a particularly high incidence in Ashkenazi Jews. Mutant CFTR proteins that are synthesized full length but fail to function normally as chloride channels are well described (class III defects), including the G551D mutation, for example.

Additional CFTR mutations exhibit partial activity of the ion channel (class IV), lower levels of CFTR mRNA (class V) or plasma membrane, and surface instability of the protein (class VI). Each of the mutants provides unique windows into the disease process that need to be addressed to achieve resolution of the initial biogenesis problem that triggers CF.

TRANSLOCATION INTO THE ER MEMBRANE

CFTR folding can be viewed as a series of sequential steps in which the nascent polypeptide is targeted to the endoplasmic reticulum (ER), oriented and inserted into the lipid bilayer, folded within the membrane, cytosol, and ER lumen, and finally assembled into its mature tertiary multidomain structure. Disruption at any one of these steps could potentially lead to misfolding of CFTR and failure to exit the ER or function properly at the cell surface. A major challenge in understanding and correcting CFTR misprocessing is therefore to define the precise point where the mutant protein deviates, either structurally or kinetically, from the normal folding pathway.

The earliest recognized stage of CFTR folding involves formation and insertion of hydrophobic helical transmembrane segments (TMs) into the lipid bilayer of the ER membrane [17,18]. In vitro reconstitution experiments reveal that ER targeting is mediated by a signal recognition particle (SRP) at a nascent chain length of about 100 amino acids as TM1 first emerges from the ribosome [19]. TM domain folding therefore begins during protein synthesis at specialized sites in the ER that contain a complex set of membrane proteins, including the Sec61 α,β,γ heterotrimer, oligosaccharyltransferase, signal peptidase, TRAM, and TRAP [20,21]. Collectively, these proteins form the translocon, a protein-conducting channel through which the nascent CFTR polypeptide chain translocates into the ER lumen and is transferred into the ER membrane. Although the precise details of membrane insertion remain a mystery, studies examining early stages of CFTR topogenesis have begun to address how and when each of its 12 TMs are oriented in the membrane and how this process can be affected by disease-related mutations [22,23], yet much remains to be done [2,24].

Models derived from relatively simple, engineered polytopic proteins demonstrate that complex topology can be generated by the action of sequential, independent topogenic determinants contained within the nascent polypeptide [25]. As these determinants pass through the ribosome exit tunnel and translocon, they control the pathway for peptide movement by opening the translocon protein-conducting pore into the ER lumen (i.e., signal sequences) or closing the pore to allow nascent chain egress into the cytosol (i.e., stop transfer sequences) [26,27]. In this manner, each TM can be properly oriented within the translocon co-translationally as it exits the ribosome. Surprisingly, CFTR and other native proteins display variations on this model in which some TM segments do not direct the expected topology independently but, rather, control the translocation pathway by nonsequential and/or cooperative mechanisms [17,23,28–30]. For example, CFTR TM1 signal sequence activity is depressed by two native polar residues, Glu92 and Lys95, in its hydrophobic core [22]. Translocation of the first extracellular loop (ECL2) is therefore facilitated by coordinated activities of both TM1 and TM2. Corresponding TMs in the second membrane spanning domain, TM7 and TM8, use a different topogenesis mechanism. Whereas TM7 functions as an efficient signal sequence that initiates translocation of ECL4 [23], another polar residue, Asp924, within TM8 decreases stop transfer activity such that TM8 extends transiently into the ER lumen and requires TM7 to terminate translocation effectively and span the membrane [23]. Furthermore, most CFTR TMs are separated by connecting loops that contain only a few residues (TM3–4, 5–6, 9–10, and 11–12) and are therefore probably inserted into the membrane as helical pairs via poorly understood mechanisms [17].

How, then, do disease-causing mutations disrupt early stages of CFTR TM folding and insertion? Although only a small subset of TM mutations has been examined to date, studies indicate that structural defects that disrupt translocation and folding are complex. Topological analysis of two class II (trafficking) mutations, G85E and G91R, demonstrate that introducing a third polar residue into the hydrophobic core of TM1 depresses signal sequence activity further but does not disrupt transmembrane topology [31]. Rather, the redundant activity of TM2 inserts the mutant polar residue into the plane of the membrane, and this in turn perturbs other aspects of CFTR folding (e.g., helical packing, domain–domain interactions, or both) that are recognized by ER sensing mechanisms [32]. CFTR folding is disrupted in a different manner by the P205S mutation in TM3. In this case, Pro205 is required to prevent CFTR misfolding by disfavoring nonnative β-structure within the helical segment [33]. Finally, a number of studies have suggested that tertiary folding of CFTR TMs can also be altered by mutations in cytosolic domains such as NBD1 (ΔF508 and short intracellular loops [12,16,34–39]). Thus, normal and pathological mechanisms of CFTR TM domain folding involve both early events of co-translational TM orientation/integration and subsequent helical packing/domain assembly that begin in association with ER biosynthetic machinery [40,41] and are completed after the nascent protein has been released into the membrane [36,37,42–45]. Understanding the rules that govern how transmembrane domain insertion is coordinated and/or potentially disrupted by mutation in these domains and cytosolic domains will shed important new insight into CF and related membrane protein folding disorders.

CO-TRANSLATIONAL FOLDING OF CFTR

Of all newly synthesized proteins that fold in the ER, those with transmembrane domains, such as CFTR, face an additional challenge: Such patients must fold domains in up to three topologically distinct spaces—the ER lumen, the membrane, and the cytosol. The domains in CFTR cannot be considered independent: They must affect each other's conformations, even across the membrane. As an example, changes in the cytosolic tails of the single transmembrane fusion proteins of HIV and influenza virus affect structural features of their ectodomains [46,47]. How mutation affects folding within a particular space and how this event is communicated across the bilayer remains a major challenge for understanding CFTR maturation.

Each compartment carries its own set of molecular chaperones that may assist the protein concurrently during its folding. Classical chaperone families of the Hsp70 and Hsp90 classes and their co-chaperones reside in both ER and cytosol [48,49], but the lectin chaperones calnexin and calreticulin are specific to the ER [50]. A wealth of folding enzymes such as oxidoreductases and proline *cis–trans* isomerases catalytically drive folding in the lumen of the ER. Given its multimembrane spanning topology, CFTR has indeed been suggested to use both lumenal chaperones (calnexin) [51] and cytosolic chaperones (extended Hsp90/70 machineries) [3,52–55]. The role of most of these components remains to be elucidated. Folding assistance within the membrane may be controlled through the transmembrane domain of calnexin or the translocon-associated J-proteins that are membrane anchored (co-chaperones of Hsp70s). The polytopic membrane protein Derlin-1, involved in degradation of misfolded CFTR (see below), associates with CFTR in a protein complex during synthesis [56,57], suggesting that folding is a challenge for both wild-type and variant protein. The increase in membrane proteins with identified chaperone function [58] will undoubtedly expand the number of putative chaperones beyond our current level of anticipation.

Communication between folding machineries in the cytosol and the ER lumen is probably disrupted in disease. This could again require folding assistants with a transmembrane domain such as Derlin-1. Evidence for this possibility came from the fact that the Hsp70 co-chaperone ERJ1 was found to have such a coordinating function in translocation. It associates with the ribosome on the cytosolic face of the ER membrane while recruiting BiP in the ER lumen to drive translocation [59]. Considering the various chaperonelike activities of the ribosome, and considering the need for co-translational folding (see above), these same J-proteins may be involved in the cross-membrane coordination of protein folding. Whereas cytosolic J proteins are important for CFTR maturation [54,60], to date none of the transmembrane ER J proteins have been implicated, but are intriguing candidates.

Proteins start to fold during synthesis and complete folding and assembly after translation, although this remains a modest controversy for CFTR, with a large amount of evidence favoring posttranslational folding in most circumstances [36,37,39,42–44]. Aside from the obvious hierarchy in time (posttranslational folding can never occur before co-translational folding), essential differences do exist between the two processes. During synthesis the C-terminus of a protein or domain is tethered to the ER membrane. This limits conformational freedom and is likely to affect folding pathways and may contribute to the inability of mutants to achieve early folding events. Freedom may be even more limited when the N-terminus is tethered to the membrane, either because an ER targeting signal peptide is not (yet) cleaved off and functions as a (transient) transmembrane domain, or because the folding domain is the C-terminal of a transmembrane domain. CFTR, with its 12 transmembrane helices, needs to fold its domains in the vicinity of the membrane, and to a large extent with N- and C-termini tethered.

During synthesis the ribosome is attached to the membrane, and for proteins such as CFTR it alternates between a sealed connection with the translocon and a partial attachment, the latter during synthesis of the cytosolic parts, perhaps a critical event disrupted in some mutants. The ribosome may not only offer chaperone activity (through its RNA or associated chaperones), but may also limit conformational freedom of the nascent chain because of steric hindrance. This occurs because the ribosome has a width of about 25 nm [61], whereas the translocon pore is an order of magnitude smaller [62]. Thus, location of mutant chain in the pore may alter the sequential links between kinetics and translation, and those of translocation/folding.

Other factors to take into account are co- and posttranslational modifications. CFTR has two N-linked glycans in the fourth extracellular loop, which is the first extracellular loop of the second transmembrane domain and the first after a long stretch of cytosolic sequence. Many other ABC-transporters also have N-linked glycans in this position, indicative of a general function for a glycan. Because N-linked glycosylation for most glycoproteins is crucial for their proper folding [50], their co-translational addition is well timed. Still, early co-translational folding can be faster than glycosylation, as illustrated by the inhibition of glycan addition by early disulfide bond formation during folding of a protein in the ER [63], although an increasing number of proteins are found to add glycans after synthesis. This competition and mutual dependence of folding and glycosylation may well determine or even regulate the outcome of the folding process, leading potentially to multiple conformations of the same protein, even if the number of glycans eventually is identical. One possibility is that because transporters such as CFTR contain suboptimal transmembrane domains, with charged and hydrophilic residues, addition of this large polar sugar moiety may prevent back-sliding of the translocated and inserted transmembrane domains. CFTR was shown to indeed need its glycans for stability, suggesting that they affect conformation [64]. How mutations affect the link between translation and N-glycan modification has not been investigated.

CFTR does not contain disulfide bonds or any other known modifications in the small lumenal parts of its sequence, but its cytosolic R region can be phosphorylated and its NBD1 and NBD2 domains can bind ATP. Whereas ATP binding is not a covalent modification, it does influence protein conformation. Phosphorylation of the R region and ATP hydrolysis have a regulatory role for channel activity, and ATP binding and phosphorylation affect R-NBD1 interaction [65] and NBD dimerization [66]. The effect of CFTR phosphorylation on its conformation and activity are unknown and probably extends to the folding stage, but this remains to be explored, particularly in mutants. We cannot exclude the fact that phosphorylation affects the folding process of CFTR or even that particular sites need to be modified to allow proper folding of the protein. The effect of ATP is less clear, although it is clear that ATP is required for CFTR folding, in particular for its release from the large ribosome–translocon-complex CFTR residues, probably in response to ATP-dependent chaperones [40]. Whereas the structure of purified NBD1 is not affected by ATP binding [67], in the context of the complete protein a direct role for ATP on CFTR conformation still cannot be excluded and indeed is likely.

Perhaps the most important determinant of co-translational folding is translation itself. Because in the absence of denaturant a protein will not remain unfolded, all newly synthesized proteins will start folding as nascent chain. As the average translation rate in a mammalian cell is four or five amino acid residues per second, a protein of about 1500 residues needs about 5.5 min to be made, although the average translation rate measured for CFTR is even lower: 2.7 residues per second implying a synthesis time of about 9 minutes for the average CFTR molecule [68]. On average, this is slow compared to the formation rate of secondary structure, but fast compared to the time it usually takes for CFTR to fold and exit the ER, which indicates that most of the protein folding occurs posttranslationally. Confounding the issue of translation rate is the fact that the ribosome has variable speed on the mRNA, due to RNA secondary structure and rare codons, which cause ribosome pausing and the piling up of ribosomes behind the pausing ribosome [69]. Pausing provides time for protein folding in the absence of downstream domains, which may decrease misfolding. More speculative but attractive is the notion that pausing sites may be present in specific positions in the mRNA to regulate the outcome of folding, possibility having an important influence on the biogenesis of CFTR.

Initial evidence for an effect of codon use and pausing on the conformation of the translated protein was found for MDR1, where a silent mutation did lead to a conformational difference and hence disease [70]. Indeed, well-timed pausing may be needed to ensure both proper insertion and proper folding. CFTR's ATP-dependent release from the ribosome–translocon complex does not happen immediately upon chain termination [40], and association with cytosolic Hdj2 and Hsp70, which started during synthesis, persists [40,54,56]. Although folding studies on CFTR followed for up to 3 hours after synthesis did not show conformational changes any longer [43], the protein did continue to mature after translation [44]. This maturation involves domain assembly [9,16], which is crucial for CFTR's function and may be the main reason for the slow exit of CFTR from the ER. Misfolding of CFTR or its mutants leads to efficient degradation by the cytosolic proteasome [71,72] (see below). The initiating event for degradation may well be co-translational, because the deletion of Phe 508 can be detected when only about one-third of the protein has been synthesized. Ubiquitination of CFTR starts during synthesis [73], the E3 ubiquitin ligase associates with CFTR nascent chains [56], but the E3 CHIP associates only after synthesis, implying steps in which both wild-type and, particularly, select mutants affect recognition for degradation of CFTR folding intermediates [56,64,74].

In summary, folding of CFTR starts during, and continues for up to an hour after, synthesis. All throughout the process, the protein associates with a series of chaperone complexes, and CFTR (mutants) can be recognized by the chaperone and degradation machineries as permanently misfolded, resulting in ubiquitination and degradation, a condition that is enhanced in many cases by disease mutations. Our understanding of these events is in its infancy.

RECOGNITION AND DEGRADATION OF MUTANT CFTR

While successful folding and maturation in the ER is a prerequisite for protein transport through the secretory pathway, it is estimated that a high percentage of translated proteins fail to be exported [75]. Retention leads to their ubiquitination and degradation by the 26S proteasome, a process denoted as ER-associated degradation (ERAD) [76,77]. The presence of a degradative pathway assures that proteins recognized as displaying highly evolutionary conserved features of "unfoldedness" are removed from the cell to avoid accumulation within the ER or the other compartments. Accumulation can lead to cell stress and the formation of toxic protein aggregates—events that trigger both lumenal and cytosolic stress responses, although neither response appears relevant to early disease, given the efficiency by which CFTR is degraded.

The selection of protein substrates for ERAD, such as CFTR, is mediated by molecular chaperones, which facilitate protein folding or degradation, depending on the conformational state of the target protein [78,79]. As noted above, the most common disease mutation that is rapidly degraded is ΔF508 [80], a class II mutation [81]. The evidence that ΔF508 is indeed a CFTR folding/processing mutation destined for ERAD comes from several fronts, and includes its temperature sensitivity, its exaggerated interactions with molecular chaperones, and the presence of intermediate conformations of the protein trapped in folding complexes, conditions under which ΔF508 CFTR appears to gain access to degradative pathways [18,82]. Although the issue of its folding in relation to translation is not settled for wild-type CFTR [43], several lines of evidence favor the concept that failure of the ΔF508 mutant to achieve appropriate domain–domain interfaces is the limiting factor in mutant protein progression [2,16,35,37,44,83,84].

Because CFTR folding is facilitated by ER-based core chaperone machinery that includes Hsp70 [53,85,86], Hsp90 [52,87], the Hsp40 co-chaperones [54,60,88,89–90], a nucleotide exchange factor [88], and calnexin [51,64,91], it is thought that these constitute a metabolic pathway to organize the polypeptide chain chemistry into an organized structure, that, in principle would be protected from degradation [3]. Indeed, many of the known chaperones have been shown to decrease NBD aggregation and improve productive CFTR folding [92]. In contrast, unstable conformations of CFTR remain bound to chaperones, resulting in their polyubiquitination and degradation [55–57,71,93,94]. Thus, intersections between the biosynthetic and degradation pathways are monitored by a complement of chaperone and co-chaperone interactions that determine the fate of CFTR. In this manner, the interactions of CFTR with folding/degradation components are determined by conformation and likely energetics of the fold [95–97]. This can be assessed indirectly by comparing the proteolytic cleavage patterns of wild-type and ΔF508 CFTR. Such studies have indicated that protease cleavage patterns are similar for the immature wild-type and ΔF508 proteins, whereas the digestion pattern of wild-type CFTR is different, reflecting a more compact, folded conformation [37,39,43]. These data support the concept that ER-retained mutants achieve intermediate conformations in the normal CFTR folding pathway but that their variant structures and interactions with chaperone or degradative systems impair passage beyond one or more critical folding steps.

One early step in detection of the mutant fold leading to degradation involves derlin-1, which initiates CFTR extraction from the ER in cooperation with the AAA-ATPase, p97 [56,57,98]. One possibility is that derlin-1 interacts with ΔF508 CFTR and some other class II mutations due to instability within their transmembrane domains. This initial step is followed by CFTR ubiquitinylation by ER resident E3 ubiquitin ligases (e.g., RMA1 and gp78) [56,99], followed by proteasome-mediated degradation. A later step involves CHIP, an Hsp70/Hsp90-interacting E3 ubiquitin ligase, which is recruited to chaperone-bound CFTR, initiating its ubiquitinylation and degradation [55]. A third step by which CFTR can leave the productive folding pathway involves the calnexin cycle and recognition by a putative lectin in the ER, EDEM [64,100], which also links Derlin and p97 to glycoprotein extraction and proteolysis [101]. Thus, CFTR faces many steps that contribute to its degradation and these contribute to disease.

Given the complexity of CFTR folding and degradation pathways, it is not surprising that novel components will continue to be discovered. A yeast expression system, coupled with a microarray analysis, has been used to select candidates for involvement in CFTR degradation. Yeast cells handle wild-type CFTR similarly to the manner in which mammalian cells degrade the ΔF508 mutant. Yeast that express CFTR are not under stress, as they grow as well as cells lacking the CFTR expression vector, and an unfolded protein response (UPR) is not evoked [102]. Therefore, the expression of specific factors required for the ERAD of CFTR might be induced in this system. Indeed, a number of factors have now been identified using this approach. More than 150 transcripts were increased more than 1.6-fold in yeast expressing CFTR, and these included genes known to be involved in CFTR biogenesis in yeast or mammalian cells. Examples include *HSP82*, which encodes the yeast Hsp90 homolog, and *FES1*, which is a nucleotide exchange factor for cytosolic Hsp70 [52,88,102,103]. Of interest is that the message encoding the small heat-shock protein (sHsp), Hsp26, increased significantly and the expression of CFTR was completely stabilized in yeast deleted for the two sHsps, Hsp26 and Hsp42. Intriguingly, the stability of other yeast ERAD substrates was not altered in strains deleted for the sHsp, and therefore the sHsps show some selectivity for CFTR.

In extending this work to mammalian cells expressing wild-type and ΔF508 CFTR, the overexpression of a mammalian sHsp, αA-crystallin, selectively accelerated the degradation of ΔF508 CFTR [74], and similar findings have now also emerged for mammalian Hsp27 (Ahner and Frizzell, unpublished observations). These findings add CFTR to a growing list of protein folding/aggregation diseases that involve interactions with sHsps [57], where the sHsps have been implicated primarily as factors that stabilize proteins during cell stress [104], and associate with partially unfolded proteins to maintain their solubility until more favorable conditions return [105–111]. It is currently thought that the sHsps tend not to interact with either native or completely denatured proteins, but with proteins of an intermediate, foldable conformation [112–114], and they can distinguish between structurally identical wild-type and mutant proteins based on small differences in unfolding free energy [115–118]. While the association of sHsps with their substrates is ATP-independent, substrate re-folding often involves the action of ATP-dependent chaperones (e.g., Hsp70, Hsp90, and Hsp104) [106,111,119–122]. This property of sHsps may contribute to their association with immature ΔF508 CFTR, which adopts an intermediate conformation that has the potential for rescue, as discussed above. Accordingly, the interaction of sHsps with CFTR may represent a branch point between degradation and folding where mutant CFTR could be recovered to the productive folding pathway [96,97]. Of more concern is the fact that these machineries are highly variable between cell types; hence, what rescues CFTR from degradation and promotes trafficking of variant CFTR in one cell or tissue type may not suffice in another [3,123,124].

CFTR STRUCTURE

An understanding of the structure of folded and misfolded variants will undoubtedly contribute significantly to our understanding of disease. Although a high-resolution structure of full-length CFTR remains elusive, recently published biochemical and structural analyses of full-length CFTR using low-resolution cryo-EM approaches [125] have illustrated key features that reflect the spatial relationship of domains probably observed in ABC transporters MsbA, BtuCD, Sal177, and Sav1866 [126–129].

The most progress has been made with understanding the fold of soluble, isolated ABC transporter NBDs and have begun to reveal the domain arrangement and mechanochemistry of this feature of CFTR structure and function. The findings indicate that nucleotide-dependent association–dissociation of the two NBDs plays a central role in harvesting the energy of ATP binding and hydrolysis [130,131]. These studies show that mutant, catalytically inactive NBDs can form stable symmetrical dimers in the presence of ATP. This dimer places two ATPs at the dimer interface, between the Walker A and B consensus sequences of one domain and the Signature sequence (LSGGQ) of the opposing domain with the two ATPs forming the bulk of the interface [131]. The Hill coefficient of 1.7 measured for the activity of the homodimeric ABC transporters is consistent with a cooperative two-site mechanism [132]. Upon hydrolysis, electrostatic mismatch at the active site and α-helical subdomain rotation within the NBDs [133,134] leads to a rapid dissociation event. This arrangement neatly solves the problem of the positive cooperativity of ATPase activity catalyzed by the isolated NBDs and the open active site and structurally remote transporter signature sequence in the monomeric NBD structures [131–135]. Recent structures of intact ABC transporters MsbA, BtuCD, Sav1866, and P-glycoproteins [171] reveal that the TMDs interact with one surface of this dimer [136–138], and thus suggest that NBD interactions can be coupled efficiently to a scissoring motion of the integral membrane solute pathway of the transporters [139].

By contrast to the homodimeric NBDs, the two NBDs of CFTR are distinct from each other in important ways. Hydrolytically active NBDs have a catalytic glutamate residue at the end of the Walker B consensus. Whereas NBD2 (the C-terminal NBD of CFTR) has a glutamate, NBD1 (the N-terminal NBD) has a serine at this position. In addition, NBD1 also lacks a critical histidine residue in the H-loop. Recent studies with the murine CFTR NBD1 indicate that it does not catalyze significant hydrolysis of ATP [67]. NBD1 contains the canonical LSGGQ in the highly conserved Signature sequence, whereas NBD2 has a more unusual LSHGH sequence. Thus, if CFTR NBD1 and NBD2 dimerize to form an ATP sandwich dimer, like the homodimeric NBDs, one of the two nucleotide sites (Walker A and B from NBD1 and the Signature from NBD2) is very unusual, perhaps consistent with CFTR's identity as a channel within a family of transporters.

Recently, structures of murine CFTR-NBD1, both wild-type and ΔF508, which is 78% identical to the human CFTR-NBD1, have been solved at high resolution [67,84]. The structure of NBD1 with AMP-PNP bound highlights both the similarities and differences between CFTR-NBD1 and the other ABC transporter NBDs solved to date and discussed above. No significant conformational changes between the six murine NBD1 structures were observed, regardless of the nucleotide bound finding. Moreover, ΔF508 showed only minor difference from wild-type. However, this has recently been found to likely result from suppressor mutations embedded in the ΔF508 sequence necessary to generate crystals [140]. The fold of the new CFTR-NBD1 structures is very similar to that of the other ATP-binding cassettes with a β-sheet subdomain, a core containing the Walker A and B sequences (similar to the RecA/F1 fold), and an α-helical subdomain that contains the signature sequence and is the site of the ΔF508 mutation. This residue is located on the surface of NBD1 at a position remote from the NBD dimer interface, but presumed to be at or near the interface for interaction with the TMDs. The location of F508 at the putative NBD–TMD interface is consistent with previous studies indicating that deletion of F508 did not affect the stability of the isolated domain significantly but destabilized the intact CFTR structure [39,141–143] and analyses of missense mutations at this position.

CFTR NBD1 contains an insertion between the first and second β-strands (residues 406 to 434) which is largely disordered in the crystals. However, when NBD1 is phosphorylated, a partial ordering of this insertion is observed which may partially occlude the nucleotide [67,84] and perhaps accounts for the very unusual nucleotide binding kinetics observed [144] (see below). Thus, by contrast to earlier predictions of the boundaries of NBD1, the first strand is composed of residues 393 to 400, and the phenylalanine at position 400 is in position to interact potentially with the nucleotide base. In all of the CFTR–NBD1 structures, the nucleotide is bound in an unusual conformation. NBD1 also has additional secondary structural elements that serve to further alter the putative dimer interface, including a helix at the extreme C-terminus that has previously been assigned to the regulatory (R) domain. Thus, formation of a canonical ATP sandwich dimer will either require reorientation of these elements or compensatory changes at the NBD2 interface to accommodate them. These initial structural characterizations of CFTR and its homologs provide new insight about the function and dysfunction of this important protein. The NBD1 domain is thought to "talk" to both the C1 and C4 cytoplasmic loops linking the various transmembrane domains. Thus, NBD1 plays a pivotal role in assembly of the full-length polypeptide, a process assisted largely by chaperones [3,16,127–129].

TRAFFICKING TO THE CELL SURFACE

The capacity of the ER to export cargo protein necessarily encompasses critical energetic relationships between protein translation, folding, degradation, and trafficking pathways [95–97,169,170,172]. Such an approach reveals an *export landscape*, defined as the "minimal export threshold" which embraces the key fundamental properties of the protein fold-folding kinetics, misfolding kinetics, and thermodynamic stability [96]. This three-dimensional landscape reveals that CFTR escapes via a highly adaptable ER environment through a diverse combination of folding-and trafficking-related pathways. Recognition that a network of interacting components may be interlinked to achieve folding and trafficking from an energetic standpoint may be the achilles heel of CF and of misfolding disease in general, allowing a broad range of both direct and indirect chemical and biological approaches to promote recovering of misfolded CFTR variants for function at the cell surface [145].

The first step in delivery of CFTR to the cell surface from the ER is its interaction with the COPII machinery [123,146]. This machinery consists largely of a small group of cytoplasmic components, including the small GTPase Sar1 and the Sec23/24 adaptor complex, which together recognize a diacidic code on the folded NBD1 domain of CFTR [147]. This interaction between CFTR and the adaptor complex results in its collection into vesicles budding from the ER surface, a process driven by the self-assembly of the cytosolic protein complex Sec13/31 [148]. The overall process can be defined as a set of linked kinetic equilibria in which the kinetics of folding, the strength of the adaptor recognition, and the kinetics of cage self-assembly, possibly catalyzed by the cargo–adaptor complexes, dictate the steady-state rate of incorporation of CFTR into COPII vesicles [123]. Moreover, CFTR exiting the ER is in competition with a large number of other cargo poised for exit. Therefore, numerous factors contribute in an as yet to be defined manner to the overall success or failure of either wild-type and mutant CFTR to evade the ERAD metabolic pathway and interact with the COPII export pathway. Like degradation and chaperone components, the levels of the COPII machinery vary widely between cell types, suggesting that CFTR wild-type and variants may face unique conditions in different cell types comprising various tissues that either robustly support or provide more limited support for export.

Following the exit from the ER, CFTR is transported through the Golgi, where its two N-linked oligosaccharides are further processed to more complex structures, followed by delivery to the cell surface. Almost nothing is known about trafficking of CFTR through the Golgi, although the small Rho-like GTPase TM10 and the protein CAL have been implicated in selection for exit from the trans Golgi network, for delivery to the cell surface, and/or for recycling from early endocytic compartments to the late compartments or the Golgi [149–151]. The physiological impact of these events on mutant function remains to be more fully clarified.

STABILITY AND TRAFFICKING AT THE CELL SURFACE

Compelling evidence indicates that ΔF508 CFTR molecules reaching the post-Golgi compartments are metabolically and functionally unstable in nonpolarized [37,45,152] and polarized [37,153–155] heterologous expression systems, as well as in primary culture of human respiratory epithelia [37]. The cell surface resident ΔF508 CFTR retains some channel activity [156]. Moreover, recent morphological and functional observations suggest that small amounts of mutant may constitutively escape the ER in primary culture, freshly isolated human and animal cells, and native tissues [157–159]. Thus, the fate of the ΔF508 CFTR in post-Golgi compartments probably influences the severity of the disease and the efficacy of therapeutic interventions in rescuing the ΔF508 CFTR as well as other variants whose cell surface activities and stability remain largely unexplored.

Reduced temperature can overcome ΔF508 failure to exit the ER by facilitating folding and/or increasing protein stability [35,45]. Intriguingly, the quasi-native conformation of the rescued ΔF508 CFTR is partially lost upon elevating the ambient temperature to 37°C, in accord with the mutant metabolic destabilization and profoundly increased ubiquitination [37]. The temperature-sensitive stability defect is associated with accelerated internalization and impaired recycling of the rescued ΔF508 CFTR back to the cell surface. This is in sharp contrast with the slow internalization and highly efficient recycling of wild-type CFTR, a prerequisite to maintaining the channel and slow metabolic turnover at the plasma membrane [37,152,153,155].

While the underlying mechanism of accelerated internalization of the rescued ΔF508 CFTR at physiological temperature remains elusive, it appears that the ubiquitination pathway plays the central role in rerouting a variety of nonnative, internalized CFTRs from recycling for lysosomal proteolysis in BHK cells [37,71], although a recycling defect could not be observed in respiratory epithelia [153]. Rescuing the peripheral folding defect attenuated the relative ubiquitination level of the mutant. Inactivation of the E1 ubiquitin-activating enzyme also stabilized ΔF508 and the C-terminal truncation Δ70 CFTR at the cell surface. These and other observations, in concert with the preferential association of ΔF508 CFTR with Ub-binding proteins (e.g., Hrs, STAM-2, and TSG101) and components of the Ub-dependent endosomal sorting machinery (ESCRT II and ESCRT III) suggest a link between ubiquitination-modification and lysosomal degradation of misfolded CFTR from the cell surface [37] and highlight potential drug targets for mutant stabilization.

A large number of additional factors have been found to associate with CFTR at the cell surface. These involve multiple components of the endocytic and recycling machineries [160], although their exact roles remained to be assessed. Moreover, the capacity of CFTR to interact with other proteins through its C-terminal PDZ domain, which, in turn, link the protein to the amiloride-sensitive epithelial Na^+ channel (ENAC) and G-protein coupled receptors and other signaling pathways [160,161], suggests that we have much to learn about CFTR as a "hub" protein in human physiology and how these linked pathways are disrupted at the onset of disease and during aging in response to CFTR misfolding.

CURRENT EFFORTS AND FUTURE OPPORTUNITIES TO CORRECT CFTR FOLDING: THE WAY FORWARD

Despite extensive past and ongoing efforts to date in the CF field, many challenges remain to fully characterize the basic biochemical and biophysical properties that direct CFTR folding and function, and to more fully understand the networks that direct both folding and function in different cell types. The concept that ΔF508 and possibly other mutants are trapped in folding intermediates implies that the ΔF508 can be rescued from intermediate conformations if the limiting steps are appropriately manipulated by adjusting the folding, trafficking, and stabilization environment associated with the ER and/or the cell surface. That this can occur is demonstrated by the fact that ΔF508 CFTR can be rescued biochemically and functionally not only by low temperature as discussed above [162] but by intragenic suppressor mutations [163,164], alteration of putative ER retention motifs [165], and/or altered chaperone activities [3,102]. In the latter case, demonstration that manipulation of the folding pathway managed by the Hsp90 chaperone through its co-chaperone Aha1 can lead to cell surface conductance supports the idea that a novel class of compounds, referred to as *proteostasis regulators* [145], can be used to adjust mutant protein folds to modify disease progression. When used independently or combined with chemical chaperones [166], or more likely, pharmacological chaperones that could bind to and stabilize the fold directly, including compounds referred to as correctors [167,168], or that activate cell surface–localized CFTR (potentiators), unanticipated avenues for disease intervention at early and late stages may result.

Interaction of mutant CFTR with local folding environment challenges the protein homeostasis or "proteostasis" program of the cell [169,170,172]. It is now apparent that an understanding of the individual contribution(s) of translation, folding, degradation, trafficking, and regulation of channel activity at the cell surface will be important in deciphering the basis for disease onset and its progression. Of central concern is that disease progression is also associated with alterations not only in surface anion and cation balance, but also inflammatory responses, alteration of carbohydrate processing, and glycan (mucin) deposition and predisposes, for example, the lung to pathogenic environments that contribute significantly to pulmonary failure. Our knowledge base of how the CFTR fold and its function contribute to these processes normally and in response to mutation is very limited. In this regard, a recent effort by the CF Foundation has led to the construction of a comprehensive "Roadmap" of current known CFTR interaction pathways using the GeneGo platform. These maps provide a baseline to begin to organize our current knowledge base, including what contributes to and/or dictates CFTR function in health and disease. Moreover, these pathways provide a starting point to integrate CFTR folding with function and therapeutics, and perhaps drive a multiple pathway-based approach to correcting this chronic early-onset disease. To accomplish these goals, it will take the combined efforts not only of the CF Consortium, but the CF community in general, to develop new tools, identify new targets, and provide new assays that are amenable to drug screening efforts that could be used to move our understanding of basic science of the folding problem into the clinic.

REFERENCES

1. Thomas, P.J., Qu, B.H., Pedersen, P.L. (1995). Defective protein folding as a basis of human disease. *Trends Biochem Sci, 20*, 456–459.
2. Riordan, J.R. (2008). CFTR function and prospects for therapy. *Annu Rev Biochem, 77*, 701–726.

3. Wang, X., Venable, J., LaPointe, P., Hutt, D.M., Koulov, A.V., Coppinger, J., Gurkan, C., Kellner, W., Matteson, J., Plutner, H., et al. (2006). Hsp90 cochaperone Aha1 downregulation rescues misfolding of CFTR in cystic fibrosis. *Cell, 127*, 803–815.

4. Xu, Y., Liu, C., Clark, J.C., Whitsett, J.A. (2006). Functional genomic responses to cystic fibrosis transmembrane conductance regulator (CFTR) and CFTR(delta508) in the lung. *J Biol Chem, 281*, 11279–11291.

5. Pollard, H.B., Eidelman, O., Jozwik, C., Huang, W., Srivastava, M., Ji, X.D., McGowan, B., Norris, C.F., Todo, T., Darling, T., et al. (2006). De novo biosynthetic profiling of high abundance proteins in cystic fibrosis lung epithelial cells. *Mol Cell Proteom, 5*, 1628–1637.

6. Knowles, M.R. (2006). Gene modifiers of lung disease. *Curr Opin Pulm Med, 12*, 416–421.

7. Quinton, P.M. (2007). Too much salt, too little soda: cystic fibrosis. *Sheng Li Xue Bao, 59*, 397–415.

8. Rowe, S.M., Miller, S., Sorscher, E.J. (2005). Cystic fibrosis. *N Engl J Med, 352*, 1992–2001.

9. Ostedgaard, L.S., Rich, D.P., DeBerg, L.G., Welsh, M.J. (1997). Association of domains within the cystic fibrosis transmembrane conductance regulator. *Biochemistry, 36*, 1287–1294.

10. King, S.A., Sorscher, E.J. (2000). R-domain interactions with distal regions of CFTR lead to phosphorylation and activation. *Biochemistry*, 39, 9868–9875.

11. Akabas, M.H., Cheung, M., Guinamard, R. (1997). Probing the structural and functional domains of the CFTR chloride channel. *J Bioenerg Biomembr, 29*, 453–463.

12. Chen, E.Y., Bartlett, M.C., Loo, T.W., Clarke, D.M. (2004). The deltaF508 mutation disrupts packing of the transmembrane segments of the cystic fibrosis transmembrane conductance regulator. *J Biol Chem, 279*, 39620–39627.

13. Chappe, V., Irvine, T., Liao, J., Evagelidis, A., Hanrahan, J.W. (2005). Phosphorylation of CFTR by PKA promotes binding of the regulatory domain. *EMBO J, 24*, 2730–2740.

14. Rich, D.P., Berger, H.A., Cheng, S.H., Travis, S.M., Saxena, M., Smith, A.E., Welsh, M.J. (1993). Regulation of the cystic fibrosis transmembrane conductance regulator Cl^- channel by negative charge in the R domain. *J Biol Chem, 268*, 20259–20267.

15. Rich, D.P., Gregory, R.J., Anderson, M.P., Manavalan, P., Smith, A.E., Welsh, M.J. (1991). Effect of deleting the R domain on CFTR-generated chloride channels. *Science, 253*, 205–207.

16. Cui, L., Aleksandrov, L., Chang, X.B., Hou, Y.X., He, L., Hegedus, T., Gentzsch, M., Aleksandrov, A., Balch, W.E., Riordan, J.R. (2007). Domain interdependence in the biosynthetic assembly of CFTR. *J Mol Biol, 365*, 981–994.

17. Sadlish, H., Skach, W.R. (2004). Biogenesis of CFTR and other polytopic membrane proteins: new roles for the ribosome–translocon complex. *J Membr Biol, 202*, 115–126.

18. Amaral, M.D., Kunzelmann. K. (2007). Molecular targeting of CFTR as a therapeutic approach to cystic fibrosis. *Trends Pharmacol Sci, 28*, 334–341.

19. Carlson, E., Skach, W.R.. (2001). Signal recognition particle mediates CFTR targeting to protease-stripped ER membranes. *Pediatr Pulmonol, S22*, 182–183.

20. Johnson, A.E., van Waes, M.A. (1999). The translocon: a dynamic gateway at the ER membrane. *Annu Rev Cell Dev Biol, 15*, 799–842.

21. Shibatani, T., David, L.L., McCormack, A.L., Frueh, K., Skach, W.R. (2005). Proteomic analysis of mammalian oligosaccharyltransferase reveals multiple subcomplexes that contain Sec61, TRAP, and two potential new subunits. *Biochemistry, 44*, 5982–5992.

22. Chen, M., Zhang, J.T. (1999). Topogenesis of cystic fibrosis transmembrane conductance regulator (CFTR): regulation by the amino terminal transmembrane sequences. *Biochemistry, 38*, 5471–5477.

23. Carveth, K., Buck, T., Anthony, V., Skach, W.R. (2002). Cooperativity and flexibility of cystic fibrosis transmembrane conductance regulator transmembrane segments participate in membrane localization of a charged residue. *J Biol Chem, 277*, 39507–39514.

24. Riordan, J.R. (2005). Assembly of functional CFTR chloride channels. *Annu Rev Physiol, 67*, 701–718.

25. Rothman, R.E., Andrews, D.W., Calayag, M.C., Lingappa, V.R. (1988). Construction of defined polytopic integral transmembrane proteins: the role of signal and stop transfer sequence permutations. *J Biol Chem, 263*, 10470–10480.

26. Rapoport, T.A., Goder, V., Heinrich, S.U., Matlack, K.E. (2004). Membrane–protein integration and the role of the translocation channel. *Trends Cell Biol, 14*, 568–575.

27. Johnson, A.E. (2005). Fluorescence approaches for determining protein conformations, interactions and mechanisms at membranes. *Traffic, 6*, 1078–1092.

28. Moss, K., Helm, A., Lu, Y., Bragin, A., Skach, W.R. (1998). Coupled translocation events generate topological heterogeneity at the endoplasmic reticulum membrane. *Mol Biol Cell, 9*, 2681–2697.

29. Goder, V., Bieri, C., Spiess, M. (1999). Glycosylation can influence topogenesis of membrane proteins and reveals dynamic reorientation of nascent polypeptides within the translocon. *J Cell Biol, 147*, 257–266.

30. Pitonzo, D., Skach, W.R. (2006). Molecular mechanisms of aquaporin biogenesis by the endoplasmic reticulum Sec61 translocon. *Biochim Biophys Acta, 1758*, 976–988.

31. Lu, Y., Xiong, X., Helm, A., Kimani, K., Bragin, A., Skach, W.R. (1998). Co- and posttranslational translocation mechanisms direct cystic fibrosis transmembrane conductance regulator N terminus transmembrane assembly. *J Biol Chem, 273*, 568–576.

32. Xiong, X., Bragin, A., Widdicombe, J.H., Cohn, J., Skach, W.R. (1997). Structural cues involved in endoplasmic reticulum degradation of G85E and G91R mutant cystic fibrosis transmembrane conductance regulator. *J Clin Invest, 100*, 1079–1088.

33. Wigley, W.C., Corboy, M.J., Cutler, T.D., Thibodeau, P.H., Oldan, J., Lee, M.G., Rizo, J., Hunt, J.F., Thomas, P.J. (2002). A protein sequence that can encode native structure by disfavoring alternate conformations. *Nat Struct Biol, 9*, 381–388.

34. Loo, T.W., Bartlett, M.C., Clarke, D.M.. (2006). Insertion of an arginine residue into the transmembrane segments corrects protein misfolding. *J Biol Chem, 281*, 29436–29440.

35. Thibodeau, P.H., Brautigam, C.A., Machius, M., Thomas, P.J. (2005). Side chain and backbone contributions of Phe508 to CFTR folding. *Nat Struct Mol Biol*, *12*, 10–16.

36. Choi, M.Y., Partridge, A.W., Daniels, C., Du, K., Lukacs, G.L., Deber, C.M. (2005). Destabilization of the transmembrane domain induces misfolding in a phenotypic mutant of cystic fibrosis transmembrane conductance regulator. *J Biol Chem*, *280*, 4968–4974.

37. Sharma, M., Pampinella, F., Nemes, C., Benharouga, M., So, J., Du, K., Bache, K.G., Papsin, B., Zerangue, N., Stenmark, H., Lukacs, G.L. (2004). Misfolding diverts CFTR from recycling to degradation: quality control at early endosomes. *J Cell Biol*, *164*, 923–933.

38. Haardt, M., Benharouga, M., Lechardeur, D., Kartner, N., Lukacs, G.L. (1999). C-terminal truncations destabilize the cystic fibrosis transmembrane conductance regulator without impairing its biogenesis. a novel class of mutation. *J Biol Chem*, *274*, 21873–21877.

39. Zhang, F., Kartner, N., Lukacs, G.L. (1998). Limited proteolysis as a probe for arrested conformational maturation of delta F508 CFTR. *Nat Struct Biol*, *5*, 180–183.

40. Oberdorf, J., Pitonzo, D., Skach, W.R. (2005). An energy-dependent maturation step is required for release of the cystic fibrosis transmembrane conductance regulator from early endoplasmic reticulum biosynthetic machinery. *J Biol Chem*, *280*, 38193–38202.

41. Oberdorf, J., Carlson, E.J., Skach, W.R. (2006). Uncoupling proteasome peptidase and ATPase activities results in cytosolic release of an ER polytopic protein. *J Cell Sci*, *119*, 303–313.

42. Lukacs, G.L., Mohamed, A., Kartner, N., Chang, X.B., Riordan, J.R., Grinstein, S. (1994). Conformational maturation of CFTR but not its mutant counterpart (delta F508) occurs in the endoplasmic reticulum and requires ATP. *EMBO J*, *13*, 6076–6086.

43. Kleizen, B., van Vlijmen, T., de Jonge, H.R., Braakman, I. (2005). Folding of CFTR is predominantly cotranslational. *Mol Cell*, *20*, 277–287.

44. Du, K., Sharma, M., Lukacs, G.L. (2005). The deltaF508 cystic fibrosis mutation impairs domain-domain interactions and arrests post-translational folding of CFTR. *Nat Struct Mol Biol*, *12*, 17–25.

45. Sharma, M., Benharouga, M., Hu, W., Lukacs, G.L. (2001). Conformational and temperature-sensitive stability defects of the delta F508 cystic fibrosis transmembrane conductance regulator in post-endoplasmic reticulum compartments. *J Biol Chem*, *276*, 8942–8950.

46. Roth, M.G., Doyle, C., Sambrook, J., Gething, M.J. (1986). Heterologous transmembrane and cytoplasmic domains direct functional chimeric influenza virus hemagglutinins into the endocytic pathway. *J Cell Biol*, *102*, 1271–1283.

47. Wyss, S., Dimitrov, A.S., Baribaud, F., Edwards, T.G., Blumenthal, R., Hoxie, J.A. (2005). Regulation of human immunodeficiency virus type 1 envelope glycoprotein fusion by a membrane-interactive domain in the gp41 cytoplasmic tail. *J Virol*, *79*, 12231–12241.

48. Young, J.C., Agashe, V.R., Siegers, K., Hartl, F.U. (2004). Pathways of chaperone-mediated protein folding in the cytosol. *Nat Rev Mol Cell Biol*, *5*, 781–791.

49. Pearl, L.H., Prodromou, C. (2006). Structure and mechanism of the Hsp90 molecular chaperone machinery. *Annu Rev Biochem*, *75*, 271–294.

50. Helenius, A., Aebi, M. (2004). Roles of N-linked glycans in the endoplasmic reticulum. *Annu Rev Biochem*, *73*, 1019–1049.

51. Pind, S., Riordan, J.R., Williams, D.B. (1994). Participation of the endoplasmic reticulum chaperone calnexin (p88, IP90) in the biogenesis of the cystic fibrosis transmembrane conductance regulator. *J Biol Chem*, *269*, 12784–12788.

52. Loo, M.A., Jensen, T.J., Cui, L., Hou, Y., Chang, X.B., Riordan, J.R. (1998). Perturbation of Hsp90 interaction with nascent CFTR prevents its maturation and accelerates its degradation by the proteasome. *EMBO J*, *17*, 6879–6887.

53. Yang, Y., Janich, S., Cohn, J.A., Wilson, J.M. (1993). The common variant of cystic fibrosis transmembrane conductance regulator is recognized by hsp70 and degraded in a pre-Golgi nonlysosomal compartment. *Proc Natl Acad Sci U S A*, *90*, 9480–9484.

54. Meacham, G.C., Lu, Z., King, S., Sorscher, E., Tousson, A., Cyr, D.M. (1999). The Hdj-2/Hsc70 chaperone pair facilitates early steps in CFTR biogenesis. *EMBO J*, *18*, 1492–1505.

55. Meacham, G.C., Patterson, C., Zhang, W., Younger, J.M., Cyr, D.M. (2001). The Hsc70 co-chaperone CHIP targets immature CFTR for proteasomal degradation. *Nat Cell Biol*, *3*, 100–105.

56. Younger, J.M., Chen, L., Ren, H.Y., Rosser, M.F., Turnbull, E.L., Fan, C.Y., Patterson, C., Cyr, D.M. (2006). Sequential quality-control checkpoints triage misfolded cystic fibrosis transmembrane conductance regulator. *Cell*, *126*, 571–582.

57. Sun, F., Zhang, R., Gong, X., Geng, X., Drain, P.F., Frizzell, R.A. (2006). Derlin-1 promotes the efficient degradation of the cystic fibrosis transmembrane conductance regulator (CFTR) and CFTR folding mutants. *J Biol Chem*, *281*, 36856–36863.

58. Lord, J.M., High, S. (2005). Polytopic proteins: preventing aggregation in the membrane. *Curr Biol*, *15*, R169–171.

59. Dudek, J., Greiner, M., Müller, A., Hendershot, L.M., Kopsch, K., Nastainczyk, W., Zimmermann, R. (2005). ERj1p has a basic role in protein biogenesis at the endoplasmic reticulum. *Nat Struct Mol Biol*, *12*, 1008–1014.

60. Zhang, H., Schmidt, B.Z., Sun, F., Condliffe, S.B., Butterworth, M.B., Youker, R.T., Brodsky, J.L., Aridor, M., Frizzell, R.A. (2006). Cysteine string protein monitors late steps in cystic fibrosis transmembrane conductance regulator biogenesis. *J Biol Chem*, *281*, 11312–11321.

61. Ban, N., Nissen, P., Hansen, J., Moore, P.B., Steitz, T.A. (2000). The complete atomic structure of the large ribosomal subunit at 2.4 A resolution. *Science*, *289*, 905–920.

62. Van den Berg, B., Clemons, W.M., Jr., Collinson, I., Modis, Y., Hartmann, E., Harrison, S.C., Rapoport, T.A. (2004). X-ray structure of a protein-conducting channel. *Nature*, *427*, 36–44.

63. Allen, S., Naim, H.Y., Bulleid, N.J. (1995). Intracellular folding of tissue-type plasminogen activator: effects of disulfide bond formation on N-linked glycosylation and secretion. *J Biol Chem*, *270*, 4797–4804.

64. Farinha, C.M., Amaral, M.D. (2005). Most F508del-CFTR is targeted to degradation at an early folding checkpoint and independently of calnexin. *Mol Cell Biol*, *25*, 5242–5252.

65. Baker, J.M., Hudson, R.P., Kanelis, V., Choy, W.Y., Thibodeau, P.H., Thomas, P.J., Forman-Kay, J.D. (2007). CFTR regulatory region interacts with NBD1 predominantly via multiple transient helices. *Nat Struct Mol Biol*, *14*, 738–745.

66. Mense, M., Vergani, P., White, D.M., Altberg, G., Nairn, A.C., Gadsby, D.C. (2006). In vivo phosphorylation of CFTR promotes formation of a nucleotide-binding domain heterodimer. *EMBO J*, *25*, 4728–4739.

67. Lewis, H.A., Buchanan, S.G., Burley, S.K., Conners, K., Dickey, M., Dorwart, M., Fowler, R., Gao, X., Guggino, W.B., Hendrickson, W.A., et al. (2004). Structure of nucleotide-binding domain 1 of the cystic fibrosis transmembrane conductance regulator. *EMBO J*, *23*, 282–293.

68. Ward, C.L., Kopito, R.R. (1994). Intracellular turnover of cystic fibrosis transmembrane conductance regulator. Inefficient processing and rapid degradation of wild-type and mutant proteins. *J Biol Chem*, *269*, 25710–25718.

69. Wolin, S.L., Walter, P. (1988). Ribosome pausing and stacking during translation of a eukaryotic mRNA. *EMBO J*, *7*, 3559–3569.

70. Kimchi-Sarfaty, C., Oh, J.M., Kim, I.W., Sauna, Z.E., Calcagno, A.M., Ambudkar, S.V., Gottesman, M.M. (2007). A "silent" polymorphism in the MDR1 gene changes substrate specificity. *Science*, *315*, 525–528.

71. Turnbull, E.L., Rosser, M.F., Cyr, D.M. (2007). The role of the UPS in cystic fibrosis. *BMC Biochem*, *8* (Suppl 1), S11.

72. Brodsky, J.L. (2007). The protective and destructive roles played by molecular chaperones during ERAD (endoplasmic-reticulum-associated degradation). *Biochem J*, *404*, 353–363.

73. Sato, S., Ward, C.L., Kopito, R.R. (1998). Cotranslational ubiquitination of cystic fibrosis transmembrane conductance regulator in vitro. *J Biol Chem*, *273*, 7189–7192.

74. Ahner, A., Nakatsukasa, K., Zhang, H., Frizzell, R.A., Brodsky, J.L. (2007). Small heat-shock proteins select delta F508-CFTR for endoplasmic reticulum–associated degradation. *Mol Biol Cell*, *18*, 806–814.

75. Schubert, U., Antón, L.C., Gibbs, J., Norbury, C.C., Yewdell, J.W., Bennink, J.R. (2000). Rapid degradation of a large fraction of newly synthesized proteins by proteasomes. *Nature*, *404*, 770–774.

76. Kostova, Z., Wolf, D.H. (2003). For whom the bell tolls: protein quality control of the endoplasmic reticulum and the ubiquitin–proteasome connection. *EMBO J*, *22*, 2309–2317.

77. McCracken, A.A., Brodsky, J.L. (2003). Evolving questions and paradigm shifts in endoplasmic-reticulum-associated degradation (ERAD). *Bioessays*, *25*, 868–877.

78. Ellgaard, L., Helenius, A. (2003). Quality control in the endoplasmic reticulum. *Nat Rev Mol Cell Biol*, *4*, 181–191.

79. Fewell, S.W., Travers, K.J., Weissman, J.S., Brodsky, J.L. (2001). The action of molecular chaperones in the early secretory pathway. *Annu Rev Genet*, *35*, 149–191.

80. Riordan, J.R., Rommens, J.M., Kerem, B., Alon, N., Rozmahel, R., Grzelczak, Z., Zielenski, J., Lok, S., Plavsic, N., Chou, J.L., et al. (1989). Identification of the

cystic fibrosis gene: cloning and characterization of complementary DNA. *Science*, *245*, 1066–1073.

81. Cheng, S.H., Gregory, R.J., Marshall, J., Paul, S., Souza, D.W., White, G.A., O'Riordan, C.R., Smith, A.E. (1990). Defective intracellular transport and processing of CFTR is the molecular basis of most cystic fibrosis. *Cell*, *63*, 827–834.
82. Amaral, M.D. (2005). Processing of CFTR: traversing the cellular maze—how much CFTR needs to go through to avoid cystic fibrosis. *Pediatr Pulmonol*, *39*, 479–491.
83. Cui, L., Aleksandrov, L., Hou, Y.X., Gentzsch, M., Chen, J.H., Riordan, J.R., Aleksandrov, A.A. (2006). The role of cystic fibrosis transmembrane conductance regulator phenylalanine 508 side chain in ion channel gating. *J Physiol*, *572*, 347–358.
84. Lewis, H.A., Zhao, X., Wang, C., Sauder, J.M., Rooney, I., Noland, B.W., Lorimer, D., Kearins, M.C., Conners, K., Condon, B., et al. (2005). Impact of the delta F508 mutation in first nucleotide-binding domain of human cystic fibrosis transmembrane conductance regulator on domain folding and structure. *J Biol Chem*, *280*, 1346–1353.
85. Choo-Kang, L.R., Zeitlin, P.L. (2001). Induction of HSP70 promotes Delta F508 CFTR trafficking. *Am J Physiol Lung Cell Mol Physiol*, *281*, L58–L68.
86. Rubenstein, R.C., Zeitlin, P.L. (2000). Sodium 4-phenylbutyrate downregulates Hsc70: implications for intracellular trafficking of deltaF508-CFTR. *Am J Physiol Cell Physiol*, *278*, C259–C267.
87. Youker, R.T., Walsh, P., Beilharz, T., Lithgow, T., Brodsky, J.L. (2004). Distinct roles for the Hsp40 and Hsp90 molecular chaperones during cystic fibrosis transmembrane conductance regulator degradation in yeast. *Mol Biol Cell*, *15*, 4787–4797.
88. Alberti, S., Böhse, K., Arndt, V., Schmitz, A., Höhfeld, J. (2004). The cochaperone HspBP1 inhibits the CHIP ubiquitin ligase and stimulates the maturation of the cystic fibrosis transmembrane conductance regulator. *Mol Biol Cell*, *15*, 4003–4010.
89. Farinha, C.M., Nogueira, P., Mendes, F., Penque, D., Amaral, M.D. (2002). The human DnaJ homologue (Hdj)-1/heat-shock protein (Hsp) 40 co-chaperone is required for the in vivo stabilization of the cystic fibrosis transmembrane conductance regulator by Hsp70. *Biochem J*, *366*, 797–806.
90. Zhang, H., Peters, K.W., Sun, F., Marino, C.R., Lang, J., Burgoyne, R.D., Frizzell, R.A. (2002). Cysteine string protein interacts with and modulates the maturation of the cystic fibrosis transmembrane conductance regulator. *J Biol Chem*, *277*, 28948–28958.
91. Okiyoneda, T., Harada, K., Takeya, M., Yamahira, K., Wada, I., Shuto, T., Suico, M.A., Hashimoto, Y., Kai, H. (2004). Delta ΔF508 CFTR pool in the endoplasmic reticulum is increased by calnexin overexpression. *Mol Biol Cell*, *15*, 563–574.
92. Strickland, E., Qu, B.H., Millen, L., Thomas, P.J. (1997). The molecular chaperone Hsc70 assists the in vitro folding of the N-terminal nucleotide-binding domain of the cystic fibrosis transmembrane conductance regulator. *J Biol Chem*, *272*, 25421–25424.

93. Jensen, T.J., Loo, M.A., Pind, S., Williams, D.B., Goldberg, A.L., Riordan, J.R. (1995). Multiple proteolytic systems, including the proteasome, contribute to CFTR processing. *Cell, 83*, 129–135.

94. Ward, C.L., Omura, S., Kopito, R.R. (1995). Degradation of CFTR by the ubiquitin–proteasome pathway. *Cell, 83*, 121–127.

95. Wiseman, R.L., Koulov, A., Powers, E., Kelly, J.W., Balch, W.E. (2007). Protein energetics in maturation of the early secretory pathway. *Curr Opin Cell Biol, 19*, 359–367.

96. Wiseman, R.L., Powers, E.T., Buxbaum, J.N., Kelly, J.W., Balch, W.E. (2007). An adaptable standard for protein export from the endoplasmic reticulum. *Cell, 131*, 809–821.

97. Sekijima, Y., Wiseman, R.L., Matteson, J., Hammarström, P., Miller, S.R., Sawkar, A.R., Balch, W.E., Kelly, J.W. (2005). The biological and chemical basis for tissue-selective amyloid disease. *Cell, 121*, 73–85.

98. Carlson, E.J., Pitonzo, D., Skach, W.R. (2006). p97 functions as an auxiliary factor to facilitate TM domain extraction during CFTR ER-associated degradation. *EMBO J, 25*, 4557–4566.

99. Morito, D., Hirao, K., Oda, Y., Hosokawa, N., Tokunaga, F., Cyr, D.M., Tanaka, K., Iwai, K., Nagata, K. (2008). Gp78 cooperates with RMA1 in endoplasmic reticulum-associated degradation of CFTRdeltaF508. *Mol Biol Cell, 19*, 1328–1336.

100. Gnann, A., Riordan, J.R., Wolf, D.H. (2004). Cystic fibrosis transmembrane conductance regulator degradation depends on the lectins Htm1p/EDEM and the Cdc48 protein complex in yeast. *Mol Biol Cell, 15*, 4125–4135.

101. Oda, Y., Okada, T., Yoshida, H., Kaufman, R.J., Nagata, K., Mori, K. (2006). Derlin-2 and Derlin-3 are regulated by the mammalian unfolded protein response and are required for ER-associated degradation. *J Cell Biol, 172*, 383–393.

102. Zhang, Y., Nijbroek, G., Sullivan, M.L., McCracken, A.A., Watkins, S.C., Michaelis, S., Brodsky, J.L. (2001). Hsp70 molecular chaperone facilitates endoplasmic reticulum-associated protein degradation of cystic fibrosis transmembrane conductance regulator in yeast. *Mol Biol Cell, 12*, 1303–1314.

103. Kabani, M., Beckerich, J.M., Brodsky, J.L. (2002). Nucleotide exchange factor for the yeast Hsp70 molecular chaperone Ssa1p. *Mol Cell Biol, 22*, 4677–4689.

104. Haslbeck, M., Miess, A., Stromer, T., Walter, S., Buchner, J. (2005). Disassembling protein aggregates in the yeast cytosol: the cooperation of Hsp26 with Ssa1 and Hsp104. *J Biol Chem, 280*, 23861–23868.

105. Biswas, A., Das, K.P. (2004). Role of ATP on the interaction of alpha-crystallin with its substrates and its implications for the molecular chaperone function. *J Biol Chem, 279*, 42648–42657.

106. Ehrnsperger, M., Graber, S., Gaestel, M., Buchner, J. (1997). Binding of non-native protein to Hsp25 during heat shock creates a reservoir of folding intermediates for reactivation. *EMBO J, 16*, 221–229.

107. Haslbeck, M., Braun, N., Stromer, T., Richter, B., Model, N., Weinkauf, S., Buchner, J. (2004). Hsp42 is the general small heat shock protein in the cytosol of *Saccharomyces cerevisiae*. *EMBO J, 23*, 638–649.

108. Haslbeck, M., Walke, S., Stromer, T., Ehrnsperger, M., White, H.E., Chen, S., Saibil, H.R., Buchner, J. (1999). Hsp26: a temperature-regulated chaperone. *EMBO J*, *18*, 6744–6751.

109. Horwitz, J. (1992). Alpha-crystallin can function as a molecular chaperone. *Proc Natl Acad Sci U S A*, *89*, 10449–10453.

110. Jakob, U., Gaestel, M., Engel, K., Buchner, J. (1993). Small heat shock proteins are molecular chaperones. *J Biol Chem*, *268*, 1517–1520.

111. Lee, G.J., Roseman, A.M., Saibil, H.R., Vierling, E. (1997). A small heat shock protein stably binds heat-denatured model substrates and can maintain a substrate in a folding-competent state. *EMBO J*, *16*, 659–671.

112. Lindner, R.A., Treweek, T.M., Carver, J.A. (2001). The molecular chaperone alpha-crystallin is in kinetic competition with aggregation to stabilize a monomeric molten-globule form of alpha-lactalbumin. *Biochem J*, *354*, 79–87.

113. Rajaraman, K., Raman, B., Rao, C.M. (1996). Molten-globule state of carbonic anhydrase binds to the chaperone-like alpha-crystallin. *J Biol Chem*, *271*, 27595–27600.

114. Treweek, T.M., Lindner, R.A., Mariani, M., Carver, J.A. (2000). The small heat-shock chaperone protein, alpha-crystallin, does not recognize stable molten globule states of cytosolic proteins. *Biochim Biophys Acta*, *1481*, 175–188.

115. Koteiche, H.A., McHaourab, H.S. (2003). Mechanism of chaperone function in small heat-shock proteins. Phosphorylation-induced activation of two-mode binding in alphaB-crystallin. *J Biol Chem*, *278*, 10361–10367.

116. McHaourab, H.S., Dodson, E.K., Koteiche, H.A. (2002). Mechanism of chaperone function in small heat shock proteins. two-mode binding of the excited states of T4 lysozyme mutants by alphaA-crystallin. *J Biol Chem*, *277*, 40557–40566.

117. Sathish, H.A., Koteiche, H.A., McHaourab, H.S. (2004). Binding of destabilized betaB2-crystallin mutants to alpha-crystallin: the role of a folding intermediate. *J Biol Chem*, *279*, 16425–16432.

118. Shashidharamurthy, R., Koteiche, H.A., Dong, J., McHaourab, H.S. (2005). Mechanism of chaperone function in small heat shock proteins: dissociation of the HSP27 oligomer is required for recognition and binding of destabilized T4 lysozyme. *J Biol Chem*, *280*, 5281–5289.

119. Cashikar, A.G., Duennwald, M., Lindquist, S.L. (2005). A chaperone pathway in protein disaggregation: Hsp26 alters the nature of protein aggregates to facilitate reactivation by Hsp104. *J Biol Chem*, *280*, 23869–23875.

120. Friedrich, K.L., Giese, K.C., Buan, N.R., Vierling, E. (2004). Interactions between small heat shock protein subunits and substrate in small heat shock protein–substrate complexes. *J Biol Chem*, *279*, 1080–1089.

121. Haslbeck, M., Franzmann, T., Weinfurtner, D., Buchner, J. (2005). Some like it hot: the structure and function of small heat-shock proteins. *Nat Struct Mol Biol*, *12*, 842–846.

122. Török, Z., Goloubinoff, P., Horváth, I., Tsvetkova, N.M., Glatz, A., Balogh, G., Varvasovszki, V., Los, D.A., Vierling, E., Crowe, J.H., Vigh, L. (2001). Synechocystis HSP17 is an amphitropic protein that stabilizes heat-stressed membranes and binds denatured proteins for subsequent chaperone-mediated refolding. *Proc Natl Acad Sci U S A*, *98*, 3098–3103.

123. Gurkan, C., Stagg, S.M., Lapointe, P., Balch, W.E. (2006). The COPII cage: unifying principles of vesicle coat assembly. *Nat Rev Mol Cell Biol*, *7*, 727–738.

124. Gurkan, C., Lapp, H., Hogenesch, J.B., Balch, W.E. (2005). Exploring trafficking GTPase function by mRNA expression profiling: use of the SymAtlas web-application and the Membrome datasets. *Methods Enzymol*, *403*, 1–10.

125. Awayn, N.H., Rosenberg, M.F., Kamis, A.B., Aleksandrov, L.A., Riordan, J.R., Ford, R.C. (2005). Crystallographic and single-particle analyses of native- and nucleotide-bound forms of the cystic fibrosis transmembrane conductance regulator (CFTR) protein. *Biochem Soc Trans*, *33*, 996–999.

126. Mendoza, J.L., Thomas, P.J. (2007). Building an understanding of cystic fibrosis on the foundation of ABC transporter structures. *J Bioenerg Biomembr*, *39*, 499–505.

127. Mornon, J.P., Lehn, P., Callebaut, I. (2008). Atomic model of human cystic fibrosis transmembrane conductance regulator: membrane-spanning domains and coupling interfaces. *Cell Mol Life Sci*, *65*, 2594–2612.

128. Serohijos, A.W., Hegedus, T., Aleksandrov, A.A., He, L., Cui, L., Dokholyan, N.V., Riordan, J.R. (2008). Phenylalanine-508 mediates a cytoplasmic-membrane domain contact in the CFTR 3D structure crucial to assembly and channel function. *Proc Natl Acad Sci U S A*, *105*, 3256–3261.

129. He, L., Aleksandrov, A.A., Serohijos, A.W., Hegedus, T., Aleksandrov, L.A., Cui, L., Dokholyan, N.V., Riordan, J.R. (2008). Multiple membrane–cytoplasmic domain contacts in cftr mediate regulation of channel gating. *J Biol Chem*, *283*, 26383–26390.

130. Chen, J., Lu, G., Lin, J., Davidson, A.L., Quiocho, F.A. (2003). A tweezers-like motion of the ATP-binding cassette dimer in an ABC transport cycle. *Mol Cell, 12*, 651–661.

131. Smith, P.C., Karpowich, N., Millen, L., Moody, J.E., Rosen, J., Thomas, P.J., Hunt, J.F. (2002). ATP binding to the motor domain from an ABC transporter drives formation of a nucleotide sandwich dimer. *Mol Cell*, *10*, 139–149.

132. Moody, J.E., Millen, L., Binns, D., Hunt, J.F., Thomas, P.J. (2002). Cooperative, ATP-dependent association of the nucleotide binding cassettes during the catalytic cycle of ATP-binding cassette transporters. *J Biol Chem*, *277*, 21111–21114.

133. Karpowich, N., Martsinkevich, O., Millen, L., Yuan, Y.R., Dai, P.L., MacVey, K., Thomas, P.J., Hunt, J.F. (2001). Crystal structures of the MJ1267 ATP binding cassette reveal an induced-fit effect at the ATPase active site of an ABC transporter. *Structure*, *9*, 571–586.

134. Yuan, Y.R., Blecker, S., Martsinkevich, O., Millen, L., Thomas, P.J., Hunt, J.F. (2001). The crystal structure of the MJ0796 ATP-binding cassette: implications for the structural consequences of ATP hydrolysis in the active site of an ABC transporter. *J Biol Chem*, *276*, 32313–32321.

135. Samanta, S., Ayvaz, T., Reyes, M., Shuman, H.A., Chen, J., Davidson, A.L. (2003). Disulfide cross-linking reveals a site of stable interaction between C-terminal regulatory domains of the two MalK subunits in the maltose transport complex. *J Biol Chem*, *278*, 35265–35271.

136. Chang, G. (2003). Structure of MsbA from *Vibrio cholera*: a multidrug resistance ABC transporter homolog in a closed conformation. *J Mol Biol*, *330*, 419–430.

137. Chang, G. (2007). Retraction of "Structure of MsbA from *Vibrio cholera*: a multidrug resistance ABC transporter homolog in a closed conformation" [*J Mol Biol* (2003) *330*, 419–430]. *J Mol Biol*, *369*, 596.

138. Locher, K.P., Lee, A.T., Rees, D.C. (2002). The *E. coli* BtuCD structure: a framework for ABC transporter architecture and mechanism. *Science*, *296*, 1091–1098.

139. Thomas, P.J., Hunt, J.F. (2001). A snapshot of nature's favorite pump. *Nat Struct Biol*, *8*, 920–923.

140. Pissarra, L.S., Farinha, C.M., Xu, Z., Schmidt, A., Thibodeau, P.H., Cai, Z., Thomas, P.J., Sheppard, D.N., Amaral, M.D. (2008). Solubilizing mutations used to crystallize one CFTR domain attenuate the trafficking and channel defects caused by the major cystic fibrosis mutation. *Chem Biol*, *15*, 62–69.

141. Qu, B.H., Strickland, E., Thomas, P.J. (1997). Cystic fibrosis: a disease of altered protein folding. *J Bioenerg Biomembr*, *29*, 483–490.

142. Qu, B.H., Strickland, E.H., Thomas, P.J. (1997). Localization and suppression of a kinetic defect in cystic fibrosis transmembrane conductance regulator folding. *J Biol Chem*, *272*, 15739–15744.

143. Qu, B.H., Thomas, P.J. (1996). Alteration of the cystic fibrosis transmembrane conductance regulator folding pathway. *J Biol Chem*, *271*, 7261–7264.

144. Basso, C., Vergani, P., Nairn, A.C., Gadsby, D.C. (2003). Prolonged nonhydrolytic interaction of nucleotide with CFTR's NH2-terminal nucleotide binding domain and its role in channel gating. *J Gen Physiol*, *122*, 333–348.

145. Balch, W.E., Morimoto, R.I., Dillin, A., Kelly, J.W. (2008). Adapting proteostasis for disease intervention. *Science*, *319*, 916–919.

146. Stagg, S.M., LaPointe, P., Balch, W.E. (2007). Structural design of cage and coat scaffolds that direct membrane traffic. *Curr Opin Struct Biol*, *17*, 221–228.

147. Wang, X., Matteson, J., An, Y., Moyer, B., Yoo, J.S., Bannykh, S., Wilson, I.A., Riordan, J.R., Balch, W.E. (2004).COPII-dependent export of cystic fibrosis transmembrane conductance regulator (CFTR) from the ER utilizes a di-acidic exit code. *J Cell Biol*, *167*, 65–74.

148. Stagg, S.M., Gürkan, C., Fowler, D.M., LaPointe, P., Foss, T.R., Potter, C.S., Carragher, B., Balch, W.E. (2006). Structure of the Sec13/31 COPII coat cage. *Nature*, *439*, 234–238.

149. Wolde, M., Fellows, A., Cheng, J., Kivenson, A., Coutermarsh, B., Talebian, L., Karlson, K., Piserchio, A., Mierke, D.F., Stanton, B.A., et al. (2007). Targeting CAL as a negative regulator of deltaF508-CFTR cell-surface expression: an RNA interference and structure-based mutagenetic approach. *J Biol Chem*, *282*, 8099–8109.

150. Cheng, J., Wang, H., Guggino, W.B. (2005). Regulation of cystic fibrosis transmembrane regulator trafficking and protein expression by a Rho family small GTPase TC10. *J Biol Chem*, *280*, 3731–3739.

151. Cheng, J., Wang, H., Guggino, W.B. (2004). Modulation of mature cystic fibrosis transmembrane regulator protein by the PDZ domain protein CAL. *J Biol Chem*, *279*, 1892–1898.

152. Gentzsch, M., Chang, X.B., Cui, L., Wu, Y., Ozols, V.V., Choudhury, A., Pagano, R.E., Riordan, J.R. (2004). Endocytic trafficking routes of wild type and

deltaF508 cystic fibrosis transmembrane conductance regulator. *Mol Biol Cell, 15*, 2684–2696.

153. Swiatecka-Urban, A., Brown, A., Moreau-Marquis, S., Renuka, J., Coutermarsh, B., Barnaby, R., Karlson, K.H., Flotte, T.R., Fukuda, M., Langford, G.M., Stanton, B.A. (2005). The short apical membrane half-life of rescued {delta}F508-cystic fibrosis transmembrane conductance regulator (CFTR) results from accelerated endocytosis of {Delta}F508-CFTR in polarized human airway epithelial cells. *J Biol Chem, 280*, 36762–36772.
154. Heda, G.D., Marino, C.R. (2000). Surface expression of the cystic fibrosis transmembrane conductance regulator mutant deltaF508 is markedly upregulated by combination treatment with sodium butyrate and low temperature. *Biochem Biophys Res Commun, 271*, 659–664.
155. Varga, K., Jurkuvenaite, A., Wakefield, J., Hong, J.S., Guimbellot, J.S., Venglarik, C.J., Niraj, A., Mazur, M., Sorscher, E.J., Collawn, J.F., Bebök, Z. (2004). Efficient intracellular processing of the endogenous cystic fibrosis transmembrane conductance regulator in epithelial cell lines. *J Biol Chem, 279*, 22578–22584.
156. Drumm, M.L., Wilkinson, D.J., Smit, L.S., Worrell, R.T., Strong, T.V., Frizzell, R.A, Dawson, D.C., Collins, F.S. (1991). Chloride conductance expressed by delta ΔF508 and other mutant CFTRs in *Xenopus* oocytes. *Science, 254*, 1797–1799.
157. Carvalho-Oliveira, I., Efthymiadou, A., Malhó, R., Nogueira, P., Tzetis, M., Kanavakis, E., Amaral, M.D., Penque, D. (2004). CFTR localization in native airway cells and cell lines expressing wild-type or F508del-CFTR by a panel of different antibodies. *J Histochem Cytochem, 52*, 193–203.
158. Bronsveld, I., Mekus, F., Bijman, J., Ballmann, M., de Jonge, H.R., Laabs, U., Halley, D.J., Ellemunter, H., Mastella, G., Thomas, S., et al. (2001). Chloride conductance and genetic background modulate the cystic fibrosis phenotype of delta F508 homozygous twins and siblings. *J Clin Invest, 108*, 1705–1715.
159. Ostedgaard, L.S., Rogers, C.S., Dong, Q., Randak, C.O., Vermeer, D.W., Rokhlina, T., Karp, P.H., Welsh, M.J. (2007). Processing and function of CFTR-delta F508 are species-dependent. *Proc Natl Acad Sci U S A, 104*, 15370–15375.
160. Guggino, W.B., Stanton, B.A. (2006). New insights into cystic fibrosis: molecular switches that regulate CFTR. *Nat Rev Mol Cell Biol, 7*, 426–436.
161. Li, C., Krishnamurthy, P.C., Penmatsa, H., Marrs, K.L., Wang, X.Q., Zaccolo, M., Jalink, K., Li, M., Nelson, D.J., Schuetz, J.D., Naren, A.P. (2007). Spatiotemporal coupling of cAMP transporter to CFTR chloride channel function in the gut epithelia. *Cell, 131*, 940–951.
162. Denning, G.M., Anderson, M.P., Amara, J.F., Marshall, J., Smith, A.E., Welsh, M.J. (1992). Processing of mutant cystic fibrosis transmembrane conductance regulator is temperature-sensitive. *Nature, 358*, 761–764.
163. DeCarvalho, A.C., Gansheroff, L.J., Teem, J.L. (2002). Mutations in the nucleotide binding domain 1 signature motif region rescue processing and functional defects of cystic fibrosis transmembrane conductance regulator delta f508. *J Biol Chem, 277*, 35896–35905.

164. Teem, J.L., Carson, M.R., Welsh, M.J. (1996). Mutation of R555 in CFTR-delta F508 enhances function and partially corrects defective processing. *Receptors Channels*, *4*, 63–72.

165. Chang, X.B., Cui, L., Hou, Y.X., Jensen, T.J., Aleksandrov, A.A., Mengos, A., Riordan, J.R. (1999). Removal of multiple arginine-framed trafficking signals overcomes misprocessing of delta ΔF508 CFTR present in most patients with cystic fibrosis. *Mol Cell*, *4*, 137–142.

166. Welch, W.J. (2004). Role of quality control pathways in human diseases involving protein misfolding. *Semin Cell Dev Biol*, *15*, 31–38.

167. Pedemonte, N., Lukacs, G.L., Du, K., Caci, E., Zegarra-Moran, O., Galietta, L.J., Verkman, A.S. (2005). Small-molecule correctors of defective deltaF508-CFTR cellular processing identified by high-throughput screening. *J Clin Invest*, *115*, 2564–2571.

168. Van Goor, F., Straley, K.S., Cao, D., González, J., Hadida, S., Hazlewood, A., Joubran, J., Knapp, T., Makings, L.R., Miller, M., et al. (2006). Rescue of deltaF508-CFTR trafficking and gating in human cystic fibrosis airway primary cultures by small molecules. *Am J Physiol Lung Cell Mol Physiol*, *290*, L1117–1130.

169. Balch, W.E., Morimoto, R.I., Dillin, A., Kelly, J.W. (2008). Adapting proteostasis for disease intervention. Review. *Science*, *319*, 916–919.

170. Powers, E.T., Morimoto, R.I., Dillin, A., Kelly, J.W., Balch, W.E. (2009). Biological and chemical approaches to diseases of proteostasis deficiency. *Annu Rev Biochem*, *78*, 959–991.

171. Aller, S.G., Yu, J., Ward, A., Weng, Y., Chittaboina, S., Zhuo, R., Harrell, P.M., Trinh, Y.T., Zhang, Q., Urbatsch, I.L., Chang, G. (2009). Structure of P-glycoprotein reveals a molecular basis for poly-specific drug binding. *Science, 323*, 1718–1722.

172. Hutt, D.M., Powers, E.T., Balch, W.E. (2009). The proteostasis boundary in misfolding diseases of membrane traffic. *FEBS Lett*, *583*, 2639–2646.

20

THIOPURINE S-METHYLTRANSFERASE PHARMACOGENOMICS: PROTEIN MISFOLDING, AGGREGATION, AND DEGRADATION

FANG LI AND RICHARD M. WEINSHILBOUM

Department of Molecular Pharmacology and Experimental Therapeutics, Mayo Clinic College of Medicine, Rochester, Minnesota

INTRODUCTION

Pharmacogenomics is the study of the role of inheritance in individual variation in drug response [1]. That variation can range from serious life-threatening adverse drug reactions at one end of the spectrum to lack of the desired therapeutic drug effect at the other. It may seem incongruous that a chapter on "pharmacogenomics" would be included in a volume devoted to protein misfolding, but it is now clear that common genetic variation in the amino acid sequences of proteins that influence drug response can result in misfolding, with resulting accelerated degradation and aggregation that can have a dramatic effect on drug response [2]. One of the most striking examples, and one of the most striking examples of the clinical importance of pharmacogenomics, involves the drug-metabolizing enzyme thiopurine S-methyltransferase (TPMT) [2]. TPMT is important not only because it is a prototypic example of the clinical impact of pharmacogenetics and pharmacogenomics—representing the first example for which the U.S. Food and Drug administration (FDA) held

Protein Misfolding Diseases: Current and Emerging Principles and Therapies,
Edited by Marina Ramirez-Alvarado, Jeffery W. Kelly, and Christopher M. Dobson
Copyright © 2010 John Wiley & Sons, Inc.

public hearings on the inclusion of pharmacogenetic information in drug labeling (www.fda.gov)—but also because there is a wealth of information with regard to mechanisms underlying the clinical impact of TPMT. Those mechanisms involve protein misfolding, aggregation, and degradation [2]. In subsequent paragraphs we briefly describe the discovery of genetic polymorphisms for *TPMT* and their clinical importance, followed by a description of mechanistic studies that have placed TPMT, and pharmacogenomics, squarely within the realm of research on protein misfolding, aggregation, and degradation.

TPMT PHARMACOGENETICS: DISCOVERY AND CLINICAL IMPORTANCE

TPMT is a cytosolic drug-metabolizing enzyme that catalyzes the methylation of sulfhydryl groups in thiopurine drugs, with *S*-adenosyl-L-methionine as the methyl donor [3,4]. Thiopurines such as 6-mercaptopurine (6-MP) and azathioprine, a prodrug that is converted to 6-MP in vivo, are used widely to treat acute lymphoblastic leukemia (ALL) of childhood, inflammatory bowel disease, and autoimmune disorders [5]. 6-MP is also a prodrug that must undergo metabolic conversion to form 6-thioguanine nucleotides (6-TGNs), followed by their incorporation into DNA to exert its anti-neoplastic and anti-inflammatory effects (see Fig. 1) [5].

Interest in TPMT emerged after the discovery of inherited variation in levels of TPMT activity in human tissue, ranging from high to virtually undetectable activity [6]. That discovery was made possible by the development of a sensitive radiochemical TPMT enzymatic assay developed specifically to test the possibility of inherited variation in the activity of this enzyme [7]. Use of this assay to measure TPMT activity in red blood cells (RBCs) demonstrated a trimodel frequency distribution of human RBC TPMT activity among randomly selected Caucasian subjects (Fig. 2) [6,7]. Segregation analysis of data from family studies established the autosomal codominant inheritance of level of RBC TPMT activity in this population, with about 10% of the subjects being heterozygous and having intermediate activity and approximately 1 in 300 being homozygous for the trait of very low TPMT activity [6]. Levels of TPMT activity in the RBC reflected relative levels of enzyme activity in other tissues, such as kidney, liver, and lymphocyte [4,8,9]. The cloning and characterization of the human TPMT cDNA and gene [10,11], made it possible to demonstrate that the variation in TPMT activity depicted graphically in Figure 2 resulted from variation in the DNA sequence of the *TPMT* gene itself: specifically two common nonsynonymous single-nucleotide polymorphisms (nsSNPs) [11]. A total of over 20 *TPMT* polymorphisms have been identified [12], many of which are associated with decreased levels of enzyme activity and/or

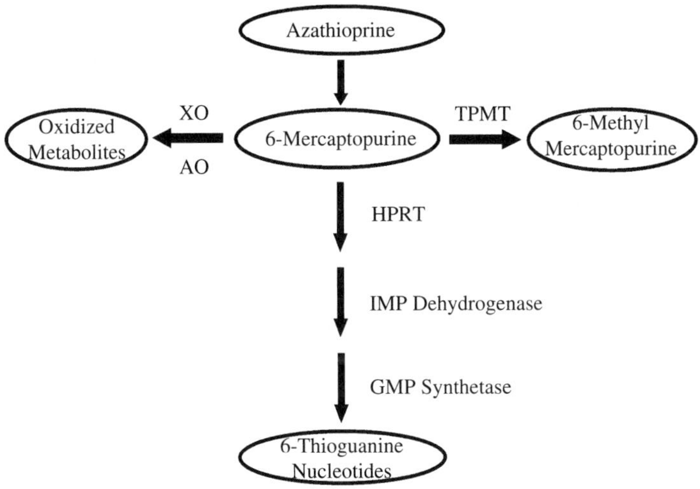

FIG. 1 Thiopurine metabolism: a simplified schematic representation of the metabolism of the thiopurine drug azathioprine, which is converted to 6-mercaptopurine (6-MP) in vivo, as well as the biotransformation of 6-MP by TPMT, xanthine oxidase (XO), and aldehyde oxidase (AO). The "metabolic activation" of 6-MP to form 6-thioguanine nucleotides occurs as a result of a series of reactions catalyzed by hypoxanthine guanine phosphoriboxyltransferase (HPRT), IMP dehydrogenase, and GMP synthetase. (From [2], with permission of Nature Publishing Company.)

thiopurine-induced toxicity [12]. Although there are many mechanisms by which variation in the DNA sequence of *TPMT* might alter enzyme activity, including the presence of a variable number of tandem repeats (VNTR) in the 5′-flanking region of the gene that may modulate transcription and SNPs that result in alterations in TPMT mRNA splicing [12–14], the majority of functionally significant variation in *TPMT* DNA sequence involves nsSNPs that alter the encoded amino acid sequence [12].

SNPs represent the most common form of DNA sequence variation, making up 90% of all known human genetic variation [15]. SNPs occur approximately once every 500 base pairs along the 3 billion–base pair human genome [16]. nsSNPs are changes in nucleotide sequence within the open reading frames of genes that result in alterations in the amino acid sequence of the encoded protein. Although, as stated previously, SNPs and other types of DNA sequence variation (e.g., insertion-deletion events) can influence function in a variety of ways, nsSNPs are a common cause of alteration in function, and the most common mechanism by which that occurs involves a change in protein quantity. TPMT illustrates these principles as well or better than any other example in pharmacogenomics [2].

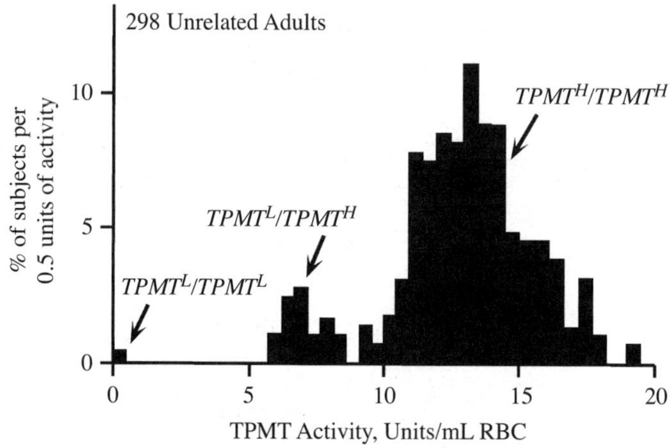

FIG. 2 The TPMT genetic polymorphism: a frequency distribution histogram of level of RBC TPMT enzyme activity in blood samples from 298 randomly selected Caucasian blood donors. Presumed genotypes for the TPMT genetic polymorphism are also indicated. These designations for high ($TPMT^H$) and low ($TPMT^L$) activity alleles were used before the molecular basis for the polymorphism was determined. Subsequently, the major allele responsible for this frequency distribution was shown to be *TPMT*3A* (see Fig. 3 and the text). (From [6], with permission of University of Chicago Press. Copyright © Elsevier 1980.)

TPMT PHARMACOGENOMICS: CLINICAL CONSEQUENCES

The clinical and functional consequence of amino acid substitutions as a result of variant *TPMT* alleles have been characterized extensively both in vitro and in vivo. *TPMT*3A* is the most common variant allele in Caucasians, with a minor allele frequency of approximately 5% [17]. This allele contains two nsSNPs that result in Ala154Thr and Tyr240Cys alterations in the encoded amino acid sequence (Fig. 3) [11,18]. *TPMT*3C*, the most common functionally relevant variant allele in East Asian populations, with a frequency of approximately 2% among Han Chinese [17], includes only the Tyr240Cys substitution, while the rare *TPMT*3B* allele includes only the Ala154Thr alteration in sequence (Fig. 3). Levels of TPMT enzyme activity and protein are virtually undetectable in the tissues of patients homozygous for *TPMT*3A* or in cultured mammalian cells transfected with this allele [11,12,19,20]. *TPMT*3B*, although rare, has functional consequences similar to those of *TPMT*3A* [12,19,20]. *TPMT*2* and **3C* do not result in functional effects that are as dramatic as those of *3A, but these alleles are also associated with significantly decreased levels of TPMT activity and protein [11,12,20,21]. Significant ethnic differences exist in the distribution and frequencies of *TPMT* variant alleles, but *TPMT*3A, *3C*, and **2* account for approximately 95% of inherited TPMT deficiency in Caucasian subjects [22]. However, the most common functionally important variant in Caucasian populations, *TPMT*3A*, has never been reported in East Asian subjects.

Thiopurine drugs are very important clinically, but like other cytotoxic agents, they have a narrow therapeutic index (i.e., the difference between therapeutic and toxic doses is relatively small) [5]. Two major reactions are involved in the catabolic inactivation of thiopurine drugs, oxidation catalyzed by xanthine oxidase (XO), and methylation catalyzed by TPMT [5] (Fig. 1). Although XO metabolizes the bulk of thiopurines, XO is not expressed in hematopoietic tissue [23]. Furthermore, it appears that TPMT is the primary factor responsible for variation in the quantity of thiopurines available for enzymatic reactions that lead to the formation of "active" thiopurine metabolites, 6-TGNs (Fig. 1). Since these metabolites are critical for the therapeutic effect of thiopurines, 6-TGN concentrations have been used as an index of the therapeutic and toxic effects of these drugs [23–25]. For example, in children taking identical doses of 6-MP there are wide interindividual variations in RBC 6-TGN concentrations [23], and high concentrations of 6-TGNs are associated with profound bone-marrow suppression, the primary life-threatening thiopurine-induced toxicity [23,25]. Clinical studies have demonstrated that genetically determined levels of RBC TPMT activity are inversely correlated with RBC 6-TGN concentrations [23,26,27], supporting the hypothesis that an inherited decrease in TPMT activity might shift the balance among thiopurine metabolic pathways toward the formation of 6-TGNs. As a result, patients homozygous for TPMT alleles associated with low TPMT activity have elevated 6-TGNs when they are treated with standard doses of thiopurines and are at greatly increased risk for the development of life-threatening bone-marrow suppression [23,26–29]. To avoid toxicity, these patients can be treated with one-tenth to one-fifteenth of the standard dose, but even then only with careful monitoring. Because of its profound clinical implications, the U.S. Food and Drug Administration (FDA) held public hearings on TPMT pharmacogenetics in 2002–2003 and, as a result, approved a label change for 6-MP (www.fda.gov). However, in addition to its striking clinical importance, the TPMT genetic polymorphism has been of equal importance for the insight that it has provided into mechanisms responsible for the functional effects of nsSNPs—the reason for the inclusion of a discussion of TPMT pharmacogenomics in this volume.

TPMT PHARMACOGENOMICS: MOLECULAR MECHANISMS

Studies of TPMT have increased our understanding of mechanisms responsible for the functional effects of nsSNPs. Thirteen variant *TPMT* alleles with nsSNPs, *2, *3A, *3B, *3C, and *5 to *13 were studied functionally by Salavaggione et al., after transient expression in COS-1 cells [12]. When constructs containing these polymorphisms were expressed transiently in these mammalian cells, recombinant variant allozymes (enzyme with altered amino

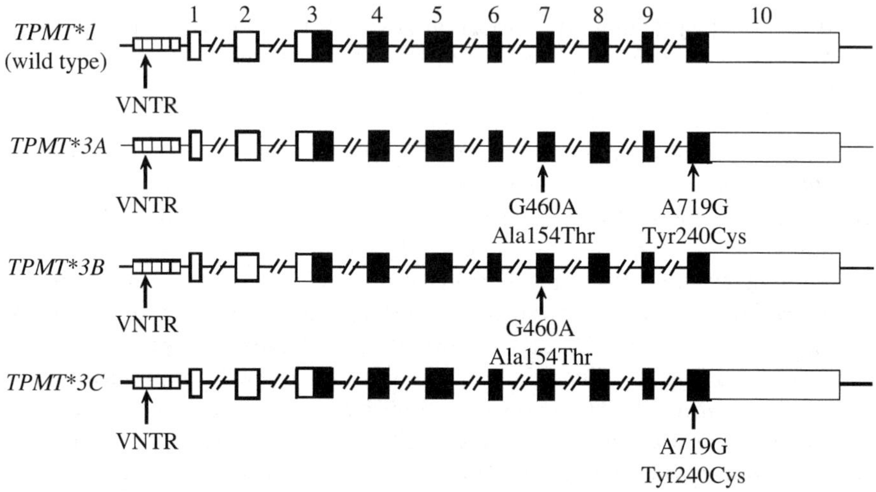

FIG. 3 TPMT alleles. *TPMT*1* is the most common allele (wild type), while *TPMT*3A* is the most common variant allele in Caucasian subjects and *TPMT*3C* is the most common variant allele in East Asian subjects. Black rectangles represent the open reading frame (ORF), and open rectangles represent 5′- and 3′-untranslated region (UTR) sequence. "VNTR" represents a GC-rich variable number of tandem repeats that is located in the 5′-flanking region of the gene. (From [2], with permission of Nature Publishing Company.)

acid sequence as a result of the translation of a variant allele) displayed levels of TPMT activity that varied from virtually undetectable to nearly 100% of that observed for the wild-type (WT) allele (Fig. 4). With the exception of TPMT*5, there was a direct correlation between decreased activity and decreased quantity of enzyme protein (Fig. 4). The alteration in amino acid sequence for TPMT*5 has been shown to affect the active site of the enzyme [30], but this mechanism is obviously the exception for this intensively studied gene. Among the allozymes studied by Salavaggione et al., TPMT*3A, *3B, *3C and *2 had the most striking functional effects, consistent with observations made over a decade ago [11,19,20,31]. For example, genotype–phenotype correlation studies performed with human liver biopsy samples at the time that the gene was originally cloned showed that the presence of **3A* allele was associated with decreased levels of TPMT protein, confirming observations made with transfected cells [11].

The association of naturally occurring nsSNPs with altered levels of protein is extremely common [32]. Altered protein level has been shown to be the primary mechanism in a series of functional genomic studies of multiple members of the sulfotransferases (SULT) gene superfamily [33], a series of methyltransferase genes other than TPMT, including catechol O-methyltransferase (COMT) [34], histamine N-methyltransferase (HNMT) [35], and phenylethanolamine N-methyltransferase (PNMT) [36], as well as glutathione S-transferases

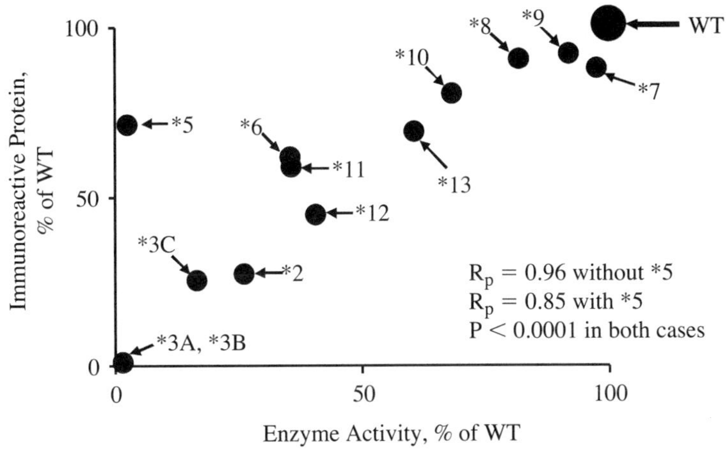

FIG. 4 Correlation of recombinant human TPMT allozyme enzyme activity and level of immunoreactive protein after expression in COS-1 cells. WT is the wild-type allozyme. The other designations refer to variant TPMT alleles that include nsSNPs. Note that the common *3A variant displays neither detectable enzyme activity nor protein. (From [12], with permission of Lippincott Williams & Wilkins.)

[37,38] and membrane-bound enzymes such as the important cytochromes P450 CYP19 (aromatase) [39]. There are many possible mechanisms that could explain how the substitution of only one or two amino acids might alter protein quantity, including decreased mRNA stability, decreased rate of protein synthesis, or accelerated protein degradation. However, accelerated degradation has been shown to be the most common mechanism when detailed studies have been performed [20,21,36,40]. Those studies have utilized a variety of experimental approaches. For example, pulse chase experiments performed after the expression of human *TPMT*3A* in yeast and COS-1 cells revealed that the TPMT*3A variant allozyme was degraded much more rapidly than was the WT enzyme [11,19–21]. Subsequently, it was shown that that process involved, at least in part, ubiquitin-proteasome-mediated degradation, without a significant change in either mRNA levels or the rate of protein synthesis [19–21]. Many of these experiments have been performed using the rabbit reticulocyte lysate (RRL), an experimental system that has been used widely to study proteasome-mediated protein degradation [41]. The RRL demonstrated the rapid degradation of the TPMT*3A allozyme, as well as the involvement of ubiquitin in that process [21]. The RRL has also been used to demonstrate the accelerated degradation of variant allozymes for SULT1A3, SULT1E1, and PNMT as well as GSTO1 and GSTO2 [36,38,42,43]. It now seems clear that naturally occurring, genetically determined alterations in encoded amino acid sequence often influence function as a result of alteration in protein quantity and that accelerated degradation is an important mechanism in that process. TPMT has also served as a valuable "model system" to pursue the details of these processes.

TPMT PHARMACOGENOMICS: MECHANISMS OF DEGRADATION

The fact that common inherited variation in amino acid sequence alters protein quantity, often as a result of accelerated degradation, raises a series of questions. One of those questions is how cells recognize an alteration in sequence that involves only one or two amino acids and then targets a variant allozyme for degradation. In the case of TPMT, molecular chaperone proteins such as the heat-shock proteins Hsp90 and Hsp70 as well as heat-shock-organizing protein (hop) are much more highly associated with the TPMT *3A variant allozyme than with the WT enzyme after expression in the RRL system [21]. Furthermore, treatment of the RRL and COS-1 cells expressing WT TPMT with the Hsp90 inhibitor geldanamycin resulted in enhanced association of Hsp90 with WT TPMT, as well as accelerated degradation of the WT protein [21]. Molecular chaperones not only play an important role in protein folding, but they are also involved in the targeting of misfolded protein for proteasome-mediated degradation [44–46]. Studies of TPMT suggest that molecular chaperones, especially Hsp90, help to mediate the balance between proper folding and degradation of TPMT, with misfolded TPMT*3A targeted for degradation [21].

TPMT*3A has also been shown to aggregate—with aggresome formation [47]. Aggresome formation is a process during which misfolded and polyubiquitinated proteins are sequestered into cytoplasmic structures around the microtubule organizing center in eukaryotic cells [48]. Aggresomes often form in response to the presence of a large quantity of misfolded protein, perhaps as a result of "overload" of the ubiquitin–proteasome system. Aggresomes form by the retrograde transport of microaggregates along microtubules, with the involvement of motor proteins such as dynein [49]. Histone deacetylase 6 (HDAC6), a protein that binds both ployubiquitinated misfolded protein and dynein, acts as an "adaptor protein" to recruit misfolded-protein cargo to dynein motors for transport to aggresomes [50]. TPMT*3A, but not the WT protein, formed aggresomes when expressed in COS-1 cells treated with proteasome inhibitors (Fig. 5) [47]. Treatment of COS-1 cells that had been transfected with *TPMT*3A* with the microtubule-destabilizing agent vinblastine, or with the HDAC6 inhibitor scriptaid, inhibited *3A aggresome formation, indicating that *3A aggresome formation was microtubule and HDAC6 dependent [47]. Although this treatment inhibited aggresome formation—as anticipated—at the same time, it dramatically increased the quantity of cytosolic microaggregates [47]. Furthermore, size-exclusion chromatography of recombinant WT, *3A, *3B, and *3C TPMT allozymes expressed in bacteria confirmed that the presence of the two nsSNPs in *TPMT*3A* resulted in protein aggregation [47]. This series of observations indicated that common nsSNPs, such as those present in *TPMT*3A*, can result in protein misfolding, leading to both accelerated degradation and aggresome formation.

TPMT PHARMACOGENOMICS: AUTOPHAGY

Studies of TPMT*3A degradation and aggregation suggested that a dynamic balance might exist among protein folding, protein degradation, and protein aggregation. In the presence of nsSNPs resulting in protein misfolding or of environmental factors such as proteasome inhibition, that balance was shifted to favor degradation and/or aggregation. This series of observations raised the question of the identity of the proteins involved in these cellular processes. In an effort to answer that question, a *Saccharomyces cerevisiae* yeast gene-deletion library was used to identify genes required for the degradation and aggregation of TPMT*3A [51]. Specifically, a "library" of 4667 yeast strains containing deletions of nonessential genes was transformed using a *TPMT*3A* expression construct that included green fluorescence protein (GFP), and the library was screened for mutants that had lost genes essential for TPMT*3A degradation and aggregation [51]. Twenty-four genes were identified that fell into several functionally related categories. The classes of genes involved included those affecting ubiquitin-dependent protein degradation (E2 ubiquitin-conjugating enzymes, E3 ubiquitin ligases, and proteasome subunits), vesicle trafficking, and vacuolar (lysosomal) degradation [51]. The presence of genes involved in vesicular transport and vacuolar degradation suggested a possible role for autophagy in TPMT*3A degradation.

Autophagy is a highly conserved eukaryotic process in which cytoplasm is sequestered into vesicles and is then delivered to vacuoles and lysosomes for degradation [52]. ATG genes, such as *ATG7* and *ATG12*, encode critical components required for the biogenesis of autophagic vesicles [53]. Therefore, it was striking that levels of aggregated TPMT*3A increased dramatically in yeast mutants for vacuolar proteases and autophagy. In addition, the expression of TPMT*3A-induced autophagy, and small interfering RNA-mediated knockdown in cultured mammalian cells of the expression of *ATG7*, an autophagy-related gene, enhanced TPMT*3A aggregation [51], indicating that autophagy is involved in TPMT*3A degradation. These results suggested that both proteasome-mediated proteolysis and autophagy participate in TPMT*3A degradation. They were also compatible with the results of studies of protein aggregation in neurodegenerative diseases.

Studies of neurodegenerative disease associated with aggregation-prone proteins such as huntingtin that contains polyglutamine expansions or mutant forms of α-synuclein have shown that both proteasome-mediated degradation and autophagy participate in the clearance of these proteins [54]. Those studies also demonstrated that dependence on autophagy for clearance is correlated with a tendency for these proteins to aggregate [54]. Furthermore, recent studies suggest a dynamic balance between the ubiquitin–proteasome system and autophagy and indicate that HDAC6 is an essential mechanistic link in that interaction [55,56]. It is uncertain whether autophagy clears only soluble species and oligomers, or also larger aggregates. In the neurons of mice with

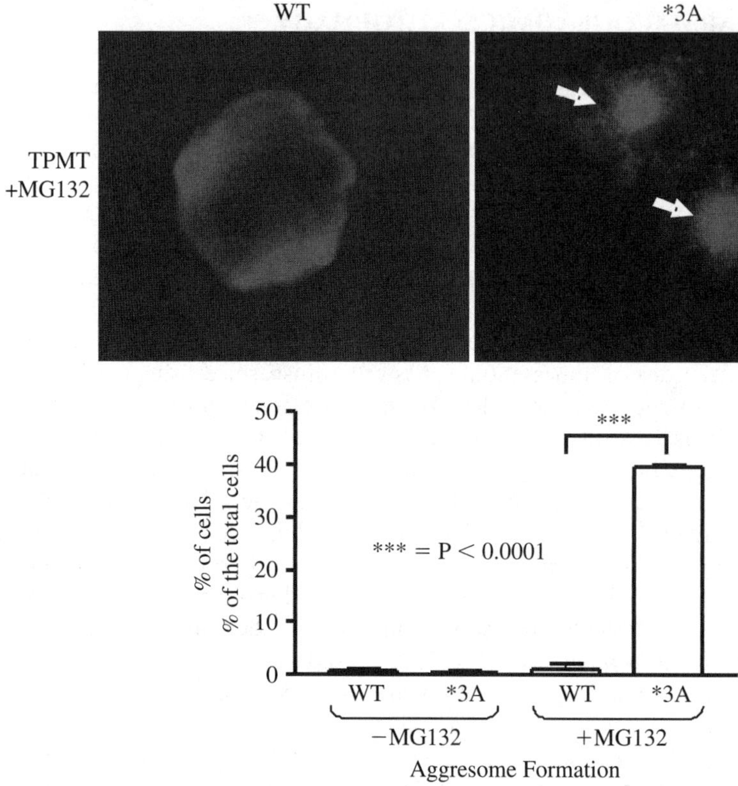

FIG. 5 TPMT aggresome formation. (A) COS-1 cells were transiently transfected with hemaggutinin (HA)-tagged WT and TPMT*3A constructs in the presence of the proteasome inhibitor MG132 and were subjected to fluorescence microscopy after immunostaining with anti-HA antibody. Arrows point to aggresomes. (B) Aggresome formation expressed as a percentage of about 200 cells counted (mean ± SEM, $N = 4$). (From [47], with permission of *Proceedings of the National Academy of Sciences*.) (*See insert for color representation of figure.*)

neuronally restricted autophagy-gene knockout, wild-type cellular proteins that are not usually aggregate-formation prone will form inclusions, suggesting that aggregates can also arise from the failure of autophagy to clear soluble proteins [57,58]. However, details of the process by which TPMT*3A is targeted for autophagy-dependent degradation remain unclear. Although autophagy is generally thought of as a nonspecific process, the autophagy-related cytoplasm-to-vacuole trafficking (CVT) pathway in yeast involves specific cargo selection [52]. Therefore, it is conceivable that specific genes encode components that mediate the selection of TPMT*3A for autophagy-dependent degradation in mammalian cells.

As mentioned previously, HDAC6, dynein, and microtubules are involved in aggresome formation [48–50], including the formation of TPMT*3A

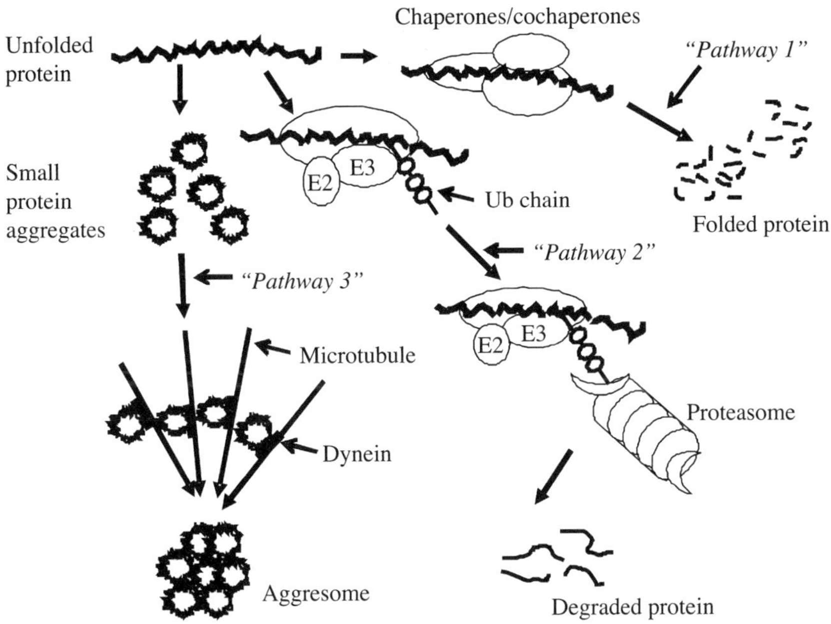

FIG. 6 Dynamic balance among protein-folding (pathway 1) proteasome-medicated degradation (pathway 2) and aggresome formation (pathway 3) for TPMT. The figure depicts various "fates" for TPMT protein, including proper folding, misfolding followed by ubiquitination, and proteasome-mediated degradation or aggregation with aggresome formation. (From [2], with permission of Nature Publishing Company.)

aggresomes [47]. Recent studies have shown that the microtubule system also plays an important role in autophagy [59–61]. HDAC6 and microtubules are required for the autophagic degradation of aggregated huntingtin [59], and dynein dysfunction retards the clearance of aggregation-prone proteins such as mutant huntingtin by reducing autophagosome–lysosome fusion [60]. It has been suggested that aggresome formation and autophagic sequestration might be cytoprotective pathways involved in the clearance of toxic aggregation-prone proteins such as those containing long polyglutamine repeats [61,62].

CONCLUSIONS

TPMT represents a prototypic example of the clinical importance of pharmacogenetics–pharmacogenomics [2]. The most common of the polymorphisms in *TPMT* involve nsSNPs, and because of their striking effect on clinical response to thiopurine drug therapy, these polymorphisms have been subjected to intense functional and mechanistic study. Those studies have demonstrated, for the first time, that protein misfolding, degradation, and aggregation represent pharmacogenomic mechanisms. They also revealed a dynamic balance among protein folding, protein degradation, and protein aggregation for TPMT*3A (Fig. 6). Under normal circumstances, WT TPMT folds properly. However, the two nsSNPs present in the common *TPMT*3A* allele result in a misfolded variant allozyme that is polyubiquinated and targeted for proteasome-mediated degradation. The misfolded TPMT *3A protein can also form microaggregates that can be translocated to aggresomes or cleared by autophagy, with the involvement of microtubules, HDAC6, and dynein. All of these processes exist in a dynamic balance that can be altered either by inherited polymorphisms or by changes in the cellular environment. In the future, the question of how this dynamic balance is regulated among the various pathways involved and how those pathways interact will have to be addressed. Therefore, studies of individual variation in thiopurine drug response led to the realization that protein misfolding—and the consequences of that process—represent important mechanisms for pharmacogenomics.

Acknowledgments

Supported in part by National Institutes of Health (NIH) grants RO1 GM28157, RO1 GM35720, and UO1 GM61388 (the Pharmacogenetics Research Network) and by a PhRMA Foundation Center of Excellence in Clinical Pharmacology award.

REFERENCES

1. Weinshilboum, R.M., Wang, L. (2006). Pharmacogenetics and pharmacogenomics: development, science, and translation. *Annu Rev Genom Hum Genet, 7*, 223–245.
2. Wang, L., Weinshilboum, R. (2006). Thiopurine S-methyltransferase pharmacogenetics: insights, challenges and future directions. *Oncogene, 25*, 1629–1638.
3. Remy, C.N. (1963). Metabolism of thiopyrimidines and thiopurines: S-methylation with *S*-adenosylmethionine transmethylase and catabolism in mammalian tissues. *J Biol Chem, 238*, 1078–1084.
4. Woodson, L.C., Weinshilboum, R.M. (1983). Human kidney thiopurine methyltransferase: purification and biochemical properties. *Biochem Pharmacol, 32*, 819–826.
5. Lennard, L. (1992). The clinical pharmacology of 6-mercaptopurine. *Eur J Clin Pharmacol, 43*, 329–339.
6. Weinshilboum, R.M., Sladek, S.L. (1980). Mercaptopurine pharmacogenetics: monogenic inheritance of erythrocyte thiopurine methyltransferase activity. *Am J Hum Genet, 32*, 651–662.
7. Weinshilboum, R.M., Raymond, F.A., Pazmino, P.A. (1978). Human erythrocyte thiopurine methyltransferase: radiochemical microassay and biochemical properties. *Clin Chim Acta, 85*, 323–333.

8. Szumlanski, C.L., Honchel, R., Scott, M.C., Weinshilboum, R.M. (1992). Human liver thiopurine methyltransferase pharmacogenetics: biochemical properties, liver-erythrocyte correlation and presence of isozymes. *Pharmacogenetics, 2*, 148–159.

9. Van Loon, J.A., Weinshilboum, R.M. (1982). Thiopurine methyltransferase biochemical genetics: human lymphocyte activity. *Biochem Genet, 20*, 637–658.

10. Honchel, R., Aksoy, I.A., Szumlanski, C., Wood, T.C., Otterness, D.M., Wieben, E.D., Weinshilboum, R.M. (1993). Human thiopurine methyltransferase: molecular cloning and expression of T84 colon carcinoma cell cDNA. *Mol Pharmacol, 43*, 878–887.

11. Szumlanski, C., Otterness, D., Her, C., Lee, D., Brandriff, B., Kelsell, D., Spurr, N., Lennard, L., Wieben, E., Weinshilboum, R. (1996). Thiopurine methyltransferase pharmacogenetics: human gene cloning and characterization of a common polymorphism. *DNA Cell Biol, 15*, 17–30.

12. Salavaggione, O.E., Wang, L., Wiepert, M., Yee, V.C., Weinshilboum, R.M. (2005). Thiopurine *S*-methyltransferase pharmacogenetics: variant allele functional and comparative genomics. *Pharmacogenet Genom, 15*, 801–815.

13. Otterness, D., Szumlanski, C., Lennard, L., Klemetsdal, B., Aarbakke, J., Park-Hah, J.O., Iven, H., Schmiegelow, K., Branum, E., O'Brien, J., Weinshilboum, R. (1997). Human thiopurine methyltransferase pharmacogenetics: gene sequence polymorphisms. *Clin Pharmacol Ther, 62*, 60–73.

14. Yan, L., Zhang, S., Eiff, B., Szumlanski, C.L., Powers, M., O'Brien, J.F., Weinshilboum, R.M. (2000). Thiopurine methyltransferase polymorphic tandem repeat: genotype–phenotype correlation analysis. *Clin Pharmacol Ther, 68*, 210–219.

15. Kruglyak, L., Nickerson, D.A. (2001). Variation is the spice of life. *Nat Genet, 27*, 234–236.

16. Guttmacher, A.E., Collins, F.S. (2002). Genomic medicine: a primer. *N Engl J Med, 347*, 1512–1520.

17. Collie-Duguid, E.S., Pritchard, S.C., Powrie, R.H., Sludden, J., Collier, D.A., Li, T., McLeod, H.L. (1999). The frequency and distribution of thiopurine methyltransferase alleles in Caucasian and Asian populations. *Pharmacogenetics, 9*, 37–42.

18. Tai, H.L., Krynetski, E.Y., Yates, C.R., Loennechen, T., Fessing, M.Y., Krynetskaia, N.F., Evans, W.E. (1996). Thiopurine S-methyltransferase deficiency: two nucleotide transitions define the most prevalent mutant allele associated with loss of catalytic activity in Caucasians. *Am J Hum Genet, 58*, 694–702.

19. Tai, H.L., Fessing, M.Y., Bonten, E.J., Yanishevsky, Y., d'Azzo, A., Krynetski, E.Y., Evans, W.E. (1999). Enhanced proteasomal degradation of mutant human thiopurine S-methyltransferase (TPMT) in mammalian cells: mechanism for TPMT protein deficiency inherited by TPMT*2, TPMT*3A, TPMT*3B or TPMT*3C. *Pharmacogenetics, 9*, 641-650.

20. Tai, H.L., Krynetski, E.Y., Schuetz, E.G., Yanishevski, Y., Evans, W.E. (1997). Enhanced proteolysis of thiopurine S-methyltransferase (TPMT) encoded by mutant alleles in humans (TPMT*3A, TPMT*2): mechanisms for the genetic polymorphism of TPMT activity. *Proc Natl Acad Sci U S A, 94*, 6444–6449.

21. Wang, L., Sullivan, W., Toft, D., Weinshilboum, R. (2003). Thiopurine *S*-methyltransferase pharmacogenetics: chaperone protein association and allozyme degradation. *Pharmacogenetics, 13*, 555–564.

22. McLeod, H.L., Krynetski, E.Y., Relling, M.V., Evans, W.E. (2000). Genetic polymorphism of thiopurine methyltransferase and its clinical relevance for childhood acute lymphoblastic leukemia. *Leukemia, 14*, 567–572.

23. Lennard, L., Van Loon, J.A., Lilleyman, J.S., Weinshilboum, R.M. (1987). Thiopurine pharmacogenetics in leukemia: correlation of erythrocyte thiopurine methyltransferase activity and 6-thioguanine nucleotide concentrations. *Clin Pharmacol Ther, 41*, 18–25.

24. Lennard, L. (1987). Assay of 6-thioinosinic acid and 6-thioguanine nucleotides, active metabolites of 6-mercaptopurine, in human red blood cells. *J Chromatogr, 423*, 169–178.

25. Lennard, L., Rees, C.A., Lilleyman, J.S., Maddocks, J.L. (1983). Childhood leukaemia: a relationship between intracellular 6-mercaptopurine metabolites and neutropenia. *Br J Clin Pharmacol, 16*, 359–363.

26. Lennard, L., Lilleyman, J.S., Van Loon, J., Weinshilboum, R.M. (1990). Genetic variation in response to 6-mercaptopurine for childhood acute lymphoblastic leukaemia. *Lancet, 336*, 225–229.

27. Lennard, L., Van Loon, J.A., Weinshilboum, R.M. (1989). Pharmacogenetics of acute azathioprine toxicity: relationship to thiopurine methyltransferase genetic polymorphism. *Clin Pharmacol Ther, 46*, 149–154.

28. Evans, W.E., Horner, M., Chu, Y.Q., Kalwinsky, D., Roberts, W.M. (1991). Altered mercaptopurine metabolism, toxic effects, and dosage requirement in a thiopurine methyltransferase-deficient child with acute lymphocytic leukemia. *J Pediatr, 119*, 985–989.

29. Schutz, E., Gummert, J., Mohr, F., Oellerich, M. (1993). Azathioprine-induced myelosuppression in thiopurine methyltransferase deficient heart transplant recipient. *Lancet, 341*, 436.

30. Peng, Y., Feng, Q., Wilk, D., Adjei, A.A., Salavaggione, O.E., Weinshilboum, R.M., Yee, V.C. (2008). Structural basis of substrate recognition in thiopurine *S*-methyltransferase. *Biochemistry, 47*, 6216–6225.

31. Krynetski, E.Y., Schuetz, J.D., Galpin, A.J., Pui, C.H., Relling, M.V., Evans, W.E. (1995). A single point mutation leading to loss of catalytic activity in human thiopurine *S*-methyltransferase. *Proc Natl Acad Sci U S A, 92*, 949–953.

32. Weinshilboum, R., Wang, L. (2004). Pharmacogenetics: inherited variation in amino acid sequence and altered protein quantity. *Clin Pharmacol Ther, 75*, 253–258.

33. Hildebrandt, M.A., Carrington, D.P., Thomae, B.A., Eckloff, B.W., Schaid, D.J., Yee, V.C., Weinshilboum, R.M., Wieben, E.D. (2007). Genetic diversity and function in the human cytosolic sulfotransferases. *Pharmacogenomics J, 7*, 133–143.

34. Shield, A.J., Thomae, B.A., Eckloff, B.W., Wieben, E.D., Weinshilboum, R.M. (2004). Human catechol O-methyltransferase genetic variation: gene resequencing and functional characterization of variant allozymes. *Mol Psychiatry, 9*, 151–160.

35. Preuss, C.V., Wood, T.C., Szumlanski, C.L., Raftogianis, R.B., Otterness, D.M., Girard, B., Scott, M.C., Weinshilboum, R.M. (1998). Human histamine N-methyltransferase pharmacogenetics: common genetic polymorphisms that alter activity. *Mol Pharmacol, 53*, 708–717.

36. Ji, Y., Salavaggione, O.E., Wang, L., Adjei, A.A., Eckloff, B., Wieben, E.D., Weinshilboum, R.M. (2005). Human phenylethanolamine N-methyltransferase

pharmacogenomics: gene resequencing and functional genomics. *J Neurochem, 95*, 1766–1776.

37. Moyer, A.M., Salavaggione, O.E., Wu, T.Y., Moon, I., Eckloff, B.W., Hildebrandt, M.A., Schaid, D.J., Wieben, E.D., Weinshilboum, R.M. (2008). Glutathione *S*-transferase p1: gene sequence variation and functional genomic studies. *Cancer Res, 68*, 4791–4801.
38. Mukherjee, B., Salavaggione, O.E., Pelleymounter, L.L., Moon, I., Eckloff, B.W., Schaid, D.J., Wieben, E.D., Weinshilboum, R.M. (2006). Glutathione *S*-transferase omega 1 and omega 2 pharmacogenomics. *Drug Metab Dispos, 34*, 1237–1246.
39. Ma, C.X., Adjei, A.A., Salavaggione, O.E., Coronel, J., Pelleymounter, L., Wang, L., Eckloff, B.W., Schaid, D., Wieben, E.D., Adjei, A.A., Weinshilboum, R.M. (2005). Human aromatase: gene resequencing and functional genomics. *Cancer Res, 65*, 11071–11082.
40. Hildebrandt, M.A., Salavaggione, O.E., Martin, Y.N., Flynn, H.C., Jalal, S., Wieben, E.D., Weinshilboum, R.M. (2004). Human SULT1A3 pharmacogenetics: gene duplication and functional genomic studies. *Biochem Biophys Res Commun, 321*, 870–878.
41. Mayer, A., Siegel, N.R., Schwartz, A.L., Ciechanover, A. (1989). Degradation of proteins with acetylated amino termini by the ubiquitin system. *Science, 244*, 1480–1483.
42. Adjei, A.A., Thomae, B.A., Prondzinski, J.L., Eckloff, B.W., Wieben, E.D., Weinshilboum, R.M. (2003). Human estrogen sulfotransferase (SULT1E1) pharmacogenomics: gene resequencing and functional genomics. *Br J Pharmacol, 139*, 1373–1382.
43. Thomae, B.A., Rifki, O.F., Theobald, M.A., Eckloff, B.W., Wieben, E.D., Weinshilboum, R.M. (2003). Human catecholamine sulfotransferase (SULT1A3) pharmacogenetics: functional genetic polymorphism. *J Neurochem, 87*, 809–819.
44. Hohfeld, J., Cyr, D.M., Patterson, C. (2001). From the cradle to the grave: molecular chaperones that may choose between folding and degradation. *EMBO Rep, 2*, 885–890.
45. Neckers, L. (2002). Hsp90 inhibitors as novel cancer chemotherapeutic agents. *Trends Mol Med, 8*, S55–61.
46. Wickner, S., Maurizi, M.R., Gottesman, S. (1999). Posttranslational quality control: folding, refolding, and degrading proteins. *Science, 286*, 1888–1893.
47. Wang, L., Nguyen, T.V., McLaughlin, R.W., Sikkink, L.A., Ramirez-Alvarado, M., Weinshilboum, R.M. (2005). Human thiopurine S-methyltransferase pharmacogenetics: variant allozyme misfolding and aggresome formation. *Proc Natl Acad Sci U S A, 102*, 9394–9399.
48. Johnston, J.A., Ward, C.L., Kopito, R.R. (1998). Aggresomes: a cellular response to misfolded proteins. *J Cell Biol, 143*, 1883–1898.
49. Johnston, J.A., Illing, M.E., Kopito, R.R. (2002). Cytoplasmic dynein/dynactin mediates the assembly of aggresomes. *Cell Motil Cytoskeleton, 53*, 26–38.
50. Kawaguchi, Y., Kovacs, J.J., McLaurin, A., Vance, J.M., Ito, A., Yao, T.P. (2003). The deacetylase HDAC6 regulates aggresome formation and cell viability in response to misfolded protein stress. *Cell, 115*, 727–738.

51. Li, F., Wang, L., Burgess, R., Weinshilboum, R. (2008) Human thiopurine S-methyltransferase (TPMT) pharmacogenomics: the role of proteasome-mediated proteolysis and autophagy in TPMT degradation, in: *The Fourth International Conference on Ubiquitin,Ubiquitin-like Proteins and Cancer*, poster 36, Houston, TXs.
52. Yorimitsu, T., Klionsky, D.J. (2005). Autophagy: molecular machinery for self-eating. *Cell Death Differ, 12* (*Suppl 2*), 1542–1552.
53. Ohsumi, Y. (2001). Molecular dissection of autophagy: two ubiquitin-like systems. *Nat Rev Mol Cell Biol, 2*, 211–216.
54. Rubinsztein, D.C. (2006). The roles of intracellular protein-degradation pathways in neurodegeneration. *Nature, 443*, 780–786.
55. Pandey, U.B., Batlevi, Y., Baehrecke, E.H., Taylor, J.P. (2007). HDAC6 at the intersection of autophagy, the ubiquitin–proteasome system and neurodegeneration. *Autophagy, 3*, 643–645.
56. Pandey, U.B., Nie, Z., Batlevi, Y., McCray, B.A., Ritson, G.P., Nedelsky, N.B., Schwartz, S.L., DiProspero, N.A., Knight, M.A., Schuldiner, O., et al. (2007). HDAC6 rescues neurodegeneration and provides an essential link between autophagy and the UPS. *Nature, 447*, 859–863.
57. Hara, T., Nakamura, K., Matsui, M., Yamamoto, A., Nakahara, Y., Suzuki-Migishima, R., Yokoyama, M., Mishima, K., Saito, I., Okano, H., Mizushima, N. (2006). Suppression of basal autophagy in neural cells causes neurodegenerative disease in mice. *Nature, 441*, 885–889.
58. Komatsu, M., Waguri, S., Chiba, T., Murata, S., Iwata, J., Tanida, I., Ueno, T., Koike, M., Uchiyama, Y., Kominami, E., Tanaka, K. (2006). Loss of autophagy in the central nervous system causes neurodegeneration in mice. *Nature, 441*, 880–884.
59. Iwata, A., Riley, B.E., Johnston, J.A., Kopito, R.R. (2005). HDAC6 and microtubules are required for autophagic degradation of aggregated huntingtin. *J Biol Chem, 280*, 40282–40292.
60. Ravikumar, B., Acevedo-Arozena, A., Imarisio, S., Berger, Z., Vacher, C., O'Kane, C.J., Brown, S.D., Rubinsztein, D.C. (2005). Dynein mutations impair autophagic clearance of aggregate-prone proteins. *Nat Genet, 37*, 771–776.
61. Webb, J.L., Ravikumar, B., Rubinsztein, D.C. (2004). Microtubule disruption inhibits autophagosome-lysosome fusion: implications for studying the roles of aggresomes in polyglutamine diseases. *Int J Biochem Cell Biol, 36*, 2541–2550.
62. Taylor, J.P., Tanaka, F., Robitschek, J., Sandoval, C.M., Taye, A., Markovic-Plese, S., Fischbeck, K.H. (2003). Aggresomes protect cells by enhancing the degradation of toxic polyglutamine-containing protein. *Hum Mol Genet, 12*, 749–757.

21

GAUCHER DISEASE

TIM EDMUNDS

Therapeutic Protein Research, Genzyme Corporation, Framingham, Massachusetts

INTRODUCTION

Gaucher disease belongs to the family of diseases known as lysosomal storage diseases (LSDs), which are a group of about 50 rare genetic disorders in the category of inborn errors of metabolism. These diseases arise from the accumulation of metabolites within the lysosome of various tissues and organs as a result of deficiencies in the activity of lysosomal hydrolases or the transport of metabolites [1–4]. Gaucher disease belongs to the subclass of neutral glycosphingolipidoses, in which the enzymatic defect results in the lysosomal accumulation of glycosphingolipids (Fig. 1). This subclass also contains two other common LSDs, Fabry and Tay–Sachs diseases. Although the enzymatic defect for all three diseases lie in the same pathway and the accumulated substrates are closely related, they differ in their pathology, clinical presentation, and age of onset of clinical symptoms.

Although the most common of the LSDs, Gaucher disease is still extremely rare, with about 6000 cases identified worldwide. Like most LSDs, Gaucher disease is an autosomal recessive genetic disorder, and although it has a panethnic distribution, it is more prevalent in the Ashkenazi Jewish population [5]. Gaucher disease is caused by a deficiency of the lysosomal enzyme acid β-glucocerebrosidase (GCase), which catabolizes the degradation of the glycolipid glycosylceramide into glucose and ceramide [6]. Deficiency of GCase leads to lysosomal accumulation of glucosylceramide in the cells of the reticuloendothelial system. Since glucosylceramide is a major component of blood cell

Protein Misfolding Diseases: Current and Emerging Principles and Therapies,
Edited by Marina Ramirez-Alvarado, Jeffery W. Kelly, and Christopher M. Dobson
Copyright © 2010 John Wiley & Sons, Inc.

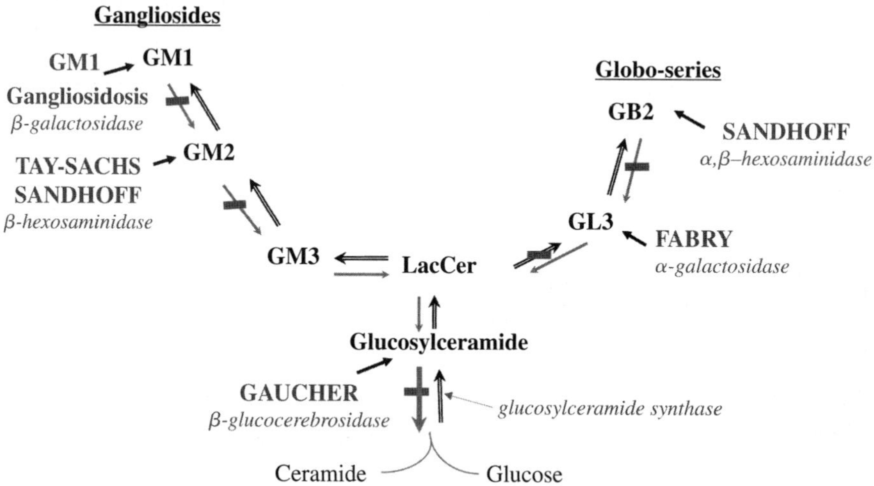

FIG. 1 Enzymes and diseases of the glycosphingolipid metabolic pathway.

membranes, accumulation is observed primarily in the macrophages of the liver and spleen, which are responsible for the breakdown of these cells [4]. The main clinical features of Gaucher disease are hepatosplenomegaly and hematological and skeletal abnormalities [7,8]. Although the increased spleen and liver size are a result of the accumulation of glucosylceramide within enlarged macrophages known as Gaucher cells, the increase in organ size cannot be accounted for by the amount of substrate that accumulates, indicating that secondary pathological events such as inflammation are involved in determining disease severity [9].

Historically, Gaucher disease has been subdivided into three clinical variants based on the age of onset and the presence or absence of neurological involvement (Table 1). In recognition of the fact that the three forms represent a continuum of severity, the disease is now classified into neuronopathic and non-neuronopathic forms. Although a handful of common mutations (Fig. 2) give rise to the majority of cases, over 200 different mutations have been identified in the GCase gene [10]. As with many LSDs, there is poor correlation between genotype and phenotype, with significant heterogeneity being observed between patients with the same phenotype. An extreme example of this heterogeneity is seen in sets of monozygotic twins where one sibling is mildly affected while the other has more severe disease [11].

GLUCOCEREBROSIDASE

Glucocerebrosidase is a membrane-associated monomeric glycoprotein; the mature protein is 497 amino acids in length and contains five potential N-linked glycosylation sites, of which four (Asn19, Asn59 Asn146, and Asn270) are

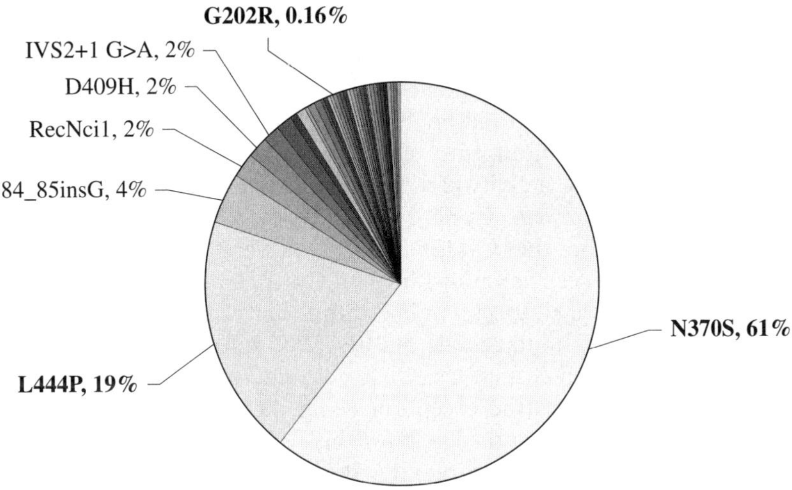

FIG. 2 Allele frequency of Gaucher mutations. The six most common mutations account for 90% of all mutations. Frequency data were calculated from 6731 alleles listed in the ICGG Gaucher Registry as of September 2007.

TABLE 1 Gaucher Disease Variants

	Non-neuronopathic	Neuronopathic	
Clinical Features	Type 1	Type 2	Type 3
Age at onset	Childhood/adulthood	Infancy	Childhood
Life span	Infancy to 80+ years	Infancy	Childhood–middle age
Hepatosplenomegaly	Yes	Yes	Yes
Skeletal disease	Yes	No	Yes
Primary central nervous in system disease	No	Yes	Yes
Predominant ethnic group	Ashkenazi Jewish	Panethnic	Norrbottnian (northern Sweden)

occupied [12–18]. The enzyme isolated from human placenta contains one oligomannose (Asn19) and three complex oligosaccharide chains. Although the presence of complex oligosaccharides are required for efficient synthesis and transport of glucocerebrosidase to the lysosome, they most likely play a role in the stability of the protein rather than in transport to the lysosome [13]. Unlike other lysosomal enzymes, which are targeted to the lysosome via the binding of mannose-6-phosphate residues to the mannose-6-phosphate receptor, GCase does not contain any mannose-6-phosphate [12,13]. Transport to the lysosome is accomplished by GCase binding to the lysosomal integral protein LIMP-2

[19]. GCase contains seven cysteine residues, the first four of which form two intrachain disulfide bonds between Cys(4–16) and Cys(18–23) while Cys126, Cys248, and Cys342 are present as free thiols [20]. Early biochemical studies demonstrated that GCase could be inactivated by thiol reagents, and activity was stabilized by reducing agents, suggesting that at least one free cysteine residue was required for activity [21]. Site-directed mutagenesis has shown that the disulfide bonds between Cys(4–16) and Cys(18–23) as well Cys342 were essential for activity but that Cys126 and Cys248 were not. These cysteines are highly conserved between species with Cys4, Cys16, Cys18, Cys23, and Cys342 being conserved across all species in line with the critical requirement for these residues identified by mutagenesis studies [22]. Interestingly, although the mutagenesis data indicated that Cys126 is not required for activity, it is highly conserved across species (the exception being the honeybee), suggesting an important but as yet unidentified role for this residue.

In vivo the activity of GCase requires the presence of the activator protein saposin C and negatively charged phospholipids [23–25]. In in vitro assays, detergents and bile salts can activate the protein in the absence of saposin C [26]. It has been proposed that the three-way interaction of saposin C, GCase, and the phospholipid membrane of the lysosome induces a conformational change in GCase, resulting in increased catalytic activity. The biological importance of this interaction is demonstrated by the fact that a deficiency of saposin C also gives rise to a very rare form of Gaucher disease [27].

The crystal structure of GCase was first solved by Dvir et al. [28], and several subsequent structures have been solved under a variety of conditions and in the presence of various inhibitors [20,22,29,31]. The basic structure consists of three domains. Domain I contains the disulfide bridges and the Asn 19 glycosylation site, domain II consists of two closely associated β-sheets which resemble an immunoglobulin fold. The third domain contains the active site residues as well as the three free cysteine residues and is a $(\beta/\alpha)_8$ TIM barrel which is consistent with the structure of other glucoside hydrolase clan members. Unfortunately, because the structures have all been solved in the absence of the activator protein saposin C and phospholipids, they do not offer any insights into the regions of GCase, which bind to these compounds.

PROTEIN FOLDING AND GAUCHER DISEASE

Early biochemical studies demonstrated that there is a poor correlation between disease severity and residual enzyme activity when measured with artificial substrates. Although the lack of correlation may be partially explained by the difficulty in replicating the in vivo environment with in vitro assay conditions, there were also differences observed in the processing of the enzyme, suggesting that the cellular distribution of protein in the more severe forms of the disease was abnormal, as was processing within the endoplasmic recticulum (ER). Ginns et al. [32] compared the protein isolated from normal fibroblasts with that

from patients with non-neuronopathic and neuronopathic forms of Gaucher disease and demonstrated a difference in the molecular species identified by SDS-PAGE analysis. Normal fibroblasts contain three species of GCase with molecular masses of 63, 61, and 56 kDa. Fibroblasts from type I patients contained only the 56-kDa form (as did normal brain tissue). Fibroblasts from type II and type III Gaucher patients contained only the high-molecular-mass species with masses of 61 and 63 kDa, suggesting that the GCase in neuronopathic patients is processed incorrectly. A pulse-chase experiment in human fibroblasts by Erickson et al. [15] showed that GCase is synthesized as a 60-kDa protein containing four oligomannose carbohydrate chains and converted within 2 hours to an intermediate 59-kDa protein with complex oligosaccharide chains. A fully processed 55-kDa protein was present after 72 hours, following further processing of the oligosaccharide chains within the lysosome (the differences in apparent molecular mass in the two reports can be explained by the different electrophoretic methods used). Additional studies by Jonsson et al. [33] also demonstrated a difference in processing between the neuronopathic and non-neuronopathic forms of Gaucher disease. In this case processing of GCase from type I fibroblasts was the same as that from normal fibroblasts, but the 59-kDa fully processed form was not observed in the fibroblasts from type II and type III patients.

These studies suggested that for neuronopathic forms of the disease, incorrect folding or processing lead to retention of a precursor form of the protein, whereas the processing of protein from type I patients was normal. Because many of these studies were conducted prior to the publication of cDNA sequence in 1985, the genotype of the patient samples used is not known.

Following the publication of the GCase cDNA sequence, over 200 disease-causing mutations have been identified in the GCase gene [10]. Most are extremely rare, with a few common mutations giving rise to the majority of Gaucher disease cases. The most common mutation is the substitution of serine for asparagine at amino acid residue 370 (N370S). The N370S mutation is predictive of the non-neuronopathic or type I disease and is the most prevalent mutation in the Ashkenazi Jewish population [5]. Patients homozygous for the N370S mutation generally have an attenuated phenotype and can present with a wide spectrum of clinical symptoms from mild to severe. In patients heterozygous for the N370S mutation, the presence of the N370S mutation on one allele is protective from the severe neuronopathic forms of Gaucher disease even when the second allele is a null allele. The second most common mutation is the substitution of proline for a leucine at amino acid residue 444 (L444P). This mutation is prevalent in the Norrbottnian (Swedish) population, and homozygosity for this mutation results in type III neuronopathic disease.

Disease-causing mutations occur throughout both the linear protein sequence and the three-dimensional structure of the protein, which, in conjunction with the high degree of homology between species, indicates that all regions of the molecule are critical for activity and/or stability. Although the crystal structure of GCase has been solved, it does not explain the mechanism by which most mutations cause disease. Biochemical data on the effect of the various mutations is also somewhat limited, focusing mainly on the major disease-causing mutations N370S and L444P. The most extensive biochemical study of mutations was carried out by Liou et al. [22], who expressed 52 mutations in an insect cell expression system. The mutations were characterized for enzyme kinetics, stability, and activator response, but this extensive study, even with the availability of crystal structure, was unable to establish a solid relationship between structure and function.

N370S MUTATION

Despite the fact that the mutant N370S GCase protein has been extensively studied from patient samples and cells as well as recombinant forms of the protein, how the mutation gives rise to Gaucher disease is not known. Although studies have implicated reduced activity [22,34,35], impaired interaction and/or activation by saposin C [36], reduced stability, and abnormal processing [37] as the defect, the majority of the evidence points to N370S being a catalytic mutation with the protein being normally processed and having normal or nearly normal stability under physiological conditions.

Several studies have investigated whether the N370S mutation leads to impaired folding or trafficking of the enzyme, but the data are inconclusive. The hypothesis that the N370S mutation results in an incorrectly folded molecule which is retained in the endoplasmic reticulum (ER) and degraded by the proteosome was first proposed by Sawkar et al. in 2002 [38]. This was based on a study in which the competitive inhibitor *N*-(*n*-nonyl)deoxynojirimycin (NN-DNJ) led to a 2-fold increase in intracellular activity of GCase when incubated with N370S fibroblasts. In addition, it was shown the NN-DNJ also protected placental GCase (Ceredase) from thermal inactivation at 48°C. In subsequent studies [39] the group demonstrated that recombinant N370S GCase expressed in insect cells has an altered thermal denaturation curve as measured by circular dichroism spectroscopy. At pH 7.0 the thermal denaturation midpoint (T_m) of N370S GCase was 4.2°C lower than for native GCase but less than 1°C lower at pH 5.3. However, the fraction of unfolded protein at 37°C was the same for the mutant and the native protein. Addition of competitive inhibitors increased the T_m value of both the N370S mutant and wild-type protein at pH 7.0 and 5.3. Decreased thermal stability and stabilization by competitive inhibitors was also observed in cell lysates from N3070S patient fibroblasts. In line with previous studies, the intracellular localization of the N370S mutant protein was similar to that of wild type, although levels were reduced. When the fibroblasts of N370S patients were grown at a reduced temperature (30°C), no effect on enzyme levels was observed in contrast to other GCase mutants, which show improved trafficking under similar conditions. Although a study of Gaucher mutations in patient fibroblasts by Ron and Horowitz [37] showed slight accumulation of incompletely processed GCase in the ER of one of three patients homozygous for the N370S mutation, there is little other evidence to indicate that the N370S mutation affects cellular distribution. In the same study, fibroblast samples from the two other patients homozygous for the N370S mutation had protein levels 70.6 and 47.3% of normal and the non-ER fractions were 84.5 and 79.8% compared to 89.8% for the normal control. The only other data supporting a possible folding or trafficking defect is a study by Schmitz et al. [40] that demonstrated delayed maturation but normal lysosomal distribution for the N370S protein.

Current evidence suggests that the N370S mutation produces a stable enzyme with reduced catalytic activity. Biochemical studies and analysis of patient fibroblasts have shown that the N370S protein has a catalytic activity between 10 and 15% of normal [22,34,35,41]. The crystal structure of GCase shows this mutation to occur at the interface between domains II and III (Fig. 3) and not to be involved directly in the catalytic mechanism of the enzyme. In addition to reduced specific activity, the biochemical studies of Liou and Ohashi [22,42] have shown that this mutant has normal stability, an increased IC_{50} for various inhibitors, normal activation by saposin C, and increased activation by phosphatidylserine. A study by Salvioli et al. [36] reported a diminished capacity of saposin C to activate the N370S mutant enzyme at low concentrations of anionic phospholipids. At high concentrations, activation was comparable to that of the normal enzyme, which may explain the apparent contradictory

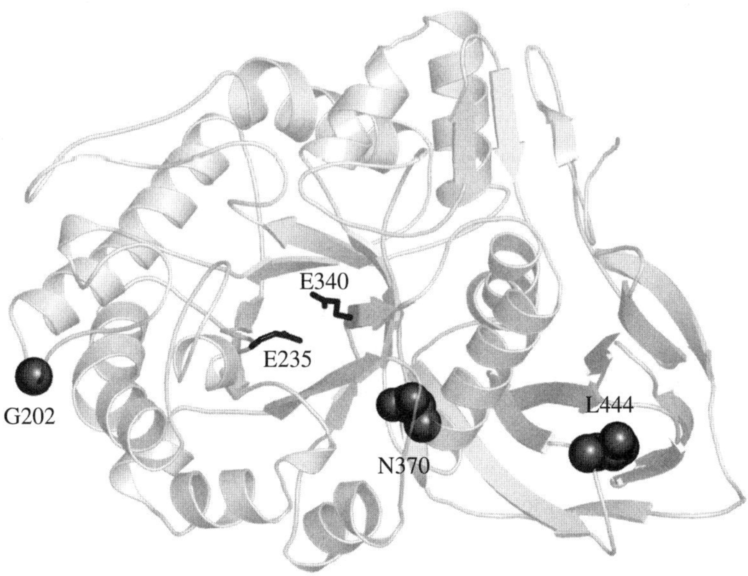

FIG. 3 Crystal structure of human glucocerebrosidase (PDB 1y7v Reference 31) with the active-site glutamic acids represented by sticks and the sites of the N360S, L444P, and G202R mutations shown as balls.

conclusions between these studies regarding the effect of the N370S mutation on saposin C activation. One interesting observation in the study by Salvioli et al. was that although the N370S GCase was present at almost normal levels and was found fully processed in the endosomal–lysosomal compartment, there was poor co-localization with saposin C compared to normal cell lines.

While the exact nature of the N370S mutation is not known, evidence supports a reduced catalytic activity, normal or near-normal lysosomal distribution, and normal stability at lysosomal pH. Whether the reduced stability at pH 7.0 is physiologically relevant is not clear, but unlike other GCase mutations, there is no direct evidence to suggest that there is a significant defect in trafficking or increased proteolysis in the endoplasmic reticulum.

L444P AND G202R MUTATIONS

The L444P and G202R mutations are both severe mutations and are associated with neuronopathic forms of Gaucher in patients homozygous for either mutation. L444P, which gives rise to type III Gaucher disease in homozygous patients, is the second most prevalent mutation after the N370S mutation and most prevalent among non-Jewish patients. Although much less common, the G202R mutation gives rise to the most severe type II form of Gaucher disease in homozygous patients.

Early pulse-chase studies of patient fibroblasts [43,44] indicated that the L444P mutation leads to an unstable protein with incompletely processed carbohydrate chains. This observation was later confirmed by recombinant expression in 3T3 and Sf9 cells [35,42,45]. The lack of stability and difficulty in expressing the protein has meant that this mutation is not as well characterized as the N370S mutation, despite its prevalence and clinical importance. The limited expression data available indicate that although unstable, the recombinant enzyme does possess a low level of residual activity and normal sensitivity to inhibitors. Immunohistochemical studies on patient fibroblasts show that cellular levels of the L444P protein are reduced significantly, precluding identification of the intracellular location of the protein, although SDS-PAGE analysis demonstrated that the protein contained incompletely processed oligosaccharide chains, indicating retention in the ER. This observation is supported further by the fact that intracellular levels can be increased by culturing patient fibroblasts at 30°C [39]. Based on the crystal structure (Fig. 3), the mutation lies in the hydrophobic core of the Ig-like domain, so the substitution of a proline for a leucine could have a significant effect on the folding of this domain.

The much rarer G202R mutation results in a protein with minimal activity (1 to 2%) and gives rise to the most severe type II Gaucher disease. This mutant GCase has very similar biochemical properties to those of the L444P variant, resulting in incompletely processed oligosaccharides and retention of the protein in the ER, which can be partially rescued by growing patients' fibroblasts at 30°C.

Unlike the N370S mutation, which is a catalytic mutation, the L444P and the G202R mutations possess many of the characteristics of a protein folding or trafficking defect. They are retained in the ER as an oligomannose containing glycoproteins, do not co-localize with lysosomal markers, and acquire a more normal cellular distribution when cells are grown at 30°C [39,46].

CURRENT THERAPIES

The current standard of care for Gaucher disease is enzyme replacement with Cerezyme (imiglucerase), a carbohydrate-modified form of recombinant human glucocerebrosidase, which is administered by intravenous infusion every 2 weeks [47]. Although the concept of enzyme replacement therapy for lysosomal storage diseases was first proposed by DeDuve in 1964 [48], it took several decades before a successful therapy was available [49]. Development was initially hampered by lack of a suitable source of enzyme, difficulties in purification due to the fact that GCase is a membrane-associated protein, and poor uptake into target cells. The supply issue was solved initially by utilizing human placentas as a source of enzyme, while the use of organic solvents and hydrophobic interaction chromatography lead to the development of an efficient purification scheme [24,50,51]. Targeting to cells was achieved by remodeling the oligosaccharide chains to expose terminal mannose residues [52]. Although placental GCase contains an oligomannose oligosaccharide at asparagine 19, the complex oligosaccharides at the other three glycosylation sites required modification to facilitate efficient uptake via the macrophage mannose receptor. These developments lead to the development of a commercial scale manufacturing process and the approval of Ceredase (alglucerase) to treat Gaucher disease in 1991. Although Ceredase was a safe and effective therapy, the number of patients that could be treated was limited by the availability of human placentas, as it required 20,000 placentas to treat a single patient for a year. In addition to supply limitations, growing concerns over the safety of a human-derived product lead to the development and approval of recombinant human glucocerebrosidase (Cerezyme) produced in a Chinese hamster ovary cell line [53]. Since approval over a decade ago, Cerezyme has been used successfully to treat more than 4500 Gaucher patients and has been shown to be extremely safe and efficacious [54].

Despite being very effective at treating type I Gaucher disease, Cerezyme does not cross the blood brain barrier and so is unable to treat the neurological complications of type II and type III Gaucher disease, although it is used to treat the visceral symptoms in these patients [55]. An alternative therapeutic approach which has the potential of addressing the neurological disease is substrate reduction therapy. The concept of substrate reduction therapy was first proposed by Radin [56], based on the fact that substrate load is a balance between synthesis and degradation. Although Cerezyme increases degradation by supplementing residual enzyme activity, a beneficial effect could also be achieved by reducing substrate synthesis through inhibition of glucosylceramide synthase (Fig. 1). This hypothesis has led to the development of *N*-butyl-1-deoxynojirimycin (Zavesca) as a second therapeutic approach to treating Gaucher disease [57]. However, lack of specificity leading to significant side effects resulted in very limited use and approval only for type I patients with mild to moderate symptoms, for whom Cerezyme infusions is not an option.

FUTURE THERAPIES

Given the success of Cerezyme, it is not surprising that there are other enzyme replacement therapies in clinical testing. One is a human GCase produced by gene activation technology in a human fibrosarcoma cell line [58]. The addition of a mannosidase inhibitor results in oligomannose oligosaccharide chains at all four sites, avoiding the need to remodel the enzyme. A second GCase variant is produced using carrot cell culture. In this case, to produce a protein with terminal mannose residues it is necessary to add cellular targeting sequences to the amino and carboxy terminals of the protein. Because these are not removed by the cell, this results in a protein with two extra amino acid residues at the amino terminus and seven at the carboxy terminus of the protein [59]. It is not known if these extra amino acids or the xylose residues that are present on the carbohydrate chains will result in increased immunogenicity relative to Cerezyme. Although different from Cerezyme, there is no biochemical or clinical data to suggest that these molecules will have any therapeutic benefits over Cerezyme, and long-term safety of these molecules still needs to be determined.

A second substrate reduction therapy is currently in clinical testing. This molecule is a ceramide analog, unlike Zavesca, which is a glucose analog, and early preclinical and clinical studies suggest that it is more potent and specific, resulting in fewer side effects [60]. If these initial results hold true through pivotal clinical testing, this would provide an alternative or complementary therapy to Cerezyme.

A third therapeutic approach, proposed by Sawkar and colleagues [38], is currently in clinical testing. This approach is based on the hypothesis that certain Gaucher mutations result in a misfolded protein that retains catalytic activity but deficient trafficking to the lysosome, and subsequent degradation in the endoplasmic reticulum leads to reduced activity. Competitive inhibitors bind to the active site and stabilize the misfolded protein in the ER sufficiently to allow transport to lysosome before degradation by the proteosome can occur. Initial studies demonstrated that incubation of fibroblasts with the immunosugar *N*-(nonyl)deoxynojirimycin (NN-DNJ) led to an increase in GCase activity. This stabilization effect was shown to occur in normal, N370S, and G202R mutant fibroblasts but not in fibroblasts from patients homozygous for the L444P mutation [38,39].

The compound currently in clinical testing is isofagomine, first identified and clinically tested as a glycocogen phosphorylase inhibitor [61,62], but which is also a potent inhibitor of GCase [63,64]. The magnitude of the effect observed in patient fibroblasts is similar to that found with NN-DNJ, but because isofagomine also inhibits several other glucosyl hydrolases, it remains to be seen whether long-term use will lead to significant side effects. An additional challenge to this approach is the fact that isofagomine is such a potent inhibitor of GCase, with a K_i value of 30 nM, that incubation of patient cells with isofagomine results in complete inhibition of intracellular activity. This then requires a washout of the inhibitor to restore activity [65], an approach which if necessary in patients could lead to a complicated dosing regiment.

Although an increase in cellular activity has been demonstrated by several groups using a variety of competitive inhibitors [66–69], the exact mechanism by which these compounds exert their effect is not firmly established. A similar increase in activity can be achieved following treatment of macrophages with NN-DNJ and uptake of Cerezyme through the mannose receptor [70]. In this case the effect is through reduction of lysosomal degradation since Cerezyme is correctly folded and delivered directly to the lysosome. That the effect may be due to decreased lysosomal degradation is supported by studies conducted with the lysosomal protease inhibitor leupeptin [33]. In these studies GCase levels increased 1.4-fold in normal fibroblasts and 2.2-fold in fibroblasts from type 1 patients following 8 days of leupeptin treatment. These results are comparable to the effect of treatment with isofagomine, which resulted in increases of 1.3-fold for normal fibroblasts and between 2.2-and 3.0-fold for N370S mutant fibroblasts, depending on the concentration and length of treatment [65].

Although chaperone therapy is a very intriguing approach that has generated a lot of interest in the last five years, it remains to be seen which (if any) patients may benefit, as there are over 200 different mutations and even more heterozygous combinations. Also, given the uncertainties around the mechanism of action, it remains to be seen whether it is possible to develop a dosing regiment that avoids intracellular inhibition of GCase and other key metabolic enzymes.

CONCLUSIONS

Like the other LSDs, Gaucher disease arises from several hundred different mutations. Some of these mutations undoubtedly result in protein folding defects with retention and/or degradation of the misfolded protein in the ER. The challenge in treating these diseases as protein folding diseases is identifying which mutations fall into this category. The most common Gaucher mutation N370S has the characteristics of a catalytic and not a folding mutation, so enzyme replacement remains the best therapeutic option currently available. The L444P and G202R mutations have the characteristics of a folding mutation, and because they give rise to neuronopathic forms of Gaucher disease, a small molecule therapy that crosses the blood–brain barrier would be ideal. Unfortunately, the reason that these mutations give rise to neuropathic disease is that the overall enzyme activity is extremely low. Thus, even if compounds can be identified that promote transport out of the ER and do not inhibit the enzyme, it is unlikely that the small increase in activity alone will have any therapeutic effect.

Acknowledgments

I would like to thank Rob Pomponio, Ronnie We and Trent Richardson for providing the figures used in this chapter.

REFERENCES

1. Neufeld, E.F. (1991). Lysosomal storage diseases. *Annu Rev Biochem, 60*, 257–280.
2. Desnick, R.J. (2004). Enzyme replacement and enhancement therapies for lysosomal diseases. *J Inherit Metab Dis, 27*, 385–410.
3. Beck, M. (2007). New therapeutic options for lysosomal storage disorders: enzyme replacement, small molecules and gene therapy. *Hum Genet, 121*, 1–22.
4. Brady, R.O. (2006). Enzyme replacement for lysosomal diseases. *Annu Rev Med, 57*, 283–296.
5. Beutler, E. (2001). Grabowski, G.A. Gaucher disease. In *The Metabolic and Molecular Bases of Inherited Disease* (Scriver, C.R., et al., Eds.), McGraw-Hill, New York, pp. 3635–3668.
6. Brady, R.O., Kanfer, J.N., Bradley, R.M., Shapiro, D. (1966). Demonstration of a deficiency of glucocerebroside-cleaving enzyme in Gaucher's disease. *J Clin Invest, 45*, 1112–1115.
7. Grabowski, G.A. (1993). Gaucher disease: enzymology, genetics, and treatment. *Adv Hum Genet, 21*, 377–441.
8. Pastores, G.M. (1997). Gaucher's disease: pathological features. *Baillieres Clin Haematol, 10*, 739–749.
9. Zhao, H., Grabowski, G.A. (2002). Gaucher disease: perspectives on a prototype lysosomal disease. *Cell Mol Life Sci, 59*, 694–707.
10. Beutler, E., Gelbart, T., Scott, C.R. (2005). Hematologically important mutations: Gaucher disease. *Blood Cells Mol Dis, 35*, 355–364.
11. Lachmann, R.H., Grant, I.R., Halsall, D., Cox, T.M. (2004). Twin pairs showing discordance of phenotype in adult Gaucher's disease. *Q J Med, 97*, 199–204.
12. Aerts, J.M., Schram, A.W., Strijland, A., van Weely, S., Jonsson, L.M., Tager, J.M., Sorrell, S.H., Ginns, E.I., Barranger, J.A., Murray G.J. (1988). Glucocerebrosidase, a lysosomal enzyme that does not undergo oligosaccharide phosphorylation. *Biochim Biophys Acta, 964*, 303–308.
13. Aerts, J.M., Brul, S., Donker-Koopman, W.E., van Weely, S., Murray, G.J., Barranger, J.A., Tager, J.M., Schram, A.W. (1986). Efficient routing of glucocerebrosidase to lysosomes requires complex oligosaccharide chain formation. *Biochem Biophys Res Commun, 141*, 452–458.

14. Edmunds, T. (1998). Cerezyme: a case study. *Dev Biol Stand, 96*, 131–140.
15. Erickson, A.H., Ginns, E.I., Barranger, J.A. (1985). Biosynthesis of the lysosomal enzyme glucocerebrosidase. *J Biol Chem, 260*, 14319–14324.
16. Sorge, J., West, C., Westwood, B., Beutler, E. (1985). Molecular cloning and nucleotide sequence of human glucocerebrosidase cDNA. *Proc Natl Acad Sci U S A, 82*, 7289–7293.
17. Takasaki, S., Murray, G.J., Furbish, F.S., Brady, R.O., Barranger, J.A., Kobata A. (1984). Structure of the *N*-asparagine-linked oligosaccharide units of human placental beta-glucocerebrosidase. *J Biol Chem, 259*, 10112–10117.
18. Tsuji, S., Choudary, P.V., Martin, B.M., Winfield, S., Barranger, J.A., Ginns, E.I. (1986). Nucleotide sequence of cDNA containing the complete coding sequence for human lysosomal glucocerebrosidase. *J Biol Chem, 261*, 50–53.
19. Reczek, D., Schwake, M., Schröder, J., Hughes, H., Blanz, J., Jin, X., Brondyk, W., Van Patten, S., Edmunds, T., Saftig, P. (2007). LIMP-2 is a receptor for lysosomal mannose-6-phosphate-independent targeting of beta-glucocerebrosidase. *Cell, 131*, 770–783.
20. Brumshtein, B., Wormald, M.R., Silman, I., Futerman, A.H., Sussman, J.L. (2006). Structural comparison of differently glycosylated forms of acid-beta-glucosidase, the defective enzyme in Gaucher disease. *Acta Crystallogr D, 62*(Pt 12), 1458–1465.
21. Berent, S.L., Radin, N.S. (1981). Mechanism of activation of glucocerebrosidase by co-beta-glucosidase (glucosidase activator protein). *Biochim Biophys Acta, 664*, 572–582.
22. Liou, B., Kazimierczuk, A., Zhang, M., Scott, C.R., Hegde, R.S., Grabowski, G.A. (2006). Analyses of variant acid beta-glucosidases: effects of Gaucher disease mutations. *J Biol Chem, 281*, 4242–4253.
23. Basu, A., Glew, R.H. (1984). Characterization of the phospholipid requirement of a rat liver beta-glucosidase. *Biochem J, 224*, 515–524.
24. Dale, G.L., Villacorte, D.G., Beutler, E. (1976). Solubilization of glucocerebrosidase from human placenta and demonstration of a phospholipid requirement for its catalytic activity. *Biochem Biophys Res Commun, 71*, 1048–1053.
25. Peters, S.P., Coyle, P., Coffee, C.J., Glew, R.H., Ho, M.W., O'Brien, J.S., Radin, N.S., Erickson J.S. (1977). Purification and properties of a heat-stable glucocerebrosidase activating factor from control and Gaucher spleen glucocerebrosidase: reconstitution of activity from macromolecular components Gaucher's disease: deficiency of "acid" -glucosidase and reconstitution of enzyme activity in vitro. *J Biol Chem*, *252*, 563–573.
26. Blonder, E., Klibansky, C., de Vries, A. (1976). Effects of detergents and choline-containing phospholipids on human spleen glucocerebrosidase. *Biochim Biophys Acta, 431*, 45–53.
27. Christomanou, H., Kleinschmidt, T., Braunitzer, G. (1987). N-Terminal amino-acid sequence of a sphingolipid activator protein missing in a new human Gaucher disease variant. *Biol Chem Hoppe Seyler, 368*, 1193–1196.
28. Dvir, H., Harel, M., McCarthy, A.A., Toker, L., Silman, I., Futerman, A.H., Sussman, J.L. (2003). X-ray structure of human acid-beta-glucosidase, the defective enzyme in Gaucher disease. *EMBO Rep, 4*, 704–709.

29. Brumshtein, B., Greenblatt, H.M., Butters, T.D., Shaaltiel, Y., Aviezer, D., Silman, I., Futerman, A.H., Sussman, J.L. (2007). Crystal structures of complexes of *N*-butyl- and *N*-nonyl-deoxynojirimycin bound to acid-beta-glucosidase: insights into the mechanism of chemical chaperone action in Gaucher disease. *J Biol Chem, 282*, 29052–29058.

30. Lieberman, R.L., Wustman, B.A., Huertas, P., Powe, A.C., Jr., Pine, C.W., Khanna, R., Schlossmacher, M.G., Ringe, D., Petsko, G.A. (2007). Structure of acid beta-glucosidase with pharmacological chaperone provides insight into Gaucher disease. *Nat Chem Biol, 3*, 101–107.

31. Premkumar, L., Sawkar, A.R., Boldin-Adamsky, S., Toker, L., Silman, I., Kelly, J.W., Futerman, A.H., Sussman, J.L. (2005). X-ray structure of human acid-beta-glucosidase covalently bound to conduritol-B-epoxide: implications for Gaucher disease. *J Biol Chem, 280*, 23815–23819.

32. Ginns, E.I., Brady, R.O., Pirruccello, S., Moore, C., Sorrell, S., Furbish, F.S., Murray, G.J., Tager, J., Barranger, J.A. (1982). Mutations of glucocerebrosidase: discrimination of neurologic and non-neurologic phenotypes of Gaucher disease. *Proc Natl Acad Sci U S A, 79*, 5607–5610.

33. Jonsson, L.M., Murray, G.J., Sorrell, S.H., Strijland, A., Aerts, J.F., Ginns, E.I., Barranger, J.A., Tager, J.M., Schram, A.W. (1987). Biosynthesis and maturation of glucocerebrosidase in Gaucher fibroblasts. *Eur J Biochem, 164*, 171–179.

34. Grace, M.E., Graves, P.N., Smith, F.I., Grabowski, G.A. (1990). Analyses of catalytic activity and inhibitor binding of human acid beta-glucosidase by site-directed mutagenesis: identification of residues critical to catalysis and evidence for causality of two Ashkenazi Jewish Gaucher disease type 1 mutations. *J Biol Chem, 265*, 6827–6835.

35. Grace, M.E., Newman, K.M., Scheinker, V., Berg-Fussman, A., Grabowski, G.A. (1994). Analysis of human acid beta-glucosidase by site-directed mutagenesis and heterologous expression. *J Biol Chem, 269*, 2283–2291.

36. Salvioli, R., Tatti, M., Scarpa, S., Moavero, S.M., Ciaffoni, F., Felicetti, F., Kaneski, C.R., Brady, R.O., Vaccaro, A.M. (2005). The N370S (Asn370 → Ser) mutation affects the capacity of glucosylceramidase to interact with anionic phospholipid-containing membranes and saposin C. *Biochem J, 390*(Pt 1), 95–103.

37. Ron, I., Horowitz, M. (2005). ER retention and degradation as the molecular basis underlying Gaucher disease heterogeneity. *Hum Mol Genet, 14*, 2387–2398.

38. Sawkar, A.R., Cheng, W.C., Beutler, E., Wong, C.H., Balch, W.E., Kelly, J.W. (2002). Chemical chaperones increase the cellular activity of N370S beta-glucosidase: a therapeutic strategy for Gaucher disease. *Proc Natl Acad Sci U S A, 99*, 15428–15433.

39. Sawkar, A.R., Schmitz, M., Zimmer, K.P., Reczek, D., Edmunds, T., Balch, W.E., Kelly, J.W. (2006). Chemical chaperones and permissive temperatures alter localization of Gaucher disease associated glucocerebrosidase variants. *ACS Chem Biol, 1*, 235–251.

40. Schmitz, M., Alfalah, M., Aerts, J.M., Naim, H.Y., Zimmer, K.P. (2005). Impaired trafficking of mutants of lysosomal glucocerebrosidase in Gaucher's disease. *Int J Biochem Cell Biol, 37*, 2310–2320.

41. van Weely, S., van den Berg, M., Barranger, J.A., Sa Miranda, M.C., Tager, J.M., Aerts, J.M. (1993). Role of pH in determining the cell-type-specific residual activity of glucocerebrosidase in type 1 Gaucher disease. *J Clin Invest, 91*, 1167–1175.

42. Ohashi, T., Hong, C.M., Weiler, S., Tomich, J.M., Aerts, J.M., Tager, J.M., Barranger, J.A. (1991). Characterization of human glucocerebrosidase from different mutant alleles. *J Biol Chem, 266*, 3661–3667.

43. Bergmann, J.E., Grabowski, G.A. (1989). Posttranslational processing of human lysosomal acid beta-glucosidase: a continuum of defects in Gaucher disease type 1 and type 2 fibroblasts. *Am J Hum Genet, 44*, 741–750.

44. Beutler, E., Kuhl, W. (1986). Glucocerebrosidase processing in normal fibroblasts and in fibroblasts from patients with type I, type II, and type III Gaucher disease. *Proc Natl Acad Sci U S A, 83*, 7472–7474.

45. Grace, M.E., Desnick, R.J., Pastores, G.M. (1997). Identification and expression of acid beta-glucosidase mutations causing severe type 1 and neurologic type 2 Gaucher disease in non-Jewish patients. *J Clin Invest, 99*, 2530–2537.

46. Zimmer, K.P., le Coutre, P., Aerts, H.M., Harzer, K., Fukuda, M., O'Brien, J.S., Naim, H.Y. (1999). Intracellular transport of acid beta-glucosidase and lysosome-associated membrane proteins is affected in Gaucher's disease (G202R mutation). *J Pathol, 188*, 407–414.

47. Weinreb, N.J., Charrow, J., Andersson, H.C., Kaplan, P., Kolodny, E.H., Mistry, P., Pastores, G., Rosenbloom, B.E., Scott, C.R., Wappner, R.S., Zimran, A. (2002). Effectiveness of enzyme replacement therapy in 1028 patients with type 1 Gaucher disease after 2 to 5 years of treatment: a report from the Gaucher Registry. *Am J Med, 113*, 112–119.

48. DeDuve, C. (1964). From cytases to lysosomes. *Fed Proc, 23*, 1045–1049.

49. Barton, N.W., Brady, R.O., Dambrosia, J.M., Di Bisceglie, A.M., Doppelt, S.H., Hill, S.C., Mankin, H.J., Murray, G.J., Parker, R.I., Argoff, C.E., et al. (1991). Replacement therapy for inherited enzyme deficiency:-macrophage-targeted glucocerebrosidase for Gaucher's disease. *N Engl J Med, 324*, 1464–1470.

50. Dale, G.L., Beutler, E., Fournier, P., Blanc, P., Liautaud, J. (1980). Large scale purification of glucocerebrosidase from human placentas. *Birth Defects Orig Artic Ser, 16*, 33–41.

51. Pentchev, P.G., Brady, R.O., Hibbert, S.R., Gal, A.E., Shapiro, D. (1973). Isolation and characterization of glucocerebrosidase from human placental tissue. *J Biol Chem, 248*, 5256–5261.

52. Furbish, F.S., Steer, C.J., Krett, N.L., Barranger, J.A. (1981). Uptake and distribution of placental glucocerebrosidase in rat hepatic cells and effects of sequential deglycosylation. *Biochim Biophys Acta, 673*, 425–434.

53. Grabowski, G.A., Barton, N.W., Pastores, G., Dambrosia, J.M., Banerjee, T.K., McKee, M.A., Parker, C., Schiffmann, R., Hill, S.C., Brady, R.O. (1995). Enzyme therapy in type 1 Gaucher disease: comparative efficacy of mannose-terminated glucocerebrosidase from natural and recombinant sources. *Ann Intern Med, 122*, 33–39.

54. Starzyk, K., Richards, S., Yee, J., Smith, S.E., Kingma, W. (2007). The long-term international safety experience of imiglucerase therapy for Gaucher disease. *Mol Genet Metab, 90*, 157–163.

55. Grabowski, G.A., Leslie, N., Wenstrup, R. (1998). Enzyme therapy for Gaucher disease: the first 5 years. *Blood Rev, 12*, 115–133.
56. Radin, N.S. (1996). Treatment of Gaucher disease with an enzyme inhibitor. *Glycoconj J, 13*, 153–157.
57. Cox, T.M., Aerts, J.M., Andria, G., Beck, M., Belmatoug, N., Bembi, B., Chertkoff, R., Vom Dahl, S., Elstein, D., Erikson, A., *et al.* (2003). The role of the iminosugar *N*-butyldeoxynojirimycin (miglustat) in the management of type I (non-neuronopathic) Gaucher disease: a position statement. *J Inherit Metab Dis, 26*, 513–526.
58. Zimran, A., Loveday, K., Fratazzi, C., Elstein, D. (2007). A pharmacokinetic analysis of a novel enzyme replacement therapy with gene-activated human glucocerebrosidase (GA-GCB) in patients with type 1 Gaucher disease. *Blood Cells Mol Dis, 39*, 115–118.
59. Shaaltiel, Y., Bartfeld, D., Hashmueli, S., Baum, G., Brill-Almon, E., Galili, G., Dym, O., Boldin-Adamsky, S.A., Silman, I., Sussman, J.L., *et al.* (2007). Production of glucocerebrosidase with terminal mannose glycans for enzyme replacement therapy of Gaucher's disease using a plant cell system. *Plant Biotechnol J, 5*, 579–590.
60. McEachern, K.A., Fung, J., Komarnitsky, S., Siegel, C.S., Chuang, W.L., Hutto, E., Shayman, J.A., Grabowski, G.A., Aerts, J.M., Cheng, S.H., *et al.* (2007). A specific and potent inhibitor of glucosylceramide synthase for substrate inhibition therapy of Gaucher disease. *Mol Genet Metab, 91*, 259–267.
61. Waagepetersen, H.S., Westergaard, N., Schousboe, A. (2000). The effects of isofagomine, a potent glycogen phosphorylase inhibitor, on glycogen metabolism in cultured mouse cortical astrocytes. *Neurochem Int, 36*, 435–440.
62. Jakobsen, P., Lundbeck, J.M., Kristiansen, M., Breinholt, J., Demuth, H., Pawlas, J., Candela, M.P., Andersen, B., Westergaard, N., Lundgren, K., Asano, N. (2001). Iminosugars: potential inhibitors of liver glycogen phosphorylase. *Bioorg Med Chem, 9*, 733–744.
63. Steet, R., Chung, S., Lee, W.S., Pine, C.W., Do, H., Kornfeld S. (2007). Selective action of the iminosugar isofagomine, a pharmacological chaperone for mutant forms of acid-beta-glucosidase. *Biochem Pharmacol, 73*, 1376–1383.
64. Oikonomakos, N.G., Tiraidis, C., Leonidas, D.D., Zographos, S.E., Kristiansen, M., Jessen, C.U., Norskov-Lauritsen, L., Agius, L. (2006). Iminosugars as potential inhibitors of glycogenolysis: structural insights into the molecular basis of glycogen phosphorylase inhibition. *J Med Chem, 49*, 5687–5701.
65. Steet, R.A., Chung, S., Wustman, B., Powe, A., Do, H., Kornfeld, S.A. (2006). The iminosugar isofagomine increases the activity of N370S mutant acid beta-glucosidase in Gaucher fibroblasts by several mechanisms. *Proc Natl Acad Sci U S A, 103*, 13813–13818.
66. Yu, L., Ikeda, K., Kato, A., Adachi, I., Godin, G., Compain, P., Martin, O., Asano, N. (2006). Alpha-1-*C*-octyl-1-deoxynojirimycin as a pharmacological chaperone for Gaucher disease. *Bioorg Med Chem, 14*, 7736–7744.
67. Yu, Z., Sawkar, A.R., Whalen, L.J., Wong, C.H., Kelly, J.W. (2007). Isofagomine- and 2,5-anhydro-2,5-imino-*D*-glucitol-based glucocerebrosidase pharmacological chaperones for Gaucher disease intervention. *J Med Chem, 50*, 94–100.

68. Lei, K., Ninomiya, H., Suzuki, M., Inoue, T., Sawa, M., Iida, M., Ida, H., Eto, Y., Ogawa, S., Ohno, K., Suzuki, Y. (2007). Enzyme enhancement activity of *N*-octyl-beta-valienamine on beta-glucosidase mutants associated with Gaucher disease. *Biochim Biophys Acta, 1772*, 587–596.

69. Compain, P., Martin, O.R., Boucheron, C., Godin, G., Yu, L., Ikeda, K., Asano, N. (2006). Design and synthesis of highly potent and selective pharmacological chaperones for the treatment of Gaucher's disease. *Chembiochem, 7*, 1356–1359.

70. Edmunds, T., Jaworski, J., Zhou, Q., Park, A., Honey, D., VanPatten, S. Lysosomal stabilization of β-glucocerebrosidase and a-galactosidase-A by competitive inhibitors: implications for chaperone therapy. Presented at the Annual Meeting of the American Society for Human Genetics, Oct. 2006, New Orleans, LA. http://www.ashg.org/genetics/ashg/annmeet/2006.

22

CATARACT AS A PROTEIN-AGGREGATION DISEASE

Yongting Wang and Jonathan A. King

Department of Biology, Massachusetts Institute of Technology, Cambridge, Massachusetts

INTRODUCTION

The most prevalent protein-deposition disease affecting vision in the human population is lens cataract. Human mature-onset cataract affects nearly 50% of the U.S. population over 75 years of age and is the leading cause of blindness in the world [1]. Although cataract is treated efficiently by surgical removal of the lens, this treatment is invasive, expensive, and is generally performed only after the cataract has caused significant loss of sight.

The eye lens functions to focus incident light on the retina and must be transparent in the visible wavelengths. The lens is composed of tightly packed onionlike layers of elongated fiber cells. These differentiate from the surface anterior layer lens epithelia, and add continuously to the outer surface of the lens with age. Within the central region (nucleus) of the lens, fiber cells enter a terminally differentiated phase. In this phase the cells lose their organelles and their ability to synthesize proteins and form elongated saclike compartments. Lens transparency depends not only on the integrity of its complex architecture but also the unique protein composition of lens crystallin proteins packed at very high concentration.

The overall protein concentration within the fibrous lens cells is very high, approaching 70% w/v. The crystallin proteins comprise 90% of the total protein content of the lens [2]. Proteins in the center of the lens are expressed

Protein Misfolding Diseases: Current and Emerging Principles and Therapies,
Edited by Marina Ramirez-Alvarado, Jeffery W. Kelly, and Christopher M. Dobson

Copyright © 2010 John Wiley & Sons, Inc.

primarily in utero and during infancy and must remain soluble and stable, without the possibility of regeneration for the entire human lifetime. The need to maintain transparency is a demanding requirement given the high protein concentration in the lens and the possibility of oxidative and radiative stress [3].

Cataracts are opaque regions of the lens categorized by clinicians based on the morphological properties and locations of the cataract. Based on its location, cataracts are classified into three types: nuclear, cortical, and subcapsular cataract. Further morphological categories include lamellar, coralliform, stellate, Coppock-like, and anterior cataract, to name a few. Morphological changes of lens fiber cells in cataractous lens have been documented [4–6], but detailed conformations of high-molecular-mass aggregates remain unknown. Biochemical differences among these different types of cataracts are not well characterized, except for a few of the rare inherited juvenile-onset cataracts [7–9].

PROPERTIES OF CATARACTS

Investigations of cataractous lenses extracted from the eye reveal the presence of insoluble aggregates of crystallin proteins [10–12]. These insoluble aggregates obstruct the passage of light through the lens, thereby preventing it from reaching the photoreceptors in the retina [13]. The conformation of crystallins in the insoluble inclusions and the detailed molecular mechanisms of mature-onset cataract formation are unknown. Mature-onset cataracts are neither precipitates nor regular protein polymers but resemble more closely the aggregates and polymers of partially folded intermediates associated with inclusion bodies and other forms of protein misfolding [14–16]. Protein inclusions of cataract can be localized in any region of the lens and by unknown mechanisms propagate across or traverse across many fiber cells and their cell membranes.

Analysis of the insoluble crystallin proteins recovered from aged and cataractous lenses reveals an accumulation of covalent modifications. The major modifications include nonnative disulfide bonds, deamidation of glutamine and asparagines, oxidation, and truncations [12,17–20]. Covalent damage to the crystallins as a result of environmental or intracellular insult may destabilize the proteins and result in partial unfolding. Such partially folded molecules may also be susceptible to further oxidative damage or other covalent modification. Such partially folded conformers may be precursors to cataractous aggregates as well as candidates for substrates of the lens chaperone protein complex α-crystallin. Aggregation from a partially folded conformation of a protein is supported by extensive investigations of inclusion body formation and protein-deposition diseases [14,21–23].

Cataract is unusual among protein-deposition diseases in that there is a direct relationship between the pathology — lens opacity and loss of sight — and the accumulation of the aggregated and polymerized crystallins that constitute the cataract. On the other hand, the mechanism of protein aggregation is poorly understood. For diseases such as sickle cell anemia, Parkinson disease, and Alzheimer disease, the structures of the polymerized states are relatively well understood. In addition, the nucleation and growth mechanisms of precursor polymerization in those cases have been well described, if not fully elucidated mechanistically. In contrast, the initial stages of aggregation and the mechanism of growth of cataracts within the lens are very poorly understood, and the gap between in vitro aggregation studies and cataract formation in situ is much greater. Nonetheless, the ability to study the unfolding and aggregation of purified crystallins in the test tube has enhanced the ability to interpret results in organ culture, animal models, and humans.

Studies using gel electrophoresis, high-performance liquid chromatography and mass spectroscopy demonstrate that modified forms of all three ubiquitous crystallins (α, β, γ) may be present in human mature-onset cataracts removed by surgery. As noted above, many covalent modifications are present in the proteins from the cataract extracted [17,20,24–27]. It is possible that some of these modifications occur during the soluble portion of the protein's lifetime, altering its stability and ultimately causing cataract formation. It is not clear whether covalent damage drives the perturbation of the native conformation or whether partially unfolded molecules are particularly susceptible to covalent damage. In either case, destabilized modified conformations of the crystallins may be at least partially responsible for cataract formation in vivo. However, it is also possible that the high-molecular-mass aggregated state traps partially folded species, so that some modifications may occur to chains already polymerized. Unfortunately, time-course data for mammalian cataract formation, which might resolve this question, are not available.

Amyloidlike protein species have been identified in situ in the mammalian lens using Congo Red and thioflavin T binding assays [28]. Additionally, in vitro studies indicate that bovine α-, β-, and γ-crystallins are capable of forming amyloid fibers under denaturing conditions [29]. Human γD (HγD)-crystallin are also found to form amyloid aggregate at low pH [30]. However, unfolding and refolding studies of HγD-crystallin identified an aggregation pathway that leads to formation of ordered fibrils distinct from amyloids [31]. It is not clear if mature-onset cataracts contain fibrillar structures that correspond to any of these described in vitro. Further studies are required to provide descriptions of the morphology of the aggregated proteins in cataract.

ETIOLOGY OF CATARACTS IN HUMANS

Mature-Onset Cataracts

Epidemiological studies have implicated a number of environmental and occupational factors in the genesis of mature-onset cataract. These include exposure to ultraviolet (UV) light from the natural environment [32–35] and occupational exposure to intense heat and infrared sources, as is seen in "glassblower's cataract" and "chain maker's cataract" [35,–37]. Accidents from the use of infrared lasers can also induce cataract [38,39].

UV Radiation. Although the cornea is a significant UV absorber, some fraction of the incident UV radiation of the wavelength absorbed by proteins (approximately 320 to 390 nm) passes through the cornea and reaches the lens [39–41]. There is substantive evidence in animal experiments for UV-induced cataracts [42]. Although there are numerous small-molecular-mass UV filters in the lens, these may not provide full protection from UV radiation [33,43,44]. The evidence for covalent modification in the insoluble fraction from cataractous lens suggests that photooxidative damage may be one of the causative agents [40]. βγ-Crystallins contain highly conserved tryptophan residues and their fluorescence is highly quenched in the native state. The quenching of Trp fluorescence in HγD-crystallin is due to Trp-to-Trp Föster resonance energy transfer and fast electron transfer through backbones [45]. The quenching of Trp fluorescence in these crystallins may be a property of crystallin fold, which has been evolved in part to enable the lens to become an effective UV filter. The efficient quenching provides an in situ mechanism to protect the tryptophans of the crystallins from photochemical degradation.

Infrared Radiation. Occupational accidents indicate that overexposure to infrared radiation (e.g., from accidents in the use of infrared lasers) are also associated with cataracts [37,39]. Occupational stress is clearly the origin of glassblower's cataract, associated with exposure to intense light and heat from glassblowing furnaces [39]. Experimental studies in rabbits demonstrated that lens damage could be due to the heating effect of radiation absorbed by the iris or pigment epithelium [46].

Oxidative Damage. Nonnative disulfide bonds are among the most prevalent covalent modifications of crystallins extracted from cataractous lenses [12]. This oxidative damage may be generated by radiation and subsequent dye-sensitized photodamage. Both patients and experimental animals exposed to hyperbaric oxygen exhibit increased incidence of cataract. Antioxidants protect rat lenses from peroxide-induced cataracts. Dillon and co-workers have suggested that one mechanism of photooxidative damage may be modification of 3-hydroxykynurinine, a compound that is believed to be in the lens to provide protection from radiation damage [47].

Sugar Cataract. Elevated glycation in diabetes is another risk factor for cataract. Prolonged exposure to uncontrolled chronic hyperglycemia in diabetes can lead to cataract and retinopathy [48,49]. The aldose reductase pathway is activated during diabetes, leading to a conversion of excess glucose to sorbitol in insulin-independent tissues like the lens [50,51]. Biochemical and epidemiological studies indicate that elevated exposure to carbohydrate may be deleterious to eye tissues [52].

Juvenile Cataracts Associated with Autosomal Dominant Mutations

Juvenile-onset cataracts from a number of the families are inherited as autosomal dominant mutations [53]. Inherited cataract is a cause of visual impairment in children. It is known to be clinically and genetically heterogeneous. Many mutations causing inherited cataract have been identified in afflicted families. Among these juvenile cataracts are cases associated with defined amino acid substitutions in γ-crystallins [54], α-crystallin [55], β-crystallins [56], aquaporin-0 [57], and connexins [58]. Table 1 summarizes known human inherited cataracts caused by mutations in crystallin genes [9,54–56,59–73].

Analysis of purified recombinant HγD-crystallin carrying human mutations [7,8] revealed probable mechanisms for their in vivo effects. The mutation R114C resulted in the presence of an extra solvent-exposed cysteine, which formed disulfide-bonded dimers with an endogenous solvent-exposed cysteine, Cys110 [7]. The disulfide linkage resulted in the formation of high-molecular-mass oligomers that precipitated out of solution in vitro. The R36S and R58S mutants

TABLE 1 Crystallin Mutations That Cause Inherited Cataracts and Their Effects on Cataract Phenotype

Protein	Mutants	Lens Phenotype	Refs.
γD	R14C	Nuclear	54
	P23T	Coralliform	59
		Cerulean	60
		Fasciculiform	62
		Lamellar	61
	R36S	Prismatic crystals	9
	R58H	Aculeiform	63
	E107A	Nuclear	64
	W156X	Central nuclear	61
	G165fs	nuclear	66
γC	T5P	Coppock-like	63
	R168W	Lamellar	61
		Nuclear	65
	52 new residues	Nuclear	67
βB1	G220X	Nuclear and cortical	72
βB2	G155X	Cerulean/nuclear/punctate/sutural	56, 68, 69
βA1/3	G91del	Lamellar	93
βA1	Splice site	Nuclear sutural/posterior sutural	70
αA	R116C	Nuclear/microcornea	55
	R49C	Nuclear	71
αB	R120G		73

formed stable native proteins. These mutant proteins had lowered barriers to nucleation of crystallization in vitro [8]. The T5P mutation of HγC-crystallin is associated with Coppock-likc cataract. This mutation causes change in the protein conformation and forms inclusion bodies during in vitro expression [74]. Despite this conformational change, the interaction between T5P HγC and αA-crystallin is not affected [75].

The juvenile-onset cataract phenotypes associated with these rare mutations differ from mature-onset cataract, which is far more frequent and forms from the wild-type protein. Below we review the structure, stability, and aggregation pathways of purified lens crystallins. We also briefly review the functions of membrane and cytoskeletal proteins of the lens. We propose models for cataract formation that draw on advances in the study of other protein aggregation disease, including Parkinson disease, Huntington disease, and light-chain amyloidosis (addressed in other chapters of this volume), as well by in vitro studies of the crystallin proteins.

STRUCTURE AND FUNCTION OF THE LENS CRYSTALLINS

Vertebrate crystallins are classified into three groups: the α-, β-, and γ-crystallins [76]. α-Crystallins, members of the small heat-shock family of chaperones, form polydisperse multimers both in vivo and in vitro. In vivo, α-crystallin is a polydisperse multimer composed of two α-crystallin subunits, αA and αB, that share about 60% sequence identity [77]. The molecular weight of α-crystallin multimer ranges from 300 to 1200 kDa, with an average of 800 kDa, corresponding to 15 to 55 monomers per complex [78–80]. αA-crystallin monomer is a 173-amino acid protein that is found primarily in the lens, while the 175-amino acid αB-crystallin has been observed in other tissues, such as heart, brain, lung, skin, kidney, and skeletal muscle [81]. The expression level of α-crystallin is variable at different developmental stages as well as between species.

Due to the polydisperse nature of the lens α-crystallins, thus far it has not been possible to determine an x-ray crystal structure of the human α-crystallins. However, x-ray structures of small heat-shock proteins (sHSPs) have been solved that contain "α-crystallin domains" linked to variable N-terminal regions [82–84]. The α-crystallin domain fold has two β-sheets arranged into a sandwich that are connected by long loops containing both α-helical and unstructured regions. In the case of wheat HSP 16.9, the monomers associate to form dimers by strand exchange, and these dimers associate further into a dodecameric structure [83].

Although the crystal structure of the human α-crystallins remains elusive, some insight into the structure of the multimeric species has been gained using cryoelectron microscopy. These analyses have led to a micellarlike structural model where the α-crystallin subunits interact to form a hollow sphere [85,86]

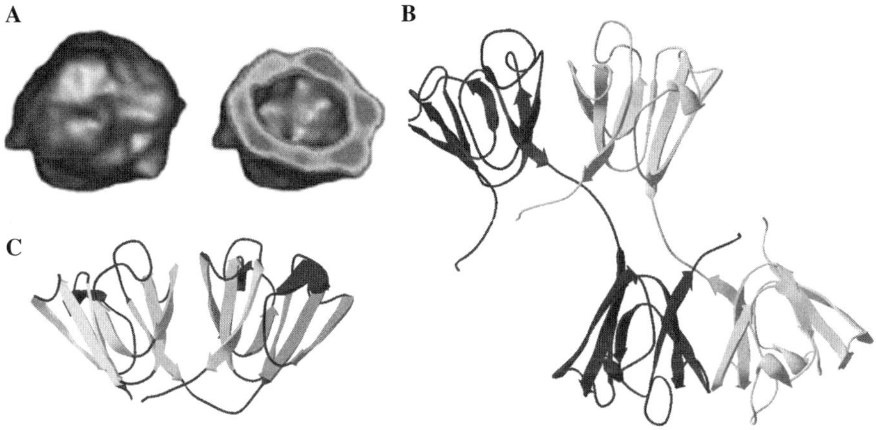

FIG. 1 Structural illustration of γ-, β-, and α-crystallins showing (A) human αB, (B) βB2, and (C) γD crystallin. (Adapted from [86, 128, 194].)

(Fig. 1A). The shell of the complex has an overall diameter of approximately 19 nm while the hollow internal cavity has a diameter of 8 nm.

α-Crystallins function both as structural proteins and as chaperones in the lens. αA- and αB-crystallin have been shown to possess molecular chaperone activity in vitro using both model substrates and lens crystallins [87–89]. The chaperone activity of α-crystallin is of intense interest for many scientists in the field of lens biology. The α-crystallins do not hydrolyze ATP and do not refold crystallins to their native states. Rather, they appear to act passively, binding and holding their substrates [78,89–91]. Brady and co-workers revealed that α-crystallin is required for maintaining lens transparency when they showed that αA crystallin knockout mice develop a cataract that starts in the nucleus and "spreads" to the entire lens with age [92].

In contrast to the chaperone activity of α-crystallins, the function of βγ-crystallins is thought to be largely structural, that is, to provide the lens with transparency and refractive power required for clear vision. The β- and γ-crystallins are small 20 to 30-kDa proteins composed primarily of antiparallel β-sheets [93]. The β- and γ-crystallins are structurally similar, both being comprised of four Greek-key motifs separated into two domains (Fig. 1B and C). The two domains are highly similar in sequence and structure and appear to be the result of a gene duplication event during evolution [94].

Sequence identities between β- and γ-crystallins are limited (approximately 30%) and the β-crystallins possess N- and C-terminal extensions, whereas the γ-crystallins do not [95]. Additionally, while the γ-crystallins studied to date behave as monomers in solution, β-crystallins exist in multimeric complexes [96,97]. For example, bovine βB2-crystallin (BβB2-Crys) forms homodimers by a quasi-domain-swapping mechanism. The dimer is formed by the packing of the N-terminal domain of one BβB2-Crys monomer against the C-terminal domain of another [98]. Despite these differences, domain topology of the β- and γ-crystallins is nearly identical.

Extensive biophysical studies have been performed in vitro on members of the βγ-crystallin superfamily [99–103]. The γ-crystallins exhibit attractive forces between molecules [104] and form micelles under appropriate conditions [13]. Although the forces between γ-crystallins assist the lens in lowering the overall osmotic pressure, they may increase susceptibility to self-association and aggregation [11].

Six of the γ-crystallin genes in humans are located in a cluster on chromosome 2 [105]. The genes encoding human γA-, γB-, γC-, and γD-crystallin are all located in the cluster and are found in appreciable levels in the human lens. The other two γ-crystallin genes in the cluster, γE and γF, are pseudogenes with in-frame termination codons and are thought to be transcriptionally inactive [105]. The gene encoding HγS-crystallin is not located in the γ-crystallin cluster and is thought to have diverged from the other γ-crystallins earlier in evolution [106,107].

In the embryonic lens, γC- and γD-crystallin comprise 90% of the γ-crystallins transcripts [105]. In the nucleus of the adult lens, the oldest region, HγC-crystallin and HγD-crystallin, continue to be the most abundant γ-crystallins. The γC- and γD-crystallins are both 173-amimo acid proteins and share 71% sequence identity.

HγS-crystallin is the major protein component of the human lens that is expressed postnatally [106]. This expression pattern results in a high concentration of HγS-crystallin in the newly divided peripheral epithelial cells and in the cortex that surrounds the aged nucleus. Both HγD- and HγS-crystallin have been cloned and expressed in bacteria [103,108]. In general, recombinant crystallins have behaved indistinguishably from native forms isolated from the lens [109].

HγS-crystallin is a 178-amino acid protein and shows 69% sequence similarity and 50% identity to HγD-crystallin. However, unlike many of the other γ-crystallins, HγS-crystallin has a three-amino acid N-terminal extension that is presumably unstructured in solution. Disulfide-bonded and covalently modified forms of HγS-crystallin have been found in aged lenses and are candidates for early precursors of cataract [110]. Covalently modified HγS-crystallin has also been recovered in protein aggregates removed from aged, cloudy lenses [111–113].

In addition to the chaperone activity of α-crystallins and the refractive function of the β,γ-crystallins, noncrystallin functions of these proteins are also an active area of research. A number of crystallin mutations have stage-specific phenotypes during development [114,115]. There is heterogeneity in the cataract phenotypes associated with similar or identical mutations in different populations. For example, an identical mutation (Q155X) in the βB2-crystallin gene in geographically disparate pedigrees leads to morphologically distinct phenotypes [56,68,69]. These data suggest that noncrystallin function is not merely a nonlens activity of a crystallin, but an essential requirement within the lens itself.

FOLDING, STABILITY, AND UNFOLDING OF CRYSTALLINS

The in vitro stabilities of the crystallin proteins have been studied intensively as a means of understanding how the proteins remain folded for a lifetime in vivo. A summary of data describing the stability of the human crystallins collected from the literature is presented in Table 2 [116–123]. Caution should be exercised when comparing these data, as stability measurements are dependent on the experimental conditions, including ionic strength, temperature, pH, and protein concentration, which may differ between laboratories. In general, these data show that the γ-crystallins are more stable than the α- and β-crystallins, with γD-crystallin being the most stable protein in the family. Additionally, some γ-crystallins have high kinetic barriers to unfolding that may aid in maintenance of their native folds [116,117,124,125].

The dissociation of protein domains is an early step in the unfolding of many multidomain proteins [126,127]. The domain interface of HγD-crystallin is a distinctive feature of the twinned Greek-key fold [93,128]. Features of the interface between the two domains are conserved among vertebrate crystallins and are likely to be important in stability, folding, and aggregation. The domain interface of HγD-crystallin is composed of a central hydrophobic cluster involving side-chain interactions of six residues: Met43, Phe56, Ile81, Val132, Leu145, and Val170. Flanking the hydrophobic cluster are peripheral interactions between Gln54 of the N-terminal domain and Gln143 of the C-terminal domain and Arg79 of the N-terminal domain and Met147 of the C-terminal domain. These two pairs of amino acids form structural boundaries between the central hydrophobic cluster and the solvent. Investigation of the contribution of these residues to the stability and folding of

TABLE 2 Stability Parameters of Human Crystallins

Protein	T_m (°C)	$C_{½, urea}$ (M)	$C_{½, GdnHCl}$ (M)	ΔG (kJ/mol)
γD	83.8		2.2, 2.8	69.4
γD-CTD	76.2		2.7	36.4
γD-NTD	64.5		1.2	15.5
γS	74.1	8.0	2.3	43.9
γS-CTD	75.1	7.6	2.3	34.3
γS-NTD	69.1	7.0	1.7	20.5
γC			2.3	36
βB1	67	5.9		
βB1-NTD		4.9		51
βB2			1,2.2	49
βA1		3.8		
βA3		3.4		
βA4		4.7		
αA				27
αB				21

HγD-crystallin revealed that the hydrophobic domain interface plays a key role in stabilizing the protein [116]. γD-Crystallin carrying interface mutations of the C-terminal domain destabilized the N-terminal domain. This suggests that the interface of the C-terminal domain acts as a template for folding of the N-terminal domain [129].

IN VITRO CRYSTALLIN AGGREGATION PATHWAYS AND CHAPERONE ACTIVITIES

HγD-crystallin is the most extensively studied crystallin with respect to its folding, unfolding, and aggregation properties. Although the pathology of juvenile-onset cataract can be explained by properties of the mutant HγD-crystallin [7,8], this observation does not account for the accumulation of wild-type HγD-crystallin in mature-onset cataracts. The most common irreversibly aggregated states of other proteins are derived from partially unfolded or partially folded species. A systematic investigation of the unfolding and refolding of wild-type HγD-crystallin [31] has been carried out at pH 7, 37°C.

At neutral pH, recombinant HγD-crystallin was resistant to denaturation by concentrations of urea up to 8 M, but could be denatured upon incubation with high concentrations of guanidine hydrochloride (GdnHCl). Equilibrium unfolding and refolding of wild-type HyD-crystallin in guanidine hydrochloride was best fit to a three-state model with transition midpoints of 2.2 and 2.8 M GdnHCl. The two transitions probably corresponded to sequential unfolding and refolding of the N-terminal domain and the C-terminal domain. Kinetic experiments revealed that the C-terminal domain refolds more rapidly than the N-terminal domain. The denatured chains could be refolded by dilution out of GdnHCl, and the reaction was reversible in the range of 1 to 5.5 M GdnHCl. This behavior is similar to that of bovine γB-crystallin, which is denatured by acid and urea and exhibits a three-stage transition, representing sequential denaturation of the C- and N-terminal domains [100,127].

HγD-crystallin displays a competing aggregation pathway during refolding out of GdnHCl in vitro at pH 7, 37°C [31]. During refolding, dilution into concentrations of GdnHCl below 1.0 M resulted in the population of an intermediate that aggregated irreversibly. Atomic force microscopy images revealed that the in vitro aggregate of HγD-crystallin had an ordered morphology. The aggregate had a filamentous appearance, as would be expected from the polymerization of structurally distinct subunits. The polymerization pathway could be resolved into stages in which small nuclei first polymerized to form long, thin protofibrils; these thin protofibrils then wound around each other, forming fiber bundles. All protofibrils were incorporated into fiber bundles within 4 hours after initiating refolding.

The aggregation pathway competed with a productive refolding pathway under the conditions described, suggesting that a partially folded intermediate, populated only in dilute GdnHCl, might be the aggregation-prone species. The partially folded species in this in vitro aggregation pathway may serve as models for the crystallin species nucleating or propagating cataract in aged lenses. Neither equilibrium nor kinetic analyses of HγD-crystallin unfolding revealed any evidence for aggregation from the native state of the protein [31].

Fibril formation by HγD-crystallin differed somewhat from other proteins in which aggregation competes with productive refolding, such as phosphoglycerate kinase and β-galactosidase [126]. For these proteins dilution to intermediate concentrations of denaturant generated an aggregating intermediate, while dilution to lower concentrations led to the recovery of the native fold.

The aggregation-prone states identified during in vitro refolding appear to be precursors in the formation of the ordered filbrillar state. Given that differential domain stability has been observed in other β- and γ-crystallins [100,102], it is possible that one domain of HγD-crystallin folds more slowly or is less stable than the other and thus lends itself to aggregation.

Even at 37°C, pH 7, both the kinetic unfolding and refolding of HγD-crystallin proceeded sufficiently slowly that the reaction course could be followed directly by the change of tryptophan fluorescence upon dilution from denaturant into buffer, or from buffer into denaturant. By using the triple tryptophan-to-phenylalanine constructs it was possible to follow the unfolding and refolding of each domain rather than the average of the four tryptophans with wild-type protein. During unfolding the N-terminal domain unfolded earlier, with the C-terminal domain unfolding more slowly. The refolding of the four mutant proteins under the same conditions showed clearly that the C-terminal domain refolded rapidly with an initial $t_{1/2}$ of 30 seconds, while the N-terminal domain folds much more slowly, with an initial $t_{1/2}$ of 190 seconds.

These results established the existence of a partially folded intermediate with the C-terminal domain folded and the N-terminal domain much less organized in both the unfolding and refolding pathways. These intermediates are observed under conditions in which the protein is refolding to its native state, the standard biophysical criteria for identification of physiologically relevant intermediates [130,131].

Bovine α-, β-, and γ-crystallins have been shown to form amyloid fibers under mild denaturing conditions [29]. Similarly, mouse congenital cataract mutations in the gene for γB-crystallin cause formation of in vivo inclusions that are stained by the amyloid-detecting dye Congo Red [115]. Recombinant proteins with these congenital mutants also formed amyloid fibers in vitro under mildly denaturing conditions [115].

Considerable evidence supports models in which α-crystallin functions as an active chaperone to the other lens crystallins [91,132,133]. Andley and co-workers described suppression of the singlet oxygen aggregation of recombinant HγD-crystallin by α-crystallin [134]. We might therefore expect α-crystallin to affect the efficiency of refolding of HγD-crystallin in vitro, or perhaps to suppress the off-pathway fibrillogenesis reaction (Acosta-Sampson and King, personal communication). In vitro studies, however, indicate that the α-crystallin may differ from the GroE class of chaperonins in that α-crystallin recognizes a more nativelike conformer of its substrate [108,135]. Such species are more likely populated in the low denaturant HγD-crystallin fibril formation pathway [31]. Although it is well documented that chaperonins distinguish partially folded intermediates from their native states, the actual mechanism of the recognition is not fully clear.

The ubiquitous nature of the heat-shock chaperonins is testimony to the problems that aggregating folding intermediates pose to cells. There is an enormous literature documenting that the physiological substrates of, for example, the Hsp60 class of chaperonins, are partially folded intermediates [14,136–138]. Fully denatured proteins in the absence of chaperonins refold—if at all—only under a very narrow set of conditions. Partially folded conformers, detected under conditions in which refolding is taking place, are prime candidates for physiologically relevant, aggregation-prone species in vivo. Such species are the best documented substrates for the heat-shock chaperones [138,139]. Much evidence indicates that this is also true for the lens α-crystallins [87,108,140].

Chaperones function by binding to partially or fully denatured polypeptide chains in order to prevent aggregation and/or facilitate correct folding. There is evidence that α-crystallin is capable of binding to nonnative conformations of crystallins in vitro [89,141]. Additionally, antibody-binding experiments have demonstrated that both β- and γ-crystallins are bound to high-molecular-mass complexes of α-crystallin extracted from lenses [142]. These results suggest that one of the roles of α-crystallin in the lens is to bind damaged and partially unfolded crystallins. Loss of lens transparency in mature-onset cataract may correlate with the saturation of α-crystallin binding, as suggested above.

If the crystallin aggregates in cataract represent an aggregation or polymerization pathway similar to any of the known cases, the prospect exists of identifying small molecules that inhibit propagation of the complex. If the precursor, nucleus, or growing surface of the polymer can be identified in vitro, there is a possibility of targeted therapeutic development [143–145].

COVALENT MODIFICATIONS OF CRYSTALLINS

Crystallins undergo a variety of irreversible covalent modifications over a long lifetime. Such modifications are caused by proteolysis, deamidation, oxidation, Maillard reactions, ultraviolet (UV) irradiation, and adduct formation, among others [146]. The critical changes for lens transparency are those that destabilize or reduce the solubility of native crystallin structures.

Several in vitro studies have addressed the effects of covalent damage on properties of the crystallins [97,118,123,147,148]. The effects of such damage ranged from formation of higher-order oligomers, increased tendency of thermal aggregation, destabilization, and partial unfolding for the

INSTITUT MEDIATHEQUE PASTEUR

βγ-crystallins [97,118,123,129,149], to changes in secondary and tertiary structure and decreased chaperone activity for α-crystallin [147,148]. These results indicate that effects of covalent damage are context dependent and that some damage may be detrimental enough to elicit in vivo changes that could cause cataract formation.

Glutamine and asparagine deamidation is a particularly pervasive form of covalent damage that has been observed in all of the major crystallin proteins recovered from cataractous lenses [20,24,150]. At the atomic level, deamidation can cause backbone isomerization and introduces a negative change at physiological pH because an amide group is replaced with a carboxyl group. Deamidation of the crystallins may cause changes in structure, stability, or solubility that could instigate or contribute to cataract formation. For example, it has been suggested that deamidation of Asnl43 in human γS-crystallin is associated specifically with mature-onset cataract formation [112]. Similarly, deamidated β-crystallins isolated from human lenses have an increased tendency to associate into noncovalent aggregates [151]. Using recombinant protein it was shown that deamidation of the HγD-crystallin domain interface glutamines Gln54 and Gln143 destabilizes the protein and lowers the kinetic barrier to unfolding [129].

MECHANISTIC MODELS FOR CATARACT FORMATION

Crystal Cataracts

The punctate juvenile cataracts removed from the patient described by Kmoch et al. [9] diffracted x-rays and were presumably crystals of mutant HγD-Crys. The R38H and R36S mutations that cause aculeiform and congenital juvenile-onset cataracts, respectively, resulted in recombinant proteins with decreased solubility and increased rates of crystal nucleation [8]. The reduced barrier to crystallization in vitro correlated with the formation of juvenile crystalline cataracts in affected persons [9].

Phase Separation

A distinctive feature of the crystallins, shared with a limited number of other proteins, is the tendency to separate spontaneously into two phases: one protein rich and the other protein poor, at high concentration [152,153]. Benedek and co-workers have systematically investigated this phenomenon and propose it as an explanation for the molecular mechanism of cataract [13,154,155]. Phase separation is sensitive to temperature and can be induced by lowering the temperature in a concentrated solution of crystallin. These can also be induced within intact lenses, yielding "cold cataracts." However, the phase separation of the crystallins is fully reversible, a behavior not found in cataracts isolated from the lens.

Association Through Nonnative Disulfide Bonds

Oxidative damage remains a major model for cataract formation [156]. Both oxidative and dye-sensitive photodamage result in the formation of disulfide bonds between protein subunits. Formation of disulfide-bonded multimers provides a pathway for formation of very large, essentially irreversible aggregates, under oxidizing conditions. Indeed, disulfide-bonded crystallins are one of the major covalently modified forms found in aging lens [12]. Disulfide-bonded forms of HγS-crystallin are present in the highest quantities of all γ-crystallins in the aging lens, suggesting that HγS-crystallin appears to be particularly susceptible to this phenomenon. Such disulfide-bonded oligomers could nucleate other polymerization reactions, leading to the development of cataract [110].

Loss of Protection from Oxidative Damage

Almost all cells contain a substantial enzymatic apparatus to protect their constituents from oxidative damage. For example, glutathione peroxidase knockout mice have increased lens opacity and cataract formation [157]. A number of observations indicate that within the lens the capacity for protection from oxidative damage is degraded with age or by exposure to UV light or agents that enhance photosensitization. For example, Linetsky and co-workers have reported that glutathione reductase in human lenses was inactivated by exposure to UVA radiation [158]. The damage was mediated by FAD at the enzyme active site and probably inactivated the catalytic site. This would suggest a two-step model of cataract formation where the protective apparatus is first inactivated, and then the crystallins are directly oxidized.

Aggregation of Partially Folded Conformers, Domain Swapping, and Loop-Sheet Insertion

The insoluble aggregated states of cataractous proteins resemble the inclusion bodies that often form during protein overexpression in bacteria in that the polypeptide chains are tightly associated and require denaturation for solubilization. In the best studied cases, inclusion bodies are formed from the polymerization of partially folded intermediates [14,159,160]. The irreversible association of protein chain within inclusion bodies is not due to disulfide bonds but rather to the same intimate interactions that keep native proteins folded [22,160,161]. This model implies that some stress on the crystallins generates partially unfolded species or otherwise conformational altered species, and these polymerize into the cataract. Similarly, studies of polypeptides conferring disease phenotypes such as α-synuclein, transthyretin, and the Aβ42 peptide have demonstrated that aggregation is not a random process, but rather, represents the sequential polymerization of distinct intermediates involving a series of specific nonnative interactions [160,162–166].

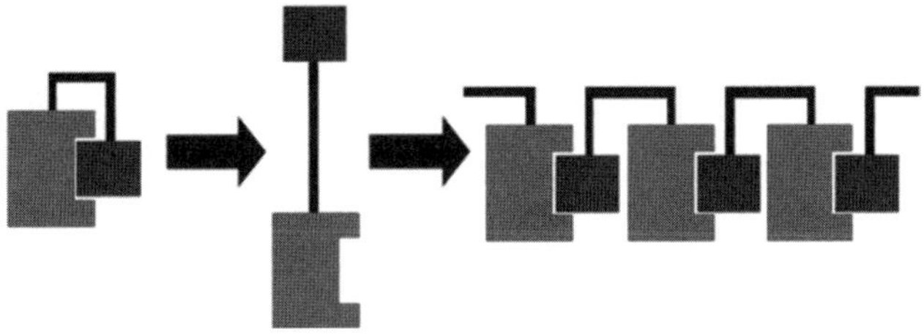

FIG. 2 Polymerization by domain swapping.

Two well-defined mechanisms of such association reactions are domain swapping and loop-sheet insertion. Domain swapping has been proposed as a mechanism of polymerization of the prion protein [167,168], and loop-sheet insertion is the mechanism of polymerization in the serpins [167,169]. Domain swapping is a particularly intriguing model of polymerization given that bovine βB2-crystallin forms dimers by a quasi-domain swapping mechanism (Fig. 2).

Chaperone Saturation

Epidemiological studies show that cataracts are very rare in humans before the age of 50 and then increase very sharply thereafter. Most of the α-crystallin chaperone molecules are synthesized very early in life, with new synthesis being limited to the lens periphery. Over a lifetime, the α-crystallin population may get saturated with damaged or partially unfolded crystallin substrates, so that later in life α-crystallin is not available to protect crystallins from aggregation [89]. Fig. 3 depicts a model where α-crystallin interacts with partially unfolded HγD crystallin to prevent its aggregation until all binding sites on α-crystallin is occupied, at which point protein aggregation propagates [170].

Membrane or Cytoskeleton Nucleation

αB-crystallin may selectively associate with cytoskeletal structures present in the lens, such as intermediate filaments, in times of stress [171]. Cobb and Petrash showed that a mutant of α-crystallin preferentially associates with membranes, and proposed that the membrane-bound formation of α-crystallin is a critical player in cataract formation [133]. An in vivo study using fluorescence resonance energy transfer suggests that all crystallins, especially αA-crystallin, interact with the membrane water channel MIP26/AQP0 [172]. One of the natural functions of α-crystallin might be to protect endogenous filamentous protein from being degraded by binding to exposed, denatured regions of the cytoskeletal protein.

Glycation Models

Protein components from aged tissues, including the lens, are often modified to form advanced glycation end products [173]. These represent proteins with glycated N-termini, or lysines. These modifications may lead to the generation

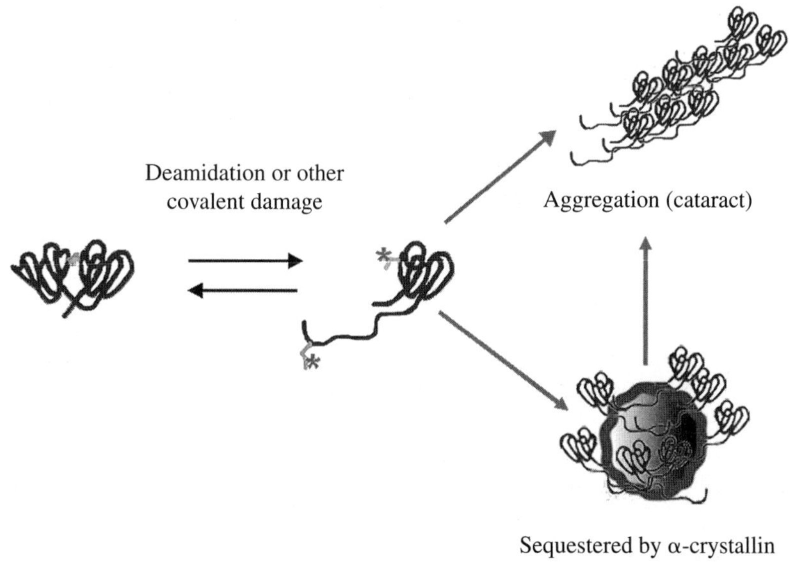

FIG. 3 Model of cataract in vitro depicting partially unfolded γ-crystallin that renders the protein aggregation-prone. Such aggregation-prone species can be rescued by α-crystallins until all the α-crystallins are consumed, at which point cataract will propagate. (From [117].)

of covalently linked proteins [174]. This pathway is generally considered most important in diabetic cataract [175].

CYTOSKELETON PROTEINS OF LENS CELLS

Lens cells possess the three major cytoskeletal elements of most eukaryotic cells: microtubules, microfilaments, and intermediate filaments. Lens cells are also hosts of a large number of cytoskeleton-associated proteins which mediate and regulate the various filament networks. Several great reviews had been written concerning the lens cytoskeleton proteins [176–179].

Microtubules are dynamic polymers of α- and β-tubulin heterodimers that add to the ends of microtubues at different rates. Microtubules are required for many important cellular functions, such as intracellular transport and chromosome movement during meiosis and mitosis. Microtubule dynamics requires high energy. In the lens, cell differentiation involves a dramatic change in cell shape, and microtubules are likely to be involved in this process. Microtubules are present in differentiating fiber cells [180] and in cultured lens epithelial cells [181]. Microtubules are rare in enucleated lens fiber cells, where ATP comes mainly from glycolysis as a consequence of the loss of mitochondria.

The expression of intermediate filament proteins is regulated developmentally and differentially. There are now over 60 intermediate filament proteins that have been identified [182]. The expression profile of each type of intermediate filament is tissue-specific. The lens fiber cells express their own specific intermediate filament proteins: filensin [183], formerly CP115 [184], and CP49 [185], sometimes called phakinin [186] or beaded filament structural protein 2 [187]. Single-point mutation as well as a deletion in CP49 have been reported to cause inherited cataract in humans [187,188].

Actin is found throughout the entire lens and is therefore expected to play a role in all stages of lens cell differentiation. Actin forms actin-membrane complexes via adhesion junctions at either side of the epithelial–fiber cell interface, designating the polarity of epithelial cells [189].

MEMBRANE PROTEINS OF LENS CELLS

Lens membrane proteins include a collection of gap junction proteins (connexins), ion pumps, water channels (aquaporins), and growth factor receptors. Membrane transport systems in the lens generate a circulating flux of ions and water, efficiently carrying nutrients into the core of the lens and returning waste products to the lens surface [190]. There is accumulating evidence that the spatial distribution of membrane transporters and channel proteins is critical to lens function [191]. Several lens membrane proteins are found to be cleaved upon fiber cell maturation. In the case of connexin 50, the cleavage abolishes the pH gating of the cell–cell channels [192]. Mutations in the genes of connexin 50, connexin 46, and aquaporin-0 have been identified as causing inherited cataract [193].

PROSPECTS FOR PROGRESS

Through the etiologies and initiation processes for mature onset cataract appear to be diverse, the loss of transparency and light scattering appear to represent aggregated states of the major lens crystallins and disruption of lens fiber architecture. The absence of tissue culture for the anucleate lens fibers retards a variety of investigations. However, continuing characterization of in vitro aggregation reactions of the lens proteins, together with continuing studies of cataract development in small animal models, may reveal key features of the mechanisms and pathways of cataract formation. These provide targets for the identification of low molecular weight inihibitors. Given the potential for therapeutic access to the eye lens by topical application, it may be feasible to develop small molecule therapies that complement cataract surgery and would offer significant public health benefit.

REFERENCES

1. *Vision Problems in the U.S.: Prevalence of Adult Vision Impairment and Age-Related Eye Disease in America*, published in conjunction with the National Eye Institute, Prevent Blindness America 2000, 38 pp.

2. Oyster, C. W. (1999). *The Human Eye: Structure and Function*, Sinauer Associates, Sunderland, MA.
3. Delaye, M., Tardieu, A. (1983). Short-range order of crystallin proteins accounts for eye lens transparency. *Nature, 302*, 415–417.
4. Hamai, Y., Fukui, H. N., Kuwabara, T. (1974). Morphology of hereditary mouse cataract. *Exp Eye Res, 18*, 537–546.
5. Hamai, Y., Kuwabara, T. (1975). Early cytologic changes of Fraser cataract: an electron microscopic study. *Invest Ophthalmol, 14*, 517–527.
6. Metlapally, S., Costello, M. J., Gilliland, K. O., Ramamurthy, B., Krishna, P. V., Balasubramanian, D., Johnsen, S. (2007). Analysis of nuclear fiber cell cytoplasmic texture in advanced cataractous lenses from Indian subjects using Debye–Bueche theory. *Exp Eye Res, 86*, 434–444.
7. Pande, A., Pande, J., Asherie, N., Lomakin, A., Ogun, O., King, J. A., Lubsen, N. H., Walton, D., Benedek, G. B. (2000). Molecular basis of a progressive juvenile-onset hereditary cataract. *Proc Natl Acad Sci U S A, 97*, 1993–1998.
8. Pande, A., Pande, J., Asherie, N., Lomakin, A., Ogun, O., King, J., Benedek, G. B. (2001). Crystal cataracts: human genetic cataract caused by protein crystallization. *Proc Natl Acad Sci U S A, 98*, 6116–6120.
9. Kmoch, S., Brynda, J., Asfaw, B., Bezouska, K., Novak, P., Rezacova, P., Ondrova, L., Filipec, M., Sedlacek, J., Elleder, M. (2000). Link between a novel human gammaD-crystallin allele and a unique cataract phenotype explained by protein crystallography. *Hum Mol Genet, 9*, 1779–1786.
10. Bellows, J. E. (1975). *Cataracts and Abnormalities of the Lens*, Grune & Stratton, New York.
11. Clark, J. (1994). *Principle and Practice of Opthamology*, Saunders College Publishing, Philadelphia.
12. Hanson, S. R., Hasan, A., Smith, D. L., Smith, J. B. (2000). The major in vivo modifications of the human water-insoluble lens crystallins are disulfide bonds, deamidation, methionine oxidation and backbone cleavage. *Exp Eye Res, 71*, 195–207.
13. Benedek, G. (1997). Cataract as a protein condensation disease: the Proctor Lecture. *Invest Ophthalmol Vis Sci, 38*, 1911–1921.
14. Mitraki, A., King, J. (1989). Protein folding intermediates and inclusion body formation. *Biotechnology, 7*, 690–697.
15. Wetzel, R. (1994). Mutations and off-pathway aggregation of proteins. *Trends Biotechnol, 12*, 193–198.
16. Betts, S., King, J. (1999). There's a right way and a wrong way: in vivo and in vitro folding, misfolding and subunit assembly of the P22 tailspike. *Structure, 7*, R131–R139.
17. Smith, J. B., Sun, Y., Smith, D. L., Green, B. (1992). Identification of the posttranslational modifications of bovine lens alpha B-crystallins by mass spectrometry. *Protein Sci, 1*, 601–608.
18. Lapko, V. N., Smith, D. L., Smith, J. B. (2003). Methylation and carbamylation of human gamma-crystallins. *Protein Sci, 12*, 1762–1774.
19. Slingsby, C., Clout, N. J. (1999). Structure of the crystallins. *Eye, 13(Pt 3b)*, 395–402.

20. Lampi, K. J., Ma, Z., Hanson, S. R., Azuma, M., Shih, M., Shearer, T. R., Smith, D. L., Smith, J. B., David, L. L. (1998). Age-related changes in human lens crystallins identified by two-dimensional electrophoresis and mass spectrometry. *Exp Eye Res, 67*, 31–43.

21. Chiti, F., Dobson, C. M. (2006). Protein misfolding, functional amyloid, and human disease. *Annu Rev Biochem, 75*, 333–366.

22. Speed, M., Wang, D., King, J. (1995). Multimeric intermediates in the pathway to the aggregated inclusion body state for P22 tailspike polypeptide chains. *Protein Sci 4*, 900–908.

23. Wetzel, R. (1996). For protein misassembly, it's the "I" decade. *Cell, 86*, 699–702.

24. Lund, A. L., Smith, J. B., Smith, D. L. (1996). Modifications of the water-insoluble human lens alpha-crystallins. *Exp Eye Res*, 63, 661–672.

25. Miesbauer, L. R., Zhou, X., Yang, Z., Yang, Z., Sun, Y., Smith, D. L., Smith, J. B. (1994). Post-translational modifications of water-soluble human lens crystallins from young adults. *J Biol Chem, 269*, 12494–12502.

26. MacCoss, M., McDonald, W., Saraf, A., Sadygov, R., Clark, J., Tasto, J., Gould, K., Wolters, D., Washburn, M., Weiss, A., et al. (2002). Shotgun identification of protein modifications from protein complexes and lens tissue. *Proc Natl Acad Sci U S A, 99*, 7900–7905.

27. Ma, Z., Hanson, S. R., Lampi, K. J., David, L. L., Smith, D. L., Smith, J. B. (1998). Age-related changes in human lens crystallins identified by HPLC and mass spectrometry. *Exp Eye Res, 67*, 21–30.

28. Frederikse, P. (2000). Amyloid-like protein structure in mammalian ocular lens. *Curr Eye Res, 20*, 462–468.

29. Meehan, S., Berry, Y., Luisi, B., Dobson, C. M., Carver, J. A., MacPhee, C. E. (2004). Amyloid fibril formation by lens crystallin proteins and its implications for cataract formation. *J Biol Chem, 279*, 3413–3419.

30. Papanikolopoulou, K., Mills-Henry, I., Thol, S. L., Wang, Y., Gross, A. A.R., Kirschner, D. A., Decatur, S. M., King, J. (2008). Formation of amyloid fibrils in vitro by human γD-crystallin and its isolated domains. *Mol Vis, 14*, 81–89.

31. Kosinski-Collins, M. S., King, J. (2003). In vitro unfolding, refolding, and polymerization of human gammaD crystallin, a protein involved in cataract formation. *Protein Sci, 12*, 480–490.

32. Hollows, F., Moran, D. (1981). Cataract: the ultraviolet risk factor. *Lancet, 2*, 1249–1250.

33. Taylor, H. R., West, S. K., Rosenthal, F. S., Munoz, B., Newland, H. S., Abbey, H., Emmett, E. A. (1988). Effect of ultraviolet radiation on cataract formation. *N Engl J Med, 319*, 1429–1433.

34. McCarty, C. A., Taylor, H. R. (2002). A review of the epidemiologic evidence linking ultraviolet radiation and cataracts. *Dev Ophthalmol, 35*, 21–31.

35. Sasaki, H., Jonasson, F., Shui, Y. B., Kojima, M., Ono, M., Katoh, N., Cheng, H. M., Takahashi, N., Sasaki, K. (2002). High prevalence of nuclear cataract in the population of tropical and subtropical areas. *Dev Ophthalmol, 35*, 60–69.

36. Lerman, S. (1962). Radiation cataractogenesis. *N Y State J Med, 62*, 3075–3085.

37. Hannah, C. (1975). Cataract of toxic etiology, In *Cataracts and Abnormalities of the Lens* (Bellows, J. G., (Ed.), Grune & Stratton, New York.

38. Wolbarsht, M. L., Orr, M. A., Yamanashi, B. S., Zigler, J. S., Matheson, I. B.C. (1977). *The Origin of Cataracts in the Lens from Infrared Laser Radiation: Annual Progress Report*, U.S. Army Medical Research and Development Command, Washington, DC.
39. Lerman, S. (1980). *Radiant Energy and the Eye*, Macmillan, New York.
40. Giblin, F. J., Leverenz, V. R., Padgaonkar, V. A., Unakar, N. J., Dang, L., Lin, L. R., Lou, M. F., Reddy, V. N., Borchman, D., Dillon, J. P. (2002). UVA light in vivo reaches the nucleus of the guinea pig lens and produces deleterious, oxidative effects. *Exp Eye Res, 75*, 445–458.
41. Sliney, D. H. (2002). How light reaches the eye and its components. *Int J Toxicol, 21*, 501–509.
42. Soderberg, P. G., Lofgren, S., Ayala, M., Dong, X., Kakar, M., Mody, V. (2002). Toxicity of ultraviolet radiation exposure to the lens expressed by maximum tolerable dose. *Dev Ophthalmol, 35*, 70–75.
43. Padayatti, P. S., Ng, A. S., Uchida, K., Glomb, M. A., Nagaraj, R. H. (2001). Argpyrimidine, a blue fluorophore in human lens proteins: high levels in brunescent cataractous lenses. *Invest Ophthalmol Vis Sci, 42*, 1299–1304.
44. Taylor, L. M., Andrew Aquilina, J., Jamie, J. F., Truscott, R. J. (2002). UV filter instability: consequences for the human lens. *Exp Eye Res, 75*, 165–175.
45. Chen, J., Flaugh, S. L., Callis, P. R., King, J. (2006). Mechanism of the highly efficient quenching of tryptophan fluorescence in human gammaD-crystallin. *Biochemistry, 45*, 11552–11563.
46. Langley, R. K., Mortimer, C. B., McCulloch, C. (1960). The experimental production of cataracts by exposure to heat and light. *Arch Ophthalmol, 63*, 473–488.
47. Dillon, J., Skonieczna, M., Mandal, K., Paik, D. (1999). The photochemical attachment of the *O*-glucoside of 3-hydroxykynurenine to alpha-crystallin: a model for lenticular aging. *Photochem Photobiol, 69*, 248–253.
48. Brownlee, M. (2001). Biochemistry and molecular cell biology of diabetic complications. *Nature, 414*, 813–820.
49. Baynes, J. W., Thorpe, S. R. (1999). Role of oxidative stress in diabetic complications: a new perspective on an old paradigm. *Diabetes, 48*, 1–9.
50. Ghahary, A., Luo, J. M., Gong, Y. W., Chakrabarti, S., Sima, A. A., Murphy, L. J. (1989). Increased renal aldose reductase activity, immunoreactivity, and mRNA in streptozocin-induced diabetic rats. *Diabetes, 38*, 1067–1071.
51. Chung, S. S., Ho, E. C., Lam, K. S., Chung, S. K. (2003). Contribution of polyol pathway to diabetes-induced oxidative stress. *J Am Soc Nephrol, 14*, S233–S236.
52. Stitt, A. W. (2005). The Maillard reaction in eye diseases. *Ann N Y Acad Sci, 1043*, 582–597.
53. Hejtmancik, J. F. (1998). The genetics of cataract: our vision becomes clearer. *Am J Hum Genet, 62*, 520–525.
54. Stephan, D. A., Gillanders, E., Vanderveen, D., Freas-Lutz, D., Wistow, G., Baxevanis, A. D., Robbins, C. M., VanAuken, A., Quesenberry, M. I., Bailey-Wilson, J., et al. (1999). Progressive juvenile-onset punctate cataracts caused by mutation of the gammaD-crystallin gene. *Proc Natl Acad Sci U S A, 96*, 1008–1012.
55. Litt, M., Kramer, P., LaMorticella, D.M., Murphey, W., Lovrien, E. W., Weleber, R. G. (1998). Autosomal dominant congenital cataract associated with

a missense mutation in the human alpha crystallin gene CRYAA. *Hum Mol Genet, 7*, 471–474.

56. Litt, M., Carrero-Valenzuela, R., LaMorticella, D. M., Schultz, D. W., Mitchell, T. N., Kramer, P., Maumenee, I. H. (1997). Autosomal dominant cerulean cataract is associated with a chain termination mutation in the human beta-crystallin gene CRYBB2. *Hum Mol Genet, 6*, 665–668.
57. Berry, V., Francis, P., Kaushal, S., Moore, A., Bhattacharya, S. (2000). Missense mutations in MIP underlie autosomal dominant "polymorphic" and lamellar cataracts linked to 12q. *Nat Genet, 25*, 15–17.
58. Shiels, A., Mackay, D., Ionides, A., Berry, V., Moore, A., Bhattacharya, S. (1998). A missense mutation in the human connexin50 gene (GJA8) underlies autosomal dominant "zonular pulverulent" cataract, on chromosome 1q. *Am J Hum Genet, 62*, 526–532.
59. Mackay, D. S., Andley, U. P., Shiels, A. (2004). A missense mutation in the gammaD crystallin gene (CRYGD) associated with autosomal dominant "coral-like" cataract linked to chromosome 2q. *Mol Vis, 10*, 155–162.
60. Nandrot, E., Slingsby, C., Basak, A., Cherif-Chefchaouni, M., Benazzouz, B., Hajaji, Y., Boutayeb, S., Gribouval, O., Arbogast, L., Berraho, A., et al. (2003). Gamma-D crystallin gene (CRYGD) mutation causes autosomal dominant congenital cerulean cataracts. *J Med Genet, 40*, 262–267.
61. Santhiya, S. T., Shyam Manohar, M., Rawlley, D., Vijayalakshmi, P., Namperumalsamy, P., Gopinath, P. M., Loster, J., Graw, J. (2002). Novel mutations in the gamma-crystallin genes cause autosomal dominant congenital cataracts. *J Med Genet, 39*, 352–358.
62. Shentu, X., Yao, K., Xu, W., Zheng, S., Hu, S., Gong, X. (2004). Special fasciculiform cataract caused by a mutation in the gammaD-crystallin gene. *Mol Vis, 10*, 233–239.
63. Heon, E., Priston, M., Schorderet, D. F., Billingsley, G. D., Girard, P. O., Lubsen, N., Munier, F. L. (1999). The gamma-crystallins and human cataracts: a puzzle made clearer. *Am J Hum Genet, 65*, 1261–1267.
64. Messina-Baas, O. M., Gonzalez-Huerta, L. M., Cuevas-Covarrubias, S. A. (2006). Two affected siblings with nuclear cataract associated with a novel missense mutation in the CRYGD gene. *Mol Vis, 12*, 995–1000.
65. Gonzalez-Huerta, L. M., Messina-Baas, O. M., Cuevas-Covarrubias, S. A. (2007). A family with autosomal dominant primary congenital cataract associated with a CRYGC mutation: evidence of clinical heterogeneity. *Mol Vis, 13*, 1333–1338.
66. Zhang, L. Y., Yam, G. H., Fan, D. S., Tam, P. O., Lam, D. S., Pang, C. P. (2007). A novel deletion variant of gammaD-crystallin responsible for congenital nuclear cataract. *Mol Vis, 13*, 2096–2104.
67. Ren, Z., Li, A., Shastry, B. S., Padma, T., Ayyagari, R., Scott, M. H., Parks, M. M., Kaiser-Kupfer, M. I., Hejtmancik, J. F. (2000). A 5-base insertion in the gammaC-crystallin gene is associated with autosomal dominant variable zonular pulverulent cataract. *Hum Genet, 106*, 531–537.
68. Gill, D., Klose, R., Munier, F. L., McFadden, M., Priston, M., Billingsley, G., Ducrey, N., Schorderet, D. F., Heon, E. (2000). Genetic heterogeneity of the Coppock-like cataract: a mutation in CRYBB2 on chromosome 22q11.2. *Invest Ophthalmol Vis Sci, 41*, 159–165.

69. Vanita, Sarhadi, V., Reis, A., Jung, M., Singh, D., Sperling, K., Singh, J. R., Burger, J. (2001). A unique form of autosomal dominant cataract explained by gene conversion between beta-crystallin B2 and its pseudogene. *J Med Genet, 38*, 392–396.
70. Burdon, K. P., Wirth, M. G., Mackey, D. A., Russell-Eggitt, I. M., Craig, J. E., Elder, J. E., Dickinson, J. L., Sale, M. M. (2004). A novel mutation in the connexin 46 gene causes autosomal dominant congenital cataract with incomplete penetrance. *J Med Genet, 41*, e106,
71. Mackay, D. S., Andley, U. P., Shiels, A. (2003). Cell death triggered by a novel mutation in the alphaA-crystallin gene underlies autosomal dominant cataract linked to chromosome 21q. *Eur J Hum Genet, 11*, 784–793.
72. Mackay, D. S., Boskovska, O. B., Knopf, H. L., Lampi, K. J., Shiels, A. (2002). A nonsense mutation in CRYBB1 associated with autosomal dominant cataract linked to human chromosome 22q. *Am J Hum Genet, 71*, 1216–1221.
73. Vicart, P., Caron, A., Guicheney, P., Li, Z., Prevost, M. C., Faure, A., Chateau, D., Chapon, F., Tome, F., Dupret, J. M., et al. (1998). A missense mutation in the alphaB-crystallin chaperone gene causes a desmin-related myopathy. *Nat Genet, 20*, 92–95.
74. Fu, L., Liang, J. J. (2002). Conformational change and destabilization of cataract gammaC-crystallin T5P mutant. *FEBS Lett, 513*, 213–216.
75. Liang, J. J. (2004). Interactions and chaperone function of alphaA-crystallin with T5P gammaC-crystallin mutant. *Protein Sci, 13*, 2476–2482.
76. Piatigorsky, J. (1998). Gene sharing in lens and cornea: facts and implications. *Prog Retin Eye Res, 17*, 145–174.
77. Bloemendal, H., de Jong, W. W. (1991). Lens proteins and their genes. *Prog Nucleic Acid Res Mol Biol, 41*, 259–281.
78. Horwitz, J. (2003). Alpha-crystallin. *Exp Eye Res, 76*, 145–153.
79. Siezen, R. J., Berger, H. (1978). The quaternary structure of bovine alpha-crystallin: size and shape studies by sedimentation, small-angle x-ray scattering and quasi-elastic light scattering. *Eur J Biochem, 91*, 397–405.
80. Veretout, F., Delaye, M., Tardieu, A. (1989). Molecular basis of eye lens transparency: osmotic pressure and x-ray analysis of alpha-crystallin solutions. *J Mol Biol, 205*, 713–728.
81. Bhat, S. P., Nagineni, C. N. (1989). Alpha B subunit of lens-specific protein alpha-crystallin is present in other ocular and non-ocular tissues. *Biochem Biophys Res Commun, 158*, 319–325.
82. Kim, K. K., Kim, R., Kim, S. H. (1998). Crystal structure of a small heat-shock protein. *Nature, 394*, 595–599.
83. van Montfort, R. L., Basha, E., Friedrich, K. L., Slingsby, C., Vierling, E. (2001). Crystal structure and assembly of a eukaryotic small heat shock protein. *Nat Struct Biol, 8*, 1025–1030.
84. Stamler, R., Kappe, G., Boelens, W., Slingsby, C. (2005). Wrapping the alpha-crystallin domain fold in a chaperone assembly. *J Mol Biol, 353*, 68–79.
85. Haley, D. A., Bova, M. P., Huang, Q. L., McHaourab, H. S., Stewart, P. L. (2000). Small heat-shock protein structures reveal a continuum from symmetric to variable assemblies. *J Mol Biol, 298*, 261–272.

86. Haley, D. A., Horwitz, J., Stewart, P. L. (1998). The small heat-shock protein, alphaB-crystallin, has a variable quaternary structure. *J Mol Biol, 277*, 27–35.

87. Horwitz, J. (1992). Alpha-crystallin can function as a molecular chaperone. *Proc Natl Acad Sci U S A, 89*, 10449–10453.

88. Rao, P. V., Huang, Q. L., Horwitz, J., Zigler, J. S., Jr. (1995). Evidence that alpha-crystallin prevents non-specific protein aggregation in the intact eye lens. *Biochim Biophys Acta, 1245*, 439–447.

89. Derham, B. K., Harding, J. J. (1999). Alpha-crystallin as a molecular chaperone. *Prog Retin Eye Res, 18*, 463–509.

90. Horwitz, J., Bova, M. P., Ding, L. L., Haley, D. A., Stewart, P. L. (1999). Lens alpha-crystallin: function and structure. *Eye, 13*(Pt 3b), 403–408.

91. Clark, J. I., Muchowski, P. J. (2000). Small heat-shock proteins and their potential role in human disease. *Curr Opin Struct Biol, 10*, 52–59.

92. Brady, J. P., Garland, D., Duglas-Tabor, Y., Robison, W. G., Jr., Groome, A., Wawrousek, E. F. (1997). Targeted disruption of the mouse alpha A-crystallin gene induces cataract and cytoplasmic inclusion bodies containing the small heat shock protein alpha B-crystallin. *Proc Natl Acad Sci U S A, 94*, 884–889.

93. Wistow, G., Turnell, B., Summers, L., Slingsby, C., Moss, D., Miller, L., Lindley, P., Blundell, T. (1983). X-ray analysis of the eye lens protein gamma-II crystallin at 1.9 Å resolution. *J Mol Biol, 170*, 175–202.

94. Piatigorsky, J., Wistow, G. (1991). The recruitment of crystallins: new functions precede gene duplication. *Science, 252*, 1078–1079.

95. Norledge, B. V., Mayr, E. M., Glockshuber, R., Bateman, O. A., Slingsby, C., Jaenicke, R., Driessen, H. P. (1996). The x-ray structures of two mutant crystallin domains shed light on the evolution of multi-domain proteins. *Nat Struct Biol, 3*, 267–274.

96. Slingsby, C., Bateman, O. A. (1990). Quaternary interactions in eye lens beta-crystallins: basic and acidic subunits of beta-crystallins favor heterologous association. *Biochemistry, 29*, 6592–6599.

97. Lampi, K. J., Oxford, J. T., Bachinger, H. P., Shearer, T. R., David, L. L., Kapfer, D. M. (2001). Deamidation of human beta B1 alters the elongated structure of the dimer. *Exp Eye Res, 72*, 279–288.

98. Bax, B., Lapatto, R., Nalini, V., Driessen, H., Lindley, P. F., Mahadevan, D., Blundell, T. L., Slingsby, C. (1990). X-ray analysis of beta B2-crystallin and evolution of oligomeric lens proteins. *Nature, 347*, 776–780.

99. Norledge, B. V., Hay, R. E., Bateman, O. A., Slingsby, C., Driessen, H. P. (1997). Towards a molecular understanding of phase separation in the lens: a comparison of the x-ray structures of two high Tc gamma-crystallins, gammaE and gammaF, with two low Tc gamma-crystallins, gammaB and gammaD. *Exp Eye Res, 65*, 609–630.

100. Rudolph, R., Siebendritt, R., Nesslauer, G., Sharma, A. K., Jaenicke, R. (1990). Folding of an all-beta protein: independent domain folding in gamma II-crystallin from calf eye lens. *Proc Natl Acad Sci USA, 87*, 4625–4629.

101. Pande, J., McDermott, M. J., Callender, R. H., Spector, A. (1991). The calf gamma crystallins: a Raman spectroscopic study. *Exp Eye Res, 52*, 193–197.

102. Mayr, E. M., Jaenicke, R., Glockshuber, R. (1997). The domains in gammaB-crystallin: identical fold-different stabilities. *J Mol Biol, 269*, 260–269.

103. Wenk, M., Herbst, R., Hoeger, D., Kretschmar, M., Lubsen, N. H., Jaenicke, R. (2000). Gamma S-crystallin of bovine and human eye lens: solution structure, stability and folding of the intact two-domain protein and its separate domains. *Biophys Chem, 86*, 95–108.

104. Tardieu, A., Veretout, F., Krop, B., Slingsby, C. (1992). Protein interactions in the calf eye lens: interactions between beta-crystallins are repulsive whereas in gamma-crystallins they are attractive. *Eur Biophys J, 21*, 1–12.

105. Siezen, R. J., Thomson, J. A., Kaplan, E. D., Benedek, G. B. (1987). Human lens gamma-crystallins: isolation, identification, and characterization of the expressed gene products. *Proc Natl Acad Sci U S A, 84*, 6088–6092.

106. Harding, J. J., Crabbe, M. J.C. (1984). The lens: development, proteins, metabolism, and cataract. In *The Eye*, Vol. 1B (Dawson, H., (Ed.), Academic Press, London, pp. 207–492.

107. Wistow, G., Wyatt, K., David, L., Gao, C., Bateman, O., Bernstein, S., Tomarev, S., Segovia, L., Slingsby, C., Vihtelic, T. (2005). GammaN-crystallin and the evolution of the betagamma-crystallin superfamily in vertebrates. *FEBS J, 272*, 2276–2291,

108. Das, K. P., Petrash, J. M., Surewicz, W. K. (1996). Conformational properties of substrate proteins bound to a molecular chaperone alpha-crystallin. *J Biol Chem, 271*, 10449–10452.

109. Hay, R. E., Andley, U. P., Petrash, J. M. (1994). Expression of recombinant bovine gamma B-, gamma C- and gamma D-crystallins and correlation with native proteins. *Exp Eye Res, 58*, 573–584.

110. Skouri-Panet, F., Bonnete, F., Prat, K., Bateman, O. A., Lubsen, N. H., Tardieu, A. (2001). Lens crystallins and oxidation: the special case of gammaS. *Biophys Chem, 89*, 65–76.

111. Lapko, V. N., Smith, D. L., Smith, J. B. (2002). S-methylated cysteines in human lens gamma S-crystallins. *Biochemistry, 41*, 14645–14651.

112. Takemoto, L., Boyle, D. (2000). Specific glutamine and asparagine residues of gamma-S crystallin are resistant to in vivo deamidation. *J Biol Chem, 275*, 26109–26112,

113. Takemoto, L., Fujii, N., Boyle, D. (2001). Mechanism of asparagine deamidation during human senile cataractogenesis. *Exp Eye Res, 72*, 559–563.

114. Loster, J., Pretsch, W., Sandulache, R., Schmitt-John, T., Lyon, M. F., Graw, J. (1994). Close linkage of the dominant cataract mutations (Cat-2) with Idh-1 and cryge on mouse chromosome 1. *Genomics, 23*, 240–242.

115. Sandilands, A., Hutcheson, A. M., Long, H. A., Prescott, A. R., Vrensen, G., Loster, J., Klopp, N., Lutz, R. B., Graw, J., Masaki, S., et al. (2002). Altered aggregation properties of mutant gamma-crystallins cause inherited cataract. *EMBO J, 21*, 6005–6014.

116. Flaugh, S. L., Kosinski-Collins, M. S., King, J. (2005). Contributions of hydrophobic domain interface interactions to the folding and stability of human gammaD-crystallin. *Protein Sci, 14*, 569–581.

117. Mills, I. A., Flaugh, S. L., Kosinski-Collins, M. S., King, J. A. (2007). Folding and stability of the isolated Greek key domains of the long-lived human lens proteins gammaD-crystallin and gammaS-crystallin. *Protein Sci, 16*, 2427–2444.

118. Kim, Y. H., Kapfer, D. M., Boekhorst, J., Lubsen, N. H., Bachinger, H. P., Shearer, T. R., David, L. L., Feix, J. B., Lampi, K. J. (2002). Deamidation, but not

truncation, decreases the urea stability of a lens structural protein, betaB1-crystallin. *Biochemistry, 41*, 14076–14084.

119. Trinkl, S., Glockshuber, R., Jaenicke, R. (1994). Dimerization of beta B2-crystallin: the role of the linker peptide and the N- and C-terminal extensions. *Protein Sci, 3*, 1392–1400.
120. Werten, P. J., Carver, J. A., Jaenicke, R., de Jong, W. W. (1996). The elusive role of the N-terminal extension of beta A3- and beta A1-crystallin. *Protein Eng, 9*, 1021–1028.
121. Bateman, O. A., Sarra, R., van Genesen, S. T., Kappe, G., Lubsen, N. H., Slingsby, C. (2003). The stability of human acidic beta-crystallin oligomers and hetero-oligomers. *Exp Eye Res, 77*, 409–422.
122. Sun, T. X., Liang, J. J. (1998). Intermolecular exchange and stabilization of recombinant human alphaA- and alphaB-crystallin. *J Biol Chem, 273*, 286–290.
123. Lampi, K. J., Amyx, K. K., Ahmann, P., Steel, E. A. (2006). Deamidation in human lens betaB2-crystallin destabilizes the dimer. *Biochemistry, 45*, 3146–3153.
124. Das, B. K., Liang, J. J. (1998). Thermodynamic and kinetic characterization of calf lens gammaF-crystallin. *Int J Biol Macromol, 23*, 191–197.
125. Jaenicke, R., Slingsby, C. (2001). Lens crystallins and their microbial homologs: structure, stability, and function. *Crit Rev Biochem Mol Biol, 36*, 435–499.
126. Ghelis, C., Yon, J. (1982). Physicochemical studies of the unfolding–folding equilibrium. In *Protein Folding* (Horecker, B., Kaplan, N. O., Marmur, J., Scheraga, H.,Eds.), Academic Press, New York.
127. Jaenicke, R. (1999). Stability and folding of domain proteins. *Prog Biophys Mol Biol, 71*, 155–241.
128. Basak, A., Bateman, O., Slingsby, C., Pande, A., Asherie, N., Ogun, O., Benedek, G. B., Pande, J. (2003). High-resolution x-ray crystal structures of human gammaD crystallin (1.25 Å) and the R58H mutant (1.15 Å) associated with aculeiform cataract. *J Mol Biol, 328*, 1137–1147.
129. Flaugh, S. L., Mills, I. A., King, J. (2006). Glutamine deamidation destabilizes human gammaD-crystallin and lowers the kinetic barrier to unfolding. *J Biol Chem, 281*, 30782–30793.
130. Kim, P. S., Baldwin, R. L. (1984). A helix stop signal in the isolated S-peptide of ribonuclease A. *Nature, 307*, 329–334.
131. Kim, P. S., Baldwin, R. L. (1990). Intermediates in the folding reactions of small proteins. *Annu Rev Biochem, 59*, 631–660.
132. Muchowski, P. J., Clark, J. I. (1998). ATP-enhanced molecular chaperone functions of the small heat shock protein human alphaB crystallin. *Proc Natl Acad Sci U S A, 95*, 1004–1009.
133. Cobb, B. A., Petrash, J. M. (2002). alpha-Crystallin chaperone-like activity and membrane binding in age-related cataracts. *Biochemistry, 41*, 483–490.
134. Andley, U. P., Mathur, S., Griest, T. A., Petrash, J. M. (1996). Cloning, expression, and chaperone-like activity of human alphaA-crystallin. *J Biol Chem, 271*, 31973–31980.
135. Das, K. P., Choo-Smith, L. P., Petrash, J. M., Surewicz, W. K. (1999). Insight into the secondary structure of non-native proteins bound to a molecular chaperone alpha-crystallin: an isotope-edited infrared spectroscopic study. *J Biol Chem, 274*, 33209–33212.

136. Ellis, R. J., van der Vies, S. M. (1991). Molecular chaperones. *Annu Rev Biochem, 60*, 321–347.

137. Horwich, A. L., Low, K. B., Fenton, W. A., Hirshfield, I. N., Furtak, K. (1993). Folding in vivo of bacterial cytoplasmic proteins: role of GroEL. *Cell, 74*, 909–917.

138. Leroux, M. R., Hartl, F. U. (2000). Cellular functions of molecular chaperones. In *Frontiers in Molecular Biology: Mechanisms of Protein Folding*, 2nd ed. (Pain R. H., Ed.), Oxford University Press, New York, pp. 364–405.

139. Hartl, F. U., Hayer-Hartl, M. (2002). Molecular chaperones in the cytosol: from nascent chain to folded protein. *Science, 295*, 1852–1858.

140. Leroux, M. R., Melki, R., Gordon, B., Batelier, G., Candido, E. P. (1997). Structure–function studies on small heat shock protein oligomeric assembly and interaction with unfolded polypeptides. *J Biol Chem, 272*, 24646–24656.

141. Sathish, H. A., Koteiche, H. A., McHaourab, H. S. (2004). Binding of destabilized betaB2-crystallin mutants to alpha-crystallin: the role of a folding intermediate. *J Biol Chem, 279*, 16425–16432.

142. Boyle, D., Takemoto, L. (1994). Characterization of the alpha-gamma and alpha-beta complex: evidence for an in vivo functional role of alpha-crystallin as a molecular chaperone. *Exp Eye Res, 58*, 9–15.

143. Soto, C., Kindy, M. S., Baumann, M., Frangione, B. (1996). Inhibition of Alzheimer's amyloidosis by peptides that prevent beta-sheet conformation. *Biochem Biophys Res Commun, 226*, 672–680.

144. Ono, K., Hasegawa, K., Naiki, H., Yamada, M. (2004). Curcumin has potent anti-amyloidogenic effects for Alzheimer's beta-amyloid fibrils in vitro. *J Neurosci Res, 75*, 742–750.

145. Ono, K., Yoshiike, Y., Takashima, A., Hasegawa, K., Naiki, H., Yamada, M. (2003). Potent anti-amyloidogenic and fibril-destabilizing effects of polyphenols in vitro: implications for the prevention and therapeutics of Alzheimer's disease. *J Neurochem, 87*, 172–181.

146. Harding, J. J. (2002). Viewing molecular mechanisms of ageing through a lens. *Ageing Res Rev, 1*, 465–479.

147. Gupta, R., Srivastava, O. P. (2004). Deamidation affects structural and functional properties of human alphaA-crystallin and its oligomerization with alphaB-crystallin. *J Biol Chem, 279*, 44258–44269.

148. Gupta, R., Srivastava, O. P. (2004). Effect of deamidation of asparagine 146 on functional and structural properties of human lens alphaB-crystallin. *Invest Ophthalmol Vis Sci, 45*, 206–214.

149. Harms, M. J., Wilmarth, P. A., Kapfer, D. M., Steel, E. A., David, L. L., Bachinger, H. P., Lampi, K. J. (2004). Laser light-scattering evidence for an altered association of beta B1-crystallin deamidated in the connecting peptide. *Protein Sci, 13*, 678–686.

150. Groenen, P. J., Merck, K. B., de Jong, W. W., Bloemendal, H. (1994). Structure and modifications of the junior chaperone alpha-crystallin: from lens transparency to molecular pathology. *Eur J Biochem, 225*, 1–19.

151. Zhang, Z., Smith, D. L., Smith, J. B. (2003). Human beta-crystallins modified by backbone cleavage, deamidation and oxidation are prone to associate. *Exp Eye Res, 77*, 259–272.

152. Broide, M. L., Berland, C. R., Pande, J., Ogun, O., Benedek, G. B. (1991). Binary-liquid phase separation of lens protein solutions. *Proc Natl Acad Sci U S A, 88*, 5660–5664.

153. Clark, J. I., Clark, J. M. (2000). Lens cytoplasmic phase separation. *Int Rev Cytol, 192*, 171–187.

154. Liu, C., Asherie, N., Lomakin, A., Pande, J., Ogun, O., Benedek, G. B. (1996). Phase separation in aqueous solutions of lens γ-crystallins: special role of γS. *Proc Natl Acad Sci U S A, 93*, 377–382.

155. Liu, C., Pande, J., Lomakin, A., Ogun, O., Benedek, G. B. (1998). Aggregation in aqueous solutions of bovine lens γ-crystallins: special role of γS. *Invest Ophthalmol Vis Sei, 39*, 1609–1619.

156. Bron, A. J., Vrensen, G. F.J. M., Koretz, J., Maraini, G., Harding, J. J. (2000). The ageing lens. *Ophthalmologica, 214*, 86–104.

157. Reddy, G. B., Das, K. P., Petrash, J. M., Surewicz, W. K. (2001). Temperature-dependent chaperone activity and structural properties of human αA- and αB-crystallins. *J Biol Chem, 275*, 4560–4570.

158. Linetsky, M., Hill, J. M.W., Chemoganskiy, V. G., Hu, F., Ortwerth, B. J. (2003). Studies on the mechanism of the UVA light-dependent loss of glutathione reductase activity in human lenses. *Invest Ophthalmol Vis Sci, 44*, 3920–3926.

159. Haase-Pettingell, C. A., King, J. (1988). Formation of aggregates from a thermolabile in vivo folding intermediate in P22 tailspike maturation: a model for inclusion body formation. *J Biol Chem, 263*, 4977–4983.

160. Speed, M. A., Wang, D. I., King, J. (1996). Specific aggregation of partially folded polypeptide chains: the molecular basis of inclusion body composition. *Nat Biotechnol, 14*, 1283–1287.

161. Speed, M. A., Morshead, T., Wang, D. I., King, J. (1997). Conformation of P22 tailspike folding and aggregation intermediates probed by monoclonal antibodies. *Protein Sci, 6*, 99–108.

162. Colon, W., Kelly, J. W. (1992). Partial denaturation of transthyretin is sufficient for amyloid fibril formation in vitro. *Biochemistry, 31*, 8654–8660.

163. Sunde, M., Blake, C. (1997). The structure of amyloid fibrils by electron microscopy and x-ray diffraction. *Adv Protein Chem, 50*, 123–159.

164. Li, J., Uversky, V. N., Fink, A. L. (2001). Effect of familial Parkinson's disease point mutations A30P and A53T on the structural properties, aggregation, and fibrillation of human alpha-synuclein. *Biochemistry, 40*, 11604–11613.

165. Lashuel, H. A., Lai, Z., Kelly, J. W. (1998). Characterization of the transthyretin acid denaturation pathways by analytical ultracentrifugation: implications for wild-type, V30M, and L55P amyloid fibril formation. *Biochemistry, 37*, 17851–17864.

166. Harper, J. D., Lansbury, P. T., Jr. (1997). Models of amyloid seeding in Alzheimer's disease and scrapie: mechanistic truths and physiological consequences of the time-dependent solubility of amyloid proteins. *Annu Rev Biochem, 66*, 385–407.

167. Liu, Y., Eisenberg, D. (2002). 3D domain swapping: as domains continue to swap. *Protein Sci, 11*, 1285–1299.

168. Knaus, K. J., Morillas, M., Swietnicki, W., Malone, M., Surewicz, W. K., Yee, V. C. (2001). Crystal structure of the human prion protein reveals a mechanism for oligomerization. *Nat Struct Biol, 8*, 770–774.

169. Carrell, R. W., Lomas, D. A. (1997). Conformational disease. *Lancet, 350*, 134–138.

170. Mills-Henry, I. A. R. (2007). Stability, unfolding and aggregation of the gamma D and gamma S humen eye crystallins. In *Biology*, Vol. Ph. D., Massachusetts Institute of Technology, Cambridge, MA.

171. Muchowski, P. J., Valdez, M. M., Clark, J. I. (1999). AlphaB-crystallin selectively targets intermediate filament proteins during thermal stress. *Invest Ophthalmol Vis Sci, 40*, 951–958.

172. Liu, B. F., Liang, J. J. (2007). Confocal fluorescence microscopy study of interaction between lens MIP26/AQP0 and crystallins in living cells. *J Cell Biochem, 104*, 51–58.

173. Zhao, H. R., Nagaraj, R. H., Abraham, E. C. (1997). The role of alpha- and epsilon-amino groups in the glycation-mediated cross-linking of gammaB-crystallin. Study of three site-directed mutants. *J Biol Chem, 272*, 14465–14469.

174. Chellan, P., Nagaraj, R. H. (2001). Early glycation products produce pentosidine cross-links on native proteins: novel mechanism of pentosidine formation and propagation of glycation. *J Biol Chem, 276*, 3895–3903.

175. Stevens, A. (1998). The contribution of glycation to cataract formation in diabetes. *J Am Optom Assoc, 69*, 519–530.

176. Quinlan, R., Prescott, A. (2004). Lens cell cytoskeleton. In *Development of the Ocular Lens* (Lovicu, F. J., Robinson, M. L., Eds.). Cambridge University Press, Cambridge, UK, pp. 173–187.

177. Perng, M. D., Quinlan, R. A. (2005). Seeing is believing! The optical properties of the eye lens are dependent upon a functional intermediate filament cytoskeleton. *Exp Cell Res, 305*, 1–9.

178. Perng, M. D., Zhang, Q., Quinlan, R. A. (2007). Insights into the beaded filament of the eye lens. *Exp Cell Res, 313*, 2180–2188.

179. Quinlan, R. A., Sandilands, A., Procter, J. E., Prescott, A. R., Hutcheson, A. M., Dahm, R., Gribbon, C., Wallace, P., Carter, J. M. (1999). The eye lens cytoskeleton. *Eye, 13*(Pt 3b), 409–416.

180. Beebe, D. C., Cerrelli, S. (1989). Cytochalasin prevents cell elongation and increases potassium efflux from embryonic lens epithelial cells: implications for the mechanism of lens fiber cell elongation. *Lens Eye Toxic Res, 6*, 589–601.

181. Lonchampt, M. O., Laurent, M., Courtois, Y., Trenchev, P., Hughes, R. C. (1976). Microtubules and microfilaments of bovine lens epithelial cells: electron microscopy and immunofluorescence staining with specific antibodies. *Exp Eye Res, 23*, 505–518.

182. Herrmann, H., Aebi, U. (2000). Intermediate filaments and their associates: multitalented structural elements specifying cytoarchitecture and cytodynamics. *Curr Opin Cell Biol, 12*, 79–90.

183. Merdes, A., Brunkener, M., Horstmann, H., Georgatos, S. D. (1991). Filensin: a new vimentin-binding, polymerization-competent, and membrane-associated protein of the lens fiber cell. *J Cell Biol, 115*, 397–410.

184. FitzGerald, P. G. (1988). Immunochemical characterization of a Mr 115 lens fiber cell-specific extrinsic membrane protein. *Curr Eye Res, 7*, 1243–1253.

185. Hess, J. F., Casselman, J. T., FitzGerald, P. G. (1993). cDNA analysis of the 49 kDa lens fiber cell cytoskeletal protein: a new, lens-specific member of the intermediate filament family? *Curr Eye Res, 12*, 77–88.

186. Merdes, A., Gounari, F., Georgatos, S. D. (1993). The 47-kD lens-specific protein phakinin is a tailless intermediate filament protein and an assembly partner of filensin. *J Cell Biol, 123*, 1507–1516.

187. Jakobs, P. M., Hess, J. F., FitzGerald, P. G., Kramer, P., Weleber, R. G., Litt, M. (2000). Autosomal-dominant congenital cataract associated with a deletion mutation in the human beaded filament protein gene BFSP2. *Am J Hum Genet, 66*, 1432–1436.

188. Conley, Y. P., Erturk, D., Keverline, A., Mah, T. S., Keravala, A., Barnes, L. R., Bruchis, A., Hess, J. F., FitzGerald, P. G., Weeks, D. E., et al. (2000). A juvenile-onset, progressive cataract locus on chromosome 3q21–q22 is associated with a missense mutation in the beaded filament structural protein-2. *Am J Hum Genet, 66*, 1426–1431.

189. Lo, W. K., Shaw, A. P., Paulsen, D. F., Mills, A. (2000). Spatiotemporal distribution of zonulae adherens and associated actin bundles in both epithelium and fiber cells during chicken lens development. *Exp Eye Res, 71*, 45–55.

190. Robinson, K. R., Patterson, J. W. (1982). Localization of steady currents in the lens. *Curr Eye Res, 2*, 843–847.

191. Kistler, J., Eckert, R., Donaldson, P. (2004) Lens cell membranes. In *Development of the Ocular Lens* (Lovicu, F. J., Robinson, M. L., Eds.), Cambridge University Press, Cambridge, UK, pp. 151–173.

192. Lin, J. S., Eckert, R., Kistler, J., Donaldson, P. (1998). Spatial differences in gap junction gating in the lens are a consequence of connexin cleavage. *Eur J Cell Biol, 76*, 246–250.

193. Reddy, M. A., Francis, P. J., Berry, V., Bhattacharya, S. S., Moore, A. T. (2004). Molecular genetic basis of inherited cataract and associated phenotypes. *Surv Ophthalmol, 49*, 300–315.

194. Jaenicke, R., Slingsby, C. (2001). Lens crystallins and their microbial homologs: structure, stability, and function. *Crit Rev Biochem Mol Biol, 36*, 435–499.

23

ISLET AMYLOID POLYPEPTIDE

ANDISHEH ABEDINI

Department of Surgery, College of Physicians and Surgeons, Columbia University, New York, New York

DANIEL P. RALEIGH

Department of Chemistry, Graduate Program in Biochemistry and Structural Biology, State University of New York at Stony Brook, Stony Brook, New York

INTRODUCTION

Amyloid deposits were found in the pancreases of patients with type 2 diabetes at the beginning of the twentieth century [1,2]. The presence of a hyaline staining substance, currently referred to as *islet amyloid*, was first described by Eugene L. Opie in 1901 [1]. The amyloid-like nature of this hyaline material was established in 1943 and later confirmed by Congo Red staining and electron microscopy. It was not until more than 80 years latter, in 1987, that a 37-residue polypeptide hormone was shown to be the principal proteinaceous component of the deposits. The polypeptide was described independently by the Westermark and Cooper groups and was named *islet amyloid polypeptide* (IAPP) and *amylin* [3,4].

IAPP has been identified in all mammalian species examined. The polypeptide is a member of the calcitionin-like family of peptides, which include calcitonin, calcitonin gene-related peptide (CGRP), and adrenomedullin [5]. IAPP is most similar to CGRP and shares some actions and receptor-binding affinities with CGRP [6,7]. Like many polypeptide hormones, IAPP is synthesized as pre-proform and undergoes processing to yield the mature hormone [8].

Protein Misfolding Diseases: Current and Emerging Principles and Therapies,
Edited by Marina Ramirez-Alvarado, Jeffery W. Kelly, and Christopher M. Dobson
Copyright © 2010 John Wiley & Sons, Inc.

Pro-insulin and pro-IAPP are processed in parallel in the trans Golgi and in the β-cell secretory granules by the prohormone convertases PC2 and PC(1/3) [9]. Defects in the processing of proIAPP, described below, have been suggested to play a key role in islet amyloid formation in vivo. Mature human IAPP is 37 residues in length, has an amidated C-terminus and a disulfide bridge between residue-2 and residue-7 (Table 1). IAPP is released in response to the same stimuli that trigger insulin release and IAPP secretion is reported to be 1 to 10% of the level of insulin secretion [10–13].

Islet amyloid is a common pathological feature of type 2 diabetes [5,14–20]. The deposition of islet amyloid is not the cause of the disease; however, it is believed to contribute to β-cell failure and the decline in insulin secretion in type 2 diabetes. Synthetic amyloid aggregates are toxic to insulin-producing β-cells, indicating that IAPP amyloid fibril formation in the pancreas may contribute to islet cell dysfunction and cell death in type 2 diabetes mellitus [14–23]. Autopsies indicate varying amounts of amyloid deposits in over 80% of patients with type 2 diabetes, although in some cases the fraction of islets affected is relatively low; however, if prefibril oligomeric species are the toxic entities, as has been postulated for a number of protein aggregation diseases, an absolute correlation between the extent of islet amyloid and the severity of β-cell loss need not be expected [24–30]. Islet amyloid deposits are not a feature of type 1 diabetes since this form of the disease is characterized by autoimmune destruction of the islet β-cells that produce insulin and IAPP.

The mechanism or mechanisms of islet amyloid formation in vivo are not known, nor is the initiation site of the amyloid formation. The mechanism of amyloid formation in vitro is also not completely understood and is certainly not nearly as well characterized as that of the Aβ peptide; however, there is a growing literature on IAPP amyloid, and a flurry of recent activity has added significantly to our knowledge base.

NORMAL PHYSIOLOGICAL ROLE OF IAPP

The exact physiological role of soluble IAPP is not completely understood, and a range of functions have been suggested. IAPP is currently thought to play an important role in maintaining glucose homeostasis by suppressing insulin-mediated glucose uptake in skeletal muscle, inhibiting glucose-stimulated insulin secretion, regulating gastric emptying and suppressing glucagon release from isolated islets [31–37]. IAPP is believed to slow exogenous (meal-derived) glucose through inhibition of gastric emptying while reducing endogenous (liver-derived) glucose as a result of suppression of glucagon release from isolated islets. IAPP has been proposed to play a role in a number of other diverse effects, including inhibition of glucose-stimulated insulin secretion; the excretion of calcium, potassium, and sodium; vasodilatation; and the regulation of energy intake and fat storage [38–41]. Many of the early experiments on

TABLE 1 Normal Processing of PreproIAPP to Mature IAPP in Islet β-Cells[a]

	1	10	20	30 37
Human	KCNTATCAT	QRLANFLVHS	SNNFGAILSS	TNVGSNTY
Monkey	KCNTATCAT	QRLANFLV*RS*	SNNFG*T*ILSS	TNVGS*D*TY
Cat	KCNTATCAT	QRLANFL*IRS*	SNN*L*GAILS*P*	TNVGSNTY
Dog	KCNTATCAT	QRLANFLV*RT*	SNN*L*GAILS*P*	TNVGSNTY
Rat	KCNTATCAT	QRLANFLV*RS*	SNN*L*G*PV*L*PP*	TNVGSNTY
Mouse	KCNTATCAT	QRLANFLV*RS*	SNN*L*G*PV*L*PP*	TNVGSNTY
Guinea pig	KCNTATCAT	QR*T*ANFLV*RS*	S*H*N*L*GAIL*P*S	*D*NVGSNTY
Hamster	KCNTATCAT	QRLANFLVHS	*N*N*L*F*PV*I*P*SS	TNVGSNTY
Degue	KCNTATCAT	QR*T*ANFLV*RS*	S*H*N*L*GAI*PP*S	NVGSNTY
Rabbit	T	QRLANFL*I*HS	SNNFGA*F*L*PP*	*S*
Hare	T	QRLANFL*I*HS	SNNFGA*F*L*PP*	T

[a]Primary sequence of IAPP from a variety of species. Residues that differ from the human peptide are indicated by *italics*. Only partial sequences are available for rabbit and hare. Human, monkeys, and cats form islet amyloid, whereas the other species listed do not. The degue does form islet amyloid, but it is derived from insulin, not IAPP.

the putative physiological role(s) of IAPP were conducted at supraphysiological concentrations and thus may not provide an accurate view of its function.

SYNTHESIS AND PROCESSING OF IAPP

The IAPP gene, which is located on the short arm of chromosome 12, is transcribed on the ribosome as an 89-residue prohormone precursor (preproIAPP) [8]. Like other secretory peptides, preproIAPP contains a signaling sequence that targets its transport to the lumen of the rough endoplasmic reticulum, where cleavage of the signaling sequence by signal peptidase forms the 67-residue prohormone proIAPP [8]. After traveling to the Golgi apparatus, proIAPP is packaged into a secretory granule, along with proinsulin, where it matures to its biologically active form [9,42,43]. Prior to secretion, both proinsulin and proIAPP are processed in parallel by the subtilisinlike proprotein convertase enzymes PC1 and PC2/PC3. Normal processing of the 67-residue proIAPP precursor depends on cleavage at two well-conserved dibasic sites both containing Lys–Arg [42,43]. PC1/ PC3 is responsible for processing proIAPP at the C-terminal dibasic site but not at the N-terminal dibasic site which is cleaved by PC2 [42,43]. Additional posttranslational modifications include formation of the disulfide bridge between cysteine residues at positions 2 and 7 in the mature peptide, and amidation of the C-terminal Tyr37, which is thought to be important for biological activity [44]. The content of IAPP in the granules is usually 1 to 2% that of insulin and is noticeably higher than that required to lead to rapid fibrillization in vitro [14–16]. This leads to an interesting question: What feature or features inhibit irreversible aggregation in the granule? pH may play a role since the intragranular environment is at lower pH (5.4) than the extracellular space, and the in vitro rate of amyloid formation by IAPP is strongly pH dependent [45–48]. In addition, several labs have shown independently that insulin is a potent inhibitor of IAPP aggregation, and this could well play an important role in maintaining the polypeptide in a soluble state suitable for secretion [49–53].

NOT ALL SPECIES FORM ISLET AMYLOID

There is correlation, although not an absolute one, between the primary sequence of IAPP derived from different species and the presence of islet amyloid. Early analysis focused on a comparison with other members of the clacitonin family. Human IAPP forms amyloid in vitro and in vivo whereas human CGRP does not. The two polypeptides differ most in the central region of the molecule, residues 20 to 29, and this leads to the hypothesis that this region was the critical determinate of amyloid formation. Further support for this idea came from sequence analysis of IAPP. Not all species form islet amyloid, and initial sequence comparisons focused attention on the region from residues 20 to 29. IAPP has been identified in all mammalian species examined, all of which have a primary sequence identity greater then 80%. The primary sequence of human IAPP is compared to that of several species in Figure 1. Despite the high sequence similarity, only humans, nonhuman primates, and cats of the species listed in Table 1 normally form islet amyloid in vivo [54–56]. In particular, humans form islet amyloid, whereas rodents do not, yet the primary sequences of rodent and human IAPP are very similar, aside from the 20–29 region. Human and rat IAPP differ at only six of 37 positions, five of which are located between residues 20 and 29. Rat IAPP contains three proline residues in this region at positions 25, 28, and 29, whereas the human sequence has none. A 10-residue peptide corresponding to the 20–29 sequence of human IAPP formed amyloid in vitro, while the related 10-residue peptide derived from the rat sequence did not [56,57]. The inability of rat IAPP to form fibrils is attributed to the proline residues, consistent with the general β-sheet disrupting effect of proline [58]. These early studies led to the view that the amyloidogenic properties of IAPP are dominated by the sequence of the 20–29 segment, and the failure to form amyloid is dictated by the presence of multiple prolines in this region. Note, however, that the dog and cat sequence are identical in this region, yet cats form islet amyloid and dogs do not. Other fragments of IAPP are capable of forming amyloid in isolation [59–61], and these studies naturally led to the question of the absolute importance of the 20–29 region. Stated differently: Does the failure of rodent IAPPs to form amyloid mean that the 20–29 sequence indeed dominates the amyloidogenic properties of IAPP, or might it simply imply that substitution with multiple prolines is sufficient to inhibit amyloid formation irrespective of whether or not they are located in the 20–29 segment? Recent work has demonstrated that multiple proline

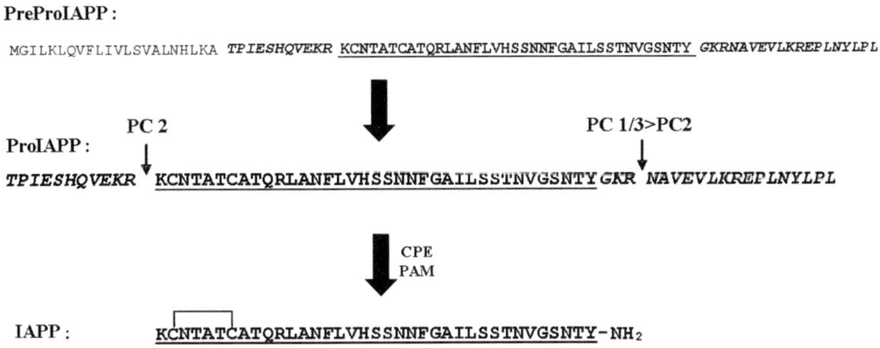

FIG. 1 The primary sequences of the 89-residue preproIAPP, the 67-residue proIAPP, and the 37-residue mature IAPP are shown. The disulfide bridge in mature IAPP is between residues 2 and 7 in the mature sequence. Cleavage of proIAPP occurs at two conserved dibasic sites; one at the N-terminus (Lys10-Arg11; numbered according to the sequence of proIAPP, in this numbering residue 12 is the first residue of mature IAPP) and the second at the C-terminus (Lys50-Arg51; numbered according to the sequence of proIAPP), indicated by arrows. Processing at the N-terminus is initiated solely by PC2. ProIAPP processing at the C-terminus is favored by PC1/PC3 but can also be mediated by PC2. Dibasic residues at the C-terminus are removed by carboxypeptidase E, and mature IAPP is formed by removal of Gly49 at the C-terminus, and amidation at this site is carried out by peptidyl amidating monooxygenase complex.

substitutions outside the 20–29 region completely abolished amyloid formation by human IAPP, thereby proving that the ability to form amyloid is not dictated solely by the 20–29 sequence [62]. Furthermore, replacement of three nonproline residues in rat IAPP, R18, L23, and V26, with the corresponding residues of the human peptide lead to a triple mutant that was able to form amyloid, albeit with a significantly reduced tendency, indicating that peptides which contain multiple prolines in the 20–29 region can still form amyloid [63].

AROMATIC INTERACTIONS AND AMYLOID FORMATION BY IAPP AND OTHER POLYPEPTIDES

There has been considerable interest in determining the factors and specific interactions that lead to amyloid formation. Aromatic–aromatic and aromatic–hydrophobic interactions are known to contribute to the stability of globular proteins and have been suggested to play a critical role in amyloid formation. Initial experiments involving alanine scanning of short peptides containing a single aromatic residue seemed to support this conjecture [64–66]. In particular, substitution of aromatic residues in short peptide fragments derived from IAPP by alanine decreased amyloid formation significantly or abolished it. Alanine is, however, a nonconservative substitution for phenylalanine, particularly with regard to amyloid formation since it is smaller, less hydrophobic, and has a smaller β-sheet propensity. In fact, aromatic–aromatic interactions are not required for amyloid formation by IAPP-derived peptides and a Leu can replace a Phe without comprising the ability to form amyloid [61]. The effects of mutating aromatic residues on the rate of amyloid formation by acyl phosphatase can also be accounted for by changes in hydrophobicity and β-sheet propensity without invoking aromatic–aromatic interactions [67]. Thus, it is clear that the original conjecture that aromatic–aromatic interactions are strictly required is incorrect. Even though they are not required for amyloid formation, they might still play a role in either helping dictate the structure of the fibril or in the kinetics of self-assembly or the stability of the fibril.

Aromatic–aromatic interactions are also not required for amyloid fibril formation by intact mature IAPP. IAPP contains two Phe's at positions 15 and 23 as well as a C-terminal tyrosine, but a triple mutant F15L, F23L, Y37L readily forms amyloid. However the rate of amyloid formation is slower for the triple mutant with an approximately five-fold-longer lag phase, and the morphology of the resulting fibrils is slightly different [68]. FRET studies have suggested that the single tyrosine of IAPP is in close proximity to one or more of the Phe's in prefibril intermediates [69]. The observation that removal of all of the aromatic residues slows amyloid formation is consistent with these interactions playing a kinetically important role.

STRUCTURAL MODELS OF THE IAPP AMYLOID PROTOFILAMENT

Structural studies of IAPP amyloid fibrils indicate that synthetic IAPP forms protofilaments that self-associate to form various fibrillar morphologies, ranging from thin single protofilaments to ropelike profilamentous bundles and twisted ribbons. Transmission electron microscopy (TEM), scanning transmission electron microscopy (STEM) mass per unit length analysis, and atomic force microscopy (AFM) reveal that the morphology of synthetic amyloid fibrils derived from human IAPP vary depending on the conditions at which they assemble [48,70,71]. In one model, fibrils that assemble from full-length IAPP predominantly contain three protofilaments which wrap around one another in a left-handed coil, while fibrils formed by the 8–37 sequence are made up primarily of two protofilaments [72]. Data from x-ray diffraction, EM, and cryo-EM have reinforced the view that IAPP amyloid fibrils, like other amyloid fibrils, consist of a cross-β arrangement of β-strands that run perpendicular to the fibril axis with the hydrogen bonds oriented parallel to the long axis of the fibril [73].

Mature IAPP has a disulfide bond between cysteines at positions 2 and 7, which is conserved in all species. Residues 1 to 7 at the N-terminus are probably not part of the amyloid fibril, due to the conformational restrictions imposed on the backbone by the disulfide bridged loop. Several atomic-level models have been proposed for the IAPP protofilament. The recent solid-state nuclear magnetic resource (NMR) structure from the National Institutes of Health (NIH) group relies on the most experimental constraints (solid-state NMR and mass per unit length measurements) and is thus, in our opinion, likely to be the most accurate. The NIH structure is strikingly similar in its overall appearance to the solid-state NMR structure of the Aβ protofilament [74]. The basic building block is a four-layer parallel β-sheet structure made up of two symmetry-related IAPP molecules. Each chain contains two β-strand regions, comprised of residues 8–17 and 28–37 with a single bend involving residues 18–27 (Fig. 2). The two β-strands are not hydrogen bonded to each other but are hydrogen bonded to the same region of other polypeptide chains, forming an intermolecular parallel β-sheet. A particularly interesting feature of this model is that a significant portion of the 20–29 region is not part of a classic β-strand but, rather, is part of the well-ordered bend that connects the two β-strands. This leads to obvious questions about the molecular origin of the sensitivity of IAPP formation to substitutions in the 20–29 region. Two other models have been proposed in recent years. A study utilizing electron paramagnetic resonance spectroscopy of spin-labeled derivatives of human IAPP reported a parallel arrangement of β-strands in the fibrillar structure [75]. Nitroxide spin-labeled sites in eight single-cysteine analogs of IAPP showed, as expected, that monomeric IAPP has a high degree of disorder and undergoes rapid motion on the subnanosecond time scale. In contrast, fibrillar IAPP was found to be highly ordered in most regions, particularly in the central 12–29 region, with the same residues from different strands in close proximity to one another, indicating a parallel arrangement of IAPP peptides within the fibril. Labeled sites within the N-terminus of the peptide were slightly less immobilized, consistent with

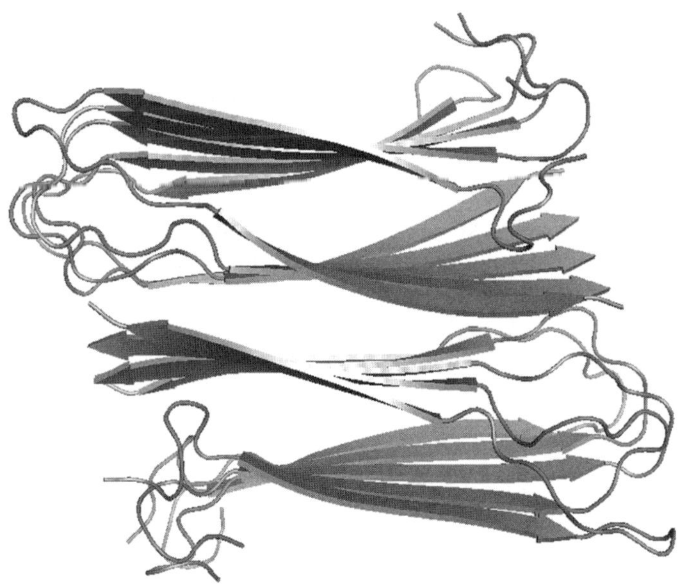

FIG. 2 Ribbon diagram of the core unit of the recently described solid-state NMR structure of the IAPP protofilament [74].

previous analyses of IAPP fragments showing that residues 1 to 7 are not required for fibril formation [59,62,72]. Based on these data, a model for the full-length IAPP fibril was proposed in which the protofibril is made up of parallel hydrogen-bonded stacks of β-strands that run from residues 12 to 37. The feature of parallel β-strands is consistent with the solid-state NMR model, but the notion of one long continuous β-strand is not. A third competing model invokes a parallel superpleated β-structure in which individual polypeptides take on a planar S-shaped or β-serpentine fold made of three β-strands spanning residues 12–17, 22–27 and 31–37 [76]. The serpentines are stacked axially and in register. This arrangement generates an array of parallel sheets in cross-β conformation. The interior is predicted to be stabilized by side-chain packing and three hydrogen-bonded "Asn ladders" created from the stacking of Asn22, Asn31 and Asn35 in successive molecules, similar to those found in β-helical conformations [77]. The first 11 residues at the N-terminus, which contains the disulfide-bonded loop, do not adopt the regular axial stacking and are not compatible with extending the serpentine. Again, the model shares some features with the solid-state NMR model but differs in others. Like the NMR structure, it argues against the chain adopting one long β-strand, and like the NMR model, it postulates parallel β-strands, but the β-serpentine model proposes three strand segments and two bends per chain. One thing to bear in mind when considering various models is that amyloid fibrils are known to be polymorphic and the various polymorphs could differ in a variety of structural aspects [78].

KINETICS OF IN VITRO AMYLOID FORMATION BY IAPP

The kinetics of in vitro amyloid formation are complex, and IAPP is no exception. Like other amyloidogeneic peptides and proteins, it displays a lag phase followed by a much more rapid growth phase (also called the *elongation phase*), and the lag phase can be abolished by seeding a solution of unaggregated peptide with a small amount of preformed fibrils. There is a rich experimental and theoretical literature on protein assembly and aggregation, and various kinetic models have been used to rationalize the time course of amyloid formation [79,80]. Other chapters in this volume address the kinetics of amyloid formation; thus, we omit an in-depth discussion here. Perhaps the classic example of a protein assembly reaction is that of actin, which is described by a nucleation-dependent mechanism; however, the model cannot explain all amyloid formation reactions. In particular, the existence of the characteristic lag phase is not well predicted [80]. Double nucleation schemes, initially developed by Eaton and co-workers during their pioneering studies of hemoglobin S polymerization, have been invoked to explain the existence of a lag time [81]. Double-nucleation models include primary nucleation steps similar to the nucleation-dependent polymerization model but they also invoke a second nucleation step which is dependent on the presence of fibrils (i.e., the fibrils can act as secondary nucleation sites). The second step is unimportant before fibrils are formed but becomes increasingly important as the amount of fibrils increase. The presence of a second, fibril-dependent nucleation step leads to a much sharper curve.

An extension to the double-nucleation model has been invoked to rationalize IAPP fibrillization kinetics [82]. The lag phase in IAPP fibrillization is only weakly dependent on concentration, which is thought to be incompatible with the double-nucleation models. In addition, the seeding behavior also differs from that predicted. The revised model, termed *dispersed phase mediated fibrillogenesis*, includes an activation step and importantly, invokes a dispersed phase that acts as a reservoir of monomers. The release of low concentrations of monomer from the dispersed phase reservoir is postulated to supply a constant concentration of momomer and is used to rationalize the concentration independence of the lag phase. The elongation phase accelerates as nucleation proceeds, due to the formation of an increasing number of fibril ends. Ultimately, elongation becomes first order when it is limited by the rate of breakdown of the dispersed-phase reservoir to soluble monomers. The dispersed-phase mechanism does an excellent job of modeling existing in vitro biophysical data and is undoubtedly the most rigorously tested kinetic scheme that has been applied to IAPP fibrillization. Essentially all in vitro experiments with IAPP involve the preparation of a stock solution in neat HFIP or other organic solvents, with fibril formation being initiated by dilution of the stock solution into aqueous buffer. The experiments that lead to the model were conducted using these protocols. The thermodynamics of HFIP–water solutions in particular, and water–fluoroalcohol solutions in general, are quite complicated, and there have been reports that HFIP forms clusters in aqueous systems, and highly fluorous compounds are usually not completely miscible in water [83,84]. Thus, the dispersed phase may be specific to this mode of sample preparation. There are other kinetic schemes that can, in principle, account for a weak dependence of the lag time on the total peptide concentration, although they have not yet been applied to IAPP. Analysis of several models shows that the concentration dependence weakens as the monomer concentration increases and can even invert [85]. The weakened concentration dependence results because the concentration of oligomers increases as the concentration of monomers is increased. Eventually, at some concentration the oligomers will be comparable in stability to the monomers, and the dependence of lag time on monomer concentration will disappear.

One difficulty in the field is that it is very difficult to compare studies conducted by different groups since the rate of amyloid formation is very sensitive to environmental parameters. A partial list of factors that have been shown to have a significant effect include the amount and type of cosolvent, pH, ionic strength, whether or not samples are agitated, the type of glass used in spectrophotometer cells, the shape and size of sample cells, and the volume of solution contained within them. The latter two factors probably reflect surface/volume ratios and the size of the air–water interface relative to the volume.

ISLET AMYLOID FORMATION IN TYPE 2 DIABETES AND IN ISLET CELL TRANSPLANTATION

Islet amyloid is a common feature in type 2 diabetes, and the evidence suggests that it plays an important role in the loss of β-cell mass and hence in the decrease in insulin production. For recent reviews, see [15–18,20]. Islet amyloid also has important implications for islet transplantation. The transplantation of pancreatic islets is potentially the ideal therapy for the treatment of type 2 diabetes, offering the prospect of complete glucose control; however, complications may arise from islet amyloid formation. For example, human islets when cultured or transplanted into nude mice develop amyloid deposits and this has raised major concerns about the impacts of amyloid formation on islet transplantation [86–88].

The mechanism or mechanisms of islet amyloid formation in vivo are not known, nor is the site of initiation of amyloid formation. Normal people produce IAPP throughout their life and yet do not form amyloid. With one exception, there is no evidence that mutations in the IAPP gene play a role in islet amyloid deposition. The single documented exception is a S20G mutation among a small Japanese cohort which leads to accelerated amyloid formation [89]. Insulin resistance in type 2 diabetes leads to abnormal levels of secretion of both insulin and IAPP, but the increase in local concentration brought about by increased insulin/IAPP secretion is not believed to be the cause of amyloid formation. Studies with transgenic mice have shown that overexpression of the human IAPP gene does not lead to amyloid formation in the absence of other factors [90]. Other components of islet amyloid include apolipoprotein E (apoE), serum amyloid P component (SAP), and the heparan sulfate proteoglycan (HSPG) perlecan [91–95]. In Alzheimer's disease there is a direct correlation between levels of apoE and extent of amyloid formation by Aβ [96,97]. This is not the case in type 2 diabetes, and the apoE knockouts do not affect islet amyloid formation [94,98,99], and there is also no correlation between the presence of SAP and islet amyloid deposition [95]. As described below, there is growing evidence which suggests that interactions with the glycosaminoglycan (GAG) component of HSPGs may play a critical role in islet amyloid formation.

Several recent papers have appeared which lay out a case that aberrant prohormone processing of proIAPP could be a crucial factor in islet amyloid formation [100–102]. Defects in the intragranular processing of prohormones and the premature secretion of immature granules occur in type 2 diabetes, leading to a significant increase in the release of incorrectly processed prohormones. Release of improperly processed insulin increases from 3% to circa 30% in type 2 diabetes, and the fraction of incompletely processed proIAPP is also increased [103,104]. Processing at the C-terminal cleavage site of proIAPP is normal, probably because this step may occur in the trans-Golgi and in the immature granule [9]. However, processing at the N-terminal cleavage site is impaired, leading to significant production of a 48-residue processing intermediate (NproIAPP) which contains the N-terminal prosequence [9,42,100–102]. Immunohistochemical investigations of islet amyloid deposits have demonstrated the presence of NproIAPP in pancreatic amyloid deposits [105,106]. Verchere and co-workers have proposed that amyloid formation is initiated by the binding of NproIAPP to the GAG portion of the proteoglycan components of the extracellular matrix (Fig. 3) [107]. The HSPG perlecan is believed to act as a scaffold for amyloid formation, whereby binding of partially processed proIAPP generates a high local concentration of aggregation-prone peptides, which could act as a seed for amyloid formation [91–93,107,108]. Perlecan is expressed on islet cells and has been implicated in virtually all the amyloidoses [91–93,108]. Biophysical studies have provided direct evidence that interactions with HSPGs promote amyloid formation by NproIAPP and demonstrated that NproIAPP fibrils can seed amyloid formation by mature IAPP [109]. Interaction with heparan sulfate was shown to lead to the rapid formation of an intermediate state, which then converted, on a slower time scale, to amyloid fibrils. The amyloid formed by the proIAPP processing intermediate in the presence and absence of heparan sulfate has the classic features of amyloid. More work remains to be done to establish that these are important interactions in vivo. One interesting feature to emerge from these studies was the observation of an intermediate with partial helical structure. Although this may seem counterintuitive given that the final fibril structure is rich in β-sheet, there is a precedent for helical intermediates in amyloid formation. Partial helical structure is induced when IAPP interacts with model vesicles, and helical structure has been postulated to play a role in the membrane-mediated association of IAPP (see below). In addition, Kirkitadze and co-workers have presented compelling evidence implicating an on-pathway helical intermediate in Aβ amyloid formation [110].

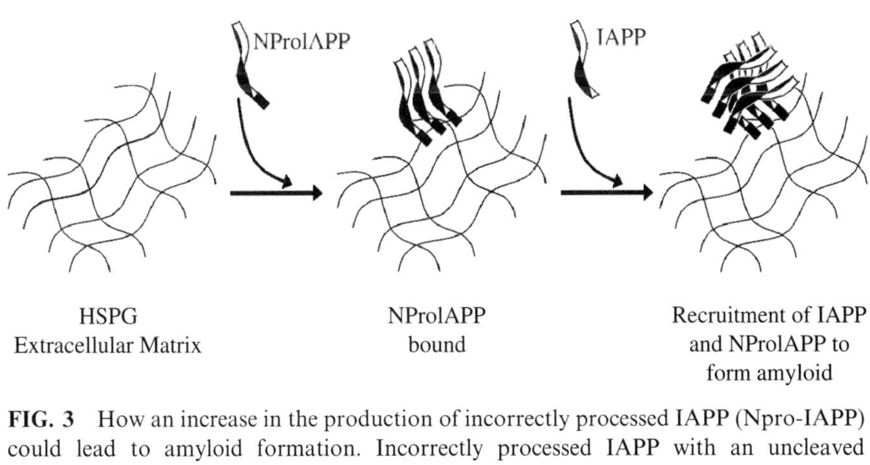

FIG. 3 How an increase in the production of incorrectly processed IAPP (Npro-IAPP) could lead to amyloid formation. Incorrectly processed IAPP with an uncleaved N-terminal extension (shaded) binds to HSPGs in the extracellular matrix, resulting in a high local concentration of the polypeptide, which in turn act as a seed for amyloid formation.

Experiments with cultured cells and with transgenic mice support an important role for processing intermediates in islet amyloid generation. Production of NproIAPP leads to amyloid formation and cell death in GH3 cells and in transgenic mice expressing human IAPP. Furthermore, inhibition of GAG synthesis prevents islet amyloid formation in islets, providing even more evidence for the importance of IAPP GAG interactions, although the study could not distinguish between effects on IAPP and the proIAPP intermediate [111].

There is also good evidence that NproIAPP and/or IAPP can form intragranule amyloid, leading to the suggestion that the release of intracellular amyloid via exocytosis could initiate extracellular amyloid formation by IAPP and NproIAPP [101]. Thus, there are two models of how processing defects could contribute to islet amyloid, one involving extracellular formation of amyloid mediated by interactions with GAGs and the second involving intracellular amyloid formation. The two processes are not mutually exclusive and they could act in parallel. The intracellular model leads to the obvious question of why NproIAPP forms intragranular amyloid whereas IAPP normally does not. One possibility is that insulin/insulin-processing intermediates are less effective in inhibiting the aggregation of NproIAPP. However, there have been no studies of the interaction of insulin or pro-insulin or its processing intermediates with NproIAPP. IAPP membrane interactions have been proposed to play a role in amyloid formation (see below), and the formal possibility exists that NproIAPP might interact more aggressively with membranes.

IAPP MEMBRANE INTERACTIONS

Polypeptide membrane interactions have been implicated as catalyzing amyloid formation and as a potential mechanism of toxicity. A wide range of amyloidogenic polypeptides have been shown to interact with model membrane systems, and IAPP is no exception [29,112–124]. It is well documented that peptide membrane interactions promote amyloid formation by IAPP. It is difficult to judge the physiological relevance of some of these studies, however, since some, but not all, used nonphysiological lipid compositions. It is also difficult to compare different studies because many used different lipid compositions or different model membrane systems (e.g., vesicles vs. planer bilayers). Electrostatic interactions involving negatively charged lipids and positively charged IAPP probably play an important role in IAPP membrane interactions, and the largest enhancement of IAPP aggregation occurs with high contents of negatively charged lipids. However, significant rate enhancements have still been observed with phosphatidylserine compositions on the order of 1 to 10%, which is in line with the reported content of phosphatidylserine in rodent islets [120,125].

The detailed molecular-level mechanism of how membranes promote IAPP aggregation is not yet known, but independent reports from two laboratories have indicated that helical intermediates play an important role [115,118]. In solution, the monomeric form of the polypeptide does not fold to a compact native state, but the region encompassing residues 5 through 20 appears to have a modest tendency to populate the helical region of the φ, ψ map transiently as judged by circular dichroism and NMR studies, and by analogy to CGRP and calcitonin [126–129]. The structures of IAPP and CGRP have been determined in mixed fluoroalcohol–water mixtures and are very similar, adopting helical structure from residues 8 to 18 in CGRP and between residues 5 and 20 in IAPP [128,129]. Interaction with negatively charged lipids induces partial helical structure, which presumably promotes association of individual polypeptides via helix–helix interactions (Figure 4). The figure is designed to summarize the results of recent studies, but is necessarily highly schematic. The spectroscopic data do not allow the location of the initial membrane-bound helical structure to be deduced, but published data on IAPP argue that it is likely to be located in the N-terminal 50 to 60% of the molecule. It is very interesting that rat IAPP also forms helical structure when it interacts with membranes, but does not assemble into amyloid [118]. Rat IAPP differs from human IAPP at only one postion between residues 1 and 23 (a His-to-Arg substitution at residue 18). The conformations of rat and human IAPP are very similar in mixed water–fluoroalcohol solutions, and rat IAPP has a tendency to populate the helical region of the φ, ψ map transiently for residues 5 to 20 in aqueous solution [127,128]. Thus, the rat peptide probably adopts a helical conformation similar to that on the membrane, but the multiple prolines in the C-terminal region inhibit β-sheet formation. Figure 4 implies that helix formation and assembly occur on the membrane surface, but it is possible that oligomerization of IAPP occurs before it interacts with membranes in vivo, or under different conditions in vitro [29]. Along these lines, there is evidence that IAPP can form helical assemblies in solution [130].

In the last five to 10 years, a number of research groups have reported that oligomeric intermediates of amyloidogenic polypeptides can permeabilize membranes. How IAPP permeabilizes membranes is not understood and, indeed, some would argue that this is true for amyloidogenic polypeptides in general [131]. Two mechanisms have been proposed; one involves the action of specific pores (channels) and the other invokes a less specific detergent-like mechanism. In the second model, surface-associated IAPP form micelle-like structures that contain peptide and lipids. Evidence in support of both models is available in the literature [29,112,114,119–124]. Again, one of the difficulties in the field is that studies have been conducted under a range of conditions, including different lipid compositions, different buffers, and different protocols for preparing IAPP–lipid mixtures. Thus, comparison of different experiments can be difficult. Along these lines, it is worth noting that studies of membrane-active host defense peptides have shown that different mechanisms can operate, depending on the lipid composition.

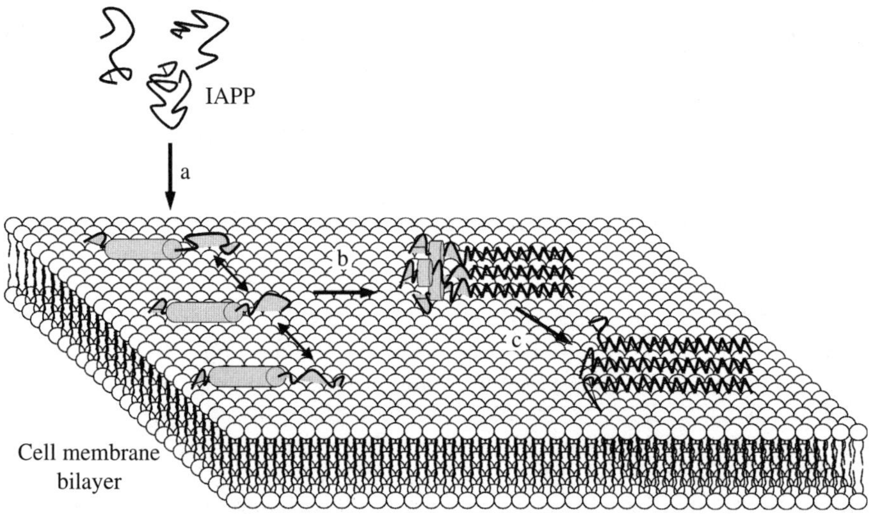

FIG. 4 How interactions between IAPP and membranes could promote amyloid formation. Interactions with membranes promote formation of partial helical structure. Association of the polypeptides via helix-mediated interactions leads to a high local concentration of the very amyloidogenic C-terminal region of IAPP. The diagram is schematic and meant to summarize recent studies of IAPP membrane interactions. IAPP may self-associate prior to interacting with membranes in vivo or in vitro under different conditions, and both oligomeric and monomeric IAPP may interact with membranes.

HELICAL INTERMEDIATES AND AMYLOID FORMATION BY IAPP: A GENERAL PHENOMENON?

Studies of NproIAPP HSPG interactions and the investigation of IAPP membrane systems outlined in the preceding subsections show that IAPP/NproIAPP populates a helical intermediate during amyloid formation in these heterogeneous systems. The membranes studied and HSPGs have little in common except that they present polyanionic surfaces. However, interactions with polyanions are not required to produce a helical intermediate. In vitro, studies of IAPP fibrillization have provided evidence for production of a helical intermediate in homogeneous solution [130]. How could the formation of helical structure promote conversion to partially ordered β-sheet structure? Helix formation and self-association are well known to be linked in many systems; classic examples include naturally occurring coiled coils and other oligomerization domains as well as peptides with a tendency to form amphiphillic helices, and various designed systems. IAPP has a predicted propensity to form an amphiphilic helix between residues 5 to 18 or 5 to 20. Thus, initial formation of IAPP oligomers might be driven by the linkage between helix formation and association. Assembly of oligomeric helical bundles would lead to a high local concentration of the highly amyloidogenic C-terminal region of IAPP, which in turn might promote intermolecular β-sheet formation [60].

INHIBITORS OF IAPP AMYLOID FORMATION

The design of inhibitors of amyloid formation and toxicity is an active area of research [132,133]. A diverse range of peptide-based and small organic ligands have been shown to act as inhibitors of IAPP fibril formation [133–141]. There is a large and growing literature on inhibitors of amyloid formation, particularly for the Aβ peptide, but also for IAPP [133]. Inhibitors of Aβ aggregation are in phase III clinical trials, and a recent clinical trial of an inhibitor of amyloid A amyoidosis is very encouraging [132,142]. However there are significant problems with IAPP inhibitors. Almost all inhibitors of fibril formation by IAPP reported to date are effective only at significant molar excess; two exceptions being a recently described point mutant of IAPP and a doubly N-methylated variant of IAPP [143,144]. Hopefully, careful studies of the mechanism of IAPP fibrillization will provide insight that leads to improved inhibitors. Inhibitors can target a variety of stages in fibril assembly, and a large fraction of existing IAPP inhibitors have little or no effect on the lag phase and, instead, affect the amount of amyloid produced (as judged by thioflavin-T binding). Such compounds are unlikely to be particularly useful if prefibril species are the toxic entities, since presumably they inhibit fibril elongation and not oligomer formation. A rarer class of inhibitors lengthen the lag phase, presumably by interfering with nucleus formation or by binding to the nucleus and sequestering it. In an ideal world, these inhibitors would also prevent fibril formation.

The development of inhibitors of Aβ aggregation is obviously an area of intense interest. IAPP and Aβ share some sequence similarity, and Aβ fibrils can cross-seed fibrillization of IAPP, suggesting that inhibitors of IAPP fibril formation might inhibit fibril formation by Aβ. Recent work has demonstrated that inhibitors of IAPP can indeed block fibril formation by Aβ (1–40) [145]. Thus, the successful design of IAPP fibril formation inhibitors may well have a major impact beyond islet amyloid formation.

ROLE OF IAPP ANALOGS IN THE TREATMENT OF TYPE 1 DIABETES

Since IAPP is produced by the pancreatic β-cells, it is deficient in people suffering from type 1 diabetes. Coadministration of IAPP with insulin helps to normalize fluctuating glucose levels; however, the extremely amyloidogenic nature of human IAPP prevents its direct use as an adjunct to insulin therapy [146,147]. An analog of human IAPP in which residues 25, 28, and 29 have been replaced by proline (the substitutions correspond to the position of prolines in rat IAPP) has been approved by the U.S. Food and Drug Administration for use in types 1 and 2 diabetes.

Acknowledgments

We thank P. Marek for critically reading the manuscript and C. Bruce Verchere for helpful discussions. We thank Robert Tycko for providing us with the coordinates of the solid-state NMR structure of the IAPP protofilament.

REFERENCES

1. Opie, E.L. (1901). The relation of diabetes mellitus to lesions of the pancreas: hyaline degeneration of the islets of Langerhans. *J Exp Med, 5*, 527–540.
2. Weichselbaum, A., Stangl, E. (1901). Zur Kenntis der feineren Veränderungen des Pankreas bei Diabetes mellitus. *Wien Klin Wochenschr, 14*, 968–972.
3. Westermark, P., Wernstedt, C., Wilander, E., Hayden, D.W., O'Brien, T.D., Johnson, K.H. (1987). Amyloid fibrils in human insulinoma and islets of Langerhans of the diabetic cat are derived from a neuropeptide-like protein also present in normal islet cells. *Proc Natl Acad Sci U S A, 84*, 3881–3885.
4. Cooper, G.J.S., Willis, A.C., Clark, A., Turner, R.C., Sim, R.B., Reid, K.B.M. (1987). Purification and characterization of a peptide from amyloid-rich pancreases of type 2 diabetic patients. *Proc Natl Acad Sci U S A, 84*, 8628–8632.
5. Cooper, G.J.S. (1994). IAPP compared with calcitonin gene-related peptide: structure, biology, and relevance to metabolic disease. *Endocrine Rev, 15*, 163–201.
6. Hay, D.L., Christopoulos, G., Christopoulos, A., Sexton, P.M. (2004). Amylin receptors: molecular composition and pharmacology. *Biochem Soc Trans, 32*, 865–867.
7. McLatchie, L.M., Fraser, N.J., Main, M.J., Wise, A., Brown, J., Thompson, N., Solari, R., Lee, M.G., Foord, S.M. (1998). RAMPS regulate the transport and ligand specificity of the calcitonin-receptor-like receptor. *Nature, 393*, 333–339.
8. Sanke, T., Bell, G.I., Sample, C., Rubenstein, A.H., Steiner, D.F. (1988). Islet amyloid peptide is derived from an 89-amino acid precursor by proteolytic processing. *J Biol Chem, 263*, 17243–17246.
9. Marzban, T., Trigo-Gonzalez, G., Verchere, C.B. (2005). Processing of pro-islet amyloid polypeptide in the constitutive and regulated secretory pathways of beta cells. *Mol Endocrinol, 19*, 2154–2163.
10. Lukinius, A., Wilander, E., Westermark, G.T., Engström, U., Westermark, P. (1989). Co-localization of islet amyloid polypeptide and insulin in the beta-cell secretory granules of the human pancreatic islets. *Diabetologia, 32*, 240–244.
11. Kahn, S.E., D'Alessio, D.A., Schwartz, M.W., Fujimoto, W.Y., Ensinck, J.W., Taborsky, G.J., Porte, D. (1990). Evidence of cosecretion of islet amyloid polypeptide and insulin by beta-cells. *Diabetes, 39*, 634–638.
12. Hanabusa, T., Kuba, K., Oki, C., Nakano, Y., Okai, K., Sanke, T., Nanjo, K. (1992). Islet amyloid polypeptide (IAPP) secretion from islet cells and its plasma concentration in patients with non-insulin-dependent diabetes mellitus. *Diabetes Res Clin Pract, 15*, 89–96.

13. Stridsberg, M., Sandler, S., Wilander, E. (1993). Cosecretion of islet amyloid polypeptide (IAPP) and insulin from isolated rat pancreatic islets following stimulation or inhibition of beta-cell function. *Regul Pept, 45*, 363–370.

14. Kahn, S.E., Andrikopoulos, S., Verchere, C.B. (1999). Islet amyloid: a long recognized but underappreciated pathological feature of type II diabetes. *Diabetes, 48*, 241–246.

15. Hull, R.L., Westermark, G.T., Westermark, P., Kahn, S.E. (2004). Islet amyloid: a critical entity in the pathogenesis of type 2 diabetes. *J Clin Endocrinol Metab, 89*, 3629–3643.

16. Jaikaran, E., Clark, A. (2001). Islet amyloid and type 2 diabetes: from molecular misfolding to islet pathophysiology. *Biochim Biophys Acta, 1537*, 179–203.

17. Phillips, L.K., Horowitz, M. (2006). Amylin. *Curr Opin Endocrinol Diabetes, 13*, 191–198.

18. Hoppener, J.W.H., Lips, C.J.M. (2006). Role of islet amyloid in type 2 diabetes mellitus. *Int J Biochem Cell Biol, 38*, 726–736.

19. Johnson, K.H., O'Brien, T.D., Betsholtz, C., Westermark, P. (1989). Islet amyloid, islet-amyloid polypeptide, and diabetes mellitus. *N Engl J Med, 321*, 513–518.

20. Clark, A., Nilsson, M.R. (2004). Islet amyloid: a complication of islet cell dysfunction or an aetiological factor in type 2 diabetes? *Diabetologia, 47*, 157–169.

21. Konarkowska, B., Aiteken, J.F., Kistler, J., Zhang, S. P., Cooper, G.J.S. (2006). The aggregation potential of human amylin determines its cytotoxicity towards islet beta-cells. *FEBS J, 273*, 3614–3624.

22. Clark, A., Wells, C.A., Buley, I.D., Cruickshank, J.K., Vanhegan, R.I., Matthews, D.R., Cooper, G.J.S., Holman, R.R., Turner, R.C. (1988). Islet amyloid, increased α-cells, reduced β-cells and exocrine fibrosis: quantitative changes in the pancreas in type 2 diabetes. *Diabetes Res Clin Pract, 9*, 151–159.

23. Lorenzo, A., Razzaboni, B., Weir, G.C., Yankner, B.A. (1994). Pancreatic islet cell toxicity of amylin associated with type 2 diabetes mellitus. *Nature, 368*, 756–760.

24. Rocken, C., Linke, R.P., Saeger, W. (1992). Immunohistology of islet amyloid polypeptide in diabetes mellitus: semiquantitative studies in a post mortem series. *Virchows Arch A, 421*, 339–344.

25. Hayden, M.R., Karuparthi, R. R., Manrique, C.M., Lastra, G. Habibi, J., Sowers, J.R. (2007). Longitudinal ultrastructure study of islet amyloid in the HIP rat model of type 2 diabetes mellitus. *Exp Biol Med, 232*, 772–779.

26. Koning, E.J., Bodkin, N.L., Hansen, B.C., Clark, A. (1993). Diabetes mellitus in macaca mulatta monkeys is characterized by islet amyloidosis and reduction in β-cell population. *Diabetologica*, *36*, 378–384.

27. Wang, F., Hull, R.L., Vidal, J., Cnop, M., Kahn, S.E. (2001). Islet amyloid develops diffusely throughout the pancreas before becoming severe and replacing endocrine cells. *Diabetes*, *50*, 2514–2520.

28. Kayed, R., Head, E., Thompson, J.L., McIntire, T.M., Milton, S.C., Cotman, C.W, Glabe, C.G. (2003). Common structure of soluble amyloid oligomers implies common mechanism of pathogenesis. *Science*, *300*, 486–489.

29. Janson, J., Ashley, R.H., Harrison, D., McIntyre, S., Butler, P.C. (1999). The mechanism of islet amyloid polypeptide toxicity is membrane disruption by intermediate-sized toxic amyloid particles. *Diabetes*, *48*, 491–498.

30. Ritzel, R.A., Meier, J.J., Lin, C.-Y., Veldhuis, J.D., Butler, P.C. (2007). Human islet amyloid polypeptide oligomers disrupt cell coupling, induce apoptosis, and impair insulin secretion in isolated human islets. *Diabetes, 56*, 65–71.

31. Butler, P.C., Chou, J., Carter, W.B., Wang, Y.N., Bu, B.H., Chang, D., Chang, J.K., Rizza, R.A. (1990). Effects of meal ingestion on plasma amylin concentration in NIDDM and nondiabetic humans. *Diabetes, 39*, 752–756.

32. Leighton, B., Cooper, G.J.S. (1988). Pancreatic amylin and calcitonin gene-related peptide cause resistance to insulin in skeletal-muscle in vitro. *Nature, 335*, 632–635.

33. Ohsawa, H., Kanatsuka, A., Yamaguchi, T., Makino, H., Yoshida, S. (1989). Islet amyloid polypeptide inhibits glucose-stimulated insulin secretion from isolated rat pancreatic islets. *Biochem Biophys Res Commun, 160*, 961–967.

34. Akesson, B., Panagiotidis, G., Westermark, P., Lundquist, I. (2003). Islet amyloid polypeptide inhibits glucagon release and exerts a dual action on insulin release from isolated islets. *Regul Pept, 111*, 55–60.

35. Wookey, P.J., Tikellis, C., Du, H.C., Qin, H.F., Sexton, P.M., Cooper, M.E. (1996). Amylin binding in rat renal-cortex, stimulation of adenylyl-cyclase, and activation of plasma-renin. *Am J Phys, 39*, F289–F294.

36. Clementi, G., Caruso, A., Cutuli, V.M.C., de Bernardis, E., Prato, A., Amico-Roxas, M. (1996). Amylin given by central or peripheral routes decreases gastric emptying and intestinal transit in the rat. *Experientia, 52*, 677–679.

37. Rushing, P.A., Hagan, M.M., Seeley, R.J., Lutz, T.A., D'Alessio, D.A., Air, E.L., Woods, S.C. (2001). Inhibition of central amylin signaling increases food intake and body adiposity in rats. *Endocrinology, 142*, 5035–5038.

38. Wookey, P.J., Xuerebi, L., Tikellis, C., Cooper, G.J.S. (2003). Amylin in the periphery. *Sci World, 3*, 163–175.

39. Riediger, T., Rauch, M., Schmid, H.A. (1999). Actions of amylin on subfornical organ neurons and on drinking behaviour in rats. *Am J Physiol, 276*, R514–R521.

40. Chin, S.Y., Hall, J.M., Brain, S.D., Morton, I.K.M. (1994). Vasodilator responses to calcitonin gene-related peptide and IAPP in the rat isolated-perfused kidney are mediated via CGRP1 receptors. *J Pharm Exp Ther, 269*, 989–992.

41. Reidelberger, R.D., Haver, A.C., Arnelo, U., Smith, D.D., Schaffert, C.S., Permert, J. (2004). Amylin receptor blockade stimulates food intake in rats. *Am J Physiol Regul Integr Comp Physiol, 287*, R568–R574.

42. Wang, J., Xu, J., Finnerty, J., Furuta, M., Steiner, D.F., Verchere, C.B. (2001). The prohormone convertase enzyme 2 (PC2) is essential for processing proislet amyloid polypeptide at the NH_2-terminal cleavage site. *Diabetes, 50*, 534–539.

43. Marzban, L., Trigo-Gonzalez, G., Zhu, X., Rhodes, C.J., Halban, P.A., Steiner, D.F., Verchere, C.B. (2004). Role of β-cell prohormone convertase PC (1/3) in processing of pro-islet amyloid polypeptide. *Diabetes, 53*, 141–148.

44. Roberts, A.N., Leighton, B., Todd, J.A., Cockburn, D., Schofield, P.N., Sutton, R., Holt, S., Boyd, Y., Day, A.J., Foot, E.A., et al. (1998). Molecular and functional characterization of amylin, a peptide associated with type 2 diabetes mellitus. *Proc Natl Acad Sci U S A, 86*, 9662–9666.

45. Hutton, J.C. (1989). The insulin secretory granule. *Diabetologia, 32*, 271–281.

46. Hutton, J.C. (1982). The internal pH and membrane potential of the insulin-secretory granule. *Biochem J, 204*, 171–178.

47. Chargé, S.B.P., de Koning, E.J.P., Clark, A. (1995). Effect of pH and insulin on fibrillogenesis of islet amyloid polypeptide in vitro. *Biochemistry, 34*, 14588–14593.
48. Abedini, A., Raleigh, D.P. (2005). The role of His-18 in amyloid formation by human islet amyloid polypeptide. *Biochemistry, 44*, 16284–16291.
49. Westermark, P., Li, Z.C., Westermark, G.T., Leckstrom, A., Steiner, D.F. (1996). Effects of beta cell granule components on human islet amyloid polypeptide fibril formation. *FEBS Lett, 379*, 203–206.
50. Janciauskiene, S., Eriksson, S., Carlemalm, E., Ahren, B. (1997). β-cell granule peptides affect human islet amyloid polypeptide (IAPP) fibril formation in vitro. *Biochem Biophys Res Commun, 236*, 580–585.
51. Jaikaran, E., Nilsson, M., Clark, A. (2004). Pancreatic β-cell granule peptides form heteromolecular complexes which inhibit islet amyloid polypeptide fibril formation. *Biochem J, 377*, 709–716.
52. Larson, J., Miranker, A.D. (2004). The mechanism of insulin action on islet amyloid polypeptide fiber formation. *J Mol Biol, 335*, 221–231.
53. Gilead, S., Wolfenson, H., Gazit, E. (2006). Molecular mapping of the recognition interface between the islet amyloid polypeptide and insulin. *Angew Chem Int Ed Engl, 45*, 6476–6480.
54. Betsholtz, C., Christmansson, L., Engstrom, U., Rorsman, F., Svensson, V., Johnson, K.H., Westermark, P. (1989). Sequence divergence in a specific region of islet amyloid polypeptide (IAPP) explains differences in islet amyloid formation between species. *FEBS Lett, 251*, 261–264.
55. Betsholtz, C., Christmanson, L., Engström, U., Rorsman, F., Jordan, K., O'Brian, T.D., Murtaugh, M., Johnson, K.H., Westermark, P. (1990). Structure of cat islet amyloid polypeptide and identification of amino acid residues of potential significance for islet amyloid formation. *Diabetes, 39*, 118–122.
56. Westermark, P., Engström, U., Johnson, K.H., Westermark, G.T., Betsholtz, C. (1990). Islet amyloid polypeptide: pinpointing amino acid residues linked to amyloid fibril formation. *Proc Natl Acad Sci U S A, 87*, 5036–5040.
57. Glenner, G.G., Eanes, E.D., Wiley, C.A. (1988). Amyloid fibrils formed from a segment of the pancreatic-islet amyloid protein. *Biochem Biophys Res Commun, 155*, 608–614.
58. Minor, D.L., Jr., Kim, P.S. (1994). Measurement of the beta-sheet-forming propensities of amino acids. *Nature, 367*, 660–663.
59. Nilsson, M.R., Raleigh, D.P. (1999). Analysis of amylin cleavage products provides new insights into the amyloidogenic region of human amylin. *J Mol Biol, 294*, 1375–1385.
60. Jaikaran, E., Higham, C.E., Serpell, L.C., Zurdo, J., Gross, M., Clark, A., Fraser, P.E. (2001). Identification of a novel human islet amyloid polypeptide β-sheet domain and factors influencing fibrillogenesis. *J Mol Biol, 308*, 515–525.
61. Tracz, S.M., Abedini, A., Driscoll, M., Raleigh, D.P. (2004). The role of aromatic interactions in amyloid formation by polypeptides: analysis of peptides derived from human amylin. *Biochemistry, 43*, 15901–15908.
62. Abedini, A., Raleigh, D.P. (2006). Destabilization of human IAPP amyloid fibrils by proline mutations outside of the putative amyloidogenic domain: is there a critical amyloidogenic domain in human IAPP? *J Mol Biol, 355*, 274–281.

63. Green, J., Goldsbury, C., Mini, T., Sunderji, S., Frey, P., Kistler, J., et al. (2003). Full-length rat amylin forms fibrils following substitution of single residues from human amylin. *J Mol Biol, 326*, 1147–1156.

64. Gazit, E. (2002). A possible role for pi-stacking in the self-assembly of amyloid fibrils. *FASEB J, 16*, 77–83.

65. Azriel, R., Gazit, E. (2001). Analysis of the minimal amyloid-forming fragment of the islet amyloid polypeptide. *J Biol Chem, 276*, 34156–34161.

66. Reches, M., Porat, Y., Gazit, E. (2002). Amyloid fibril formation by pentapeptide and tetrapeptide fragments of human calcitonin. *J Biol Chem, 277*, 35475–35480.

67. Bemporad, F., Taddei, N. Steffani, M., Chiti, F. (2006). Assessing the role of aromatic residues in the amyloid aggregation of human muscle acylphosphatase. *Protein Sci,15*, 862–870.

68. Marek, P., Abedini, A., Song, B., Kanungo, M., Johnson, M.E., Gupta, R., Zaman, W., Wong, S.S., Raleigh, D.P. (2007). Aromatic interactions are not required for amyloid fibril formation by islet amyloid polypeptide but do influence the rate of fibril formation and fibril morphology. *Biochemistry, 46*, 3255–3261.

69. Padrick, S.B., Miranker, A.D. (2001). Islet amyloid polypeptide: identification of long-range contacts and local order on the fibrillogenesis pathway. *J Mol Biol, 308*, 783–794.

70. Goldsbury, C.S., Cooper, G.J.S., Goldie, K.N., Muller, S.A., Saafi, E.L., Gruijters, W.T.M., Misur, M.P. (1997). Polymorphic fibrillar assembly of human amylin. *J Struct Biol, 119*, 17–27.

71. Goldsbury, C., Kistler, J., Aebi,U., Arvinte, T., Cooper, G.J.S. (1999). Watching amyloid fibrils grow by timelapse atomic force microscopy. *J Mol Biol, 285*, 33–39.

72. Goldsbury, C., Goldie, K., Pellaud, J., Seelig, J., Frey, P., Muller, S.A., Kistler, J., Cooper, G.J.S., Aebi, U. (2000). Amyloid fibril formation from full-length and fragments of amylin. *J Struct Biol, 130*, 352–362.

73. Makin, S.O., Serpell, L.C. (2004). Structural characterization of islet amyloid polypeptide fibrils. *J Mol Biol, 335*, 1279–1288.

74. Luca, S., Yau, W.-M., Leapman, R., Tycko, R. (2007). Peptide conformation and supramolecular organization in amylin fibrils: constraints from solid-state NMR. *Biochemistry, 46*, 13505–13522.

75. Jayasinghe, S.A., Langen, R. (2004). Identifying structural features of fibrillar islet amyloid polypeptide using site-directed spin labeling. *J Biol Chem, 279*, 48420–48425.

76. Kajava, A.V., Aebi, U., Steven, A.C. (2005). The parallel superpleated beta structure as a model for amyloid fibrils of human amylin. *J Mol Biol, 348*, 247–252.

77. Yoder, M.D., Keen, N.T., Jurnak, F. (1993). New domain motif: the structure of pectate lyase C, a secreted plant virulence factor. *Science, 260*, 1503–1507.

78. Gordon, D.J., Balbach, J.J., Tycko, R., Meredith, S.C. (2004). Increasing the amphiphilicity of an amyloidogenic peptide changes the beta-sheet structure in the fibrils from antiparallel to parallel. *Biophys J, 86*, 428–434.

79. Oosawa, F., Asakura, S. (1975). *Thermodynamics of the Polymerization of Protein*, Academic Press, New York.

80. Ferrone, F. (2006). Nucleation: the connections between equilibrium and kinetic behavior. *Methods Enzymol, 412*, 285–299.

81. Mozzarelli, A., Hofrichter, J., Eaton, W.A. (1987). Delay time of hemoglobin-S polymerization prevents most cells from sickling in vivo. *Science*, 237, 500–506.
82. Padrick, S.B., Miranker, A.D. (2002). Islet amyloid: phase partitioning and secondary nucleation are central to the mechanism of fibrillogenesis. *Biochemistry, 41*, 4694–4703.
83. Hong, D.P., Hoshino M., Kuboi R., Goto Y. (1999). Clustering of fluorine-substituted alcohols as a factor responsible for their marked effects on proteins and peptides. *J Am Chem Soc, 121*, 8427–8433.
84. Yoshida, K., Yamaguchi, T., Adachi, T., Otomo, T., Matsuo, D., Takamuku, T. Nishi, N. (2003). Structure and dynamics of hexafluoroisopropanol–water mixtures by x-ray diffraction, small-angle neutron scattering, NMR spectroscopy, and mass spectrometry. *J Chem Phys, 119*, 6132–6142.
85. Powers, E.T., Powers, D.L. (2006). The kinetics of nucleated polymerizations at high concentrations: amyloid fibril formation near and above the "supercritical concentration." *Biophys J, 91*, 122–132.
86. Westermark, P., Eizirik, D.L., Pipeleers, D.G., Hellersrom, C., Anderson, A. (1994). Rapid deposition of amyloid in human islets transplanted into nude mice. *Diabetologia, 38*, 543–549.
87. Westermark, P., Andersson, A., Westermark, G. (2005). Is aggregated IAPP a cause of beta-cell failure in transplanted human pancreatic islets? *Cur Diabetes Rep*, *5*, 185–188.
88. Westermark, G.T., Westermark, P., Nordin A., Tornelius E., Andersson, A. (2003). Formation of amyloid in human pancreatic islets transplanted to the liver and spleen of nude mice. *Upsala J Med Sci*, *108*, 193–203.
89. Ma, Z., Westermark, G.T., Sakagashira, S., Sanke, T., Gustavsson, A., Sakamoto, H., Engstrom, U., Nanjo, K., Westermark, P. (2001). Enhanced in vitro production of amyloid-like fibrils from mutant (S20G) islet amyloid polypeptide. *Amyloid*, *8*, 242–249.
90. Matveyenko, A.V., Butler, P.C. (2007). Islet amyloid polypeptide (IAPP) transgenic rodents as models for type 2 diabetes. *ILAR J, 47*, 225–233.
91. Young, I.D., Ailles, L., Narindrasorasak, S., Tan, R., Kisilevsky, R. (1992). Localization of the basement membrane heparan sulfate proteoglycan in islet amyloid deposits in type II diabetes mellitus. *Arch Pathol Lab Med, 116*, 951–954.
92. Conde-Knape, K. (2001). Heparan sulfate proteoglycans in experimental models of diabetes: a role for perlecan in diabetes complications. *Diabetes Metab Res Rev, 17*, 412–421.
93. Potter-Perigo, S., Hull, R.L., Tsoi, C., Braun, K.R., Andrikopoulos, S., Teague, J., Verchere, C.B., Kahn, S.E., Wright, T.N. (2003). Proteoglycans synthesized and secreted by pancreatic islet β– cells bind amylin. *Arch Biochem Biophys, 413*, 182–190.
94. Vidal, J., Verchere, C.B., Andrikopoulos, S., Wang, F., Hull, R.L., Cnop, M., Olin, K.L., LeBoeuf, R.C., O'Brian, K.D., Chait, A., Khan, S.E. (2003). The effect of apolipoprotein E deficiency on islet amyloid deposition in human islet amyloid polypeptide transgenic mice. *Diabetologica, 46*, 71–79.
95. Pepys, M.B., Booth, D.R., Hutchinson, W.L., Gallimore, J.R., Collins, P.M., Hohenester, E. (1997). Amyloid P component: a critical review. *Amyloid: Int J Exp Clin Invest, 4*, 274–295.

96. Schmechel, D.E., Saunders, A.M., Strittmatter, W.J., Crain, B.J., Hulette, C.M., Joo, S.H., Pericak-Vance, M.A., Goldgaber, D., Roses, A.D. (1993). Increased amyloid β-peptide deposition in cerebral cortex as a consequence of apolipoprotein E genotype in late-onset Alzheimer disease. *Proc Natl Acad Sci U S A, 90*, 9649–9653.

97. Bales, K.R., Verina, T., Cummins, D.J., Du, Y., Dodel, R.C., Saura, J., Fishman, C.E., DeLong, C.A., Piccardo, P., Petegnief, V., et al. (1997). Lack of apolipoprotein E dramatically reduces amyloid β-peptide deposition. *Nat Genet, 17*, 263–264.

98. Eto, M., Watanabe, K., Makino, I., Ishii, K. (1991). Apolipoprotein E allele frequencies in non-insulin-dependent diabetes mellitus with hypertriglyceridemia (type IIb, III, IV, and V hyperlipoproteinemia). *Metabolism, 40*, 776–780.

99. Inamdar, P.A., Kelkar, S.M., Devasagayam, T.P., Bapat, M.M. (2000). Apolipoprotein E polymorphism in non-insulin-dependent diabetes of Mumbai, India and its effect on plasma lipids and lipoproteins. *Diabetes Res Clin Pract, 47*, 217–223.

100. Paulsson, J.F., Westermark, G.T. (2005). Aberrant processing of human proislet amyloid polypeptide results in increased amyloid formation. *Diabetes, 54*, 2117–2125.

101. Paulsson, J.F., Andersson, A., Westermark, P., Westermark, G.T. (2006). Intracellular amyloid-like deposits contain unprocessed pro-islet amyloid polypeptide (proIAPP) in beta cells of transgenic mice overexpressing the gene for human and transplanted human islets. *Diabetologia, 49*, 1237–1246.

102. Marzban, L., Rhodes, C., Steiner, D., Haataja, L., Halban, P., Verchere, C.B. (2006). Impaired NH_2-terminal processing of human proislet amyloid polypeptide by the prohormone convertase PC2 leads to amyloid formation and cell death. *Diabetes, 55*, 2192–2201.

103. Kahn, S.E. (1997). Release of incompletely processed proinsulin is the cause of the disproportionate proinsulinemia of NIDDM. *Diabetes, 46*, 1725–1732.

104. Zhu, X., Orci, L., Carroll, R., Norrbom, C., Ravazzola, M., Steiner, D.F. (2002). Severe block in processing of proinsulin to insulin accompanied by elevation of des-64, 65 proinsulin intermediates in islets of mice lacking prohormone convertase PC1/3. *Proc Natl Acad Sci U S A, 99*, 10299–10304.

105. Westermark, G.T., Steiner, D.F., Gebre-Medhin, S., Engström, U., Westermark, P. (2000). Proislet amyloid polypeptide immunoreactivity in the islets of Langerhans. *Upsala J Med Sci, 105*, 97–106.

106. Westermark, P., Engström, U., Westermark, G.T., Johnson, K.H., Permerth, J., Betsholtz, C. (1989). Islet amyloid polypeptide (IAPP) and pro-IAPP immunoreactivity in human islets of Langerhans. *Diabetes Res Clin Pract, 7*, 219–226.

107. Park, K., Verchere, C.B. (2001). Identification of a heparin binding domain in the N-terminal cleavage site of pro-islet amyloid polypeptide. *J Biol Chem, 276*, 16611–16616.

108. Ancsin, J.B. (2003). Amyloidogenesis: historical and modern observations point to heparan sulfate proteoglycans as a major culprit. *Amyloid, 10*, 67–79.

109. Meng, F., Abedini, A., Raleigh, D. P. (2007). Amyloid formation by pro-islet amyloid polypeptide processing intermediates: examination of the role of protein

heparan sulfate interactions and implications for islet amyloid formation in type 2 diabetes. *Biochemistry, 46*, 12091–12099.

110. Kirkitadze, M.D., Condron, M.M., Teplow, D.B. (2001). Identification and characterization of key kinetic intermediates in amyloid β-protein fibrillogenesis. *J Mol Biol, 312*, 1103–1109.

111. Hull, R., Zraika, S., Udayasankar, J., Kisilevsky, R., Szarek, W., Wight, T., Kahn, S.E. (2007). Inhibition of glucosaminoglycan synthesis and protein glycosylation with WAS-406 and azaserine result in reduced islet amyloid formation in vitro. *Amer J Phys Cell Physiol, 293*, C1586–C1593.

112. Mirzabekov, T.A., Lin, M.C., Kagan, B.L. (1996). Pore formation by the cytotoxic islet amyloid peptide amylin. *J Biol Chem, 271*, 1988–1992.

113. Harroun, T.A., Bradshaw, J.P., Ashley, R.H. (2001). Inhibitors can arrest the membrane activity of human islet amyloid polypeptide independently of amyloid formation. *FEBS Lett, 507*, 200–204.

114. Anguiano, M., Nowak, R.J., Lansbury, P.T. (2002). Protofibrillar islet amyloid polypeptide permeabilizes synthetic vesicles by a pore-like mechanism that may be relevant to type II diabetes. *Biochemistry, 41*, 11338–11343.

115. Jayasinghe, S.A., Langen, R. (2005). Lipid membranes modulate the structure of islet amyloid polypeptide. *Biochemistry, 44*, 12113–12119.

116. Balali-Mood, K., Ashley, R.H., Hauss, T., Bradshaw, J.P. (2005). Neutron diffraction reveals sequence-specific membrane insertion of pre-fibrillar islet amyloid polypeptide and inhibition by rifampicin. *FEBS Lett, 579*, 1143–1148.

117. Engel, M.F.M., Yigittop, H., Elgersma, R.C., Rijkers, D.T.S., Liskamp, R.M.J., De Kruijff, B., Hoeppener, J.W.M., Antoinette Killian, J. (2006). Islet amyloid polypeptide inserts into phospholipid monolayers as monomer. *J Mol Biol, 356*, 783–789.

118. Knight, J.D., Hebda, J.A., Miranker, A.D. (2006). Conserved and cooperative assembly of membrane bound α-helical states of islet amyloid polypeptide. *Biochemistry, 45*, 9496–9508.

119. Lopes, D.H.J., Meister, A., Gohlke, A., Hauser, A., Blume, A., Winter, R. (2007). Mechanism of islet amyloid polypeptide fibrillation at lipid interfaces studied by infrared reflection absorption spectroscopy. *Biophys J, 93*, 3132–3141.

120. Jayasinghe, S.A., Langen, R. (2007). Membrane interactions of islet amyloid polypeptide. *Biochim Biophys Acta, 1768*, 2002–2009.

121. Sparr, E., Engel, M.F.M., Sakharov, D.V., Sprong, M., Jacobs, J., de Kruijff, B., Hoppener, J.W.M., Antoinette Killian, J. (2004). Islet amyloid polypeptide-induced membrane leakage involves uptake of lipids by forming amyloid fibers. *FEBS Lett, 577*, 117–120.

122. Porat, Y., Kolusheva, S., Jelinek, R., Gazit, E. (2003). The human islet amyloid polypeptide forms transient membrane-active prefibrillar assemblies. *Biochemistry, 42*, 10971–10977.

123. Green, J.D., Kreplan, L., Goldsbury, C., Blatter, X.L., Stolz, M., Cooper, G.J.S., Seelig, A., Kistler, J., Aebi, U. (2004). Atomic force microscopy reveals defects within mica supported lipid bilayers induced by the amyloidogenic human amylin peptide. *J Mol Biol, 342*, 877–887.

124. Kayed, R., Sokolov, Y., Edmonds, B., McIntire, T.M., Milton, S.C., Hall, J.E., Glabe, C.G. (2004). Permeabilization of lipid bilayers is a common conformation-dependent activity of soluble amyloid oligomers in protein misfolding disease. *J Biol Chem, 279*, 46363–46366.

125. Rustenbeck, I., Matthies, A., Lenzen, S. (1994). Lipid composition of glucose-stimulated pancreatic islets and insulin-secreting tumor cells. *Lipids, 29*, 685–692.

126. Hubbard, J., Martin, S., Chaplin, L.C., Bose, C., Kelly, S.M., Prince, N.C. (1991). Solution structures of calcitonin-gene-related-peptide analogs of calcitonin-gene-related peptide and amylin. *Biochem J, 275*, 785–788.

127. Williamson, J.A., Miranker, A.D. (2007). Direct detection of transient α-helical states in islet amyloid polypeptide. *Protein Sci, 16*, 1–8.

128. Cort, J., Liu, Z., Lee, G., Harris, S.M., Prickett, K.S., Gaeta, S.L., Andersen, N.H. (1994). β-Structure in human amylin and two designer β-peptides: CD and NMR spectroscopic comparisons suggests soluble β-oligomers and the absence of significant population of β-strand dimers. *Biochem Biophys Res Commun, 204*, 1088–1095.

129. Breeze, A.L., Harvey, T.S., Bazzo, R., Campbell, I.D. (1991). Solution structure of human calcitonin gene-related peptide by H-1-NMR and distance geometry with restrained molecular-dynamics. *Biochemistry, 30*, 575–582.

130. Abedini, A., Raleigh, D.P. (2009). A role for helical intermediates in amyloid formation by natively unfolded polypeptides? *Phys Biol, 6*, 15005.

131. Eliezer, D. (2006). Amyloid ion channels: a porous argument or a thin excuse? *J Gen Physiol, 128*, 631–636.

132. Lansbury, P.T., Lashuel, H.A. (2006). A century-old debate on protein aggregation and neurodegeneration enters the clinic. *Nature, 443*, 774–779.

133. Rochet, J. (2007). Novel therapeutic strategies for the treatment of protein-misfolding diseases. *Expert Rev Mol Med, 9*, 1–31.

134. Kapurniotu, A., Schmauder, A., Tenidis, K. (2002). Structure-based design and study of non-amyloidogenic, double *N*-methylated IAPP amyloid core sequences as inhibitors of IAPP amyloid formation and cytotoxicity. *J Mol Biol, 315*, 339–350.

135. Rijkers, D.T.S., Hoppener, J.W.M., Posthuma, G., Lips, C.J.M., Kiskamp, R.M.J. (2002). Inhibition of amyloid fibril formation of human amylin by N-alkylated amino acid and alpha-hydroxy acid residue containing peptides. *Chemistry, 8*, 4285–4291.

136. Porat, Y., Mazor, Y., Efrat, S., Gazit, E. (2004). Inhibition of islet amyloid polypeptide fibril formation: a possible role for hetero-aromatic interactions. *Biochemistry, 43*, 14454–14462.

137. Scrocchi, L.A., Chen, Y., Waschuk, S., Wang, F., Cheung, S., Darabie, A.A., McLaurin, J., Fraser, P.E. (2002). Design of peptide-based inhibitors of human islet amyloid polypeptide fibrillogenesis. *J Mol Biol, 318*, 697–706.

138. Cohen, T., Frydman-Marom, A., Rechter, M., Gazit, E. (2006). Inhibition of amyloid fibril formation and cytotoxicity by hydroxyindole derivatives. *Biochemistry, 45*, 4727–4735.

139. Scrocchi, L.A., Chen, Y., Wang, F., Han, K., Ha, K., Fraser, P.E. (2003). Inhibitors of islet amyloid polypeptide fibrillogenesis, and the treatment of type-2 diabetes. *Lett Pept Sci, 10*, 545–551.

140. Aiteken, J.F., Lommes, K.M., Konarkowska, B., Cooper, G.J.S. (2003). Suppression by polycyclic compounds of the conversion of human amylin into insoluble amyloid. *Biochem J, 374*, 779–784.

141. Gazit, E. (2005). Mechanisms of amyloid fibril self-assembly and inhibition. *FASEB J, 272*, 5971–5978.

142. Dember, L.M., Hawkins, P.N., Hazenberg, B.P., Gorevic, P.D., Merlini, G., Butrimiene, I., Livneh, A., Lesnyak, O., Puéchal, X., Lachmann, H.J., et al., Eprodisate for AA Amyloidosis Trial Group. (2007). Eprodisate for the treatment of renal disease in AA amyloidosis. *N Engl J Med, 356*, 2349–2359.

143. Abedini, A., Meng, F., Raleigh, D.P. (2007). A single-point mutation converts the highly amyloidogenic human islet amyloid polypeptide into a potent fibrillization inhibitor. *J Am Chem Soc, 129*, 11300–11301.

144. Yan, L.M., Tatarek-Nossol, M., Velkova, A., Kazantzis, A., Kapurniotu, A. (2006). Design of a mimic of nonamyloidogenic and bioactive human islet amyloid polypeptide (IAPP) as a nanomolar affinity inhibitor of IAPP cytotoxic fibrillogenesis. *Proc Natl Acad Sci U S A, 103*, 2046–2051.

145. Yan, L.M, Velkova, A., Tatarek-Nossol, M., Andreetto, E., Kapurniotu, A. (2007). IAPP mimic blocks Aβ cytotoxic self-assembly: cross-suppression of amyloid toxicity of Aβ and IAPP suggests a molecular link between Alzheimer's disease and type II diabetes. *Angew Chem Int Ed Engl, 46*, 1246–1252.

146. Ratner R.E., Dickey R., Fineman M., Maggs D.G., Shen L., Strobel S.A., Weyer C., Kolterman O.G. (2004) Amylin replacement with pramlintide as an adjunct to insulin therpy improves long-term glycaemic and weight control in type-1 diabetes mellitus: a 1-year, randomized controlled trial. *Diabetes Med, 21*, 1204–1212.

147. Allard, F.D., Wallace, A.E., Greenbaum, G.J. (2005). Emerging therapies: going beyond insulin in treating individuals with type-I diabetes mellitus. *Curr Opin Endocrinol Diabetes, 12*, 303–308.

PART III

ROLE OF ACCESSORY MOLECULES AND RISK FACTORS

24

ROLE OF METALS IN ALZHEIMER DISEASE

BLAINE R. ROBERTS

Mental Health Research Institute of Victoria, Parkville, Victoria, Australia; Department of Pathology, University of Melbourne, Parkville, Australia

ASHLEY I. BUSH

Mental Health Research Institute of Victoria, Parkville, Victoria, Australia; Department of Pathology, University of Melbourne, Parkville, Victoria, Australia; Genetics and Aging Research Unit, Massachusetts General Hospital, Charlestown, Massachusetts

INTRODUCTION

Senile plaques, neurofibrillary tangles, and Lewy bodies are all pathological hallmarks that result from misfolded proteins in neurodegenerative diseases. Parkinson disease (PD), amyotrophic lateral sclerosis (ALS), Huntington disease (HD), and Alzheimer disease (AD) all exhibit these forms of misfolded proteins. Each disease results in the selective loss of specific neuronal cells, and each disease has a different protein that is implicated by genetic mutation in the disease process. α-Synuclein is implicated in PD, huntingtin in HD, copper–zinc superoxide dismutase in ALS, and the amyloid beta peptide in AD. How these proteins are involved in the disease process or how mutations of these proteins cause loss of a specific set of neurons remains uncertain. Additionally, the role of these proteins in sporadic cases is far from clear. For these diseases it is thought that the misfolding of the proteins and associated deposition are

Protein Misfolding Diseases: Current and Emerging Principles and Therapies,
Edited by Marina Ramirez-Alvarado, Jeffery W. Kelly, and Christopher M. Dobson
Copyright © 2010 John Wiley & Sons, Inc.

closely related to pathogenesis. In this chapter we discuss the role of amyloid beta (Aβ) in AD and current therapeutic approaches.

ROLE OF AMYLOID BETA IN AD

In AD, Aβ is regarded as the principal toxic species. However, the mechanism by which Aβ exerts toxic effects remains an issue of debate and a major challenge to the development of AD-modifying drugs. The amyloid cascade hypothesis [1–3] suggests that a disturbance in the processing of the amyloid precursor protein is the initiating event that leads to the aggregation of Aβ. Although there are problems with the amyloid cascade hypothesis, the majority of the research into drug development targets the removal of Aβ from the brain. Indeed, genetic studies do implicate Aβ in the biochemistry of the disease [4]; however, the mechanism of how Aβ accumulates or induces dementia remains uncertain. For example, Aβ is produced as a normal part of metabolism, and in vitro Aβ is neurotoxic at nonphysiological (micromolar) concentrations and neurotrophic at physiological (nanomolar) concentrations [5]. The role of metals in neurobiology is a developing area, and increasing amounts of evidence suggest that iron, copper, and zinc have important roles in modulating toxicity in neurodegenerative diseases. Alzheimer disease is the most common neurodegenerative disease, and in this chapter we focus mainly on the role of metals in AD.

Aβ is a soluble component of all biological fluids [6,7] but in AD Aβ is enriched in amyloid deposits [8] and elevated in some familial AD-linked mutations [9]. The length of Aβ is considered to be an important factor that mediates the toxicity of the peptide. Indeed, synthetic Aβ 1–42 does have a greater propensity to self-aggregate than Aβ 1–40 [10,11]. However, not all forms of soluble Aβ are toxic, as healthy persons have soluble Aβ in all their biological fluids, including the brain [6]. What is different about the Aβ in AD cases? There is growing evidence that the formation of soluble Aβ oligomers, but not fibrils, correlate with tangles and the neuritic changes associated with AD [12–14].

Current therapeutic approaches are based largely on the model that Aβ is the principal toxic species and have either attempted to prevent Aβ production (secretase inhibitors) or to remove Aβ (immunotherapy). Although the involvement of Aβ in the pathogenesis of AD is almost indisputable, there is evidence that other neurochemical reactions and processes apart from Aβ production must contribute to the amyloid formation in AD. For example, if Aβ levels were responsible for the initiation of amyloid, it would be difficult to explain why amyloid deposits are not uniform and tend to relate to synapses and the cerebrovascular lamina media. Additionally, the presence of Aβ 1–42 alone is insufficient, as this peptide is a normal component of cerebrospinal fluid (CSF) of healthy persons [6]. Finally, the biggest risk factor of AD is age, and if Aβ levels do not rise with age, other age-related neurochemical changes would have an essential role in the reaction that causes Aβ to accumulate. An age-dependent increase in oxidative damage is a common observation, and in AD the oxidative damage of neuronal cells precedes Aβ deposition and thus may act as a trigger for amyloid deposition. [15–17]

Since the discovery that Aβ becomes amyloidogenic upon binding of Zn^{2+} and Cu^{2+} [18,19], it has become clear that Aβ is a metalloprotein [16,20]. Based on the ability of copper, zinc, and potentially iron to modulate the toxicity of Aβ, a new therapeutic strategy to target Aβ and metals was developed.

A point of confusion regarding the role of metals in neurodegenerative diseases is that both deficiency and excess can contribute to severe pathology. All transition metals are potentially toxic, yet are essential for many biochemical functions. Constitutively, the brain has more than enough endogenous transition metal ions to mediate toxicity when homeostasis is lost. For example, during neurotransmission the concentrations of zinc reach levels (ca. 300 μM) that are more than sufficient to be toxic to neuronal cells in culture [21]. Thus, there must be very efficient processes to regulate the distribution of transition metals in the cell. The ability of the cell to sequester and chaperone zinc, copper, and iron is critical for preventing toxicity or aberrant redox activity from occurring. In Alzheimer disease, disruption of the homeostasis of copper and zinc in the central nervous system results in a twofold insult with respect to Aβ. First the binding of copper to Aβ results in the production of hydrogen peroxide. Second, the binding of copper or zinc by Aβ increases the aggregation of the peptide [19]. The synapse is a place of highly regulated release and uptake of metals, with zinc being a principal metal released during neurotransmission. The release of zinc at the synapse may explain why amyloid deposits are related to the synapse and the neocortex, where synaptic activity is high. The impact of ZnT3-mediated zinc ion release in the glutamatergic synapse is discussed later.

The aggregation properties of Aβ are increased dramatically by zinc, copper or, iron. Aβ has both a low and a high affinity-binding site for either copper or zinc [18,22–24]. Zinc rapidly precipitates Aβ, and both Cu^{2+} and Fe^{3+}, under mildly acidic conditions (e.g., pH 6.8 to 7.0), markedly induce Aβ aggregation [18,24,25]. The K_d value of Aβ for zinc or copper has been an issue of debate. Originally, the high affinity binding was reported to be about 100 nM and about 5 μM for the low affinity-binding site [18,24], but now it is understood that the buffer conditions (e.g., the presence of NaCl [26]), the aggregation state of the peptide [18,27,28], and how the bound and free metal ions are assayed [29] are critical for the values observed. Despite the difficulties in determining the exact K_d for metal binding to Aβ, it is generally agreed that the micromolar concentrations of zinc and copper that are released at the cortical synapse are sufficient to induce Aβ aggregation [25,28–30]. Interestingly, the high affinity Cu^{2+} apparent binding constant for Aβ is different between Aβ(1–42) (7.0×10^{-18} M) and Aβ(1–40) (5.0×10^{-11} M) but may reflect a perturbed equilibrium brought about by the reaction of Cu^{2+} with the peptide [23]. The greater affinity of Aβ(1–42) for Cu^{2+} is associated with the enhanced precipitation of Aβ(1–42) by Cu^{2+} [31,32], increased formation of SDS-resistant dimerization of Aβ(1–42) [31] and with the increased redox activity and toxicity of copper bound to Aβ [33].

At pH 7.4, Aβ binds equimolar amounts of copper and zinc; however, under slightly more acidic conditions (pH 6.6) copper completely displaces the zinc from Aβ [23]. The stoichiometry of either copper or zinc binding to Aβ is 2.5 equivalents; the fractional binding of metals suggests that the formation of Aβ oligomers provides an additional metal binding site [23]. Interestingly, isoforms of apolipoprotein E can prevent the metal mediated aggregation of Aβ in a manner that that is inversely proportional to their disease risk (e.g., ApoE4 prevents aggregation the least) [34]. When Cu^{2+} or Fe^{3+} bind to Aβ, they become reduced and can produce hydrogen peroxide by subsequent reduction of molecular oxygen [16,35–37]. The redox chemistry that is facilitated by the binding of Cu^{2+} or Fe^{3+} to Aβ is critical to the oxidative stress-induced toxicity that is observed in cell culture that is partially mediated by methionine 35 [38,39] and tyrosine 10 [40,41]. Copper or iron bound to Aβ redox cycle to produce hydrogen peroxide in the presence of biological reducing agents. The most likely reductants are cholesterol and long-chain fatty acids [16,36, 40–44], consistent with the toxicity of Aβ being associated with the membrane [39] and with the generation of lipid oxidation products, oxysterols, and 4-hydroxynonenal (HNE), which are found elevated in brain tissue in AD and in amyloid-β protein precursor transgenic mice [16,36,41,43,44]. HNE can covalently modify the histidines of Aβ, which results in a HNE-Aβ species that has an increased affinity for lipid membranes and an enhanced ability to aggregate [42]. This provides a mechanism where the pro-oxidant activity of Aβ and copper leads to its own modification and accelerated amyloidogenesis.

The redox activity of copper bound to Aβ is important because there is a large body of evidence which shows that oxidative injury, mediated by hydrogen peroxide, has a significant role in AD. Generation of the oxidative markers observed in AD can stem from the reaction of hydrogen peroxide and reduced iron or copper (Fe^{2+}, Cu^{+}) to generate hydroxyl radicals by Fenton chemistry. Hydroxyl radicals are highly chemically reactive and in turn generate lipid peroxidation products, protein carbonyl modifications, and nucleic acid adducts such as 8-hydroxyguanosine, all of which characterize AD neuropathology [45–47]. The length of the Aβ species is regarded as an important factor in AD pathogenesis, due largely to its enrichment in amyloid deposits [8,48,49] despite its relatively low abundance in biological fluids [6]. Interestingly, the redox activity of Aβ is greatest for $A\beta42_{human} > A\beta40_{human} \gg A\beta40_{mouse} \approx 0$ [50]; this relationship is also relevant because Aβ(1–42) is overproduced in some familial forms of AD. The relationship of redox activity and the length of Aβ also correlate with the neurotoxicity observed in neuronal cultures, which is largely mediated by the Cu^{2+}–Aβ interaction [16,50]. The interaction of Aβ with the cell membrane is promoted by the binding of copper or zinc. Chelators such as TETA and cliquinol (CQ) prevent the redox activity of copper or iron and block the toxicity of Aβ in cell culture [43,51].

The redox activity of copper–Aβ can also lead to the oxidation of Aβ side chains. Mass spectrometry has shown that copper can promote the addition of oxygen atoms, with the formation of methionine sulfoxide and methionine sulfone being the most likely products [38]. In addition to methionine oxidation, a number of other amino acid modifications can result from copper-mediated redox activity. For example, lysine residues can be modified by aldehydes produced from lipid peroxidation [52], and multiple tyrosine modifications, including dityrosine and nitrotyrosine [53,54]. Oxidized products of histidine and N^3-pyroglutamate have been isolated from AD plaques. Interestingly, the positron emission tomography imaging ligand PIB has a particularly high affinity for the pyroglutamate-modified form of Aβ [55]. Many different species of Aβ from dimer to fibril have been proposed to be the toxic form of Aβ. However, due to the difficult nature of aggregated or fibrillized proteins, the various "toxic" species have remained poorly characterized. There is a growing interest in soluble oligomers, which appear to be especially toxic [56,57], as are soluble dimers of Aβ. Although the mechanism for the formation of dimeric Aβ in vivo remains to be shown, it is possible that the formation of dityrosine cross-linked Aβ dimers could be a result of the redox activity of copper bound to Aβ [32].

Aside from the redox activity of Aβ–metal complexes, both copper and zinc can induce Aβ aggregation in vivo. Analysis of plaques and congophilic angiopathy from both AD and APP transgenic mice have demonstrated that zinc, copper, and iron are highly enriched in these structures [20,58–64]. In the plaques, the levels of Cu (390 μM), Zn (1055 μM), and Fe (940 μM) are elevated severalfold compared to age-matched control neuropil (Cu [70 μM], Zn [350 μM], and Fe [340 μM]) [60], and Aβ coordinates copper and zinc directly, but not iron [16,20]. Iron is found complexed with ferritin in the plaque-associated neuritic processes [65].

In addition to the presence of copper and zinc in AD plaques, genetic manipulation of the release of presynaptic zinc also demonstrates a primary role for metal ions in the aggregation of Aβ. The evidence for this comes from experiments in ZnT3 knockout mice crossed with Tg2576 APP transgenic mice. ZnT3 is responsible for concentrating zinc into glutamatergic synaptic vesicles and ZnT3 knockout mice have 15% less zinc in their cortex [66]. Characterization of the progeny from the cross of the ZnT3 knockout with Tg2576 APP transgenic mice demonstrated that the ablation of zinc markedly inhibits amyloid pathology and congophilic angiopathy [58,67] and actually increased the concentration of soluble Aβ [67]. The increase in soluble Aβ is consistent with the idea that Aβ plaques represent a store of zinc and Aβ that is trapped in a dissociable equilibrium with soluble Aβ and soluble zinc [68]. These findings are also consistent with the hypothesis that the release of zinc into the synaptic cleft can trigger the formation of amyloid plaques and with the observation that the synapse is the location of the first deposits of amyloid in AD. The glutamatergic synapse is probably the most important site for dysfunction in AD because it is the site of long-term potentiation, a requirement for memory formation [69]. Interestingly, ZnT3 is only found in membrane of glutamatergic vesicles, which may explain why amyloid deposits are not found in the peripheral tissue despite Aβ being expressed throughout the body.

Recently, it was reported that postsynaptic NMDA neurites release copper upon NMDA activation [70]. Copper has been reported to be neuroprotective against NMDA-mediated excitotoxicity in a manner that is dependent on endogenous nitric oxide [71]. Despite accumulating in amyloid plaques, in AD parenchymal tissue copper levels decrease with advanced pathology [72]. Additionally, both genetic and dietary manipulations that increase copper in the brain ameliorate amyloid pathology in APP transgenic mice [73,74]. However, there are also reports that high-fat diets in combination with exposure to copper can increase the risk for AD [75]. Opposite to the levels of copper, zinc levels increase with AD, correlating with Aβ burden in humans but not in APP transgenic mice [72]. In advanced age, nutritional deficiencies in zinc are common and a recent report indicates that zinc deficiency in APP transgenic mice increases the volume of amyloid plaques [62]. Overall, these data indicate the complexities of the metal metabolism in AD. The consensus is that copper and zinc are enriched in amyloid where they coordinate Aβ, iron is enriched in tissue and neuritic pathology, and there is evidence of functional copper deficiency. Thus, a pharmacotherapy that targets aberrant Aβ metallation and that is capable of making the metal bioavailable would have the potential to release the metals trapped by Aβ and return them to normal metabolism.

CURRENT THERAPEUTIC APPROACHES TO AD

A common misconception in the envisioning of therapies to treat the disordered metabolism of metals is the belief that chelation, or the removal of metals, is the obvious intervention. However, the maldistribution of metals in AD is complex with metals being pooled in plaques and the neighboring cells being relatively deficient. Thus, a simple metal chelator may fix one problem while worsening the other. Additionally, there is the risk of the removal of essential metal ions with subsequent side effects (e.g., iron-deficiency anemia). Although some of the side effects of general chelation can be addressed, a compound with more sophisticated properties (e.g., ionophores) is being explored. Several metal-complexing agents have been tested. The lipophilic chelator DP109 reduced the levels of insoluble Aβ and conversely increased soluble Aβ forms in a mouse model [59]. Treatment of APP transgenic mice with nicotine resulted in a decrease amyloidosis through a mechanism that appears to involve the modulation of copper and zinc homeostasis [76]. Additionally, there have been two clinical trials of metal chelators: desferrioxamine and *d*-penicillamine. Desferrioxamine treatment over a two-year period led to a significant reduction in the rate of decline of daily living skills [77]. With *d*-penicillamine there was a decrease in oxidative markers over a six-month period but no change in cognitive decline [78].

An orally administered ionophore, clioquinol (CQ, 5-chloro-7-iodo-8-hydroxyquinoline), in Tg2576 mice resulted in a 49% reduction of cortical amyloid deposits, with improvement or stability in the health and weight parameters compared to untreated mice [79]. CQ is able to cross the blood–brain barrier (BBB) and increase brain copper and zinc levels in treated mice. The affinity of CQ for Cu^{2+} and Zn^{2+} is modest (nanomolar) [17], but sufficient to facilitate the dissociation of these metals from the low affinity-binding site of Aβ. CQ is capable of crossing the BBB in Tg2576 mice and forming a complex to amyloid plaques, as well as coordinating zinc-metallated Aβ from postmortem AD affected brain specimens [17]. In a phase II human clinical trial, CQ administered orally for 36 weeks slowed the rate of cognitive decline and caused a reduction in plasma Aβ 1–42 levels compared to placebo controls [80].

Recently, a double-blind, placebo-controlled phase II clinical trial in 78 subjects over 12 weeks for the treatment of early AD was completed on an 8-OH quinoline analog of CQ, known as PBT2. The results showed that CSF levels of Aβ were significantly lowered at the 250-mg dose at 12 weeks and that there was significant improvement in performance on executive tests of the Neuropsychological Test Battery (NTB) at 12 weeks [81]. Further clinical testing of what may be a disease-modifying drug based on the metal hypothesis is planned.

The treatment of APP transgenic mice with PBT2 results in an impressive improvement in the animals' cognitive abilities within days of beginning treatment [82]. The mode of action for CQ or PBT2 remains to be confirmed; however, it is understood that CQ enters the brain and combines with metallated Aβ plaques and potentially diffuse Aβ plaques [17]. Also, CQ is not simply acting as a chelator because treatment of transgenic mice increases the levels of copper and zinc in the brain [79], and in a phase II clinical trial in AD patients, the plasma levels of zinc were increased significantly [80]. In cell culture, CQ–copper complexes markedly decrease the secretion of Aβ by a mechanism involving the elevation of intracellular copper and the up-regulation of metalloprotease-2 (MMP)-2 and MMP-3 [83]. We hypothesize that the mechanism of action for CQ and PBT2 involves the entry of these compounds into the brain, where they are attracted to the pool of metals that are in a dissociable equilibrium in amyloid. After binding copper or zinc, the drug–metal complex enters the cell and activates MMP, which in turn facilitates the clearance of Aβ in the synapse. Concomitantly, the drug blocks the oxidative oligomerization of Aβ and the toxic redox activity of Aβ–copper complexes.

The overall goal of drugs derived from the metal hypothesis for AD is to relocate copper and zinc from where they do harm (i.e., bound to Aβ) and to return them into the pool of copper and zinc that are needed for normal neuronal function. CQ has also been shown to have efficacy in animal models of both Parkinson [84] and Huntington disease [85], which have been associated with iron or copper overload [86], leading to oxidative stress.

As the search for disease-modifying therapeutics continues, it seems likely that the pharmacological treatment of AD or other neurodegenerative diseases will involve combination therapies. In this regard there is no barrier to the potential combination of metal-complexing drugs with other potential interventions, such as secretase inhibitors or immunotherapy.

REFERENCES

1. Hardy, J., Allsop, D. (1991). Amyloid deposition as the central event in the aetiology of Alzheimer's disease. *Trends Pharmacol Sci, 12*, 383–388.
2. Hardy, J., Selkoe, D.J. (2002). The amyloid hypothesis of Alzheimer's disease: progress and problems on the road to therapeutics. *Science, 297*, 353–356.
3. Tanzi, R.E., Bertram, L. (2005). Twenty years of the Alzheimer's disease amyloid hypothesis: a genetic perspective. *Cell, 120*, 545–555.
4. Goate, A., Chartier-Harlin, M.C., Mullan, M., Brown, J., Crawford, F., Fidani, L., Giuffra, L., Haynes, A., Irving, N., James, L., et al. (1991). Segregation of a missense mutation in the amyloid precursor protein gene with familial Alzheimer's disease. *Nature, 349*, 704–706.
5. Luo, Y., Sunderland, T., Roth, G.S., Wolozin, B. (1996). Physiological levels of beta-amyloid peptide promote PC12 cell proliferation. *Neurosci Lett, 217*, 125–128.
6. Vigo-Pelfrey, C., Lee, D., Keim, P., Lieberburg, I., Schenk, D.B. (1993). Characterization of β-amyloid peptide from human cerebrospinal fluid. *J Neurochem, 61*, 1965–1968.
7. Seubert, P., Vigo-Pelfrey, C., Esch, F., Lee, M., Dovey, H., Davis, D., Sinha, S., Schlossmacher, M., Whaley, J., Swindlehurst, C., et al. (1992). Isolation and quantification of soluble Alzheimer's β-peptide from biological fluids. *Nature, 359*, 325–327.
8. Masters, C.L., Multhaup, G., Simms, G., Pottgiesser, J., Martins, R.N., Beyreuther, K. (1985). Neuronal origin of a cerebral amyloid: neurofibrillary tangles of Alzheimer's disease contain the same protein as the amyloid of plaque cores and blood vessels. *EMBO J, 4*, 2757–2763.
9. Citron, M., Westaway, D., Xia, W., Carlson, G., Diehl, T., Levesque, G., Johnson-Wood, K., Lee, M., Seubert, P., Davis, A., *et al.* (1997). Mutant presenilins of Alzheimer's disease increase production of 42-residue amyloid β-protein in both transfected cells and transgenic mice. *Nat Med, 3*, 67–72.
10. Jarrett, J.T., Berger, E.P., Lansbury, P.T., Jr. (1993). The carboxy terminus of the β amyloid protein is critical for the seeding of amyloid formation: implications for the pathogenesis of Alzheimer's disease. *Biochemistry, 32*, 4693–4697.
11. Hilbich, C., Kisters-Woike, B., Reed, J., Masters, C.L., Beyreuther, K. (1991). Aggregation and secondary structure of synthetic amyloid β A4 peptides of Alzheimer's disease. *J Mol Biol, 218*, 149–163.
12. Maynard, C.J., Cappai, R., Volitakis, I., Cherny, R.A., White, A.R., Beyreuther, K., Masters, C.L., Bush, A.I., Li, Q.-X. (2002). Overexpression of Alzheimer's disease β-amyloid opposes the age-dependent elevations of brain copper and iron levels, *J Biol Chem, 277*, 44670–44676.
13. Melov, S., Adlard, P.A., Morten, K., Johnson, F., Golden, T.R., Hinerfeld, D., Schilling, B., Mavros, C., Masters, C.L., Volitakis, I., *et al.* (2007). Mitochondrial oxidative stress causes hyperphosphorylation of tau. *PLoS ONE, 2*, e536.

14. Yamamoto, A., Shin, R.W., Hasegawa, K., Naiki, H., Sato, H., Yoshimasu, F., Kitamoto, T. (2002). Iron(III) induces aggregation of hyperphosphorylated tau and its reduction to iron(II) reverses the aggregation: implications in the formation of neurofibrillary tangles of Alzheimer's disease. *J Neurochem, 82*, 1137–1147.

15. Nunomura, A., Perry, G., Pappolla, M.A., Wade, R., Hirai, K., Chiba, S., Smith, M.A. (1999). RNA oxidation is a prominent feature of vulnerable neurons in Alzheimer's disease. *J Neurosci, 19*, 1959–1964.

16. Opazo, C., Huang, X., Cherny, R.A., Moir, R.D., Roher, A.E., White, A.R., Cappai, R., Masters, C.L., Tanzi, R.E., Inestrosa, N.C., Bush, A.I. (2002). Metalloenzyme-like activity of Alzheimer's disease beta-amyloid. Cu-dependent catalytic conversion of dopamine, cholesterol, and biological reducing agents to neurotoxic H(2)O(2). *J Biol Chem, 277*, 40302–40308.

17. Opazo, C., Luza, S., Villemagne, V.L., Volitakis, I., Rowe, C., Barnham, K.J., Strozyk, D., Masters, C.L., Cherny, R.A., Bush, A.I. (2006). Radioiodinated clioquinol as a biomarker for beta-amyloid: Zn complexes in Alzheimer's disease. *Aging Cell, 5*, 69–79.

18. Bush, A.I., Pettingell, W.H., Jr., Paradis, M.D., Tanzi, R.E. (1994). Modulation of Aβ adhesiveness and secretase site cleavage by zinc. *J Biol Chem, 269*, 12152–12158.

19. Bush, A.I., Pettingell, W.H., Multhaup, G., Paradis, M.D., Vonsattel, J.P., Gusella, J.F., Beyreuther, K., Masters, C.L., Tanzi, R.E. (1994). Rapid induction of Alzheimer Aβ amyloid formation by zinc. *Science, 265*, 1464–1467.

20. Dong, J., Atwood, C.S., Anderson, V.E., Siedlak, S.L., Smith, M.A., Perry, G., Carey, P.R. (2003). Metal binding and oxidation of amyloid-beta within isolated senile plaque cores: Raman microscopic evidence. *Biochemistry, 42*, 2768–2773.

21. Frederickson, C.J. (1989). Neurobiology of zinc and zinc-containing neurons. *Int Rev Neurobiol, 31*, 145–328.

22. Atwood, C.S., Moir, R.D., Huang, X., Bacarra, N.M.F., Scarpa, R.C., Romano, D.M., Hartshorn, M.A., Tanzi, R.E., Bush, A.I. (1998). Dramatic aggregation of Alzheimer Aβ by Cu(II) is induced by conditions representing physiological acidosis. *J Biol Chem, 273*, 12817–12826.

23. Atwood, C.S., Scarpa, R.C., Huang, X., Moir, R.D., Jones, W.D., Fairlie, D.P., Tanzi, R.E., Bush, A.I. (2000). Characterization of copper interactions with Alzheimer Aβ peptides: identification of an attomolar affinity copper binding site on Aβ1–42. *J Neurochem, 75*, 1219–1233.

24. Bush, A.I., Pettingell, W.H., Multhaup, G., Paradis, M.D, Vonsattel, J.P., Gusella, J.F., Beyreuther, K., Masters, C.L., Tanzi, R.E. (1994). Rapid induction of Alzheimer A beta amyloid formation by zinc. *Science, 265*, 1464–1467.

25. Ha, C., Ryu, J., Park, C.B. (2007). Metal ions differentially influence the aggregation and deposition of Alzheimer's beta-amyloid on a solid template. *Biochemistry, 46*, 6118–6125.

26. Huang, X., Cuajungco, M.P., Atwood, C.S., Hartshorn, M.A., Tyndall, J., Hanson, G.R., Stokes, K.C., Leopold, M., Multhaup, G., Goldstein, L.E., et al. (1999). Cu(II) potentiation of Alzheimer Aβ neurotoxicity: correlation with cell-free hydrogen peroxide production and metal reduction. *J Biol Chem, 274*, 37111–37116.

27. Garai, K., Sengupta, P., Sahoo, B., Maiti, S. (2006). Selective destabilization of soluble amyloid beta oligomers by divalent metal ions. *Biochem Biophys Res Commun, 345*, 210–215.

28. Jun, S., Saxena, S. (2007). The aggregated state of amyloid-beta peptide in vitro depends on Cu^{2+} ion concentration, *Angew Chem Int Ed Engl, 46*, 3959–3961.

29. Tougu, V., Karafin, A., Palumaa, P. (2008). Binding of zinc(II) and copper(II) to the full-length Alzheimer's amyloid-beta peptide. *J Neurochem, 104*, 1249–1259.

30. Stellato, F., Menestrina, G., Dalla Serra, M., Potrich, C., Tomazzolli, R., Meyer-Klaucke, W., Morante, S. (2006). Metal binding in amyloid β-peptides shows intra- and inter-peptide coordination modes. *Eur Biophys J, 35*, 340–351.

31. Atwood, C.S., Moir, R.D., Huang, X., Scarpa, R.C., Bacarra, N.M., Romano, D.M., Hartshorn, M.A., Tanzi, R.E., Bush, A.I. (1998). Dramatic aggregation of Alzheimer Abeta by Cu(II) is induced by conditions representing physiological acidosis, *J Biol Chem, 273*, 12817–12826.

32. Atwood, C.S., Perry, G., Zeng, H., Kato, Y., Jones, W.D., Ling, K.Q., Huang, X., Moir, R.D., Wang, D., Sayre, L.M., *et al.* (2004). Copper mediates dityrosine cross-linking of Alzheimer's amyloid-beta. *Biochemistry, 43*, 560–568.

33. Huang, X., Atwood, C.S., Hartshorn, M.A., Multhaup, G., Goldstein, L.E., Scarpa, R.C., Cuajungco, M.P., Gray, D.N., Lim, J., Moir, R.D., *et al.* (1999). The Aβ peptide of Alzheimer's disease directly produces hydrogen peroxide through metal ion reduction. *Biochemistry, 38*, 7609–7616.

34. Moir, R.D., Atwood, C.S., Romano, D.M., Laurans, M.H., Huang, X., Bush, A.I., Smith, J.D., Tanzi, R.E. (1999). Differential effects of apolipoprotein E isoforms on metal-induced aggregation of Aβ using physiological concentrations. *Biochemistry, 38*, 4595–4603.

35. Dikalov, S.I., Vitek, M.P., Mason, R.P. (2004). Cupric-amyloid beta peptide complex stimulates oxidation of ascorbate and generation of hydroxyl radical, *Free Radic Biol Med, 36*, 340–347.

36. Nelson, T.J., Alkon, D.L. (2005). Oxidation of cholesterol by amyloid precursor protein and beta-amyloid peptide. *J Biol Chem, 280*, 7377–7387.

37. Tabner, B.J., Turnbull, S., El-Agnaf, O.M., Allsop, D. (2002). Formation of hydrogen peroxide and hydroxyl radicals from A(beta) and alpha-synuclein as a possible mechanism of cell death in Alzheimer's disease and Parkinson's disease. *Free Radic Biol Med, 32*, 1076–1083.

38. Ali, F.E., Separovic, F., Barrow, C.J., Cherny, R.A., Fraser, F., Bush, A.I., Masters, C.L., Barnham, K.J. (2005). Methionine regulates copper/hydrogen peroxide oxidation products of Abeta. *J Pept Sci, 11*, 353–360.

39. Ciccotosto, G.D., Tew, D., Curtain, C.C., Smith, D., Carrington, D., Masters, C.L., Bush, A.I., Cherny, R.A., Cappai, R., Barnham, K.J. (2004). Enhanced toxicity and cellular binding of a modified amyloid beta peptide with a methionine to valine substitution. *J Biol Chem, 279*, 42528–42534.

40. Barnham, K.J., Haeffner, F., Ciccotosto, G.D., Curtain, C.C., Tew, D., Mavros, C., Beyreuther, K., Carrington, D., Masters, C.L., Cherny, R.A., *et al.* (2004). Tyrosine gated electron transfer is key to the toxic mechanism of Alzheimer's disease beta-amyloid. *FASEB J, 18*, 1427–1429.

41. Haeffner, F., Smith, D.G., Barnham, K.J., Bush, A.I. (2005). Model studies of cholesterol, and ascorbate oxidation by copper complexes: relevance to Alzheimer's disease beta-amyloid metallochemistry. *J Inorg Biochem, 99*, 2403–2422.

42. Murray, I.V., Sindoni, M.E., Axelsen, P.H. (2005). Promotion of oxidative lipid membrane damage by amyloid beta proteins. *Biochemistry, 44*, 12606–12613.

43. Puglielli, L., Friedlich, A.L., Setchell, K.D.R., Nagano, S., Opazo, C., Cherny, R.A., Barnham, K.J., Wade, J.D., Melov, S., Kovacs, D.M., Bush, A.I. (2005). Cholesterol oxidase mimetic activity of Alzheimer's disease β-amyloid. *J Clin Invest, 115*, 2556–2563.

44. Smith, D.P., Smith, D.G., Curtain, C.C., Boas, J.F., Pilbrow, J.R., Ciccotosto, G.D., Lau, T.L., Tew, D.J., Perez, K., Wade, J.D., *et al.* (2006). Copper-mediated amyloid-beta toxicity is associated with an intermolecular histidine bridge. *J Biol Chem, 281*, 15145–15154.

45. Markesbery, W.R., Lovell, M.A. (2007). Damage to lipids, proteins, DNA, and RNA in mild cognitive impairment. *Arch Neurol, 64*, 954–956.

46. Smith, M.A., Perry, G., Richey, P.L., Sayre, L.M., Anderson, V.E., Beal, M.F., Kowall, N. (1996). Oxidative damage in Alzheimer's. *Nature, 382*, 120–121.

47. Smith, M.A., Richey Harris, P.L., Sayre, L.M., Beckman, J.S., Perry, G. (1997). Widespread peroxynitrite-mediated damage in Alzheimer's disease. *J.Neurosci, 17*, 2653–2657.

48. Kang, J., Lemaire, H.G., Unterbeck, A., Salbaum, J.M., Masters, C.L., Grzeschik, K.H., Multhaup, G., Beyreuther, K., Muller-Hill, B. (1987). The precursor of Alzheimer's disease amyloid A4 protein resembles a cell-surface receptor. *Nature, 325*, 733–736

49. Masters, C.L., Simms, G., Weinman, N.A., Multhaup, G., McDonald, B.L., Beyreuther, K. (1985). Amyloid plaque core protein in Alzheimer disease and Down syndrome. *Proc Natl Acad Sci U S A, 82*, 4245–4249.

50. Huang, X., Cuajungco, M.P., Atwood, C.S., Hartshorn, M.A., Tyndall, J.D., Hanson, G.R., Stokes, K.C., Leopold, M., Multhaup, G., Goldstein, L.E., et al. (1999). Cu(II) potentiation of Alzheimer Abeta neurotoxicity: correlation with cell-free hydrogen peroxide production and metal reduction, *J Biol Chem, 274*, 37111–37116.

51. Abramov, A.Y., Canevari, L., Duchen, M.R. (2003). Changes in intracellular calcium and glutathione in astrocytes as the primary mechanism of amyloid neurotoxicity. *J Neurosci, 23*, 5088–5095.

52. Chen, K., Kazachkov, M., Yu, P.H. (2007). Effect of aldehydes derived from oxidative deamination and oxidative stress on beta-amyloid aggregation; pathological implications to Alzheimer's disease. *J Neural Transm, 114*, 835–839.

53. Hensley, K., Maidt, M.L., Yu, Z., Sang, H., Markesbery, W.R., Floyd, R.A. (1998). Electrochemical analysis of protein nitrotyrosine and dityrosine in the Alzheimer brain indicates region-specific accumulation. *J Neurosci, 18*, 8126–8132.

54. Reynolds, M.R., Berry, R.W., Binder, L.I. (2005). Site-specific nitration differentially influences tau assembly in vitro. *Biochemistry, 44*, 13997–14009.

55. Maeda, J., Ji, B., Irie, T., Tomiyama, T., Maruyama, M., Okauchi, T., Staufenbiel, M., Iwata, N., Ono, M., Saido, T.C., *et al.* (2007). Longitudinal, quantitative assessment of amyloid, neuroinflammation, and anti-amyloid treatment in a living mouse model of Alzheimer's disease enabled by positron emission tomography. *J Neurosci, 27*, 10957–10968.

56. Shankar, G.M., Li, S., Mehta, T.H., Garcia-Munoz, A., Shepardson, N.E., Smith, I., Brett, F.M., Farrell, M.A., Rowan, M.J., Lemere, C.A., *et al.* (2008). Amyloid-beta protein dimers isolated directly from Alzheimer's brains impair synaptic plasticity and memory. *Nat Med, 14*, 837–842.

57. Lesne, S., Koh, M.T., Kotilinek, L., Kayed, R., Glabe, C.G., Yang, A., Gallagher, M., Ashe, K.H. (2006). A specific amyloid-beta protein assembly in the brain impairs memory. *Nature, 440*, 352–357.

58. Friedlich, A.L., Lee, J.Y., van Groen, T., Cherny, R.A., Volitakis, I., Cole, T.B., Palmiter, R.D., Koh, J.Y., Bush, A.I. (2004). Neuronal zinc exchange with the blood vessel wall promotes cerebral amyloid angiopathy in an animal model of Alzheimer's disease. *J Neurosci, 24*, 3453–3459.

59. Lee, J.Y., Friedman, J.E., Angel, I., Kozak, A., Koh, J.Y. (2004). The lipophilic metal chelator DP-109 reduces amyloid pathology in brains of human beta-amyloid precursor protein transgenic mice. *Neurobiol Aging, 25*, 1315–1321.

60. Lovell, M.A., Robertson, J.D., Teesdale, W.J., Campbell, J.L., Markesbery, W.R. (1998). Copper, iron and zinc in Alzheimer's disease senile plaques. *J Neurol Sci, 158*, 47–52.

61. Miller, L.M., Wang, Q., Telivala, T.P., Smith, R.J., Lanzirotti, A., Miklossy, J. (2006). Synchrotron-based infrared and x-ray imaging shows focalized accumulation of Cu and Zn co-localized with beta-amyloid deposits in Alzheimer's disease. *J Struct Biol, 155*, 30–37.

62. Stoltenberg, M., Bruhn, M., Sondergaard, C., Doering, P., West, M.J., Larsen, A., Troncoso, J.C., Danscher, G. (2005). Immersion autometallographic tracing of zinc ions in Alzheimer beta-amyloid plaques. *Histochem Cell Biol, 123*, 605–611.

63. Stoltenberg, M., Bush, A.I., Bach, G., Smidt, K., Larsen, A., Rungby, J., Lund, S., Doering, P., Danscher, G. (2007). Amyloid plaques arise from zinc-enriched cortical layers in APP/PS1 transgenic mice and are paradoxically enlarged with dietary zinc deficiency. *Neuroscience, 150*, 357–369.

64. Suh, S.W., Jensen, K.B., Jensen, M.S., Silva, D.S., Kesslak, P.J., Danscher, G., Frederickson, C.J. (2000). Histochemically-reactive zinc in amyloid plaques, angiopathy, and degenerating neurons of Alzheimer's diseased brains. *Brain Res, 852*, 274–278.

65. Grundke-Iqbal, I., Fleming, J., Tung, Y.C., Lassmann, H., Iqbal, K., Joshi, J.G. (1990). Ferritin is a component of the neuritic (senile) plaque in Alzheimer dementia. *Acta Neuropathol, 81*, 105–110.

66. Cole, T.B., Wenzel, H.J., Kafer, K.E., Schwartzkroin, P.A., Palmiter, R.D. (1999). Elimination of zinc from synaptic vesicles in the intact mouse brain by disruption of the ZnT3 gene. *Proc Natl Acad Sci U S A, 96*, 1716–1721.

67. Lee, J.-Y., Cole, T.B., Palmiter, R.D., Suh, S.W., Koh, J.-Y. (2002). Contribution by synaptic zinc to the gender-disparate plaque formation in human Swedish mutant APP transgenic mice. *Proc Natl Acad Sci U S A, 99*, 7705–7710.

68. Bush, A.I., Tanzi, R.E. (2002). The galvanization of beta-amyloid in Alzheimer's disease. *Proc Natl Acad Sci U S A, 99*, 7317–7319.

69. Terry, R.D., Masliah, E., Salmon, D.P., Butters, N., DeTeresa, R., Hill, R., Hansen, L.A., Katzman, R. (1991). Physical basis of cognitive alterations in

Alzheimer's disease: synapse loss is the major correlate of cognitive impairment. *Ann Neurol, 30*, 572–580.

70. Schlief, M.L., Craig, A.M., Gitlin, J.D. (2005). NMDA receptor activation mediates copper homeostasis in hippocampal neurons. *J Neurosci, 25*, 239–246.
71. Schlief, M.L., West, T., Craig, A.M., Holtzman, D.M., Gitlin, J.D. (2006). Role of the Menkes copper-transporting ATPase in NMDA receptor-mediated neuronal toxicity. *Proc Natl Acad Sci U S A, 103*, 14919–14924.
72. Religa, D., Strozyk, D., Cherny, R.A., Volitakis, I., Haroutunian, V., Winblad, B., Naslund, J., Bush, A.I. (2006). Elevated cortical zinc in Alzheimer disease. *Neurology, 67*, 69–75.
73. Bayer, T.A., Schafer, S., Simons, A., Kemmling, A., Kamer, T., Tepests, R., Eckert, A., Schussel, K., Eikenberg, O., Sturchler-Pierrat, C., *et al.* (2003). Dietary Cu stabilizes brain superoxide dismutase 1 activity and reduces amyloid A{beta} production in APP23 transgenic mice, *Proc Natl Acad Sci U S A, 100*, 14187–14192.
74. Phinney, A.L., Drisaldi, B., Schmidt, S.D., Lugowski, S., Coronado, V., Liang, Y., Horne, P., Yang, J., Sekoulidis, J., Coomaraswamy, J., *et al.* (2003). In vivo reduction of amyloid-{beta} by a mutant copper transporter. *Proc Natl Acad Sci U S A, 100*, 14193–14198.
75. Morris, M.C., Evans, D.A., Tangney, C.C., Bienias, J.L., Schneider, J.A., Wilson, R.S., Scherr, P.A. (2006). Dietary copper and high saturated and trans fat intakes associated with cognitive decline. *Arch Neurol, 63*, 1085–1088.
76. Zhang, J., Liu, Q., Chen, Q., Liu, N.Q., Li, F.L., Lu, Z.B., Qin, C., Zhu, H., Huang, Y.Y., He, W., Zhao, B.L. (2006). Nicotine attenuates beta-amyloid-induced neurotoxicity by regulating metal homeostasis. *FASEB J, 20*, 1212–1214.
77. Crapper McLachlan, D.R., Dalton, A.J., Kruck, T.P.A., Bell, M.Y., Smith, W.L., Kalow, W., Andrews, D.F. (1991). Intramuscular desferrioxamine in patients with Alzheimer's disease. *Lancet, 337*, 1304–1308.
78. Squitti, R., Rossini, P.M., Cassetta, E., Moffa, F., Pasqualetti, P., Cortesi, M., Colloca, A., Rossi, L., Finazzi Agro, A. (2002). D-Penicillamine reduces serum oxidative stress in Alzheimer's disease patients. *Eur J Clin Invest, 32*, 51–59.
79. Cherny, R.A., Atwood, C.S., Xilinas, M.E., Gray, D.N., Jones, W.D., McLean, C.A., Barnham, K.J., Volitakis, I., Fraser, F.W., *et al.* (2001). Treatment with a copper–zinc chelator markedly and rapidly inhibits β-amyloid accumulation in Alzheimer's disease transgenic mice. *Neuron, 30*, 665–676.
80. Ritchie, C.W., Bush, A.I., Mackinnon, A., Macfarlane, S., Mastwyk, M., MacGregor, L., Kiers, L., Cherny, R., Li, Q.X., Tammer, A., *et al.* (2003). Metal–protein attenuation with iodochlorhydroxyquin (clioquinol) targeting Abeta amyloid deposition and toxicity in Alzheimer disease: a pilot phase 2 clinical trial. *Arch Neurol, 60*, 1685–1691.
81. Lannfelt, L., Blennow, K., Zetterberg, H., Batsman, S., Ames, D., Harrison, J., Masters, C.L., Targum, S., Bush, A.I., Murdoch, R., *et al.* (2008). Safety, efficacy, and biomarker findings of PBT2 in targeting Abeta as a modifying therapy for Alzheimer's disease: a phase IIa, double-blind, randomised, placebo-controlled trial. *Lancet Neurol, 7*, 779–786.

82. Adlard, P.A., Cherny, R.A., Finkelstein, D.I., Gautier, E., Robb, E., Cortes, M., Volitakis, I., Liu, X., Smith, J.P., Perez, K., *et al.* (2008). Rapid restoration of cognition in Alzheimer's transgenic mice with 8-hydroxy quinoline analogs is associated with decreased interstitial Abeta. *Neuron, 59*, 43–55.

83. White, A.R., Du, T., Laughton, K.M., Volitakis, I., Sharples, R.A., Xilinas, M.E., Hoke, D.E., Holsinger, R.M., Evin, G., Cherny, R.A., *et al.* (2006). Degradation of the Alzheimer disease amyloid beta-peptide by metal-dependent up-regulation of metalloprotease activity. *J Biol Chem, 281*, 17670–17680.

84. Kaur, D., Yantiri, F., Kumar, J., Mo, J.O., Rajagopalan, S., Viswanath, V., Boonplueang, R., Jacobs, R., Yang, L., Beal, M.F., *et al.* (2003). Genetic or pharmacological iron chelation prevents MPTP-induced neurotoxicity in vivo: a novel therapy for Parkinson's disease. *Neuron, 37*, 923–933.

85. Nguyen, T., Hamby, A., Massa, S.M. (2005). Clioquinol down-regulates mutant huntingtin expression in vitro and mitigates pathology in a Huntington's disease mouse model. *Proc Natl Acad Sci U S A, 102*, 11840–11845.

86. Fox, J.H., Kama, J.A., Lieberman, G., Chopra, R., Dorsey, K., Chopra, V., Volitakis, I., Cherny, R.A., Bush, A.I., Hersch, S. (2007). Mechanisms of copper ion mediated Huntington's disease progression. *PLoS ONE, 2*, e334.

25

WHY STUDY THE ROLE OF HEPARAN SULFATE IN IN VIVO AMYLOIDOGENESIS?

ROBERT KISILEVSKY

Department of Pathology and Molecular Medicine and Department of Biochemistry, Queen's University, Kingston, Ontario, Canada; Syl and Molly Apps Research Center, Kingston General Hospital Kingston, Ontario, Canada

JOHN ANCSIN

Department of Pathology and Molecular Medicine, Queen's University, Kingston, Ontario, Canada; Syl and Molly Apps Research Center, Kingston General Hospital, Kingston, Ontario, Canada

INTRODUCTION

It is rare in science to have the opportunity to write about what one would *like* to know instead of what one does know about a scientific problem. Within specific topics, scientific writing focuses predominantly on what has been established, on recent discoveries, and on the immediate implications of the novel results within existing paradigms. Occasionally, accumulated data overturn such paradigms and we are forced to speculate. Despite this, most scientific writing avoids going beyond what is supported by available, or imminently anticipated, experimental data.

Referees of journal submissions frown on speculative papers; submissions are expected to conform to a set format that begins with a review of the pertinent literature, states the issues emerging from the review, and then focuses on the investigators' particular interests and the reasons for such interests.

Protein Misfolding Diseases: Current and Emerging Principles and Therapies,
Edited by Marina Ramirez-Alvarado, Jeffery W. Kelly, and Christopher M. Dobson
Copyright © 2010 John Wiley & Sons, Inc.

These are matters that can usually be addressed technically, and they require detailed descriptions of the methodology. Results and a discussion of the results follow, and novelty is explained within existing and competing hypotheses. Questions that cannot be presently addressed technically are seldom raised because supporting data cannot be provided for proposed answers. Within this format there is little opportunity to reflect on what one would like to know or what we might call "cognitive desiderata." Yet in our own minds and in informal discussions, this is exactly what we, and our colleagues, do. We raise bold conjectures and venture fanciful hypotheses—but then work on the answerable questions instead of those we still cannot address.

Contrary to established practice, in this chapter I focus on the *verboten;* on what I would like to know instead of what I do know about the role of the extracellular matrix (ECM), particularly heparan sulfate (HS), and its influence on amyloidogenesis in vivo. The hope is to engage in and to prompt some productive speculation and experimentation.

WHAT DO I KNOW ABOUT HEPARAN SULFATE AND ITS ROLE IN AMYLOIDOGENESIS IN VIVO?

Our current knowledge of the role played by HS during in vivo amyloidogenesis has been reviewed and referenced extensively recently [1] and is summarized below. Much of what is known has come from an examination of the prototype of in vivo amyloidogenesis, AA amyloid, and the structure of AA amyloid *in situ.*

The very name *amyloid,* coined by Virchow in 1854, was based on the chemical behavior of tissue amyloid deposits when exposed to acidified iodine. The blue–black reaction product indicated the presence of a complex carbohydrate, erroneously thought to be starch or cellulose. As early as the 1920s the carbohydrate had been identified as a glycosaminoglycan (GAG) (previously called a mucopolysaccharide), and by 1942 amyloid-laden tissues, such as spleen containing the AA form, were known to be a rich source of HS for study, a clear clue that HS may be involved in tissue amyloid deposition. Subsequent data obtained in the early 1980s demonstrated that the deposition of the GAG (now clearly identified as HS) and the AA peptide in spleen were coincident temporally and anatomically. The mRNAs for splenic perlecan, the HS proteoglycan (HSPG) deposited in AA amyloid, and several additional extracellular matrix proteins (e.g., type IV collagen and several laminin chains) increased in spleen coincidently with, or shortly before, amyloid was detectable in tissue. Histochemistry and immunohistochemistry revealed that the HSPG was a structural component of amyloid fibrils in situ, which was then substantiated by immunoelectron and high-resolution electron microscopy. In addition, the in vivo kinetics of HS and AA amyloid deposition were similar and consistent with a process that "seeds" fibrillogenesis [2,3].

In vitro studies provided details of the interaction of HS with serum amyloid A (SAA, the AA peptide precursor). These revealed that only the combination of HS and the fibrillogenic isoform of mouse SAA (SAA1.1) resulted in a conformational change of SAA1.1. This interaction significantly increased SAA1.1's β-sheet structure, the characteristic folding state of amyloid proteins. No other isoform of SAA underwent this change in folding on binding to HS, and no other GAG interacting with SAA1.1 induced this change in SAA1.1. These data indicated that (1) this property of HS *vis-à-vis* SAA1.1 was in part due to molecular charge, but not simply one of charge, since other equivalently charged GAGs, such as chondroitin sulfate, failed to prompt the conformational change in SAA1.1, and (2) there were specificities of GAG structure necessary for the interaction of HS and SAA1.1. The nature of these specificities remains to be determined and is one of the aspects of HS structure that I would like to understand. Furthermore, two HS binding domains on SAA1.1 were identified [4,5], one of which inhibits AA amyloidogenesis in cell culture [6] and a second that enhances this process [7]. Recognizing that AA amyloidogenesis in vivo is an interactive process between, at least, HS and the amyloidogenic form of SAA raised the possibility or opportunity that interference with this interaction could perturb this process and provide (1) additional understanding of the in vivo amyloidogenic process concerned, and (2) molecular clues to potential therapies. Molecular analogs of the complementary binding faces of HS–SAA proved to be remarkably effective in inhibiting AA amyloidogenesis in culture and in vivo [6,8], as did inhibitors of HS biosynthesis [9]. The latter also proved to be effective in an in vivo model of Aβ and a culture model of AIAPP amyloid deposition [1,10], indicating that the HS–protein interaction was not restricted to AA amyloid but also operated in amyloidogenesis of Alzheimer disease and type 2 diabetes. Not surprisingly, transgenic mice overexpressing human heparanase proved to be resistant to AA amyloid deposition, but remarkably only in those organs expressing the transgene [11]. In the same mouse, the spleen, which did not express the gene or the heparanase protein, continued to deposit amyloid, leaving little doubt that HS is not simply an accessory molecule but an integral player in the entire process of AA amyloid deposition in vivo, and probably other forms of in vivo amyloidogenesis. As discussed previously [1], this process is not taking place at sites of preexisting physiological HSPG deposits (e.g., basement membranes) but at sites of de novo basement membrane protein synthesis.

The interactions between HS and amyloidogenic proteins represents a special case, or subset, of HS–protein interactions/binding, a topic that has been of interest to investigators in diverse areas of research. HS–protein interaction/binding plays a role in a host of biological, physiological, and pathological processes [12–21]: embryogenesis, cell proliferation, cell differentiation and migration, cell–cell adhesion, cell–extracellular matrix interactions, wound repair, angiogenesis, lipid metabolism, kidney function, blood coagulation, viral infections, and amyloidogenesis. The numbers of proteins and peptides involved are myriad and include, but are not limited to, fibroblast growth factors, chemokines, cytokines, extracellular matrix components, adhesion molecules, angiogenic factors, viral capsid proteins, malarial circumsporozoite proteins, and lipoproteins, among which is SAA.

WHAT WOULD I LIKE TO KNOW ABOUT THE ROLE OF HEPARAN SULFATE IN IN VIVO AMYLOIDOGENESIS?

Two general aspects must be understood to appreciate the manner and mechanism by which a protein binds to or interacts with HS. The first concerns the structure (from primary to quaternary) of the HS-binding site on the protein, and the second concerns the complementary structure (primary to quaternary) of the protein-binding site on HS. Significantly more progress has been made addressing the former than the latter. One of the major reasons for this difference is that the techniques for purifying a protein to virtual homogeneity and establishing its amino acid sequence, or synthesizing a sufficiently large peptide of known sequence, are now well established. When combined with NMR, x-ray diffraction, crystallography, mass spectrometry, molecular mechanics, molecular modeling, and amino acid substitutions, substantial understanding of the various levels of the protein structure can be achieved in the absence, and in the presence, of a model of HS: namely, heparin. However, analogous techniques to achieve this level of information for specific HS polysaccharides are either nonexistent or at a very early stage of development. Although purification of HSPGs, and even a specific HSPG, can be accomplished, the HS polysaccharide chains prepared following these purification steps remain heterogeneous because there is usually more than one HS chain per HSPG molecule (e.g., perlecan or syndecan). This situation with HS is analogous to making a partially purified protein extract from a tissue and attempting to determine the amino acid sequence of the protein preparation. The results would be incomprehensible! An absolute requirement is the purification of a protein to virtual homogeneity, the first step to elucidating a protein's structure. Analogously, this defines for us the first technical hurdle that must be overcome to determine if (1) within a given tissue or cell type, and (2) at a specific serine locus on a defined proteoglycan protein backbone, there is a consistent/specific sulfation and epimerization pattern along the linear structure of the relevant HS chain. If such a pattern exists for a specific HS polysaccharide chain (everything proposed below is predicated on this assumption as otherwise there is no point in addressing this question), such information can form a basis on which to determine the higher-order structural organization of the HS polysaccharide concerned. This is turn will open opportunities to define how the specific HS interacts and influences the conformation of its protein partner(s), including in vivo amyloidogenesis.

A partial solution to this problem has been to study heparin, a surrogate for HS, and the manner in which it interacts with proteins (e.g., antithrombin III) [20,22]. Like HS, heparin (Fig. 1) is a linear polysaccharide composed of alternating *N*-acetylglucosamine (GlcNac) and glucuronic acid (GlcA) sugars.

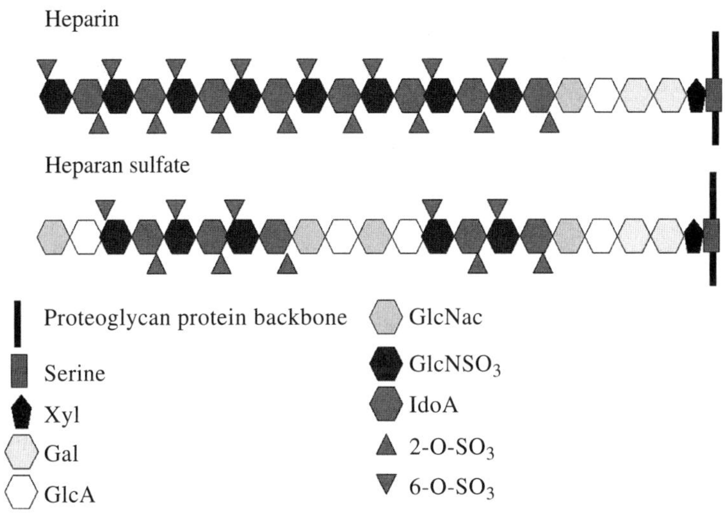

FIG. 1 Comparison of the linear structures of heparin and heparan sulfate. Note that beyond the tetrasaccharide linkage region heparin has a consistent pattern of repeating disaccharides. On occasion a 3-O-SO_3 is present in the $GlcNSO_3$ (not shown). By contrast, heparan sulfate has short stretches of heparinlike structure separated by regions that are poorly sulfated within which is GlcA rather than sulfated IdoA. (*See insert for color representation of figure.*)

The majority of the GlcA has undergone epimerization about the C-5 carbon [generating iduronate (IdoA)], and sulfation has occurred at the 2-O position. The GlcNac has been modified by N-deacetylation and N-sulfation, and sulfation has occurred at the 6-O position and occasionally at the 3-O position. This relatively uniform sulfation and structural pattern along the entire length of the heparin molecule has made it amenable to conformational analysis, how it binds to a protein, and what influence it has on protein structure [22]. Nevertheless, the results obtained with heparin, although well studied and useful because of its clinical utility, are not sufficiently relevant because physiologically it is found predominantly in mast cells. Heparan sulfate is the form found ubiquitously in the body, and it is this form that is less well understood.

Heparan sulfate (Fig. 1) is also a linear polysaccharide composed of alternating GlcNac and GlcA sugars; however, the virtually consistent sulfation and epimerization pattern seen along the length of a heparin molecule is not maintained in HS. There are stretches of heparinlike structure of variable length that are separated by regions in which the GlcNac has not been deacetylated or sulfated, and since the latter modification is necessary for epimerization and 2-O-sulfation of adjacent GlcAs, 2-O-sulfated IdoA is also lacking in these regions. Nevertheless, the introduction of these modifications does not appear to be random. Random introduction of sulfation and epimerization would result in HSs that have overall similarity in their disaccharide compositions, regardless of the tissue, cell, or type of HSPG examined. The degree to which these modifications take place in HS is, however, consistent for each organ, cell type, and type of HSPG found at these locations. Furthermore, there are aspects of embryogenesis indicating that specific modifications are necessary for specific organ development [23]. Where does this HS specificity lie, and is there a linear structural pattern that is maintained for a HS at a given serine locus on a specific HSPG? For example, there are three HS chains on perlecan, each linked to a specific but separate serine residue in the amino acid sequence of the perlecan protein backbone. Is the linear structural pattern the same for each of these HS chains at the different serine loci, or are they different? Within a specific tissue, is the linear structural pattern at each serine locus the same from one perlecan molecule to another? Such information will be fundamental in determining the higher-order structural organization of the specific HS polysaccharide and how it interacts with proteins of interest, including those that are amyloidogenic. Preparation of synthetic libraries of short oligosaccharides (8 to 10 sugars) of defined "sequence" has been attempted, but how these molecules interact with known protein factors has yielded results of variable quality [24–28]. How, then, might one obtain a sufficient quantity of a HSPG that contains but a single HS chain at a specific locus on the proteoglycan backbone? Phrased differently, if such a chain exists, how does one purify such a HS chain to virtual homogeneity and in sufficient quantity for analysis?

One may perhaps address this challenge in the following way:

1. Choose a cell line that expresses a small HSPG that has HS chains at defined proteoglycan serine positions [e.g. syndecan-4 (Syn4) is well studied, its base sequence and amino acid sequence are known, it has a molecular mass of approximately 22,000 and possesses three HS chains] [29].
2. Create a Syn4 "knockout" of the cell line concerned (so that no wild-type Syn4 can contaminate subsequent isolation procedures). Reintroduce into this cell line a Syn4 gene that has been altered by site-directed mutagenesis so that only one of the three specific serine residue that would normally carry HS is retained [e.g., Syn4 possessing only the serine 1 locus (Syn4Ser1)], the remaining two serine coding loci being altered to code for alanine, a residue similar in structure to serine but lacking the OH group necessary for the initiation of GAG synthesis. Theoretically, one should be able to prepare three separate cell lines, each retaining a specific serine HS attachment locus (Syn4Ser1, 2, or 3) and then compare the structures of the HS chains at each of these loci.
3. After culturing the Syn4Ser1 cell line, the cells can be lysed, the nuclei removed, the postnuclear preparation dissolved in mild detergent and guanidine HCl, the proteoglycans isolated by ion-exchange chromatography, and the Syn4Ser1-HS subsequently isolated with a Syn4

antibody. The HS chains can then be released by β-elimination (alkali and sodium borohydride reduction). If for a defined cell type a specific HS linear structural pattern really exists at a specific proteoglycan serine locus, the foregoing preparation would be the one likely to contain such a pattern. How might one uncover this pattern?

Recent publications have shown that a 4-deoxy analog of GlcNac can truncate HS biosynthesis in cell culture, and such an analog can inhibit AA, Aβ, and AIAPP amyloid deposition in culture and in vivo [1,9,10]. Using the Syn4Ser1 cell line described above, exposure of the cell line to the 4-deoxy analog of GlcNac (or a different analog modified at the C-4 position) will abruptly terminate HS elongation, generating a population of nested HS polysaccharides of varying length (Fig. 2). Each of these will begin with the

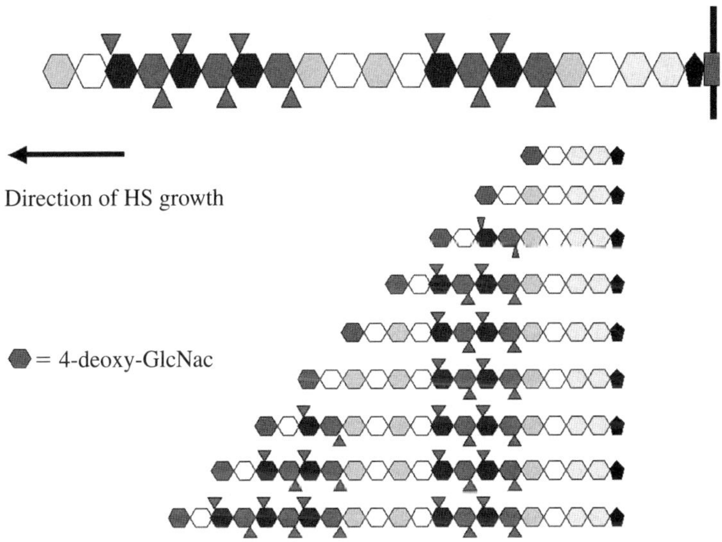

FIG. 2 Nested series of HS polysaccharides that would be generated by 4-deoxy-GlcNac if there is a specific HS linear structural pattern (e.g., top of figure) that exists at a specific proteoglycan serine locus. Each of these will begin with the Xyl–Gal–Gal–GlcA tetrasaccharide linkage region that serves as a bridge between the elongating HS chain and the relevant proteoglycan serine residue and terminate with 4-deoxy-GlcNac. These unique structures "tag" each end of the polysaccharide. Because HS grows by the alternating addition of GlcNac and GlcA, the shortest polysaccharide within the isolated HS preparation will be a pentasaccharide consisting of Xyl–Gal–Gal–GlcA–4-deoxy-GlcNac. Each larger HS within the isolated polysaccharides population will increase in size by a disaccharide consisting of GlcA–4-deoxy-GlcNac and a MW of approximately 400. Sulfation occurring in the disaccharide immediately behind the truncation site may increase the MW by an additional 200 to 250 mass units. With unique tags at each end of these polysaccharides, the intervening sugar modifications should be amenable to mass spectrometric analysis to determine the consistent linear structural pattern, if present. (*See insert for color representation of figure.*)

xylose–galactose–galactose–glucuronate (Xyl-Gal-Gal-GlcA) tetrasaccharide linkage region that serves as the bridge between the elongating HS chain and the relevant proteoglycan serine residue and terminate with 4-deoxy-GlcNac. These unique structures "tag" each end of the polysaccharide. Furthermore, because HS grows by the alternating addition of GlcNac and GlcA, the shortest polysaccharide within the isolated HS preparation will be a pentasaccharide consisting of Xyl–Gal–Gal–GlcA–4-deoxy-GlcNac. Each larger HS within the isolated polysaccharides population will increase in size by a disaccharide consisting of GlcA–4-deoxy-GlcNac and an MW of approximately 400. Oligosaccharides ranging in size from 4 to 20 sugars and differing by one disaccharide unit (i.e., 4, 6, 8, 10, etc. sugars long) can be separated by liquid or thin-layer chromatography [30]. Somewhat larger oligosaccharides may be separated by capillary electrophoresis [31–33], and these different sizes can be subjected to mass spectrometry [34–38]. With unique "tags" at each end of these polysaccharides, the intervening sugar modifications should be amenable to analysis to determine a consistent structural pattern, if present. The data will either support, or deny, the existence of a specific HS linear structural pattern at a specific proteoglycan serine locus. A comparison of the linear structural pattern in each of the three different cell lines will confirm, or deny, whether Syn4Ser1, 2, and 3 carry HS chains with similar or different linear structural patterns. Why would this information be important?

A failure to demonstrate the existence of a specific HS linear structural pattern at a specific proteoglycan serine locus would indicate that the specific manner in which HS interacts with a protein will be a much more intractable problem than is presently appreciated. On the other hand, the ability to obtain, and demonstrate, a HS preparation with a defined linear structural pattern would allow the application of techniques such as NMR, x-ray diffraction, mass spectrometry, molecular mechanics- and molecular modeling to determine the higher, ordered structure, and in turn, to have immense value for the investigators in the diverse areas of research in which HS–protein interaction/binding in known to play a role. Such information will form the basis on which to identify the specific HS structures (linear to quaternary) that bind or interact with complementary structural domains in specific proteins, and the design and synthesis of specific compounds to inhibit such interactions for the purposes of (1) perturbing the biological systems to reveal more of their biochemical mechanisms, and (2) identifying lead compounds for therapy. From the perspective of amyloidogenesis, one may determine the structural specificity of HS necessary for its interaction with the amyloidogenic protein concerned and thus call on rational design for the development of antiamyloid compounds.

Acknowledgments

This work has been supported primarily by grant MOP-3153 from the Canadian Institutes for Health Research and grant 200301 from the Institute for the Study of Aging in New York. The technical assistance of Ruth Tan, Lee Boudreau, and Elizabeth Hearn is gratefully acknowledged. A thank you is also in order to Sherry Taylor, David Lillicrap, John Stone, and Carlos Prado for helpful discussions.

REFERENCES

1. Kisilevsky, R., Ancsin, J.B., Szarek, W.A., Petanceska, S. (2007). Heparan sulfate as a therapeutic target in amyloidogenesis: prospects and possible complications. *Amyloid, 14*, 21–32.
2. Kisilevsky, R., Boudreau, L. (1983). The kinetics of amyloid deposition: I. The effect of amyloid enhancing factor and splenectomy. *Lab Invest, 48*, 53–59.
3. Snow, A.D., Kisilevsky, R., Stephens, C., Anastassiades, T. (1987). Characterization of tissue and plasma glycosaminoglycans during experimental AA amyloidosis and acute inflammation: quantitative and qualitative analysis. *Lab Invest, 56*, 665–675.
4. Ancsin, J.B., Kisilevsky, R. (1999). The heparin/heparan sulfate-binding site on apo-serum amyloid A: implications for the therapeutic intervention of amyloidosis. *J Biol Chem, 274*, 7172–7181.
5. Ancsin, J.B., Elimova, E., Kisilevsky, R. (2001). Serum amyloid A has two heparin/heparan sulfate binding sites with different pH optima. *Amyloid J Prot Fold Disord, 8* Suppl 2, 40.
6. Elimova, E., Kisilevsky, R., Szarek, W.A., Ancsin, J.B. (2004). Amyloidogenesis recapitulated in cell culture: a peptide inhibitor provides direct evidence for the role of heparan sulfate and suggests a new treatment strategy. *FASEB J, 18*, 1749–1751.
7. Elimova, E., Kisilevsky, R., Ancsin, J.B. (2005). A serum amyloid A peptide contains a low-pH heparin binding site which promotes AA-amyloidogenesis in cell culture. In *Amyloid and Amyloidosis: Proceedings of the 10th International Symposium on Amyloidosis* (Grateau, G., Kyle, R.A., Skinner, M., Eds.) CRC Press, Boca Raton, FL, pp. 194–196.
8. Kisilevsky, R., Lemieux, L.J., Fraser, P.E., Kong, X.Q., Hultin, P.G., Szarek, W.A. (1995). Arresting amyloidosis in vivo using small-molecule anionic sulphonates or sulphates: implications for Alzheimer's disease. *Nat Med, 1*, 143–148.
9. Kisilevsky, R., Szarek, W.A., Ancsin, J.B., Elimova, E., Marone, S., Bhat, S., Berkin, A. (2004). Inhibition of amyloid A amyloidogenesis in vivo and in tissue culture by 4-deoxy analogues of peracetylated 2-acetamido-2-deoxy-α- and β-D-glucose: implications for the treatment of various amyloidoses. *Am J Pathol, 164*, 2127–2137.
10. Hull, R., Zraika, S., Udayasankar, J., Kisilevsky, R., Szarek, W.A., Wight, T.N., Kahn, S.E. (2007). Inhibition of glycosaminoglycan synthesis and protein glycosylation with WAS-406 and azaserine result in reduced islet amyloid formation in vitro. *Am J Physiol Cell Physiol, 293*, C1586–1593.
11. Li, J.P., Galvis, M.L., Gong, F., Zhang, X., Zcharia, E., Metzger, S., Vlodavsky, I., Kisilevsky, R., Lindahl, U. (2005). In vivo fragmentation of heparan sulfate by heparanase overexpression renders mice resistant to amyloid protein A amyloidosis. *Proc Natl Acad Sci U S A, 102*, 6473–6477.

12. Sugahara, K., Kitagawa, H. (2000). Recent advances in the study of the biosynthesis and functions of sulfated glycosaminoglycans. *Curr Opin Struct Biol*, *10*, 518–527.
13. David, G., Bernfield, M. (1998). The emerging roles of cell surface heparan sulfate proteoglycans. *Matrix Biol*, *17*, 461–463.
14. Bernfield, M., Gotte, M., Park, P.W., Reizes, O., Fitzgerald, M.L., Lincecum, J., Zako, M. (1999). Functions of cell surface heparan sulfate proteoglycans. *Annu Rev Biochem*, *68*, 729–777.
15. Tumova, S., Woods, A., Couchman, J.R. (2000). Heparan sulfate proteoglycans on the cell surface: versatile coordinators of cellular functions. *Int J Biochem Cell Biol*, *32*, 269–288.
16. Park, P.W., Reizes, O., Bernfield, M. (2000). Cell surface heparan sulfate proteoglycans: selective regulators of ligand-receptor encounters. *J Biol Chem*, *275*, 29923–29926.
17. Iozzo, R.V. (2001). Heparan sulfate proteoglycans: intricate molecules with intriguing functions. *J Clin Invest*, *108*, 165–167.
18. Shriver, Z., Liu, D.F., Sasisekharan, R. (2002). Emerging views of heparan sulfate glycosaminoglycan structure/activity relationships modulating dynamic biological functions. *Trends Cardiovasc Med*, *12*, 71–77.
19. Gesslbauer, B., Rek, A., Falsone, F., Rajkovic, E., Kungl, A.J. (2007). Proteoglycanomics: tools to unravel the biological function of glycosaminoglycans. *Proteomics*, *16*, 2870–2880.
20. Munoz, E.M., Linhardt, R.J. (2004). Heparin-binding domains in vascular biology. *Arterioscler Thromb Vasc Biol*, *24*, 1549–1557.
21. Lindahl, U. (2007). Heparan sulfate–protein interactions: a concept for drug design? *Thromb Haemost*, *98*, 109–115.
22. Mulloy, B., Forster, M.J. (2000). Conformation and dynamics of heparin and heparan sulfate. *Glycobiology*, *10*, 1147–1156.
23. Li, J.P., Gong, F., Hagner-McWhirter, A., Forsberg, E., Abrink, M., Kisilevsky, R., Zhang, X., Lindahl, U. (2003). Targeted disruption of a murine glucuronyl C5-epimerase gene results in heparan sulfate lacking L-iduronic acid and in neonatal lethality. *J Biol Chem*, *278*, 28363–28366.
24. Jemth, P., Kreuger, J., Kusche-Gullberg, M., Sturiale, L., Gimenez-Gallego, G., Lindahl, U. (2002). Biosynthetic oligosaccharide libraries for identification of protein-binding heparan sulfate motifs: exploring the structural diversity by screening for FGF1 and FGF2 binding. *J Biol Chem*, *277*, 30567–30573.
25. Kreuger, J., Prydz, K., Pettersson, R.F., Lindahl, U., Salmivirta, M. (1999). Characterization of fibroblast growth factor 1 binding to heparan sulfate domain. *Glycobiology*, *9*, 723–729.
26. Kreuger, J., Salmivirta, M., Sturiale, L., Gimenez-Gallego, G., Lindahl, U. (2001). Sequence analysis of heparan sulfate epitopes with graded affinities for fibroblast growth factors 1 and 2. *J Biol Chem*, *276*, 30744–30752.
27. Kreuger, J., Jemth, P., Sanders-Lindberg, E., Eliahu, L., Ron, D., Basilico, C., Salmivirta, M., Lindahl, U. (2005). Fibroblast growth factors share binding sites in heparan sulfate. *Biochem J*, *389*, 145–150.
28. Kreuger, J., Spillmann, D., Lindahl, U. (2006). Interactions between heparan sulfate and proteins, the concept of specificity. *J Cell Biol*, *174*, 323–327.

29. Tkachenko, E., Rhodes, J.M., Simons, M. (2005). Syndecans: new kids on the signaling block. *Circ Res*, *96*, 488–500.
30. Zhang, Z., Xie, J., Zhang, F., Linhardt, R.J. (2007). Thin-layer chromatography for the analysis of glycosaminoglycan oligosaccharides. *Anal Biochem*, *371*, 118–120.
31. Grimshaw, J. (1997). Analysis of glycosaminoglycans and their oligosaccharide fragments by capillary electrophoresis. *Electrophoresis*, *18*, 2408–2414.
32. Karamanos, N.K., Hjerpe, A. (1998). A survey of methodological challenges for glycosaminoglycan/proteoglycan analysis and structural characterization by capillary electrophoresis. *Electrophoresis*, *19*, 2561–2571.
33. Militsopoulou, M., Lamari, F., Karamanos, N.K. (2003). Capillary electrophoresis: a tool for studying interactions of glycans/proteoglycans with growth factors. *J Pharm Biomed Anal*, *32*, 823–828.
34. Zaia, J., Costello, C.E. (2001). Compositional analysis of glycosaminoglycans by electrospray mass spectrometry. *Anal Chem, 73*, 233–239.
35. Zaia, J., Chan, S.Y., Costello, C.E. (2003). Tandem mass spectrometric strategies for determination of sulfation positions and uronic acid epimerization in chondroitin sulfate oligosaccharides. *J Am Soc Mass Spectrom, 14*, 1270–1281.
36. Saad, O.M., Leary, J.A. (2003). Compositional analysis and quantification of heparin and heparan sulfate by electrospray ionization ion trap mass spectrometry. *Anal Chem,75*, 2985–2995.
37. Lawrence, R., Kuberan, B., Lech, M., Beeler, D.L., Rosenberg, R.D. (2004). Mapping critical biological motifs and biosynthetic pathways of heparan sulfate. *Glycobiology, 14*, 467–479.
38. Wu, Z.L., Lech, M., Beeler, D.L., Rosenberg, R.D. (2004). Determining heparan sulfate structure in the vicinity of specific sulfotransferase recognition sites by mass spectrometry. *J Biol Chem 279*, 1861–1866.

26

SERUM AMYLOID P COMPONENT

Simon Kolstoe and Steve Wood
Centre for Amyloidosis and Acute Phase Proteins, University College London Medical School, London, UK

INTRODUCTION

Amyloid fibers associated with a particular disease condition are made up of essentially one protein type and are believed to grow by a seeded nucleation mechanism, fed by misfolded forms of the protein [1]. Discussion of these processes is available elsewhere in this volume. Pathological fibers are deposited in a closely associated mesh along with glycosaminoglycans and a variety of accessory proteins. The most prominent of these accessory proteins is the plasma glycoprotein serum amyloid P component (SAP), making up 15% of the mass of amyloid deposits in vivo [2]. About 20 unrelated proteins have been described as dominant fiber constituents, yet all known amyloid fibers are bound by SAP. This specific recognition has been exploited in the development of whole-body scintigraphy using radiolabeled SAP as a diagnostic tool in the clinical management of amyloidosis [3]. SAP binding appears to be a remarkably specific identifier of amyloid and must involve recognition of some common protein fiber motif, since fibers produced in vitro from purified proteins retain the ability to bind SAP [4]. This motif is certainly not defined by any clear sequence homology among fibrillogenic proteins and is not displayed by the natively folded form of the proteins, as SAP does not bind to them. Most likely, the motif is defined by some structural feature of amyloid. In the absence of a detailed molecular model, our best guess at the nature of the

Protein Misfolding Diseases: Current and Emerging Principles and Therapies,
Edited by Marina Ramirez-Alvarado, Jeffery W. Kelly, and Christopher M. Dobson
Copyright © 2010 John Wiley & Sons, Inc.

interaction comes from what we know of the structures and properties of the isolated partners.

Defining this SAP–amyloid interaction is important to both the fundamental science of the recognition process and also in the context of drug development aiming to assist amyloid clearance or prevent its formation. Both are plainly important issues in the systemic amyloidoses where massive fiber deposition can occur with physical implications for tissue structure and organ function. Pharmacological depletion of SAP has been linked to amyloid regression in animals [8] and is thus viewed as a viable therapy, certainly for treatment of the systemic amyloidoses. However, work implicating a cytotoxic species composed of soluble oligomeric Aβ(1–42) in Alzheimers disease [5] suggests that fiber formation might be protective in sequestering such toxic precursors. On the other hand, the considerable stability of the fiber form of prion proteins may contribute to transmission of the associated diseases. Evidence from in vitro experiments that implicate SAP as a stabilizer of amyloid fibers against proteolysis and phagocytosis [6], along with experiments suggesting that SAP may interact with protein folding intermediates [7], may be especially relevant to these issues.

STRUCTURE OF AMYLOID FIBERS

Although a limited subset of proteins and their fragments are found in clinically relevant amyloid deposits, the observation that many proteins and peptides can be persuaded to adopt a fibrous structure reminiscent of pathological amyloid has allowed a somewhat relaxed terminology to penetrate the field [9]. This is partly due to difficulties in applying classical structural techniques such as x-ray crystallography to such large aggregates, necessitating a search for experimentally tractable samples. However, such approaches have highlighted the central issue that many polypeptides relax to a common low-energy form of β-structure when heated at low pH or destabilized in other ways. Indeed, in vitro studies of some small fragments of proteins have shown that a fibril state can be adopted without undue provocation and have provided very detailed descriptions of the atomic arrangements likely to exist in amyloid fibers [10]. This may reflect the fact that in some fibers only a small part of the polypeptide chain is involved in the cross-β structure or that the extreme propensity of these fragments fulfill some seeding or template role in fiber extension. These ideas suggest that diseases of protein misfolding may in part be an expression of a thermodynamic propensity for polypeptides to organize in a primordial fold that is distinct from the native fold and prone to oligomerization [11]. Destabilization of the native fold by mutation or proteolysis [12,13] apparently removes an energy barrier and allows access to the fiber state.

The basic form of amyloid fibers was initially established by negative stain transmission electron microscopy of ex vivo materials, revealing unbranched structures up to 16,000 Å in length and 100 Å in width [14]. Staining with

Congo Red dye provided a material that appeared green when viewed through the crossed polars of a light microscope. The birefringence was thought to derive from a particular mode of dye binding that implicated long-range order in fibers. The nature of this order was subsequently revealed by x-ray fiber diffraction with 10-Å and 4.8-Å repeats, indicating the presence of cross-β structure in the fibers [15,16], where polypeptide chains run at right angles to the fiber axis. The structure was thus similar to that first described by Astbury in 1935, modeled by Linus Pauling in 1955 [17], and subsequently proposed for the egg-stalk silk of the *Cryosopa* moth [18]. Models suggest that the 4.8-Å meridional reflections derive from hydrogen-bonded β-strands with the second, more diffuse 10-Å equatorial reflection corresponding to wider spacing between individual β-sheets [19]. However, exactly how Congo Red binds such fibers is still not clear. Spectroscopic effects have been interpreted as demonstrating dye molecules aligned with the fiber axis [20], but crystallographic investigations of insulin with bound Congo Red showed the dye closely associated with the antiparallel β-sheet of the insulin dimer interface [21]. Congo Red is a flat aromatic compound that one might expect to be an intercalator, slipping between the β-sheets of the fiber in a similar manner to ethidium bromide in DNA. Indeed, stacking of dye molecules has been suggested as a source of the cotton effect observed spectroscopically. Such a binding mode would be in accord with the wide varieties of amyloid detectable with the dye. Contemporary electron microscopic (EM) methods employing vitrified unstained samples and image processing methods have produced more detailed models of fiber organization, suggesting that fibers are composed of multiple intertwined chains of filaments [22,23]. Progress is also being made in defining the conformation of polypeptide chains within such filaments.

Blake and co-workers used synchrotron radiation to investigate fiber diffraction from an ex vivo transthyretin fiber and suggested that the pattern could be explained by a continuous twisted β-helix of antiparallel β-strands [24]. This arrangement could be generated by the end-to-end stacking of transthyretin (TTR) dimers, but where conformational adjustments enabled the formation of a new intermolecular β-structure [25]. This model was extended by spin label studies that implicated the participation of strands A and B in the contact [26]. These strands become exposed when the C/D strand region of TTR is displaced, as described in the crystallographic structure of the G53S/E54D/L55S triple mutant where a three-residue "β-slip" destabilizes strand D [27]. Further hydrogen/deuterium, (H/D) exchange nuclear magnetic resonance (NMR) experiments of TTR molecules released by fiber breakdown confirmed a pattern of protected hydrogen bonds within the fibers, consistent with a H-bonding network related to native TTR subunits and inclusive of the C/D strand displacement [28]. This work also highlighted the likelihood of substantial regions of flexible chain termini protruding from the fiber surface (Fig. 1). The existence of novel intermolecular contacts in the TTR fibers is consistent with evidence that tetramer disassembly is a prerequisite of fiber formation. The strand C/D displacement violates the protein design principle of

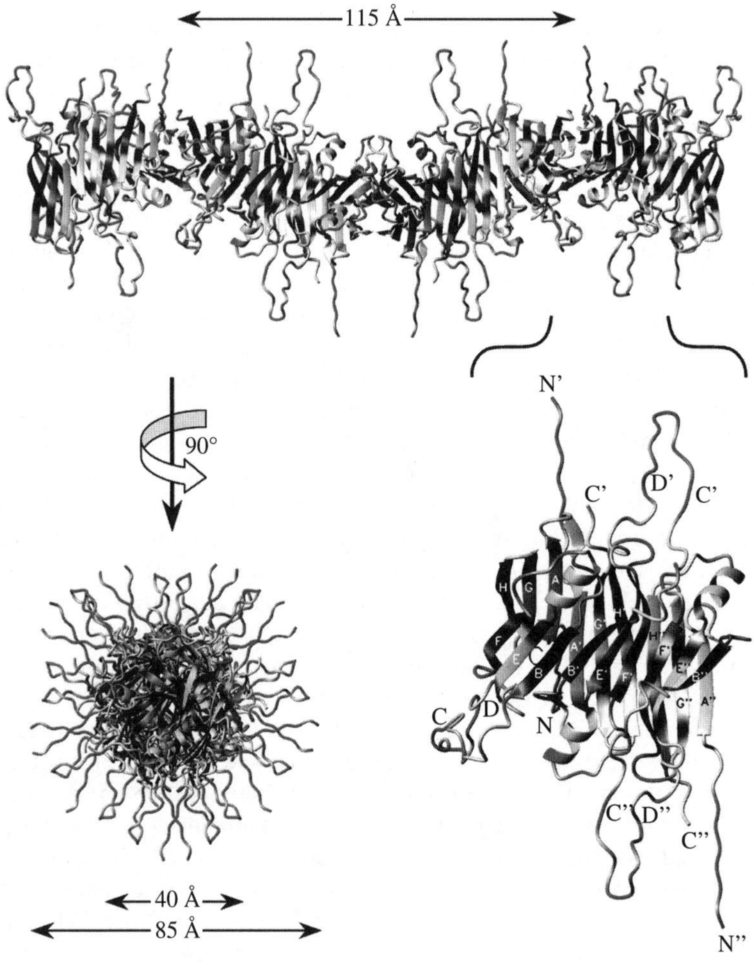

FIG. 1 Proposed structure of an TTR fiber based on deuterium exchange protection NMR and electron spin resonance data. Destabilization of strands C and D along with small modifications of the native fold expose edge strands A and B, leading to fiber formation and propagation. Loops between strands along with the N and C termini are exposed on the surface of the fiber, providing potential ligands for SAP. (From [28], with permission. Copyright © 2004 American Society for Biochemistry and Molecular Biology.) (*See insert for color representation of figure.*)

concealing regular exposed strands [29] and enables fiber formation. Stabilization of TTR tetramers is being pursued as one avenue of chemotherapeutic inhibition of fiber growth [30].

In conclusion our view of the TTR fiber involves a close-packed network of antiparallel β-strands twisted into an extended spiral with a general topology not far removed from native TTR dimers linked end to end via a new β network. Probably four of these superhelices entwine, with disordered loops and chain termini decorating the surface. It seems unlikely that SAP can access the compact cross-β core of the fiber. Turns interconnecting strands and disordered regions populating the fiber surface must contain the motif seen by SAP. The turns may not be very different from those in native TTR, which does not bind SAP, but the oligomeric form of the fiber means that they are present at high density.

Fibers of Aβ(1–42), the polypeptide component of senile plaques in Alzheimer's disease, have been studied using EM over many years, with the evidence suggesting multiple entwined filaments composed of much flatter layers of polypeptide [31]. This is in keeping with more recent models proposed from solid-state NMR methods by Tycko and co-workers [32,33]. These models consist of fibers constructed from stacked dimers of Aβ with each chain arranged as a hairpin between residues 10 and 40, with polar residues 25–29 forming the turn. The extended strands do not hydrogen bond, but their interaction is stabilized by hydrophobic interactions. The dimer is similarly stabilized by hydrophobic interaction between twofold-axis-related regions 30 to 40. The hydrogen-bonding network of the cross-β structure takes place between the stacked sets of four extended strands. The strands are all parallel, in register, and propagate with a twist as observed in sheets in globular proteins (Fig. 2). A broadly similar model has been proposed from H/D exchange NMR measurements with dimetheyl sulfoxide dispersed fibers of Aβ(1–42), although this model does not involve dimers and places the turn between residues 26 and 30 [34]. In this model the in-register strands generate stripes of acidic residues down the fiber face that are either exposed to solvent or involved in stabilizing adjacent chains. However, both models allow for results, suggesting that the N-terminal regions of each chain can be released from the fiber by proteolysis, and thus are likely to comprise an additional nonfibrous section of the deposit [35]. Interestingly, this region has a high affinity for copper ions and may be ordered with bound metal [36,37]. In conclusion, it seems likely that the fiber surface seen by approaching SAP molecules will be populated by exposed turns, stripes of acidic residues, and more disordered regions perhaps incorporating metal ions.

These two amyloid systems provide us with the greatest structural detail of intact fibers that are relevant to disease, although many other models and mechanisms of fiber formation involving strand displacement and domain swapping have been discussed [38]. Subtle reorganization of the TTR fold may be all that is required following tetramer disassembly to initiate fiber growth. On the other hand, the Aβ sequence, originating as an α-helical transmembrane region of amyloid precursor protein (APP), is not unexpectedly metastable and prone to aggregation into β-structured forms in an aqueous environment [39]. These fiber structures provide rather similar types of surface features potentially recognized by SAP, even though the form of the cross-β component varies substantially.

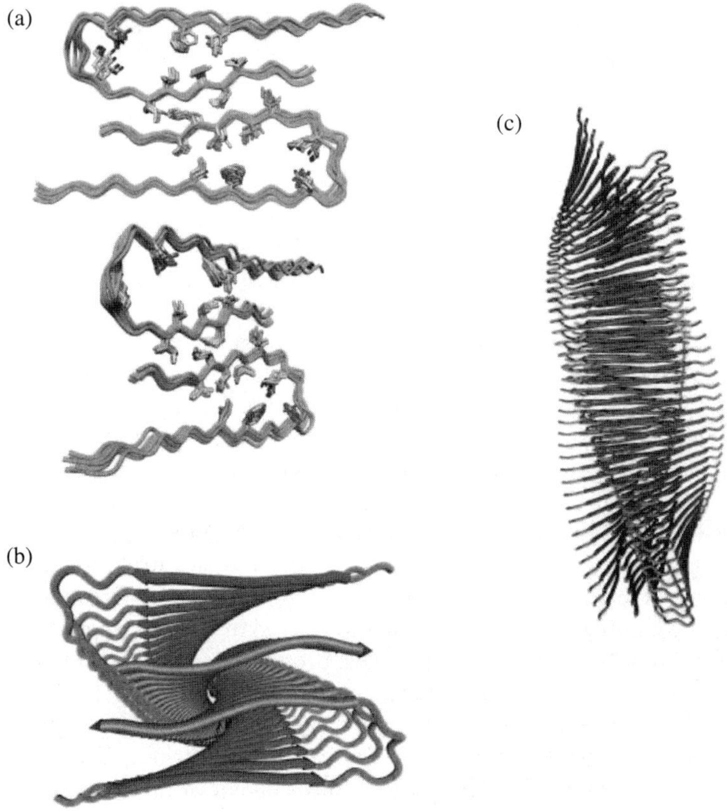

FIG. 2 Proposed structure of Aβ(1–40) amyloid. (a) Aβ dimers interacting via hydrophobic residues 30 to 40. Dimers can be layered (b) with (c) illustrating how stacking can lead to fibril morphology. (Adapted from [33].) (*See insert for color representation of figure.*)

STRUCTURE OF SAP

The crystal structure of SAP was solved in 1994 to 2-Å resolution [40] and showed five subunits, each adopting a distinctive flattened β-jellyroll structure similar to that of the legume lectins. The subunits pack into a radially symmetric pentameric disk approximately 100-Å in diameter, 35 Å in depth, with a 20-Å pore in the center. The x-ray structure confirmed earlier electron microscopy studies responsible for coining the name *pentraxin* for the larger protein family, derived from the Greek word for five (*penta*) and berries (*ragos*) [41]. Each SAP subunit consists of 204 amino acids in a single polypeptide chain with a disulfide bridge linking Cys36 and Cys95, a complex oligosaccharide N-glycosylation site on Asn32, and a distinctive metal-binding site containing two calcium ions. SAP is highly resistant to proteolytic cleavage in the presence of calcium ions, probably because of the short loops interconnecting β

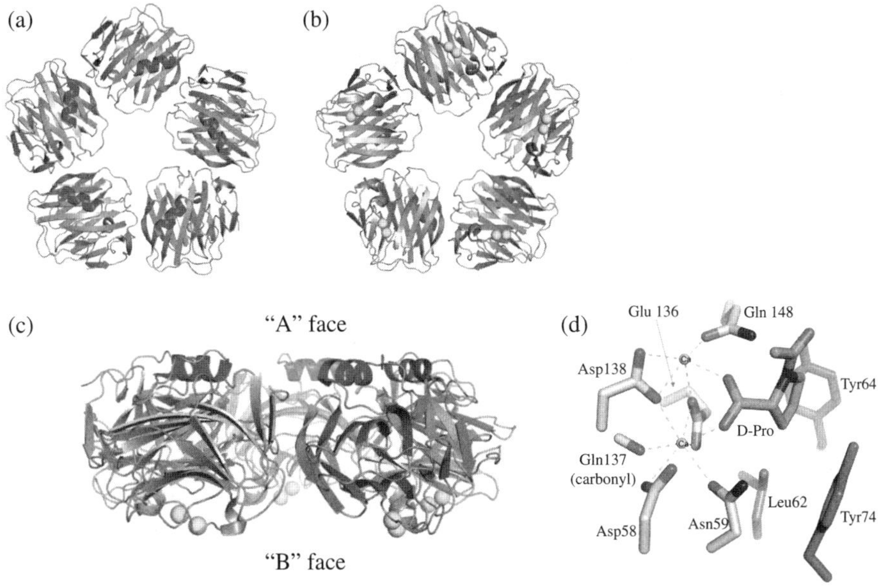

FIG. 3 Three orientations of the SAP pentamer: (a) the five α-helixes on the A face of the protein, (b) the five double calcium binding sites on the B face of the protein with calcium atoms in yellow; and (c) side view of the pentamer (made from the coordinates 1SAC [40] and using the program PyMOL [61]); (d) calcium-binding site of SAP with the ligand D-proline. Residues making up the hydrophobic pocket are shown in blue, calcium coordinating residues in gray, and calcium atoms in yellow and the ligand green (prepared using PyMOL [61]). (*See insert for color representation of figure.*)

strands [42], while its protomers are tightly associated, requiring strong denaturants to separate them. One of the two antiparallel β-sheets making up the subunit carries a 10-residue α-helix on its face and is termed the A face. The calcium-binding site is located on the other sheet on the opposite side of the molecule and is termed the B face (Fig. 3a). The calcium-binding sites consist of polar residues from loops assembled at the concave face of this twisted sheet with one calcium ion coordinated by Asp58, Asn59, Glu136, Asp138, the main chain carbonyl oxygen of Gln137, and a carboxylate or phosphate group of a ligand. The second calcium ion is coordinated by Glu136, Asp138, Gln148, two water molecules, and a contribution from a carboxylate or phosphate group of a ligand (Fig. 3d). This second site is more open, due to fewer coordinating groups and has been found to lose calcium fairly easily. This suggests that other metals, such as copper or zinc, might be accommodated preferentially at this site. Both calcium ions sit in an electrostatically strained arrangement about 4 Å apart, with the charge balance deriving from the carboxylate ligands to the metals. A small hydrophobic pocket created by the side chains of Leu62, Tyr64, and Tyr74 is located close by. The radial fivefold symmetry generates a polar pentamer with five prominent α-helices on the A face and 10 calcium ions distributed in five double binding sites on the B face.

The interaction of SAP with amyloid fibers is known to be calcium dependent as small-molecule SAP ligands binding at the double calcium site compete for the interaction [43,44]. Several of these ligands were identified in early preparative biochemical procedures, where they served as affinity supports. Most notably, the calcium-dependent binding of SAP to agarose was discovered to be due to the presence of the R isomer of the pyruvate acetal of β-D-galactose in the polymer [45]. SAP binds even more tightly to phosphoethanolamine (PE) [46], a property that is utilized for SAP affinity purification. Crystal structures of SAP cocrystallized with these ligands have confirmed that the prominent feature of interactions involve the carboxylate group of MOβDG and the phosphate group of PE coordinating directly with the calcium ions of each subunit and completing the pentagonal bipyrramidal ligand field of the metal atoms [47]. Both ligands also send atoms into the hydrophobic pocket toward Leu62. For MOβDG a single methyl group from the pyruvate acetal points into the pocket, whereas it is the methylene groups of PE that dip into the pocket. More recently, D-proline has been identified as an even higher affinity ligand, with its pyrrolidine ring fitting especially well into the hydrophobic pocket [8]. MOβDG, PE, and D-proline bind SAP with μM dissociation constants with apparently independent binding to each subunit. A remarkable enhancement in ligand affinity has been observed when bivalent ligands cross-link pairs of SAP subunits in a B face-to-B-face arrangement. This phenomenon was first observed with the ligand dAMP, where the phosphate group of the nucleotide is bound to the calcium ions and decamerization of two SAP pentamers is stabilized by stacking of the adenine rings [48]. The decamerization resurfaced again in high-throughput screens for new SAP ligands that identified the tight-binding palindromic compound CPHPC, comprising two D-proline residues linked through their imino nitrogens via a six-carbon alkyl chain. In this case the two D-proline residues binding into all of the metal sites and hydrophobic pockets of twofold-axis-related pentamers enhanced the apparent affinity of the proline head group for SAP by three orders of magnitude [8]. A similar affinity enhancement was subsequently reported for cross-linked MOβDG molecules, although the linker employed allowed a 20° relative rotation of the pentamers about their common fivefold axis, staggering the subunit pattern [49]. In the context of the current discussion, this cross-linking work suggests that a multipoint mode of binding to amyloid fibrils could be important in the tight binding of SAP. This further implies that the fiber motifs recognized by SAP might be distributed in a way that at least in part satisfies the fivefold symmetry of the binding partner. The current understanding we have of fiber structure does not reveal an obvious source for such symmetry.

Further clues as to the role of symmetry in SAP–amyloid recognition come from structural studies of C-reactive protein (CRP). This second acute-phase plasma protein member of the pentraxin family is closely related to SAP, with an amino acid sequence that is 63% homologous. The polypeptide chain length is approximately the same, and the three-dimensional structure of the molecule reveals that it is organized in a very similar way, including the presence of a double calcium binding site on each subunit [50]. CRP also binds small-molecule ligands with an acidic group coordinating the calcium ions. Although the tightest-binding monomeric ligand known for CRP is phosphocholine (PC), PE is bound sufficiently tightly to enable CRP to be bound by an immobilized PE affinity column such as SAP. Despite this similarity in ligand specificity, CRP does not bind to amyloid fibers with high affinity [43]. Similarly, it does not share SAP's molecular chaperone properties [7]. One major difference between SAP and CRP that could contribute to this issue concerns the precise organization of the double calcium sites on CRP. Rotation of the subunits of CRP by 20° toward the fivefold axis relative to their positions in SAP means that the ligand-binding sites are positioned at a different and wider radius from the rotation axis [51]. A second major difference concerns the shape of the hydrophobic pocket adjacent to the calcium-binding site, where instead of one of the two tyrosines found in SAP, CRP has a much smaller threonine residue. This leads to a far more open topology that is able to incorporate the bulkier methyl groups on PC, which cannot fit into the smaller hydrophobic pocket on SAP. Although it is conceivable that this second difference may prevent CRP from recognizing the amyloid motif, it is probably the former, geometric difference in the arrangement of the five binding sites, that disables the binding cooperativity and thus prevents CRP from binding amyloid. This hypothesis implies some strict symmetry relations in amyloid motifs.

Another hint of symmetry-driven cooperative binding derive from observations of the way in which SAP molecules interact with each other. In a number of crystal forms of SAP, an exposed glutamate 167 residue displayed prominently on the A-face helix of SAP docks with the double calcium site of another molecule to generate a crystal contact [40]. Electron microscopy shows that SAP can form long strings of stacked pentamers [52] and plausible models of a stack can be generated where Glu167 side chains from each subunit plug into the double calcium sites of adjacent pentamers.

As already noted, we have no evidence suggesting any fivefold symmetry elements in protein fiber structure, but there is a strong likelihood that multisite binding is important for SAP recognition of amyloid. This apparent contradiction might be resolved if SAP was able to bind to sets of five ligands selected from a crowded surface of acidic ligands with no explicit symmetry. As discussed previously, the condensed fiber structure is likely to have a surface populated by disordered loops and chain termini as well as turns between β-strands that together are likely to be populated by negatively charged amino acids. A parallel view has been postulated to explain the way in which CRP might bind to multiple phosphocholine head groups of phospholipids when it attaches to lipid bilayers [51].

One remaining issue of interest is at what point during fibrillogenesis is SAP binding supported. It is not entirely clear that all fibers extend by simple monomer addition. This possibility has been cited as a mode of prion protein conformational isomerization [53]. Similarly, in the case of Aβ(1–42), distinct oligomeric but soluble forms have been identified as possible fiber precursors [54]. SAP has been reported to interact with the fiber-forming process in vitro, depending on its state of calcification. Calcium-free SAP interferes with Aβ fiber assembly [55], whereas the presence of calcium is reported to enhance fiber formation and tangling [56]. This suggests that sites other than the double calcium site of SAP may participate, although these experiments are prone to technical artifacts due to structural changes in SAP and its self-association in the presence of calcium ions. However, these observations do share conclusions with experiments showing that SAP can interact with folding intermediates of other proteins in a manner reminiscent of molecular chaperones [7] in the absence of calcium.

There are plainly many outstanding questions as to the role of SAP in amyloid that remain to be resolved. Other accessory proteins, such as apolipoprotein E, are also reported to interact with fibers [57], and similarly, the role of fiber-associated glycosaminoglycans remains to be resolved [58]. SAP clearly binds to dermatan sulfate, which would sustain multipoint binding [59]. SAP has been shown to bind in vivo to apoptotic cells, surface blebs bearing chromatin fragments, and nuclear material in skin biopsies. In vitro SAP binds to double-stranded DNA [60]. It is therefore a reasonable hypothesis that SAP normally functions as a scavenging protein able to recognize nuclear cell debris released during apoptotic and necrotic cell death and masking them from the immune system. The recognition of amyloid by SAP may be related to this, perhaps through its more general anti-opsonin function proving beneficial in masking low levels of amyloid deposition, which appear to be a common feature of aging. Alternatively, the recognition could be the result of a pathological accident, perhaps related to SAP's more general chaperone like function.

REFERENCES

1. Chiti, F., Dobson, C.M. (2006). Protein misfolding, functional amyloid, and human disease. *Annu Rev Biochem*, *75*, 333–366.
2. Pepys, M.B., Booth, D.R., Hutchinson, W.L., Gallimore, J.R., Collins, P.M., Hohenester, E. (1997). Amyloid P component: a critical review. *Amyloid: Int J Exp Clin Invest*, *4*, 274–295.
3. Hawkins, P.N., Lavender, J.P., Pepys, M.B. (1990). Evaluation of systemic amyloidosis by scintigraphy with ^{123}I-labeled serum amyloid P component. *N Engl J Med*, *323*, 508–513.
4. Pepys, M.B., Baltz, M.L., de Beer, F.C., Dyck, R.F., Holford, S., Breathnach, S.M., Black, M.M., Tribe, C.R., Evans, D.J., Feinstein, A. (1982). Biology of serum amyloid P component. *Ann N Y Acad Sci*, *389*, 286–298.

5. Kirkitadze, M.D., Bitan, G., Teplow, D.B. (2002). Paradigm shifts in Alzheimer's disease and other neurodegenerative disorders: the emerging role of oligomeric assemblies. *J Neurosci Res*, *69*, 567–577.
6. Tennent, G.A., Lovat, L.B., Pepys, M.B. (1995). Serum amyloid P component prevents proteolysis of the amyloid fibrils of Alzheimer disease and systemic amyloidosis. *Proc Natl Acad Sci U S A*, *92*, 4299–4303.
7. Coker, A.R., Purvis, A., Baker, D., Pepys, M.B., Wood, S.P. (2000). Molecular chaperone properties of serum amyloid P component. *FEBS Lett*, *473*, 199–202.
8. Pepys, M.B., Herbert, J., Hutchinson, W.L., Tennent, G.A., Lachmann, H.J., Gallimore, J.R., Lovat, L.B., Bartfai, T., Alanine, A., Hertel, C., et al. (2002). Targeted pharmacological depletion of serum amyloid P component for treatment of human amyloidosis. *Nature*, *417*, 254–259.
9. Pepys, M.B. (2006). Amyloidosis. *Annu Rev Med*, *57*, 223–241.
10. Nelson, R., Sawaya, M.R., Balbirnie, M., Madsen, A.O., Riekel, C., Grothe, R., Eisenberg, D. (2005). Structure of the cross-beta spine of amyloid-like fibrils. *Nature*, *435*, 773–778.
11. Fandrich, M., Dobson, C.M. (2002). The behaviour of polyamino acids reveals an inverse side chain effect in amyloid structure formation. *EMBO J*, *21*, 5682–5690.
12. Booth, D.R., Sunde, M., Bellotti, V., Robinson, C.V., Hutchinson, W.L., Fraser, P.E., Hawkins, P.N., Dobson, C.M., Radford, S.E., Blake, C.C., Pepys, M.B. (1997). Instability, unfolding and aggregation of human lysozyme variants underlying amyloid fibrillogenesis. *Nature*, *385*, 787–793.
13. Kelly, J.W. (1996). Alternative conformations of amyloidogenic proteins govern their behavior. *Curr Opin Struct Biol*, *6*, 11–17.
14. Sipe, J.D., Cohen, A.S. (2000). Review: history of the amyloid fibril. *J Struct Biol*, *130*, 88–98.
15. Bonar, L., Cohen, A.S., Skinner, M.M. (1969). Characterization of the amyloid fibril as a cross-beta protein. *Proc Soc Exp Biol Med*, *131*, 1373–1375.
16. Eanes, E.D., Glenner, G.G. (1968). X-ray diffraction studies on amyloid filaments. *J Histochem Cytochem*, *16*, 673–677.
17. Marsh, R.E., Corey, R.B., Pauling, L. (1955). An investigation of the structure of silk fibroin. *Biochim Biophys Acta*, *16*, 1–34.
18. Geddes, A.J., Parker, K.D., Atkins, E.D., Beighton, E. (1968). "Cross-beta" conformation in proteins. *J Mol Biol*, *32*, 343–358.
19. Sunde, M., Serpell, L.C., Bartlam, M., Fraser, P.E., Pepys, M.B., Blake, C.C.F. (1997). Common core structure of amyloid fibrils by synchrotron x-ray diffraction. *J Mol Biol*, *273*, 729–739.
20. Uversky, V.N., Fink, A.L. (2004). Conformational constraints for amyloid fibrillation: the importance of being unfolded. *Biochim Biophys Acta*, *1698*, 131–153.
21. Turnell, W.G., Finch, J.T. (1992). Binding of the dye Congo Red to the amyloid protein pig insulin reveals a novel homology amongst amyloid-forming peptide sequences. *J Mol Biol*, *227*, 1205–1223.
22. Jimenez, J.L., Nettleton, E.J., Bouchard, M., Robinson, C.V., Dobson, C.M., Saibil, H.R. (2002). The protofilament structure of insulin amyloid fibrils. *Proc Natl Acad Sci U S A*, *99*, 9196–9201.

23. Jimenez, J.L., Tennent, G., Pepys, M., Saibil, H.R. (2001). Structural diversity of ex vivo amyloid fibrils studied by cryo-electron microscopy. *J Mol Biol, 311*, 241–247.
24. Blake, C., Serpell, L. (1996). Synchrotron x-ray studies suggest that the core of the transthyretin amyloid fibril is a continuous beta-sheet helix. *Structure, 4*, 989–998.
25. Sunde, M., Blake, C.C. (1998). From the globular to the fibrous state: protein structure and structural conversion in amyloid formation. *Q Rev Biophys, 31*, 1–39.
26. Serag, A.A., Altenbach, C., Gingery, M., Hubbell, W.L., Yeates, T.O. (2002). Arrangement of subunits and ordering of beta-strands in an amyloid sheet. *Nat Struct Biol, 9*, 734–739.
27. Eneqvist, T., Andersson, K., Olofsson, A., Lundgren, E., Sauer-Eriksson, A.E. (2000). The beta-slip: a novel concept in transthyretin amyloidosis. *Mol Cell, 6*, 1207–1218.
28. Olofsson, A., Ippel, J.H., Wijmenga, S.S., Lundgren, E., Ohman, A. (2004). Probing solvent accessibility of transthyretin amyloid by solution NMR spectroscopy. *J Biol Chem, 279*, 5699–5707.
29. Richardson, J.S., Richardson, D.C. (2002). Natural beta-sheet proteins use negative design to avoid edge-to-edge aggregation. *Proc Natl Acad Sci U S A, 99*, 2754–2759.
30. Johnson, S.M., Wiseman, R.L., Sekijima, Y., Green, N.S., Adamski-Werner, S.L., Kelly, J.W. (2005). Native state kinetic stabilization as a strategy to ameliorate protein misfolding diseases: a focus on the transthyretin amyloidoses. *Acc Chem Res, 38*, 911–921.
31. Serpell, L.C. (2000). Alzheimer's amyloid fibrils: structure and assembly. *Biochim Biophys Acta, 1502*, 16–30.
32. Petkova, A.T., Ishii, Y., Balbach, J.J., Antzutkin, O.N., Leapman, R.D., Delaglio, F., Tycko, R. (2002). A structural model for Alzheimer's beta-amyloid fibrils based on experimental constraints from solid state NMR. *Proc Natl Acad Sci U S A, 99*, 16742–16747.
33. Petkova, A.T., Yau, W.M., Tycko, R. (2006). Experimental constraints on quaternary structure in Alzheimer's beta-amyloid fibrils. *Biochemistry, 45*, 498–512.
34. Luhrs, T., Ritter, C., Adrian, M., Riek-Loher, D., Bohrmann, B., Dobeli, H., Schubert, D., Riek, R. (2005). 3D structure of Alzheimer's amyloid-beta(1-42) fibrils. *Proc Natl Acad Sci U S A, 102*, 17342–17347.
35. Kheterpal, I., Chen, M., Cook, K.D., Wetzel, R. (2006). Structural differences in Abeta amyloid protofibrils and fibrils mapped by hydrogen exchange-mass spectrometry with on-line proteolytic fragmentation. *J Mol Biol, 361*, 785–795.
36. Stellato, F., Menestrina, G., Serra, M.D., Potrich, C., Tomazzolli, R., Meyer-Klaucke, W., Morante, S. (2006). Metal binding in amyloid beta-peptides shows intra- and inter-peptide coordination modes. *Eur Biophys J, 35*, 340–351.
37. Syme, C.D., Nadal, R.C., Rigby, S.E., Viles, J.H. (2004). Copper binding to the amyloid-beta (Abeta) peptide associated with Alzheimer's disease: folding, coordination geometry, pH dependence, stoichiometry, and affinity of Abeta-(1–28): insights from a range of complementary spectroscopic techniques. *J Biol Chem, 279*, 18169–18177.
38. Nelson, R., Eisenberg, D. (2006). Recent atomic models of amyloid fibril structure. *Curr Opin Struct Biol, 16*, 260–265.

39. Tomaselli, S., Esposito, V., Vangone, P., van Nuland, N.A., Bonvin, A.M., Guerrini, R., Tancredi, T., Temussi, P.A., Picone, D. (2006). The alpha-to-beta conformational transition of Alzheimer's Abeta-(1–42) peptide in aqueous media is reversible: a step by step conformational analysis suggests the location of beta conformation seeding. *Chembiochem*, *7*, 257–267.

40. Emsley, J., White, H.E., O'Hara, B.P., Oliva, G., Srinivasan, N., Tickle, I.J., Blundell, T.L., Pepys, M.B., Wood, S.P. (1994). Structure of pentameric human serum amyloid P component. *Nature*, *367*, 338–345.

41. Osmand, A.P., Friedenson, B., Gewurz, H., Painter, R.H., Hofmann, T., Shelton, E. (1977). Characterization of C-reactive protein and the complement subcomponent C1t as homologous proteins displaying cyclic pentameric symmetry (pentraxins). *Proc Natl Acad Sci U S A*, *74*, 739–743.

42. Kinoshita, C.M., Gewurz, A.T., Siegel, J.N., Ying, S.C., Hugli, T.E., Coe, J.E., Gupta, R.K., Huckman, R., Gewurz, H. (1992). A protease-sensitive site in the proposed Ca(2+)-binding region of human serum amyloid P component and other pentraxins. *Protein Sci*, *1*, 700–709.

43. Pepys, M.B., Dyck, R.F., de Beer, F.C., Skinner, M., Cohen, A.S. (1979). Binding of serum amyloid P-component (SAP) by amyloid fibrils. *Clin Exp Immunol*, *38*, 284–293.

44. Hind, C.R., Collins, P.M., Caspi, D., Baltz, M.L., Pepys, M.B. (1984). Specific chemical dissociation of fibrillar and non-fibrillar components of amyloid deposits. *Lancet*, *2*, 376–378.

45. Hind, C.R., Collins, P.M., Renn, D., Cook, R.B., Caspi, D., Baltz, M.L., Pepys, M.B. (1984). Binding specificity of serum amyloid P component for the pyruvate acetal of galactose. *J Exp Med*, *159*, 1058–1069.

46. Schwalbe, R.A., Dahlback, B., Coe, J.E., Nelsestuen, G.L. (1992). Pentraxin family of proteins interact specifically with phosphorylcholine and/or phosphorylethanolamine. *Biochemistry*, *31*, 4907–4915.

47. Thompson, D., Pepys, M.B., Tickle, I., Wood, S. (2002). The structures of crystalline complexes of human serum amyloid P component with its carbohydrate ligand, the cyclic pyruvate acetal of galactose. *J Mol Biol*, *320*, 1081–1086.

48. Hohenester, E., Hutchinson, W.L., Pepys, M.B., Wood, S.P. (1997). Crystal structure of a decameric complex of human serum amyloid P component with bound dAMP. *J Mol Biol*, *269*, 570–578.

49. Ho, J.G., Kitov, P.I., Paszkiewicz, E., Sadowska, J., Bundle, D.R., Ng, K.K. (2005). Ligand-assisted aggregation of proteins: dimerization of serum amyloid P component by bivalent ligands. *J Biol Chem*, *280*, 31999–32008.

50. Shrive, A.K., Cheetham, G.M., Holden, D., Myles, D.A., Turnell, W.G., Volanakis, J.E., Pepys, M.B., Bloomer, A.C., Greenhough, T.J. (1996). Three dimensional structure of human C-reactive protein. *Nat Struct Biol*, *3*, 346–354.

51. Thompson, D., Pepys, M.B., Wood, S.P. (1999). The physiological structure of human C-reactive protein and its complex with phosphocholine. *Structure*, *7*, 169–177.

52. Pinteric, L., Painter, R.H. (1979). Electron microscopy of serum amyloid protein in the presence of calcium; alternative forms of assembly of pentagonal molecules in two-dimensional lattices. *Can J Biochem*, *57*, 727–736.

53. Caughey, B. (2001). Interactions between prion protein isoforms: the kiss of death? *Trends Biochem Sci*, *26*, 235–242.

54. Goldsbury, C.S., Wirtz, S., Muller, S.A., Sunderji, S., Wicki, P., Aebi, U., Frey, P. (2000). Studies on the in vitro assembly of a beta 1–40: implications for the search for a beta fibril formation inhibitors. *J Struct Biol*, *130*, 217–231.

55. Janciauskiene, S., Garcia de Frutos, P., Carlemalm, E., Dahlback, B., Eriksson, S. (1995). Inhibition of Alzheimer beta-peptide fibril formation by serum amyloid P component. *J Biol Chem*, *270*, 26041–26044.

56. MacRaild, C.A., Stewart, C.R., Mok, Y.F., Gunzburg, M.J., Perugini, M.A., Lawrence, L.J., Tirtaatmadja, V., Cooper-White, J.J., Howlett, G.J. (2004). Non fibrillar components of amyloid deposits mediate the self-association and tangling of amyloid fibrils. *J Biol Chem*, *279*, 21038–21045.

57. Gunzburg, M.J., Perugini, M.A., Howlett, G.J. (2007). Structural basis for the recognition and cross-linking of amyloid fibrils by human apolipoprotein E. *J Biol Chem*, *282*, 35831–35841.

58. Inoue, S., Kuroiwa, M., Saraiva, M.J., Guimaraes, A., Kisilevsky, R. (1998). Ultrastructure of familial amyloid polyneuropathy amyloid fibrils: examination with high-resolution electron microscopy. *J Struct Biol*, *124*, 1–12.

59. Hamazaki, H. (1987). Ca^{2+}-mediated association of human serum amyloid P component with heparan sulfate and dermatan sulfate. *J Biol Chem*, *262*, 1456–1460.

60. Pepys, M.B., Booth, S.E., Tennent, G.A., Butler, P.J., Williams, D.G. (1994). Binding of pentraxins to different nuclear structures: C-reactive protein binds to small nuclear ribonucleoprotein particles, serum amyloid P component binds to chromatin and nucleoli. *Clin Exp Immunol*, *97*, 152–157.

61. DeLano, W.L. (2002). *The PyMOL Molecular Graphics System*, DeLano Scientific, Palo Alto, *CA*.

27

ROLE OF OXIDATIVELY STRESSED LIPIDS IN AMYLOID FORMATION AND TOXICITY

Paul H. Axelsen and Hiroaki Komatsu

Department of Pharmacology, University of Pennsylvania School of Medicine, Philadelphia, Pennsylvania

LIPIDS AND ALZHEIMER DISEASE

The evidence linking lipids to the formation of amyloid β (Aβ) protein fibrils and to the pathogenesis of Alzheimer disease (AD) is largely circumstantial. However, the amount of evidence pointing to some type of link is overwhelming. The evidence trail begins with the origin of Aβ proteins in a transmembrane protein, the amyloid precursor protein (APP). The 40-residue Aβ protein (Aβ40) and the 42-residue Aβ protein (Aβ42) are both produced by cleavage of the transmembrane segment of AAP within the lipid membrane. As a consequence, the C-terminal residues of Aβ proteins that were once part of the APP transmembrane segment are uniformly hydrophobic. Following their cleavage from APP, Aβ proteins have been observed to remain associated with detergent-resistant lipid membrane domains in the brain [1] or membrane-anchored APP [2]. The general conclusion reached by most investigators is that Aβ proteins have little affinity for neutral lipid membranes, and that electrostatic interactions with anionic lipids tend to induce β structure [3–16]. α-Helical structure is observed in detergent micelles [17] and in association with lipid/membranes at low lipid/protein ratios [18], high cholesterol concentrations [19], or in conjunction with metal ions [11,20,21].

Protein Misfolding Diseases: Current and Emerging Principles and Therapies,
Edited by Marina Ramirez-Alvarado, Jeffery W. Kelly, and Christopher M. Dobson
Copyright © 2010 John Wiley & Sons, Inc.

Aβ proteins appear to have special relationships or interactions with specific membrane components. For example, the ganglioside G_{M1} is bound together with Aβ proteins in diffuse amyloid plaques [22], and G_{M1}-containing membranes accelerate fibril formation in vitro [23–33]. Our own work shows that oxidatively damaged polyunsaturated lipids are potent promoters of Aβ fibril formation [34] and that the fibrillogenic effects of oxidatively damaged lipids can be mimicked by 4-hydroxy-2-nonenal (HNE) [35], a well known product of ω-6 lipid oxidation which modifies the three His residues of Aβ proteins [36]. Others have shown that modification of the Lys residues in Aβ40 with HNE analogs can increase the tendency to aggregate and the toxicity of Aβ40 [37].

The relative abundance of various lipid classes is altered in AD [38–41], altering membrane composition protects PC12 cells from toxic effects of Aβ proteins [42], and membrane binding by Aβ proteins appears to mediate some types of neurotoxicity [21,43,44]. Plasma membranes isolated from human brain accelerate Aβ fibrillogenesis [45], while fibrillizing Aβ proteins disrupt the structure of membranes formed from both synthetic lipids [9,10] and whole-brain lipid extracts [46]. Ultrastructural studies suggest that fibril formation tends to occur first in portions of diffuse deposits that are closest to membranes [47–49]. Aβ40 with the E22Q mutation (as occurs in hereditary cerebral hemorrhage with amyloidosis–Dutch type) will fibrillize on the surface membrane of human cerebrovascular smooth muscle cells [50]. Membranes and fatty acids can promote the aggregation/multimerization of synucleins in a pathological process that exhibits interesting parallels to Aβ proteins in AD [51,52].

Our understanding of the role of lipids in the pathogenesis of AD tends to lag behind that of other chemical classes because lipids introduce many experimental challenges. Lipids are chemically diverse, and naturally occurring membranes are heterogeneous mixtures of lipids, proteins, and other compounds. Lipids are also physically diverse. Few lipid molecules exist as monomeric species in solution; the vast majority form micelles and vesicles, and the latter may be unilammelar or multilammelar. Thus, lipid suspensions generally have two or more phases and complex interphase equilibria. This physical heterogeneity complicates most types of physicochemical analysis, even when using pure synthetic lipid preparations, and it makes some others outright impossible. It also impedes the measurement of fibril formation. The measurement of fibril formation by turbidity, for example, is confounded by ambiguity over whether the species causing turbidity is protein or lipid. Lipids markedly increase the fluorescence of the thioflavin T apart from fibril formation, while assays based on Congo Red absorbance ratios entail practical restrictions on protein and lipid concentrations that limit the flexibility of this assay.

OXIDATIVE STRESS AND LIPIDS

More than 3500 papers related to oxidative stress and AD were published in the past five years. As with lipids and AD, the evidence for a relationship between oxidative stress and AD is overwhelming, although there is no consensus and *even fewer* details about the nature of this relationship. In general, oxidative damage is difficult to quantify and characterize because the chemical products are diverse [53–58]. Numerous associations have been described between AD and oxidative changes in various cell components [59–64]. The brain in AD appears to sustain more oxidative damage than does normal brain [65–67], and it exhibits an increased susceptibility to oxidative stress [68–70], perhaps as a consequence of relatively low levels of naturally occurring antioxidants [71] such as α-tocopherol [72,73].

Polyunsaturated fatty acyl chains within lipid membranes are likely to be the prime victims of oxidative damage in the brain. Their susceptibility to peroxidation is well known, and at least in principle they may undergo radical-mediated chain reactions. It may be significant that precautions against spontaneous air oxidation of lipids are not mentioned in most studies of the interactions between Aβ proteins and lipids (e.g., solutions are not deoxygenated, they are exposed to air, antioxidants are not added, etc.). Without vigorous precautions, some of the distinctions made between the properties of membranes from different sources may be specious.

Evidence that lipid peroxidation may be involved in the pathogenesis of AD has been reviewed extensively [59,64,74,75]. Lipid oxidation products and the susceptibility of lipids to oxidative damage are both increased in AD [76–79]. Lipid peroxides undergo spontaneous (nonenzymatic) decomposition to yield aldehydes such as 4-oxo-2-nonenal and HNE [80,81] and eicosanoids such as isoprostanes [82,83]. HNE concentrations in human ventricular fluid are approximately 15 μM and are elevated in AD [68,76,84]. HNE is highly reactive, it has a well-known propensity to form adducts with the side chains of various amino acid residues, and HNE–protein adducts have been used as biomarkers of oxidative stress [85]. In contrast, isoprostanes are chemically stable, although they have also been used as biomarkers of oxidative stress [86–88], and they are elevated specifically in AD [89–91].

Some investigators have focused on the role of acrolein in the pathogenesis of AD. Acrolein is a well-known environmental toxin and a recently recognized product of lipid peroxidation [92–94]. Acrolein reacts with Lys residue side chains, forming a cyclic *N*ε-(3-formyl-3,4-dehydropiperidino) derivative. Acrolein is elevated in AD brain, it appears to be more neurotoxic than HNE, it causes elevated intracellular calcium levels [95,96], and it appears to inactivate flippase, inducing a breakdown of lipid membrane asymmetry and apoptosis [97,98]. Acrolein-modified proteins have also been considered a biomarker for oxidative stress in AD [99].

Aβ PROTEINS AND OXIDATIVE STRESS

Both prooxidant and antioxidant properties have been ascribed to Aβ proteins, and both properties have been linked to the affinity of Aβ proteins for redox-active metals. Several laboratories have reported direct antioxidant effects of Aβ proteins [100–105]. The density of plaques containing Aβ protein correlates inversely with markers of oxidative damage [106], and in Down syndrome, cortical deposition of Aβ proteins correlates with reduced oxidative damage [107]. Moreover, Aβ proteins prevent lipoprotein oxidation [108] and metal-induced neuronal death in culture [109].

On the other hand, other investigators have found that regions of the brain rich in Aβ proteins also have increased levels of protein oxidation [110]. The overexpression of Aβ proteins in transgenic mice, in *Caenorhabditis elegans*, and in cell culture results in an increase in biomarkers of oxidative stress and in HNE production [91,111,112]. Aβ proteins exhibit direct prooxidant activity by producing H_2O_2 and oxidizing compounds such as dopamine, phospholipids, and cholesterol [113–123].

Amyloid plaques bind redox-active transition metals that are likely to be the actual site of redox activity [11,101], and some structural information is available for complexes between Aβ proteins and various metal ions [11,124–127]. Copper levels are significantly increased in AD [128], and metal depletion promotes dissolution of Aβ fibrils [129]. Treating mice with antibiotics that chelate Cu^{2+} and Zn^{2+} ions appears to inhibit Aβ deposition in brain tissue [130]. Cu(II) potentiates the neurotoxicity of Aβ42 > Aβ40 in embryonic rodent neurons, and its effect is mediated by H_2O_2 [113,131]. Zinc(II) is redox-inert and is able to protect/rescue human cells in tissue culture from Aβ and Cu(II) toxicity [106]. Aβ proteins may also be the substrate for oxidative damage [132].

The production of H_2O_2 by Aβ appears to involve the reduction of Fe(III) or Cu(II) [114]. The H_2O_2 produced may then react via the Fenton reaction with Fe(II) or Cu(I) to produce highly toxic hydroxyl radicals ($^{\cdot}OH$). However, the role of redox-active metal ions in the toxicity of Aβ proteins is far from clear. For example, investigations into the role of Met35 in mediating Aβ toxicity in which the three histidine residues of Aβ42 were replaced with tyrosine to eliminate Cu(II) binding. This altered neither fibril formation by Aβ nor its neurotoxic properties [133]. These investigators nonetheless proposed that aggregated Aβ peptides, perhaps in concert with redox metal ions, form a secondary structure conducive to sulfur-initiated oxidative stress and subsequent neurotoxicity. This secondary structure is disturbed by an I31P mutation, which is interpreted as depriving Met35 of a critical interaction with the Ile31 backbone [134]. By itself, oxidation of Met35 inhibits fibril formation [124,135–137].

Because of the potential for Aβ proteins to mediate redox reactions and the well-known susceptibility of unsaturated lipids to oxidation, it may be expected that membrane-associated Aβ proteins promote oxidative damage to lipids in a membrane. Indeed, Aβ42 does increase HNE modification of a glutamate transporter in synaptosomes [138]. Our own work with Aβ42 demonstrated that its prooxidant activity toward polyunsaturated lipids could be neutralized by lipophilic antioxidants, chelation of metal ions, anaerobic conditions, mutation of His13 or His14 to Ala, or modification of the Met35 side chain [121]. Furthermore, the prooxidant activity of Aβ42 increases HNE production, which, in turn, chemically modifies Aβ and causes it to form fibrils [36,139–141] or inhibit proteasome function [142]. Together with the observation that immunoreactivity of antibodies to HNE-modified His residues localizes to amyloid plaques [143,144], these observations suggest that Aβ proteins not only promote lipid oxidation, but that there may also be a mechanistic link between the lipid oxidation products formed during oxidative stress and Aβ misfolding [145].

The neurotoxic properties of amyloid β proteins may depend on redox reactions and highly reactive oxygen species. Aβ proteins cause H_2O_2 and lipid peroxides to accumulate in cells [112]. Catalase protects cells from Aβ toxicity, and cell lines selected for resistance to Aβ toxicity also become resistant to the cytotoxic action of H_2O_2. However, significant methodologic differences exist among laboratories, and it is inherently difficult to quantify oxidative damage. For example, one study concluding that the primary mechanism of Aβ toxicity does *not* involve oxidative pathways measured TBARS to assess lipid peroxidation [146]. Assays for TBARS, however, measure only a small subset of the diverse oxidation products generated.

It has been suggested that Aβ proteins split into fragments and that these fragments are both neurotoxic and able to generate additional oxygen radicals [147,148]. Support for a causal link between Aβ fibril formation and free radical–mediated toxicity was provided by electron paramagnetic resonance spectroscopy showing that Aβ42 generates free radicals in solution and that there is a strong correlation between the intensity of radical generation by Aβ and neurotoxicity [147,149]. In these studies, preincubation of Aβ to form fibrils increased its toxicity. In contrast, replacing the redox-active sulfur atom in residue Met35 with methylene ($-CH_2-$) resulted in a peptide that formed fibrillar structures but had no demonstrable toxicity toward cultured hippocampal neurons. In the same experimental system, vitamin E (presumably acting as an antioxidant) neutralized the neurotoxicity of Aβ but had no effect on its ability to form fibrils [150]. However, the suggestion that Aβ proteins split into fragments bearing free radicals has been strongly refuted [151]. Another study concluding that free radicals and lipid peroxidation do not mediate Aβ-induced neurotoxicity used ginkgolides (the antioxidant component of *Ginkgo biloba* leaves) and vitamin E to inhibit oxidation [152]. Neither of these agents rescued PC12 nerve cells from Aβ-induced apoptosis and cell death. It should be noted that the agents used in these studies certainly attenuated, but probably did not completely halt, oxidative damage.

LIPIDS AND THE THERMODYNAMICS OF FIBRIL FORMATION

The concentration of monomeric Aβ40 in equilibrium with fibrillar Aβ40 has been estimated at 6 to 9 μM [153], ~15 μM [154], and 0.7 to 1.0 μM [155]. These measurements are affected by the concentration of extendable fibril ends with which to aggregate, and this is a difficult concentration to measure and control. Surface plasmon resonance (SPR) techniques can circumvent that difficulty, although somewhat varied results are also reported with this technique. For example, a single-stage K_d value of 11 μM has been reported, with Cu^{2+} ions decreasing this value to 1 μM [156], while two-stage K_d values of 20 and 200 nM [157] and three-stage K_d values of 122 μM, 69 mM, and 93 mM [158] have been reported by others. Numerous other SPR studies of Aβ aggregation have been performed [159], in most cases for the purpose of screening inhibitors [160–162] or assessing membrane affinity [14,163,164].

At sufficiently high concentrations, Aβ proteins undergo nucleated polymerization, characterized by a lag period in which a high-energy barrier is crossed with kinetics that are time and concentration dependent [153,165–168]. Oligomeric seeds are formed, becoming low-energy binding sites for protein monomers. Fibril formation proceeds rapidly with fibril extension kinetics that are first order in monomer concentration [169,170]. Fundamentally similar processes are probably also occurring when immobilized Aβ proteins are prepared as templates, and "docked" Aβ monomers settle over time into a thermodynamically hyperstable "locked" configuration [171,172].

Nevertheless, in vitro thermodynamics and kinetics studies of fibril formation by Aβ proteins have not been able to explain why fibrils form in vivo. The concentrations of Aβ proteins in human spinal fluid are several orders of magnitude lower than the equilibrium monomer concentrations mentioned above [173], so one does not expect or see spontaneous aggregation in this body compartment. Aβ proteins are subject to remarkably fast turnover [174], suggesting that the tissue concentrations of monomeric Aβ proteins are determined by the relative rates of production and elimination, as well as an equilibrium between mobile low-molecular-mass species and immobile fibrils.

In considering possible explanations for in vivo aggregation, it has been suggested that there is a tenuous metabolic balance in brain tissue such that a small disturbance in the rates of Aβ protein production, aggregation, or elimination, applied over sufficient time, produces the pathological amyloid deposition observed in AD. Such a disturbance may be caused by the chemical modification of Aβ proteins, especially by lipid oxidation products [36,139,140,145,175], cross-linking [117,176–178], or metal complex formation [179,180] forming heterogeneous seeds for fibril formation. As with homogeneous seeds formed at high concentrations of Aβ protein, the creation of heterogeneous seeds may be viewed as the creation of new binding sites for monomeric protein with a much lower thermodynamic energy state that did not originally exist apart from the seed. Fibril formation occurs when each monomer added to such a seed becomes part of this "template" and makes a similar low-energy state available for another monomer.

THE PROTEOMICS OF OXIDATIVE STRESS IN AD

Numerous studies of AD brain using the tools of proteomics have been described. The results may be sorted into relatively few groups. Studies of unmodified proteins comprise the first group [181,182–183]. Remarkably, none of these studies identified Aβ proteins in AD brains. Studies focused on proteins damaged by oxidation, nitration, and other reactive substances comprise a second group [55,62,184–194]. In most cases, reactions with protein carbonyl groups is taken for evidence of damage by oxidation, and the nature of the protein modification is not defined explicitly. Two studies have examined the ability of Aβ proteins to induce oxidative modifications in other proteins [195]. In any case, it is difficult to imagine how oxidative damage to proteins, especially Aβ proteins, would lower the aqueous solubility of Aβ proteins and promote fibril formation.

A third group used laser-capture microdissection to isolate amyloid plaques for proteomics analysis. Two such papers report that hundreds of proteins co-localize with plaques [196,197], whereas a third paper reports that only Aβ proteins were found [198]. It is not clear that any of these three studies, however, would have identified HNE-modified or carbonylated Aβ proteins [199].

Finally, there is a group of studies focused on protein modification by HNE, using anti-HNE antibodies to identify the modification [96,200,201]. Of note in these studies, insoluble pellets of aggregated protein were excluded from analysis; only soluble supernatants were examined. As in other studies, Aβ proteins were not identified. Efforts to get more specific information from mass spectrometry about the nature of HNE–protein adducts are hindered by their insolubility and difficulties with ionization [141,202]. However, approaches involving digestion with AspN and MALDI ionization have shown that HNE modifies the three His residues of Aβ40 by Michael addition [36].

LIPOPROTEIN E AND OXIDATIVE STRESS

Apolipoprotein E (apoE) is a 299-amino acid protein involved in lipid transport and cholesterol homeostasis that occurs as three common isoforms [203]. ApoE3 is the most common (77% of the alleles) and is therefore considered to be the wild type. ApoE2 has an R158C substitution, while apoE4 has a C112R substitution [204]. Persons with one apoE4 allele also have a threefold increased incidence of AD, while persons with two copies of the gene have an eightfold increased incidence. One apoE2 allele reduces the incidence of AD by 60% [205–207].

In addition to this epidemiological evidence, the *experimental* evidence linking apoE to the pathogenesis of AD is manifold: First, the apoE4 allele is associated with increased Aβ deposits in the brain, and a distinct neuropathological phenotype [206]. Second, the characteristic amyloid plaques of AD immunostain for apoE [208], apoE and Aβ exist as bound complexes within amyloid plaques [209], and apoE copurifies with Aβ from amyloid plaques [210]. Third, apoE4 is able to accelerate fibril formation by

Aβ40 under some conditions [211–213]. Under other conditions, apoE4 appears to inhibit amyloid nucleation, but dimeric forms of apoE3 are much more efficient inhibitors of this process [214]. Circulating apoE3 is therefore much more effective at inhibiting nucleation, because approximately 27% of it is a disulfide-linked dimer (apoE4 cannot form disulfide bonded dimers due to its C112R substitution) [215]. Fourth, apoE2 and apoE3 bind Aβ to form circulating complexes, but not apoE4 [216]. ApoE3 complexes with Aβ that is already in β conformation; it dramatically accelerates folding into β conformation and fibrillogenesis for Aβ that is in random coil conformation. Fifth, even though the presence of apoE alters neither transcription or translation of the APP(V717F) transgene nor its processing to Aβ in the mouse, it appears that apoE promotes both the deposition and fibrillization of Aβ, ultimately affecting the clearance of protease-resistant Aβ/apoE aggregates [217]. Conversely, a lack of apoE reduces Aβ deposition in mice [218].

Despite all of these observations, the reason that different isoforms of ApoE have different levels of risk for AD is not known. Our own recently published study demonstrated no isoform-dependent difference in the structure of lipoprotein particles [219]. The case for isoform-specific direct interactions between apoE and Aβ proteins has been made [220], but clear conclusions are elusive due to problems with protein purity, aggregation and denaturation of the apoE protein, and indirect assay methods [209,221–226]. Others have suggested that risk is related to differences in antioxidant activity [227–229]. Specifically, apoE4 has no Cys residues and hence no free thiol groups. ApoE3 has one Cys, while apoE2 has two. Free thiol groups have significant antioxidant activity via mechanisms that differ from those of glutathione [230–232]. Variants of apoA-I with a free Cys residue side chain exhibit significantly enhanced antioxidant activity [233]. Therefore, it is intruiging to speculate that apoE4 confers increased risk of AD, while apoE2 confers decreased risk, because Cys residue side chains protect lipids in lipoprotein E particles against oxidative stress.

MOUSE MODELS OF AD AND OXIDATIVE STRESS

A partial list of the means that have been tested in mouse models for the treatment of AD includes antioxidants [234–237] as well as immunization [238,239], ω-3 fatty acids [240,241], small-molecule inhibitors of β/γ-secretases or fibril formation [162,242–244], and oligopeptides [161,245–254]. Although the TG2576 mouse became available in 1996, investigations of oxidative stress in this model have only been reported relatively recently. Studies in this model system of hereditary AD may not model the cause of sporadic AD in humans, but they can help delineate the effect of oxidative lipid damage on plaque formation and the development of neurological deficits. The available studies have found elevated levels of oxidative and nitrative stress in proteins

[192,255,256] and lipids [122]. Given that this model differs from background only in expressing human APP_{Swe}, it suggests that the presence of this protein contributes to oxidative stress. Consistent with this conclusion, the intracerebral injection of fibrillar Aβ40 into rats causes inflammation and oxidative stress [257]. Some of the data from mouse models suggests that certain inflammatory mediators are potent drivers of the disease; however, there is still no conclusive evidence that inflammation triggers or promotes Alzheimer disease [258,259]. Consistent with this, clinical trials of cyclooxygenase inhibitors in AD have been disappointing [260].

The spice curcumin and ethyl esters of the related compound ferulic acid appear to reduce the prooxidant effects of Aβ proteins [234,237,261,262]. Dietary supplementation with omega-3 fatty acids reduces the amyloid burden in TG2576 mice [240], while a deficiency depletes significant protein markers and increases caspase protein fragments [241]. Acrolein (along with HNE) has been implicated as a mediator of the oxidative stress induced by PGJ2 prostaglandins [263].

Taken altogether, the data from mouse models expressing human Aβ proteins suggest that there is a bidirectional relationship between oxidative stress and amyloid plaque formation: Aβ proteins increase signs of oxidative stress, while reductions of oxidative stress alleviate Aβ protein deposition into plaque. However, it has been difficult to translate this appreciation for the role of oxidative stress in mouse models of AD to therapy for humans with AD. For example, mouse models repeatedly point to a beneficial effect of α-tocopherol [235,236,264], but this approach has not had any demonstrable effect in humans [265,266].

VICIOUS CYCLES INVOLVING LIPIDS AND AD

The effects described above of Aβ proteins on lipids and, conversely, the effects of lipids on Aβ proteins, suggest that a vicious cycle may be involved in producing amyloid fibrils in the brain under conditions in which they would not otherwise form, and in causing the neurotoxic effects associated with amyloid formation. The scheme illustrated in Figure 1 has components in common with "vicious" cycles implicated by others in the pathogenesis of AD [267,268], and every step has been observed either in vitro or in vivo.

As cited above, there is abundant (albeit circumstantial) evidence suggesting that redox-active cations and reactive oxygen species are involved in mediating the neurotoxicity of Aβ proteins. These proteins bind metal ions and create reactive oxygen species that activate inflammatory processes, deleterious signaling pathways, or apoptotic processes. However, the scheme in Figure 1 is only intended to suggest how oxidative stress and amyloid fibril formation may be mechanistically linked by lipid membranes. The role of Aβ fibrils or prefibrillar intermediate forms of Aβ in actually causing the neuronal loss observed in AD remains unclear.

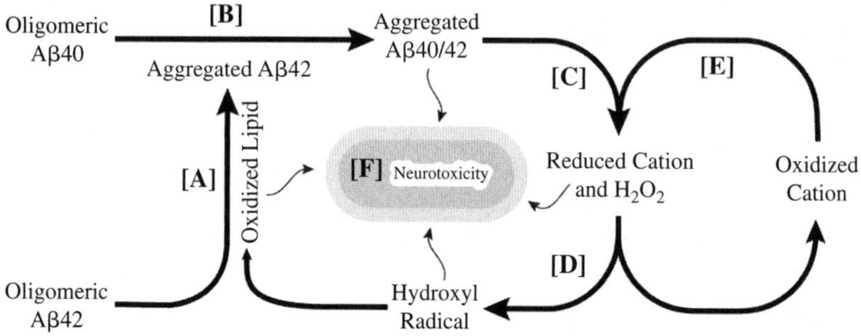

FIG. 1 Possible relationships between lipid oxidation and amyloid formation based on in vitro observations. [A] Accelerated accumulation and folding of oligomeric Aβ42 into β structure by oxidatively damaged lipids as described [34]. ApoE complexes may provide the oxidatively damaged lipid for this reaction [221,222,269]. [B] Aβ42 that has assumed β structure on membranes containing oxidatively damaged lipids accelerates the accumulation and folding of oligomeric Aβ40 into β structure. [C] Aβ42pi > Aβ40 produces H_2O_2 through a mechanism involving cation reduction [114]. [D] The production of hydroxyl radicals by H_2O_2 and Fe(II) is known as the Fenton reaction. This reaction proceeds similarly with Cu(I) [270]. [E] Bound redox-active metals may be reduced by various electron donors: for example, ascorbate, which is present in brain tissue at 50 to 100 μM [271]. Metals appear to be involved in the neurotoxicity of Aβ [101], and this proposed feedback mechanism is consistent with the reported potentiation of Aβ toxicity by Cu(II) [113]. Alternatively, oxidized cations such as Cu(II) may accelerate fibrillogenesis [130,272], although this point is controversial [131]. [F] Many of the components involved in these mechanisms may be neurotoxic.

SUMMARY

The problem of sporadic AD in the human population is large, it is increasing in size, we do not know the cause, and we have no effective therapy. We have a vast array of disparate clues about the cause, and it seems prudent at this point to explore the implications of mechanistic links between these clues rather than championing one and excluding others as the fundamental cause of sporadic AD. An understanding of lipid interactions with Aβ proteins and other common components of the extracellular compartment in brain tissue points to a strong hypothesis about a mechanistic link between two of the most consistent features of AD: amyloid plaque formation and oxidative stress. The operation of this mechanism, in turn, has the potential to cause widespread death and loss of neurons in AD.

REFERENCES

1. Lee, S.J., Liyanage, U., Bickel, P.E., Xia, W.M., Lansbury, P.T., Kosik, K.S. (1998). A detergent-insoluble membrane compartment contains A beta in vivo. *Nat Med, 4,* 730–734.

2. Lorenzo, A., Yuan, M.L., Zhang, Z.H., Paganetti, P.A., Sturchler-Pierrat, C., Staufenbiel, M., Mautino, J., Sol, V., Sommer, B., Yankner, B.A. (2000). Amyloid beta interacts with the amyloid precursor protein: a potential toxic mechanism in Alzheimer's disease. *Nat Neurosci, 3*, 460–464.

3. Terzi, E., Holzemann, G., Seelig, J. (1994). Alzheimer beta-amyloid aeptide 25–35: electrostatic interactions with phospholipid membranes. *Biochemistry, 33*, 7434–7441.

4. Terzi, E., Holzemann, G., Seelig, J. (1995). Self-association of beta-amyloid peptide (1–40) in solution and binding to lipid membranes. *J Mol Biol, 252*, 633–642.

5. Hertel, C., Terzi, E., Hauser, N., Jakob-Rotne, R., Seelig, J., Kemp, J.A. (1997). Inhibition of the electrostatic interaction between beta-amyloid peptide and membranes prevents beta-amyloid-induced toxicity. *Proc Natl Acad Sci USA, 94*, 9412–9416.

6. McLaurin, J., Chakrabartty, A. (1997). Characterization of the interactions of Alzheimer beta-amyloid peptides with phospholipid membranes. *Eur J Biochem, 245*, 355–363.

7. Martinez-Senac, M.D., Villalain, J., Gomez-Fernandez, J.C. (1999). Structure of the Alzheimer beta-amyloid peptide (25–35) and its interaction with negatively charged phospholipid vesicles. *Eur J Biochem, 265*, 744–753.

8. Chauhan, A., Ray, I., Chauhan, V.P.S. (2000). Interaction of amyloid beta-protein with anionic phospholipids: possible involvement of Lys(28) and C-terminus aliphatic amino acids. *Neurochem Res, 25*, 423–429.

9. Kremer, J.J., Pallitto, M.M., Sklansky, D.J., Murphy, R.M. (2000). Correlation of beta-amyloid aggregate size and hydrophobicity with decreased bilayer fluidity of model membranes. *Biochemistry, 39*, 10309–10318.

10. Kremer, J.J., Sklansky, D.J., Murphy, R.M. (2001). Profile of changes in lipid bilayer structure caused by beta-amyloid peptide. *Biochemistry, 40*, 8563–8571.

11. Curtain, C.C., Ali, F., Volitakis, I., Cherny, R.A., Norton, R.S., Beyreuther, K., Barrow, C.J., Masters, C.L., Bush, A.I., Barnham, K.J. (2001). Alzheimer's disease amyloid-beta binds copper and zinc to generate an allosterically ordered membrane-penetrating structure containing superoxide dismutase-like subunits. *J Biol Chem, 276*, 20466–20473.

12. McLaurin, J., Darabie, A.A., Morrison, M.R. (2002). Cholesterol, a modulator of membrane-associated A beta-fibrillogenesis. *Ann N Y Acad Sci, 977*, 376–383.

13. Yip C.M., Darabie, A.A., McLaurin J. (2002). A beta 42-peptide assembly on lipid bilayers. *J Mol Biol, 318*, 97–107.

14. Kremer, J.J., Murphy, R.M. (2003). Kinetics of adsorption of beta-amyloid peptide A beta(1–40) to lipid bilayers. *J Biochem Biophys Methods, 57*, 159–169.

15. Bokvist, M., Lindstrom, F., Watts, A., Grobner, G. (2004). Two types of Alzheimer's beta-amyloid (1–40) peptide membrane interactions: aggregation preventing transmembrane anchoring versus accelerated surface fibril formation. *J Mol Biol, 335*, 1039–1049.

16. Ege, C., Lee, K.Y.C. (2004). Insertion of Alzheimer's A beta 40 peptide into lipid monolayers. *Biophys J, 87*, 1732–1740.

17. Shao, H., Jao, S., Ma, K., Zagorski, M.G. (1999). Solution structures of micelle-bound amyloid beta-(1–40) and beta-(1–42) peptides of Alzheimer's disease. *J Mol Biol, 285*, 755–773.

18. Terzi, E., Holzemann, G., Seelig, J. (1997). Interaction of Alzheimer beta-amyloid peptide(1–40) with lipid membranes. *Biochemistry, 36*, 14845–14852.
19. Ji, S.R., Wu, Y., Sui, S.F. (2002). Cholesterol is an important factor affecting the membrane insertion of beta-amyloid peptide (A beta 1–40), which may potentially inhibit the fibril formation. *J Biol Chem, 277*, 6273–6279.
20. Curtain, C.C., Ali, F.E., Smith, D.G., Bush, A.I., Masters, C.L., Barnham, K.J. (2003). Metal ions, pH, and cholesterol regulate the interactions of Alzheimer's disease amyloid-beta peptide with membrane lipid. *J Biol Chem, 278*, 2977–2982.
21. Barnham, K.J., Ciccotosto, G.D., Tickler, A.K., Ali, F.E., Smith, D.G., Williamson, N.A., Lam, Y.H., Carrington, D., Tew, D., Kocak, G., et al. (2003). Neurotoxic, redox-competent Alzheimer's beta-amyloid is released from lipid membrane by methionine oxidation. *J Biol Chem, 278*, 42959–42965.
22. Yanagisawa, K., Odaka, A., Suzuki, N., Ihara, Y. (1995). GM1 ganglioside-bound amyloid beta-protein (A beta): a possible form of preamyloid in Alzheimer's disease. *Nat Med, 1*, 1062–1066.
23. Choo-Smith, L.P., Surewicz, W.K. (1997). The interaction between Alzheimer amyloid beta(1–40) peptide and ganglioside GM1-containing membranes. *FEBS Lett, 402*, 95–98.
24. Choo-Smith, L.P., Garzon-Rodriguez, W., Glabe, C.G., Surewicz, W.K. (1997). Acceleration of amyloid fibril formation by specific binding of Abeta-(1–40) peptide to ganglioside-containing membrane vesicles. *J Biol Chem, 272*, 22987–22990.
25. Matsuzaki, K., Horikiri, C. (1999). Interactions of amyloid beta-peptide (1–40) with ganglioside-containing membranes. *Biochemistry, 38*, 4137–4142.
26. Kakio, A., Nishimoto, S., Yanagisawa, K., Kozutsumi, Y., Matsuzaki, K. (2001). Cholesterol-dependent formation of GM1 ganglioside-bound amyloid beta-protein, an endogenous seed for Alzheimer amyloid. *J Biol Chem, 276*, 24985–24990.
27. Kakio, A., Nishimoto, S., Yanagisawa, K., Kozutsumi, Y., Matsuzaki, K. (2002). Interactions of amyloid beta-protein with various gangliosides in raft-like membranes: importance of GM1 ganglioside-bound form as an endogenous seed for Alzheimer amyloid. *Biochemistry, 41*, 7385–7390.
28. Kakio, A., Nishimoto, S.I., Kozutsumi, Y., Matsuzaki, K. (2003). Formation of a membrane-active form of amyloid beta-protein in raft-like model membranes. *Biochem Biophys Res Commun, 303*, 514–518.
29. Hayashi, H., Kimura, N., Yamaguchi, H., Hasegawa, K., Yokoseki, T., Shibata, M., Yamamoto, N., Michikawa, M., Yoshikawa, Y., Terao, K., et al. (2004). A seed for Alzheimer amyloid in the brain. *J Neurosci, 24*, 4894–4902.
30. Kakio, A., Yano, Y., Takai, D., Kuroda, Y., Matsumoto, O., Kozutsumi, Y., Matsuzaki, K. (2004). Interaction between amyloid beta-protein aggregates and membranes. *J Pept Sci, 10*, 612–621.
31. Wakabayashi, M., Okada, T., Kozutsumi, Y., Matsuzaki, K. (2005). GM1 ganglioside-mediated accumulation of amyloid beta-protein on cell membranes. *Biochem Biophys Res Commun, 328*, 1019–1023.
32. Matsuzaki, K., Noguch, T., Wakabayashi, M., Ikeda, K., Okada, T., Ohashi, Y., Hoshino, M., Naiki, H. (2007). Inhibitors of amyloid beta-protein aggregation mediated by GM1-containing raft-like membranes. *Biochim Biophys Acta, 1768*, 122–130.

33. Wakabayashi, M., Matsuzaki, K. (2007). Formation of amyloids by a beta-(1–42) on NGF-differentiated PC12 cells: roles of gangliosides and cholesterol. *J Mol Biol, 371*, 924–933.

34. Koppaka, V., Axelsen, P.H. (2000). Accelerated accumulation of amyloid beta proteins on oxidatively damaged lipid membranes. *Biochemistry, 39*, 10011–10016.

35. Koppaka, V., Paul, C., Murray, I.V.J., Axelsen P.H. (2003). Early synergy between Abeta42 and oxidatively damaged membranes in promoting amyloid fibril formation by Abeta40. *J Biol Chem, 278*, 36277–36284.

36. Murray, I.V.J., Liu, L., Komatsu, H., Uryu, K., Xiao, G., Lawson, J.A., Axelsen, P.H. (2007). Membrane mediated amyloidogenesis and the promotion of oxidative lipid damage by amyloid beta proteins. *J Biol Chem, 282*, 9335–9345.

37. Qahwash, I.M., Boire, A., Lanning, J., Krausz, T., Pytel, P., Meredith, S.C. (2007). Site-specific effects of peptide lipidation on beta-amyloid aggregation and cytotoxicity. *J Biol Chem, 282*, 36987–36997.

38. Prasad, M.R., Lovell, M.A., Yatin, M., Dhillon, H., Markesbery, W.R. (1998). Regional membrane phospholipid alterations in Alzheimer's disease. *Neurochem Res, 23*, 81–88.

39. Pettegrew, J.W., Panchalingam, K., Hamilton, R.L., McClure, R.J. (2001). Brain membrane phospholipid alterations in Alzheimer's disease. *Neurochem Res, 26*, 771–782.

40. Han, X.L., Holtzman, D.M., McKeel, D.W., Kelley, J., Morris, J.C. (2002). Substantial sulfatide deficiency and ceramide elevation in very early Alzheimer's disease: potential role in disease pathogenesis. *J Neurochem, 82*, 809–818.

41. Mulder, C., Wahlund, L.O., Teerlink, T., Blomberg, M., Veerhuis, R., van Kamp, G.J., Scheltens, P., Scheffer, P.G. (2003). Decreased lysophosphatidylcholine/phosphatidylcholine ratio in cerebrospinal fluid in Alzheimer's disease. *J Neural Trans, 110*, 949 955.

42. Wang, S.S.S., Rymer, D.L., Good, T.A. (2001). Reduction in cholesterol and sialic acid content protects cells from the toxic effects of beta-amyloid peptides. *J Biol Chem, 276*, 42027–42034.

43. Ciccotosto, G.D., Tew, D., Curtain, C.C., Smith, D., Carrington, D., Masters, C.L., Bush, A.I., Cherny, R.A., Cappai, R., Barnham, K.J. (2004). Enhanced toxicity and cellular binding of a modified amyloid beta peptide with a methionine to valine substitution. *J Biol Chem, 279*, 42528–42534.

44. Tickler, A.K., Smith, D.G., Ciccotosto, G.D., Tew, D.J., Curtain, C.C., Carrington, D., Masters, C.L., Bush, A.I., Cherny, R.A., Cappai, R., et al. (2005). Methylation of the imidazole side chains of the Alzheimer disease amyloid-beta peptide results in abolition of superoxide dismutase-like structures and inhibition of neurotoxicity. *J Biol Chem, 280*, 13355–13363.

45. Waschuk, S.A., Elton, E.A., Darabie, A.A., Fraser, P.E., McLaurin, J. (2001). Cellular membrane composition defines A beta-lipid interactions. *J Biol Chem, 276*, 33561–33568.

46. Yip, C.M., McLaurin, J. (2001). Amyloid-beta peptide assembly: a critical step in fibrillogenesis and membrane disruption. *Biophys J, 80*, 1359–1371.

47. Natte, R., Yamaguchi, H., Maat-Schieman, M.L.C., Prins, F.A., Neeskens, P., Roos, R.A.C., van Duinen, S.G. (1999). Ultrastructural evidence of early

non-fibrillar A beta 42 in the capillary basement membrane of patients with hereditary cerebral hemorrhage with amyloidosis, Dutch type. *Acta Neuropathol, 98*, 577–582.

48. Yamaguchi, H., Maat-Schieman, M.L.C., van Duinen, S.G., Prins, F.A., Neeskens, P., Natte, R., Roos, R.A.C. (2000). Amyloid beta protein (A beta) starts to deposit as plasma membrane-bound form in diffuse plaques of brains from hereditary cerebral hemorrhage with amyloidosis-Dutch type, Alzheimer disease and non-demented aged subjects. *J Neuropathol Exp Neurol, 59*, 723–732.
49. Torp, R., Head, E., Milgram, N.W., Hahn, F., Ottersen, O.P., Cotman, C.W. (2000). Ultrastructural evidence of fibrillar beta-amyloid associated with neuronal membranes in behaviorally characterized aged dog brains. *Neuroscience, 96*, 495–506.
50. Van Nostrand, W.E., Melchor, J.P., Ruffini, L. (1998). Pathologic amyloid beta-protein cell surface fibril assembly on cultured human cerebrovascular smooth muscle cells. *J Neurochem, 70*, 216–223.
51. Narayanan, V., Scarlata, S. (2001). Membrane binding and self-association of alpha-synucleins. *Biochemistry, 40*, 9927–9934.
52. Perrin, R.J., Woods, W.S., Clayton, D.F., George, J.M. (2001). Exposure to long chain polyunsaturated fatty acids triggers rapid multimerization of synucleins. *J Biol Chem, 276*, 41958–41962.
53. Smith, M.A., Taneda, S., Richey, P.L., Miyata, S., Yan, S.D., Stern, D., Sayre, L.M., Monnier, V.M., Perry, G. (1994). Advanced Maillard reaction end-products are associated with Alzheimer-disease pathology. *Proc Natl Acad Sci USA, 91*, 5710–5714.
54. Smith, M.A., Perry, G., Richey, P.L., Sayre, L.M., Anderson, V.E., Beal, M.F., Kowall, N. (1996). Oxidative damage in Alzheimer's. *Nature, 382*, 120–121.
55. Smith, M.A., Richey Harris, P.L., Sayre, L.M., Beckman, J.S., Perry, G. (1997). Widespread peroxynitrite-mediated damage in Alzheimer's disease. *J Neurosci, 17*, 2653.
56. Sayre, L.M., Zelasko, D.A., Harris, P.L.R., Perry, G., Salomon, R.G., Smith, M.A. (1997). 4-Hydroxynonenal-derived advanced lipid peroxidation end products are increased in Alzheimer's disease. *J Neurochem, 68*, 2092–2097.
57. Castellani, R.J., Perry, G., Harris, P.L.R., Cohen, M.L., Sayre, L.M., Salomon, R.G., Smith, M.A. (1998). Advanced lipid peroxidation end-products in Alexander's disease. *Brain Res, 787*, 15–18.
58. Rottkamp, C.A., Nunomura, A., Raina, A.K., Sayre, L.M., Perry, G., Smith, M.A. (2000). Oxidative stress, antioxidants, and Alzheimer disease. *Alzheimer Dis Assoc Disord, 14*, S62–S66.
59. Markesbery, W.R. (1997). Oxidative stress hypothesis in Alzheimer's disease. *Free Rad Biol Med, 23*, 134–147.
60. Markesbery, W.R., Carney, J.M. (1999). Oxidative alterations in Alzheimer's disease. *Brain Pathol, 9*, 133–146.
61. Pratico, D., Delanty, N. (2000). Oxidative injury in diseases of the central nervous system: focus on Alzheimer's disease. *Am J Med, 109*, 577–585.
62. Butterfield, D.A., Lauderback, C.M. (2002). Lipid peroxidation and protein oxidation in Alzheimer's disease brain: potential causes and consequences involving

INSTITUT MEDIATHEQUE PASTEUR

amyloid beta-peptide-associated free radical oxidative stress. *Free Radic Biol Med, 32*, 1050–1060.

63. Drake, J., Link, C.D., Butterfield, D.A. (2003). Oxidative stress precedes fibrillar deposition of Alzheimer's disease amyloid beta-peptide (1–42) in a transgenic *Caenorhabditis elegans* model. *Neurobiol Aging, 24*, 415–420.
64. Butterfield, D.A., Boyd-Kimball, D. (2004). Amyloid beta peptide(1–42) contributes to the oxidative stress and neurodegeneration found in Alzheimer disease brain. *Brain Pathol, 14*, 426–432.
65. McIntosh, L.J., Trush, M.A., Troncoso, J.C. (1997). Increased susceptibility of Alzheimer's disease temporal cortex to oxygen free radical–mediated processes. *Free Radic Biol Med, 23*, 183–190.
66. Marcus, D.L., Thomas, C., Rodriguez, C., Simberkoff, K., Tsai, J.S., Strafaci, J.A., Freedman, M.L. (1998). Increased peroxidation and reduced antioxidant enzyme activity in Alzheimer's disease. *Exp Neurol, 150*, 40–44.
67. Varadarajan, S., Yatin, S., Aksenova, M., Butterfield, D.A. (2000). Alzheimer's amyloid beta-peptide-associated free radical oxidative stress and neurotoxicity. *J Struct Biol, 130*, 184–208.
68. Lovell, M.A., Ehmann, W.D., Mattson, M.P., Markesbery, W.R. (1997). Elevated 4-hydroxynonenal in ventricular fluid in Alzheimer's disease. *Neurobiol Aging, 18*, 457–461.
69. Schippling, S., Kontush, A., Arlt, S., Buhmann, C., Sturenburg, H.J., Mann, U., Muller-Thomsen, T., Beisiegel, U. (2000). Increased lipoprotein oxidation in Alzheimer's disease. *Free Radic Biol Med, 28*, 351–360.
70. Montine, T.J., Neely, M.D., Quinn, J.F., Beal, M.F., Markesbery, W.R., Roberts, L.J., Morrow, J.D. (2002). Lipid peroxidation in aging brain and Alzheimer's disease. *Free Radic Biol Med, 33*, 620–626.
71. Knight, J.A. (1998). Free radicals: their history and current status in aging and disease. *Ann Clin Lab Sci, 28*, 331–346.
72. Jeandel, C., Nicolas, M.B., Dubois, F., Nabet-Belleville, F., Penin, F., Cuny, G. (1989). Lipid peroxidation and free radical scavengers in Alzheimer's disease. *Gerontology, 35*, 275–282.
73. Zaman, Z., Roche, S., Fielden, P., Frost, P.G., Niriella, D.C., Cayley, A.C. (1992). Plasma concentrations of vitamins A and E and carotenoids in Alzheimer's disease. *Age Ageing, 21*, 91–94.
74. Arlt, S., Beisiegel, U., Kontush, A. (2002). Lipid peroxidation in neurodegeneration: new insights into Alzheimer's disease. *Curr Opin Lipidol, 13*, 289–294.
75. Butterfield, D.A., Castegna, A., Lauderback, C.M., Drake, J. (2002). Evidence that amyloid beta-peptide-induced lipid peroxidation and its sequelae in Alzheimer's disease brain contribute to neuronal death. *Neurobiol Aging, 23*, 655–664.
76. McGrath, L.T., McGleenon, B.M., Brennan, S., McColl, D., McIlroy, S., Passmore, A.P. (2001). Increased oxidative stress in Alzheimer's disease as assessed with 4-hydroxynonenal but not malondialdehyde. *Q J Med, 94*, 485–490.
77. Sultana, R., Butterfield, D.A. (2004). Oxidatively modified GST and MRP1 in Alzheimer's disease brain: implications for accumulation of reactive lipid peroxidation products. *Neurochem Res, 29*, 2215–2220.

78. Galbusera, C., Facheris, M., Magni, F., Galimberti, G., Sala, G., Tremolada, L., Isella, V., Guerini, F.R., Appollonio, I., Galli-Kienle, M., Ferrarese, C. (2004). Increased susceptibility to plasma lipid peroxidation in Alzheimer disease patients. *Curr Alzheimer Res, 1*, 103–109.

79. Cutler, R.G., Kelly, J., Storie, K., Pedersen, W.A., Tammara, A., Hatanpaa, K., Troncoso, J.C., Mattson, M.P. (2004). Involvement of oxidative stress-induced abnormalities in ceramide and cholesterol metabolism in brain aging and Alzheimer's disease. *Proc Natl Acad Sci USA, 101*, 2070–2075.

80. Esterbauer, H., Benedetti, A., Lang, J., Fulceri, R., Fauler, G., Comporti, M. (1986). Studies on the mechanism of formation of 4-hydroxynonenal during microsomal lipid-peroxidation. *Biochim Biophys Acta, 876*, 154–166.

81. Esterbauer, H., Schaur, R.J., Zollner, H. (1991). Chemistry and biochemistry of 4-hydroxynonenal, malonaldehyde and related aldehydes. *Free Radic Biol Med, 11*, 81–128.

82. Morrow, J.D., Awad, J.A., Boss, H.J., Blair, I.A., Roberts, L.J., II (1992). Non-cyclooxygenase-derived prostanoids (F2-isoprostanes) are formed in situ on phospholipids. *Proc Natl Acad Sci USA, 89*, 10721–10725.

83. Morrow, J.D., Minton, T.A., Mukundan, C.R., Campbell, M.D., Zackert, W.E., Daniel, V.C., Badr, K.F., Blair, I.A., Roberts, L.J., II (1994). Free radical–induced generation of isoprostanes in vivo: evidence for the formation of D-ring and E-ring isoprostanes. *J Biol Chem, 269*, 4317–4326.

84. Markesbery, W.R., Lovell, M.A. (1998). Four-Hydroxynonenal, a product of lipid peroxidation, is increased in the brain in Alzheimer's disease. *Neurobiol Aging, 19*, 33–36.

85. Uchida K. (2003). 4-Hydroxy-2-nonenal: a product and mediator of oxidative stress. *Prog Lipid Res, 42*, 318–343.

86. Roberts, L.J., II, Morrow J.D. (1994). Isoprostanes: novel markers of endogenous lipid peroxidation and potential mediators of oxidant injury. *Ann N Y Acad Sci, 744*, 237–242.

87. Patrono, C., FitzGerald, G.A. (1997). Isoprostanes: potential markers of oxidant stress in atherothrombotic disease. *Arterioscler Thromb Vasc Biol, 17*, 2309–2315.

88. Lawson, J.A., Rokach, J., FitzGerald, G.A. (1999). Isoprostanes: formation, analysis and use as indices of lipid peroxidation in vivo. *J Biol Chem, 274*, 24441–24444.

89. Pratico, D., Lee, V.M.Y., Trojanowski, J.Q., Rokach, J., FitzGerald, G.A. (1998). Increased F-2-isoprostanes in Alzheimer's disease: evidence for enhanced lipid peroxidation in vivo. *FASEB J, 12*, 1777–1783.

90. Pratico, D., Clark, C.M., Lee, V.M.Y., Trojanowski, J.Q., Rokach, J., FitzGerald, G.A. (2000). Increased 8,12-iso-iPF(2 alpha)-VI in Alzheimer's disease: correlation of a noninvasive index of lipid peroxidation with disease severity. *Ann Neurol, 48*, 809–812.

91. Pratico, D., Uryu, K., Leight, S., Trojanowswki, J.Q., Lee, V.M.Y. (2001). Increased lipid peroxidation precedes amyloid plaque formation in an animal model of Alzheimer amyloidosis. *J Neurosci, 21*, 4183–4187.

92. Uchida, K., Kanematsu, M., Sakai, K., Matsuda, T., Hattori, N., Mizuno, Y., Suzuki, D., Miyata, T., Noguchi, N., Niki, E., Osawa, T. (1998). Protein-bound

acrolein: potential markers for oxidative stress. *Proc Natl Acad Sci USA, 95*, 4882–4887.

93. Uchida, K., Kanematsu, M., Morimitsu, Y., Osawa, T., Noguchi, N., Niki, E. (1998). Acrolein is a product of lipid peroxidation reaction: formation of free acrolein and its conjugate with lysine residues in oxidized low density lipoproteins. *J Biol Chem, 273*, 16058–16066.
94. Uchida, K. (1999). Current status of acrolein as a lipid peroxidation product. *Trends Cardiovasc Med, 9*, 109–113.
95. Lovell, M.A., Xie, C.S., Markesbery, W.R. (2001). Acrolein is increased in Alzheimer's disease brain and is toxic to primary hippocampal cultures. *Neurobiol Aging, 22*, 187–194.
96. Williams, T.I., Lynn, B.C., Markesbery, W.R., Lovell, M.A. (2006). Increased levels of 4-hydroxynonenal and acrolein, neurotoxic markers of lipid peroxidation, in the brain in mild cognitive impairment and early Alzheimer's disease. *Neurobiol Aging, 27*, 1094–1099.
97. Castegna, A., Lauderback, C.M., Mohammad-Abdul, H., Butterfield, D.A. (2004). Modulation of phospholipid asymmetry in synaptosomal membranes by the lipid peroxidation products, 4-hydroxynonenal and acrolein: implications for Alzheimer's disease. *Brain Res, 1004*, 193–197.
98. Abdul, H.M., Butterfield, D.A. (2005). Protection against amyloid beta-peptide (1–42)-induced loss of phospholipid asymmetry in synaptosomal membranes by tricyclodecan-9-xanthogenate (D609) and ferulic acid ethyl ester: implications for Alzheimer's disease. *Biochim Biophys Acta, 1741*, 140–148.
99. Calingasan, N.Y., Uchida, K., Gibson, G.E. (1999). Protein-bound acrolein: a novel marker of oxidative stress in Alzheimer's disease. *J Neurochem, 72*, 751–756.
100. Berthon, G. (2000). Does human beta A4 exert a protective function against oxidative stress in Alzheimer's disease? *Med Hypotheses, 54*, 672–677.
101. Sayre, L.M., Perry, G., Harris, P.L.R., Liu, Y.H., Schubert, K.A., Smith, M.A. (2000). In situ oxidative catalysis by neurofibrillary tangles and senile plaques in Alzheimer's disease: a central role for bound transition metals. *J Neurochem, 74*, 270–279.
102. Kontush, A., Donarski, N., Beisiegel, U. (2001). Resistance of human cerebrospinal fluid to in vitro oxidation is directly related to its amyloid-beta content. *Free Radic Res, 35*, 507–517.
103. Kontush, A. (2001). Amyloid-[beta]: an antioxidant that becomes a pro-oxidant and critically contributes to Alzheimer's disease. *Free Radic Biol Med, 31*, 1120–1131.
104. Smith, M.A., Casadesus, G., Joseph, J.A., Perry, G. (2002). Amyloid-beta and tau serve antioxidant functions in the aging and Alzheimer brain. *Free Radic Biol Med, 33*, 1194–1199.
105. Atwood, C.S., Obrenovich, M.E., Liu, T.B., Chan, H., Perry, G., Smith, M.A., Martins, R.N. (2003). Amyloid-beta: a chameleon walking in two worlds: a review of the trophic and toxic properties of amyloid-beta. *Brain Res Rev, 43*, 1–16.
106. Cuajungco, M.P., Goldstein, L.E., Nunomura, A., Smith, M.A., Lim, J.T., Atwood, C.S., Huang, X., Farrag, Y.W., Perry, G., Bush, A.I. (2000). Evidence that the beta-amyloid plaques of Alzheimer's disease represent the redox-silencing and entombment of abeta by zinc. *J Biol Chem, 275*, 19439–19442.

107. Nunomura, A., Perry, G., Pappolla, M.A., Friedland, R.P., Hirai, K., Chiba, S., Smith, M.A. (2000). Neuronal oxidative stress precedes amyloid-beta deposition in Down syndrome. *J Neuropathol Exp Neurol, 59*, 1011–1017.

108. Kontush, A., Berndt, C., Weber, W., Akopyan, V., Arlt, S., Schippling, S., Beisiegel, U. (2001). Amyloid-beta is an antioxidant for lipoproteins in cerebrospinal fluid and plasma. *Free Radic Biol Med, 30*, 119–128.

109. Zou, K., Gong, J.S., Yanagisawa, K., Michikawa, M. (2002). A novel function of monomeric amyloid beta-protein serving as an antioxidant molecule against metal-induced oxidative damage. *J Neurosci, 22*, 4833–4841.

110. Hensley, K., Hall, N., Subramaniam, R., Cole, P., Harris, M., Aksenov, M., Aksenova, M., Gabbita, S.P., Wu, J.F., Carney, J.M., et al. (1995). Brain regional correspondence between Alzheimer's disease histopathology and biomarkers of protein oxidation. *J Neurochem, 65*, 2146–2156.

111. Wu, Y., Luo, Y. (2005). Transgenic *C. elegans* as a model in Alzheimer's research. *Curr Alzheimer Res, 2*, 37–47.

112. Behl, C., Davis, J.B., Lesley, R., Schubert, D. (1994). Hydrogen peroxide mediates amyloid beta protein toxicity. *Cell, 77*, 817–827.

113. Huang, X.D., Cuajungco, M.P., Atwood, C.S., Hartshorn, M.A., Tyndall, J.D.A., Hanson, G.R., Stokes, K.C., Leopold, M., Multhaup, G., Goldstein, L.E., et al. (1999). Cu(II) potentiation of Alzheimer A beta neurotoxicity: correlation with cell-free hydrogen peroxide production and metal reduction. *J Biol Chem, 274*, 37111–37116.

114. Huang, X.D., Atwood, C.S., Hartshorn, M.A., Multhaup, G., Goldstein, L.E., Scarpa, R.C., Cuajungco, M.P., Gray, D.N., Lim, J., Moir, R.D., et al. (1999). The A beta peptide of Alzheimer's disease directly produces hydrogen peroxide through metal ion reduction. *Biochemistry, 38*, 7609–7616.

115. Opazo, C., Huang, X.D., Cherny, R.A., Moir, R.D., Roher, A.E., White, A.R., Cappai, R., Masters, C.L., Tanzi, R.E., Inestrosa, N.C., Bush, A.I. (2002). Metalloenzyme-like activity of Alzheimer's disease beta-amyloid–Cu-dependent catalytic conversion of dopamine, cholesterol, and biological reducing agents to neurotoxic H2O2. *J Biol Chem, 277*, 40302–40308.

116. Nelson, T.J., Alkon, D.L. (2005). Oxidation of cholesterol by amyloid precursor protein and beta-amyloid peptide. *J Biol Chem, 280*, 7377–7387.

117. Barnham, K.J., Haeffner, F., Ciccotosto, G.D., Curtain, C.C., Tew, D., Beyreuther, K., Carrington, D., Masters, C.L., Cherny, R.A., Cappai, R., Bush, A.I. (2004). Tyrosine gated electron transfer is key to the toxic mechanism of Alzheimer's disease β-amyloid. *FASEB J, 18*, 1427–1429.

118. Butterfield, D.A., Kanski, J. (2002). Methionine residue 35 is critical for the oxidative stress and neurotoxic properties of Alzheimer's amyloid beta-peptide 1–42 [Review]. *Peptides, 23*, 1299–1309.

119. Butterfield, D.A. (2002). Amyloid beta-peptide (1–42)-induced oxidative stress and neurotoxicity: implications for neurodegeneration in Alzheimer's disease brain: a review. *Free Radic Res, 36*, 1307–1313.

120. Butterfield, D.A., Bush, A.I. (2004). Alzheimer's amyloid beta-peptide (1–42): involvement of methionine residue 35 in the oxidative stress and neurotoxicity properties of this peptide. *Neurobiol Aging, 25*, 563–568.

121. Murray, I.V.J., Sindoni, M.E., Axelsen, P.H. (2005). Promotion of oxidative lipid membrane damage by amyloid beta proteins. *Biochemistry, 44*, 12606–12613.

122. Puglielli, L., Friedlich, A.L., Setchell, K.D.R., Nagano, S., Opazo, C., Cherny, R.A., Barnham, K.J., Wade, J.D., Melov, S., Kovacs, D.M., Bush, A.I. (2005). Alzheimer disease {beta}-amyloid activity mimics cholesterol oxidase. *J Clin Invest, 115*, 2556–2563.

123. Yoshimoto, N., Tasaki, M., Shimanouchi, T., Umakoshi, H., Kuboi, R. (2005). Oxidation of cholesterol catalyzed by amyloid beta-peptide (A beta)-Cu complex on lipid membrane. *J Biosci Bioeng, 100*, 455–459.

124. Hou, L.M., Shao, H.Y., Zhang, Y.B., Li, H., Menon, N.K., Neuhaus, E.B., Brewer, J.M., Byeon, I.J.L., Ray, D.G., Vitek, M.P., et al. (2004). Solution NMR studies of the A beta(1–40) and A beta(1–42) peptides establish that the met35 oxidation state affects the mechanism of amyloid formation. *J Am Chem Soc, 126*, 1992–2005.

125. Zirah, S., Kozin, S.A., Mazur, A.K., Blond, A., Cheminant, M., Segalas-Milazzo, I., Debey, P., Rebuffat, S. (2006). Structural changes of region 1–16 of the Alzheimer disease amyloid beta-peptide upon zinc binding and in vitro aging. *J Biol Chem, 281*, 2151–2161.

126. Dong, J., Canfield, J.M., Mehta, A.K., Shokes, J.E., Tian, B., Childers, W.S., Simmons, J.A., Mao, Z., Scott, R.A., Warncke, K., Lynn, D.G. (2007). Engineering metal ion coordination to regulate amyloid fibril assembly and toxicity. *Proc Natl Acad Sci USA*, 0702669104.

127. Nakamura, M., Shishido, N., Nunomura, A., Smith, M.A., Perry, G., Hayashi, Y., Nakayama, K., Hayashi, T. (2007). Three histidine residues of amyloid-beta peptide control the redox activity of copper and iron. *Biochemistry, 46*, 12737–12743.

128. Strausak, D., Mercer, J.F.B., Dieter, H.H., Stremmel, W., Multhaup, G. (2001). Copper in disorders with neurological symptoms: Alzheimer's, Menkes, and Wilson diseases. *Brain Res Bull, 55*, 175–185.

129. Cherny, R.A., Legg, J.T., Mclean, C.A., Fairlie, D.P., Huang, X.D., Atwood, C.S., Beyreuther, K., Tanzi, R.E., Masters, C.L., Bush, A.I. (1999). Aqueous dissolution of Alzheimer's disease A beta amyloid deposits by biometal depletion. *J Biol Chem, 274*, 23223–23228.

130. Cherny, R.A., Atwood, C.S., Xilinas, M.E., Gray, D.N., Jones, W.D., Mclean, C.A., Barnham, K.J., Volitakis, I., Fraser, F.W., Kim, Y.S., et al. (2001). Treatment with a copper–zinc chelator markedly and rapidly inhibits beta-amyloid accumulation in Alzheimer's disease transgenic mice. *Neuron, 30*, 665–676.

131. Yoshiike, Y., Tanemura, K., Murayama, O., Akagi, T., Murayama, M., Sato, S., Sun, X.Y., Tanaka, N., Takashima, A. (2001). New insights on how metals disrupt amyloid beta-aggregation and their effects on amyloid-beta cytotoxicity. *J Biol Chem, 276*, 32293–32299.

132. Atwood, C.S., Huang, X.D., Khatri, A., Scarpa, R.C., Kim, Y.S., Moir, R.D., Tanzi, R.E., Roher, A.E., Bush, A.I. (2000). Copper catalyzed oxidation of alzheimer A beta. *Cell Mol Biol (Noisy-Le-Grand), 46*, 777–783.

133. Butterfield, D.A., Boyd-Kimball, D. (2005). The critical role of methionine 35 in Alzheimer's amyloid-beta-peptide (1–42)-induced oxidative stress and neurotoxicity. *Biochim Biophys Acta, 1703*, 149–156.

134. Kanski, J., Aksenova, M., Schoneich, C., Butterfield, D.A. (2002). Substitution of isoleucine-31 by helical-breaking proline abolishes oxidative stress and neurotoxic properties of Alzheimer's amyloid beta-peptide (1–42). *Free Radic Biol Med, 32*, 1205–1211.

135. Hou, L.M., Kang, I., Marchant, R.E., Zagorski, M.G. (2002). Methionine 35 oxidation reduces fibril assembly of the amyloid A beta-(1–42) peptide of Alzheimer's disease. *J Biol Chem, 277*, 40173–40176.

136. Palmblad, M., Westlind-Danielsson, A., Bergquist, J. (2002). Oxidation of methionine 35 attenuates formation of amyloid beta-peptide 1–40 oligomers. *J Biol Chem, 277*, 19506–19510.

137. Johansson, A.S., Bergquist, J., Volbracht, C., Paivio, A., Leist, M., Lannfelt, L., Westlind-Danielsson, A. (2007). Attenuated amyloid-beta aggregation and neurotoxicity owing to methionine oxidation. *Neuroreport, 18*, 559–563.

138. Lauderback, C.M., Hackett, J.M., Huang, F.F., Keller, J.N., Szweda, L.I., Markesbery, W.R., Butterfield, D.A. (2001). The glial glutamate transporter, GLT-1, is oxidatively modified by 4-hydroxy-2-nonenal in the Alzheimer's disease brain: the role of A beta 1–42. *J Neurochem, 78*, 413–416.

139. Bieschke, J., Zhang, Q., Powers, E.T., Lerner, R.A., Kelly, J.W. (2005). Oxidative metabolites accelerate Alzheimer's amyloidogenesis by a two-step mechanism, eliminating the requirement for nucleation. *Biochemistry, 44*, 4977–4983.

140. Zhang, Q.H., Powers, E.T., Nieva, J., Huff, M.E., Dendle, M.A., Bieschke, J., Glabe, C.G., Eschenmoser, A., Wentworth, P., Lerner, R.A., Kelly, J.W. (2004). Metabolite-initiated protein misfolding may trigger Alzheimer's disease. *Proc Natl Acad Sci USA, 101*, 4752–4757.

141. Magni, F., Galbusera, C., Tremolada, L., Ferrarese, C., Kienle, M.G. (2002). Characterisation of adducts of the lipid peroxidation product 4-hydroxy-2-nonenal and amyloid beta-peptides by liquid chromatography/electrospray ionisation mass spectrometry. *Rapid Commun Mass Spectrom, 16*, 1485–1493.

142. Shringarpure, R., Grune, T., Sitte, N., Davies, K.J.A. (2000). 4-Hydroxynonenal-modified amyloid-beta peptide inhibits the proteasome: possible importance in Alzheimer's disease. *Cell Mol Life Sci, 57*, 1802–1809.

143. Ando, Y., Brannstrom, T., Uchida, K., Nyhlin, N., Nasman, B., Suhr, O., Yamashita, T., Olsson, T., El Salhy, M., Uchino, M., Ando, M. (1998). Histochemical detection of 4-hydroxynonenal protein in Alzheimer amyloid. *J Neurol Sci, 156*, 172–176.

144. Rofina, J.E., Singh, K., Skoumalova-Vesela, A., van Ederen, A.M., van Asten, A.J.A.M., Wilhelm, J., Gruys, E. (2004). Histochemical accumulation of oxidative damage products is associated with Alzheimer-like pathology in the canine. *Amyloid, 11*, 90–100.

145. Komatsu, H., Liu, L., Murray, I.V., Axelsen, P.H. (2007). A mechanistic link between oxidative stress and membrane mediated amyloidogenesis revealed by infrared spectroscopy. *Biochim Biophys Acta, 1768*, 1913–1922.

146. Pike, C.J., RamezanArab, N., Cotman, C.W. (1997). Beta-amyloid neurotoxicity in vitro: evidence of oxidative stress but not protection by antioxidants. *J Neurochem, 69*, 1601–1611.

147. Hensley, K., Carney, J.M., Mattson, M.P., Aksenova, M., Harris, M., Wu, J.F., Floyd, R.A., Butterfield, D.A. (1994). A model for beta-amyloid aggregation and

neurotoxicity based on free radical generation by the peptide: relevance to Alzheimer disease. *Proc Natl Acad Sci USA, 91*, 3270–3274.

148. Butterfield, D.A., Hensley, K., Harris, M., Mattson, M.P., Carney, J.M. (1994). Beta-amyloid peptide free radical fragments initiate synaptosomal lipoperoxidation in a sequence-specific fashion: implications to Alzheimer's disease. *Biochem Biophys Res Commun, 200*, 710–715.
149. Monji, A., Utsumi, H., Ueda, T., Imoto, T., Yoshida, I., Hashioka, S., Tashiro, K., Tashiro, N. (2001). The relationship between the aggregational state of the amyloid-beta peptides and free radical generation by the peptides. *J Neurochem, 77*, 1425–1432.
150. Varadarajan, S., Yatin, S., Butterfield, D.A. (2000). Alzheimer's amyloid beta peptide (1–42) fibrils are not always neurotoxic. *Alzheimer's Rep, 3*, 71–76.
151. Dikalov, S.I., Vitek, M.P., Maples, K.R., Mason, R.P. (1999). Amyloid beta peptides do not form peptide-derived free radicals spontaneously, but can enhance metal-catalyzed oxidation of hydroxylamines to nitroxides. *J Biol Chem, 274*, 9392–9399.
152. Yao, Z.X., Drieu, K., Szweda, L.I., Papadopoulos, V. (1999). Free radicals and lipid peroxidation do not mediate beta-amyloid-induced neuronal cell death. *Brain Res, 847*, 203–210.
153. Jarrett, J.T., Berger, E.P., Lansbury, P.T. (1993). The carboxy terminus of the beta amyloid protein is critical for the seeding of amyloid formation: implications for the pathogenesis of Alzheimer's disease. *Biochemistry, 32*, 4693–4697.
154. Sengupta, P., Garai, K., Sahoo, B., Shi, Y., Callaway, D.J.E., Maiti, S. (2003). The amyloid beta peptide (A beta(1–40)) is thermodynamically soluble at physiological concentrations. *Biochemistry, 42*, 10506–10513.
155. O'Nuallain, B., Shivaprasad, S., Kheterpal, I., Wetzel, R. (2005). Thermodynamics of A beta(1–40) amyloid fibril elongation. *Biochemistry, 44*, 12709–12718.
156. Hu, W.P., Chang, G.L., Chen, S.J., Kuo, Y.M. (2006). Kinetic analysis of beta-amyloid peptide aggregation induced by metal ions based on surface plasmon resonance biosensing. *J Neurosci Methods, 154*, 190–197.
157. Hasegawa, K., Ono, K., Yamada, M., Naiki, H. (2002). Kinetic modeling and determination of reaction constants of Alzheimer's beta-amyloid fibril extension and dissociation using surface plasmon resonance. *Biochemistry, 41*, 13489–13498.
158. Cannon, M.J., Williams, A.D., Wetzel, R., Myszka, D.G. (2004). Kinetic analysis of beta-amyloid fibril elongation. *Anal Biochem, 328*, 67–75.
159. Aguilar, M.I., Small, D.H. (2005). Surface plasmon resonance for the analysis of beta-amyloid interactions and fibril formation in Alzheimer's disease research. *Neurotox Res, 7*, 17–27.
160. Bohrmann, B., Tjernberg, L.O., Kuner, P., Poli, S., Levet-Trafit, B., Naslund, J., Richards, G., Huber, W., Dobeli, H., Nordstedt, C. (1999). Endogenous proteins controlling amyloid beta-peptide polymerization: possible implications for beta-amyloid formation in the central nervous system and in peripheral tissues. *J Biol Chem, 274*, 15990–15995.
161. Cairo, C.W., Strzelec, A., Murphy, R.M., Kiessling, L.L. (2002). Affinity-based inhibition of beta-amyloid toxicity. *Biochemistry, 41*, 8620–8629.

162. Kakuyama, H., Soderberg, L., Horigome, K., Winblad, B., Dahlqvist, C., Naslund, J., Tjernberg, L.O. (2005). CLAC binds to aggregated A beta and A beta fragments, and attenuates fibril elongation. *Biochemistry, 44*, 15602–15609.
163. Subasinghe, S., Unabia, S., Barrow, C.J., Mok, S.S., Aguilar, M.I., Small, D.H. (2003). Cholesterol is necessary both for the toxic effect of A beta peptides on vascular smooth muscle cells and for A beta binding to vascular smooth muscle cell membranes. *J Neurochem, 84*, 471–479.
164. Devanathan, S., Salamon, Z., Lindblom, G., Grobner, G., Tollin, G. (2006). Effects of sphingomyelin, cholesterol and zinc ions on the binding, insertion and aggregation of the amyloid A beta(1–40) peptide in solid-supported lipid bilayers. *FEBS J, 273*, 1389–1402.
165. Jarrett, J.T., Lansbury, P.T. (1992). Amyloid fibril formation requires a chemically discriminating nucleation event: studies of an amyloidogenic sequence from the bacterial protein OsmB. *Biochemistry, 31*, 12345–12352.
166. Jarrett, J.T., Lansbury, P.T. (1993). Seeding "one-dimensional crystallization" of amyloid: a pathogenic mechanism in Alzheimer's disease and scrapie? *Cell, 73*, 1055–1058.
167. Harper, J.D., Lansbury, P.T. (1997). Models of amyloid seeding in Alzheimer's disease and scrapie: mechanistic truths and physiological consequences of the time-dependent solubility of amyloid proteins. *Annu Rev Biochem, 66*, 385–407.
168. Lomakin, A., Teplow, D.B., Kirschner, D.A., Benedek, G.B. (1997). Kinetic theory of fibrillogenesis of amyloid beta-protein. *Proc Natl Acad Sci USA, 94*, 7942–7947.
169. Esler, W.P., Stimson, E.R., Ghilardi, J.R., Vinters, H.V., Lee, J.P., Mantyh, P.W., Maggio, J.E. (1996). In vitro growth of Alzheimer's disease beta-amyloid plaques displays first-order kinetics. *Biochemistry, 35*, 749–757.
170. Lomakin, A., Chung, D.S., Benedek, G.B., Kirschner, D.A., Teplow, D.B. (1996). On the nucleation and growth of amyloid beta-protein fibrils: detection of nuclei and quantitation of rate constants. *Proc Natl Acad Sci USA, 93*, 1125–1129.
171. Esler, W.P., Stimson, E.R., Jennings, J.M., Vinters, H.V., Ghilardi, J.R., Lee, J.P., Mantyh, P.W., Maggio, J.E. (2000). Alzheimer's disease amyloid propagation by a template-dependent dock–lock mechanism. *Biochemistry, 39*, 6288–6295.
172. Nguyen, P.H., Li, M.S., Stock, G., Straub, J.E., Thirumalai, D. (2007). Monomer adds to preformed structured oligomers of A beta-peptides by a two-stage dock–lock mechanism. *Proc Natl Acad Sci USA, 104*, 111–116.
173. Oe, T., Ackermann, B.L., Inoue, K., Berna, M.J., Garner, C.O., Gelfanova, V., Dean, R.A., Siemers, E.R., Holtzman, D.M., Farlow, M.R., Blair, I.A. (2006). Quantitative analysis of amyloid beta peptides in cerebrospinal fluid of Alzheimer's disease patients by immunoaffinity purification and stable isotope dilution liquid chromatography/negative electrospray ionization tandem mass spectrometry. *Rapid Commun Mass Spectrom, 20*, 3723–3735.
174. Bateman, R.J., Munsell, L.Y., Morris, J.C., Swarm, R., Yarasheski, K.E., Holtzman, D.M. (2006). Human amyloid-beta synthesis and clearance rates as measured in cerebrospinal fluid in vivo. *Nat Med, 12*, 856–861.
175. Siegel, S.J., Bieschke, J., Powers, E.T., Kelly, J.W. (2007). The oxidative stress metabolite 4-hydroxynonenal promotes Alzheimer protofibril formation. *Biochemistry, 46*, 1503–1510.

176. Atwood, C.S., Perry, G., Zeng, H., Kato, Y., Jones, W.D., Ling, K.Q., Huang, X.D., Moir, R.D., Wang, D.D., Sayre, L.M., et al. (2004). Copper mediates dityrosine cross-linking of Alzheimer's amyloid-beta. *Biochemistry, 43*, 560–568.

177. Smith, D.P., Smith, D.G., Curtain, C.C., Boas, J.F., Pilbrow, J.R., Ciccotosto, G.D., Lau, T.L., Tew, D.J., Perez, K., Wade, J.D., et al. (2006). Copper mediated amyloid-beta toxicity is associated with an intermolecular histidine bridge. *J Biol Chem, 281*, 15145–15154.

178. Smith, D.P., Ciccotosto, G.D., Tew, D.J., Fodero-Tavoletti, M.T., Johanssen, T., Masters, C.L., Barnham, K.J., Cappai, R. (2007). Concentration dependent Cu^{2+} induced aggregation and dityrosine formation of the Alzheimer's disease amyloid-beta peptide. *Biochemistry, 46*, 2881–2891.

179. Ali, F.E., Separovic, F., Barrow, C.J., Yao, S.G., Barnham, K.J. (2006). Copper and zinc mediated oligomerisation of A beta peptides. *Int J Pept Res Ther, 12*, 153–164.

180. Dai, X.L., Sun, Y.X., Jiang, Z.F. (2006). Cu(II) potentiation of Alzheimer Abeta1–40 cytotoxicity and transition on its secondary structure. *Acta Biochim Biophys Sin (Shanghai)*, *38*, 765–772.

181. Schonberger, S.J., Edgar, P.F., Kydd, R., Faull, R.L.M., Cooper, G.J.S. (2001). Proteomic analysis of the brain in Alzheimer's disease: molecular phenotype of a complex disease process. *Proteomics, 1*, 1519–1528.

182. Shiozaki, A., Tsuji, T., Kohno, R., Kawamata, J., Uemura, K., Teraoka, H., Shimohama, S. (2004). Proteome analysis of brain proteins in Alzheimer's disease: subproteomics following sequentially extracted protein preparation. *J Alzheimer's Dis, 6*, 257–268.

183. Cottrell, B.A., Galvan, V., Banwait, S., Gorostiza, O., Lombardo, C.R., Williams, T., Schilling, B., Peel, A., Gibson, B., Koo, E.H., et al. (2005). A pilot proteomic study of amyloid precursor interactors in Alzheimer's disease. *Ann Neurol, 58*, 277–289.

184. Tien, M., Berlett, B.S., Levine, R.L., Chock, P.B., Stadtman, E.R. (1999). Peroxynitrite-mediated modification of proteins at physiological carbon dioxide concentration: pH dependence of carbonyl formation, tyrosine nitration, and methionine oxidation. *Proc Natl Acad Sci USA, 96*, 7809–7814.

185. Dalle-Donne, I., Rossi, R., Giustarini, D., Gagliano, N., Lusini, L., Milzani, A., Di Simplicio, P., Colombo R. (2001). Actin carbonylation: from a simple marker of protein oxidation to relevant signs of severe functional impairment. *Free Radic Biol Med, 31*, 1075–1083.

186. Castegna, A., Aksenov, M., Aksenova, M., Thongboonkerd, V., Klein, J.B., Pierce, W.M., Booze, R., Markesbery, W.R., Butterfield, D.A. (2002). Proteomic identification of oxidatively modified proteins in Alzheimer's disease brain: I. Creatine kinase BB, glutamine synthase, and ubiquitin carboxy-terminal hydrolase L-1. *Free Radic Biol Med, 33*, 562–571.

187. Castegna, A., Aksenov, M., Thongboonkerd, V., Klein, J.B., Pierce, W.M., Booze, R., Markesbery, W.R., Butterfield, D.A. (2002). Proteomic identification of oxidatively modified proteins in Alzheimer's disease brain: II. dihydropyrimidinase-related protein 2, alpha-enolase and heat shock cognate 71. *J Neurochem, 82*, 1524–1532.

188. Castegna, A., Thongboonkerd, V., Klein, J.B., Lynn, B., Markesbery, W.R., Butterfield, D.A. (2003). Proteomic identification of nitrated proteins in Alzheimer's disease brain. *J Neurochem, 85*, 1394–1401.

189. Ghezzi, P., Bonetto, V. (2003). Redox proteomics: identification of oxidatively, modified proteins. *Proteomics, 3*, 1145–1153.

190. Butterfield, D.A., Boyd-Kimball, D., Castegna, A. (2003). Proteomics in Alzheimer's disease: insights into potential mechanisms of neurodegeneration. *J Neurochem, 86*, 1313–1327.

191. Butterfield, D.A. (2004). Proteomics: a new approach to investigate oxidative stress in Alzheimer's disease brain. *Brain Res, 1000*, 1–7.

192. Shin, S.J., Lee, S.E., Boo, J.H., Kim, M., Yoon, Y.D., Kim, S.I., Mook-Jung, I. (2004). Profiling proteins related to amyloid deposited brain of Tg2576 mice. *Proteomics, 4*, 3359–3368.

193. Pamplona, R., Dalfo, E., Ayala, V.R., Bellmunt, M.J., Prat, J., Ferrer, I., Portero-Otin, M. (2005). Proteins in human brain cortex are modified by oxidation, glycoxidation, and lipoxidation: effects of Alzheimer disease and identification of lipoxidation targets. *J Biol Chem, 280*, 21522–21530.

194. Butterfield, D.A., Reed, T., Newman, S.F., Sultana, R. (2007). Roles of amyloid beta-peptide-associated oxidative stress and brain protein modifications in the pathogenesis of Alzheimer's disease and mild cognitive impairment. *Free Radic Biol Med, 43*, 658–677.

195. Boyd-Kimball, D., Castegna, A., Sultana, R., Poon, H.F., Petroze, R., Lynn, B.C., Klein, J.B., Butterfield, D.A. (2005). Proteomic identification of proteins oxidized by A beta(1–42) in synaptosomes: implications for Alzheimer's disease. *Brain Res, 1044*, 206–215.

196. Liao, L., Cheng, D., Wang, J., Duong, D.M., Losik, T.G., Gearing, M., Rees, H.D., Lah, J.J., Levey, A.I., Peng, J. (2004). Proteomic characterization of postmortem amyloid plaques isolated by laser capture microdissection. *J Biol Chem, 279*, 37061–37068.

197. Gozal, P.M., Cheng, D.M., Duong, D.M., Lah, J.J., Levey, A.I., Peng, J.M. (2006). Merger of laser capture microdissection and mass spectrometry: a window into the amyloid plaque proteome. *Amyloid Prions Other Protein Aggregates, 412* (Pt B), 77–93.

198. Söderberg, L., Bogdanovic, N., Axelsson, B., Winblad, B., Näslund, J., Tjernberg, L.O. (2006). Analysis of single Alzheimer solid plaque cores by laser capture microscopy and nanoelectrospray/tandem mass spectrometry. *Biochemistry, 45*, 9849–9856.

199. Boyd-Kimball, D., Sultana, R., Poon, H.F., Lynn, B.C., Casamenti, F., Pepeu, G., Klein, J.B., Butterfield, D.A. (2005). Proteomic identification of proteins specifically oxidized by intracerebral injection of amyloid beta-peptide (1–42) into rat brain: implications for Alzheimer's disease. *Neuroscience, 132*, 313–324.

200. Perluigi, M., Poon, H.F., Hensley, K., Pierce, W.M., Klein, J.B., Calabrese, V., De Marco, C., Butterfield, D.A. (2005). Proteomic analysis of 4-hydroxy-2-nonenal-modified proteins in G93A-SOD1 transgenic mice: a model of familial amyotrophic lateral sclerosis. *Free Radic Biol Med, 38*, 960–968.

201. Butterfield, D.A., Reed, T., Perluigi, M., De Marco, C., Coccia, R., Cini, C., Sultana, R. (2006). Elevated protein-bound levels of the lipid peroxidation product, 4-hydroxy-2-nonenal, in brain from persons with mild cognitive impairment. *Neurosci Lett, 397*, 170–173.

202. Carini, M., Aldini, G., Facino, R.M. (2004). Mass spectrometry for detection of 4-hydroxy-*trans*-2-nonenal (HNE) adducts with peptides and proteins. *Mass Spectrom Rev, 23*, 281–305.

203. Huang, Y.D. (2006). Apolipoprotein E and Alzheimer disease. *Neurology, 66*, S79–S85.

204. Rosseneu, M., Labeur, C. (1995). Physiological significance of apolipoprotein mutants. *FASEB J, 9*, 768–776.

205. Corder, E.H., Saunders, A.M., Strittmatter, W.J., Schmechel, D.E., Gaskell, P.C., Small, G.W., Roses, A.D., Haines, J.L., Pericakvance, M.A. (1993). Gene dose of apolipoprotein E type 4 allele and the risk of Alzheimer's disease in late onset families. *Science, 261*, 921–923.

206. Schmechel, D.E., Saunders, A.M., Strittmatter, W.J., Crain, B.J., Huletter, C.M., Joo, S.H., Pericakvance, M.A., Goldgaber, D., Roses, A.D. (1993). Increased amyloid beta-peptide deposition in cerebral cortex as a consequence of apolipoprotein E genotype in late-onset Alzheimer disease. *Proc Natl Acad Sci USA, 90*, 9649–9653.

207. Saunders, A.M., Schmader, K., Breitner, J.C.S., Benson, M.D., Brown, W.T., Goldfarb, L., Goldgaber, D., Manwaring, M.G., Szymanski, M.H., McCown, N., et al. (1993). Apolipoprotein E epsilon 4 allele distributions in late-onset Alzheimer's disease and in other amyloid-forming diseases. *Lancet, 342*, 710–711.

208. Aizawa, Y., Fukatsu, R., Takamaru, Y., Tsuzuki, K., Chiba, H., Kobayashi, K., Fujii, N., Takahata, N. (1997). Amino-terminus truncated apolipoprotein E is the major species in amyloid deposits in Alzheimer's disease affected brains: a possible role for apolipoprotein E in Alzheimer's disease. *Brain Res, 768*, 208–214.

209. Permanne, B., Perez, C., Soto, C., Frangione, B., Wisniewski, T. (1997). Detection of apolipoprotein E dimeric soluble amyloid beta complexes in Alzheimer's disease brain supernatants. *Biochem Biophys Res Commun, 240*, 715–720.

210. Baumann, M.H., Kallijarvi, J., Lankinen, H., Soto, C., Haltia, M. (2000). Apolipoprotein E includes a binding site which is recognized by several amyloidogenic polypeptides. *Biochem J, 349*, 77–84.

211. Sanan, D.A., Weisgraber, K.H., Russell, S.J., Mahley, R.W., Huang, D., Saunders, A., Schmechel, D., Wisniewski, T., Frangione, B., Roses, A.D., Strittmatter, W.J. (1994). Apolipoprotein E associates with beta amyloid peptide of Alzheimer's disease to form novel monofibrils: isoform apoE4 associates more efficiently than apoE3. *J Clin Invest, 94*, 860–869.

212. Wisniewski, T., Castano, E.M., Golabek, A., Vogel, T., Frangione, B. (1994). Acceleration of Alzheimer's fibril formation by apolipoprotein E in vitro. *Am J Pathol, 145*, 1030–1035.

213. Ma, J.Y., Yee, A., Brewer, H.B., Das, S., Potter, H. (1994). Amyloid-associated proteins alpha 1-antichymotrypsin and apolipoprotein E promote assembly of Alzheimer beta-protein into filaments. *Nature, 372*, 92–94.

214. Evans, K.C., Berger, E.P., Cho, C.G., Weisgraber, K.H., Lansbury, P.T. (1995). Apolipoprotein E is a kinetic but not a thermodynamic inhibitor of amyloid formation: implications for the pathogenesis and treatment of Alzheimer disease. *Proc Natl Acad Sci USA, 92*, 763–767.

215. Weisgraber, K.H., Shinto, L.H. (1991). Identification of the disulfide-linked homodimer of apolipoprotein E3 in plasma. Impact on receptor binding activity. *J Biol Chem, 266*, 12029–12034.

216. Yang, D.S., Smith, J.D., Zhou, Z.M., Gandy, S.E., Martins, R.N. (1997). Characterization of the binding of amyloid-beta peptide to cell culture–derived native apolipoprotein E2, E3, and E4 isoforms and to isoforms from human plasma. *J Neurochem, 68*, 721–725.

217. Bales, K.R., Verina, T., Cummins, D.J., Du, Y.S., Dodel, T.C., Saura, J., Fishman, C.E., DeLong, C.A., Piccardo, P., Petegnief, V., et al. (1999). Apolipoprotein E is essential for amyloid deposition in the APP(V717F) transgenic mouse model of Alzheimer's disease. *Proc Natl Acad Sci USA, 96*, 15233–15238.

218. Bales, K.R., Verina, T., Dodel, R.C., Du, Y.S., Altstiel, L., Bender, M., Hyslop, P., Johnstone, E.M., Little, S.P., Cummins, D.J., et al. S.M. (1997). Lack of apolipoprotein E dramatically reduces amyloid beta-peptide deposition. *Nat Genet, 17*, 263–264.

219. Schneeweis, L.A., Koppaka, V., Lund-Katz, S., Phillips, M.C., Axelsen, P.H. (2005). Structural analysis of lipoprotein E particles. *Biochemistry, 44*, 12525–12534.

220. Namba, Y., Tomonaga, M., Kawasaki, H., Otomo, E., Ikeda, K. (1991). Apolipoprotein E immunoreactivity in cerebral amyloid deposits and neurofibrillary tangles in Alzheimer's disease and kuru plaque amyloid in Creutzfeldt–Jakob disease. *Brain Res, 541*, 163–166.

221. Strittmatter, W.J., Weisgraber, K.H., Huang, D.Y., Dong, L.M., Salvesen, G.S., Pericakvance, M., Schmechel, D., Saunders, A.M., Goldgaber, D., Roses, A.D. (1993). Binding of human apolipoprotein E to synthetic amyloid beta peptide: isoform-specific effects and implications for late-onset Alzheimer disease. *Proc Natl Acad Sci USA, 90*, 8098–8102.

222. Ladu, M.J., Falduto, M.T., Manelli, A.M., Reardon, C.A., Getz, G.S., Frail, D.E. (1994). Isoform-specific binding of apolipoprotein-E to beta-amyloid. *J Biol Chem, 269*, 23403–23406.

223. Ladu, M.J., Pederson, T.M., Frail, D.E., Reardon, C.A., Getz, G.S., Falduto, M.T. (1995). Purification of apolipoprotein E attenuates isoform-specific binding to beta-amyloid. *J Biol Chem, 270*, 9039–9042.

224. Aleshkov, S., Abraham, C.R., Zannis, V.I. (1997). Interaction of nascent ApoE2, ApoE3, and ApoE4 isoforms expressed in mammalian cells with amyloid peptide beta (1–40). Relevance to Alzheimer's disease. *Biochemistry, 36*, 10571–10580.

225. Manelli, A.M., Stine, W.B., Van Eldik, L.J., Ladu, M.J. (2004). ApoE and Abeta1–42 interactions: effects of isoform and conformation on structure and function. *J Mol Neurosci, 23*, 235–246.

226. Stratman, N.C., Castle, C.K., Taylor, B.M., Epps, D.E., Melchior, G.W., Carter, D.B. (2005). Isoform-specific interactions of human apolipoprotein E to an intermediate conformation of human Alzheimer amyloid-beta peptide. *Chem Phys Lipids, 137*, 52–61.

227. Miyata, M., Smith, J.D. (1996). Apolipoprotein E allele–specific antioxidant activity and effects on cytotoxicity by oxidative insults and beta-amyloid peptides. *Nat Genet, 14*, 55–61.

228. Tamaoka, A., Miyatake, F., Matsuno, S., Ishii, K., Nagase, S., Sahara, N., Ono, S., Mori, H., Wakabayashi, K., Tsuji, S., et al. (2000). Apolipoprotein E allele-dependent antioxidant activity in brains with Alzheimer's disease. *Neurology, 54*, 2319–2321.

229. Lauderback, C.M., Kanski, J., Hackett, J.M., Maeda, N., Kindy, M.S., Butterfield, D.A. (2002). Apolipoprotein E modulates Alzheimer's A beta(1–42)-induced oxidative damage to synaptosomes in an allele-specific manner. *Brain Res, 924*, 90–97.

230. Giles, G.I., Jacob, C. (2002). Reactive sulfur species: an emerging concept in oxidative stress. *Biol Chem, 383*, 375–388.

231. Wahrle, S.E., Jiang, H., Parsadanian, M., Hartman, R.E., Bales, K.R., Paul, S.M., Holtzman, D.M. (2005). Deletion of Abca1 increases A beta deposition in the PDAPP transgenic mouse model of Alzheimer disease. *J Biol Chem, 280*, 43236–43242.

232. Zerbinatti, C.V., Wahrle, S.E., Kim, H., Cam, J.A., Bales, K., Paul, S.M., Holtzman, D.M., Bu, G.J. (2006). Apolipoprotein E and low density lipoprotein receptor-related protein facilitate intraneuronal A beta 42 accumulation in amyloid model mice. *J Biol Chem, 281*, 36180–36186.

233. Bielicki, J.K., Oda, M.N. (2002). Apolipoprotein A-I-Milano and apolipoprotein A-I-Paris exhibit an antioxidant activity distinct from that of wild-type apolipoprotein A-I. *Biochemistry, 41*, 2089–2096.

234. Lim, G.P., Chu, T., Yang, F.S., Beech, W., Frautschy, S.A., Cole, G.M. (2001). The curry spice curcumin reduces oxidative damage and amyloid pathology in an Alzheimer transgenic mouse. *J Neurosci, 21*, 8370–8377.

235. Sung, S., Yao, Y.M., Uryu, K., Yang, H.X., Lee, V.M.Y., Trojanowski, J.Q., Pratico, D. (2003). Early Vitamin E supplementation in young but not aged mice reduces A beta levels and amyloid deposition in a transgenic model of Alzheimer's disease. *FASEB J, 17*, 323–+.

236. Conte, V., Uryu, K., Fujimoto, S., Yao, Y., Rokach, J., Longhi, L., Trojanowski, J.Q., Lee, V.M.Y., McIntosh, T.K., Pratico, D. (2004). Vitamin E reduces amyloidosis and improves cognitive function in Tg2576 mice following repetitive concussive brain injury. *J Neurochem, 90*, 758–764.

237. Yang, F.S., Lim, G.P., Begum, A.N., Ubeda, O.J., Simmons, M.R., Ambegaokar, S.S., Chen, P.P., Kayed, R., Glabe, C.G., Frautschy, S.A., Cole, G.M. (2005). Curcumin inhibits formation of amyloid beta oligomers and fibrils, binds plaques, and reduces amyloid in vivo. *J Biol Chem, 280*, 5892–5901.

238. Janus, C., Pearson, J., McLaurin, J., Mathews, P.M., Jiang, Y., Schmidt, S.D., Chishti, M.A., Horne, P., Heslin, D., French, J., et al. (2000). A beta peptide immunization reduces behavioural impairment and plaques in a model of Alzheimer's disease. *Nature, 408*, 979–982.

239. Klyubin, I., Walsh, D.M., Lemere, C.A., Cullen, W.K., Shankar, G.M., Betts, V., Spooner, E.T., Jiang, L.Y., Anwyl, R., Selkoe, D.J., Rowan, M.J. (2005). Amyloid beta protein immunotherapy neutralizes A beta oligomers that disrupt synaptic plasticity in vivo. *Nat Med, 11*, 556–561.

240. Lim, G.P., Calon, F., Morihara, T., Yang, F.S., Teter, B., Ubeda, O., Salem, N., Frautschy, S.A., Cole, G.M. (2005). A diet enriched with the omega-3 fatty acid docosahexaenoic acid reduces amyloid burden in an aged Alzheimer mouse model. *J Neurosci, 25*, 3032–3040.

241. Calon, F., Lim, G.P., Morihara, T., Yang, F.S., Ubeda, O., Salem, N., Frautschy, S.A., Cole, G.M. (2005). Dietary *n*-3 polyunsaturated fatty acid depletion activates caspases and decreases NMDA receptors in the brain of a transgenic mouse model of Alzheimer's disease. *Eur J Neurosci, 22*, 617–626.

242. Bohrmann, B., Adrian, M., Dubochet, J., Kuner, P., Muller, F., Huber, W., Nordstedt, C., Dobeli, H. (2000). Self-assembly of beta-amyloid 42 is retarded by small molecular ligands at the stage of structural intermediates. *J Struct Biol, 130*, 232–246.

243. McLaurin, J., Kierstead, M.E., Brown, M.E., Hawkes, C.A., Lambermon, M.H.L., Phinney, A.L., Darabie, A.A., Cousins, J.E., French, J.E., Lan, M.F., et al. (2006). Cyclohexanehexol inhibitors of A beta aggregation prevent and reverse Alzheimer phenotype in a mouse model. *Nat Med, 12*, 801–808.

244. Townsend, M., Cleary, J.P., Mehta, T., Hofmeister, J., Lesne, S., O'Hare, E., Walsh, D.M., Selkoe, D.J. (2006). Orally available compound prevents deficits in memory caused by the Alzheimer amyloid-beta oligomers. *Ann Neurol, 60*, 668–676.

245. Tjernberg, L.O., Naslund, J., Lindqvist, F., Johansson, J., Karlstrom, A.R., Thyberg, J., Terenius, L., Nordstedt, C. (1996). Arrest of beta-amyloid fibril formation by a pentapeptide ligand. *J Biol Chem, 271*, 8545–8548.

246. Tjernberg, L.O., Lilliehook, C., Callaway, D.J.E., Naslund, J., Hahne, S., Thyberg, J., Terenius, L., Nordstedt, C. (1997). Controlling amyloid beta-peptide fibril formation with protease-stable ligands. *J Biol Chem, 272*, 12601–12605.

247. Soto, C., Sigurdsson, E.M., Morelli, L., Kumar, R.A., Castano, E.M., Frangione, B. (1998). Beta-sheet breaker peptides inhibit fibrillogenesis in a rat brain model of amyloidosis: implications for Alzheimer's therapy [see comments]. *Nat Med, 4*, 822–826.

248. Findeis, M.A., Musso, G.M., Arico-Muendel, C.C., Benjamin, H.W., Hundal, A.M., Lee, J.J., Chin, J., Kelley, M., Wakefield, J., Hayward, N.J., Molineaux, S.M. (1999). Modified-peptide inhibitors of amyloid beta-peptide polymerization. *Biochemistry, 38*, 6791–6800.

249. Pallitto, M.M., Ghanta, J., Heinzelman, P., Kiessling, L.L., Murphy, R.M. (1999). Recognition sequence design for peptidyl modulators of beta-amyloid aggregation and toxicity. *Biochemistry, 38*, 3570–3578.

250. Lowe, T.L., Strzelec, A., Kiessling, L.L., Murphy, R.M. (2001). Structure–function relationships for inhibitors of beta-amyloid toxicity containing the recognition sequence KLVFF. *Biochemistry, 40*, 7882–7889.

251. Watanabe, K., Nakamura, K., Akikusa, S., Okada, T., Kodaka, M., Konakahara, T., Okuno, H. (2002). Inhibitors of fibril formation and cytotoxicity of beta-amyloid peptide composed of KLVFF recognition element and flexible hydrophilic disrupting element. *Biochem Biophys Res Commun, 290*, 121–124.

252. Gordon, D.J., Tappe, R., Meredith, S.C. (2002). Design and characterization of a membrane permeable *N*-methyl amino acid–containing peptide that inhibits A beta(1–40) fibrillogenesis. *J Pept Res, 60*, 37–55.

253. Gibson, T.J., Murphy, R.M. (2005). Design of peptidyl compounds that affect beta-amyloid aggregation: importance of surface tension and contex. *Biochemistry, 44*, 8898–8907.

254. Rzepecki, P., Schrader, T. (2005). Beta-sheet ligands in action: KLVFF recognition by aminopyrazole hybrid receptors in water. *J Am Chem Soc, 127*, 3016–3025.

255. Apelt, J., Bigl, M., Wunderlich, P., Schliebs, R. (2004). Aging-related increase in oxidative stress correlates with developmental pattern of beta-secretase activity

and beta-amyloid plaque formation in transgenic Tg2576 mice with Alzheimer-like pathology. *Int J Dev Neurosci, 22*, 475–484.

256. Manczak, M., Anekonda, T.S., Henson, E., Park, B.S., Quinn, J., Reddy, P.H. (2006). Mitochondria are a direct site of A beta accumulation in Alzheimer's disease neurons: implications for free radical generation and oxidative damage in disease progression. *Hum Mol Genet, 15*, 1437–1449.
257. Rosales-Corral, S., Tan, D.X., Reiter, R.J., Valdivia-Velázquez, M., Acosta-Martinez, J.P., Ortiz, G.G. (2004). Kinetics of the neuroinflammation-oxidative stress correlation in rat brain following the injection of fibrillar amyloid-beta onto the hippocampus in vivo. *J Neuroimmunol, 150*, 20–28.
258. Wyss-Coray, T. (2006). Inflammation in Alzheimer disease: Driving force, bystander or beneficial response? *Nat Med, 12*, 1005–1015.
259. Zipp, F., Aktas, O. (2006). The brain as a target of inflammation: common pathways link inflammatory and neurodegenerative diseases. *Trends Neurosci, 29*, 518–527.
260. Firuzi, O., Pratico, D. (2006). Coxibs and Alzheimer's disease: Should they stay or should they go? *Ann Neurol, 59*, 219–228.
261. Sultana, R., Ravagna, A., Mohammad-Abdul, H., Calabrese, V., Butterfield, D.A. (2005). Ferulic acid ethyl ester protects neurons against amyloid beta-peptide(1–42)-induced oxidative stress and neurotoxicity: relationship to antioxidant activity. *J Neurochem, 92*, 749–758.
262. Perluigi, M., Joshi, G., Sultana, R., Calabrese, V., De Marco, C., Coccia, R., Cini, C., Butterfield, D.A. (2006). In vivo protective effects of ferulic acid ethyl ester against amyloid-beta peptide 1–42-induced oxidative stress. *J Neurosci Res, 84*, 418–426.
263. Kondo, M., Oya-Ito, T., Kumagai, T., Osawa, T., Uchida, K. (2001). Cyclopentenone prostaglandins as potential inducers of intracellular oxidative stress. *J Biol Chem, 276*, 12076–12083.
264. Nishida, Y., Yokota, T., Takahashi, T., Uchihara, T., Jishage, K.I., Mizusawa, H. (2006). Deletion of vitamin E enhances phenotype of Alzheimer disease model mouse. *Biochem Biophys Res Commun, 350*, 530–536.
265. Meagher, E.A., Barry, O.P., Lawson, J.A., Rokach, J., FitzGerald, G.A. (2001). Effects of vitamin E on lipid peroxidation in healthy persons. *JAMA, 285*, 1178–1182.
266. Petersen, R.C., Thomas, R.G., Grundman, M., Bennett, D., Doody, R., Ferris, S., Galasko, D., Jin, S., Kaye, J., Levey, A., et al., Alzheimer's Disease Cooperative Study Group. (2005). Vitamin E and donepezil for the treatment of mild cognitive impairment. *N Engl J Med, 352*, 2379–2388.
267. Bayer, T.A., Schafer, S., Breyhan, H., Wirths, O., Treiber, C., Multhaup, G. (2006). A vicious circle: role of oxidative stress, intraneuronal A beta and Cu in Alzheimer's disease. *Clin Neuropathol, 25*, 163–171.
268. Standridge, J.B. (2006). Vicious cycles within the neuropathophysiologic mechanisms of Alzheimer's disease. *Curr Alzheimer Res, 3*, 95–107.
269. Jolivalt, C., Leininger-Muller, B., Bertrand, P., Herber, R., Christen, Y., Siest, G. (2000). Differential oxidation of apolipoprotein E isoforms and interaction with phospholipids. *Free Radic Biol Med, 28*, 129–140.

270. Halliwell, B., Gutteridge, J.M.C. (1999). *Free Radicals in Biology and Medicine*, 3rd ed., Oxford University Press, Oxford, UK.

271. Rice, M.E. (2000). Ascorbate regulation and its neuroprotective role in the brain. *Trends Neurosci, 23*, 209–216.

272. Atwood, C.S., Moir, R.D., Huang, X., Scarpa, R.C., Bacarra, N.M., Romano, D.M., Hartshorn, M.A., Tanzi, R.E., Bush, A.I. (1998). Dramatic aggregation of Alzheimer Abeta by Cu(II) is induced by conditions representing physiological acidosis. *J Biol Chem, 273*, 12817–12826.

28

ROLE OF OXIDATIVE STRESS IN PROTEIN MISFOLDING AND/OR AMYLOID FORMATION

JOHANNA C. SCHEINOST, DANIEL P. WITTER, AND GRANT E. BOLDT

Department of Biochemistry, The Scripps–Oxford Laboratory, University of Oxford, Oxford, UK

PAUL WENTWORTH, JR.

Department of Biochemistry, The Scripps–Oxford Laboratory, University of Oxford, Oxford, UK; Department of Chemistry, The Scripps Research Institute, La Jolla, California

INTRODUCTION

Although considerable knowledge of the pathology, onset, and progression of protein misfolding or amyloid-based syndromes, such as Alzheimer disease (AD) and Parkinson disease, has been obtained from the familial incidence of such diseases, such information is limited clinically because genetic predisposition typically constitutes less than 10% of the cases observed. The vast majority of protein misfolding diseases are sporadic in nature and, as such, are associated with amyloidogenesis of native polypeptides. In such cases, environmental factors are being considered as the trigger(s) for disease onset. Thus, either the local environment of the protein may be different in affected persons or an unusual posttranslational modification of the protein may lead to misfolding. It is the determination of these factors that are at the forefront of our understanding of how to slow or prevent progression of the sporadic protein amyloidoses.

Protein Misfolding Diseases: Current and Emerging Principles and Therapies,
Edited by Marina Ramirez-Alvarado, Jeffery W. Kelly, and Christopher M. Dobson
Copyright © 2010 John Wiley & Sons, Inc.

Oxidative stress arises when the body has a compromised ability to destroy reactive oxygen species (ROS) that are generated during normal in vivo metabolic processes. Human beings have evolved a number of enzymatic systems by which these ROS are detoxified, such as superoxide dismutase (SOD), catalase, glutathione peroxidase, and thioredoxin reductase. Thus, a delicate balance exists between the body's utilization of oxygen and its ability to defend against the damage produced by the by-products of this process, and it is this balance that becomes perturbed as aging progresses. Any cellular damage that occurs as the result of an imbalance in these processes is referred to globally as *oxidative stress*.

Oxidative stress results in a number of chemical consequences, including DNA damage [1], lipid [2] and protein modification [3], and ultimately cell death, via both necrosis [4,5] and apoptosis [6]. Critically, numerous studies have linked oxidative stress to age-related disorders [7–12], and oxidative stress is considered a major contributory factor in aging and age-related diseases. This focused review details a new aspect of how ROS are being linked to misfolding and amyloid diseases. Specifically, we will focus on how ROS-generated lipid peroxidation products are being studied as potential triggers of protein misfolding and amyloidogenesis.

ROS are generated primarily during the biological processes of oxidative phosphorylation in mitochondria and the oxidative burst of activated phagocytes [13], and can also be produced by exposure to ionizing radiation [14] (Table 1). Nitric oxide (•NO) and its derivatives, reactive nitrogen species (RNS), are generated by NO-synthases and subsequent chemical reduction/reaction. The cascade of oxidative damage begins initially with the single electron reduction of ground-state oxygen (which is a triplet biradical $^3O_2$) to form superoxide anion ($\bullet O_2^-$) which is then further reduced (one electron) to form hydrogen peroxide (HOOH). Hydrogen peroxide may then be converted into hypohalous acids (HOCl/HOBr) by the action of leukocyte myeloperoxidase enzymes, or reduced further (Fenton reduction) to form the hydroxyl radical (•OH). In addition, HOOH has been shown to react with hypohalous acids to generate the short-lived low-energy singlet state of molecular oxygen ($^1O_2$) in biological systems.

All these reactive species interact with neighboring biomolecules directly, but their damage is ultimately increased by the toxicity associated with stable products of their reaction with these biomolecules, such as 8-*oxo*-guanine (generated by nucleic acid oxidation) or 4-hydroxynonenal (4-HNE) (generated by oxidation of esterified or free unsaturated fatty acids) [15–17].

Whether by interaction with free radicals or with lipid peroxidation products, proteins are prime targets for oxidative damage. Direct oxidation of proteins can result in irreparable protein carbonylation, a modification closely linked with aging [18,19]. In addition to carbonylation, or because of it, protein oxidation can lead to a variety of pathological consequences: enhanced proteolysis of the oxidized protein [20,21], interruption of signaling pathways [22], decreased enzyme activity [23–25], and misfolding of proteins [26,27], which can lead to the formation of amyloid.

TABLE 1 Biologically Relevant Reactive Oxygen and Nitrogen Species

Hydroxyl radical	•OH
Hydrogen peroxide	HOOH
Hypohalous acids	HOCl/HOBr
Nitric oxide/nitroxyl anion	•NO/−NO
Peroxynitrite	ONO_2^-
Singlet dioxygen	$^1O_2$
Superoxide anion/hydroperoxyl radical	$•O_2^-/•O_2H$

LIPID-DERIVED ALDEHYDES: CHEMICAL AND BIOLOGICAL ORIGINS

Lipid peroxidation in biological systems is always accompanied by formation of aldehyde by-products. Volatile aldehydes formed by autoxidation processes in fats, oils, and foods have long intrigued food chemists because they are intrinsic to the taste and aroma of food and beverages. However, more insidiously it has become evident that some of these aldehydic lipid peroxidation products are biologically very active and can produce a number of pathological effects in vivo. The reactivity of such aldehydes when generated in vivo is so high that it has been suggested that such lipidic aldehydes should not be seen as end products of lipid peroxidation processes but, rather, as "second toxic messengers" for the primary free radicals that spawn them. For example, 4-HNE, generated from the peroxidation of the ω-6 polyunsaturated fatty acids linoleic and arachadonic acid, is cytotoxic, genotoxic, mutagenic, and hepatotoxic [28] (Fig. 1). 4-HNE is the product of oxidation of the polyunsaturated fatty acid chains, and it has been widely studied ever since for its cytotoxic effects [29–34] and more recently by us for its ability to induce protein misfolding [35–39]. Clinically, 4-HNE has been associated with both Alzheimer and Parkinson diseases [40]. This aldehyde exists at 0.1 to 1.0 mM under normal circumstances [15], but its levels are elevated in the brain of persons with Alzheimer disease relative to age-matched controls [41], and it has been detected histochemically in Alzheimer amyloid plaques [42] and parkinsonian Lewy bodies [43,44].

Recently, we have discovered a new family of lipid aldehydes in vivo that are derived from cholesterol oxidation [45]. These cholesterol *seco*-sterol aldehydes, termed the *atheronals*, are possible mediators of the known link among inflammation, hypercholesterolemia, and atherosclerosis progression (Fig. 1). The atheronals are present in the plaque material and plasma of patients with atherosclerosis [45]. The origin of these cholesterol oxidation products, whether chemical, biochemical, or environmental, is still unknown; however, it has been shown that ex vivo activation of residual macrophages within human atheroma leads to a significant increase in the levels of these *seco*-sterols ($p < 0.05$). The formation of these cholesterol metabolites has been proposed to be linked to the antibody-catalyzed water oxidation pathway (ACWOP), which itself is

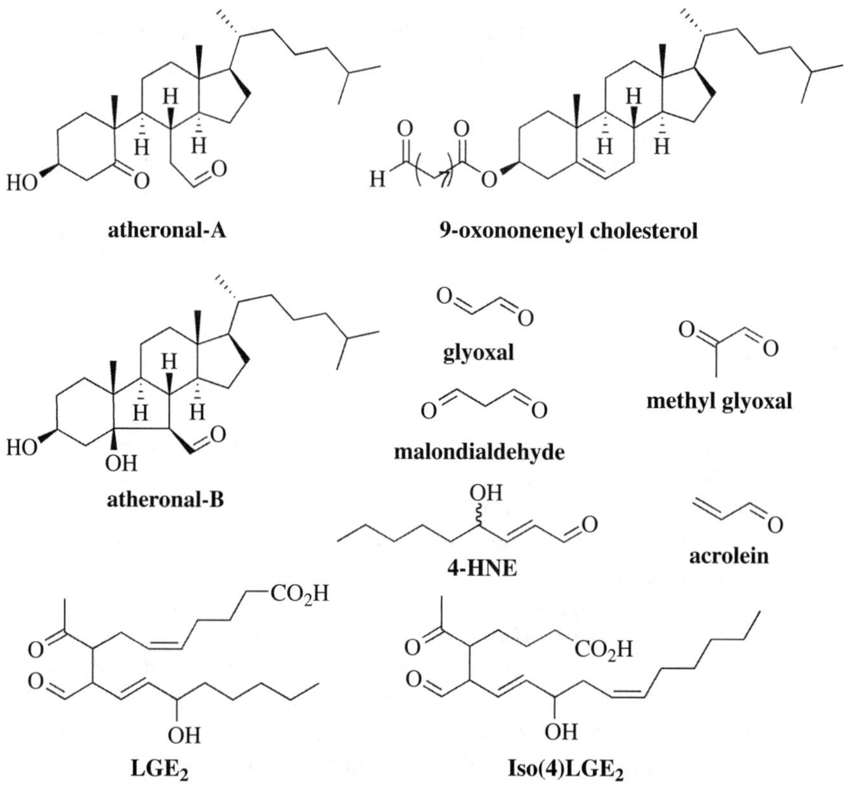

FIG. 1 Various biologically relevant lipid oxidation products.

activated by singlet dioxygen ($^1O_2$), a reactive oxygen species generated by activated leukocytes, such as polymorphonuclear neutrophils (PMNs) and monocytes [45–48].

The main mechanism of formation of aldehydes from lipid hydroperoxides is hemolytic scission (β-cleavage) of the two C–C bonds adjacent to the nascent alkyl hydroperoxy group generated by oxidation of the double bond with, for example, superoxide anion. This scission reaction proceeds via a lipid alkoxl radical and is accelerated strongly by traces of reduced forms of transition metal ions such as Fe^{2+}, Cu^{+}, and Co^{2+}. The principal polyunsaturated fatty acids (PUFA) in mammalian tissues and cells are linoleic acid (18 : 2), arachidonic acid (20 : 4), and docosahexanoic acid (22 : 6). To grasp a sense of the number and diversity of lipidic aldehydes that occur in vivo, when lipid peroxidation of these acids occurs, they are converted into their positional isomeric hydroperoxides, the number (n) of which is given by $2x - 2$, where x is the number of double bonds in the fatty acid. Therefore, two, six, and 10 different hydroperoxides result from 18 : 2, 20 : 4, and 22 : 6, respectively. Upon β-cleavage, each of these hydroperoxides gives an aldehyde with either a methyl or a carboxylate terminus. The aldehydes linked to the carboxylates are important, because

PUFAs in biological systems are bound to phospholipids, cholesterol, and triglycerides. Thus, when the aldehydes are on the same fragment as the carboxylate, is the structural diversity multiplied when complexed with such biological headgroups.

LIPID ALDEHYDES AND PROTEIN MISFOLDING OR AMYLOIDOGENESIS

The amyloid cross-β-sheet quaternary structures, found in amyloidosis patients, represent an alternative intermolecular fold available to peptides and proteins that are either natively unfolded or folded. Historically, these cross-β-sheet structures, wherein the peptide chains are oriented perpendicular to the long axis of the fibril, have been characterized either by low-resolution analytical methods such as Fourier transfer–infrared and x-ray diffraction (4.7- and 10-Å lattice spacing) [49] or more grossly by their resistance to sodium dodecyl sulfate denaturation, selective Congo Red binding, and green birefringence [typically used for thick (>1 μm) protein deposits] [50].

The ability of lipid-derived aldehydes to induce protein misfolding, first shown by us to affect apoB_{100} misfolding in low-density lipoprotein particles in the context of atherosclerosis, is now a rapidly growing field of study and has been linked to the misfolding of proteins involved in Lewy body dementia, Alzheimer disease, and most recently, antibody light-chain (LC) amyloidosis [51–54]. The amyloidoses are all age-related diseases and often share the risk factors of oxidative stress and a protein that is known to misfold; thus, the link between lipid oxidation products and protein aggregation may have clinical relevance in such processes. We consider that the lipid aldehydes induce misfolding of proteins by an initial covalent adduction followed by a "facilitated" nucleation process that may ultimately lead to the production of the entire family of prefibrillar and fibrillar amyloid forms [55]. According to the amyloid hypothesis of disease, amyloidogenesis occurs as a series of steps. The aggregating protein undergoes nucleation or seeding, after which it forms small, round, oligomeric protofibrils, which typically remain soluble. These protofibrillar seeds induce further aggregation until long mature fibrils are formed. These insoluble fibrils are the primary component of amyloid and are the insoluble proteinaceous deposits that form in various amyloid diseases. These amyloid deposits are also referred to as plaques. Misfolded proteins characteristically exhibit β-sheet structure at every stage of the amyloid cascade. Toxicity has been linked to protein aggregates at both stages prefibrillar oligomers and amyloid plaques [3,56]. Thus, the nature and cause of the misfolding are of considerable interest.

Mechanistically, 4-HNE and atheronals are considered to behave differently in their ability to modify proteins covalently. Both undergo initial Schiff base formation between the aldehyde moiety of the lipid aldehyde and one of the amino groups of the protein, either the ε-amino group of a lysine side chain, or the α-amino group of the N-terminal amino acid. It is thought that the local change in hydrophobicity caused by this adduction is sufficient to induce conformational changes in the protein that direct misfolding (Fig. 2A). However, in addition to forming a Schiff base, 4-HNE can also undergo 1,4-conjugate addition with the nucleophilic side chains of histidine or cysteine [57] (Fig. 2B, and C). Because of this additional reactivity, 4-HNE has the potential to cause the cross-linking of proteins both intramolecularly and intermolecularly [35] (Fig. 2D).

The nucleation and aggregation of Aβ peptides, Aβ(1–40) and Aβ(1–42), into neurotoxic oligomers is considered a primary event in AD pathogenesis [58,59]. The 40-amino acid peptide, Aβ(1–40) comprises about 90% of the Aβ found in vivo with Aβ(1–42) making up the majority of the remainder. With two extra hydrophobic amino acids added to an already hydrophobic C-terminus, Aβ(1–42) is far more prone to aggregation and is thought to be the primary species involved in initiating the amyloid cascade [60]. As with most protein misfolding diseases, some cases (ca.10%) of AD can be traced back to genetic factors (i.e., mutations in the amyloid precursor protein or processing proteins) [61,62]. However, the vast majority of AD occurs sporadically (>85%) [63]. Therefore, intensive research is dedicated to classify these in vivo triggers of AD onset that ultimately can be traced back to factors that facilitate nucleation and aggregation of native Aβ peptide. Numerous risk factors for AD have been identified, including atherosclerosis [64], hypercholesterolemia [65], lipid peroxidation [66,67], and oxidative stress [68].

The ability of lipid-derived aldehydes such as 4-HNE and the atheronals to induce the aggregation of proteins has been studied most widely with the amyloid-β peptide (Aβ). As described above, a possible link between 4-HNE and AD has long been thought to exist [69]. Levels of 4-HNE are elevated significantly in the brains of AD patients [41], and immunohistochemical analysis has revealed 4-HNE in AD plaques [42]. 4-HNE has also been shown to bind Aβ covalently via Schiff base formation [39].

Siegel et al. [70] have studied the nature of the interaction between 4-HNE and examined the mechanisms behind the binding of 4-HNE to Aβ(1–40) and the nature of the aggregates that are formed. Samples of Aβ(1–40) were incubated with 4-HNE and the aggregates analyzed by the environment-sensitive fluorophore thioflavin T (ThT). It was observed that the initial rate of HNE-induced amyloidogenesis is proportional to the concentration of 4-HNE in the sample. Atomic force microscopy (AFM) analysis of 4-HNE-adducted Aβ aggregates has shown them to be small and round, resembling prefibrillar amyloid forms. A combination of MALDI-TOF mass spectroscopy and dot blot analysis, using an antibody to 4-HNE-protein adducts, was performed to analyze the interaction between the aldehyde and protein. These studies revealed that 1,4-conjugate (Michael) addition of 4-HNE to Aβ is occurring, followed by putative interpeptide cross-linking via Schiff base formation. As the nature of the Aβ aggregates that give rise to neurotoxicity is currently under speculation, the formation of prefibrillar aggregates by 4-HNE-Aβ adducts without the subsequent formation of long, straight fibrils is potentially a very important observation.

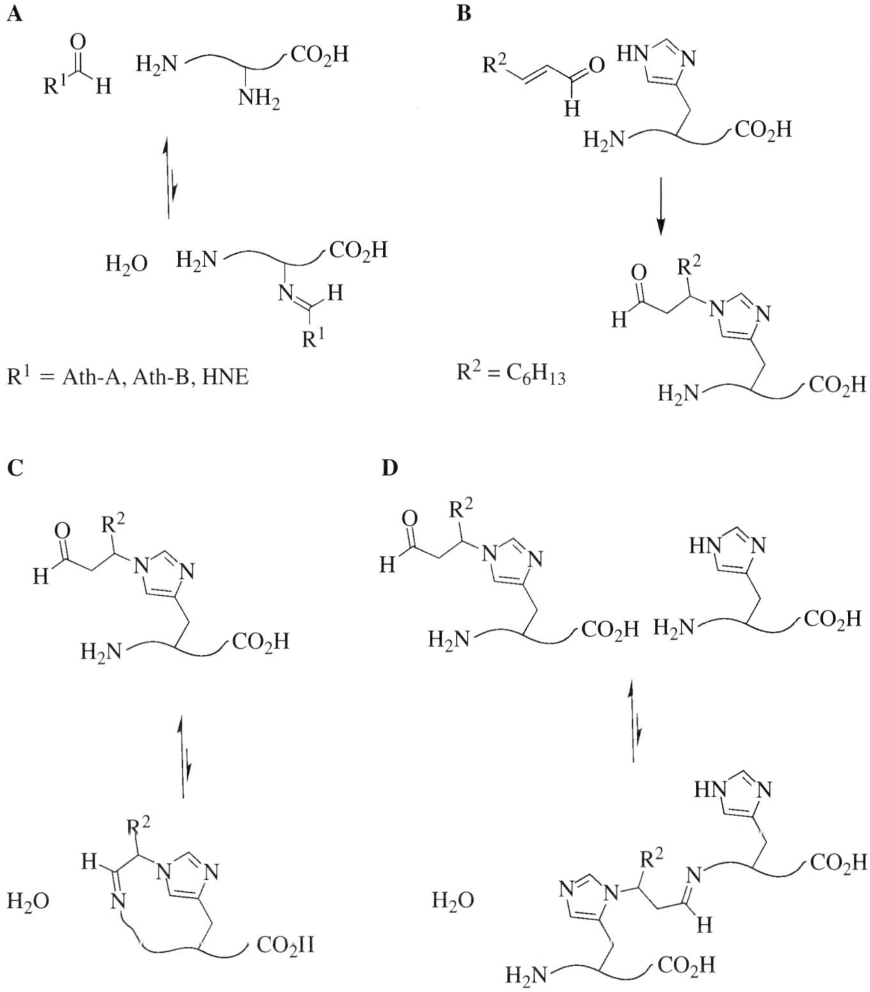

FIG. 2 (A) Schiff base formation between aldehyde (4-HNE or atheronals) and a peptide amino group (N-terminus or side chain); (B) 1,4-conjugate addition of 4-HNE to histidine; (C) intramolecular cross-linking: 1,4-conjugate addition of 4-HNE to histidine followed by Schiff base formation with an amino group of the same peptide; (D) intermolecular cross-linking: 1,4-conjugate addition of 4-HNE to histidine followed by Schiff base formation with an amino group of a second peptide.

A subsequent study by Liu et al. [71] investigated how adduction of 4-HNE to Aβ results in misfolding and where this covalent binding occurs. In contrast to the preceding study, the association of 4-HNE and Aβ was examined on or in the vicinity of synthetic lipid membranes. Thus, 4-HNE-induced Aβ(1–40) and Aβ(1–42) fibrillization was studied on monolayer lipid membranes. FTIR revealed that 1,4-conjugate addition of 4-HNE occurs at all three histidines of Aβ(1–42) and that this adduction greatly increases the protein's hydrophobicity.

It is this elevation in hydrophobicity that leads to fibrillar aggregation of Aβ(1–40) on monolayer lipid membranes. Binding of Congo Red to 4-HNE-induced aggregates of both Aβ(1–40) and Aβ(1–42) in lipid bilayers was then studied. This study revealed that even at subphysiological levels of 4-HNE, this aldehyde had a significant effect on the initiation of aggregation of both Aβ(1–40) and Aβ(1–42). Finally, surface plasmon resonance studies revealed that 4-HNE binding of Aβ probably occurs in solution rather than on the lipid membrane, but once bound together, the 4-HNE-Aβ adduct has increased affinity for neutral lipid membranes, where it initiates the formation of amyloid fibrils. Evidence from this study supported the mechanism proposed in the Siegel study: namely, that 4-HNE and Aβ are initially joined together by 1,4-conjugate addition on the Aβ side chain followed by Schiff base formation and cross-linking of proteins. Interestingly, however, the evidence evinced for cross-linking was more suggestive of intrapeptide binding than of interpeptide binding. The fibrillar nature of the aggregates in this study is also somewhat different from the prefibrillar structure observed by Siegel et al. This is probably due to differences in the in vitro experimental design, as the conformation that peptides adapt during folding or misfolding is highly dependent on the surrounding conditions [72].

Incubation of seed-free Aβ(1–40) and Aβ(1–42) with the cholesterol *seco*-sterols atheronal-A and atheronal-B dramatically increases the propensity of Aβ(1–40) and Aβ(1–42) to misfold. The aggregation kinetics of Aβ(1–40) and Aβ(1–42) change from a nucleated polymerization to a thermodynamically favored downhill polymerization in the presence of the atheronals [39] (Fig. 3A). AFM data of Aβ incubated in the presence of atheronals without agitations reveals that condensed protofibrils are formed, not full-length fibrils. However, if agitation occurs during the aggregation protocol or 1% of fibrillar seeds are added to the protofibrils formed by atheronal-induced aggregates of Aβ, full-length fibrils are formed (Fig. 3B). The acceleration process occurs in an aldehyde concentration-dependent manner and reduces the critical concentration of Aβ(1–40) needed for oligomerization to <90 nM, more than 100-fold lower than the critical concentration for Aβ alone (10 to 40 μM) [39,73]. This remarkable decrease in the critical concentration was a first indication as to how the presence of atheronals could trigger misfolding of the native peptide in sporadic AD at physiological concentrations of Aβ.

Atheronal concentrations in the frontal cortex tissue of AD patients are not elevated significantly to persons with no AD histopathology (0.44 pmol/mg in AD patients; 0.35 pmol/mg in controls) [74]. However, it is rationalized that a transient local increase in cholesterol *seco*-sterols may be sufficient to initiate the aggregation of Aβ, and in such a case the total cortex levels may not reflect this change. An intriguing aspect of atheronal-induced misfolding of Aβ peptides is that the reaction can be traceless. Atheronals initiate only in the early stages of amyloidogenesis (i.e., the energetically unfavored formation of seeds); subsequent propagation of fibril formation is fast and does not require further metabolite interaction.

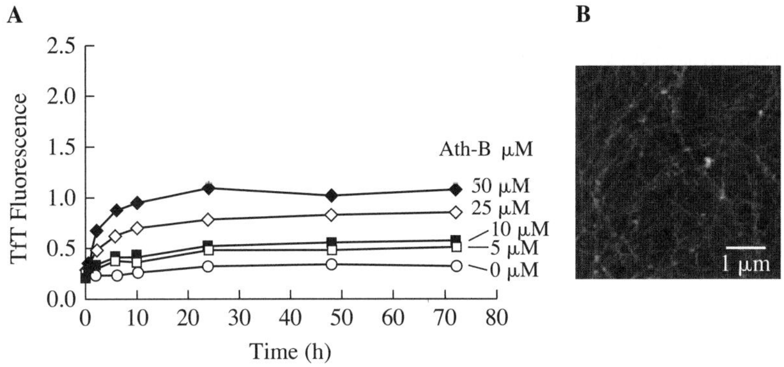

FIG. 3 (A) Concentration-dependent instantaneous acceleration of Aβ(1–40) misfolding in the presence of atheronal-B (3). In the absence of Ath-B, no detectable aggregation of Aβ is observed within the time shown. (B) Fibrils formed from Aβ with atheronal-B upon agitation adopt a classic amyloidlike network of fibrils as observed by AFM. (From [39], with permission. Copyright © Wiley-VCH Verlag GmbH & Co. KGaA.)

An important question regarding lipid–aldehyde-induced protein misfolding is the nature of the interaction between the aldehyde and the protein and how this association facilitates protein aggregation [35]. As described above, it is expected that the aldehyde moiety of the atheronals reacts with the primary amines of the peptide to form an initial Schiff base (Fig. 2). High-performance liquid chromatography and mass spectrometry analysis revealed that the nature of the interaction between atheronal-B and Aβ(1–40) is indeed covalent [39]. Aβ incubated in the presence of atheronal-B was treated with $NaBH_4$, and the Aβ aggregates contained a single modification. Moreover, small aldehyde scavenging drugs such as aminoguanidine and carnosine were shown to reduce the effect of atheronals at low concentrations on Aβ peptides, confirming the importance of the aldehyde group and the resulting covalent interactions between the peptide and the metabolite [75].

To investigate whether adduction of atheronals to any of the three primary amines present on the amyloid-β peptide is sufficient to trigger protein aggregation, or whether adduction to a specific locus is required, Scheinost et al. [76] have used a panel of synthetic mono-, bis-, and tris-*N*,*N*-dimethylamine-containing Aβ(1–40) protein sequences **2b** to **2f** (Fig. 4). These peptides are characterized by the presence of one to three tertiary amines replacing primary amines at the side chains of Lys16 and Lys28 and at the N-terminal amine which are present in native Aβ(1–40) (**2a**), therefore rendering the peptide unable to form the critical, putative site-specific Schiff base with aldehydes. Kinetic analyses using ThT fluorescence and far-ultraviolet circular dichroism (CD) reveal that no initiation in oligomerization is observed when atheronal-B adducts to either the ε-amino group of Lys28 or the α-amino group amine of Asp1. However, there is clear acceleration in fibrillization of Aβ by atheronal-B if the ε-amino group of Lys16 is available for covalent modification. This observation is the most dramatic in the case of peptide **2b** (K*16) in which the ε-amino group of Lys-28 and the α-amino group of Asp1 are both available for adduction, but there is no initiation in fibrillization upon incubation of 2b with atheronal-B (Fig. 4D and J).

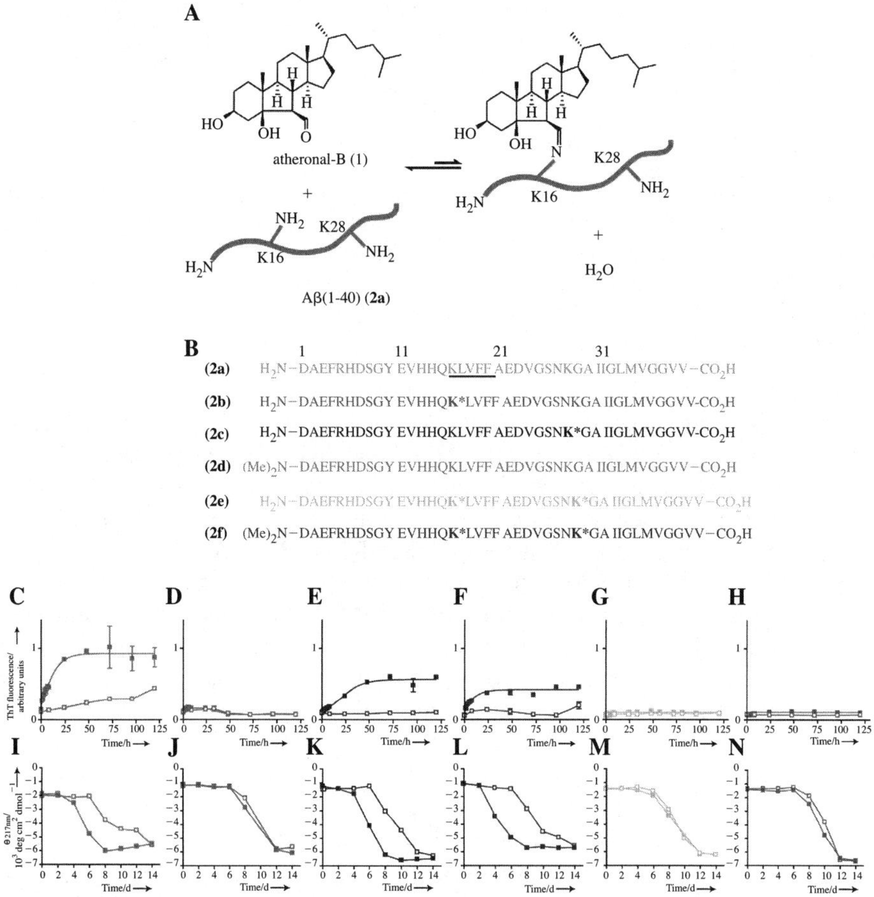

FIG. 4 (A) Schiff base formation between atheronal-B and primary amines on Aβ; (B) sequences of mono-, bis-, and tris-dimethylated peptide analogs (**2b** to **2f**) to native Aβ (1–40) (**2a**); (C–N) kinetics of atheronal-B-induced aggregation of amyloid-β peptides **2a** to **2f**. (C–H) ThT analyses, ex: 440 nm and em: 485 nm reported as mean ± SD; (I–N) far-UV CD analyses (reported as an average of three scans) of mean residue ellipticity (Θ) at 217 nm, of peptides **2a** (green, wild-type); **2b** (red, **K*16**); **2c** (blue, **K*28**); **2d** (purple, Me$_2$N-D1); **2e** (orange, **K*16**, **K*28**); **2f** (brown, Me$_2$N-D1, **K*16**, **K*28**). In each case, the peptide (100 μM) in phosphate-buffered saline, pH 7.4, is incubated quiescently in the presence (filled squares) or absence (open squares) of aldehdye 1 (100 μM) at 37°C. (From [76], with permission. Copyright © Wiley-VCH Verlag GmbH & Co. KGaA.) (*See insert for color representation of figure.*)

In line with current thinking on how hydrophobic peptide mutations contribute to peptide fibrillization via burial of hydrophobic surface, we had initially speculated that covalent modification of the ε-amino group of Lys16, Lys28, and the α-amino group of Asp1 with the hydrophobic aldehyde 1 would be sufficient to trigger amyloidogenesis. The hydrophobic effect may still affect amyloidogenesis once the atheronal is adducted to Aβ, but what is clear is that this process is specific to Lys16. Adduction and increase of local hydrophobicity at Lys28 and Asp1 are not sufficient to trigger fibrillization.

Lys16 sits at the N-terminus of the central hydrophobic cluster (CHC), which has been suggested to be a cholesterol-binding domain of Aβ (Fig. 4B). The binding of cholesterol by Aβ has been linked to a role of membrane stabilization. Given that atheronal-B and cholesterol share structural simile, it seems plausible that upon adduction to Lys16 binding of atheronal-B, *seco*-sterol motif in the CHC may occur. We investigated such a hypothesis by studying the effect of cholesterol on atheronal-B-induced fibrillization of Aβ(1–40). Cholesterol was found to exhibit a concentration-dependent reduction of the ability of atheronal-B to induce fibrillization of Aβ(1–40), with an EC_{50} value an effective (concentration that reduces the maximum ThT-positive aggregates to 50% of the untreated material) of about 30 μM. This is the first example of inhibition of atheronal-induced Aβ aggregation by a molecule that does not trap the aldehyde of atheronal-B with a nucleophile. Cholesterol could be affecting aldehyde-induced fibrillization of Aβ by competing with atheronal B for the CHC domain or, alternatively, by binding in the CHC domain cholesterol may block adduction of atheronal-B to Lys16. This is thought to be due to the proximity of the putative cholesterol-binding site (residues 17 to 21) [77].

CONCLUSIONS

Oxidative stress either in vivo or ex vivo is always accompanied by lipid peroxidation, in a process that is purely chemical in nature and is not enzyme dependent. Once peroxidized, lipids are present in biological systems, and formation of aldehydic products such as 4-HNE, atheronal-A, and atheronal-B occurs spontaneously. The exact mechanism by which such ROS products interact with proteins remains poorly understood, particularly in the context of protein function after a covalent or noncovalent interaction between the lipid oxidation product and the protein has occurred. In the future, it will be critically important to delineate the precise mechanism by which lipid oxidation products such as atheronal-A and atheronal-B accelerate and/or induce misfolding of amyloidagenic proteins into insoluble fibrils. A greater understanding of such interactions will provide valuable insights to the consequences of inflammation and the toxic species that arise from this process, leading to potentially new therapeutic avenues.

REFERENCES

1. Hammond, E.M., Kaufmann, M.R., Giaccia, A.J. (2007). Oxygen sensing and the DNA-damage response. *Curr Opin Cell Biol, 19*, 680–684.
2. Fruhwirth, G.O., Loidl, A., Hermetter, A. (2007). Oxidized phospholipids: from molecular properties to disease. *Biochim Biophys Acta, 1772*, 718–736.
3. Cecarini, V., Gee, J., Fioretti, E., Amici, M., Angeletti, M., Eleuteri, A.M., Keller, J.N. (2007). Protein oxidation and cellular homeostasis: emphasis on metabolism. *Biochim Biophys Acta, 1773*, 93–104.
4. Lee, Y.J., Shacter, E. (1999). Oxidative stress inhibits apoptosis in human lymphoma cells. *J Biol Chem, 274*, 19792–19798.
5. Lelli, J.L., Jr., Becks, L.L., Dabrowska, M.I., Hinshaw, D.B. (1998). ATP converts necrosis to apoptosis in oxidant-injured endothelial cells. *Free Radic Biol Med, 25*, 694–702.
6. Genestra, M. (2007). Oxyl radicals, redox-sensitive signalling cascades and antioxidants. *Cell Signal, 19*, 1807–1819.
7. Ames, B.N., Shigenaga, M.K., Hagen, T.M. (1993). Oxidants, antioxidants, and the degenerative diseases of aging. *Proc Natl Acad Sci U S A, 90*, 7915–7922.
8. Beckman, K.B., Ames, B.N. (1998). The free radical theory of aging matures. *Physiol Rev, 78*, 547–581.
9. Carney, J.M., Smith, C.D., Carney, A.M., Butterfield, D.A. (1994). Aging- and oxygen-induced modifications in brain biochemistry and behavior. *Ann N Y Acad Sci, 738*, 44–53.
10. Cortopassi, G., Liu, Y., Hutchin, T. (1996). Degeneration of human oncogenes and mitochondrial genes occurs in cells that exhibit age-related pathology. *Exp Gerontol, 31*, 253–265.
11. Gilchrest, B.A., Bohr, V.A. (1997). Aging processes, DNA damage, and repair. *FASEB J, 11*, 322–330.
12. Muller, F.L., Lustgarten, M.S., Jang, Y., Richardson, A., Van Remmen, H. (2007). Trends in oxidative aging theories. *Free Radic Biol Med, 43*, 477–503.
13. Krinsky, N.I. (1974). Singlet excited oxygen as a mediator of the antibacterial action of leukocytes. *Science, 186*, 363–365.
14. Zhao, W., Diz, D.I., Robbins, M.E. (2007). Oxidative damage pathways in relation to normal tissue injury. *Br J Radiol, 80* Spec No 1, S23–S31.
15. Esterbauer, H., Schaur, R.J., Zollner, H. (1991). Chemistry and biochemistry of 4-hydroxynonenal, malonaldehyde and related aldehydes. *Free Radic Biol Med, 11*, 81–128.
16. Fessel, J.P., Porter, N.A., Moore, K.P., Sheller, J.R., Roberts, L.J., 2nd. (2002). Discovery of lipid peroxidation products formed in vivo with a substituted tetrahydrofuran ring (isofurans) that are favored by increased oxygen tension. *Proc Natl Acad Sci U S A, 99*, 16713–16718.
17. Petersen, D.R., Doorn, J.A. (2004). Reactions of 4-hydroxynonenal with proteins and cellular targets. *Free Radic Biol Med, 37*, 937–945.
18. Stadtman, E.R., Levine, R.L. (2000). Protein oxidation. *Ann N Y Acad Sci, 899*, 191–208.

19. Smith, C.D., Carney, J.M., Starke-Reed, P.E., Oliver, C.N., Stadtman, E.R., Floyd, R.A., Markesbery, W.R. (1991). Excess brain protein oxidation and enzyme dysfunction in normal aging and in Alzheimer disease. *Proc Natl Acad Sci U S A, 88*, 10540–10543.
20. Agarwal, S., Sohal, R.S. (1994). Aging and proteolysis of oxidized proteins. *Arch Biochem Biophys, 309*, 24–28.
21. Grune, T., Klotz, L.O., Gieche, J., Rudeck, M., Sies, H. (2001). Protein oxidation and proteolysis by the nonradical oxidants singlet oxygen or peroxynitrite. *Free Radic Biol Med, 30*, 1243–1253.
22. Droge, W., Schipper, H.M. (2007). Oxidative stress and aberrant signaling in aging and cognitive decline. *Aging Cell, 6*, 361–370.
23. Aksenova, M.V., Aksenov, M.Y., Carney, J.M., Butterfield, D.A. (1998). Protein oxidation and enzyme activity decline in old brown Norway rats are reduced by dietary restriction. *Mech Ageing Dev, 100*, 157–168.
24. Floyd, R.A. (1990). Role of oxygen free radicals in carcinogenesis and brain ischemia. *FASEB J, 4*, 2587–2597.
25. Hensley, K., Hall, N., Subramaniam, R., Cole, P., Harris, M., Aksenov, M., Aksenova, M., Gabbita, S.P., Wu, J.F., Carney, J.M., et al. (1995). Brain regional correspondence between Alzheimer's disease histopathology and biomarkers of protein oxidation. *J Neurochem, 65*, 2146–2156.
26. Rakhit, R., Cunningham, P., Furtos-Matei, A., Dahan, S., Qi, X.F., Crow, J.P., Cashman, N.R., Kondejewski, L.H., Chakrabartty, A. (2002). Oxidation-induced misfolding and aggregation of superoxide dismutase and its implications for amyotrophic lateral sclerosis. *J Biol Chem, 277*, 47551–47556.
27. Ursini, F., Davies, K.J., Maiorino, M., Parasassi, T., Sevanian, A. (2002). Atherosclerosis: another protein misfolding disease? *Trends Mol Med, 8*, 370–374.
28. Benedetti, A., Comporti, M., Esterbauer, H. (1980). Identification of 4-hydroxynonenal as a cytotoxic product originating from the peroxidation of liver microsomal lipids. *Biochim Biophys Acta, 620*, 281–296.
29. Cheng, J.Z., Singhal, S.S., Sharma, A., Saini, M., Yang, Y., Awasthi, S., Zimniak, P., Awasthi, Y.C. (2001). Transfection of mGSTA4 in HL-60 cells protects against 4-hydroxynonenal-induced apoptosis by inhibiting JNK-mediated signaling. *Arch Biochem Biophys, 392*, 197–207.
30. He, N., Singhal, S.S., Awasthi, S., Zhao, T., Boor, P.J. (1999). Role of glutathione S-transferase 8-8 in allylamine resistance of vascular smooth muscle cells in vitro. *Toxicol Appl Pharmacol, 158*, 177–185.
31. Kinter, M., Roberts, R.J. (1996). Glutathione consumption and glutathione peroxidase inactivation in fibroblast cell lines by 4-hydroxy-2-nonenal. *Free Radic Biol Med, 21*, 457–462.
32. Renes, J., de Vries, E.E., Hooiveld, G.J., Krikken, I., Jansen, P.L., Muller, M. (2000). Multidrug resistance protein MRP1 protects against the toxicity of the major lipid peroxidation product 4-hydroxynonenal. *Biochem J, 350 (Pt 2)*, 555–561.
33. Spitz, D.R., Malcolm, R.R., Roberts, R.J. (1990). Cytotoxicity and metabolism of 4-hydroxy-2-nonenal and 2-nonenal in H_2O_2-resistant cell lines. Do aldehydic byproducts of lipid peroxidation contribute to oxidative stress? *Biochem J, 267*, 453–459.

34. Sullivan, S.J., Oberley, T.D., Roberts, R.J., Spitz, D.R. (1992). A stable O_2-resistant cell line: role of lipid peroxidation byproducts in O_2-mediated injury. *Am J Physiol, 262*, L748–L756.
35. Bieschke, J., Zhang, Q., Bosco, D.A., Lerner, R.A., Powers, E.T., Wentworth, P., Jr., Kelly, J.W. (2006). Small molecule oxidation products trigger disease-associated protein misfolding. *Acc Chem Res, 39*, 611–619.
36. Chen, K., Maley, J., Yu, P.H. (2006). Potential inplications of endogenous aldehydes in beta-amyloid misfolding, oligomerization and fibrillogenesis. *J Neurochem, 99*, 1413–1424.
37. Janue, A., Olive, M., Ferrer, I. (2007). Oxidative stress in desminopathies and myotilinopathies: a link between oxidative damage and abnormal protein aggregation. *Brain Pathol, 17*, 377–388.
38. Stewart, C.R., Wilson, L.M., Zhang, Q., Pham, C.L., Waddington, L.J., Staples, M.K., Stapleton, D., Kelly, J.W., Howlett, G.J. (2007). Oxidized cholesterol metabolites found in human atherosclerotic lesions promote apolipoprotein C-II amyloid fibril formation. *Biochemistry, 46*, 5552–5561.
39. Zhang, Q., Powers, E.T., Nieva, J., Huff, M.E., Dendle, M.A., Bieschke, J., Glabe, C.G., Eschenmoser, A., Wentworth, P., Jr., Lerner, R.A., Kelly, J.W. (2004). Metabolite-initiated protein misfolding may trigger Alzheimer's disease. *Proc Natl Acad Sci U S A, 101*, 4752–4757.
40. Zarkovic, K. (2003). 4-Hydroxynonenal and neurodegenerative diseases. *Mol Aspects Med, 24*, 293–303.
41. Sayre, L.M., Zelasko, D.A., Harris, P.L., Perry, G., Salomon, R.G., Smith, M.A. (1997). 4-Hydroxynonenal-derived advanced lipid peroxidation end products are increased in Alzheimer's disease. *J Neurochem, 68*, 2092–2097.
42. Ando, Y., Brannstrom, T., Uchida, K., Nyhlin, N., Nasman, B., Suhr, O., Yamashita, T., Olsson, T., El Salhy, M., Uchino, M., Ando, M. (1998). Histochemical detection of 4-hydroxynonenal protein in Alzheimer amyloid. *J Neurol Sci, 156*, 172–176.
43. Castellani, R.J., Perry, G., Siedlak, S.L., Nunomura, A., Shimohama, S., Zhang, J., Montine, T., Sayre, L.M., Smith, M.A. (2002). Hydroxynonenal adducts indicate a role for lipid peroxidation in neocortical and brainstem Lewy bodies in humans. *Neurosci Lett, 319*, 25–28.
44. Yoritaka, A., Hattori, N., Uchida, K., Tanaka, M., Stadtman, E.R., Mizuno, Y. (1996). Immunohistochemical detection of 4-hydroxynonenal protein adducts in Parkinson disease. *Proc Natl Acad Sci U S A, 93*, 2696–2701.
45. Babior, B.M., Takeuchi, C., Ruedi, J., Gutierrez, A., Wentworth, P., Jr. (2003). Investigating antibody-catalyzed ozone generation by human neutrophils. *Proc Natl Acad Sci U S A, 100*, 3031–3034.
46. Datta, D., Vaidehi, N., Xu, X., Goddard, W.A., 3rd. (2002). Mechanism for antibody catalysis of the oxidation of water by singlet dioxygen. *Proc Natl Acad Sci U S A, 99*, 2636–2641.
47. Wentworth, P., Jr., Jones, L.H., Wentworth, A.D., Zhu, X., Larsen, N.A., Wilson, I.A., Xu, X., Goddard, W.A., 3rd, Janda, K.D., Eschenmoser, A., Lerner, R.A. (2001). Antibody catalysis of the oxidation of water. *Science, 293*, 1806–1811.

48. Wentworth, P., Jr., McDunn, J.E., Wentworth, A.D., Takeuchi, C., Nieva, J., Jones, T., Bautista, C., Ruedi, J.M., Gutierrez, A., Janda, K.D., et al. (2002). Evidence for antibody-catalyzed ozone formation in bacterial killing and inflammation. *Science, 298*, 2195–2199.
49. Eanes, E.D., Glenner, G.G. (1968). X-ray diffraction studies on amyloid filaments. *J Histochem Cytochem, 16*, 673 677.
50. Glenner, G.G., Wong, C.W., Quaranta, V., Eanes, E.D. (1984). The amyloid deposits in Alzheimer's disease: their nature and pathogenesis. *Appl Pathol, 2*, 357–369.
51. Liu, Q., Raina, A.K., Smith, M.A., Sayre, L.M., Perry, G. (2003). Hydroxynonenal, toxic carbonyls, and Alzheimer disease. *Mol Aspects Med, 24*, 305–313.
52. Picklo, M.J., Montine, T.J., Amarnath, V., Neely, M.D. (2002). Carbonyl toxicology and Alzheimer's disease. *Toxicol Appl Pharmacol, 184*, 187–197.
53. Shigematsu, S., Takahashi, N., Hara, M., Yoshimatsu, H., Saikawa, T. (2007). Increased incidence of coronary in-stent restenosis in type 2 diabetic patients is related to elevated serum malondialdehyde-modified low-density lipoprotein. *Circ J, 71*, 1697–1702.
54. Uchida, K. (2003). 4-Hydroxy-2-nonenal: a product and mediator of oxidative stress. *Prog Lipid Res, 42*, 318–343.
55. Hardy, J.A., Higgins, G.A. (1992). Alzheimer's disease: the amyloid cascade hypothesis. *Science, 256*, 184–185.
56. Kayed, R., Head, E., Thompson, J.L., McIntire, T.M., Milton, S.C., Cotman, C.W., Glabe, C.G. (2003). Common structure of soluble amyloid oligomers implies common mechanism of pathogenesis. *Science, 300*, 486–489.
57. Doorn, J.A., Petersen, D.R. (2003). Covalent adduction of nucleophilic amino acids by 4-hydroxynonenal and 4-oxononenal. *Chem Biol Interact, 143–144*, 93–100.
58. Haass, C., Steiner, H. (2001). Protofibrils, the unifying toxic molecule of neurodegenerative disorders? *Nat Neurosci, 4*, 859–860.
59. Klein, W.L., Stine, W.B., Jr., Teplow, D.B. (2004). Small assemblies of unmodified amyloid beta-protein are the proximate neurotoxin in Alzheimer's disease. *Neurobiol Aging, 25*, 569–580.
60. Hardy, J. (1997). Amyloid, the presenilins and Alzheimer's disease. *Trends Neurosci, 20*, 154–159.
61. Teplow, D.B., Lomakin, A., Kirschner, D.A., Walsh, D.M. (1997). Effects of beta-protein mutations on amyloid fibril nucleation and elongation. In *Alzheimer's Disease: Biology, Diagnosis, and Therapeutics* (Iqbak, K., (Ed.), Wiley, Chichester, UK, pp. 311–319.
62. Tsubuki, S., Takaki, Y., Saido, T.C. (2003). Dutch, Flemish, Italian, and Arctic mutations of APP and resistance of Abeta to physiologically relevant proteolytic degradation. *Lancet, 361*, 1957–1958.
63. Chiti, F., Dobson, C.M. (2006). Protein misfolding, functional amyloid, and human disease. *Annu Rev Biochem, 75*, 333–366.
64. Casserly, I., Topol, E. (2004). Convergence of atherosclerosis and Alzheimer's disease: inflammation, cholesterol, and misfolded proteins. *Lancet, 363*, 1139–1146.

65. Hartmann, T. (2001). Cholesterol, A beta and Alzheimer's disease. *Trends Neurosci, 24*, S45–S48.
66. Patrico, D., Uryu, K., Leight, S., Trojanoswki, J.Q., Lee, V.M.-Y. (2001). Increased lipid peroxidation precedes amyloid plaque formation in an animal model of Alzheimer amyloidosis. *J Neurosci, 21*, 4183–4187.
67. Sayre, L.M., Smith, M.A., Perry, G. (2001). Chemistry and biochemistry of oxidative stress in neurodegenerative disease. *Curr Med Chem, 8*, 721–738.
68. Markesbery, W.R., Carney, J.M. (1999). Oxidative alterations in Alzheimer's disease. *Brain Pathol, 9*, 133–146.
69. Mark, R.J., Lovell, M.A., Markesbery, W.R., Uchida, K., Mattson, M.P. (1997). A role for 4-hydroxynonenal, an aldehydic product of lipid peroxidation, in disruption of ion homeostasis and neuronal death induced by amyloid beta-peptide. *J Neurochem, 68*, 255–264.
70. Siegel, S.J., Bieschke, J., Powers, E.T., Kelly, J.W. (2007). The oxidative stress metabolite 4-hydroxynonenal promotes Alzheimer protofibril formation. *Biochemistry, 46*, 1503–1510.
71. Liu, L., Komatsu, H., Murray, I.V., Axelsen, P.H. (2008). Promotion of amyloid beta protein misfolding and fibrillogenesis by a lipid oxidation product. *J Mol Biol, 377*, 1236–1250.
72. Chiti, F., Webster, P., Taddei, N., Clark, A., Stefani, M., Ramponi, G., Dobson, C.M. (1999). Designing conditions for in vitro formation of amyloid protofilaments and fibrils. *Proc Natl Acad Sci U S A, 96*, 3590–3594.
73. Harper, J.D., Lansbury, P.T., Jr. (1997). Models of amyloid seeding in Alzheimer's disease and scrapie: mechanistic truths and physiological consequences of the time-dependent solubility of amyloid proteins. *Annu Rev Biochem, 66*, 385–407.
74. Bosco, D.A., Fowler, D.M., Zhang, Q., Nieva, J., Powers, E.T., Wentworth, P., Jr., Lerner, R.A., Kelly, J.W. (2006). Elevated levels of oxidized cholesterol metabolites in Lewy body disease brains accelerate alpha-synuclein fibrilization. *Nat Chem Biol, 2*, 249–253.
75. Bieschke, J., Zhang, Q., Powers, E.T., Lerner, R.A., Kelly, J.W. (2005). Oxidative metabolites accelerate Alzheimer's amyloidogenesis by a two-step mechanism, eliminating the requirement for nucleation. *Biochemistry, 44*, 4977–4983.
76. Scheinost, J.C., Wang, H., Boldt, G.E., Offer, J., Wentworth, P., Jr. (2008). Cholesterol *seco*-sterol-induced aggregation of methylated amyloid-beta peptides: insights into aldehyde-initiated fibrillization of amyloid-beta. *Angew Chem Int Ed Engl, 47*, 3919–3922.
77. Nelson, T.J., Alkon, D.L. (2005). Oxidation of cholesterol by amyloid precursor protein and beta-amyloid peptide. *J Biol Chem, 280*, 7377–7387.

29

AGING AND AGGREGATION-MEDIATED PROTEOTOXICITY

EHUD COHEN AND ANDREW DILLIN

Molecular and Cell Biology Laboratory, The Salk Institute for Biological Studies, La Jolla, California

INTRODUCTION

Aging is the major risk factor for the development of human neurodegenerative disorders such as prion maladies, Huntington (HD), Parkinson (PD), and Alzheimer (AD) diseases [1], all tightly linked to the aggregation, accumulation, and deposition of aberrantly folded proteins [2,3]. In AD, a dual proteolytic digestion of the amyloid precursor protein (APP) releases aggregation-prone peptides, termed Aβ, which form fibrillar structures of various sizes, and initiate disease [4]. Similarly, α-synuclein aggregation is associated with the emergence of PD [5] and the aggregation of mutant huntingtin, bearing abnormally long polyglutamine (polyQ) stretches, initiates HD [6]. The detailed mechanisms that lead to the development of these disorders are unknown; however, a large body of data suggests that high-molecular-weight aggregates are not causative, but rather, small, intermediate aggregating structures are the major toxic species that initiate neurodegenerative disorders (reviewed by caughey and Lansbury [7]).

Neurodegenerations manifest in three distinct fashions: sporadically, as mutation-linked familial diseases or, uniquely to prion maladies, as transmissible disorders [8,9]. AD and PD generally appear sporadically and rarely as familial diseases [2], whereas HD is solely inherited [6]. The majority of all sporadic cases emerge during the seventh decade of the patient's life or later,

Protein Misfolding Diseases: Current and Emerging Principles and Therapies,
Edited by Marina Ramirez-Alvarado, Jeffery W. Kelly, and Christopher M. Dobson
Copyright © 2010 John Wiley & Sons, Inc.

while most people who carry disease-linked mutations develop clinical signs during their fifth decade [1]. Why different neurodegenerations emerge late in life, in a relatively narrow age range, are key unsolved enigmas. Two basic models can explain the late-onset phenomenon; one suggests that the buildup of protein aggregation-mediated toxicity is a very slow process that requires many years to initiate disease. The other model predicts that the aging process plays an active role in enabling these syndromes to onset. Recently, the latter model was strongly supported as studies suggest that aggregation-mediated proteotoxicity is not a stochastic process but depends on the aging process [10–12]. Aging is suspected to negatively regulate cellular counter-proteotoxicity mechanisms, a process that enables constitutive aggregation to become toxic late in life.

THE REGULATION OF LIFE SPAN AND AGING

At least three distinguishable pathways regulate life span and aging: dietary intake, mitochondrial respiration, and the insulin/IGF-1 signaling (IIS) pathway (reviewed by Wolff and Dillin [13]). The IIS, perhaps the most prominently studied aging regulatory pathway, controls the life span and stress resistance of worms, flies [14], and mice [15]. In the nematode *Caenorhabditis elegans*, the sole insulin/IGF-1 receptor, DAF-2, mediates the phosphorylation of its downstream forkheadlike transcription factor, DAF-16, prevents its nuclear localization, compromises its target genes expression, shortens the life span, and elevates stress sensitivity [16,17]. Thus, *daf-2* knockdown hyperactivates DAF-16, creating long-lived, youthful, stress-resistant worms [14,18]. The ability of reduced IIS to promote longevity in worms is completely dependent on the single FOXO family member in *C. elegans*, DAF-16, as all long-lived *daf-2* mutated strains exhibit a short life span when *daf-16* is either mutated or reduced by RNAi [17–19]. Similarly, in mice, deletion of one copy of *igf1-R*, the murine *daf-2* ortholog, increases longevity and stress resistance [15]. Additionally, tissue-specific knockout of the insulin receptor in adipose tissue [20], as well as brain-specific knockout of the downstream insulin receptor substrate (IRS-2) [21], increased mice life span.

Heat-shock factor 1 (HSF-1), a highly conserved [22] leucine zipper containing [23] transcription factor that plays critical roles in stress response [24] and innate immunity [25], is also vital for worm life span extension facilitated by compromised IIS. First, long-lived worms, expressing mutated, weak *daf-2* allele, and wild-type worms had similar, exceptionally short life spans when developed and grown on bacteria expressing *hsf-1* dsRNA. (In worms, gene knockdown can be achieved by feeding worms bacteria expressing dsRNA of the gene of interest [26].) Second, worms that express an additional copy of the *hsf-1* gene live longer and are more stress resistant than their wild-type counterparts [11,27]. Thus, although it is not yet clear how HSF-1 is linked mechanistically to the IIS, it is required for the IIS regulation of longevity and stress resistance.

IIS AND TOXIC PROTEIN AGGREGATION

Studies in worm models indicate that manipulation of the IIS affects the onset of aggregation-mediated proteotoxicity. The Morimoto lab created a series of transgenic worm strains, each expressing a polyQ stretch of different length, fused to the yellow fluorescent protein (YFP) [12]. Exploiting the transparency of *C. elegans*, the researchers visualized the aggregative fluorescent polypeptides within living worms and found that at least 40 glutamine repeats are required for efficient aggregation in young (day 2 of adulthood) worms. Interestingly, the threshold number of repeats needed for aggregation decreased as the animals aged. In worms, expressing polyQ lengths of 35, polyQ35-YFP, aggregates were observable by day 4 of adulthood, while polyQ29-YFP aggregates could not be detected earlier than day 9 of adulthood (midlife for a worm). Since in this model polyQ-YFP aggregation impairs motility, Morley and colleagues examined this toxic effect. In accordance with aggregation, worms expressing 33 to 35 glutamine repeats exhibited no motility impairment at young ages but succumbed to toxicity later in life. The researchers also presented data indicating that the RNAi-mediated reduction of *age-1* (a component of the IIS that when inactivated results in long-lived animals comparable to *daf-2* reduction) protects worm embryos from the aggregation of polyQ82-YFP. Accordingly, *age-1* RNAi reduced the motility impairment of young polyQ82-YFP worms. These protective effects were *daf-16* dependent, as RNAi toward *daf-16* abolished them. Analogous observations were reported by Parker et al. [28], who found that in worms, neuronal polyQ-mediated toxicity is mitigated by Resveratrol, a sirtuin activator known to extend the life span of yeast [29], worms, flies [30], and fish [31] in a *daf-16*-dependent fashion. The sirtuin, *sir-2*, is thought to act immediately upstream of *daf-16* to help regulate the expression of DAF-16 longevity genes [32]). Importantly, this study indicated that reduced IIS plays a role in protecting neurons from proteotoxicity.

The link between aging and the onset of polyQ aggregation was further established by Hsu and colleagues [11], who found that in worms, HSF-1 is essential not only for life span extension by reduced IIS but also for the mitigation of polyQ40-YFP aggregation. This aggregation was elevated when the expression of *daf-16* was compromised. This study also points to small heat-shock proteins, members of the crystalline family that are transcriptionally regulated by HSF-1, as important players in the antiaggregation effect of reduced IIS. Similar conclusions were suggested by Morley and Morimoto in a following study [27], which confirmed the critical role of HSF-1 and of chaperones in the compromised IIS-mediated life span extension of worms. Interestingly, in this study, additional chaperones, members of the hsp70 family, were found to be vital for the reduced IIS-mediated life span extension.

The results described above established the idea that reduction of the IIS pathway can protect worms from polyQ proteotoxicity; however, key questions were left unanswered. First, can reduced IIS counter the proteotoxicity of other disease-linked aggregation-prone proteins? What are the mechanistic details of this reduced IIS-mediated protective effect? To address these and other questions, we employed model worms that express the AD-linked human Aβ(1–42) in their body-wall muscles. This expression and the subsequent Aβ aggregation result in progressive paralysis of the animals [33]. Analogous to its effect on polyQ-YFP toxicity, decreased IIS significantly reduced the Aβ(1–42)-mediated proteotoxicity [10]. Using genetic, biochemical, in vitro, and microscopic techniques (Fig. 1), we found that the amounts of high-molecular-weight Aβ(1–42) aggregates do not correlate with toxicity, as (1) reduction of the IIS (*daf-2* RNAi) resulted in reduced toxicity but in no reduction of high-molecular-weight Aβ(1–42) aggregates; (2) reduction of DAF-16 (*daf-16* RNAi) reduced the amount of high-molecular-mass Aβ(1–42) aggregates but enhanced toxicity; and (3) *hsf-1* RNAi treatment led to elevation of both toxicity and the amount of high-molecular-weight Aβ(1–42) aggregates.

What is the nature of the toxic Aβ structures? Recent studies point to small Aβ aggregates (also termed *protofibrils*) as the key source of toxicity. First, Cleary and colleagues injected soluble fractions containing secreted Aβ dimers and trimers into rat brains and found that learning functions were disrupted significantly [34]. Secreted fractions containing Aβ trimers also exhibit the most efficient interference with the long-term potentiation (LTP) that follows electrical excitation of cultured cells [35], a hallmark of brain slices recovered from AD rodent models. Consistent with these studies, we found that in our worm system, Aβ trimers correlated in vivo with toxicity [10]. Moreover, Aβ protofibrils close to 56 kDa (the approximate size of a 12-mer of Aβ monomers) correlated with memory impairment in AD model mice [36]. In addition, soluble Aβ structures termed amyloid beta – derived diffusible ligands were found to correlate with toxicity [37]. Small aggregative structures were also shown to be the major toxic species in neurodegenerative diseases other than AD [7,38,39].

Similar to the polyQ-based studies, our observations support the apparent mechanistic link between aging and the onset of proteotoxicity in general and the involvement of the IIS in regulating protective activities in particular. Based on these findings, we proposed a model suggesting that the IIS regulates both *hsf-1* and *daf-16* negatively and, subsequently, compromises the counter-proteotoxic mechanisms, disaggregation and active aggregation, which both play roles in Aβ protofibril detoxification (Fig. 2).

However, in contrast to the current polyQ-based studies discussed above, our data argue that there is no direct correlation between high-molecular-weight aggregates and toxicity. This apparent contradiction might be a result of the possibility that hyperaggregation is protective and that an animal forms large aggregates before succumbing to toxicity. Thus, the large polyQ aggregates observed might be a marker of proteotoxicity, but not its source. Consistent with this hypothesis, the Finkbeiner group discovered that cell survival of individual cultured neurons expressing expanded polyQ correlated with the appearance of inclusion bodies containing huntingtin and not soluble molecules [40].

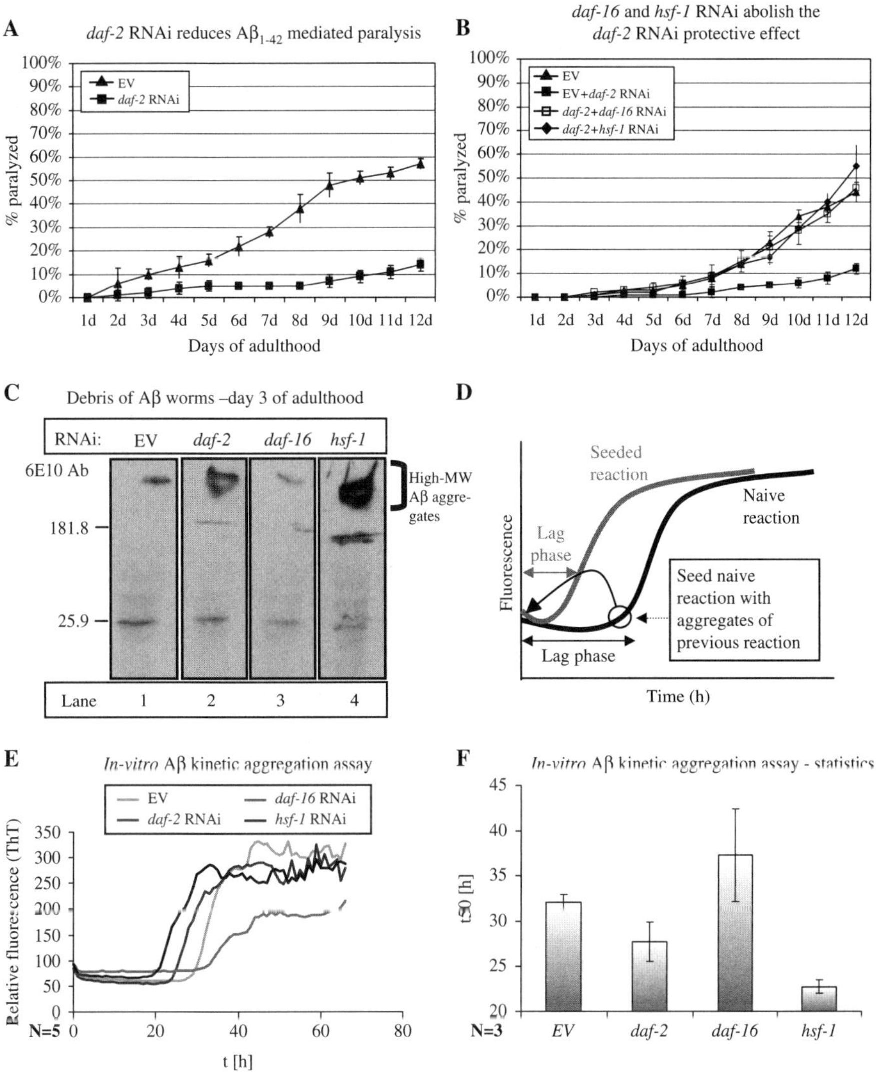

FIG. 1 Lack of correlation between high-molecular-mass Aβ aggregates and toxicity. (A) *daf-2* RNAi protects Aβ worms from the paralysis phenotype associated with Aβ expression. (B) The *daf-2* RNAi-mediated protective effect is *daf-16* and *hsf-1* dependent. *daf-2* RNAi protected worms from paralysis when diluted with EV bacteria but not when mixed with either *daf-16* or *hsf-1* RNAi bacteria. (C) Aβ worms were grown on RNAi bacteria as indicated. At day 3 of adulthood the worms were homogenized, spun, and debris was separated from the soluble fraction. Aβ contents in worm debris were analyzed using Western blot and 6E10 antibody. (From [10], with permission of AAAS.) (D) In vitro kinetic aggregation assay. The typical lag phase that is associated with in vitro aggregation of Aβ can be shortened by seeding of the reaction with previously aggregated Aβ. This technique has been exploited to measure Aβ aggregate content in worm samples. (E) Aβ seed contents of worms grown on RNAi bacteria (as indicated) were evaluated using kinetic aggregation assay. (From [10], with permission of AAAS.) (F) Quantification of three independent in vitro kinetic aggregation assays [as in (E)]. (*See insert for color representation of figure.*)

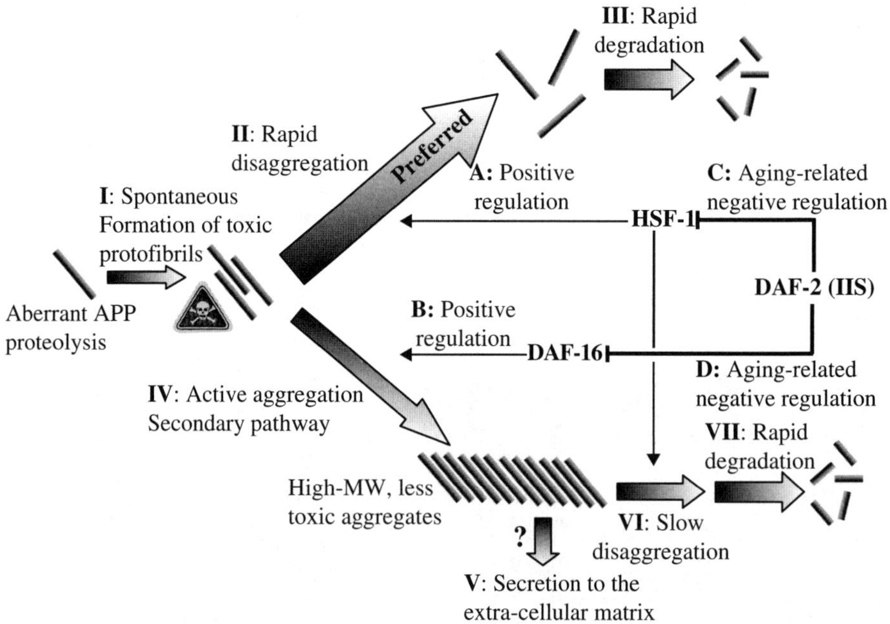

FIG. 2 The insulin/IGF-1 signaling pathway links aging and proteotoxicity. (I) Serial digestion of the amyloid precursor protein (APP) releases the aggregative-prone peptides Aβ, which spontaneously form small toxic aggregates. (II) HSF-1-regulated (A) disaggregation machinery disrupts the Aβ aggregates and prepares them for (III) degradation. (IV) When the disaggregation pathway is overtaxed, a DAF-16-regulated (B) active aggregation apparatus creates high-molecular-mass aggregates of lower toxicity which are possibly secreted from the cell (V). (VI) The high-molecular-mass aggregates undergo slow disaggregation and subsequent degradation (VII). DAF-2 compromises the activity of both protective mechanisms in an age-dependent manner by negatively regulating HSF-1 (C) and DAF-16 (D). (From [10], with permission of AAAS.)

An alternative explanation suggests that Aβ and polyQ differ in their aggregation and toxicity properties. Although the worm Aβ and huntingtin studies do not appear to support a single toxic species of a distinct size, they do indicate that IIS reduction protects worms from proteotoxicity.

Cell culture–based reports and studies performed in rodent models appear to disagree with the idea that counter-proteotoxicity activities are mediated by reduced IIS. Humbert et al. [41] reported that activation of IGF/AKT pathway promotes neuroprotection from specific polyQ-mediated toxicity. This protection involved the phosphorylation of huntingtin by AKT, a kinase positively regulated by the IIS, and its clearance by autophagy. Consistent with this observation, Yamamoto and colleagues [42] found that autophagy-mediated clearance of huntingtin is triggered by the activation of insulin receptor substrate 2 (IRS-2), a protein that mediates the signaling of growth factors such as insulin and insulin-like growth factor 1 (IGF-1). However, unlike

Humbert et al. Yamamoto and colleagues reported that the counter polyQ toxicity effect is AKT independent, suggesting that signals downstream of IRS-2 must diverge to counter proteotoxicity in this setting.

Similarly, Carro et al. [43] reported that increased levels of circulating IGF-1 induces the clearance of Aβ in rat brains. They infused IGF-1 into the brains of old rats and found that Aβ levels in the animal's brains were reduced to these observed in young rat brains. The researchers concluded that hyperactivation of the IIS, by injection of IGF-1, was neuroprotective. Recently, the same group reported that IGF-1 injection to AD-model mice reduced the typical behavioral impairments associated with increased Aβ and reduced the Aβ in their brains [44]. These results, generated in cell cultures and rodent models, appear to contradict the data obtained from *Caenorhabditis elegans* polyQ and Aβ proteotoxicity models. However, the possibility that IGF-1, AKT, and IRS-2 play complex roles in signaling pathways other than the IIS should not be excluded. An additional possible explanation to reconcile the results from mammalian and worm experiments is that local, acute up-regulation of the IIS pathway results in a prolonged dampening of IIS signaling. In this model, spike activation of IIS signaling, by injecting large amounts of insulin or IGF-1, results in future compensation of the pathway to reduce further stimulation of the IIS pathway. This idea might be supported by the discovery of a positive feedback loop that controls the IIS activity in worms [45]. At least two key experiments need to be performed to resolve this discrepancy. The first, reduction of IIS in a bona fide mammalian model of AD, needs evaluation. The second is the question of whether ectopic, acute stimulation of the IIS pathway results in later down-regulation and compensation of the IIS pathway in these mammalian models of AD. Nonetheless, the fact that aging is the major risk factor for the development of neurodegenerations and that reduced IIS slows aging of mice [15] seriously challenges the idea that activated IIS indeed protects animals from proteotoxicity.

BIOLOGICAL COUNTER-PROTEOTOXICITY ACTIVITIES

Countering proteotoxicity involves the detoxification and clearance of harmful protein aggregates by biological activities such as disaggregation [46], degradation [47,48], and perhaps active protein aggregation [49,50]. The involvement of the IIS in regulating proteotoxicity raise several key questions: Do protective activities decline with age? What are the cellular components that promote these activities? Does the IIS play a role in regulating the expression and stability of these components and, subsequently, these protective activities?

The disassembly of monomeric polypeptides and their separation from large aggregates appears to be required to enable their efficient degradation. In yeast, the chaperone HSP104 mediates disaggregation activity in concert with other heat-shock proteins [51]. No obvious *hsp104* orthologs can be identified in mammalian systems or *C. elegans;* however, disaggregation activity is present in these systems, as disaggregation has been observed in mammalian cells [52] and in worm homogenates [10].

To date it is unclear what cellular components promote disaggregation activity; however, HSF-1 is involved in the regulation of this activity [10]. This finding suggests that disaggregation might be compromised by the aging process. Perhaps the apparent involvement of the crystalline chaperones in countering polyQ aggregation and their regulation by HSF-1 and/or DAF-16 [11] points to these chaperones as possible mediators of disaggregation. In support of this idea, overexpression of hsp-16.2 delayed the onset of proteotoxicity in the worm Aβ model [53].

Proteolysis is an additional critical counter-proteotoxicity activity. Two cellular degradation mechanisms are known to digest abnormally processed, unfolded, and damaged proteins: the ubiquitin-proteasome system (UPS) and lysosomes. Proteasomes possess the capability to degrade polyubiquitin-tagged proteins (reviewed by Ciechanover and Brundin [54]). Proteasomes were found within the vicinity of potentially toxic protein aggregates [55] and proposed to be involved in polyQ aggregate clearance [56]. However, several studies indicate that proteases, not proteasomes, mediate the degradation of misfolded aggregative proteins, perhaps through a mechanism termed *chaperone-mediated autophagy* (CMA) [57]. Proteases including neprilysin [48] and the insulin degrading enzyme (IDE) [58] have been also shown to be involved directly in the degradation of Aβ. It was suggested that the aging process in general and the IIS in particular play roles in the regulation of proteolysis [59]. Accordingly, members of the autophagic pathway have been found to play a role in the increased longevity and decreased Aβ proteotoxicity of IIS-reduced worms [60].

The third protective activity to be discussed here is active aggregation. Traditionally, protein aggregation was thought to be an uncontrolled, sporadic, toxic process that living cells sought to prevent. This view has been challenged with indications that cells actively aggregate proteins under certain conditions, as large aggregates have been proposed to bear lower toxicity than that of their smaller counterparts. For example, the yeast chaperone Hsp104 can catalyze aggregation of the yeast prion-like protein Sup35 [50]. This chaperone activity was observed when Sup35 was present in high concentration and was abolished upon dilution. It is also shown in this study that Hsp104 disrupts large aggregates; therefore, the authors suggest that the Hsp104 machinery possesses opposing activities in an aggregate-concentration-dependent manner. This raises the intriguing possibility that in higher organisms, the distinct and differentially regulated disaggregation and aggregation activities evolved from the primordial Hsp104 functional ortholog.

Similarly, the cytosolic chaperonin TRiC promotes the aggregation of polyQ stretches [49]. This activity was found to be associated with the Hsp70/Hsp40 machinery and to be cell protective, as the larger aggregates were less toxic than their smaller counterparts. Analogous to Hsp104, TRiC active aggregation activity is concentration dependent, as overexpression of TRiC resulted in the reduction of polyQ aggregation. The idea that active aggregation is protective was supported by our study [10], which indicated that this activity is regulated by the transcription factor DAF-16 and thus probably declines with age.

Collectively, the studies reviewed herein argue that the aging process compromises counter-proteotoxicity mechanisms. Plausibly, this aging-associated decline enables constitutive protein aggregation to become toxic and to initiate neurodegeneration late in life.

INFLUENCES ON THE AGE OF ONSET OF NEURODEGENERATION

When will a certain person succumb to accumulating toxic aggregates and develop neurodegeneration? The answer to this essential question probably depends on the balance between toxic protein aggregation and the efficiency of detoxification activities. Therefore, it is plausible that a threshold amount of toxic species is required to initiate disease. Early in life the aggregate clearance capabilities exceed the rate of aggregation, and toxic species do not accumulate to reach the disease-initiation threshold. As the person ages, the aging-associated decline in protective activities (disaggreation, active aggregation, proteolysis, export) enables the accumulation of toxic structures to the levels required to begin neurodegeneration. In this regard, an interesting study

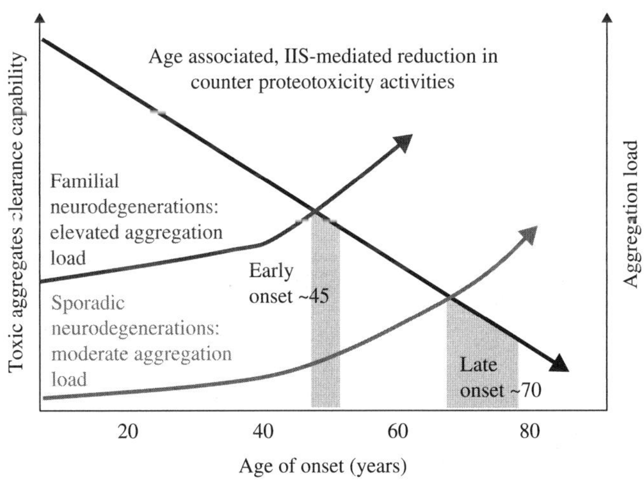

FIG. 3 The balance between the production of toxic protein aggregates and the aging-associated reduction in counter-proteotoxicity activities determines the age at which the amount of toxic aggregates required for disease onset will cross the threshold level. A higher aggregation load and lower protective activities will lead to early onset, while a lower aggregation load and higher protective capabilities will postpone the disease age of onset. This model proposes that the similar ages of onset of distinct neurodegenerations stem from one phenomenon: the age-related decline in the natural counter-proteotoxicity activities. (From *Nature Reviews Neuroscience*, *9*, 759–767.)

suggests that a single perturbation of the proteome by an uncontrolled aggregation can initiate disease by affecting the folding of other proteins [61].

Apparently, as a population, human beings age at similar rates; thus, it is plausible that their capabilities to clear proteotoxic species decline at similar rates. This might explain why distinct neurodegenerations emerge at similar ages. The rates of potentially toxic aggregate formation probably vary among individuals depending on their specific genetic background and environmental conditions. Nevertheless, it is clear that patients harboring neurodegeneration-linked mutated genes produce relatively high amounts of aggregation-prone toxic polypeptides and are more likely to develop disease early in life. People who do not carry neurodegeneration-linked mutated genes produce fewer toxic aggregates and therefore develop disease later in life, if at all (Fig. 3).

Many aspects of the mechanisms that link aging and neurodegenerative diseases are still obscure; nevertheless, current understanding points toward the manipulation of aging and the consequent maintenance of counter-aggregation activities as a possible exciting avenue toward the development of future anti-neurodegeneration treatments.

REFERENCES

1. Amaducci, L., Tesco, G. (1994). Aging as a major risk for degenerative diseases of the central nervous system. *Curr Opin Neurol, 7*, 283–286.
2. Selkoe, D.J. (2003). Folding proteins in fatal ways. *Nature, 426*, 900–904.
3. Kopito, R.R., Ron, D. (2000). Conformational disease. *Nat Cell Biol, 2*, E207–E209.
4. Haass, C., Selkoe, D.J. (2007). Soluble protein oligomers in neurodegeneration: lessons from the Alzheimer's amyloid beta-peptide. *Nat Rev Mol Cell Biol, 8*, 101–112.
5. Lee, V.M., Trojanowski, J.Q. (2006). Mechanisms of Parkinson's disease linked to pathological alpha-synuclein: new targets for drug discovery. *Neuron, 52*, 33–38.
6. Bates, G. (2003). Huntingtin aggregation and toxicity in Huntington's disease. *Lancet, 361*, 1642–1644.
7. Caughey, B., Lansbury, P.T. (2003). Protofibrils, pores, fibrils, and neurodegeneration: separating the responsible protein aggregates from the innocent bystanders. *Annu Rev Neurosci, 26*, 267–298.
8. Aguzzi, A., Polymenidou, M. (2004). Mammalian prion biology: one century of evolving concepts. *Cell, 116*, 313–327.
9. Gajdusek, D.C., Gibbs, C.J., Jr. (1968). Slow, latent and temperate virus infections of the central nervous system. *Res Publ Assoc Res Nerv Ment Dis, 44*, 254–280.
10. Cohen, E., Bieschke, J., Perciavalle, R.M., Kelly, J.W., Dillin, A. (2006). Opposing activities protect against age-onset proteotoxicity. *Science, 313*, 1604–1610,
11. Hsu, A.L., Murphy, C.T., Kenyon, C. (2003). Regulation of aging and age-related disease by DAF-16 and heat-shock factor. *Science, 300*, 1142–1145.

12. Morley, J.F., Brignull, H.R., Weyers, J.J., Morimoto, R.I. (2002). The threshold for polyglutamine-expansion protein aggregation and cellular toxicity is dynamic and influenced by aging in *Caenorhabditis elegans*. *Proc Natl Acad Sci U S A, 99*, 10417–10422.
13. Wolff, S., Dillin, A. (2006). The trifecta of aging in *Caenorhabditis elegans*. *Exp Gerontol, 41*, 894–903.
14. Kenyon, C. (2005). The plasticity of aging: insights from long-lived mutants. *Cell, 120*, 449–460.
15. Holzenberger, M., Dupont, J., Ducos, B., Leneuve, P., Géloën, A., Even, P.C., Cervera, P., Le Bouc, Y. (2003). IGF-1 receptor regulates lifespan and resistance to oxidative stress in mice. *Nature, 421*, 182–187.
16. Henderson, S.T., Johnson, T.E. (2001). *daf-16* integrates developmental and environmental inputs to mediate aging in the nematode *Caenorhabditis elegans*. *Curr Biol, 11*, 1975–1980.
17. Lee, R.Y., Hench, J., Ruvkun, G. (2001). Regulation of *C. elegans* DAF-16 and its human ortholog FKHRL1 by the *daf-2* insulin-like signaling pathway. *Curr Biol, 11*, 1950–1957.
18. Kenyon, C., Chang, J., Gensch, E., Rudner, A., Tabtiang, R. (1993). A *C. elegans* mutant that lives twice as long as wild type. *Nature, 366*, 461–464.
19. Tissenbaum, H.A., Ruvkun, G. (1998). An insulin-like signaling pathway affects both longevity and reproduction in *Caenorhabditis elegans*. *Genetics, 148*, 703–717.
20. Bluher, M., Kahn, B.B., Kahn, C.R. (2003). Extended longevity in mice lacking the insulin receptor in adipose tissue. *Science, 299*, 572–574.
21. Taguchi, A., Wartschow, L.M., White, M.F. (2007). Brain IRS2 signaling coordinates life span and nutrient homeostasis. *Science, 317*, 369–372.
22. Liu, X.D., Liu, P.C., Santoro, N., Thiele, D.J. (1997). Conservation of a stress response: human heat shock transcription factors functionally substitute for yeast HSF. *EMBO J, 16*, 6466–6477.
23. Rabindran, S.K., Haroun, R.I., Clos, J., Wisniewski, J., Wu, C. (1993). Regulation of heat shock factor trimer formation: role of a conserved leucine zipper. *Science, 259*, 230–234.
24. Sarge, K.D., Murphy, S.P., Morimoto, R.I. (1993). Activation of heat shock gene transcription by heat shock factor 1 involves oligomerization, acquisition of DNA-binding activity, and nuclear localization and can occur in the absence of stress. *Mol Cell Biol, 13*, 1392–1407.
25. Singh, V., Aballay, A. (2006). Heat-shock transcription factor (HSF)-1 pathway required for *Caenorhabditis elegans* immunity. *Proc Natl Acad Sci U S A, 103*, 13092–13097.
26. Timmons, L., Fire, A. (1998). Specific interference by ingested dsRNA. *Nature, 395*, 854.
27. Morley, J.F., Morimoto, R.I. (2004). Regulation of longevity in *Caenorhabditis elegans* by heat shock factor and molecular chaperones. *Mol Biol Cell, 15*, 657–664.
28. Parker, J.A., Arango, M., Abderrahmane, S., Lambert, E., Tourette, C., Catoire, H., Néri, C. (2005). Resveratrol rescues mutant polyglutamine cytotoxicity in nematode and mammalian neurons. *Nat Genet, 37*, 349–350.

29. Howitz, K.T., Bitterman, K.J., Cohen, H.Y., Lamming, D.W., Lavu, S., Wood, J.G., Zipkin, R.E., Chung, P., Kisielewski, A., Zhang, L.L., *et al.* (2003). Small molecule activators of sirtuins extend *Saccharomyces cerevisiae* lifespan. *Nature, 425*, 191–196.

30. Wood, J.G., Rogina, B., Lavu, S., Howitz, K., Helfand, S.L., Tatar, M., Sinclair, D. (2004). Sirtuin activators mimic caloric restriction and delay ageing in metazoans. *Nature, 430*, 686–689.

31. Valenzano, D.R., Terzibasi, E., Genade, T., Cattaneo, A., Domenici, L., Cellerino, A. (2006). Resveratrol prolongs lifespan and retards the onset of age-related markers in a short-lived vertebrate. *Curr Biol, 16*, 296–300.

32. Tissenbaum, H.A., Guarente, L. (2001). Increased dosage of a *sir-2* gene extends lifespan in *Caenorhabditis elegans. Nature, 410*, 227–230.

33. Link, C. (1995). Expression of human beta-amyloid peptide in transgenic *Caenorhabditis elegans. Proc Natl Acad Sci U S A, 92*, 9368–9372.

34. Cleary, J.P., Walsh, D.M., Hofmeister, J.J., Shankar, G.M., Kuskowski, M.A., Selkoe, D.J., Ashe, K.H. (2005). Natural oligomers of the amyloid-beta protein specifically disrupt cognitive function. *Nat Neurosci, 8*, 79–84.

35. Townsend, M., Shankar, G.M., Mehta, T., Walsh, D.M., Selkoe, D.J. (2006). Effects of secreted oligomers of amyloid beta-protein on hippocampal synaptic plasticity: a potent role for trimers. *J Physiol, 572*, 477–492.

36. Lesné, S., Koh, M.T., Kotilinek, L., Kayed, R., Glabe, C.G., Yang, A., Gallagher, M., Ashe, K.H. (2006). A specific amyloid-beta protein assembly in the brain impairs memory. *Nature, 440*, 352–357.

37. Klein, W.L. (2002). Abeta toxicity in Alzheimer's disease: globular oligomers (ADDLs) as new vaccine and drug targets. *Neurochem Int, 41*, 345–352.

38. Silveira, J.R., Raymond, G.J., Hughson, A.G., Race, R.E., Sim, V.L., Hayes, S.F., Caughey, B. (2005). The most infectious prion protein particles. *Nature, 437*, 257–261.

39. Chesebro, B., Trifilo, M., Race, R., Meade-White, K., Teng, C., LaCasse, R., Raymond, L., Favara, C., Baron, G., Priola, S., *et al.* (2005). Anchorless prion protein results in infectious amyloid disease without clinical scrapie. *Science, 308*, 1435–1439.

40. Arrasate, M., Mitra, S., Schweitzer, E.S., Segal, M.R., Finkbeiner, S. (2004). Inclusion body formation reduces levels of mutant huntingtin and the risk of neuronal death. *Nature, 431*, 805–810.

41. Humbert, S., Bryson, E.A., Cordelières, F.P., Connors, N.C., Datta, S.R., Finkbeiner, S., Greenberg, M.E., Saudou, F. (2002). The IGF-1/Akt pathway is neuroprotective in Huntington's disease and involves huntingtin phosphorylation by Akt. *Dev Cell, 2*, 831–837.

42. Yamamoto, A., Cremona, M.L., Rothman, J.E. (2006). Autophagy-mediated clearance of huntingtin aggregates triggered by the insulin-signaling pathway. *J Cell Biol, 172*, 719–731.

43. Carro, E., Trejo, J.L., Gomez-Isla, T., LeRoith, D., Torres-Aleman, I. (2002). Serum insulin-like growth factor I regulates brain amyloid-beta levels. *Nat Med, 8*, 1390–1397.

44. Carro, E., Trejo, J.L., Gerber, A., Loetscher, H., Torrado, J., Metzger, F., Torres-Aleman, I. (2006). Therapeutic actions of insulin-like growth factor I on APP/PS2 mice with severe brain amyloidosis. *Neurobiol Aging, 27*, 1250–1257.
45. Murphy, C.T., McCarroll, S.A., Bargmann, C.I., Fraser, A., Kamath, R.S., Ahringer, J., Li, H., Kenyon, C. (2003). Genes that act downstream of DAF-16 to influence the lifespan of *Caenorhabditis elegans*. *Nature, 424*, 277–283.
46. Bosl, B., Grimminger, V., Walter, S. (2006). The molecular chaperone Hsp104: a molecular machine for protein disaggregation. *J Struct Biol, 156*, 139–148.
47. Bandhyopadhyay, U., Cuervo, A.M. (2007). Chaperone-mediated autophagy in aging and neurodegeneration: lessons from alpha-synuclein. *Exp Gerontol, 42*, 120–128.
48. Iwata, N., Tsubuki, S., Takaki, Y., Shirotani, K., Lu, B., Gerard, N.P., Gerard, C., Hama, E., Lee, H.J., Saido, T.C. (2001). Metabolic regulation of brain Abeta by neprilysin. *Science, 292*, 1550–1552.
49. Behrends, C., Langer, C.A., Boteva, R., Böttcher, U.M., Stemp, M.J., Schaffar, G., Rao, B.V., Giese, A., Kretzschmar, H., Siegers, K., Hartl, F.U. (2006). Chaperonin TRiC promotes the assembly of polyQ expansion proteins into nontoxic oligomers. *Mol Cell, 23*, 887–897.
50. Shorter, J., Lindquist, S. (2004). Hsp104 catalyzes formation and elimination of self-replicating Sup35 prion conformers. *Science, 304*, 1793–1797.
51. Glover, J.R., Lindquist, S. (1998). Hsp104, Hsp70, and Hsp40: a novel chaperone system that rescues previously aggregated proteins. *Cell, 94*, 73–82.
52. Nollen, E.A., Salomons, F.A., Brunsting, J.F., Want, J.J., Sibon, O.C., Kampinga, H.H. (2001). Dynamic changes in the localization of thermally unfolded nuclear proteins associated with chaperone-dependent protection. *Proc Natl Acad Sci U S A, 98*, 12038–12043.
53. Fonte, V., Kipp, D.R., Yerg, J., 3rd, Merin, D., Forrestal, M., Wagner, E., Roberts, C.M., Link, C.D. (2008). Suppression of in vivo beta-amyloid peptide toxicity by overexpression of the HSP-16.2 small chaperone protein. *J Biol Chem, 283*, 784–791.
54. Ciechanover, A., Brundin, P. (2003). The ubiquitin proteasome system in neuro-degenerative diseases: sometimes the chicken, sometimes the egg. *Neuron, 40*, 427–446.
55. Cohen, E., Taraboulos, A. (2003). Scrapie-like prion protein accumulates in aggresomes of cyclosporin A–treated cells. *EMBO J, 22*, 404–417.
56. Bennett, E.J., Bence, N.F., Jayakumar, R., Kopito, R.R. (2005). Global impairment of the ubiquitin-proteasome system by nuclear or cytoplasmic protein aggregates precedes inclusion body formation. *Mol Cell, 17*, 351–365.
57. Cuervo, A.M., Stefanis, L., Fredenburg, R., Lansbury, P.T., Sulzer, D. (2004). Impaired degradation of mutant alpha-synuclein by chaperone-mediated autophagy. *Science, 305*, 1292–1295.
58. Leissring, M.A., Farris, W., Chang, A.Y., Walsh, D.M., Wu, X., Sun, X., Frosch, M.P., Selkoe, D.J. (2003). Enhanced proteolysis of beta-amyloid in APP transgenic mice prevents plaque formation, secondary pathology, and premature death. *Neuron, 40*, 1087–1093.
59. Massey, A.C., Zhang, C., Cuervo, A.M. (2006). Chaperone-mediated autophagy in aging and disease. *Curr Top Dev Biol, 73*, 205–235.

60. Florez-McClure, M.L., Hohsfield, L.A., Fonte, G., Bealor, M.T., Link, C.D. (2007). Decreased insulin-receptor signaling promotes the autophagic degradation of beta-amyloid peptide in *C. elegans*. *Autophagy, 3*, 569–580.

61. Gidalevitz, T., Ben-Zvi, A., Ho, K.H., Brignull, H.R., Morimoto, R.I. (2006). Progressive disruption of cellular protein folding in models of polyglutamine diseases. *Science, 311*, 1471–1474.

PART IV

MEDICAL ASPECTS OF DISEASE: DIAGNOSIS AND CURRENT THERAPIES

30

IMAGING OF MISFOLDED PROTEINS

Harry LeVine, III

Department of Molecular and Cellular Biochemistry, University of Kentucky, Lexington, Kentucky

INTRODUCTION

Protein misfolding is being recognized as a primary etiology in an increasing number of diseases. Mutation is well known to destabilize protein structure, leading to the affected protein aggregating in the endoplasmic reticulum, accumulating intracellularly, and activating the unfolded-protein response. What is less intuitive are the sporadic diseases defined pathologically by misfolded-protein deposits that are not associated with mutation or endoplasmic reticulum disturbance as the primary insult. Age is the primary risk factor and many of the diseases involve the brain. The proteins involved are from a recently recognized class of proteins containing lengthy sequences described variously as natively disordered, intrinsically disordered, or natively unfolded [33,116]. While the flexibility of these disordered regions is thought to allow low affinity but specific interactions upon binding to target sites on multiple different proteins or subcellular components, a liability is their potential for also forming intra- and intermolecular cross-β-sheet structure [14,36,69], causing aggregation, insolubility, and resistance to degradation and impaired clearance.

Neurons of the brain are particularly vulnerable to the accumulation of these materials because once they mature and differentiate, they are postmitotic. Mutations in these minimally stable proteins can enhance their β-sheet propensity, as has been observed for forms of familial Alzheimer disease (Aβ), and for

Protein Misfolding Diseases: Current and Emerging Principles and Therapies,
Edited by Marina Ramirez-Alvarado, Jeffery W. Kelly, and Christopher M. Dobson
Copyright © 2010 John Wiley & Sons, Inc.

Parkinson disease (α-synuclein), dementia with Lewy bodies (α-synuclein), as well as frontotemporal dementias (tau) and other tauopathies [20,121,126].

The pathologies of these diseases are currently defined by the postmortem observation of amorphous masses and various forms of amyloid fibrils in characteristic regions of the brain. Staining with a standard set of histological dyes (Congo Red with birefringence, thioflavin S, and Gallyas silver) reflects similarity on a macromolecular scale, although the amino acid sequences differ. Although there is honest disagreement about whether the proximal etiologic agent is a soluble oligomeric form of misfolded protein or the insoluble deposits, the two species are probably linked by a molecular equilibrium. The general temporal sequence of pathology development followed by the clinical manifestations of disease has suggested to many observers that detecting incipient preclinical pathology might provide enough lead time to interfere in a meaningful way with disease progression in these otherwise incurable diseases. Specific and robust diagnostic biochemical markers for these diseases have thus far proven elusive. Could an answer be to identify early brain deposit pathology that defines the disease in the living subject brain?

WHY IMAGING?

Biomarkers of disease are important for diagnosing disease and for determining the efficacy of therapeutic interventions. Many of these involve changes in biochemical parameters of body fluids, usually blood or urine. Chronic neurodegenerative diseases such as Alzheimer disease (AD) or Parkinson disease (PD) are defined clinically by specific functional losses reflecting pathology that develops in particular regions of the brain responsible for the functions affected. Early predictive biochemical markers for these events have been difficult to establish. The pathologies involve distinct proteins affecting different neurotransmitter systems in a variety of brain regions; AD produces dementia in specific cognitive domains, while PD results in lesions of motor control. Yet both result from misfolding events of proteins (Aβ, tau, α-synuclein) leading to neuronal dysfunction and death. Definitive diagnosis of AD has traditionally required postmortem histological and morphological analysis of the brain. Modern techniques of brain imaging can now detect changes in brain regional anatomy and functional properties in the living subject before overt clinical signs develop. Refinement of imaging methods and the development of ligands for imaging the pathological lesions have the potential to identify biomarkers signaling ever-earlier preclinical changes in the brain [19].

EARLY DETECTION

The key to treatment of chronic neurodegenerative diseases is to identify the early signs, because by the time clinical symptoms appear, irreparable damage has occurred. The presence of disease is signaled by changes in body chemistry, which is often reflected in the blood or other body fluids. Specific changes that can be correlated with the presence of a particular disease process are termed *biomarkers*. Familiar examples are cholesterol bound to a particular type of lipoprotein particle in the blood for cardiovascular disease, and prostatic serum antigen (PSA) overproduced by prostatic cancer cells for prostatic cancer. Monitoring of these biomarkers permits early diagnosis of disease and assessment of the efficacy of therapy.

The search for a serum biomarker in AD, the most common chronic neurodegenerative disease, has been punctuated with reports of selectivity and specificity comparable to clinical evaluation which lose significance in larger studies [25,108,113]. Groups of markers are now being proposed for use in AD, as the likelihood of the discovery of a single predictive marker is considered extremely low. A panel of 18 cytokines has recently been proposed to detect an early stage of AD called mild cognitive impairment (MCI) [96], but the performance of this test remains to be confirmed by other groups and in a larger cohort of subjects. In AD as well as other chronic neurodegenerative diseases, progression is prolonged and insidious for most of the disease course. Informative early biomarkers could provide improved opportunity for therapeutic intervention. Support for early intervention comes from retrospective epidemiological studies showing significant reduction in the incidence of AD in populations that were receiving drug treatments for other conditions, such as nonsteroidal anti-inflammatories [3,112] or cholesterol-lowering statins [129]. By contrast, direct trials demonstrated little curative effect. This suggested that the positive effects of these treatments in the retrospective epidemiological studies could be attributed to delaying the onset of symptoms rather than affecting established disease [97,109].

Since the Aβ-containing amyloid plaques and neurofibrillary tangles (NFTs) comprised of hyperphosphorylated tau are the defining lesions of AD, and definitive earlier markers have not been established, perhaps these pathologies can be employed as reference biomarkers for the disease. The progression of the pathology described by Braak and Braak [9,10] and others showed that the lesions developed in appropriate brain regions prior to clinical dysfunction. Although the NFT pathology appears more closely related to neuronal dysfunction, Aβ pathology preceded tau pathology in a triple transgenic mouse model (APP, PS1, tau) and could drive the tau pathology while the reverse was not observed [84]. Tau pathology, although morphologically distinguishable, also occurs in a distinct set of neurodegenerative diseases, the tauopathies, represented by frontotemporal dementia [11]. Thus, in AD, Aβ appears to be a prime mover. This provides support for the idea that imaging ligands that are sensitive and relatively specific for Aβ pathology could have the potential to detect preclinical pathology which could serve as a biomarker of disease.

HISTOLOGY

The standard for determining protein deposition pathology before in vivo imaging was histology, the staining of biological tissue samples with organic dyes or metal ion treatments to reveal pathological features. Protocols were devised to highlight specific structures. In the mid-nineteenth century, before AD pathology was described by Alois Alzheimer, an abnormal structure was recognized in diseases where masses of protein accumulated in the spleen, heart, and kidney. These waxy deposits stained with iodine in a way that reminded Rudolph Virchow of cellulose, so he called them amyloid, believing them to be made up of carbohydrate [117]. They also stained with the cotton dye Congo Red, and the regular fibrillar structure of these deposits was revealed by their metachromasy and birefringence under polarized light. This green Congo Red birefringence became a definitive method for detecting fibrillar amyloid protein deposits. Assessment of amyloid content required representative biopsy as well as experience in histology and pathological interpretation, as Congo Red binding and even birefringence (collagen fibrils) are not restricted to amyloid fibrils. Originally, the term *amyloid* was an adjective to describe a particular type of fibrils, but it is now officially applied to both intracellular and extracellular protein deposits. According to the Nomenclature Committee of the International Society of Amyloidosis, amyloid now refers to "an in vivo deposited material, which can be distinguished from non-amyloid deposits by characteristic fibrillar electron microscopic appearance, typical x-ray diffraction pattern and histological staining reactions, particularly affinity for the dye Congo Red with resulting green birefringence" [127]. The biochemical literature is somewhat confused, as the nomenclature for amyloid-related structures and structural intermediates remains in flux [20,31].

The regular assemblies of amyloid fibrils formed from a variety of proteins of different primary amino acid sequence [106] were shown by George Glenner [37,38] to be rich in highly similar cross-β-sheet secondary structure. Common x-ray fiber diffraction patterns further emphasized the similar organization of the macromolecular fibril structure [101,102]. Dyes differing in structure from Congo Red, such as the benzothiazoles, the cationic thioflavine T and thioflavine S, also show relative selectivity for amyloid fibril structures. Thioflavine T and thioflavine S, but not other benzothiazoles such as BTA-1, change their fluorescent properties in characteristic ways when they bind to amyloid fibrils. This has been useful for in vitro studies of amyloid assembly [67,77]. Congo Red and the thioflavines were starting points for the development of amyloid imaging ligands. The size, charge, and metabolic liabilities of the chemical structures of the classical dyes prevented them from being used directly for in vivo brain imaging.

LABELING OF AD PATHOLOGY IN VIVO

Whereas histological staining specifically delineating the different components of AD pathology could readily be performed postmortem on tissue sections, achieving similar specificity in the living brain introduced orders of magnitude more complexity. Irrespective of the mode of detection, there are issues such as pharmacokinetics and pharmacodynamics which govern whether the label will reach the intended target at an appropriate concentration and be retained long enough to measure. This is overlaid on the specificity and selectivity for the target under prevailing metabolic conditions. In the living subject there is no opportunity to "clean up" nonspecific interactions. In vitro tissue sections are washed and chemically fixed or cross-linked with aldehydes before they are stained with high concentrations of ligand and frequently differentiated with alcohols to remove nonspecifically bound ligand. Sometimes the sections are subjected to antigen retrieval by heating or treatment with concentrated formic acid to remove interfering material or to expose binding sites. Obviously, none of these manipulations can be performed in vivo; thus, in vivo histology now enters the realm of receptor-binding pharmacology, where specific and nonspecific binding have to be defined and differentiated. A similar challenge of in vivo versus in vitro was encountered and answered by Paul Ehrlich, who is credited with articulating the concept of targeted chemotherapy and its successful clinical application. Ehrlich suggested that dyes that specifically stained pathogenic microorganisms bound to unique "chemoreceptors" and that derivatives of these dyes containing an added "toxicophore" might be therapeutic. In 1907, after testing 605 compounds, he identified salvarsan (diaminodioxyarsenobenzene). "606" was a miracle drug for syphilis marketed by Hoechst in 1910.

Following an analogous process, ligands for AD plaque pathology were derived from Congo Red [24,51,52,99,110]. A series of modifications substituted the metabolically labile and potentially carcinogenic azo linkages and the charged sulfonate groups [52]. The evolution of the structures can be followed in Figure 1. The planar core structure and its molecular dimensions were essential for the binding while the "spinach" modifications promoted solubility and acceptable biodistribution. The new molecules penetrated the blood–brain barrier, survived first-pass metabolism, and showed improved affinity for fibrillar amyloid deposits.

Other dyes commonly used to stain AD pathology are the fluorescent thioflavines S and T. Thioflavine S is preferred by pathologists for staining tissue sections, while thioflavine T is favored for in vitro studies of amyloid fibril formation [65–67, 77]. While Congo Red stains neuritic amyloid plaques, the thioflavines bind to both neuritic Aβ deposits and to neurofibrillary tangles as well as Lewy bodies containing α-synuclein. The tangles are comprised of

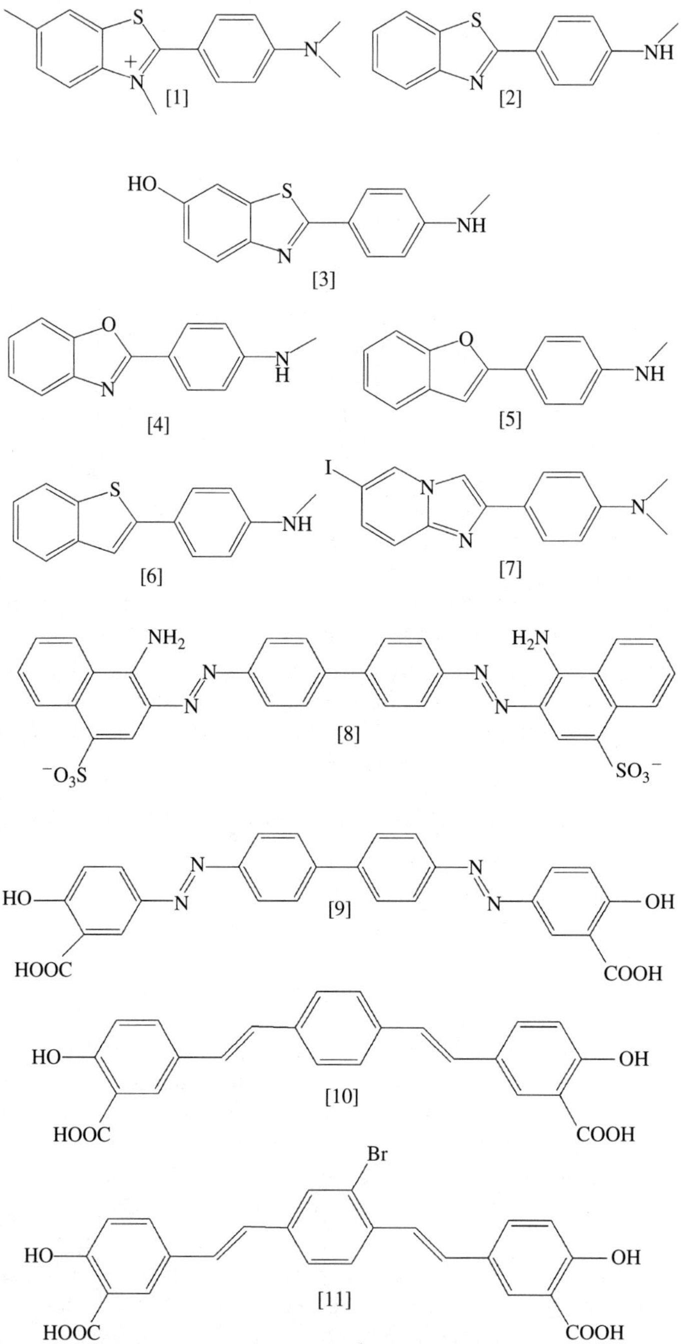

FIG. 1 Chemical structures of amyloid ligands.

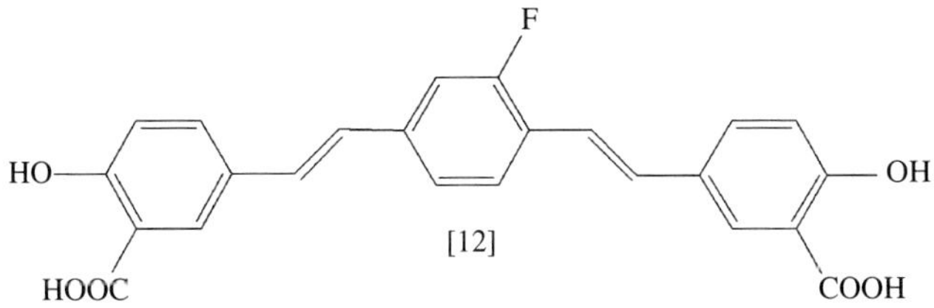

[12]

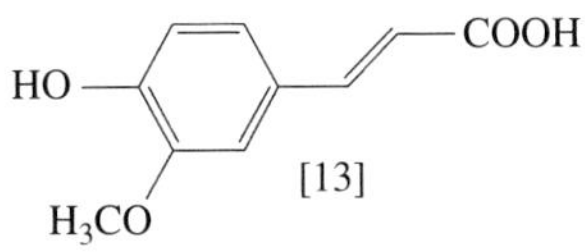

[13]

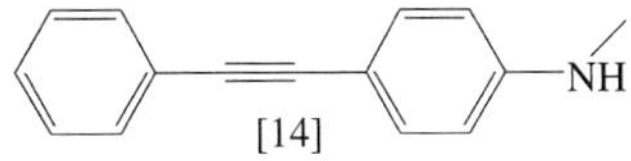

[14]

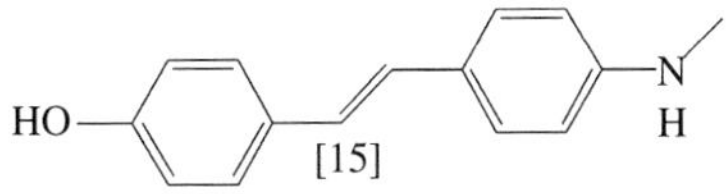

[15]

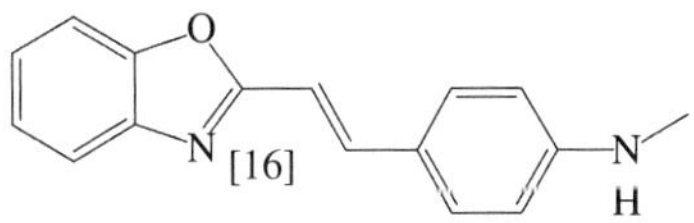

[16]

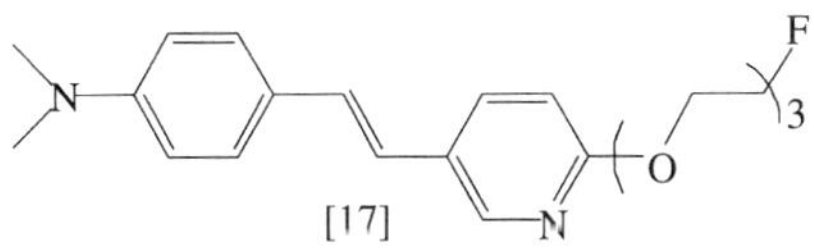

[17]

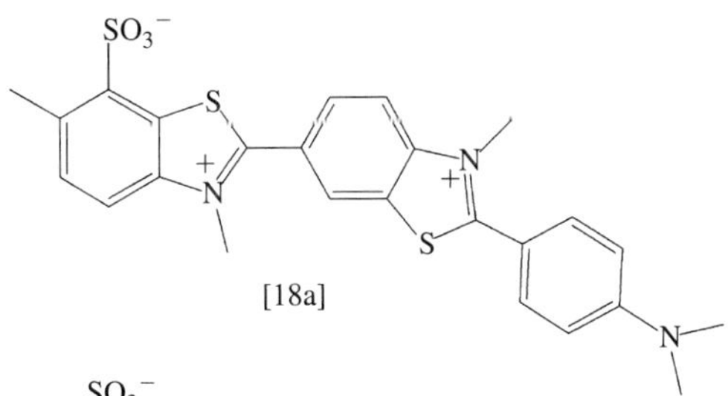

[18a]

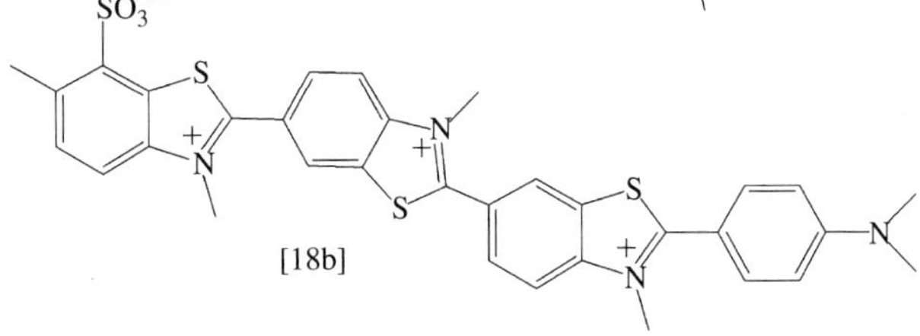

[18b]

FIG. 1 *(Continued)*

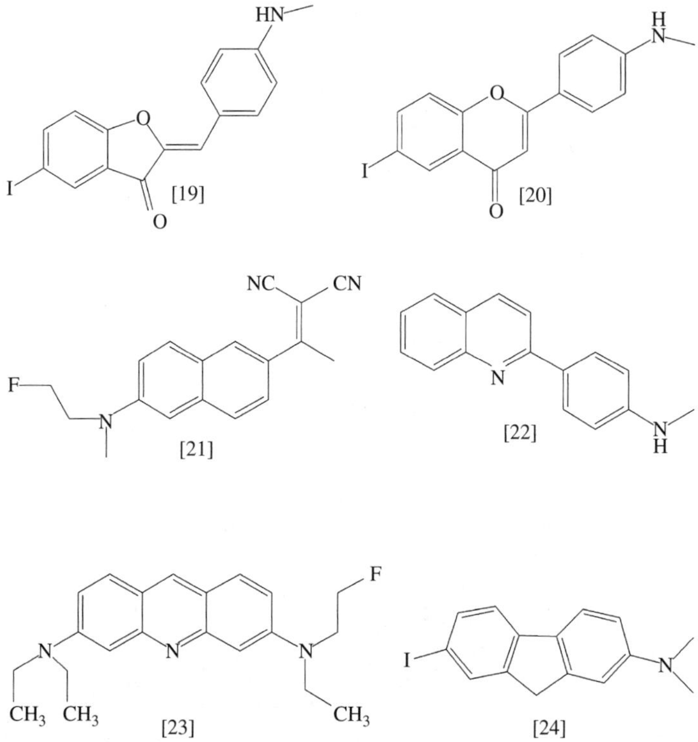

FIG. 1 ***(Continued)***. The structure name for the ligand is followed by the reference citation. 1, Thioflavin T [77]; 2, BTA-1 [56]; 3, PIB [53]; 4, benzoxazole [18] ; 5, benzofuran [86]; 6, benzothiophene [18]; 7, IMPY [61]; 8, Congo Red [52]; 9, Chrysamine G [52]; 10, X-34 [110]; 11, bromostyrylbenzene (BSB) [99]; 12, fluorostyrylbenzene (FSB) [98]; 13, ferulic acid [13]; 14, diphenylacetylene [17]; 15, styrylbenzene SB-13 [60]; 16, styrylbenzoazole [43]; 17, styrylpyridine [95]; 18a, 18b, thioflavine-S components [125]; 19, aurone [87]; 20, flavone [88]; 21, FDDNP [1]; 22, benzoquinoline BF-158 [85]; 23, aminoacridine BF-108 [111]; 24, fluorene [58].

polymers of hyperphosphorylated microtubule protein tau that are distributed differently temporally and regionally from neuritic Aβ deposits. This was originally seen as a potential confounding complication due to frequent comorbidity of tangles and Lewy bodies in AD as well as tangle presence in the absence of Aβ in the frontal temporal dementias and Lewy bodies in dementia with Lewy bodies and Parkinson disease. It turned out to be much less of an issue than was anticipated. For unexplained reasons, engineering in selectivity for Aβ binding over tau tangles has been accomplished readily in all the chemical series of potential imaging agents. Thioflavines S and T both contain the benzothiazole nucleus. Thioflavine T has received the most developmental attention because it is smaller, easily synthesized and modified, and is a single molecular entity, unlike thioflavine S, which is a mixture of larger primuline-like compounds (Fig. 1, 18a and 18b) [125].

A great variety of thioflavine-T analogs have been prepared (Fig. 1). This has facilitated the medicinal chemical exploration of the structure–activity relationship (SAR) of Aβ fibril binding. Applying analyses akin to those used for ligand–receptor interactions, synthetic fibril preparations and human and transgenic mouse brain homogenates have been characterized with respect to the affinity and stoichiometry of binding sites for the ligands. There are multiple, at least three, types of benzothiazole binding sites as defined by their affinity and by their position-dependent tolerance for modification of the benzothiazole nucleus [70]. Some of them appear to overlap with the binding site for Congo Red and thioflavine S [29]. This type of analysis remains in flux because of site heterogeneity and the existence of Aβ fibril polymorphism [91]. There are important ramifications of fibril polymorphism for comparison of synthetic Aβ fibrils and transgenic mouse model plaques to disease-associated pathology in the AD brain. This crucial topic is considered in a later section.

IN VIVO IMAGING

In vivo imaging of acetylcholine, opiate, benzodiazepine, or dopamine neurotransmitter binding has been interpreted successfully with the help of sophisticated models that take into account ligand availability, blood flow and diffusion in interstitial fluid, and binding kinctics. It is tempting to treat in vivo ligand binding to AD plaque and/or tangle pathology similarly. However, the situation is more complex for misfolded protein deposits because of the unknown contribution of the deposit microenvironment, the density of deposits, perturbation of local perfusion, and penetration of ligand into heterogeneous deposits [74], which are less of a problem for receptors [21]. Quantitative interpretation of images—in particular, changes that occur within a person during disease progression or with therapeutic intervention—are subject to significant uncertainty. Modeling schemes continue to be developed [103–105] to be able to deal with these issues. A detailed accounting of the parameters involved is a subfield in itself.

In vivo brain imaging of radioligands employs either SPECT (single-photon-emission computed tomography) or PET (positron-emission tomography). Briefly, these are relatively low-resolution methodologies (7 mm for SPECT; 1 to 2 mm for PET) that employ short-half-life gamma emissions and relatively short scan times (important for agitated demented subjects) to generate maps of the regional distribution of labeled ligand. Individual plaques (10 to 250 μm) cannot be resolved. SPECT instruments and the longer-lived isotopes (^{123}I:13.13 h) are less sensitive but cheaper and more readily available than those for PET, whose very short half-lived isotopes (^{11}C:20.3 min; ^{18}F:109.8 min) require a nearby cyclotron and rapid synthesis or purification of the labeled ligand. Because of its sensitivity, speed of acquisition, resolution, and utility in quantitative and temporal studies, PET with either ^{11}C or ^{18}F is used for most amyloid imaging studies, and it is effective for both structural studies and metabolic functional measurements such as glucose uptake with [^{18}F]fluorodexoyglucose (FDG) [22]. ^{123}I SPECT is used for proteins and peptides or small organic molecules such as IMPY (Fig. 1) which are readily iodinated.

Benzothiazoles

The most extensively in vivo studied ligand is Pittsburgh compound B (PIB) [2-(4′-methylaminophenyl)-6-hydroxybenzothiazole], derived from the structure of thioflavine T. Binding of tritiated 2-(4′-methylaminophenyl)benzothiazole (BTA-1) to AD brain homogenates was shown to be dominated by the amyloid component [55,56]. A number of changes were incorporated to improve the physicochemical properties of the ligand and to increase its affinity to the nanomolar (nM) level, which is a rule of thumb for successful imaging ligands [124]. The charge on the ring quaternary nitrogen of thioflavine T was removed by omitting the ring *N*-methyl group, the phenyl dimethylamine was converted to a secondary amine, and the ring methyl group was replaced with a hydroxyl. These changes optimized the affinity, brain uptake, retention, and washout characteristics. The ring hydroxyl was a compromise between lipophilicity and metabolism. Modeling of the interactions of PIB with comparison to data acquired with human subjects [94] is being used to validate simpler means of quantifying amyloid load [130]. IMPY is another thioflavine T analog that can be labeled for SPECT [62,78] or PET [15,16,100], which has been used in APP-overexpressing transgenic mice [59].

PIB has been used in humans since 2004 [53]. In multiple reports PIB binding has been able to distinguish AD from age-matched nondemented brain [35,49]. One of the clinical hallmarks of AD dementia that distinguishes it from dysfunction due to an acute insult such as a stroke is the lengthy and progressive decline of cognitive function in AD. The progression of AD has been divided into stages based on clinical signs. A specific type of mild cognitive impairment, (MCI), amnestic MCI, is suggested to be a precursor to AD because a larger proportion of MCI patients progress to the dementia of AD than directly from normal to AD [8,64,90]. The hope is that PIB binding during or before the clinical predementia early stages of AD will distinguish those who will not become demented from those who will become demented. A potential complication is that progression to AD dementia is not entirely a simple one-way street, at least in the early stages. Some amnestic MCI patients revert to normal; others cycle back and forth between normal and MCI. It is not clear at this time whether the pathology is reversible or whether there are compensating mechanisms in some people. PIB binding detected amyloid deposition in 50 to 60% of MCI patients, and binding was higher in those who converted to AD at later follow-up [35,49]. Larger patient numbers are needed to confirm the finding that MCI patients with higher PIB binding are at risk to convert at an accelerated rate to frank AD.

Naphthylmalononitriles

Another small-molecule ligand used for imaging amyloid deposits spawned a series of derivatives of naphthylmalononitriles. Originally used as microenvironment viscosity reporters [1,92], these ligands bind with sub-nM to nM affinity to a low-abundance site (1:3.5 $\times$ 10^4 Aβ) in a relatively hydrophobic region of Aβ fibrils that is still exposed to solvent. FDDNP binds to NFT pathology but more weakly than to Aβ plaques, a characteristic shared with PIB. An ^{18}F-labeled fluoroethyl derivative of DDNP, FDDNP, (Fig. 1) has been used for both in vitro binding and in living human subjects for in vivo imaging in AD [50,107]. Binding of FDDNP to prion aggregates in vivo is weaker than for PIB [7,12]. Competition for FDDNP binding is observed with (*S*)-naproxen and (*R*)- and (*S*)-ibuprofen [2], nonsteroidal anti-inflammatories being investigated as potential prophylactic treatments for AD. The lack of competition for Congo Red analogs or benzothiazole derivatives [68,70,128] suggests distinct binding sites on fibrils for the naphthylmalononitriles and the traditional amyloid dyes.

The cross-reactivity of DDNP analogs for both senile plaques and NFTs at the concentrations employed for imaging in vivo could be either an asset or a detriment. This cross-reactivity with NFTs is more pronounced for the DDNP series than for the thioflavine- or Congo Red–like molecules. Depending on the intended use, dual specificity either complicates interpretation of images or gives a more complete picture of overall pathology. It is likely that higher selectivity could be achieved by modifying the DDNP molecular structure. Tau pathology-selective quinoline and benzimidazole compounds (BF-126, BF-158, BF-170) (Fig. 1) recognize the paired helical NFTs in tissue sections but, interestingly, not the straight tau fibrils in Pick disease or the globose tangles and glial fibrillary inclusions in progressive supranuclear palsy (PSP) [85]. Derivatives of bis-styrylbenzenes also vary in their relative selectivity for tau and Aβ [34]. These compounds have the potential to be developed as imaging agents, as they are brain-penetrant. Relatively little has been published on tau-specific imaging reagents. Imaging agent selectivity is also likely to be achievable for other misfolded proteins. Small molecules that can distinguish among Aβ, tau, and α-synuclein have been found by screening libraries of synthetic compounds [42].

Antibodies and Peptides

Antibodies have been employed successfully to target therapeutics or as diagnostic imaging agents outside the brain. The blood–brain barrier (BBB) imposed by the endothelial cells of the cerebral blood vessels and the brain astrocytes efficiently excludes large molecules such as proteins unless they are transported specifically. Nevertheless, a small fraction of 1% of injected antibodies enter the brain by unknown mechanisms [5]. Reduction of the size of the antibodies to Fab fragments or conversion to single-chain antibodies also improves uptake [39]. Temporary opening of the BBB induced by mannitol osmotic stress increases large-molecule penetration, but it could be a risky procedure in AD subjects since some of them already have a compromised BBB. Chemical modification of the agent by coupling of cationic polyamines such as putresceine increases uptake without compromising the BBB. This approach has permitted brain targeting of a variety of ligands, including anti-Aβ antibodies radioisotopically labeled with ^{123}I for SPECT imaging or magnetic contrast labeled with gadolinium for MRI (see the section on MRI).

The exquisite specificity of amyloid fibril growth for the cognate peptide sequence [83] has been used selectively on fibrillar Aβ plaques and polyglutamine inclusions [89] for radiolabeling [73,114] or magnetic contrast tagging [46]. Aβ peptides are surprisingly permeable in both directions through the BBB utilizing the receptor for advanced glycation end products (RAGE) and the lipoprotein-receptor-like receptor LPRLR [26]. These systems have not received the attention accorded the small-molecule ligands, for a number of reasons. These include issues of sensitivity as well as potential immunogenicity and its implications for follow-up imaging. Diffusion of these larger molecules in the interstitial spaces and penetration into the interior of deposits is unknown but will probably be restricted to the periphery of the deposit structures, which complicates quantification.

Individual Plaque Detection: microNMR Imaging

The imaging methodology with the highest spatial resolution is nuclear magnetic resonance imaging. Commonly referred to as MRI because early proponents of the technology felt that the public would fear that radioactivity was involved, the water molecules of the tissue are the origin of the signals. No ionizing radiation is involved, but the subject is confined in the field of a powerful magnet, and a radiofrequency signal is used to excite the magnetic dipoles of the protons in the water nuclei. The environment of the water molecules influences the dipoles, and when the signal is processed, an "image" corresponding to those different environments can be created. Because this technique produces such exquisite structural detail of soft tissues such as the brain, MRI images are often used to provide coordinates to register the lower-resolution SPECT and PET images, which detect only the radiolabeled ligand. This is a well-developed and vibrant field with many studies, and a large body of literature using MRI to quantify structural changes in the brain. Brain regions such as the hippocampus atrophy, the cortex thins, and the ventricles enlarge as the neuronal loss in AD progresses. Functional brain parameters relevant to the pathology of AD can also be determined through magnetic resonance measures sensitive to blood flow, and cerebral metabolic use of oxygen. Since this chapter is concerned with ligand-based imaging of misfolded proteins, the focus in this section is on ligand-induced changes in MRI parameters of plaque-containing regions.

BRAIN PATHOLOGY

Magnetic resonance imaging of AD plaque pathology is attractive because it is relatively high resolution, less expensive than PET imaging, requires either no labeling agents or uses reagents that are stable in storage, and does not require exposure of subjects to radioactive substances. It does require immobilization of the subject under claustrophobic conditions for relatively long times (hours), which are issues for AD subjects and many nondemented persons. MR microimaging (μMRI) studies of plaques to date have been restricted to APP transgenic mice for technology development. Plaques in a postmortem AD brain were seen only after 20 hours of imaging [6] and were not detected in isolated AD brain tissue using higher field strength and resolution [27].

Nonimaging magnetic resonance studies use a highly homogeneous magnetic field to bring all nuclear spins into resonance in a band of excitation frequencies excited simultaneously by a radio-frequency pulse. The result is an average of the signal over the volume within the receiver coils. Details of the molecular structures of the molecules bearing the spins can be determined with high resolution as well as information about their environment and motional characteristics. No information about the location of spins within the sample volume is available from this type of measurement. Images are produced in MRI by controlled variation of the magnetic field in space (a field gradient) to bring separate depths of the sample into resonance at different times. Data processing results in a series of "slices" that are combined for tomographic (three-dimensional) spatial resolution. Resolution is determined both by the quality of the field gradient and the signal intensity within a volumetric element (voxel). Individual plaques 100 to 250 μm in diameter have been resolved in vivo in transgenic mice [30], albeit with lengthy signal acquisitions for typical field strengths. A variety of radio- frequency pulse schemes have been implemented to improve plaque detection by suppressing background. Higher magnetic fields (9.4 and 17.6 tesla) provide greater sensitivity but also compress the T2 transverse relaxation difference between plaques and neuropil [30]. These higher magnetic fields have also not yet been approved for use with humans.

Plaque detection requires contrast between plaque elements and background tissue signal. Unlabeled plaques are observable because the water structure around/in them is different from the surrounding tissue. Aβ plaques also accumulate iron and possibly other paramagnetic metals that alter the water signal [28] which can be processed to make them stand out. Targeting agents were developed to increase sensitivity and distinguish conditions other than the presence of plaques (calcium or metal-ion deposits, nonamyloid lesions) that produce artifactual signals otherwise indistinguishable from plaques. Antibodies or cognate amyloid peptides were conjugated to gadolinium chelates or superparamagnetic iron oxide nanoparticles to act as MRI contrast agents or to otherwise alter the water signal in the vicinity of the targeting molecule bound to a plaque [44–46,93].

Dynamics of Plaques

Looking at the pathology in postmortem brain sections gives a static two-dimensional picture of misfolded-protein deposits and their effects on surrounding neuronal structures. A series of elegant studies by Hyman and collaborators using laser confocal microscopy to study the structure of plaques in AD brain tissue [23] and in transgenic mouse brain tissue [115] revealed plaques with a porous three-dimensional network substructure. Clearly not all portions of that network are going to be equally accessible to all ligands. It is also unknown whether the plaque pathology in early stages of the disease will have the same structure and accessibility as later stages. Subsequent multi-wavelength three-dimensional multiphoton imaging within the brains of living transgenic mice through transparent windows in the skull emphasized the dynamic nature of plaque structure and labeling with ligands. They were also able to demonstrate the enhanced production of free radicals around neuritic plaques and neuronal responses to anti-Aβ antibody administration [4,71,74]. While some of the dynamism may reflect the generally more mutable pathology seen in transgenic mice due to lack of cross-linking and other modifications that accumulate over years in humans, the observations emphasize that some changes may occur rapidly, even in human brains, especially in early stages of the disease. Variability should be expected, possibly accounting for some of the back-and-forth transitions in MCI.

What the Ligands Can Tell Us

Misfolded-protein diseases or proteopathies [120–122] are all characterized by protein deposits in affected tissues. Each disease involves a distinct protein containing natively disordered sequences expressed in tissues or organs other than those in which they are deposited. The ultimate key to the disease mechanism is probably tied to the mystery surrounding the selectivity of the pathology. Since both genetic and sporadic forms of these proteopathies occur, differing mainly in the age of onset, conformational stability is a key property. For the spectrum of prion diseases, infectious proteopathies [118,123], multiple conformational misfolded states of the same protein, have been identified and linked to characteristic clinical presentation with histologically distinguishable protein deposits in particular brain regions. AD appears to be a dual proteopathy involving both Aβ and tau depositing in temporal sequence [9,10]. The Aβ peptide deposits in two main forms, diffuse and senile or neuritic plaques, distinguishable by the presence or lack of amyloid dye staining, birefringence of Congo Red staining, and thioflavine fluorescence. This dichotomy may be oversimplified, as other reagents, such as anionic and cationic polythiophenes, reveal heterogeneity within the organization of individual plaques in AD brain [41,79–81]. Although considerable genetic and biochemical evidence implicates Aβ in the etiology of clinical AD, there is discordance between the regional location of Aβ deposits and the disrupted clinical modality, including the senile plaques, which are most closely correlated with disease.

A potential explanation for the mismatch between visible pathology and regional dysfunction consistent with the involvement of Aβ in the disease is that the particular Aβ assembly responsible for the effects is not being distinguished from other forms [76]. Soluble toxic Aβ oligomers are currently a leading candidate, causing synaptic dysfunction leading to neurodegeneration [32,48]. This species of Aβ represents a small proportion of the already low concentration of soluble Aβ in brain [40], which makes them challenging to detect even with available oligomer conformation-specific antibodies [47,63]. Currently there are no small-molecule ligands published that are specific for oligomers *and* have high enough affinity (= 1 nM) for in vivo imaging.

A key observation with the PIB amyloid imaging ligand may lead to a better understanding of the etiology of AD and possibly its use as a biomarker for development of the disease. Klunk et al. [54] noted the discrepancy between the high-affinity binding of PIB to affected regions of AD brain and the lack of high-affinity binding to plaque-rich regions of brain in several APP transgenic mouse models depositing human sequence Aβ. High-affinity specific binding was undetectable in unaffected brain regions in AD brain or in any region of non-AD control brain. The lack of binding to the plaque-rich APP transgenic mouse brain has been taken by some investigators to question the relevance of PIB binding to disease. Abundant plaques in the mice are visible with Aβ antibodies and with the high concentrations of unlabeled dye used for histologically staining tissue sections. Subsequent labeling with nM ^{11}C-PIB with tenfold-higher specific radioactivity detected binding, but the stoichiometry calculated was 1 PIB per 5000 or 10,000 Aβ peptides [72]. This stoichiometry is similar to that for synthetic Aβ fibrils [54]. The stoichiometry calculated in affected regions of AD brain is much higher, on the order of 1 PIB per 2 Aβ peptides [54]. Interestingly, the binding stoichiometry of Congo Red and its derivatives is similar in synthetic Aβ fibrils and AD brain (ca. 1:1) [52,68], which is consistent with distinct localization of the high-affinity PIB site and the Congo Red site.

The high-affinity PIB binding remains associated with the insoluble Aβ upon fractionation of AD brain [54]. Whether the phenomenon of high-affinity PIB binding is causative or results from the AD disease process, the physical basis for the binding needs to be understood. A reasonable hypothesis is that it represents a disease-associated specific conformation of Aβ or an as-yet undefined specific Aβ complex [54]. There is precedence from solid-state NMR studies for hydrogen-bonding polymorphisms in synthetic Aβ fibrils [91]. Amyloid fibril assemblies are known to grow with exquisite conformational fidelity [57,82,83], a property reflected in the retention of species barriers and strain-specific properties of prions. AD brain extracts readily seed Aβ fibril deposition in the brains of APP-overexpressing mice, while synthetic fibrils are much less efficient when the same amount of Aβ peptide is infused [75,119,123]. Hence Aβ assembly polymorphism may be the key to resolving the conundrum of the amyloid hypothesis and offers the potential for the development of predictive biomarkers.

CONCLUSIONS

Imaging of misfolded-protein pathology in diseases such as AD and PD is beginning to provide a window into early presymptomatic stages of chronic neurodegenerative disease. This is important because studies indicate that early intervention has the greatest impact on delaying the onset of clinical dementia. Much remains to be done. Ligands selective for other misfolded-protein pathologies need to be developed, since those diseases could also benefit from early detection. α-Synuclein is frequently comorbid in AD with Lewy bodies. Tau pathology is common to multiple neurodegenerative diseases (tauopathies) and in some form is a marker for neuronal dysfunction. Effective use of imaging will need to be combined with other diagnostic modalities, including more sensitive clinical testing, measurements of regional brain blood flow to provide a more detailed understanding of the early neurodegenerative process.

Acknowledgments

The author would like to thank Lary C. Walker at the Center for Neurodegenerative Disease, Emory University, Atlanta, Georgia, for helpful discussions during the writing of this chapter. Funding was provided by the Sanders-Brown Center on Aging and the Chandler Medical Center of the University of Kentucky.

REFERENCES

1. Agdeppa, E.D., Kepe, V., Liu, J., Small, G.W., Huang, S.C., Petric, A., Satyamurthy, N., Barrio, J.R. (2003). 2-Dialkylamino-6-acylmalononitrile substituted naphthalenes (DDNP analogs): novel diagnostic and therapeutic tools in Alzheimer's disease. *Mol Imaging Biol*, 5, 404–417.
2. Agdeppa, E.D., Kepe, V., Petri, A., Satyamurthy, N., Liu, J., Huang, S.C., Small, G.W., Cole, G.M., Barrio, J.R. (2003). In vitro detection of (*S*)-naproxen and ibuprofen binding to plaques in the Alzheimer's brain using the positron emission tomography molecular imaging probe 2-(1-[6-[(2-[(18)F]fluoroethyl)(methyl) amino]-2-naphthyl]ethylidene)malono nitrile. *Neuroscience, 117*, 723–730.
3. Anthony, J.C., Breitner, J.C., Zandi, P.P., Meyer, M.R., Jurasova, I., Norton, M.C., Stone, S.V. (2000). Reduced prevalence of AD in users of NSAIDs and H_2 receptor antagonists: the Cache County study. *Neurology, 54*, 2066–2071.
4. Bacskai, B.J., Hickey, G.A., Skoch, J., Kajdasz, S.T., Wang, Y., Huang, G.F., Mathis, C.A., Klunk, W.E., Hyman, B.T. (2003). Four-dimensional multiphoton imaging of brain entry, amyloid binding, and clearance of an amyloid-{beta} ligand in transgenic mice. *Proc Natl Acad Sci U S A, 100*, 12462–12467.
5. Banks, W.A., Terrell, B., Farr, S.A., Robinson, S.M., Nonaka, N., Morley, J.E. (2002). Passage of amyloid beta protein antibody across the blood–brain barrier in a mouse model of Alzheimer's disease. *Peptides, 23*, 2223–2226.

6. Benveniste, H., Einstein, G., Kim, K.R., Hulette, C., Johnson, G.A. (1999). Detection of neuritic plaques in Alzheimer's disease by magnetic resonance microscopy. *Proc Natl Acad Sci U S A, 96*, 14079–14084.
7. Boxer, A.L., Rabinovici, G.D., Kepe, V., Goldman, J., Furst, A.J., Huang, S.C., Baker, S.L., O'Neil J.P., Chui, H., Geschwind, M.D., et al. (2007). Amyloid imaging in distinguishing atypical prion disease from Alzheimer disease. *Neurology*, 69, 283–290.
8. Boyle, P.A., Wilson, R.S., Aggarwal, N.T., Tang, Y., Bennett, D.A. (2006). Mild cognitive impairment: risk of Alzheimer disease and rate of cognitive decline. *Neurology, 67*, 441–445.
9. Braak, H., Braak, E. (1991). Neuropathological stageing of Alzheimer-related changes. *Acta Neuropathol (Berl), 82*, 239–259.
10. Braak, H., Braak, E. (1995). Staging of Alzheimer's disease-related neurofibrillary changes. *Neurobiol Aging, 16*, 271–278; discussion, 278–284.
11. Brandt, R., Hundelt, M., Shahani, N. (2005). Tau alteration and neuronal degeneration in tauopathies: mechanisms and models. *Biochim Biophys Acta, 1739*, 331–354.
12. Bresjanac, M., Smid, L.M., Vovko, T.D., Petric, A., Barrio, J.R., Popovic, M. (2003). Molecular-imaging probe 2-(1-[6-[(2-fluoroethyl)(methyl) amino]-2-naphthyl] ethylidene) malononitrile labels prion plaques in vitro. *J Neurosci, 23*, 8029–8033.
13. Byeon, S.R., Jin, Y.J., Lim, S.J., Lee, J.H., Yoo, K.H., Shin, K.J., Oh, S.J., Kim, D.J. (2007). Ferulic acid and benzothiazole dimer derivatives with high binding affinity to beta-amyloid fibrils. *Bioorg Med Chem Lett, 17*, 4022–4025.
14. Caflisch, A. (2006). Computational models for the prediction of polypeptide aggregation propensity. *Curr Opin Chem Biol, 10*, 437–444.
15. Cai, L., Chin, F.T., Pike, V.W., Toyama, H., Liow, J.S., Zoghbi, S.S., Modell, K., Briard, E., Shetty, H.U., Sinclair, K., et al. (2004). Synthesis and evaluation of two (18)F-labeled 6-Iodo-2-(4′-*N*,*N*-dimethylamino)phenylimidazo[1,2-α]pyridine derivatives as prospective radioligands for beta-amyloid in Alzheimer's disease. *J Med Chem, 47*, 2208–2218.
16. Cai, L., Liow, J.S., Zoghbi, S.S., Cuevas, J., Baetas, C., Hong, J., Shetty, H.U., Seneca, N.M., Brown, A.K., Gladding, R., et al. (2008). Synthesis and evaluation of *N*-methyl and *S*-methyl (11)C-labeled 6-methylthio-2-(4′-*N*,*N*-dimethylamino)phenylimidazo[1,2-α]pyridines as radioligands for imaging beta-amyloid plaques in Alzheimer's disease. *J Med Chem, 51*, 148–158.
17. Chandra, R., Oya, S., Kung, M.P., Hou, C., Jin, L.W., Kung, H.F. (2007). New diphenylacetylenes as probes for positron emission tomographic imaging of amyloid plaques. *J Med Chem, 50*, 2415–2423.
18. Chang, Y.S., Jeong, J.M., Lee, Y.S., Kim, H.W., Ganesha, R.B., Kim, Y.J., Lee, D.S., Chung, J.K., Lee, M.C. (2006). Synthesis and evaluation of benzothiophene derivatives as ligands for imaging beta-amyloid plaques in Alzheimer's disease. *Nucl Med Biol, 33*, 811–820.
19. Chertkow, H., Black, S. (2007). Imaging biomarkers and their role in dementia clinical trials. *Can J Neurol Sci, 34* Suppl 1, S77–S83.
20. Chiti, F., Dobson, C.M. (2006). Protein misfolding, functional amyloid, and human disease. *Annu Rev Biochem, 75*, 333–366.

21. Cohen, R.M. (2007). The application of positron-emitting molecular imaging tracers in Alzheimer's disease. *Mol Imaging Biol, 9*, 204–216.

22. Coimbra, A., Williams, D.S., Hostetler, E.D. (2006). The role of MRI and PET/SPECT in Alzheimer's disease. *Curr Top Med Chem, 6*, 629–647.

23. Cruz, L., Urbanc, B., Buldyrev, S.V., Christie, R., Gomez-Isla, T., Havlin, S., MaNamara, M., Stanley, H.E., Hyman, B.T. (1997). Aggregation and disaggregation of senile plaques in Alzheimer disease. *Proc Natl Acad Sci U S A, 97*, 7612–7616.

24. Crystal, A.S., Giasson, B.I., Crowe, A., Kung, M.P., Zhuang, Z.P., Trojanowski, J.Q., Lee, V.M. (2003). A comparison of amyloid fibrillogenesis using the novel fluorescent compound K114. *J Neurochem, 86*, 1359–1368.

25. de Jong, D., Kremer, B.P., Olde Rikkert, M.G., Verbeek, M.M. (2007). Current state and future directions of neurochemical biomarkers for Alzheimer's disease. *Clin Chem Lab Med, 45*, 1421–1434.

26. Deane, R., Zlokovic, B.V. (2007). Role of the blood–brain barrier in the pathogenesis of Alzheimer's disease. *Curr Alzheimer Res, 4*, 191–197.

27. Dhenain, M., Privat, N., Duyckaerts, C., Jacobs, R.E. (2002). Senile plaques do not induce susceptibility effects in T2*-weighted MR microscopic images. *NMR Biomed, 15*, 197–203.

28. El Tayara Nel, T., Volk, A., Dhenain, M., Delatour, B. (2007). Transverse relaxation time reflects brain amyloidosis in young APP/PS1 transgenic mice. *Magn Reson Med, 58*, 179–184.

29. Elhaddaoui, A., Pigorsch, E., Delacourte, A., Turrell, S. (1995). Competition of Congo Red and thioflavin S binding to amyloid sites in Alzheimer's diseased brain. *Biospectroscopy, 1*, 351–356.

30. Faber, C., Zahneisen, B., Tippmann, F., Schroeder, A., Fahrenholz, F. (2007). Gradient-echo and CRAZED imaging for minute detection of Alzheimer plaques in an APP(V717I) × ADAM10-dn mouse model. *Magn Reson Med, 57*, 696–703.

31. Fandrich, M. (2007). On the structural definition of amyloid fibrils and other polypeptide aggregates. *Cell Mol Life Sci, 64*, 2066–2078.

32. Ferreira, S.T., Vieira, M.N., De Felice, F.G. (2007). Soluble protein oligomers as emerging toxins in Alzheimer's and other amyloid diseases. *IUBMB Life, 59*, 332–345.

33. Fink, A.L. (2005). Natively unfolded proteins. *Curr Opin Struct Biol, 15*, 35–41.

34. Flaherty, D.P., Walsh, S.M., Kiyota, T., Dong, Y., Ikezu, T., Vennerstrom, J.L. (2007). Polyfluorinated bis-styrylbenzene beta-amyloid plaque binding ligands. *J Med Chem, 50*, 4986–4992.

35. Forsberg, A., Engler, H., Almkvist, O., Blomquist, G., Hagman, G., Wall, A., Ringheim, A., Langstrom, B., Nordberg, A. (2008). PET imaging of amyloid deposition in patients with mild cognitive impairment. *Neurobiol Aging, 29*, 1456–1465.

36. Galzitskaya, O.V., Garbuzynskiy, S.O., Lobanov, M.Y. (2006). Prediction of amyloidogenic and disordered regions in protein chains. *PLoS Comput Biol, 2*, e177.

37. Glenner, G.G. (1980). Amyloid deposits and amyloidosis. the beta-fibrilloses (first of two parts). *N Engl J Med, 302*, 1283–1292.

38. Glenner, G.G. (1980). Amyloid deposits and amyloidosis: the beta-fibrilloses (second of two parts). *N Engl J Med, 302*, 1333–1343.

39. Habicht, G., Haupt, C., Friedrich, R.P., Hortschansky, P., Sachse, C., Meinhardt, J., Wieligmann, K., Gellermann, G.P., Brodhun, M., Gotz, J., et al. (2007). Directed selection of a conformational antibody domain that prevents mature amyloid fibril formation by stabilizing A protofibrils. *Proc Natl Acad Sci U S A, 104*, 19232 19237.

40. Haes, A.J., Chang, L., Klein, W.L., Van Duyne, R.P. (2005). Detection of a biomarker for Alzheimer's disease from synthetic and clinical samples using a nanoscale optical biosensor. *J Am Chem Soc, 127*, 2264–2271.

41. Herland, A., Nilsson, K.P., Olsson, J.D., Hammarstrom, P., Konradsson, P., Inganas, O. (2005). Synthesis of a regioregular zwitterionic conjugated oligoelectrolyte, usable as an optical probe for detection of amyloid fibril formation at acidic pH. *J Am Chem Soc, 127*, 2317–2323.

42. Honson, N.S., Johnson, R.L., Huang, W., Inglese, J., Austin, C.P., Kuret, J. (2007). Differentiating Alzheimer disease–associated aggregates with small molecules. *Neurobiol Dis, 28*, 251–260.

43. Ishikawa, K., Kudo, Y., Nishida, N., Suemoto, T., Sawada, T., Iwaki, T., Doh-Ura, K. (2006). Styrylbenzoazole derivatives for imaging of prion plaques and treatment of transmissible spongiform encephalopathies. *J Neurochem, 99*, 198–205.

44. Jack, C.R., Jr., Marjanska, M., Wengenack, T.M., Reyes, D.A., Curran, G.L., Lin, J., Preboske, G.M., Poduslo, J.F., Garwood, M. (2007). Magnetic resonance imaging of Alzheimer's pathology in the brains of living transgenic mice: a new tool in Alzheimer's disease research. *Neuroscientist, 13*, 38–48.

45. Kandimalla, K.K., Curran, G.L., Holasek, S.S., Gilles, E.J., Wengenack, T.M., Ramirez-Alvarado, M., Poduslo, J F. (2006). Physiological and biophysical factors that influence Alzheimer's disease amyloid plaque targeting of native and putrescine modified human amyloid {beta} 40. *J Pharmacol Exp Ther, 318*, 17–25.

46. Kandimalla, K.K., Wengenack, T.M., Curran, G.L., Gilles, E.J., Poduslo, J.F. (2007). Pharmacokinetics and amyloid plaque targeting ability of a novel peptide-based magnetic resonance contrast agent in wild type and Alzheimer's disease transgenic mice. *J Pharmacol Exp Ther, 318*, 17–25.

47. Kayed, R., Glabe, C.G. (2006). Conformation-dependent anti-amyloid oligomer antibodies. *Methods Enzymol, 413*, 326–344.

48. Kayed, R., Head, E., Thompson, J.L., McIntire, T.M., Milton, S.C., Cotman, C.W., Glabe, C.G. (2003). Common structure of soluble amyloid oligomers implies common mechanism of pathogenesis. *Science, 300*, 486–489.

49. Kemppainen, N.M., Aalto, S., Wilson, I.A., Nagren, K., Helin, S., Bruck, A., Oikonen, V., Kailajarvi, M., Scheinin, M., Viitanen, M., et al. (2007). PET amyloid ligand [^{11}C]PIB uptake is increased in mild cognitive impairment. *Neurology, 68*, 1603–1606.

50. Kepe, V., Huang, S.C., Small, G.W., Satyamurthy, N., Barrio, J.R. (2006). Visualizing pathology deposits in the living brain of patients with Alzheimer's disease. *Methods Enzymol, 412*, 144–160.

51. Klunk, W.E., Debnath, M.L., Pettegrew, J.W. (1994). Development of small molecule probes for the beta-amyloid protein of Alzheimer's disease. *Neurobiol Aging, 15*, 691–698.

52. Klunk, W.E., Debnath, M.L., Pettegrew, J.W. (1995). Chrysamine-G binding to Alzheimer and control brain: autopsy study of a new amyloid probe. *Neurobiol Aging, 16*, 541–548.

53. Klunk, W.E., Engler, H., Nordberg, A., Wang, Y., Blomqvist, G., Holt, D.P., Bergstrom, M., Savitcheva, I., Huang, G.F., Estrada, S., et al. (2004). Imaging brain amyloid in Alzheimer's disease with Pittsburgh compound-B. *Ann Neurol, 55*, 306–319.

54. Klunk, W.E., Lopresti, B.J., Ikonomovic, M.D., Lefterov, I.M., Koldamova, R.P., Abrahamson, E.E., Debnath, M.L., Holt, D.P., Huang, G.F., Shao, L., et al. (2005). Binding of the positron emission tomography tracer Pittsburgh compound-B reflects the amount of amyloid-beta in Alzheimer's disease brain but not in transgenic mouse brain. *J Neurosci, 25*, 10598–10606.

55. Klunk, W.E., Wang, Y., Huang, G.F., Debnath, M.L., Holt, D.P., Mathis, C.A. (2001). Uncharged thioflavin-T derivatives bind to amyloid-beta protein with high affinity and readily enter the brain. *Life Sci, 69*, 1471–1484.

56. Klunk, W.E., Wang, Y., Huang, G.F., Debnath, M.L., Holt, D.P., Shao, L., Hamilton, R.L., Ikonomovic, M.D., DeKosky, S.T., Mathis, C.A. (2003). The binding of 2-(4′-methylaminophenyl)benzothiazole to postmortem brain homogenates is dominated by the amyloid component. *J Neurosci, 23*, 2086–2092.

57. Kodali, R., Wetzel, R. (2007). Polymorphism in the intermediates and products of amyloid assembly. *Curr Opin Struct Biol, 17*, 48–57.

58. Kung, H.F., Kung, M.P., Zhuang, Z.P., Hou, C., Lee, C.W., Plossl, K., Zhuang, B., Skovronsky, D.M., Lee, V.M., Trojanowski, J.Q. (2003). Iodinated tracers for imaging amyloid plaques in the brain. *Mol Imaging Biol, 5*, 418–426.

59. Kung, M.P., Hou, C., Zhuang, Z.P., Cross, A.J., Maier, D.L., Kung, H.F. (2004). Characterization of IMPY as a potential imaging agent for beta-amyloid plaques in double transgenic PSAPP mice. *Eur J Nucl Med Mol Imaging, 31*, 1136–1145.

60. Kung, M.P., Hou, C., Zhuang, Z.P., Skovronsky, D., Kung, H.F. (2004). Binding of two potential imaging agents targeting amyloid plaques in postmortem brain tissues of patients with Alzheimer's disease. *Brain Res, 1025*, 98–105.

61. Kung, M.P., Hou, C., Zhuang, Z.P., Zhang, B., Skovronsky, D., Trojanowski, J.Q., Lee, V.M., Kung, H.F. (2002). IMPY: an improved thioflavin-T derivative for in vivo labeling of beta-amyloid plaques. *Brain Res, 956*, 202–210.

62. Kung, M.P., Zhuang, Z.P., Hou, C., Kung, H.F. (2004). Development and evaluation of iodinated tracers targeting amyloid plaques for SPECT imaging. *J Mol Neurosci*, 24, 49–54.

63. Lambert, M.P., Velasco, P.T., Chang, L., Viola, K.L., Fernandez, S., Lacor, P.N., Khuon, D., Gong, Y., Bigio, E.H., Shaw, P., et al. (2007). Monoclonal antibodies that target pathological assemblies of Abeta. *J Neurochem, 100*, 23–35.

64. Levey, A., Lah, J., Goldstein, F., Steenland, K., Bliwise, D. (2006). Mild cognitive impairment: an opportunity to identify patients at high risk for progression to Alzheimer's disease. *Clin Ther, 28*, 991–1001.

65. LeVine, H., III. (1993). Thioflavine T interaction with synthetic Alzheimer's disease beta-amyloid peptides: detection of amyloid aggregation in solution. *Protein Sci, 2*, 404–410.

66. LeVine, H., III. (1995). Thioflavine T interaction with amyloid β-sheet structures. *Amyloid: Int J Exp Clin Invest, 2*, 1–6.

67. LeVine, H., III. (1999). Quantification of beta-sheet amyloid fibril structures with thioflavin T. *Methods Enzymol, 309*, 274–284.

68. LeVine, H., III. (2005). Multiple ligand binding sites on A beta(1–40) fibrils. *Amyloid, 12*, 5–14.

69. Linding, R., Schymkowitz, J., Rousseau, F., Diella, F., Serrano, L. (2004). A comparative study of the relationship between protein structure and beta-aggregation in globular and intrinsically disordered proteins. *J Mol Biol, 342*, 345–353.

70. Lockhart, A., Ye, L., Judd, D.B., Merritt, A.T., Lowe, P.N., Morgenstern, J.L., Hong, G., Gee, A.D., Brown, J. (2005). Evidence for the presence of three distinct binding sites for the thioflavin T class of Alzheimer's disease PET imaging agents on beta-amyloid peptide fibrils. *J Biol Chem, 280*, 7677–7684.

71. Lombardo, J.A., Stern, E.A., McLellan, M.E., Kajdasz, S.T., Hickey, G.A., Bacskai, B.J., Hyman, B.T. (2003). Amyloid-beta antibody treatment leads to rapid normalization of plaque-induced neuritic alterations. *J Neurosci, 23*, 10879–10883.

72. Maeda, J., Ji, B., Irie, T., Tomiyama, T., Maruyama, M., Okauchi, T., Staufenbiel, M., Iwata, N., Ono, M., Saido, T.C., et al. (2007). Longitudinal, quantitative assessment of amyloid, neuroinflammation, and anti-amyloid treatment in a living mouse model of Alzheimer's disease enabled by positron emission tomography. *J Neurosci, 27*, 10957–10968.

73. Marshall, J.R., Stimson, E.R., Ghilardi, J.R., Vinters, H.V., Mantyh, P.W., Maggio, J.E. (2002). Noninvasive imaging of peripherally injected Alzheimer's disease type synthetic A beta amyloid in vivo. *Bioconjug Chem, 13*, 276–284.

74. McLellan, M.E., Kajdasz, S.T., Hyman, B.T., Bacskai, B.J. (2003). In vivo imaging of reactive oxygen species specifically associated with thioflavine *S*–positive amyloid plaques by multiphoton microscopy. *J Neurosci, 23*, 2212–2217.

75. Meyer-Luehmann, M., Coomaraswamy, J., Bolmont, T., Kaeser, S., Schaefer, C., Kilger, E., Neuenschwander, A., Abramowski, D., Frey, P., Jaton, A.L., et al. (2006). Exogenous induction of cerebral beta-amyloidogenesis is governed by agent and host. *Science, 313*, 1781–1784.

76. Montalto, M.C., Farrar, G., Hehir, C.T. (2007). Fibrillar and oligomeric beta-amyloid as distinct local biomarkers for Alzheimer's disease. *Ann N Y Acad Sci, 1097*, 239–258.

77. Naiki, H., Higuchi, K., Hosokawa, M., Takeda, T. (1989). Fluorometric determination of amyloid fibrils in vitro using the fluorescent dye, thioflavin T1. *Anal Biochem, 177*, 244–249.

78. Newberg, A.B., Wintering, N.A., Plossl, K., Hochold, J., Stabin, M.G., Watson, M., Skovronsky, D., Clark, C.M., Kung, M.P., Kung, H.F. (2006). Safety, biodistribution, and dosimetry of 123I-IMPY: a novel amyloid plaque-imaging agent for the diagnosis of alzheimer's disease. *J Nucl Med, 47*, 748–754.

79. Nilsson, K.P., Aslund, A., Berg, I., Nystrom, S., Konradsson, P., Herland, A., Inganas, O., Stabo-Eeg, F., Lindgren, M., Westermark, G.T., et al. (2007). Imaging distinct conformational states of amyloid-beta fibrils in Alzheimer's disease using novel luminescent probes. *ACS Chem Biol, 2*, 553–560.

80. Nilsson, K.P., Hammarstrom, P., Ahlgren, F., Herland, A., Schnell, E.A., Lindgren, M., Westermark, G.T., Inganas, O. (2006). Conjugated polyelectrolytes:conformation-sensitive optical probes for staining and characterization of amyloid deposits. *Chembiochem, 7*, 1096–1104.

81. Nilsson, K.P., Herland, A., Hammarstrom, P., Inganas, O. (2005). Conjugated polyelectrolytes: conformation-sensitive optical probes for detection of amyloid fibril formation. *Biochemistry, 44*, 3718–3724.

82. O'Nuallain, B., Shivaprasad, S., Kheterpal, I., Wetzel, R. (2005). Thermodynamics of abeta(1–40) amyloid fibril elongation. *Biochemistry, 44*, 12709–12718.

83. O'Nuallain, B., Williams, A.D., Westermark, P., Wetzel, R. (2004). Seeding specificity in amyloid growth induced by heterologous fibrils. *J Biol Chem, 279*, 17490–17499.

84. Oddo, S., Caccamo, A., Kitazawa, M., Tseng, B.P., LaFerla, F.M. (2003). Amyloid deposition precedes tangle formation in a triple transgenic model of Alzheimer's disease. *Neurobiol Aging, 24*, 1063–1070.

85. Okamura, N., Suemoto, T., Furumoto, S., Suzuki, M., Shimadzu, H., Akatsu, H., Yamamoto, T., Fujiwara, H., Nemoto, M., Maruyama, M., et al. (2005). Quinoline and benzimidazole derivatives: candidate probes for in vivo imaging of tau pathology in Alzheimer's disease. *J Neurosci, 25*, 10857–10862.

86. Ono, M., Kawashima, H., Nonaka, A., Kawai, T., Haratake, M., Mori, H., Kung, M.P., Kung, H.F., Saji, H., Nakayama, M. (2006). Novel benzofuran derivatives for PET imaging of beta-amyloid plaques in Alzheimer's disease brains. *J Med Chem, 49*, 2725–2730.

87. Ono, M., Maya, Y., Haratake, M., Ito, K., Mori, H., Nakayama, M. (2007). Aurones serve as probes of beta-amyloid plaques in Alzheimer's disease. *Biochem Biophys Res Commun, 361*, 116–121.

88. Ono, M., Yoshida, N., Ishibashi, K., Haratake, M., Arano, Y., Mori, H., Nakayama, M. (2005). Radioiodinated flavones for in vivo imaging of beta-amyloid plaques in the brain. *J Med Chem, 48*, 7253–7260.

89. Osmand, A.P., Berthelier, V., Wetzel, R. (2006). Imaging polyglutamine deposits in brain tissue. *Methods Enzymol, 412*, 106–122.

90. Petersen, R.C. (2004). Mild cognitive impairment as a diagnostic entity. *J Intern Med, 256*, 183–194.

91. Petkova, A.T., Leapman, R.D., Guo, Z., Yau, W.M., Mattson, M.P., Tycko, R. (2005). Self-propagating, molecular-level polymorphism in Alzheimer's beta-amyloid fibrils. *Science, 307*, 262–265.

92. Petric, A., Jacobson, A.F., Barrio, J.R. (1998). Functionalization of a viscosity-sensitive fluorophore for probing of biological systems. *Bioorg Med Chem Lett, 8*, 1455–1460.

93. Poduslo, J.F., Ramakrishnan, M., Holasek, S.S., Ramirez-Alvarado, M., Kandimalla, K.K., Gilles, E.J., Curran, G.L., Wengenack, T.M. (2007). In vivo targeting of antibody fragments to the nervous system for Alzheimer's disease immunotherapy and molecular imaging of amyloid plaques. *J Neurochem, 102*, 420–433.

94. Price, J.C., Klunk, W.E., Lopresti, B.J., Lu, X., Hoge, J.A., Ziolko, S.K., Holt, D.P., Meltzer, C.C., Dekosky, S.T. Mathis, C.A. (2005). Kinetic modeling of amyloid binding in humans using PET imaging and Pittsburgh compound-B. *J Cereb Blood Flow Metab, 25*, 1528–1547.

95. Qu, W., Kung, M.P., Hou, C., Benedum, T.E., Kung, H.F. (2007). Novel styrylpyridines as probes for SPECT imaging of amyloid plaques. *J Med Chem, 50*, 2157–2165.

96. Ray, S., Britschgi, M., Herbert, C., Takeda-Uchimura, Y., Boxer, A., Blennow, K., Friedman, L.F., Galasko, D.R., Jutel, M., Karydas, A., et al. (2007). Classification and prediction of clinical Alzheimer's diagnosis based on plasma signaling proteins. *Nat Med, 13*, 1359–1362.

97. Rockwood, K. (2006). Epidemiological and clinical trials evidence about a preventive role for statins in Alzheimer's disease. *Acta Neurol Scand Suppl, 185*, 71–77.

98. Sato, K., Higuchi, M., Iwata, N., Saido, T.C., Sasamoto, K. (2004). Fluoro-substituted and (13)C-labeled styrylbenzene derivatives for detecting brain amyloid plaques. *Eur J Med Chem, 39*, 573–578.

99. Schmidt, M.L., Schuck, T., Sheridan, S., Kung, M.P., Kung, H., Zhuang, Z.P., Bergeron, C., Lamarche, J.S., Skovronsky, D., Giasson, B.I., et al. (2001). The fluorescent Congo Red derivative, (*trans*, *trans*) - 1-bromo-2,5-bis-(3-hydroxycarbonyl-4-hydroxy)styrylbenzene (BSB), labels diverse beta-pleated sheet structures in postmortem human neurodegenerative disease brains. *Am J Pathol, 159*, 937–943.

100. Seneca, N., Cai, L., Liow, J.S., Zoghbi, S.S., Gladding, R.L., Hong, J., Pike, V.W., Innis, R.B. (2007). Brain and whole-body imaging in nonhuman primates with [(11)C]MeS-IMPY, a candidate radioligand for beta-amyloid plaques. *Nucl Med Biol, 34*, 681–689.

101. Serpell, L.C., Fraser, P.E., Sunde, M. (1999). X-ray fiber diffraction of amyloid fibrils. *Methods Enzymol, 309*, 526–536.

102. Serpell, L.C., Sunde, M., Benson, M.D., Tennent, G.A., Pepys, M.B., Fraser, P.E. (2000). The protofilament substructure of amyloid fibrils. *J Mol Biol, 300*, 1033–1039

103. Shoghi-Jadid, K., Barrio, J.R., Kepe, V., Huang, S.C. (2006). Exploring a mathematical model for the kinetics of beta-amyloid molecular imaging probes through a critical analysis of plaque pathology. *Mol Imaging Biol, 8*, 151–162.

104. Shoghi-Jadid, K., Barrio, J.R., Kepe, V., Wu, H.M., Small, G.W., Phelps, M.E., Huang, S.C. (2005). Imaging beta-amyloid fibrils in Alzheimer's disease: a critical analysis through simulation of amyloid fibril polymerization. *Nucl Med Biol, 32*, 337–351.

105. Simmons, M.K., Manjeshwar, R., Agdeppa, E.D., Mattheyses, R.M., Kiehl, T.R., Montalto, M.C. (2005). A computational positron emission tomography simulation model for imaging beta-amyloid in mice. *Mol Imaging Biol, 7*, 69–77.

106. Sipe, J.D., Cohen, A.S. (2000). Review: history of the amyloid fibril. *J Struct Biol, 130*, 88–98.

107. Small, G.W., Kepe, V., Ercoli, L.M., Siddarth, P., Bookheimer, S.Y., Miller, K.J., Lavretsky, H., Burggren, A.C., Cole, G.M., Vinters, H.V., et al. (2006). PET of brain amyloid and tau in mild cognitive impairment. *N Engl J Med, 355*, 2652–2663.

108. Sonnen, J.A., Keene, C.D., Montine, K.S., Li, G., Peskind, E.R., Zhang, J., Montine, T.J. (2007). Biomarkers for Alzheimer's disease. *Expert Rev Neurother, 7*, 1021–1028.

109. Sparks, D.L., Sabbagh, M., Connor, D., Soares, H., Lopez, J., Stankovic, G., Johnson-Traver, S., Ziolkowski, C., Browne, P. (2006). Statin therapy in Alzheimer's disease. *Acta Neurol Scand Suppl, 185*, 78–86.
110. Styren, S.D., Hamilton, R.L., Styren, G.C., Klunk, W.E. (2000). X-34, a fluorescent derivative of Congo Red: a novel histochemical stain for Alzheimer's disease pathology. *J Histochem Cytochem, 48*, 1223–1232.
111. Suemoto, T., Okamura, N., Shiomitsu, T., Suzuki, M., Shimadzu, H., Akatsu, H., Yamamoto, T., Kudo, Y., Sawada, T. (2004). In vivo labeling of amyloid with BF-108. *Neurosci Res, 48*, 65–74.
112. Szekely, C.A., Thorne, J.E., Zandi, P.P., Ek, M., Messias, E., Breitner, J.C., Goodman, S.N. (2004). Nonsteroidal anti-inflammatory drugs for the prevention of Alzheimer's disease: a systematic review. *Neuroepidemiology, 23*, 159–169.
113. Thal, L.J., Kantarci, K., Reiman, E.M., Klunk, W.E., Weiner, M.W., Zetterberg, H., Galasko, D., Pratico, D., Griffin, S., Schenk, D., Siemers, E. (2006). The role of biomarkers in clinical trials for Alzheimer disease. *Alzheimer Dis Assoc Disord, 20*, 6–15.
114. Tseng, B.P., Esler, W.P., Clish, C.B., Stimson, E.R., Ghilardi, J.R., Vinters, H.V., Mantyh, P.W., Lee, J.P., Maggio, J.E. (1999). Deposition of monomeric, not oligomeric, Abeta mediates growth of Alzheimer's disease amyloid plaques in human brain preparations. *Biochemistry, 38*, 10424–10431.
115. Urbanc, B., Cruz, L., Buldyrev, S.V., Havlin, S., Irizarry, M.C., Stanley, H.E., Hyman, B.T. (1999). Dynamics of plaque formation in Alzheimer's disease. *Biophys J, 76*, 1330–1334.
116. Uversky, V.N. (2002). Natively unfolded proteins: a point where biology waits for physics. *Protein Sci, 11*, 739–756.
117. Virchow, R. (1854). Uber eine im Gehirn und Ruchenmark des Menschen aufgefundene Substanz mit der chemischen Reaktion der cellulose. *Virchows Arch Pathol Anat, 6*, 135–138.
118. Walker, L., LeVine, H., Jucker, M. (2006). Koch's postulates and infectious proteins. *Acta Neuropathol, 112*, 1–4.
119. Walker, L.C., Bian, F., Callahan, M.J., Lipinski, W.J., Durham, R.A., LeVine, H. (2002). Modeling Alzheimer's disease and other proteopathies in vivo: Is seeding the key? Review article. *Amino Acids, 23*, 87–93.
120. Walker, L.C., LeVine, H. (2000). The cerebral proteopathies. *Neurobiol Aging, 21*, 559–561.
121. Walker, L.C., LeVine, H. (2000). The cerebral proteopathies: neurodegenerative disorders of protein conformation and assembly. *Mol Neurobiol, 21*, 83–95.
122. Walker, L.C., LeVine, H., 3rd. (2002). Proteopathy: the next therapeutic frontier? *Curr Opin Investig Drugs, 3*, 782–787.
123. Walker, L.C., Levine, H., 3rd, Mattson, M.P., Jucker, M. (2006). Inducible proteopathies. *Trends Neurosci, 29*, 438–443.
124. Wang, Y., Klunk, W.E., Debnath, M.L., Huang, G.F., Holt, D.P., Shao, L., Mathis, C.A. (2004). Development of a PET/SPECT agent for amyloid imaging in Alzheimer's disease. *J Mol Neurosci, 24*, 55–62.

125. Wei, J., Wu, C., Lankin, D., Gulrati, A., Valyi-Nagy, T., Cochran, E., Pike, V.W., Kozikowski, A., Wang, Y. (2005). Development of novel amyloid imaging agents based upon thioflavin S. *Curr Alzheimer Res, 2*, 109–114.

126. Westermark, P. (2005). Aspects on human amyloid forms and their fibril polypeptides. *FEBS J, 272*, 5942–5949.

127. Westermark, P., Benson, M.D., Buxbaum, J.N., Cohen, A.S., Frangione, B., Ikeda, S., Masters, C.L., Merlini, G., Saraiva, M.J., Sipe, J.D. (2007). A primer of amyloid nomenclature. *Amyloid, 14*, 179–183.

128. Ye, L., Morgenstern, J.L., Gee, A.D., Hong, G., Brown, J., Lockhart, A. (2005). Delineation of PET imaging agent binding sites on beta-amyloid peptide fibrils. *J Biol Chem, 280*, 23599–23604.

129. Zandi, P.P., Sparks, D.L., Khachaturian, A.S., Tschanz, J., Norton, M., Steinberg, M., Welsh-Bohmer, K.A., Breitner, J.C. (2005). Do statins reduce risk of incident dementia and Alzheimer disease? The Cache County Study. *Arch Gen Psychiatry, 62*, 217–224.

130. Ziolko, S.K., Weissfeld, L.A., Klunk, W.E., Mathis, C.A., Hoge, J.A., Lopresti, B.J., Dekosky, S.T., Price, J.C. (2006). Evaluation of voxel-based methods for the statistical analysis of PIB PET amyloid imaging studies in Alzheimer's disease. *Neuroimage, 33*, 94–102.

31

DIAGNOSIS OF SYSTEMIC AMYLOID DISEASES

Morie A. Gertz
Department of Medicine, Division of Hematology, Mayo Clinic College of Medicine, Rochester, Minnesota

INTRODUCTION

Amyloidosis is a group of diseases characterized by deposition of protein fibrils with a β-sheet structure in one or more organ systems. Signs and symptoms of amyloidosis vary considerably among the different disease types, and the type of disease determines the treatment [1]. Amyloidosis is classified by the structural units of the amyloid protein. Specifically, immunoglobulin light-chain proteins indicate primary amyloidosis (AL), amyloid A proteins indicate secondary amyloidosis (AA), and transthyretin and other proteins indicate familial amyloidosis (AF) and senile systemic amyloidosis [2]. After the diagnosis of amyloidosis is established, the extent of systemic involvement is evaluated by functional assays and specialized assessment of the myocardium to establish the overall prognosis. When available, scintigraphy with serum amyloid P (SAP) component is also performed [3].

CONGO RED STAIN

Identification of amyloid deposits in biopsy specimens is the only method of confirming the diagnosis of amyloidosis. The β-sheet structure of amyloid proteins gives them affinity for Congo Red stain. The first use of Congo Red as

Protein Misfolding Diseases: Current and Emerging Principles and Therapies,
Edited by Marina Ramirez-Alvarado, Jeffery W. Kelly, and Christopher M. Dobson
Copyright © 2010 John Wiley & Sons, Inc.

a specific stain for detection of amyloid deposits was reported in 1922 [4], and observations of green birefringence under polarized light, the sine qua non of amyloid protein identification, was first reported in 1927 [5].

Use of Congo Red stain may be challenging. For example, overfixation of biopsy specimens may result in poor staining, and trapping of the Congo Red dye may result in false-positive results. Occasionally, fibrin and elastin in skin and fat tissues bind Congo Red, although they do not show green birefringence. For inexperienced pathologists, this may be a difficult distinction to make and could result in false-positive diagnoses. Phenolic Congo Red stain may be superior to the conventional alkaline Congo Red stain. In one study of 10 patients [6], phenolic Congo Red showed amyloid deposits for all patients, whereas the alkaline Congo Red stain showed positive results for only seven.

Over the years, multiple attempts were made to improve the sensitivity and specificity of Congo Red. Potassium permanganate sensitivity once was used to distinguish AA from other forms of amyloidosis. After incubation in potassium permanganate, amyloid deposits of AA lost affinity for the Congo Red stain (permanganate sensitive), whereas AL and AF were permanganate-resistant forms of amyloidosis. However, this technique has largely been abandoned in favor of more sensitive techniques [7]. Performate was also used to distinguish AA (performate sensitive) from other forms of amyloidosis that remain positively birefringent [8].

Congo Red birefringence has been combined with immunocytochemistry and Congo Red fluorescence to improve the diagnostic value of the technique. Congo Red fluorescence, which is visualized with ultraviolet light instead of polarized light, is more sensitive than conventional Congo Red birefringence and may detect minute amyloid deposits with greater accuracy. Furthermore, it is reportedly the most sensitive method for direct diagnosis of amyloidosis from tissue sections [9] and does not interfere with immunohistochemical stains.

Congo Red fluorescence has been applied to frozen kidney biopsy specimens. In a prospective study of 15 patients with amyloidosis, no false-positive or false-negative results were observed [10]. When 146 renal biopsy specimens previously stained with Congo Red were reevaluated for Congo Red fluorescence, 87 Congo Red–positive cases were confirmed by fluorescence and one additional positive case was identified. Congo Red fluorescence is simple to perform and has high specificity and sensitivity; amyloid deposits are more pronounced and therefore easier to evaluate than with the traditional alkaline Congo Red stain. Although sulfated Alcian Blue is used to stain endomyocardial biopsy specimens and crystal violet is used for screening nerve biopsy sections, Congo Red stain remains a requirement for the diagnosis of amyloidosis.

BIOPSY DIAGNOSIS OF AMYLOID

Because amyloidosis is a widespread deposition disorder, it may be diagnosed by performing a biopsy at virtually any site. In clinical practice, biopsies are highly sensitive because they are usually directed to the clinically affected organ: the liver in hepatomegaly (in excess of 10 cm below the right costal margin), the kidney in nephrotic syndrome, and so on. Endomyocardial biopsy for cardiac amyloidosis is 100% sensitive when sampling error is eliminated by obtaining a minimum of four samples [11]. In a study of 36 renal biopsies, all biopsy specimens showed characteristic fibrillary deposits of amyloid with electron microscopy and stained positively with Congo Red or thioflavin T [12]. Amyloidosis may be diagnosed with fine-needle aspirates; this technique has been applied to liver diagnoses, thereby reducing the risks of performing core-needle biopsies [13].

Although biopsy of the affected tissue remains the diagnostic standard for amyloidosis, alternative methods are available [14]. An endoscopic biopsy of the upper digestive tract or duodenum may be preferred for the diagnosis of renal amyloidosis. The frequency of amyloid deposition in biopsy specimens is 100% for the duodenum, 95% for the stomach, 91% for the colorectum, and 72% for the esophagus. Endoscopy may be preferred over renal biopsy because it generally is better tolerated and is a safer technique for the diagnosis of systemic amyloidosis [15,16]. Rectal biopsy, skin biopsy, and labial salivary gland biopsy may also be performed. Because labial salivary gland biopsy is minimally invasive, it has been used to diagnose amyloidosis in patients presenting with polyneuropathy [17]. When small amyloid deposits are anticipated, a kidney biopsy may be preferable because renal tissue is amenable to electron microscopic study and provides a greater amount of amyloid for accurate classification with a panel of antibodies [18].

In established amyloidosis treatment centers, the most commonly used technique is the subcutaneous fat aspiration [19]. In one of the first studies of the technique [20], the sensitivity was 84%. It was proposed as the diagnostic procedure of choice because it requires no specialty consultation or technical expertise and causes minimal patient discomfort. In a blinded, controlled analysis of subcutaneous fat aspiration, the procedure was performed on 82 patients with amyloidosis and 72 disease-free adult volunteers. The sensitivity was 72%, the specificity was 99%, and a diagnosis could be established within 24 hours. Two pathologists, blinded to the results, showed concordance for 95% of samples. The authors noted that equivocally positive stains should be interpreted with caution because weak and nonspecific histologic staining may occur.

Several methods have been used to identify and classify amyloidosis from fat biopsy specimens. An enzyme-linked immunosorbent assay successfully characterized the type of amyloidosis in 14 of 15 positive fat biopsy specimens [21]. Another study, using a similar method, showed a specificity of 100% and a positive predictive value of 100% [22]. Sensitivity, however, was low (58%), and the percentage of inadequate specimens was high (11%). Difficulties in confirming amyloidosis included ambiguous interpretation of the Congo Red stain (pale-stained amyloid fibrils and collagen birefringence). Congo Red fluorescence, as described above, has been used in the analysis of amyloid fat pad aspirates. This technique remains useful when examining archival slides of previously stained aspirates [23]. Immunoelectron microscopy also has been used to characterize amyloidosis in fat biopsy specimens in patients with cardiac amyloidosis [24], and immunohistochemical classification of the amyloid deposits in subcutaneous fat samples may be determined by Western blot analysis with specific antibodies against amyloid fibril proteins [25].

Fine-needle aspiration biopsy may be used to diagnose amyloidosis. In a study of 91 patients who underwent fine-needle aspiration biopsy of fat at Johns Hopkins Hospital [26], findings were positive for 22%, negative for 68%, insufficient for 9%, and equivocal for 1%. Of the 62 patients with negative findings, follow-up biopsies were performed for 19, and 5 were positive for amyloidosis when samples were stained with Congo Red. In this series, 55% of patients with positive findings and 31% with negative findings underwent a second biopsy. The sensitivity and specificity were 75% and 92%, respectively. A conclusive positive result helped exclude other underlying conditions.

In a study of 120 patients with established systemic amyloidosis [38 with AA, 70 with AL, and 12 with transthyretin-related amyloidosis (ATTR)], all underwent subcutaneous fat biopsy [27]. Thoroughly examined smears were positive for 93% of patients. Specificity was 100% for 45 control samples, and the clinical utility of fat tissue aspiration was greater than that of biopsy of the rectum. The authors concluded that subcutaneous fat aspiration was the preferred method for detecting amyloidosis; when one pathologist examined three fat smears, the sensitivity was 80%, but when two pathologists examined three smears, sensitivity exceeded 90%. The additional value of a subsequent rectal biopsy was negligible, and a biopsy of the organ affected was recommended.

In summary, the preferred technique for the diagnosis of suspected systemic amyloidosis is subcutaneous fat biopsy. Specimens should be examined with Congo Red birefringence.

DISTINGUISHING BETWEEN LOCALIZED AND SYSTEMIC AMYLOIDOSIS

Before proceeding with therapeutic intervention, it is critical to classify the disease as systemic or localized amyloidosis. Patients with localized amyloidosis may present with hematuria, respiratory difficulties, and visual disturbances, which may easily be confused with signs and symptoms of systemic amyloidosis. A localized amyloidosis syndrome may be suspected by the pattern of amyloid deposition, which may include the skin, the tracheobronchial tree [28], or the bladder and ureters. The production of localized amyloid deposits may be related to plasma cells at the deposition site that produce insoluble light-chain proteins. Patients with pulmonary amyloidosis may have a systemic process or a localized pulmonary process such as nodular pulmonary amyloidosis [29]. Localized amyloidosis may involve the vocal cords and lead to hoarseness. Ureterovesicular amyloid deposits nearly always are localized, and patients present with hematuria. Lichen and macular forms of cutaneous amyloidosis invariably are localized and are relatively innocuous conditions. Nodular amyloidosis may be associated with systemic AL and may be a clue to an underlying, life-threatening process. Carpal tunnel syndrome may be observed with systemic or localized amyloidosis. Localized amyloidosis has also been reported in the conjunctiva and orbits and is managed surgically or expectantly. Trace amounts of localized amyloid deposits may be seen in hip cartilage and knee arthroplasty specimens [30].

CLASSIFICATION OF AMYLOIDOSIS ON BIOPSY TISSUE SPECIMENS

After amyloidosis has been established by a Congo Red–stained tissue biopsy specimen showing green birefringence, and after it is clear that the amyloidosis is systemic, it is critical to identify the specific amyloidosis syndrome. Treatments for AL, AA, AF, and senile systemic amyloidosis vary dramatically, and reports have described cases with initially incorrect classification of amyloidosis. In one study, fibrinogen A-α chain renal amyloidosis and ATTR were confirmed by genetic testing when the initial clinical diagnosis was AL [31]. Another study described three patients with an incidental monoclonal gammopathy and a hereditary variant of systemic amyloidosis [32]. These patients could easily have been treated with chemotherapy, an inappropriate approach for the management of hereditary amyloidosis. One case report described a patient, initially thought to have renal amyloidosis, who ultimately received a diagnosis of Waldenström macroglobulinemia and minimal change disease [33]

One of the difficulties in characterizing the type of amyloidosis from biopsy specimens arises from the small amount of amyloid proteins available for analysis, despite a large amount of biopsy tissue. Micromethods have been developed for the extraction, purification, and amino acid sequencing of amyloid proteins contained in minute specimens. Amyloid proteins may also be extracted and sequenced from formalin-fixed tissue specimens, and exact identification of the protein in fibrillar amyloid deposits is possible [34]. Amyloid proteins that are extracted and purified by micro techniques also may be used for immunochemical and chemical characterization [35].

The analytic techniques used to identify amyloid deposits include enzyme-linked immunosorbent assays, Western blot, amino acid sequencing, and mass spectroscopy [36]. Of these techniques, immunohistochemical assays are used most commonly, although considerable controversy exists with regard to the value of this technique in classifying amyloidosis. Immunoperoxidase techniques were the first methods used to identify amyloid deposits in tissue section, and they were initially reported to be quite sensitive [37].

In an autopsy series of 43 patients [38], a panel of noncommercial antibodies was used and identified AA in 21, ATTR in 11, and AL in 10. One patient had more than one type of amyloid protein that was deposited systematically.

For patients with AA or AL, the kidney was involved most frequently; for patients with ATTR, the heart and lungs were involved most frequently. This series showed that for most patients, amyloidosis could be classified using specific antibodies against five major amyloid fibril proteins. Micro extracts of amyloid proteins from 11 fat aspirates have been analyzed immunochemically with concordant results in three of four patients with κ light-chain AL, five of six with λ light-chain AL, and one patient with AA [39].

Polyclonal antibodies against synthetic peptides have been developed; they correspond to amino acid positions 118 to 134 of λ light chains and positions 116 to 133 of κ light chains. These synthetic peptides were useful for the classification of amyloidosis from formalin-fixed, paraffin-embedded tissue specimens [40]. Anti-λ antiserum reacted with samples from 18 of 19 patients with A-λ amyloidosis, and anti-κ antiserum reacted with samples from 9 of 10 patients with A-κ amyloidosis. Immunofluorescence staining of kidney biopsy specimens for κ and λ light-chain proteins was shown to be unreliable; one study described negative immunofluorescence results for 12 patients with plasma cell dyscrasia (35.3% of study patients) [12]. Because the sensitivity of immunofluorescence microscopy for the detection of AL in the kidney was low, additional diagnostic studies were required.

In a series of 169 biopsies from 121 patients [41], 12 patients with amyloidosis could not be classified using immunohistochemical techniques. In 32% of biopsies, amyloid deposits were not unequivocally stained for λ or κ light chains; this is indicative of the limited sensitivity of immunohistochemical techniques for the detection of light-chain amyloidosis. Presumably, some antibodies do not recognize light-chain epitopes when they are in amyloid configuration, or proteolysis involved in amyloid subunit production deletes the necessary epitopes. The authors concluded, as is frequently noted, that immunohistochemical classification of amyloidosis still poses a problem, particularly when commercial anti-AL antibodies are used. Although classification of non-AL deposits is relatively straightforward, the classification of AL and rare forms of non-ATTR hereditary amyloidosis often cannot be made with certainty when conclusive clinical information is not available.

A recent article summarized mistakes in immunohistochemical assays that have resulted in misdiagnosis of amyloidosis [42]. Errors included inconsistent immunolabeling, nonspecific background staining, and inconsistent reactions. A recently developed antibody against the λ light-chain peptide has been suitable for immunohistochemical classification of AL amyloid proteins [43]. It specifically stained proteins in Western blots and formalin-fixed and paraffin-embedded tissue sections. Formic acid extraction has been combined with immunochemical and biochemical characterization to classify amyloidosis accurately; in one study that used this method, constant-region sequences of the immunoglobulin light chain were observed instead of the expected variable-chain sequence [36].

Mass spectrometry has been applied recently to the diagnosis of systemic amyloidosis. A microtechnique may be used to extract fibrillar deposits from formalin-fixed biopsy specimens, and the chemical composition of the deposits can then be determined through amino acid sequencing or mass spectroscopy [44]. Tandem mass spectrometry of material extracted from formalin-fixed, amyloid-containing tissue biopsy specimens can identify the type of amyloid deposits precisely [45].

A unique molecular profile in light-chain amyloidosis has been identified through functional gene expression analysis of clonal plasma cells [46]. Class prediction analysis shows a subset of 12 genes that can be used to discriminate amyloid of the AL type from other amyloid proteins. Molecular profiling of clonal plasma cells may also provide insights into the pathogenesis of light-chain amyloidosis.

After amyloidosis has been diagnosed, it is essential to attempt classification of the type from tissue biopsy specimens. The recently introduced immunoglobulin free light-chain assay has been used to further classify the type of amyloidosis. Abnormalities of the immunoglobulin free light-chain ratio may be specific for AL, but nonreactivity of individual monoclonal free light-chain proteins has been reported and thus may result in false-negative findings [47]. Pooled human immunoglobulin contains antibodies that specifically recognize fibrils formed from light-chain proteins associated with AA and ATTR. A peptide purified from human immunoglobulin is reactive against all forms of amyloid deposits, including those of AA, ATTR, and islet amyloidosis. These antibodies immunostain human amyloid tissue deposits, and fibril affinity-purified intravenous immunoglobulin has potential as a diagnostic agent for patients with amyloid-associated disease [48].

RADIONUCLIDE IMAGING OF AMYLOID DEPOSITS

Use of imaging to specifically identify amyloid deposits began some 25 years ago. In one study, seven patients with myocardial amyloidosis underwent scintigraphy with technetium 99m-labeled pyrophosphate [49]. The myocardial uptake of tracer was negligible and could not be used to diagnose amyloidosis, but it did show that four had reduced systolic function and six had impaired diastolic function. Others have indicated that myocardial uptake of technetium-labeled pyrophosphate was useful for diagnosing amyloid polyneuropathy [50].

Thallium-201 scintigraphic studies have been performed in patients with cardiac amyloidosis. Washout rates during rest and during delayed thallium imaging may reflect the severity of amyloidosis in the myocardium, and mean washout rates of the entire heart are higher in patients with amyloidosis than in control patients. Researchers showed a very high washout rate in four of five patients with amyloidosis, and all died in less than a year [51].

Gallium-67 and thallium-201 single-photon emission tomography has been used to study dialysis amyloidosis [52]. This technique was compared to imaging with technetium-labeled methylene diphosphonate. Technetium-labeled methylene diphosphonate whole-body bone scans were able to detect active and inactive (preexisting) deposits of dialysis amyloid. Gallium and thallium scans were also helpful when differentiating between active and inactive deposits and when evaluating therapeutic effects of interventions.

Technetium-labeled N_2S_2 conjugates of chrysamine G have been used to image amyloid deposits. This agent localizes specifically to amyloid deposits in human kidney tissue, suggesting its applicability as a specific targeting agent for diagnostic purposes [53]. The use of technetium-99m 3,3-diphosphono-1,2-propanodicarboxylic acid in eight patients with ATTR was highly sensitive, showing whole-body tracer retention of 80% in patients with familial amyloidotic polyneuropathy compared with 56% in controls 3 hours after injection [54]. Increased uptake of the technetium-labeled bone tracer molecule hydroxymethylene diphosphonate has been reported for patients with cardiac amyloidosis [55]. Technetium-labeled aprotinin has also been studied as a specific marker of amyloidosis because it appears to be a fairly sensitive and specific diagnostic molecule. In a group of 23 patients, 22 had focal accumulations of technetium-labeled aprotinin in different organs. Findings for 20 were confirmed by biopsy or autopsy. The diagnostic accuracy of technetium-99m 3,3-diphosphono-1,2-propanodicarboxylic acid was investigated in cardiac amyloidosis [56]. The accuracy for differentiating between ATTR and AL was 100%, and uptake was absent in controls. In patients with ATTR, sensitivity and specificity were 100%. In patients with AL, sensitivity was 0% and specificity was 100%. Therefore, technetium-99m 3,3-diphosphono-1,2-propanodicarboxylic acid was useful for differentiating between ATTR and AL of the heart [56].

Quantitative high-resolution microradiographic imaging of amyloid deposits has been developed in a murine model; it combines radio-iodinated SAP component and single-photon emission tomography imaging. The use of radiography to discern the extent of amyloid burden was beneficial for quantitating the total-body burden of amyloid and for evaluation of the therapeutic efficacy of pharmacologic compounds [57].

SAP COMPONENT SCAN

A highly specific, safe quantitative tracer molecule that identifies amyloid deposits is radio-iodinated SAP component. This nonfibrillar pentraxin plasma protein undergoes specific calcium-dependent binding to amyloid fibrils and thus is always present in amyloid deposits. In 1990, scintigraphy after the injection of ^{123}I-labeled SAP component was reported to accurately diagnose, locate, and monitor the extent of systemic amyloidosis [58]. The uptake of [^{123}I] SAP is proportional to the quantity of amyloid protein in different tissues, and 24-hour retention levels are abnormal in all patients with known AL [59].

The scan has been used to show regression of visceral amyloid deposits after liver transplantation for patients with ATTR [60]. Using this scan, hepatic amyloid deposits were identified in 54% of patients with AL and 18% of patients with AA but only 2% of patients with ATTR [61]. Results of hepatic SAP scans were in complete concordance with those of liver biopsy. Liver involvement was always associated with amyloid deposits in other organ systems. The SAP scan also is a noninvasive method of monitoring patients for amyloid deposits after renal transplantation for end-stage renal disease [62]. ^{131}I-labeled amyloid P has been used for diagnostic scans. However, its imaging characteristics are unfavorable compared with ^{123}I-labeled SAP [63].

Systemic amyloidosis is characterized by accelerated initial clearance of ^{123}I-labeled SAP from the plasma and increased interstitial exchange rate and extravascular retention. These findings indicate reversible binding of radiolabeled SAP and provide clinically useful information when monitoring the effects of therapy [64]. Clinically useful observations include organ distribution of the amyloid proteins, evidence of amyloidosis in sites unavailable for biopsy, and evidence of progression or regression of the amyloid deposits in different organs [65,66].

ECHOCARDIOGRAPHIC ASSESSMENT OF CARDIAC AMYLOIDOSIS

Echocardiography has been used to establish the diagnosis of cardiac amyloidosis and to differentiate cardiac amyloidosis from other forms of hypertrophic cardiomyopathy [67]. Echocardiographic features in amyloidosis are well described and include left ventricular fractional shortening reduction, a transmitral flow velocity compatible with abnormal relaxation, right ventricular systolic dysfunction, and ventricular wall thickening [68]. Pulsed tissue Doppler imaging adds value to two-dimensional imaging studies for evaluation of systolic and diastolic left ventricular function [69]. Right ventricular dysfunction in cardiac amyloidosis has been characterized using the Tei index, which is increased markedly for patients with cardiac amyloidosis compared with control patients [70]. Cardiac amyloidosis may be diagnosed using the velocity profile in the hypertrophied left ventricular wall. The myocardial velocity profile in the ventricular septum and left ventricular posterior wall shows a distinctive serrated pattern that may be related to amyloid deposition in the myocardium. Color-coded tissue Doppler imaging also provides diagnostic information [71]. Low-voltage patterns, pseudoinfarction patterns, increased myocardial thickness, and a speckled appearance of the myocardium on the echocardiogram are associated with biopsy-proven cardiac amyloidosis. Multivariate analysis showed that a combination of low voltage and measures of myocardial thickness produced the most statistically powerful models. The combination of low voltage and increased intraventricular septal thickness also is a useful diagnostic tool [72].

Although AA uncommonly causes cardiac symptoms, asymptomatic patients may have systolic and diastolic dysfunction. Tissue Doppler imaging is a more reliable method for early detection of cardiac relaxation abnormalities for these patients with AA [73]. The recent development of strain rate imaging to assess myocardial function has added greatly to tissue Doppler evaluation of patients with cardiac amyloidosis. Because it sensitively detects the earliest degrees of cardiac dysfunction, it is now used routinely in the assessment of patients with cardiac amyloidosis at Mayo Clinic [74]. The combination of pulsed tissue Doppler imaging and myocardial strain rate imaging is capable of recognizing signs of infiltrative cardiac amyloidosis, even for patients in early stages of ATTR [75].

MAGNETIC RESONANCE IMAGING ASSESSMENT OF AMYLOIDOSIS

Magnetic resonance imaging is a noninvasive diagnostic method for amyloidosis; it may be used to identify typical morphologic markers and suggest the presence of infiltrative disease by tissue characterization [76]. Two types of lesions typically are noted: (1) capsular and tendon lesions (thickened, hypointense lesions that are enhanced by gadolinium on T1-weighted images and hyperintense on T2-weighted images), and (2) tumor-forming periarticular and osseous lesions (hypointense on T1- and T2-weighted imaging and not enhanced by gadolinium). Both are useful findings for the diagnosis of amyloidosis arthropathy [77].

A number of independent groups have verified the value of cardiovascular magnetic resonance imaging for cardiac amyloidosis [78–80]. Patients with cardiac amyloidosis have a characteristic pattern of global, subendocardial, late enhancement coupled with abnormal gadolinium kinetics in the myocardium and blood pools. These findings are in concordance with the transmural histologic distribution of amyloid protein.

CONCLUSIONS

All patients with amyloidosis must have Congo Red–stained deposits showing green birefringence under polarized light. Subcutaneous fat is the easiest source of such tissue. Amyloid deposits must be characterized to establish systemic or localized disease, and all forms of systemic amyloidosis must be further classified using immunohistochemistry, immunofluorescence, or mass spectroscopy techniques. When available, SAP component imaging may be used to quantify the extent of amyloid deposition. Both echocardiography and magnetic resonance imaging are important tools for assessing the extent of cardiac amyloidosis and establishing the prognosis for patients with this disease.

REFERENCES

1. Hazenberg, B.P., van Gameren, I.I., Bijzet, J., Jager, P.L., van Rijswijk, M.H. (2004). Diagnostic and therapeutic approach of systemic amyloidosis. *Neth J Med, 62*, 121–128.

2. Gertz, M.A. (2002). Diagnosing primary amyloidosis. *Mayo Clin Proc, 77*, 1278–1279.
3. Hachulla, E., Grateau, G. (2002). Diagnostic tools for amyloidosis. *Joint Bone Spine, 69*, 538–545.
4. Bennhold, H. (1922). Eine spezifische amyloidfärbung mit. *MMW Munch Med Wochenschr, 69*, 1537.
5. Divry, P., Florkin, M. (1927). Sur les propriétés optiques de l'amyloïde. *C Soc Biol, 97*, 1808–1810.
6. Ishii, W., Matsuda, M., Nakamura, N., Katsumata, S., Toriumi, H., Suzuki, A., Ikeda, S. (2003). Phenol Congo Red staining enhances the diagnostic value of abdominal fat aspiration biopsy in reactive AA amyloidosis secondary to rheumatoid arthritis. *Intern Med, 42*, 400–405.
7. Srinivasan, R., Nijhawan, R., Grautam, U., Bambery, P. (1994). Potassium permanganate resistant amyloid in fine-needle aspirate of the liver. *Diagn Cytopathol, 10*, 383–384.
8. Bely, M., Apathy, A. (2000). Histochemical and immunohistochemical differential diagnosis of amyloidosis: a brief illustrated essay and personal experience with Romhanyi's method. *Amyloid, 7*, 212–217.
9. Linke, R.P. (2000). Highly sensitive diagnosis of amyloid and various amyloid syndromes using Congo Red fluorescence. *Virchows Arch, 436*, 439–448.
10. Sen, S., Basdemir, G. (2003). Diagnosis of renal amyloidosis using Congo Red fluorescence. *Pathol Int, 53*, 534–538.
11. Pellikka, P.A., Holmes, D.R., Jr, Edwards, W.D., Nishimura, R.A., Tajik, A.J., Kyle, R.A. (1988). Endomyocardial biopsy in 30 patients with primary amyloidosis and suspected cardiac involvement. *Arch Intern Med, 148*, 662–666.
12. Novak, L., Cook, W.J., Herrera, G.A., Sanders, P.W. (2004). AL-amyloidosis is underdiagnosed in renal biopsies. *Nephrol Dial Transplant, 19*, 3050–3053.
13. Gangane, N., Anshu, A., Shivkumar, V.B., Sharta, S. (2006). Cytodiagnosis of hepatic amyloidosis by fine needle aspiration cytology: a case report. *Acta Cytol, 50*, 574–576.
14. Kristen, A.V., Dengler, T.J., Katus, H.A. (2007). Suspected cardiac amyloidosis: endomyocardial biopsy remains the diagnostic gold-standard. *Am J Hematol, 82*, 328.
15. Taillan, B., Fuzibet, J.G., Vinti, H., Pesce, A., Dujardin, P. (1990). AL amyloid deposits in temporal artery mimicking giant cell arteritis. *Clin Rheumatol, 9*, 256.
16. Zumrutdal, A., Ozelsancak, R., Bolat, F., Sezerl, S., Ozdemir, F.N. (2005). Duodenal biopsy may be preferred first for diagnosis of renal amyloidosis in patients with nephrotic syndrome. *Clin Nephrol, 64*, 242–243.
17. Lechapt-Zalcman, E., Authier, F.J., Creange, A., Voisin, M.C., Gherardi, R.K. (1999). Labial salivary gland biopsy for diagnosis of amyloid polyneuropathy. *Muscle Nerve, 22*, 105–107.
18. Picken, M.M. (2007). Immunoglobulin light and heavy chain amyloidosis AL/AH: renal pathology and differential diagnosis. *Contrib Nephrol, 153*, 135–155.
19. Duston, M.A., Skinner, M., Shirahama, T., Cohen, A.S. (1987). Diagnosis of amyloidosis by abdominal fat aspiration: analysis of four years' experience. *Am J Med, 82*, 412–414.

20. Gertz, M.A., Li, C.Y., Shirahama, T., Kyle, R.A. (1988). Utility of subcutaneous fat aspiration for the diagnosis of systemic amyloidosis (immunoglobulin light chain). *Arch Intern Med, 148*, 929–933.

21. Olsen, K.E., Slennen, K., Westermark, P. (1999). The use of subcutaneous fat tissue for amyloid typing by enzyme-linked immunosorbent assay. *Am J Clin Pathol, 111*, 355–362.

22. Guy, C.D., Jones, C.K. (2001). Abdominal fat pad aspiration biopsy for tissue confirmation of systemic amyloidosis: specificity, positive predictive value, and diagnostic pitfalls. *Diagn Cytopathol, 24*, 181–185.

23. Giorgadze, T.A., Shiina, N., Baloch, Z.W., Tomaszewski, J.E., Gupta, P.K. (2004). Improved detection of amyloid in fat pad aspiration: an evaluation of Congo Red stain by fluorescent microscopy. *Diagn Cytopathol, 31*, 300–306.

24. Arbustini, E., Verga, L., Concardi, M., Palladini, G., Obici, L., Merlini, G. (2002). Electron and immuno-electron microscopy of abdominal fat identifies and characterizes amyloid fibrils in suspected cardiac amyloidosis. *Amyloid, 9*, 108–114.

25. Westermark, P., Davey, E., Lindbom, K., Enqvist, S. (2006). Subcutaneous fat tissue for diagnosis and studies of systemic amyloidosis. *Acta Histochem, 108*, 209–213.

26. Ansari-Lari, M.A., Ali, S.Z. (2004). Fine-needle aspiration of abdominal fat pad for amyloid detection: a clinically useful test? *Diagn Cytopathol, 30*, 178–181.

27. van Gameren, I.I., Hazenberg, B.P., Bijzet, J., van Rijswijk, M.H. (2006). Diagnostic accuracy of subcutaneous abdominal fat tissue aspiration for detecting systemic amyloidosis and its utility in clinical practice. *Arthritis Rheum, 54*, 2015–2021.

28. Siddachari, R.C., Chaukar, D.A., Pramesh, C.S., Naresh, K.N., de Souza, C.E., Dcruz, A.K. (2005). Laryngeal amyloidosis. *J Otolaryngol, 34*, 60–63.

29. Shah, P.L., Gillmore, J.D., Copley, S.J., Collins, J.V., Wells, A.U., du Bois, R.M., Hawkins, P.N., Nicholson, A.G. (2002). The importance of complete screening for amyloid fibril type and systemic disease in patients with amyloidosis in the respiratory tract. *Sarcoidosis Vasc Diffuse Lung Dis, 19*, 134–142.

30. Biewend, M.L., Menke, D.M., Calamia, K.T. (2006). The spectrum of localized amyloidosis: a case series of 20 patients and review of the literature. *Amyloid, 13*, 135–142.

31. Lachmann, H.J., Booth, D.R., Booth, S.E., Bybee, A., Gilbertson, J.A., Gillmore, J.D., Pepys, M.B., Hawkins, P.N. (2002). Misdiagnosis of hereditary amyloidosis as AL (primary) amyloidosis. *N Engl J Med, 346*, 1786–1791.

32. Comenzo, R.L., Zhou, P., Fleisher, M., Clark, B., Teruya-Feldstein, J. (2006). Seeking confidence in the diagnosis of systemic AL (Ig light-chain) amyloidosis: patients can have both monoclonal gammopathies and hereditary amyloid proteins. *Blood, 107*, 3489–3491.

33. Terrier, B., Buzyn, A., Hummel, A., Deroure, B., Bollee, G., Joblonski, M., de Serre, N.P., Noel, L.H., Fakhouri, F. (2006). Serum monoclonal component and nephrotic syndrome: It is not always amyloidosis. Diagnosis: WM complicated by retroperitoneal and renal infiltration and associated with a minimal change disease. *Nephrol Dial Transplant, 21*, 3327–3329.

34. Kaplan, B., Hrncic, R., Murphy, C.L., Gallo, G., Weiss, D.T., Solomon, A. (1999). Microextraction and purification techniques applicable to chemical characterization of amyloid proteins in minute amounts of tissue. *Methods Enzymol, 309*, 67–81.

35. Kaplan, B., Shtrasberg, S., Pras, M. (2003). Micropurification techniques in the analysis of amyloid proteins. *J Clin Pathol, 56*, 86–90.
36. Kaplan, B., Martin, B.M., Livneh, A., Pras, M., Gallo, G.R. (2004). Biochemical subtyping of amyloid in formalin-fixed tissue samples confirms and supplements immunohistologic data. *Am J Clin Pathol, 121*, 794–800.
37. Levo, Y., Livni, N., Laufer, A. (1982). Diagnosis and classification of amyloidosis by an immuno-histological method. *Pathol Res Pract, 175*, 373–379.
38. Strege, R.J., Saeger, W., Linke, R.P. (1998). Diagnosis and immunohistochemical classification of systemic amyloidoses: report of 43 cases in an unselected autopsy series. *Virchows Arch, 433*, 19–27.
39. Kaplan, B., Vidal, R., Kumar, A., Ghiso, J., Gallo, G. (1999). Immunochemical microanalysis of amyloid proteins in fine-needle aspirates of abdominal fat. *Am J Clin Pathol, 112*, 403–407.
40. Hoshii, Y., Setoguchi, M., Iwata, T., Ueda, J., Cui, D., Kawano, H., Gondo, T., Takahashi, M., Ishihara, T. (2001). Useful polyclonal antibodies against synthetic peptides corresponding to immunoglobulin light chain constant region for immunohistochemical detection of immunoglobulin light chain amyloidosis. *Pathol Int, 51*, 264–270.
41. Kebbel, A., Rocken, C. (2006). Immunohistochemical classification of amyloid in surgical pathology revisited. *Am J Surg Pathol, 30*, 673–683.
42. Linke, R.P., Oos, R., Wiegel, N.M., Nathrath, W.B. (2006). Classification of amyloidosis: misdiagnosing by way of incomplete immunohistochemistry and how to prevent it. *Acta Histochem, 108*, 197–208.
43. Kuci, H., Ebert, M.P., Rocken, C. (2007). Anti-lambda-light chain-peptide antibodies are suitable for the immunohistochemical classification of AL amyloid. *Histol Histopathol, 22*, 379–387.
44. Murphy, C.L., Eulitz, M., Hrncic, R., Sletten, K., Westermark, P., Williams, T., Macy, S.D., Wooliver, C., Wall, J., Weiss, D.T., Solomon, A. (2001). Chemical typing of amyloid protein contained in formalin-fixed paraffin-embedded biopsy specimens. *Am J Clin Pathol, 116*, 135–142.
45. Murphy, C.L., Wang, S., Williams, T., Weiss, D.T., Solomon, A. (2006). Characterization of systemic amyloid deposits by mass spectrometry. *Methods Enzymol, 412*, 48–62.
46. Abraham, R.S., Ballman, K.V., Dispenzieri, A., Grill, D.E., Manske, M.K., Price-Troska, T.L., Paz, N.G., Gertz, M.A., Fonseca, R. (2005). Functional gene expression analysis of clonal plasma cells identifies a unique molecular profile for light chain amyloidosis. *Blood, 105*, 794–803.
47. Tate, J.R., Mollee, P., Dimeski, G., Carter, A.C., Gill, D. (2007). Analytical performance of serum free light-chain assay during monitoring of patients with monoclonal light-chain diseases. *Clin Chim Acta, 376*, 30–36.
48. O'Nuallain, B., Hrncic, R., Wall, J.S., Weiss, D.T., Solomon, A. (2006). Diagnostic and therapeutic potential of amyloid-reactive IgG antibodies contained in human sera. *J Immunol, 176*, 7071–7078.
49. Hartmann, A., Frenkel, J., Hopf, R., Baum, R.P., Hor, G., Schneider, M., Kaltenbach, M. (1990). Is technetium-99m-pyrophosphate scintigraphy valuable in the diagnosis of cardiac amyloidosis? *Int J Card Imaging, 5*, 227–231.

50. Fournier, C., Grimon G., Rinaldi, J.P., Terral, A., Boujon, B., Adams, D., Desgrez, A., Blondeau, M. (1993). Usefulness of technetium-99m pyrophosphate myocardial scintigraphy in amyloid polyneuropathy and correlation with echocardiography. *Am J Cardiol, 72*, 854–857.

51. Kodama, K., Hamada, M., Kuwahara, T., Nakamura, M., Shigematsu, Y., Hiwada, K., Iwata, T., Hoshii, Y., Ishihara, T. (1999). Rest-redistribution thallium-201 myocardial scintigraphic study in cardiac amyloidosis. *Int J Card Imaging, 15*, 371–378.

52. Yen, T.C., Tzen, K.Y., Chen, K.S., Tsai, C.J. (2000). The value of gallium-67 and thallium-201 whole-body and single-photon emission tomography images in dialysis-related beta 2-microglobulin amyloid. *Eur J Nucl Med, 27*, 56–61.

53. Dezutter, N.A., Sciot, R.M., de Groot, T.J., Bormans, G.M., Verbruggen, A.M. (2001). In vitro affinity of ^{99}Tcm-labelled N2S2 conjugates of chrysamine G for amyloid deposits of systemic amyloidosis. *Nucl Med Commun, 22*, 553–558.

54. Puille, M., Altland, K., Linke, R.P., Steen-Muller, M.K., Kiett, R., Steiner, D., Bauer, R. (2002). ^{99m}Tc-DPD scintigraphy in transthyretin-related familial amyloidotic polyneuropathy. *Eur J Nucl Med Mol Imaging, 29*, 376–379.

55. Kulhanek, J., Movahed, A. (2003). Uptake of technetium 99m HDP in cardiac amyloidosis. *Int J Cardiovasc Imaging, 19*, 225–227.

56. Perugini, E., Guidalotti, P.L., Salvi, F., Cooke, R.M., Pettinato, C., Riva, L., Leone, O., Farsad, M., Ciliberti, P., Bacchi-Reggiani, L., et al. (2005). Noninvasive etiologic diagnosis of cardiac amyloidosis using ^{99m}Tc-3, 3-diphosphono-1,2-propanodicarboxylic acid scintigraphy. *Am J Cardiol, 46*, 1076–1084.

57. Wall, J.S., Kennel, S.J., Paulus, M.J., Gleason, S., Gregor, J., Baba, J., Schell, M., Richey, T., O'Nuallain, B., Donnell, R., et al (2005). Quantitative high-resolution microradiographic imaging of amyloid deposits in a novel murine model of AA amyloidosis. *Amyloid, 12*, 149–156.

58. Hawkins, P.N., Lavender, J.P., Pepys, M.B. (1990). Evaluation of systemic amyloidosis by scintigraphy with ^{123}I-labeled serum amyloid P component. *N Engl J Med, 323*, 508–513.

59. Hachulla, E., Maulin, L., Deveaux, M., Facon, T., Bletry, O., Vanhille, P., Wechsler, B., Godeau, P., Levesque, H., Hatron, P.Y., et al (1996). Prospective and serial study of primary amyloidosis with serum amyloid P component scintigraphy: from diagnosis to prognosis. *Am J Med, 101*, 77–87.

60. Rydh, A., Suhr, O., Hietala, S.O., Alstrom, K.R., Pepys, M.B., Hawkins, P.N. (1998). Serum amyloid P component scintigraphy in familial amyloid polyneuropathy: regression of visceral amyloid following liver transplantation. *Eur J Nucl Med, 25*, 709–713.

61. Lovat, L.B., Persey, M.R., Madhoo, S., Pepys, M.B., Hawkins, P.N. (1998). The liver in systemic amyloidosis: insights from ^{123}I serum amyloid P component scintigraphy in 484 patients. *Gut, 42*, 727–734.

62. Gillmore, J.D., Madhoo, S., Pepys, M.B., Hawkins, P.N. (2000). Renal transplantation for amyloid end-stage renal failure: insights from serial serum amyloid P component scintigraphy. *Nucl Med Commun, 21*, 735–740.

63. Hawkins, P.N., Aprile, C., Capri, G., Vigano, L., Munzone, E., Gianni, L., Pepys, M.B., Merlini, G. (1998). Scintigraphic imaging and turnover studies with iodine-131

labelled serum amyloid P component in systemic amyloidosis. *Eur J Nucl Med, 25*, 701–708.

64. Jager, P.L., Hazenberg, B.P., Franssen, E.J., Limburg, P.C., van Rijswijk, M.H., Piers, D.A. (1998). Kinetic studies with iodine-123-labeled serum amyloid P component in patients with systemic AA and AL amyloidosis and assessment of clinical value. *J Nucl Med, 39*, 699–706.
65. Pepys, M.B. (2001). Pathogenesis, diagnosis and treatment of systemic amyloidosis. *Philos Trans R Soc Lond B, 356*, 203–210.
66. Hawkins, P.N. (2002). Serum amyloid P component scintigraphy for diagnosis and monitoring amyloidosis. *Curr Opin Nephrol Hypertens, 11*, 649–655.
67. Eriksson, P., Backman, C., Eriksson, A., Eriksson, S., Karp, K., Olofsson, B.O. (1987). Differentiation of cardiac amyloidosis and hypertrophic cardiomyopathy: a comparison of familial amyloidosis with polyneuropathy and hypertrophic cardiomyopathy by electrocardiography and echocardiography. *Acta Med Scand, 221*, 39–46.
68. Moyssakis, I., Triposkiadis, F., Rallidis, L., Hawkins, P., Kyriakidis, M., Nihoyannopoulos, P. (1999). Echocardiographic features of primary, secondary and familial amyloidosis. *Eur J Clin Invest, 29*, 484–489.
69. Koyama, J., Ray-Sequin, P.A., Falk, R.H. (2003). Longitudinal myocardial function assessed by tissue velocity, strain, and strain rate tissue Doppler echocardiography in patients with AL (primary) cardiac amyloidosis. *Circulation, 107*, 2446–2452.
70. Kim, W.H., Otsuji, Y., Yuasa, T., Minagoe, S., Seward, J.B., Tei, C. (2004). Evaluation of right ventricular dysfunction in patients with cardiac amyloidosis using Tei index. *J Am Soc Echocardiogr, 17*, 45–49.
71. Oki, T., Tanaka, H., Yamada, H., Tabata, T., Oishi, Y., Ishimoto, T., Nagase, N., Shinohara, H., Sakabe, K., Fukuda, N. (2004). Diagnosis of cardiac amyloidosis based on the myocardial velocity profile in the hypertrophied left ventricular wall. *Am J Cardiol, 93*, 864–869.
72. Rahman, J.E., Helou, E.F., Gelzer-Bell, R., Thompson, R.E., Kuo, C., Rodriguez, E.R., Hare, J.M., Baughman, K.L., Kasper, E.K. (2004). Noninvasive diagnosis of biopsy-proven cardiac amyloidosis. *J Am Coll Cardiol, 43*, 410–415.
73. Demir, M., Paydas, S., Cayli, M., Akpinar, O., Balal, M., Acarturk, E. (2005). Tissue Doppler is a more reliable method in early detection of cardiac dysfunction in patients with AA amyloidosis. *Ren Fail, 27*, 415–420.
74. Sallach, J.A., Klein, A.L. (2004). Tissue Doppler imaging in the evaluation of patients with cardiac amyloidosis. *Curr Opin Cardiol, 19*, 464–471.
75. Lindqvist, P., Olofsson, B.O., Backman, C., Suhr, O., Waldenstrom, A. (2006). Pulsed tissue Doppler and strain imaging discloses early signs of infiltrative cardiac disease: a study on patients with familial amyloidotic polyneuropathy. *Eur J Echocardiogr, 7*, 22–30.
76. Fattori, R., Rocchi, G., Celletti, F., Bertaccini, P., Rapezzi, C., Gavelli, G. (1998). Contribution of magnetic resonance imaging in the differential diagnosis of cardiac amyloidosis and symmetric hypertrophic cardiomyopathy. *Am Heart J, 136*, 824–830.
77. Miyata, M., Sato, N., Watanabe, H., Kumakawa, H., Saito, A., Funabashi, H., Iwatsuki, K., Hashimoto, Y., Sato, Y., Kasukawa, R. (2000). Magnetic resonance

imaging findings in primary amyloidosis-associated arthropathy. *Intern Med, 39*, 313–319.

78. Crean, A., Merchant, N. (2004). Role of cardiac magnetic resonance imaging in identification of amyloid cardiomyopathy. *Indian Heart J, 56*, 682–683.
79. Kwong, R.Y., Falk, R.H. (2005). Cardiovascular magnetic resonance in cardiac amyloidosis. *Circulation, 111*, 122–124.
80. Maceira, A.M., Joshi, J., Prasad, S.K., Moon, J.C., Perugini, E., Harding, I., Sheppard, M.N., Poole-Wilson, P.A., Hawkins, P.N., Pennell, D.J. (2005). Cardiovascular magnetic resonance in cardiac amyloidosis. *Circulation, 111*, 186–193.

32

IDENTIFICATION OF BIOMARKERS FOR DIAGNOSIS OF AMYLOID DISEASES: QUANTITATIVE FREE LIGHT-CHAIN ASSAYS

Jerry A. Katzmann

Department of Laboratory Medicine and Pathology, Mayo Clinic, Rochester, Minnesota

INTRODUCTION

Primary systemic amyloidosis (AL) is a plasma cell proliferative disorder. The plasma cell proliferative disorders are characterized by a clonal expansion of plasma cells and can be subgrouped into malignant diseases (multiple myeloma, plasma cell leukemia, plasmacytoma, Waldenström's macroglobulinemia), premalignant syndromes (monoclonal gammopathy of undetermined significance, smoldering myeloma), and protein conformation disorders (AL, light-chain deposition disease, cryoglobulinemia). Because of the clonal plasma cell proliferation, most of these diseases are also characterized by the presence of a monoclonal immunoglobulin and are therefore often described as monoclonal gammopathies. This monoclonal protein (Fig. 1) serves a surrogate marker for the clonal plasma cell expansion and was one of the first (and remains one of the best) clonal cell markers, or tumor markers. In Figure 1B the immunofixation electrophoresis (IFE) identifies an IgGκ monoclonal gammopathy. The identification of this monoclonal IgGκ is not diagnostic for any of the specific diseases listed above, but it is diagnostic for a plasma cell proliferative disorder. In addition, changes in the quantitation of the

Protein Misfolding Diseases: Current and Emerging Principles and Therapies,
Edited by Marina Ramirez-Alvarado, Jeffery W. Kelly, and Christopher M. Dobson
Copyright © 2010 John Wiley & Sons, Inc.

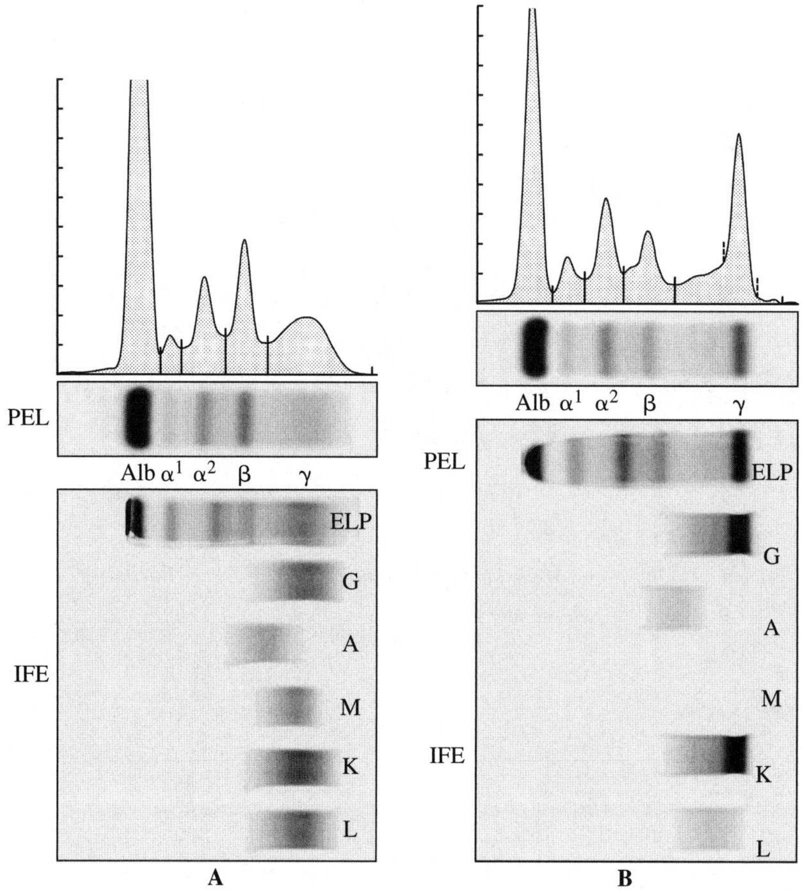

FIG. 1 (A) Protein electrophoresis (PEL) and immunoelectrophoresis (IFE) of normal serum: The proteins in the γ fraction on the PEL gel and the precipitated immunoglobulins on the IFE gel exhibit a smooth Gaussian distribution, indicating a polyclonal distribution. (B) Protein electrophoresis (PEL) and immunoelectrophoresis (IFE) of a serum containing a monoclonal IgG κ : The proteins in the γ fraction on the PEL gel and the precipitated immunoglobulins on the IFE gel exhibit a discrete band indicating a monoclonal protein. The uppermost graph is a scan of the PEL gel, and the area under the monoclonal protein spike is used in conjunction with the serum total protein concentration to quantitate the amount of monoclonal immunoglobulin.

monoclonal protein reflect changes in the size of the plasma cell clone and can be used to monitor disease or response to therapy. In Figure 1B the electropherogram of the protein electrophoresis gel (PEL) allows quantitation of the amount of the monoclonal protein and can be used as a marker for changes in the size of the plasma cell clone. Not all plasma cell proliferative disorders, however, produce a detectable or quantifiable monoclonal protein. Nonsecretory multiple myeloma, for example, does not secrete an easily detectable

TABLE 1 Distribution (%) of Plasma Cell Proliferative Disorders in Clinical Practice[a]

Monoclonal gammopathy of undetermined significance	61
Multiple myeloma	17
Amyloidosis AL	9
Lymphoproliferative disease	3
Smoldering myeloma	4
Solitary or extramedullary plasmacytoma	2
Macroglobulinemia	2
Other	2

Source: Mayo Clinic Dysproteinemia Data Base, 1960–2003.
[a]$n = 31{,}479$.

monoclonal protein even though the bone marrow may be packed with malignant plasma cells containing cytoplasmic monoclonal immunoglobulin light chain. In AL patients with significant disease, it may also be difficult to detect a monoclonal protein and often impossible to quantitate.

The distribution of monoclonal gammopathies in our clinical practice is listed in Table 1. By far the most common plasma cell proliferative disorder is monoclonal gammopathy of undetermined significance (MGUS). In the normal population MGUS has a 3% prevalence in people older than 50 years [1]. The prevalence increases with age, such that between ages 50 and 60 the prevalence is 1.7% and in ages greater than 80 years the prevalence is 6.6%. This very common premalignant disorder has no clinical symptoms, but these patients have a 1% per year progression to diseases such as multiple myeloma or AL [2]. The diagnosis of a plasma cell proliferative disorder as MGUS is dependent on the absence of related organ or tissue impairment (e.g., no end-organ damage or bone lesions), less than 10% clonal bone marrow plasma cells, and less that 3 g/dL of serum monoclonal protein. The diagnosis of multiple myeloma requires greater than 10% clonal bone marrow plasma cells as well as clinical symptoms such as hypercalcemia, renal disease, anemia, and/or bone lesions due to the clonal plasma cell proliferation. Multiple myeloma has an incidence of approximately 40 per million per year in Caucasians, and the incidence is two to threefold higher in black populations [3]. There are approximately 13,000 new cases each year in the United States.

By contrast, AL is a relatively rare plasma cell proliferative disorder. It is one-fifth as common as multiple myeloma, with approximately 2500 new patients diagnosed in the United States every year. Like all forms of amyloid, diagnosis requires histologic staining with Congo Red for the identification of amyloid fibrils in tissue [4]. Because AL is a monoclonal gammopathy, the identification of a monoclonal protein is a useful initial screen for patients with AL. The 3% prevalence of MGUS in the population over 50 years, however, means that some of these coincident findings will be unrelated. Immunohistochemical staining or some other direct characterization of the amyloid fibrils may be necessary to confirm the type of amyloidosis [5].

The fibrils in AL are derived from intact or fragments of monoclonal immunoglobulin light chains. These patients often have free monoclonal immunoglobulin light chains detected in the serum and/or urine. The free light chains (FLCs) are secreted from the plasma cells without being bound to heavy chain in an intact immunoglobulin molecule, and the monoclonal light chains or fragments are precursor components for the amyloid fibrils. Unlike patients with multiple myeloma, AL patients may have a very small population of clonal plasma cells in the bone marrow. The continuous long-term secretion of the amyloidogenic monoclonal light chain results in amyloid fibril deposition and eventual organ failure. The presence of small clonal plasma cell populations and the small amount of secreted monoclonal light chains may make it difficult to document the monoclonal gammopathy by serum and/or urine immunofixation electrophoresis (IFE) [6]. A newer laboratory approach uses antisera that distinguishes free from bound κ and λ light chains and nephelometry for quantitation of the serum concentration.

FREE LIGHT-CHAIN QUANTITATION

The introduction of automated assays for the quantitation of immunoglobulin FLCs has given clinical laboratories a new tool for evaluating AL. The FLC assays have increased the diagnostic sensitivity for identification of light-chain diseases such as AL, and in addition they have improved disease monitoring and prognosis. The FLC assay was described by Bradwell and colleagues in 2001 [7]. The method is an automated nephelometric assay that uses a commercially available reagent set of polyclonal antibodies to quantitate both κ and λ FLCs by immunonephelometry (Freelite, the Binding Site, San Diego, California). The antibodies show no reactivity by immunoelectrophoresis or by Western blots to light chains contained in intact immunoglobulin. Sensitive hemagglutination assays showed reactivity to the appropriate FLCs at dilutions above 1 : 16,000 and no reactivity to light chains contained within intact Ig at antisera dilutions >1 : 2. The assay reagents therefore appear to have a minimum of a 10,000-fold difference in reactivity to FLCs compared to light chain contained within intact Ig. This high specificity allows κ and λ FLCs to be quantitated in the presence of a large excess of serum IgG, IgA, and IgM. Furthermore, labs can perform the assays on a number of automated laboratory instruments, including the Dade Behring (Deerfield, Illinois) BN II and BN ProSpec; Beckman Coulter (Brea, California) IMMAGE; Roche/Hitachi (Indianapolis, Indiana) 911, 912, 917, and module P; and the Olympus (Melville, New York) AU series.

Because the prevalence of monoclonal gammopathies increases with increasing age, a clinical laboratory reference range study was performed with sera from healthy donors aged 21 to 90 years [8]. Sera were screened by protein electrophoresis and IFE to exclude samples with a monoclonal protein (e.g., unknown MGUS patients), and κ FLC, λ FLC, and cystatin C were quantitated. In addition, the κ/λ FLC ratio was calculated. This reference range study revealed an apparent effect of age on the two FLC concentrations, but the κ/λ FLC ratio showed no age dependence (Table 2). Cystatin C showed the same apparent age dependence, and the increase in serum FLC concentrations with age is most likely due to reduced renal clearance. The use of the κ/λ FLC ratio normalized the effects of reduced clearance.

Abnormal κ and λ FLC concentrations may result from a number of clinical situations, including immune suppression, immune stimulation, reduced renal clearance, or monoclonal plasma cell proliferative disorders. Sera from patients with either polyclonal hypergammaglobulinemia or renal impairment often have elevated κ and λ FLCs, due to increased synthesis or reduced renal clearance, respectively. The κ/λ FLC ratio, however, usually remains normal in these conditions [8]. A significantly abnormal κ/λ FLC ratio should only be due to a plasma cell proliferative disorder that secretes excess free light chain and disturbs the normal balance between κ and λ secretion. An abnormally high ratio suggests expansion of a κ-producing plasma cell clone, whereas an abnormally low ratio suggests expansion of a λ-producing plasma cell clone. Normal reference ranges are usually defined by a 95% reference range, and therefore, by definition, the 2.5% tails of the population distribution of the κ/λ FLC ratio are abnormal. Because the usual 95% reference range defines 5% of the population as abnormal, a diagnostic range for the κ/λ FLC ratio was defined by 100% of the normal sample study (Table 3).

TABLE 2 κ FLC, λ FLC, κ/λ FLC Ratio[a]

Age (years)	κ FLC (mg/dL)	λ FLC (mg/dL)	κ/λ FLC Ratio
20–29	4.1	12.6	0.38
30–39	5.0	12.3	0.38
40–49	5.2	12.8	0.38
50–59	5.4	14.2	0.39
60–69	5.8	16.3	0.39
70–79	6.9	19.3	0.39
80+	9.2	23.2	0.39

[a]Median values by decade.

TABLE 3 FLC Quantitation: Serum FLC Reference Intervals and Diagnostic Range

	95% Reference Interval (Normal Range)	Diagnostic Range
κ FLC	0.33–1.94 mg/dL	
λ FLC	0.57–2.63 mg/dL	
κ/λ FLC ratio	0.3–1.2	0.26–1.65

DIAGNOSTIC SENSITIVITY FOR IDENTIFICATION OF MONOCLONAL FLC

The concept of using the alteration of the FLC ratio as a sensitive marker of clonal FLC synthesis was not obvious. In a study of 28 patients with nonsecretory multiple myeloma, however, Drayson and colleagues found that although some patients had no serum or urine monoclonal protein detectable by protein electrophoresis or IFE, 19 (68%) had abnormal κ/λ FLC ratios [9]. In another study of 262 AL patients, Lachmann and colleagues detected abnormal serum FLC in 98% of the AL patients, whereas the serum or urine IFE was positive in only 79% [10]. This increase in diagnostic sensitivity of the serum FLC assay for monoclonal light-chain diseases such as AL was an unexpected diagnostic advance and has been confirmed by other authors [11,12]. In a prospective study to further evaluate the diagnostic performance in clinical practice, we have confirmed the increased sensitivity in a series of all patients with newly diagnosed monoclonal gammopathies that were seen in our practice in one calendar year [13]. In 110 untreated AL patients, the FLC assay was more sensitive (91%) than the serum or urine IFE assay (69% and 83% sensitivity, respectively). In addition, the IFE and FLC assays were complementary for the detection of monoclonal FLCs in AL patients. If both the serum IFE and FLC assay was performed, 109 of the 110 AL patients (99%) had an abnormal result in at least one of the two tests (Table 4). The inclusion of urine testing did not increase the diagnostic sensitivity. This enhanced ability to detect monoclonal free light chains does not address whether the light chains are amyloidogenic, but supports the need to search for amyloid deposits as well as the classification of tissue-confirmed amyloid.

The high diagnostic sensitivity for AL when using serum IFE and FLC assays suggests that the recommended diagnostic screening algorithm for monoclonal gammopathies may not be the best approach. The recommended laboratory testing for patients suspected of having MM, AL, or related disorders requires protein electrophoresis and IFE of both serum and urine.

TABLE 4 Diagnostic Sensitivity (%) in AL[a]

Assay	Sensitivity
κ/λ FLC ratio	91
Serum IFE	69
Urine IFE	83
Serum IFE ± urine IFE	95
FLC κ/λ ratio + urine IFE	91
FLC κ/λ ratio + serum IFE	99
All three assays	99%

[a] $n = 110$.

In initial laboratory evaluations, however, urine is submitted for testing in only approximately 30% of cases. The absence of urine reduces the diagnostic sensitivity most significantly in AL, light-chain deposition disease, and non-secretory MM. Because of this increased sensitivity of the serum FLC assays, it is not apparent that urine studies are still necessary as part of the diagnostic screening algorithm. In a study of 224 patients with light-chain multiple myeloma (LCMM) it was reported that serum IFE and FLC identified 100% of the patients and that urine IFE added no additional information [14]. Similarly, in the study described above, it was found that 109 of 110 patients with AL had abnormal results in either serum IFE or FLC assays and that urine studies also added no diagnostic information [13]. Because of these reports suggesting that urine studies may not add diagnostic information to serum IFE and FLC assays, the inconvenience in routinely obtaining a 24-hour urine sample, and the additional patients detected by serum FLC assays, it is reasonable to replace urine IFE studies with serum FLC assays. The danger of using these studies to dismiss the 24-hour urine IFE as part of the screen for monoclonal gammopathies is that none of these studies addressed the question of diagnostic screening. We therefore used a large cohort of patients with an assortment of plasma cell proliferative diagnoses and a monoclonal urine protein who had urine PEL and IFE as well as serum PEL, IFE, and FLC performed [15]. The study was designed to determine which of these patients would have been undiagnosed in the absence of urine studies. There were 428 patients who had positive urine studies and also had serum PEL, IFE, and FLC performed. These patients had diagnoses of MM (n = 148), AL (n = 123), monoclonal gammopathy of undetermined significance (n = 69), smoldering multiple myeloma (n = 59), solitary plasmacytomas (n = 5), and other less frequently detected monoclonal gammopathies. All 428 had a monoclonal urine protein (by definition of the cohort), and in the serum 86% had an abnormal serum-free light-chain κ/λ ratio, 91% had an abnormal serum protein electrophoresis, and 93% had an abnormal serum immunofixation. However, there were only two cases that were normal in all the serum assays. Both of these cases had monoclonal gammopathy (idiopathic Bence Jones proteinuria). Discontinuation of urine studies and reliance on a diagnostic algorithm using solely serum studies (protein electrophoresis, immunofixation, and free light-chain quantitation), missed 0.5 % of the 428 monoclonal gammopathies with urinary monoclonal proteins, and these two cases required no medical intervention. All 123 cases of AL in this cohort were identified using serum IFE and FLC, and although the urine studies were positive, urine assays were not required for the diagnostic screen.

MONITORING DISEASE ACTIVITY

The enhanced sensitivity for detecting monoclonal-free light chains does not, of course, address whether the light chains are amyloidogenic, but supports the need to search for amyloid deposits. Once the diagnosis of AL is confirmed, the FLC serum assay provides a quantitation of the monoclonal FLC involved and therefore a way to monitor the plasma cell clone analogous to a tumor marker [16]. This allows an assessment of the hematologic response to therapy in addition to organ-based response criteria. Lachmann and colleagues detected abnormal serum FLC ratios in 98% of AL patients tested, but only 79% of the same patients showed an abnormal IFE [10]. Equally interesting, however, was the observation that 46% of the 262 AL patients had no serum or urine M-spike which could be used to monitor treatment. For AL, therefore, the serum FLC assay provided not only a more sensitive diagnostic tool but also a quantitative assessment for monitoring disease activity. Just as a 50% reduction in M-spike values is used as a response criterion when monitoring MM, a 50% reduction in the monoclonal FLC indicated a therapeutic response and was predictive of a significant survival advantage in AL. Dispenzieri et al. have reported that normalization of the FLC level after stem cell transplant predicted organ response in AL patients [12].

The quantitation of immunoglobulin-free light chains has proved to be a useful biomarker in the monoclonal gammopathies in general and in AL in particular. The diagnostic sensitivity of the serum FLC assay in conjunction with serum IFE defines a sensitive diagnostic screen for AL and reduces the need for urine in the screening algorithm. In addition, the concentration of the FLC involved provides a simple way to monitor the disease process and hematologic response.

REFERENCES

1. Kyle, R.A., Therneau, T.M., Rajkumar, S.V., Orford, J.R., Larson, D.R., Plevak, M.F., Melton, L.J. (2006). Prevalence of monoclonal gammopathy of undetermined significance. *N Engl J Med, 354*, 1362–1369.
2. Kyle, R.A., Therneau, T.M., Rajkumar, S.V., Larson, D.R., Plevak, M.F., Orford, J.R., Dispenzieri, A., Katzmann, J.A., Melton, L.J. (2002). A long-term study of prognosis in monoclonal gammopathy of undetermined significance. *N Engl J Med, 346*, 564–569.
3. Landgren, O., Gridley, G., Turesson, I., Caporaso, N.E., Goldin, L.R., Baris, D., Fears, T.R., Hoover, R.N., Linet, M.S. (2006). Risk of monoclonal gammopathy of undetermined significance (MGUS) and subsequent multiple myeloma among African American and white veterans in the United States. *Blood, 107*, 904–906.
4. Steensma, D.P. (2001). "Congo" Red: out of Africa? *Arch Pathol Lab Med, 125*, 250–252.
5. Strege, R.J., Saeger, W., Linke, R.P. (1998). Diagnosis and immunohistochemical classification of systemic amyloidoses: report of 43 cases in an unselected autopsy series. *Virchows Arch, 433*, 19–27.
6. Gertz, M.A., Lacy, M.Q., Dispenzieri, A. (1999). Amyloidosis: recognition, confirmation, prognosis, and therapy. *Mayo Clin Proc, 74*, 490–494.

7. Bradwell, A.R., Carr-Smith, H.D., Mead, G.P., Tang, L.X., Showell, P.J., Drayson, M.T., Drew, R. (2001). Highly sensitive automated immunoassay for immunoglobulin free light chains in serum and urine. *Clin Chem*, *47*, 637–680.

8. Katzmann, J.A., Clark, R.J., Abraham, R.S., Bryant, S., Lymp, J.F., Bradwell, A.R., Kyle, R.A. (2002). Serum reference intervals and diagnostic ranges for free kappa and free lambda immunoglobulin light chains: relative sensitivity for detection of monoclonal light chains. *Clin Chem*, *48*, 1437–1444.

9. Drayson, M., Tang, L.X., Drew, R., Mead, G.P., Carr-Smith, H., Bradwell, A.R. (2001). Serum free light-chain measurements for identifying and monitoring patients with nonsecretory multiple myeloma. *Blood*, *97*, 2900–2902.

10. Lachmann, H.J., Gallimore, R., Gillmore, J.D., Carr-Smith, H.D., Bradwell, A.R., Pepys, M.B., Hawkins, P.N. (2003). Outcome in systemic AL amyloidosis in relation to changes in concentration of circulating free immunoglobulin light chains following chemotherapy. *Br J Haematol*, *122*, 78–84.

11. Abraham, R.S., Katzmann, J.A., Clark, R.J., Bradwell, A.R., Kyle, R.A., Gertz, M.A. (2003). Quantitative analysis of serum free light chains: a new marker for the diagnostic evaluation of primary systemic amyloidosis. *Am J Clin Pathol*, *119*, 274–278.

12. Dispenzieri, A., Lacy, M.Q., Katzmann, J.A., Rajkumar, S.V., Abraham, R.S., Hayman, S.R., Kumar, S.K., Clark, R., Kyle, R.A., Litzow, M.R., et al. (2006). Absolute values of immunoglobulin free light chains are prognostic in patients with primary systemic amyloidosis undergoing peripheral blood stem cell transplantation. *Blood*, *107*, 3378–3383.

13. Katzmann, J.A., Abraham, R.S., Dispenzieri, A., Lust, J.A., Kyle, R.A. (2005). Diagnostic performance of quantitative kappa and lambda free light chain assays in clinical practice. *Clin Chem*, *51*, 878–881.

14. Bradwell, A.R., Carr-Smith, H.D., Mead, G.P., Harvey, T.C., Drayson, M.T. (2003). Serum test for assessment of patients with Bence Jones myeloma. *Lancet*, *361*, 489–491.

15. Katzmann, J.A., Dispenzieri, A., Kyle, R.A., Snyder, M.R., Plevak, M.F., Larson, D.R., Abraham, R.S., Lust, J.A., Melton, L.J., Rajkumar, S.V. (2006). Elimination of the need for urine studies in the screening algorithm for monoclonal gammopathies by using serum immunofixation and free light chain assays. *Mayo Clin Proc*, *81*(12), 1575–1578.

16. Gertz, M.A., Comenzo, R., Falk, R.H., Fermand, J.P., Hazenberg, B.P., Hawkins, P.N., Merlini, G., Moreau, P., Ronco, P., Sanchorawala, V., et al. (2005). Definition of organ involvement and treatment response in immunoglobulin light chain amyloidosis (AL): a consensus opinion from the 10th International Symposium on Amyloid and Amyloidosis, Tours, France, 18–22 April 2004. *Am J Hematol*, *79*, 319–328.

INSTITUT
MEDIATHEQUE
PASTEUR

33

REAL-TIME OBSERVATION OF AMYLOID-β FIBRIL GROWTH BY TOTAL INTERNAL REFLECTION FLUORESCENCE MICROSCOPY

TADATO BAN AND YUJI GOTO
Institute for Protein Research, Osaka University, Osaka, Japan

INTRODUCTION

In Alzheimer disease, amyloid β (Aβ) peptide forms amyloid fibrils that deposit in the extracellular space of the brain as senile amyloid plaques, pathological hallmarks of Alzheimer's disease, and also in the walls of cerebral blood vessels [1–5]. The formation of Aβ amyloid fibrils is considered to be a nucleation-dependent process in which Aβ peptides slowly associate to form a nucleus, which then grows via an extension reaction involving the sequential incorporation of Aβ peptides, producing rigid and straight morphology consisting of several layers of cross-β sheets [6,7]. This process is influenced by several factors: peptide concentration, pH, ionic strength, and interactions with other components [8,9]. The interactions with lipid membranes in particular have received attention because the membrane surface might be responsible for both neurotoxicity and senile plaque formation [10–12].

For several proteins including Aβ, amyloid fibrils prepared on the solid substrates such as mica or quartz produce radial assemblies [13–15]. Considering that in several neurodegenerative diseases, radial and spherical aggregates of amyloid fibrils are found in tissue deposits [5,16], surface interaction may play dominant roles in the formation of amyloid fibrils and their deposition in vivo

Protein Misfolding Diseases: Current and Emerging Principles and Therapies,
Edited by Marina Ramirez-Alvarado, Jeffery W. Kelly, and Christopher M. Dobson
Copyright © 2010 John Wiley & Sons, Inc.

[7,17]. Moreover, amyloid deposits are found in a specific tissue region, suggesting that a specific surface chemistry is involved in the fibril formation and deposition processes in general. However, the behavior of amyloid fibrils on solid surfaces is still far from clear. To obtain further insight into the mechanism of fibril growth and deposition, direct observation of individual fibrils is essential.

TOTAL INTERNAL REFLECTION FLUORESCENCE MICROSCOPY

We developed a new technique for the real-time observation of amyloid fibrils using total internal reflection fluorescence microscopy (TIRFM) combined with thioflavin T (ThT) fluorescence (Fig. 1) [18–23]. TIRFM has been useful for monitoring single molecules by effectively reducing the background fluorescence under the evanescent field formed on the surface of a quartz slide [24–26]. When a laser is incident on the interface between a quartz slide (high reflection index) and an aqueous solution (low reflection index) at the critical angle for total internal reflection, the evanescent field is produced beyond the interface in the solution. Because the evanescent field is produced with a penetration depth of about 150 nm, the illumination is restricted to fluorophores either bound to the quartz slide surface or located close by, resulting in highly reduced background fluorescence. Furthermore, with the careful selection of optical elements, the background fluorescence can be reduced 2000-fold compared with ordinary epi-fluorescence microscopy. On the other hand, ThT is a reagent known to become strongly fluorescent upon binding to amyloid fibrils [27], so that one can detect the fibrils specifically without covalent modification. Importantly, because the evanescent field formed by the total internal reflection of laser light penetrates to a depth of 150 nm one can selectively monitor fibrils lying along the slide glass within 150 nm and so can obtain the exact length of the fibrils. By combining amyloid fibril-specific ThT fluorescence and TIRFM, it would be possible to observe the amyloid fibrils and the process by which they form without introducing any fluorescence reagent bound covalently to the protein molecule.

REAL-TIME OBSERVATION OF Aβ(1–40) FIBRIL GROWTH

Real-time observation of the growth of individual fibrils following seed-dependent extension was carried out on the surface of quartz slides (Fig. 2) [19]. The growth of fibrils occurred concomitantly at many seeds. Although several fibrils often developed from apparently one seed, it is likely that the clustered seeds produced such a radial pattern. Once started, unidirectional growth continued, producing remarkably long fibrils more than 15 μm in length. Considering that TIRFM selectively monitors fibrils lying along the slide within 150 nm, the interaction of fibrils with the quartz surface caused the lateral growth. In addition, the combination of relatively rapid fibril

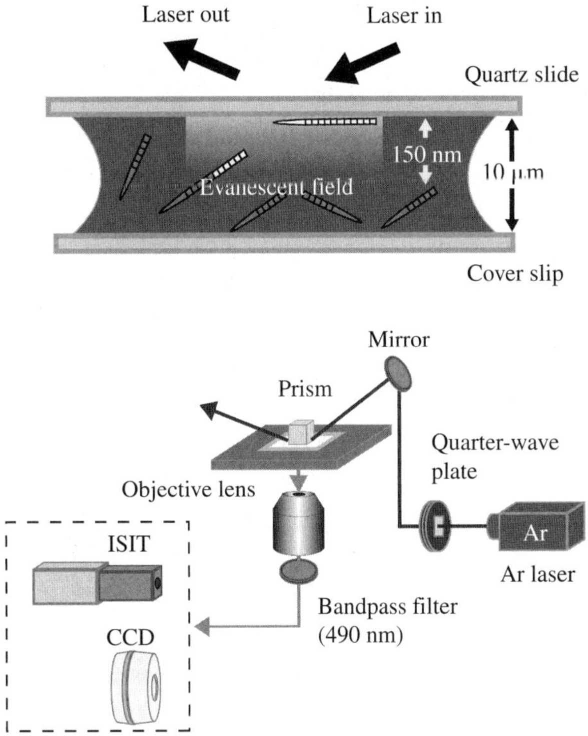

FIG. 1 Schematic representation of amyloid fibrils revealed by total internal reflection fluorescence microscopy. The penetration depth of the evanescent field formed by the total internal reflection of laser light is about 150 nm for a laser light at 455 nm, so that only amyloid fibrils lying in parallel with the slide glass surface were observed. (From [20], with permission of Elsevier.)

growth and less aggregation of fibrils weakly fixed on the quartz surface enabled the formation of remarkably long fibrils.

Intriguingly, we occasionally observed growth to be aligned in a similar direction (e.g., vertically aligned growth) (Fig. 3a), implying that the interaction with an ordered quartz surface significantly affected the direction of growth. We sometimes observed a swinging motion of the growing head, resulting in a shift in the direction of growth and thus producing rugged fibrils (Fig. 3c). In addition, real-time observation revealed important images, suggesting a transient loss of cooperativity in fibril growth (Fig. 3b). After the cooperative growth with a blunt end (images from 6 to 10 min), the end frayed into three thinner filaments (image at 12 min). In the next step, braiding of the three filaments recovered the blunt end (image at 14 min), implying that the mature fibril is made of three protofilaments.

The remarkable length of the fibrils enabled an exact analysis of the rate of growth of individual fibrils [19]. The growth at the early and middle stages

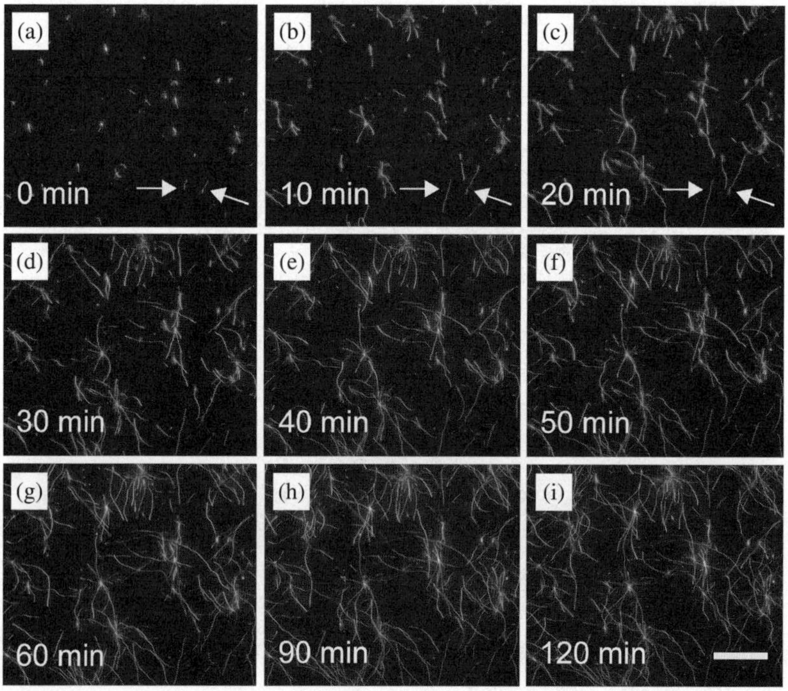

FIG. 2 Direct observation of Aβ(1–40) amyloid fibril growth by TIRFM. Real-time monitoring of fibril growth on glass slides. Arrows indicate the unidirectional growth of Aβ from a single seed fibril. The scale bar represents 10 μm. (From [19], with permission of Elsevier.)

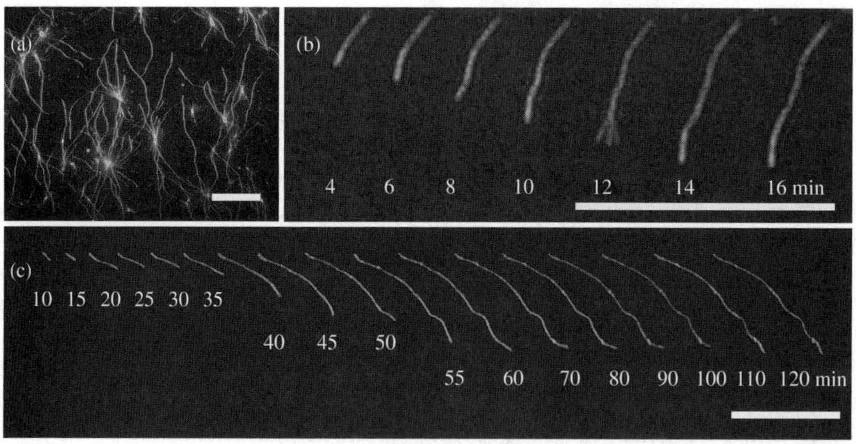

FIG. 3 Characteristic images of Aβ(1–40) amyloid fibril growth revealed by TIRFM: (a) vertically aligned image of fibrils; (b) growth with transient fraying of the growing end at 12 minutes; (c) growth with a swinging head producing a rugged fibril. The scale bars are 10 μm. (From [19], with permission of Elsevier.)

seems to occur in an all-or-none manner: When the fibril extends, the rate is almost constant (ca.0.3 μm/min), independent of fibril species. There were cases where the growth paused briefly, possibly because of physical obstacles or local depletion of monomers. When the growth restarted, however, a similar rate of 0.3 μm/min was regained. Similar discontinuous growth, termed the *stop-and-run mechanism*, was also observed during the growth of α-synuclein protofibrils monitored by AFM in situ [28].

EFFECTS OF VARIOUS SURFACES ON THE GROWTH OF Aβ FIBRILS

There is some evidence that surfaces may be crucial for amyloid fibril formation. The size and the shape of fibrils, as well as kinetics of formation, are dependence on the physicochemical nature of the surface [13–15]. We focused on the effects of the physicochemical properties of surface on the growth of amyloid fibrils of Aβ. Using specific chemical modifications, it is possible to modify the properties of the quartz surface, in terms of both net charge and hydrophobicity. We observed the seed-dependent formation of Aβ (1–40) fibrils on the surface of various chemically modified substrates that were created either by alternative adsorption of polyelectrolytes or with self-assembled monolayer of silanes, as described [22]. We observed the seed-dependent formation of Aβ(1–40) fibrils on the surface of various chemically modified substrates which were created either by alternative adsorption of polyelectrolytes or with self-assembled monolayers of silanes. The results are compiled in Figure 4. In the presence of the Aβ(1–40) seed fibrils, enhanced fibril formation was observed on negatively charged surfaces, including quartz and polyethyleneimine (PEI)/polyvinylsulfonate (PVS). On quartz, intense growth led to remarkably long fibrils, as reported previously [19]. We often observed radial growth patterns, suggesting the presence of clustered seeds. Extensive fibril formation was generally observed on surfaces with negative charges, regardless of whether they were modified by a polyelectrolyte or silane (Fig. 4a–f). In contrast, fibril growth was largely suppressed on positively charged or hydrophobic surfaces (Fig. 4g and i). The image obtained with a hydrophobic surface suggested that the binding efficiency of seed fibrils is less than it is with other surfaces. Thus, the efficiency of seed adsorption may be an important factor determining the fibril growth.

FORMATION OF Aβ SPHERULITIC STRUCTURES

Fibril growth was especially prominent on surfaces covered with PEI/PVS, highly negatively charged and hydrophilic polyelectrolytes (Fig. 4d). Initially we presumed that the growth of fibrils on the PEI/PVS initiated from large clustered seeds attached to the surface. However, the real-time observation

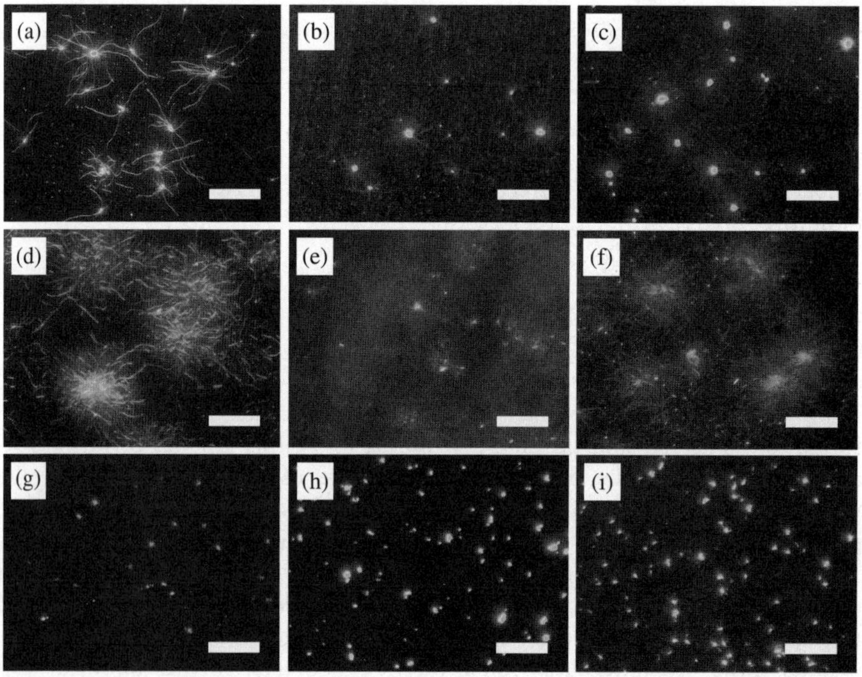

FIG. 4 Surface-dependent growth of Aβ(1–40) amyloid fibrils. Seed-dependent growth was performed on various surfaces: (a) quartz; (b) negatively charged CDDS; (c) negatively charged APTS/PVS; (d) negatively charged PEI/PVS; (e) negatively charged PEI/PSS; (f) negatively charged PEI/PAA; (g) hydrophobic OTS; (h) positively charged APTS; (i) positively charged PEI. Concentrations of Aβ(1–40) monomers, seeds, and ThT were 50 μM, 5 μg/mL, and 5 μM, respectively. The scale bar represents 10 μm. Extensive fibril growth was observed on the surfaces with negative charges (a–f), while the formation of fibrils was suppressed on hydrophobic (g) or positively charged (h, i) surfaces. (From [22], with permission. Copyright © 2006 American Society for Biochemistry and Molecular Biology.) (*See insert for color representation of figure.*)

revealed striking images of fibril growth, producing huge spherical assemblies with a densely packed radial pattern (Fig. 5). Importantly, no branching of the growing ends was observed, as on quartz. In the spherical assemblies, fibril growth continued linearly as on quartz, when length was plotted against time, until the depletion of monomers [19]. Incidentally, the rates of growth were similar (ca. 0.3 μm/min) for both types of fibrils [19]. This result supports the morphological similarity of individual fibrils. Thus, once fibril growth started, the rate of fibril growth might be less affected by the physicochemical properties of the surface, although the extent of seed clustering and adsorption depend significantly on the surface.

Considering that TIRFM illumination has a depth of penetration of about 150 nm and the depth of focus on the objective lens is about 100 nm, the large

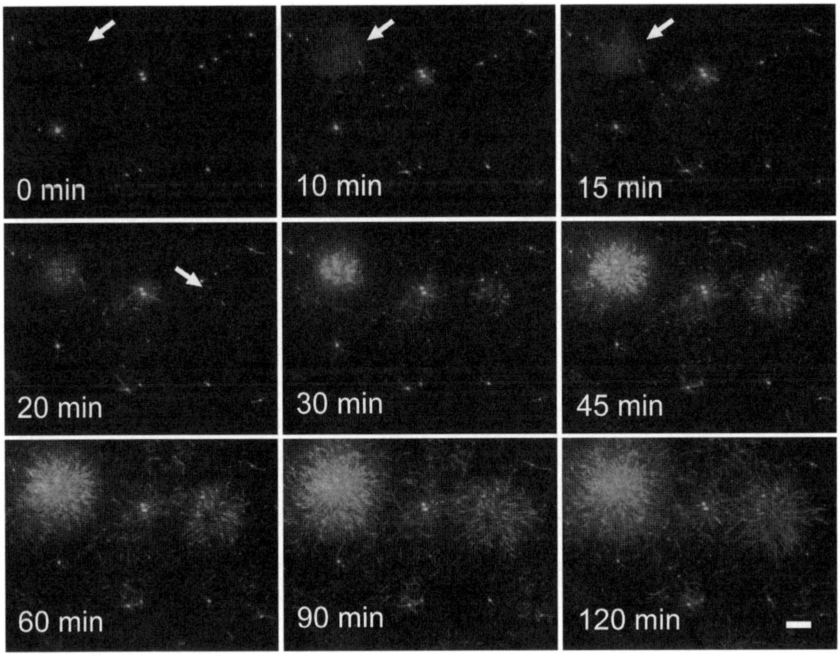

FIG. 5 Real-time observations of the formation of Aβ(1–40) spherulite. Real-time observations of Aβ(1–40) amyloid fibril growth on PEI/PVS at pH 7.5 and 37°C. Concentrations of Aβ(1–40) monomers, seeds, and ThT were 50 μM, 5 μg/Ml, and 5 μM, respectively. White arrows in panels of 0 to 20 minutes indicate the hazy area detected before clear images of spherical amyloid fibrils were obtained. At time zero, large clusters were not observed on the surface. At 10 minutes, hazy globular objects were identified. At 15 minutes, fibrils emerged. Fibrils grew both in size and number with time, forming huge spherical amyloid assemblies with a radius of more than 20 μm at 120 minutes. (From [22], with permission Copyright © 2006 American Society for Biochemistry and Molecular Biology.) (*See insert for color representation of figure.*)

clusters of seeds formed at first *in solution* and were not in contact with the substrate. The hazy areas observed at the initial stages, as indicated by arrows in Figure 5, may represent the clustered seeds or aggregated intermediates formed in solution. Since the thickness of the water medium estimated from the fine focus stroke between the quartz slide and coverslip is about 10 μm, the spherical assemblies observed here are, in fact, flattened spheres. The surface used for TIRFM observation was located on the upper side of the cell, so the clustered fibrils on the surface are not deposited by gravitational force.

Most important, these spherulitic structures resemble the amyloid core of senile plaques observed in the central cortices of patients suffering from Alzheimer disease [5]. Similar spherical amyloid deposits are observed in a mouse model of Alzheimer disease [16], in patients with Creutzfeldt–Jakob disease [17], and in several other neurodegenerative diseases [7], indicating that they are a common architectural feature of fibrils. Furthermore, spherulites were observed in vitro in many systems, including natural and synthetic polymers (see Krebs et al. [29] for details): for example, in insulin [29,30], pathogenic immunoglobulin chains [31], β-lactoglobulin [32], and synthetic peptides [33], indicating that they are a common architectural feature of fibers. We consider that the senile plaquelike spherical objects observed here correspond to *spherulites*, a higher-order spherical assembly of amyloid fibrils ranging in diameter from 10 to 150 μm. In a polarizing light microscope, spherulites exhibit a typical Maltese-cross extinction pattern [29].

The similarity of the amyloid core of senile plaques and the spherulitic assemblies observed in the present study suggests that senile plaques in patients also develop through the cooperative growth and association of amyloid fibrils as visualized here, together with interactions between Aβ monomers or fibrils and various biological molecules. During concurrent fibril growth from clumped seeds, moderate repulsion between the fibrils may keep the growing ends separated and thus active. At the same time, moderate attraction between growing fibrils is presumably important in maintaining the spherical shape. Thus, the formation of senile plaques is likely to be governed both by the physicochemical properties of Aβ amyloid fibrils per se and by the molecular environment in situ.

CONCLUSIONS

Our observations with TIRFM clearly characterized the behavior of amyloid fibrils on solid surfaces. The surface properties have crucial roles in both promoting and suppressing fibril growth and interactions. By controlling surface properties, we reproduced the senile-plaque-like spherulitic assemblies of Aβ(1–40). The real-time images at the single fibrillar level revealed that a balance of attractive and repulsive interactions coupled with intense growth without branching produces a huge spherical object. On the other hand, on a quartz surface, intense growth led to remarkably long fibrils, implying similarity to the deposition on vascular basement membranes. These observations argue that the physicochemical properties of amyloid fibrils per se play a dominant role for the formation of senile plaques, as well as deposition on vascular basement membranes, giving insights into creating the strategies preventing Alzheimer disease. Importantly, because this approach using ThT can be applied to various amyloid fibrils, it will be useful for developing a new diagnosis of amyloid diseases.

Acknowledgments

We would like to acknowledge Hironobu Naiki (Fukui University), Tetsuichi Wazawa (Tohoku University), and Daizo Hamada (Research Institute and Osaka Medical Center for Maternal and Child Health) for their support and encouragement. This work was supported by grants in aid from the Japanese

Ministry of Education, Culture, Sports, Science and Technology, and by the Japan Society for Promotion of Science Research Fellowships for Young Scientists to T.B.

REFERENCES

1. Chiti, F. Dobson, C.M. (2006). Protein misfolding, functional amyloid, and human disease. *Annu Rev Biochem, 75*, 333–366.
2. Lansbury, P.T., Jr. (2004). Back to the future: the "old-fashioned" way to new medication for neurodegeneration. *Nat Med, 10*, 13709–13715.
3. Hardy, J., Selkoe, D. (2002). The amyloid hypothesis of Alzheimer's disease: progress and problems on the road to therapeutics. *Science, 297*, 353–356.
4. Mattson, M.P. (2004). Pathways toward and away from Alzheimer's disease. *Nature, 430*, 631–639.
5. Jin, L., Claborn, K., Kurimoto, M., Geday, M., Maezawa, I., Sohraby, F., Estrada, M., Kaminsky, W., Kahr, B. (2003). Imaging linear birefringence and dichroism in cerebral amyloid pathologies. *Proc Natl Acad Sci U S A, 100*, 15294–15298.
6. Naiki, H., Nakakuki, K. (1996). First-order kinetic model of Alzheimer's β-amyloid fibril extension in vitro. *Lab Invest, 74*, 374–383.
7. Harper, J.D., Lansbury, P.T., Jr. (1997). Models of amyloid seeding in Alzheimer's disease and scrapie. mechanistic truth and physiological consequence of the time-dependent solubility of amyloid proteins. *Annu Rev Biochem, 66*, 385–407.
8. McLaurin, J., Yang, D.S., Yip, C.M., Fraser, P.E. (2000). Review: modulationg factors in amyloid-β ibril formation. *J Struct Biol, 130*, 259–270.
9. Petkova, A.T., Leapman, R.D., Guo, A., Yau, W.M., Mattson, M.P., Tycko, R. (2005). Self-propagating, molecular-level polymorphism in Alzheimer's β-amyloid fibrils. *Science, 307*, 262–265.
10. Kakio, A., Nishimoto, S., Yanagisawa, K., Kozutsumi, Y., Matsuzaki, K. (2002). Interaction of amyloid β-protein with various gangliosides in raft-like membrane: importance of GM1 ganglioside-bound form as an endogenous seed for Alzheimer amyloid. *Biochemistry, 41*, 7385–7390.
11. Gellermann, G.P., Appel, T.R., Tannert, A., Radestock, A., Hortschansky, P., Schroeckh, V., Leisner, C., Lütkepohl, T., Shtrasburg, S., Röcken, C., et al. (2005). Raft lipid as common components of human extracellular amyloid fibrils. *Proc Natl Acad Sci U S A, 102*, 6297–6302.
12. Yip, C.M., Darabie, A.A., McLaurin, J. (2002). Aβ42-peptide assembly on lipid bilayers. *J Miol Biol, 318*, 97–107.
13. Blackley, H.K.L., Sanders, G.H.W., Davies, M.C., Roberts, C.J., Tendler, S.J.B., Wilkinson, M.J. (2000). In situ atomic force microscopy study of β-amyloid fibrillization. *J Mol Biol, 298*, 833–840.
14. Kowalewski, T., Holtzman, D.M. (1999). In situ atomic microscopy study of Alzheimer's β-amyloid peptide on different substrates: new insights into mechanism of β-sheet formation. *Proc Natl Acad Sci U S A, 96*, 3688–3693.
15. Zhu, M., Souillac, P.O., Ionescu-Zanetti, C., Carter, S.A., Fink, A.L. (2002). Annular oligomeric amyloid intermediates observed by in situ atomic force microscopy. *J Biol Chem, 277*, 50914–50922.

16. Hsiao, K., Chapman, P., Nilsen, S., Eckman, C., Harigawa, Y., Younkin, S., Yang, F., Cole, G. (1996). Correlative memory deficits, Aβ elevation, and amyloid plaques in transgenic mice. *Science, 274*, 99–102.
17. Manuelidis, L., Fritch, W., Xi, Y.-G. (1997). Evolution of a strain of CJD that induces BSE-like plaques. *Science, 277*, 94–98.
18. Ban, T., Hamada, D., Hasegawa, K., Naiki, H., Goto, Y. (2003). Direct observation of amyloid fibril growth monitored by ThT fluorescence. *J Biol Chem, 278*, 16462–16465.
19. Ban, T., Hoshino, M., Takahashi, S., Hamada, D., Hasegawa, K., Naiki, H., Goto, Y. (2004). Direct observation of Aβ amyloid fibril growth and inhibition. *J Mol Biol, 344*, 757–767.
20. Ban, T., Goto, Y. (2006). Direct observation of amyloid growth monitored by total internal reflection fluorescence microscopy. *Methods Enzymol, 413*, 91–102.
21. Ban, T., Yamaguchi, K., Goto, Y. (2006). Direct observation of amyloid fibril growth, propagation, and adaptation. *Acc Chem Res, 39*, 663–670.
22. Ban, T., Morigaki, K., Yagi, H., Kawasaki, T., Kobayashi, A., Yuba, S., Naiki, H., Goto, Y. (2006). Real-time and single fibril observation of the formation of amyloid β spherulitic structures. *J Biol Chem, 281*, 33677–33683.
23. Yagi, H., Ban, T., Morigaki, K., Naiki, N., Goto, Y. (2007). Visualization and classification of amyloid β supramolecular assemblies. *Biochemistry, 46*, 15009–15017
24. Funatsu, T., Harada, Y., Tokunaga, M., Saito, K., Yanagida, T. (1995). Imaging of single fluorescent molecules and individual ATP turnovers by single myosin molecules in aqueous solution. *Nature, 374*, 555–559.
25. Yamasaki, R., Hoshino, M., Wazawa, T., Ishii, Y., Yanagida, T., Kawata, Y., Higurashi, T., Sakai, K., Nagai, J., Goto, Y. (1999). Single molecular observation of the interaction of GroEL with substrate proteins. *J Mol Biol, 292*, 965–972.
26. Wazawa, T., Ueda, M. (2005). Total internal reflection fluorescence microscopy in single molecule nanobioscience. *Adv Biochem Eng Biotechonol, 95*, 77–106.
27. Naiki, H., Higuchi, K., Hosokawa, M., Takeda, T. (1989). Fluorometric determination of amyloid fibrils in vitro using the fluorescent dye, thioflavin T. *Anal Biochem, 177*, 244–249.
28. Hoyer, W., Cherny, D., Subramaniam, V., Jovin T.M. (2004). Rapid self-assembly of a-synuclein observed by in situ atomic force microscopy. *J Mol Biol, 340*, 127–139.
29. Krebs, M.R.H., MacPhee, C.E., Miller, A.F., Dunlop, I.E., Dobson, C.M., Donald, A.M. (2004). The formation of spherulites by amyloid fibrils of bovine insulin. *Proc Natl Acad Sci U S A, 101*, 14420–14424.
30. Roger, S.S., Krebs, M.R.H., Bromley, E.H.C., van der Linden E., Donald, A.M. (2006). Optical microscopy of growing insulin amyloid spherulites on surfaces in vitro. *Biophys J, 90*, 1043–1054.
31. Raffen, R., Dieckman, L.J., Szpunar, M., Wunschl, C., Pokkuluri, P., Dave, P., Stevens, P.W., Cai, X., Schiffer, M., Stevens, F.J. (1999). Physicochemical consequences of amino acid variations that contribute to fibril formation by immunoglobulin light chains. *Protein Sci, 8*, 509–517.

32. Sagis, L.M.C., Veerman, C., van der Linden, E. (2004). Mesoscopic properties of semiflexible amyloid fibrils. *Langmuir, 20*, 924–927.

33. Fezoui, Y., Martley, D.M., Walsh, D.M., Selkoe, D.J., Osterhout, J.J., Teplow, D.B. A. (2000). A de novo designed helix-turn-helix peptide forms nontoxic amyloid fibrils. *Nat Struct Biol*, *7*, 1095–1099.

34

CURRENT AND FUTURE THERAPIES FOR ALZHEIMER DISEASE

PARAMITA CHAKRABARTY, PRITAM DAS, AND TODD E. GOLDE
Department of Neuroscience, College of Medicine, Mayo Clinic Florida, Jacksonville, Florida

INTRODUCTION

Alzheimer disease (AD) patients typically present with forgetfulness, behavioral changes, and disturbed short-term memory. As the dementia progresses, patients suffer from long-term memory deficits, aphasia, and deficits in face and object recognition; finally, they become apraxic and moribund, requiring constant caregiver attention. AD is a multifactorial disease. Whereas, the infrequent early-onset forms (<60 years of age) have a much stronger genetic component, genetics, environment and co-morbidities contribute to the risk of developing the late onset sporadic disease (LOAD), which accounts for the majority of AD patients. Irrespective of early or late presentation, both familial and sporadic AD are characterized by similar underlying pathological phenotypes, characterized by accumulation of extracellular neuritic plaques composed of Aβ peptide and intracellular deposits or neurofibrillary tangles composed of hyperphosphorylated tau protein.

There is evidence that AD pathology results from the inexorable long-term accumulation of pathological lesions in the brain. Bioimaging studies [positron-emission tomography (PET) or magnetic resonance imaging (MRI)] have hinted at age-progressive buildup of amyloid, neuronal dysfunction, and compromised metabolic markers years to decades prior to the onset of symptoms [1–4]. Longitudinal studies also show that neuropsychological

Protein Misfolding Diseases: Current and Emerging Principles and Therapies,
Edited by Marina Ramirez-Alvarado, Jeffery W. Kelly, and Christopher M. Dobson
Copyright © 2010 John Wiley & Sons, Inc.

abnormalities, such as delayed recall, precede an actual diagnosis by up to 10 years [5]. Indeed, these data all indicate that even the earliest clinical diagnosis of AD occurs long after the initial onset of brain pathology [6].

Medications that are presently being used in AD patients fall into two classes: selective N-methyl D-aspartic acid (NMDA) receptor antagonist (Memantine) and acetyl cholinesterase (AChE) inhibitors (Galantamine, Donepezil, Rivastigmine). Memantine is reported to alleviate pathological glutamate-induced excitotoxicity resulting from excessive activity of the NMDA receptors without inhibiting the normal levels of NMDA receptor activity [7]. AChE inhibitors rescue the cholinergic transmission deficits present in AD patients by increasing acetylcholine levels in the synapse [8]. These are purely symptomatic therapies that have been used successfully to delay cognitive decline in patients; however, these do not ameliorate the actual neuropathological hallmarks of AD or delay progression from mild cognitive impairment (MCI) to full-blown AD (reviewed in [9–11]). Additional neurotransmitter based drugs are currently in early-stage clinical trials. These include a 5-hydroxytryptamine$_{1A}$ antagonist (phase II, Wyeth), a 5-HT$_6$ antagonist (phase II, GlaxoSmithKline), and a histamine H$_3$ antagonist (phase II, Abbott). These drugs aim to improve on the cholinergic-based therapy currently available [12]. In addition, several novel ACh receptor agonists also being tested (phase II trial) for symptomatic management are AZD340 (AstraZeneca/Targecept), MEM3454 (Roche/Memory Pharmaceutical), and GTS-21 (CoMentis).

In contrast to the approved symptomatic therapies and palliative actions of many neurotransmitter-based therapies in development, to eradicate AD it is necessary to identify therapies that work to prevent or delay the onset of disease. Although eradication is a lofty goal, it is theoretically achievable. Modest delays in disease onset can have a huge impact on prevalence: delaying AD onset by five years or 10 years reduces the overall prevalence by 50% or 75%, respectively [13]. Of course, there are limitations to disease-modifying therapies that may show little or no immediate symptomatic improvements. Indeed, once an advanced threshold in disease progression and disability is passed, it is far less compelling, and perhaps unethical, to intervene to slow the underlying pathologic progression and thereby prolong the suffering of the patient.

AD is characterized by selective neuronal and synaptic loss, extracellular neuritic plaques with an amyloid core that is composed largely of aggregated Aβ peptides and intracellular neurofibrillary tangles of hyperphosphorylated tau protein (Fig. 1). The amyloid cascade hypothesis, formulated in the early 1990s, implicated Aβ aggregation as an initiating event in the neuropathology of AD, although currently there is a debate regarding whether the oligomeric or the fibrillar form of Aβ is the neurotoxic entity [14,15]. The prevalent theory is that altered homeostasis of Aβ caused by (1) altered production of the amyloid-β protein precursor (APP), (2) altered clearance of Aβ, or (3) some combination of these events leads to accumulation of aggregated Aβ in the brain, which then triggers a complex pathological cascade leading to AD. Genetic studies have linked the APP gene to autosomal early-onset familial AD (FAD) and Down

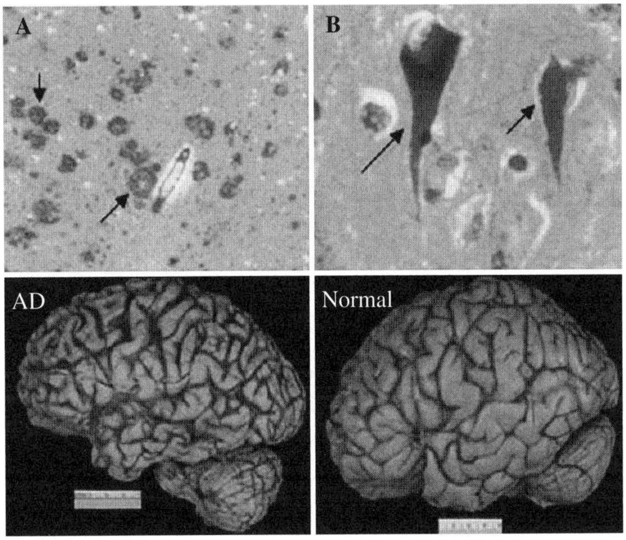

FIG. 1 Typical neuropathological lesions in Alzheimer disease brain. Immunohistochemical staining reveals the presence of extracellular senile plaques (top panel, A, arrows) and intracellular neurofibrillary tangles (top panel, B, arrows) in a typical human AD patient brain section embedded in paraffin. These constitute the defining neuropathological benchmarks of the disease. The bottom panel shows the gross anatomical changes in the brain from an AD patient (bottom panel, left) compared to a brain from an age matched normal individual (bottom panel, right). Notice the very apparent decrease in brain volume and remarkable frontal and temporal lobe atrophy in the AD patient brain. (Courtesy of Dennis Dickson, Mayo Clinic, Jacksonville, Florida.) (*See insert for color representation of figure.*)

syndrome patients. Studies of the biological effects of these genetic alterations have provided the core evidence that increased total Aβ levels, selective increases in Aβ42, or altered aggregation properties induced by mutations in the primary sequence of Aβ trigger AD [16]. Simultaneous genetic and mouse modeling studies of mutant tau protein [from a related dementia condition called fronto-temporal dementia with parkinsonism linked to chromosome 17 (FTDP-17)] demonstrated that accumulation of abnormal hyperphosphorylated tau could lead to formation of paired helical filaments (PHFs) and frank neurofibrillary tangles (NFTs) in mice which are highly reminiscent of the neuritic tau pathology seen in AD [17]. Identification of Aβ and tau and their role in disease pathology has led the way to the development of AD disease-modifying therapies that focus on preventing the abnormal misfolding, aggregation, and accumulation of Aβ and/or tau.

Currently, numerous disease-modifying therapies are being developed against AD. In the following sections we describe the current status of potentially disease-modifying therapies targeting Aβ, tau, and other metabotropic pathways (see Table 1). In this chapter, we will, for the sake of clarity,

TABLE 1 Disease-Modifying Therapies Presently in Use for AD[a]

Strategy	Example	Mechanism	Refs.
Preclinical Therapies			
α-Secretase activator	Bryostatin1, M1 agonists	α-Secretase activator, PKC activator—prevents Aβ production	42–46, 51
γ-Secretase inhibitor	LY-411, 575 and others	Modulates Aβ42/40 production; prevents Aβ production	72, 76–78
Aβ and amyloid clearance	IDE, Neprilysin	Enzymatically degrades Aβ	135, 136
Aβ aggregation inhibitor	β-Sheet breaker peptides	Inhibits fibrillogenesis, inhibits Aβ-ApoE4 interaction	116, 118, 120, 121
Removal via "peripheral sink"	LRP/anti-Aβ antibodies	Sequestration of Aβ in the periphery	171, 172, 199
GPCR antagonist	ICI 118,551	Reduces α-secretase activity	87, 88
Cholesterol-lowering treatment	BM15.766	Reduces amyloidogenic APP cleavage	306, 316, 317
Kinase inhibitors	Lithium	GSK3β inhibitor-inhibits tau phosphorylation	242, 243
Caspase inhibitors	Anthraquinones	Inhibits tau NFT formation	262, 266
Tau degradation	hsp 90, 70	Induces proteasomal removal of tau	257, 258
Taxol	Paclitaxel	Microtubule stabilization/anti-Aβ effects	190, 259
Cell transplantation/stem cell therapy	Stem cells, siRNA therapy	Induces neurogenesis, down-regulate-specific tau alleles or APP	380, 386
Clinical Therapies			
Active Aβ immunotherapy	AN1792 (halted), ACC-001	Removes Aβ by immunological clearance; second-generation vaccine	200–206, 207–211
Passive Aβ immunization	3D6, IVIg	Removes Aβ by immunological clearance	169

γ-Secretase inhibitor/ modulator	LY450139, Flurbiprofen	Reduces "amyloidogenic" cleavage of APP, produces shorter antiamyloidogenic Aβ fragments	68, 74, 84, 85
β-Secretase inhibitor	CTS21166 (CoMentis)	Reduces "amyloidogenic" cleavage of APP	www.Comentis.com
Muscarinic agonists	Talsaclidine	Induces α-secretase cleavage	49
Statin	Simvastatin	Lowers ApoE levels, unknown?	304–309
Metal-chelating compounds	Clioquinol, PBT-2	Inhibits Aβ fibrillization	107
APP transcription modulators	Phenserine	Lowers Aβ levels	94, 386
Aβ aggregation inhibitor	Tramiprosate, *Gingko*	Inhibits Aβ aggregation; glycosaminoglycan mimetic	102, 103, 355
Astrocytic modulators	Arundic acid	Lowers astrocytic S100b production	295, 296
Anti-inflammatory reagents	NSAID, etanercept doxycycline	Inhibits neuroinflammation; inhibits Aβ fibrillization	84, 115, 286, 290
PPARγ agonists	Pioglitazone, rosiglitazone	Anti-inflammatory/MAO inhibitor	301, 361
Ampakine	Amrakine	AMPA receptor agonist	360, Cortex Pharma
NMDA receptor antagonist	Memantine	Ameliorates excitotoxicity	7
Acetyl cholinesterase inhibitors	Donepezil	Enhances cholinergic transmission	8, 12
Antioxidants	EGCG, vitamin E, Curcumin	Antioxidative potential; pleiotropic modality	301, 351, 355
Growth factor therapy	Cerebrolysin, NGF	Neurotrophic activity	374, 384
Estrogen replacement therapy	Estrogen	Trophic activity (?)	328, 329
Tau aggregation inhibitor	Methylene Blue	Inhibits tau aggregation	227, 228
Homocysteine-lowering therapy	Vitamins B_{12}, B_6, and folic acid	Lowers serum homocysteine	341, 344, 345

[a]Potential AD disease-modifying therapies being investigated in preclinical studies and in clinical trials for the treatment or prevention of AD are listed.

divide these into three broad categories: anti-amyloid therapy, anti-tau therapy, and therapies that have pleiotropic modes of action. The first category details antiamyloid therapies that have been designed to prevent Aβ production, decrease its aggregation, or improve Aβ aggregate clearance mechanisms. The second category describes anti-tau therapies aimed largely at preventing hyperphosphorylation events that lead to misfolding of tau and formation of pathologic aggregates. The third category consists of a variety of neuroprotective therapies focused on ameliorating the synaptic and neuronal loss presumably triggered by metabolic dysfunction arising from protein aggregation and misfolding. These therapies have multiple modalities of action, in the sense that they may alter either Aβ or tau dysfunction or cognitive deficits by acting on different cellular signaling pathways. In the following sections we highlight validation of various targets thus obtained and also the specific concerns and drawbacks of each. Finally, the need for improvements in clinical trial designs and suggestions for streamlining future trials are discussed.

MOUSE MODELS OF AD

To understand the basic biology leading to development of the neuropathologic hallmarks in AD and to test putative drug targets, studies on transgenic mouse models have been invaluable. Modeling AD in rodents was initiated following the characterization of the APP gene as the origin of Aβ peptide and identification of the mutant APP gene in causing FAD; the mutated APP gene promotes Aβ aggregation and deposition in neuritic plaques. The mutations in the APP gene flank the Aβ region and affect the target sites where proteolytic cleavage occurs, leading to increased production of the amyloidogenic Aβ42, an APP-derived peptide of 42 amino acids (Fig. 2). Some mutations that alter the Aβ coding sequence increase the propensity of the peptide to aggregate. More recently, it was shown that APP gene duplication resulting in increased gene dosage can also lead to familial AD [18]. The discovery of APP mutations led to the creation of the PDAPP transgenic mice line followed by the Tg2576 and APP23 transgenic lines, among others [19–21]. The PDAPP transgenic mouse expresses human APP carrying the Indiana familial AD mutation (V717F/G) driven by the platelet-derived growth factor β promoter, whereas both the Tg2576 and APP23 mice express human APP with the Swedish mutation (K670N/M671) driven by the hamster prion protein promoter and the murine Thy-1 promoter, respectively. Later, it was established that mutations in the presinilin 1 gene (PS1) are the most common cause of familial AD. Involvement of the presenilin (PS) genes in AD spurred efforts to create mutant PS1 and PS2 transgenic mice lines (reviewed in [22]) and double transgenics expressing both APP and presenilin mutants [23,24]. Presenilins were identified to be part of the muticomponent γ-secretase complex that is involved in APP processing (Fig. 2) [25]. The PS mutations generally alter the cleavage site in APP, generating the more amyloidogenic Aβ42 peptide. The

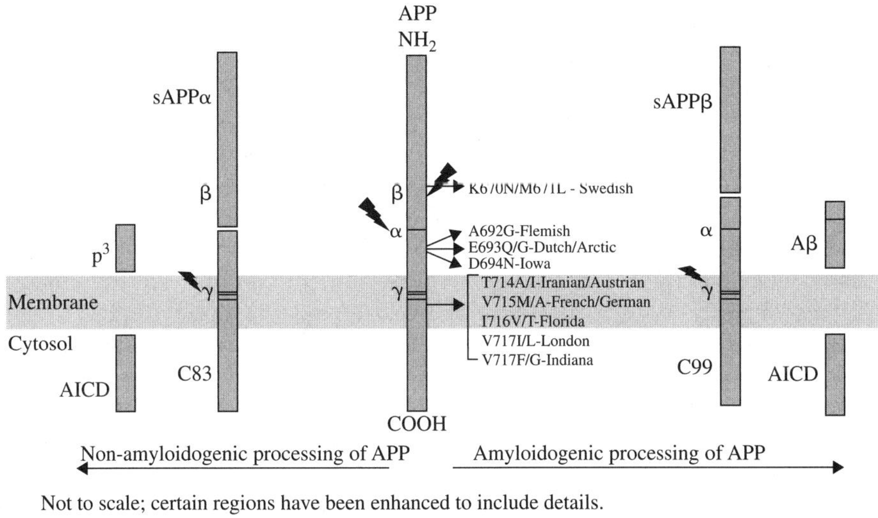

FIG. 2 Schematic representation of APP processing and location of different familial mutations associated with early-onset AD. In the non-amyloidogenic pathway, full-length APP is cleaved sequentially by α-secretase in the middle of the Aβ sequence followed by γ-secretase, which yields the soluble 3-kDa P3 fragment and the AICD (APP intracellular domain) whose function is unclear. In the amyloidogenic pathway, β-secretase cleavage leads to secretion of the ectodomain, sAPPβ; the membrane stub, CTFβ or C99, is then cleaved by the γ-secretase complex, liberating the Aβ and AICD. γ-Secretase can cleave at three sites in the transmembrane domain—the cleavage sites are referred to as γ-, ζ-, and ε-sites (from amino to carboxy terminal). The γ-site is thus variable and can occur after amino acids 38, 40 or 42, leading to creation of either Aβ38, 40 or 42 which have different aggregation properties. Some of the mutations associated with early-onset AD are mentioned along with their position in the APP sequence. These mutations can alter APP processing and are named after the nationality or the location of the first family in which it was identified (e.g., Swedish, Florida, etc). For a complete list of APP mutations associated with AD, refer to Price et al, [36]. The preclinical studies for characterization of disease-modifying therapies in AD have been done in transgenic mice overexpressing one or more of these mutated APP sequences.

major limitation of these mouse models is that none develop frank NFTs or overt cell death, two major hallmarks of AD.

Recent evidence implicated that tau, either by itself or triggered by Aβ accumulation, may be functionally very closely linked to neuronal dystrophy. Identification of tau mutants that cause FTDP-17, a disease with very extensive NFTs, enabled the development of transgenic mice models expressing these mutant tau proteins [17,26–28]. In contrast to APP mice models, some of these tau mice (JNPL3 and rTg4510 transgenic lines) show extensive neurofibrillary pathology and neurodegeneration, but these mice do not develop extracellular plaques [17,29]. Crossing these two transgenic lines (mutant APP and tau) or injection of Aβ peptide into JNPL3 mice produced a mouse model that exhibited both amyloid and tau abnormalities [30,31]. Surprisingly, tau knockout mice, by themselves, have relatively mild phenotypes [32–34]. Subsequently, a triple transgenic model overexpressing mutant tau, APP, and PS1 was created by microinjection of mutant tau and mutant APP into PS1 mouse embryo, ensuring that tau and APP would be segregated together. These transgenic mice (3XTg-AD) recapitulate multiple features of AD pathology, including neuritic pathology and cognitive deficits [35].

Current AD transgenic mice models are clearly very useful for target and therapeutic validation, but they cannot substitute for clinical trials in humans. A major limitation of the APP mice and wild-type tau mice is that they do not fully recapitulate one of the prime features of AD pathology: neuronal loss [36]. On the other hand, although the mutant tau mice show pronounced neurodegeneration, there is no plaque pathology in these mice. Despite these caveats, the importance of having good preclinical models to test proof of concept therapeutic paradigms can never be overstated; these studies will eventually lead to the identification of the druggable targets, drug efficacy, route of drug administration, dose–response relationship, drug bioavailability, and determination of the window of opportunity for efficacious treatment.

APP AND Aβ-SPECIFIC THERAPEUTIC TARGETS

As Aβ aggregation into amyloid and other smaller aggregates is a concentration-dependent phenomenon, Aβ production by β- and γ-secretase cleavage of APP is one of the most pivotal points for therapeutic intervention in AD. A tremendous amount of progress has been achieved in developing preclinical therapeutic paradigms that prevent Aβ production, block its aggregation, lower its effectual concentration in the brain, and disassemble preformed plaques (Fig. 3). In the following sections, we describe these potential anti-Aβ therapies and their efficacy in preclinical and clinical settings.

Targeting Aβ Production: Role of α-, β-, and γ-Secretases

APP, the precursor of Aβ, is an integral membrane protein with a signal sequence, a large extracellular domain, a transmembrane domain, and a small cytosolic C terminal domain. Sequential cleavage of APP by β- and γ-secretase leads to the formation and secretion of Aβ, a physiological and constitutive process [16]. Aβ can normally be found in plasma, cerebrospinal fluid (CSF), and brain of normal healthy persons [37].

α-Secretase cleavage of APP produces a large (ca. 110 to 135 kDa) secreted fragment (sAPPα), and a small (ca. 10 kDa) membrane-bound stub (C83 or CTFα), effectively cutting Aβ into two fragments (Fig. 2). α-Secretases belong to the family of membrane-bound metalloprotease disintegrins (e.g., ADAM10 and ADAM17) [38–40]. Thus, inducing α-secretase cleavage over β- or γ-secretase activities may theoretically not only reduce Aβ production but

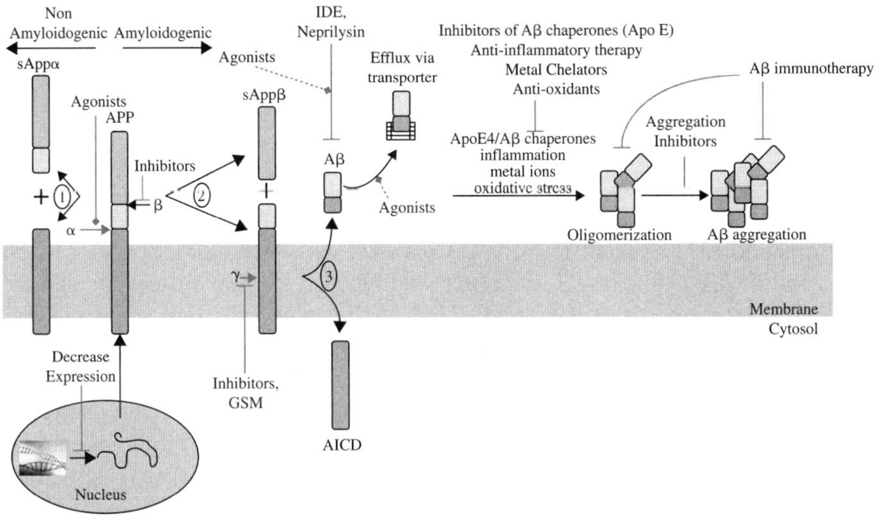

FIG. 3 Potential Aβ therapeutic targets. Full-length APP is sequentially processed by β- and γ-secretases to produce the amyloidogenic Aβ peptide (steps marked 2 and 3). Action of the α-secretase precludes formation of Aβ production since its cleavage site is within the peptide sequence (step marked 1). Potential AD disease-modifying therapies (preclinical as well as clinical studies) include targeting the secretase activities by inducing α-secretase activity or inhibiting β- and γ-secretase activities. GSMs that can modulate γ-secretase activity to produce shorter Aβ peptides have also been investigated. Other putative therapies include up-regulating the activity of Aβ-degrading enzymes such as neprilysin and IDE and enabling the efflux of the amyloidogenic peptides by pharmacological intervention. Decreasing the overall cellular concentration of APP by utilizing translational inhibitors has also been done. In addition, since the propensity of Aβ to self-assemble into oligomers and fibrillar forms has been shown to be neurotoxic, various therapies, such as a Aβ immunotherapy and pharmacological aggregation inhibitors that can perturb this fibrillization pathways, are also being tested for their efficacies. For a detailed description, refer to the text and to Table 1. α, β, and γ mark secretase cleavage sites on APP.

also increase the production of sAPPα, a potentially neuroprotective form of APP. Genetic up-regulation of the α-secretase, ADAM10 in an APP transgenic mouse increased sAPPα, decreased Aβ production, reduced plaque formation, and alleviated cognitive deficits providing proof of concept for this strategy [41]. Also, the polyphenol (−)-epigallocatechin-3-gallate (EGCG, a constituent of green tea) has been shown to activate ADAM10, an effect that may contribute to its efficacy in attenuating Aβ deposition in APP mouse models [42–44]. However, there are concerns that pharmacologic activation of these enzymes might produce off-target effects, as ADAM10 and ADAM17 participate in multiple signaling pathways.

Several groups have shown that indirect pharmacologic activation of α-secretase by various agents, including muscarinic agonists, serotonin, glutamate, and estrogens, can reduce Aβ production (reviewed in [45]). M1 muscarinic ACh receptor agonists have been hypothesized not only for symptomatic treatment but also for therapeutic treatment of AD [46]. A 10-week intraperitoneal treatment of the 3XTg-AD mice with the M1 agonist AF267B could reduce both tau and amyloid pathology as well as ameliorate cognition [47]. The muscarinic M1 agonist Talsaclidine also shifts APP cleavage toward the α-secretase-mediated nonamyloidogenic pathway [48], and in a 4-week-long double-blind randomized trial on AD patients with Talsaclidine, a significant decrease in CSF Aβ42 levels was observed [49]. A third M1 muscarinic agonist that can activate α-secretase, NGX267, is in phase I trials (TorreyPines Therapeutics). Although these preclinical data are noteworthy, there are lingering concerns that muscarinic agonists may not be well tolerated for long-term use because of lack of receptor selectivity and potential cholinergic-based toxicity.

Other pharmacologic α-secretase agonists alter trafficking and subcellular distribution of APP or α-secretase or both by more directly activating PKC-mediated signaling pathways [50,51]. An example of a PKC-mediated α-secretase inducer is Bryostatin1, which is a macrolide lactone isolated from a bryozoan [52]. Bryostatin is presently involved in a multitude of clinical trials as an anticancer agent. More interestingly, Bryostatin 1 has been shown to cause a significant increase in sAPPα and reduction in brain Aβ40 and Aβ42 levels in APP/PS1 mice. Bryostatin 1 also leads to an induction of proteins involved in long-term potentiation (LTP; a measure of learning and memory) and seems to decrease aberrant tau hyperphosphorylation, making it an attractive candidate for clinical trials [53].

The membrane-bound aspartyl protease, β-secretase (BACE1 or memapsin), is the rate-limiting enzyme in Aβ production and thus is an attractive target for therapies aimed at reducing Aβ production [54–57]. The BACE1 knockout mice are viable, have no gross behavioral anomalies, and have dramatically lowered Aβ levels [58,59]. However, BACE1 has been shown to have other substrates, including APP-like proteins (APLP1 and 2), P-selectin glycoprotein ligand-1, lipoprotein receptor–related protein (LRP), and neuregulin [60–64], raising the possibility that inhibition of BACE1 may have unexpected toxicities. Indeed, a recent report suggests that BACE1 knockout may have a harmful phenotype, raising concerns over the potential safety of β-secretase inhibitors [65,66]. BACE1 has also been shown to regulate the development of the myelin sheath by its cleavage of neuregulin; thus, inhibiting BACE1 may cause accumulation of unprocessed neuregulin and consequent demyelination and neuropathy [64]. Whether pharmacological BACE1 inhibitors can inhibit neuregulin cleavage in adults and whether this leads to changes in myelination patterns has not been determined. In addition, BACE1 seems to have a large catalytic pocket, making the design of a small and thus metabolically stable and brain penetrant compound extremely challenging. Despite current difficulties in developing successful BACE1 inhibitors, many pharmaceutical companies are committed to developing brain penetrant, orally bioavailable, and selective BACE1 inhibitors. Among the BACE1 inhibitors currently being tested is one by CoMentis Ltd This inhibitor (CTS-21166) is an orally bioavailable small-molecule BACE1 inhibitor. Currently, this is in phase I clinical trials and the phase II trial is expected to be initiated in 2008.

Potent γ-secretase inhibitors that target the multicomponent enzyme have been under development and have entered into phase I and II human clinical trials [67,68]. These inhibitors appear to target PS1 and PS2, the catalytic components of the γ-secretase complex [69–71]. Among the more potent small-molecule γ-secretase inhibitors is LY411,575 (Eli Lilly), which has shown promising results in trials on Tg2576 mice, lowering both brain and plasma Aβ [72]. However, it has been shown by others that this compound leads to abnormalities in lymphocyte development and intestinal cell differentiation [73]. Another γ-secretase inhibitor, LY450,139 (Eli Lilly), was shown to be reduce plasma Aβ (but not CSF Aβ) in a 6-week phase II trial [74]. Although this trial did not show any cognitive improvement, the decrease in Aβ42 levels is a very promising lead that is worth follow-up. A caveat of this strategy is that use of such APP-selective γ-secretase inhibitors can lead to the accumulation of Aβ that is bearing fragments of APP that may be neurotoxic [75].

Adenosine triphosphate (ATP) can activate the generation of Aβ by purified γ-secretase without similarly affecting Notch cleavage in vitro. Thus, in principle it may be possible to decrease γ-secretase cleavage of APP by interfering with its ATP docking site using kinase inhibitors [76,77]. Several companies are in the process of developing such APP-selective γ-secretase inhibitors, but the precise targets of these compounds are unknown.

Another approach to targeting γ-secretase may be to modulate its cleavage sites to produce less of Aβ42 compared to the other Aβ peptide species [78,79]. A subset of NSAIDs (nonsteroidal anti-inflammatory compounds) was identified to act as GSMs (γ-secretase modulators) in vitro. It is now established that the inhibition of γ-secretase by the NSAIDs is not mediated by cyclooxygenase inhibition but rather by a more direct effect on γ-secretase itself. Aβ42-lowering GSMs alter the cleavage site specificity of γ-secretase by shifting APP cleavage from Aβ42 to shorter Aβ peptides. Notably, GSMs do not result in substrate accumulation or functional impairment of other γ-secretase substrates [78,80–82]. Recent data that Aβ40 and possibly other shorter Aβ derivatives may inhibit Aβ deposition into neuritic plaques indicates that shifting the balance toward producing shorter peptides may be neuroprotective [83]. A selective Aβ42-lowering NSAID, tarenflurbil (formerly called *R*-flurbiprofen, Flurizan, MPC-7869, Myriad Pharmaceuticals, Ltd) is currently being evaluated in two phase III human trials. In a short-term placebo-controlled double-blind trial, tarenflurbil was shown to be well tolerated across a wide dosage range by healthy individuals [84]. A one-year randomized, placebo-controlled, double-blind phase II study of tarenflurbil in 207 subjects with mild to moderate AD showed promising benefits in measures of daily activities and overall function, but not in measures of memory and thinking skills in subjects taking the highest dose only [85] (see Note Added in Proof).

In vitro data had shown that there is a selective increase in adrenergic receptors in the cerebellum of AD patients [86]. In a follow-up study, in an effort to dissect a correlation between adrenergic receptor activation and Aβ pathology, daily oral treatment with selective β_2 adrenergic antagonist ICI 118,551 was shown to ameliorate plaque pathology in APPswe/PS1 ΔE9 mice [87]; an opposite effect was seen following acute treatment with the β_2 adrenergic agonist clenbuterol. The increase in Aβ production induced by activation of the adrenergic receptors was associated with an increased γ-secretase activity, once again showing the potential efficacy of GPCR (G protein-coupled receptor, e.g., adrenergic receptors) targeting as a means of altering Aβ production indirectly. An angiotensin receptor blocker, valsartan, which is currently being used in antihypertensive therapy, has also been shown to inhibit Aβ aggregation in Tg2576 mice [88]. Interestingly, a recent epidemiological study revealed that there was a suggestive trend associated with the use of antihypertensive drugs (β_2 adrenergic antagonists) and decreased incidence of AD [89].

There has been limited enthusiasm regarding the use of nonselective secretase inhibitors in AD therapy. Other than APP, several biological targets for γ-secretase have been identified (e.g., Notch, LRP, ErbB-4, nectrin receptor) (reviewed in [90]). In addition, transgenic mice lacking PS 1 and 2 have a severe developmental phenotype [91]. Thus, AD-modifying therapies encompassing γ-secretase (or, for that matter, β-secretase) inhibitors must take into account possible adverse effects arising out of mistargeting of physiologically relevant secretase-modified pathways. Nevertheless, even partial inhibition of these secretases may be sufficient to lower Aβ production to a level that may delay the onset and progression of AD without producing unwanted side effects.

Decreasing APP Expression

Decreasing APP expression at the transcriptional or translational level can be achieved using small molecules. Metal cheating agents, such as EGCG (see the section "Targeting Aβ Production: Role of α-, β-, and γ-Secretases"), as a disease-modifying factor for AD came into focus because of their effects on the iron responsive element (IRE) in the 5′ untranslated region (UTR) of the APP gene. A cross-sectional study (Tsurugaya project) found a positive association between the consumption of green tea and a higher retention of cognitive function in elderly Japanese subjects [92]. Interestingly, in addition to metal chelation, these polyphenol compounds (often occurring naturally, like EGCG and curcumin) have anti-inflammatory properties, can induce α-secretase activity and can effectively scavenge free radicals thus conferring neuroprotection at multiple levels. Recently, several U.S. Food and Drug Administration (FDA)–approved drugs that can potentially target APP expression via this particular IRE have been identified from a drug screen and tested successfully in mice [93]. Another compound that acts through this IRE is phenserine, developed initially as a noncompetitive AChE inhibitor. Phenserine interferes with translation of APP and can decrease Aβ production [94]. However, the effective dose at which the inhibitory effects on APP translation are observed is much higher than the dose needed to elicit the anti-AChE activity, thus limiting the clinical potential of the compound as a dual-action anti-AChE anti-Aβ therapy. A phase III trial of phenserine failed to show efficacy in AD, although the trial design may have been suboptimal. Another new approach has been the identification of APP 5′ UTR-directed drugs and RNA-directed agents to lower APP translation and Aβ production [95]. An interesting preclinical study reported that some of these potential small-molecule RNA-directed agents could lower Aβ in transgenic mice, and more proof-of-concept studies are warranted to investigate whether these are feasible future therapeutics [96].

Targeting Aβ Aggregation

Small-Molecule Aggregation Inhibitors. Since Aβ aggregation per se has been shown to be sufficient for disease progression phenotype [97], several small-molecule inhibitors of Aβ aggregation have been identified and some of these are presently in clinical trials [98]. One such compound, homotaurine (NC758/Alzhemed, tramiprosate, 3-aminopropane-1-sulfonic acid or 3-APS), a low-molecular-mass compound that inhibits the interaction of Aβ with glycosaminoglycans (GAG) and prevents β-sheet formation, is currently in a phase III AD trial [99–101]. Preclinical studies showed that homotaurine could lower plaque number as well as plaque size in APP mice in a dose-dependent manner [102]. In a three-month randomized, double-blind, placebo-controlled phase II study (Neurochem Inc., Canada) with mild-to-moderate AD patients, this drug was well tolerated in human subjects. Although a significant dose-dependent reduction of CSF Aβ42 levels was reported, the net change in CSF Aβ42 was less then 10% of the average total levels of Aβ42, raising issues of biological significance [103]. The U.S. Alzhemed trial was ruled inconclusive by the FDA. More recently, scyllo-inositol (AZD-103), a membrane glycolipid which can inhibit Aβ aggregation in vitro, was shown to ameliorate behavioral deficits in a mouse model of AD [104] and has been used in a phase I human trial (Elán and Transition Therapeutics Ltd, Canada). Caprospinol (SP-233), a 22*R*-hydroxycholesterol derivative that can prevent oligomerization of Aβ will be entering phase I trials (Samaritan Pharmaceuticals) after promising preclinical results that this compound could reduce Aβ load in animal studies ([105]; Neurochem Ltd, Canada). A metal-chelating compound, clioquinol, which is postulated to both inhibit Aβ fibril formation and also disaggregate preformed fibrils by binding zinc and copper, had been in a phase II clinical trial; however, the trial was recently halted, due to the inability to remove a contaminant during purification of the compound [106,107]. A second-generation clioquinol derivative called PBT 2 (8-hydroxyquinoline) with improved pharmacokinetic properties is currently in phase IIa clinical trials (Prana Pharmaceuticals). Similarly, curcumin, a natural constituent of the spice turmeric, is an Aβ-aggregation inhibitor with a rich pharmacology that includes antioxidant, metal-chelating, cholesterol lowering, and anti-inflammatory properties

[108–110]. Based on its efficacy in reducing Aβ deposition in mouse models of AD, it is being tested as a potential AD therapeutic, although poor oral bioavailability has been an issue [111,112]. Another interesting small-molecule inhibitor is the rifampicin and tetracycline family of antibiotics. Doxycycline, a tetracycline analog, was found to not only inhibit fibrillar Aβ formation but also to destabilize preformed fibrils in vitro [113,114]. A three month randomized controlled clinical trial using rifampicin and doxycycline was reported to reduce cognitive decline in mild to moderate AD patients, although the exact mechanism of action is not clarified [115].

Although preclinical and clinical trials with these aggregation inhibitors show promise, one important issue that needs to be addressed is the effective concentration needed in situ. The in vitro assays of these compounds are performed with micromolar concentrations of Aβ; however, the steady-state levels of Aβ in the human brain are in the nanomolar range. In addition, many of these compounds act on multiple pathways; as an example, curcumin can act as an antioxidant, metal chelator, and antiaggregation compound. Also, an important point to consider would be that there are multiple, perhaps competing, Aβ ligands present in the human brain (e.g., GAG, ApoE). Therefore, the pharmacokinetic optima of these compounds need to be established before we can expect striking human clinical trial results.

Peptide-Based Aggregation Inhibitors. Peptide inhibitors have been proposed to be more efficient than small pharmacologic compounds in inhibiting Aβ aggregation, as the former can potentially interact with extended regions of Aβ. A number of such peptide-based Aβ aggregation, inhibitors have been identified, and several have been shown to effectively decrease deposition of Aβ in mouse models by either inhibiting the aggregation process or by disaggregating preformed Aβ fibrils [116,117]. For example, using β-sheet breaker peptides, several groups have shown that it is possible to block Aβ aggregation in vitro and in vivo [118–120]. Another interesting study showed that a truncated Aβ peptide (Aβ 12–28; the ApoE binding domain) could bind to ApoE and prevent it from inducing Aβ aggregation both in vitro and in APP and APP/PS1 mice [121,122]. However, the biggest challenge for such strategies remains that of targeted delivery in situ.

Enhancing Aβ and Amyloid Clearance

Aβ-Degrading Enzymes. Aggregated Aβ is highly resistant to proteolyis and when aggregated into amyloid may have an indefinite half-life [123–125]. In contrast, soluble Aβ is rapidly metabolized in the brain ($t_{1/2} \sim 2$ hours) and periphery ($t_{1/2}$ in plasma < 10 minutes) [126]. Multiple proteases [e.g., insulin-degrading enzyme (IDE), neprilysin, endothelin-converting enzyme (ECE), and angiotensin-converting enzyme (ACE)] have been identified that can cleave monomeric Aβ, and thus their agonists can potentially help prevent Aβ deposition [127–133]. Validation of this theory comes from transgenic mouse studies where IDE or neprilysin overexpression decreases Aβ deposition in APP mice brain [135,136]. Among the multiple proteases studied, plasmin is unique in that it is the only one characterized that can degrade preformed Aβ fibrils in vitro. Aβ can stimulate the tissue plasminogen activator (tPA), which subsequently activates plasminogen and generates plasmin [137–139]. However, in vivo studies in mice show that knocking out plasmin does not alter brain Aβ levels [140,141]. In addition, tPA by itself can cause Aβ neurotoxicity and tau phosphorylation via activation of ERK1/2 [142]. Moreover since none of these proteases are exclusively Aβ-selective, there might be detrimental off-target reactions to therapeutics based on in vivo activation of these proteases.

Neprilysin has been shown to undergo age-progressive decline; thus, its deficiency may be one cause of the age-progressive rise in Aβ levels observed in humans [143]. In transgenic mice, high neprilysin activity prior to deposition caused a delay in the onset of Aβ deposition; however, once the plaques have formed, neprilysin activity does not cause any significant removal of the aggregated Aβ [135,144]. Therefore, as with many other therapeutics, neprilysin-based therapy should be optimized in a "window of opportunity" period. Interestingly, somatostatin, a peptide neurotransmitter that acts as a physiological inhibitor of human growth hormone, up-regulates neprilysin activity. Thus, agonist-mediated somatostatin activity modulation may constitute a pharmacologic intervention step for AD therapy [145].

Insulin degrading enzyme (IDE) is also a bona fide Aβ-degrading enzyme; thus, there might be a link between hyperinsulinemia, increased Aβ levels, and AD [146,147]. Interestingly, when intranasal insulin was administered to humans, there were improvements in memory tests without appreciable effects on blood sugar or insulin levels [148,149].

Aβ Chaperones. A number of cofactors (e.g., ApoE, ACT, clusterin/ApoJ, HSPG) that are co-deposited as a nonfibrillar component of the amyloid plaques, can modify aggregated Aβ deposition [150–153]. Despite the promising potential in developing these pathologic chaperones as potential drug targets for AD, very little effort is presently being undertaken in the clinic. Apolipoprotein E or ApoE is a critical mediator of amyloid aggregate formation and has been linked genetically to the development of AD [153,154]. There are three structurally divergent ApoE alleles in humans: ApoE2, 3, and 4. Overexpression of ApoE3 allele has been reported to decrease Aβ deposition in mice [155,156]. The ApoE4 allele, on the other hand, has definitely been linked to increased risk of getting sporadic AD and cerebral amyloid angiopathy (CAA), thus suggesting that it can possibly modify Aβ structure, clearance, and neurotoxicity in vivo. Neutralizing the chaperone effect of ApoE4 may have an ameliorating effect on Aβ deposition, as has been shown in APP/ApoE$^{-/-}$ double transgenic mice [157,158]. ApoE4 has been shown to bind Aβ12–28 and form insoluble micellar structures. Interestingly, this inhibitor peptide, Aβ12–28P can inhibit Aβ aggregation and reduce Aβ plaques by 60% in mice [121]. However, there is ambiguity of whether the effects observed are due to direct inhibition of Aβ aggregation or indirect inhibition of Aβ–ApoE4 interaction.

Another lipoprotein, ApoJ (clusterin), is also implicated as an amyloid-associated protein that increases the propensity of Aβ to aggregate into thiofavin S–positive plaques in vivo; however, in vitro data shows that ApoJ can actually inhibit Aβ aggregation by binding directly to Aβ [153,159,160]. Another "pathologic chaperone" that promotes Aβ aggregation is the acute-phase protein, α1 antichymotrypsin (ACT). Astrocytic overexpression of ACT in APP transgenic mice leads to increase in age-dependent Aβ plaque deposition and impaired cognitive functions [150,161,162].

Aβ can also bind to proteoglycans, and this has been suggested to alter the conformation of soluble Aβ to a more pro-amyloidogenic state. Heparan sulfate proteoglycans (HSPGs) bind with APP via the carbohydrate moieties on the HSPGs, whereas Aβ binds to the core protein [163,164]. Thus, there might be specific interactions between APP and Aβ with the HSPGs during the progressive nucleation stages of amyloid formation. Heparan sulfate can accelerate Aβ aggregation in vitro [164]. Small-molecule anionic sulfonates (e.g., 1,3-propanediol disulfates) inhibit this process and can also disassemble preformed fibrils in vitro [165]. A low-molecular-mass structural analog of heparin sulfate, enoxaparin, could also reduce the Aβ load and associated neuroinflammation in the APP mouse [166].

Although not technically defined as an Aβ chaperone, Pin1 prolyl isomerase has been shown to bind APP and modulate APP isomerization to promote its processing into the neurotrophic sAPPα fragment [167]. Interestingly, Pin1 promoter polymorphisms appear to associate with reduced Pin1 levels and increased risk for LOAD [168]. Previously, its knockout was found to result in tauopathy and neurodegeneration (see the section on "Tau Chaperones"). In the most recent study, it was shown that Pin1 can stabilize the *trans* form of phospho-threonine 668-APP in vivo; in addition, it can also potentially modulate the secretase activity by stabilizing one conformation of APP over another [167]. In such a scenario, Pin1 seems like an attractive target for AD therapeutics. Although Pin1 does seem to integrate two underlying dysfunctions seen in AD (tau and Aβ), any therapeutic endeavor concerning Pin1 should consider the fact that it is a highly pleitropic molecule and regulates cell-cycle control, cell proliferation, and apoptosis, among other things.

Aβ Binding Human Antibodies. A recent report has described updates from an open-label add-on trial using human IVIg (intravenous immunoglobulin), which showed, a slight improvement in cognition following administration to human patients [169]. IVIgs are purified human polyclonal antibodies originating from the plasma of blood donors and is in present use as an FDA-approved treatment for immune inflammatory diseases. The mixture of natural antibodies in IVIg has been shown to dissociate Aβ fibrils and enhance Aβ clearance [170]. Presently, a 24-subject, double-blind, placebo-controlled phase

II trial is being conducted with a follow-up phase III trial through the Alzheimer Disease Cooperative Study group.

Promoting Efflux of Brain Aβ

Aβ can also bind to *transporter proteins*, which can assist its clearance from the brain to the blood via the blood–brain barrier. Thus, promoting or enhancing Aβ efflux from the brain can be utilized as a potential therapeutic strategy for AD [171,172]. A few transporter proteins characterized are LRP (lipoprotein receptor), ABCA1 (ATP-binding cassette), and P-glycoprotein transporters [171,173]. Interestingly, P glycoprotein [also known as multidrug resistance (Mdr1)] is a major pharmacologic target in chemotherapy and other preventive regimens because of its regulatory role in distribution and bioavailability of different drugs. Chronic treatment with drug regimens can alter P-glycoprotein function and affect its Aβ transporting function. Genetic knockout or inhibition of these transporters results in increased Aβ deposition [174–179]. More recently, a construct expressing one of the domains of LRP, LRP-IV, was shown to be sufficient in effectively sequestering Aβ from AD patient–derived plasma and also in APP mice following intravenous administration [180]. LRP-IV was also demonstrated to be brain impermeant, and thus the authors concluded that it could bind and sequester plasma Aβ and subsequently act as a "peripheral sink" by enabling brain Aβ to be drawn out.

Aβ Immunotherapy

Schenk and colleagues at Elan Pharmaceuticals first reported that active immunization with aggregated Aβ42 in APP transgenic mice, initiated before the manifestation of significant AD-like pathology, resulted in markedly decreased plaque deposition [181]. Subsequent studies have confirmed the ability of active immunization to decrease brain Aβ deposition and related behavioral deficits in several mouse models of AD [182–186]. Passive immunization, in which anti-Aβ antibodies have been administered peripherally or infused directly into the brain, can also result in attenuation of Aβ plaque pathology and cognitive deficits [172,187–191]. Interestingly, peripheral administration of anti-Aβ antibodies can induce a rapid improvement in cognition in old APP transgenic mice (with abundant plaque pathology) without any clear effect on plaque load [192,193], suggesting that certain forms of soluble Aβ species may account for ongoing neuronal dysfunction in APP transgenic mice. The efficiency of the treatment is greatly influenced by several factors; the forms of Aβ used for the immunization, the specificity of the anti-Aβ antibody, the mode of delivery, the mouse model used, and the extent of AD-like pathology present when immunization is initiated (reviewed in [194]). For example, in PDAPP mice different types of Aβ plaques are more easily altered by immunization compared to Tg2576 mice, wherein more dense-cored plaques are deposited predominantly and not readily altered by Aβ immunizations [195]. Immunization of older transgenic mice with significant preexisting plaque pathology also results consistently in reduced efficacies following immunotherapy [192]. Intact antibodies do not appear to be required for efficacy; treatment with Fab fragments as well as scFv fragments of anti-Aβ antibodies have been shown to work well [196,197]. Similarly, Aβ-binding proteins and gangliosides or, more recently, sLRP (soluble LRP) was shown to effectively reduce Aβ plaques burdens, suggesting that the binding of Aβ by antibody or binding proteins and not the effector functions of the antibody are required for efficacy [198,199].

Given the impressive preclinical data of the Aβ vaccination studies [198], the first attempt to use an active immunization approach in a human AD clinical trial (AN-1792) was initiated by Elan Pharmaceuticals. However, the phase II trial had to be terminated prematurely, due to aseptic meningoencephalitis in about 6% of the persons vaccinated [200–204]. Many of the patients who were vaccinated with Aβ42 in the initial phase of the study developed increased anti-Aβ42 antibody serum titers [200]. In follow-up studies, subjects who developed high levels of anti-Aβ antibodies following Aβ42 immunization showed a slower rate of cognitive decline, particularly patients who had mild to moderate AD [200,205,206]. Autopsy of three immunized patients, one year after the trial was halted, showed evidence of Aβ clearance in specific regions of the brain [202]. However, two of these patients also showed signs of encephalitis, including the presence of T-cell infiltrates in the brain [201]. Because there was no correlation between anti-Aβ antibody titers and the development of encephalitis, it has been postulated that the encephalitis was due to an enhanced T-cell response to Aβ, although the exact mechanism underlying the development of encephalitis remains unclear. Brain atrophy due to neuronal loss was also monitored in patients receiving Aβ immunization. Volumetric magnetic resonance imaging analysis was used to determine whether the progression of brain atrophy was altered in immunized patients [205]. Surprisingly, immunized persons with increased anti-Aβ antibody titers demonstrated a larger decrease in brain volume and a greater increase in ventricular enlargement than did patients receiving placebo. The reason for this decrease in brain volume in immunized individuals is unclear, but it has been speculated that clearance of Aβ amyloid and/or fluid shifts from brain parenchyma to cerebrospinal fluid may be involved.

Although the AN-1792 immunization trial resulted in unexpected complications, there may have been positive improvements on certain memory tasks and reduced plaque pathology in a small subset of patients, so Aβ-targeted immunotherapies are still actively pursued as a viable treatment strategy for AD. Indeed, to circumvent the adverse side effects of active immunization, passive immunization with humanized Aβ monoclonal antibodies are currently being evaluated. Elan Pharmaceuticals in collaboration with Wyeth Pharmaceuticals is testing the anti-Aβ monoclonal antibody, AAB-001, for passive immunization in patients with mild to moderate AD. Similarly, Eli Lilly is testing LY206430, a humanized version of m266 antibody that recognizes Aβ(6–23). However, one of the significant issues with passive immunization using humanized antibodies is that it may prove to be too costly to be widely available to the public. Thus, alternative active immunization strategies are also being investigated. To prevent possible T-cell involvement, vaccination with Aβ fragments containing B-cell epitopes [Aβ(1–16)] and lacking T-cell epitopes [Aβ(16–42)] have been shown to ameliorate Aβ pathology and improve behavioral deficits in an APP mouse model without appreciable T-cell responses [207–211]. For example, intranasal immunization with dendrimeric Aβ(1–15) or a tandem repeat of Aβ(1–15) was shown to ameliorate Aβ pathology and improve behavioral deficits in an APP mouse model [208,212]. Based on these initial experiments, at least two additional clinical trails have been initiated with active immunization using Aβ(1–16) (B-cell epitopes). ACC-001, derived from the Aβ(1–7) B-cell epitope region, is in phase I clinical trials (Elan). Another active vaccination strategy in clinical trials (CAD106, Immunodrug; Novartis), utilizes the first six N-terminal amino acids of Aβ, attached to a viruslike particle. This vaccine has been shown to elicit primarily a B-cell response and thus may avoid T-cell activation. Although such altered vaccination strategies have been tested successfully in mice, given the heterogeneity and complexity of the human immune response, it is difficult to predict whether such strategies will prevail in human trials.

Although a wealth of data exists regarding Aβ immunotherapies in several mouse models tested to date, the exact mechanisms underlying the immune-mediated clearance of Aβ from the brains of transgenic mice remain a mystery. Both the active and passive immunization paradigms may act through a direct "Control Nervous System (CNS) clearance" mechanism, whereby the antibodies cross the blood–brain barrier, enter the brain, and either help in dissolution of the fibrils or prevent formation of Aβ fibrils [213–216]. Anti-Aβ antibodies can activate microglial cells to clear plaques through Fc-receptor-mediated phagocytosis and subsequent peptide degradation [195,217]. However, it has also been demonstrated that Fc-independent processes can remove Aβ equally efficiently in both active and passive immunization paradigms [215,218]. Several studies have provided evidence to support Aβ removal via the "peripheral sink" hypothesis by showing that plasma Aβ levels increase dramatically following both active and passive Aβ immunotherapy, and that at least some of the Aβ in the plasma is complexed to the anti-Aβ antibodies [172,184,219,220]. Although such studies are intriguing, the precise mechanism is still contested, as other studies suggest that there is simply not enough anti-Aβ antibody present following active or passive immunization to directly influence bulk sequestration of brain Aβ in the periphery [221].

One of the critical issues that needs further characterization is the exact binding properties of the individual antibodies to their ligand (i.e., whether binding to monomeric or fibrillar or some intermediate toxic forms of Aβ is the most efficacious). Identifying the optimal target epitope may affect future active as well as passive anti-Aβ immunotherapy studies. Despite all these uncertainties, Aβ immunotherapy remains a promising therapeutic approach for AD, and ongoing studies with respect to a precise mechanism of action will help to achieve the most efficacy in future clinical trials.

TAU-SPECIFIC THERAPEUTIC TARGETS

Accumulation of hyperphosphorylated tau as paired helical filaments (PHFs) within neurofibrillary tangles (NFTs), although considered an event secondary to neuritic plaques, is a hallmark of AD pathogenesis and contributes to neurodegeneration in AD [222,223]. Although tau dysfunction is an integral part of AD neuropathologic lesions, it has remained sidelined as a therapeutic target.

The normal function of tau is to stabilize the microtubules that form the structural framework of neurons. It is surmised that in AD, an imbalance of kinases and phosphatases lead to phosphorylation of tau at multiple sites (Fig. 4). Such hyperphosphorylated tau gets detached from microtubules leading to disruption of the microtubules and neuronal transport mechanisms. Within the last few years, efforts have been dedicated to identify and validate

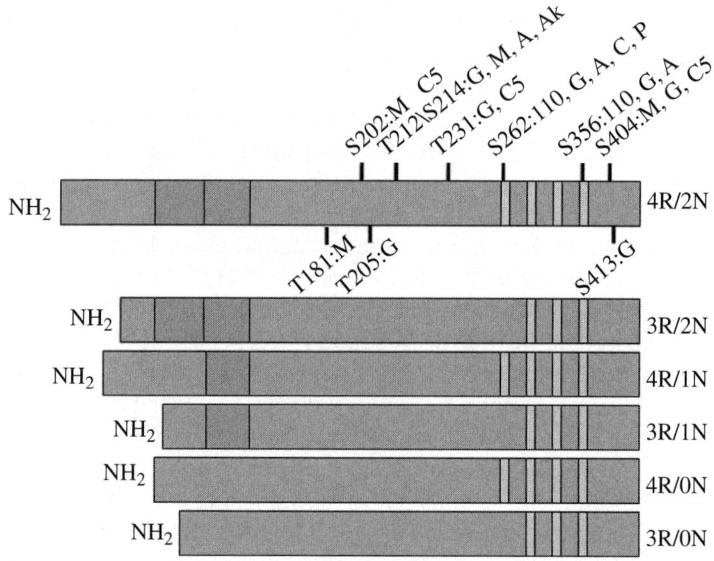

FIG. 4 The domain organization and some phosphorylation sites of tau isoforms expressed in adult humans. The basic structure of tau consists of a highly acidic 29-amino acid stretch (varying from 0N to 2N; shown in orange) in the N terminus or the projection domain and repeated stretches of tubulin binding domain in the C terminus (varying from 1R to 4R; shown in blue). The C terminus is responsible for microtubule binding and stabilization as well as for PHF assembly. Between these two domains is a proline-rich domain which provides additional points of contact with the microtubule surface. The different isoforms result from alternative splicing. Some of the putative phosphorylation sites on tau (serine, S or threonine, T) are marked. Kinases that can act on these sites are abbreviated as follows: A, protein kinase A; Ak, Akt or protein kinase B; C, protein kinase C; C5, CDK5; G, GSK-3; M, MAP kinase, and P, phosphorylase kinase. (*See insert for color representation of figure.*)

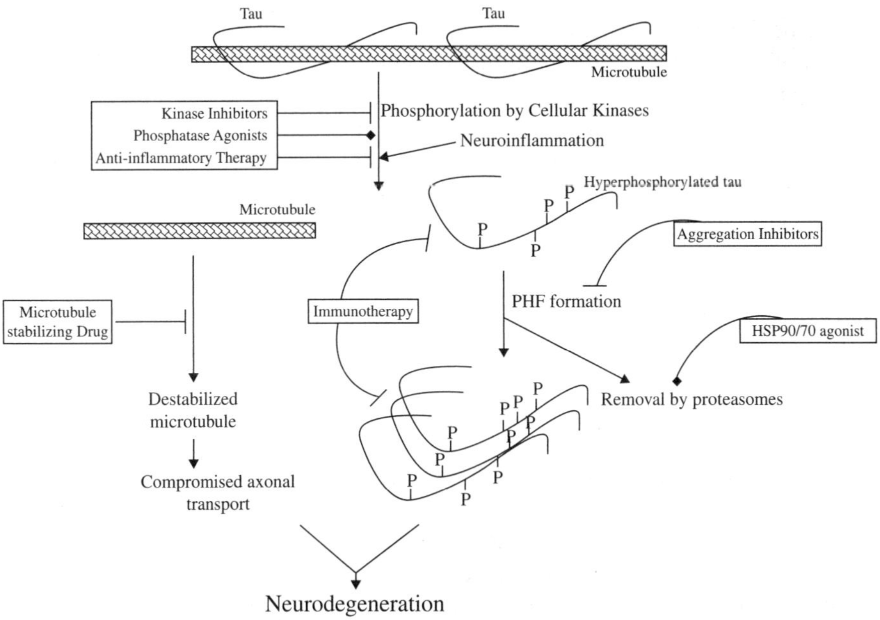

FIG. 5 Potential tau therapeutic targets. Tau binds to and stabilizes microtubules, enabling maintenance of the cytoskeletal network and vesicular transport in the neurons. Hyperphosphorylation of tau leads to its dissociation from the microtubules, leading to the disruption of the microtubule network. Hyperphosphorylated tau can form highly insoluble intracellular PHFs, which along with compromised microtubule functioning, contribute to the neuropathology of AD. Glial neuroinflammation is also postulated to exacerbate the pathology. General principles of tau-based therapies include inhibition of tau hyperphosphorylation and chronic inflammation as well as using aggregation inhibitors. In addition, inducing the clearance of pathologic tau, either by immunotherapy or proteasomal machinery and stabilization of microtubules by pharmacologic intervention, is also being tested for therapeutic efficacy.

novel tau-specific targets (Fig. 5). In the following sections we describe recent preclinical and clinical therapeutic developments designed against tau.

Targeting Tau Production

AD therapies targeting tau are still in early-stage development. Nevertheless, recent advances in the pathobiology of tau have spurred renewed interest in tau therapeutics. It has long been established that NFTs correlate well with neuronal loss and severity of dementia, suggesting they may be closely linked to the underlying pathology. However, recent evidence from mutant tau-overexpressing transgenic mice (rTg4510) suggest that cognitive deficits could be rescued significantly by abrogating tau production even after significant neurofibrillary pathology has set in [29]. This finding has provided even more rationale for anti-tau therapy and raises hope that targeting tau even in later stages of AD may have therapeutic benefit.

Genetic studies with the FTDP-17 tau mutation led to an interesting observation that an imbalance between exon10+ (4repeat or 4R) and exon 10– (3repeat or 3R) forms of tau arising out of splice site mutation might favor tangle pathology [26]. Importantly, the tangles found in human patients are made up almost exclusively of the 4R form (Fig 4). Recent efforts to identify factors that might regulate exon10 splicing have turned up some interesting candidates: hTRA2-β1, thyroid hormone, and CLK-2 [224–226]. Thus, compounds that can modulate the expression or in vivo activity of any of these can, in theory, be employed in the design of tau-based therapeutics.

Altering Tau Aggregation

Several tau aggregation inhibitors have been identified from in vitro screening assays. One of these compounds, a phenothiazine, methylene blue, is currently being evaluated in a human trial [227,228]. Others have also utilized high–throughput screens to characterize aggregation inhibitors in vitro for related neurodegenerative diseases [229]. None of these small-molecule inhibitors have, however, been tested rigorously in animals for efficacy and safety.

It has been hypothesized that a misfolded and hyperphosphorylated soluble form of tau, rather than aggregated tau tangles, may be the real neurotoxic entity. Thus, there are some concerns that compounds that block tau aggregation (and not misfolding) might increase toxicity by accumulating misfolded intermediates. Recent studies in transgenic mice and *Drosophila* overexpressing the wild-type or mutant tau has shown that neurodegeneration can occur independent of NFT formation and that accumulating NFTs do not irrevocably cause neuronal death [29,230,231]. Other independent studies have also shown in mice that cell death and cognitive impairment preceded formation of NFTs [232–234]. However, two studies from Mandelkow and co-workers showed that in cell culture and in transgenic mice, mutant tau that cannot aggregate is less toxic than aggregatable tau [228,235]. Based on the data presently available, an attractive hypothesis is that tau aggregates represent a protective mechanism to sequester toxic forms of abnormal tau, a theory reminiscent of the huntingtin protein aggregates serving a neuroprotective function [236,237]; however, presently there is no direct evidence to support this hypothesis.

Altering Tau Phosphorylation

Increased hyperphosphorylation of tau leading to tangle formation and resulting destabilization of the microtubule network is thought to lead to neurodegeneration. This has led to the examination of the role of kinase inhibitors in ameliorating tau aggregation. In the human brain, tau is phosphorylated at over 38 residues by one or more of the following kinases: GSK-3, cdk5, CK-1, PKA, CaMKII, MAPK, ERK1 and 2, and SAPK (Fig. 4).

Several tau kinase transgenic mouse models (e.g., GSK-3β, CDK5) showed age-progressive accumulation of hyperphosphorylated tau in neurons and neurodegeneration [238–240]. Recently, a number of interesting experiments have demonstrated that the use of small-molecule inhibitors to inhibit tau phosphorylation and reinstate physiological functions is feasible. Most of these kinase inhibitors are ATP-competitive. Intrathecal administration of a calpain inhibitor following spinal cord hemisection in rats could prevent cdk5 phosphorylation and tau phosphorylation and improve neurological function [241]. Two other notable preclinical studies also suggest that inhibiting GSK3β-induced tau phosphorylation can potentially ameliorate the disease outcome. Lithium, a dual-action GSK3β inhibitor, and AR-A014418, another GSK3β inhibitor, reduced tau phosphorylation, pathologic tau accumulation, and axonal degeneration in tau mice [242,243] and has also been shown to modulate Aβ production, although the mechanistics are still debated [244,245]. A GSK3β inhibitor, valproic acid, is in phase II clinical trial against progressive supranuclear palsy (a tauopathy with gait abnormalities and dementia); a combination therapy of lithium and Divalproex (depakote/valproic acid) is also in phase II trials on AD patients. Another study showed beneficial effects of a small-molecule wide-spectrum kinase inhibitor, SRN-003-556, in a mouse model of tauopathy [246]. This indolocarbazole derivative can potentially inhibit ERK2, CDK1, GSK3β, PKA, and PKC, and its use reduced soluble aggregated hyperphosphorylated tau and delayed motor phenotype impairment in vivo. The redundance of intracellular kinase pathways and the nonselective action of kinase inhibitors on multiple related kinases potentially compromises the therapeutic feasibility of these inhibitors. In addition, using kinase inhibitors can lead to off-target effects, as many of these kinases are required to maintain cellular homeostasis. For example, GSK3β roles in the Wnt pathway may mean that its chronic inhibition may be oncogenic. There are similar problems of nonspecific effects of most cdk5 inhibitors; encouragingly, a new class of compounds called the *indolinones* have proved to be very specific in inhibiting cdk5 without affecting cdk2, Jnk, or MAPK activity [247,248]. A feasible therapy may entail partial inhibition of multiple kinases using oligospecific inhibitors rather than full inhibition of either one of them with a monospecific inhibitor. This might help in achieving a high degree of inhibition of tau phosphorylation without complete suppression of individual kinase functions. In addition, the fact that several well-characterized kinase inhibitors are already in clinical trials for cancer provides an opportunity for tau-based therapies to bypass the initial safety and selectivity issues.

Another approach for inhibiting cdk5 is using a peptide known as the cdk5-inhibitory peptide (CIP) [249,250]. CIP, derived from residues 154 to 279 of p35, binds to cdk5 and inhibits its phosphorylation, thereby inhibiting its cleavage to the pathologic p25 form. Neuronal transfection of CIP suppresses cdk5-mediated tau phosphorylation by specifically inhibiting the pathologic cdk5–p25 complex. Because of the exclusive specificity of CIP for cdk5, this remains to be investigated as a potential therapeutic target.

Along with the possible aberrant activation of kinases, it has been shown that there is a 20 to 30% decrease in phosphatase activities (PP2A and PP1) in the AD brain [251]. Restoring the phosphatase activity might help in maintaining phospho-tau homeostasis; indeed, memantine (an NMDA receptor antagonist) could restore okadaic acid–inhibited PP2A activity in in vitro culture systems [252]. However, memantine can enter the cell only under excitotoxic conditions, which limits its degree of efficaciousness in up-regulating PP2A activity under normal circumstances. Whether such PP2A up-regulation by other bioavailable compounds is a valid in vivo therapeutic target remains to be seen.

Targeting Tau Chaperones

Other potential intracellular druggable targets that can impair pathologic tau hyperphosphorylation are a group of intracellular proteins that can either regulate the aggregation and folding of tau or mediate clearance of the misfolded and aggregated tau. One such protein is Pin1, a peptidylprolyl isomerase, which is reduced in AD patients. Pin1 isomerizes phosphorylated tau and restores the ability of phosphorylated tau to bind microtubules. It also promotes dephosphorylation of hyperphosphorylated tau by PP2A phosphatase [167,253,254]. Interestingly, Pin1 has also been shown to modulate Aβ production by stabilizing a phospho-APP isomeric form (see the section on "Enhancing Aβ and Amyloid Clearance"). A plausible hypothesis is that up-regulation of Pin1 could be beneficial in AD by specifically targeting hyperphosphorylated tau and APP; however, this pathway and the pharmacokinetics of its modulators remain to be investigated.

It has been theorized that soluble intermediates in the tau aggregation pathway might be toxic species in the mouse model of tauopathy [29]. Additionally, in a *Drosophila* taoupathy model, it was shown that expression of mutant human tau protein resulted in intracellular accumulation of hyperphosphorylated tau, neurodegeneration, and death without appearance of frank NFTs [231]. Because the protein degradation machinery normally removes the toxic misfolded tau, a possibility exists that overburdening or malfunctioning of this system may result in intracellular accumulation of tau aggregates. This led to an investigation of the role of at least three different proteins of the ubiquitin proteasome system that can polyubiquinate and degrade pathologic tau: CHIP, hsp70, and hsp90 [255,256]. CHIP knockout mice accumulate phospho-tau in many areas of the brain, suggesting that ubiquitination may play a protective role in removing the toxic pretangle tau species [257,258]. Additionally, other members of the proteasome machinery [e.g., hsp70 and hsp90 (heat-shock proteins)], also reduce phospho-tau levels and restore tau association with microtubules [255]. Induction of chaperones such as hsp90 can reduce tau phosphorylation at certain sites via induction of a heat-shock response in vitro [257]. Small-molecule HSP90 inhibitors (e.g., geldanamycin analogs) are currently being tested in human trials as anticancer agents, suggesting that modulation of CHIP and the ubiquitin proteasome system to alter tau pathology may be a feasible disease-modifying therapy in the near future.

Miscellaneous Strategies Targeting Tau

As hyperphosphorylated tau gets sequestered in intracellular inclusions, its microtubule stabilizing function is compromised. Based on this hypothesis, microtubule stabilizing drugs have been used in preclinical studies to determine their ability to rescue the neuropathological outcome. Paclitaxel (Taxol), in use clinically as a potent anticancer drug, is one such drug. Paclitaxel could restore fast axonal transport in spinal axons, increase microtubule number, and ameliorate motor impairments in tau transgenic mice [190]. Recently, paclitaxel has been also shown to inhibit Aβ-induced neuronal death in vitro at sublethal concentrations [259]. However, a caveat in these studies is that Taxol is blood–brain barrier impermeant, so newer taxoid derivatives with better neuroprotective functions need to be designed.

A recent report of active immunization with a phosphorylated tau epitope reported significant decrease in aggregated tau in mutant tau transgenic mice brain. Eventually such immunotherapy could ameliorate tangle-related behavioral phenotype in P301L mice [192]. A concern for this study is the presence of high levels of anti-tau autoantibodies in even the control-vaccinated mice, as is also the fact that the study did not address the specific tau isoforms that was targeted by this immunotherapy regimen. In the absence of validation from other research groups, this preclinical study nonetheless represents a novel and promising therapeutic strategy.

The potential role of proteolysis in controlling seeding of tau aggregation has also been explored. Caspase cleavage of tau may result in production of a highly fibrillogenic tau species [260–265]. Thus, caspase inhibitors may be an attractive target for attenuating tau aggregation. A host of other likely candidates that can inhibit formation of PHF tangles was identified from a high–throughput screen; these compounds belonging to the anthraquinone family did not interfere with microtubule stabilization, thus making them physiologically safe. Interestingly, some of the compounds identified in this screen (e.g., daunomycin, transthyretin) have been found in other screens designed for the inhibition of Aβ fibrillation [266,267]. Other compounds (e.g., M1 receptor agonists, statins, curcumin), discussed previously as anti-Aβ therapies, might also work directly or indirectly to modulate tau pathology [268,269].

TARGETING PLEITROPIC FACTORS THAT MODULATE AD NEUROPATHOLOGY

Although abnormalities in Aβ and tau are associated with AD onset and progression, the pathological cascades leading to neurodegeneration are still ambiguous [222,270,271]. Secondary factors such as chronic neuroinflammation, oxidative stress, cholesterol metabolism dysfunction, lack of trophic support, and receptor dysfunction probably contribute to the degenerative process and modify the neuropathology [272]. In this section we describe how multiple pleiotropic signaling pathways can influence the neuropathologic hallmarks of AD and the recent advances made in identifying various metabolic druggable targets that can potentially ameliorate disease outcome by targeting such pathways.

Targeting Neuroinflammation in AD

Microglia are involved in regulating neuronal survival by release of trophic factors and engaging in tissue repair and neurogenesis [109,273]. However, deregulation and misactivation of microglia have recently been recognized as key players in neurodegeneration [274]. Indeed, in AD patients, volumetric MRI and PET scans have demonstrated an upsurge in activated microglia in the entorhinal, temporoparietal, and cingulate cortices, even in early stages of the disease. A number of polymorphisms in cytokine and chemokine genes and their promoters (e.g., IL1α, and β, TNFα, IL6, IL10) have been associated with an increased risk of developing AD [275]. In the triple transgenic mouse model, 3XTg-AD, chronic neuroinflammation was shown to induce tau hyperphosphorylation at cdk5 specific sites [276]. In addition, induction of pro-inflammatory cytokines was seen to accompany overexpression of the mutant P301S tau in mice [277]. In APP mice, cored Aβ plaques are surrounded by microglia and astrocytes immunoreactive for various cytokines [272,278]. In addition, several preclinical reports have shown that alterations in microglial activity (either by knocking off cytokine receptors or by overexpression of inflammatory mediators) lead to changes in Aβ deposition [279]. Since microglial activation may potentially exacerbate tangle and amyloid neuropathology, using anti-inflammatory therapy may be beneficial in such cases. However, the role of neuroinflammation in development of the neuropathological lesions is still hotly debated: Is microglial activation beneficial in removing toxic debris such as Aβ aggregates, or does microglial activation cause neuronal toxicity and neurodegeneration, or perhaps both? [279].

Available results in transgenic mice mostly demonstrate that hyperactivation of microglia and astrocyte does not precede but usually follows the appearance of Aβ plaques and tau tangles; however, an intriguing piece of data that neuronal BACE1 increases within localized foci of activated astrocytes and microglia and precedes plaque formation raises the notion that chronic neuroinflammation may increase the local production of Aβ by altering APP metabolism [280]. In addition, different cytokines have been shown to increase APP levels in vitro either by increased transcription or stabilization of transcripts. For example, IL1, TNFα, and TGFβ may cause increased APP transcription, whereas IFNγ can lead to increased BACE RNA levels [281–284].

The role of chronic neuroinflammation in modifying the pathologic hallmarks in AD is supported indirectly by observational retrospective and prospective studies showing that long-term NSAID use led to a preventive effect against AD [285–287]. However, given that some NSAIDs may act as GSMs and may also block fibril formation, it is not possible to attribute this apparent protective effect definitively to their anti-inflammatory properties [78,80,288]. Besides NSAIDs, another class of anti-inflammatory drugs, the aminopyridazines, have also been shown to be effective in reducing glial inflammation and Aβ-induced neuronal toxicity in an Aβ infusion model in rats [289]. More recently, an open trial using etanercept (TNFα antagonist) for six months on mild to severe AD patients showed evidence for significant cognitive improvement [290]. Bioavailable TNFa antagonists have previously been used successfully in ameliorating neuropathology in a mouse model of Parkinson disease [291]. A T-cell activator, glatiramer acetate, which promotes Aβ clearance efficiently through an IGF-1-dependent pathway, has been proposed to act by inducing microglial activity [292–294]. However provocative the role of IGF-1 may seem from this mouse model, it is worthwhile mentioning that a recent human trial with an IGF-1-inducing agent (MK-0677, Merck) failed to produce any improvement in AD patients, despite a 60% increase in serum IGF-1 levels.

Reactive astrocytosis has been shown to be responsible for increased neurotoxicity in vivo; especially, S100B produced by activated astrocytes has been correlated with progression of brain damage following stroke and brain lesions. Arundic acid [(*R*)-(−)-2-propyloctanoic acid, ONO-2506] negatively regulates S100B production from astrocytes and has been shown to reduce reactive astrogliosis as well as Aβ plaque load in APP mice [295]. The oral formulation of arundic acid (Cereact) is in phase I trials in the UK for AD and PD (Parkinson's disease) [296]. In the United States, a randomized double-blind phase II clinical trial to test its efficacy in AD patients was concluded in July, 2007.

In addition to a variety of cellular binding partners of Aβ, much as P glycoprotein, a cell surface receptor that binds advanced glycation end products (RAGE) has been shown to bind Aβ [297]. RAGE is a member of the immunoglobulin superfamily of cell surface molecules expressed on endothelial cells and microglia/macrophages. Aβ, by binding to RAGE, have been shown to up-regulate glial inflammatory response. In addition, such an interaction leads to increased levels of oxidative stress, increased Aβ influx at the blood–brain barrier, and vascular dysfunction. Recently, TTP488, an orally bioavailable small-molecule RAGE antagonist has entered phase IIa trials (TransTech Pharma/Pfizer). It works by blocking interaction between Aβ and RAGE and in preclinical studies has been shown to reduce amyloid burden.

Another approach to controlling neuritic plaque-associated inflammation has been to use PPARγ agonists. PPARγ agonists are already being used in clinical treatment of type 2 diabetes [298]. PPARγ is a nuclear receptor whose activation can lead to transcriptional inhibition of pro-inflammatory genes [299]. An acute weekly regimen of a PPARγ agonist, pioglitazone (Actos; Takeda/Eli Lilly), could alleviate inflammatory landmarks in APP transgenic mice when administered orally [300]. In addition to lowering COX-2 and iNOS levels, pioglitazone could also decrease BACE and lower Aβ42 plaque load in these mice. Interestingly, a recent AD trial for rosiglitazone (another PPARγ agonist) was deemed as ineffective; however; following restratification of patients on the basis of their ApoE genotype, non-ApoE4 patients showed significant improvement in their cognition compared with placebo-treated patients [301]. However, a caveat of such global immune inhibition strategies is that this might not always be beneficial, as microglia may also serve to clear Aβ from plaques and thus have neuroprotective functions in vivo. For example, Fiala and co-workers found that an active but minor ingredient of curcumin, bisdesmethoxycurcumin, can induce Aβ phagocytosis by inducing the phagocytic activity of microglia [302]. This is presumably mediated via transcriptional up-regulation of microglial *N*-acetylglucosaminyltransferase III (GlcNAc-TIII) and Toll-like receptor genes. This opens up a new therapeutic potential of curcumin through rescue of the functional and transcriptional deficits of AD macrophages.

Targeting the Cholesterol Pathway in AD

Cholesterol is an essential component of the cell membrane, helping to maintain its structure and fluidity. A human postmortem study revealed that neuropathologically diagnosed AD patients show greater elevation of low-density lipoprotein cholesterol, apolipoprotein B, and lower levels of HDL-C compared to controls [303]. In addition, use of statins (to lower cholesterol) seems to be associated with a decreased risk of dementia [304,305]. The effect of simvastatin (but not lovastatin or atorvastatin) was particularly robust in patients, leading to a 55% chance of being less likely to develop dementia, an effect that remained significant even after adjusting for covariates. Interestingly, the effect of statins may be mediated by a concurrent decrease in ApoE levels, as evident from studies on transgenic mouse models [306,307]. Two Finnish studies found that high cholesterol levels in midlife correlated with higher incidence of AD when scored 20 years later [308,309]. In contrast, two other studies with more than 25 years of follow-up—one on Japanese-American elderly and the other on the Framingham Heart Study cohort—did not find any significant correlation between the two [310,311]. However, the discrepancy of these results may be dependent on the variable timeline of examination and follow-through, use of single timepoints over time-averaged cholesterol measurements, inherent discrepancy in gender-related cholesterol levels, variations of cholesterol levels with age, longitudinal vs. cross-sectional sample sets, and so on.

In vitro cell culture experiments showed that cholesterol levels modulate APP metabolism (e.g., cholesterol depletion by lovastatin resulted in lowered Aβ as well as decreased β-CTF production by BACE1) [312]. In addition, a high-cholesterol diet exacerbated the amyloid pathology in APP mice, a phenotype that could be rescued by treatment of these mice with a cholesterol synthesis inhibitor, BM15.766 [306,313]. It is worthwhile to remember that ApoE, a significant risk factor for AD, is a cholesterol transport protein; persons who are homozygous for the ApoE4 allele have elevated serum cholesterol and also increased risk of developing AD [314,315].

Recent evidence also ties APP processing with cholesterol; cholesterol-rich lipid rafts are the preferential sites for cellular β- and γ-secretase action [177]. In contrast, the enzymes of the α-secretase pathway reside in the phospholipid-rich domain of the plasma membrane and show increased activity when cellular cholesterol content is lowered [316]. Thus, alterations in cholesterol levels may potentially alter the distribution of APP-cleaving enzymes complex within the membrane and shift the Aβ production levels from the "protective" to the "pathogenic" mode. Cross-linking of APP and BACE at the cell surface, either using antibodies or by adding a GPI anchor to BACE as a tether, show that these proteins co-localize with known raft markers; in addition, such cross-linking also increased Aβ production and up-regulation of sAPPα, indicating that cleavage of APP by BACE occurs more efficiently in cholesterol-rich microdomains of the cell membrane [317–319]. A likely hypothesis is that during aging or in patients with a high cholesterol level, the proportion of the membrane containing lipid rafts will be increased and this may allow BACE and APP to be more available in closer apposition to each other for efficient interaction. However plausible this hypothesis may seem, there has been little or no effort therapeutically to inhibit the amyloidogenic processing of APP by specifically targeting the lipid rafts.

Following the efficacy of an acetyl-CoA:cholesterol acetyl transferase inhibitor (ACAT inhibitor; CP-113,818) in reducing Aβ production in cell culture assays, in vivo experiments in APP mutant mice using this compound showed a reduction in Aβ levels and plaque load as well [320]. CP-113,818 could reduce APP processing selectively without affecting β- or γ-secretase activity. Follow-up studies with Avasimibe (Pfizer), which is a small-molecule ACAT inhibitor, showed that it could reduce plaque load in APP/PS1 mice by 60%. Avasimibe had already made it to phase III trials for cardiovascular disease, but Pfizer discontinued its development a few years ago for reasons unconnected with its safety or efficacy. The potential for Avasimibe in treating AD patients is awaited in further clinical trials.

Targeting the Androgenic Hormonal Pathways

Recent work on the triple transgenic mice (3XTg-AD) showed that both testosterone and estrogen depletion can lead to increased Aβ load [321,322]. Previous work had shown that estrogen could prevent Aβ-associated neurotoxicity in PC12 and neuroblastoma cells as well as in primary hippocampal neurons [323–325]. Epidemiological studies showed that menopausal women are at a greater risk of developing AD [326,327] relative to men. In addition, estrogen replacement therapy has been shown to have a negative association with risk for AD [328,329], improving cognitive function and/or delaying disease progression [330,331]. Interestingly, estradiol deprivation due to ovariectomy in APP and PS/APP mice increases the levels of soluble Aβ42 and Aβ40, an effect that can be reversed by exogenous estrogen administration [332,333]. It was also shown that estradiol therapy was accompanied by an increase in sAPPα levels with no changes in holo APP, conferring neuroprotection. Although the recently concluded National Institutes of Health–led Women's Health Initiative (2007) study found that hormone replacement therapy and dementia are positively correlated, another recently conducted longitudinal study found that premature ovariectomy increases the risk of cognitive decline by 1.5 times [334]. The answer might lie in the complex pathways (anti-inflammatory, proliferative, and neuroprotective pathways) that are sensitive to different levels of activation of Erα (estrogen receptor α) and Erβ (estrogen receptor β) receptors. Interestingly, it was shown that an FDA-approved antiestrogenic compound (ICI 182,780), which antagonizes the ability of estrogen to drive proliferation in reproductive tissues, has estrogen receptor agonist activity in neurons [335,336]. This calls for the development of "selective estrogen receptor modulators" that can offer neuroprotection by way of functioning exclusively as an antiamyloidogenic in neurons.

Environmental Enrichment and Lifestyle: Role in AD Therapy

Epidemiological data from large cohorts of aged persons show that higher levels of education and occupational attainment, along with regular participation in cognitively stimulating tasks, reduce the risk of developing sporadic AD [337,338–339]. The "cognitive reserve" hypothesis formulates that enriched lifestyles result in the establishment of more functionally efficient cognitive networks, building up a cognitive reserve that can delay the onset of clinical symptoms [340]. Interestingly, both exercise and environmental stimuli can increase neprilysin activity and thus decrease Aβ. In addition to this, certain diet choices have been shown to have a modulatory effect on disease onset; for example, omega-3 fatty acids, vitamins E, B_6, and B_{12}, and moderate intake of red wine are protective, while the risk factors include high calorie intake [341–343]. Vitamins B_6, B_{12} and folic acid have been implicated to reduce the plasma levels of homocysteine in a recent open label trial study [344]; homocysteine has previously been shown to be associated with an increased risk of AD in epidemiological studies [345]. Homocysteine can also impair DNA repair, which in postmitotic neurons have been hypothesized to lead to neuronal death [346]. In the APP mouse model, long-term environmental enrichment can lead to cognitive improvement, without a reduction in Aβ burden [347]. On the other hand, Lazarov et al. reported that in APP/PS1 mice, enrichment actually led to a decrease in cerebral Aβ deposits, concominant with increased neprilysin levels and induction of various other immediate early genes [348].

Oxidative stress has been reported to be an age-associated event that may aggravate AD pathology (reviewed in [349]). Vitamin E is a fat-soluble antioxidant that has been touted widely as a preventive therapeutic against ageing and oxidative stress. Vitamin E–supplemented food significantly reduced lipid peroxidation adducts in young and old T2576 mice; however, an Aβ-lowering effect was seen only when administered in young mice [350]. Given this, however, the failure of antioxidants such as vitamins C and E in clinical trials have been very disappointing [351]. Another compound, *Gingko biloba*, when supplemented in the drinking water of APP mice, could rescue their behavioral deficits [352]. The bioactive ingredients in *Gingko* were flavonoids and terpenoids; however, despite behavioral improvement, there was no significant alteration in either Aβ levels or plaque burden following its administration. Another study found that *Gingko* extract Egb761 could prevent Aβ fibrillogenesis in cell culture and in the transgenic *Caenorhabditis elegans* and mice models of Aβ deposition [353,354]. A randomized double-blind phase IV clinical trial of Egb761 (Tanakan; Ipsen Pharma) in human patients is ongoing; in addition, a National Institutes of Health–supported GEM (Ginkgo Evaluation of Memory) study in the United States and a GuidAge study in Europe are under way to evaluate Egb 761 as a preventive drug [355]. To date, treatments targeting these processes (e.g., non-Aβ42-lowering NSAIDs, vitamin E, among others) have been inconclusive in human trials. Nevertheless, omega-3-fatty acids, curcumin, *Gingko*, and statins might influence multiple pathologic pathways, such as inflammation and oxidative stress, in addition to influencing Aβ or tau metabolism [108,356,357]. Given the relative safety of such compounds, further studies in humans are certainly warranted to reach a state of optimal dosage and effectivity in vivo.

Targeting Glucose Metabolism

A physiological manifestation of aging in mammals is dysfunction of glucose metabolism, and this has been purported to contribute to cognitive deficits. There is evidence that plasma ketone bodies can form an effective alternative source of energy substrate for the brain and alleviate cognitive decline [358]. A compound called AC-1202 and trademarked Ketasyn (Accera), which can increase serum ketone bodies, was used in a double-blind placebo-controlled study in an aged dog model of dementia and more recently in a phase IIb study on mild to moderate AD patients. The study showed that patients with ApoE2 or 3 alleles benefited the most (in terms of significant cognitive improvement), whereas those with ApoE4 allele did not. This mirrors the rosiglitazone trial and opens up new challenges in the area of pharmacogenomics where a person's genetic makeup modifies the response to a drug.

Targeting Trophic Factor and Receptor-Mediated Pathways in AD Therapy

Recent studies with AMPA (ionotropic glutaminergic receptor agonist) receptor agonists (e.g., LY404187) had shown promising leads in enhancing synaptic function and ameliorating cognitive deficits in animal models [359]. LY404187 has been shown to enhance glutaminergic synaptic transmission directly by potentiation of AMPA receptor function, as well as an indirect recruitment of voltage-dependent NMDA receptor activity. AMPA agonists or ampakines, through expression of BDNF (brain-derived neurotrophic factor), were shown to reverse regional, age-related deficits in LTP in rats [360]. One such AMPA agonist, Ampakine CX-717 (Cortex Pharmaceuticals), is currently being used in clinical trials on AD patients in the United States.

Other compounds that confer neuroprotection through trophic support and mitochondrial up-regulation are selegiline and rasagiline [361]. Rasagiline (Azilect, Eisai Pharmaceuticals) is a potent, selective, irreversible monoamine oxidase (MAO) type B inhibitor and is presently being used in the symptomatic treatment of Parkinson disease. Rasagiline increases glial cell–derived neurotrophic factor, NGF (nerve growth factor), and BDNF levels, leading to increased neuroplasticity and improved LTP. It is presently involved in a randomized trial in a combination therapy with Aricept in mild to moderate AD patients.

Recently, another interesting eight-amino acid peptide fragment (NAP/AL-108; Allon Therapeutics) derived from the activity-dependent neuroprotective protein (ADNP) has entered phase II trials. ADNP is up-regulated during brain injury and the NAP peptide has been shown to have neuroprotective function by stimulating astrocyte-mediated neurite growth [362]. Specifically, NAP binds to tubulin and promotes microtubule assembly and stabilization. Interestingly, this peptide has been shown to abrogate not only Aβ-mediated learning deficits but also neuronal injury in mouse models of head injury, fetal alcohol syndrome, and stroke [363].

Among other receptor-based intervention strategies for management of AD symptoms is the recently discovered endogenous cannabinoid system (reviewed in [364]). Encouraging data have emerged that Dronabinol, an oil-based Δ^9-THC, improves disturbed behavior in AD patients [365,366]. The proposed site of action of cannabinoids are the NMDA receptors, possibly via inhibition of presynaptic Ca^{2+} entry and subsequent inhibition of excessive glutaminergic synaptic activity [367,368]. In addition, like memantine, cannabinoids are also capable of increasing BDNF to confer protection against excitotoxicity [369]. In vitro Aβ aggregation studies demonstrated that cannabinol could prevent AChE-induced Aβ aggregation, a key neuropathological hallmark of AD-type dementia [370]. In a mouse model, cannabidiol could ameliorate Aβ-induced neuroinflammation by suppressing IL1β and iNOS [371] showing that cannabinoids can, in principle, regulate Aβ-mediated neuropathology. However, the use of cannabinoids is fraught with medical and ethical dilemmas, unless it is derivatized to a nonaddictive formulation.

Recently, other neuroprotective agent, such as cerebrolysin, have been used as disease-modifying therapy against AD [372]. Cerebrolysin is a peptide-based drug that mimics NGF-like neurotrophic activity. Treatment of APP mice with cerebrolysin ameliorated performance deficits, along with lowered Aβ pathology, inhibition of APP maturation, and reduction in GSK3β and cdk5 kinase levels [373]. In a randomized clinical trial (Ebewe Pharmaceutical), it was reported to confer cognitive improvement in AD patients [374].

Gene Therapy and Regenerative Tissue Implantation in AD Therapy

Several nonpharmacologic approaches to reduce Aβ load, including using siRNAs (small interfering RNAs) and antibodies, have been carried out in a preclinical setting. Using viral vectors, small hairpin RNAs (shRNA) have been used to target disease pathology in ALS (amyotrophic lateral sclerosis), HD (Huntington disease), and APP mouse models (reviewed in [375]). Proof of concept that in vitro APP silencing and allele-specific silencing of tau by siRNAs could be achieved has been reported [134,376]. However, expression of shRNAs in neurons can potentially interfere with dendritic spine structure and function, and can result in decreases in synapse number in mice, regardless of the sequence targeted [377]. Also, a recent report showed that the administration of viral vectors that produce shRNAs can be fatal to mice by competing with the cellular machinery that produces endogenous microRNAs needed for cellular homeostasis and survival [378].

Recently, antisense oligonucleotides have been used successfully to ameliorate ALS type pathology in mice models by using osmotic pumps for intracerebroventricular delivery [379]. A phase I trial on humans using such antisense oligos is in the works, and if this proves to be safe, targeting of presenilins or kinases such as GSK3β seems to be a prospective AD modifying therapy.

Using an antibody-based approach it was shown that it is possible to inhibit BACE by blocking its cleavage site of APP [380]. Although such approaches are feasible, the challenge lies in targeting these to the brain; thus, developing optimal gene delivery vectors (e.g., AAV2 used for gene delivery to PD patients [381] and Duchene dystrophic patients [382] need to be undertaken to realize the full potential of such approaches [383]. Recent technological improvements in intracerebral infusion devices have been very encouraging, but the high cost and invasive nature of this type of therapeutic approach is a major deterrent.

Ex vivo NGF gene delivery to patients with mild AD was done by implanting autologous fibroblasts expressing human NGF [384]. Long-term follow-up showed that AD patients who received bilateral injections of NGF-releasing cells maintained stable cognitive status as well as the ability to live independently. Patients who received injections into only one side of their brain declined on these scores. Recently, Ceregene Ltd has filed an investigational new drug application with the FDA for a phase I trial for this treatment regimen (reviewed in [385]). The main concern about these strategies is their safety and effectiveness; additionally, the challenge with such studies is the invasiveness of the procedure.

The use of stem cell therapy in AD and other progressive degenerative diseases is hotly debated. In a recent study, it was shown that transplanted human neural stem cells in phenserine (that can posttranscriptionally attenuate APP production)-treated APP23 mice can lead to these neurons differentiating and migrating to the cortex and hippocampus, two brain areas that undergo neurodegeneration in AD [386]. In theory, stem cell transplantation could lead to the recovery of cognitive deficits caused by the degeneration of the basal cholinergic neurons. However, given the widespread and progressive damage in AD patients, the long-term effectivity of such approaches remains doubtful. On the whole, stem cell and tissue transplantation therapies remain experimental and far from effective clinical application.

FUTURE CHALLENGES IN DEVELOPMENT OF CLINICALLY RELEVANT THERAPIES

Although there have been tremendous advances in preclinical science relevant to AD, the design of clinical trials has not changed substantially in the last several decades, leading to a somewhat frustrating gap between the bench and the bedside. A cautionary note from the prematurely abandoned AN1792 trial has to be remembered in this era of translational research. Although rodents are the model of choice for the design and development of new drugs, the fact that there are basic metabolic, physiological, and anatomical differences between humans and the available disease models means that nothing substitutes for a human trial.

Regulatory hurdles and the sheer cost and complexity of these AD trials have made them even more challenging. Given the slow and variable rate of decline, especially among patients with mild AD or MCI, a single 12- to18-month well-powered double-blind placebo-controlled phase III trial requiring thousands of patients typically costs upward of $50 million. An important issue is thus prioritization of preclinical studies that could be expected to have the best outcome in phased clinical trials; in addition, standardization of the criteria needed to conclude a particular trial successfully and for extending follow-up trials needs to be established. Presently, the primary endpoint measures of disease-modifying therapies in AD are ADAS-Cog, ADCS-ADI, and SIB, three different measures of cognitive ability and day-to-day independent functioning. However, a set of standardized and objectively defined endpoint biomarkers need to be established not only to measure the efficacy of different treatment modalities but also to stratify recruited patients into different stages of the disease. In addition, biomarkers should be able to predict progression of MCI or presymptomatic patients into full-blown dementia. The need to establish such biomarkers is of prime importance. It will be critical to develop reproducible and minimally invasive biomarker and/or imaging technique(s) that will enable clinical trials to be completed and analyzed in a time- and cost-effective manner. Coupled with identification of genetic risk factors, better imaging agents, and biomarker tools to assess disease progression status, one can hope to target the population that could benefit the most from such trials [6,13,79,387–389]. Using such risk stratification, it may not only be possible to identify much smaller cohorts of patients at increased risk for disease and need for a particular preventive drug, but such strategies may make future trials more affordable and conclusive. One can hope that the advent of a postgenomic era of systems biology will help in better clustering of biomarker and genetic profiling for patients, which may lead to reliable preclinical diagnosis and prediction of patient-specific clinical outcome.

An important point to consider is that AD in humans, unlike the structured rodent disease models, is a complex set of pathologic symptoms; thus, individual responses to a treatment regimen can result from variations in disease stage, age, nutritional and dietetic status, underlying medical conditions, and other medications administered. In such cases, it might be more effective to follow either a "cocktail" or a multimodal therapy to affect the clinical course of AD significantly. For example, a cocktail of therapies that target tau, Aβ, inflammation, and cognitive symptoms may be more efficacious than any monotherapy. In addition, while targeting phospho-specific pathways, it may be more advantageous to suboptimally target multiple tau kinases rather than inhibit one single kinase; this might help reduce the off-target effects of these kinase inhibitors. Even the employment of kinase of inhibitor monotherapies with lithium or muscarinic M1 agonists that can multiple pathways simultaneously may be more useful in the long run [47,243,244,280,390].

SUMMARY

Despite seminal discoveries in the field of AD and related dementia, we still do not have a tangible end in sight. Will anti-Aβ vaccines be the magic bullet, or do we need to have a multipronged combination therapy of anti-amyloid, anti-tau, and anti-inflammatory drugs? In the United States alone, the annual socioeconomic burden from central nervous system (CNS) disorders such as stroke, schizophrenia, and AD is currently estimated to be $250 billion and rising. With the current global thrust in the effort and resources dedicated to developing disease-modifying therapies for AD, it is hoped that better and more methodologically sound clinical trials will pave the way to effective treatment in the near future.

Acknowledgments

T.E.G and P.D. are supported by grants from the Alzheimer's Association, AFAR, and the National Institutes of Health. P.C. acknowledges the Robert H. and Clarice Smith Foundation postdoctoral fellowship.

NOTE ADDED IN PROOF

The phase III clinical study evaluated 800 mg *R*-flurbiprofen twice-daily versus placebo for 18 months in 1800 patients with mild Alzheimer disease. However, in 2008, Myriad Genetics concluded that the drug did not significantly improve either cognition or the ability of patients to carry out daily activities. On June 30, 2008, Myriad announced that it will no longer be developing Flurizan (Abstract O3-04-01 presented at ICAD 2008: Alzheimer's Association International Conference on Alzheimer's Disease. July 29, 2008).

REFERENCES

1. Fox, N.C., Crum, W.R., Scahill, R.I., Stevens, J.M., Janssen, J.C., Rossor, M.N. (2001). Imaging of onset and progression of Alzheimer's disease with voxel-compression mapping of serial magnetic resonance images. *Lancet, 358*, 201–205.
2. Leenders, K.L., Oertel, W.H. (2001). Parkinson's disease: clinical signs and symptoms, neural mechanisms, positron emission tomography, and therapeutic interventions. *Neural Plast, 8*, 99–110.
3. Small, G.W., Mazziotta, J.C., Collins, M.T., Baxter, L.R., Phelps, M.E., Mandelkern, M.A., Kaplan, A., La Rue, A., Adamson, C.F., Chang, L., et al. (1995). Apolipoprotein E type 4 allele and cerebral glucose metabolism in relatives at risk for familial Alzheimer disease. *JAMA, 273*, 942–947.
4. Fagan, A.M., Mintun, M.A., Mach, R.H., Lee, S.Y., Dence, C.S., Shah, A.R., LaRossa, G.N., Spinner, M.L., Klunk, W.E., Mathis, C.A., et al. (2005). Inverse relation between in vivo amyloid imaging load and cerebrospinal fluid Abeta(42) in humans. *Ann Neurol, 59*, 512–519.
5. Lin, K.N., Wang, P.N., Chen, C., Chiu, Y.H., Kuo, C.C., Chuang, Y.Y., Liu, H.C. (2003).The three-item clock-drawing test: a simplified screening test for Alzheimer's disease. *Eur Neurol, 49*, 53–58.
6. Golde, T.E. (2003). Alzheimer disease therapy: Can the amyloid cascade be halted? *J Clin Invest, 111*, 11–18.
7. Lipton, S.A. (2005). The molecular basis of memantine action in Alzheimer's disease and other neurologic disorders: low-affinity, uncompetitive antagonism. *Curr Alzheimer Res, 2*, 155–165.
8. Holzgrabe, U., Kapková, P., Alptüzün, V., Scheiber, J., Kugelmann, E. (2007). Targeting acetylcholinesterase to treat neurodegeneration. *Expert Opin Ther Targets, 11*, 161–179.
9. Sonkusare, S.K., Kaul, C.L., Ramarao, P. (2005). Dementia of Alzheimer's disease and other neurodegenerative disorders: memantine, a new hope. *Pharmacol Res, 51*, 1–17.
10. Scarpini, E., Scheltens, P., Feldman, H. (2003). Treatment of Alzheimer's disease: current status and new perspectives. *Lancet Neurol, 2*, 539–547.
11. Raschetti, R., Albanese, E., Vanacore, N., Maggini, M. (2007). Cholinesterase inhibitors in mild cognitive impairment: a systematic review of randomised trials. *PLoS Med, 4*, e338.
12. Jacobsen, J.S., Reinhart, P., Pangalos, M.N. (2005). Current concepts in therapeutic strategies targeting cognitive decline and disease modification in Alzheimer's disease. *NeuroRx, 2*, 612–626.
13. Thal, L.J., Carta, A., Doody, R., Leber, P., Mohs, R., Schneider, L., Shimohama, S., Silber, C. (1997). Prevention protocols for Alzheimer disease: Position paper

from the International Working Group on Harmonization of Dementia Drug Guidelines. *Alzheimer Dis Assoc Disord, 11* (Suppl 3), 46–49.

14. Hardy, J.A., Higgins, G.A. (1992). Alzheimer's disease: the amyloid cascade hypothesis. *Science, 256*, 184–185.
15. Haass, C., Selkoe, D.J. (2007). Soluble protein oligomers in neurodegeneration: lessons from the Alzheimer's amyloid beta-peptide. *Nat Rev Mol Cell Biol, 8*, 101–112.
16. Golde, T.E., Eckman, C.B., Younkin, S.G. (2000). Biochemical detection of Abeta isoforms: implications for pathogenesis, diagnosis, and treatment of Alzheimer's disease. *Biochim Biophys Acta, 1502*, 172–187.
17. Lewis, J., McGowan, E., Rockwood, J., Melrose, H., Nacharaju, P., Van Slegtenhorst, M., Gwinn-Hardy, K., Paul Murphy, M., Baker, M., Yu, X., et al. (2000). Neurofibrillary tangles, amyotrophy and progressive motor disturbance in mice expressing mutant (P301L) tau protein. *Nat Genet, 25*, 402–405.
18. Rovelet-Lecrux, A., Hannequin, D., Raux, G., Le Meur, N., Laquerrière, A., Vital, A., Dumanchin, C., Feuillette, S., Brice, A., Vercelletto, M., et al. (2006). APP locus duplication causes autosomal dominant early-onset Alzheimer disease with cerebral amyloid angiopathy. *Nat Genet, 38*, 24–26.
19. Games, D., Adams, D., Alessandrini, R., Barbour, R., Berthelette, P., Blackwell, C., Carr, T., Clemens, J., Donaldson, T., Gillespie, F., et al. (1995). Alzheimer-type neuropathology in transgenic mice overexpressing V717F beta-amyloid precursor protein. *Nature, 373*, 523–527.
20. Hsiao, K., Chapman, P., Nilsen, S., Eckman, C., Harigaya, Y., Younkin, S., Yang, F., Cole, G. (1996). Correlative memory deficits, Abeta elevation, and amyloid plaques in transgenic mice. *Science, 274*, 99–102.
21. Sturchler-Pierrat, C., Abramowski, D., Duke, M., Wiederhold, K.H., Mistl, C., Rothacher, S., Ledermann, B., Bürki, K., Frey, P., Paganetti, P.A., et al. (1997). Two amyloid precursor protein transgenic mouse models with Alzheimer disease-like pathology. *Proc Natl Acad Sci U S A, 94*, 13287–13292.
22. Newman, M., Musgrave, I.F., Lardelli, M. (2007). Alzheimer disease: amyloidogenesis, the presenilins and animal models. *Biochim Biophys Acta, 1772*, 285–297.
23. Holcomb, L., Gordon, M.N., McGowan, E., Yu, X., Benkovic, S., Jantzen, P., Wright, K., Saad, I., Mueller, R., Morgan, D., et al. (1998). Accelerated Alzheimer-type phenotype in transgenic mice carrying both mutant amyloid precursor protein and presenilin 1 transgenes. *Nat Med, 4*, 97–100.
24. Richards, J.G., Higgins, G.A., Ouagazzal, A.M., Ozmen, L., Kew, J.N., Bohrmann, B., Malherbe, P., Brockhaus, M., Loetscher, H., Czech, C., et al. (2003). PS2APP transgenic mice, coexpressing hPS2mut and hAPPswe, show age-related cognitive deficits associated with discrete brain amyloid deposition and inflammation. *J Neurosci, 23*, 8989–9003.
25. Li, Y.M., Lai, M.T., Xu, M., Huang, Q., DiMuzio-Mower, J., Sardana, M.K., Shi, X.P., Yin, K.C., Shafer, J.A., Gardell, S.J. (2000). Presenilin 1 is linked with gamma-secretase activity in the detergent solubilized state. *Proc Natl Acad Sci U S A, 97*, 6138–6143.
26. Hutton, M., Lendon, C.L., Rizzu, P., Baker, M., Froelich, S., Houlden, H., Pickering-Brown, S., Chakraverty, S., Isaacs, A., Grover, A., et al. (1998).

Association of missense and 5′-splice-site mutations in tau with the inherited dementia FTDP-17. *Nature, 393*, 702–705.

27. Spillantini, M.G., Murrell, J.R., Goedert, M., Farlow, M.R., Klug, A., Ghetti, B. (1998). Mutation in the tau gene in familial multiple system tauopathy with presenile dementia. *Proc Natl Acad Sci U S A, 95*, 7737–7741.
28. Lee, V.M., Goedert, M., Trojanowski, J.Q. (2001). Neurodegenerative tauopathies. *Annu Rev Neurosci, 24*, 1121–1159.
29. Santacruzm, K., Lewis, J., Spires, T., Paulson, J., Kotilinek, L., Ingelsson, M., Guimaraes, A., DeTure, M., Ramsden, M., McGowan, E., et al. (2005). Tau suppression in a neurodegenerative mouse model improves memory function. *Science, 309*, 476–481.
30. Lewis, J., Dickson, D.W., Lin, W.L., Chisholm, L., Corral, A., Jones, G., Yen, S.H., Sahara, N., Skipper, L., Yager, D., et al. (2001). Enhanced neurofibrillary degeneration in transgenic mice expressing mutant tau and APP. *Science, 293*, 1487–1491.
31. Götz, J., Chen, F., Barmettler, R., Nitsch, R.M. (2001). Tau filament formation in transgenic mice expressing P301L tau. *J Biol Chem, 276*, 529–534.
32. Harada, A., Oguchi, K., Okabe, S., Kuno, J., Terada, S., Ohshima, T., Sato-Yoshitake, R., Takei, Y., Noda, T., Hirokawa, N. (1994). Altered microtubule organization in small-calibre axons of mice lacking tau protein. *Nature, 369*, 488–491.
33. Takei, Y., Teng, J., Harada, A., Hirokawa, N. (2000). Defects in axonal elongation and neuronal migration in mice with disrupted tau and map1b genes. *J Cell Biol, 150*, 989–1000.
34. Ikegami, S., Harada, A., Hirokawa, N. (2000). Muscle weakness, hyperactivity, and impairment in fear conditioning in tau-deficient mice. *Neurosci Lett, 279*, 129–132.
35. Oddo, S., Caccamo, A., Shepherd, J.D., Murphy, M.P., Golde, T.E., Kayed, R., Metherate, R., Mattson, M.P., Akbari, Y., LaFerla, F.M. (2003). Triple-transgenic model of Alzheimer's disease with plaques and tangles: intracellular Abeta and synaptic dysfunction. *Neuron, 39*, 409–421.
36. Price, D.L., Sisodia, S.S. (1998). Mutant genes in familial Alzheimer's disease and transgenic models. *Annu Rev Neurosci, 21*, 479–505.
37. Gravina, S.A., Ho, L., Eckman, C.B., Long, K.E., Otvos, L., Jr., Younkin, L.H., Suzuki, N., Younkin, S.G. (1995). Amyloid beta protein (A beta) in Alzheimer's disease brain. Biochemical and immunocytochemical analysis with antibodies specific for forms ending at A beta 40 or A beta 42(43). *J Biol Chem, 270*, 7013–7016.
38. Kojro, E., Fahrenholz, F. (2005). The non-amyloidogenic pathway: structure and function of alpha-secretases. *Subcell Biochem, 38*, 105–127.
39. Buxbaum, J.D., Liu, K.N., Luo, Y., Slack, J.L., Stocking, K.L., Peschon, J.J., Johnson, R.S., Castner, B.J., Cerretti, D.P., Black, R.A. (1998). Evidence that tumor necrosis factor alpha converting enzyme is involved in regulated alpha-secretase cleavage of the Alzheimer amyloid protein precursor. *J Biol Chem, 273*, 27765–27767.
40. Fahrenholz, F., Gilbert, S., Kojro, E., Lammich, S., Postina, R. (2000). Alpha-secretase activity of the disintegrin metalloprotease ADAM 10. influences of domain structure. *Ann N Y Acad Sci, 920*, 215–222.

41. Postina, R., Schroeder, A., Dewachter, I., Bohl, J., Schmitt, U., Kojro, E., Prinzen, C., Endres, K., Hiemke, C., Blessing, M., et al. (2004). A disintegrin-metalloproteinase prevents amyloid plaque formation and hippocampal defects in an Alzheimer disease mouse model. *J Clin Invest, 113*, 1456–1464.

42. Levites, Y., Amit, T., Mandel, S., Youdim, M.B. (2003). Neuroprotection and neurorescue against Abeta toxicity and PKC-dependent release of nonamyloidogenic soluble precursor protein by green tea polyphenol (−)-epigallocatechin-3-gallate. *FASEB J, 17*, 952–954.

43. Obregon, D.F., Rezai-Zadeh, K., Bai, Y., Sun, N., Hou, H., Ehrhart, J., Zeng, J., Mori, T., Arendash, G.W., Shytle, D., et al. (2006). ADAM10 activation is required for green tea (−)-epigallocatechin-3-gallate-induced alpha-secretase cleavage of amyloid precursor protein. *J Biol Chem, 281*, 16419–16427.

44. Rezai-Zadeh, K., Shytle, D., Sun, N., Mori, T., Hou, H., Jeanniton, D., Ehrhart, J., Townsend, K., Zeng, J., Morgan, D., et al. (2005). Green tea epigallocatechin-3-gallate (EGCG) modulates amyloid precursor protein cleavage and reduces cerebral amyloidosis in Alzheimer transgenic mice. *J Neurosci, 25*, 8807–8814.

45. Vardy, E.R., Catto, A.J., Hooper, N.M. (2005). Proteolytic mechanisms in amyloid-beta metabolism: therapeutic implications for Alzheimer's disease. *Trends Mol Med, 11*, 464–472.

46. Fisher, A., Michaelson, D.M., Brandeis, R., Haring, R., Chapman, S., Pittel, Z. (2000). M1 muscarinic agonists as potential disease-modifying agents in Alzheimer's disease. Rationale and perspectives. *Ann N Y Acad Sci, 920*, 315–320.

47. Caccamo, A., Oddo, S., Billings, L.M., Green, K.N., Martinez-Coria, H., Fisher, A., LaFerla, F.M. (2006). M1 receptors play a central role in modulating AD-like pathology in transgenic mice. *Neuron, 49*, 671–682.

48. Müller, D.M., Mendla, K., Farber, S.A., Nitsch, R.M. (1997). Muscarinic M1 receptor agonists increase the secretion of the amyloid precursor protein ectodomain. *Life Sci, 60*, 985–991.

49. Hock, C., Maddalena, A., Raschig, A., Müller-Spahn, F., Eschweiler, G., Hager, K., Heuser, I., Hampel, H., Müller-Thomsen, T., Oertel, W., et al. (2003). Treatment with the selective muscarinic m1 agonist talsaclidine decreases cerebrospinal fluid levels of A beta 42 in patients with Alzheimer's disease. *Amyloid, 10*, 1–6.

50. Gandy, S., Greengard, P. (1994). Regulated cleavage of the Alzheimer amyloid precursor protein: molecular and cellular basis. *Biochimie, 76*, 300–303.

51. Koo, E.H. (1997). Phorbol esters affect multiple steps in beta-amyloid precursor protein trafficking and amyloid beta-protein production. *Mol Med, 3*, 204–211.

52. Etcheberrigaray, R., Tan, M., Dewachter, I., Kuipéri, C., Van der Auwera, I., Wera, S., Qiao, L., Bank, B., Nelson, T.J., Kozikowski, A.P., et al. (2004). Therapeutic effects of PKC activators in Alzheimer's disease transgenic mice. *Proc Natl Acad Sci U S A, 101*, 11141–11146.

53. Sun, M.K., Alkon, D.L. (2006). Bryostatin-1: pharmacology and therapeutic potential as a CNS drug. *CNS Drug Rev, 12*, 1–8.

54. Vassar, R., Bennett, B.D., Babu-Khan, S., Kahn, S., Mendiaz, E.A., Denis, P., Teplow, D.B., Ross, S., Amarante, P., Loeloff, R., et al. (1999). Beta-secretase celavage of Alzheimer's amyloid precursor protein by the transmembrane aspartic protease BACE. *Science, 286*, 735–740.

55. Sinha, S., Anderson, J.P., Barbour, R., Basi, G.S., Caccavello, R., Davis, D., Doan, M., Dovey, H.F., Frigon, N., Hong, J., et al. (1999). Purification and cloning of amyloid precursor protein beta-secretase from human brain. *Nature, 402*, 537–540.

56. Hussain, I., Powell, D., Howlett, D.R., Tew, D.G., Meek, T.D., Chapman, C., Gloger, I.S., Murphy, K.E., Southan, C.D., Ryan, D.M., et al. (1999). Identification of a novel aspartic protease (Asp 2) as beta-secretase. *Mol Cell Neurosci, 14*, 419–427.

57. Yan, R., Bienkowski, M.J., Shuck, M.E., Miao, H., Tory, M.C., Pauley, A.M., Brashier, J.R., Stratman, N.C., Mathews, W.R., Buhl, A.E., et al. (1999). Membrane-anchored aspartyl protease with Alzheimer's disease beta- secretase activity. *Nature, 402*, 533–537.

58. Luo, Y., Bolon, B., Kahn, S., Bennett, B.D., Babu-Khan, S., Denis, P., Fan, W., Kha, H., Zhang, J., Gong, Y., et al. (2001). Mice deficient in BACE1, the Alzheimer's beta-secretase, have normal phenotype and abolished beta-amyloid generation. *Nat Neurosci, 4*, 231–232.

59. Cai, H., Wang, Y., McCarthy, D., Wen, H., Borchelt, D.R., Price, D.L., Wong, P.C. (2001). BACE1 is the major beta-secretase for generation of Abeta peptides by neurons. *Nat Neurosci, 4*, 233–234.

60. Li, Q., Sudhof, T.C. (2004). Cleavage of amyloid-beta precursor protein and amyloid-beta precursor-like protein by BACE 1. *J Biol Chem, 279*, 10542–10550.

61. Pastorino, L., Ikin, A.F., Lamprianou, S., Vacaresse, N., Revelli, J.P., Platt, K., Paganetti, P., Mathews, P.M., Harroch, S., Buxbaum, J.D. (2004). BACE (beta-secretase) modulates the processing of APLP2 in vivo. *Mol Cell Neurosci, 25*, 642–649.

62. Lichtenthaler, S.F., Dominguez, D.I., Westmeyer, G.G., Reiss, K., Haass, C., Saftig, P., De Strooper, B., Seed, B. (2003). The cell adhesion protein P-selectin glycoprotein ligand-1 is a substrate for the aspartyl protease BACE1. *J Biol Chem, 278*, 48713–48719.

63. von Arnim, C.A., Kinoshita, A., Peltan, I.D., Tangredi, M.M., Herl, L., Lee, B.M., Spoelgen, R., Hshieh, T.T., Ranganathan, S., Battey, F.D., et al. (2005). The low density lipoprotein receptor-related protein (LRP) is a novel beta-secretase (BACE1) substrate. *J Biol Chem, 280*, 17777–17785.

64. Willem, M., Garratt, A.N., Novak, B., Citron, M., Kaufmann, S., Rittger, A., DeStrooper, B., Saftig, P., Birchmeier, C., Haass, C. (2006). Control of peripheral nerve myelination by the beta-secretase BACE1. *Science, 314*, 664–666.

65. Laird, F.M., Cai, H., Savonenko, A.V., Farah, M.H., He, K., Melnikova, T., Wen, H., Chiang, H.C., Xu, G., Koliatsos, V.E., et al. (2005). BACE1, a major determinant of selective vulnerability of the brain to amyloid-beta amyloidogenesis, is essential for cognitive, emotional, and synaptic functions. *J Neurosci, 25*, 11693–11709.

66. Dominguez, D., Tournoy, J., Hartmann, D., Huth, T., Cryns, K., Deforce, S., Serneels, L., Camacho, I.E., Marjaux, E., Craessaerts, K., et al. (2005). Phenotypic and biochemical analyses of BACE1- and BACE2-deficient mice. *J Biol Chem, 280*, 30797–30806.

67. Golde, T.E., Eckman, C.B. (2003). Physiologic and pathologic events mediated by intramembranous and juxtamembranous proteolysis. *Sci STKE, 2003*, RE4.

68. Siemers, E., Skinner, M., Dean, R.A., Gonzales, C., Satterwhite, J., Farlow, M., Ness, D., May, P.C. (2005). Safety, tolerability, and changes in amyloid beta

concentrations after administration of a gamma-secretase inhibitor in volunteers. *Clin Neuropharmacol, 28*, 126–132.

69. Li, Y.M., Xu, M., Lai, M.T., Huang, Q., Castro, J.L., DiMuzio-Mower, J., Harrison, T., Lellis, C., Nadin, A., Neduvelil, J.G., et al. (2000). Photoactivated gamma-secretase inhibitors directed to the active site covalently label presenilin 1. *Nature, 405*, 689–694.
70. Esler, W.P., Kimberly, W.T., Ostaszewski, B.L., Diehl, T.S., Moore, C.L., Tsai, J.Y., Rahmati, T., Xia, W., Selkoe, D.J., Wolfe, M.S. (2000). Transition-state analogue inhibitors of gamma-secretase bind directly to presenilin-1. *Nat Cell Biol, 2*, 428–434.
71. Seiffert, D., Bradley, J.D., Rominger, C.M., Rominger, D.H., Yang, F., Meredith, J.E., Jr., Wang, Q., Roach, A.H., Thompson, L.A., Spitz, S.M., et al. (2000). Presenilin-1 and -2 are molecular targets for gamma-secretase inhibitors. *J Biol Chem, 275*, 34086–34091.
72. Lanz, T.A., Himes, C.S., Pallante, G., Adams, L., Yamazaki, S., Amore, B., Merchant, K.M. (2003). The gamma-secretase inhibitor *N*-[*N*-(3,5-difluorophenacetyl)-*L*-alanyl]-*S*-phenylglycine *t*-butyl ester reduces A beta levels in vivo in plasma and cerebrospinal fluid in young (plaque-free) and aged (plaque-bearing) Tg2576 mice. *J Pharmacol Exp Ther, 305*, 864–871.
73. Wong, G.T., Manfra, D., Poulet, F.M., Zhang, Q., Josien, H., Bara, T., Engstrom, L., Pinzon-Ortiz, M., Fine, J.S., Lee, H.J., et al. (2004). Chronic treatment with the gamma-secretase inhibitor LY-411,575 inhibits Abeta production and alters lymphopoiesis and intestinal cell differentiation. *J Biol Chem, 279*, 12876–12882.
74. Siemers, E.R., Quinn, J.F., Kaye, J., Farlow, M.R., Porsteinsson, A., Tariot, P., Zoulnouni, P., Galvin, J.E., Holtzman, D.M., Knopman, D.S., et al. (2006). Effects of a gamma-secretase inhibitor in a randomized study of patients with Alzheimer disease. *Neurology, 66*, 602–604.
75. Yankner, B.A., Dawes, L.R., Fisher, S., Villa-Komaroff, L., Oster-Granite, M.L., Neve, R.L. (1989). Neurotoxicity of a fragment of the amyloid precursor associated with Alzheimer's disease. *Science, 245*, 417–420.
76. Netzer, W.J., Dou, F., Cai, D., Veach, D., Jean, S., Li, Y., Bornmann, W.G., Clarkson, B., Xu, H., Greengard, P. (2003). Gleevec inhibits beta-amyloid production but not Notch cleavage. *Proc Natl Acad Sci U S A, 100*, 12444–12449.
77. Fraering, P.C., Ye, W., LaVoie, M.J., Ostaszewski, B.L., Selkoe, D.J., Wolfe, M.S. (2005). Gamma-secretase substrate selectivity can be modulated directly via interaction with a nucleotide-binding site. *J Biol Chem, 280*, 41987–41996.
78. Eriksen, J.L., Sagi, S.A., Smith, T.E., Weggen, S., Das, P., McLendon, D.C., Ozols, V.V., Jessing, K.W., Zavitz, K.H., Koo, E.H., Golde, T.E. (2003). NSAIDs and enantiomers of flurbiprofen target gamma-secretase and lower Abeta 42 in vivo. *J Clin Invest, 112*, 440–449.
79. Lleó, A., Berezovska, O., Herl, L., Raju, S., Deng, A., Bacskai, B.J., Frosch, M.P., Irizarry, M., Hyman, B.T. (2004). Nonsteroidal anti-inflammatory drugs lower Abeta42 and change presenilin 1 conformation. *Nat Med, 10*, 1065–1066.
80. Weggen, S., Eriksen, J.L., Das, P., Sagi, S.A., Wang, R., Pietrzik, C.U., Findlay, K.A., Smith, T.E., Murphy, M.P., Bulter, T., et al. (2001). A subset of NSAIDs

lower amyloidogenic Abeta42 independently of cyclooxygenase activity. *Nature, 414*, 212–216.

81. Weggen, S., Eriksen, J.L., Sagi, S.A., Pietrzik, C.U., Ozols, V., Fauq, A., Golde, T.E., Koo, E.H. (2003). Evidence that nonsteroidal anti-inflammatory drugs decrease amyloid beta 42 production by direct modulation of gamma-secretase activity. *J Biol Chem, 278*, 31831–31837.
82. Kukar, T., Murphy, M.P., Eriksen, J.L., Sagi, S.A., Weggen, S., Smith, T.E., Ladd, T., Khan, M.A., Kache, R., Beard, J., et al. (2005). Diverse compounds mimic Alzheimer disease–causing mutations by augmenting Abeta42 production. *Nat Med, 11*, 545–550.
83. Kim, J., Onstead, L., Randle, S., Price, R., Smithson, L., Zwizinski, C., Dickson, D.W., Golde, T., McGowan, E. (2007). Abeta40 inhibits amyloid deposition in vivo. *J Neurosci, 27*, 627–633.
84. Galasko, D.R., Graff-Radford, N., May, S., Hendrix, S., Cottrell, B.A., Sagi, S.A., Mather, G., Laughlin, M., Zavitz, K.H., Swabb, E., et al. (2007). Safety, tolerability, pharmacokinetics, and Abeta levels after short-term administration of *R*-flurbiprofen in healthy elderly individuals. *Alzheimer Dis Assoc Disord, 21*, 292–299.
85. Geerts, H. (2007). Drug evaluation: (*R*)-flurbiprofen—an enantiomer of flurbiprofen for the treatment of Alzheimer's disease. *IDrugs, 10*, 121–133.
86. Russo-Neustadt, A., Cotman, C.W. (1997). Adrenergic receptors in Alzheimer's disease brain: selective increases in the cerebella of aggressive patients. *J Neurosci, 17*, 5573–5580.
87. Ni, Y., Zhao, X., Bao, G., Zou, L., Teng, L., Wang, Z., Song, M., Xiong, J., Bai, Y., Pei, G. (2006). Activation of beta2-adrenergic receptor stimulates gamma-secretase activity and accelerates amyloid plaque formation. *Nat Med, 12*, 1390–1396.
88. Wang, J., Ho, L., Chen, L., Zhao, Z., Zhao, W., Qian, X., Humala, N., Seror, I., Bartholomew, S., Rosendorff, C., Pasinetti, G.M. (2007). Valsartan lowers brain beta-amyloid protein levels and improves spatial learning in a mouse model of Alzheimer disease. *J Clin Invest, 117*, 3393–3402.
89. Khachaturian, A.S., Zandi, P.P., Lyketsos, C.G., Hayden, K.M., Skoog, I., Norton, M.C., Tschanz, J.T., Mayer, L.S., Welsh-Bohmer, K.A., Breitner, J.C. (2006). Antihypertensive medication use and incident Alzheimer disease: the Cache County Study. *Arch Neurol, 63*, 686–692.
90. Cheng, H., Vetrivel, K.S., Gong, P., Meckler, X., Parent, A., Thinakaran, G. (2007). Mechanisms of disease: new therapeutic strategies for Alzheimer's disease—targeting APP processing in lipid rafts. *Nat Clin Pract Neurol, 3*, 374–382.
91. De Strooper, B., Saftig, P., Craessaerts, K., Vanderstichele, H., Guhde, G., Annaert, W., Von Figura, K., Van Leuven, F. (1998). Deficiency of presenilin-1 inhibits the normal cleavage of amyloid precursor protein. *Nature, 391*, 387–390.
92. Kuriyama, S., Hozawa, A., Ohmori, K., Shimazu, T., Matsui, T., Ebihara, S., Awata, S., Nagatomi, R., Arai, H., Tsuji, I. (2006). Green tea consumption and cognitive function: a cross-sectional study from the Tsurugaya Project 1. *Am J Clin Nutr, 83*, 355–361.
93. Utsuki, T., Yu, Q.S., Davidson, D., Chen, D., Holloway, H.W., Brossi, A., Sambamurti, K., Lahiri, D.K., Greig, N.H., Giordano, T. (2006). Identification of

novel small molecule inhibitors of amyloid precursor protein synthesis as a route to lower Alzheimer's disease amyloid-beta peptide. *J Pharmacol Exp Ther, 318*, 855–862.

94. Greig, N.H., Ruckle, J., Comer, P., Brownell, L., Holloway, H.W., Flanagan, D.R., Jr., Canfield, C.J., Burford, R.G. (2005). Anticholinesterase and pharmacokinetic profile of phenserine in healthy elderly human subjects. *Curr Alzheimer Res, 2*, 483–492.
95. Rogers, J.T., Randall, J.D., Eder, P.S., Huang, X., Bush, A.I., Tanzi, R.E., Venti, A., Payton, S.M., Giordano, T., Nagano, S., et al. (2002). Alzheimer's disease drug discovery targeted to the APP mRNA 5′ untranslated region. *J Mol Neurosci, 19*, 77–82.
96. Tucker, S., Ahl, M., Cho, H.H., Bandyopadhyay, S., Cuny, G.D., Bush, A.I., Goldstein, L.E., Westaway, D., Huang, X., Rogers, J.T. (2006). RNA therapeutics directed to the non coding regions of APP mRNA, in vivo anti-amyloid efficacy of paroxetine, erythromycin, and *N*-acetyl cysteine. *Curr Alzheimer Res, 3*, 221–227.
97. Walsh, D.M., Selkoe, D.J. (2004). Deciphering the molecular basis of memory failure in Alzheimer's disease. *Neuron, 44*, 181–193.
98. Soto, C., Estrada, L. (2005). Amyloid inhibitors and beta-sheet breakers. *Subcell Biochem, 38*, 351–364.
99. McLaurin, J., Franklin, T., Zhang, X., Deng, J., Fraser, P.E. (1999). Interactions of Alzheimer amyloid-beta peptides with glycosaminoglycans effects on fibril nucleation and growth. *Eur J Biochem, 266*, 1101–1110.
100. Huang, T.H., Yang, D.S., Plaskos, N.P., Go, S., Yip, C.M., Fraser, P.E., Chakrabartty, A. (2000). Structural studies of soluble oligomers of the Alzheimer beta-amyloid peptide. *J Mol Biol, 297*, 73–87.
101. Huang, T.H., Yang, D.S., Fraser, P.E., Chakrabartty, A. (2000). Alternate aggregation pathways of the Alzheimer beta-amyloid peptide. an in vitro model of preamyloid. *J Biol Chem, 275*, 36436–36440.
102. Gervais, F., Paquette, J., Morissette, C., Krzywkowski, P., Yu, M., Azzi, M., Lacombe, D., Kong, X., Aman, A., Laurin, J., et al. (2007). Targeting soluble Abeta peptide with Tramiprosate for the treatment of brain amyloidosis. *Neurobiol Aging, 28*, 537–547.
103. Aisen, P.S., Saumier, D., Briand, R., Laurin, J., Gervais, F., Tremblay, P., Garceau, D. (2006). A phase II study targeting amyloid-beta with 3APS in mild-to-moderate Alzheimer disease. *Neurology, 67*, 1757–1763.
104. McLaurin, J., Yang, D., Yip, C.M., Fraser, P.E. (2000). Review: modulating factors in amyloid-beta fibril formation. *J Struct Biol, 130*, 259–270.
105. Yao, Z.X., Brown, R.C., Teper, G., Greeson, J., Papadopoulos, V. (2002). 22*R*-Hydroxycholesterol protects neuronal cells from beta-amyloid-induced cytotoxicity by binding to beta-amyloid peptide. *J Neurochem, 8*, 1110–1119.
106. Curtain, C.C., Ali, F., Volitakis, I., Cherny, R.A., Norton, R.S., Beyreuther, K., Barrow, C.J., Masters, C.L., Bush, A.I., Barnham, K.J. (2001). Alzheimer's disease amyloid-beta binds copper and zinc to generate an allosterically ordered membrane-penetrating structure containing superoxide dismutase-like subunits. *J Biol Chem, 276*, 20466–20473.
107. Cherny, R.A., Atwood, C.S., Xilinas, M.E., Gray, D.N., Jones, W.D., McLean, C.A., Barnham, K.J., Volitakis, I., Fraser, F.W., Kim, Y., et al. (2001). Treatment

with a copper–zinc chelator markedly and rapidly inhibits beta-amyloid accumulation in Alzheimer's disease transgenic mice. *Neuron, 30*, 665–676.

108. Ringman, J.M., Frautschy, S.A., Cole, G.M., Masterman, D.L., Cummings, J.L. (2005). A potential role of the curry spice curcumin in Alzheimer's disease. *Curr Alzheimer Res, 2*, 131–136.

109. Davalos, D., Grutzendler, J., Yang, G., Kim, J.V., Zuo, Y., Jung, S., Littman, D.R., Dustin, M.L., Gan, W.B. (2005). ATP mediates rapid microglial response to local brain injury in vivo. *Nat Neurosci, 8*, 752–758.

110. Frautschy, S.A., Hu, W., Kim, P., Miller, S.A., Chu, T., Harris-White, M.E., Cole, G.M. (2001). Phenolic anti-inflammatory antioxidant reversal of Abeta-induced cognitive deficits and neuropathology. *Neurobiol Aging, 22*, 993–1005.

111. Lim, G.P., Chu, T., Yang, F., Beech, W., Frautschy, S.A., Cole, G.M. (2001). The curry spice curcumin reduces oxidative damage and amyloid pathology in an Alzheimer transgenic mouse. *J Neurosci, 21*, 8370–8377.

112. McLaurin, J., Kierstead, M.E., Brown, M.E., Hawkes, C.A., Lambermon, M.H., Phinney, A.L., Darabie, A.A., Cousins, J.E., French, J.E., Lan, M.F., et al. (2006). Cyclohexanehexol inhibitors of Abeta aggregation prevent and reverse Alzheimer phenotype in a mouse model. *Nat Med, 12*, 801–808.

113. Forloni, G., Colombo, L., Girola, L., Tagliavini, F., Salmona, M. (2001). Anti-amyloidogenic activity of tetracyclines: studies in vitro. *FEBS Lett, 487*, 404–407.

114. Tomiyama, T., Shoji, A., Kataoka, K., Suwa, Y., Asano, S., Kaneko, H., Endo, N. (1996). Inhibition of amyloid beta protein aggregation and neurotoxicity by rifampicin: its possible function as a hydroxyl radical scavenger. *J Biol Chem, 271*, 6839–6844.

115. Loeb, M.B., Molloy, D.W., Smieja, M., Standish, T., Goldsmith, C.H., Mahony, J., Smith, S., Borrie, M., Decoteau, E., Davidson, W., et al. (2004). A randomized, controlled trial of doxycycline and rifampin for patients with Alzheimer's disease. *J Am Geriatr Soc, 52*, 381–387.

116. Permanne, B., Adessi, C., Saborio, G.P., Fraga, S., Frossard, M.J., Van Dorpe, J., Dewachter, I., Banks, W.A., Van Leuven, F., Soto, C. (2002). Reduction of amyloid load and cerebral damage in a transgenic mouse model of Alzheimer's disease by treatment with a beta-sheet breaker peptide. *FASEB J, 16*, 860–862.

117. Sigurdsson, E.M., Permanne, B., Soto, C., Wisniewski, T., Frangione, B. (2000). In vivo reversal of amyloid-beta lesions in rat brain. *J Neuropathol Exp Neurol, 59*, 11–17.

118. Gordon, D.J., Sciarretta, K.L., Meredith, S.C. (2001). Inhibition of beta-amyloid(40) fibrillogenesis and disassembly of beta-amyloid(40) fibrils by short beta-amyloid congeners containing *N*-methyl amino acids at alternate residues. *Biochemistry, 40*, 8237–8245.

119. Soto, C., Sigurdsson, E.M., Morelli, L., Kumar, R.A., Castaño, E.M., Frangione, B. (1998). Beta-sheet breaker peptides inhibit fibrillogenesis in a rat brain model of amyloidosis: implications for Alzheimer's therapy. *Nat Med, 4*, 822–826.

120. Chacán, M.A., Barría, M.I., Soto, C., Inestrosa, N.C. (2004). Beta-sheet breaker peptide prevents Abeta-induced spatial memory impairments with partial reduction of amyloid deposits. *Mol Psychiatry, 9*, 953–961.

121. Sadowski, M., Pankiewicz, J., Scholtzova, H., Ripellino, J.A., Li, Y., Schmidt, S.D., Mathews, P.M., Fryer, J.D., Holtzman, D.M., Sigurdsson, E.M., Wisniewski, T. (2004). A synthetic peptide blocking the apolipoprotein E/beta-amyloid binding mitigates beta-amyloid toxicity and fibril formation in vitro and reduces beta-amyloid plaques in transgenic mice. *Am J Pathol, 165*, 937–948.

122. Sadowski, M.J., Pankiewicz, J., Scholtzova, H., Mehta, P.D., Prelli, F., Quartermain, D., Wisniewski, T. (2006). Blocking the apolipoprotein E/amyloid-beta interaction as a potential therapeutic approach for Alzheimer's disease. *Proc Natl Acad Sci U S A, 103*, 18787–18792.

123. Glenner, G.G. (1988). Alzheimer's disease: its proteins and genes. *Cell, 52*, 307–308.

124. Masters, C.L., Simms, G., Weinman, N.A., Multhaup, G., McDonald, B.L., Beyreuther, K. (1985). Amyloid plaque core protein in Alzheimer disease and Down syndrome. *Proc Natl Acad Sci U S A, 82*, 4245–4249.

125. Jankowsky, J.L., Slunt, H.H., Gonzales, V., Savonenko, A.V., Wen, J.C., Jenkins, N.A., Copeland, N.G., Younkin, L.H., Lester, H.A., Younkin, S.G., Borchelt, D.R. (2005). Persistent amyloidosis following suppression of Abeta production in a transgenic model of Alzheimer disease. *PLoS Med, 2*, e355.

126. Bateman, R.J., Munsell, L.Y., Morris, J.C., Swarm, R., Yarasheski, K.E., Holtzman, D.M. (2006). Human amyloid-beta synthesis and clearance rates as measured in cerebrospinal fluid in vivo. *Nat Med, 12*, 856–861.

127. Qiu, W.Q., Walsh, D.M., Ye, Z., Vekrellis, K., Zhang, J., Podlisny, M.B., Rosner, M.R., Safavi, A., Hersh, L.B., Selkoe, D.J. (1998). Insulin-degrading enzyme regulates extracellular levels of amyloid beta-protein by degradation. *J Biol Chem, 273*, 32730–32738.

128. Iwata, N., Tsubuki, S., Takaki, Y., Shirotani, K., Lu, B., Gerard, N.P., Gerard, C., Hama, E., Lee, H.J., Saido, T.C. (2001). Metabolic regulation of brain Abeta by neprilysin. *Science, 292*, 1550–1552.

129. Takaki, Y., Iwata, N., Tsubuki, S., Taniguchi, S., Toyoshima, S., Lu, B., Gerard, N.P., Gerard, C., Lee, H.J., Shirotani, K., Saido, T.C. (2000). Biochemical identification of the neutral endopeptidase family member responsible for the catabolism of amyloid beta peptide in the brain. *J Biochem, 128*, 897–902.

130. Iwata, N., Tsubuki, S., Takaki, Y., Watanabe, K., Sekiguchi, M., Hosoki, E., Kawashima-Morishima, M., Lee, H.J., Hama, E., Sekine-Aizawa, Y., Saido, T.C. (2000). Identification of the major Abeta1-42-degrading catabolic pathway in brain parenchyma: suppression leads to biochemical and pathological deposition. *Nat Med, 6*, 143–150.

131. Eckman, E.A., Reed, D.K., Eckman, C.B. (2001). Degradation of the Alzheimer's amyloid beta peptide by endothelin-converting enzyme. *J Biol Chem, 276*, 24540–24548.

132. Eckman, E.A., Watson, M., Marlow, L., Sambamurti, K., Eckman, C.B. (2003). Alzheimer's disease beta-amyloid peptide is increased in mice deficient in endothelin-converting enzyme. *J Biol Chem, 278*, 2081–2084.

133. Shirotani, K., Tsubuki, S., Iwata, N., Takaki, Y., Harigaya, W., Maruyama, K., Kiryu-Seo, S., Kiyama, H., Iwata, H., Tomita, T., et al. (2001). Neprilysin degrades both amyloid beta peptides 1–40 and 1–42 most rapidly and efficiently

among thiorphan- and phosphoramidon-sensitive endopeptidases. *J Biol Chem, 276*, 21895–21901.

134. Miller, V.M., Xia, H., Marrs, G.L., Gouvion, C.M., Lee, G., Davidson, B.L., Paulson, H.L. (2003). Allele-specific silencing of dominant disease genes. *Proc Natl Acad Sci U S A, 100*, 7195–7200.
135. Mohajeri, M.H., Kuehnle, K., Li, H., Poirier, R., Tracy, J., Nitsch, R.M. (2004). Anti-amyloid activity of neprilysin in plaque-bearing mouse models of Alzheimer's disease. *FEBS Lett, 562*, 16–21.
136. Leissring, M.A., Farris, W., Chang, A.Y., Walsh, D.M., Wu, X., Sun, X., Frosch, M.P., Selkoe, D.J. (2003). Enhanced proteolysis of beta-amyloid in APP transgenic mice prevents plaque formation, secondary pathology, and premature death. *Neuron, 40*, 1087–1093.
137. Tucker, H.M., Kihiko-Ehmann, M., Wright, S., Rydel, R.E., Estus, S. (2000). Tissue plasminogen activator requires plasminogen to modulate amyloid-beta neurotoxicity and deposition. *J Neurochem, 75*, 2172–2177.
138. Tucker, H.M., Kihiko, M., Caldwell, J.N., Wright, S., Kawarabayashi, T., Price, D., Walker, D., Scheff, S., McGillis, J.P., Rydel, R.E., Estus, S. (2000). The plasmin system is induced by and degrades amyloid-beta aggregates. *J Neurosci, 20*, 3937–3946.
139. Ledesma, M.D., Da Silva, J.S., Crassaerts, K., Delacourte, A., De Strooper, B., Dotti, C.G. (2000). Brain plasmin enhances APP alpha-cleavage and Abeta degradation and is reduced in Alzheimer's disease brains. *EMBO Rep, 1*, 530–535.
140. Ertekin-Taner, N., Ronald, J., Feuk, L., Prince, J., Tucker, M., Younkin, L., Hella, M., Jain, S., Hackett, A., Scanlin, L., et al. (2005). Elevated amyloid beta protein (Abeta42) and late onset Alzheimer's disease are associated with single nucleotide polymorphisms in the urokinase-type plasminogen activator gene. *Hum Mol Genet, 14*, 447–460.
141. Tucker, H.M., Simpson, J., Kihiko-Ehmann, M., Younkin, L.H., McGillis, J.P., Younkin, S.G., Degen, J.L., Estus, S. (2004). Plasmin deficiency does not alter endogenous murine amyloid beta levels in mice. *Neurosci Lett, 368*, 285–289.
142. Medina, M.G., Ledesma, M.D., Domínguez, J.E., Medina, M., Zafra, D., Alameda, F., Dotti, C.G., Navarro, P. (2005). Tissue plasminogen activator mediates amyloid-induced neurotoxicity via Erk1/2 activation. *EMBO J, 24*, 1706–17016.
143. Russo, R., Borghi, R., Markesbery, W., Tabaton, M., Piccini, A. (2005). Neprylisin decreases uniformly in Alzheimer's disease and in normal aging. *FEBS Lett, 579*, 6027–6030.
144. Mohajeri, M.H., Wollmer, M.A., Nitsch, R.M. (2002). Abeta 42-induced increase in neprilysin is associated with prevention of amyloid plaque formation in vivo. *J Biol Chem, 277*, 35460–35465.
145. Saito, T., Iwata, N., Tsubuki, S., Takaki, Y., Takano, J., Huang, S.M., Suemoto, T., Higuchi, M., Saido, T.C. (2005). Somatostatin regulates brain amyloid beta peptide Abeta42 through modulation of proteolytic degradation. *Nat Med, 11*, 434–439.
146. Craft, S. (2005). Insulin resistance syndrome and Alzheimer's disease: age- and obesity-related effects on memory, amyloid, and inflammation. *Neurobiol Aging, 26*, 65–69.

147. Watson, G.S., Peskind, E.R., Asthana, S., Purganan, K., Wait, C., Chapman, D., Schwartz, M.W., Plymate, S., Craft, S. (2003). Insulin increases CSF Abeta42 levels in normal older adults. *Neurology, 60*, 1899–1903.

148. Reger, M.A., Watson, G.S., Green, P.S., Wilkinson, C.W., Baker, L.D., Cholerton, B., Fishel, M.A., Plymate, S.R., Breitner, J.C., DeGroodt, W., et al. (2008). Intranasal insulin improves cognition and modulates {beta}-amyloid in early AD. *Neurology, 70*, 440–448.

149. Benedict, C., Hallschmid, M., Schmitz, K., Schultes, B., Ratter, F., Fehm, H.L., Born, J., Kern, W. (2007). Intranasal insulin improves memory in humans: superiority of insulin aspart. *Neuropsychopharmacology, 32*, 239–243.

150. Potter, H., Wefes, I.M., Nilsson, L.N. (2001). The inflammation-induced pathological chaperones ACT and apo-E are necessary catalysts of Alzheimer amyloid formation. *Neurobiol Aging, 22*, 923–930.

151. DeMattos, R.B., O'Dell, M.A., Parsadanian, M., Taylor, J.W., Harmony, J.A., Bales, K.R., Paul, S.M., Aronow, B.J., Holtzman, D.M. (2002). Clusterin promotes amyloid plaque formation and is critical for neuritic toxicity in a mouse model of Alzheimer's disease. *Proc Natl Acad Sci U S A, 99*, 10843–10848.

152. Wisniewski, T., Frangione, B. (2005). Immunological and anti-chaperone therapeutic approaches for Alzheimer disease. *Brain Pathol, 15*, 72–77.

153. Holtzman, D.M. (2004). In vivo effects of ApoE and clusterin on amyloid-beta metabolism and neuropathology. *J Mol Neurosci, 23*, 247–254.

154. Strittmatter, W.J., Roses, A.D. (1995). Apolipoprotein E and Alzheimer disease. *Proc Natl Acad Sci U S A, 92*, 4725–4727.

155. Koistinaho, M., Lin, S., Wu, X., Esterman, M., Koger, D., Hanson, J., Higgs, R., Liu, F., Malkani, S., Bales, K.R., Paul, S.M. (2004). Apolipoprotein E promotes astrocyte colocalization and degradation of deposited amyloid-beta peptides. *Nat Med, 10*, 719–726.

156. Dodart, J.C., Marr, R.A., Koistinaho, M., Gregersen, B.M., Malkani, S., Verma, I.M., Paul, S.M. (2005). Gene delivery of human apolipoprotein E alters brain Abeta burden in a mouse model of Alzheimer's disease. *Proc Natl Acad Sci U S A, 102*, 1211–1216.

157. Holtzman, D.M., Fagan, A.M., Mackey, B., Tenkova, T., Sartorius, L., Paul, S.M., Bales, K., Ashe, K.H., Irizarry, M.C., Hyman, B.T. (2000). Apolipoprotein E facilitates neuritic and cerebrovascular plaque formation in an Alzheimer's disease model. *Ann Neurol, 47*, 739–747.

158. Holtzman, D.M., Bales, K.R., Wu, S., Bhat, P., Parsadanian, M., Fagan, A.M., Chang, L.K., Sun, Y., Paul, S.M. (1999). Expression of human apolipoprotein E reduces amyloid-beta deposition in a mouse model of Alzheimer's disease. *J Clin Invest, 103*, R15–R21.

159. Matsubara, E., Frangione, B., Ghiso, J. (1995). Characterization of apolipoprotein J-Alzheimer's A beta interaction. *J Biol Chem, 270*, 7563–7567.

160. Oda, T., Wals, P., Osterburg, H.H., Johnson, S.A., Pasinetti, G.M., Morgan, T.E., Rozovsky, I., Stine, W.B., Snyder, S.W., Holzman, T.F., et al. (1995). Clusterin (apoJ) alters the aggregation of amyloid beta-peptide (A beta 1–42) and forms slowly sedimenting A beta complexes that cause oxidative stress. *Exp Neurol, 136*, 22–31.

161. Abraham, C.R., Potter, H. (1989). Alpha 1-antichymotrypsin in brain aging and disease. *Prog Clin Biol Res, 317*, 1037–1048.

162. Nilsson, L.N., Bales, K.R., DiCarlo, G., Gordon, M.N., Morgan, D., Paul, S.M., Potter, H. (2001). Alpha-1-antichymotrypsin promotes beta-sheet amyloid plaque deposition in a transgenic mouse model of Alzheimer's disease. *J Neurosci, 21*, 1444–1451.

163. Narindrasorasak, S., Lowery, D., Gonzalez-DeWhitt, P., Poorman, R.A., Greenberg, B., Kisilevsky, R. (1991). High affinity interactions between the Alzheimer's beta-amyloid precursor proteins and the basement membrane form of heparan sulfate proteoglycan. *J Biol Chem, 266*, 12878–12883.

164. Snow, A.D., Sekiguchi, R., Nochlin, D., Fraser, P., Kimata, K., Mizutani, A., Arai, M., Schreier, W.A., Morgan, D.G. (1994). An important role of heparan sulfate proteoglycan (Perlecan) in a model system for the deposition and persistence of fibrillar A beta-amyloid in rat brain. *Neuron, 12*, 219–234.

165. Fraser, P.E., Nguyen, J.T., Chin, D.T., Kirschner, D.A. (1992). Effects of sulfate ions on Alzheimer beta/A4 peptide assemblies: implications for amyloid fibril–proteoglycan interactions. *J Neurochem, 59*, 1531–1540.

166. Bergamaschini, L., Rossi, E., Storini, C., Pizzimenti, S., Distaso, M., Perego, C., De Luigi, A., Vergani, C., De Simoni, M.G. (2004). Peripheral treatment with enoxaparin, a low molecular weight heparin, reduces plaques and beta-amyloid accumulation in a mouse model of Alzheimer's disease. *J Neurosci, 24*, 4181–4186.

167. Pastorino, L., Sun, A., Lu, P.J., Zhou, X.Z., Balastik, M., Finn, G., Wulf, G., Lim, J., Li, S.H., Li, X., et al. (2006). The prolyl isomerase Pin1 regulates amyloid precursor protein processing and amyloid-beta production. *Nature, 440*, 528–534.

168. Segat, L., Pontillo, A., Annoni, G., Trabattoni, D., Vergani, C., Clerici, M., Arosio, B., Crovella, S. (2007). PIN1 promoter polymorphisms are associated with Alzheimer's disease. *Neurobiol Aging, 28*, 69–74.

169. Dodel, R.C., Du, Y., Depboylu, C., Hampel, H., Frölich, L., Haag, A., Hemmeter, U., Paulsen, S., Teipel, S.J., Brettschneider, S., et al. (2004). Intravenous immunoglobulins containing antibodies against beta-amyloid for the treatment of Alzheimer's disease. *J Neurol Neurosurg Psychiatry, 75*, 1472–1474.

170. Istrin, G., Bosis, E., Solomon, B. (2006). Intravenous immunoglobulin enhances the clearance of fibrillar amyloid-beta peptide. *J Neurosci Res, 84*, 434–443.

171. Zlokovic, B.V. (2004). Clearing amyloid through the blood–brain barrier. *J Neurochem, 89*, 807–811.

172. DeMattos, R.B., Bales, K.R., Cummins, D.J., Dodart, J.C., Paul, S.M., Holtzman, D.M. (2001). Peripheral anti-A beta antibody alters CNS and plasma A beta clearance and decreases brain A beta burden in a mouse model of Alzheimer's disease. *Proc Natl Acad Sci U S A, 98*, 8850–8855.

173. Lam, F.C., Liu, R., Lu, P., Shapiro, A.B., Renoir, J.M., Sharom, F.J., Reiner, P.B. (2001). Beta-amyloid efflux mediated by *p*-glycoprotein. *J Neurochem, 76*, 1121–1128.

174. Zlokovic, B.V., Deane, R., Sallstrom, J., Chow, N., Miano, J.M. (2005). Neurovascular pathways and Alzheimer amyloid beta-peptide. *Brain Pathol, 15*, 78–83.

175. Deane, R., Wu, Z., Zlokovic, B.V. (2004). RAGE (yin) versus LRP (yang) balance regulates alzheimer amyloid beta-peptide clearance through transport across the blood–brain barrier. *Stroke, 35*, 2628–2631.

176. Brendza, R.P., Bacskai, B.J., Cirrito, J.R., Simmons, K.A., Skoch, J.M., Klunk, W.E., Mathis, C.A., Bales, K.R., Paul, S.M., Hyman, B.T., Holtzman, D.M. (2005). Anti-Abeta antibody treatment promotes the rapid recovery of amyloid-associated neuritic dystrophy in PDAPP transgenic mice. *J Clin Invest, 115*, 428–433.

177. Wahrle, S., Das, P., Nyborg, A.C., McLendon, C., Shoji, M., Kawarabayashi, T., Younkin, L.H., Younkin, S.G., Golde, T.E. (2002). Cholesterol-dependent gamma-secretase activity in buoyant cholesterol-rich membrane microdomains. *Neurobiol Dis, 9*, 11–23.

178. Koldamova, R., Staufenbiel, M., Lefterov, I. (2005). Lack of ABCA1 considerably decreases brain ApoE level and increases amyloid deposition in APP23 mice. *J Biol Chem, 280*, 43224–43235.

179. Hirsch-Reinshagen, V., Maia, L.F., Burgess, B.L., Blain, J.F., Naus, K.E., McIsaac, S.A., Parkinson, P.F., Chan, J.Y., Tansley, G.H., Hayden, M.R., et al. (2005). The absence of ABCA1 decreases soluble ApoE levels but does not diminish amyloid deposition in two murine models of Alzheimer disease. *J Biol Chem, 280*, 43243–43256.

180. Sagare, A., Deane, R., Bell, R.D., Johnson, B., Hamm, K., Pendu, R., Marky, A., Lenting, P.J., Wu, Z., Zarcone, T., et al. (2007). Clearance of amyloid-beta by circulating lipoprotein receptors. *Nat Med, 13*, 1029–1031.

181. Schenk, D., Barbour, R., Dunn, W., Gordon, G., Grajeda, H., Guido, T., Hu, K., Huang, J., Johnson-Wood, K., et al. (1999). Immunization with amyloid-beta attenuates Alzheimer-disease-like pathology in the PDAPP mouse. *Nature, 400*, 173–177.

182. Janus, C., Pearson, J., McLaurin, J., Mathews, P.M., Jiang, Y., Schmidt, S.D., Chishti, M.A., Horne, P., Heslin, D., French, J., et al. (2000). A beta peptide immunization reduces behavioural impairment and plaques in a model of Alzheimer's disease. *Nature, 408*, 979–982.

183. Sigurdsson, E.M., Scholtzova, H., Mehta, P.D., Frangione, B., Wisniewski, T. (2001). Immunization with a nontoxic/nonfibrillar amyloid-beta homologous peptide reduces Alzheimer's disease-associated pathology in transgenic mice. *Am J Pathol, 159*, 439–447.

184. Lemere, C.A., Spooner, E.T., Leverone, J.F., Mori, C., Iglesias, M., Bloom, J.K., Seabrook, T.J. (2003). Amyloid-beta immunization in Alzheimer's disease transgenic mouse models and wildtype mice. *Neurochem Res, 28*, 1017–1027.

185. Frenkel, D., Katz, O., Solomon, B. (2000). Immunization against Alzheimer's beta-amyloid plaques via EFRH phage administration. *Proc Natl Acad Sci U S A, 97*, 11455–11459.

186. Lavie, V., Becker, M., Cohen-Kupiec, R., Yacoby, I., Koppel, R., Wedenig, M., Hutter-Paier, B., Solomon, B. (2004). EFRH-phage immunization of Alzheimer's disease animal model improves behavioral performance in Morris water maze trials. *J Mol Neurosci, 24*, 105–113.

187. McLaurin, J., Cecal, R., Kierstead, M.E., Tian, X., Phinney, A.L., Manea, M., French, J.E., Lambermon, M.H., Darabie, A.A., Brown, M.E., et al. (2002).

Therapeutically effective antibodies against amyloid-beta peptide target amyloid-beta residues 4–10 and inhibit cytotoxicity and fibrillogenesis. *Nat Med, 8*, 1263–1269.

188. Kotilinek, L.A., Bacskai, B., Westerman, M., Kawarabayashi, T., Younkin, L., Hyman, B.T., Younkin, S., Ashe, K.H. (2002). Reversible memory loss in a mouse transgenic model of Alzheimer's disease. *J Neurosci, 22*, 6331–6335.

189. Oddo, S., Billings, L., Kesslak, J.P., Cribbs, D.H., LaFerla, F.M. (2004). Abeta immunotherapy leads to clearance of early, but not late, hyperphosphorylated tau aggregates via the proteasome. *Neuron, 43*, 321–332.

190. Lee, E.B., Leng, L.Z., Zhang, B., Kwong, L., Trojanowski, J.Q., Abel, T., Lee, V.M. (2006). Targeting Abeta oligomers by passive immunization with a conformation selective monoclonal antibody improves learning and memory in APP transgenic mice. *J Biol Chem, 281*, 4292–4299.

191. Lee, E.B., et al. (2006). Targeting amyloid-beta peptide (Abeta) oligomers by passive immunization with a conformation-selective monoclonal antibody improves learning and memory in Abeta precursor protein (APP) transgenic mice. *J Biol Chem, 281*, 4292–4299.

192. Asuni, A.A., Boutajangout, A., Quartermain, D., Sigurdsson, E.M. (2007). Immunotherapy targeting pathological tau conformers in a tangle mouse model reduces brain pathology with associated functional improvements. *J Neurosci, 27*, 9115–9129.

193. Das, P., Murphy, M.P., Younkin, L.H., Younkin, S.G., Golde, T.E. (2001). Reduced effectiveness of Abeta1-42 immunization in APP transgenic mice with significant amyloid deposition. *Neurobiol Aging, 22*, 721–727.

194. Solomon, B. (2007). Clinical immunologic approaches for the treatment of Alzheimer's disease. *Expert Opin Investig Drugs, 16*, 819–828.

195. Bard, F., Barbour, R., Cannon, C., Carretto, R., Fox, M., Games, D., Guido, T., Hoenow, K., Hu, K., Johnson-Wood, K., et al. (2003). Epitope and isotype specificities of antibodies to beta-amyloid peptide for protection against Alzheimer's disease-like neuropathology. *Proc Natl Acad Sci U S A, 100*, 2023–2028.

196. Fukuchi, K., Accavitti-Loper, M.A., Kim, H.D., Tahara, K., Cao, Y., Lewis, T.L., Caughey, R.C., Kim, H., Lalonde, R. (2006). Amelioration of amyloid load by anti-Abeta single-chain antibody in Alzheimer mouse model. *Biochem Biophys Res Commun, 344*, 79–86.

197. Levites, Y., Das, P., Price, R.W., Rochette, M.J., Kostura, L.A., McGowan, E.M., Murphy, M.P., Golde, T.E. (2006). Anti-Abeta42- and anti-Abeta40-specific mAbs attenuate amyloid deposition in an Alzheimer disease mouse model. *J Clin Invest, 116*, 193–201.

198. Matsuoka, Y., Saito, M., LaFrancois, J., Saito, M., Gaynor, K., Olm, V., Wang, L., Casey, E., Lu, Y., Shiratori, C., et al. (2003). Novel therapeutic approach for the treatment of Alzheimer's disease by peripheral administration of agents with an affinity to beta-amyloid. *J Neurosci, 23*, 29–33.

199. Vasilevko, V., Xu, F., Previti, M.L., Van Nostrand, W.E., Cribbs, D.H. (2007). Experimental investigation of antibody-mediated clearance mechanisms of amyloid-beta in CNS of Tg-SwDI transgenic mice. *J Neurosci, 27*, 13376–13383.

200. Gilman, S., Koller, M., Black, R.S., Jenkins, L., Griffith, S.G., Fox, N.C., Eisner, L., Kirby, L., Rovira, M.B., Forette, F., Orgogozo, J.M; AN1792(QS-21)-201

Study Team. (2005). Clinical effects of A{beta} immunization (AN1792) in patients with AD in an interrupted trial. *Neurology, 64*, 1553–1562.

201. Orgogozo, J.M., Gilman, S., Dartigues, J.F., Laurent, B., Puel, M., Kirby, L.C., Jouanny, P., Dubois, B., Eisner, L., Flitman, S., et al. C. (2003). Subacute meningoencephalitis in a subset of patients with AD after Abeta42 immunization. *Neurology, 61*, 46–54.

202. Nicoll, J.A., Wilkinson, D., Holmes, C., Steart, P., Markham, H., Weller, R.O. (2003). Neuropathology of human Alzheimer disease after immunization with amyloid-beta peptide: a case report. *Nat Med, 9*, 448–452.

203. Ferrer, I., Boada Rovira, M., Sánchez Guerra, M.L., Rey, M.J., Costa-Jussá, F. (2004). Neuropathology and pathogenesis of encephalitis following amyloid-beta immunization in Alzheimer's disease. *Brain Pathol, 14*, 11–20.

204. Masliah, E., Hansen, L., Adame, A., Crews, L., Bard, F., Lee, C., Seubert, P., Games, D., Kirby, L., Schenk, D. (2005). Abeta vaccination effects on plaque pathology in the absence of encephalitis in Alzheimer disease. *Neurology, 64*, 129–131.

205. Fox, N.C., Black, R.S., Gilman, S., Rossor, M.N., Griffith, S.G., Jenkins, L., Koller, M.; AN1792(QS-21)-201 Study Team. (2005). Effects of A{beta} immunization (AN1792) on MRI measures of cerebral volume in Alzheimer disease. *Neurology, 64*, 1563–1572.

206. Hock, C., Konietzko, U., Streffer, J.R., Tracy, J., Signorell, A., Müller-Tillmanns, B., Lemke, U., Henke, K., Moritz, E., Garcia, E., et al. (2003). Antibodies against beta-amyloid slow cognitive decline in Alzheimer's disease. *Neuron, 38*, 547–554.

207. Agadjanyan, M.G., Ghochikyan, A., Petrushina, I., Vasilevko, V., Movsesyan, N., Mkrtichyan, M., Saing, T., Cribbs, D.H. (2005). Prototype Alzheimer's disease vaccine using the immunodominant B cell epitope from beta-amyloid and promiscuous T cell epitope pan HLA DR-binding peptide. *J Immunol, 174*, 1580–1586.

208. Maier, M., Seabrook, T.J., Lazo, N.D., Jiang, L., Das, P., Janus, C., Lemere, C.A. (2006). Short amyloid-beta (Abeta) immunogens reduce cerebral Abeta load and learning deficits in an Alzheimer's disease mouse model in the absence of an Abeta-specific cellular immune response. *J Neurosci, 26*, 4717–4728.

209. Lemere, C.A., Maron, R., Selkoe, D.J., Weiner, H.L. (2001). Nasal vaccination with beta-amyloid peptide for the treatment of Alzheimer's disease. *DNA Cell Biol, 20*, 705–711.

210. Seabrook, T.J., Jiang, L., Thomas, K., Lemere, C.A. (2006). Boosting with intranasal dendrimeric Abeta1-15 but not Abeta1-15 peptide leads to an effective immune response following a single injection of Abeta1-40/42 in APP-tg mice. *J Neuroinflammation, 3*, 14.

211. Nikolic, W.V., Bai, Y., Obregon, D., Hou, H., Mori, T., Zeng, J., Ehrhart, J., Shytle, R.D., Giunta, B., Morgan, D., et al. (2007). Transcutaneous beta-amyloid immunization reduces cerebral beta-amyloid deposits without T cell infiltration and microhemorrhage. *Proc Natl Acad Sci U S A, 104*, 2507–2512.

212. Seabrook, T.J., Thomas, K., Jiang, L., Bloom, J., Spooner, E., Maier, M., Bitan, G., Lemere, C.A. (2007). Dendrimeric Abeta1-15 is an effective immunogen in wildtype and APP-tg mice. *Neurobiol Aging, 28*, 813–823.

213. Solomon, B., Koppel, R., Frankel, D., Hanan-Aharon, E. (1997). Disaggregation of Alzheimer beta-amyloid by site-directed mAb. *Proc Natl Acad Sci U S A, 94*, 4109–4112.

214. Frenkel, D., Balass, M., Solomon, B. (1998). N-terminal EFRH sequence of Alzheimer's beta-amyloid peptide represents the epitope of its anti-aggregating antibodies. *J Neuroimmunol, 88*, 85–90.

215. Bacskai, B.J., Kajdasz, S.T., McLellan, M.E., Games, D., Seubert, P., Schenk, D., Hyman, B.T. (2002). Non-Fc-mediated mechanisms are involved in clearance of amyloid-beta in vivo by immunotherapy. *J Neurosci, 22*, 7873–7878.

216. Klyubin, I., Walsh, D.M., Lemere, C.A., Cullen, W.K., Shankar, G.M., Betts, V., Spooner, E.T., Jiang, L., Anwyl, R., Selkoe, D.J., Rowan, M.J. (2005). Amyloid beta protein immunotherapy neutralizes Abeta oligomers that disrupt synaptic plasticity in vivo. *Nat Med, 11*, 556–561.

217. Bard, F., Cannon, C., Barbour, R., Burke, R.L., Games, D., Grajeda, H., Guido, T., Hu, K., Huang, J., Johnson-Wood, K., et al. (2000). Peripherally administered antibodies against amyloid beta-peptide enter the central nervous system and reduce pathology in a mouse model of Alzheimer disease. *Nat Med, 6*, 916–919.

218. Das, P., Howard, V., Loosbrock, N., Dickson, D., Murphy, M.P., Golde, T.E. (2003). Amyloid-beta immunization effectively reduces amyloid deposition in FcRgamma$^{-/-}$knock-out mice. *J Neurosci, 23*, 8532–8538.

219. DeMattos, R.B., Bales, K.R., Cummins, D.J., Paul, S.M., Holtzman, D.M. (2002). Brain to plasma amyloid-beta efflux: a measure of brain amyloid burden in a mouse model of Alzheimer's disease. *Science, 295*, 2264–2267.

220. Sigurdsson, E.M., Knudsen, E., Asuni, A., Fitzer-Attas, C., Sage, D., Quartermain, D., Goni, F., Frangione, B., Wisniewski, T. (2004). An attenuated immune response is sufficient to enhance cognition in an Alzheimer's disease mouse model immunized with amyloid-beta derivatives. *J Neurosci, 24*, 6277–6282.

221. Levites, Y., Smithson, L.A., Price, R.W., Dakin, R.S., Yuan, B., Sierks, M.R., Kim, J., McGowan, E., Reed, D.K., Rosenberry, T.L., et al. (2006). Insights into the mechanisms of action of anti-Abeta antibodies in Alzheimer's disease mouse models. *FASEB J, 20*, 2576–2578.

222. Selkoe, D.J. (2001). Alzheimer's disease: genes, proteins, and therapy. *Physiol Rev, 81*, 741–766.

223. Lee, V.M.-Y., Trojanowski, J.Q. (1992). The disordered neuronal cytoskeleton in Alzheimer's disease. *Curr Opin Neurobiol, 2*, 653–656.

224. Glatz, D.C., Rujescu, D., Tang, Y., Berendt, F.J., Hartmann, A.M., Faltraco, F., Rosenberg, C., Hulette, C., Jellinger, K., Hampel, H., et al. (2006). The alternative splicing of tau exon 10 and its regulatory proteins CLK2 and TRA2-BETA1 changes in sporadic Alzheimer's disease. *J Neurochem, 96*, 635–644.

225. Cuadrado, A., García-Fernández, L.F., Imai, T., Okano, H., Muñoz, A. (2002). Regulation of tau RNA maturation by thyroid hormone is mediated by the neural RNA-binding protein musashi-1. *Mol Cell Neurosci, 20*, 198–210.

226. Donahue, J.E., Flaherty, S.L., Johanson, C.E., Duncan, J.A., 3rd, Silverberg, G.D., Miller, M.C., Tavares, R., Yang, W., Wu, Q., Sabo, E., et al. (2006). RAGE, LRP-1, and amyloid-beta protein in Alzheimer's disease. *Acta Neuropathol, 112*, 405–415.

227. Wischik, C.M., Edwards, P.C., Lai, R.Y., Roth, M., Harrington, C.R. (1996). Selective inhibition of Alzheimer disease-like tau aggregation by phenothiazines. *Proc Natl Acad Sci U S A, 93*, 11213–11218.

228. Khlistunova, I., Biernat, J., Wang, Y., Pickhardt, M., von Bergen, M., Gazova, Z., Mandelkow, E., Mandelkow, E.M. (2006). Inducible expression of Tau repeat domain in cell models of tauopathy: aggregation is toxic to cells but can be reversed by inhibitor drugs. *J Biol Chem, 281*, 1205–1214.

229. Masuda, M., Suzuki, N., Taniguchi, S., Oikawa, T., Nonaka, T., Iwatsubo, T., Hisanaga, S., Goedert, M., Hasegawa, M. (2006). Small molecule inhibitors of alpha-synuclein filament assembly. *Biochemistry, 45*, 6085–6094.

230. Andorfer, C., Acker, C.M., Kress, Y., Hof, P.R., Duff, K., Davies, P. (2005). Cell-cycle reentry and cell death in transgenic mice expressing nonmutant human tau isoforms. *J Neurosci, 25*, 5446–5454.

231. Wittmann, C.W., Wszolek, M.F., Shulman, J.M., Salvaterra, P.M., Lewis, J., Hutton, M., Feany, M.B. (2001). Tauopathy in *Drosophila:* neurodegeneration without neurofibrillary tangles. *Science, 293*, 711–714.

232. Yoshiyama, Y., Higuchi, M., Zhang, B., Huang, S.M., Iwata, N., Saido, T.C., Maeda, J., Suhara, T., Trojanowski, J.Q., Lee, V.M. (2007). Synapse loss and microglial activation precede tangles in a P301S tauopathy mouse model. *Neuron, 53*, 337–351.

233. Spires, T.L., Orne, J.D., SantaCruz, K., Pitstick, R., Carlson, G.A., Ashe, K.H., Hyman, B.T. (2006). Region-specific dissociation of neuronal loss and neurofibrillary pathology in a mouse model of tauopathy. *Am J Pathol, 168*, 1598–1607.

234. Billings, L.M., Oddo, S., Green, K.N., McGaugh, J.L., LaFerla, F.M. (2005). Intraneuronal Abeta causes the onset of early Alzheimer's disease–related cognitive deficits in transgenic mice. *Neuron, 45*, 675–688.

235. Eckermann, K., Mocanu, M.M., Khlistunova, I., Biernat, J., Nissen, A., Hofmann, A., Schönig, K., Bujard, H., Haemisch, A., Mandelkow, E., et al. (2007). The beta-propensity of tau determines aggregation and synaptic loss in inducible mouse models of tauopathy. *J Biol Chem, 282*, 31755–31765.

236. Arrasate, M., Mitra, S., Schweitzer, E.S., Segal, M.R., Finkbeiner, S. (2004). Inclusion body formation reduces levels of mutant huntingtin and the risk of neuronal death. *Nature, 431*, 805–810.

237. Yamamoto, A., Lucas, J.J., Hen, R. (2000). Reversal of neuropathology and motor dysfunction in a conditional model of Huntington's disease. *Cell, 101*, 57–66.

238. Noble, W., Olm, V., Takata, K., Casey, E., Mary, O., Meyerson, J., Gaynor, K., LaFrancois, J., Wang, L., Kondo, T., et al. (2003). Cdk5 is a key factor in tau aggregation and tangle formation in vivo. *Neuron, 38*, 555–565.

239. Plattner, F., Angelo, M., Giese, K.P. (2006). The roles of cyclin-dependent kinase 5 and glycogen synthase kinase 3 in tau hyperphosphorylation. *J Biol Chem, 281*, 25457–25465.

240. Engel, T., Lucas, J.J., Gómez-Ramos, P., Moran, M.A., Avila, J., Hernández, F. (2006). Cooexpression of FTDP-17 tau and GSK-3beta in transgenic mice induce tau polymerization and neurodegeneration. *Neurobiol Aging, 27*, 258–268.

241. Hung, K.S., Hwang, S.L., Liang, C.L., Chen, Y.J., Lee, T.H., Liu, J.K., Howng, S.L., Wang, C.H. (2005). Calpain inhibitor inhibits p35–p25–cdk5 activation,

decreases tau hyperphosphorylation, and improves neurological function after spinal cord hemisection in rats. *J Neuropathol Exp Neurol, 64*, 15–26.

242. Nakashima, H., Ishihara, T., Suguimoto, P., Yokota, O., Oshima, E., Kugo, A., Terada, S., Hamamura, T., Trojanowski, J.Q., Lee, V.M., Kuroda, S. (2005). Chronic lithium treatment decreases tau lesions by promoting ubiquitination in a mouse model of tauopathies. *Acta Neuropathol, 110*, 547–556.
243. Noble, W., Planel, E., Zehr, C., Olm, V., Meyerson, J., Suleman, F., Gaynor, K., Wang, L., LaFrancois, J., Feinstein, B., et al. (2005). Inhibition of glycogen synthase kinase-3 by lithium correlates with reduced tauopathy and degeneration in vivo. *Proc Natl Acad Sci U S A, 102*, 6990–6995.
244. Phiel, C.J., Wilson, C.A., Lee, V.M., Klein, P.S. (2003). GSK-3alpha regulates production of Alzheimer's disease amyloid-beta peptides. *Nature, 423*, 435–439.
245. Feyt, C., Kienlen-Campard, P., Leroy, K., N'Kuli, F., Courtoy, P.J., Brion, J.P., Octave, J.N. (2005). Lithium chloride increases the production of amyloid-beta peptide independently from its inhibition of glycogen synthase kinase 3. *J Biol Chem, 280*, 33220–33227.
246. Le Corre, S., Klafki, H.W., Plesnila, N.,. Hübinger, G., Obermeier, A., Sahagún, H., Monse, B., Seneci, P., Lewis, J., Eriksen, J., et al. (2006). An inhibitor of tau hyperphosphorylation prevents severe motor impairments in tau transgenic mice. *Proc Natl Acad Sci U S A, 103*, 9673–9678.
247. Weishaupt, J.H., Kussmaul, L., Grötsch, P., Heckel, A., Rohde, G., Romig, H., Bähr, M., Gillardon, F. (2003). Inhibition of CDK5 is protective in necrotic and apoptotic paradigms of neuronal cell death and prevents mitochondrial dysfunction. *Mol Cell Neurosci, 24*, 489–502.
248. Gillardon, F., Schrattenholz, A., Sommer, B. (2005). Investigating the neuroprotective mechanism of action of a CDK5 inhibitor by phosphoproteome analysis. *J Cell Biochem, 95*, 817–826.
249. Amin, N.D., Albers, W., Pant, H.C. (2002). Cyclin-dependent kinase 5 (cdk5) activation requires interaction with three domains of p35. *J Neurosci Res, 67*, 354–362.
250. Zheng, Y.L., Kesavapany, S., Gravell, M., Hamilton, R.S., Schubert, M., Amin, N., Albers, W., Grant, P., Pant, H.C. (2005). A cdk5 inhibitory peptide reduces tau hyperphosphorylation and apoptosis in neurons. *EMBO J, 24*, 209–220.
251. Gong, C.X., Shaikh, S., Wang, J.Z., Zaidi, T., Grundke-Iqbal, I., Iqbal, K. (1995). Phosphatase activity toward abnormally phosphorylated tau: decrease in Alzheimer disease brain. *J Neurochem, 65*, 732–738.
252. Chohan, M.O., Khatoon, S., Iqbal, I.G., Iqbal, K. (2006). Involvement of I2PP2A in the abnormal hyperphosphorylation of tau and its reversal by Memantine. *FEBS Lett, 580*, 3973–3979.
253. Lu, P.J., Wulf, G., Zhou, X.Z., Davies, P., Lu, K.P. (1999). The prolyl isomerase Pin1 restores the function of Alzheimer-associated phosphorylated tau protein. *Nature, 399*, 784–788.
254. Zhou, X.Z., Kops, O., Werner, A., Lu, P.J., Shen, M., Stoller, G., Küllertz, G., Stark, M., Fischer, G., Lu, K.P. (2000). Pin1-dependent prolyl isomerization regulates dephosphorylation of cdc25C and tau proteins. *Mol Cell, 6*, 873–883.
255. Petrucelli, L., Dickson, D., Kehoe, K., Taylor, J., Snyder, H., Grover, A., De Lucia, M., McGowan, E., Lewis, J., Prihar, G., et al. (2004). CHIP and

Hsp70 regulate tau ubiquitination, degradation and aggregation. *Hum Mol Genet, 13*, 703–714.

256. Dickey, C.A., Kamal, A., Lundgren, K., Klosak, N., Bailey, R.M., Dunmore, J., Ash, P., Shoraka, S., Zlatkovic, J., Eckman, C.B., et al. (2007). The high-affinity HSP90-CHIP complex recognizes and selectively degrades phosphorylated tau client proteins. *J Clin Invest, 117*, 648–658

257. Dickey, C.A., Yue, M., Lin, W.L., Dickson, D.W., Dunmore, J.H., Lee, W.C., Zehr, C., West, G., Cao, S., Clark, A.M., et al. (2006). Deletion of the ubiquitin ligase CHIP leads to the accumulation, but not the aggregation, of both endogenous phospho- and caspase-3-cleaved tau species. *J Neurosci, 26*, 6985–6996.

258. Shimura, H., Schwartz, D., Gygi, S.P., Kosik, K.S. (2004). CHIP–Hsc70 complex ubiquitinates phosphorylated tau and enhances cell survival. *J Biol Chem, 279*, 4869–4876.

259. Michaelis, M.L., Ansar, S., Chen, Y., Reiff, E.R., Seyb, K.I., Himes, R.H., Audus, K.L., Georg, G.I. (2005). {beta}-Amyloid-induced neurodegeneration and protection by structurally diverse microtubule-stabilizing agents. *J Pharmacol Exp Ther, 312*, 659–668.

260. Rissman, R.A., Poon, W.W., Blurton-Jones, M., Oddo, S., Torp, R., Vitek, M.P., LaFerla, F.M., Rohn, T.T., Cotman, C.W. (2004). Caspase-cleavage of tau is an early event in Alzheimer disease tangle pathology. *J Clin Invest, 114*, 121–130.

261. Newman, J., Rissman, R.A., Sarsoza, F., Kim, R.C., Dick, M., Bennett, D.A., Cotman, C.W., Rohn, T.T., Head, E. (2005). Caspase-cleaved tau accumulation in neurodegenerative diseases associated with tau and alpha-synuclein pathology. *Acta Neuropathol, 110*, 135–144.

262. Kang, H.J., Yoon, W.J., Moon, G.J., Kim, D.Y., Sohn, S., Kwon, H.J., Gwag, B.J. (2005). Caspase-3-mediated cleavage of PHF-1 tau during apoptosis irrespective of excitotoxicity and oxidative stress: an implication to Alzheimer's disease. *Neurobiol Dis, 18*, 450–458.

263. Cotman, C.W., Poon, W.W., Rissman, R.A., Blurton-Jones, M. (2005). The role of caspase cleavage of tau in Alzheimer disease neuropathology. *J Neuropathol Exp Neurol, 64*, 104–112.

264. Guillozet-Bongaarts, A.L., Garcia-Sierra, F., Reynolds, M.R., Horowitz, P.M., Fu, Y., Wang, T., Cahill, M.E., Bigio, E.H., Berry, R.W., Binder, L.I. (2005). Tau truncation during neurofibrillary tangle evolution in Alzheimer's disease. *Neurobiol Aging, 26*, 1015–1022.

265. Horowitz, P.M., Patterson, K.R., Guillozet-Bongaarts, A.L., Reynolds, M.R., Carroll, C.A., Weintraub, S.T., Bennett, D.A., Cryns, V.L., Berry, R.W., Binder, L.I. (2004). Early N-terminal changes and caspase-6 cleavage of tau in Alzheimer's disease. *J Neurosci, 24*, 7895–7902.

266. Howlett, D.R., Perry, A.E., Godfrey, F., Swatton, J.E., Jennings, K.H., Spitzfaden, C., Wadsworth, H., Wood, S.J., Markwell, R.E. (1999). Inhibition of fibril formation in beta-amyloid peptide by a novel series of benzofurans. *Biochem J, 340*, 283–289.

267. Cardoso, I., Merlini, G., Saraiva, M.J. (2003). 4′-Iodo-4′-deoxydoxorubicin and tetracyclines disrupt transthyretin amyloid fibrils in vitro producing noncytotoxic species: screening for TTR fibril disrupters. *FASEB J, 17*, 803–809.

268. Narlawar, R., Pickhardt, M., Leuchtenberger, S., Baumann, K., Krause, S., Dyrks, T., Weggen, S., Mandelkow, E., Schmidt, B. (2008). Curcumin-derived pyrazoles and isoxazoles: Swiss Army knives or blunt tools for Alzheimer's disease? *ChemMedChem, 3*, 165–172.

269. Ohm, T.G., Meske, V. (2006). Cholesterol, statins and tau. *Acta Neurol Scand Suppl, 185*, 93–101.

270. Small, D.H., Mok, S.S., Bornstein, J.C. (2001). Alzheimer's disease and Abeta toxicity: from top to bottom. *Nat Rev Neurosci, 2*, 595–598.

271. Price, D.L., Tanzi, R.E., Borchelt, D.R., Sisodia, S.S. (1998). Alzheimer's disease: genetic studies and transgenic models. *Annu Rev Genet, 32*, 461–493.

272. Akiyama, H., Barger, S., Barnum, S., Bradt, B., Bauer, J., Cole, G.M., Cooper, N.R., Eikelenboom, P., Emmerling, M., Fiebich, B.L., et al. (2000). Inflammation and Alzheimer's disease. *Neurobiol Aging, 21*, 383–421.

273. Polazzi, E., Contestabile, A. (2002). Reciprocal interactions between microglia and neurons: from survival to neuropathology. *Rev Neurosci, 13*, 221–242.

274. Block, M.L., Zecca, L., Hong, J.S. (2007). Microglia-mediated neurotoxicity: uncovering the molecular mechanisms. *Nat Rev Neurosci, 8*, 57–69.

275. Bertram, L., McQueen, M.B., Mullin, K., Blacker, D., Tanzi, R.E. (2007). Systematic meta-analyses of Alzheimer disease genetic association studies: the AlzGene database. *Nat Genet, 39*, 17–23.

276. Kitazawa, M., Oddo, S., Yamasaki, T.R., Green, K.N., LaFerla, F.M. (2005). Lipopolysaccharide-induced inflammation exacerbates tau pathology by a cyclin-dependent kinase 5-mediated pathway in a transgenic model of Alzheimer's disease. *J Neurosci, 25*, 8843–8853.

277. Bellucci, A., Westwood, A.J., Ingram, E., Casamenti, F., Goedert, M., Spillantini, M.G. (2004). Induction of inflammatory mediators and microglial activation in mice transgenic for mutant human P301S tau protein. *Am J Pathol, 165*, 1643–1652.

278. Frautschy, S.A., Yang, F., Irrizarry, M., Hyman, B., Saido, T.C., Hsiao, K., Cole, G.M. (1998). Microglial response to amyloid plaques in APPsw transgenic mice. *Am J Pathol, 152*, 307–317.

279. Wyss-Coray, T. (2006). Inflammation in Alzheimer disease: driving force, bystander or beneficial response? *Nat Med, 12*, 1005–1015.

280. Heneka, M.T., Sastre, M., Dumitrescu-Ozimek, L., Dewachter, I., Walter, J., Klockgether, T., Van Leuven, F. (2005). Focal glial activation coincides with increased BACE1 activation and precedes amyloid plaque deposition in APP[V717I] transgenic mice. *J Neuroinflammation, 2*, 22.

281. Amara, F.M., Junaid. A., Clough, R.R., Liang, B. (1999). TGF-beta(1), regulation of alzheimer amyloid precursor protein mRNA expression in a normal human astrocyte cell line: mRNA stabilization. *Brain Res Mol Brain Res, 71*, 42–49.

282. Cho, H.J., Kim, S.K., Jin, S.M., Hwang, E.M., Kim, Y.S., Huh, K., Mook-Jung, I. (2007). IFN-gamma-induced BACE1 expression is mediated by activation of JAK2 and ERK1/2 signaling pathways and direct binding of STAT1 to BACE1 promoter in astrocytes. *Glia, 55*, 253–262.

283. Rogers, J.T., Leiter, L.M., McPhee, J., Cahill, C.M., Zhan, S.S., Potter, H., Nilsson, L.N. (1999). Translation of the Alzheimer amyloid precursor protein

mRNA is up-regulated by interleukin-1 through 5′-untranslated region sequences. *J Biol Chem, 274*, 6421–6431.

284. Ge, Y.W., Lahiri, D.K. (2002). Regulation of promoter activity of the APP gene by cytokines and growth factors: implications in Alzheimer's disease. *Ann N Y Acad Sci, 973*, 463–467.
285. McGeer, P.L., Schulzer, M., McGeer, E.G. (1996). Arthritis and anti-inflammatory agents as possible protective factors for Alzheimer's disease: a review of 17 epidemiologic studies. *Neurology, 47*, 425–432.
286. Stewart, W.F., Kawas, C., Corrada, M., Metter, E.J. (1997). Risk of Alzheimer's disease and duration of NSAID use. *Neurology, 48*, 626–632.
287. in 't Veld, B.A., Ruitenberg, A., Hofman, A., Launer, L.J., van Duijn, C.M., Stijnen, T., Breteler, M.M., Stricker, B.H. (2001). Nonsteroidal antiinflammatory drugs and the risk of Alzheimer's disease. *N Engl J Med, 345*, 1515–1521.
288. Hirohata, M., Ono, K., Naiki, H., Yamada, M. (2005). Non-steroidal anti-inflammatory drugs have anti-amyloidogenic effects for Alzheimer's beta-amyloid fibrils in vitro. *Neuropharmacology, 49*, 1088–1099.
289. Ralay Ranaivo, H., Craft, J.M., Hu, W., Guo, L., Wing, L.K., Van Eldik, L.J., Watterson, D.M. (2006). Glia as a therapeutic target: selective suppression of human amyloid-beta-induced upregulation of brain proinflammatory cytokine production attenuates neurodegeneration. *J Neurosci, 26*, 662–670.
290. Tobinick, E., Gross, H., Weinberger, A., Cohen, H. (2006). TNF-alpha modulation for treatment of Alzheimer's disease: a 6-month pilot study. *MedGenMed, 8*, 25.
291. Steed, P.M., Tansey, M.G., Zalevsky, J., Zhukovsky, E.A., Desjarlais, J.R., Szymkowski, D.E., Abbott, C., Carmichael, D., Chan, C., Cherry, L., et al. (2003). Inactivation of TNF signaling by rationally designed dominant-negative TNF variants. *Science, 301*, 1895–1898.
292. Weiner, H.L., Frenkel, D. (2006). Immunology and immunotherapy of Alzheimer's disease. *Nat Rev Immunol, 6*, 404–416.
293. Monsonego, A., Imitola, J., Petrovic, S., Zota, V., Nemirovsky, A., Baron, R., Fisher, Y., Owens, T., Weiner, H.L. (2006). Abeta-induced meningoencephalitis is IFN-gamma-dependent and is associated with T cell-dependent clearance of Abeta in a mouse model of Alzheimer's disease. *Proc Natl Acad Sci U S A, 103*, 5048–5053.
294. Frenkel, D., Maron, R., Burt, D.S., Weiner, H.L. (2005). Nasal vaccination with a proteosome-based adjuvant and glatiramer acetate clears beta-amyloid in a mouse model of Alzheimer disease. *J Clin Invest, 115*, 2423–2433.
295. Mori, T., Town, T., Tan, J., Yada, N., Horikoshi, Y., Yamamoto, J., Shimoda, T., Kamanaka, Y., Tateishi, N., Asano, T. (2006). Arundic acid ameliorates cerebral amyloidosis and gliosis in Alzheimer transgenic mice. *J Pharmacol Exp Ther, 318*, 571–578.
296. de Paulis, T. (2003). ONO-2506. Ono. *Curr Opin Investig Drugs, 4*, 863–867.
297. Yan, S.D., Chen, X., Fu, J., Chen, M., Zhu, H., Roher, A., Slattery, T., Zhao, L., Nagashima, M., Morser, J., et al. (1996). RAGE and amyloid-beta peptide neurotoxicity in Alzheimer's disease. *Nature, 382*, 685–691.
298. Pascual, G., Ricote, M., Hevener, A.L. (2007). Macrophage peroxisome proliferator activated receptor gamma as a therapeutic target to combat type 2 diabetes. *Expert Opin Ther Targets, 11*, 1503–1520.

299. Ricote, M., Glass, C.K. (2007). PPARs and molecular mechanisms of transrepression. *Biochim Biophys Acta, 1771*, 926–935.

300. Heneka, M.T., Sastre, M., Dumitrescu-Ozimek, L., Hanke, A., Dewachter, I., Kuiperi, C., O'Banion, K., Klockgether, T., Van Leuven, F., Landreth, G.E. (2005). Acute treatment with the PPARgamma agonist pioglitazone and ibuprofen reduces glial inflammation and Abeta1–42 levels in APPV717I transgenic mice. *Brain, 128*, 1442–1453.

301. Roses, A.D., Saunders, A.M., Huang, Y., Strum, J., Weisgraber, K.H., Mahley, R.W. (2007). Complex disease-associated pharmacogenetics: drug efficacy, drug safety, and confirmation of a pathogenetic hypothesis (Alzheimer's disease). *Pharmacogenom J, 7*, 10–28.

302. Fiala, M., Liu, P.T., Espinosa-Jeffrey, A., Rosenthal, M.J., Bernard, G., Ringman, J.M., Sayre, J., Zhang, L., Zaghi, J., Dejbakhsh, S., et al. (2007). Innate immunity and transcription of MGAT-III and Toll-like receptors in Alzheimer's disease patients are improved by bisdemethoxycurcumin. *Proc Natl Acad Sci U S A, 104*, 12849–12854.

303. Kuo, Y.M., Emmerling, M.R., Bisgaier, C.L., Essenburg, A.D., Lampert, H.C., Drumm, D., Roher, A.E. (1998). Elevated low-density lipoprotein in Alzheimer's disease correlates with brain abeta 1–42 levels. *Biochem Biophys Res Commun, 252*, 711–715.

304. Wolozin, B., Kellman, W., Ruosseau, P., Celesia, G.G., Siegel, G. (2000). Decreased prevalence of Alzheimer disease associated with 3-hydroxy-3- methyglutaryl coenzyme A reductase inhibitors. *Arch Neurol, 57*, 1439–1443.

305. Jick, H., Zornberg, G.L., Jick, S.S., Seshadri, S., Drachman, D.A. (2000). Statins and the risk of dementia. *Lancet, 356*, 1627–1631.

306. Refolo, L.M., Pappolla, M.A., LaFrancois, J., Malester, B., Schmidt, S.D., Thomas-Bryant, T., Tint, G.S., Wang, R., Mercken, M., Petanceska, S.S., Duff, K.E. (2001). A cholesterol-lowering drug reduces beta-amyloid pathology in a transgenic mouse model of Alzheimer's disease. *Neurobiol Dis, 8*, 890–899.

307. Refolo, L.M., Eckman, C., Prada, C.M., Yager, D., Sambamurti, K., Mehta, N., Hardy, J., Younkin, S.G. (1999). Antisense-induced reduction of presenilin 1 expression selectively increases the production of amyloid beta42 in transfected cells. *J Neurochem, 73*, 2383–2388.

308. Notkola, I.L., Sulkava, R., Pekkanen, J., Erkinjuntti, T., Ehnholm, C., Kivinen, P., Tuomilehto, J., Nissinen, A. (1998). Serum total cholesterol, apolipoprotein E epsilon 4 allele, and Alzheimer's disease. *Neuroepidemiology, 17*, 14–20.

309. Kivipelto, M., Helkala, E.L., Hänninen, T., Laakso, M.P., Hallikainen, M., Alhainen, K., Soininen, H., Tuomilehto, J., Nissinen, A. (2001). Midlife vascular risk factors and late-life mild cognitive impairment: a population-based study. *Neurology, 56*, 1683–1689.

310. Kalmijn, S., Foley, D., White, L., Burchfiel, C.M., Curb, J.D., Petrovitch, H., Ross, G.W., Havlik, R.J., Launer, L.J. (2000). Metabolic cardiovascular syndrome and risk of dementia in Japanese-American elderly men. The Honolulu–Asia aging study. *Arterioscler Thromb Vasc Biol, 20*, 2255–2260.

311. Tan, Z.S., Seshadri, S., Beiser, A., Wilson, P.W., Kiel, D.P., Tocco, M., D'Agostino, R.B., Wolf, P.A. (2003). Plasma total cholesterol level as a risk

factor for Alzheimer disease: the Framingham Study. *Arch Intern Med, 163*, 1053–1057.

312. Frears, E.R., Stephens, D.J., Walters, C.E., Davies, H., Austen, B.M. (1999). The role of cholesterol in the biosynthesis of beta-amyloid. *Neuroreport, 10*, 1699–1705.
313. Refolo, L.M., Malester, B., LaFrancois, J., Bryant-Thomas, T., Wang, R., Tint, G.S., Sambamurti, K., Duff, K., Pappolla, M.A. (2000). Hypercholesterolemia accelerates the Alzheimer's amyloid pathology in a transgenic mouse model. *Neurobiol Dis, 7*, 321–331.
314. Sparks, D.L. (1997). Coronary artery disease, hypertension, ApoE, and cholesterol: a link to Alzheimer's disease? *Ann N Y Acad Sci, 826*, 128–146.
315. Wolozin, B. (2004). Cholesterol and the biology of Alzheimer's disease. *Neuron, 41*, 7–10.
316. Kojro, E., Gimpl, G., Lammich, S., Marz, W., Fahrenholz, F. (2001). Low cholesterol stimulates the nonamyloidogenic pathway by its effect on the alpha-secretase ADAM 10. *Proc Natl Acad Sci U S A, 98*, 5815–5820.
317. Ehehalt, R., Keller, P., Haass, C., Thiele, C., Simons, K. (2003). Amyloidogenic processing of the Alzheimer beta-amyloid precursor protein depends on lipid rafts. *J Cell Biol, 160*, 113–123.
318. Cordy, J.M., Hussain, I., Dingwall, C., Hooper, N.M., Turner, A.J. (2003). Exclusively targeting beta-secretase to lipid rafts by GPI-anchor addition up-regulates beta-site processing of the amyloid precursor protein. *Proc Natl Acad Sci U S A, 100*, 11735–11740.
319. Cordy, J.M., Hooper, N.M., Turner, A.J. (2006).The involvement of lipid rafts in Alzheimer's disease. *Mol Membr Biol, 23*, 111–122.
320. Hutter-Paier, B., Huttunen, H.J., Puglielli, L., Eckman, C.B., Kim, D.Y., Hofmeister, A., Moir, R.D., Domnitz, S.B., Frosch, M.P., Windisch, M., Kovacs, D.M. (2004). The ACAT inhibitor CP-113,818 markedly reduces amyloid pathol ogy in a mouse model of Alzheimer's disease. *Neuron, 44*, 227–238.
321. Rosario, E.R., Carroll, J.C., Oddo, S., LaFerla, F.M., Pike, C.J. (2006). Androgens regulate the development of neuropathology in a triple transgenic mouse model of Alzheimer's disease. *J Neurosci, 26*, 13384–13389.
322. Carroll, J.C., Rosario, E.R., Chang, L., Stanczyk, F.Z., Oddo, S., LaFerla, F.M., Pike, C.J. (2007). Progesterone and estrogen regulate Alzheimer-like neuropathology in female 3xTg-AD mice. *J Neurosci, 27*, 13357–13365.
323. Zhang, L., Rubinow, D.R., Xaing, G., Li, B.S., Chang, Y.H., Maric, D., Barker, J.L., Ma, W. (2001). Estrogen protects against beta-amyloid-induced neurotoxicity in rat hippocampal neurons by activation of Akt. *Neuroreport, 12*, 1919–1923.
324. Green, P.S., Bishop, J., Simpkins, J.W. (1997). 17 Alpha-estradiol exerts neuroprotective effects on SK-N-SH cells. *J Neurosci, 17*, 511–515.
325. Chae, H.S., Bach, J.H., Lee, M.W., Kim, H.S., Kim, Y.S., Kim, K.Y., Choo, K.Y., Choi, S.H., Park, C.H., Lee, S.H., et al. (2001). Estrogen attenuates cell death induced by carboxy-terminal fragment of amyloid precursor protein in PC12 through a receptor-dependent pathway. *J Neurosci Res, 65*, 403–407.
326. Zhang, M.Y., Katzman, R., Salmon, D., Jin, H., Cai, G.J., Wang, Z.Y., Qu, G.Y., Grant, I., Yu, E., Levy, P., et al. (1990). The prevalence of dementia and

Alzheimer's disease in Shanghai, China: impact of age, gender, and education. *Ann Neurol, 27*, 428–437.

327. Bachman, D.L., Wolf, P.A., Linn, R., Knoefel, J.E., Cobb, J., Belanger, A., D'Agostino, R.B., White, L.R. (1992). Prevalence of dementia and probable senile dementia of the Alzheimer type in the Framingham Study. *Neurology, 42*, 115–119.
328. Henderson, V.W., Paganini-Hill, A., Emanuel, C.K., Dunn, M.E., Buckwalter, J.G. (1994). Estrogen replacement therapy in older women: comparisons between Alzheimer's disease cases and nondemented control subjects. *Arch Neurol, 51*, 896–900.
329. Kawas, C., Resnick, S., Morrison, A., Brookmeyer, R., Corrada, M., Zonderman, A., Bacal, C., Lingle, D.D., Metter, E. (1997). A prospective study of estrogen replacement therapy and the risk of developing Alzheimer's disease: the Baltimore Longitudinal Study of Aging. *Neurology, 48*, 1517–1521.
330. Ohkura, T., Isse, K., Akazawa, K., Hamamoto, M., Yaoi, Y., Hagino, N. (1994). Evaluation of estrogen treatment in female patients with dementia of the Alzheimer type. *Endocrine J, 41*, 361–371.
331. Doraiswamy, P.M., Bieber, F., Kaiser, L., Krishnan, K.R., Reuning-Scherer, J., Gulanski, B. (1997). The Alzheimer's Disease Assessment Scale: patterns and predictors of baseline cognitive performance in multicenter Alzheimer's disease trials. *Neurology, 48*, 1511–1517.
332. Zheng, H., Xu, H., Uljon, S.N., Gross, R., Hardy, K., Gaynor, J., Lafrancois, J., Simpkins, J., Refolo, L.M., Petanceska, S., et al. (2002). Modulation of A(beta) peptides by estrogen in mouse models. *J Neurochem, 80*, 191–196.
333. Levin-Allerhand, J.A., Lominska, C.E., Wang, J., Smith, J.D. (2002). 17Alpha-estradiol and 17beta-estradiol treatments are effective in lowering cerebral amyloid-beta levels in AbetaPPSWE transgenic mice. *J Alzheimer's Dis, 4*, 449–457.
334. Rocca, W.A., Bower, J.H., Maraganore, D.M., Ahlskog, J.E., Grossardt, B.R., de Andrade, M., Melton, L.J., 3rd. (2007). Increased risk of cognitive impairment or dementia in women who underwent oophorectomy before menopause. *Neurology, 69*, 1074–1083.
335. Zhao, L., O'Neill, K., Brinton, R.D. (2006). Estrogenic agonist activity of ICI 182,780 (Faslodex) in hippocampal neurons: implications for basic science understanding of estrogen signaling and development of estrogen modulators with a dual therapeutic profile. *J Pharmacol Exp Ther, 319*, 1124–1132.
336. Tiwari-Woodruff, S., Morales, L.B., Lee, R., Voskuhl, R.R. (2007). Differential neuroprotective and antiinflammatory effects of estrogen receptor (ER)alpha and ERbeta ligand treatment. *Proc Natl Acad Sci U S A, 104*, 14813–14818.
337. Stern, Y., Gurland, B., Tatemichi, T.K., Tang, M.X., Wilder, D., Mayeux, R. (1994). Influence of education and occupation on the incidence of Alzheimer's disease. *JAMA, 271*, 1004–1010.
338. Friedland, R.P., Fritsch, T., Smyth, K.A., Koss, E., Lerner, A.J., Chen, C.H., Petot, G.J., Debanne, S.M. (2001). Patients with Alzheimer's disease have reduced activities in midlife compared with healthy control-group members. *Proc Natl Acad Sci U S A, 98*, 3440–3445.
339. Wilson, R.S., Bennett, D.A., Bienias, J.L., Aggarwal, N.T., Mendes De Leon, C.F., Morris, M.C., Schneider, J.A., Evans, D.A. (2002). Cognitive activity and

incident AD in a population-based sample of older persons. *Neurology, 59*, 1910–1914.

340. Whalley, L.J., Deary, I.J., Appleton, C.L., Starr, J.M. (2004). Cognitive reserve and the neurobiology of cognitive aging. *Ageing Res Rev, 3*, 369–382.

341. Mattson, M.P., Shea, T.B. (2003). Folate and homocysteine metabolism in neural plasticity and neurodegenerative disorders. *Trends Neurosci, 26*, 137–146.

342. McDowell, I. (2001). Alzheimer's disease: insights from epidemiology. *Aging (Milano), 13*, 143–162.

343. Luchsinger, J.A., Tang, M.X., Shea, S., Mayeux, R. (2002). Caloric intake and the risk of Alzheimer disease. *Arch Neurol, 59*, 1258–1263.

344. Aisen, P.S., Egelko, S., Andrews, H., Diaz-Arrastia, R., Weiner, M., DeCarli, C., Jagust, W., Miller, J.W., Green, R., Bell, K., Sano, M. (2003). A pilot study of vitamins to lower plasma homocysteine levels in Alzheimer disease. *Am J Geriatr Psychiatry, 11*, 246–249.

345. McIlroy, S.P., Dynan, K.B., Lawson, J.T., Patterson, C.C., Passmore, A.P. (2002). Moderately elevated plasma homocysteine, methylenetetrahydrofolate reductase genotype, and risk for stroke, vascular dementia, and Alzheimer disease in Northern Ireland. *Stroke, 33*, 2351–2356.

346. Mattson, M.P., Mattson, E.P. (2002). Amyloid peptide enhances nail rusting: novel insight into mechanisms of aging and Alzheimer's disease. *Ageing Res Rev, 1*, 327–330.

347. Arendash, G.W., Garcia, M.F., Costa, D.A., Cracchiolo, J.R., Wefes, I.M., Potter, H. (2004). Environmental enrichment improves cognition in aged Alzheimer's transgenic mice despite stable beta-amyloid deposition. *Neuroreport, 15*, 1751–1754.

348. Lazarov, O., Robinson, J., Tang, Y.P., Hairston, I.S., Korade-Mirnics, Z., Lee, V.M., Hersh, L.B., Sapolsky, R.M., Mirnics, K., Sisodia, S.S. (2005). Environmental enrichment reduces Abeta levels and amyloid deposition in transgenic mice. *Cell, 120*, 701–713.

349. Mattson, M.P. (2004). Pathways towards and away from Alzheimer's disease. *Nature, 430*, 631–639.

350. Sung, S., Yao, Y., Uryu, K., Yang, H., Lee, V.M., Trojanowski, J.Q., Praticò, D. (2004). Early vitamin E supplementation in young but not aged mice reduces Abeta levels and amyloid deposition in a transgenic model of Alzheimer's disease. *FASEB J, 18*, 323–325.

351. Boothby, L.A., Doering, P.L. (2005). Vitamin C and vitamin E for Alzheimer's disease. *Ann Pharmacother, 39*, 2073–2080.

352. Stackman, R.W., Eckenstein, F., Frei, B., Kulhanek, D., Nowlin, J., Quinn, J.F. (2003). Prevention of age-related spatial memory deficits in a transgenic mouse model of Alzheimer's disease by chronic *Ginkgo biloba* treatment. *Exp Neurol, 184*, 510–520.

353. Luo, Y., Smith, J.V., Paramasivam, V., Burdick, A., Curry, K.J., Buford, J.P., Khan, I., Netzer, W.J., Xu, H., Butko, P. (2002). Inhibition of amyloid-beta aggregation and caspase-3 activation by the *Ginkgo biloba* extract EGb761. *Proc Natl Acad Sci U S A, 99*, 12197–12202.

354. Wu, Y., Wu, Z., Butko, P., Christen, Y., Lambert, M.P., Klein, W.L., Link, C.D., Luo, Y. (2006). Amyloid-beta-induced pathological behaviors are suppressed by

Ginkgo biloba extract EGb 761 and ginkgolides in transgenic *Caenorhabditis elegans*. *J Neurosci, 26*, 13102–13113.

355. DeKosky, S.T., Fitzpatrick, A., Ives, D.G., Saxton, J., Williamson, J., Lopez, O.L., Burke, G., Fried, L., Kuller, L.H., Robbins, J., et al.; GEMS Investigators. (2006). The Ginkgo Evaluation of Memory (GEM) Study: design and baseline data of a randomized trial of *Ginkgo biloba* extract in prevention of dementia. *Contemp Clin Trials, 27*, 238–253.
356. Calon, F., Lim, G.P., Yang, F., Morihara, T., Teter, B., Ubeda, O., Rostaing, P., Triller, A., Salem, N., Jr., Ashe, K.H., et al. (2004). Docosahexaenoic acid protects from dendritic pathology in an Alzheimer's disease mouse model. *Neuron, 43*, 633–645.
357. Wolozin, B. (2001). A fluid connection: cholesterol and Abeta. *Proc Natl Acad Sci U S A, 98*, 5371–5373.
358. Reger, M.A., Henderson, S.T., Hale, C., Cholerton, B., Baker, L.D., Watson, G.S., Hyde, K., Chapman, D., Craft, S. (2004). Effects of beta-hydroxybutyrate on cognition in memory-impaired adults. *Neurobiol Aging, 25*, 311–314.
359. Quirk, J.C., Nisenbaum, E.S. (2002). LY404187: a novel positive allosteric modulator of AMPA receptors. *CNS Drug Rev, 8*, 255–282.
360. Rex, C.S., Lauterborn, J.C., Lin, C.Y., Kramár, E.A., Rogers, G.A., Gall, C.M., Lynch, G. (2006). Restoration of long-term potentiation in middle-aged hippocampus after induction of brain-derived neurotrophic factor. *J Neurophysiol, 96*, 677–685.
361. Youdim, M.B., Bar Am, O., Yogev-Falach, M., Weinreb, O., Maruyama, W., Naoi, M., Amit, T. (2005). Rasagiline: neurodegeneration, neuroprotection, and mitochondrial permeability transition. *J Neurosci Res, 79*, 172–179.
362. Gozes, I., Spivak-Pohis, I. (2006). Neurotrophic effects of the peptide NAP: a novel neuroprotective drug candidate. *Curr Alzheimer Res, 3*, 197–199.
363. Gozes, I., Zaltzman, R., Hauser, J., Brenneman, D.E., Shohami, E., Hill, J.M. (2005). The expression of activity-dependent neuroprotective protein (ADNP) is regulated by brain damage and treatment of mice with the ADNP derived peptide, NAP, reduces the severity of traumatic head injury. *Curr Alzheimer Res, 2*, 149–153.
364. Campbell, V.A., Gowran, A. (2007). Alzheimer's disease; taking the edge off with cannabinoids? *Br J Pharmacol, 152*, 655–662.
365. Volicer, L., Stelly, M., Morris, J., McLaughlin, J., Volicer, B.J. (1997). Effects of dronabinol on anorexia and disturbed behavior in patients with Alzheimer's disease. *Int J Geriatr Psychiatry, 12*, 913–919.
366. Walther, S., Mahlberg, R., Eichmann, U., Kunz, D. (2006). Delta-9-tetrahydrocannabinol for nighttime agitation in severe dementia. *Psychopharmacology (Berl), 185*, 524–528.
367. Shen, M., Thayer, S.A. (1998). The cannabinoid agonist Win55,212-2 inhibits calcium channels by receptor-mediated and direct pathways in cultured rat hippocampal neurons. *Brain Res, 783*, 77–84.
368. Takahashi, K.A., Castillo, P.E. (2006). The CB1 cannabinoid receptor mediates glutamatergic synaptic suppression in the hippocampus. *Neuroscience, 139*, 795–802.

369. Khaspekov, L.G., Brenz Verca, M.S., Frumkina, L.E., Hermann, H., Marsicano, G., Lutz, B. (2004). Involvement of brain-derived neurotrophic factor in cannabinoid receptor-dependent protection against excitotoxicity. *Eur J Neurosci, 19*, 1691–1698.

370. Eubanks, L.M., Rogers, C.J., Beuscher, A.E. 4th, Koob, G.F., Olson, A.J., Dickerson, T.J., Janda, K.D. (2006). A molecular link between the active component of marijuana and Alzheimer's disease pathology. *Mol Pharm, 3*, 773–777.

371. Esposito, G., Scuderi, C., Savani, C., Steardo, L., Jr., De Filippis, D., Cottone, P., Iuvone, T., Cuomo, V., Steardo, L. (2007). Cannabidiol in vivo blunts beta-amyloid induced neuroinflammation by suppressing IL-1beta and iNOS expression. *Br J Pharmacol, 151*, 1272–1279.

372. Hashimoto, M., Rockenstein, E., Masliah, E. (2003). Transgenic models of alpha-synuclein pathology: past, present, and future. *Ann N Y Acad Sci, 991*, 171–188.

373. Rockenstein, E., Torrance, M., Mante, M., Adame, A., Paulino, A., Rose, J.B., Crews, L., Moessler, H., Masliah, E. (2006). Cerebrolysin decreases amyloid-beta production by regulating amyloid protein precursor maturation in a transgenic model of Alzheimer's disease. *J Neurosci Res, 83*, 1252–1261.

374. Alvarez, X.A., Cacabelos, R., Laredo, M., Couceiro, V., Sampedro, C., Varela, M., Corzo, L., Fernandez-Novoa, L., Vargas, M., Aleixandre, M., et al. (2006). A 24-week, double-blind, placebo-controlled study of three dosages of cerebrolysin in patients with mild to moderate Alzheimer's disease. *Eur J Neurol, 13*, 43–54.

375. Farah, M.H. (2007). RNAi silencing in mouse models of neurodegenerative diseases. *Curr Drug Deliv, 4*, 161–167.

376. Miller, V.M., Gouvion, C.M., Davidson, B.L., Paulson, H.L. (2004). Targeting Alzheimer's disease genes with RNA interference: an efficient strategy for silencing mutant alleles. *Nucleic Acids Res, 32*, 661–668.

377. Alvarez, V.A., Ridenour, D.A., Sabatini, B.L. (2006). Retraction of synapses and dendritic spines induced by off-target effects of RNA interference. *J Neurosci, 26*, 7820–7825.

378. Grimm, D., Streetz, K.L., Jopling, C.L., Storm, T.A., Pandey, K., Davis, C.R., Marion, P., Salazar, F., Kay, M.A. (2006). Fatality in mice due to oversaturation of cellular microRNA/short hairpin RNA pathways. *Nature, 441*, 537–541.

379. Smith, R.A., Miller, T.M., Yamanaka, K., Monia, B.P., Condon, T.P., Hung, G., Lobsiger, C.S., Ward, C.M., McAlonis-Downes, M., Wei, H., et al. (2006). Antisense oligonucleotide therapy for neurodegenerative disease. *J Clin Invest, 116*, 2290–2296.

380. Arbel, M., Yacoby, I., Solomon, B. (2005). Inhibition of amyloid precursor protein processing by beta-secretase through site-directed antibodies. *Proc Natl Acad Sci U S A, 102*, 7718–7723.

381. Fiandaca, M., Forsayeth, J., Bankiewicz, K. (2007). Current status of gene therapy trials for Parkinson's disease. *Exp Neurol, 1*, 51–57.

382. Athanasopoulos, T., Graham, I.R., Foster, H., Dickson, G. (2004). Recombinant adeno-associated viral (rAAV) vectors as therapeutic tools for Duchenne muscular dystrophy (DMD). *Gene Ther, 11*, S109–S121.

383. Singer, O., Marr, R.A., Rockenstein, E., Crews, L., Coufal, N.G., Gage, F.H., Verma, I.M., Masliah, E. (2005). Targeting BACE1 with siRNAs ameliorates Alzheimer disease neuropathology in a transgenic model. *Nat Neurosci, 8*, 1343–1349.

384. Tuszynski, M.H., Thal, L., Pay, M., Salmon, D.P., U, H.S., Bakay, R., Patel, P., Blesch, A., Vahlsing, H.L., Ho, G., et al. (2005). A phase 1 clinical trial of nerve growth factor gene therapy for Alzheimer disease. *Nat Med, 11*, 551–555.

385. Tuszynski, M.H. (2007). Nerve growth factor gene delivery: animal models to clinical trials. *Dev Neurobiol, 67*, 1204–1215.

386. Marutle, A., Ohmitsu, M., Nilbratt, M., Greig, N.H., Nordberg, A., Sugaya, K. (2007). Modulation of human neural stem cell differentiation in Alzheimer (APP23) transgenic mice by phenserine. *Proc Natl Acad Sci U S A, 104*, 12506–12511.

387. Klunk, W.E., Engler, H., Nordberg, A., Wang, Y., Blomqvist, G., Holt, D.P., Bergström, M., Savitcheva, I., Huang, G.F., Estrada, S., et al. (2004). Imaging brain amyloid in Alzheimer's disease with Pittsburgh compound-B. *Ann Neurol, 55*, 306–319.

388. Blennow, K. (2004). Cerebrospinal fluid protein biomarkers for Alzheimer's disease. *NeuroRx, 1*, 213–225.

389. Rosenberg, R.N. (2005). Translational research on the way to effective therapy for Alzheimer disease. *Arch Gen Psychiatry, 62*, 1186–1192.

390. Sastre, M., Dewachter, I., Landreth, G.E., Willson, T.M., Klockgether, T., van Leuven, F., Heneka, M.T. (2003). Nonsteroidal anti-inflammatory drugs and peroxisome proliferator-activated receptor-gamma agonists modulate immunostimulated processing of amyloid precursor protein through regulation of beta-secretase. *J Neurosci, 23*, 9796–9804.

35

CURRENT THERAPIES FOR LIGHT-CHAIN AMYLOIDOSIS

Angela Dispenzieri and Shaji Kumar
Division of Hematology, Mayo Clinic, Rochester, Minnesotea

INTRODUCTION

An accurate diagnosis of immunoglobulin light-chain amyloidosis (AL) is the first and most important step in managing the disorder. Misclassification of amyloidosis occurs more frequently than anticipated [1,2] and assures treatment failure. Once the amyloid is typed, a thorough assessment of the extent of involvement is required to develop a treatment plan [3]. Virtually all cases of systemic AL will require therapy; in contrast, not all cases of localized AL will. Once a determination of which organs are and are not involved, one should consider the extent of organ involvement in order to estimate the risk associated with different treatment modalities and to modulate expectations for the patient. Finally, one should consider which parameters will be followed to modulate ongoing treatment decisions. This last recommendation cannot be overemphasized; patients with AL are complex, often very ill, and have multiple disease parameters to follow, making it difficult to track a patient's progress in a busy clinical practice.

Virtually all patients with systemic disease will require immediate treatment. In contrast, patients with localized disease may not. The designation *localized* applies to those cases of AL in which the precursor protein (the immunoglobulin light chain) is made at the site of amyloid deposition [4] and is typically not associated with a detectable circulating monoclonal protein in the serum or urine. The classic examples of localized AL are tracheobronchial, urinary tract,

Protein Misfolding Diseases: Current and Emerging Principles and Therapies,
Edited by Marina Ramirez-Alvarado, Jeffery W. Kelly, and Christopher M. Dobson
Copyright © 2010 John Wiley & Sons, Inc.

cutaneous, lymph node, and nodular cutaneous involvement. The bone marrows of these patients do not have clonal plasma cells, but rather, in the case of tracheobronchial or urinary tract AL, the mucosal tissues that have the amyloid deposited within them also harbor the clonal plasma cells immediately adjacent to the amyloid. It is not unusual for the pathologist to overlook these plasma cells, because they are typically few in number and overwhelmed by the surrounding amyloid. Although it is not understood why the plasma cells in these unusual presentations stay specific to their chosen tissue type, the disease does not characteristically generalize to other tissue types, and the management strategy of choice is localized, palliative therapy as required (see below).

In contrast, a patient diagnosed with AL who has clonal bone marrow plasmacytosis producing their circulating amyloidogenic precursor protein (i.e., systemic disease) will almost always require treatment rather than observation. The goal in these patients is to achieve a reduction in the precursor protein (monoclonal protein), ideally complete disappearance of it; however, for some patients a mere 50% reduction of the monoclonal protein is sufficient to both halt the progression of organ damage and to allow for significant improvement of organ function [5]. A priori, it is unclear which patients will require complete hematologic response or partial response. Therefore, the ultimate challenge in managing these patients is balancing treatment-related toxicity and efficacy.

ASSIGNING A PROGNOSTIC CATEGORY TO PATIENTS WITH SYSTEMIC AL

Many prognostic factors identified in these patients, including β_2-microglobulin and level of circulating immunoglobulin free light chains [6,7]. One of the most commonly applied prognostic factors applied in the past decade has been the number of organs involved [8]. Although useful, more important than number is the extent of the organs involved. The most important determinant of survival has always been the extent of cardiac involvement [9]. Efforts at characterizing cardiac involvement have employed standard echocardiography [10], but these measurements do not detect early involvement and there are issues with interobserver variability. More recently, cardiac biomarkers [troponin T, troponin I, N-terminal brain naturietic factor (NT-BNP), and brain naturietic factor (BNP)] that can be measured in the blood have been shown to be excellent predictors of prognosis (Fig.1) [11–14]. These routine laboratory tests are reproducible and relatively inexpensive. Additional work is being done using Doppler myocardial imaging, and systolic strain of left ventricular segments may have a role to play in assessing risk in these patients [15,16]. In addition, MRI of the heart may play a role [17].

Another important factor to consider in reviewing clinical trial data is that the diagnosis of AL is not always straightforeword, and patients are frequently quite ill at the time of diagnosis with an ECOG performance status greater than 2. Most trials require that its participants have a performance status of

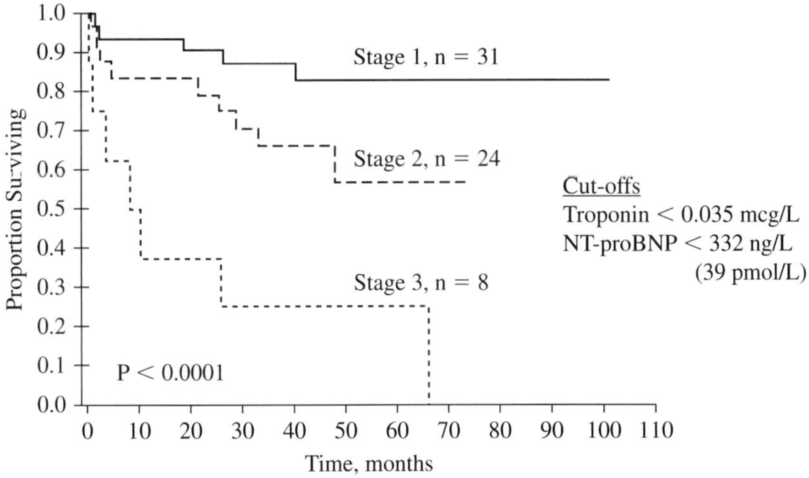

FIG. 1 Survival based on baseline cardiac biomarkers.

2 or less for inclusion, thereby skimming off patients who a priori will have a better prognosis than an unselected group of presenting patients. If we examine our experience with the randomized melphalan-based clinical trials conducted at the Mayo Clinic, we find that the median survival for those patients was 18 to 29 months [18–20], whereas the median survival of contemporaneous patients presenting to Mayo Clinic within 30 days of their diagnosis was 12 months. Another example are those eligible for a HSCT trial (but not undergoing transplant) versus those not eligible [21]. Only 16% of all comers would have been eligible for transplant, and this group had a median survival of 42 months. These types of observations drive home the importance of well-designed, well-powered randomized trials and the uniform incorporation of powerful prognostic factors.

TREATMENT OF SYSTEMIC AL

Therapy for AL is in evolution and there is a paucity of randomized clinical trials to direct the best therapies. Box 1 and Figure 2 illustrate the current treatment strategy employed for patients with systemic AL seen at the Mayo Clinic. The three most important questions relating to therapy of AL concern (1) the role of high-dose chemotherapy with peripheral blood stem cell support (HSCT), (2) the role of novel therapies, and (3) the duration of therapy. In routine practice, the first question we ask is whether a patient is a candidate for high-dose chemotherapy with HSCT. As discussed below, we base this decision on our experience with treatment-related morbidity and mortality in select populations. Risk factors include physiologic age, performance status, cardiac function as determined by serum troponin T and functional class, and numbers of organs involved [12,22,23]. During the course of therapy, the next vital

BOX 1 Hematopoietic Stem Cell Transplantation Eligible

- Physiologic age = 70 years
- Eastern Cooperative Oncology Group performance score = 2
- Troponin T< 0.06 ng/mL
- Creatinine clearance = 30 mL/min[a]
- New York Heart Association class I/II[b]
- No more than two organs *significantly* involved

[a]Selected chronic dialysis patients may become eligible for PBSCT with renal transplantation.

[b]Selected severely affected cardiac amyloidosis patients may become eligible for PBSCT with cardiac transplantation.

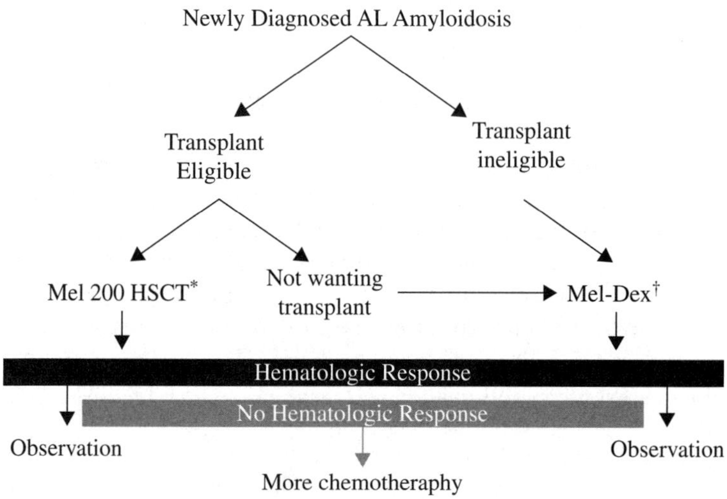

FIG. 2 Mayo Clinic treatment algorithm. *Consider second-line therapy if hematological partial response not achieved at day +100 or organ progression at six months. †Treat to maximum response +2 (no more than 10 cycles). Consider second-line therapy if hematological minimal response not seen after four cycles or organ progression at six months.

question is whether they have achieved an adequate hematological response. Finally, with emerging therapies luch as lenalidomide and bortezomib for the treatment of myeloma, one must ask how and when these treatments fit into the therapeutic armamentarium.

High-Dose Chemotherapy with Stem Cell Transplant

Capitalizing on the improved overall survival observed with patients with myeloma undergoing high-dose chemotherapy with peripheral blood stem cell transplant, stem cell transplantation has been widely applied to patients with AL, and over 1000 patients have been reported in the literature. The most common conditioning regimen is melphalan 200 mg/m^2. Hematologic responses have been reported anywhere from 32 to 68% and complete hematologic responses from 16 to 50% [24–29]. Organ response is time dependent and a median time to respond can take up to one year, but organ responses, the most important outcome, range anywhere from 31 to 64%. Patients with the deepest hematologic responses are more likely to have long-term overall survival [5]. The treatment-related mortality is quoted from 6 to 27% at day 100, with the majority of deaths due to cardiac causes, but patients also succumb to multiorgan failure, hepatic insufficiency, hemorrhage, and sepsis [24–29]. At Mayo Clinic, the day-100 mortality is 11%, and the median survival of patients with three-organ involvement is 30 months, two-organ involvement 68 months, and 98 months for patients with one-organ involvement [27]. Overall survival of 337 patients is a projected median of 82 months.

A case–control study comparing overall survival of 63 AL patients undergoing HSCT with 63 patients not undergoing transplantation was performed [30]. According to design, there was no difference between the groups with respect to gender (57% males), age (median, 53 years), left ventricular ejection fraction (65%), number of patients with peripheral nerve involvement (17%), cardiac interventricular septal wall thickness (12 mm), serum creatinine (1.1 mg/dL), and bone marrow plasmacytosis (8%). For HSCT and control groups, respectively, the one-, two-, and four-year overall survival rates were 89 and 71%; 81 and 55%; and 71 and 41%. In contrast, a small, prospective randomized controlled trial comparing melphalan and dexamethasone to HSCT did not demonstrate any improvement in overall survival for the HSCT arm over the conventional chemotherapy arm (see below) [28].

Finally, allogeneic bone marrow transplantation is feasible in patients with amyloidosis; however, treatment-related mortality and morbidity is high. The European Group for Blood and Marrow Transplantation (EBMT) registry reported 19 patients with AL who underwent allogeneic (n = 15) or syngeneic (n = 4) hematopoietic stem cell transplantation between 1991 and 2003 [31]. For allo-SCT, full-intensity conditioning was used in seven patients and reduced-intensity conditioning in eight patients. With a median follow-up time of 19 months, overall and progression-free survival was 60 and 53% at one year, respectively. Overall, 40% of patients died of transplant-related mortality.

Standard Chemotherapy Options

The successful use of cytotoxic chemotherapy to produce regression of AL was reported 30 years ago [32,33]. The use of alkylating agents to suppress the plasma cell clone in the bone marrow of AL patients followed the recognized success of these agents in the management of multiple myeloma. In the case of intermittent low-dose chemotherapy, it can be difficult to distinguish those patients who are destined to respond but need longer exposure to therapy from those who are destined to fail therapy and should be offered an alternative therapeutic approach. In vitro studies of the plasma cells from AL patients demonstrate aberrant light-chain synthesis. In one study, the cytoplasm of the plasma cells contained light-chain tetramers [34]. After successful alkylating agent-based chemotherapy, light-chain synthesis was suppressed, and a clinical response was demonstrated.

Role of Low-Dose Melphalan as a Treatment for AL

In a prospective study of melphalan, prednisone, and placebo, 55 patients with biopsy-proven AL were randomized to either placebo or melphalan and prednisone in a double-blind fashion [35]. There was no survival difference between the two groups, but patients who received melphalan and prednisone for a longer time and received larger doses had superior survival. Nephrotic syndrome disappeared in two patients, and proteinuria decreased by greater than 50% in an additional eight patients. There were no such responses in the placebo arm. Of the 13 AL patients who received 12 months of therapy with melphalan and prednisone, six improved, three were stable, and four progressed.

A subsequent randomized crossover study of melphalan and prednisone versus colchicine was reported [19]. The 101 eligible patients were stratified by age and their dominant clinical manifestation: heart failure, neuropathy, nephrotic syndrome, and other. Of the 49 patients who received melphalan and prednisone, eigth crossed over to colchicine. Of the 52 randomized to colchicine, 35 crossed over to melphalan and prednisone at six months due to progression. There was no difference in survival. Upon subgroup analysis, there were significant differences that favored melphalan and prednisone when patients receiving only one regimen were analyzed. Also, the time to death or to crossover favored those patients randomized to melphalan and prednisone.

There have been two prospective, randomized noncrossover studies evaluating the value of melphalan and prednisone in AL [18,36] (Table 1). In one, 219 patients received (1) daily colchicine (n = 72), (2) melphalan and prednisone for 7 days every six weeks (n = 70), or (3) regimens 1 and 2 combined (n = 69). Stratification was by age, gender, and clinical manifestation. Half had nephrotic-range proteinuria, and 20% had heart failure. The median survival was

TABLE 1 Standard Chemotherapy for AL

	N	Hematologic Response (%)	Organ Response (%)	Median Survival (months)
MP [18,19,35,36]	~200	28	20–30	18–29
VBMCP [20]	49	29	31	29
Melphalan-Dex [28,41]	96	52–67	39–48	57–60
Dex [48,49]	77	—	15–35	12–21
Dex-IFN [52]	93	33	—	29
VAD [40,43–47]	~100	42–50		

significantly superior in the melphalan- and prednisone-containing groups compared to the colchicine group: 17 versus 8.5 months.

The second study randomized 100 patients [36] to colchicine or a combination of melphalan, prednisone, and colchicine (Table 1). Stratification was a function of gender, time from diagnosis to study entry, and dominant organ system involvement. The overall survival of the patient group was 6.7 months in the colchicine group and 12.2 months in the melphalan group. A survival advantage was seen particularly for those patients presenting with either peripheral neuropathy or "other" (nonrenal, noncardiac, nonhepatic). In a multivariate analysis, melphalan had a significant impact on survival when heart failure was not present.

The earliest reports of patient factors that predicted lack of response to melphalan and prednisone were cardiac involvement, lambda light-chain restriction, and low β_2 microglobulin [37]. Serum creatinine levels over 3 mg/dL had also been associated with low organ response rates and poorer overall survival [38]. With melphalan and prednisone, nephrotic-only patients with a normal serum creatinine enjoyed an organ response rate of 39%, whereas patients with cardiomyopahthy had organ response rates of only 15%. Although only 18% of patients responded to melphalan and prednisone, their median survival was 89 months, and the five-year survival of this small subgroup was 78%. The median time to response was one year. Nonresponders had a median survival of 15 months.

Because of the putative beneficial role of VBMCP (vincristine, carmustine, melphalan, cyclophosphamide, and prednisone) over melphalan and prednisone in multiple myeloma, a prospective randomized study comparing these regimens in patients with AL was performed (Table 1) [20]. One hundred one patients were randomized and were stratified by age, clinical manifestation, and the presence or absence of heart failure. The median overall survival for the entire group was 29 months. There was no difference between the arms. The most common cause of death while receiving dialysis was intractable hypotension related to cardiac involvement. Myelodysplasia was documented in eight and all have died, including one who died after nonmyeloablative allogeneic transplant following diagnosis of myelodysplasia.

In a separate retrospective study of 153 amyloid patients receiving melphalan, cytogenetic abnormalities consistent with secondary myelodysplasia were recognized in 10 [39]. Overall, bone marrow damage consistent with alkylator-induced toxicity was seen in 7% of the total patient population, but the actuarial risk for developing myelodysplasia or acute leukemia at 42 months after initiating therapy was 21%. Eight of these 10 died of pancytopenia and one of progressive renal amyloid and one was alive at the time of the report. Morphologically, four had acute leukemia and five had myelodysplasia. The median survival following the diagnosis of leukemia or myelodysplasia was eight months.

Infusional low-dose melphalan has also been used off-study in 24 patients with AL. Doses of 25 mg/m^2 every 4 to 6 weeks for two to six courses have provided serum immunoglobulin-free light-chain responses in approximately 50% of patients, with a median overall survival approaching five years [40].

In 2004, Palladini et al. reported their experience treating 41 patients who were not transplant candidates with melphalan and dexamethasone. They showed that by merely exchanging dexamethasone for prednisone, they could achieve hematological response rates of 67%, including 33% complete responses and organ response rates of 48% [41]. In a five-year update, these patients had an overall median survival of 5.1 years and a progression-free survival of 3.8 years [42]. A total of 21 patients—13 nonresponders and 8 responders—died after a median of 1.6 years (range 0.1 to 5.7 years). Death was not amyloid-related in four patients who responded to melphalan and dexamethasone, whereas it was due to progressive cardiac amyloidosis in the remaining cases. There was only one case of myelodysplasia. The authors attribute that to the modest total dose of melphalan administered (median, 288 mg; range, 48 to 912 mg), even considering the additional cycles delivered in four relapsing patients.

Finally, a prospective randomized study of 100 patients randomized to autologous stem cell transplant with melphalan compared to oral melphalan and dexamethasone showed no difference between the two arms for hematologic responses, and the landmark analysis performed to correct for early mortality associated with transplant also showed no difference in overall survival [28]. On an intention-to-treat basis, the median survival for melphalan and dexamethasone was 57 months versus. 22 months for the stem cell transplant arm. This important study is limited by its very small size for a disease that is very heterogeneous in its extent. Among the 50 patients randomized to receive HSCT, only 37 actually made it to transplant, and then nine of those died within 100 days, a 24% treatment-related mortality, leaving only 28 patients for their landmark analysis. In contrast, of the 50 patients randomized to melphalan and dexamethasone, 43 patients received three or more cycles of therapy.

Corticosteroid-Based Therapy for AL

Infusional vincristine, doxorubicin, and dexamethasone (VAD) over 96 hours has been reported in just over 100 patients from multiple small series (Table 1) [40,43–47]. The overall hematological response rate was about 42 to 50%. The use of infusional VAD is an option for patients with AL, but the use of vincristine in patients with amyloid neuropathy and doxorubicin in patients with cardiomyopathy may limit its applicability.

Variable degrees of success have been reported with single-agent dexamethasone (Table 1). When we treated 19 patients with newly diagnosed AL with high-dose dexamethasone (40 mg on days 1 to 4, 9 through 12, and 17 through 20 every 5 weeks), only 3 of 19 showed an objective organ response [48]. The median survival of the entire group was 11.2 months. When cardiac amyloid was not present, high-dose dexamethasone did produce a benefit in some patients with AL. We also tested dexamethasone in melphalan and prednisone failures [49]. Twenty-five patients received high-dose dexamethasone; three objective responses with organ-specific improvement were seen. The median survival of the entire group was 13.8 months. In our experience, toxicity can be formidable, including fluid retention, gastrointestinal bleeding, and colonic perforation. In contrast, Palladini et al. treated 23 patients with an attenuated schedule consisting of dexamethasone 40 mg on days 1 to 4 every 21 days for up to eight cycles and achieved a 35% organ response rate at a median time of four months (range two to six months) without significant toxicity [50].

Dhodopkar added interferon to the high-dose dexamethasone regimen in a pilot of nine consecutive patients who he treated with dexamethasone alone, given 40 mg on days 1 to 4, 9 through 12, and 17 through 20 every 5 weeks for three to six cycles, followed by maintenance interferon in a dose of 3 to 6 million units three times per week. Three of these patients received maintenance dexamethasone at 40 mg on days 1 through 4 monthly for one year. AL organ improvement was reported in eight of nine patients. Of seven with nephrotic-range proteinuria, six had a 50% reduction in proteinuria with a median time to response of only four months. Organ function improvement was reported in amyloid neuropathy, hepatic involvement, and gastrointestinal involvement. Neither patient with heart failure improved [51]. In a subsequent study 93 patients were treated between 1996 and 2003 in a prospective U.S. national cooperative group trial (Table 1) [52]. Hematologic complete remissions were observed in 24%, and improvement in AL amyloidosis-related organ dysfunction occurred in 45% of patients evaluable for response. Median survival of the entire cohort was 31 months, with an estimated two-year overall survival (OS) and event-free survival (EFS) of 60 and 52%, respectively. The presence of congestive heart failure and an increased level of serum β_2 microglobulin $\geq$3.5 mg/L were dominant predictors of adverse outcome.

Therapies for AL That Have Not Shown Promise

Fifteen patients were treated with single-agent interferon α_2. None of the patients showed any objective regression of their disease; the median survival of the entire group was 26.3 months [53]. Sixteen patients were treated with vitamin E because of an animal study suggesting that vitamin E could inhibit amyloid production [54]. There were no objective responses in this cohort. The median survival was 19 months. A cohort of 15 patients was treated with subcutaneous injections of interferon α-2b three times a week. No responses were seen in this cohort. Interferon is not a useful agent in the treatment of AL [53]. 4′-Iodo-4′-deoxydoxorubicin has been reported in the treatment of AL [55,56]. Patients with visceral amyloid deposits have a lower response rate than those who have soft tissue amyloid. Of 45 patients treated with deoxydoxorubicin, the response rate was 15%. Production of the drug has ceased, and no further trials are possible at present.

Novel Therapies for AL

New therapies for amyloidosis involving incorporation of novel agents have also been described. Thalidomide, as a single agent, has a heightened toxicity in patients with amyloidosis, and no hematologic or organ responses have been reported (Table 2) [57–59]. In contrast, in combination with dexamethasone, 48% of 31 patients achieved hematologic response, with eight (26%) organ responses. Median time to response was 3.6 months (range, 2.5 to 8.0 months). Treatment-related toxicity was frequent (65%), and symptomatic bradycardia was a common (26%) adverse reaction [60]. Wechalekar et al. have treated 75 patients with cyclophosphamide, thalidomide, and dexamethasone combination [61]. A hematological response occurred in 48 (74%) of 65 evaluable patients, including complete hematological responses in 14 (21%). With a median follow-up of 22 months, median estimated overall survival from the beginning of treatment was 41 months.

Lenalidomide has been combined with dexamethasone for the treatment of AL. In a trial of 23 patients, one patient had a hematologic and organ response with lenalidomide alone [62]. Eleven additional patients received dexamethasone, and overall response rates of 12 who completed more than three cycles of therapy were 10/12 with 9/12 hematologic and 5/12 organ responses. Of interest, 10 patients did not complete three cycles of therapy, and analysis demonstrated that the cardiac troponin T level was highly predictive of patients being able to complete protocolized therapy. In a similar study, with more restrictive eligibility entry criteria, of 24 evaluable patients, the overall hematologic response rate was 67%, including a 29% hematologic complete response [63].

Bortezomib has been studied in a phase I study with a tolerated dose of 1.6 mg/m^2 days 1, 8, 15, and 22 every 5 weeks. Grade 3 to 4 adverse events were seen in 40% of patients, congestive heart failure in 10%, and infection in 15%. Of 19 evaluable patients, nine (47%) had an objective response, including five complete hematologic responses and four partial hematologic responses; 10 patients (53%) had stable disease. Organ responses were reported in 10 patients (50%) [64].

Using Prognostic Factors to Make Treatment Decisions

Among 271 patients undergoing stem cell transplantation, troponin T was a powerful predictor of treatment-related mortality. Patients with troponin levels of 0.06 mg/L or higher had a day-100 all-cause mortality rate of 28% [23] compared to patients with troponin levels of less than 0.06 mg/L, who had a day-100 all-cause mortality rate of 7% ($p = 0.001$), despite risk-adapted dose modification of melphalan. Among the 40 patients with troponin levels of 0.06 mg/L or higher, the conditioning dosage of melphalan was reduced for 31 patients (78%). In addition, we have noted that the 52% of patients who gained more than 2% body weight during mobilization had a poorer outcome

TABLE 2 Immune Modulatory Derivatives and Proteosome Inhibitors in Patients with AL[a]

Regimen	N	No Prior Rx (%)	Two Organs (%)	Cardiac	Heme Response	Organ Response	Median Follow-up (months)	Grade 3–4 AE (%)
Thal 200–800 mg [58]	16	6	31	25	25	0	NR	50
Thal/Dex [60]	31	42	61	38	48	26	32	65
Thal 200–800 mg [87]	12	58	67	42	0	11	2.3[b]	58
Thal 50–200 mg [59]	18	28	50	67	0	11	5.6[b]	75
CTX/Thal/Dex								
Len ± Dex [62]	22	43	57	65	43	26	17	83
Len ± Dex [63]	34				47	21		>35
Bortez ± Dex [64]	18	61	61	84	94 (44CR)	28	11	NR

[a] Rx, treatment; AE, adverse events; Cardiac, cardiac involvement Dex, dexamethasone; Thal, thalidomide; Len, lenalidomide.

[b] Median time on treatment.

[c] Induction phase.

[d] Maintenance phase.

following HDM-SCT. First-year mortality was significantly higher in those with more than 2% weight gain (33.9% versus 9.8%, $p = 0.002$) [65].

Zhou et al. have recently shown that high levels of expression of plasma cell calreticulin predicts for better response to high-dose melphalan [66]. Calreticulin is a pleiotropic calcium-binding protein found in the endoplasmic reticulum and the nucleus, whose overexpression is associated with increased sensitivity to apoptotic stimuli. Although this prognostic factor is not ready to be applied in clinical practice, it may play an important role in the future.

Because the immunoglobulin free light chain is an important prognostic factor in AL patients in terms of both baseline value and extent of reduction [7,40,67], the group at Memorial Sloan Kettering have looked at treating those patients who did not achieve a complete hematological response at three months after their HSCT with adjuvant thalidomide and dexamethasone. Thirty-one patients began adjuvant therapy, with 16 (52%) completing nine months of treatment and 13 (42%) achieving an improvement in hematological response. By intention-to-treat, the overall hematological response rate was 71% (36% complete response), with 44% having organ responses. With a median follow-up of 31 months, two-year survival was 84% (95% confidence interval: 73%, 94%) [68].

TREATING LOCALIZED AMYLOIDOSIS

The location of the amyloid is an important clue in recognizing the amyloid as being localized. The most frequent sites of localized amyloid are respiratory tract, genitourinary tract, and skin [69]. Pulmonary amyloid can be subdivided into nodular, tracheobronchial, or diffuse interstitial. Only the third represents a manifestation of systemic AL [70,71]. The diagnosis of tracheobronchial amyloidosis is via bronchoscopy while evaluating a patient with obstruction, cough, dyspnea, wheezing, or hemoptysis. The usual treatment is yttrium–aluminum–garnet (YAG) laser resection of the tissue. Tracheobronchial amyloid deposits are derived from immunoglobulin light chains [72]. Tracheobronchial amyloidosis has been difficult to treat, due to the limitations of treatment, recurrence, and complications. EBRT appears to be safe and can provide symptomatic as well as objective improvement [73]. The nodular form of amyloid presents as solitary pulmonary nodules or multiple nodules. This does not represent the systemic form of AL [74]. These nodules are not calcified and often require resection to exclude a diagnosis of malignancy. The diagnosis is made at thoracotomy or a video thoracoscopic surgical procedure. Amyloid can involve the vocal cords and false vocal cords, causing traction on the structures, leading to hoarseness. This form of laryngeal amyloid is always localized [75]. Obstructive ureterovesicular amyloidosis is always localized. Patients present with hematuria or obstruction [76]. The prebiopsy diagnosis is cancer. Amyloid is found when cystoscopic biopsies are performed. Eighty-five percent of patients present with hematuria. Partial cystectomy, fulguration, and transurethral resection have all been used [77]. Dimethyl sulfoxide instillation in the bladder has been reported to improve these deposits [78]. Colchicine has also been reported to be beneficial.

Amyloid involving the renal pelvis or ureter is a localized amyloid syndrome [79,80]. These patients present with colic due to obstruction or hematuria. The deposits are found at surgery. Nephrectomy is commonly performed because of the concern that this would represent a transitional cell malignancy. The recognition of amyloid avoids nephrectomy. Amyloidosis can involve the urethra and present with dysuria and hematuria. The preoperative diagnosis is usually a urethral malignancy. Resection is the treatment of choice.

There are three forms of amyloidosis recognized in the skin. The lichen and macular forms are localized [81]. Nodular amyloidosis is associated with AL [82,83]. The generated keratin fibrils are the source of macular and papular amyloid [84]. Dermabrasion and other forms of local therapy are adequate for control. Lichen and macular amyloids usually are associated with a history of local skin trauma or inflammation. Lichen or macular amyloid is an innocuous condition. The nodular form, however, can be an important clinical clue to an underlying life-threatening process.

Carpal tunnel amyloidosis can be seen in systemic AL and AF and may be localized as well [85]. If a patient presents with carpal tunnel syndrome as the only manifestation of amyloid, the median survival is 12 years. Virtually all these patients have localized disease, and only two patients with localized carpal tunnel amyloid developed AL [86]. The amino acid composition of localized carpal tunnel amyloid is transthyretin (TTR). The conjunctiva and orbits are sites where localized amyloidosis is also seen [69]. The best treatment is surgical excision.

Acknowledgments

This work was supported in part by the Hematologic Malignancies Fund.

REFERENCES

1. Lachmann, H.J., Booth, D.R., Booth, S.E., Bybee, A., Gilbertson, J.A., Gillmore, J.D., Pepys, M.B., Hawkins, P.N. (2002). Misdiagnosis of hereditary amyloidosis as AL (primary) amyloidosis. *N Engl J Med*, *346*, 1786–1791.
2. Comenzo, R.L., Zhou, P., Fleisher, M., Clark, B., Teruya-Feldstein, J. (2006). Seeking confidence in the diagnosis of systemic AL (Ig light-chain) amyloidosis: patients can have both monoclonal gammopathies and hereditary amyloid proteins. *Blood*, *107*, 3489–3491.
3. Gertz, M.A., Lacy, M.Q., Dispenzieri, A., Hayman, S.R. (2005). Amyloidosis: diagnosis and management. *Clin Lymphoma Myeloma*, *6*, 208–219.
4. Hamidi Asl, K., Liepnieks, J.J., Nakamura, M., Benson, M.D. (1999). Organ-specific (localized) synthesis of Ig light chain amyloid. *J Immunol*, *162*, 5556–5560.

5. Gertz, M.A., Lacy, M.Q., Dispenzieri, A., Hayman, S.R., Kumar, S.K., Leung, N., Gastineau, D.A. (2007). Effect of hematologic response on outcome of patients undergoing transplantation for primary amyloidosis: importance of achieving a complete response. *Haematologica, 92*, 1415–1418.
6. Gertz, M.A., Kyle, R.A., Greipp, P.R., Katzmann, J.A., O'Fallon, W.M. (1990). Beta 2-microglobulin predicts survival in primary systemic amyloidosis. *Am J Med, 89*, 609–614.
7. Dispenzieri, A., Lacy, M.Q., Katzmann, J.A., Rajkumar, S.V., Abraham, R.S., Hayman, S.R., Kumar, S.K., Clark, R., Kyle, R.A., Litzow, M.R., *et al.* (2006). Absolute values of immunoglobulin free light chains are prognostic in patients with primary systemic amyloidosis undergoing peripheral blood stem cell transplantation. *Blood, 107*, 3378–3383.
8. Moreau, P., Leblond, V., Bourquelot, P., Facon, T., Huynh, A., Caillot, D., Hermine, O., Attal, M., Hamidou, M., Nedellec, G., *et al.* (1998). Prognostic factors for survival and response after high-dose therapy and autologous stem cell transplantation in systemic AL amyloidosis: a report on 21 patients. *Br J Haematol, 101*, 766–769.
9. Gertz, M.A., Lacy, M.Q., Dispenzieri, A., Hayman, S.R. (2005). Amyloidosis. *Best Pract Res Clin Haematol, 18*, 709–727.
10. Cueto-Garcia, L., Reeder, G.S., Kyle, R.A., Wood, D.L., Seward, J.B., Naessens, J., Offord, K.P., Greipp, P.R., Edwards, W.D., Tajik, A.J. (1985). Echocardiographic findings in systemic amyloidosis: spectrum of cardiac involvement and relation to survival. *J Am Coll Cardiol, 6*, 737–743.
11. Dispenzieri, A., Kyle, R.A., Gertz, M.A., Therneau, T.M., Miller, W.L., Chandrasekaran, K., McConnell, J.P., Burritt, M.F., Jaffe, A.S. (2003). Survival in patients with primary systemic amyloidosis and raised serum cardiac troponins. *Lancet, 361*, 1787–1789.
12. Dispenzieri, A., Gertz, M.A., Kyle, R.A., Lacy, M.Q., Burritt, M.F., Therneau, T.M., McConnell, J.P., Litzow, M.R., Gastineau, D.A., Tefferi, A., *et.al.* (2004). Prognostication of survival using cardiac troponins and N-terminal pro-brain natriuretic peptide in patients with primary systemic amyloidosis undergoing peripheral blood stem cell transplantation. *Blood, 104*, 1881–1887.
13. Dispenzieri, A., Gertz, M.A., Kyle, R.A., Lacy, M.Q., Burritt, M.F., Therneau, T.M., Greipp, P.R., Witzig, T.E., Lust, J.A., Rajkumar, S.V.,*et al.* (2004). Serum cardiac troponins and N-terminal pro-brain natriuretic peptide: a staging system for primary systemic amyloidosis. *J Clin Oncol, 22*, 3751–3757.
14. Palladini, G., Campana, C., Klersy, C., Balduini, A., Vadacca, G., Perfetti, V., Perlini, S., Obici, L., Ascari, E., d'Eril, G.M., et al. (2003). Serum N-terminal pro-brain natriuretic peptide is a sensitive marker of myocardial dysfunction in AL amyloidosis. *Circulation, 107*, 2440–2445.
15. Koyama, J., Ray-Sequin, P.A., Falk, R.H. (2003). Longitudinal myocardial function assessed by tissue velocity, strain, and strain rate tissue Doppler echocardiography in patients with AL (primary) cardiac amyloidosis. *Circulation, 107*, 2446–2452.
16. Bellavia, D., Abraham, T.P., Pellikka, P.A., Al-Zahrani, G.B., Dispenzieri, A., Oh, J.K., Bailey, K.R., Wood, C.M., Novo, S., Miyazaki, C., Miller, F.A., Jr. (2007). Detection of left ventricular systolic dysfunction in cardiac amyloidosis with strain rate echocardiography. *J Am Soc Echocardiogr, 20*, 1194–1202.

17. Hosch, W., Bock, M., Libicher, M., Ley, S., Hegenbart, U., Dengler, T.J., Katus, H.A., Kauczor, H.U., Kauffmann, G.W., Kristen, A.V. (2007). MR-relaxometry of myocardial tissue: significant elevation of T1 and T2 relaxation times in cardiac amyloidosis. *Invest Radiol*, *42*, 636–642.

18. Kyle, R.A., Gertz, M.A., Greipp, P.R., Witzig, T.E., Lust, J.A., Lacy, M.Q., Therneau, T.M. (1997). A trial of three regimens for primary amyloidosis: colchicine alone, melphalan and prednisone, and melphalan, prednisone, and colchicine [see comments]. *N Engl J Med*, *336*, 1202–1207.

19. Kyle, R.A., Greipp, P.R., Garton, J.P., Gertz, M.A. (1985). Primary systemic amyloidosis. Comparison of melphalan/prednisone versus colchicine. *Am J Med*, *79*, 708–716.

20. Gertz, M.A., Lacy, M.Q., Lust, J.A., Greipp, P.R., Witzig, T.E., Kyle, R.A. (1999). Prospective randomized trial of melphalan and prednisone versus vincristine, carmustine, melphalan, cyclophosphamide, and prednisone in the treatment of primary systemic amyloidosis. *J Clin Oncol*, *17*, 262–267.

21. Dispenzieri, A., Lacy, M.Q., Kyle, R.A., Therneau, T.M., Larson, D.R., Rajkumar, S.V., Fonseca, R., Greipp, P.R., Witzig, T.E., Lust, J.A., Gertz, M.A. (2001). Eligibility for hematopoietic stem-cell transplantation for primary systemic amyloidosis is a favorable prognostic factor for survival. *J Clin Oncol*, *19*, 3350–3356.

22. Gertz, M.A., Lacy, M.Q., Dispenzieri, A., Ansell, S.M., Elliott, M.A., Gastineau, D.A., Inwards, D.J., Micallef, I.N., Porrata, L.F., Tefferi, A., Litzow, M.R. (2004). Risk-adjusted manipulation of melphalan dose before stem cell transplantation in patients with amyloidosis is associated with a lower response rate. *Bone Marrow Transplant*, *34*, 1025–1031.

23. Gertz, M.A., et al. (2007). Should troponin level be an exclusion criterion for stem cell transplantation in primary amyloidosis? *ASH Annu Meet Abstr*, *110*, 3003.

24. Moreau, P., Milpied, N., de Faucal, P., Petit, T., Herbouiller, P., Bataille, R., Harousseau, J.L. (1996). High-dose melphalan and autologous bone marrow transplantation for systemic AL amyloidosis with cardiac involvement. *Blood*, *87*, 3063–3064.

25. Comenzo, R.L., Vosburgh, E., Falk, R.H., Sanchorawala, V., Reisinger, J., Dubrey, S., Dember, L.M., Berk, J.L., Akpek, G., LaValley, M., et al. (1998). Dose-intensive melphalan with blood stem-cell support for the treatment of AL (amyloid light-chain) amyloidosis: survival and responses in 25 patients. *Blood*, *91*, 3662–3670.

26. Skinner, M., Sanchorawala, V., Seldin, D.C., Dember, L.M., Falk, R.H., Berk, J.L., Anderson, J.J., O'Hara, C., Finn, K.T., Libbey, C.A., *et.al.* (2004). High-dose melphalan and autologous stem-cell transplantation in patients with AL amyloidosis: an 8-year study. *Ann Intern Med*, *140*, 85–93.

27. Gertz, M.A., Lacy, M.Q., Dispenzieri, A., Hayman, S.R., Kumar, S. (2007). Transplantation for amyloidosis. *Curr Opin Oncol*, *19*, 136–141.

28. Jaccard, A., Moreau, P., Leblond, V., Leleu, X., Benboubker, L., Hermine, O., Recher, C., Asli, B., Lioure, B., Royer, B., et al.; Myélome Autogreffe (MAG) and Intergroupe Francophone du Myélome (IFM) Intergroup. (2007). High-dose melphalan versus melphalan plus dexamethasone for AL amyloidosis. *N Engl J Med*, *357*, 1083–1093.

29. Vesole, D.H., Pérez, W.S., Akasheh, M., Boudreau, C., Reece, D.E., Bredeson, C.N.; Plasma Cell Disorders Working Committee of the Center for International Blood and Marrow Transplant Research. (2006). High-dose therapy and

autologous hematopoietic stem cell transplantation for patients with primary systemic amyloidosis: a Center for International Blood and Marrow Transplant Research Study. *Mayo Clin Proc*, *81*, 880–888.

30. Dispenzieri, A., Kyle, R.A., Lacy, M.Q., Therneau, T.M., Larson, D.R., Plevak, M.F., Rajkumar, S.V., Fonseca, R., Greipp, P.R., Witzig, T.E., *et.al.* (2004). Superior survival in primary systemic amyloidosis patients undergoing peripheral blood stem cell transplantation: a case–control study. *Blood*, *103*, 3960–3963.
31. Schönland, S.O., Lokhorst, H., Buzyn, A., Leblond, V., Hegenbart, U., Bandini, G., Campbell, A., Carreras, E., Ferrant, A., Grommisch, L., *et al.;* Chronic Leukemia Working Party; Myeloma Subcommittee of the European Cooperative Group for Blood and Marrow Transplantation. (2006). Allogeneic and syngeneic hematopoietic cell transplantation in patients with amyloid light-chain amyloidosis: a report from the European Group for Blood and Marrow Transplantation. *Blood*, *107*, 2578–2584.
32. Jones, N.F., Hilton, P.J., Tighe, J.R., Hobbs, J.R. (1972). Treatment of "primary" renal amyloidosis with melphalan. *Lancet*, *2*, 616–619.
33. Cohen, H.J., Lessin, L.S., Hallal, J., Burkholder, P. (1975). Resolution of primary amyloidosis during chemotherapy: studies in a patient with nephrotic syndrome. *Ann Intern Med*, *82*, 466–473.
34. Buxbaum, J.N., Hurley, M.E., Chuba, J., Spiro, T. (1979). Amyloidosis of the AL type: clinical, morphologic and biochemical aspects of the response to therapy with alkylating agents and prednisone. *Am J Med*, *67*, 867–878.
35. Kyle, R.A., Greipp, P.R. (1978). Primary systemic amyloidosis: comparison of melphalan and prednisone versus placebo. *Blood*, *52*, 818–827.
36. Skinner, M., Anderson, J., Simms, R., Falk, R., Wang, M., Libbey, C., Jones, L.A., Cohen, A.S. (1996). Treatment of 100 patients with primary amyloidosis: a randomized trial of melphalan, prednisone, and colchicine versus colchicine only. *Am J Med*, *100*, 290–298.
37. Fielder, K., Durie, B.G. (1986). Primary amyloidosis associated with multiple myeloma: predictors of successful therapy. *Am J Med*, *80*, 413–418.
38. Gertz, M.A., Kyle, R.A., Greipp, P.R. (1991). Response rates and survival in primary systemic amyloidosis. *Blood*, *77*, 257–262.
39. Gertz, M.A., Kyle, R.A. (1990). Acute leukemia and cytogenetic abnormalities complicating melphalan treatment of primary systemic amyloidosis. *Arch Intern Med*, *150*, 629–633.
40. Lachmann, H.J., Gallimore, R., Gillmore, J.D., Carr-Smith, H.D., Bradwell, A.R., Pepys, M.B., Hawkins, P.N. (2003). Outcome in systemic AL amyloidosis in relation to changes in concentration of circulating free immunoglobulin light chains following chemotherapy. *Br J Haematol*, *122*, 78–84.
41. Palladini, G., Perfetti, V., Obici, L., Caccialanza, R., Semino, A., Adami, F., Cavallero, G., Rustichelli, R., Virga, G., Merlini, G. (2004). Association of melphalan and high-dose dexamethasone is effective and well tolerated in patients with AL (primary) amyloidosis who are ineligible for stem cell transplantation. *Blood*, *103*, 2936–2938.
42. Palladini, G., Russo, P., Nuvolone, M., Lavatelli, F., Perfetti, V., Obici, L., Merlini, G. (2007). Treatment with oral melphalan plus dexamethasone produces long-term remissions in AL amyloidosis. *Blood*, *110*, 787–788.

43. Levy, Y., Belghiti-Deprez, D., Sobel, A. (1988). [Treatment of AL amyloidosis without myeloma]. *Ann Med Interne (Paris)*, *139*, 190–193.

44. Wardley, A.M., Jayson, G.C., Goldsmith, D.J., Venning, M.C., Ackrill, P., Scarffe, J.H. (1998). The treatment of nephrotic syndrome caused by primary (light chain) amyloid with vincristine, doxorubicin and dexamethasone. *Br J Cancer*, *78*, 774–776.

45. van Gameren I.I., Hazenberg, B.P., Jager, P.L., Smit, J.W., Vellenga, E. (2002). AL amyloidosis treated with induction chemotherapy with VAD followed by high dose melphalan and autologous stem cell transplantation. *Amyloid*, *9*, 165–174.

46. Ichida, M., Imagawa, S., Ohmine, K., Komatsu, N., Hatake, K., Ozawa, K., Miura, Y. (2000). Successful treatment of multiple myeloma–associated amyloidosis by interferon-alpha, dimethyl sulfoxide, and VAD (vincristine, adriamycin, and dexamethasone). *Int J Hematol*, *72*, 491–493.

47. Gono, T., Matsuda, M., Shimojima, Y., Ishii, W., Koyama, J., Sakashita, K., Koike, K., Hoshii, Y., Ikeda, S. (2004). VAD with or without subsequent high-dose melphalan followed by autologous stem cell support in AL amyloidosis: Japanese experience and criteria for patient selection. *Amyloid*, *11*, 245–256.

48. Gertz, M.A., Lacy, M.Q., Lust, J.A., Greipp, P.R., Witzig, T.E., Kyle, R.A. (1999). Phase II trial of high-dose dexamethasone for untreated patients with primary systemic amyloidosis. *Med Oncol*, *16*, 104–109.

49. Gertz, M.A., Lacy, M.Q., Lust, J.A., Greipp, P.R., Witzig, T.E., Kyle, R.A. (1999). Phase II trial of high-dose dexamethasone for previously treated immunoglobulin light-chain amyloidosis. *Am J Hematol*, *61*, 115–119.

50. Palladini, G., Anesi, E., Perfetti, V., Obici, L., Invernizzi, R., Balduini, C., Ascari, E., Merlini, G. (2001). A modified high-dose dexamethasone regimen for primary systemic (AL) amyloidosis. *Br J Haematol*, *113*, 1044–1046.

51. Dhodapkar, M.V., Jagannath, S., Vesole, D., Munshi, N., Naucke, S., Tricot, G., Barlogie, B. (1997). Treatment of AL-amyloidosis with dexamethasone plus alpha interferon. *Leuk Lymphoma*, *27*, 351–356.

52. Dhodapkar, M.V., Hussein, M.A., Rasmussen, E., Solomon, A., Larson, R.A., Crowley, J.J., Barlogie, B.; United States Intergroup Trial Southwest Oncology Group. (2004). Clinical efficacy of high-dose dexamethasone with maintenance dexamethasone/alpha interferon in patients with primary systemic amyloidosis: results of United States Intergroup Trial Southwest Oncology Group (SWOG) S9628. *Blood*, *104*, 3520–3526.

53. Gertz, M.A., Kyle, R.A. (1993). Phase II trial of recombinant interferon alfa-2 in the treatment of primary systemic amyloidosis. *Am J Hematol*, *44*, 125–128.

54. Gertz, M.A.,. Kyle, R.A (1990). Phase II trial of alpha-tocopherol (vitamin E) in the treatment of primary systemic amyloidosis. *Am J Hematol*, *34*, 55–58.

55. Merlini, G., Ascari, E., Amboldi, N., Bellotti, V., Arbustini, E., Perfetti, V., Ferrari, M., Zorzoli, I., Marinone, M.G., Garini, P., et al. (1995). Interaction of the anthracycline 4′-iodo-4′-deoxydoxorubicin with amyloid fibrils: inhibition of amyloidogenesis. *Proc Natl Acad Sci U S A*, *92*, 2959–2963.

56. Gertz, M.A., Lacy, M.Q., Dispenzieri, A., Cheson, B.D., Barlogie, B., Kyle, R.A., Palladini, G., Geyer, S.M., Merlini, G. (2002). A multicenter phase II trial of 4′-iodo-4′deoxydoxorubicin (IDOX) in primary amyloidosis (AL). *Amyloid*, *9*, 24–30.

57. Dispenzieri, A., et al. (2002). High doses of thalidomide are not well tolerated in patients with primary systemic amyloidosis. *Blood, 98*(11), Abstract 1545.
58. Seldin, D.C., Choufani, E.B., Dember, L.M., Wiesman, J.F., Berk, J.L., Falk, R.H., O'Hara, C., Fennessey, S., Finn, K.T., Wright, D.G., *et. al.* (2003). Tolerability and efficacy of thalidomide for the treatment of patients with light chain–associated (AL) amyloidosis. *Clin Lymphoma, 3*, 241–246.
59. Dispenzieri, A., et al. (2004). Low dose single agent thalidomide is tolerated in patients with primary systemic amyloidosis, but responses are limited. *ASH Annu Meet Abstr, 104*(11), 4920–.
60. Palladini, G., Perfetti, V., Perlini, S., Obici, L., Lavatelli, F., Caccialanza, R., Invernizzi, R., Comotti, B., Merlini, G. (2005). The combination of thalidomide and intermediate-dose dexamethasone is an effective but toxic treatment for patients with primary amyloidosis (AL). *Blood, 105*, 2949–2951.
61. Wechalekar, A.D., Goodman, H.J., Lachmann, H.J., Offer, M., Hawkins, P.N., Gillmore, J.D. (2007). Safety and efficacy of risk-adapted cyclophosphamide, thalidomide, and dexamethasone in systemic AL amyloidosis. *Blood, 109*, 457–464.
62. Dispenzieri, A., Lacy, M.Q., Zeldenrust, S.R., Hayman, S.R., Kumar, S.K., Geyer, S.M., Lust, J.A., Allred, J.B., Witzig, T.E., Rajkumar, S.V., et al. (2007). The activity of lenalidomide with or without dexamethasone in patients with primary systemic amyloidosis. *Blood, 109*, 465–470.
63. Sanchorawala, V., Wright, D.G., Rosenzweig, M., Finn, K.T., Fennessey, S., Zeldis, J.B., Skinner, M., Seldin, D.C.. (2007). Lenalidomide and dexamethasone in the treatment of AL amyloidosis: results of a phase 2 trial. *Blood, 109*, 492–496.
64. Kastritis, E., Anagnostopoulos, A., Roussou, M., Toumanidis, S., Pamboukas, C., Migkou, M., Tassidou, A., Xilouri, I., Delibasi, S., Psimenou, E., et al. (2007). Treatment of light chain (AL) amyloidosis with the combination of bortezomib and dexamethasone. *Haematologica, 92*, 1351–1358.
65. Leung, N., Leung, T.R., Cha, S.S., Dispenzieri, A., Lacy, M.Q., Gertz, M.A. (2005). Excessive fluid accumulation during stem cell mobilization: a novel prognostic factor of first-year survival after stem cell transplantation in AL amyloidosis patients. *Blood, 106*, 3353–3357.
66. Zhou, P., Teruya-Feldstein, J., Lu, P., Fleisher, M., Olshen, A., Comenzo, R.L. (2008). Calreticulin expression in the clonal plasma cells of patients with systemic light-chain (AL-) amyloidosis is associated with response to high-dose melphalan. *Blood, 111*, 549–557.
67. Sanchorawala, V., et al. (2005). Early serum free light chain responses following high-dose melphalan and stem cell transplantation for AL amyloidosis predict treatment outcomes. *ASH Annu Meet Abstr, 106*(11), 1160–.
68. Cohen, A.D., Zhou, P., Chou, J., Teruya-Feldstein, J., Reich, L., Hassoun, H., Levine, B., Filippa, D.A., Riedel, E., Kewalramani, T., et al. (2007). Risk-adapted autologous stem cell transplantation with adjuvant dexamethasone +/− thalidomide for systemic light-chain amyloidosis: results of a phase II trial. *Br J Haematol, 139*, 224–233.
69. Biewend, M.L., Menke, D.M., Calamia, K.T. (2006). The spectrum of localized amyloidosis: a case series of 20 patients and review of the literature. *Amyloid, 13*, 135–142.

70. Utz, J.P., Gertz, M.A., Kalra, S. (2001). External-beam radiation therapy in the treatment of diffuse tracheobronchial amyloidosis. *Chest*, *120*, 1735–1738.

71. Pitz, M.W., Gibson, I.W., Johnston, J.B. (2006). Isolated pulmonary amyloidosis: case report and review of the literature. *Am J Hematol*, *81*, 212–213.

72. Miura, K., Shirasawa, H. (1993). Lambda III subgroup immunoglobulin light chains are precursor proteins of nodular pulmonary amyloidosis. *Am J Clin Pathol*, *100*, 561–566.

73. Neben-Wittich, M.A., Foote, R.L., Kalra, S. (2007). External beam radiation therapy for tracheobronchial amyloidosis. *Chest*, *132*, 262–267.

74. Dundore, P.A., Aisner, S.C., Templeton, P.A., Krasna, M.J., White, C.S., Seidman, J.D. (1993). Nodular pulmonary amyloidosis: diagnosis by fine-needle aspiration cytology and a review of the literature. *Diagn Cytopathol*, *9*, 562–564.

75. Michaels, L., Hyams, V.J. (1979). Amyloid in localised deposits and plasmacytomas of the respiratory tract. *J Pathol*, *128*, 29–38.

76. Mariani, A.J., Barrett, D.M., Kurtz, S.B., Kyle, R.A. (1978). Bilateral localized amyloidosis of the ureter presenting with anuria. *J Urol*, *120*, 757–759.

77. Shittu, O.B., Weston, P.M. (1994). Localised amyloidosis of the urinary bladder: a case report and review of treatment. *West Afr J Med*, *13*, 252–253.

78. Malek, R.S., Wahner-Roedler, D.L., Gertz, M.A., Kyle, R.A. (2002). Primary localized amyloidosis of the bladder: experience with dimethyl sulfoxide therapy. *J Urol*, *168*, 1018–1020.

79. Gepi-Attee, S., Gingell, J.C., Rigby, H.S. (1992). Urethral amyloid. *Br J Urol*, *69*, 431–432.

80. Hayashi, T., Kojima, S., Sekine, H., Mizuguchi, K. (1998). Primary localized amyloidosis of the ureter. *Int J Urol*, *5*, 383–385.

81. Breathnach, S.M. (1988). Amyloid and amyloidosis. *J Am Acad Dermatol*, *18*, 1–16.

82. Moon, A.O., Calamia, K.T., Walsh, J.S. (2003). Nodular amyloidosis: review and long-term follow-up of 16 cases. *Arch Dermatol*, *139*, 1157–1159.

83. Kakani, R.S., Goldstein, A.E., Meisher, I., Hoffman, C. (2001). Nodular amyloidosis: case report and literature review. *J Cutan Med Surg*, *5*, 101–104.

84. Inoue, K., Takahashi, M., Hamamoto, Y., Muto, M., Ishihara, T. (2000). An immunohistochemical study of cytokeratins in skin-limited amyloidosis. *Amyloid*, *7*, 259–265.

85. Bastian, F.O. (1974). Amyloidosis and the carpal tunnel syndrome. *Am J Clin Pathol*, *61*, 711–717.

86. Kyle, R.A., Eilers, S.G., Linscheid, R.L., Gaffey, T.A. (1989). Amyloid localized to tenosynovium at carpal tunnel release. Natural history of 124 cases. *Am J Clin Pathol*, *91*, 393–397.

87. Dispenzieri, A., Lacy, M.Q., Rajkumar, S.V., Geyer, S.M., Witzig, T.E., Fonseca, R., Lust, J.A., Greipp, P.R., Kyle, R.A., Gertz, M.A. (2003). Poor tolerance to high doses of thalidomide in patients with primary systemic amyloidosis. *Amyloid*, *10*, 257–261.

36

FAMILIAL AND SENILE AMYLOIDOSIS CAUSED BY TRANSTHYRETIN

STEVEN R. ZELDENRUST

Division of Hematology, Mayo Clinic College of Medicine, Rochester, Minnesota

MERRILL D. BENSON

Department of Pathology and Laboratory Medicine, Indiana University School of Medicine, Indianapolis, Indiana

INTRODUCTION

Transthyretin-associated amyloidosis (ATTR) and senile systemic amyloidosis (SSA) are two different forms of systemic amyloidosis that result in the formation of amyloid deposits derived from the same plasma protein, transthyretin (TTR). Despite their common origin, the clinical features of the diseases vary considerably, resulting in markedly different outcomes. Patients with ATTR typically experience a progressive, fatal form of the disease with extensive morbidity, requiring aggressive treatment approaches. Patients with SSA have a more benign course, and supportive care is generally the preferred approach. We will highlight herein the clinical characteristics for each form of the disease and discuss what is known about the pathogenesis of transthyretin-derived amyloidosis and the impact of various treatment approaches.

Protein Misfolding Diseases: Current and Emerging Principles and Therapies,
Edited by Marina Ramirez-Alvarado, Jeffery W. Kelly, and Christopher M. Dobson
Copyright © 2010 John Wiley & Sons, Inc.

TRANSTHYRETIN-ASSOCIATED AMYLOIDOSIS

Background

To date, over 100 different mutations in the TTR protein have been reported, most in association with ATTR [1]. There are a wide variety of clinical syndromes in affected individuals, representing the varied target organs of amyloid deposition (Table 1). Depending on the predominant organ involved, patients may present with congestive heart failure, peripheral and/or autonomic neuropathy, nephrotic-range proteinuria, malabsorption, and even intracranial bleeding.

Clinical Characteristics

The initial descriptions of ATTR focused on the progressive peripheral sensory neuropathy present in many cases. The nervous system remains the most commonly affected organ, with up to 90% of affected persons developing peripheral nerve involvement at some point in the disease course [2]. Affected persons typically present with a progressive peripheral parasthesia, in a stocking-glove pattern of distribution. Pain and temperature impairment is also common, with many experiencing painful dysesthesias. Bilateral carpal tunnel syndrome is seen in nearly half of patients with ATTR. More advanced involvement leads to motor weakness and gait instability. Autonomic involvement is less common, affecting a little over a third of patients [3]. Patients may present with dyshidrosis, impotence, orthostatic hypotension, or urinary retention. Alternating diarrhea and constipation are also frequently indicative of autonomic failure. More rarely, intestinal pseudo-obstruction develops, resulting in persistent nausea, vomiting, and malnutrition.

More recently, reports of central nervous system (CNS) involvement in ATTR have been published [4–7]. These people may exhibit cerebral infarction and hemorrhage, ataxia, seizures, and even dementia. Of note, although TTR has been implicated in Alzheimer disease, a form of CNS amyloid, no parenchymal amyloid deposits have been reported in ATTR [8,9]. The relationship between leptomeningeal amyloid deposits and the clinical symptoms seen in these patients remains unclear. It has been suggested that aberrant metabolism of TTR produced locally by the choroid plexus, rather than circulating plasma protein, is responsible for the CNS deposits [10]. This theory has been substantiated by the finding of new leptomeningeal deposits in recipients of orthotopic liver transplantation, in which the circulating variant form of the protein is undetectable and only wild-type protein is synthesized by the liver [11].

Cardiac deposits occur less frequently, with only a fourth having evidence of cardiomyopathy at diagnosis [2]. As with nerve deposition, clinically apparent heart involvement becomes more common during the course of the disease, with nearly two-thirds ultimately developing cardiac amyloid. Deposition of amyloid within the myocardium is not always uniform, but classic features are

TABLE 1 Transthyretin Amyloidosis

Mutation	Codon Change	Clinical Features[a]	Geographic Kindreds
Cys10Arg	TGT–CGT	Heart, eye, PN	United States (PA)
Leu12Pro	CTG–CCG	LM	UK
Asp18Glu	GAT–GAA	PN	South America, United States
Asp18Gly	GAT–GGT	LM	Hungary
Asp18Asn	GAT–AAT	Heart	United States
Val20Ile	GTC–ATC	Heart, CTS	Germany, United States
Ser23Asn	AGT–AAT	Heart, PN, eye	United States
Pro24Ser	CCT–TCT	Heart, CTS, PN	United States
Ala25Ser	GCC–TCC	Heart, CTS, PN	United States
Ala25Thr	GCC–ACC	LM, PN	Japan
Val28Met	GTG–ATG	PN, AN	Portugal
Val30Met	GTG–ATG	PN, AN, eye, LM	Portugal, Japan, Sweden, United States (FAP I)
Val30Ala	GTG–GCG	Heart, AN	United States
Val30Leu	GTG–CTG	PN, heart	Japan
Val30Gly	GTG–GGG	LM, eye	United States
Val32Ala	GTG–GCG	PN	Israel
Phe33Ile	TTC–ATC	PN, eye	Israel
Phe33Leu	TTC–CTC	PN, heart	United States
Phe33Val	TTC–GTC	PN	UK, Japan, China
Phe33Cys	TTC–TGC	CTS, heart, eye, kidney	United States
Arg34Thr	AGA–ACA	PN, heart	Italy
Arg34Gly	AGA–GGA	Eye	UK
Lys35Asn	AAG–AAC	PN, AN, heart	France
Lys35Thr	AAG–ACG	Eye	United States
Ala36Pro	GCT–CCT	Eye, CTS	United States
Asp38Ala	GAT–GCT	PN, heart	Japan
Trp41Leu	TGG–TTG	Eye, PN	United States
Glu42Gly	GAG–GGG	PN, AN, heart	Japan, United States, Russia
Glu42Asp	GAG–GAT	Heart	France
Phe44Ser	TTT–TCT	PN, AN, heart	United States
Ala45Thr	GCC–ACC	Heart	United States
Ala45Asp	GCC–GAC	Heart, PN	United States
Ala45Ser	GCC–TCC	Heart	Sweden
Gly47Arg	GGG–CGG/AGG	PN, AN	Japan
Gly47Ala	GGG–GCG	Heart, AN	Italy, France
Gly47Val	GGG–GTG	CTS, PN, AN, heart	Sri Lanka
Gly47Glu	GGG–GAG	Heart, PN, AN	Turkey, United States, Germany
Thr49Ala	ACC–GCC	Heart, CTS	France, Italy
Thr49Ile	ACC–ATC	PN, heart	Japan, Spain
Thr49Pro	ACC–CCC	Heart, PN	United States

(Continued)

TABLE 1 ***(Continued)***

Mutation	Codon Change	Clinical Features[a]	Geographic Kindreds
Ser50Arg	AGT–AGG	AN, PN	Japan, France/Italy, United States
Ser50Ile	AGT–ATT	Heart, PN, AN	Japan
Glu51Gly	GAG–GGG	Heart	United States
Ser52Pro	TCT–CCT	PN, AN, heart, kidney	UK
Gly53Glu	GGA–GAA	LM, heart	Basque, Sweden
Glu54Gly	GAG–GGG	PN, AN, eye	UK
Glu54Lys	GAG–AAG	PN, AN, heart, eye	Japan
Glu54Leu	GAG – CTG		UK
Leu55Pro	CTG–CCG	Heart, AN, eye	United States, Taiwan
Leu55Arg	CTG–CGG	LM	Germany
Leu55Gln	CTG–CAG	Eye, PN	United States
Leu55Glu	CTG–CAG	Heart, PN, AN	Sweden
His56Arg	CAT–CGT	Heart	United States
Gly57Arg	GGG–AGG	Heart	Sweden
Leu58His	CTC–CAC	CTS, heart	United States (MD) (FAP II)
Leu58Arg	CTC–CGC	CTS, AN, eye	Japan
Thr59Lys	ACA–AAA	Heart, PN, AN	Italy, United States (Chinese)
Thr60Ala	ACT–GCT	Heart, CTS	United States (Appalachian)
Glu61Lys	GAG–AAG	PN	Japan
Glu61Gly	GAG–GGG	Heart, PN	United States
Phe64Leu	TTT–CTT/ TTG	PN, CTS, heart	United States, Italy
Phe64Ser	TTT–TCT	LM, PN, eye	Canada, UK
Ile68Leu	ATA–TTA	Heart	Germany
Tyr69His	TAC–CAC	Eye, LM	Canada, United States
Tyr69Ile	TAC–ATC[b]	Heart, CTS, AN	Japan
Lys70Asn	AAA–AAC	Eye, CTS, PN	United States
Val71Ala	GTG–GCG	PN, eye, CTS	France, Spain
Ile73Val	ATA–GTA	PN, AN	Bangladesh
Ser77Tyr	TCT–TAT	Kidney	United States (IL, TX), France
Ser77Phe	TCT–TTT	PN, AN, heart	France
Tyr78Phe	TAC–TTC	PN, CTS, skin	France
Ala81Thr	GCA–ACA	Heart	United States
Ala81Val	GCA–GTA	Heart	UK
Ile84Ser	ATC–AGC	Heart, CTS, eye	United States (IN), Hungary (FAP II)
Ile84Asn	ATC–AAC	Heart, eye	United States
Ile84Thr	ATC–ACC	Heart, PN	Germany, UK
His88Arg	CAT–CGT	Heart	Sweden
Glu89Gln	GAG–CAG	PN, heart	Italy
Glu89Lys	GAG–AAG	PN, heart	United States
His90Asp	CAT–GAT	Heart	UK
Ala91Ser	GCA–TCA	PN, CTS, heart	France
Glu92Lys	GAG–AAG	Heart	Japan

(Continued)

INSTITUT MEDIATHEQUE PASTEUR

TABLE 1 ***(Continued)***

Mutation	Codon Change	Clinical Features[a]	Geographic Kindreds
Val94Ala	GTA–GCA	Heart, PN, AN, kidney	Germany, United States
Ala97Gly	GCC–GGC	Heart, PN	Japan
Ala97Ser	GCC–TCC	PN, heart	Taiwan, United States
Ile107Val	ATT–GTT	Heart, CTS, PN	United States
Ile107Met	ATT–ATG	PN, heart	Germany
Ile107Phe	ATT–TTT	PN, AN	UK
Ala109Ser	GCC–TCC	PN, AN	Japan
Leu111Met	CTG–ATG	Heart	Denmark
Ser112Ile	AGC–ATC	PN, heart	Italy
Tyr114Cys	TAC–TGC	PN, AN, eye, LM	Japan, United States
Tyy114His	TAC–CAC	CTS, skin	Japan
Tyr116Ser	TAT–TCT	PN, CTS, AN	France
Ala120Ser	GCT–TCT	Heart	Afro-Caribbean
Val122Ile	GTC–ATC	Heart	United States
ΔVal122	GTC–ΔΔΔ	Heart, PN	United States (Ecuadoran), Spain
Val122Ala	GTC–GCC	Heart, eye, PN	United States

[a] AN, autonomic neuropathy; CTS, carpal tunnel syndrome; eye, vitreous deposits; LM, leptomeningeal; PN, peripheral neuropathy.

[b] Double nucleotide substitution.

symmetric thickening of the intraventricular septum and posterior ventricular wall. The electrocardiogram may show typical low-voltage changes or the pseudoinfarction pattern due to loss of anterolateral or inferior forces. The echocardiogram remains the gold standard, showing concentric hypertrophy and diastolic dysfunction as the classic features. A granular, sparkling pattern of the myocardium is often observed. Systolic function, measured by ejection fraction, is typically intact until late in the course of the disease. Cardiac biomarkers such as troponin I (TnI) and brain natriuretic peptide (BNP) have been shown to be highly sensitive and predictive of survival in light-chain amyloidosis (AL) [12]. No such studies have been reported in familial forms of the disease to date. More advanced imaging methods include cardiac magnetic resonance imaging to detect delayed gadolinium enhancement, which has been shown to be informative in three-fourths of patients with cardiac amyloidosis [13]. Scintigraphy using ^{99m}Tc-labeled phosphonates has also been reported to have a high sensitivity and specificity for cardiac involvement and may be helpful in distinguishing the different forms of amyloid [14]. Retention of these compounds in the myocardium noted during a routine bone scan is occasionally the first evidence of TTR amyloidosis with cardiac involvement (Fig. 1).

Patients with cardiac amyloid involvement typically present with dyspnea on exertion or orthopnea, reflecting the diastolic dysfunction seen early in the disease. More advanced cardiac involvement leads to systolic failure and conduction system disruption. Conduction delay can result in isolated bundle

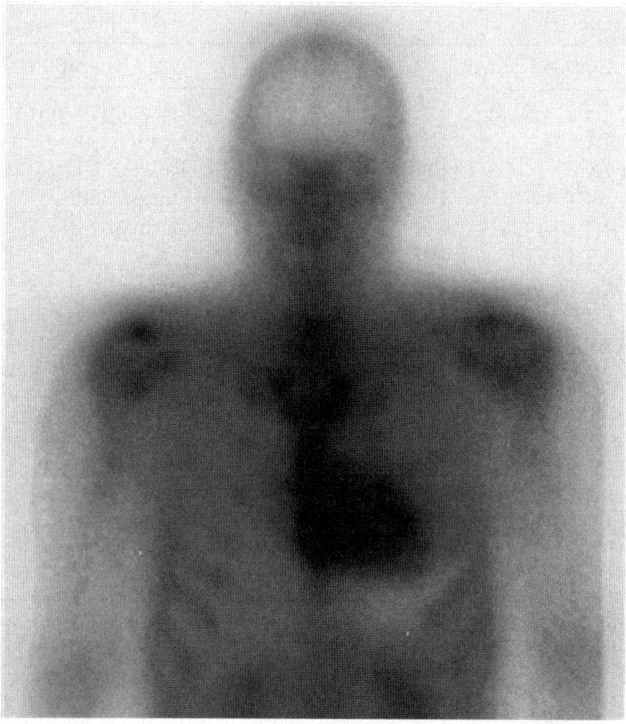

FIG. 1 Cardiac uptake of ^{99m}Tc PYP seen on a bone scan of a patient with TTR cardiac amyloidosis. This finding in nearly always associated with TTR-associated amyloidosis.

branch block, atrioventricular nodal block, tachyarrhythmias, or atrial fibrillation. Sudden death remains a common cause of mortality. More rarely, patients may present with classical anginal symptoms probably due to subendocardial ischemia resulting from vascular amyloid deposition.

Of particular note in regard to cardiac involvement in ATTR is the V122I mutation found in a significant number of African-Americans. Carriers of this mutation have a high incidence of symptomatic cardiac amyloidosis. The high allele frequency makes this one of the most common pathologic genetic mutations in the United States and may explain the increased incidence of cardiovascular disease in elderly blacks. Unfortunately, recognition of the disease is frequently poor, leading to decreased reporting and appropriate treatment in many cases.

Ocular involvement, although not a fatal consequence, remains a significant source of morbidity in ATTR patients [15]. TTR is known to be synthesized locally by the pigmented retinal epithelium [16]. Vitreous amyloid accumulation leads to vision loss and frequently requires vitrectomy, which is helpful in restoring vision. Most patients will develop recurrent deposits and require additional procedures. Secondary glaucoma may result, requiring more aggressive surgical treatment [17]. The scalloped pupil deformity seen in both Japanese and Swedish patients with ATTR is another ocular manifestation, probably related to ciliary nerve involvement [18].

Renal involvement, while common in AL, is less frequent in ATTR patients with up to one-third of patients showing evidence of significant proteinuria at some stage of the disease [19]. Patients may develop glomerular deposition with resulting nephrotic-range proteinuria, leading to significant hypoalbuminemia and edema. More rarely, vascular deposits have been observed, resulting in a decrease in glomerular filtration and eventual end-stage renal disease requiring dialysis.

Direct involvement of the gastrointestinal tract is a frequent finding in some kindreds, but frank malabsorption is rare [20]. More often, gastrointestinal symptoms are the result of significant autonomic neuropathy, as described above. Mucosal friability due to amyloid infiltration can lead to gastrointestinal bleeding in rare cases [2]. Given the prevalence of gastrointestinal symptoms, endoscopy with biopsy of the stomach, colon, or small bowel is often the initial diagnostic procedure.

The wide variety and variable severity of clinical manifestations of ATTR often results in a substantial delay in accurate diagnosis. Peripheral neuropathy and renal injury are frequently attributed to diabetes and cardiomyopathy to ischemic injury. A positive family history is often obtained only after the diagnosis has been established. In one series, the time from initial onset of symptoms to the histologic diagnosis of amyloidosis ranged from 2 to 100 months, with a median of nearly three years [2]. Even when the diagnosis of amyloidosis is established, patients can be misclassified as AL or AA and treated inappropriately with cytotoxic chemotherapy. In a retrospective analysis of 350 patients diagnosed with AL at a major amyloid treatment center in the UK, nearly 4% were found to have ATTR [21].

Survival in ATTR is difficult to predict in individual cases, due to the considerable clinical heterogeneity seen in different kindreds. As a group, patients with ATTR fare considerably better than those with AL. The median survival in ATTR is between 5 and 15 years, depending on the population being studied, compared to a median of 20 months for AL [2,3]. The site of predominant clinical involvement is critical, with the presence of cardiomyopathy conferring a median survival of 3.4 years [2,3]. Once again, this compares favorably to AL patients with cardiomyopathy, who have a median survival of less than one year. This suggests that the rate of amyloid deposition and/or toxic effects of fibrils derived from TTR are less than that of light-chain-derived amyloid. Age, gender (male), and the presence of peripheral or autonomic neuropathy and carpal tunnel syndrome have also been shown to affect survival. Death typically results from progressive heart failure, inanition related to progressive neuropathy, wasting from malabsorption, and infection. Sudden cardiac death is not uncommon. End-stage renal failure and resulting complications related to hemodialysis are also reported.

Pathogenesis

Although a great deal has been discovered about the underlying cause of ATTR in the last 25 years, including the genetic lesions responsible and resulting changes in the TTR protein, the mechanism by which amyloid fibrils of any type exert their toxic effects is not known. It has been clearly shown that intact TTR amyloid fibrils themselves are nontoxic in vitro, but the formation of fibrils from a soluble precursor results in apoptosis [22]. These data suggest that it is an intermediate formed during the conversion of TTR from its normal soluble state into the mature fibril that is the toxic species. The exact nature of this intermediate remains elusive, but studies have implicated monomeric and dimeric forms of TTR, as well as soluble oligomers and nonfibrillar aggregates, as potential candidates [23–25]. A similar phenomenon has been observed in the toxicity of the Aβ protein in Alzheimer disease, in which oligomeric or protofibrillar species have been identified as potent toxins [26,27]. Thus, it appears that the mature amyloid deposits seen at the time of biopsy may represent a final endpoint on the path of toxicity, with the damage being done much earlier in the course of fibril formation.

It is clear, however, that amyloid deposits do result in direct effects on end-organ function. This is particularly evident in the effects of amyloid on the heart. Early deposition results primarily in diastolic dysfunction, resulting from impaired relaxation [28]. The thickening of the myocardium as a result of interstitial amyloid buildup results in significant mass effect on the heart, with an increase in size from 350 g to over 1000 g in many cases. Contractility, as measured by systolic ejection fraction, is maintained until the wall thickness reaches a critical threshold. Early systolic impairment can be seen through the use of more sensitive methods of detection, such as strain rate Doppler analysis echocardiography [29]. This effect presumably represents a direct physical effect of the amyloid deposits rather than a toxic effect of the fibrils or intermediates on the cardiac myocytes themselves.

Deposition of amyloid within the vitreous of the eye is clearly a direct effect in which the fibrils interrupt the path of light to the retina. This results in decreased visual acuity. Removal of the deposits by vitrectomy immediately restores vision, confirming the lack of a deleterious effect on the retina itself. Vitreous deposits ultimately recur and diminished vision develops, resulting in the need for repeat extraction, in most cases.

Gastrointestinal involvement can result in malabsorption through a variety of means. Dysmotility from autonomic neuropathy can cause gastric dumping and altered digestion of crucial nutrients. Extensive submucosal deposits can also have direct effects by interfering with the transit of fatty acids and amino acids across the intestinal epithelium. This results in steatorrhea and high-volume diarrhea. This disruption of the normal function of the gastrointestinal tract is yet another example of direct effects of amyloid deposits themselves.

The presence of amyloid deposits in peripheral nerves results in axonal degeneration and demyelination [30]. Fibers near amyloid deposits demonstrate distortion of the myelin sheath, segmental demyelination, and Wallerian degeneration [31]. Due to the proximity of these effects to the deposits themselves, direct compression of the nerve as a result of amyloid deposition has been postulated to play a role in the development of peripheral neuropathy [32]. In addition, the presence of amyloid within endoneurial blood vessels suggests that ischemia is another mechanism of direct neurotoxicity of the deposits. In support of this theory, interstitial edema of the endoneurium adjacent to deposits of amyloid in the sciatic nerves and brachial plexuses of FAP patients have been reported [33]. Altered permeability resulting from vascular amyloid involvement may produce severe endoneurial edema, leading directly to ischemic injury of peripheral nerves.

Leptomeningeal amyloidosis is yet another example of vascular involvement leading directly to tissue injury. Patients with specific TTR variants associated with leptomeningeal deposits exhibit cerebral hemorrhage, dementia, ataxia, cerebral infarction, and seizures [5,34–36]. Pathologic examination in these patients reveals extensive leptomeningeal and vascular wall involvement with no discernible parenchymal deposits. Symptoms in these patients have been linked to direct impairment of cortical blood flow resulting from the amyloid itself. Cerebral hemorrhage results from breakdown of vascular integrity, also a direct effect of the amyloid deposits themselves. Hydrocephalus has been reported in some cases, requiring ventriculoperitoneal shunt placement [37]. It is possible that the presence of amyloid within the leptomeninges results in a chronic basal arachnoiditis, a well-known cause of hydrocephalus, representing yet another direct effect of the deposits.

A second mystery is why so many different mutations within the TTR gene result in the same tendency to form amyloid. It has been shown that amyloidogenic mutations destabilize the native structure of TTR, leading to conformational changes that favor reassembly into fibrils [38,39]. A wide variety of altered conditions can induce dissociation of TTR into alternative monomeric intermediates, including temperature, ionic strength, pH, and protein concentration, promoting the formation of soluble aggregates [40,41]. A correlation between the thermodynamic stability of TTR variants and their ability to unfold and form aggregates has been demonstrated [42,43]. Wild-type TTR has been shown to undergo similar conformational changes at increased temperatures at physiologic pH and under hydrostatic pressure, resulting in increased fibrillogenesis [44,45].

The means through which the toxic form of TTR exerts its effect on the cell remains equally obscured. Multiple pathways have been implicated in studies of TTR toxicity, including activation of the receptor for advanced glycation end products (RAGE), mitogen-activated protein kinases (MAPK), endoplasmic reticulum stress, oxidative stress, and calcium homeostasis [24,46–50]. The activation of RAGE is an attractive hypothesis, as it has been shown to be involved in many pathways of cellular toxicity, such as the regulation of nuclear factor κ-B, MAPK, and Jun-N terminal kinase signaling. A greater understanding of the toxicity of amyloid deposits is clearly needed, since treatment for these patients is less effective than desired in many cases.

Treatment of ATTR

The goal of treatment for all forms of systemic amyloidosis is to reduce or remove the supply of the precursor protein, thereby preventing further amyloid deposition. Since the synthesis of circulating TTR is almost exclusively hepatic, orthotopic liver transplantation (OLT) has been offered as a potentially curative treatment for ATTR since 1990 [51]. This approach has been widely adopted, and data from the Familial Amyloidotic Polyneuropathy World Transplant Registry (FAPWTR) documents over 1300 OLTs performed worldwide as of 2006 (unpublished data). Given the normal synthetic function of the liver in ATTR, surgical outcomes are excellent in ATTR patients, leading to decreased surgical mortality, shorter hospital stays, and decreased need for blood products. The overall survival of transplanted ATTR patients is also good, at 77% at five years [52].

The benefit of OLT in this patient population is predominantly subjective improvement in neurologic symptoms. Of the 149 evaluable patients, 42% reported an improvement in their sensory neuropathy. Patients with duration of symptoms for less than two years and relatively mild impairment prior to transplantation were more likely to result in improvement after OLT. The severity of the neurologic deficit prior to transplant was shown to be similarly predictive of outcome in a report of 25 French patients [53]. In this study, no improvement was noted in any patient, and 40% showed progression of their neurologic deficit following OLT. Similar results of stable peripheral neuropathy with lack of objective improvement has been reported in a series of patients from the UK [54]. In a series of 15 patients treated with OLT in the United States, only 35% showed stable or improved peripheral neuropathy symptoms following transplant [55]. Overall, despite encouraging results from some centers, the benefit of OLT appears to be greatest in patients with shorter duration and milder presenting peripheral neuropathy.

Autonomic neuropathy is reported by the FAPWTR to improve in up to half of patients following OLT, as determined by improvement in gastrointestinal symptoms [52]. Improvement in autonomic neuropathy was noted to be the earliest sign of improvement following OLT in a series of nine patients from the United States, with most showing an improvement in gastrointestinal symptoms, orthostasis, and anhidrosis [56]. Other series have shown a lack of improvement in symptoms such as orthostatic hypotension, but recovery of nutritional status has been reported in patients who were malnourished due to significant autonomic neuropathy prior to OLT [57].

The benefit of OLT for ATTR patients with cardiac involvement is less clear. A high rate of death due to cardiovascular causes occurs following OLT, often within 90 days of surgery [52]. Despite early reports of stable or improvement in cardiac status of ATTR patients undergoing OLT, subsequent studies have shown that it is common for cardiac involvement by echocardiographic findings to progress in some patients [55,58–60]. Similarly, cardiac conduction abnormalities become more frequent and can even be life-threatening following OLT [61]. Progressive cardiac amyloidosis following OLT is not limited to those with preoperative cardiomyopathy or specific TTR mutations.

It has been noted that some patients appear to show a parodoxical increase in the rate of amyloid deposition following OLT [60]. Since the source of variant TTR production leading to amyloid fibril is removed at the time of OLT, the accumulation of new amyloid deposits is completely unexpected. Subsequent evaluation of the amyloid deposits has shown that wild-type TTR is present in increased amounts in the heart following OLT [62,63]. Thus, it appears that wild-type TTR can continue to form amyloid deposits, particularly in the heart, in patients with ATTR following OLT. This may also explain why patients with nerve involvement show progressive symptoms following OLT, as wild-type TTR has also been shown to contribute to peripheral nerve amyloid deposits [64].

Progressive vitreous amyloid deposits have similarly been reported to occur following OLT [65]. Since TTR is known to be synthesized by the pigmented retinal epithelium, it is likely that continued production of variant TTR is responsible for ongoing amyloid formation in these cases. Similar results of de novo leptomeningeal amyloid deposition have been reported in two Japanese patients following OLT, further documenting that variant TTR produced by the choroid plexus can form progressive amyloid [11].

Since the synthetic function of the liver is intact in ATTR patients, it has become common practice to perform a domino transplant, in which the liver harvested from the ATTR patient is subsequently transplanted into another person [66–68]. A total of 532 domino transplantations are reported from 47 centers in 16 counties by the FAPWTR. The recipients of the ATTR livers tended to be somewhat older than most liver transplant candidates and most often had either primary or metastatic malignancy as the indication for transplant. Survival does not appear adversely affected, although tumor recurrence and infection remain frequent causes of death. Interestingly, reports of amyloid developing in the peripheral nerves and gastric mucosa of recipients of ATTR livers have now surfaced after extended periods [69,70]. This has necessitated a second liver transplant in at least one case [69].

Given the variable benefit of OLT in ATTR and the fact that many patients are either too old or too ill to be considered candidates for a major surgery at the time of diagnosis, alternative forms of treatment have been explored. These studies have focused on alternative means of suppressing variant TTR production by the liver and stabilization of the TTR tetramer.

Efforts to prevent or diminish the formation of amyloid fibrils through the use of small molecules that stabilize the TTR tetramer have been pursued based on several lines of evidence. The first was the finding that Portuguese patients carrying both an amyloidogenic V30M-mutated TTR gene and a nonpathogenic T119M allele had a milder clinical course than those with the V30M mutation alone [71]. It was subsequently shown that hybrid tetramers containing both the V30M and T119M variants showed reduced amyloid-forming potential than that off tetramers containing only V30M [72]. This decrease in amyloidogenicity was linked to increased stabilization of the tetramer, decreasing the pool of free monomer able to undergo conformational shifts leading to amyloid formation.

Further work in this area has demonstrated that small molecules that bind to the thyroxine-binding site of the tetramer can have a similar stabilizing effect, resulting in a decreased rate of tetramer dissociation [73]. One of these compounds is the commercially available anti-inflammatory drug diflunisal. Although not the most potent stabilizer of TTR, the safety and tolerability of diflunisal make it an attractive choice for testing in clinical trials in humans. Testing performed in vivo have demonstrated that orally administered diflunisal provides sufficient serum levels to stabilize TTR tetramers against dissociation [74]. This work has led to a first-in-class randomized international phase III clinical trial of diflunisal for the prevention of ATTR in patients with peripheral neuropathy that is currently under way at multiple centers. An additional novel compound with much more potent tetramer stabilizing activity is also currently being tested in a phase II clinical trial. The results of these trials are eagerly awaited as proof of concept that stabilization of the TTR tetramer will result in clinical benefits.

The finding that TTR knockout mice, which produce absolutely no detectable TTR, have no phenotypic consequences, has led to the investigation of additional treatment approaches designed to diminish or abolish variant TTR production without the need for liver transplantation. Recently published data using antisense oligonucleotides in mice transgenic for the I84S human TTR gene have shown that hepatic TTR synthesis can be effectively suppressed for substantial periods [75]. Other groups have shown that suppression of TTR synthesis by ribozymes may be a feasible approach for treatment [76]. In addition, gene therapy in the form of targeted conversion of the variant allele has been demonstrated to be possible both in vitro and in vivo [77]. Although these approaches have not yet been tested in clinical trials, they offer hope for patients that are either not candidates for liver transplantation or those who develop progressive disease after liver transplantation.

SENILE SYSTEMIC AMYLOIDOSIS

Background

Senile systemic amyloidosis (SSA) is the most common form of systemic amyloidosis. In autopsy series, up to 28% of persons over 80 years old were found to have cardiac amyloid deposits composed of TTR [78,79]. Early studies identified TTR as the main component of the amyloid fibrils is these patients [80]. Subsequent sequencing of the TTR protein extracted from patients with SSA showed only wild-type protein, confirming the lack of an inherited mutation as seen in ATTR [81]. Sequencing of the cDNA from a patient with SSA confirmed the lack of a mutation within the TTR gene and altered transcription as a potential contributor [82]. Some overlap in the clinical picture and age of onset in people with the V122I mutation led to initial confusion whether all patients with SSA had an inherited form of the disease [83]. It is now recognized that although there is distinct similarity in the phenotype of V122I ATTR and SSA, wild-type TTR is solely responsible for the amyloid in SSA.

Clinical Characteristics

The clinical picture of SSA is much different from that in ATTR. Cardiac involvement is seen in virtually all cases, and congestive heart failure is the dominating presenting symptom, which led to the initial use of the term *senile cardiac amyloidosis* for this disease [84]. More detailed examination has revealed that deposits are not limited to the heart, but frequently involve the aorta, lung, gastrointestinal tract, liver, and kidney, prompting the adoption of senile systemic amyloidosis (SSA) as being more accurate [78]. Men are affected far more often than women, for unclear reasons. Despite the widespread appearance of amyloid deposits, patients rarely present with any symptoms attributable to disease outside the heart. Carpal tunnel syndrome may be an exception, as it is not uncommon to see localized amyloid deposits at the time of surgical release [85,86]. Patients may have hepatomegaly, but this is almost always secondary to severe congestive heart failure. Renal failure is uncommon, but may be significant in some cases.

Cardiac involvement is often extensive at the time of diagnosis of SSA. The severity of symptoms is typically less than one would expect for a patient with a similar amount of AL deposits. Patients with SSA have significantly greater septal and ventricular wall thickness at diagnosis then to patients with AL [87]. This may be due to less impairment of the conduction system, although atrial fibrillation is commonly seen due to atrial enlargement. Low voltage on electrocardiogram, a classic finding in AL patients with cardiac involvement, is seen in only about a third of SSA patients [87]. However, conduction abnormalities, including complete heart block, can be seen in a significant number of patients. Rarely, septal thickening can lead to left ventricular tract outflow obstruction, as seen more commonly in hypertrophic obstructive cardiomyopathy [88]. Thickening of the cardiac valves is occasionally seen, due to direct infiltration of the valve leaflets by amyloid (Fig. 2). The impact on cardiac function is predominantly a restrictive cardiomyopathy with resulting diastolic dysfunction. Survival for SSA is vastly superior to that of patients with AL cardiac involvement, with a median of 60 to 75 months versus 6 to 11 months [87,89]. Progressive heart failure is the most common cause of death.

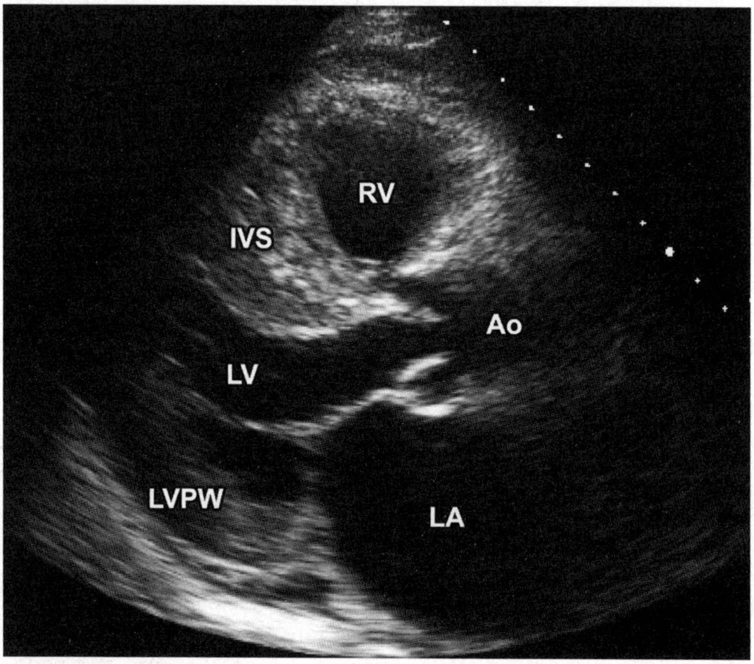

FIG. 2 Two-dimensional echocardiogram for an 80-year-old man with senile cardiac amyloidosis. Classic features of amyloid cardiomyopathy are thickened intraventricular septum (IVS), thickened left ventricular posterior wall (LVPW), and enlarged left atrium (LA). Right ventricle (RV) and aorta (Ao) are also labeled. This long-axis view is displayed in midsystole to demonstrate thickening of aortic valve leaflets, which may occur but is relatively uncommon.

Treatment of SSA

Unlike ATTR, in which aggressive therapy is warranted immediately upon diagnosis of symptomatic disease, the treatment of SSA focuses on supportive care. Since the amyloidogenic protein in SSA is not altered in any detectable fashion, liver transplantation does not have a role in the management of this disease. Diuretics are the mainstay of treatment, but careful attention must be paid to fluid status and daily weight to prevent a significant loss of filling pressure. Calcium channel blockers have been reported to cause clinical worsening in amyloid heart disease, presumably due to their negative inotropic effects [90]. Beta-adrenergic receptor blockers are avoided for similar concerns. Implantation of a permanent pacemaker is helpful in symptom relief if indicated clinically, but has not shown a survival benefit [91].

Although the mean age at presentation is over 70 years, rare patients may present earlier. Of 18 patients diagnosed with SSA at the Mayo Clinic between 1984 and 1992, three were younger than 70 [89]. In these younger patients with significant symptoms, consideration should be given to orthotopic heart transplantation, which has been successful in restoring quality of life in some cases [92].

Based on the accumulating data on TTR tetramer stabilization, it is reasonable to consider SSA patients for clinical trials of compounds designed to prevent dissociation of TTR into the monomeric or oligomeric intermediates. Indeed, the wild-type TTR protein has been shown to undergo conformational changes and resulting fibril formation similar to amyloidogenic variants under denaturing conditions [38]. Studies of wild-type TTR undergoing heat denaturation have shown that several key changes occur during the process that are not completely reversed on renaturation, leading to the hypothesis that these permanent conformational events may promote subsequent fibril formation [44]. Kinetic stabilization by small molecules has been shown to stabilize the wild-type TTR tetramer and prevent or slow the formation of amyloid fibrils in vitro [93]. Thus, these compounds have significant promise in preventing the progression of SSA, in which patients are generally far older than ATTR patients. Methods of suppression of TTR synthesis, such as antisense oligonucleotides, hold similar hope.

In summary, significant advances have been made in understanding the molecular and pathologic consequences of genetic alterations in the TTR protein leading to systemic amyloidosis. With greater understanding, new therapeutic methods are being employed to prevent the inevitable fatal progression of these diseases. It is an exciting time in the study of these devastating diseases, and we look forward to the future, where a cure seems possible for the first time.

REFERENCES

1. Connors, L.H., Lim, A., Prokaeva, T., Roskens, V.A., Costello, C.E. (2003). Tabulation of human transthyretin (TTR) variants, 2003. *Amyloid, 10*, 160–184.
2. Gertz, M.A., Kyle, R.A., Thibodeau, S.N. (1992). Familial amyloidosis: a study of 52 North American-born patients examined during a 30-year period. *Mayo Clin Proc, 67*, 428–440.
3. Ando, Y., Nakamura, M., Araki, S. (2005). Transthyretin-related familial amyloidotic polyneuropathy. *Arch Neurol, 62*, 1057–1062.
4. Ellie, E., Camou, F., Vital, A., Rummens, C., Grateau, G., Delpech, M., Valleix, S. (2001). Recurrent subarachnoid hemorrhage associated with a new transthyretin variant (Gly53Glu). *Neurology, 57*, 135–137.
5. Goren, H., Steinberg, M.C., Farboody, G.H. (1980). Familial oculoleptomeningeal amyloidosis. *Brain, 103*, 473–495.
6. Petersen, R.B., Goren, H., Cohen, M., Richardson, S.L., Tresser, N., Lynn, A., Gali, M., Estes, M., Gambetti, P. (1997). Transthyretin amyloidosis: a new mutation associated with dementia. *Ann Neurol, 41*, 307–313.
7. Ushiyama, M., Ikeda, S., Yanagisawa, N. (1991). Transthyretin-type cerebral amyloid angiopathy in type I familial amyloid polyneuropathy. *Acta Neuropathol* (Berl), *81*, 524–528.

8. Riisoen, H. (1988). Reduced prealbumin (transthyretin) in CSF of severely demented patients with Alzheimer's disease. *Acta Neurol Scand, 78*, 455–459.
9. Serot, J.M., Christmann, D., Dubost, T., Couturier, M. (1997). Cerebrospinal fluid transthyretin: aging and late onset Alzheimer's disease. *J Neurol Neurosurg Psychiatry, 63*, 506–508.
10. Hammarstrom, P., Sekijima, Y., White, J.T., Wiseman, R.L., Lim, A., Costello, C.E., Altland, K., Garzuly, F., Budka, H., Kelly, J.W. (2003). D18G transthyretin is monomeric, aggregation prone, and not detectable in plasma and cerebrospinal fluid: a prescription for central nervous system amyloidosis? *Biochemistry, 42*, 6656–6663.
11. Ando, Y., Terazaki, H., Nakamura, M., Ando, E., Haraoka, K., Yamashita, T., Ueda, M., Okabe, H., Sasaki, Y., Tanihara, H., et al. (2004). A different amyloid formation mechanism: de novo oculoleptomeningeal amyloid deposits after liver transplantation. *Transplantation, 77*, 345–349.
12. Dispenzieri, A., Gertz, M.A., Kyle, R.A., Lacy, M.Q., Burritt, M.F., Therneau, T.M., Greipp, P.R., Witzig, T.E., Lust, J.A., Rajkumar, S.V., et al. (2004). Serum cardiac troponins and N-terminal pro-brain natriuretic peptide: a staging system for primary systemic amyloidosis. *J Clin Oncol, 22*, 3751–3757.
13. Perugini, E., Rapezzi, C., Piva, T., Leone, O., Bacchi-Reggiani, L., Riva, L., Salvi, F., Lovato, L., Branzi, A., Fattori, R. (2006). Non-invasive evaluation of the myocardial substrate of cardiac amyloidosis by gadolinium cardiac magnetic resonance. *Heart, 92*, 343–349.
14. Perugini, E., Guidalotti, P.L., Salvi, F., Cooke, R.M.T., Pettinato, C., Riva, L., Leone, O., Farsad, M., Ciliberti, P., Bacchi-Reggiani, L., et al. (2005). Noninvasive etiologic diagnosis of cardiac amyloidosis using ^{99m}Tc-3,3-diphosphono-1,2-propanodicarboxylic acid scintigraphy. *J Am Coll Cardiol, 46*, 1076–1084.
15. Falls, H.F., Jackson, J., Carey, J.H., Rukavina, J.G., Block, W.D. (1955). Ocular manifestations of hereditary primary systemic amyloidosis. *AMA Arch Opthalmol, 54*, 660–664.
16. Ong, D.E., Davis, J.T., O'Day, W.T., Bok, D. (1994). Synthesis and secretion of retinol-binding protein and transthyretin by cultured retinal pigment epithelium. *Biochemistry, 33*, 1835–1842.
17. Tsukahara, S., Matsuo, T. (1977). Secondary glaucoma accompanied with primary familial amyloidosis. *Ophthalmologica, 175*, 250–262.
18. Benson, M.D., Kincaid, J.C. (2007). The molecular biology and clinical features of amyloid neuropathy. *Muscle Nerve, 36*, 411–423.
19. Lobato, L. (2003). Portuguese-type amyloidosis (transthyretin amyloidosis, ATTR V30M). *J Nephrol, 16*, 438–442.
20. Takahashi, K., Yi, S., Kimura, Y., Araki, S. (1991). Familial amyloidotic polyneuropathy type 1 in Kumamoto, Japan: a clinicopathologic, histochemical, immunohistochemical, and ultrastructural study. *Hum Pathol, 22*, 519–527.
21. Lachmann, H.J., Booth, D.R., Booth, S.E., Bybee, A., Gilbertson, J.A., Gillmore, J.D., Pepys, M.B., Hawkins, P.N. (2002). Misdiagnosis of hereditary amyloidosis as AL (primary) amyloidosis [see comment]. *N Engl J Med, 346*, 1786–1791.
22. Andersson, K., Olofsson, A., Nielsen, E.H., Svehag, S.E., Lundgren, E. (2002). Only amyloidogenic intermediates of transthyretin induce apoptosis. *Biochem Biophys Res Commun, 294*, 309–314.

23. Matsubara, K., Mizuguchi, M., Igarashi, K., Shinohara, Y., Takeuchi, M., Matsuura, A., Saitoh, T., Mori, Y., Shinoda, H., Kawano, K. (2005). Dimeric transthyretin variant assembles into spherical neurotoxins. *Biochemistry, 44*, 3280–3288.
24. Reixach, N., Deechongkit, S., Jiang, X., Kelly, J.W., Buxbaum, J.N. (2004). Tissue damage in the amyloidoses: Transthyretin monomers and nonnative oligomers are the major cytotoxic species in tissue culture. *Proc Natl Acad Sci U S A, 101*, 2817–2822.
25. Sousa, M.M., Cardoso, I., Fernandes, R., Guimaraes, A., Saraiva, M.J. (2001). Deposition of transthyretin in early stages of familial amyloidotic polyneuropathy: evidence for toxicity of nonfibrillar aggregates. *Am J Pathol, 159*, 1993–2000.
26. Lambert, M.P., Barlow, A.K., Chromy, B.A., Edwards, C., Freed, R., Liosatos, M., Morgan, T.E., Rozovsky, I., Trommer, B., Viola, K.L., et al. (1998). Diffusible, nonfibrillar ligands derived from Abeta1–42 are potent central nervous system neurotoxins. *Proc Natl Acad Sci U S A, 95*, 6448–6453.
27. Walsh, D.M., Hartley, D.M., Kusumoto, Y., Fezoui, Y., Condron, M.M., Lomakin, A., Benedek, G.B., Selkoe, D.J., Teplow, D.B. (1999). Amyloid beta-protein fibrillogenesis: structure and biological activity of protofibrillar intermediates. *J Biol Chem, 274*, 25945–25952.
28. Klein, A.L., Hatle, L.K., Burstow, D.J., Seward, J.B., Kyle, R.A., Bailey, K.R., Luscher, T.F., Gertz, M.A., Tajik, A.J. (1989). Doppler characterization of left ventricular diastolic function in cardiac amyloidosis. *J Am Coll Cardiol, 13*, 1017–1026.
29. Sallach, J.A., Klein, A.L. (2004). Tissue Doppler imaging in the evaluation of patients with cardiac amyloidosis. *Curr Opin Cardiol, 19*, 464–471.
30. Vital, C., Vital, A., Bouillot-Eimer, S., Brechenmacher, C., Ferrer, X., Lagueny, A. (2004). Amyloid neuropathy: a retrospective study of 35 peripheral nerve biopsies. *J Peripher Nerv Syst, 9*, 232–241.
31. Said, G., Ropert, A., Faux, N. (1984). Length-dependent degeneration of fibers in Portuguese amyloid polyneuropathy: a clinicopathologic study. *Neurology, 34*, 1025–1032.
32. Cohen, A., Benson, M. (1975). Amyloid neuropathy. In *Peripheral Neuropathy* (Dyck, P., Thomas, P., Lambert, E., Eds.), W.B. Saunders, Philadelphia, pp. 1067–1091.
33. Hanyu, N., Ikeda, S., Nakadai, A., Yanagisawa, N., Powell, H.C. (1989). Peripheral nerve pathological findings in familial amyloid polyneuropathy: a correlative study of proximal sciatic nerve and sural nerve lesions. *Ann Neurol, 25*, 340–350.
34. Brett, M., Persey, M.R., Reilly, M.M., Revesz, T., Booth, D.R., Booth, S.E., Hawkins, P.N., Pepys, M.B., Morgan-Hughes, J.A. (1999). Transthyretin Leu12Pro is associated with systemic, neuropathic and leptomeningeal amyloidosis. *Brain, 122*, 183–190.
35. Nakamura, M., Yamashita, T., Ueda, M., Obayashi, K., Sato, T., Ikeda, T., Washimi, Y., Hirai, T., Kuwahara, Y., Yamamoto, M.T., et al. (2005). Neuroradiologic and clinicopathologic features of oculoleptomeningeal type amyloidosis. *Neurology, 65*, 1051–1056.
36. Uitti, R.J., Donat, J.R., Rozdilsky, B., Schneider, R.J., Koeppen, A.H. (1988). Familial oculoleptomeningeal amyloidosis: report of a new family with unusual features. *Arch Neurol, 45*, 1118–1122.

37. Jin, K., Sato, S., Takahashi, T., Nakazaki, H., Date, Y., Nakazato, M., Tominaga, T., Itoyama, Y., Ikeda, S. (2004). Familial leptomeningeal amyloidosis with a transthyretin variant Asp18Gly representing repeated subarachnoid haemorrhages with superficial siderosis. *J Neurol Neurosurg Psychiatry, 75*, 1463–1466.

38. Cardoso, I., Goldsbury, C.S., Muller, S.A., Olivieri, V., Wirtz, S., Damas, A.M., Aebi, U., Saraiva, M.J. (2002). Transthyretin fibrillogenesis entails the assembly of monomers: a molecular model for in vitro assembled transthyretin amyloid-like fibrils. *J Mol Biol, 317*, 683–695.

39. Kelly, J.W., Colon, W., Lai, Z., Lashuel, H.A., McCulloch, J., McCutchen, S.L., Miroy, G.J., Peterson, S.A. (1997). Transthyretin quaternary and tertiary structural changes facilitate misassembly into amyloid. *Adv Protein Chem, 50*, 161–181.

40. Quintas, A., Saraiva, M.J., Brito, R.M. (1999). The tetrameric protein transthyretin dissociates to a non-native monomer in solution: a novel model for amyloidogenesis. *J Biol Chem, 274*, 32943–32949.

41. Quintas, A., Saraiva, M.J., Brito, R.M. (1997). The amyloidogenic potential of transthyretin variants correlates with their tendency to aggregate in solution. *FEBS Lett, 418*, 297–300.

42. Quintas, A., Vaz, D.C., Cardoso, I., Saraiva, M.J., Brito, R.M. (2001). Tetramer dissociation and monomer partial unfolding precedes protofibril formation in amyloidogenic transthyretin variants. *J Biol Chem, 276*, 27207–27213.

43. Shnyrov, V.L., Villar, E., Zhadan, G.G., Sanchez-Ruiz, J.M., Quintas, A., Saraiva, M.J., Brito, R.M. (2000). Comparative calorimetric study of non-amyloidogenic and amyloidogenic variants of the homotetrameric protein transthyretin. *Biophys Chem, 88*, 61–67.

44. Chung, C.M., Connors, L.H., Benson, M.D., Walsh, M.T. (2001). Biophysical analysis of normal transthyretin: implications for fibril formation in senile systemic amyloidosis. *Amyloid, 8*, 75–83.

45. Ferrao-Gonzales, A.D., Palmieri, L., Valory, M., Silva, J.L., Lashuel, H., Kelly, J.W., Foguel, D. (2003). Hydration and packing are crucial to amyloidogenesis as revealed by pressure studies on transthyretin variants that either protect or worsen amyloid disease. *J Mol Biol, 328*, 963–974.

46. Fiszman, M.L., Di Egidio, M., Ricart, K.C., Repetto, M.G., Borodinsky, L.N., Llesuy, S.F., Saizar, R.D., Trigo, P.L., Riedstra, S., Costa, P.P., et al. (2003). Evidence of oxidative stress in familial amyloidotic polyneuropathy type 1. *Arch Neurol, 60*, 593–597.

47. Hou, X., Parkington, H.C., Coleman, H.A., Mechler, A., Martin, L.L., Aguilar, M.-I., Small, D.H. (2007). Transthyretin oligomers induce calcium influx via voltage-gated calcium channels. *J Neurochem, 100*, 446–457.

48. Monteiro, F.A., Sousa, M.M., Cardoso, I., do Amaral, J.B., Guimaraes, A., Saraiva, M.J. (2006). Activation of ERK1/2 MAP kinases in familial amyloidotic polyneuropathy. *J Neurochem, 97*, 151–161.

49. Sousa, M.M., Du Yan, S., Fernandes, R., Guimaraes, A., Stern, D., Saraiva, M.J. (2001). Familial amyloid polyneuropathy: receptor for advanced glycation end products-dependent triggering of neuronal inflammatory and apoptotic pathways. *J Neurosci, 21*, 7576–7586.

50. Teixeira, P.F., Cerca, F., Santos, S.D., Saraiva, M.J. (2006). Endoplasmic reticulum stress associated with extracellular aggregates: evidence from transthyretin deposition in familial amyloid polyneuropathy. *J Biol Chem, 281*, 21998–22003.

51. Holmgren, G., Steen, L., Ekstedt, J., Groth, C.G., Ericzon, B.G., Eriksson, S., Andersen, O., Karlberg, I., Norden, G., Nakazato, M., et al. (1991). Biochemical effect of liver transplantation in two Swedish patients with familial amyloidotic polyneuropathy (FAP-met30). *Clin Genet, 40*, 242–246.

52. Herlenius, G., Wilczek, H.E., Larsson, M., Ericzon, B.-G., Familial Amyloidotic Polyneuropathy World Transplant Registry. (2004). Ten years of international experience with liver transplantation for familial amyloidotic polyneuropathy: results from the Familial Amyloidotic Polyneuropathy World Transplant Registry. *Transplantation, 77*, 64–71.

53. Adams, D., Samuel, D., Goulon-Goeau, C., Nakazato, M., Costa, P.M., Feray, C., Plante, V., Ducot, B., Ichai, P., Lacroix, C., et al. (2000). The course and prognostic factors of familial amyloid polyneuropathy after liver transplantation. *Brain, 123*, 1495–1504.

54. Stangou, A.J., Hawkins, P.N. (2004). Liver transplantation in transthyretin-related familial amyloid polyneuropathy. *Curr Opin Neurol, 17*, 615–620.

55. Sharma, P., Perri, R.E., Sirven, J.E., Zeldenrust, S.R., Brandhagen, D.J., Rosen, C.B., Douglas, D.D., Mulligan, D.C., Rakela, J., Wiesner, R.H., Balan, V. (2003). Outcome of liver transplantation for familial amyloidotic polyneuropathy. *Liver Transpl, 9*, 1273–1280.

56. Bergethon, P.R., Sabin, T.D., Lewis, D., Simms, R.W., Cohen, A.S., Skinner, M. (1996). Improvement in the polyneuropathy associated with familial amyloid polyneuropathy after liver transplantation. *Neurology, 47*, 944–951.

57. Suhr, O.B. (2003). Impact of liver transplantation on familial amyloidotic polyneuropathy (FAP) patients' symptoms and complications. *Amyloid, 10* (Suppl 1), 77–83.

58. Olofsson, B.-O., Backman, C., Karp, K., Suhr, O.B. (2002). Progression of cardiomyopathy after liver transplantation in patients with familial amyloidotic polyneuropathy, Portuguese type. *Transplantation, 73*, 745–751.

59. Pomfret, E.A., Lewis, W.D., Jenkins, R.L., Bergethon, P., Dubrey, S.W., Reisinger, J., Falk, R.H., Skinner, M. (1998). Effect of orthotopic liver transplantation on the progression of familial amyloidotic polyneuropathy. *Transplantation, 65*, 918–925.

60. Stangou, A.J., Hawkins, P.N., Heaton, N.D., Rela, M., Monaghan, M., Nihoyannopoulos, P., O'Grady, J., Pepys, M.B., Williams, R. (1998). Progressive cardiac amyloidosis following liver transplantation for familial amyloid polyneuropathy: implications for amyloid fibrillogenesis. *Transplantation, 66*, 229–233.

61. Hornsten, R., Wiklund, U., Olofsson, B.-O., Jensen, S.M., Suhr, O.B. (2004). Liver transplantation does not prevent the development of life-threatening arrhythmia in familial amyloidotic polyneuropathy, Portuguese-type (ATTR Val30Met) patients. *Transplantation, 78*, 112–116.

62. Liepnieks, J.J., Benson, M.D. (2007). Progression of cardiac amyloid deposition in hereditary transthyretin amyloidosis patients after liver transplantation. *Amyloid, 14*, 277–282.

63. Yazaki, M., Mitsuhashi, S., Tokuda, T., Kametani, F., Takei, Y.I., Koyama, J., Kawamorita, A., Kanno, H., Ikeda, S.I. (2007). Progressive wild-type transthyretin

deposition after liver transplantation preferentially occurs onto myocardium in FAP patients. *Am J Transplant, 7*, 235–242.

64. Yazaki, M., Liepnieks, J.J., Kincaid, J.C., Benson, M.D. (2003). Contribution of wild-type transthyretin to hereditary peripheral nerve amyloid. *Muscle Nerve, 28*, 438–442.
65. Munar-Ques, M., Salva-Ladaria, L., Mulet-Perera, P., Sole, M., Lopez-Andreu, F.R., Saraiva, M.J. (2000). Vitreous amyloidosis after liver transplantation in patients with familial amyloid polyneuropathy: ocular synthesis of mutant transthyretin. *Amyloid, 7*, 266–269.
66. Azoulay, D., Samuel, D., Castaing, D., Adam, R., Adams, D., Said, G., Bismuth, H. (1999). Domino liver transplants for metabolic disorders: experience with familial amyloidotic polyneuropathy. *J Am Coll Surg, 189*, 584–593.
67. Furtado, A., Tome, L., Oliveira, F.J., Furtado, E., Viana, J., Perdigoto, R. (1997). Sequential liver transplantation. *Transplant Proc, 29*, 467–468.
68. Schmidt, H.H., Nashan, B., Propsting, M.J., Nakazato, M., Flemming, P., Kubicka, S., Boker, K., Pichlmayr, R., Manns, M.P. (1999). Familial amyloidotic polyneuropathy: domino liver transplantation. *J Hepatol, 30*, 293–298.
69. Stangou, A.J., Heaton, N.D., Hawkins, P.N. (2005). Transmission of systemic transthyretin amyloidosis by means of domino liver transplantation. *N Engl J Med, 352*, 2356.
70. Takei, Y.-I., Gono, T., Yazaki, M., Ikeda, S.-I., Ikegami, T., Hashikura, Y., Miyagawa, S.-I., Hoshii, Y. (2007). Transthyretin-derived amyloid deposition on the gastric mucosa in domino recipients of familial amyloid polyneuropathy liver.[see comment]. *Liver Transpl, 13*, 215–218.
71. Coelho, T., Carvalho, M., Saraiva, M.J., Alves, I., Almeida, M.R., Costa, P.P. (1993). A strikingly benign evolution of FAP in an individual compound heterozygote for two TTR mutations: TTR Met30 and TTR Met119. *J Rheumatol, 20*, 179.
72. Hammarstrom, P., Schneider, F., Kelly, J.W. (2001). Trans-suppression of misfolding in an amyloid disease. *Science, 293*, 2459–2462.
73. Hammarstrom, P., Wiseman, R.L., Powers, E.T., Kelly, J.W. (2003). Prevention of transthyretin amyloid disease by changing protein misfolding energetics. *Science, 299*, 713–716.
74. Sekijima, Y., Dendle, M.A., Kelly, J.W. (2006). Orally administered diflunisal stabilizes transthyretin against dissociation required for amyloidogenesis. *Amyloid, 13*, 236–249.
75. Benson, M.D., Kluve-Beckerman, B., Zeldenrust, S.R., Siesky, A.M., Bodenmiller, D.M., Showalter, A.D., Sloop, K.W. (2006). Targeted suppression of an amyloidogenic transthyretin with antisense oligonucleotides. *Muscle Nerve, 33*, 609–618.
76. Tanaka, K., Yamada, T., Ohyagi, Y., Asahara, H., Horiuchi, I., Kira, J. (2001). Suppression of transthyretin expression by ribozymes: a possible therapy for familial amyloidotic polyneuropathy. *J Neurol Sci, 183*, 79–84.
77. Nakamura, M., Ando, Y., Nagahara, S., Sano, A., Ochiya, T., Maeda, S., Kawaji, T., Ogawa, M., Hirata, A., Terazaki, H., et al. (2004). Targeted conversion of the transthyretin gene in vitro and in vivo. *Gene Ther, 11*, 838–846.
78. Cornwell, G.G., 3rd, Murdoch, W.L., Kyle, R.A., Westermark, P., Pitkanen, P. (1983). Frequency and distribution of senile cardiovascular amyloid: a clinicopathologic correlation. *Am J Med, 75*, 618–623.

79. Westermark, P., Johansson, B., Natvig, J.B. (1979). Senile cardiac amyloidosis: evidence of two different amyloid substances in the ageing heart. *Scand J Immunol, 10*, 303–308.
80. Cornwell, G.G., 3rd, Sletten, K., Johansson, B., Westermark, P. (1988). Evidence that the amyloid fibril protein in senile systemic amyloidosis is derived from normal prealbumin. *Biochem Biophys Res Commun, 154*, 648–653.
81. Westermark, P., Sletten, K., Johansson, B., Cornwell, G.G., 3rd. (1990). Fibril in senile systemic amyloidosis is derived from normal transthyretin. *Proc Natl Acad Sci U S A, 87*, 2843–2845.
82. Christmanson, L., Betsholtz, C., Gustavsson, A., Johansson, B., Sletten, K., Westermark, P. (1991). The transthyretin cDNA sequence is normal in transthyretin-derived senile systemic amyloidosis. *FEBS Lett, 281*, 177–180.
83. Gorevic, P.D., Prelli, F.C., Wright, J., Pras, M., Frangione, B. (1989). Systemic senile amyloidosis: identification of a new prealbumin (transthyretin) variant in cardiac tissue: immunologic and biochemical similarity to one form of familial amyloidotic polyneuropathy. *J Clin Invest, 83*, 836–843.
84. Pomerance, A. (1965). Senile cardiac amyloidosis. *Br Heart J, 27*, 711–718.
85. Kyle, R.A., Gertz, M.A., Linke, R.P. (1992). Amyloid localized to tenosynovium at carpal tunnel release immunohistochemical identification of amyloid type. *Am J Clin Pathol, 97*, 250–253.
86. Takei, Y.-I., Hattori, T., Gono, T., Tokuda, T., Saitoh, S., Hoshii, Y., Ikeda, S.-I. (2002). Senile systemic amyloidosis presenting as bilateral carpal tunnel syndrome. *Amyloid, 9*, 252–255.
87. Ng, B., Connors, L.H., Davidoff, R., Skinner, M., Falk, R.H. (2005). Senile systemic amyloidosis presenting with heart failure: a comparison with light chain–associated amyloidosis [see comment]. *Arch Intern Med, 165*, 1425–1429.
88. Mookadam, F., Haley, J.H., Olson, L.J., Cikes, M., Mookadam, M. (2006). Dynamic left ventricular outflow tract obstruction in senile cardiac amyloidosis. *Eur J Echocardiogr, 7*, 465–468.
89. Kyle, R.A., Spittell, P.C., Gertz, M.A., Li, C.Y., Edwards, W.D., Olson, L.J., Thibodeau, S.N. (1996). The premortem recognition of systemic senile amyloidosis with cardiac involvement. *Am J Med, 101*, 395–400.
90. Gertz, M.A., Falk, R.H., Skinner, M., Cohen, A.S., Kyle, R.A. (1985). Worsening of congestive heart failure in amyloid heart disease treated by calcium channel-blocking agents. *Am J Cardiol, 55*, 1645.
91. Mathew, V., Olson, L.J., Gertz, M.A., Hayes, D.L. (1997). Symptomatic conduction system disease in cardiac amyloidosis. *Am J Cardiol, 80*, 1491–1492.
92. Fuchs, U., Zittermann, A., Suhr, O., Holmgren, G., Tenderich, G., Minami, K., Koerfer, R. (2005). Heart transplantation in a 68-year-old patient with senile systemic amyloidosis. *Am J Transplant, 5*, 1159–1162.
93. Johnson, S., Wiseman, R., Sekijima, Y., Green, N., Adamski-Werner, S., Kelly, J. (2005). Native state kinetic stabilization as a strategy to ameliorate protein misfolding diseases: a focus on the transthyretin amyloidoses. *Acc Chem Res, 38*, 911–921.

37

IDENTIFYING TARGETS IN α-SYNUCLEIN METABOLISM TO TREAT PARKINSON DISEASE AND RELATED DISORDERS

JULIANNA TOMLINSON
Division of Neuroscience, Ottawa Health Research Institute, University of Ottawa, Ottawa, Ontario, Canada

VALERIE CULLEN
LINK Medicine, Cambridge, Massachusetts

MICHAEL G. SCHLOSSMACHER
Division of Neuroscience, Ottawa Health Research Institute, University of Ottawa, Ottawa, Ontario, Canada

PARKINSON DISEASE

Parkinson disease (PD) is the second most common neurodegenerative disorder [after Alzheimer disease (AD) (see Chapter 34)] and is estimated to affect over 1 million people in North America [1]. AD and PD share advanced age as an important risk factor. Patients with parkinsonism suffer from neurological abnormalities that include slowness of movements, stiffness of limbs and trunk, rest tremor, and postural reflex loss. These motoric deficits result from dopamine deficiency in the striatum of the forebrain. This loss is due to the progressive reduction in neuronal cell bodies that are located in the substantia

Protein Misfolding Diseases: Current and Emerging Principles and Therapies,
Edited by Marina Ramirez-Alvarado, Jeffery W. Kelly, and Christopher M. Dobson
Copyright © 2010 John Wiley & Sons, Inc.

nigra of the midbrain and their projecting axons [2]. Typical PD patients may also develop autonomic changes and, following years of progression, cognitive impairment due to the involvement of nondopaminergic cell populations throughout the nervous system [1,3].

Akin to the many different forms of dementia, parkinsonism is understood as a complex syndrome with distinct etiologies. Approximately 75 to 80%of PD cases are considered idiopathic; they are characterized by symptom onset at age >55 years, usually present in a nonheritable form and show distinct features at autopsy [1,4]. Since 1997 an increasing number of pedigrees have been published that are afflicted by rare, familial PD. These studies have revealed unprecedented genetic insights that led, as of today, to the discovery of 12 chromosomal loci and the identification of five genes, *SNCA, Parkin, Dj-1, Pink1*, and *LRRK2*, which are linked to heritable PD [5–7]. Nevertheless the question as to how these susceptibility loci and five genes interact with one another, the environment, and with the aging process to promote PD pathogenesis remains unknown.

Currently, there is no cure for PD. The movement-disorder community is in urgent need of validated treatment options for three related indications: to delay onset of PD in at-risk persons (neuroprevention), to protect remaining cells after the clinical diagnosis has been established (neuroprotection), and to reverse neuronal loss (neurorestoration; [8]). Currently available interventions by either pharmacological or surgical means are limited to symptom relief and aimed at the improvement of motor dysfunction. Notably, these therapies do not change the course of PD, do not treat its many nonmotoric symptoms, and do not address the root causes of parkinsonism. The progressive nature of PD and the absence of bona fide disease-modifying drugs highlight the importance of exploratory programs within industry and academia that pursue novel targets. Ideally, such targets will be inspired by and related mechanistically to the genetics of PD (reviewed by Farrer [5]). In this chapter we discuss the evidence that recent insights into the metabolism of α-synuclein (αSyn) which is encoded by the PD-linked *SNCA* gene, have begun to reveal potential targets for novel drug-screening programs. For a review of other PD-linked genes, we refer the reader to recent, well-written articles in the literature (e.g., [5,9]).

RELATED SYNUCLEINOPATHY DISORDERS OF THE BRAIN

Increased *SNCA* gene expression due to gene multiplication events is causally linked to rare forms of familial PD. The intracellular accumulation of its encoded protein, αSyn, represents one of the hallmark findings of PD and of several related disorders that are collectively referred to as *synucleinopathies* [10]. This descriptive term was coined by neuropathologists and refers to the detection of the neuronal or oligodendroglial accumulation of αSyn aggregates at autopsy in the form of Lewy bodies (Fig. 1) and glial cytoplasmic inclusions, respectively. These lesions can be found in a wide spectrum of incurable

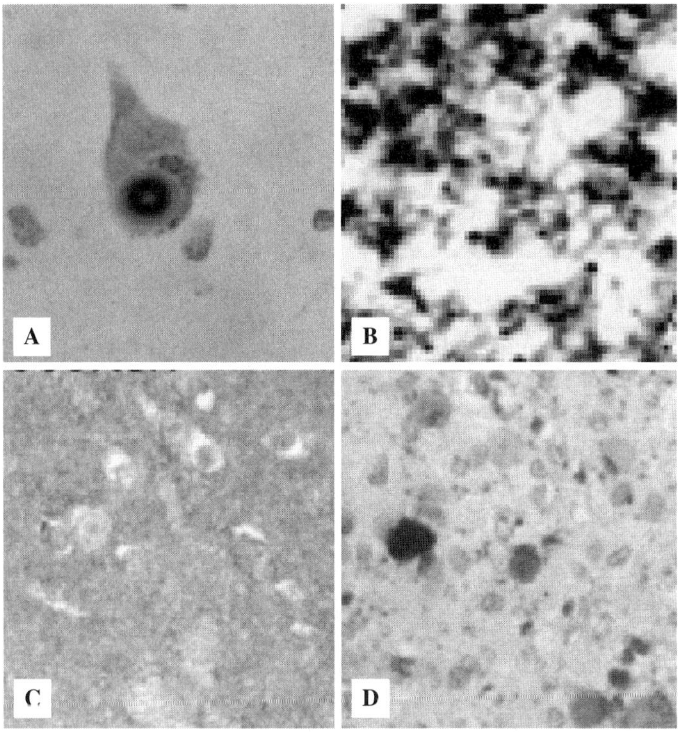

FIG. 1 Neuropathological findings of two human brain disorders associated with synucleinopathy. Immunohistochemical (A,C,D) and immunoelectron microscopy (B) images were obtained from midbrain (A,B) and frontal cortex (C,D) sections from patients with Parkinson disease (A,B), a normal infant (C), and an infant with a fatal lysosomal storage disease (D). Sections were probed with anti-αSyn antibodies, developed and either counterstained with hematoxyline (blue; A,C,D), or developed with an immunogold particle-coated secondary antibody (B). In (A), a reduced number of dopamine neurons, reactive gliosis and the presence of a round, classical Lewy body inclusion (brown) located in the cytoplasm of a single remaining neuron can be seen. The α-synuclein-containing Lewy body (diameter, 6 μm) is partially surrounded by clusters of physiological neuromelanin and located in proximity to the nucleus. In (B), round gold particles (black; diameter, 12 nm) decorate αSyn fibrils (gray) in an affinity-enriched Lewy body. Healthy human cortex shown in (C) contains high levels of soluble α-synuclein seen diffusely throughout the neuropil (brown). In the absence of cathepsin D expression (D), the neuronal architecture of the cortex is disrupted, αSyn signals are reduced in the neuropil and instead begin to aggregate (diameter, 1 to 6 μm) within the cytoplasm and neurites. (Image (B) provided by Wei Ping Gai, Flinders University, Australia; image (D) provided by Jaana Tyynelae, Helsinki University, Finland; panels (C) and (D) from [80].) (*See insert for color representation of figure.*)

disorders that clinically present with dementia or parkinsonism (or both; Table 1); their post mortem detection by immunohistochemistry has significantly enhanced our ability to better classify several disorders of the human nervous system [11]. Synucleinopathies also have a biochemical signature that underlies their microscopic abnormalities: namely, the formation of remarkably insoluble, higher-molecular-mass aggregates (oligomers) that can be detected by serial fractionation, SDS/PAGE, and proteinase K digestion [12,13].

Synucleinopathy disorders can be grouped according to the neural cell type that is predominantly affected (neurons vs. glia), but also according to whether microscopic evidence for αSyn mishandling is consistently present ("invariable synucleinopathies"), for example in typical PD, or only sometimes ("variable"), such as in parkinsonism caused by a neurovisceral storage disorder (Fig. 1). This observation raises the possibility that in the former category misprocessing of αSyn is an event that is proximal to the disease process. By inference, it may be a rather secondary event in the category of variable synucleinopathies, possibly related to the fact that αSyn is expressed abundantly in vivo. Table 1 lists the currently known human conditions that are associated with microscopic evidence of αSyn mishandling. The sheer number of patients afflicted by one of the many incurable synucleinopathy disorders is motivation enough to study the metabolism of the protein involved and the mechanisms by which it may confer neurotoxicity [14]. Furthermore, these intracellular αSyn-related pathologies cannot yet be visualized in living subjects without brain biopsy; however, the extracellular quantification of αSyn, such as in peripheral blood or cerebrospinal fluid, is being explored aggressively for the purpose of laboratory marker development (e.g., [15–17]).

ALPHA-SYNUCLEIN: A PROTEIN PRONE TO MISFOLDING

α-Synuclein is a 140-amino acid long protein. It is expressed in the central and peripheral nervous system, where it is believed to be involved in the maturation of presynaptic vesicles and the fatty acid composition of membranes. In neurons, αSyn acts as a co-regulator of neurotransmission [18,19]. The amino terminus of the protein contains six to seven imperfect repeats, which are predicted to function as lipid-binding domains. The protein encodes a centrally located hydrophobic domain that contains the non-Aβ component of amyloid plaque precursor protein (NAC) domain, so named because the NAC portion had first been isolated from human AD brain, and a carboxyl-terminal acidic domain (Fig. 2). Biochemical and structural studies suggest that in its native state, αSyn is cytosolic, highly soluble, and assumes an unfolded structure. However, upon binding to phospholipids, the amino terminus of the protein adopts an amphipathic α-helical conformation (Fig. 2). At higher cellular concentrations (*molecular crowding*), αSyn has the propensity to multimerize, leading to the formation of initially soluble prefibrillar oligomers or protofibrils. The accumulation of protofibrils is thought to lead eventually to insoluble amyloid fibrils which deposit in the brain as Lewy bodies (Fig. 1), Lewy neurites, and glial cytoplasmic inclusions [10,20,21].

TABLE 1 Neurological Syndromes with Microscopic Evidence of Synucleinopathy

A. *Invariable* synucleinopathy of neurons in:

- Parkinsonism
  - Sporadic Parkinson disease
  - Familial, *SNCA*-linked Parkinson disease
- Dementia
  - Sporadic, pure dementia with Lewy bodies
  - Familial, *SNCA*-linked dementia with Lewy bodies
  - Lewy body variant of Alzheimer disease
  - Familial, *APP*-linked Alzheimer disease
  - Down syndrome
  - Familial, *Presenilin1*-linked Alzheimer disease
  - Familial, *Presenilin2*-linked Alzheimer disease
- Neuroaxonal dystrophy
  - Sporadic neurodegeneration with brain iron accumulation (i.e., Hallervorden–Spatz disease)
  - Familial, *PANK2*-linked neurodegeneration with brain iron accumulation
- Lysosomal storage disease
  - Familial, *CTSD*-linked neuronal ceroid lipofuscinosis
- Pure autonomic failure
- Incidental Lewy body disease (in the absence of known neurological deficits)

B. *Invariable* synucleinopathy of glia in:

- Multiple-system atrophy
  - Parkinsonism variant (i.e., striatonigral degeneration)
  - Ataxia variant (i.e., olivopontocerebellar degeneration)
  - Autonomic variant (i.e., Shy–Drager syndrome)

C. *Variable* synucleinopathy in:

- Parkinsonism
  - Familial, *LRRK2*-linked Parkinson disease
  - Familial, *Parkin*-linked Parkinson disease
- Dementia with parkinsonism
  - Pick disease
  - Progressive supranuclear palsy (i.e., Steele–Richardson syndrome)
  - Prion disease (e.g., Creutzfeldt–Jakob disease)
- Neurovisceral storage disease
  - Familial, *GBA*-linked Gaucher disease
  - Familial, *NP1C*-linked Niemann–Pick disease
- Other condition
  - Juvenile-onset neuroaxonal dystrophy
  - Amyotrophic lateral sclerosis
  - Traumatic brain injury

Source: Modified from [11,89].

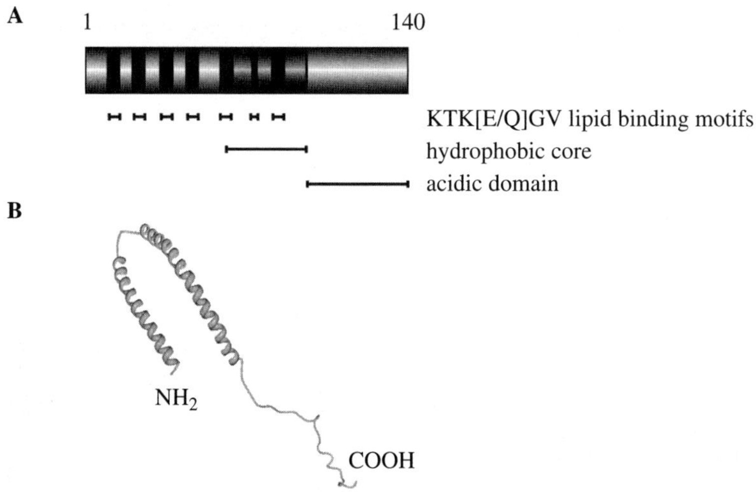

FIG. 2 Graphic representation of human α-synuclein. (A) Schematic of human αSyn, highlighting the six to seven lipid-binding motifs, the hydrophobic core, and the acid carboxyl terminus of the protein. In its native state, αSyn is unfolded, highly soluble, and cytoplasmic. (B) Ribbon diagram representing the tertiary structure of micelle-bound αSyn derived from solution NMR spectroscopy. (From the PDB database, accession number 1XQ8 [102].)

As the second of currently four known members of the synuclein protein family, β-synuclein (βSyn) is of relevance because of its significant homology to αSyn; it is expressed in highly similar regions of the central nervous system and equally enriched in presynaptic terminals. Human βSyn is encoded by *SNCB* and confers neuroprotection in αSyn-mediated toxicity models both in vitro and in vivo [22]. To date, the *SNCB* gene has not been linked convincingly to a neurological disease phenotype. In contrast, rare familial forms of early-onset PD with autosomal dominant inheritance have been linked to three missense point mutations within the *SNCA* gene, which encode Ala30Pro, Ala53Thr, and Glu46Lys substitutions. Humans who are heterozygous for one of these three-point mutations develop a severe form of PD with a high penetrance rate, cognitive impairment, and often with significant autonomic dysfunction ([23]; reviewed in [5]). Initial biochemical studies suggested that these mutations in αSyn alter the physiochemical properties of the protein in a manner that increases the propensity to oligomerize in vitro [24,25] and promote fibril formation and neurodegeneration in vivo. However, so far, no shared uniform gain-of-function mechanism has emerged for these three point mutants of human αSyn that underlies their interchangeable clinical phenotypes, even after a decade of studies conducted in multiple laboratories using physicochemical, biochemical, cellular, and in vivo modeling (e.g., [26–29]). Although the specific identity of *the* most neurotoxic αSyn species remains elusive and represents a major focus of ongoing research efforts worldwide (e.g., [30, 31]), the discovery of these missense mutations demonstrated that dysregulated αSyn metabolism can function as the primary event in PD pathogenesis.

ALPHA-SYNUCLEIN METABOLISM AS THE FOCUS OF RESEARCH

There is strong genetic, experimental, and pathological evidence that increased intracellular concentrations of αSyn contribute to the development of PD and related synucleinopathies. Most notable is the genetic evidence that *SNCA* gene dosage is linked to PD. Triplication and duplication of the *SNCA* gene locus, which have been demonstrated to lead to increased *SNCA* mRNA and αSyn production, cause PD [32]. Importantly, *SNCA* gene dosage correlates closely with the penetrance, age of onset, and severity of the ensuing synucleinopathy (Fig. 3; Table 1).

Recent data suggest that *SNCA* dosage is not only limited to rare familial cases of PD, as individuals with sporadic PD carrying a *SNCA* duplication event have been identified [33]. Furthermore, an estimated 3% of people with sporadic, late-onset PD carry a polymorphism within the *SNCA* promoter termed Rep-1, which has been postulated to increase *SNCA* expression and thus PD susceptibility [34]. Additional single nucleotide polymorphisms (SNPs) spanning the *SNCA* locus have been shown to correlate with altered susceptibility to sporadic PD (reviewed by Farrer [5]). Experimentally, overexpression of human wild type αSyn in human dopaminergic cells, yeast, *Caenorhabditis elegans*, *Drosophila melanogaster*, and rodent models is toxic and promotes neuronal death; paradoxically, a still poorly understood neuroprotective effect of αSyn has been documented as well in several experimental paradigms [35,36].

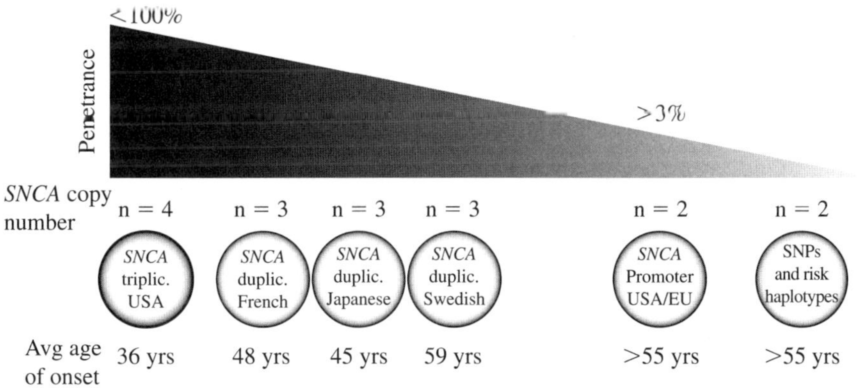

FIG. 3 Effects of *SNCA* gene dosage on Parkinson disease penetrance and its age of onset. Gene multiplication events lead to increased αSyn expression and have been linked to familial cases of early-onset PD. Gene dosage appears to correlate inversely with the age of onset and severity of parkinsonism in affected individuals. *n* represents the *SNCA* allele copy number. The *SNCA* promoter represents the identified Rep-1 polymorphism within the 5′ region that has been linked to increased PD susceptibility. Relevant references are included within the body of the text. (Adapted from [103].)

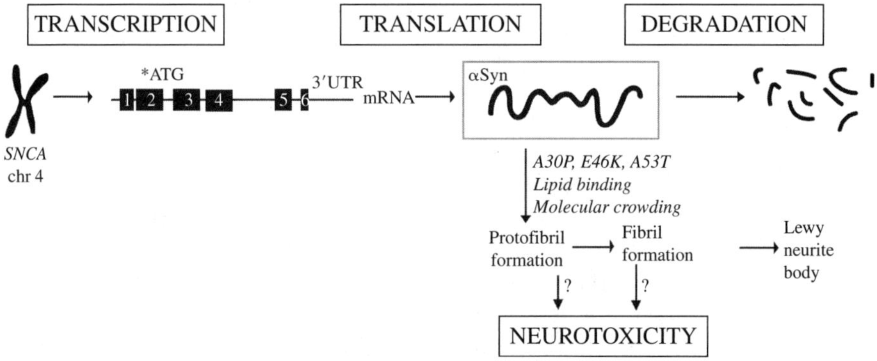

FIG. 4 Schematic representation of α-synuclein metabolism. Steady-state levels of soluble αSyn are maintained through equilibrium between αSyn production (processes of transcription and translation), protein degradation, and the generation of insoluble oligomeric species. Our current understanding of the molecular mechanisms and pathways that regulate protein expression and degradation are discussed in the text. Soluble αSyn is prone to formation of oligomeric species mediated through its central hydrophobic core. The formation of higher-order structures is modulated by lipid binding and molecular crowding and is enhanced by point mutations that are linked to autosomal dominant heritable PD (A30P, E46K, A53T), and may be regulated by posttranslational modifications, such as Ser129 phosphorylation. The relative neurotoxicity of the insoluble protofibril and fibril species remains an important question in understanding the pathogenesis of synucleinopathies.

Taken together with the presence of αSyn aggregates in Lewy inclusions, the overwhelming neurogenetic evidence suggests that a sustained increase in the concentration of neural αSyn is central in promoting PD pathology in aging humans. Therefore, several research teams, including ours, postulate that reducing the pool of total αSyn in vivo represents a plausible focus in target identification efforts. Given the strong correlation between *SNCA* gene dosage and the penetrance, age of onset, and severity of parkinsonism, we hypothesize that a modest decrease of αSyn protein concentration within neural cells could confer neuropreventive, neuroprotective and neurorestorative effects in vivo.

Steady-state levels of αSyn are maintained through equilibrium between de novo protein synthesis and protein degradation (Fig. 4). Our mechanistic understanding of the molecular pathways that govern αSyn expression and its turnover are relatively limited. These pathways are currently understudied in PD; however, their characterization is critical for both target identification and validation efforts to pharmacologically decrease αSyn protein levels in the future.

DE NOVO PROTEIN SYNTHESIS: TRANSCRIPTION AND TRANSLATION

Protein synthesis is a two-step process consisting of gene transcription and translation. There are two broad approaches to targeting these processes to reduce αSyn expression. In the first instance, one could target specific regulatory molecules such as transcription factors or their accessory factors to either inhibit or enhance their activity. In this regard, either the function or the expression of these factors themselves could be modulated therapeutically. Alternatively, one could attenuate gene expression by blocking translation using various nucleic acid molecule–based therapeutic agents that are designed to interfere with the process. Both of these approaches have been used successfully as therapeutic interventions, and they will be discussed in further detail as potential approaches to treating PD.

MODULATORS OF *SNCA* GENE TRANSCRIPTION

The transcriptional activation of a gene is tightly regulated and is mediated by a network of transcriptional regulatory factors. These include the *trans*-acting factors and *cis*-acting regulatory elements that can either enhance or silence transcription. Repression of *SNCA* transcription would lower production of αSyn protein. A prerequisite of this therapeutic approach is that it requires mechanistic insight into the identity of these regulatory proteins and the pathways in which they function to regulate αSyn expression. The identities of the molecular players that potentiate *SNCA* transcription are just beginning to emerge.

Rep-1 Allele

As previously described, an estimated 3% of people with idiopathic PD carry a polymorphism at the Rep-1 allele within the *SNCA* promoter [34]. The Rep-1 allele is a polymorphic microsatellite repeat located approximately 10 kb upstream of the *SNCA* transcription start site [37,38]. In 2006, a large-scale collaborative analysis of 2692 PD cases and 2652 controls firmly established that an expanded Rep-1 allele length is associated with an increased risk of sporadic PD, without affecting the age of disease onset [34]. The Rep-1 allele is hypothesized to be a *cis*-acting transcriptional regulatory element that modulates expression of αSyn. The regulatory potential of this allele was demonstrated using neuroblastoma cell–based transcription studies, which showed that expanded variants of the Rep-1 allele promoted up to three-fold elevation in transcription [39].

Rep-1 allelic variation has been linked to αSyn levels in human blood [40]. Until recently however, the in vivo evidence of the association between the pathogenic Rep-1 allele and increased αSyn production in the brain was lacking. This was addressed by studies performed by Chiba-Falek and colleagues in which they engineered transgenic mice lines that carried the human *SNCA* locus with three distinct Rep-1 variants: an expanded Rep1 261-bp allele, a protective shorter Rep1 259-bp allele, and a Rep-1 deletion mutant. Mice carrying the risk-associated Rep-1 allele had a 1.7-fold increase in human *SNCA* mRNA and protein in the brain compared with mice carrying the Rep1 259-bp allele. The brains of mice carrying the Rep-1 deletion had the lowest levels of human αSyn among the three transgenic lines [41]. These studies provide the first in vivo evidence that Rep-1 acts as an enhancer element to increase the transcriptional regulation of *SNCA* expression in the brain. The up-regulation of human αSyn in these animals is therefore strikingly similar to the 1.5-fold increase recorded in persons carrying the duplication of the *SNCA* locus. The expected modest chronic increase in αSyn production conferred by Rep-1 expansion could therefore explain the associated risk modulation linked to this allele.

Having established that Rep-1 allele functions as a transcriptional regulatory element in vivo, it now remains necessary to elucidate the molecular mechanisms by which it mediates its effects. Studies have shown specific binding of poly(ADP-ribose) transferase/polymerase-1 (PARP-1) to Rep-1 and that PARP-1 down-regulated transcriptional activity from the *SNCA* promoter in cell-based transcription assays [42]. Further defining the role of PARP-1 in this regulatory pathway, identifying potential co-regulatory factors that may bind to the Rep-1 element and determining how it interacts with other transcriptional regulatory elements on the *SNCA* promoter will provide insight into potential therapeutic targets which could interfere with its enhancer function and ultimately down-regulate *SNCA* transcription.

GATA Transcription Factors

The first insight into the transcription factors that regulate *SNCA* expression emerged recently with the identification of GATA-1 and GATA-2 as sequence-specific transcription factors that are targeted to the *SNCA* gene regulatory region [43]. Stemming from the initial observation that αSyn is highly abundant in human erythroid cells, Scherzer and colleagues identified a tight correlation between *SNCA* transcription and three heme metabolism genes, all of which were co-induced by the transcription factor GATA-1. They demonstrated that GATA-1 and the related family member GATA-2 selectively occupied a GATA response element located within the first intron of the *Snca* gene in a murine hematopoietic cell line. Furthermore, GATA-1 and GATA-2 expression correlated positively with *Snca* mRNA abundance and protein expression in murine hematopoietic cells and human dopaminergic cells, respectively. However, whereas GATA-1 expression was restricted to erythroid cells, GATA-2 was abundantly expressed in dopaminergic neuronal cells and brain regions affected by PD, including the substantia nigra and frontal cortex of human tissue [43].

GATA-1 and GATA-2, along with GATA-3, are closely related members of a family of transcription factors (GATA-1 to GATA-6 in vertebrates) which are involved in the regulation of developmental processes and tissue-specific functions. GATA proteins bind as monomers to conserved DNA motifs, (A/T)GATA(A/G), within the transcriptional regulatory regions of responsive genes and differentially modulate gene expression in GATA isoform-, promoter-, and cell-type-specific contexts [44]. Further explorations are warranted to evaluate the potential role of other GATA family members in the regulation of *SNCA* transcription in regions of the brain most susceptible to PD.

Importantly, these studies provide the first mechanistic insight into the regulation of *SNCA* transcription. They also identify GATA protein-encoding genes and those of their associated regulatory factors, such as Friend of GATA (Fog) proteins, as possible susceptibility loci for PD and other synucleinopathies.

Other Factors Contributing to Transcriptional Regulation of *SNCA*

There is evidence that *SNCA* is induced in response to prolonged stimulation with NGF and bFGF in PC12 cells in culture and that this induction is mediated by MAPK signaling. This inducibility maps to a region within the first intron of the gene [45]. However, the mechanism underlying this potentiation is unknown. Of interest, the activity of GATA transcription factors can be modulated by MAPK-dependent phosphorylation [46]. Separate studies have shown that endogenous αSyn is up-regulated in response to stimulation with liver X receptor (LXR) ligands in human neuronal and oligodendrocyte cell models in culture. LXR is a ligand-activated transcription factor that functions as a heterodimer with its obligate binding partner, 9-*cis*-retinoic acid receptor (RXR). Taken together with the identification of two putative LXR response elements in the human *SNCA* promoter, these results suggest that *SNCA* may be a target of LXR/RXR transcription factors. Further studies are required to determine if this regulation is direct or indirect [47]. Finally, putative DNA-binding elements for the CCAAT/enhancer binding protein β (C/EBPβ), GADD 153, N-Myc, and Sp1 have been identified within the proximal promoter region of *SNCA* by computational methods [39]. However, the functionality and significance of these elements have yet to be investigated.

Transcription factors, their co-regulatory proteins, and the signaling pathways that modulate their activity all represent potential therapeutic targets that could lower the production of αSyn. Further work is required to identify these targets more clearly and how they interact to drive *SNCA* transcription. Therapeutically, these factors could be targeted in two ways: either by modulating their activity or their expression. The regulatory potential of these factors could be altered using small-molecule antagonists or inhibitors or by applying nucleic acid molecule–based strategies, including oligonucleotide decoys and aptamers [48,49].

Targeting transcription factors as a therapeutic intervention is not without precedent. Steroid hormones, including glucocorticoids, mineralocorticoids, and PPAR agonists, are prescribed routinely as anti-inflammatory medications, and for autonomic as well as for metabolic diseases. These hormones exert their effects by binding to cognate nuclear hormone receptors which are ligand-dependent transcription factors. Of particular relevance to *SNCA* transcription, GATA-3 is presently being explored for treating allergic diseases, including asthma. GATA-3 expression has been reduced using both RNA interference and antisense oligonucleotide approaches with some success in rodent models [50].

Alternatively, targeting the p38 mitogen-activated protein kinase (p38) pathway, which phosphorylates GATA-3, thus increasing its transcriptional potential (and which is independently linked to multiple system atrophy-type synucleinopathy [51] (Table 1)), has been used to reduce cytokine production in murine asthma models [50].

MODULATING *SNCA* mRNA TRANSLATION

A more common approach to gene-specific silencing is to block the translation of the mRNA into protein. Several nucleic acid molecule-derived strategies are used to inhibit translation. They include antisense oligonucleotides, RNA interference (RNAi), including small interfering RNAs (siRNAs), short hairpin RNAs (shRNAs), and micro RNAs (miRNAs), ribozymes, and DNAzymes. All of these molecules share the common feature that they are based on complementary sequences to the targeted nucleic acid sequence. They differ in the mechanisms through which they inhibit translation. In general, they function either to inhibit the translational process or to cause mRNA to be more readily degraded, thus reducing protein production. Numerous clinical trials are under way, evaluating the use of these molecules as therapeutic agents, and some have been approved and others under consideration for approval by the U.S. Food and Administration (FDA), including ribozymes, antisense oglionucleotides, and siRNA-based drugs [48,49].

Reducing αSyn expression using RNAi approaches is currently being explored by Alynylam Pharmaceuticals, which has patented an RNAi approach targeted against human and rodent *SNCA* [52]. When tested in animal models, direct infusion of this RNAi molecule into the brain led to decreased *SNCA* mRNA and αSyn protein production in multiple brain regions. This and related approaches hold great promise for *SNCA* targeted treatments in preclinical models.

In addition to SNPs located in the 5′ regulatory region of the SNCA gene, which includes the Rep-1 allele, several other SNPs spanning the *SNCA* locus have been identified that are linked to sporadic PD [53–55]. How these SNPs alter *SNCA* regulation remains unknown. They may affect translational efficiency, mRNA splicing, mRNA stability, encode either an additional enhancer or a silencing element, or could affect a miRNA target sequence. Further characterization of how these sequences contribute to the regulation of *SNCA* could provide molecular insight into additional therapeutic targets.

DEGRADATION OF NEURAL α-SYNUCLEIN

Two Routes to αSyn Degradation: Proteasome and Lysosome

Given that αSyn inclusions are a prerequisite feature of synucleinopathies, the processing of αSyn has been examined extensively in in vitro, ex vivo and in vivo models. These investigations have focused either on posttranslational modifications of αSyn (e.g., [12]) (a complete review of which would exceed the scope of this chapter) or on mechanisms of degradation. The issue of monitoring the total pool of intracellular αSyn is complicated by two factors: its relatively long half-life (estimated to be >50 hours) [56] and its overall abundance in mammalian brain (ca. 0.1% of its proteome) [17]. Initially, a key role had been postulated for the ubiquitin proteasome pathway (UPP) in the degradation of αSyn, because mutations in two UPP-related genes, *Parkin* and *UchL-1*, have been shown to influence PD risk [57–60] and because biochemical, cellular, and animal studies linked these genes to UPP-dependent processing of αSyn [61–63]. However, mounting evidence has indicated that the lysosome, as well as the proteasome, can mediate degradation of αSyn [64,65]. In general, proteins are sequestered within lysosomes by one of three known processes: macroautophagy, microautophagy, and chaperone-mediated autophagy (CMA) (reviewed by Ventruti and Cucruo [66]); it appears that αSyn can be processed by macroautophagy and CMA [67–70].

Searching for Enzymes with Synuclein-Degrading Activity

It is assumed that upon entry into a lysosomal compartment αSyn undergoes rapid degradation by a proteolytic enzyme (or enzyme complexes, which we refer to here as *synucleinases*). Several research teams, including ours, have recently begun to focus on the potential of specific lysosomal proteases to degrade αSyn in various experimental paradigms. Cathepsins represent a class of lysosomal proteases whose enzymatic activity is conferred by critical residues (*e.g.*, serine, cysteine, or aspartic acid). Of these, cathepsin D (CathD) is a major lysosomal aspartyl protease, which is composed of two disulfide-linked polypeptide chains, both produced from a single protein precursor [71]; it shares structural features with pepsin. Interestingly, CathD deficiency and its enzymatic inactivation in mammals results in an early-onset, progressive, and ultimately fatal neurodegenerative disease, which has been classified as a neuronal ceroid lipofuscinosis (NCL) [72–76]. Early in vitro experiments by Hossain et al. indicated that the treatment of recombinant αSyn with CathD resulted in partial αSyn proteolysis [77], and this was confirmed more recently by Sevlever et al. [78].

Cathepsin D Function Promotes Synucleinase Activity

Parallel experiments conducted in three independent laboratories focused on neuroblastoma cell culture models and the processing of αSyn in the context of CathD function [78–80]. *CTSD* gene expression (encoding CathD) reduced αSyn concentrations in a manner that was dependent on the level of newly synthesized CathD [78,80]. The effect of *CTSD* cDNA appeared specific because overexpression of cathepsin L (CathL) or cathepsin E (CathE) under the same conditions did not cause a reduction of aSyn concentrations [80].

The simplest explanation for the effects of *CTSD* expression on αSyn observed would be that CathD-mediated synucleinase activity within lysosomes was enhanced (alternative explanations are discussed below). For this explanation to be valid, αSyn must have entered the lysosome by either macroautophagy or CMA, or both, given that CathD is thought to be active only at low pH.

Several investigators have shown that the Asp98 and Gln99 residues of αSyn are involved in the recognition by Hsc70 chaperone and subsequent binding of the αSyn/Hsc70 complex to the Lamp2a receptor preceding CMA [68,70,81]. However, in the cell culture system employed by Cullen et al., the D98A/Q99A double mutant of αSyn was also readily degradable by CathD. Furthermore, according to Cuervo et al., the Ala30Pro and Ala53Thr mutants of αSyn are able to bind to isolated lysosomal membranes but are unable to translocate into the lysosomal lumen for degradation [70], whereas these PD-linked mutants were also efficiently degraded by CathD in the cell-culture system employed by Cullen et al. These collective findings suggested that CathD, at least in some cell types, does not rely entirely on CMA as a means of transporting αSyn into lysosomes, and that, rather, macroautophagy must have been as important as, or even more important than, CMA [68,81]. Intriguingly, Sarkar et al. recently demonstrated that small-molecule activators of mammalian autophagy, acting through mechanisms that are both dependent on and independent of the classical rapamycin-linked pathway, conferred protection in several mutant αSyn-induced toxicity models [82].

Of note, changing the phosphorylation state of human αSyn at Ser129 did not appear to alter the synucleinase activity exhibited by exogenous CathD [80]. The regulation of αSyn phosphorylation has been of intense research interest recently, because essentially all human synucleinopathies examined to date feature a pathological phosphorylation state of αSyn at Ser129 in post mortem studies (e.g., [12]). Specifically, antibodies raised to detect phosphorylated Ser129 (but not unphosphorylated αSyn) routinely decorate all the hallmark lesions of PD and related disorders; furthermore, modulation of Ser129 phosphorylation in *D. melanogaster* suggested a pivotal role in its neurotoxicity as well as inclusion formation [83]. Therefore, in pursuit of better defining a potential drug target in αSyn metabolism, Inglis and colleagues recently discovered in polo-like kinase 2 the key enzyme for the phosphorylation at Ser129 in mammalian brain [84]. However, the most recent validation effort for the relevance of modification at Ser129 in a rat model of PD cast doubt on the concept that preventing the phosphorylation of αSyn at that site represents a promising target for drug screening [85].

Aside from a possible direct enzyme/substrate-like relation inferred above, various indirect effects could underlie the interaction observed between CathD and αSyn. For example, *CTSD* gene expression could activate another cathepsin or noncathepsin protease, and this second enzyme may process αSyn for degradation. Another possible mechanism for their interaction is a change in lysosomal contents downstream of CathD protein function. Perturbations in the activities of several lysosomal sphingolipid-metabolizing enzymes have been documented in CathD-mutant sheep and dogs [74], while *ctsd*$^{-/-}$ mice accumulate phospho-and glycol-sphingolipids, including gangliosides [86]. Since αSyn is a lipid-and ganglioside-binding protein [87–89], and since the composition and amount of lipids coregulate the oligomerization state of αSyn in cells and in vivo [90,91], it is plausible that dysregulated lipid metabolism represents the indirect mechanism by which CathD deficiency leads to aSyn misprocessing. CathD is also known to exhibit activities that are unrelated to its enzymatic function [92–94], and it is therefore possible that nonproteolytic effects of CathD are essential in αSyn processing.

CTSD and *SNCA* Genes Interact in Human and Mouse Brain

To extend these cellular findings and to examine the relevance of endogenous CathD expression on endogenous αSyn processing in vivo, two of the three research teams that had initially discovered the cellular effects of CathD on αSyn processing independently examined brain tissue from homozygous *ctsd*$^{-/-}$ mice [95]. Both groups detected a consistent decrease (rather than the expected increase) in *soluble* αSyn proteins in CathD-deficient mouse brains versus those of wild-type littermates using a variety of techniques. However, this apparent discrepancy between cellular findings and initial brain biochemistry was reconciled by immunohistochemical experiments; these demonstrated that CathD-deficient mouse brains exhibited marked reduction in normal αSyn levels throughout the neuropil [80]. Since Tyynelae et al. had previously shown that CathD-deficient mice show a decrease in presynaptic markers [76] and because αSyn is a predominantly pre-synaptic protein, it was surmised that the reduction of soluble αSyn concentration recorded in the *ctsd*$^{-/-}$ mice occurred as a result of presynaptic abnormalities and overall thinning of the cortex. Importantly, in parallel to the loss of *soluble* αSyn throughout the neuropil, Qiao et al. and Cullen et al. also detected αSyn aggregates in neurons of many brain cortical and subcortical regions of *ctsd*$^{-/-}$ mice. These findings were substantiated by the observation that increased levels of *insoluble* species of *oligomeric* αSyn can be detected in formic acid extracts of *ctsd*$^{-/-}$ mouse brain [80]. These complementary findings by several research labs suggested that the presence of CathD is important for the prevention of αSyn misprocessing in the developing nervous system and that its absence facilitates the formation of αSyn aggregation in vivo. These CathD-deficient mice [95] therefore also represent the first mouse model that replicates select features of human synucleinopathy and is based on the misprocessing of endogenous murine αSyn in the absence of a human transgene and occurs without the effects of an environmental toxin.

Human subjects suffering from a congenital lysosomal storage disorder, such as a neuronal ceroid lipofuscinosis (NCL), are often affected at birth and die prematurely. Two mutational events affecting the human *CTSD* gene have been described to underlie a subset of congenital NCL cases [72,96]. The brains of these infants show extreme cortical atrophy (in excess of that seen in mice), loss of neurons and myelin, and display generalized gliosis [72,96,97]. Cullen et al. examined brain specimens of affected siblings from a NCL pedigree linked to homozygous mutations in the *CTSD* gene; there, anti-αSyn staining of the neuropil was reduced markedly throughout the brain (Fig. 1). However, prominent intracellular accumulation of αSyn was detected in cortical and thalamic populations of neurons. These intracellular neuronal inclusions morphologically resembled early Lewy body formation in cortical synucleinopathies (Table 1). However, in contrast to PD and Lewy body–positive dementias, these brain lesions were negative for ubiquitin and did not stain with anti-phospho Ser129 αSyn antibodies [80]. Although the morphological changes seen in *CTSD* gene-inactivated brain specimens did not match all the characteristics of classical synucleinopathies, these collective findings nevertheless identified in CathD the first protease that promotes the proper degradation of neuronal αSyn in human brain.

Up-Regulating Specific Enzyme Activities: A Possible Approach to Treating Synucleinopathies

With the characterization of the first enzyme promoting synucleinase activity in vivo comes the mandate to pursue further validation studies and to consider screening programs. The next phase of target validation will need to: establish the mechanisms by which the observed effect is mediated (directly or indirectly); provide proof of concept by demonstrating that elevation in human CathD expression can ameliorate human αSyn-induced pathology in the nervous system of a suitable animal model; and delineate that the up-regulation of CathD is both specific in its desired therapeutic effect and safe for the host.

Provided that CathD emerges as a fully validated target from further study, general considerations are due as to how up-regulation of its activity could be achieved in the human brain. One avenue would be to enhance the pool of available CathD enzyme in the central nervous system using a *recombinant protein-based approach* after overcoming structural impediments, such as the blood–brain barrier and cell membrane integrity. This approach underlies the concept of enzyme replacement therapy that has been pursued successfully by Genzyme since 1991 in the treatment of classical lysosomal storage disorders such as Gaucher disease, where a critical, glucosylceramide-metabolizing enzyme encoded by the glucocerebrosidase gene (*GBA*) is missing or defective. Intraparenchymal therapy either with a recombinant enzyme or through delivery of its cDNA *by a viral approach* [98] may be useful to elevate CathD levels in vivo.

An alternative approach might be to chaperone a larger pool of newly synthesized CathD to its correct lysosomal compartment than is the case during the normal aging process. For example, Jung et al. [99] have shown that while CathD is exclusively found within lysosomes in the cerebral cortex of young rats, in aged rats, CathD becomes prominently cytosolic in nature, and that this altered distribution is paralleled by cellular degeneration. This redistribution could be due partially to reduced efficiency in lysosomal targeting of

the enzyme of interest; by inference, enhancing the proper routing of target enzymes through the *use of pharmacological chaperones* may represent a useful remedy. Of note, Amicus Therapeutics has explored such a chaperone-based strategy to pursue novel therapeutics for those lysosomal storage disorders where defective processing of the relevant enzyme played a role in disease pathogenesis. Specifically, in the case of Gaucher disease, Lieberman et al. recently provided proof of concept for enhanced enzymatic activity of both wild-type (!) and mutant GBA proteins through chaperone treatment in preclinical ex vivo studies [100]. Finally, *increasing the shuttling of αSyn to the lysosome*, such as through increased binding to cognate heat-shock proteins, has the potential to promote a proteolytic effect of lysosomal CathD on αSyn. The induction and augmentation (or both) of the autophagic pathway by virtue of small-molecule therapy is an approach that a number of groups are currently pursuing actively [82,101].

CONCLUDING REMARKS

The currently approved treatment strategies for PD and related synucleinopathies are symptomatic at best, are not yet driven by pathogenetic insights, and are invariably of limited duration. To date, progress in cause-directed treatment of synucleinopathies has been prevented mostly by our limited knowledge of the key regulators that control αSyn steady state. Because researchers have uncovered only a few of the molecular underpinnings that regulate *SNCA* transcription and αSyn degradation and have not yet elucidated the mechanisms of αSyn oligomer-induced toxicity, the PD field follows any progress in lead discovery with great anticipation. We speculate that a reduction of about 50% in total αSyn levels may be sufficient for neuroprevention and neuroprotection in *SNCA*-linked and sporadic PD given that gene duplication and triplication events led to a modest 1.5-to 2-fold increase in αSyn levels, respectively. This assumption is supported by the fact that the partial and complete absence of *snca* gene expression in mice is well tolerated and does not lead to any detectable organ and nervous system dysfunction during their two-year life span, as determined by several investigators (e.g., [18]). Further exploration of the mechanisms by which human αSyn concentrations can be safely down-regulated in the nervous system, ideally during decades of treatment, holds great promise for caregivers and patients alike. Our shared vision is ultimately to treat PD and its many related, devastating disorders at one of the root causes.

REFERENCES

1. Lang, A.E., Lozano, A.M. (1998). Parkinson's disease: first of two parts. *N Engl J Med, 339*, 1044–1053.

2. Bernheimer, H., Birkmayer, W., Hornykiewicz, O., Jellinger, K., Seitelberger, F. (1973). Brain dopamine and the syndromes of Parkinson and Huntington. clinical, morphological and neurochemical correlations. *J Neurol Sci, 20*, 415–455.
3. Braak, H., Del Tredici, K. (2008). Invited Article: Nervous system pathology in sporadic Parkinson disease. *Neurology, 70*, 1916–1925.
4. Klein, C., Schlossmacher, M.G. (2006). The genetics of Parkinson disease: implications for neurological care. *Nat Clin Pract Neurol, 2*, 136–146.
5. Farrer, M.J. (2006). Genetics of Parkinson disease: paradigm shifts and future prospects. *Nat Rev Genet, 7*, 306–318.
6. Klein, C., Schlossmacher, M.G. (2007). Parkinson disease, 10 years after its genetic revolution: multiple clues to a complex disorder. *Neurology, 69*, 2093–2104.
7. Pankratz, N., Wilk, J.B., Latourelle, J.C., DeStefano, A.L., Halter, C., Pugh, E.W., Doheny, K.F., Gusella, J.F., Nichols, W.C., Foroud, T., Myers, R.H. (2009). Genomewide association study for susceptibility genes contributing to familial Parkinson disease. *Hum Genet, 124*, 593–605.
8. Horstink, M., Tolosa, E., Bonuccelli, U., Deuschl, G., Friedman, A., Kanovsky, P., Larsen, J.P., Lees, A., Oertel, W., Poewe, W., et al. (2006). Review of the therapeutic management of Parkinson's disease: report of a joint task force of the European Federation of Neurological Societies and the Movement Disorder Society-European Section. Part I: Early (uncomplicated) Parkinson's disease. *Eur J Neurol, 13*, 1170–1185.
9. Cookson, M.R. (2005). The biochemistry of Parkinson's disease. *Annu Rev Biochem, 74*, 29–52.
10. Spillantini, M.G., Schmidt, M.L., Lee, V.M., Trojanowski, J.Q., Jakes, R., Goedert, M. (1997). Alpha-synuclein in Lewy bodies. *Nature, 388*, 839–840.
11. Galvin, J.E., Lee, V.M., Trojanowski, J.Q. (2001). Synucleinopathies: clinical and pathological implications. *Arch Neurol, 58*, 186–190.
12. Anderson, J.P., Walker, D.E., Goldstein, J.M., de Laat, R., Banducci, K., Caccavello, R.J., Barbour, R., Huang, J., Kling, K., Lee, M., et al. (2006). Phosphorylation of Ser-129 is the dominant pathological modification of alpha-synuclein in familial and sporadic Lewy body disease. *J Biol Chem, 281*, 29739–29752.
13. Neumann, M., Muller, V., Kretzschmar, H.A., Haass, C., Kahle, P.J. (2004). Regional distribution of proteinase K–resistant alpha-synuclein correlates with Lewy body disease stage. *J Neuropathol Exp Neurol, 63*, 1225–1235.
14. Goldberg, M.S., Lansbury, P.T., Jr. (2000). Is there a cause-and-effect relationship between alpha-synuclein fibrillization and Parkinson's disease? *Nat Cell Biol, 2*, E115–E119.
15. El-Agnaf, O.M., Salem, S.A., Paleologou, K.E., Curran, M.D., Gibson, M.J., Court, J.A., Schlossmacher, M.G., Allsop, D. (2006). Detection of oligomeric forms of alpha-synuclein protein in human plasma as a potential biomarker for Parkinson's disease. *FASEB J, 20*, 419–425.
16. Tokuda, T., Salem, S.A., Allsop, D., Mizuno, T., Nakagawa, M., Qureshi, M.M., Locascio, J.J., Schlossmacher, M.G., El-Agnaf, O.M. (2006). Decreased alpha-synuclein in cerebrospinal fluid of aged individuals and subjects with Parkinson's disease. *Biochem Biophys Res Commun, 349*, 162–166.

17. Mollenhauer, B., Cullen, V., Kahn, I., Krastins, B., Outeiro, T.F., Pepivani, I., Ng, J., Schulz-Schaeffer, W., Kretzschmar, H.A., McLean, P.J., et al. (2008). Direct quantification of CSF alpha-synuclein by ELISA and first cross-sectional study in patients with neurodegeneration. *Exp Neurol, 213*, 315–325.

18. Abeliovich, A., Schmitz, Y., Farinas, I., Choi-Lundberg, D., Ho, W.H., Castillo, P.E., Shinsky, N., Verdugo, J.M., Armanini, M., Ryan, A., et al. (2000). Mice lacking alpha-synuclein display functional deficits in the nigrostriatal dopamine system. *Neuron, 25*, 239–252.

19. Sharon, R., Bar-Joseph, I., Mirick, G.E., Serhan, C.N., Selkoe, D.J. (2003). Altered fatty acid composition of dopaminergic neurons expressing alpha-synuclein and human brains with alpha-synucleinopathies. *J Biol Chem, 278*, 49874–49881.

20. Gai, W.P., Power, J.H., Blumbergs, P.C., Blessing, W.W. (1998). Multiple-system atrophy: a new alpha-synuclein disease? *Lancet, 352*, 547–548.

21. El-Agnaf, O.M., Jakes, R., Curran, M.D., Middleton, D., Ingenito, R., Bianchi, E., Pessi, A., Neill, D., Wallace, A. (1998). Aggregates from mutant and wild-type alpha-synuclein proteins and NAC peptide induce apoptotic cell death in human neuroblastoma cells by formation of beta-sheet and amyloid-like filaments. *FEBS Lett, 440*, 71–75.

22. Park, J.Y., Lansbury, P.T., Jr. (2003). Beta-synuclein inhibits formation of alpha-synuclein protofibrils: a possible therapeutic strategy against Parkinson's disease. *Biochemistry, 42*, 3696–3700.

23. Polymeropoulos, M.H., Lavedan, C., Leroy, E., Ide, S.E., Dehejia, A., Dutra, A., Pike, B., Root, H., Rubenstein, J., Boyer, R., et al. (1997). Mutation in the alpha-synuclein gene identified in families with Parkinson's disease. *Science, 276*, 2045–2047.

24. El-Agnaf, O.M., Jakes, R., Curran, M.D., Wallace, A. (1998). Effects of the mutations Ala30 to Pro and Ala53 to Thr on the physical and morphological properties of alpha-synuclein protein implicated in Parkinson's disease. *FEBS Lett, 440*, 67–70.

25. Conway, K.A., Harper, J.D., Lansbury, P.T., Jr. (2000). Fibrils formed in vitro from alpha-synuclein and two mutant forms linked to Parkinson's disease are typical amyloid. *Biochemistry, 39*, 2552–2563.

26. Feany, M.B., Bender, W.W. (2000). A *Drosophila* model of Parkinson's disease. *Nature, 404*, 394–398.

27. Lee, M.K., Stirling, W., Xu, Y., Xu, X., Qui, D., Mandir, A.S., Dawson, T.M., Copeland, N.G., Jenkins, N.A., Price, D.L. (2002). Human alpha-synuclein–harboring familial Parkinson's disease–linked Ala-53→Thr mutation causes neurodegenerative disease with alpha-synuclein aggregation in transgenic mice. *Proc Natl Acad Sci U S A, 99*, 8968–8973.

28. Outeiro, T.F., Lindquist, S. (2003). Yeast cells provide insight into alpha-synuclein biology and pathobiology. *Science, 302*, 1772–1775.

29. Fredenburg, R.A., Rospigliosi, C., Meray, R.K., Kessler, J.C., Lashuel, H.A., Eliezer, D., Lansbury, P.T., Jr. (2007). The impact of the E46K mutation on the properties of alpha-synuclein in its monomeric and oligomeric states. *Biochemistry, 46*, 7107–7118.

30. Lansbury, P.T., Lashuel, H.A. (2006). A century-old debate on protein aggregation and neurodegeneration enters the clinic. *Nature, 443*, 774–779.

31. Lashuel, H.A., Hartley, D., Petre, B.M., Walz, T., Lansbury, P.T., Jr. (2002). Neurodegenerative disease: amyloid pores from pathogenic mutations. *Nature*, *418*, 291.

32. Ross, O.A., Braithwaite, A.T., Skipper, L.M., Kachergus, J., Hulihan, M.M., Middleton, F.A., Nishioka, K., Fuchs, J., Gasser, T., Maraganore, D.M., et al. (2008). Genomic investigation of alpha-synuclein multiplication and parkinsonism. *Ann Neurol, 63*, 743–750.

33. Theuns, J., Van Broeckhoven, C. (2008). alpha-Synuclein gene duplications in sporadic Parkinson disease. *Neurology, 70*, 7–9.

34. Maraganore, D.M., de Andrade, M., Elbaz, A., Farrer, M.J., Ioannidis, J.P., Kruger, R., Rocca, W.A., Schneider, N.K., Lesnick, T.G., Lincoln, S.J., et al. (2006). Collaborative analysis of alpha-synuclein gene promoter variability and Parkinson disease. *JAMA, 296*, 661–670.

35. Manning-Bog, A.B., McCormack, A.L., Purisai, M.G., Bolin, L.M., Di Monte, D.A. (2003). Alpha-synuclein overexpression protects against paraquat-induced neurodegeneration. *J Neurosci, 23*, 3095–3099.

36. Chandra, S., Gallardo, G., Fernandez-Chacon, R., Schluter, O.M., Sudhof, T.C. (2005). Alpha-synuclein cooperates with CSPalpha in preventing neurodegeneration. *Cell, 123*, 383–396.

37. Xia, Y., Rohan de Silva, H.A., Rosi, B.L., Yamaoka, L.H., Rimmler, J.B., Pericak-Vance, M.A., Roses, A.D., Chen, X., Masliah, E., DeTeresa, R., et al. (1996). Genetic studies in Alzheimer's disease with an NACP/alpha-synuclein polymorphism. *Ann Neurol, 40*, 207–215.

38. Touchman, J.W., Dehejia, A., Chiba-Falek, O., Cabin, D.E., Schwartz, J.R., Orrison, B.M., Polymeropoulos, M.H., Nussbaum, R.L. (2001). Human and mouse alpha-synuclein genes: comparative genomic sequence analysis and identification of a novel gene regulatory element. *Genome Res, 11*, 78–86.

39. Chiba-Falek, O., Nussbaum, R.L. (2001). Effect of allelic variation at the NACP-Rep1 repeat upstream of the alpha-synuclein gene (SNCA) on transcription in a cell culture luciferase reporter system. *Hum Mol Genet, 10*, 3101–3109.

40. Fuchs, J., Tichopad, A., Golub, Y., Munz, M., Schweitzer, K.J., Wolf, B., Berg, D., Mueller, J.C., Gasser, T. (2008). Genetic variability in the SNCA gene influences alpha-synuclein levels in the blood and brain. *FASEB J, 22*, 1327–1334.

41. Cronin, K.D., Ge, D., Manninger, P., Linnertz, C., Rossosheck, A., Orrisson, B.M., Bernard, D., El-Agnaf, O.M.A., Schlossmacher, M.G., Nussbaum, R.L., Chiba-Falek, O. (2009). Expansion of the Parkinson disease–associated SNCA Rep1 allele up-regulates human alpha-synuclein in transgenic mouse brain. *Hum Mol Genet, 18*, 3274–3285.

42. Chiba-Falek, O., Kowalak, J.A., Smulson, M.E., Nussbaum, R.L. (2005). Regulation of alpha-synuclein expression by poly (ADP ribose) polymerase-1 (PARP-1) binding to the NACP-Rep1 polymorphic site upstream of the SNCA gene. *Am J Hum Genet, 76*, 478–492.

43. Scherzer, C.R., Grass, J.A., Liao, Z., Pepivani, I., Zheng, B., Eklund, A.C., Ney, P.A., Ng, J., McGoldrick, M., Mollenhauer, B., et al. (2008). GATA transcription

factors directly regulate the Parkinson's disease–linked gene alpha-synuclein. *Proc Natl Acad Sci U S A, 105*, 10907–10912.

44. Patient, R.K., McGhee, J.D. (2002). The GATA family (vertebrates and invertebrates). *Curr Opin Genet Dev, 12*, 416–422.
45. Clough, R.L., Stefanis, L. (2007). A novel pathway for transcriptional regulation of alpha-synuclein. *FASEB J, 21*, 596–607.
46. Yu, Y.L., Chiang, Y.J., Chen, Y.C., Papetti, M., Juo, C.G., Skoultchi, A.I., Yen, J.J. (2005). MAPK-mediated phosphorylation of GATA-1 promotes Bcl-XL expression and cell survival. *J Biol Chem, 280*, 29533–29542.
47. Cheng, D., Kim, W.S., Garner, B. (2008). Regulation of alpha-synuclein expression by liver X receptor ligands in vitro. *Neuroreport, 19*, 1685–1689.
48. Bhindi, R., Fahmy, R.G., Lowe, H.C., Chesterman, C.N., Dass, C.R., Cairns, M.J., Saravolac, E.G., Sun, L.Q., Khachigian, L.M. (2007). Brothers in arms: DNA enzymes, short interfering RNA, and the emerging wave of small-molecule nucleic acid–based gene-silencing strategies. *Am J Pathol, 171*, 1079–1088.
49. Rayburn, E.R., Zhang, R. (2008). Antisense, RNAi, and gene silencing strategies for therapy: mission possible or impossible? *Drug Discov Today, 13*, 513–521.
50. Cousins, D.J., McDonald, J., Lee, T.H. (2008). Therapeutic approaches for control of transcription factors in allergic disease. *J Allergy Clin Immunol, 121*, 803–809; quiz, 810–801.
51. Wenning, G.K., Stefanova, N., Jellinger, K.A., Poewe, W., Schlossmacher, M.G. (2008). Multiple system atrophy: a primary oligodendrogliopathy. *Ann Neurol, 64*, 239–246.
52. Lewis, J., Melrose, H., Bumcrot, D., Hope, A., Zehr, C., Lincoln, S., Braithwaite, A., He, Z., Ogholikhan, S., Hinkle, K., et al. (2008). In vivo silencing of alpha-synuclein using naked siRNA. *Mol Neurodegener, 3*, 19.
53. Mueller, J.C., Fuchs, J., Hofer, A., Zimprich, A., Lichtner, P., Illig, T., Berg, D., Wullner, U., Meitinger, T., Gasser, T. (2005). Multiple regions of alpha-synuclein are associated with Parkinson's disease. *Ann Neurol, 57*, 535–541.
54. Winkler, S., Hagenah, J., Lincoln, S., Heckman, M., Haugarvoll, K., Lohmann-Hedrich, K., Kostic, V., Farrer, M., Klein, C. (2007). Alpha-synuclein and Parkinson disease susceptibility. *Neurology, 69*, 1745–1750.
55. Mizuta, I., Satake, W., Nakabayashi, Y., Ito, C., Suzuki, S., Momose, Y., Nagai, Y., Oka, A., Inoko, H., Fukae, J., et al. (2006). Multiple candidate gene analysis identifies alpha-synuclein as a susceptibility gene for sporadic Parkinson's disease. *Hum Mol Genet, 15*, 1151–1158.
56. Okochi, M., Walter, J., Koyama, A., Nakajo, S., Baba, M., Iwatsubo, T., Meijer, L., Kahle, P.J., Haass, C. (2000). Constitutive phosphorylation of the Parkinson's disease associated alpha-synuclein. *J Biol Chem, 275*, 390–397.
57. Leroy, E., Anastasopoulos, D., Konitsiotis, S., Lavedan, C., Polymeropoulos, M.H. (1998). Deletions in the Parkin gene and genetic heterogeneity in a Greek family with early onset Parkinson's disease. *Hum Genet, 103*, 424–427.
58. Kitada, T., Asakawa, S., Hattori, N., Matsumine, H., Yamamura, Y., Minoshima, S., Yokochi, M., Mizuno, Y., Shimizu, N. (1998). Mutations in the parkin gene cause autosomal recessive juvenile parkinsonism. *Nature, 392*, 605–608.

59. Carmine Belin, A., Westerlund, M., Bergman, O., Nissbrandt, H., Lind, C., Sydow, O., Galter, D. (2007). S18Y in ubiquitin carboxy-terminal hydrolase L1 (UCH-L1) associated with decreased risk of Parkinson's disease in Sweden. *Parkinsonism Relat Disord, 13*, 295–298.

60. Maraganore, D.M., Lesnick, T.G., Elbaz, A., Chartier-Harlin, M.C., Gasser, T., Kruger, R., Hattori, N., Mellick, G.D., Quattrone, A., Satoh, J., et al. (2004). UCHL1 is a Parkinson's disease susceptibility gene. *Ann Neurol, 55*, 512–521.

61. Shimura, H., Schlossmacher, M.G., Hattori, N., Frosch, M.P., Trockenbacher, A., Schneider, R., Mizuno, Y., Kosik, K.S., Selkoe, D.J. (2001). Ubiquitination of a new form of alpha-synuclein by parkin from human brain: implications for Parkinson's disease. *Science, 293*, 263–269.

62. Liu, Y., Fallon, L., Lashuel, H.A., Liu, Z., Lansbury, P.T., Jr. (2002). The UCH-L1 gene encodes two opposing enzymatic activities that affect alpha-synuclein degradation and Parkinson's disease susceptibility. *Cell, 111*, 209–218.

63. Lo Bianco, C., Deglon, N., Pralong, W., Aebischer, P. (2004). Lentiviral nigral delivery of GDNF does not prevent neurodegeneration in a genetic rat model of Parkinson's disease. *Neurobiol Dis, 17*, 283–289.

64. Webb, J.L., Ravikumar, B., Atkins, J., Skepper, J.N., Rubinsztein, D.C. (2003). Alpha-synuclein is degraded by both autophagy and the proteasome. *J Biol Chem, 278*, 25009–25013.

65. Shin, Y., Klucken, J., Patterson, C., Hyman, B.T., McLean, P.J. (2005). The co-chaperone carboxyl terminus of Hsp70-interacting protein (CHIP) mediates alpha-synuclein degradation decisions between proteasomal and lysosomal pathways. *J Biol Chem, 280*, 23727–23734.

66. Ventruti, A., Cuervo, A.M. (2007). Autophagy and neurodegeneration. *Curr Neurol Neurosci Rep, 7*, 443–451.

67. Sarkar, S., Davies, J.E., Huang, Z., Tunnacliffe, A., Rubinsztein, D.C. (2007). Trehalose, a novel mTOR-independent autophagy enhancer, accelerates the clearance of mutant huntingtin and alpha-synuclein. *J Biol Chem, 282*, 5641–5652.

68. Vogiatzi, T., Xilouri, M., Vekrellis, K., Stefanis, L. (2008). Wild type alpha-synuclein is degraded by chaperone-mediated autophagy and macroautophagy in neuronal cells. *J Biol Chem, 283*, 23542–23556.

69. Lee, H.J., Khoshaghideh, F., Patel, S., Lee, S.J. (2004). Clearance of alpha-synuclein oligomeric intermediates via the lysosomal degradation pathway. *J Neurosci, 24*, 1888–1896.

70. Cuervo, A.M., Stefanis, L., Fredenburg, R., Lansbury, P.T., Sulzer, D. (2004). Impaired degradation of mutant alpha-synuclein by chaperone-mediated autophagy. *Science, 305*, 1292–1295.

71. Hasilik, A., Neufeld, E.F. (1980). Biosynthesis of lysosomal enzymes in fibroblasts. Synthesis as precursors of higher molecular weight. *J Biol Chem, 255*, 4937–4945.

72. Siintola, E., Partanen, S., Stromme, P., Haapanen, A., Haltia, M., Maehlen, J., Lehesjoki, A.E., Tyynela, J. (2006). Cathepsin D deficiency underlies congenital human neuronal ceroid-lipofuscinosis. *Brain, 129*, 1438–1445.

73. Tyynela, J., Sohar, I., Sleat, D.E., Gin, R.M., Donnelly, R.J., Baumann, M., Haltia, M., Lobel, P. (2000). A mutation in the ovine cathepsin D gene causes a congenital

lysosomal storage disease with profound neurodegeneration. *EMBO J, 19*, 2786–2792.

74. Awano, T., Katz, M.L., O'Brien, D.P., Taylor, J.F., Evans, J., Khan, S., Sohar, I., Lobel, P., Johnson, G.S. (2006). A mutation in the cathepsin D gene (CTSD) in American bulldogs with neuronal ceroid lipofuscinosis. *Mol Genet Metab, 87*, 341–348.
75. Koike, M., Nakanishi, H., Saftig, P., Ezaki, J., Isahara, K., Ohsawa, Y., Schulz-Schaeffer, W., Watanabe, T., Waguri, S., Kametaka, S., et al. (2000). Cathepsin D deficiency induces lysosomal storage with ceroid lipofuscin in mouse CNS neurons. *J Neurosci, 20*, 6898–6906.
76. Partanen, S., Haapanen, A., Kielar, C., Pontikis, C., Alexander, N., Inkinen, T., Saftig, P., Gillingwater, T.H., Cooper, J.D., Tyynela, J. (2008). Synaptic changes in the thalamocortical system of cathepsin D–deficient mice: a model of human congenital neuronal ceroid-lipofuscinosis. *J Neuropathol Exp Neurol, 67*, 16–29.
77. Hossain, S., Alim, A., Takeda, K., Kaji, H., Shinoda, T., Ueda, K. (2001). Limited proteolysis of NACP/alpha-synuclein. *J Alzheimer's Dis, 3*, 577–584.
78. Sevlever, D., Jiang, P., Yen, S.H. (2008). Cathepsin D is the main lysosomal enzyme involved in the degradation of alpha-synuclein and generation of its carboxy-terminally truncated species. *Biochemistry, 47*, 9678–9687.
79. Qiao, L., Hamamichi, S., Caldwell, K.A., Caldwell, G.A., Yacoubian, T.A., Wilson, S., Xie, Z.L., Speake, L.D., Parks, R., Crabtree, D., et al. (2008). Lysosomal enzyme cathepsin D protects against alpha-synuclein aggregation and toxicity. *Mol Brain, 1*, 17.
80. Cullen, V., Lindfors, M., Ng, J., Paetau, A., Swinton, E., Kolodzeij, P., Boston, H., Saftig, P., Woulfe, J., Feany, M.B., et al. (2009). Cathepsin D expression level affects alpha-synuclein processing, aggregation, and toxicity in vivo. *Mol Brain, 2*, 5.
81. Xilouri, M., Vogiatzi, T., Vekrellis, K., Stefanis, L. (2008). Alpha-Synuclein degradation by autophagic pathways: a potential key to Parkinson's disease pathogenesis. *Autophagy, 4*, 917–919.
82. Sarkar, S., Perlstein, E.O., Imarisio, S., Pineau, S., Cordenier, A., Maglathlin, R.L., Webster, J.A., Lewis, T.A., O'Kane, C.J., Schreiber, S.L., Rubinsztein, D.C. (2007). Small molecules enhance autophagy and reduce toxicity in Huntington's disease models. *Nat Chem Biol, 3*, 331–338.
83. Chen, L., Feany, M.B. (2005). Alpha-synuclein phosphorylation controls neurotoxicity and inclusion formation in a *Drosophila* model of Parkinson disease. *Nat Neurosci, 8*, 657–663.
84. Inglis, K.J., Chereau, D., Brigham, E.F., Chiou, S.S., Schobel, S., Frigon, N.L., Yu, M., Caccavello, R.J., Nelson, S., Motter, R., et al. (2009). Polo-like kinase 2 (PLK2) phosphorylates {alpha}-Synuclein at serine 129 in central nervous system. *J Biol Chem, 284*, 2598–2602.
85. Azeredo da Silveira, S., Schneider, B.L., Cifuentes-Diaz, C., Sage, D., Abbas-Terki, T., Iwatsubo, T., Unser, M., Aebischer, P. (2009). Phosphorylation does not prompt, nor prevent, the formation of alpha-synuclein toxic species in a rat model of Parkinson's disease. *Hum Mol Genet, 18*, 872–887.
86. Jabs, S., Quitsch, A., Kakela, R., Koch, B., Tyynela, J., Brade, H., Glatzel, M., Walkley, S., Saftig, P., Vanier, M.T., Braulke, T. (2008). Accumulation of

bis(monoacylglycero)phosphate and gangliosides in mouse models of neuronal ceroid lipofuscinosis. *J Neurochem, 106*, 1415–1425.

87. Perrin, R.J., Woods, W.S., Clayton, D.F., George, J.M. (2000). Interaction of human alpha-synuclein and Parkinson's disease variants with phospholipids: structural analysis using site-directed mutagenesis. *J Biol Chem, 275*, 34393–34398.
88. Cole, N.B., Murphy, D.D., Grider, T., Rueter, S., Brasaemle, D., Nussbaum, R.L. (2002). Lipid droplet binding and oligomerization properties of the Parkinson's disease protein alpha-synuclein. *J Biol Chem, 277*, 6344–6352.
89. Schlossmacher, M.G., Cullen, V., Muething, J. (2005). The glucocerebrosidase gene and Parkinson's disease in Ashkenazi Jews. *N Engl J Med, 352*, 728–731; author reply, 728–731.
90. Sharon, R., Bar-Joseph, I., Frosch, M.P., Walsh, D.M., Hamilton, J.A., Selkoe, D.J. (2003). The formation of highly soluble oligomers of alpha-synuclein is regulated by fatty acids and enhanced in Parkinson's disease. *Neuron, 37*, 583–595.
91. Sharon, R., Goldberg, M.S., Bar-Josef, I., Betensky, R.A., Shen, J., Selkoe, D.J. (2001). Alpha-synuclein occurs in lipid-rich high molecular weight complexes, binds fatty acids, and shows homology to the fatty acid-binding proteins. *Proc Natl Acad Sci U S A, 98*, 9110–9115.
92. Glondu, M., Coopman, P., Laurent-Matha, V., Garcia, M., Rochefort, H., Liaudet-Coopman, E. (2001). A mutated cathepsin-D devoid of its catalytic activity stimulates the growth of cancer cells. *Oncogene, 20*, 6920–6929.
93. Tardy, C., Tyynela, J., Hasilik, A., Levade, T., Andrieu-Abadie, N. (2003). Stress-induced apoptosis is impaired in cells with a lysosomal targeting defect but is not affected in cells synthesizing a catalytically inactive cathepsin D. *Cell Death Differ, 10*, 1090–1100.
94. Benes, P., Vetvicka, V., Fusek, M. (2008). Cathepsin D: many functions of one aspartic protease. *Crit Rev Oncol Hematol, 68*, 12–28.
95. Saftig, P., Hetman, M., Schmahl, W., Weber, K., Heine, L., Mossmann, H., Koster, A., Hess, B., Evers, M., von Figura, K., et al. (1995). Mice deficient for the lysosomal proteinase cathepsin D exhibit progressive atrophy of the intestinal mucosa and profound destruction of lymphoid cells. *EMBO J, 14*, 3599–3608.
96. Fritchie, K., Siintola, E., Armao, D., Lehesjoki, A.E., Marino, T., Powell, C., Tennison, M., Booker, J.M., Koch, S., Partanen, S., et al. (2009). Novel mutation and the first prenatal screening of cathepsin D deficiency (CLN10). *Acta Neuropathol, 117*, 201–208.
97. Garborg, I., Torvik, A., Hals, J., Tangsrud, S.E., Lindemann, R. (1987). Congenital neuronal ceroid lipofuscinosis: a case report. *Acta Pathol Microbiol Immunol Scand A, 95*, 119–125.
98. Marshall, J., McEachern, K.A., Kyros, J.A., Nietupski, J.B., Budzinski, T., Ziegler, R.J., Yew, N.S., Sullivan, J., Scaria, A., van Rooijen, N., et al. (2002). Demonstration of feasibility of in vivo gene therapy for Gaucher disease using a chemically induced mouse model. *Mol Ther, 6*, 179–189.
99. Jung, H., Lee, E.Y., Lee, S.I. (1999). Age-related changes in ultrastructural features of cathepsin B-and D-containing neurons in rat cerebral cortex. *Brain Res, 844*, 43–54.

100. Lieberman, R.L., Wustman, B.A., Huertas, P., Powe, A.C., Jr., Pine, C.W., Khanna, R., Schlossmacher, M.G., Ringe, D., Petsko, G.A. (2007). Structure of acid beta-glucosidase with pharmacological chaperone provides insight into Gaucher disease. *Nat Chem Biol, 3*, 101–107.

101. Wyttenbach, A., Hands, S., King, M.A., Lipkow, K., Tolkovsky, A.M. (2008). Amelioration of protein misfolding disease by rapamycin: translation or autophagy? *Autophagy*, *4*, 542–545.

102. Ulmer, T.S., Bax, A., Cole, N.B., Nussbaum, R.L. (2005). Structure and dynamics of micelle-bound human alpha-synuclein. *J Biol Chem, 280*, 9595–9603.

103. Klein, C., Lohmann-Hedrich, K., Rogaeva, E., Schlossmacher, M.G., Lang, A.E. (2007). Deciphering the role of heterozygous mutations in genes associated with parkinsonism. *Lancet Neurol, 6*, 652–662.

38

EMERGING MOLECULAR TARGETS IN THE THERAPY OF DIALYSIS-RELATED AMYLOIDOSIS

GENNARO ESPOSITO

Dipartimento di Scienze e Tecnologie Biomediche, Università di Udine, Udine, Italy

VITTORIO BELLOTTI

Dipartimento di Biochimica, Università di Pavia, Pavia, Italy

INTRODUCTION

In recent years new categories of diseases coming from abnormal protein folding processes have been described. In particular, a folding-linked origin has been recognized in a group of diseases, called *amyloidoses*, where incorrect protein folding is responsible for the formation of fibrillar aggregates exhibiting the cross-β structure, a particularly stable generic fold of the polypeptide chain, accessible under specific conditions in vitro and in vivo, despite sequence and corresponding native fold diversity [1,2]. In amyloid fibrils the aggregating polypeptide chain can be represented by globular proteins or, alternatively, by short peptides corresponding to truncated forms of protein precursors. Short peptides should acquire a certain level of secondary structure for the occurrence of an aggregation process. The conversion of globular proteins into insoluble fibrillar aggregates requires, instead, partial unfolding of the native geometry that entails significant conformational changes such as the partial loss of tertiary and quaternary interactions and/or conversion into β secondary structure. For many of these proteins, amyloid fibril formation is facilitated by amino acid

Protein Misfolding Diseases: Current and Emerging Principles and Therapies,
Edited by Marina Ramirez-Alvarado, Jeffery W. Kelly, and Christopher M. Dobson
Copyright © 2010 John Wiley & Sons, Inc.

mutations that destabilize the native state and confer a structural flexibility to the molecule, but other proteins are amyloidogenic in the wild-type form [3].

Following the wealth of studies on the fibril formation process, some general guidelines to interpret amyloidogenesis have emerged. It is widely accepted that fibril formation occurs via a nucleated growth mechanism [4–8]. Prefibrillar oligomeric species that can be considered structured protofibrils have been observed repeatedly [9–15]. The morphology of these oligomers is certainly related to the monomer structure, which, in turn, depends on the precursor protein rearrangement, except for the de novo adopted conformations of unstructured amyloidogenic peptides. It is generally accepted that aggregation of globular proteins into oligomeric species occurs via partial unfolding, a rearrangement that may constitute an alternative to proper folding—misfolding—through which partially folded intermediates are driven to an aggregate state with favorable overall free energy [3,16,17]. The conformational "distance" of these partially unfolded intermediates from the native state may not be very large. Especially in the initial oligomerization steps, nativelike structures should be involved [18,19].

In recent years many research efforts have been devoted to the identification and structural definition of the intermediates, along the same general line as the protein folding studies. The structural characterization of the folding–misfolding intermediates, from which the fibril formation is primed, should give very useful clues to prevent pathological misfolding. To date, more than 30 different proteins or fragments have been described to cause amyloidosis, such as Alzheimer and Parkinson diseases, type 2 diabetes, and familial amyloid neuropathies [3]. Of those proteins, some 15 are responsible for systemic amyloidoses in humans, where the pathogen is a plasma protein that is transported as a soluble product and forms fibrils at the deposition sites by partially or totally unclear mechanisms. Several proteins involved in the systemic amyloid diseases belong to the immunoglobulin superfamily. Among them, one of the most representative is β_2-microglobulin (β_2m), responsible for dialysis-related amyloidosis (DRA). β_2m is the nonpolymorphic light chain of the class I major histocompatibility complex (MHC-I). It consists of 99 residues, with a single disulfide bridge between the two Cys residues of the sequence, at positions 25 and 80, and folds into the classical β-sandwich motif of the immunoglobulin constant domains [20]. As mentioned, the deposition of β_2m fibrils is associated with DRA [21], a disease that arises in people with chronic renal failure, as the inescapable complication of long-term hemodialysis. Because of renal failure, in fact, the clearance of β_2m following dissociation from MHC-I proves heavily compromised, resulting in an increase in the circulating protein up to 60-fold, with concentrations reaching the micromolar level in long-term hemodialyzed patients. More than 90% of patients undergoing dialysis for about 10 years develop symptoms such as destructive arthropathy, bone fractures, and carpal tunnel. Because of the ensuing physical disabilities and the widespread occurrence of renal failure, this type of amyloidosis can be regarded as a high-cost social disease.

DRA can be considered an iatrogenic amyloidosis. The introduction in the clinical practice of hemodialytic procedure by Nils Alwall in September 1946 (see http://www.med.lu.se/) has saved millions of patients from death caused by renal failure, but because of the inability to replace completely the filtration selectivity of the kidney, has posed the conditions for an abnormal catabolism of β_2m. The fact that part of the bone disease associated with chronic hemodialysis could be due to amyloid deposition was discovered in 1985 by a Japanese team leaded by Gejyo [21]. The hemodialytic procedure has improved enormously over the last six decades, and several side effects of the dyalitic procedures, such as iperparathyroidism and anemia, have been minimized by increasing the biocompatibility of the membranes and the sterility of water and by mimicking, with drugs and synthetic hormones, the endocrine function of the kidney [22,23]. The frequency and severity of the amyloid complication have also been minimized by continuous improvement of the hemodialytic procedures. Floeges group documented such achievements by comparing the population of patients under hemodialysis in 1988 and in 1996 [24]. The frequency of amyloid complication decreased by about 80% in eight years, and effective expression of "vanishing complication" was employed [24] for describing the favorable trend of this *drawback*. We can say that successful therapeutic achievements were reached in this amyloidosis before the scientific community could understand the molecular basis of the process of amyloid fibril formation of globular proteins such as β_2m. A better dialytic procedure reduces the amyloid risk directly simply by reducing the concentration of circulating β_2m, because a high concentration of the protein is a prerequisite for the amyloid deposition.

However, it seems that a high β_2m concentration is not the only factor that influences the formation of amyloid fibrils. There are other factors that cannot be invoked so intuitively, but could modulate the genesis of this amyloidosis. The interest in discovering the multiple factors implied in the disease and in describing, at the molecular level, the dynamics of structural conversions, as determined for other amyloidogenic proteins [3], is justified by the perception that the tendency to vanish is not complete and the incidence of the disease through the years assumes the profile of an asymptotic curve. In the context of the global phenomenon of population aging and of an unprecedented number of individuals for which the right to health will hopefully be recognized, new approaches to effectively prevent this complication of dialysis should be pursued. We think that for curing β_2m amyloidosis new hints could come from the body of biochemical and biophysical information obtained in the last decade by researchers who have studied the amyloidogenesis of β_2m at the molecular level. This protein has been investigated for its capacity to adapt its polypeptide chain to a monomeric globular shape 2×4 nm (Fig. 1) or to an element of the polymeric structure of amyloid fibril with the same biochemical and biophysical methods as those used to solve the similar problem of other amyloidogenic proteins, such as lysozyme or transthyretin.

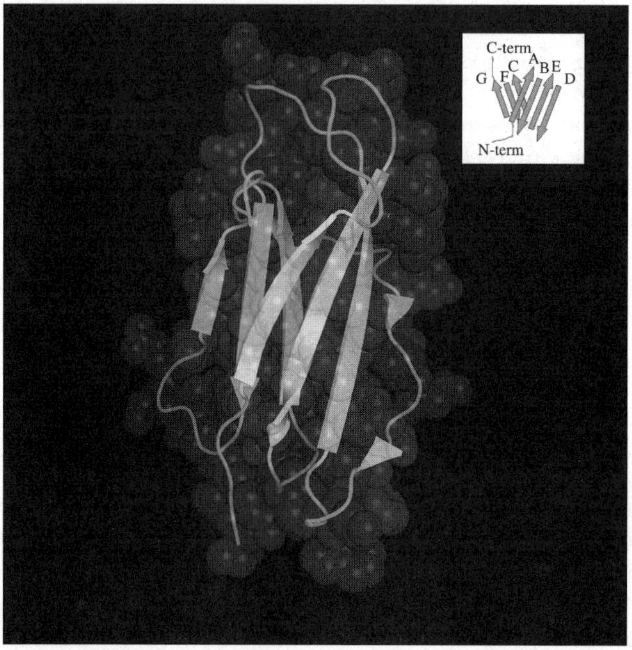

FIG. 1 Wild-type β_2m structure in MHC-I [20]. The strand segments are colored in yellow, the loop and bulge segments in green. The van der Waals surface of the molecule is shown in transparency. The strand naming scheme according to this representation is reported. (*See insert for color representation of figure.*)

A peculiar pathological property of DRA consists in the preferential growth of the amyloid fibrils at the level of joint tissues such as cartilage, capsule synovial membrane, and tendons. Localization in these sites cannot be explained on the basis of a local production of the protein, but probably is caused by the presence in these districts of a molecular environment and cofactors that assist globular β_2m conversion into fibrils. At the present state of knowledge, modulation or inhibition of β_2m amyloidogenesis in vivo can be envisaged to direct toward three different targets:

1. Reduction of circulating β_2m
2. Identification of β_2m interactors that stabilize the globular state against the fibrillar conformation
3. Modification of the local environment in which constitutive factors interact with β_2m favoring a fibrillogenic pathway

In the following the details of the structural aspects outlined above will be analyzed to highlight the emerging molecular targets in the therapy of DRA. Novel therapeutic strategies cannot, however, neglect the role of cofactors that should be addressed as a further chance to defeat the disease.

REDUCTION OF CIRCULATING β_2m

β_2m has a molecular weight of 11,740 Da, is not glycosylated, and circulates in monomeric form in the blood. Size-exclusion chromatography indicates that approximately 10% of the protein is dimeric in plasma. The protein derives from the continuous shedding of type I MHC from mononucleate cells and has become one of the parameters for the characterization of the efficiency of the dialytic procedure with regard to the removal in vivo of medium-sized molecules [25]. The attempt to remove β_2m from plasma becomes a valuable objective of the dialytic procedure after the demonstration that a high β_2m concentration is the main responsible for DRA. In the early 1990s some retrospective studies were carried out in which the concentration of serum β_2m was compared between a group of patients dialyzed with high-flux polyacrylonitrile membranes and a matching group treated with low-flux cuprophane membranes [26]. These types of studies have shown that high-flux membranes remove significant β_2m from plasma, and this effect correlates with a delay in the onset and progression of DRA. Some dialysis membranes eliminate β_2m mainly through absorption of the protein on their surface [27]. The absorptive process reaches saturation, of course, after a certain treatment duration. The efficiency of β_2m removal in high-flux hemodialysis is determined by the efficiency of convective transport and internal filtration. The efficiency in β_2m removal has been reported to range approximately between 40 and 60%. The attempt to improve the removal of β_2m through more effective filtration of medium-sized proteins with specific dialytic procedures such as hemofiltration [28] and hemodiafiltration [29] should always be confronted with the risk of loss of important components such as albumin.

Selective absorption of β_2m could fulfill the objective of very efficient removal of this protein without major losses of other important molecules. A Japanese company has developed the first system of selective absorption of β_2m, called Lixelle, consisting of a hemoperfusion-type absorbent column that uptakes β_2m through hydrophobic interactions with an hexadecyl group linked covalently to a cellulose surface. To avoid the loss of albumin and other proteins that would be bound equally by the hydrophobic groups, the system is equipped with a sieving section that prevents contact with the hydrophobic surface for proteins larger than β_2m. The capacity of β_2m removal is extremely high; the total amount of protein that can be eliminated is up to 300 mg, corresponding to about a 70% reduction in serum content. Determination of the co-absorbed proteins shows that Lixelle columns also bind peptides and proteins within a range of molecular weights from 4000 to 20,000 Da, including lysozyme, transthyretin, retinol-binding protein, and insulin.

It is worth noting that several cytokines, such as IL1 and IL6, are also captured by the hydrophobic groups linked to the cellulose stationary phase, and the removal of inflammatory mediators could also contribute to counteract the amyloid damage. In 2004, Gejyo group reported on the effectiveness of Lixelle in modifying symptoms and pathological evidence of DRA [30].

Those clinicians stated, in conclusion, that treatment with Lixelle improves the activity of daily living significantly, and in particular reduces stiffness, joint pain, and makes quite rare the formation of new bone cysts. Regression of amyloid was not investigated with specific tracers, but radiologic evidence would suggest simply the lack of progression of amyloid deposition. Apparently, treatment with Lixelle was able to affect the symptoms of the disease before any regression of amyloid could be obtained. The concomitance of β_2m reduction and symptom regression does not imply per se a cause–effect relationship because other mediators of inflammation, such as cytokines, can be responsible. A temporal dissociation between the rapid regression of muscle–skeletal symptoms of DRA and the slow reabsorption of the amyloid deposits is well established in patients receiving a kidney transplant that is associated to rapid normalization of serum β_2m. This observation cannot be explained easily at the molecular level and constitutes one of the paradigmatic but obscure features of the complex physiopathology of this amyloid disease.

IDENTIFICATION OF β_2m INTERACTORS: MOLECULAR OVERVIEW

Analysis of the material extracted from DRA patients have shown that only full-length wild-type β_2m and proteolysis products thereof occur in fibrils [31–33], along with some glycation [34] and oxidation derivatives [35] and auxiliary proteins (apoE, serum amyloid P component) [34,36]. The most abundant species (25 to 30%) among the truncation products of β_2m is a form devoid of the six N-terminal residues (ΔN6β_2m). This species was shown to have a higher tendency to self-aggregate than the full-length protein, being extremely amyloidogenic even at neutral pH, and to fail forming a fully folded native state at the end of the refolding procedure [33,37]. A comparative investigation of full-length β_2m and ΔN6β_2m by ^{1}H nuclear magnetic resonance (NMR), electrospray mass spectrometry, and limited proteolysis led to the establishment of the extent and location of the deep structural modifications undergone by β_2m upon removal of the N-terminal hexapeptide [37]. In a subsequent study on the fibrils of β_2m and ΔN6β_2m, overlapping patterns of limited proteolysis were observed, with an additional cleavage for the full-length protein fibrils leading to the truncated species [38]. The implication of ΔN6β_2m as fibrillogenesis-initiating species in vivo remains a possibility that still escapes conclusive proof. The stability hierarchy of the secondary structure elements in β_2m that determines ΔN6β_2m residual structure was also confirmed independently by other groups [39,40].

We believe that ΔN6β_2m is the closest model available to date for the conformation of β_2m in fibrils. Even minor deviations from the structure that β_2m adopts in MHC-I crystal may significantly affect its stability, as shown by the three-dimensional solution structure determined by NMR [41] (PDB code 1JNJ). The most important rearrangements of this structure were observed for strands D and E, the intervening loop and strand A, including the N-terminal segment. These modifications can be considered as prodromes of the amyloid transition. The latter was proposed to start with unpairing of strand A, leading to polymerization and precipitation into fibrils or amorphous aggregates [8,37,41]. The same mechanism was proposed [41] to account for the partial unfolding and fiber formation subsequent to Cu^{2+} binding [42], which was shown to occur primarily at His31 [41,43]. The role of His31 is crucial for the charge state of the loop between strands B and C, which in turn is very close to the N-terminal strand (Fig. 2). This area of the molecule is a key region for the stability of the entire structure. According to our expectations, the mutant His31Tyr, where a tyrosine residue replaces His31, should be more stable than a wild-type sequence because it is devoid of the partial charge of the histidine imidazole group, which could generate electrostatic repulsion with the guanidinium group of Arg3 [44].

Stopped-flow kinetic analysis [45] did not succeed in identifying any metastable unfolding intermediate of full-length β_2m, but could clearly reveal the occurrence of a long-lived intermediate, I_2, along the refolding pathway. This metastable intermediate was subsequently shown to possess higher propensity to aggregate than the fully folded conformation and to occur in equilibrium with the latter [46], suggesting that the amiloidogenicity of β_2m could arise from the fraction of the protein in this conformation. Real-time NMR measurements, performed during refolding of wild-type β_2m, showed that in solution a minor species exists whose concentration decays at a rate similar to that I_2 kinetics, while the natively folded species increases simultaneously [8]. By considering the type of resonances that are seen to decrease, it was concluded that I_2 is a conformer with a high degree of tertiary structure, very closely related to the fully folded protein. This transient conformation, apparently characterized by packing differences in the inner hydrophobic cluster compared to the folded species, converts slowly (tens of minutes at ambient temperature) into the most stable form. According to Chiti and co-workers [45,46] the slow refolding phase that populates the I_2 intermediate should be attributed to *trans–cis* isomerization of Pro32, an interpretation reinforced by more recent reports with additional NMR and spectroscopic evidence [47,48]. Over the last years β_2m has been investigated extensively in several laboratories. The molecule represents, in fact, an interesting paradigm of amyloidogenic protein that accumulates with tissue specificity. A nucleation-dependent mechanism has been proposed for β_2m in vitro fibrillogenesis [4,8], but no satisfactory interpretation is yet available for in vivo amyloidogenesis.

The structural rigidity and packing in fibrils were addressed deeply along with the kinetics and thermodynamics of fibril growth [5,49]. These studies led to rationalization into the general framework based on the basis of the main-chain-directed theory of fibril structuration and the template-dependent propagation and maturation of amyloids [50]. Fibrillogenesis occurring in the presence of glycosaminoglycans, organic solvents, collagen, and other factors [51–53] was also analyzed to assess the relevance of various physiological

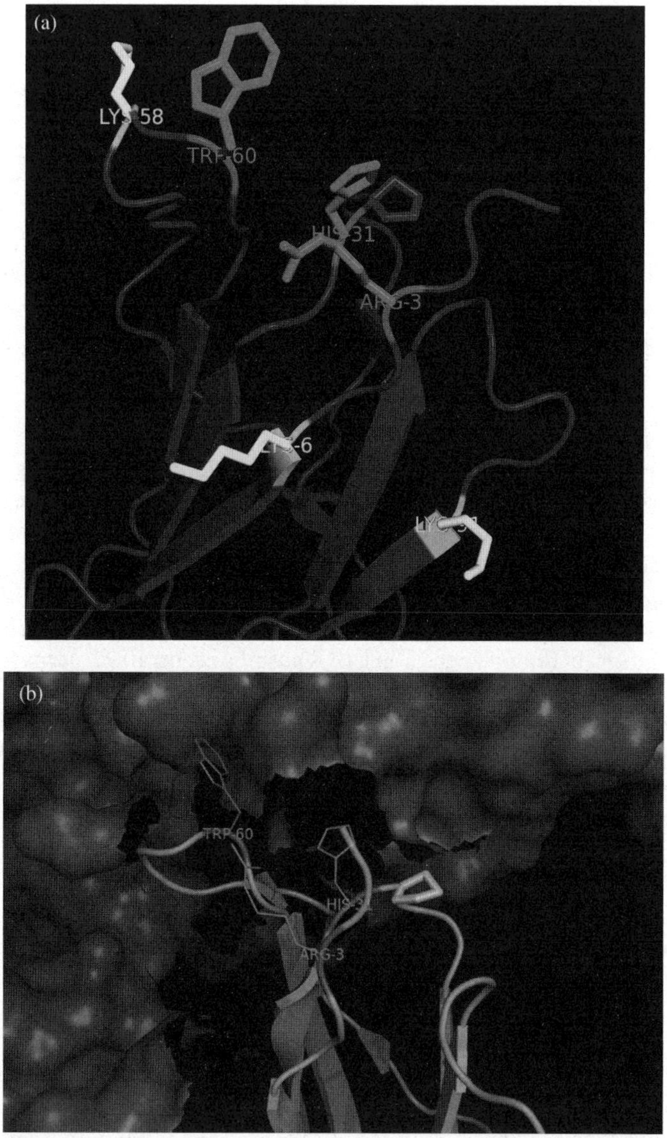

FIG. 2 (a) View of the functionally relevant apical region in β_2m solution structure [41]. The backbone and side chain of Arg3, His31, and Trp60 are shown. The exposed and positively charged side chains of Lys6, Lys58, and Lys91, along with the corresponding backbone positions, are also shown. The side chain of Pro32 is also drawn. (b) Same apical region of β_2m in MHC-I [20] with Arg3, His31, and Trp60 highlighted. The heavy-chain binding surface is shown.

conditions [48,54]. Recently, a novel approach was undertaken to stress the peculiar depositional properties of β_2m that occurs specifically on collagen substrates. The interaction was investigated in the heterogeneous phase, first by measuring the affinity of collagen for different β_2m conformations and derivatives [55] and then by interpreting the atomic force microscopy results in terms of the electrostatics that lead to fibril deposition onto a collagen or polylysine-coated solid surface [56].

The aggregation process that eventually leads to fibrils starts with the formation of dynamic oligomers formed by nativelike molecules. For β_2m, following the seminal work of Naiki and colleagues [4], this phase was first addressed by NMR observations [8], subsequently by laser light scattering [54] and mass spectrometry [15], and finally by molecular dynamics simulations performed with a cluster of 27 protein monomers and about 150,000 water molecules [57]. Analysis of the dynamics trajectories highlighted the important role of establishing intermolecular contacts of a β_2m tryptophan residue, Trp60 located in the DE loop, by the same apical region containing the N-terminal segment and the BC loop, where His31 and Pro32 are also located (Fig. 2). The role of Trp60 has been investigated experimentally by expression and characterization of specific mutants ([58,59]. The results confirm the relevance of a bulky aromatic sidechain in position 60 for the aggregation properties of β_2m.

IDENTIFICATION OF β_2m INTERACTORS: LEARNING FROM RATIONAL MUTANT DESIGN

The rationale that led us to recognize the prodromes of β_2m amyloid transition was put to work to design mutants and variants of the molecule. In particular, the crucial role of His31 for stability (Fig. 2) prompted us to engineer a mutant that, while keeping the aromatic character of the side chain, could not be charged around neutral pH. According to expectations, the His31Tyr mutant, where a tyrosine replaces histidine, was found to be more stable than wild-type species by 1.5 kcal/mol, whereas dropping the aromatic side chain by substituting a serine for histidine stabilized by only 0.6 kcal/mol [8]. Also the aggregation properties of the His31Tyr mutant appeared remarkably different with respect to wild-type protein [54], and fibrillogenesis failed under either mild (seeding and TFE) or extreme acidic (pH 2.5) conditions [59]. The His31Tyr mutant structure, determined by x-ray and NMR [44], shows (1) the conformational closeness of His31Tyr β_2m and wild-type protein, when the latter is either complexed in MHC-I [20] or isolated in solution [41], but not in the crystal [60], and (2) the heterogeneity in His31Tyr β_2m crystal, which also includes an alternative minor conformation where the N-terminal strand (strand A) is displaced from its native location, in agreement with the proposed evolution toward the amyloidogenic conformation.

Besides predicting more stable mutants based on the critical residue His31, the structural information on the solution structure of β_2m could be used to predict less stable mutants than natural sequence. Suppression of the unfavorable electrostatic interaction between residues 3 and 31 does not ensure engineering a more stable β_2m mutant. Indeed, the Arg3Ala mutant is less stable than wild type by about ½ kcal/mol [8]. This highlights the relevance of a bulky and positively charged side chain in position 3. Arg3 was proposed earlier to contribute, with Lys6, Lys58, and Lys91, to the overall stability of the molecule by breaking the extension of an exposed hydrophobic region [37]. The NMR solution structure of the Arg3Ala mutant, however, looks still very similar to that of the parent sequence, with some dispersion increase within the conformational ensemble at DE and FG loops [8]. While suppression of the positive charge in position 31 for His31Tyr mutant affects mainly the electrostatic contribution to folding free energy, suppression of the positive charge in position 3 for the Arg3Ala mutant affects mostly the corresponding solvation contribution [8]. Similarly, solvation free energy should contribute most of the 1.9 kcal/mol stability decrease of ΔN3β_2m, the truncated β_2m variant devoid of the N-terminal tripeptide, although in this case the actual lack of a sequence fragment impairs direct comparison between full-length sequence and truncated species. In addition, the presence in ΔN3β_2m of a N-terminal methionyl residue, coming from expression in *Escherichia coli*, places the backbone N-terminus ammonium in position 3 and hence does not suppress the positive charge at that location. This may account for the substantial conservation of the tertiary folding in ΔN3β_2m inferred from the relative invariance of the NMR chemical shifts with respect to full-length β_2m, at variance with the extensively destructured ΔN6β_2m [37], which also proves the most destabilized variant of β_2m (2.5 kcal/mol).

As mentioned, the search for folding/misfolding intermediates of β_2m and variants led to identifying a slowly converting form on the pathway of refolding named I_2 [8,45,46] The kinetic lability of I_2 was shown to be inversely related to the thermodynamic stability of the fully folded final product. Based on NMR evidence, in general the equilibrium population of I_2 was rather low, except for the truncated sequence ΔN3β_2m. In this case, I_2 accounts for some 25% of the overall protein concentration at 298 K [8,61].

The β_2m refolding intermediate I_2 was formerly regarded as an ensemble of species [45] and later as a single species [46–48]. Following the earlier interpretation of Chiti and co-workers [45,46], it has been shown that this intermediate contains a nonnative *trans*-peptide bond between His31 and Pro32 that slowly converts into *cis* form during the final refolding step [47]. The isomerization occurs with rearrangement of the protein toward native structure from a differently packed [8] or nativelike state [47,48]. The experimental data of Jahn et al. [48] suggest that a *trans* geometry for the ω_{31} angle can be accommodated in the native form of β_2m, based on the folding kinetics and the NMR pattern of the mutant Pro32Gly. This mutant protein with a *trans*-peptide bond between Gly32 and the peptidyl group of the preceding chain, however, does not form fibrils (i.e. an intermediate other than native state is required) [48]. Thus, the occurrence of a *trans*-peptide bond between residues

31 and 32 in the native structure of β_2m does not necessarily imply fibrillar aggregation. Rather, it is a proper conformational flexibility that appears necessary for fibrillogenesis. In fact, the mutation Pro32Gly and the elimination of *cis* peptide bond 31–32 should destabilize the apical region, leading to an increase in the population of local conformers that can be considered as the pool, including the fibril-competent intermediate(s). This interpretation is consistent with the resonance intensity reduction that was observed for residues of strand A and loops BC, DE, and FG, and was ascribed to exchange broadening [48]. Accordingly, the increased fibrillogenic potential does not arise from the presence of a nonnative *trans*-peptide bond but from the increased extent of unfolding in critical regions. Among these critical regions there is loop DE. In particular, the conformational flexibility of loop DE containing Trp60 could play a key role in the early steps of fibrillogenesis of β_2m. This conclusion arises from the entire body of experimental evidence obtained by comparing wild-type and Trp60Gly β_2m. The interest in this single-point mutant was first stimulated by inspection of molecular dynamics trajectories at the early steps of β_2m aggregation: the intermolecular contacts captured by the simulation snapshots suggested that Trp60 should play a most relevant role, together with nearby N-terminal residues [57].

The Trp60Gly mutant was then expressed and characterized structurally and functionally. The replacement of Trp60 with a Gly led to an increase of 0.8 kcal/mol in the thermodynamic stability of the protein, unexpectedly at first glance for a molecular position totally exposed to solvent that should contribute little difference between folded and denatured states. Although the presence of a Gly in position 60 should somehow alleviate the conformational strain of the local a_L secondary structure due to the unique Ramachandran profile of an amino acid without a side chain, nearly the same stabilization (0.7 kcal/mol) had, intriguingly, been reported for the Trp60Phe mutant [58]. In general, however, comparisons among protein stabilities should always be considered with some caution because the specific denatured states are largely unknown.

The most striking feature of Trp60Gly β_2m was its failure to form fibrils under mild conditions (i.e., 20% TFE neutral aqueous solution in the presence of seeds), whereas the fibrillogenic potential was fully retained at low pH (2.5). In contrast, the β_2m mutants addressed by Kihara et al. [58], with Phe substitutions for Trp60 or Trp95, formed fibrils under either set of conditions, which could suggest the relevance of an aromatic residue in position 60 or 95.

These results could be related to the structural features of the Gly mutants of the evolutionarily conserved Trp residues. The solution structure of Trp95Gly β_2m mutant proved remarkably different with respect to wild type, as evident from circular dichroison and NMR patterns. The latter was consistent with a statistically disordered structure. Also, the Trp95Gly mutant did not form fibrils in the presence of seeds and 20% TFE, despite the massive precipitation into amorphous aggregates. Overall, the evidence is consistent with the absence of fibril-competent conformations within the distribution of Trp95Gly, not

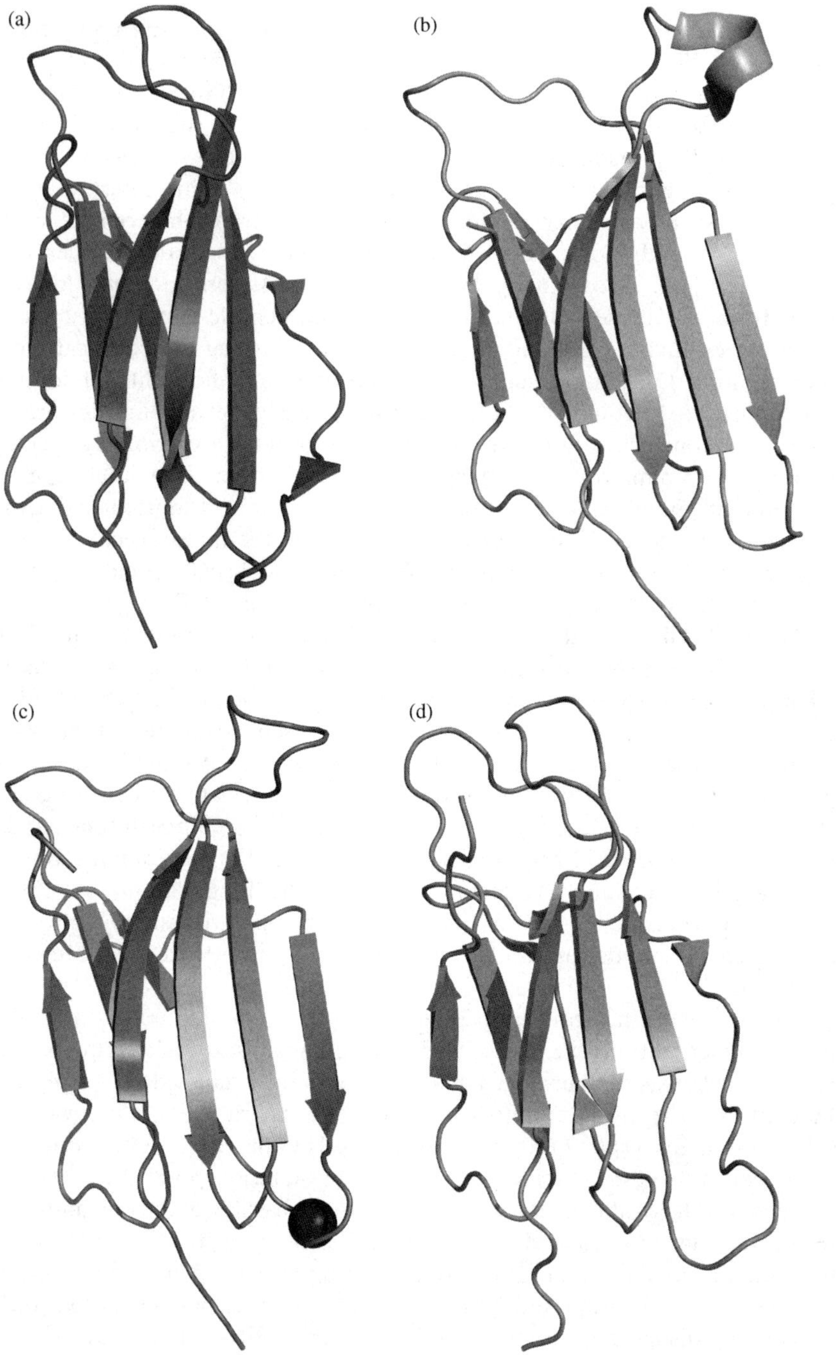

FIG. 3 Comparison between the crystal structures of (a) MHC-I-bound β_2m [20], (b) isolated β_2m [60], (c) Trp60Gly β_2m where the sphere indicates the mutation position [59], and (d) the solution structure of wild-type protein [41]. β_2m strand D (the rightmost

only in neutral aqueous solutions, but also under conditions that generally promote fibrillogenesis (i.e., pH 2.5). For the Trp60Gly mutant, in addition to the NMR solution structure, an x-ray structure could be obtained because of successful crystallization [59]. Based on the NMR results, Trp60Gly β_2m was seen to maintain in solution the same general folding as that of wild-type β_2m. The crystal structure, on the other hand, proved similar to the conformation already reported for the isolated β_2m [60]: an extended regular D strand devoid of a bulge at residue 53, and an outward, rather than inward orientation of the AB loop (Fig. 3). These two features are the most relevant deviations with respect to the structure adopted by the β_2m in MHC-I [20] and in solution [41]. In particular, the lack of a bulge in the D strand was considered to be the hallmark of fibril competence in isolated β_2m [60]: the rare event involving the crucial edge strand required for intermolecular amyloid aggregation [62]. The failure to form fibrils of Trp60Gly β_2m under mild conditions, despite the continuous D strand observed in its crystal structure, disproves this interpretation and suggests that the conformation of β_2m D strand in the crystalline state is determined by interactions within the specific lattice.

The behavior of Trp60Gly mutant is much better accounted for by the structural dynamics determinations obtained by NMR. The ^{15}N relaxation rate measurements clearly indicate that the extent of association of the soluble protein is reduced for the mutant Trp60Gly β_2m in comparison to wild type, as inferred from the lower mean t_m value. In addition, but most significantly, several residues in the region around Trp60 (D–E loop) are observed to undergo a slow-time-scale conformational exchange in wild-type β_2m that proves substantially absent in Trp60Gly β_2m. Taken together, fibrillogenesis failure, reduced association, and increased DE loop rigidity of Trp60Gly β_2m call for the role of an aromatic ring within the latter loop to promote intermolecular contacts, as expected from molecular dynamics simulations [57]. These contacts represent early events in the oligomerization process, during which the entire population of β_2m conformers (i.e., both the folded and partially unstructured forms) experience dynamic association that should, after nucleation and elongation, lead eventually to fibrils, according to an NCC (nucleated conformational conversion) type mechanism [8]. Thus, the role of the conformational flexibility to achieve fibril-competent arrangements assumes a complementary relevance with respect to the oligomerization equilibria in the search for a nucleated assembly [8,15,54].

strand in all panels) is split into two substrands by a bulge centered at Asp53 in MHC-I (a), but only one of these substrands survives in solution (d). On the other hand, a regular and continuous β-strand D is observed when the protein is crystallized without the heavy chain (b), along with an outward orientation of the AB loop (refer to Fig. 1 for strand naming). Both features also occur in the crystal structure of the Trp60Gly mutant (c), which does not form fibrils when seeded in 20% TFE. Hence, a continuous D strand can not be considered the hallmark of β_2m fibril-competent conformation [60].

Given the fibrillogenesis attitudes observed, it was argued that the "risky" conformational flexibility of the natural β_2m sequence may be a necessary compromise to enforce proper affinity with the α-chain of class I MHC, at the price of partial unfolding danger. This unavoidable condition should eventually be responsible for pathological aggregation of β_2m at nonphysiological high concentrations. To challenge this speculation, two parallel binding experiments, using wild-type and Trp60Gly β_2m, respectively, were carried out to isolate a ternary complex (heavy chain, β_2m, and synthetic nonapeptide epitope) of the type involved in cell-mediated immunity. The results showed that mutant Trp60Gly β_2m has a 20-fold lower affinity for the heavy chain than has wild-type protein. Besides the loss of Trp60 indole-specific interactions, conformational rigidity may also conceivably determine the affinity decrease observed.

The consistency of different lines of experimental data suggests that our interpretation of β_2m conformational plasticity may provide new insights for DRA therapies.

MODIFICATION OF THE FIBRILLOGENIC ENVIRONMENT

Does the tissue environment where amyloid is formed represent a therapeutic target? The answer to this question recapitulates the characteristics of DRA. In the perspective of a global therapeutic strategy against DRA, the mechanisms related to the peculiar tissue specificity of this amyloidosis should not be ignored. In other amyloidoses the deposition of amyloid fibrils in specific tissues or districts of an organism can largely be explained by the local production of a highly amyloidogenic precursor. This is the case in medullary amyloidosis or Alzheimer disease. However, selective tissue specificity can also occur in other amyloidoses in which the tissue where the amyloid is deposited is different from that at the site of production of the precursor, and this process calls for a convincing molecular explanation. As mentioned above, the anatomic selectivity of DRA is extremely high and certainly cannot be explained by the local production levels of the protein.

In the last three years our group has scrutinized the hypothesis, first raised by Homma et al. [63], about the role of collagen in determining the affinity of β_2m amyloid fibrils for the muscle–skeletal system. The affinity of β_2m for type I collagen was evaluated by band-shift electrophoresis and surface plasmon resonance [55]. This last technique suggests that the affinity constant is quite low ($K_d \sim 0.2$ mM), which itself cannot justify the stable absorption and accumulation of circulating β_2m onto collagen. The affinity raises when the constant decreases to about 5 μM if the binding is measured for the truncated species of β_2m lacking six residues at the N-terminal end (ΔN6β_2m). The occurrence of ΔN6β_2m in natural fibrils was first reported in 1986 [64], and our group demonstrated its high amyloidogenic propensity nearly 15 years later [37]. The hypothesis that collagen could play a role in vivo in the genesis of

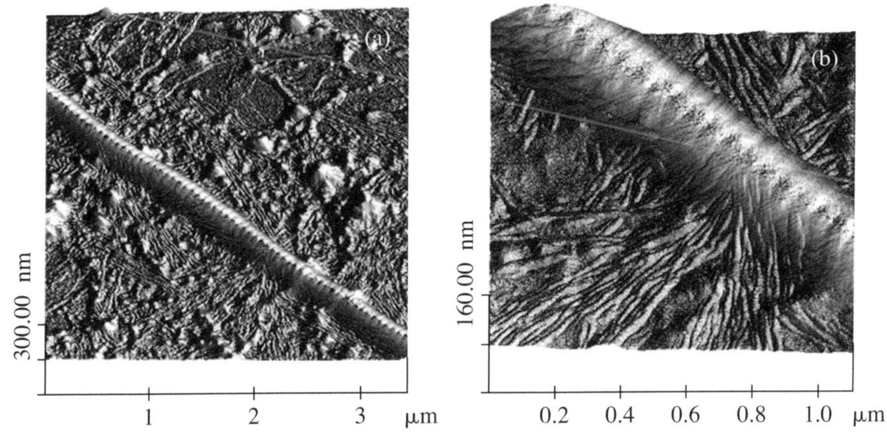

FIG. 4 Surface plots of a TM-AFM image (height data) of β_2m ex vivo amyloid material obtained from the femoral head of a patient affected by DRA: (a) amyloid fibrils surrounding a collagen fibril; (b) higher-resolution image of a portion of panel (a) showing amyloid fibrils crowding around the collagen fibril. (From [56], with permission.)

amyloid fibrils came from the observation of images of natural fibrils obtained by atomic force microscopy by Relini and Gliozzi group [56]. The image, reproduced in Figure 4, allows one to appreciate distinctly the surface of a collagen fiber, clearly identifiable from the characteristic periodic pattern and the large diameter, and many thinner amyloid fibrils that emerge from collagen and are apparently anchored and grown orthogonally to the fiber axis.

The pattern observed in vivo can surprisingly be reproduced in vitro through the incubation of synthetic collagen fibres with recombinant wild-type β_2m. The formation of fibrils occurs in conditions in which β_2m conserves a native structure (0.1M phosphate buffer, pH 6.4 and 37°C), and this finding cannot easily be explained because all the methods of β_2m fibrillogenesis reported previously invariably required a significant perturbation of the native structure. The kinetics of collagen-driven amyloidogenesis was influenced favorably by raising the temperature from 37°C to 40°C, and was affected negatively by a change of pH from 6.4 to 7.4 or by preliminary filtration of β_2m solution with small (20-nm) pore filters [56]. It is possible that all these effects can be ascribed to the profibrillogenic role of β_2m oligomers [54] and to the molecular effects of pH and temperature [8,41]. The oligomer concentration in solution is higher at pH 6.4 than 7.4, and is higher at 40°C than at 37°C. The soluble oligomeric aggregates can be removed temporarily by extreme filtration (i.e., using 20-nm threshold filters), resulting in oligomer-poor solutions with a slow rate of new formation that can be accelerated by seeds [54]. A specific model of β_2m accumulation onto surfaces with charge arrays such as collagen, which includes the molecular effects of pH and temperature on the monomeric protein or its oligomeric assembly, was proposed [56].

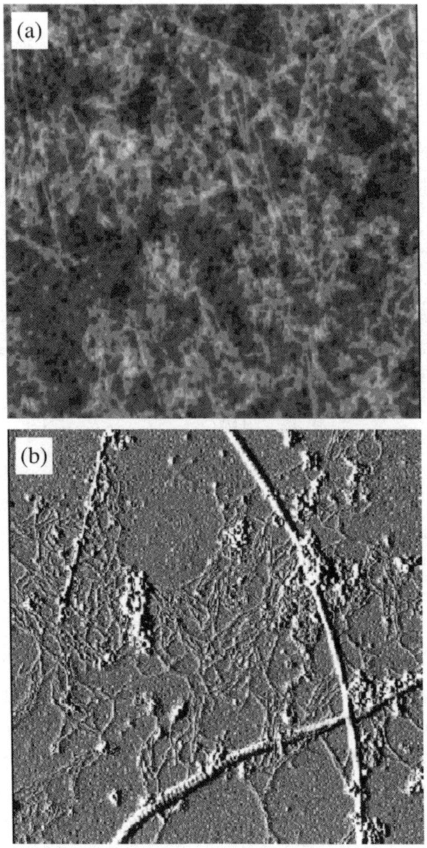

FIG. 5 TM-AFM images of β_2m incubated at 37°C and pH 6.4 in the presence of fibrillar collagen and heparin. (a) After 17 hours of incubation, thin filaments are clearly visible on a background of nonfibrillar material. Height data: scan size, 1.2 μm, Z range, 8.0 nm. (b) Fibril network connecting isolated collagen fibrils, observed after 24 hours of incubation. Nonfibrillar aggregates are also present. Amplitude data: scan size, 5.7 μm. (From [65], with permission.)

We have recently discovered that the fibrillogenesis on collagen is increased further by heparin (Fig. 5) at a concentration that can probably be reached in plasma throughout the hemodialytic procedures, where heparin is the necessary anticoagulant [65]. Glycosaminoglycans (GAGs) are a constitutive component of amyloid deposits [66], including those associated with DRA. The potent and generic pro-amyloidogenic effect of glycosaminoglycans (GAGs) has long been established for different proteins [67]. β_2m binds heparin and other GAGs with a K_d value of approximately 0.1 mM [68], similar to the affinity measured between β_2m and collagen. However, despite the low affinity, GAGs strongly favor the in vitro fibrillogenesis of wild-type β_2m when fibrils are obtained by methods in which the native structure is perturbed [49,51–53]. Moreover, fibrils can be formed in vitro under physiological conditions if the truncated $\Delta N6\beta_2m$ is incubated with GAGs.

In vitro studies have shown that GAGs enhance the nucleation of amyloid β peptides [69] and favor fibril formation and stabilization [70]. Heparin and, to a lesser extent, heparan sulfate were reported to increase significantly the rate of fibrillation of α-synuclein [71] and gelsolin [72]. Extensive experimental evidence shows that, among GAGs, heparin is particularly effective in accelerating both fibril formation and extension [51], and it has been proposed that such behavior is highly dependent on the sulfate groups present in this GAG. Furthermore, in vivo studies have demonstrated the coincident deposition of amyloidogenic protein and GAGs [73], whereas in a mouse model of AA amyloidosis a resistance against the amyloid deposition can be achieved through the overexpression of heparanase, which degrades heparan sulfate [74]. A key role of GAGs in causing amyloid deposition is also proved by the effectiveness of therapeutic agents able to displace GAGs from the amyloid fibrils [75]. Despite the role of GAGs in the amyloid deposition of DRA that is clearly emerging from several and different experimental approaches, these sugars have so far not been considered as a possible therapeutic target. Heparin is even used regularly in dialysis, and apparently at the moment there is no alternative in terms of efficacy to prevent blood clotting. The possibility of replacing the infusion of heparin with other anticoagulant approaches is under evaluation [76] not for the amyloid risk, but because heparin has well-known side effects, such as the induction of lipolysis and osteoporosis, and can provoke hemorragic thrombocytopenia.

The molecular mechanisms involved in the enhancement of protein aggregation by heparin and other GAGs are still unknown. Calamai et al. [77] proposed an interesting model for explaining the effect of heparin on oligomerization and amyloidogenesis of acylphosphatase. Heparin is a strongly charged polyanion that interacts with the positive charges located on the surface of the protein and could minimize the repulsive forces acting in maintaining the single molecules at the monomeric state.

CONCLUSIONS

In dialysis-related amyloidosis, an effective and decisive therapeutic approach fortunately exists and consists of kidney transplantation, which normalize the β_2m concentration and rapidly eliminate symptoms of the disease. Unfortunately, in forthcoming years we should expect an increasing number of subjects who, in the presence of severe renal failure, will be treated with hemodialytic procedures and will not have any chance to receive a renal transplant. For patients treated with a hemodialytic procedure, the possibility of avoiding amyloidosis caused by the hemodiaysis will possibly be based on three complementary strategies: (1) development of devices able to deplete blood

β_2m, (2) identification of drugs that interact with β_2m and inhibit its intrinsic amyloidogenic propensity, and (3) manipulation of the interaction between β_2m and natural cofactors such as collagen and glycosaminoglycans that are probably essential in local deposition of the fibrils. The therapeutic approaches for this amyloidosis are not dissimilar from those used in other amyloidoses where reduction of the precursor can represent a primary therapeutic target, as in AL amyloidosis, or the inhibition of misfolding and aggregation that can be obtained by stabilization of the native fold of the precursor through a potent ligand, as in the case of transthyretin. In DRA, a quite peculiar therapeutic target might be the environment of the tissue where the amyloid grows. In these tissues we have discovered two constitutive elements, collagen and GAGs, that might play an important pathogenic role and could become complementary pharmaceutical targets.

Acknowledgments

This work was supported by MIUR (2006058958, RBNE03PX83) and the European Union (LSHM-CT-2005-037525). The suggestions of A. Makek are acknowledged.

REFERENCES

1. Dobson, C.M. (1999). Protein misfolding, evolution and disease. *Trends Biochem Sci, 24*, 329–332.
2. Sunde, M., Blake, C.C. (1998). From the globular to the fibrous state: protein structure and structural conversion in amyloid formation. *Q Rev Biophys, 31*, 1–39.
3. Chiti, F., Dobson, C.M. (2006). Protein misfolding, functional amyloid, and human disease. *Annu Rev Biochem, 75*, 333–366.
4. Naiki, H., Hashimoto, N., Suzuki, S., Kimura, H., Nakakuki, K., Gejyo, F. (1997). Establishment of a kinetic model of dialysis-related amyloid fibril extension in vitro. *Amyloid: Int J Exp Clin Invest, 4*, 223–232.
5. Naiki, H., Yamamoto, S., Hasegawa, K., Yamaguchi, I., Goto, Y., Gejyo, F. (2005). Molecular interactions in the formation and deposition of beta2-microglobulin-related amyloid fibrils. *Amyloid, 12*, 15–25.
6. Serio, T.R., Cashikar, A.G., Kowal, A.S., Sawicki, G.J., Moslehi, J.J., Serpell, L., Arnsdorf, M.F., Lindquist, S.L. (2000). Nucleated conformational conversion and the replication of conformational information by a prion determinant. *Science, 289*, 1317–1321.
7. Uversky, V.N., Li, J., Souillac, P., Millett, I.S., Doniach, S., Jakes, R., Goedert, M., Fink, A.L. (2002). Biophysical properties of the synucleins and their propensities to fibrillate: inhibition of alpha-synuclein assembly by beta-and gamma-synucleins. *J Biol Chem, 277*, 11970–11978.
8. Corazza, A., Pettirossi, F., Viglino, P., Verdone, G., Garcia, J., Dumy, P., Giorgetti, S., Mangione, P., Raimondi, S., Stoppini, M., et al. (2004). Properties of some

variants of human beta2-microglobulin and amyloidogenesis. *J Biol Chem, 279*, 9176–9189.

9. Harper, J.D., Wong, S.S., Lieber, C.M., Lansbury, P.T. (1997). Observation of metastable Abeta amyloid protofibrils by atomic force microscopy. *Chem Biol, 4*, 119–125.
10. Walsh, D.M., Lomakin, A., Benedek, G.B., Condron, M.M., Teplow, D.B. (1997). Amyloid beta-protein fibrillogenesis. Detection of a protofibrillar intermediate. *J Biol Chem, 272*, 22364–22372.
11. Ionescu-Zanetti, C., Khurana, R., Gillespie, J.R., Petrick, J.S., Trabachino, L.C., Minert, L.J., Carter, S.A., Fink, A.L. (1999). Monitoring the assembly of Ig light-chain amyloid fibrils by atomic force microscopy. *Proc Natl Acad Sci U S A, 96*, 13175–13179.
12. Quintas, A., Vaz, D.C., Cardoso, I., Saraiva, M.J., Brito, R.M. (2001). Tetramer dissociation and monomer partial unfolding precedes protofibril formation in amyloidogenic transthyretin variants. *J Biol Chem, 276*, 27207–27213.
13. Kayed, R., Sokolov, Y., Edmonds, B., McIntire, T.M., Milton, S.C., Hall, J.E., Glabe, C.G. (2004). Permeabilization of lipid bilayers is a common conformation-dependent activity of soluble amyloid oligomers in protein misfolding diseases. *J Biol Chem, 279*, 46363–46366.
14. Borysik, A.J., Radford, S.E., Ashcroft, A.E. (2004). Co-populated conformational ensembles of beta2-microglobulin uncovered quantitatively by electrospray ionization mass spectrometry. *J Biol Chem, 279*, 27069–27077.
15. Smith, A.M., Jahn, T.R., Ashcroft, A.E., Radford, S.E. (2006). Direct observation of oligomeric species formed in the early stages of amyloid fibril formation using electrospray ionisation mass spectrometry. *J Mol Biol, 364*, 9–19.
16. Pande, V.S., Grosberg, A., Tanaka, T., Rokhsar, D.S. (1998). Pathways for protein folding: Is a new view needed? *Curr Opin Struct Biol, 8*, 68–79.
17. Dobson, C.M., Karplus, M. (1999). The fundamentals of protein folding: bringing together theory and experiment. *Curr Opin Struct Biol, 9*, 92–101.
18. Esposito, G., Corazza, A., Viglino, P., Verdone, G., Pettirossi, F., Fogolari, F., Makek, A., Giorgetti, S., Mangione, P., Stoppini, M., Bellotti, V. (2005). Solution structure of beta(2)-microglobulin and insights into fibrillogenesis. *Biochim Biophys Acta, 1753*, 76–84.
19. Plakoutsi, G., Bemporad, F., Calamai, M., Taddei, N., Dobson, C.M., Chiti, F. (2005). Evidence for a mechanism of amyloid formation involving molecular reorganisation within native-like precursor aggregates. *J Mol Biol, 351*, 910–922.
20. Bjorkman, P.J., Saper, M.A., Samraoui, B., Bennett, W.S., Strominger, J.L., Wiley, D.C. (1987). Structure of the human class I histocompatibility antigen, HLA-A2. *Nature, 329*, 506–512.
21. Gejyo, F., Yamada, T., Odani, S., Nakagawa, Y., Arakawa, M., Kunitomo, T., Kataoka, H., Suzuki, M., Hirasawa, Y., Shirahama, T., et al. (1985). A new form of amyloid protein associated with chronic hemodialysis was identified as beta 2-microglobulin. *Biochem Biophys Res Commun, 129*, 701–706.
22. Faull, R.J. (2007). Managing bone parameters in dialysis patients: international guideline conflicts. *Semin Dial, 20*, 191–194.

23. Cases, A. (2007). The latest advances in kidney diseases and related disorders. *Drug News Perspect, 20*, 647–654.

24. Schwalbe, S., Holzhauer, M., Schaeffer, J., Galanski, M., Koch, K.M., Floege, J. (1997). Beta 2-microglobulin associated amyloidosis: a vanishing complication of long-term hemodialysis? *Kidney Int, 52*, 1077–1083.

25. European Best Practices Guidelines Expert Group on Hemodialysis ERA. (2002). Haemodialysis adequacy. *Nephrol Dial Transplant*, Suppl 7, 16–31.

26. van Ypersele de Strihou, C., Jadoul, M., Malghem, J., Maldague, B., Jamart, J. (1991). Effect of dialysis membrane and patient's age on signs of dialysis-related amyloidosis. The Working Party on Dialysis Amyloidosis. *Kidney Int, 39*, 1012–1019.

27. Krieter, D.H., Lemke, H.D., Canaud, B., Wanner, C. (2005). Beta(2)-microglobulin removal by extracorporeal renal replacement therapies. *Biochim Biophys Acta, 1753*, 146–153.

28. Kramer, B.K., Pickert, A., Hohmann, C., Liebich, H.M., Muller, G.A., Hablitzel, M., Risler, T. (1992). In vivo clearance and elimination of nine marker substances during hemofiltration with different membranes. *Int J Artif Organs, 15*, 408–412.

29. Leber, H.W., Wizemann, V., Goubeaud, G., Rawer, P., Schutterle, G. (1978). Hemodiafiltration: a new alternative to hemofiltration and conventional hemodialysis. *Artif Organs, 2*, 150–153.

30. Gejyo, F., Kawaguchi, Y., Hara, S., Nakazawa, R., Azuma, N., Ogawa, H., Koda, Y., Suzuki, M., Kaneda, H., Kishimoto, H., et al. (2004). Arresting dialysis-related amyloidosis: a prospective multicenter controlled trial of direct hemoperfusion with a beta2-microglobulin adsorption column. *Artif Organs, 28*, 371–380.

31. Linke, R.P., Hampl, H., Bartel-Schwarze, S., Eulitz, M. (1987). Beta 2-microglobulin, different fragments and polymers thereof in synovial amyloid in long-term hemodialysis. *Biol Chem Hoppe Seyler, 368*, 137–144.

32. Linke, R.P., Hampl, H., Lobeck, H., Ritz, E., Bommer, J., Waldherr, R., Eulitz, M. (1989). Lysine-specific cleavage of beta 2-microglobulin in amyloid deposits associated with hemodialysis. *Kidney Int, 36*, 675–681.

33. Bellotti, V., Stoppini, M., Mangione, P., Sunde, M., Robinson, C., Asti, L., Brancaccio, D., Ferri, G. (1998). Beta2-microglobulin can be refolded into a native state from ex vivo amyloid fibrils. *Eur J Biochem, 258*, 61–67.

34. Miyata, T., Oda, O., Inagi, R., Iida, Y., Araki, N., Yamada, N., Horiuchi, S., Taniguchi, N., Maeda, K., Kinoshita, T. (1993). Beta 2-microglobulin modified with advanced glycation end products is a major component of hemodialysis-associated amyloidosis. *J Clin Invest, 92*, 1243–1252.

35. Capeillere-Blandin, C., Delaveau, T., Descamps-Latscha, B. (1991). Structural modifications of human beta 2 microglobulin treated with oxygen-derived radicals. *Biochem J, 277* (Pt 1), 175–182.

36. Pepys, M.B., Booth, D.R., Hutchinson, W.L., Gallimore, J.R., Collins, P.M., Hohenester, E. (1997). Amyloid P component: a critical review. *Amyloid: Int J Exp Clin Invest, 4*, 274–295.

37. Esposito, G., Michelutti, R., Verdone, G., Viglino, P., Hernandez, H., Robinson, C.V., Amoresano, A., Dal Piaz, F., Monti, M., Pucci, P., et al. (2000). Removal of the

N-terminal hexapeptide from human beta2-microglobulin facilitates protein aggregation and fibril formation. *Protein Sci, 9*, 831–845.

38. Monti, M., Principe, S., Giorgetti, S., Mangione, P., Merlini, G., Clark, A., Bellotti, V., Amoresano, A., Pucci, P. (2002). Topological investigation of amyloid fibrils obtained from beta2-microglobulin. *Protein Sci, 11*, 2362–2369.
39. Hoshino, M., Katou, H., Hagihara, Y., Hasegawa, K., Naiki, H., Goto, Y. (2002). Mapping the core of the beta(2)-microglobulin amyloid fibril by H/D exchange. *Nat Struct Biol, 9*, 332–336.
40. McParland, V.J., Kalverda, A.P., Homans, S.W., Radford, S.E. (2002). Structural properties of an amyloid precursor of beta(2)-microglobulin. *Nat Struct Biol, 9*, 326–331.
41. Verdone, G., Corazza, A., Viglino, P., Pettirossi, F., Giorgetti, S., Mangione, P., Andreola, A., Stoppini, M., Bellotti, V., Esposito, G. (2002). The solution structure of human beta2-microglobulin reveals the prodromes of its amyloid transition. *Protein Sci, 11*, 487–499.
42. Morgan, C.J., Gelfand, M., Atreya, C., Miranker, A.D. (2001). Kidney dialysis-associated amyloidosis: a molecular role for copper in fiber formation. *J Mol Biol, 309*, 339–345.
43. Villanueva, J., Hoshino, M., Katou, H., Kardos, J., Hasegawa, K., Naiki, H., Goto, Y. (2004). Increase in the conformational flexibility of beta 2-microglobulin upon copper binding: a possible role for copper in dialysis-related amyloidosis. *Protein Sci, 13*, 797–809.
44. Rosano, C., Zuccotti, S., Mangione, P., Giorgetti, S., Bellotti, V., Pettirossi, F., Corazza, A., Viglino, P., Esposito, G., Bolognesi, M. (2004). Beta2-microglobulin H31Y variant 3D structure highlights the protein natural propensity towards intermolecular aggregation. *J Mol Biol, 335*, 1051–1064.
45. Chiti, F., Mangione, P., Andreola, A., Giorgetti, S., Stefani, M., Dobson, C.M., Bellotti, V., Taddei, N. (2001). Detection of two partially structured species in the folding process of the amyloidogenic protein beta 2-microglobulin. *J Mol Biol, 307*, 379–391.
46. Chiti, F., De Lorenzi, E., Grossi, S., Mangione, P., Giorgetti, S., Caccialanza, G., Dobson, C.M., Merlini, G., Ramponi, G., Bellotti, V. (2001). A partially structured species of beta 2-microglobulin is significantly populated under physiological conditions and involved in fibrillogenesis. *J Biol Chem, 276*, 46714–46721.
47. Kameda, A., Hoshino, M., Higurashi, T., Takahashi, S., Naiki, H., Goto, Y. (2005). Nuclear magnetic resonance characterization of the refolding intermediate of beta 2-microglobulin trapped by non-native prolyl peptide bond. *J Mol Biol, 348*, 383–397.
48. Jahn, T.R., Parker, M.J., Homans, S.W., Radford, S.E. (2006). Amyloid formation under physiological conditions proceeds via a native-like folding intermediate. *Nat Struct Mol Biol, 13*, 195–201.
49. Yamaguchi, K., Takahashi, S., Kawai, T., Naiki, H., Goto, Y. (2005). Seeding-dependent propagation and maturation of amyloid fibril conformation. *J Mol Biol, 352*, 952–960.
50. Chatani, E., Goto, Y. (2005). Structural stability of amyloid fibrils of beta(2)-microglobulin in comparison with its native fold. *Biochim Biophys Acta, 1753*, 64–75.

51. Yamamoto, S., Hasegawa, K., Yamaguchi, I., Tsutsumi, S., Kardos, J., Goto, Y., Gejyo, F., Naiki, H. (2004). Low concentrations of sodium dodecyl sulfate induce the extension of beta 2-microglobulin-related amyloid fibrils at a neutral pH. *Biochemistry, 43*, 11075–11082.
52. Yamamoto, S., Yamaguchi, I., Hasegawa, K., Tsutsumi, S., Goto, Y., Gejyo, F., Naiki, H. (2004). Glycosaminoglycans enhance the trifluoroethanol-induced extension of beta 2-microglobulin-related amyloid fibrils at a neutral pH. *J Am Soc Nephrol, 15*, 126–133.
53. Myers, S.L., Jones, S., Jahn, T.R., Morten, I.J., Tennent, G.A., Hewitt, E.W., Radford, S.E. (2006). A systematic study of the effect of physiological factors on beta2-microglobulin amyloid formation at neutral pH. *Biochemistry, 45*, 2311–2321.
54. Piazza, R., Pierno, M., Iacopini, S., Mangione, P., Esposito, G., Bellotti, V. (2006). Micro-heterogeneity and aggregation in beta2-microglobulin solutions: effects of temperature, pH, and conformational variant addition. *Eur Biophys J, 35*, 439–445.
55. Giorgetti, S., Rossi, A., Mangione, P., Raimondi, S., Marini, S., Stoppini, M., Corazza, A., Viglino, P., Esposito, G., Cetta, G., et al. (2005). Beta2-microglobulin isoforms display an heterogeneous affinity for type I collagen. *Protein Sci, 14*, 696–702.
56. Relini, A., Canale, C., De Stefano, S., Rolandi, R., Giorgetti, S., Stoppini, M., Rossi, A., Fogolari, F., Corazza, A., Esposito, G., et al. (2006). Collagen plays an active role in the aggregation of beta2-microglobulin under physiopathological conditions of dialysis-related amyloidosis. *J Biol Chem, 281*, 16521–16529.
57. Fogolari, F., Corazza, A., Viglino, P., Zuccato, P., Pieri, L., Faccioli, P., Bellotti, V., Esposito, G. (2007). Molecular dynamics simulation suggests possible interaction patterns at early steps of beta2-microglobulin aggregation. *Biophys J, 92*, 1673–1681.
58. Kihara, M., Chatani, E., Iwata, K., Yamamoto, K., Matsuura, T., Nakagawa, A., Naiki, H., Goto, Y. (2006). Conformation of amyloid fibrils of beta2-microglobulin probed by tryptophan mutagenesis. *J Biol Chem, 281*, 31061–31069.
59. Esposito, G., Ricagno, S., Corazza, A., Rennella, E., Gumral, D., Mimmi, M.C., Betto, E., Pucillo, C.E., Fogolari, F., Viglino, P., et al. (2008). The controlling roles of Trp60 and Trp95 in beta2-microglobulin function, folding and amyloid aggregation properties. *J Mol Biol, 378*, 885–895.
60. Trinh, C.H., Smith, D.P., Kalverda, A.P., Phillips, S.E., Radford, S.E. (2002). Crystal structure of monomeric human beta-2-microglobulin reveals clues to its amyloidogenic properties. *Proc Natl Acad Sci U S A, 99*, 9771–9776.
61. Mimmi, M.C., Jorgensen, T.J., Pettirossi, F., Corazza, A., Viglino, P., Esposito, G., De Lorenzi, E., Giorgetti, S., Pries, M., Corlin, D.B., et al. (2006). Variants of beta-microglobulin cleaved at lysine-58 retain the main conformational features of the native protein but are more conformationally heterogeneous and unstable at physiological temperature. *FEBS J*, 273, 2461–2474.
62. Richardson, J.S., Richardson, D.C. (2002). Natural beta-sheet proteins use negative design to avoid edge-to-edge aggregation. *Proc Natl Acad Sci U S A, 99*, 2754–2759.
63. Homma, N., Gejyo, F., Isemura, M., Arakawa, M. (1989). Collagen-binding affinity of beta-2-microglobulin, a preprotein of hemodialysis-associated amyloidosis. *Nephron, 53*, 37–40.

64. Linke, R.P., Bommer, J., Ritz, E., Waldherr, R., Eulitz, M. (1986). Amyloid kidney stones of uremic patients consist of beta 2-microglobulin fragments. *Biochem Biophys Res Commun, 136*, 665–671.

65. Relini, A., De Stefano, S., Torrassa, S., Cavalleri, O., Rolandi, R., Gliozzi, A., Giorgetti, S., Raimondi, S., Marchese, L., Verga, L., et al. (2008). Heparin strongly enhances the formation of beta2-microglobulin amyloid fibrils in the presence of type I collagen. *J Biol Chem, 283*, 4912–4920,

66. Lindahl, U. (2007). Heparan sulfate–protein interactions: a concept for drug design? *Thromb Haemost, 98*, 109–115.

67. Kisilevsky, R. (2000). Review: amyloidogenesis–unquestioned answers and unanswered questions. *J Struct Biol, 130*, 99–108.

68. Ohashi, K., Kisilevsky, R., Yanagishita, M. (2002). Affinity binding of glycosaminoglycans with beta(2)-microglobulin. *Nephron, 90*, 158–168.

69. McLaurin, J., Franklin, T., Zhang, X., Deng, J., Fraser, P.E. (1999). Interactions of Alzheimer amyloid-beta peptides with glycosaminoglycans effects on fibril nucleation and growth. *Eur J Biochem, 266*, 1101–1110.

70. Castillo, G.M., Lukito, W., Wight, T.N., Snow, A.D. (1999). The sulfate moieties of glycosaminoglycans are critical for the enhancement of beta-amyloid protein fibril formation. *J Neurochem, 72*, 1681–1687.

71. Cohlberg, J.A., Li, J., Uversky, V.N., Fink, A.L. (2002). Heparin and other glycosaminoglycans stimulate the formation of amyloid fibrils from alpha-synuclein in vitro. *Biochemistry, 41*, 1502–1511.

72. Suk, J.Y., Zhang, F., Balch, W.E., Linhardt, R.J., Kelly, J.W. (2006). Heparin accelerates gelsolin amyloidogenesis. *Biochemistry, 45*, 2234–2242.

73. Snow, A.D., Kisilevsky, R. (1985). Temporal relationship between glycosaminoglycan accumulation and amyloid deposition during experimental amyloidosis: a histochemical study. *Lab Invest, 53*, 37–44.

74. Li, J.P., Galvis, M.L., Gong, F., Zhang, X., Zacharia, E., Metzger, S., Vlodavsky, I., Kisilevsky, R., Lindahl, U. (2005). In vivo fragmentation of heparan sulfate by heparanase overexpression renders mice resistant to amyloid protein A amyloidosis. *Proc Natl Acad Sci U S A, 102*, 6473–6477.

75. Kisilevsky, R., Lemieux, L.J., Fraser, P.E., Kong, X., Hultin, P.G., Szarek, W.A. (1995). Arresting amyloidosis in vivo using small-molecule anionic sulphonates or sulphates: implications for Alzheimer's disease. *Nat Med, 1*, 143–148.

76. Evenepoel, P., Dejagere, T., Verhamme, P., Claes, K., Kuypers, D., Bammens, B., Vanrenterghem, Y. (2007). Heparin-coated polyacrylonitrile membrane versus regional citrate anticoagulation: a prospective randomized study of 2 anticoagulation strategies in patients at risk of bleeding. *Am J Kidney Dis, 49*, 642–649.

77. Calamai, M., Kumita, J.R., Mifsud, J., Parrini, C., Ramazzotti, M., Ramponi, G., Taddei, N., Chiti, F., Dobson, C.M. (2006). Nature and significance of the interactions between amyloid fibrils and biological polyelectrolytes. *Biochemistry, 45*, 12806–12815.

39

FAMILIAL AMYLOIDOSIS CAUSED BY LYSOZYME

Mireille Dumoulin
Centre d'Ingénierie des Protéines, Institut de Chimie, Université de Liège, Liège, Belgium

INTRODUCTION

Lysozyme is a 130-residue protein which is highly expressed in hematopoietic cells and is found in granulocytes, monocytes, and macrophages as well as in their bone marrow precursors [1]. It is widely distributed in a variety of tissues and body fluids, including liver, articular cartilage, saliva, and tears. It is a bacteriolytic enzyme that hydrolyzes preferentially the β-1,4 glycosidic linkages between the *N*-acetylmuramic acid and *N*-acetylglucosamine groups that occur in the peptidoglycan cell wall structure of certain microorganisms [2]. In the last 15 years, six variants of human lysozyme (i.e., I56T, F57I, F57I/T70N, W64R, D67H, and T70N/W112R) have been found to be associated with a nonneuropathic systemic amyloidosis [3–6]. This rare autosomal-dominant disease, which has been reported so far to affect nine unrelated families, involves fibrillar deposits found to accumulate in a wide range of tissues, including the liver, spleen, and kidneys [3–7]. Moreover, two other natural variants, T70N and W112R, have been found in 3.5 to 12% of control populations [4,6,8]. Individually, neither of these two mutations are associated with disease; however, they become pathogenic when found in combination, suggesting that both mutations are effective in disease manifestation [4]. On the contrary,

Protein Misfolding Diseases: Current and Emerging Principles and Therapies,
Edited by Marina Ramirez-Alvarado, Jeffery W. Kelly, and Christopher M. Dobson
Copyright © 2010 John Wiley & Sons, Inc.

in patients having both the T70N and pathogenic F57I alleles, the mutation at position 70 has not detectable effects on the clinical phenotype [6].

The wild type, as well as the I56T, D67H, and T70N variants, have been expressed successively in large quantities in heterologous organisms [9–12] permitting extensive studies of their properties, such as activity, stability, folding, dynamics, and aggregation. Characterization of the more recently discovered F57I and W64R variants have also been initiated [13]. Comparison of the properties of this series of natural lysozyme variants, only some of which are associated with clinical pathologies, with those of the wild-type protein have made it possible to explore in detail how specific mutations can lead to amyloid diseases.

In this chapter we outline the clinical manifestations associated with human lysozyme amyloidoses and the therapeutic approaches that have been used so far. We summarize the results of in vitro biophysical and biochemical characterization of the lysozyme variants, and we discuss the insights into the molecular mechanism of amyloid fibril formation that these studies have provided. Finally, we discuss results of recent studies on the in vitro inhibition of fibril formation and how they could be used for the rational design of new therapeutic strategies.

LYSOZYME AMYLOIDOSIS: CLINICAL MANIFESTATIONS AND CURRENT THERAPIES

The implication of two variants of human lysozyme (I56T and D67H) in amyloidosis was first described in 1993 [3]. Since then, four new amyloidogenic lysozyme variants have been discovered: F57I, W64R, F57I/T70N, and T70N/W112R [4–6]. In all cases, the patients are heterozygous, the disease being transmitted through an autosomal dominant mechanism. Human lysozyme (ALys) amyloidosis can affect the viscera, with severe involvement when located in the kidneys and liver (Fig. 1A). Renal dysfunction [3,5–7] and massive hepatic hemorrhage [14,15] constitute two severe visceral clinical features. Gastrointestinal involvement associated with bleeding is another common characteristic of ALys [4,16], and dermatological manifestations such as sicca syndrome [5], pepechiae, and purpura [3] have also been observed. The age at which amyloid deposits appear, their distribution in tissue, and their clinical effects are very variable both within and between families [3–7]. Such variability has also been reported for other systemic amyloidosis, including the most common such disorder that is associated with transthyretin, but its causes remain largely unknown [17].

As for other systemic amyloidosis, ALys causes serious morbidity and is usually fatal. The mechanism by which the amyloid deposits damage tissues and compromise organ function are very poorly understood. The systematic extracellular amyloid fibrils, and in particular their smaller, more diffusible precursor assemblies (oligomers and protofibrils), might act in an amphipathic

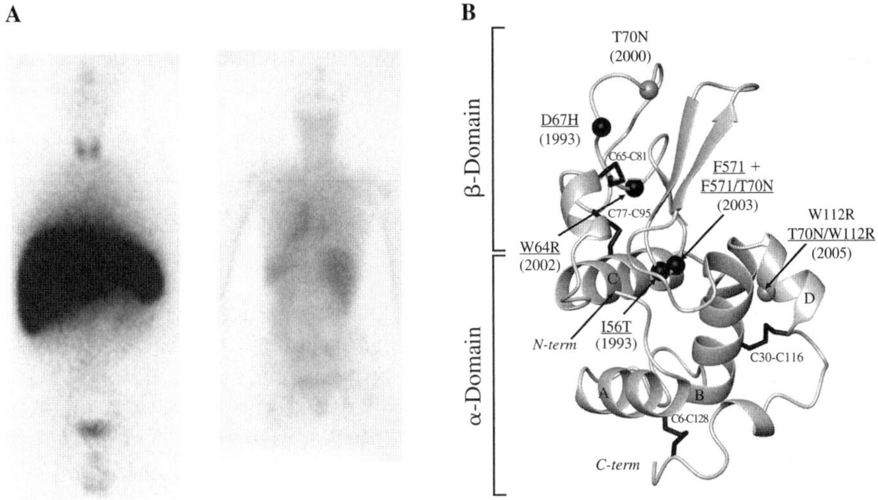

FIG. 1 (A) Posterior whole-body scintigraphic images after intraventous injection of [^{123}I]human SAP (serum amyloid protein, which binds selectively to amyloid deposits) showing a patient with lysozyme amyloidosis (left) and a normal subject (right). In the patient with the disease, there is heavy amyloid deposition in the liver, spleen, and kidneys, whereas in the patient without the disease no protein deposits are observed and the SAP is distributed throughout the blood pool. (Adapted from [7]). (B) Ribbon diagram of the structure of human wild-type lysozyme (WT) showing the locations of the known natural mutations, along with the year of their discovery. The α-helices are labeled A to D; the four disulfide bonds and the N- and C-termini are also indicated. The single point mutants, T70N and W112R, are not associated with disease; they are, however, pathogenic when found in combination. (Adapted from [39].)

fashion to bind and perturb multiple cell-surface receptors and/or channels rather nonspecifically [18]. Clearly, however, the presence of massive deposits of fibrils, which may involve kilograms of protein, is structurally disruptive and incompatible with the normal biological functions of any organs in which they are located [17]. Organs that are heavily infiltrated by amyloids can fail precipitously, with little or no warning, often leading to death. So far, only palliative treatments are available; they consist mainly of careful maintenance of the impaired organ functions and replacement of end-stage organs. At least five ALys patients have received renal transplants. One of them died from postoperative gastric bleeding, whereas the others have been well many years after the transplantation [5–7]. Moreover, two emergency liver transplantations following spontaneous liver rupture due to the accumulation of large quantities of lysozyme fibrils have been reported [14,15]. As mentioned above, lysozyme is a ubiquitous protein that is produced diffusely within the body, so liver transplant is not a curative treatment, and amyloid deposition in other tissues may continue after liver transplantation. Although the long-term outcome of liver transplantation for ALys amyloidosis remains unclear, patients have clearly benefited from this type of treatment [14,15].

INSIGHT INTO THE MOLECULAR MECHANISM OF LYSOZYME FIBRIL FORMATION FROM EXAMINATION OF EX VIVO FIBRILS

Analysis of ex vivo amyloid fibrils from the tissue of patients with amyloidosis associated with the I56T and D67H lysozyme variants has revealed that these aggregates are made up only of the variant proteins [3,9]. Thus, although wild-type lysozyme as well as the amyloidogenic variants are produced in these patients, the former does not convert into fibrillar structures. The I56T and D67H proteins isolated form ex vivo fibrils were found to be full length [3,9]. Moreover, the D67H variant extracted from ex vivo fibrils is able under appropriate conditions to refold to its native state. In addition, it was shown that all four disulfide bonds are present in the form of the D67H protein found in the deposits [9]. These results suggest that neither cleavage of the polypeptide chain nor reduction of disulfide bonds is required for, or result from, fibril formation; the D67H variant must therefore fold correctly in the cell prior to deposition in tissue. This is probably also the case with the I56T variant [9].

Ex vivo amyloid fibrils from patients carrying the T70N/W112R mutations were shown to be made up of the full-length protein but also from fragments of lysozyme of 12.1, 10.0, and 6.3 kDa [4]. Extraction and characterization of the protein in amyloid deposits from patients having the F57I, F57I/T70N, or W64R mutations has not yet been reported. Interestingly, attempts to identify the W64R variant in urine and plasma of affected patients by a combination of chromatographic separation and mass spectrometry have revealed no traces of the pathogenic variant protein, although wild-type lysozyme is readily detected [5]. This finding could result from selective and highly efficient incorporation of the W64R protein into the amyloid deposits or from its high level of intra-or extracellular degradation. All together, these reports suggest that the effects of the W64R and T70N/W112R mutations are such that the corresponding variant proteins are more susceptible to proteolysis.

When observed by electron microscopy, after negative staining with uranylacetate, most of the ex vivo D67H lysozyme fibrils are "wavy" in nature and their diameters range from 8 to 13 nm [19]. X-ray fiber diffraction of such fibrils showed a meridional reflection at 4.6 to 4.8 Å and a broad equatorial reflection at 8 to 14 Å characteristic of the amyloid cross-β structure [20]. No reflections could be attributed to helical structure, suggesting that if helices persist after transformation of the soluble protein to the fibrillar form, they are not regularly ordered. Cryoelectron microscopy analyses of ex vivo D67H fibrils suggest that the fibrils are made of five or six protofilaments and that they have a hollow core [19,21].

INSIGHT INTO THE MOLECULAR MECHANISM OF LYSOZYME FIBRIL FORMATION FROM BIOCHEMICAL AND BIOPHYSICAL CHARACTERIZATION OF THE VARIANT PROTEINS

Effects of the Mutations on the Structure of Lysozyme

The native structure of wild-type human lysozyme consists of two closely interacting domains: the α-helical domain (residues 1 to 40 and 83 to 100) made up of 4-helices and two 3_{10}-helices, and the β-domain (residues 41 to 82), made up of a triple-stranded β-sheet, a 3_{10}-helix, and a long loop (Fig. 1B). The protein contains four disulfide bonds two of which are located in the α-domain, one of which is located in the long loop of the β-domain, and one that connects the two domains. The active site is located in the cleft that is formed between the two domains. All the natural mutations discovered so far are located in the β-domain with the exception of the W112R mutation, which is in the D helix of the α-domain; the latter mutation has, however, only been found to be associated with amyloidosis in conjunction with the β-domain mutation T70N [4]. The I56T [9], D67H [9], F57I (unpublished data), and T70N [10] have been found to be functional with specific activities very similar to those of wild-type lysozyme. On the other hand, the W64R protein is much less active (unpublished data), suggesting that the W64R mutation may significantly affect the structure of the protein; this observation is consistent with the proposition that the W64R variant is highly susceptible to proteolysis (see above).

X-ray crystallographic data have shown that the I56T, D67H, and T70N variants have the same overall native fold as the wild-type protein, with all the disulfide bonds correctly formed [9,12]. They show, however, that both amyloidogenic mutations result in the loss of a number of long-range interactions that bridge the α/β domain interface, suggesting that the interface region of the I56T and D67H variants is significantly destabilized, either directly or indirectly, relative to the situation in the wild-type protein [9]. In contrast, the nonpathogenic T70N mutation does not appear to perturb significantly the domain interface region [12]. All together, these findings imply that disruption of long-range interactions between the two structural domains of lysozyme could be an important factor in their amyloidogenic properties.

Effects of the Mutations on the Folding of Lysozyme

Folding of the wild-type lysozyme in vitro from both the reduced and oxidized states is characterized by the existence of multiple pathways, involving a series of metastable intermediates [10,11,22–25]. The I56T and D67H variants refold in a manner qualitatively similar to that the wild-type protein, although the rate at which they do so differs. Only the most important findings are summarized here since these studies have been described extensively in previous review articles [26,27].

The unfolding kinetics of the I56T, D67H, and T70N variants and wild-type lysozyme in high concentrations of guanidinium chloride under nonreducing conditions have all been found to fit well to single exponential functions. However, the T70N, I56T, and D67H variants unfold, respectively, about 3, 30, and 160 times faster than do wild-type lysozyme [10,22].

Upon refolding from its guanidinium chloride denatured state under nonreducing conditions (i.e., in the presence of its native disulfide bonds), the majority of the wild-type lysozyme molecules follow a slow-track, the first step of which involves stabilization of the A and B helices and the C-terminal 3_{10}-helix in a locally cooperative manner [22,23]. This step is followed by the cooperative folding of helices C and D, leading to an intermediate in which the α-domain has a nativelike structure, and finally by the folding of the β-domain [22,23]. As found for the homologous hen egg-white lysozyme, it is likely that a final step involving the docking of the two domains is required to generate the native close-packed structure with a functional active site [22,23]. About 10% of the molecules, however, fold along a fast track that arises from a population of molecules able to form the native state more efficiently than that the majority of molecules. For these rapidly folding molecules, the β-domain becomes structured concomitantly with the formation of the α-domain. The refolding of both the I56T and D67H variants also occurs via multiple parallel pathways and through a series of well-defined intermediates [22]. The refolding of the D67H variant is virtually identical to that of the wild-type protein, whereas for the I56T variant the coalescence of the β-domain on to the folded α-domain is about 10 time slower [22]. These findings can be attributed to the fact that residue 67 is located in a loop region within the β-domain, while residue 56 is located at the crucial interface between the α and β domains.

Unfolded lysozyme under reducing conditions (i.e., with all of its eight cysteine residues in the thiol form) corresponds to ensembles of unfolded conformers, and in the initial stages of folding these species collapse rapidly via a large number of parallel pathways to form a multitude of relatively unstructured intermediates having one or two disulfide bonds [25]. The majority of these species then fold to form a nativelike three-disulfide intermediate lacking the 77–95 bond. The final and slowest step involves a conformational rearrangement requiring at least local unfolding of the latter species in order to allow the remaining free thiol groups to form the fourth disulfide linkage and hence to generate the fully oxidized native protein. Although the I56T and D67H variants refold in a manner qualitatively similar to that of wild-type lysozyme, they fold faster by a factor of 2 and 3, respectively [25]. This finding can be attributed primarily to the fact that the lower stabilities of the native-like intermediates of the variants compared with those of the wild-type protein facilitates the conformational rearrangements associated with the final folding step.

Effects of the Mutations on the Stability and Global Cooperativity of Lysozyme

The thermostability of both the I56T and D67H variants monitored by far-ultraviolet circular dichroism (CD) at pH values between 2.5 and 7.0 is decreased by about 10 to 12°C relative to that of the wild-type protein [9,11,24, 28–30]. For example, at pH 5.0 the temperature of midtransition (T_m) of both the I56T and D67H variants is 68 ± 1°C versus 80 ± 1°C for the wild-type protein (Fig. 2A). Moreover, the thermal unfolding of the I56T and D67H proteins is associated with a large increase in 8-anilino-1-naphthalene-sulfonic acid (ANS) fluorescence observed near the denaturation temperature (Fig. 2A), indicating that a partially unfolded intermediate with one or more hydrophobic regions exposed to the solvent is populated significantly during unfolding. The wild-type protein also binds ANS near the temperature of the midpoint of its unfolding transition at pH 5.0. The intensity of ANS binding is, however, only about 37% of that observed for the amyloidogenic I56T and D67H variants (Fig. 2A).

The T_m value of the nonamyloidogenic T70N variant at pH 5.0 is 4°C lower than that of the wild-type but 8°C higher than that of the I56T and D67H variants (Fig. 2A)[10,12]. The thermostability of the T70N variant is therefore intermediate between that of the amyloidogenic and wild-type lysozymes. Most important, its thermal unfolding is substantially more cooperative than that of the I56T and D67H proteins but less than that of the wild-type protein, as revealed by ANS binding experiments (Fig. 2A).

The more recently discovered F57I and W64R proteins display thermostabilities that are reduced to a degree similar to that of the I56T and D67H proteins; their T_m values at pH 6.0 in the presence of 0.5 M NaCl are 60.4 ± 1.1°C and 61.7 ± 1.0°C, respectively [13]. Moreover, for both F57I and W64R proteins, a significant ANS fluorescence increase was observed upon thermal unfolding, indicating that like the I56T and D67H proteins, both variants populate partially unfolded species with increased exposure of their hydrophobic regions relative to that of the wild-type protein [13].

All together these results indicate that the ability to populate an intermediate species upon thermal unfolding is an intrinsic property of the lysozyme fold and is not restricted to the amyloidogenic variants. Most important, however, they suggest that the significant reduction in cooperativity, and hence the ability to form intermediate species much more readily than the wild-type protein, is a common property of the amyloidogenic variants and could therefore be a feature linked to their amyloidogenicity. To further investigate the effects of the mutations on the global cooperativity under more physiologically relevant conditions, hydrogen–deuterium exchange properties of the labile amide and side-chain hydrogens was monitored using a combination of electrospray ionization mass spectrometry (ESI-MS) and nuclear magnetic resonance (NMR) spectroscopy [12,28,31]. These experiments have clearly shown that at 35 to 37°C and pH 5.0 to 8.0, the I56T and D67H variants populate at a significant level a partially folded intermediate that is in dynamic equilibrium

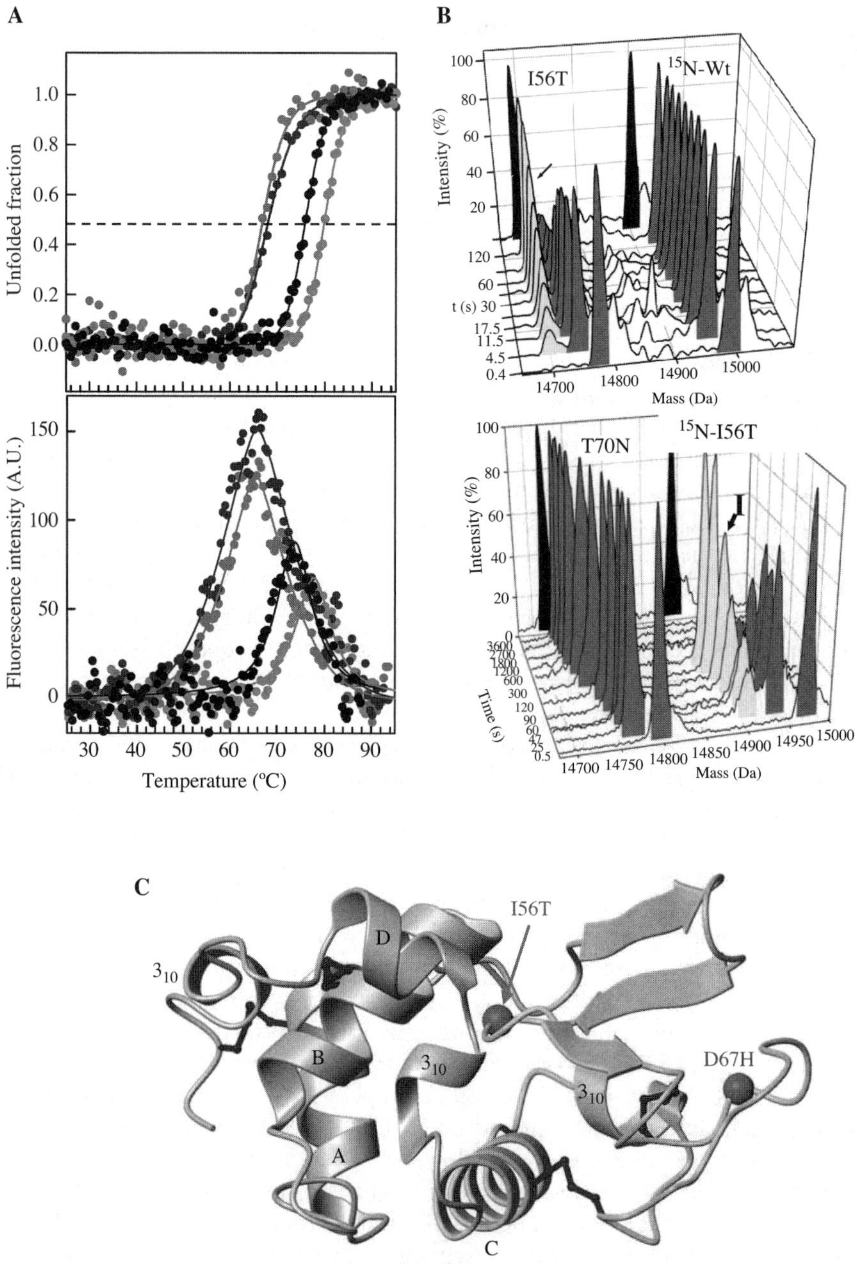

FIG. 2 Effects of the mutations on the stability and global cooperativity of lysozyme. (A) Thermal unfolding monitored by far-UV CD at 228 nm (upper panel) and by ANS fluorescence (lower panel) of the I56T (red), the D67H (blue), the T70N (black), and the wild-type (green) lysozymes. The lysozyme concentration was 0.2 mg/mL and 0.035 mg/mL in 0.1 M sodium acetate buffer (pH 5.0) for CD and ANS fluorescence measurements, respectively. The ANS concentration was 315 μM. (B) Electrospray mass spectra of an equimolar mixture of ^{14}N-labeled 156T variant and ^{15}N-labeled wild-type lysozyme

with the native protein (Fig. 2B) [28,31]. The regions of the protein that are unfolded in this partially structured species have been found to be remarkably similar for both variants, despite the different locations of the mutations [28]. They correspond to the regions of the protein that form the β-domain and the adjacent C-helix in the native state, and they are referred to as *amylotope* (Fig. 2C) [28,31]. Interestingly, the intermediate that is occasionally sampled in this way by the variant proteins under equilibrium conditions resembles that populated in the normal refolding of the protein from a highly denatured state, emphasizing the close link between normal and aberrant folding behavior [22].

No detectable population of partially unfolded species has been detected for the T70N and the wild-type protein under physiologically relevant conditions (Fig. 2B). At higher temperatures, however, the T70N (47°C) and wild-type protein (57°C) both undergo partial unfolding in a locally cooperative manner that is essentially identical to that seen at lower temperature for the amyloidogenic I56T and D67H variants [12].

All together, these observations demonstrate that the ability to form partially unfolded species in which the C-helix and the β-domain are simultaneously unfolded is an intrinsic property of the lysozyme fold rather than the results of specific mutations. The experimental conditions required to populate the partially unfolded species readily depends, however, on particular mutations and seems somehow correlated with the reduction in thermostability induced by the mutations. The amyloidogenic mutations, I56T and D67H, destabilize the native state of lysozyme so that the variant proteins can access the partially folded species under physiologically relevant conditions, whereas the nonamyloidogenic T70N lysozyme is not sufficiently destabilized to allow the population of the amyloidogenic intermediate under such conditions. On this evidence it has been suggested that the ability to access partially

(upper panel) and of ^{15}N-labeled I56T and ^{14}N-labeled T70N variant (lower panel). Mass spectra were recorded following exposure to hydrogen exchange conditions at pH 8.0 and 37°C for periods of time ranging from 0.4 to 3600 s. The four proteins were exposed to D_2O initially to replace all the labile hydrogens with deuterium atoms; the exchange process therefore involved subsequent replacement of these deuterium atoms with hydrogen atoms from the solvent H_2O. The peaks observed in spectra of control samples recorded after complete H/D exchange are shown in black. The peaks colored red (I56T and WT) and green (T70N) arise from the gradual loss of deuterium during the course of exchange that occurs via an EX2 mechanism due to local fluctuations [12,28,31]. The peaks colored yellow were observed in the spectra of the I56T variant but not in that of the T70N and the wild-type lysozyme. They result from a transient and cooperative unfolding event that gives rise to exchange by an EX1 mechanism. Note that in these experiments, the D67H variant has been found to behave like the I56T variant [31]. (C) Ribbon diagram of the lysozyme, showing in green regions of the protein involved in the transient cooperative unfolding event observed for the I56T and D67H variants as determined by real-time H/D exchange experiments analyzed by MNR [28, 31]. The 3_{10}-and α-helixes are labeled A to D, and the four disulfide bonds are shown in blue. (*See insert for color representation of figure.*)

unfolded species under physiologically relevant conditions is the primary origin of the amyloidogenic character of the I56T and D67H proteins and that this transient, locally cooperative unfolding process is a crucial step in the events that lead to aggregation and amyloid formation. Interestingly, the fact that the T70N mutation is not associated with ALys despite its destabilization relative to the wild-type suggests that a fine threshold exists between a benign mutation and one that induces amyloid disease [10,12].

Effects the Mutations on the In Vitro Aggregation Propensity of Lysozyme

The I56T, D67H, and T70N variants, and indeed wild-type human lysozyme, are able to form amyloid fibrils in vitro under conditions where a significant fraction of the protein molecules are at least partially unfolded, such as low pH, high temperature, a moderate concentration of denaturant, or following the application of high pressure [26,27]. However, as a result of their lower stability and global cooperativity, and therefore their higher ability to populate partially unfolded states (Fig. 2A and B), the amyloidogenic variants form fibrils in vitro much more easily than does the T70N variant, especially the wild-type protein (Fig. 3A) [12,28–30]. The fibrils formed in vitro from the various lysozyme species resemble those formed in vivo, and their amyloid character has been shown by a wide range of techniques, some of which are illustrated in Figure 3. The kinetics of aggregation are sigmoidal, comprising a lag phase followed by an exponential growth phase (Fig. 3A) [12,28–30]. Moreover, fibril formation by I56T, D67H, and wild-type lysozyme can be greatly accelerated by seeding the solution with preformed fibrils from either the I56T variant or the wild-type protein [30]. These results are consistent with a nucleation-dependent growth process that has been suggested as a common mechanism of fibril formation [32].

To identify the region involved in the core of the fibrils, fibrils formed from full-length wild-type lysozyme obtained by incubation at pD 1.5 and 45°C were subjected to limited proteolysis, and the resulting protease-resistant protein core was identified by mass spectrometry and N-terminal amino-acid sequencing [33]. The cross-β structure core of most of fibrils was found to be composed of residues 32 to 108, whereas the rest of the chain is largely unstructured and/or weakly packed. Interestingly, the pepsin-resistant sequence 32 to 108 encompasses the entire region corresponding to the amylotope (i.e., the β-domain and the C-helix). These results suggest that the region of the protein involved in the cooperative unfolding of the amylotope is a key factor in determining the specific structure of the lysozyme fibrils [33].

MECHANISM OF LYSOZYME AMYLOID FIBRIL FORMATION

As a result of the findings from the various biochemical and biophysical studies described above, a mechanism of lysozyme amyloid fibril formation can be proposed, at least for the I56T and D67H variants (Fig. 4). Because the

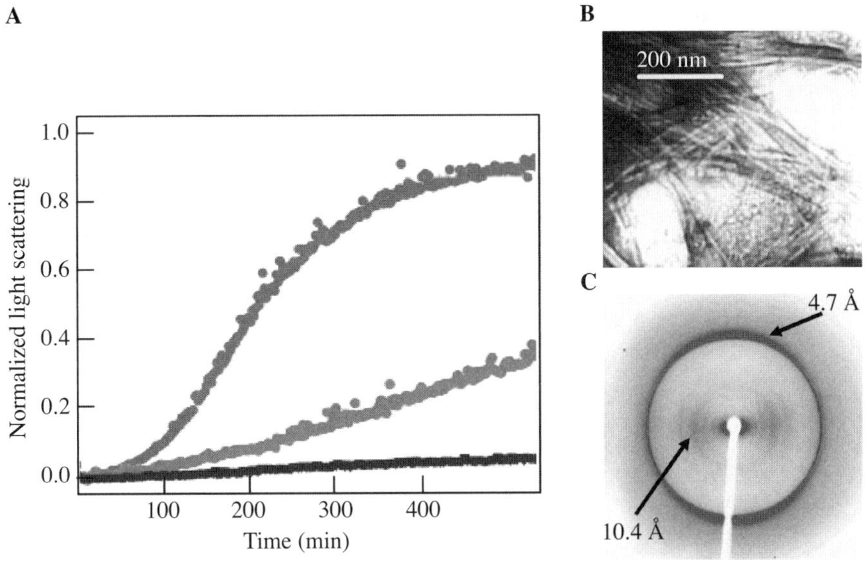

FIG. 3 (A) Time course of the aggregation of the I56T (upper curve), the T70N (middle curve), and the wild-type (lower curve) lysozyme, as monitored by light scattering. Lysozyme solutions (0.1 mg/mL) were incubated at 48°C in 0.1 M citrate buffer pH 5.5 containing 3 M urea. Note that in these experiments, the D67H variant behaves like the I56T protein. (Adapted from [12].) (B) Representative image of fibrils formed from the D67H variant as produced by transmission electron microscopy. (C) X-ray diffraction pattern of the same type of fibril showing a prominent meridional reflection at 4.7 Å and an equatorial reflection at 10.4 Å, features typical of the cross-β structure of amyloid fibril. (Adapted from [29].)

pathological fibrillar deposits are predominantly extracellular and the ex vivo fibrils from patients carrying one of these two mutations are made of the full-length protein, it is likely that I56T and D67H variants are able to fold correctly and to be secreted in the extracellular space where they normally function. The mutations, however, destabilize the native state sufficiently that the proteins are able under physiological conditions to populate transiently a partially unfolded intermediate in which the α-domain and the C-helix are unfolded simultaneously, whereas the rest of the α-domain (helices A, B, and D) largely maintains its nativelike structure. Intermolecular interactions through the mutually unfolded regions of at least two lysozyme molecules are likely to initiate the aggregation phenomenon that ultimately leads to the formation of amyloid fibril. Further reorganization of the A, B, and D helices is likely to occur upon the formation of small oligomers and mature fibrils.

Ex vivo fibrils from patients carrying the T70N/W112R mutation are in part made up of the full-length variant protein, suggesting that fibril formation could occur following a mechanism similar to that describing the I56T and

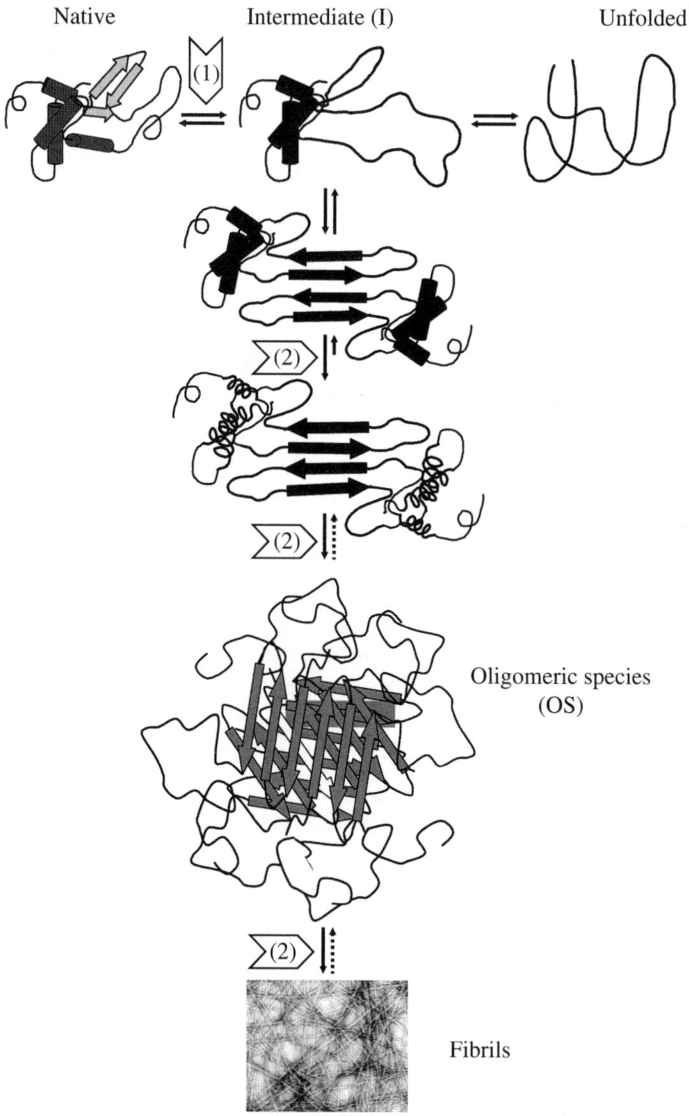

FIG. 4 Schematic mechanism for amyloid fibril formation by the I56T and D67H variants of lysozyme. Under physiological relevant conditions, the variants populate transiently an intermediate species (I) in which the α-domain and the C-helix are cooperatively unfolded, whereas the remainder of the α-domain remains nativelike. The formation of intermolecular interactions between the regions that are unfolded in this intermediate species can generate dimers; further rearrangement is likely to occur in the remainder of the structure prior to the formation of oligomeric species (OS) that are subsequently able to nucleate amyloid fibril growth. This figure is not intended to represent any defined secondary structural type, except that of the native protein (N). Note also that disulfide bridges, although not represented in this scheme, are present in the fibrils. Arrows (1) and (2) indicate the targets of camelid antibody fragments and clusterin, respectively.

D67H variants. However, some of these fibrils contain proteolytic fragments, suggesting that another parallel aggregation pathway could take place for T70N/W112R protein. This might also be the case for the W64R protein, which seems highly prone to degradation. A deeper characterization of these lysozyme variants is, however, necessary to draw definitive conclusions.

INHIBITION OF IN VITRO LYSOZYME AMYLOID FIBRIL FORMATION

Inhibition by Small Molecules

A 27-μM solution of wild-type lysozyme aggregates into amyloid fibril in about a week when incubated at pH 2 and 57°C; the presence of 200 μM of 2-amino-4-chlorophenol (2A4CP), however, completely inhibits amyloid fibril formation [34]. Moreover, the 2A4CP molecule is able effectively to disaggregate human lysozyme preformed fibrils [34]. This small-molecule compound was also found to prevent and reverse fibril formation from the Aβ-amyloid peptide; it could therefore serve as a prototype for the development of drugs for the prevention and treatment of a series of amyloidosis-including ALys.

Inhibition by Clusterin

The in vitro aggregation of the I56T amyloidogenic variant into amyloid fibrils is inhibited significantly by clusterin, a human extracellular molecular chaperone distributed widely throughout the body [35]. Remarkably, this inhibitory effect was observed even at clusterin/lysozyme ratios as low as 1 : 80 (i.e., one clusterin molecule per 80 lysozyme molecules). Under conditions where inhibition of aggregation occurs, clusterin was found not to bind detectably to the native or fibrillar states of lysozyme or to the monomeric transient intermediate thought to be a key species in the aggregation reaction (Fig. 4). Rather, it seems to interact with oligomeric species that are present at low concentrations during the nucleation phase of the aggregation reaction. Thus, clusterin probably prevents lysozyme aggregation by sequestering these species, thus preventing their concentration from reaching the threshold that leads to fibril formation. These findings suggest that clusterin, and perhaps other extracellular chaperones, could play a key role in limiting the potentially pathogenic effects of the misfolding and aggregation of proteins that, like lysozyme, are secreted into the extracellular environment [35].

Inhibition by Camelid Antibody Fragments

Binding an equimolar amount of cAb-HuL6, the antigen-binding domain of a camelid heavy-chain antibody specific for human lysozyme [36] to the I56T and D67H variant proteins restores to both proteins the thermostability and global

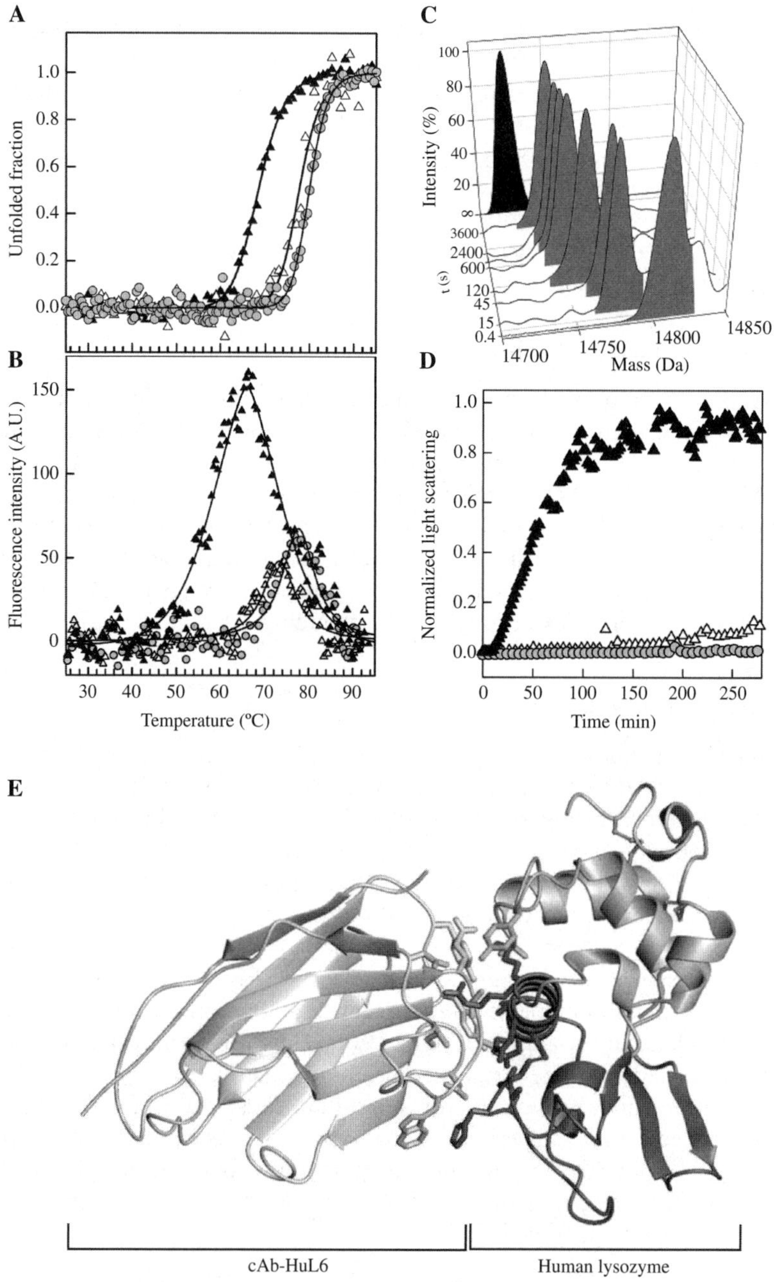

FIG. 5 Effects of the binding of cAb-HuL6 on the stability and global cooperativity of the D67H variant. (A,B) Thermal unfolding monitored by far-UV CD at 228 nm and by ANS fluorescence, respectively, of the D67H variant (filled triangles), the D67H variant complexed to an equimolar amount of cAb-HuL6 (open triangles), and wild-type lysozyme (open circles). The lysozyme concentration was 0.2 mg/mL and 0.035 mg/mL in 0.1 M sodium acetate buffer (pH 5.0) for CD and ANS fluorescence measurements,

cooperativity characteristic of wild-type protein (Fig. 5A to C) and, as a result, inhibits dramatically in vitro fibril formation by both variants (Fig. 5D) [28,29]. These results therefore provide further evidence that the formation of a partially unfolded species with a high propensity to aggregate, resulting from the locally cooperative unfolding of the amylotope, is the critical event that triggers the aggregation process that occurs in the absence of the antibody fragment. X-ray crystallographic data of the complex between the wild-type protein and cAb-HuL6 reveal that the epitope on the lysozyme molecule includes neither the site of the mutation nor most of the residues in the region of the protein structure that is destabilized by the mutations (Fig. 5E) [29]. Thus, the effects of binding are not simply to mask the entire region of the protein destabilized by the mutation and hence prevent its unfolding from the remainder of the structure. Analysis of the NMR chemical shift changes of lysozyme resulting from binding of the antibody fragment shows that restoration of the global cooperativity occurs, at least in part, through the transmission of long-range conformational effects to the interface between the two structural domains [28,29].

Preliminary studies with cAb-HuL22, the antigen-binding domain of another camelid heavy-chain antibody specific of human lysozyme, show that inhibition of amyloid fibril formation through the stabilization of the native

respectively. The ANS concentration was 315 μM. In the presence of an equimolar amount of cAb-HuL6, the stability of the D67H lysozyme is raised by about 10°C, so that when complexed to the antibody fragment, the D67H lysozyme is almost as stable as the wild type protein. Moreover, in the presence of the antibody fragment, the ANS fluorescence intensity is similar to that observed with the wild-type protein. (C) Electrospray mass spectra of a mixture of the D67H lysozyme in the presence of an equimolar amount of cAb-HuL6. A single peak, whose mass deceases with the length of time for which the exchange was allowed to proceed, is observed. The peaks of the species of lower-mass species observed in the spectra of the free D67H variant (which is similar to that observed with the I56T protein in Figure 2B) [31], which result from a locally cooperative unfolding of the amylotope, are therefore not observed in the spectra of the D67H protein in the presence of the antibody fragment. This result indicates that the binding of the antibody fragment to the D67H variant restores the stability and global cooperativity that is characteristic of wild-type lysozyme [29]. Similar results have been obtained for the I56T variant [28]. Thus, in the presence of the antibody fragment, the variant proteins do not populate the partially folded intermediate (species I in Fig. 4), which otherwise can initiate the aggregation process. The result of antibody binding is therefore to prevent the ready conversion of the lysozyme variants into their aggregated states as shown on panel D. (D) Time course of the aggregation of D67H variant (filled triangles), the D67H variant complexed to equimolar amount of cAb-HuL6 (open triangles), and the wild-type lysozyme (open circles) as monitored by light scattering. Lysozyme solutions (0.1 mg/mL) were incubated in 0.1 M acetate buffer pH 5.0 under 65°C. (E) X-ray structure of the complex between wild-type human lysozyme and cAb-HuL6. The side-chain residues constituting the paratope and the epitope are shown. The amylotope is shown in dark gray on the lysozyme structure.

state of the amyloidogenic lysozyme variants can be obtained by targeting the active site of the enzyme (Chan et al., unpublished data).

These findings suggest that new therapeutic strategies could be based on the use of ligands to stabilize the native state of the amyloidogenic variants. One of the unique properties of the antigen-binding domain of camelid heavy-chain antibodies is the relative simplicity of their paratope, which is restricted to three CDRs (complementary-determining regions) instead of six for conventional antibody fragments [37]. These loops therefore constitute a rather simple molecular scaffold for the design of small molecules capable of binding the antigen in a manner analogous to that of the parent antibody [38].

CONCLUSIONS AND PERSPECTIVES

Investigations of the effects of the I56T, D67H, and T70N mutations on the structure, folding, stability, dynamics, and in vitro aggregating behavior of lysozyme have allowed considerable progress to be made in understanding the molecular events that lead to its conversion into amyloid fibrils. These studies have shown that the reduction in stability, and especially in global cooperativity, is the major factor underlying the amyloidogenicity of pathogenic lysozyme mutations. Further detailed in vitro studies of the more recently discovered natural variants (W64R, F57I, F57I/T70N, T70N/W112R, and W112R) should undoubtedly further enhance our understanding as to how particular changes in the primary structure of an otherwise nonamyloidogenic protein contributes to amyloidogenicity. In this context, the W112R mutation is of special interest since it is the only natural mutation known so far that is located in the α-domain of the protein. Moreover, the detail in which the process of lysozyme amyloid fibril formation is understood has allowed the investigation at the molecular level of the effects on the aggregation pathway of a series of ligands. The results of these studies should enable new rational therapeutic strategies for lysozyme amyloidosis to be investigated.

Acknowledgments

The author expresses her deepest gratitude to Christopher Dobson for his continuous support. She is grateful to the European Commission, the BBRSC, the Belgian Government under the framework of Interuniversity Attraction Poles (I.A.P. P6/19), and the FRS-FNRS for their support of those parts of her own research that are described in this chapter.

REFERENCES

1. Reitamo, S., Klockars, M., Adinolfi, M., Osserman, E.F. (1978). Human lysozyme (origin and distribution in health and disease). *Ric Clini Lab*, *8*, 211–231.
2. Jolles, P., Jolles, J. (1984). What's new in lysozyme research? Always a model system, today as yesterday. *Mol Cell Biochem*, *63*, 165–189.

3. Pepys, M.B., Hawkins, P.N., Booth, D.R., Vigushin, D.M., Tennent, G.A., Soutar, A.K., Totty, N., Nguyen, O., Blake, C.C., Terry, C.J., et al. (1993). Human lysozyme gene mutations cause hereditary systemic amyloidosis. *Nature*, *362*, 553–557.

4. Rocken, C., Becker, K., Fandrich, M., Schroeckh, V., Stix, B., Rath, T., Kahne, T., Dierkes, J., Roessner, A., Albert, F.W. (2006). ALys amyloidosis caused by compound heterozygosity in exon 2 (Thr70Asn) and exon 4 (Trp112Arg) of the lysozyme gene. *Hum Mutat*, *27*, 119–120.

5. Valleix, S., Drunat, S., Philit, J.B., Adoue, D., Piette, J.C., Droz, D., MacGregor, B., Canet, D., Delpech, M., Grateau, G. (2002). Hereditary renal amyloidosis caused by a new variant lysozyme W64R in a French family. *Kidney Int*, *61*, 907–912.

6. Yazaki, M., Farrell, S.A., Benson, M.D. (2003). A novel lysozyme mutation Phe57Ile associated with hereditary renal amyloidosis. *Kidney Int*, *63*, 1652–1657.

7. Gillmore, J.D., Booth, D.R., Madhoo, S., Pepys, M.B., Hawkins, P.N. (1999). Hereditary renal amyloidosis associated with variant lysozyme in a large English family. *Nephrol Dial Transplant*, *14*, 2639–2644.

8. Booth, D.R., Pepys, M.B., Hawkins, P.N. (2000). A novel variant of human lysozyme (T70N) is common in the normal population. *Hum Mutat*, *16*, 180.

9. Booth, D.R., Sunde, M., Bellotti, V., Robinson, C.V., Hutchinson, W.L., Fraser, P.E., Hawkins, P.N., Dobson, C.M., Radford, S.E., Blake, C.C., Pepys, M.B. (1997). Instability, unfolding and aggregation of human lysozyme variants underlying amyloid fibrillogenesis. *Nature*, *385*, 787–793.

10. Esposito, G., Garcia, J., Mangione, P., Giorgetti, S., Corazza, A., Viglino, P., Chiti, F., Andreola, A., Dumy, P., Booth, D., et al. (2003). Structural and folding dynamics properties of T70N variant of human lysozyme. *J Biol Chem*, *278*, 25910–25918.

11. Funahashi, J., Takano, K., Ogasahara, K., Yamagata, Y., Yutani, K. (1996). The structure, stability, and folding process of amyloidogenic mutant human lysozyme. *J Biochem* (*Tokyo*), *120*, 1216–1223.

12. Johnson, R.J., Christodoulou, J., Dumoulin, M., Caddy, G.L., Alcocer, M.J., Murtagh, G.J., Kumita, J.R., Larsson, G., Robinson, C.V., Archer, D.B., et al. (2005). Rationalising lysozyme amyloidosis: insights from the structure and solution dynamics of T70N lysozyme. *J Mol Biol*, *352*, 823–836.

13. Kumita, J.R., Johnson, R.J., Alcocer, M.J., Dumoulin, M., Holmqvist, F., McCammon, M.G., Robinson, C.V., Archer, D.B., Dobson, C.M. (2006). Impact of the native-state stability of human lysozyme variants on protein secretion by Pichia pastoris. *FEBS J*, *273*, 711–720.

14. Harrison, R.F., Hawkins, P.N., Roche, W.R., MacMahon, R.F., Hubscher, S.G., Buckels, J.A. (1996). 'Fragile' liver and massive hepatic haemorrhage due to hereditary amyloidosis. *Gut*, *38*, 151–152.

15. Loss, M., Ng, W.S., Karim, R.Z., Strasser, S.I., Koorey, D.J., Gallagher, P.J., Verran, D.J., McCaughan, G.W. (2006). Hereditary lysozyme amyloidosis: spontaneous hepatic rupture (15 years apart) in mother and daughter. Role of emergency liver transplantation. *Liver Transpl*, *12*, 1152–1155.

16. Granel, B., Serratrice, J., Valleix, S., Grateau, G., Droz, D., Lafon, J., Sault, M.C., Chaudier, B., Disdier, P., Laugier, R., et al. (2002). A family with gastrointestinal amyloidosis associated with variant lysozyme. *Gastroenterology*, *123*, 1346–1349.
17. Pepys, M. B. (2001). Pathogenesis, diagnosis and treatment of systemic amyloidosis. *Philos Trans R Soc Lond B, 356*, 203–210.
18. Selkoe, D.J. (2003). Folding proteins in fatal ways. *Nature*, *426*, 900–904.
19. Jimenez, J.L., Tennent, G., Pepys, M., Saibil, H.R. (2001). Structural diversity of ex vivo amyloid fibrils studied by cryo-electron microscopy. *J Mol Biol*, *311*, 241–247.
20. Sunde, M., Serpell, L.C., Bartlam, M., Fraser, P.E., Pepys, M.B., Blake, C.C.F. (1997). Common core structure of amyloid fibrils by synchrotron x-ray diffraction. *J Mol Biol*, *273*, 729–739.
21. Serpell, L.C., Sunde, M., Benson, M.D., Tennent, G.A., Pepys, M.B. Fraser, P.E. (2000). The protofilament substructure of amyloid fibrils. *J Mol Biol*, *300*, 1033–1039.
22. Canet, D., Sunde, M., Last, A.M., Miranker, A., Spencer, A., Robinson, C.V., Dobson, C.M. (1999). Mechanistic studies of the folding of human lysozyme and the origin of amyloidogenic behavior in its disease-related variants. *Biochemistry*, *38*, 6419–6427.
23. Hooke, S.D., Radford, S.E., Dobson, C.M. (1994). The refolding of human lysozyme: a comparison with the structurally homologous hen lysozyme. *Biochemistry*, *33*, 5867–5876.
24. Takano, K., Funahashi, J., Yutani, K. (2001). The stability and folding process of amyloidogenic mutant human lysozymes. *Eur J Biochem*, *268*, 155–159.
25. Wain, R., Smith, L.J., Dobson, C.M. (2005). Oxidative refolding of amyloidogenic variants of human lysozyme. *J Mol Biol*, *351*, 662–671.
26. Dumoulin, M., Bellotti, V., Dobson, C.M. (2005). In *The Beta Pleated Sheet Conformation and Diseases*, Vol. II. (Sipe, J.D., (Ed.), VCH Verlag, Weinheim, Germany, pp. 635–656.
27. Dumoulin, M., Johnson, J.R.K., Bellotti, V., Dobson, C.M. (2007). In *Protein Misfolding, Aggregation, and Conformational Diseases*, Vol. 6 Uversky, V.N., Fink, A., Eds.), Springer, New York, pp. 285–308.
28. Dumoulin, M., Canet, D., Last, A.M., Pardon, E., Archer, D.B., Muyldermans, S., Wyns, L., Matagne, A., Robinson, C.V., Redfield, C., Dobson, C.M. (2005). Reduced global cooperativity is a common feature underlying the amyloidogenicity of pathogenic lysozyme mutations. *J Mol Biol*, *346*, 773–788.
29. Dumoulin, M., Last, A.M., Desmyter, A., Decanniere, K., Canet, D., Larsson, G., Spencer, A., Archer, D.B., Sasse, J., Muyldermans, S., *et al.* (2003). A camelid antibody fragment inhibits the formation of amyloid fibrils by human lysozyme. *Nature*, *424*, 783–788.
30. Morozova-Roche, L.A., Zurdo, J., Spencer, A., Noppe, W., Receveur, V., Archer, D.B., Joniau, M., Dobson, C.M. (2000). Amyloid fibril formation and seeding by wild-type human lysozyme and its disease-related mutational variants. *J Struct Biol*, *130*, 339–351.
31. Canet, D., Last, A.M., Tito, P., Sunde, M., Spencer, A., Archer, D.B., Redfield, C., Robinson, C.V., Dobson, C.M. (2002). Local cooperativity in the unfolding of an amyloidogenic variant of human lysozyme. *Nat Struct Biol*, *9*, 308–315.

32. Harper, J.D., Lansbury, P.T., Jr. (1997). Models of amyloid seeding in Alzheimer's disease and scrapie: mechanistic truths and physiological consequences of the time-dependent solubility of amyloid proteins. *Annu Rev Biochem*, *66*, 385–407.

33. Frare, E., Mossuto, M.F., Polverino de Laureto, P., Dumoulin, M., Dobson, C.M., Fontana, A. (2006). Identification of the core structure of lysozyme amyloid fibrils by protcolysis. *J Mol Biol*, *361*, 551–561.

34. Vieira, M.N., Figueroa-Villar, J.D., Meirelles, M.N., Ferreira, S.T., De Felice, F.G. (2006). Small molecule inhibitors of lysozyme amyloid aggregation. *Cell Biochem Biophys*, *44*, 549–553.

35. Kumita, J.R., Poon, S., Caddy, G.L., Hagan, C.L., Dumoulin, M., Yerbury, J.J., Stewart, E.M., Robinson, C.V., Wilson, M.R., Dobson, C.M. (2007). The extra-cellular chaperone clusterin potently inhibits human lysozyme amyloid formation by interacting with prefibrillar species. *J Mol Biol*, *369*, 157–167.

36. Dumoulin, M., Conrath, K., Van Meirhaeghe, A., Meersman, F., Heremans, K., Frenken, L.G., Muyldermans, S., Wyns, L., Matagne, A. (2002). Single-domain antibody fragments with high conformational stability. *Protein Sci*, *11*, 500–515.

37. Muyldermans, S. (2001). Single domain camel antibodies: current status. *J Biotechnol*, *74*, 277–302.

38. Marquardt, A., Muyldermans, S., Przybylski, M. (2006). A synthetic camel anti-lysozyme peptide antibody (peptibody) with flexible loop structure identified by high-resolution affinity mass spectrometry. *Chemistry*, *12*, 1915–1923.

39. Dumoulin, M., Kumita, J.R., Dobson, C.M. (2006). Normal and aberrant biological self-assembly: Insights from studies of human lysozyme and its amyloidogenic variants. *Acc Chem Res*, *39*, 603–610.

40

THERAPEUTIC PROSPECTS FOR POLYGLUTAMINE DISEASE

MARIA PENNUTO
Department of Neuroscience, Italian Institute of Technology, Genoa, Italy

KENNETH H. FISCHBECK
Neurogenetics Branch, National Institute of Neurological Disorders and Stroke, National Institutes of Health, Bethesda, Maryland

INTRODUCTION

Polyglutamine diseases are a family of inherited, late-onset, progressive neurodegenerative disorders. To date, at least nine different diseases have been included in this family: Huntington disease (HD), spinobulbar muscular atrophy (SBMA), dentatorubral-pallidoluysian atrophy (DRPLA), and six types of spinocerebellar ataxia (SCA 1, 2, 3, 6, 7, and 17). All these disorders are inherited in an autosomal-dominant fashion, except for SBMA, which is X-linked and gender-specific, as full disease manifestations occur only in males. Polyglutamine diseases are caused by expansion of CAG repeats, encoding polyglutamine tracts, in nine different genes. One interesting aspect of polyglutamine diseases is that in each of the diseases, specific populations of neurons are vulnerable despite widespread and overlapping expression of the disease protein. Although the mutations may cause both loss and gain of protein function, there is genetic evidence that expansion of the polyglutamine tract causes disease through a toxic gain of function. Since identification in the early 1990s of the expansion of the polyglutamine tract as a major determinant for late-onset neurodegeneration, finding an effective

Protein Misfolding Diseases: Current and Emerging Principles and Therapies,
Edited by Marina Ramirez-Alvarado, Jeffery W. Kelly, and Christopher M. Dobson
Copyright © 2010 John Wiley & Sons, Inc.

treatment remains a challenge. Here we discuss recent and promising therapeutic strategies developed in animal models for these disorders.

REDUCING POLYGLUTAMINE PROTEIN EXPRESSION

Development of SCA 1 and HD conditional mouse models of polyglutamine disease has shown that the disease manifestations can be reversed by blocking mutant protein expression [1,2]. Reduction of mutant protein expression may therefore be effective in postsymptomatic as well as in presymptomatic stages of disease. One strategy to down-regulate protein expression is the antisense technique. Oligonucleotides directed against huntingtin (htt) exon 1 have been shown to decrease htt protein expression both in cell-free systems and in cells expressing mutant htt [3,4]. Another strategy used successfully to reduce htt expression in mammalian cells has been to use an oligonucleotide with RNA-cleaving enzymatic activity [5]. However, the efficacy of oligonucleotides in animal models is very low. A similar but more promising method is to use short-hairpin RNA (shRNA) and small interfering RNA (siRNA) to produce double-stranded RNA, which is degraded by the RNA-interference (RNAi) mechanism. RNAi has been shown to decrease expression and reduce the cytotoxicity of the mutant androgen receptor (AR) in SBMA [6]. Injection of adeno-associated virus type 1 expressing shRNA targeted to ataxin 1 into the cerebellum of SCA1 transgenic mice reduced protein accumulation in the nuclei of Purkinje cells and improved motor coordination [7]. Liposome-mediated delivery of siRNA targeted against exon 1 just 5′ to the region encoding the htt polyglutamine tract has been shown to reduce mutant protein expression, nuclear inclusions, and disease manifestations in the R6/2 mouse model of HD [8]. Adeno-associated virus-mediated transduction of shRNA targeting htt in the brain of HD mice resulted in decreased protein expression and rescue of some aspects of pathology in HD-N171-82Q mice [9] and R6/1 mice [10]. These findings show that siRNA has therapeutic promise. However, before application of this approach to patients, it is important to consider that the shRNA and siRNA used in mouse were directed against the mutant human transgene and thus did not suppress expression of the endogenous mouse gene. To achieve the same result in humans, it may be necessary to design siRNA specifically targeted toward the product of the mutant allele. One potential strategy would be that of making use of single-nucleotide polymorphisms and constructing siRNAs that target only the mutant allele. In this case, siRNAs may need to be designed and used in a patient-specific manner. A further degree of complexity comes from the observation that the ability of siRNAs to suppress mutant htt expression may be cell-type specific [11].

TARGETING POLYGLUTAMINE PROTEIN FOR DEGRADATION

Accumulation of unfolded proteins leads to neuronal dysfunction and death. Cells dispose of unwanted proteins by proteolysis through the ubiquitin–proteasome system (UPS) and the autophagy–lysosomal degradation system (Fig. 1). In the UPS, proteins are first polyubiquitinated in a three-step reaction executed by the E1 ubiquitin-activating enzyme, an E2 ubiquitin carrier, and an E3 ubiquitin ligase, and then recognized by the proteasome for degradation (reviewed by Hegde and Upadhya [12]). In autophagy, proteins are delivered to lysosomes for pH-dependent degradation (reviewed by Shacka et al. [13]). UPS and autophagy are coupled genetically and functionally. Proteasome dysfunction activates autophagy [14], in a process mediated by the histone deacetylase HDAC 6 [15]. Expanded polyglutamine causes impaired UPS function [16], and UPS dysfunction may be an early event in HD [17]. Genetic inhibition of autophagy leads to accumulation of ubiquitin-positive aggregates and neurodegeneration [18,19]. Therefore, therapy to enhance the UPS and autophagy might be beneficial for the polyglutamine diseases. However, enhancement of proteasome activity might be problematic for at least two reasons. One is that degradation of oligomeric and aggregated polyglutamine proteins by the proteasome is probably inefficient, due to the inability of the expanded polyglutamine proteins to enter the proteasome pore. The other is that many short-lived proteins that regulate important aspects of cellular homeostasis are substrates of the UPS, and up-regulation of the UPS may have a deleterious effect on such regulators. An alternative approach is to increase protein clearance through the proteasome by

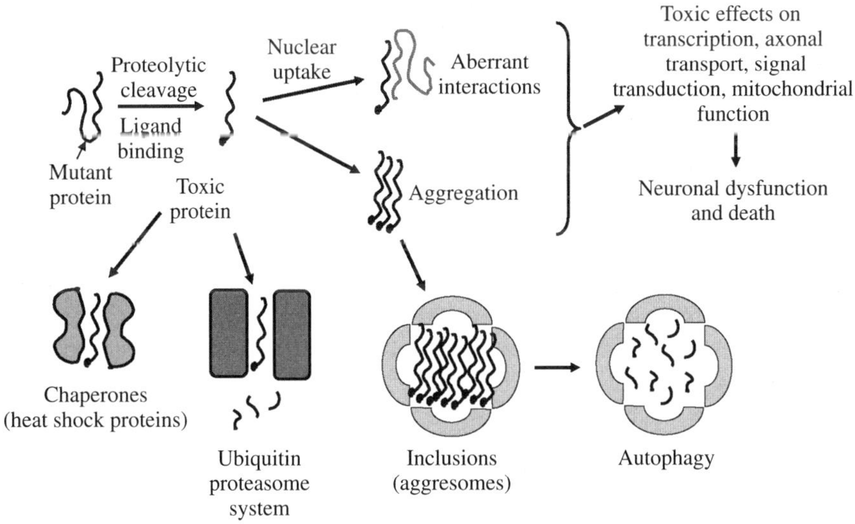

FIG. 1 Expanded polyglutamine toxicity. Expansion of the polyglutamine tract results in protein cleavage, nuclear uptake, and generation of aggregates. The cells cope with the accumulation of aberrant species, trying to increase protein folding through chaperone activity and protein degradation through the ubiquitin–proteasome system and autophagy. If this process is inefficient, the aberrant protein accumulates in the cell, affecting several cellular processes, such as transcription, axonal transport, signal transduction, and mitochondrial function. This leads to neuronal dysfunction and eventually to death.

enhancing chaperone activity. By promoting protein folding, molecular chaperones can increase the amount of monomeric polyglutamine protein that is a substrate of the proteasome. Consistent with this, overexpression of the molecular chaperone heat-shock protein (Hsp) 40 has been shown to protect cells from the toxicity of mutant ataxin 1 [20] and ataxin 3 [21]. Hsp27 suppressed toxicity of mutant htt in cells [22], and Hsp105 was protective in an SBMA cell culture model [23]. Overexpression of Hsp70 was neuroprotective in SBMA [24] and SCA1 mouse models [25], and overexpression of Hsp104 protected mice from mutant htt [26]. The Hsp70 interacting protein CHIP (C terminus of Hsp70 interacting protein) represents the molecular link between protein refolding mediated by chaperones and degradation carried out by the UPS. Overexpression of CHIP has been shown to ameliorate disease manifestations in animal models of SBMA [27], SCA1 [28], and HD [29]. How can these findings be converted into therapy? Overexpression of Hsps can be induced pharmacologically. Oral administration of geranylgeranylacetone, an acyclic isoprenoid molecule known to induce Hsps, to SBMA mice induced expression of Hsp70, Hsp90, and Hsp105 and reduced neuromuscular degeneration [30]. Geldanamycin has been shown to induce expression of Hsp90, 70, and 40, and to decrease mutant htt aggregation and toxicity in cells [31]. Unfortunately, its liver toxicity precludes its use in patients. 17-Allyl-aminogeldanamycin (17-AAG) is a geldanamycin derivative that has been shown to reduce motor neuron degeneration and increase survival in SBMA mice [32]. Recently, a novel derivate of geldanamycin, 17-DMAG, has been shown to be more potent than 17-AAG in cells [33]. It remains to be determined whether this compound is effective in animal models of polyglutamine disease and suitable for human testing.

Induction of autophagy is another approach to reducing accumulation of polyglutamine proteins in cells. A negative regulator of autophagy is target of rapamycin (TOR). Therefore, autophagy can be up-regulated by rapamycin, which inhibits TOR. Treatment of cell, fly, and mouse models of HD with rapamycin resulted in increased autophagy, and decreased aggregates and toxicity of this mutant protein [34]. Importantly, rapamycin had effects not only in HD, but also in other polyglutamine disorders, including SCA 3 [35] and SBMA [15]. Rapamycin is already used in patients to prevent transplant rejection and for treatment of gliomas. Unfortunately, prolonged use of rapamycin in humans has undesirable side effects, including immunosuppression and altered wound healing. A search for alternative compounds to induce autophagy has led to the identification of small molecules that attenuate HD manifestations [36]. These new compounds may be used as a basis for a future polyglutamine disease therapy.

HISTONE DEACETYLASE INHIBITION

Expression of polyglutamine proteins alters gene transcription. This is probably due to depletion of nuclear factors and chromatin remodeling. Chromatin is remodeled through various histone modifications, including acetylation. Increased histone acetylation is generally associated with increased gene transcription, and deacetylation is associated with reduced transcription. Histone acetylation is regulated by the activity of histone acetylase (HAT) and deacetylase (HDAC) enzymes. A tightly regulated equilibrium between the activities of these enzymes helps to ensure proper gene expression. Histone acetylation is altered by expanded polyglutamine htt [37]. Polyglutamine proteins sequester and deplete HATs, including CREB-binding protein (CBP). Overexpression of CBP rescued histone acetylation and neurodegeneration in fly and mouse models of polyglutamine disease [38–40]. This research suggests that inhibition of HDAC activity may be of therapeutic value. Administration of HDAC inhibitors to HD flies reduced neurodegeneration and extended life-span [37]. The HDAC inhibitor sodium butyrate improved body weight and motor function in HD [41], DRPLA [42], and SBMA [43] mice. The HDAC inhibitor suberoylanilide hydroxamic acid (vorinostat) protected HD mice from neurodegeneration [44]. Administration of the less potent HDAC inhibitor phenylbutyrate in HD mice was beneficial even when treatment was started after the onset of symptoms [45]. Several HDAC inhibitors are approved for treatment of other disorders. Vorinostat is approved as treatment for cutaneous T-cell lymphomas, and phenylbutyrate is currently used for treatment of urea cycle defects and hemoglobinopathy and is in a phase II clinical trial for HD. Thus, HDAC inhibition is worth pursuing as therapy for polyglutamine disorders.

CASPASE INHIBITION

Apoptosis or programmed cell death is a process that leads to nuclear condensation, DNA fragmentation, cell shrinkage, and eventually cell death. At least two pathways of apoptosis exist, one extrinsic and the other intrinsic. The extrinsic pathway is initiated by activation of cell-surface death receptors, which transmit a signal that leads to activation of the initiator caspases 8 and 10. The intrinsic pathway involves release of cytochrome c from mitochondria and activation of caspase 9. Activation of the initiator caspases triggers activation of the executioner caspase 3, 6, and 7. Caspase 8 is activated in HD brain specimens and cell culture [46], and caspase 1 and 3 are activated in the brains of R6/2 mice [47]. Evidence that caspase activation plays an important role in the pathogenesis of HD came from experiments performed in mouse models of HD. Expression of a dominant negative form of caspase 1 in the brain of R6/2 mice and intraventricular administration of the caspase inhibitor z-VAD-fmk improved motor weakness and delayed the onset of symptoms and death [47]. Intraperitoneal injection of minocycline, a tetracycline derivative that inhibits caspases, in R6/2 mice delayed the onset of disease and prolonged survival [48]. An important consequence of caspase activation is that polyglutamine proteins are cleaved by caspases, thus generating a polyglutamine-containing fragment that is probably more toxic than the full-length protein [49]. Caspase cleavage of polyglutamine proteins may be a common step in the pathogenesis of polyglutamine disease [50]. Proteolytic cleavage of ataxin 3 and 7 correlates with protein aggregation [51,52]. Cleavage-resistant mutant versions of AR [53], atrophin 1 [54], htt [55], and ataxin 7 [56] have reduced cytotoxicity. Htt is also a substrate of calpain [57]. Mutant calpain-resistant versions of htt have reduced cytotoxicity, with decreased protein aggregation and nuclear accumulation. Calpain activity is increased in HD mice. Calpains are proteases activated by calcium. Alteration of calcium homeostasis in HD has been attributed to enhanced calcium influx through the *N*-methyl-D-aspartate (NMDA) receptors [58]. Treatment of a rodent model of striatal degeneration mimicking HD with the NMDA antagonist memantine protected striatal neurons from death and reduced calpain activity and htt proteolysis [59]. Therefore, inhibition of protease activity can be of therapeutic value, due to a double effect, one to reduce or block programmed cell death triggered by expression of mutant htt and the other to decrease proteolytic cleavage of polyglutamine htt, thereby reducing cytotoxicity.

NEUROTROPHIC FACTORS

Neurotrophic factors (neurotrophins) promote neuronal survival by triggering signaling pathways that prevent initiation of programmed cell death, and this generally occurs through activation of the stress kinase pathway and the phosphatidylinositol 3-kinase/Akt signaling pathway. Altered neurotrophin signaling has been reported in several neurodegenerative disorders, including the polyglutamine diseases, where this may contribute to disease manifestations. Neurotrophin levels are decreased in a mouse model of SBMA [60], and Akt signaling is altered in patients and animal models of HD [61,62]. Several neurotrophins have been shown to protect striatal neurons from degeneration in rodent models of HD. Virus-mediated transfer or transplantation of cells secreting brain-derived neurotrophic factors (BDNF), neurotrophin 3, neurotrophin 4/5, and human ciliary neurotrophic factor (CNTF) into brain attenuated neurodegeneration in rats lesioned with quinolinate to mimic HD [63–66]. Another class of neurotrophic factors that has been shown to prevent striatal neuron degeneration in animal models of HD is the glial-derived neurotrophic factor (GDNF) family of ligands, which includes GDNF and neurturin [67,68]. Although this research shows a beneficial effect of neurotrophins, it is important to note that in all the experiments described, neurotrophins were administered in presymptomatic animals. Therefore, it remains to be established whether neurotrophin treatment is also effective when started after the onset of symptoms.

In addition to neurotrophins, growth factors have been shown to protect against the toxicity of polyglutamine proteins. Vascular endothelial growth factor [69] and insulin-like growth factor 1 (IGF-1) [62,70,71] protected neurons from polyglutamine toxicity. Intranasal administration of IGF-1 in SCA1 mice rescued motor deficits and prevented Purkinje cell degeneration [72]. The effect of IGF-1 on polyglutamine toxicity depends on activation of pro-survival ERK and Akt signaling.

TRANSGLUTAMINASE INHIBITION

Polyglutamine proteins, including htt [73], AR [74], and ataxin 1 [75], are cross-linked by transglutaminases, enzymes that generate an isopeptide bond between glutamine and lysine residues. Multiple lines of evidence suggest a role for transglutaminase in polyglutamine disease. Tissue transglutaminase and transglutaminase activity are increased in striatal neurons of HD patients [76,77] and in mouse models of HD [78,79], SCA 1 [75], and SBMA [74]. Ablation of the tissue transglutaminase gene in R6/1 and R6/2 HD mice improved motor performance and prolonged life span [80]. This indicates that transglutaminases may represent another therapeutic target for polyglutamine diseases. Support for this came first from the observation that cells are protected from mutant atrophin 1–induced polyglutamine toxicity by treatment with the transglutaminase inhibitor cystamine [81]. Subsequently, administration of cystamine to R6/2 HD mice was shown to inhibit transglutaminase activity in the brain and improve survival and motor performance [79] and to ameliorate striatal pathology, albeit with no effect on motor performance, in YAC128 HD mice [78]. Interestingly, cystamine has been shown to protect cells from polyglutamine-induced toxicity not only by inhibiting transglutaminase activity, but also by preventing caspase 3 activation [82]. Although further research is required for therapy optimization, based on these results, inhibition of transglutaminase activity by cystamine has potential as a treatment for polyglutamine disease.

DISEASE-SPECIFIC TREATMENT

HD

Gsk 3β has been implicated in a variety of neurodegenerative disorders, including HD. Because lithium is an inhibitor of Gsk 3β and therefore potentially neuroprotective, it has been tested in rodent and cell models of HD. In the rat, treatment with lithium before induction of striatal lesions by quinolinic acid protected neurons from degeneration and improved motor performance [83]. In HD cell models, the effect of lithium has been shown to occur through inhibition of Gsk 3β [84]. However, treatment of R6/2 mice before and after appearance of disease manifestations has given inconsistent results [85]. Therefore, further research is necessary to establish the potential of lithium as treatment for HD.

Another important feature of HD is mitochondrial dysfunction, as indicated by altered lactate levels and mitochondrial-complex activity in HD patients, leading to oxidative damage [86,87]. Compounds such as creatine, coenzyme Q_{10}, and idebenone, which stabilize mitochondria and protect against oxidative tissue damage, have been shown to protect HD mice from neurodegeneration [88]. Importantly, creatine was effective even when administered after the onset of disease manifestations [89].

SBMA

In contrast to the other polyglutamine diseases, SBMA is ligand-dependent. In humans, only males are fully symptomatic, and females homozygous for the mutation are only mildly affected [90]. Mouse and fly models of SBMA have shown that males develop disease manifestations because of higher levels of testosterone in the serum. Testosterone reduction by castration in SBMA transgenic mice restored motor function [91,92], and testosterone administration to female transgenic mice induced an SBMA phenotype [91]. Consistent with the testosterone-dependent toxicity of the mutant AR in SBMA, treatment of transgenic mice with the antiandrogen leuprorelin extended survival and rescued behavioral deficits [93], and gave promising results in phase 2 clinical trial [94]. Leuprorelin is a lutenizing hormone–releasing hormone agonist, which decreases serum testosterone levels and thus is a potential therapy for SBMA.

Another approach to treating SBMA is to reduce interaction of AR with critical nuclear cofactors. This approach has recently been taken by Chang's lab [95]. Treatment of SBMA mice with the curcumin-related compound 5-hydroxy-1,7-bis[3,4-dimethoxyphenyl]-1,4,6-heptatrien-3-one (ASC-J9) reduced interaction of AR with ARA70, increased AR degradation, decreased nuclear aggregation, prolonged survival, and ameliorated disease manifestations.

SCA 1

As mentioned above, lithium may have neuroprotective effects. Recently, Watase and colleagues have shown that treatment of a mouse model of SCA 1 with lithium leads to motor alterations, reduces dendritic pathology in the hippocampus, and rescues alteration in gene transcription [96]. Importantly, treatment of postsymptomatic mice with lithium was effective. This observation makes lithium a potentially promising therapy for SCA 1 patients.

Acknowledgments

This work was supported by intramural NINDS funds, National Institutes of Health, and Telethon-Italy (GFP04005), Muscular Dystrophy Association and Kennedy's Disease Association funds to M.P.

REFERENCES

1. Zu, T., Duvick, L.A., Kaytor, M.D., Berlinger, M.S., Zoghbi, H.Y., Clark, H.B., Orr, H.T. (2004). Recovery from polyglutamine-induced neurodegeneration in conditional SCA1 transgenic mice. *J Neurosci, 24*, 8853–8861.
2. Yamamoto, A., Lucas, J.J., Hen, R. (2000). Reversal of neuropathology and motor dysfunction in a conditional model of Huntington's disease. *Cell, 101*, 57–66.
3. Boado, R.J., Kazantsev, A., Apostol, B.L., Thompson, L.M., Pardridge, W.M. (2000). Antisense-mediated down-regulation of the human huntingtin gene. *J Pharmacol Exp Ther, 295*, 239–243.
4. Nellemann, C., Abell, K., Nørremølle, A., Løkkegaard, T., Naver, B., Röpke, C., Rygaard, J., Sørensen, S.A., Hasholt, L. (2000). Inhibition of Huntington synthesis by antisense oligodeoxynucleotides. *Mol Cell Neurosci, 16*, 313–323.
5. Yen, L., Strittmatter, S.M., Kalb, R.G. (1999). Sequence-specific cleavage of Huntingtin mRNA by catalytic DNA. *Ann Neurol, 46*, 366–373.
6. Caplen, N.J., Taylor, J.P., Statham, V.S., Tanaka, F., Fire, A., Morgan, R.A. (2002). Rescue of polyglutamine-mediated cytotoxicity by double-stranded RNA-mediated RNA interference. *Hum Mol Genet, 11*, 175–184.
7. Xia, H., Mao, Q., Eliason, S.L., Harper, S.Q., Martins, I.H., Orr, H.T., Paulson, H.L., Yang, L., Kotin, R.M., Davidson, B.L. (2004). RNAi suppresses polyglutamine-induced neurodegeneration in a model of spinocerebellar ataxia. *Nat Med, 10*, 816–820.
8. Wang, Y.L., Liu, W., Wada, E., Murata, M., Wada, K., Kanazawa, I. (2005). Clinico-pathological rescue of a model mouse of Huntington's disease by siRNA. *Neurosci Res, 53*, 241–249.
9. Harper, S.Q., Staber, P.D., He, X., Eliason, S.L., Martins, I.H., Mao, Q., Yang, L., Kotin, R.M., Paulson, H.L., Davidson, B.L. (2005). RNA interference improves motor and neuropathological abnormalities in a Huntington's disease mouse model. *Proc Natl Acad Sci U S A, 102*, 5820–5825.
10. Rodriguez-Lebron, E., Denovan-Wright, E.M., Nash, K., Lewin, A.S., Mandel, R.J. (2005). Intrastriatal rAAV-mediated delivery of anti-huntingtin shRNAs induces partial reversal of disease progression in R6/1 Huntington's disease transgenic mice. *Mol Ther, 12*, 618–633.
11. Gary, D.S., Davidson, A., Milhavet, O., Slunt, H., Borchelt, D.R. (2007). Investigation of RNA interference to suppress expression of full-length and fragment human huntingtin. *Neuromolecular Med, 9*, 145–155.
12. Hegde, A.N., Upadhya, S.C. (2007). The ubiquitin–proteasome pathway in health and disease of the nervous system. *Trends Neurosci, 30*, 587–595.
13. Shacka, J.J., Roth, K.A., Zhang, J. (2008). The autophagy–lysosomal degradation pathway, role in neurodegenerative disease and therapy. *Front Biosci, 13*, 718–736.
14. Ding, W.X., Ni, H.M., Gao, W., Yoshimori, T., Stolz, D.B., Ron, D., Yin, X.M. (2007). Linking of autophagy to ubiquitin–proteasome system is important for the regulation of endoplasmic reticulum stress and cell viability. *Am J Pathol, 171*, 513–524.
15. Pandey, U.B., Nie, Z., Batlevi, Y., McCray, B.A., Ritson, G.P., Nedelsky, N.B., Schwartz, S.L., DiProspero, N.A., Knight, M.A., Schuldiner, O., Padmanabhan, R.,

Hild, M., Berry, D.L., Garza, D., Hubbert, C.C., Yao, T.P., Baehrecke, E.H., Taylor, J.P. (2007). HDAC6 rescues neurodegeneration and provides an essential link between autophagy and the UPS. *Nature, 447*, 859–863.

16. Bence, N.F., Sampat, R.M., Kopito, R.R. (2001). Impairment of the ubiquitin-proteasome system by protein aggregation. *Science, 292*, 1552–1555.
17. Bennett, E.J., Shaler, T.A., Woodman, B., Ryu, K.Y., Zaitseva, T.S., Becker, C.H., Bates, G.P., Schulman, H., Kopito, R.R. (2007). Global changes to the ubiquitin system in Huntington's disease. *Nature, 448*, 704–708.
18. Komatsu, M., Waguri, S., Chiba, T., Murata, S., Iwata, J., Tanida, I., Ueno, T., Koike, M., Uchiyama, Y., Kominami, E., Tanaka, K. (2006). Loss of autophagy in the central nervous system causes neurodegeneration in mice. *Nature, 441*, 880–884.
19. Hara, T., Nakamura, K., Matsui, M., Yamamoto, A., Nakahara, Y., Suzuki-Migishima, R., Yokoyama, M., Mishima, K., Saito, I., Okano, H., Mizushima, N. (2006). Suppression of basal autophagy in neural cells causes neurodegenerative disease in mice. *Nature, 441*, 885–889.
20. Cummings, C.J., Mancini, M.A., Antalffy, B., DeFranco, D.B., Orr, H.T., Zoghbi, H.Y. (1998). Chaperone suppression of aggregation and altered subcellular proteasome localization imply protein misfolding in SCA1. *Nat Genet, 19*, 148–154.
21. Chai, Y., Koppenhafer, S.L., Bonini, N.M., Paulson, H.L. (1999). Analysis of the role of heat shock protein (Hsp). molecular chaperones in polyglutamine disease. *J Neurosci, 19*, 10338–10347.
22. Wyttenbach, A., Sauvageot, O., Carmichael, J., Diaz-Latoud, C., Arrigo, A.P., Rubinsztein, D.C. (2002). Heat shock protein 27 prevents cellular polyglutamine toxicity and suppresses the increase of reactive oxygen species caused by huntingtin. *Hum Mol Genet, 11*, 1137–1151.
23. Ishihara, K., Yamagishi, N., Saito, Y., Adachi, H., Kobayashi, Y., Sobue, G., Ohtsuka, K., Hatayama, T. (2003). Hsp105alpha suppresses the aggregation of truncated androgen receptor with expanded CAG repeats and cell toxicity. *J Biol Chem, 278*, 25143–25150.
24. Adachi, H., Katsuno, M., Minamiyama, M., Sang, C., Pagoulatos, G., Angelidis, C., Kusakabe, M., Yoshiki, A., Kobayashi, Y., Doyu, M., Sobue, G. (2003). Heat shock protein 70 chaperone overexpression ameliorates phenotypes of the spinal and bulbar muscular atrophy transgenic mouse model by reducing nuclear-localized mutant androgen receptor protein. *J Neurosci, 23*, 2203–2211.
25. Cummings, C.J., Sun, Y., Opal, P., Antalffy, B., Mestril, R., Orr, H.T., Dillmann, W.H., Zoghbi, H.Y. (2001). Over-expression of inducible HSP70 chaperone suppresses neuropathology and improves motor function in SCA1 mice. *Hum Mol Genet, 10*, 1511–1518.
26. Vacher, C., Garcia-Oroz, L., Rubinsztein, D.C. (2005). Overexpression of yeast hsp104 reduces polyglutamine aggregation and prolongs survival of a transgenic mouse model of Huntington's disease. *Hum Mol Genet, 14*, 3425–3433.
27. Adachi, H., Waza, M., Tokui, K., Katsuno, M., Minamiyama, M., Tanaka, F., Doyu, M., Sobue, G. (2007). CHIP overexpression reduces mutant androgen receptor protein and ameliorates phenotypes of the spinal and bulbar muscular atrophy transgenic mouse model. *J Neurosci, 27*, 5115–5126.

28. Al-Ramahi, I., Lam, Y.C., Chen, H.K., de Gouyon, B., Zhang, M., Pérez, A.M., Branco, J., de Haro, M., Patterson, C., Zoghbi, H.Y., Botas, J. (2006). CHIP protects from the neurotoxicity of expanded and wild-type ataxin-1 and promotes their ubiquitination and degradation. *J Biol Chem, 281*, 26714–26724.

29. Miller, V.M., Nelson, R.F., Gouvion, C.M., Williams, A., Rodriguez-Lebron, E., Harper, S.Q., Davidson, B.L., Rebagliati, M.R., Paulson, H.L. (2005). CHIP suppresses polyglutamine aggregation and toxicity in vitro and in vivo. *J Neurosci, 25*, 9152–9161.

30. Katsuno, M., Sang, C., Adachi, H., Minamiyama, M., Waza, M., Tanaka, F., Doyu, M., Sobue, G. (2005). Pharmacological induction of heat-shock proteins alleviates polyglutamine-mediated motor neuron disease. *Proc Natl Acad Sci U S A, 102*, 16801–16806.

31. Sittler, A., Lurz, R., Lueder, G., Priller, J., Lehrach, H., Hayer-Hartl, M.K., Hartl, F.U., Wanker, E.E. (2001). Geldanamycin activates a heat shock response and inhibits huntingtin aggregation in a cell culture model of Huntington's disease. *Hum Mol Genet, 10*, 1307–1315.

32. Waza, M., Adachi, H., Katsuno, M., Minamiyama, M., Sang, C., Tanaka, F., Inukai, A., Doyu, M., Sobue, G.. (2005). 17-AAG, an Hsp90 inhibitor, ameliorates polyglutamine-mediated motor neuron degeneration. *Nat Med, 11*, 1088–1095.

33. Herbst, M., Wanker, E.E. (2007). Small molecule inducers of heat-shock response reduce polyglutamine-mediated huntingtin aggregation: a possible therapeutic strategy. *Neurodegener Dis, 4*, 254–260.

34. Ravikumar, B., Vacher, C., Berger, Z., Davies, J.E., Luo, S., Oroz, L.G., Scaravilli, F., Easton, D.F., Duden, R., O'Kane, C.J., Rubinsztein, D.C. (2004). Inhibition of mTOR induces autophagy and reduces toxicity of polyglutamine expansions in fly and mouse models of Huntington disease. *Nat Genet, 36*, 585–595.

35. Berger, Z., Ravikumar, B., Menzies, F.M., Oroz, L.G., Underwood, B.R., Pangalos, M.N., Schmitt, I., Wullner, U., Evert, B.O., O'Kane, C.J., Rubinsztein, D.C. (2006). Rapamycin alleviates toxicity of different aggregate-prone proteins. *Hum Mol Genet, 15*, 433–442.

36. Sarkar, S., Perlstein, E.O., Imarisio, S., Pineau, S., Cordenier, A., Maglathlin, R.L., Webster, J.A., Lewis, T.A., O'Kane, C.J., Schreiber, S.L., Rubinsztein, D.C. (2007). Small molecules enhance autophagy and reduce toxicity in Huntington's disease models. *Nat Chem Biol, 3*, 331–338.

37. Steffan, J.S., Bodai, L., Pallos, J., Poelman, M., McCampbell, A., Apostol, B.L., Kazantsev, A., Schmidt, E., Zhu, Y.Z., Greenwald, M., Kurokawa, R., Housman, D.E., Jackson, G.R., Marsh, J.L., Thompson, L.M. (2001). Histone deacetylase inhibitors arrest polyglutamine-dependent neurodegeneration in *Drosophila*. *Nature, 413*, 739–743.

38. Taylor, J.P., Taye, A.A., Campbell, C., Kazemi-Esfarjani, P., Fischbeck, K.H., Min, K.T. (2003). Aberrant histone acetylation, altered transcription, and retinal degeneration in a *Drosophila* model of polyglutamine disease are rescued by CREB-binding protein. *Genes Dev, 17*, 1463–1468.

39. Nucifora, F.C. Jr., Sasaki, M., Peters, M.F., Huang, H., Cooper, J.K., Yamada, M., Takahashi, H., Tsuji, S., Troncoso, J., Dawson, V.L., Dawson, T.M., Ross, C.A.

(2001). Interference by huntingtin and atrophin-1 with cbp-mediated transcription leading to cellular toxicity. *Science, 291*, 2423–2428.

40. McCampbell, A., Taylor, J.P., Taye, A.A., Robitschek, J., Li, M., Walcott, J., Merry, D., Chai, Y., Paulson, H., Sobue, G., Fischbeck, K.H. (2000). CREB-binding protein sequestration by expanded polyglutamine. *Hum Mol Genet, 9*, 2197–2202.
41. Ferrante, R.J., Kubilus, J.K., Lee, J., Ryu, H., Beesen, A., Zucker, B., Smith, K., Kowall, N.W., Ratan, R.R., Luthi-Carter, R., Hersch, S.M. (2003). Histone deacetylase inhibition by sodium butyrate chemotherapy ameliorates the neurodegenerative phenotype in Huntington's disease mice. *J Neurosci, 23*, 9418–9427.
42. Ying, M., Xu, R., Wu, X., Zhu, H., Zhuang, Y., Han, M., Xu, T. (2006). Sodium butyrate ameliorates histone hypoacetylation and neurodegenerative phenotypes in a mouse model for DRPLA. *J Biol Chem, 281*, 12580–12586.
43. Minamiyama, M., Katsuno, M., Adachi, H., Waza, M., Sang, C., Kobayashi, Y., Tanaka, F., Doyu, M., Inukai, A., Sobue, G. (2004). Sodium butyrate ameliorates phenotypic expression in a transgenic mouse model of spinal and bulbar muscular atrophy. *Hum Mol Genet, 13*, 1183–1192.
44. Hockly, E., Richon, V.M., Woodman, B., Smith, D.L., Zhou, X., Rosa, E., Sathasivam, K., Ghazi-Noori, S., Mahal, A., Lowden, P.A., Steffan, J.S., Marsh, J.L., Thompson, L.M., Lewis, C.M., Marks, P.A., Bates, GP. (2003). Suberoylanilide hydroxamic acid, a histone deacetylase inhibitor, ameliorates motor deficits in a mouse model of Huntington's disease. *Proc Natl Acad Sci U S A, 100*, 2041–2046.
45. Gardian, G., Browne, S.E., Choi, D.K., Klivenyi, P., Gregorio, J., Kubilus, J.K., Ryu, H., Langley, B., Ratan, R.R., Ferrante, R.J., Beal, M.F. (2005). Neuroprotective effects of phenylbutyrate in the N171-82Q transgenic mouse model of Huntington's disease. *J Biol Chem, 280*, 556–563.
46. Sánchez, I., Xu, C.J., Juo, P., Kakizaka, A., Blenis, J., Yuan, J. (1999). Caspase-8 is required for cell death induced by expanded polyglutamine repeats. *Neuron, 22*, 623–633.
47. Ona, V.O., Li, M., Vonsattel, J.P., Andrews, L.J., Khan, S.Q., Chung, W.M., Frey, A.S., Menon, A.S., Li, X.J., Stieg, P.E., Yuan, J., Penney, J.B., Young, A.B., Cha, J.H., Friedlander, R.M. (1999). Inhibition of caspase-1 slows disease progression in a mouse model of Huntington's disease. *Nature, 399*, 263–267.
48. Chen, M., Ona, V.O., Li, M., Ferrante, R.J., Fink, K.B., Zhu, S., Bian, J., Guo, L., Farrell, L.A., Hersch, S.M., Hobbs, W., Vonsattel, J.P., Cha, J.H., Friedlander, R.M. (2000). Minocycline inhibits caspase-1 and caspase-3 expression and delays mortality in a transgenic mouse model of Huntington disease. *Nat Med, 6*, 797–801.
49. Hackam, A.S., Singaraja, R., Wellington, C.L., Metzler, M., McCutcheon, K., Zhang, T., Kalchman, M., Hayden, M.R. (1998). The influence of huntingtin protein size on nuclear localization and cellular toxicity. *J Cell Biol, 141*, 1097–1105.
50. Wellington, C.L., Ellerby, L.M., Hackam, A.S., Margolis, R.L., Trifiro, M.A., Singaraja, R., McCutcheon, K., Salvesen, G.S., Propp, S.S., Bromm, M., Rowland, K.J., Zhang, T., Rasper, D., Roy, S., Thornberry, N., Pinsky, L., Kakizuka, A., Ross, C.A., Nicholson, D.W., Bredesen, D.E., Hayden, M.R. (1998). Caspase

INSTITUT MEDIATHEQUE PASTEUR

cleavage of gene products associated with triplet expansion disorders generates truncated fragments containing the polyglutamine tract. *J Biol Chem, 273*, 9158–9167.

51. Haacke, A., Broadley, S.A., Boteva, R., Tzvetkov, N., Hartl, F.U., Breuer, P. (2006). Proteolytic cleavage of polyglutamine-expanded ataxin-3 is critical for aggregation and sequestration of non-expanded ataxin-3. *Hum Mol Genet, 15*, 555–568.
52. Garden, G.A., Libby, R.T., Fu, Y.H., Kinoshita, Y., Huang, J., Possin, D.E., Smith, A.C., Martinez, R.A., Fine, G.C., Grote, S.K., Ware, C.B., Einum, D.D., Morrison, R.S., Ptacek, L.J., Sopher, B.L., La Spada, A.R. (2002). Polyglutamine-expanded ataxin-7 promotes non-cell-autonomous Purkinje cell degeneration and displays proteolytic cleavage in ataxic transgenic mice. *J Neurosci, 22*, 4897–4905.
53. Ellerby, L.M., Hackam, A.S., Propp, S.S., Ellerby, H.M., Rabizadeh, S., Cashman, N.R., Trifiro, M.A., Pinsky, L., Wellington, C.L., Salvesen, G.S., Hayden, M.R., Bredesen, D.E. (1999). Kennedy's disease, caspase cleavage of the androgen receptor is a crucial event in cytotoxicity. *J Neurochem, 72*, 185–195.
54. Ellerby, L.M., Andrusiak, R.L., Wellington, C.L., Hackam, A.S., Propp, S.S., Wood, J.D., Sharp, A.H., Margolis, R.L., Ross, C.A., Salvesen, G.S., Hayden, M.R., Bredesen, D.E. (1999). Cleavage of atrophin-1 at caspase site aspartic acid 109 modulates cytotoxicity. *J Biol Chem, 274*, 8730–8736.
55. Graham, R.K., Deng, Y., Slow, E.J., Haigh, B., Bissada, N., Lu, G., Pearson, J., Shehadeh, J., Bertram, L., Murphy, Z., Warby, S.C., Doty, C.N., Roy, S., Wellington, C.L., Leavitt, B.R., Raymond, L.A., Nicholson, D.W., Hayden, M.R. (2006). Cleavage at the caspase-6 site is required for neuronal dysfunction and degeneration due to mutant huntingtin. *Cell, 125*, 1179–1191.
56. Young, J.E., Gouw, L., Propp, S., Sopher, B.L., Taylor, J., Lin, A., Hermel, E., Logvinova, A., Chen, S.F., Chen, S., Bredesen, D.E., Truant, R., Ptacek, L.J., La Spada, A.R., Ellerby, L.M. (2007). Proteolytic cleavage of ataxin-7 by caspase-7 modulates cellular toxicity and transcriptional dysregulation. *J Biol Chem, 282*, 30150–30160.
57. Gafni, J., Hermel, E., Young, J.E., Wellington, C.L., Hayden, M.R., Ellerby, L.M. (2004). Inhibition of calpain cleavage of huntingtin reduces toxicity, accumulation of calpain/caspase fragments in the nucleus. *J Biol Chem, 279*, 20211–20220.
58. Zeron, M.M., Hansson, O., Chen, N., Wellington, C.L., Leavitt, B.R., Brundin, P., Hayden, M.R., Raymond, L.A. (2002). Increased sensitivity to *N*-methyl-D-aspartate receptor-mediated excitotoxicity in a mouse model of Huntington's disease. *Neuron, 33*, 849–860.
59. Lee, S.T., Chu, K., Park, J.E., Kang, L., Ko, S.Y., Jung, K.H., Kim, M. (2006). Memantine reduces striatal cell death with decreasing calpain level in 3-nitropropionic model of Huntington's disease. *Brain Res, 1118*, 199–207.
60. Yu, Z., Dadgar, N., Albertelli, M., Gruis, K., Jordan, C., Robins, D.M., Lieberman, A.P. (2006). Androgen-dependent pathology demonstrates myopathic contribution to the Kennedy disease phenotype in a mouse knock-in model. *J Clin Invest, 116*, 2663–2672.
61. Colin, E., Régulier, E., Perrin, V., Dürr, A., Brice, A., Aebischer, P., Déglon, N., Humbert, S., Saudou, F. (2005). Akt is altered in an animal model of Huntington's disease and in patients. *Eur J Neurosci, 21*, 1478–1488.

62. Humbert, S., Bryson, E.A., Cordelières, F.P., Connors, N.C., Datta, S.R., Finkbeiner, S., Greenberg, M.E., Saudou, F. (2002). The IGF-1/Akt pathway is neuroprotective in Huntington's disease and involves Huntingtin phosphorylation by Akt. *Dev Cell, 2*, 831–837.

63. de Almeida, L.P., Zala, D., Aebischer, P., Déglon, N. (2001). Neuroprotective effect of a CNTF-expressing lentiviral vector in the quinolinic acid rat model of Huntington's disease. *Neurobiol Dis, 8*, 433–446.

64. Pérez-Navarro, E., Canudas, A.M., Akerund, P., Alberch, J., Arenas, E. (2000). Brain-derived neurotrophic factor, neurotrophin-3, and neurotrophin-4/5 prevent the death of striatal projection neurons in a rodent model of Huntington's disease. *J Neurochem, 75*, 2190–2199.

65. Bemelmans, A.P., Horellou, P., Pradier, L., Brunet, I., Colin, P., Mallet, J. (1999). Brain-derived neurotrophic factor-mediated protection of striatal neurons in an excitotoxic rat model of Huntington's disease, as demonstrated by adenoviral gene transfer. *Hum Gene Ther, 10*, 2987–2997.

66. Emerich, D.F., Lindner, M.D., Winn, S.R., Chen, E.Y., Frydel, B.R., Kordower, J.H. (1996). Implants of encapsulated human CNTF-producing fibroblasts prevent behavioral deficits and striatal degeneration in a rodent model of Huntington's disease. *J Neurosci, 16*, 5168–5181.

67. Ramaswamy, S., McBride, J.L., Herzog, C.D., Brandon, E., Gasmi, M., Bartus, R.T., Kordower, J.H. (2007). Neurturin gene therapy improves motor function and prevents death of striatal neurons in a 3-nitropropionic acid rat model of Huntington's disease. *Neurobiol Dis, 26*, 375–384.

68. McBride, J.L., Ramaswamy, S., Gasmi, M., Bartus, R.T., Herzog, C.D., Brandon, E.P., Zhou, L., Pitzer, M.R., Berry-Kravis, E.M., Kordower, J.H. (2006). Viral delivery of glial cell line–derived neurotrophic factor improves behavior and protects striatal neurons in a mouse model of Huntington's disease. *Proc Natl Acad Sci U S A*, 103, 9345–9350.

69. Sopher, B.L., Thomas, P.S. Jr., LaFevre-Bernt, M.A., Holm, I.E., Wilke, S.A., Ware, C.B., Jin, L.W., Libby, R.T., Ellerby, L.M., La Spada, A.R. (2004). Androgen receptor YAC transgenic mice recapitulate SBMA motor neuronopathy and implicate VEGF164 in the motor neuron degeneration. *Neuron, 41*, 687–699.

70. Palazzolo, I., Burnett, B.G., Young, J.E., Brenne, P.L., La Spada, A.R., Fischbeck, K.H., Howell, B.W., Pennuto, M. (2007). Akt blocks ligand binding and protects against expanded polyglutamine androgen receptor toxicity. *Hum Mol Genet, 16*, 1593–1603.

71. Palazzolo, I., Stack, C., Kong, L., Musaro, A., Adachi, H., Katsuno, K., Sobue, G., Taylor, J.P., Sumner, C.J., Fischbeck, K.H., Pennuto, M. (2009). Overexpression of IGF-1 in muscle attenuates disease in a mouse model of spinal and bulbar muscular atrophy. *Neuron, 63*, 316–328.

72. Vig, P.J., Subramony, S.H., D'Souza, D.R., Wei, J., Lopez, M.E. (2006). Intranasal administration of IGF-I improves behavior and Purkinje cell pathology in SCA1 mice. *Brain Res Bull, 69*, 573–579.

73. Kahlem, P., Green, H., Djian, P. (1998). Transglutaminase action imitates Huntington's disease, selective polymerization of Huntingtin containing expanded polyglutamine. *Mol Cell, 1*, 595–601.

74. Mandrusiak, L.M., Beitel, L.K., Wang, X., Scanlon, T.C., Chevalier-Larsen, E., Merry, D.E., Trifiro, M.A. (2003). Transglutaminase potentiates ligand-dependent proteasome dysfunction induced by polyglutamine-expanded androgen receptor. *Hum Mol Genet, 12*, 1497–1506.

75. D'Souza, D.R., Wei, J., Shao, Q., Hebert, M.D., Subramony, S.H., Vig, P.J. (2006). Tissue transglutaminase crosslinks ataxin-1, possible role in SCA1 pathogenesis. *Neurosci Lett, 409*, 5–9.

76. Lesort, M., Chun, W., Johnson, G.V., Ferrante, R.J. (1999). Tissue transglutaminase is increased in Huntington's disease brain. *J Neurochem, 73*, 2018–2027.

77. Karpuj, M.V., Garren, H., Slunt, H., Price, D.L., Gusella, J., Becher, M.W., Steinman, L. (1999). Transglutaminase aggregates huntingtin into nonamyloidogenic polymers, and its enzymatic activity increases in Huntington's disease brain nuclei. *Proc Natl Acad Sci U S A, 96*, 7388–7393.

78. Van Raamsdonk, J.M., Pearson, J., Bailey, C.D., Rogers, D.A., Johnson, G.V., Hayden, M.R., Leavitt, B.R. (2005). Cystamine treatment is neuroprotective in the YAC128 mouse model of Huntington disease. *J Neurochem, 95*, 210–220.

79. Dedeoglu, A., Kubilus, J.K., Jeitner, T.M., Matson, S.A., Bogdanov, M., Kowall, N.W., Matson, W.R., Cooper, A.J., Ratan, R.R., Beal, M.F., Hersch, S.M., Ferrante, R.J. (2002). Therapeutic effects of cystamine in a murine model of Huntington's disease. *J Neurosci, 22*, 8942–8950.

80. Mastroberardino, P.G., Iannicola, C., Nardacci, R., Bernassola, F., De Laurenzi, V., Melino, G., Moreno, S., Pavone, F., Oliverio, S., Fesus, L., Piacentini, M. (2002). 'Tissue' transglutaminase ablation reduces neuronal death and prolongs survival in a mouse model of Huntington's disease. *Cell Death Differ, 9*, 873–880.

81. Igarashi, S., Koide, R., Shimohata, T., Yamada, M., Hayashi, Y., Takano, H., Date, H., Oyake, M., Sato, T., Sato, A., Egawa, S., Ikeuchi, T., Tanaka, H., Nakano, R., Tanaka, K., Hozumi, I., Inuzuka, T., Takahashi, H., Tsuji, S. (1998). Suppression of aggregate formation and apoptosis by transglutaminase inhibitors in cells expressing truncated DRPLA protein with an expanded polyglutamine stretch. *Nat Genet, 18*, 111–117.

82. Lesort, M., Lee, M., Tucholski, J., Johnson, G.V. (2003). Cystamine inhibits caspase activity. Implications for the treatment of polyglutamine disorders. *J Biol Chem, 278*, 3825–3830.

83. Wei, H., Qin, Z.H., Senatorov, V.V., Wei, W., Wang, Y., Qian, Y., Chuang, D.M. (2001). Lithium suppresses excitotoxicity-induced striatal lesions in a rat model of Huntington's disease. *Neurosci, 106*, 603–612.

84. Carmichael, J., Sugars, K.L., Bao, Y.P., Rubinsztein, D.C. (2002). Glycogen synthase kinase-3beta inhibitors prevent cellular polyglutamine toxicity caused by the Huntington's disease mutation. *J Biol Chem, 277*, 33791–33798.

85. Wood, N.I., Morton, A.J. (2003). Chronic lithium chloride treatment has variable effects on motor behaviour and survival of mice transgenic for the Huntington's disease mutation. *Brain Res Bull, 61*, 375–383.

86. Gu, M., Gash, M.T., Mann, V.M., Javoy-Agid, F., Cooper, J.M., Schapira, A.H. (1996). Mitochondrial defect in Huntington's disease caudate nucleus. *Ann Neurol, 39*, 385–389.

87. Browne, S.E., Bowling, A.C., MacGarvey, U., Baik, M.J., Berger, S.C., Muqit, M.M., Bird, E.D., Beal, M.F. (1997). Oxidative damage and metabolic dysfunction in Huntington's disease, selective vulnerability of the basal ganglia. *Ann Neurol, 41*, 646–653.

88. Ferrante, R.J., Andreassen, O.A., Jenkins, B.G., Dedeoglu, A., Kuemmcrle, S., Kubilus, J.K., Kaddurah-Daouk, R., Hersch, S.M., Beal, M.F. (2000). Neuroprotective effects of creatine in a transgenic mouse model of Huntington's disease. *J Neurosci, 20*, 4389–4397.

89. Dedeoglu, A., Kubilus, J.K., Yang, L., Ferrante, K.L., Hersch, S.M., Beal, M.F., Ferrante, R.J. (2003). Creatine therapy provides neuroprotection after onset of clinical symptoms in Huntington's disease transgenic mice. *J Neurochem, 85*, 1359–1367.

90. Schmidt, B.J., Greenberg, C.R., Allingham-Hawkins, D.J., Spriggs, E.L. (2002). Expression of X-linked bulbospinal muscular atrophy (Kennedy disease) in two homozygous women. *Neurology, 59*, 770–772.

91. Katsuno, M., Adachi, H., Kume, A., Li, M., Nakagomi, Y., Niwa, H., Sang, C., Kobayashi, Y., Doyu, M., Sobue, G. (2002). Testosterone reduction prevents phenotypic expression in a transgenic mouse model of spinal and bulbar muscular atrophy. *Neuron, 35*, 843–854.

92. Chevalier-Larsen, E.S., O'Brien, C.J., Wang, H., Jenkins, S.C., Holder, L., Lieberman, A.P., Merry, D.E. (2004). Castration restores function and neurofilament alterations of aged symptomatic males in a transgenic mouse model of spinal and bulbar muscular atrophy. *J Neurosci, 24*, 4778–4786.

93. Katsuno, M., Adachi, H., Doyu, M., Minamiyama, M., Sang, C., Kobayashi, Y., Inukai, A., Sobue, G. (2003). Leuprorelin rescues polyglutamine-dependent phenotypes in a transgenic mouse model of spinal and bulbar muscular atrophy. *Nat Med, 9*, 768–773.

94. Banno, H., Katsuno, M., Suzuki, K., Takeuchi, Y., Kawashima, M., Suga, N., Takamori, M., Ito, M., Nakamura, T., Matsuo, K., Yamada, S., Oki, Y., Adachi, H., Minamiyama, M., Waza, M., Atsuta, N., Watanabe, H., Fujimoto, Y., Nakashima, T., Tanaka, F., Doyu, M., Sobue, G. (2009). Phase 2 trial of leuprorelin in patients with spinal and bulbar muscular atrophy. *Annual Neurology, 65*, 140–150.

95. Yang, Z., Chang, Y.J., Yu, I.C., Yeh, S., Wu, C.C., Miyamoto, H., Merry, D.E., Sobue, G., Chen, L.M., Chang, S.S., Chang, C. (2007). ASC-J9 ameliorates spinal and bulbar muscular atrophy phenotype via degradation of androgen receptor. *Nat Med, 13*, 348–353.

96. Watase, K., Gatchel, J.R., Sun, Y., Emamian, E., Atkinson, R., Richman, R., Mizusawa, H., Orr, H.T., Shaw, C., Zoghbi, H.Y. (2007). Lithium therapy improves neurological function and hippocampal dendritic arborization in a spinocerebellar ataxia type 1 mouse model. *PLoS Med, 4*, e182.

PART V

APPROACHES FOR NEW AND EMERGING THERAPIES

41

CHEMISTRY AND BIOLOGY OF AMYLOID INHIBITION

Mark A. Findeis

Satori Pharmaceuticals Incorporated, Cambridge, Massachusetts

INTRODUCTION

Amyloid and amyloidlike diseases are characterized by the pathological accumulation of aggregated, insoluble, and fibrillar protein deposits associated with cellular and organ dysfunction. These deposits are classically distinguished by their tinctorial properties, including the birefringent red–green staining produced with Congo Red and the fluorescent dye shifts observed with thioflavins [1]. The most extensively studied of these diseases is Alzheimer disease (AD), which is associated with the aggregation and deposition of amyloid beta-peptide (A-beta, Aβ). Early efforts to identify disease-modifying therapeutic approaches to treat these diseases focused on the idealized goal of preventing the formation of the pathological deposits. Progress in understanding the process of amyloidogenic peptide formation, followed by subsequent aggregation and deposition, has resulted in a refined focus on earlier events in aggregation in which still soluble forms of aggregated amyloidogenic peptide and protein are mediating cellular toxicity [2,3]. These observations demonstrating the pathological significance of soluble amyloid oligomers emphasize their importance when considering how to screen for and characterize potential inhibitors of amyloidosis.

Therapeutic approaches to mechanism-based disease-modifying intervention in amyloid diseases include at least four broad approaches: block the production of the amyloidogenic peptide or protein, block its "misfolding" or

Protein Misfolding Diseases: Current and Emerging Principles and Therapies,
Edited by Marina Ramirez-Alvarado, Jeffery W. Kelly, and Christopher M. Dobson
Copyright © 2010 John Wiley & Sons, Inc.

transformation from a nonpathogenic monomer or low-oligomer to toxic oligomers and polymers, block the toxic effects of amyloid, or modulate an auxiliary cellular pathway in a manner that affects beneficially one or more of the foregoing approaches. Depending on the target disease and the level of knowledge about the production, processing, and clearance of the target peptide or protein, each of the approaches above will have varying levels of tractability. In the case of AD and Aβ, sufficient research has been accomplished to demonstrate all of these approaches at least at the cellular level, and the same is being done for other diseases.

Because the amyloid diseases are diseases in which a normal and known, or presumably, beneficial (if not fully understood) peptide in the monomeric or low-oligomeric state is transformed through conformational change and polymerization into toxic high oligomers and pathological deposits, they are a toxic gain-of-function process. The most specific intervention in the amyloid cascade would thus be to prevent the conversion of normal nonpathological forms of peptide to those that are toxic. In this chapter we discuss these approaches, with a particular emphasis on Aβ, where at present there is the largest and most varied body of research in these methods.

MACROMOLECULAR INHIBITORS OF AMYLOID FORMATION

A large variety of macromolecules have been described that modulate the aggregation of amyloidogenic proteins. The complex biological milieu in vivo provides many different macromolecules that have been demonstrated to interact with one or more amyloidogenic proteins. Many proteins are observed associated with amyloid deposits. Whether these observations reflect an interaction that happens before aggregation and deposition, and whether the interaction is inhibitory or promoting with respect to aggregation, is not always clear. For proteins associated with deposits, even if a specific interaction, we do not typically have firm data on whether the additional protein has any influence on pathology. With this general caveat in mind, there are three types of macromolecules that have received considerable attention as modulators of amyloid formation: antibodies against Aβ, glycoproteins, and apolipoprotein E (ApoE). In each case the molecular-level details of how these macromolecules interact with their target are not well defined. But, in turn, new opportunities have emerged that have added to efforts to discover small-molecule antagonists of amyloidosis.

ANTIBODIES AND IMMUNOTHERAPY

The most advanced disease-modifying approach to treating a major amyloid disease is that of using immunotherapy for AD. This approach has progressed to clinical trials in humans based on compelling results in animal models. Early research demonstrated that antibodies against Aβ could inhibit its aggregation in vitro [4,5]. Subsequently, active vaccination of transgenic animals with Aβ in various forms and passive vaccination (administration of previously made antibody) were shown to provide apparent benefits in animals with respect to amyloid deposition and pathology [6] as well as cognitive performance [7,8]. Progress in the clinic has been slowed due to side effects observed in the first trial, but many variations of how to design vaccines and engineer antibodies have been the subject of ongoing research, and revised immunotherapies will continue to enter development.

From a mechanistic point of view, the antagonism of Aβ amyloidosis by antibodies thoroughly validates the concept that misfolding of amyloidogenic peptides can be inhibited by the binding of a structural element (e.g., the relatively large antibody) that sterically interferes with conformational changes and/or subsequent polymerization processes. In addition, binding of antibodies is expected to promote clearance of its target peptide through preservation of its solubility and prevention from deposition.

APOLIPOPROTEIN E

The first significant risk factor identified for acquisition of sporadic late-onset AD, in addition to age, was related to allelic variation of ApoE. Individuals homozygous for ApoE4 were found to be more likely to acquire AD than those with genes for ApoE2 and ApoE3 [9]. Upon observing that ApoE genetics were related to risk of AD, ApoE was found to interact with Aβ [10]. Further, ApoE was observed to be an inhibitor of Aβ aggregation, with ApoE4 being less effective than ApoE3 [11]. From this point of view, ApoE4 may not be so much a positive contributor to risk for acquiring AD as a less effective protectant from AD. Whether direct interactions of ApoE with Aβ, or its role in lipid metabolism, represent phenomena that can be exploited for treatment of AD remains to be determined. ApoE and a number of other proteins are also associated with interactions with Aβ and APP processing that have been studied and may present additional opportunities for drug discovery [12].

SMALL-MOLECULE INHIBITORS OF AMYLOIDOSIS

A variety of relatively low molecular weight inhibitors of amyloid aggregation have been described. For the purposes of the present discussion, these compounds can be described as derived from a range of molecular types, that including peptides, carbohydrates, and glycoproteins (and related compounds), and organic compounds.

Peptides

Early experiments in which site-specific changes in the sequence of Aβ were made using chemical synthesis identified sites in the Aβ sequence that appeared to be quite important in promoting or allowing folding of the peptide into a form that would allow amyloidosis [13–15]. In particular, the Phe–Phe dipeptide at position 19–20 in Aβ was identified as important. Long Aβ peptides in which these two residues were changed to smaller nonaromatic residues were found to be less competent to form amyloid fibrils and also to act as inhibitors of the fibrillization of wild-type Aβ. The suggestion was made that such nonamyloidogenic long peptides might even have potential therapeutic use [14]. Subsequent experiments by Tjernberg et al., [16] used an iterative process of spot synthesis of short peptides and binding to Aβ-identified short peptides based on Aβ(16–21), KLVFF, with high affinity for Aβ and the ability to act as inhibitors. Soto developed the concept of "beta-sheet breaker" peptides based on the inclusion of a prolyl residue which was reported to improve inhibitory activity ([17]; and see Chapter 43), although this series of compounds has been cited as less active than related nonprolyl peptides in some assays [18–20]. Somewhat related approaches were undertaken by others using α, α-dialkyl amino acids [21] and *N*-methyl peptides to achieve similar effects [18,22]. In the same time frame, the Praecis Pharmaceuticals group had already undertaken systematic studies of these and other approaches to the adaptation of small peptides as inhibitors of amyloidosis and identified alternative preferred approaches. Initially their main approaches were based on organic-modified peptides of various lengths [23–26], and with further optimization, the identification of sequence-modified "enhanced" Aβ peptide derivatives with potent antifibrillization activity [19,20,27]. An extensive series of inverso (all-D) and retro-inverso (all-D reverse sequence) peptides included a compound numbered PPI-1019, *N*-methyl-D-Leu-D-Val-D-Phe-D-Phe-D-Leu-amide, which progressed through phase 1 clinical trials [27]. While the intention to begin phase II trials in AD patients was announced by Praecis, this effort was put on hold, apparently due to business-restructuring issues.

Glycoprotein Mimetics

Identification of glycoproteins and glycosaminoglycans (GAGs) in AD plaques in association with Aβ prompted further studies that found that these proteins and polysaccharides could bind to Aβ with high affinity [28]. These observations led to the development GAG-mimetics which inhibit amyloidosis [29,30]. The most advanced examples of this approach are experimental drugs being developed by Neurochem, including tramiprosate (3-aminopropanesulfonic acid) for AD and eprodisate (propanedisulfonic acid) for amyloid A amyloidosis. Both compounds have been the subject of phase III clinical trials and are being considered for marketing approval. Less than robust clinical data may require further study to gain approval but do suggest that aggregation inhibitors have clinical promise.

In other work based on the study of phosphatidylinositols, neutral inositols were found to inhibit aggregation [31]. Quite notably, while peptide inhibitors are reported to maintain Aβ in a monomeric state, inositols were found to stabilize an oligomeric form of Aβ which was nontoxic to cells. Similarly to the GAG mimetics, these sorts of inhibitors does not appear to be as potent as the high-affinity peptide inhibitors, but appear to be very safe, very soluble, and therefore acceptable in profile for clinical development [32,33].

A related polyol, the disaccharide trehalose ((1-α-D-glucopyranosyl-1-β-D-glucopyranoside) is also reported to inhibit Aβ aggregation [34]. Careful study of this inhibitor revealed that it was less potent for Aβ42 relative to the less amyloidogenic Aβ40. Importantly, inhibition was retained for mixtures of Aβ40 and Aβ42.

Small Organics

Early screening efforts to identify inhibitors of Aβ aggregation from libraries of organic compounds appear to have been undertaken by a number of pharmaceutical companies as well as academic labs, starting in the early 1990s if not earlier. A general result of these efforts is that a variety of compounds (reviewed elsewhere, and beyond the scope of this chapter) with activity were described, but none proved robust enough in potency or drugability to reach clinical development. A particular challenge in early screens was the rather high concentration of Aβ used in assays and a resulting difficulty in discerning whether apparently potent compounds were merely "good" or "excellent" and supportive of further work. Persistence on the part of experienced investigators and those new to the field have provided more recent progress, however, often from interesting and divergent starting points. Among the compounds that have been reported with direct or indirect inhibitory activities are curcumin [35] and the structurally related ferulic acid [36], methylene blue [37], aryl amino acid derivatives [38], naphthalene sulfonates [39], and metal chelators [40]. In addition to histological dyes such as Congo Red and the thioflavins T and S, a general structural theme among these compounds is some degree of aromaticity and hydroxylation or other heteroatomic elaboration [41] presumably to support a mixture of hydrophobic interactions and hydrogen bonding, whose specificity is not yet elucidated.

MECHANISMS OF ACTION

Attempts to understand the mechanisms of action of any of these inhibitors are limited by the level of understanding amyloid structure, particularly in the context of protein–inhibitor complexes. Sophisticated structural studies of Aβ under varying solvent conditions or polymer state reveal very different structures (Fig.1) [42–45]. The only clear detail in this regard is that from studies of transthyretin in which a stable native tetramer provides a cavity in which stabilizers of the native state can be studied by protein structural methods, including crystallography (see Chapter 45). For those amyloidoses where a stable native state is not available for study, one is left to infer about

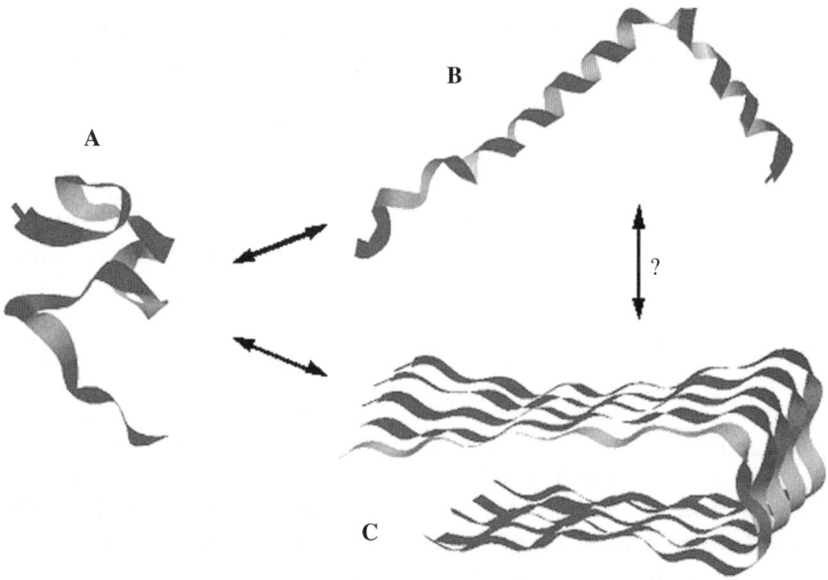

FIG. 1 Dynamic structure of amyloid beta peptide in different states of environment and self-assembly (PDB file number, reference): (A) monomer as collapsed coil in water (1HZ3, [42]); (B) highly helical monomer in 80% hexafluoro isopropano l–20%water (1IYT, [43]); (C) fibrillar Aβ (2BEG, [44]; see also [45]). Not shown is a representation of toxic soluble oligomeric amyloid.

the mechanism based on structure–activity relationships (SARs) of different related compounds and differences in the effects of inhibitors at different stages of the amyloid cascade.

In the case of the peptide-based inhibitors of Aβ aggregation, there is a decent level of experimental data to suggest that a particular site of binding is the hydrophobic core of Aβ at positions 17 to 21, Leu–Val–Phe–Phe–Ala or (LVFFA). The affinities between synthetic peptides of varying sequence and Aβ [16] and the SAR of "enhanced" inverso and retro-inverso peptidic inhibitors of Aβ aggregation ([27] and references therein) support a model in which peptidic inhibitors bind to an amyloidogenic target in which the target has already (or, concomitantly) adopted a beta-structured state but is not yet in a toxic oligomeric or fibrillar form. Studies of the solubility and conformational properties of mixtures of D-and L-polyamino acids which show low solubility [46] and transition to beta-sheet-rich structure [47] are consistent with these observations. More recent studies suggest that the thermodynamic driving force for the stability of "diastereomeric" beta-structured peptide complexes is a greater reduction of hydration and a corresponding savings in entropy [48].

Evidence for a strong interaction of inhibitors with beta structure is seen in the use of a fibril-binding assay which can document subnanomolar dissociation constants of potent peptidic inhibitors [19,20]. Given the efficacy of such compounds to block fibrillogenesis and maintain Aβ in a monomeric or low-oligomeric state [26], it suggests that even in a prefibrillar nontoxic state, a completely or substantially beta-structured binding site for the inhibitor may exist.

Attention to both biophysical studies and cellular toxicity are important measures to be used together to evaluate the properties of potential inhibitors. As noted above, compounds such as the inositols can stabilize a non toxic but oligomeric form of Aβ. By some screening approaches, the presence of oligomer would disqualify a compound as not useful. By also examining cellular toxicity, an uncommon pairing of properties was noted. By contrast, naphthalene sulfonates such as bis-ANS also stabilize low oligomers of Aβ, but these oligomers are toxic [39]. Other compounds are able to inhibit fibrillization but not oligomerization [49]. Although such compounds might slow accumulation of amyloid deposits, they might not have any impact on acute toxicity associated with soluble oligomers and in fact might exacerbate it. In the case of trehalose, its potency-inhibiting Aβ40 aggregation contrasts with it much reduced potency against Aβ42. Whether the lesser potency against Aβ42 represents a liability that would keep such a compound from working therapeutically remains to be determined. In the case of AD and Aβ, a question such as this becomes quite important. When mixed with Aβ42, Aβ40 inhibits the aggregation of Aβ42 [50]. If an inhibitor were somehow to be selective in a manner that reduced the ratio of Aβ40 to Aβ42 (e.g., by a mechanism involving binding and facilitated clearance), such an effect might actually be promoting of disease [12].

SUMMARY

A variety of inhibitors of aggregation and fibrillization of amyloidogenic proteins have been described. Limited molecular detail is available to fully describe the mechanism of action of these compounds, except in special cases. Fully profiling the characteristics of compounds through SAR studies and multiple forms of biophysical, cellular, and in vivo assays is essential to fully correlate interactions of a compound with its target and any associated potential therapeutic benefit. Past and current progress is now bringing amyloid inhibitors into clinical trials and the ability of discovery techniques to provide drug candidates that will benefit patients is being addressed.

REFERENCES

1. LeVine, H., 3rd. (1993). Thioflavine T interaction with synthetic Alzheimer's disease beta-amyloid peptides: detection of amyloid aggregation in solution. *Protein Sci, 2*, 404–410.
2. Klein, W.L., Krafft, G.A., Finch, C.E. (2001). Targeting small Abeta oligomers: the solution to an Alzheimer's disease conundrum? *Trends Neurosci, 24*, 219–224.

3. Walsh, D.M., Klyubin, I., Fadeeva, J.V., Cullen, W.K., Anwyl, R., Wolfe, M.S., Rowan, M.J., Selkoe, D.J. (2002). Naturally secreted oligomers of amyloid beta protein potently inhibit hippocampal long-term potentiation in vivo. *Nature, 416*, 535–539.
4. Solomon, B., Koppel, R., Hanan, E., Katzav, T. (1996). Monoclonal antibodies inhibit in vitro fibrillar aggregation of the Alzheimer beta-amyloid peptide. *Proc Natl Acad Sci U S A, 93*, 452–455.
5. Solomon, B., Koppel, R., Frankel, D., Hanan-Aharon, E. (1997). Disaggregation of Alzheimer beta-amyloid by site-directed mAb. *Proc Natl Acad Sci U S A, 94*, 4109–4112.
6. Schenk, D., Barbour, R., Dunn, W., Gordon, G., Grajeda, H., Guido, T., Hu, K., Huang, J., Johnson-Wood, K., Khan, K., et al. (1999). Immunization with amyloid-beta attenuates Alzheimer-disease-like pathology in the PDAPP mouse. *Nature, 400*, 173–177.
7. Janus, C., Pearson, J., McLaurin, J., Mathews, P.M., Jiang, Y., Schmidt, S.D., Chishti, M.A., Horne, P., Heslin, D., French, J., et al. (2000). A beta peptide immunization reduces behavioural impairment and plaques in a model of Alzheimer's disease. *Nature, 408*, 979–982.
8. Morgan, D., Diamond, D.M., Gottschall, P.E., Ugen, K.E., Dickey, C., Hardy, J., Duff, K., Jantzen, P., DiCarlo, G., Wilcock, D., et al. (2000). A beta peptide vaccination prevents memory loss in an animal model of Alzheimer's disease. *Nature, 408*, 982–985.
9. Strittmatter, W.J., Saunders, A.M., Schmechel, D., Pericak-Vance, M., Enghild, J., Salvesen, G.S., Roses, A.D. (1993). Apolipoprotein E: high-avidity binding to beta-amyloid and increased frequency of type 4 allele in late-onset familial Alzheimer disease. *Proc Natl Acad Sci U S A, 90*, 1977–1981.
10. Wisniewski, T., Golabek, A., Matsubara, E., Ghiso, J., Frangione, B. (1993). Apolipoprotein E: binding to soluble Alzheimer's beta-amyloid. *Biochem Biophys Res Commun, 192*, 359–365.
11. Evans, K.C., Berger, E.P., Cho, C.G., Weisgraber, K.H., Lansbury, P.T., Jr. (1995). Apolipoprotein E is a kinetic but not a thermodynamic inhibitor of amyloid formation: implications for the pathogenesis and treatment of Alzheimer disease. *Proc Natl Acad Sci U S A, 92*, 763–767.
12. Findeis, M.A. (2007). The role of amyloid beta peptide 42 in Alzheimer's disease. *Pharmacol Ther, 116*, 266–286.
13. Hilbich, C., Kisters-Woike, B., Reed, J., Masters, C.L., Beyreuther, K. (1991). Aggregation and secondary structure of synthetic amyloid beta A4 peptides of Alzheimer's disease. *J Mol Biol, 218*, 149–163.
14. Hilbich, C., Kisters-Woike, B., Reed, J., Masters, C.L., Beyreuther, K. (1992). Substitutions of hydrophobic amino acids reduce the amyloidogenicity of Alzheimer's disease beta A4 peptides. *J Mol Biol, 228*, 460–473.
15. Wood, S.J., Wetzel, R., Martin, J.D., Hurle, M.R. (1995). Prolines and amyloidogenicity in fragments of the Alzheimer's peptide beta/A4. *Biochemistry, 34*, 724–730.
16. Tjernberg, L.O., Naslund, J., Lindqvist, F., Johansson, J., Karlstrom, A.R., Thyberg, J., Terenius, L., Nordstedt, C. (1996). Arrest of beta-amyloid fibril formation by a pentapeptide ligand. *J Biol Chem, 271*, 8545–8548.

17. Soto, C., Kindy, M.S., Baumann, M., Frangione, B. (1996). Inhibition of Alzheimer's amyloidosis by peptides that prevent beta-sheet conformation. *Biochem Biophys Res Commun, 226*, 672–680.

18. Kokkoni, N., Stott, K., Amijee, H., Mason, J.M., Doig, A.J. (2006). N-Methylated peptide inhibitors of beta-amyloid aggregation and toxicity. Optimization of the inhibitor structure. *Biochemistry, 45*, 9906–9918.

19. Findeis, M.A., Phillips, K., Olson, G.L., Self, C. (2003). Modulators of beta-amyloid peptide aggregation. U.S. patent 6,610,658, Praecis Pharmaceuticals Incorporated.

20. Findeis, M.A., Phillips, K. (2004). Modulators of beta-amyloid peptide aggregation. U.S. patent 6,831,066, Praecis Pharmaceuticals Incorporated.

21. Etienne, M.A., Aucoin, J.P., Fu, Y., McCarley, R.L., Hammer, R.P. (2006). Stoichiometric inhibition of amyloid beta-protein aggregation with peptides containing alternating alpha,alpha-disubstituted amino acids. *J Am Chem Soc, 128*, 3522–3523.

22. Gordon, D.J., Tappe, R., Meredith, S.C. (2002). Design and characterization of a membrane permeable N-methyl amino acid–containing peptide that inhibits Abeta1-40 fibrillogenesis. *J Pept Res, 60*, 37–55.

23. Findeis, M.A., Molineaux, S.M. (1998). Discovery and characterization of peptidoorganic inhibitors of amyloid beta-peptide polymerization. In *Advances in Behavioral Biology,* Vol. 49 (Fisher, A., Hanin, I., Yoshida, M., Eds.), Plenum Press, New York, pp. 191–195.

24. Findeis, M.A., Molineaux, S.M. (1999). Design and testing of inhibitors of fibril formation. In *Methods in Enzymology,* Vol. 309 (Wetzel, R., Ed.), Academic Press, San Diego, CA, pp. 476–488.

25. Findeis, M.A., Musso, G.M., Arico-Muendel, C.C., Benjamin, H.W., Hundal, A.M., Lee, J.J., Chin, J., Kelley, M., Wakefield, J., Hayward, N.J., Molineaux, S.M. (1999). Modified-peptide inhibitors of amyloid beta-peptide polymerization. *Biochemistry, 38*, 6791–6800.

26. Findeis, M.A., Lee, J.J., Kelley, M., Wakefield, J.D., Zhang, M.H., Chin, J., Kubasek, W., Molineaux, S.M. (2001). Characterization of cholyl-leu-val-phe-phe-ala-OH as an inhibitor of amyloid beta-peptide polymerization. *Amyloid, 8*, 231–241.

27. Findeis, M.A. (2002). Peptide inhibitors of beta amyloid aggregation. *Curr Top Med Chem, 2*, 417–423.

28. Narindrasorasak, S., Lowery, D., Gonzalez-DeWhitt, P., Poorman, R.A., Greenberg, B., Kisilevsky, R. (1991). High affinity interactions between the Alzheimer's beta-amyloid precursor proteins and the basement membrane form of heparan sulfate proteoglycan. *J Biol Chem, 266*, 12878–12883.

29. Kisilevsky, R., Lemieux, L.J., Fraser, P.E., Kong, X., Hultin, P.G., Szarek, W.A. (1995). Arresting amyloidosis in vivo using small-molecule anionic sulphonates or sulphates: implications for Alzheimer's disease. *Nat Med, 1*, 143–148.

30. Kisilevsky, R., Szarek, W., Weaver, D. (1999). Method using anionic group-containing compounds for treating amyloidosis. U.S. patent 5,972,328, Queen's University, Canada.

31. McLaurin, J., Golomb, R., Jurewicz, A., Antel, J.P., Fraser, P.E. (2000). Inositol stereoisomers stabilize an oligomeric aggregate of Alzheimer amyloid beta peptide and inhibit abeta-induced toxicity. *J Biol Chem, 275*, 18495–18502.

32. McLaurin, J., Kierstead, M.E., Brown, M.E., Hawkes, C.A., Lambermon, M.H., Phinney, A.L., Darabie, A.A., Cousins, J.E., French, J.E., Lan, M.F., et al. (2006). Cyclohexanehexol inhibitors of Abeta aggregation prevent and reverse Alzheimer phenotype in a mouse model. *Nat Med, 12*, 801–808.
33. Fenili, D., Brown, M., Rappaport, R., McLaurin, J. (2007). Properties of scyllo-inositol as a therapeutic treatment of AD-like pathology. *J Mol Med, 85*, 603–611.
34. Liu, R., Barkhordarian, H., Emadi, S., Park, C.B., Sierks, M.R. (2005). Trehalose differentially inhibits aggregation and neurotoxicity of beta-amyloid 40 and 42. *Neurobiol Dis, 20*, 74–81.
35. Yang, F., Lim, G.P., Begum, A.N., Ubeda, O.J., Simmons, M.R., Ambegaokar, S.S., Chen, P.P., Kayed, R., Glabe, C.G., Frautschy, S.A., Cole, G.M. (2005). Curcumin inhibits formation of amyloid beta oligomers and fibrils, binds plaques, and reduces amyloid in vivo. *J Biol Chem, 280*, 5892–5901.
36. Ono, K., Hirohata, M., Yamada, M. (2005). Ferulic acid destabilizes preformed beta-amyloid fibrils in vitro. *Biochem Biophys Res Commun, 336*, 444–449.
37. Necula, M., Breydo, L., Milton, S., Kayed, R., van der Veer, W.E., Tone, P., Glabe, C.G. (2007). Methylene blue inhibits amyloid Abeta oligomerization by promoting fibrillization. *Biochemistry, 46*, 8850–8860.
38. Cohen, T., Frydman-Marom, A., Rechter, M., Gazit, E. (2006). Inhibition of amyloid fibril formation and cytotoxicity by hydroxyindole derivatives. *Biochemistry, 45*, 4727–4735.
39. Ferrao-Gonzales, A.D., Robbs, B.K., Moreau, V.H., Ferreira, A., Juliano, L., Valente, A.P., Almeida, F.C., Silva, J.L., Foguel, D. (2005). Controlling {beta}-amyloid oligomerization by the use of naphthalene sulfonates: trapping low molecular weight oligomeric species. *J Biol Chem, 280*, 34747–34754.
40. Cuajungco, M.P., Frederickson, C.J., Bush, A.I. (2005). Amyloid-beta metal interaction and metal chelation. *Subcell Biochem, 38*, 235–254.
41. Findeis, M.A. (2000). Approaches to discovery and characterization of inhibitors of amyloid beta-peptide polymerization. *Biochim Biophys Acta, 1502*, 76–84.
42. Zhang, S., Iwata, K., Lachenmann, M.J., Peng, J.W., Li, S., Stimson, E.R., Lu, Y., Felix, A.M., Maggio, J.E., Lee, J.P. (2000). The Alzheimer's peptide a beta adopts a collapsed coil structure in water. *J Struct Biol, 130*, 130–141.
43. Crescenzi, O., Tomaselli, S., Guerrini, R., Salvadori, S., D'Ursi, A.M., Temussi, P.A., Picone, D. (2002). Solution structure of the Alzheimer amyloid beta-peptide (1-42) in an apolar microenvironment. Similarity with a virus fusion domain. *Eur J Biochem, 269*, 5642–5648.
44. Luhrs, T., Ritter, C., Adrian, M., Riek-Loher, D., Bohrmann, B., Dobeli, H., Schubert, D., Riek, R. (2005). 3D structure of Alzheimer's amyloid-beta(1–42) fibrils. *Proc Natl Acad Sci U S A, 102*, 17342–17347.
45. Petkova, A.T., Yau, W.M., Tycko, R. (2006). Experimental constraints on quaternary structure in Alzheimer's beta-amyloid fibrils. *Biochemistry, 45*, 498–512.
46. Yoshida, T., Sakurai, S., Okuda, T., Takagi, Y. (1962). A racemic compound of alpha-helical polypeptide. *J Am Chem Soc, 84*, 3590–3591.
47. Fuhrhop, J.-H., Krull, M., Bueldt, G. (1987). Precipitates with beta -pleated sheet structure by mixing aqueous solutions of helical poly(D-lysine) and poly(L-lysine). *Angew Chem, 99*, 707–708.

48. Dzwolak, W., Ravindra, R., Nicolini, C., Jansen, R., Winter, R. (2004). The diastereomeric assembly of polylysine is the low-volume pathway for preferential formation of beta-sheet aggregates. *J Am Chem Soc, 126*, 3762–3768.

49. Necula, M., Kayed, R., Milton, S., Glabe, C.G. (2007). Small molecule inhibitors of aggregation indicate that amyloid beta oligomerization and fibrillization pathways are independent and distinct. *J Biol Chem, 282*, 10311–10324.

50. Snyder, S.W., Ladror, U.S., Wade, W.S., Wang, G.T., Barrett, L.W., Matayoshi, E.D., Huffaker, H.J., Krafft, G.A., Holzman, T.F. (1994). Amyloid-beta aggregation: selective inhibition of aggregation in mixtures of amyloid with different chain lengths. *Biophys J, 67*, 1216–1228.

42

IMMUNOTHERAPY IN SECONDARY AND LIGHT-CHAIN AMYLOIDOSIS

JONATHAN WALL

Human Immunology and Cancer Program, Preclinical and Diagnostic Molecular Imaging Laboratory, University of Tennessee Graduate School of Medicine, Knoxville, Tennessee

INTRODUCTION

The goal of immunotherapy is to modulate the immune system to achieve a prophylactic or therapeutic response to pathogens (virus or bacteria), toxins (tetanus or diphtheria), or cancerous tumors [1,2]. With respect to amyloidosis, the goal is generally to effect amyloid removal and thereby facilitate organ recovery. Active immunotherapy (or immunization) stimulates the host's intrinsic immune system to mount a response to antigenic material that will ideally prevent or ameliorate the effects of the pathogenic, toxic, or cancerous insult. In contrast, passive (or adoptive) immunotherapy is achieved by transferring either antibodies or lymphocytes as a means of providing transient protection against the target.

Immunotherapy has been used historically to provide prophylactic protection against acute infectious diseases caused by pathogens or their secreted toxins; however, these approaches have now been applied effectively to the treatment of cancer [2] and the removal of Aβ-associated amyloid plaques from patients with Alzheimer disease (AD) [3]. Indeed, both passive and active immunization have been evaluated in clinical trials as methods to remove cerebral Aβ amyloid in patients with AD. Although these trials have experienced setbacks associated with a fatal T-cell-mediated cerebral inflammation in 6% of the participants,

Protein Misfolding Diseases: Current and Emerging Principles and Therapies,
Edited by Marina Ramirez-Alvarado, Jeffery W. Kelly, and Christopher M. Dobson
Copyright © 2010 John Wiley & Sons, Inc.

they have also provided encouraging data, in that many patients achieved a reduction in amyloid burden and a concomitant increase in organ function (i.e., cognitive ability) [4,5].

The aim of this chapter is to extend the discussion of immunotherapy for amyloid disease beyond AD and to introduce novel strategies and reagents that have been developed and evaluated preclinically as interventions for the most common forms of peripheral (noncerebral) amyloidosis. Light-chain (AL) and secondary (AA) amyloidoses are systemic diseases associated with the deposition of immunoglobulin light chains (LCs) and serum amyloid protein A (sAA), respectively. Both are characterized by generally widespread (multiorgan), often massive deposits of amyloid within the viscera as well as high concentrations of precursor protein in the circulation [6–8]. These features, which are common to many of the peripheral amyloidoses and define systemic forms of the disease, provide unique challenges in the design and implementation of effective immunotherapies.

IMMUNOTHERAPY FOR AMYLOID DISEASES

Amyloid deposits are generally extracellular matrices that consist of fibrils derived from normally innocuous, functional proteins and accessory molecules such as serum amyloid P component, apolipoproteins, and heparan sulfate proteoglycans [9–11]. Despite the structural and functional heterogeneity of peripheral amyloid precursor proteins, amyloid fibrils exhibit a remarkable ultrastructural homology irrespective of the precursor protein from which they are formed. For example, they produce similar x-ray diffraction patterns, bind certain dyes in a well-structured manner that results in birefringence and novel fluorescence spectra [10], and present common epitopes that are recognized by pan-fibril reactive antibodies [12]. In systemic amyloidosis the precursor proteins generally circulate at high concentrations within the blood (e.g., 14 g/L and about 24 mg/L are median values for LC and sAA reported in patients with AL and AA, respectively) [13,14].

As was the case in patients with AD, the current goal of a successful immunotherapy for AL or AA is to elicit a humoral immune response to the amyloid fibrils that will facilitate antibody-mediated dissolution of the deposits presumably by the host's immune cells, such as polymorphonuclear leukocytes (PMN) or macrophages. In contrast to the AD experience, however, the immune response to AL and AA amyloid deposits would ideally be limited to the amyloid fibril and not involve the free, circulating precursor protein.

ANTIGENS FOR IMMUNOTHERAPY OF AL AND AA

Opsonization of amyloid deposits with antibody (Ab) derived either by active immunization or passive administration is a prerequisite to their recognition and destruction by the cellular immune system. Both passive delivery and active induction of Aβ-reactive Abs have reduced the amount of amyloid in the brains of patients with AD and led to modest recoveries in organ function [5]. In the case of AL and AA, however, the high serum concentrations of LC and sAA precursor proteins provide constraints on the antibody specificity. Ideally, the opsonizing Abs would bind specifically to the fibrillar form of the protein and not the free precursor protein. Active immunization leading to a humoral response that included Abs to the soluble precursor protein might be expected to result in renal deposition of complexes and possibly immune complex glomerular nephiritis. For this reason, passive immunization is appealing, as the specificity of the therapeutic mAb can be well defined and the immune response well regulated. It is therefore worth considering, in general terms, how producing a fibril-specific humoral immune response may be achieved before talking about specific examples that have been developed and assayed experimentally.

The conversion of soluble, natively folded proteins into amyloid fibrils rich in β-sheet secondary structure requires, at the very least, a partial unfolding of the molecule but may also involve primary structural changes, including proteolytic degradation. The fibrillogenesis of LC and sAA probably involves both these processes to some degree [15–18].

The thermodynamic folding stability of LC proteins is critically important in determining their pathogenic and amyloidogenic propensity. Using recombinant LC variable domains (VLs) produced in bacteria as well as isolated LC proteins, it has been well established that structurally unstable LC proteins, exhibit a greater propensity for amyloid formation [10,18–20]. In addition, amyloidogenic LC proteins undergo proteolytic degradation, and as a result, AL fibrils are composed predominantly of the VL domain (with a limited number of adjoining constant domain residues). Similarly, sAA which circulates bound to high-density lipoprotein during the acute-phase reaction, and is predicted to be predominantly α-helical in structure [21], undergoes unfolding and proteolytic degradation resulting in the loss of up to about 70 C-terminal amino acids [22]. Therefore, the truncated and misfolded proteins comprising amyloid fibrils are morphologically quite distinct from their circulating counterparts. Consequently, the fibrils present unique antigenic determinants that can be targeted to generate fibril-specific Ab reagents [12,23]. A complete description of the many theoretical neoepitopes associated with fibril formation is presented elsewhere [24].

PRECLINICAL IMMUNOTHERAPY FOR AA AND AL AMYLOIDOSIS

Passive Immunization with the 11-1F4 mAb

We have developed and evaluated the therapeutic efficacy of an IgGκ1 mAb-designated 11-1F4 in a preclinical murine model of AL amyloidoma.

This reagent was generated using heat-denatured, fibrillar κ4 LC Len [25–27], and its ability to bind AL amyloid extracts and synthetic fibrils irrespective of the LC isotype and subgroup was recognized during enzyme-linked immunosorbant (ELISA) screening. The original immunogen used to generate the 11-1F4 was shown to bind the amyloidophilic fluorescent dye thioflavin T and produce fluorescence excitation and emission wavelength shifts to 450 and 490 nm, respectively, indicative of the presence of amyloidlike fibrils [28].

The binding of the 11-1F4 mAb to fibrils was specific, as little reactivity with the native, soluble precursor LC proteins of any subgroup was observed [29,30]. If, however, LC were partially denatured (e.g., by surface adsorption to the wells of a microtiter plate), the mAb bound with an estimated K_d value of about 0.3 nM (for κ4 LC proteins). The specificity for an epitope present on fibrils and partially denatured LC proteins was confirmed using competitive EuLISA (europium-linked immunsorbent assays) in which a 50- or 100-fold molar excess of soluble κ4 LC protein was unable to compete for surface-adsorbed κ4 LC in the microplate well. In contrast, heat-aggregated and fibrillar κ4 LC proteins were found to be competitive inhibitors in this assay. Thus, the epitope recognized by the 11-1F4 mAb was not present on the soluble LC precursor proteins but was "generated" and exposed on LC in amyloid and synthetic fibrils. These are important, possibly essential criteria for a candidate immunotherapeutic mAb for the treatment of AL amyloidosis [29–31]. The presence of the 11-1F4 mAb epitope in patient amyloid samples was assessed both quantitatively by EuLISA using ex vivo human AL amyloid extracts as the substrate and qualitatively by immunohistochemical testing of AL amyloid-containing, formalin-fixed, paraffin-embedded human tissue sections. Consistent with our ELISA data, the 11-1F4 mAb bound both κ and λ AL amyloid extracts and tissue deposits with varying degrees of relative affinity (10 to 100 nM), implying that the epitope may be conformational and was common to AL amyloid fibrils but that the affinity exhibited some dependence on amino acid sequence.

The epitope bound by the 11-1F4 mAb was identified within the N-terminal 15 amino acids of the κ4 LC Len using synthetic peptide mapping, alanine-scanning mutagenesis, and peptide phage display techniques [29,30]. It was readily apparent that the integrity of the binding site was critically dependent on the prolyl residue at position 8. Alanine substitution of Pro8 resulted in a complete loss of 11-1F4 mAb binding to a synthetic κ4 peptide spanning residues 1 to 18. Similar but less significant disruption of the binding occurred when residues within the 1–4 and 11–15 positions were substituted for alanine. The criticality of the proline at position 8 was further evidenced in peptide phage display epitope mapping experiments in which the majority of peptides (n = 60) isolated using the 11-1F4 mAb as a selecting ligand contained at least one centrally positioned proline in the 12-mer sequence [67]. Based on these findings, we have hypothesized that the conformational epitope recognized by the 11-1F4 mAb consists of a proline-anchored type VI β-turn that juxtaposes residues 1–4 and 11–15, which constitute the Ab-interacting amino acids. This turn is induced in the N-terminal of LC proteins as a result of denaturation

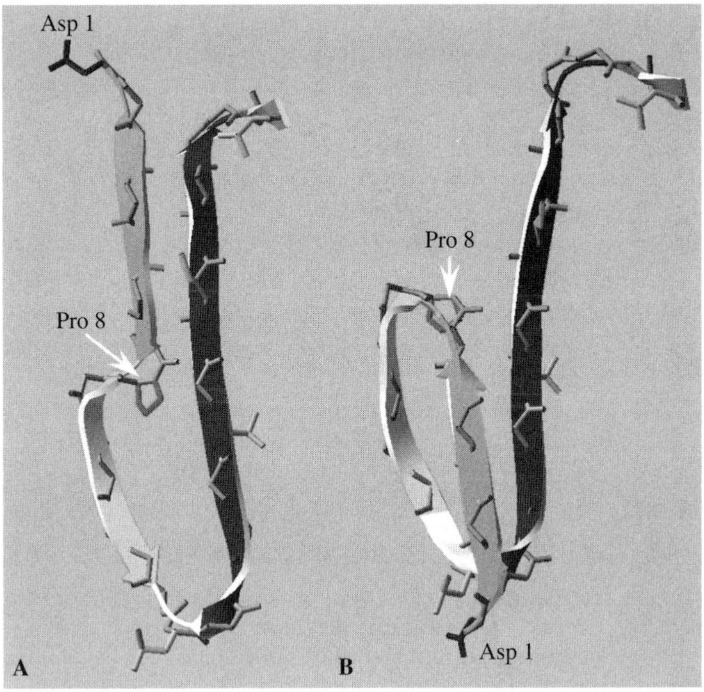

FIG. 1 Hypothetical N-terminal loop-flip conformation of fibril-incorporated LC proteins: (A) structure of κ4 Len (1–22) from x-ray crystallographic studies; (B) hypothetical type VI β-turn anchored by the proline at position 8 that juxtaposes residues 1–4 and 11–15, which constitute the epitope recognized by the 11-1F4 mAb. (From [29].)

either by surface adsorption or, more important by incorporation into amyloid fibrils (Fig. 1).

The diagnostic potential of the 11-1F4 mAb was demonstrated in vivo using a murine model of AL amyloidoma in which human AL amyloid extract was injected subcutaneously between the scapulae. Mice that received an intravenous bolus of ^{125}I-labeled 11-1F4 mAb were imaged by single-photon-emission computed tomography (SPECT) and the specific activity in the amyloid and healthy tissues measured [32–34]. The amyloidoma was readily visible in the SPECT images confirming the reaction of [^{125}I]11-1F4 mAb with the amyloid in vivo (Fig. 2). Approximately 15 to 45 % injected 11-1F4 mAb per gram of tissue was observed in the amyloid, with greater accumulation in κ-type amyloidomas with respect to those composed of λ LC [33].

In further studies it was found that subsequent to the binding of 11-1F4 mAb, cell-mediated resolution of the AL amyloidoma (amyloidolysis) was accelerated relative to its behavior in untreated mice. Without the 11-1F4 mAb therapy, mice mounted a humoral response to the xenogeneic material, and the mass resolved spontaneously by about 25 to 40 days for ALκ and ALλ

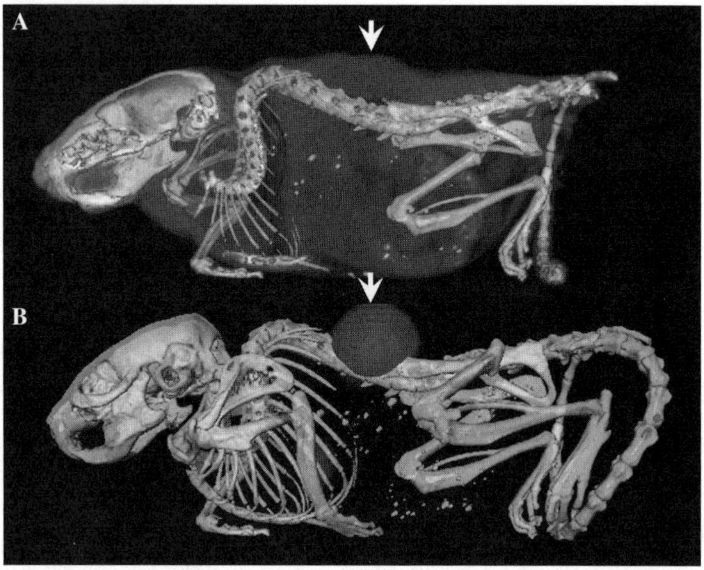

FIG. 2 Immunoimaging of human AL amyloidoma in mice using radioiodinated 11-1F4 mAb: (A) x-ray CT image showing the subcutaneous AL amyloidoma (arrow); (B) the [^{125}I]11-1F4 accumulated in the intrascapular human AL amyloidoma 3 days after injection, as evidenced in co-registered SPECT/CT images. (From [33].)

amyloidomas, respectively [35]. Opsonization of the amyloid by the 11-1F4 mAb led to the recruitment of macrophages and PMNs as evidenced histologically in sections of amyloidoma stained with α-naphthyl acetate and naphthol AS-D chloroacetate esterases, respectively [35]. Treatment of amyloidoma-bearing mice with 11-1F4 mAb either intravenously or subcutaneously expedited the clearance of the mass, which was totally resolved in as few as 5 days after initiating mAb therapy [35]. Dissolution of the amyloid mass probably occurred via enzymatic proteolysis and oxidative degradation, although the precise mechanisms remain unknown. This is functionally similar to the mAb-mediated removal of Aβ amyloid plaques in the brains of experimental animal models of AD that required microglial activation [36].

This was the first demonstration that opsonized human AL amyloid was susceptible to cell-mediated and complete dissolution and that the process could be hastened in vivo using a fibril-binding mAb administered parenterally. However, to translate this passive immunotherapeutic approach to the clinic, modifications of the murine mAb are necessary. To permit long-term administration to patients and minimize the risk of adverse events arising from human anti-mouse Ab activity, a chimeric (c) form of the 11-1F4 reagent has been prepared and its reactivity with amyloid extracts and LC fibrils has been shown to be equivalent to the murine form [37]. As a precursor to use of 11-1F4 in human therapy, an Exploratory Investigational New Drug (E-IND) application has been approved by the U.S. Food and Drug Administration (FDA) to evaluate the biodistribution and amyloid-targeting capabilities of an ^{124}I (positron-emitting)–conjugated form of murine 11-1F4 in patients with AL amyloidosis using positron-emission tomography (PET) imaging. In addition, the c11-1F4 Ab is being prepared under the auspices of the Rapid Access to Intervention Development (RAID) program at the National Cancer Institute for future studies to test its immunotherapeutic efficacy in patients with AL amyloidosis.

Passive Intravenous Immunization with Enriched Immune Globulin

Passive and active immunotherapies have been widely explored as methods to effect the removal of Aβ-containing amyloid plaques from the brain, and these are now being evaluated clinically in patients with AD [3]. One incarnation of the passive approach is the use of intravenous immune globulin (IGIV) such as Gamunex (Talecris Biotherapeutics, North Carolina) or Gammagard (Baxter, Illinois), which are 10% w/v liquid suspensions of purified human IgG [38]. Numerous studies have reported both in vitro [39,40] and in vivo efficacy of IGIV treatment in murine models and patients with AD [41–44]. In addition, a single report has documented the treatment of a patient with transthyretin (TTR) amyloidosis, which usually involves the heart and peripheral nerves, but in this case the amyloid deposits, composed of the Ala25Thr TTR mutant, were confined to the leptomeninges [45]. In this unique case study, hearing loss, sensory ataxia, and asymmetrical polyneuropathy all improved dramatically following high-dose IGIV therapy.

In our laboratory we have found that IGIV contains fibril-specific reactive IgG that bind amyloid composed of LC, TTR, islet amyloid precursor polypeptide (IAPP), and Aβ(1–40) [23]. Furthermore, these polyclonal Abs can be isolated from the commercially available preparations by affinity chromatography using synthetic Aβ or LC fibrils as the target. The enriched IGIV fraction retained the pan-fibril reactivity. Notably, fibril-enriched IGIV does not bind to nonfibrillar, soluble forms of LC, TTR, IAPP, or Aβ(1–40) as evidenced using a competition EuLISA similar to that described above for the 11-1F4 mAb. When labeled with ^{125}I, enriched IGIV co-localized with subcutaneous AL and ATTR amyloidomas, as evidenced by SPECT imaging (Fig. 3). When the affinity-purified IGIV was used in therapy experiments in the mouse amyloidoma model, this interaction resulted in about a 82% and 70% reduction in the masses relative to untreated mice [23].

The origin of the fibril-reactive Abs in IGIV and the nature of the initial antigenic challenge remain enigmatic; however, it is possible that "natural" fibrils such as silk [46,47], curli protein produced by *E. coli* [48], apolipoprotein A1 fibrils present in atherosclerotic plaques [49], or fibrils in food such as pâté de fois gras [50] may provide the antigenic stimulus for their production.

Irrespective of the source, the fibril-specific pan-amyloid-reactive Abs contained in normal human serum and concentrated in commercial IGIV products provide a novel opsonizing immunotherapeutic for use in patients with AL, AA, and other peripheral amyloidoses as well as those with AD.

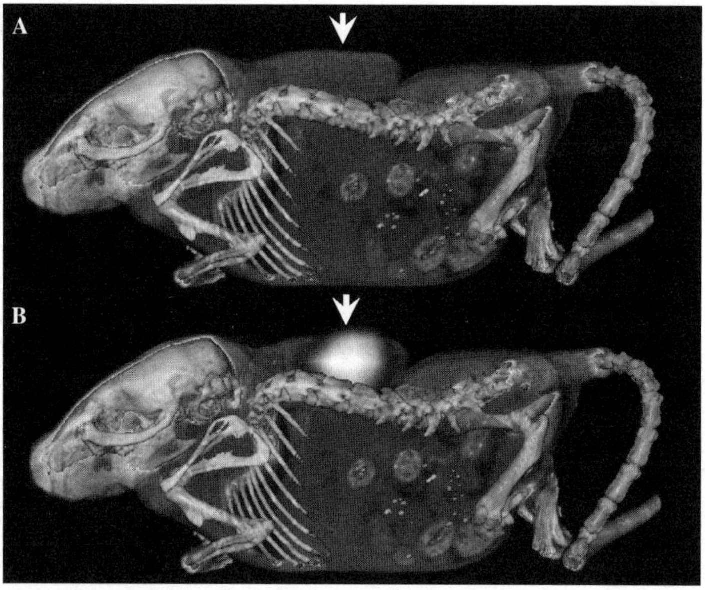

FIG. 3 Immunoimaging of human AL amyloidoma in mice using radioiodinated-enriched IGIV Ab: (A) x-ray CT image showing the subcutaneous AL amyloidoma (arrow); (B) the [^{125}I]IGIV accumulated in the human amyloidomas, as evidenced in co-registered SPECT/CT images.

Active Immunization with Heterogeneous Fibrils

The goal of active immunization is to elicit an immune response to an immunogen that will provide protection or ameliorate the infection or disease. In contrast to passive immunotherapy using mAb infusions, immunization requires a competent immune system capable of mounting a humoral or cellular response and retaining immunologic memory. This approach was shown efficacious at removing Aβ-containing amyloid plaques from mice and in patients with mild AD [51–53]. Unfortunately, clinical evaluation of the inaugural Aβ fibril vaccine (AN1792) was halted because cerebral bleeding was observed in 6% of the patients. Such an adverse event was not predicted from the preclinical murine studies or from the stringent safety studies performed in phase I [3,54,55,]. Thus, immunization clearly provides an effective method for amyloid removal; however, the path forward must be taken with care.

Ideally, an immunogen used for peripheral amyloidoses will produce a humoral response resulting in the production of opsonizing mAbs that bind the amyloid fibril but not the large pool of soluble precursor protein in the circulation, which could lead to immune complex disease and autoimmunity. Unfortunately, the immune response to fibrillar immunogens most often results in what appears to be a T-cell-independent IgM response [12] and/or IgG that reacts with the fibril as well as the soluble precursor protein [56,57]. We have therefore considered an alternative approach, heterogeneous immunization, that uses as the immunogen synthetic fibrils composed of heterogeneous proteins (i.e., unlike those deposited as amyloid in the host). The goal is to generate an Ab response that includes reactivity with fibril-specific conformational epitopes as well as avoiding precursor-reactive Abs. As a model system, we have tested the efficacy of synthetic λ6 LC fibrils compounded with an aluminum-based adjuvant (Alum) as an active immunotherapy for AA amyloidosis in which the fibrillar protein is derived from sAA, using a transgenic murine model of the pathology. Mice that constitutively express the human interleukin-6 (hIL-6) transgene under the control of the mouse metallothionein promoter develop systemic AA amyloidosis at five months of age which progresses rapidly over the next three months, at which time every major organ is involved and the mice show signs of morbidity [58]. Amyloidogenesis can be greatly accelerated in these animals by an intravenous injection of amyloid-enhancing factor (AEF) containing AA fibrils, which when administered to 8-week-old mice results in amyloid-related morbidity within 8 to 10 weeks post-injection [58]. We have found that mice immunized with synthetic fibrils composed of recombinant λ6VL domains formulated with Alum adjuvant will elicit a humoral immune response after 5 intraperitoneal injections. When these mice are then administered AEF intravenously to initiate AA fibrillogenesis, the extent of the resulting pathology, the hepatic function, and the survival rate are all improved compared to these factors in nonimmunized mice [59].

Although the splenic amyloid burden was statistically similar in the immunized and untreated groups (for reasons that are not yet known), the hepatic deposits were decreased significantly (by about 80%) in the λ6 VL-treated group (Fig. 4A). In addition, hepatocyte functionality, assessed histologically by staining formalin-fixed tissue sections with periodic acid Schiff and naphthol AS-D chloroacetateesterase, was partially salvaged in the immunized mice relative to the untreated group (Fig. 4B). Most notable was the increased longevity of the immunized group, which had a median survival of 13 weeks, compared to only 3.5 weeks in the untreated mice (Fig. 5). Indeed, at 16 weeks post-AEF injection (the end of the monitoring period), 50% of the immunized mice were thriving. These experiments demonstrated the feasibility of the heterogeneous immunization approach and validated the hypothesis that structural (conformational) epitopes expressed by amyloid fibrils can give rise to pan-fibril reactive Abs capable of affording therapeutic protection against peripheral amyloidosis.

Immunization with heterogeneous fibrils theoretically circumvents a major obstacle to the use of active immunization as a treatment for peripheral amyloidosis: the development of immune complex disease and/or other complications arising from reactivity of Abs with the soluble (or cell surface-associated) amyloid precursor protein. At present, this approach to immunization is experimental, and further studies are required to characterize the immune response to fibrils and to identify the most suitable fibrillar immunogen.

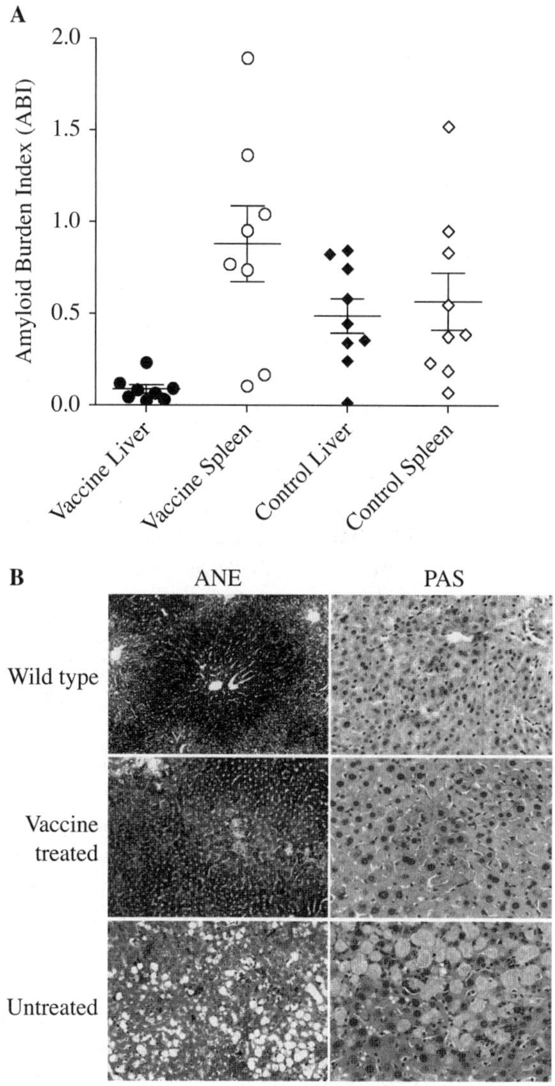

FIG. 4 Manifestation of heterogeneous immunization in mice with AA amyloidosis. (A) Amyloid burden in the liver and spleen of immunized and untreated mice. ABI was calculated by measuring the area occupied by Congo Red birefringent material in formalin fixed tissue sections using Image Pro-Plus visualization software. (B) Histological evaluation of hepatocyte integrity using naphthol AS-D chloroacetate esterase (ANE) and periodic acid Schiff (PAS). Original magnification × 160.

Plasma Cell–Directed Immunotherapy for AL

AL is unique among the systemic amyloidoses in that the monoclonal plasma population secreting the aberrant LC proteins can be targeted and destroyed without sustaining a critical loss of physiological functionality, as would be

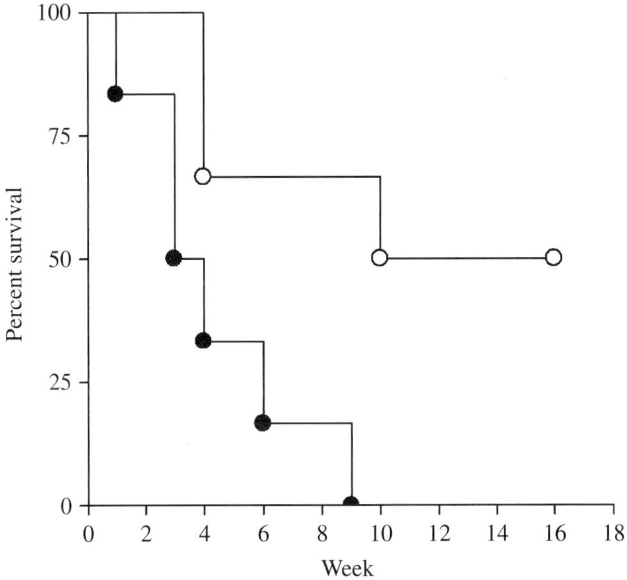

FIG. 5 Survival of hIL-6 transgenic mice with AA following immunization using λ6 VL fibrils. Immunized (open circles) and untreated (filled circles) mice were followed for 16 weeks post-AEF injection. The median survival for immunized and untreated mice was 13 and 3.5 weeks, respectively ($p < 0.02$). Data were analyzed by a Kaplan–Meier plot.

associated with the destruction of hepatocytes (the source of many amyloidogenic precursor proteins) [60]. Recent reports have described experimental immunotherapeutic approaches that target the monoclonal plasma cell population in patients with AL.

Cytotoxic T-cells (CTLs) can be induced and targeted to mediate specific cellular destruction using major histocompatibility (MHC) class I–restricted peptide vaccination. This method has been employed successfully to vaccinate against cancer [61–63]. When peptides derived from a human λ6 LC protein and the plasma cell–associated B-lymphocyte-induced maturation protein-1 (BLIMP-1) molecule injected into mice expressing human HLA-A*0201 allele, the mice mounted a CTL response as evidenced by the in vitro killing of λ6 LC-transfected human lymphoma cells [64]. A similar cytotoxic effect was observed using the BLIMP-1-derived peptide sequence.

Plasma cell killing can also be mediated by targeted Ab therapy, but as with amyloid fibrils, this requires the presentation of specific cell-associated biomarkers or epitopes. Zhou and co-workers have identified one such molecule that is present on 99% of $CD138^+$ clonal plasma cells isolated from patients with AL [65]. The low-affinity IgG Fcγ-receptor (CD32b) is found on B lymphocytes but is highly expressed on the surface of plasma cells from patients with AL and B-cell lymphomas and can therefore be targeted using mAb [66]. Chimeric and humanized forms of an anti-CD32b mAb have been generated, and approval of an IND to evaluate the therapeutic efficacy of the first of these reagents in patients with B-cell lymphoproliferative disorders was expected in 2008.

These novel immunotherapeutic approaches that target the cause of AL offer new hope to patients with this devastating disease. As these or similar treatments enter the clinic we will have the opportunity to study in great detail the complex relationship between LC precursor toxicity, amyloid deposition, and organ dysfunction, which will undoubtedly provide a quantum leap in our understanding of the etiopathology of AL amyloidosis.

PROSPECTS AND PROBLEMS WITH IMMUNOTHERAPY FOR PERIPHERAL AMYLOIDOSES

Immunotherapy is the most promising approach currently in clinical evaluation to effect removal of amyloid in AD. Both passive and active immunization protocols have demonstrated amyloidolytic efficacy in this population. When applied to the treatment of peripheral amyloidoses such as AL and AA, there are significant complications that must be overcome; for example, the high concentrations of soluble precursor protein in the circulation, the potential for adverse inflammatory reactions in visceral organs, and the potential for eliciting autoimmune disease all need to be considered. To this end, the identification of Abs that bind specifically to neo- or cryptic epitopes present on amyloid fibrils and not the soluble precursor protein provide hope and promise for new therapeutic amyloid-opsonizing reagents. As novel immunotherapeutic approaches for AL and AA enter clinical trials, we will be able to evaluate the risks and benefits of these mAbs and immunizations that facilitate amyloid dissolution and also determine whether in the peripheral amyloid diseases, removal of the fibrillar deposits will result in increased organ functionality, which remains the ultimate goal of all treatments for amyloidosis.

Acknowledgments

I wish to thank Alan Solomon, Maria Schell, Rudi Hrncic, Stephen Kennel, Tina Richey, Brian O'Nuallain, Sallie Macy, Craig Wooliver, and Denny Wolfenbarger for their many contributions over the years that have led to our evaluation and increased understanding of immunotherapy and amyloidosis.

REFERENCES

1. Waldmann, T.A. (2003). Immunotherapy: past, present and future. *Nat Med*, *9*, 269–277.
2. Waldmann, T.A., Morris, J.C. (2006). Development of antibodies and chimeric molecules for cancer immunotherapy. *Adv Immunol, 90*, 83–131.

3. Schenk, D.B., Seubert, P., Grundman, M., Black, R. (2005). A beta immunotherapy: lessons learned for potential treatment of Alzheimer's disease. *Neurodegener Dis, 2*, 255–260.
4. Gilman, S., Koller, M., Black, R.S., Jenkins, L., Griffith, S.G., Fox, N.C., Eisner, L., Kirby, L., Rovira, M.B., Forette, F., Orgogozo, J.M.; AN1792(QS-21)-201 Study Team. (2005). Clinical effects of Abeta immunization (AN1792) in patients with AD in an interrupted trial. *Neurology, 64*, 1553–1562.
5. Bombois, S., Maurage, C.A., Gompel, M., Deramecourt, V., Mackowiak-Cordoliani, M.A., Black, R.S., Lavielle, R., Delacourte, A., Pasquier, F. (2007). Absence of beta-amyloid deposits after immunization in Alzheimer disease with Lewy body dementia. *Arch Neurol, 64*, 583–587.
6. Lachmann, H.J., Hawkins, P.N. (2006). Systemic amyloidosis. *Curr Opin Pharmacol, 6*, 214–220.
7. Bellotti, V., Mangione, P., Merlini, G. (2000). Review: immunoglobulin light chain amyloidosis: the archetype of structural and pathogenic variability. *J Struct Biol, 130*, 280–289.
8. Rocken, C., Shakespeare, A. (2002). Pathology, diagnosis and pathogenesis of AA amyloidosis. *Virchows Arch, 440*, 111–122.
9. Pepys, M.B. (2006). Amyloidosis. *Annu Rev Med, 57*, 223–241.
10. Merlini, G., Bellotti, V. (2003). Molecular mechanisms of amyloidosis. *N Engl J Med, 349*, 583–596.
11. Pepys, M.B. (2001). Pathogenesis, diagnosis and treatment of systemic amyloidosis. *Philos Trans R Soc Lond B, 356*, 203–210; discussion, 210–211.
12. O'Nuallain, B., Wetzel, R. (2002). Conformational Abs recognizing a generic amyloid fibril epitope. *Proc Natl Acad Sci U S A, 99*, 1485–1490.
13. Kyle, R.A., Gertz, M.A. (1995). Primary systemic amyloidosis: clinical and laboratory features in 474 cases. *Semin Hematol, 32*, 45–59.
14. Dember, L.M., Hawkins, P.N., Hazenberg, B.P., Gorevic, P.D., Merlini, G., Butrimiene, I., Livneh, A., Lesnyak, O., Puéchal, X., Lachmann, H.J., et al.; Eprodisate for AA Amyloidosis Trial Group. (2007). Eprodisate for the treatment of renal disease in AA amyloidosis. *N Engl J Med, 356*, 2349–2360.
15. Novak, J., Moldoveanu, Z., Renfrow, M.B., Yanagihara, T., Suzuki, H., Raska, M., Hall, S., Brown, R., Huang, W.Q., Goepfert, A., et al. (2007). IgA nephropathy and Henoch–Schoenlein purpura nephritis: aberrant glycosylation of IgA1, formation of IgA1-containing immune complexes, and activation of mesangial cells. *Contrib Nephrol, 157*, 134–138.
16. Wall, J.S., Gupta, V., Wilkerson, M., Schell, M., Loris, R., Adams, P., Solomon, A., Stevens, F., Dealwis, C. (2004). Structural basis of light chain amyloidogenicity: comparison of the thermodynamic properties, fibrillogenic potential and tertiary structural features of four Vlambda6 proteins. *J Mol Recognit, 17*, 323–331.
17. Stevens, F.J. (2000). Four structural risk factors identify most fibril-forming kappa light chains. *Amyloid, 7*, 200–211.
18. Myatt, E.A., Westholm, F.A., Weiss, D.T., Solomon, A., Schiffer, M., Stevens, F.J. (1994). Pathogenic potential of human monoclonal immunoglobulin light chains: relationship of in vitro aggregation to in vivo organ deposition. *Proc Natl Acad Sci U S A, 91*, 3034–3038.

19. Wall, J., Schell, M., Murphy, C., Hrncic, R., Stevens, F.J., Solomon, A. (1999). Thermodynamic instability of human lambda 6 light chains: correlation with fibrillogenicity. *Biochemistry, 38*, 14101–14108.

20. Raffen, R., Dieckman, L.J., Szpunar, M., Wunschl, C., Pokkuluri, P.R., Dave, P., Wilkins Stevens, P., Cai, X., Schiffer, M., Stevens, F.J. (1999). Physicochemical consequences of amino acid variations that contribute to fibril formation by immunoglobulin light chains. *Protein Sci, 8*, 509–517.

21. Stevens, F.J. (2004). Hypothetical structure of human serum amyloid A protein. *Amyloid, 11*, 71–80.

22. Röcken, C., Menard, R., Bühling, F., Vöckler, S., Raynes, J., Stix, B., Krüger, S., Roessner, A., Kähne, T. (2005). Proteolysis of serum amyloid A and AA amyloid proteins by cysteine proteases: cathepsin B generates AA amyloid proteins and cathepsin L may prevent their formation. *Ann Rheum Dis, 64*, 808–815.

23. O'Nuallain, B., Hrncic, R., Wall, J.S., Weiss, D.T., Solomon, A. (2006). Diagnostic and therapeutic potential of amyloid-reactive IgG antibodies contained in human sera. *J Immunol, 176*, 7071–7078.

24. Dumoulin, M., Dobson, C.M. (2004). Probing the origins, diagnosis and treatment of amyloid diseases using antibodies. *Biochimie, 86*, 589–600.

25. Solomon, A., Weiss, D.T., Macy, S.D., Antonucci, R.A. (1990). Immunocytochemical detection of kappa and lambda light chain V region subgroups in human B-cell malignancies. *Am J Pathol, 137*, 855–862.

26. Abe, M., Goto, T., Kennel, S.J., Wolfenbarger, D., Macy, S.D., Weiss, D.T., Solomon, A. (1993). Production and immunodiagnostic applications of antihuman light chain monoclonal antibodies. *Am J Clin Pathol, 100*, 67–74.

27. Abe, M., Ozaki, S., Wolfenbarger, D., deBram-Hart, M., Weiss, D.T., Solomon, A. (1994). Variable-region subgroup distribution among lambda-type immunoglobulins in normal human serum. *J Clin Lab Anal, 8*, 4–9.

28. Wall, J., Murphy, C.L., Solomon, A. (1999). In vitro immunoglobulin light chain fibrillogenesis. *Methods Enzymol, 309*, 204–217.

29. O'Nuallain, B., Allen, A., Kennel, S.J., Weiss, D.T., Solomon, A., Wall, J.S. (2007). Localization of a conformational epitope common to non-native and fibrillar immunoglobulin light chains. *Biochemistry, 46*, 1240–1247.

30. O'Nuallain, B., et al. (2005). The amyloid-reactive monoclonal antibody 11-1F4 binds a cryptic epitope on fibrils and partially denatured immunoglobulin light chains and inhibits fibrillogenesis. In *Amyloid and Amyloidosis: Proceedings of the Xth International Symposium on Amyloidosis*, Tours, France, CRC Press, Boca Raton, FL.

31. Solomon, A., Weiss, D.T., Wall, J.S. (2003). Immunotherapy in systemic primary (AL) amyloidosis using amyloid-reactive monoclonal antibodies. *Cancer Biother Radiopharm, 18*, 853–860.

32. Wall, J.S., et al. (2005). Radioimaging of primary (AL) amyloidosis with an amyloid-reactive monoclonal antibody. In *Amyloid and Amyloidosis: Proceedings of the Xth International Symposium on Amyloidosis*, Tours, France, CRC Press, Boca Rato, FL.

33. Wall, J.S., Kennel, S.J., Paulus, M., Gregor, J., Richey, T., Avenell, J., Yap, J., Townsend, D., Weiss, D.T., Solomon, A. (2006). Radioimaging of light chain amyloid with a fibril-reactive monoclonal antibody. *J Nucl Med, 47*, 2016–2024.

34. Wall, J.S., Paulus, M.J., Gleason, S., Gregor, J., Solomon, A., Kennel, S.J. (2006). Micro-imaging of amyloid in mice. *Methods Enzymol, 412*, 161–182.
35. Hrncic, R., Wall, J., Wolfenbarger, D.A., Murphy, C.L., Schell, M., Weiss, D.T., Solomon, A. (2000). Antibody-mediated resolution of light chain–associated amyloid deposits. *Am J Pathol, 157*, 1239–1246.
36. Schenk, D., Barbour, R., Dunn, W., Gordon, G., Grajeda, H., Guido, T., Hu, K., Huang, J., Johnson-Wood, K., Khan, K., et al. (1999). Immunization with amyloid-beta attenuates Alzheimer-disease-like pathology in the PDAPP mouse. *Nature, 400*, 173–177.
37. Solomon, A., Weiss, D.T., Wall, J.S. (2003). Therapeutic potential of chimeric amyloid-reactive monoclonal antibody 11-1F4. *Clin Cancer Res, 9*, 3831S–3838S.
38. Newcombe, C., Newcombe, A.R. (2007). Antibody production: polyclonal-derived biotherapeutics. *J Chromatogr B, 848*, 2–7.
39. Solomon, B. (2007). Intravenous immunoglobulin and Alzheimer's disease immunotherapy. *Curr Opin Mol Ther, 9*, 79–85.
40. Istrin, G., Bosis, E., Solomon, B. (2006). Intravenous immunoglobulin enhances the clearance of fibrillar amyloid-beta peptide. *J Neurosci Res, 84*, 434–443.
41. Weksler, M.E. (2004). The immunotherapy of Alzheimer's disease. *Immun Ageing, 1*, 2.
42. Dodel, R.C., Du, Y., Depboylu, C., Hampel, H., Frölich, L., Haag, A., Hemmeter, U., Paulsen, S., Teipel, S.J., Brettschneider, S., et al. (2004). Intravenous immunoglobulins containing antibodies against beta-amyloid for the treatment of Alzheimer's disease. *J Neurol Neurosurg Psychiatry, 75*, 1472–1474.
43. Du, Y., Wei, X., Dodel, R., Sommer, N., Hampel, H., Gao, F., Ma, Z., Zhao, L., Oertel, W.H., Farlow, M. (2003). Human anti-beta-amyloid antibodies block beta-amyloid fibril formation and prevent beta-amyloid-induced neurotoxicity. *Brain, 126*, 1935–1939.
44. Dodel, R., Hampel, H., Depboylu, C., Lin, S., Gao, F., Schock, S., Jäckel, S., Wei, X., Buerger, K., Höft, C., et al. (2002). Human antibodies against amyloid beta peptide: a potential treatment for Alzheimer's disease. *Ann Neurol, 52*, 253–256.
45. Shimizu, Y., Takeuchi, M., Matsumura, M., Tokuda, T., Iwata, M. (2006). A case of biopsy-proven leptomeningeal amyloidosis and intravenous Ig-responsive polyneuropathy associated with the Ala25Thr transthyretin gene mutation. *Amyloid, 13*, 37–41.
46. Lundmark, K., Westermark, G.T., Olsén, A., Westermark, P. (2005). Protein fibrils in nature can enhance amyloid protein A amyloidosis in mice: cross-seeding as a disease mechanism. *Proc Natl Acad Sci U S A, 102*, 6098–6102.
47. Kisilevsky, R., Lemieux, L., Boudreau, L., Yang, D.S., Fraser, P. (1999). New clothes for amyloid enhancing factor (AEF): silk as AEF. *Amyloid, 6*, 98–106.
48. Cherny, I., Rockah, L., Levy-Nissenbaum, O., Gophna, U., Ron, E.Z., Gazit, E. (2005). The formation of *Escherichia coli* curli amyloid fibrils is mediated by prion-like peptide repeats. *J Mol Biol, 352*, 245–252.
49. Röcken, C., Tautenhahn, J., Bühling, F., Sachwitz, D., Vöckler, S., Goette, A., Bürger, T. (2006). Prevalence and pathology of amyloid in atherosclerotic arteries. *Arterioscler Thromb Vasc Biol, 26*, 676–677.
50. Solomon, A., Richey, T., Murphy, C.L., Weiss, D.T., Wall, J.S., Westermark, G.T., Westermark, P. (2007). Amyloidogenic potential of foie gras. *Proc Natl Acad Sci U S A, 104*, 10998–11001.

51. Nicoll, J.A., Barton, E., Boche, D., Neal, J.W., Ferrer, I., Thompson, P., Vlachouli, C., Wilkinson, D., Bayer, A., Games, D., et al. (2006). Abeta species removal after abeta42 immunization. *J Neuropathol Exp Neurol, 65*, 1040–1048.

52. Masliah, E., Hansen, L., Adame, A., Crews, L., Bard, F., Lee, C., Seubert, P., Games, D., Kirby, L., Schenk, D. (2005). Abeta vaccination effects on plaque pathology in the absence of encephalitis in Alzheimer disease. *Neurology, 64*, 129–131.

53. Schenk, D., Hagen, M., Seubert, P. (2004). Current progress in beta-amyloid immunotherapy. *Curr Opin Immunol, 16*, 599–606.

54. Schenk, D. (2004). Hopes remain for an Alzheimer's vaccine. *Nature, 431*, 398.

55. Schenk, D. (2002). Amyloid-beta immunotherapy for Alzheimer's disease: the end of the beginning. *Nat Rev Neurosci, 3*, 824–828.

56. Frenkel, D., Kariv, N., Solomon, B. (2001). Generation of auto-antibodies towards Alzheimer's disease vaccination. *Vaccine, 19*, 2615–2619.

57. Solomon, B., Koppel, R., Hanan, E., Katzav, T. (1996). Monoclonal antibodies inhibit in vitro fibrillar aggregation of the Alzheimer beta-amyloid peptide. *Proc Natl Acad Sci U S A, 93*, 452–455.

58. Solomon, A., Weiss, D.T., Schell, M., Hrncic, R., Murphy, C.L., Wall, J., McGavin, M.D., Pan, H.J., Kabalka, G.W., Paulus, M.J. (1999). Transgenic mouse model of AA amyloidosis. *Am J Pathol, 154*, 1267–1272.

59. Schell, M., et al. (2001). Prevention of AA-amyloidosis by active immunotherapy. In *Amyloid and Amyloidosis: Proceedings of the IXth International Symposium on Amyloidosis*, Budapest, Hungary.

60. Westermark, P., Benson, M.D., Buxbaum, J.N., Cohen, A.S., Frangione, B., Ikeda, S., Masters, C.L., Merlini, G., Saraiva, M.J., Sipe, J.D. (2007). A primer of amyloid nomenclature. *Amyloid, 14*, 179–183.

61. Moron, G., Dadaglio, G., Leclerc, C. (2004). New tools for antigen delivery to the MHC class I pathway. *Trends Immunol, 25*, 92–97.

62. Durrant, L.G., Ramage, J.M. (2005). Development of cancer vaccines to activate cytotoxic T lymphocytes. *Expert Opin Biol Ther, 5*, 555–563.

63. Gross, G., Margalit, A. (2007). Targeting tumor-associated antigens to the MHC class I presentation pathway. *Endocr Metab Immune Disord Drug Targets, 7*, 99–109.

64. Flies, A., Sherr, D.H. (2006). A peptide-based vaccine elicits CTL responses towards AL amyloid-related immunoglobulin light chain. In *XIth International Symposium on Amyloidosis* (Skinner, M., et al., Eds.), CRC Press, Boca Raton, FL, pp. 355–357.

65. Zhou, P., et al. (2006). The inhibitory Fcgamma-receptor IIB (CD32B) is highly expressed on clonal plasma cells from patients with systemic light-chain (AL) amyloidosis and provides a target for monoclonal antibody therapy. In *XIth International Symposium on Amyloidosis* (Skinner, M., et al., Eds.), CRC Press, Boca Raton, FL, pp. 371–373.

66. Rankin, C.T., Veri, M.C., Gorlatov, S., Tuaillon, N., Burke, S., Huang, L., Inzunza, H.D., Li, H., Thomas, S., Johnson, S., et al. (2006). CD32B, the human inhibitory Fc-gamma receptor IIB, as a target for monoclonal antibody therapy of B-cell lymphoma. *Blood, 108*, 2384–2391.

67. O'Nuallain, B., Allen, A., Ataman, D., Weiss, D.T., Solomon, A., Wall, J.S. (2008). Peptide mimetics of a conformational fibril epitope identified by phage display. *Biochemistry, 46*, 13049–13058.

43

ANTI-MISFOLDING AND ANTI-FIBRILLIZATION THERAPIES FOR PROTEIN MISFOLDING DISORDERS

ZANE MARTIN AND CLAUDIO SOTO

Departments of Neurology, Neuroscience and Cell Biology, and Biochemistry and Molecular Biology, George and Cynthia Mitchell Center for Neurodegenerative Diseases, University of Texas Medical Branch, Galveston, Texas

INTRODUCTION

Protein misfolding disorders (PMDs) make up a plethora of diseases that have a common mechanism involving structural rearrangement between a monomeric native protein that is composed primarily of unordered or α-helical structures to a β-sheet-rich higher-ordered conformation [1]. Neurodegenerative diseases are a subset of PMDs that are extremely debilitating, affecting cognition and/or movement. This group of disorders is composed of clinically diverse diseases, including Alzheimer disease (AD), Parkinson disease (PD), Huntington disease (HD), amyotrophic lateral sclerosis (ALS), and the prion diseases [2]. Even though these diseases are caused by different proteins, the molecular mechanism of pathogenesis is similar and involves the accumulation of large deposits, termed *amyloid.*

When amyloid was first discovered, it was mistakenly thought to be starch, based on crude iodine-staining techniques, hence given the Latin name *amylum,*

Protein Misfolding Diseases: Current and Emerging Principles and Therapies,
Edited by Marina Ramirez-Alvarado, Jeffery W. Kelly, and Christopher M. Dobson
Copyright © 2010 John Wiley & Sons, Inc.

meaning "starch". Through time, however, it has been determined that amyloid is actually proteinaceous [3]. Amyloid aggregates are composed of diverse peptides or proteins that have common features such that they form 6-to 10-nm cross β-sheet structures that are parallel and in register. These structures can be recognized through birefringence when stained with Congo Red using polarized light [4]. Amyloids form by a seeding/nucleation pathway in which after the initial misfolding, a set of intermediate structures are formed, including oligomers, protofibrils, and fibrils (Fig. 1).

There has been extensive debate as to which of these species is most toxic. When the amyloid hypothesis was first proposed to explain the pathogenesis of AD [5,6], it was thought that fibrils were the toxic species causing neuronal death. However, several findings did not fit completely with this proposal. First, there was no good correlation between disease progression and quantity of amyloid plaques [7]. Second, transgenic mice have been shown to develop behavioral problems before developing plaque deposits [8]. As a result, the focus turned to oligomeric and protofibrillar intermediates as the putative toxic species [9]. In parallel, the initial perception that neuronal death was the cause of brain malfunction and clinical disease symptoms has also been changing, since at the early stages of the disease in humans there is no significant neuronal loss, and transgenic mice show clear memory dysfunction without nerve cell death [10]. The current version of the molecular mechanism for neurodegenerative diseases is that soluble oligomers induce extensive synaptic dysfunction, leading to brain damage and disease [10].

Soluble oligomers are small assemblies of misfolded proteins that are present in the buffer-soluble fraction of brain extract and include structures ranging in size from dimers to 24-mers [11]. These oligomers then aggregate into protofibrils, shown by electron microscopy (EM) to be curvilinear structures of 4 to 11 nm in diameter and less then 200 nm long [12]. Protofibrils increase in size with increased time and protein concentration and are elongated by growth on their ends [13]. Annular protofibrils are porelike assemblies that form in the cell membrane and may contribute to cell death [14]. Protofibrils then aggregate into fibrils, which have been shown to be the most stable species [15]. Evidence indicates that these intermediates as well as the monomeric protein and the fibrillar aggregates are all present simultaneously and in a dynamic equilibrium between each other [16,17].

Drug discovery looks at this amyloid pathway to develop therapeutics that interfere with amyloid formation or increase degradation of misfolded aggregates. Proteins, peptides, small molecules, and immunotherapy have been designed to inhibit and/or reverse the conformational changes that result in formation of the pathological protein conformer and its sequential aggregation. In later sections we go into further detail regarding each of the approaches under development. Other approaches, including the use of gene therapy, silence RNA, or stem cells, are not discussed in this chapter but have reviewed been recently [18–22].

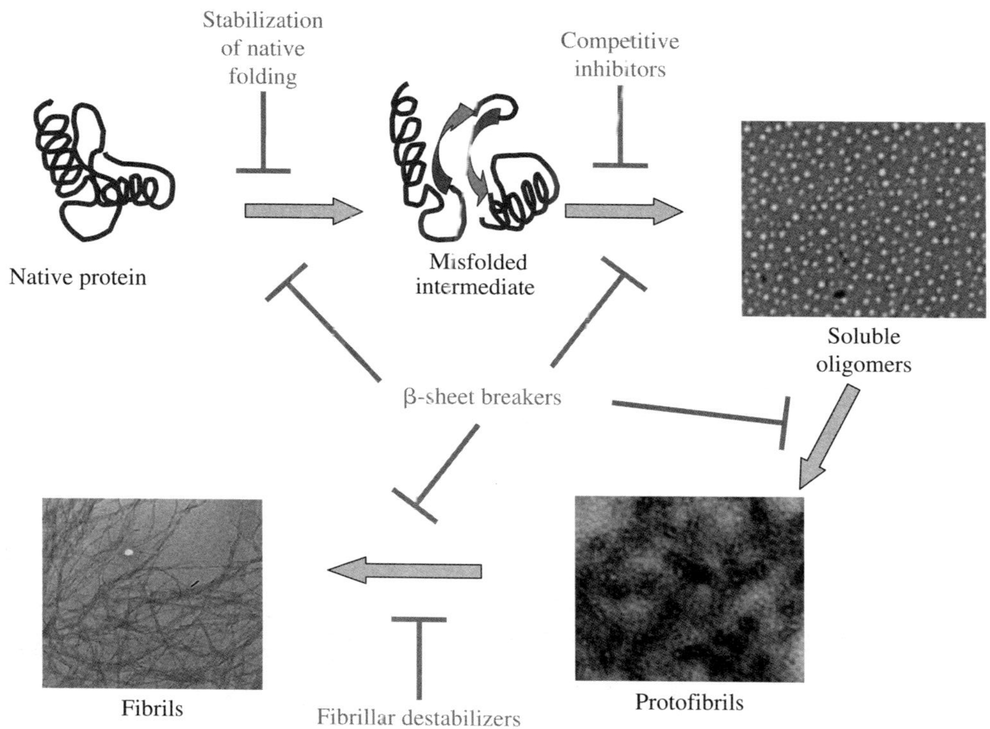

FIG. 1 Protein misfolding and fibrillogenesis pathway and targets for intervention. The aggregation of proteins in PMD follows a seeding and nucleation model in which a series of misfolded intermediates are formed. Currently, it is not know which of these conformational species is the most toxic. Various strategies have been proposed to prevent and reverse this pathway for therapeutic purposes.

AMYLOID-BINDING PROTEINS AS NATURAL INHIBITORS OF PROTEIN MISFOLDING AND AGGREGATION

The first approach focuses on the natural partners for misfolded amyloid aggregates in the body. These endogenous factors can modulate aggregation either as inhibitors of aggregation or by promoting fibrillar formation. Several proteins have been identified immunohistochemically or biochemically to be associated with amyloid deposits, including apolipoproteins E and J, serum amyloid-P, a1-antichymotrypsin, laminin, transthyretin, and proteoglycans [23,24]. The problem with this approach, though, is that different studies give contradictory results with respect to the effect of these proteins in aggregate formation. For example, the interaction between some of these proteins and the amyloidogenic protein has been shown to stabilize the native folding of protein and prevent its aggregation [25–31]. In contrast, under different conditions the same proteins have been shown to bind and promote fibrillogenesis in vitro [32–34]. The reason for this discrepancy could be the distinct concentration of proteins used or differences in assays utilized to measure the effect. These results indicate that more needs to be elucidated in determining the mechanistic interaction of these proteins with amyloids. Nevertheless, one approach aimed to prevent the interaction of the amyloidogenic protein with one of its natural partners (proteoglycans) is currently in phase III clinical trials for AD treatment. This compound is Alzhemed, a glycosaminoglycan mimetic able to displace the interaction between Aβ and heparan sulfate proteoglycans and prevent Aβ accumulation in vitro and in vivo [35]. However, the results of a recent phase III human trial were disappointing.

Another approach of protein-based therapy for PMDs takes advantage of molecular chaperones, which represent the cellular "professional" machinery to prevent and correct protein misfolding. The accumulation of misfolded proteins induces a coordinated adaptive program termed the *unfolded-protein response* (UPR) [36]. Activation of the UPR affects the expression of many proteins and, in particular, results in the up-regulation of several chaperone proteins. Molecular chaperones are ubiquitous stress-induced proteins, and some of them have been found to be effective in preventing misfolding of different disease-causing proteins [37].

ANTIBODIES AND VACCINES TO PREVENT AND REMOVE MISFOLDED AGGREGATES

Another approach that has been used successfully in several PMDs is active and passive immunization to prevent amyloid aggregation. This was first used in AD by Schenk and co-workers [38]. They used active immunization with a synthetic Aβ(1–42) peptide administered to 6-week-old mice along with an adjuvant. Vaccination prevented development of Aβ plaques, neuritic dystrophy, and astrogliosis. They also immunized 11-month-old mice, and the results showed a reduction in the extent and progression of the pathology [38]. This pioneer study was then repeated and confirmed by many groups, showing that immunization also prevents memory deficits in transgenic mice [39,40]. Passive immunization has also been done by intraperitoneal injection of antibodies against Aβ, which reduced pathology in the brain [41]. These results can be explained by considering that the antibodies entered the central nervous system (CNS) where they bind to plaques labeling them for microglial removal or by the *peripheral sink hypothesis,* which that postulates that antibodies bind to soluble peptide in the peripheral circulation, displacing the equilibrium between brain and blood Aβ [42]. Another study using old APP transgenic mice reported a decrease in amyloid plaques and cognitive improvement. However, cerebral microhemorrhages were detected, indicating a potential side effect of the therapy [43]. The promising results with immunotherapy in AD prompted investigators to apply similar strategies for other PMDs. Indeed, positive results with diverse passive or active immunization paradigms have been reported in diverse diseases, including PD, prion diseases, and systemic amyloidosis [40]. The immunization approach showed such great promise that it moved quickly to human clinical trials. However, a phase II trial in humans was halted because 6% of the patients treated developed severe meningoencephalitis [44]. Even though the clinical trial was terminated, postmortem histological examination of several patients receiving the treatment showed an unusually low number of amyloid deposits [45], suggesting that the treatment may have been effective in reducing amyloid pathology in the brain. This correlated with further follow-up of 30 patients that demonstrated a significant slowdown of cognitive decline [46]. Immunotherapy is currently regarded as one of the most promising strategies for treatment if research can reduce side effects maintaining efficacy. Several modifications are currently under development, including the use of shorter or modified peptides as immunogens and DNA immunization among others [39,40].

SMALL-MOLECULE INHIBITORS OF AMYLOID FORMATION

A traditional approach in drug discovery is using small chemical molecules because they offer the best druglike properties. Pharmaceutical companies usually use high-throughput screening (HTS) to identify potential candidates. This allows researchers to test hundreds of thousands of compounds very quickly. There are many small chemical compounds that have been reported to inhibit protein misfolding and aggregation, which have been identified either serendipitously, from HTS, or because epidemiological studies have suggested that they may be active. Examples of small molecules that that have been reported to prevent the misfolded and aggregation of proteins involved in PMDs include Congo red and derivatives, curcumin and rosmarinic acid, small sulfonated anions, wine polyphenols and tannic acid, melatonin, nicotine, estrogen, hexadecyl-*N*-methylpiperidinium bromide, benzofuran-based compounds, amphiphilic surfactants such as di-C6-PC and di-C7-PC, the disaccharide trehalose, the anti-leprosy drugs dapsone and rifampicin, anthracyclines, thyroxine, flufenamic acid, N-substituted anthranilic acids, *N*-phenylphenoxazines, nitrophenols, diclofenac analog, inositol, nordihydroguaiaretic acid, β-cyclodextrins, apomorphine and analog, tetracyclines, quinacrine, branched polyamines, acridine and phenotiazine derivatives, porphyrins and phtalocyanines, 4′-iodo-4′-deoxydoxorubicin, pentosane polysulfate, amphotericin B, suramin, and "chemical chaperones" (e.g., glycerol, dimethyl sulfoxide, trimethylamine-*N*-oxide). (A more comprehensive list of compounds, including the specific references, may be found in [47–49].) These compounds result in a net reduction of misfolded aggregates but reach this goal acting through diverse mechanisms (Fig. 1). Some of the compounds act by stabilizing the native conformation of the protein. Other molecules work as competitive inhibitors of protein–protein interaction needed for aggregation. Some of them prevent fibrillogenesis by acting on partially folded intermediates of the folding process as well as on soluble oligomers that populate the initial phase of fibril formation. Others are able to prevent and stabilize the abnormal β-sheet structure of the proteins (β-sheet breakers) and thus can attack the pathway at several different stages. Yet others compounds interact with mature fibrillar aggregates, destabilizing the structure of the polymer, resulting in disassembly of the aggregate (fibril destabilizers). Finally, some drugs may displace fundamental cofactors involved in the process of protein misfolding and aggregation. It is also possible that some of the small-molecule inhibitors may act by more than one of these mechanisms. The usefulness of these small molecules is compromised by their lack of specificity and their unclear mechanism of action in most of the cases. In addition, many of them are highly toxic, making it difficult to use them in humans.

PEPTIDE INHIBITORS OF PROTEIN MISFOLDING

A strategy that has been successful in identifying hit compounds is the rational development of specific inhibitors based on the use of short peptides targeting the protein region needed for protein–protein interaction [50,51]. Although the approaches to come out with peptide inhibitors are different, they usually consist of synthesizing short peptides combining a self-recognition motif with a β-sheet-disrupting element. The self-recognition domain is typically the region of the protein implicated in early misfolding and protein–protein interaction. As disrupting elements, different chemical groups have been used in distinct strategies [52], including the use of a bulky group (e.g., cholyl) that sterically inhibits protèin aggregation; N-methylations (or N-alkylations) to generate peptides having a blocking face; β-sheet breaker amino acids to disrupt β-sheet conformation; and addition of charged residues to reduce the hydrophic interactions that trigger protein aggregation.

Because of the high prevalence of AD in the population and the lack of an efficient treatment to alter the progression of the disease, most of the effort to develop peptide inhibitors of protein misfolding and aggregation has focused on this disease. However, many of the approaches can be (and in some cases already have been) extrapolated to design compounds beneficial for the treatment of other PMDs. Identification of the region involved in protein–protein interactions is crucial in the rational design of specific peptide inhibitors. Several pieces of evidence have shown that in AD the central region (residues 16 to 20) of Aβ is essential for self-recognition. Tjernberg et al. have shown that Aβ(16–20) is able to bind to full-length Aβ and inhibit the formation of amyloid fibrils [53]. Because these peptides have the ability to form aggregates by themselves and to incorporate into amyloid-fibrils, they cannot be used as therapeutic inhibitors. Therefore, several groups began to modify this sequence to produce peptide derivatives containing the self-recognition motif but, at the same time, a disrupting element enhancing their inhibitory activity and preventing them from aggregating [52]. For example, one group tried adding charged residues to the end of the recognition motif because it is well known that the major force driving Aβ aggregation is hydrophobicity [54,55]. They demonstrated that at least three lysines are required as an appropriate disrupting element. The compound (KLVFFKKKK) showed activity in altering fibril morphology and reducing cellular toxicity in vitro. The anionic disrupting compound KLVFFEEEE had similar effects, whereas the neutral compound KLVFFSSSS was ineffective, suggesting that the charged nature of the disrupting element is critical [55]. A different approach to design aggregation inhibitors was to add a bulky group to the recognition sequence, creating an inhibitor through steric hindrance [56,57]. The incorporation of N-methyl amino acids into peptides as disrupting elements has also been used by several investigators [58,59]. Having N-methyl groups placed in alternate positions in the peptide backbone creates a "blocking face," avoiding hydrogen bonding. The other side constitutes a "complementary" face to the protein, which allows binding [51]. An additional advantage of this strategy is that N-methylated peptides are more resistant than regular peptides to proteolysis.

Our approach using β-sheet breaker amino acids was the first to lead to modified peptides with inhibitory activity [60]. Residues 17 to 21 of Aβ were used as a self-recognition motif, but a key residue for β-sheet formation was replaced by an amino acid thermodynamically unable to fit in the β-sheet structure [60,61]. A residue known to promote the β-sheet folding of Aβ (valine at position 18) was replaced by a proline, an amino acid thermodynamically unable to fit in a β-sheet structure [62]. Several β-sheet breaker peptides were generated and tested in vitro [60]. One β-sheet breaker of five residues, termed iAβ5, was shown to bind to Aβ with high affinity and to inhibit Aβ misfolding and aggregation [63]. iAβ5 also induces the disassembly of preformed fibrils in vitro and prevent neuronal death in neuroblastoma cultures [63]. The ability of iAβ5 to inhibit and disassembly amyloid fibrils was also demonstrated in vivo, using three different animal models of AD, including transgenic mice that develop many of pathological hallmarks of AD [63–66].

Peptide inhibitors based on similar principles have also been reported to prevent misfolding and aggregation of several proteins involved in other PMDs, including prion diseases, PD, HD, and type 2 diabetes [47]. The major advantage of the peptide approach is that highly potent and specific compounds can be produced which usually are not overly toxic. However, the peptide nature of these molecules imposes serious problems for administration and delivery, especially for compounds needed to act on the brain [67]. In addition, peptides are rapidly degraded, resulting in the need for frequent administration and the use of large doses. Nevertheless, the great advance in peptide chemistry permits the use of a variety of strategies to minimize these weaknesses [67].

REFERENCES

1. Makin, O.S., Serpell, L.C. (2002). Examining the structure of the mature amyloid fibril. *Biochem Soc Trans*, *30*, 521–525.
2. Soto, C. (2003). Unfolding the role of protein misfolding in neurodegenerative diseases. *Nat Rev Neurosci*, *4*, 49–60.
3. Osborne, B.M., Butler, J.J., Mackay, B. (1979). Proteinaceous lymphadenopathy with hypergammaglobulinemia. *Am J Surg Pathol*, *3*, 137–145.
4. Wolman, M., Bubis, J.J. (1965). The cause of the green polarization color of amyloid stained with Congo Red. *Histochemie*, *4*, 351–356.
5. Hardy, J. (2001). Genetic dissection of primary neurodegenerative diseases. *Biochem Soc Symp*, 51–57.
6. Selkoe, D.J. (1994). Alzheimer's disease: a central role for amyloid. *J Neuropathol Exp Neurol*, *53*, 438–447.
7. Katzman, R. (1988). Alzheimer's disease as an age-dependent disorder. *Ciba Found Symp*, *134*, 69–85.
8. Moechars, D., Dewachter, I., Lorent, K., Reverse, D., Baekelandt, V., Naidu, A., Tesseur, I., Spittaels, K., Haute, C.V., Checler, F., Horowitz, P., et al. (1999). Early phenotypic changes in transgenic mice that overexpress different mutants of amyloid precursor protein in brain. *J Biol Chem*, *274*, 6483–6492.
9. Walsh, D.M., Selkoe, D.J. (2004). Oligomers on the brain: the emerging role of soluble protein aggregates in neurodegeneration. *Protein Pept Lett*, *11*, 213–228.
10. Selkoe, D.J. (2002). Alzheimer's disease is a synaptic failure. *Science*, *298*, 789–791.
11. Chromy, B.A., Nowak, R.J., Lambert, M.P., Viola, K.L., Chang, L., Velasco, P.T., Jones, B.W., Fernandez, S.J., Lacor, P.N., Horowitz, P., et al. (2003). Self-assembly of Abeta(1–42) into globular neurotoxins. *Biochemistry*, *42*, 12749–12760.
12. Walsh, D.M., Hartley, D.M., Kusumoto, Y., Fezoui, Y., Condron, M.M., Lomakin, A., Benedek, G.B., Selkoe, D.J., Teplow, D.B. (1999). Amyloid beta-protein fibrillogenesis. Structure and biological activity of protofibrillar intermediates. *J Biol Chem*, *274*, 25945–25952.

13. Harper, J.D., Wong, S.S., Lieber, C.M., Lansbury, P.T., Jr. (1999). Assembly of A beta amyloid protofibrils: an in vitro model for a possible early event in Alzheimer's disease. *Biochemistry*, *38*, 8972–8980.

14. Lin, Y.M., Raffen, R., Zhou, Y., Cassidy, C.S., Flavin, M.T., Stevens, F.J. (2001). Amyloid fibril formation in microwell plates for screening of inhibitors. *Amyloid*, *8*, 182–193.

15. Modler, A.J., Gast, K., Lutsch, G., Damaschun, G. (2003). Assembly of amyloid protofibrils via critical oligomers: a novel pathway of amyloid formation. *J Mol Biol*, *325*, 135–148.

16. Teplow, D.B. (1998). Structural and kinetic features of amyloid beta-protein fibrillogenesis. *Amyloid*, *5*, 121–142.

17. Levine, H., III. (1995). Soluble multimeric Alzheimer beta(1–40) pre-amyloid complexes in dilute solution. *Neurobiol Aging*, *16*, 755–764.

18. Cardone, M. (2007). Prospects for gene therapy in inherited neurodegenerative diseases. *Curr Opin Neurol*, *20*, 151–158.

19. Boudreau, R.L., Davidson, B.L. (2006). RNAi therapy for neurodegenerative diseases. *Curr Top Dev Biol*, *75*, 73–92.

20. Koutsilieri, E., Rethwilm, A., Scheller, C. (2007). The therapeutic potential of siRNA in gene therapy of neurodegenerative disorders. *J Neural Transm Suppl*, 43–49.

21. Kan, I., Melamed, E., Offen, D. (2007). Autotransplantation of bone marrow-derived stem cells as a therapy for neurodegenerative diseases. *Handb Exp Pharmacol*, 219–242.

22. Sonntag, K.C., Sanchez-Pernaute, R. (2006). Tailoring human embryonic stem cells for neurodegenerative disease therapy. *Curr Opin Investi Drugs*, *7*, 614–618.

23. Golabek, A., Marques, M.A., Lalowski, M., Wisniewski, T. (1995). Amyloid beta binding proteins in vitro and in normal human cerebrospinal fluid. *Neurosci Lett*, *191*, 79–82.

24. Soto, C., Ghiso, J., Frangione, B. (1997). Alzheimer's amyloid-β aggregation is modulated by the interaction of multiple factors. *Alzheimer's Res*, *3*, 215–222.

25. Evans, K.C., Berger, E.P., Cho, C.G., Weisgraber, K.H., Lansbury, P.T., Jr. (1995). Apolipoprotein E is a kinetic but not a thermodynamic inhibitor of amyloid formation: implications for the pathogenesis and treatment of Alzheimer disease. *Proc Natl Acad Sci U S A*, *92*, 763–767.

26. Matsubara, E., Soto, C., Governale, S., Frangione, B., Ghiso, J. (1996). Apolipoprotein J and Alzheimer's amyloid beta solubility. *Biochem J*, *316* (Pt 2), 671–679.

27. Schwarzman, A.L., Gregori, L., Vitek, M.P., Lyubski, S., Strittmatter, W.J., Enghilde, J.J., Bhasin, R., Silverman, J., Weisgraber, K.H., Coyle, P.K. (1994). Transthyretin sequesters amyloid beta protein and prevents amyloid formation. *Proc Natl Acad Sci U S A*, *91*, 8368–8372.

28. Janciauskiene, S., Garcia, D.F., Carlemalm, E., Dahlback, B., Eriksson, S. (1995). Inhibition of Alzheimer beta-peptide fibril formation by serum amyloid P component. *J Biol Chem*, *270*, 26041–26044.

29. Janciauskiene, S., Rubin, H., Lukacs, C.M., Wright, H.T. (1998). Alzheimer's peptide Abeta1–42 binds to two beta-sheets of alpha1-antichymotrypsin and transforms it from inhibitor to substrate. *J Biol Chem*, *273*, 28360–28364.

30. Castillo, G.M., Lukito, W., Peskind, E., Raskind, M., Kirschner, D.A., Yee, A.G., Snow, A.D. (2000). Laminin inhibition of beta-amyloid protein (Abeta) fibrillogenesis and identification of an Abeta binding site localized to the globular domain repeats on the laminin a chain. *J Neurosci Res*, *62*, 451–462.
31. Bronfman, F.C., Garrido, J., Alvarez, A., Morgan, C., Inestrosa, N.C. (1996). Laminin inhibits amyloid-beta-peptide fibrillation. *Neurosci Lett*, *218*, 201–203.
32. Castano, E.M., Prelli, F., Wisniewski, T., Golabek, A., Kumar, R.A., Soto, C., Frangione, B. (1995). Fibrillogenesis in Alzheimer's disease of amyloid beta peptides and apolipoprotein E. *Biochem J*, *306* (Pt 2), 599–604.
33. Ma, J., Yee, A., Brewer, H.B., Jr., Das, S., Potter, H. (1994). Amyloid-associated proteins alpha 1-antichymotrypsin and apolipoprotein E promote assembly of Alzheimer beta-protein into filaments. *Nature*, *372*, 92–94.
34. Hamazaki, H. (1995). Amyloid P component promotes aggregation of Alzheimer's beta-amyloid peptide. *Biochem Biophys Res Commun*, *211*, 349–353.
35. Geerts, H. (2004). NC-531 (Neurochem). *Curr Opin Invest Drugs*, *5*, 95–100.
36. Hetz, C.A., Soto, C. (2006). Stressing out the ER: a role of the unfolded protein response in prion-related disorders. *Curr Mol Med*, *6*, 37–43.
37. Chaudhuri, T.K., Paul, S. (2006). Protein misfolding diseases and chaperone-based therapeutic approaches. *FEBS J*, *273*, 1331–1349.
38. Schenk, D., Barbour, R., Dunn, W., Gordon, G., Grajeda, H., Guido, T., Hu, K., Huang, J., Johnson-Wood, K., Khan, K., et al. (1999). Immunization with amyloid-beta attenuates Alzheimer-disease-like pathology in the PDAPP mouse. *Nature*, *400*, 173–177.
39. Schenk, D., Hagen, M., Seubert, P. (2004). Current progress in beta-amyloid immunotherapy. *Curr Opin Immunol*, *16*, 599–606.
40. Sigurdsson, E.M. (2006). Immunotherapy for conformational diseases. *Curr Pharm Des*, *12*, 2569–2585.
41. Bard, F., Cannon, C., Barbour, R., Burke, R.L., Games, D., Grajeda, H., Guido, T., Hu, K., Huang, J., Johnson-Wood, K., et al. (2000). Peripherally administered antibodies against amyloid beta-peptide enter the central nervous system and reduce pathology in a mouse model of Alzheimer disease. *Nat Med*, *6*, 916–919.
42. DeMattos, R.B., Bales, K.R., Cummins, D.J., Dodart, J.C., Paul, S.M., Holtzman, D.M. (2001). Peripheral anti-A beta antibody alters CNS and plasma A beta clearance and decreases brain A beta burden in a mouse model of Alzheimer's disease. *Proc Natl Acad Sci U S A*, *98,* 8850–8855.
43. Pfeifer, M., Boncristiano, S., Bondolfi, L., Stalder, A., Deller, T., Staufenbiel, M., Mathews, P.M., Jucker, M. (2002). Cerebral hemorrhage after passive anti-Abeta immunotherapy. *Science*, *298*, 1379.
44. Orgogozo, J.M., Gilman, S., Dartigues, J.F., Laurent, B., Puel, M., Kirby, L.C., Jouanny, P., Dubois, B., Eisner, L., Flitman, S., et al. (2003). Subacute meningoencephalitis in a subset of patients with AD after Abeta42 immunization. *Neurology*, *61*, 46–54.
45. Spinney, L. (2004). Update on Elan vaccine for Alzheimer's disease. *Lancet Neurol*, *3*, 5.

46. Hock, C., Konietzko, U., Streffer, J.R., Tracy, J., Signorell, A., Muller-Tillmanns, B., Lemke, U., Henke, K., Moritz, E., Garcia, E., et al. (2003). Antibodies against beta-amyloid slow cognitive decline in Alzheimer's disease. *Neuron*, *38*, 547–554.

47. Estrada, L.D., Yowtak, J., Soto, C. (2006). Protein misfolding disorders and rational design of antimisfolding agents. *Methods Mol Biol*, *340*, 277–293.

48. De Lorenzi, E., Giorgetti, S., Grossi, S., Merlini, G., Caccialanza, G., Bellotti, V. (2004). Pharmaceutical strategies against amyloidosis: old and new drugs in targeting a "protein misfolding disease." *Curr Med Chem*, *11*, 1065–1084.

49. Cohen, F.E., Kelly, J.W. (2003). Therapeutic approaches to protein misfolding diseases. *Nature*, *426*, 905–909.

50. Soto, C. (2001). Protein misfolding and disease; protein refolding and therapy. *FEBS Lett*, *498*, 204–207.

51. Mason, J.M., Kokkoni, N., Stott, K., Doig, A.J. (2003). Design strategies for anti-amyloid agents. *Curr Opin Struct Biol*, *13*, 526–532.

52. Estrada, L.D., Soto, C. (2006). Inhibition of protein misfolding and aggregation by small rationally-designed peptides. *Curr Pharm Des*, *12*, 2557–2567.

53. Tjernberg, L.O., Naslund, J., Lindqvist, F., Johansson, J., Karlstrom, A.R., Thyberg, J., Terenius, L., Nordstedt, C. (1996). Arrest of beta-amyloid fibril formation by a pentapeptide ligand. *J Biol Chem*, *271*, 8545–8548.

54. Ghanta, J., Shen, C.L., Kiessling, L.L., Murphy, R.M. (1996). A strategy for designing inhibitors of beta-amyloid toxicity. *J Biol Chem*, *271*, 29525–29528.

55. Pallitto, M.M., Ghanta, J., Heinzelman, P., Kiessling, L.L., Murphy, R.M. (1999). Recognition sequence design for peptidyl modulators of beta-amyloid aggregation and toxicity. *Biochemistry*, *38*, 3570–3578.

56. Findeis, M.A., Musso, G.M., Arico-Muendel, C.C., Benjamin, H.W., Hundal, A.M., Lee, J.J., Chin, J., Kelley, M., Wakefield, J., Hayward, N.J., Molineaux, S.M. (1999). Modified-peptide inhibitors of amyloid beta-peptide polymerization. *Biochemistry*, *38*, 6791–6800.

57. Findeis, M.A., Lee, J.J., Kelley, M., Wakefield, J.D., Zhang, M.H., Chin, J., Kubasek, W., Molineaux, S.M. (2001). Characterization of cholyl-leu-val-phe-phe-ala-OH as an inhibitor of amyloid beta-peptide polymerization. *Amyloid*, *8*, 231–241.

58. Hughes, E., Burke, R.M., Doig, A.J. (2000). Inhibition of toxicity in the beta-amyloid peptide fragment beta-(25–35) using N-methylated derivatives: a general strategy to prevent amyloid formation. *J Biol Chem*, *275*, 25109–25115.

59. Gordon, D.J., Sciarretta, K.L., Meredith, S.C. (2001). Inhibition of beta-amyloid(40) fibrillogenesis and disassembly of beta-amyloid(40) fibrils by short beta-amyloid congeners containing *N*-methyl amino acids at alternate residues. *Biochemistry*, *40*, 8237–8245.

60. Soto, C., Kindy, M.S., Baumann, M., Frangione, B. (1996). Inhibition of Alzheimer's amyloidosis by peptides that prevent beta-sheet conformation. *Biochem Biophys Res Commun*, *226*, 672–680.

61. Soto, C. (1999). Plaque busters: strategies to inhibit amyloid formation in Alzheimer's disease. *Mol Med Today*, *5*, 343–350.

62. Wood, S.J., Wetzel, R., Martin, J.D., Hurle, M.R. (1995). Prolines and amyloidogenicity in fragments of the Alzheimer's peptide beta/A4. *Biochemistry*, *34*, 724–730.

63. Soto, C., Sigurdsson, E.M., Morelli, L., Kumar, R.A., Castano, E.M., Frangione, B. (1998). Beta-sheet breaker peptides inhibit fibrillogenesis in a rat brain model of amyloidosis: implications for Alzheimer's therapy. *Nat Med*, *4*, 822–826.

64. Sigurdsson, E.M., Permanne, B., Soto, C., Wisniewski, T., Frangione, B. (2000). In vivo reversal of amyloid-beta lesions in rat brain. *J Neuropathol Exp Neurol*, *59*, 11–17.

65. Permanne, B., Adessi, C., Fraga, S., Frossard, M.J., Saborio, G.P., Soto, C. (2002). Are beta-sheet breaker peptides dissolving the therapeutic problem of Alzheimer's disease? *J Neural Transm Suppl*, 293–301.

66. Chacon, M.A., Barria, M.I., Soto, C., Inestrosa, N.C. (2004). b-sheet breaker peptide prevents Abeta-induced spatial memory impairments with partial reduction of amyloid deposits. *Mol Psychiatry*, *9*, 953–961.

67. Adessi, C., Soto, C. (2002). Converting a peptide into a drug: strategies to improve stability and bioavailability. *Curr Med Chem*, *9*, 963–978.

44

THERAPIES AIMED AT CONTROLLING GENE EXPRESSION, INCLUDING UP-REGULATING A CHAPERONE OR DOWN-REGULATING AN AMYLOIDOGENIC PROTEIN

GREGOR P. LOTZ

Gladstone Institute of Neurological Disease, University of California, San Francisco, California

PAUL J. MUCHOWSKI

Gladstone Institute of Neurological Disease, and Departments of Biochemistry and Biophysics, and Neurology, University of California, San Francisco, California

INTRODUCTION

Many protein misfolding diseases are characterized by the accumulation of amyloid, an ordered protein aggregate that contains high levels of a cross β-sheet structure. Amyloid formation is a complex process that may involve oligomeric and protofibrillar intermediate species. It is still unclear how misfolded proteins mediate cytotoxicity in diseases, but there is substantial evidence that amyloidogenic proteins are pathogenic. It has long been thought that the expression of molecular chaperones, proteins that mediate the proper

Protein Misfolding Diseases: Current and Emerging Principles and Therapies,
Edited by Marina Ramirez-Alvarado, Jeffery W. Kelly, and Christopher M. Dobson
Copyright © 2010 John Wiley & Sons, Inc.

folding and assembly of other proteins, is impaired in protein misfolding diseases. Molecular chaperones are emerging as critical modulators of protein aggregation and toxicity in a variety of protein misfolding diseases, and the induction of chaperone expression appears to confer neuroprotection.

In this chapter we provide a comprehensive overview of two novel strategies to control gene expression as a treatment for protein misfolding diseases. The first strategy seeks to up-regulate the expression of molecular chaperones by direct or indirect modulation of the heat-shock response. The second strategy seeks to down-regulate the expression of amyloidogenic proteins in Alzheimer disease (AD), Parkinson disease (PD), Huntington disease (HD), and familial amyotrophic lateral sclerosis (FALS) by RNA interference (RNAi).

MOLECULAR CHAPERONES: FUNCTION IN FOLDING AND MISFOLDING

Organisms have evolved a variety of quality control mechanisms to prevent the assembly of proteins into toxic conformations [1]. One major protective mechanism is mediated by a class of proteins called *molecular chaperones*. Chaperones represent several different structural families that are organized by molecular size and function, including Hsp100, Hsp90, Hsp70, Hsp60, Hsp40, and small heat-shock protein (sHsp) families [2]. They do not change the native conformation of a protein, nor are they part of the native structure. Chaperones transiently shield aggregation-prone surfaces from improper interactions until the protein can reach its native fold [3,4]. Typically, this is achieved by multiple rounds of binding and release of substrate polypeptide in an ATP-regulated manner. Thus, chaperones provide an environment in which the folding equilibrium is shifted away from unintended off-pathway reactions toward intramolecular interactions that are productive for proper folding [5].

Molecular chaperones are important for facilitating protein folding under normal conditions, but they are even more critical when cells are subjected to stress, such as extreme temperature or pH alterations. Under stressful conditions, the native conformations of folded proteins are destabilized, and aggregation-prone hydrophobic surfaces that are normally buried in the native state become exposed. To combat the potentially dangerous effects of misfolded proteins, cells have adapted a cellular response system, the heat-shock response (HSR) [6–8], which results in up-regulation of protective factors, including molecular chaperones [6,9]. This response is activated by many types of environmental stress [10], by pathologic conditions such as ischemia, inflammation, and infection, and by misfolded mutant proteins associated with genetic and sporadic diseases.

Aging is associated with a decrease in the functional capacity of protein quality control systems, and the late onset of many protein misfolding diseases correlates with age-dependent deficits in this system. For example, molecular chaperones and other components of the protein quality control system are often co-localized with fibrillar deposits, highly organized fibrils of disease-related misfolded proteins. This observation suggests a critical role for molecular chaperones in the aggregation pathway of pathogenic, amyloidogenic proteins.

Numerous studies have analyzed the effects of chaperones in cellular and animal models of protein misfolding diseases. For example, the overexpression of chaperones in an adenoviral-based system in primary neurons rescued intracellular Aβ42-mediated neurotoxicity [11]. Additionally, down-regulation of heat-shock factor 1 (Hsf1), the major transcription factor of the HSR that induces the expression of several chaperones, increased Aβ42 protein aggregation and accelerated paralysis in the worm model of AD [12]. Although these studies used artificial models in which the concentration of Aβ42 was intentionally increased in the cytoplasm, the results suggest that intracellular accumulation of soluble, prefibrillar forms of Aβ42 cause neuronal dysfunction and death and that molecular chaperones may alter the conformation of cytoplasmic Aβ42 in a way that prevents the accumulation of toxic Aβ42 species or enhances their degradation.

The HSR and molecular chaperones have also been implicated in several other protein misfolding diseases, including inherited disorders caused by CAG/polyglutamine expansion (polyQ) as occurs in HD and the spinocerebellar ataxias. In a worm model of HD, imaging experiments in live cells showed that the Hsp70 system actively regulates the conformation of misfolded huntingtin proteins [13]. Hsp70 exhibits rapid kinetics of association and dissociation with polyQ inclusion bodies that are similar to its interactions with unfolded substrates [13], and mutations in the substrate-binding domain of Hsp70 reduce its co-localization with inclusion bodies [13]. Interestingly, the modulation of polyQ inclusion body formation by Hsp70 depends on the ratio of its cofactors [14]. The cofactor Chip (carboxy terminus of Hsc70-interacting protein), which binds Hsp70 and acts as an E3-ligase to facilitate the transfer of a polyubiquitin chain to misfolded substrates, can induce degradation of polyQ-containing proteins in an Hsp70-dependent manner [15–17]. Overexpression of Hsp70 and Hsp40 inhibited the conversion of a mutant huntingtin fragment into spherical and ringlike oligomers by binding and sequestering the monomeric protein [18]. In animal models of spinocerebellar ataxias, overexpression of molecular chaperones potently suppresses neurodegeneration. In *Drosophila melanogaster*, ocular expression of a truncated form of ataxin-3 that contains a polyQ tract expansion (MJDtr-Q78) causes severe neurodegeneration. This phenotype is exacerbated by a dominant-negative Hsp70 and is rescued by coexpression of human Hsp70 [19,20].

The critical role of molecular chaperones in protein misfolding diseases has been the subject of several recent reviews [1,21–23]. Collectively, the studies described in those reviews all suggest that increasing levels of molecular chaperones suppress the aggregation and toxicity of amyloidogenic proteins. Therefore, pharmacological regulation of the HSR and identification of novel drugs to up-regulate the expression of molecular chaperones are of broad potential interest for the treatment of protein misfolding diseases [7,9].

REGULATION OF THE HSR

The HSR is controlled by the binding and activation of a family of heat-shock transcription factors (HSFs) to heat-shock elements (HSEs) on heat-shock promoters [7,9]. Numerous elegant studies have described the mechanisms and machinery that mediate the HSR. The best characterized heat-shock transcription factor is heat-shock factor 1 (HSF1). A variety of stress signals induce trimerization and accumulation of HSF1 in the nucleus. HSF1 trimers bind with high affinity to the HSEs and regulate genes encoding Hsp70, Hsp90, and sHsp [7]. Furthermore, posttranslational modifications of HSF1 (e.g., phosphorylation and sumoylation) may play a critical role in regulation of the HSR [24–26]. In mammals, the transcription factors HSF2 and HSF4 may participate in the HSR, although they have not been characterized extensively [27,28]. Accumulation of misfolded protein in the cytoplasm or treatment with heavy metals quickly activates the HSR, suggesting that a cellular sensing mechanism exists to detect misfolded proteins [9]. However, the signals that initiate the HSR to the accumulation of misfolded proteins are not well understood.

The most common model for the HSR describes a negatively regulated feedback loop of HSF1 through the multi chaperone machinery of Hsp70 and Hsp90 [9]. Hsp70 and Hsp90 sequester HSF1 in the cytoplasm. Environmental or physiological stress leads to the accumulation of misfolded proteins, thereby increasing the substrate levels for Hsp90 and Hsp70. The consequence is that the chaperones release HSF1, resulting in higher concentrations of cytosolic and nuclear HSF1. As a trimer, HSF1 binds the DNA and activates the expression of protective proteins such as chaperones through a multistep process. Thus, induction of the HSR by several stress signaling pathways and the complex mechanisms of HSF1 activation (e.g., posttranslational modification, interaction with cytosolic chaperones, trimerization of HSF1 and binding to DNA) could be targets for compounds regulating the HSR [7,9].

PHARMACOLOGICAL INDUCTION OF THE HSR

A number of recent studies have described small molecules that manipulate the HSR via HSF1. These small molecules activate or induce HSF1 directly or induce HSF1 activation indirectly by inhibiting Hsp90 (and causing HSF1 dissociation). Next, we discuss a diverse set of chemically unrelated compounds that modulate the HSR (Table 1).

Nonsteroidal Anti-inflammatory Drugs

Nonsteroidal anti-inflammatory drugs–(NSAIDs) such as acetylsalicylic acid and indomethacin display potent anti-inflammatory properties by inhibiting cyclooxygenases, which convert arachidonic acid to prostaglandins [29]. At high concentrations, NSAIDS also activate the HSR, and this mechanism of

TABLE 1 Compounds That Modulate HSR

Compound	Mechanism of HSR Modulation	Cell/Animal Model of Protein Misfolding Disease Tested	Refs.
Puromycin, azetidine	Protein synthesis inhibitors	None tested	40
MG132, lactacystin	Proteasome inhibitors	None tested	41
DCIC, TPCK, TLCK	Serin protease inhibitors	None tested	42
Celastrol	Coinduction of HSF1	PD mouse model; preclinical (ALS)	57, 59, 60
Curcumin	?	AD mouse model; phase II (AD, cancer)	59, 62–65
Paeniflorin	?	None tested	61
Hydroxylamine derivatives: arimoclomol, bimoclomol	Coinduction of HSF1	ALS mouse model; phase IIb ongoing (ALS)	43–52
NSAIDS: indomethacin, sodium salycylate	Coinduction of HSF1	None tested	32–39
Cyclopentenone prostaglandines	Inflammatory mediators	None tested	30
Phospholipase A2	Inflammatory mediators	None tested	29
Acetyl-l-carnitine	Coinduction of HSF1	AD cortical cell model	56
Radicicol	Hsp90 inhibitors	HD cell model	70
Geldanamycin	Hsp90 inhibitors	HD + PD fly and cell model	69–72
17-AAG	Hsp90 inhibitors	SBMA mouse model; Phase II + III (cancer)	73, 74
Novobiocin analog	Hsp90 inhibitors	Cellular model of AD	75
Low-molecular-mass Hsp90 inhibitors	Hsp90 inhibitors	Cellular model of tauopathy	77,78

Source: Modified from [7].

action may contribute to their protective effects, such as antitumor activity [30–32]. The molecular mechanism through which NSAIDS modulate the HSR is unclear but involves the translocation of HSF1 to the nucleus and induction of heat-shock proteins [33–36].

Metabolism Inhibitors

Another class of molecules that regulate the HSR are a protein synthesis inhibitor [37], proteasome inhibitors [38], and serine protease inhibitors [39] (Table 1). All of these compounds increase the abundance of misfolded proteins in cells, thereby activating HSF1 and the expression of chaperones [7]. However, intentionally increasing the levels of misfolded proteins is not likely to be a beneficial strategy for treating diseases.

Hydroxylamine Derivatives

Bimoclomol and arimoclomol are nontoxic hydroxylamine derivatives that co-induce the HSR in many cell types and stimulate the expression of chaperones under stress by enhanced HSF1 activation [40–42]. The functional mechanism for how they work is not well established, but a few studies reported enhanced nuclear transport and binding of active HSF1 to DNA in the presence of these compounds [40–42]. Another potential signal to induce the HSR is alteration of the physical and chemical properties of lipid membranes, such as the density of domains and changes in fluidity, during stress. Interestingly, bimoclomol affected the plasma membrane in a manner similar to membrane alteration during stress [43,44]. Targeting lipid membranes as a novel therapeutic strategy for induction of chaperones has been the subject of excellent reviews [43,45].

Treatment with arimoclomol rescues motor neurons in vivo from injury-induced cell death [46]. Arimoclomol does not induce the HSR in motor neurons itself but in glial cells surrounding injured motor neurons [46]. These data suggest that an enhanced HSR in astroglia contributes to an increased survival of motor neurons to cytotoxic effects of nerve cells [47]. Remarkably, administration of arimoclomol in a transgenic mouse model of inherited amyotrophic lateral sclerosis (ALS) increased survival significantly and dramatically. Arimoclomol increased HSF1 phosphorylation in the spinal cord, followed by the increased expression of HSPs in both astroglial cells and motor neurons [42]. Arimoclomol is being tested in clinical trial as a treatment for ALS. It was recently administrated orally at different dosages to ALS patients over 12 weeks; it appears to be safe and well tolerated and crosses the blood–brain barrier [48,49].

Acetyl-L-Carnitine

AD brains are under significant oxidative stress [50–52]. Acetyl-L-carnitine is a mitochondrial membrane molecule involved in maintaining stable energetics and decreasing oxidative stress. In a study of cortical neurons in vitro, addition of acetyl-L-carnitine increased the level of HSPs and attenuated Aβ42-mediated toxicity and protein oxidation [53]. These characteristics make acetyl-L-carnitine an intriguing potential candidate for treating neurological disorders; however, studies have not been performed in animal models, and a direct molecular link to HSR activation has not been established.

HSR-Inducing Compounds Found in Herbal Medicine

In a high-throughput screen for induction of the HSR by small molecules, celastrol was identified as a compound that activates chaperone expression [54]. The pentacyclic triterpene celastrol (also known as tripterine) is an active component in a Chinese herbal medicine extracted from the plant *Tripterygium wilfordii*, a member of the Celastraceae family. Celastrol has anti-inflammatory activity in animal models of arthritis, lupus, and protein misfolding diseases such as ALS and AD [55], and antiproliferative effects in cancer cell lines. Its mechanism of action has not been eludicated in detail, but it modulates gene expression [55], proteasome inhibition, and heat-shock activation [56,57]. Remarkably, celastrol activates HSF1, leading to expression of chaperones with kinetics similar to those observed for the HSR. Therefore, celastrol provides cytoprotection during exposure to several stresses by activating HSF1. Thus, celastrol may be a promising new class of pharmacologically active regulators of HSR pathways [54]. However, data on the safety and pharmacokinetics of celastrol in humans are lacking, and it is not clear that this drug can cross the blood–brain barrier [48].

Other ingredients in herbal medicines also induce the HSR. Paeoniflorin enhanced phosphorylation and acquisition of the DNA-binding ability of HSF1 and conferred thermotolerance in cells [58]. The polyphenol curcumin of *Curcuma longa* showed anti-inflammatory, immunomodulatory, antimalaria, and anticancer effects [56,59]. In transformed cell lines, curcumin also induced the HSR [43,60], which may be responsible for its general therapeutic effects. Additionally, it inhibited formation of amyloid oligomers and fibrils and reduced amyloid in a transgenic mouse model of AD [61]. Curcumin was well tolerated up to 8000 mg/day in a phase I clinical trial and is being tested in phase II trials for the treatment of cancer, psoriasis, and AD [56,62].

Hsp90 Inhibitors

Initially identified in a screen for anticancer drugs with antiproliferative effects, the fungal antibiotics radicicol and the benzoquinone ansamycin (herbimycin A, geldanamycin, and macbecin) bind Hsp90, block ATP hydrolysis by Hsp90, and cause degradation of client proteins, including oncogenic proteins. Although radicicol is structurally distinct from geldanamycin, they both inhibit the ATPase activity of Hsp90. The geldanamycin derivatives 17-AAG (tanespimycin) and 17-DMAG were developed for cancer therapies and are now in phase II clinical trials and, in combination with other anticancer drugs, in phase

III trials. Hsp90 inhibitors increase levels of free HSF1 in the cytosol, thereby promoting HSF1 activation [63].

In cancer models, treatment with geldanamycin and its derivatives induced the expression of Hsp70 and Hsp27 [64,65], suggesting that Hsp90 inhibitors may be useful for treating diseases caused by protein misfolding. Indeed, geldanamycin inhibits huntingtin aggregation in a cell culture model of HD [66]. Geldanamycin and radicicol suppressed mutant huntingtin toxicity and aggregation in organotypic slice cultures derived from a mouse model [67]. In cultured cells and in a fly model of PD, geldanamycin protected against α-synuclein-mediated toxicity [68,69]. Recent studies demonstrated that administration of 17-AAG improved motor deficits in a transgenic mouse model of spinal and bulbar muscular atrophy (SBMA), a polyQ disorder caused by a CAG repeat expansion in the androgen receptor, which itself is an Hsp90 client protein [70]. 17-AAG apparently decreased fibril formation of the mutant androgen receptor by inducing Hsp70 and Hsp40. Interestingly, 17-AAG caused no detectable toxicity in these mice [71].

Taken together, these data suggest that inhibition of Hsp90 may prevent the formation of toxic protein intermediates by up-regulating other chaperones (i.e., Hsp70). However, a potential toxic side effect of increasing Hsp70 levels by inhibiting Hsp90 results from decreasing the levels of critical Hsp90 client proteins [72,73]. These results suggest that only a small therapeutic window is available for treating neurodegenerative diseases by inhibiting Hsp90 directly. Small-molecule agents that pass the blood–brain barrier must be developed to induce the HSR at lower concentrations [73]. Along these lines, novobiocin analogs that bind to the C-terminus of Hsp90 and not directly to the ATP-binding domain induce Hsp90 and Hsp70 overexpression at nanomolar concentrations. Moreover, these compounds suppressed Aβ-induced toxicity and protected neurons in a cellular model of AD [72].

Other novel low-molecular mass Hsp90 inhibitors were identified that induce Hsp70, Hsp40, and Hsp27 expression in a cell model of human tauopathy: diseases characterized by intracellular fibrillar tangles of misfolded, phosphorylated protein tau. These inhibitors reduce the levels of specific forms of phosphorylated tau (p-tau) by selective clearance [74,75]. Although direct evidence is still missing that increased chaperone levels are linked to clearance of tau, the selectivity of these compounds is presumably achieved through chaperone-mediated degradation of aberrant p-tau species by the proteasome. These Hsp90 inhibitors facilitated the degradation of aberrant p-tau species in a humanized tau transgenic mouse [75]. Another novel small-molecule Hsp90 inhibitor (STA-9090) is being tested in a phase I clinical trial against multiple myeloma (http://www.syntapharma.com/). This compound is 10 to 100 times more potent than the geldanamycin family of inhibitors. Although the mechanism of Hsp90 inhibition has not been described, the development of low-molecular-mass Hsp90 inhibitors may offer sufficient blood–brain barrier penetration to make systemic administration for neurological disorders feasible.

PROSPECTS

The HSR has great potential as a therapeutic target in the treatment of diseases caused by the accumulation of misfolded proteins. Of the small number of compounds that target the HSR, the majority act through HSF1 or Hsp90. The complex cascade of HSR activation offers many additional entry points to manipulate Hsp expression in vivo. The discovery of novel molecular targets may make it possible to generate compounds with increased potency and selectivity for inducing the HSR. However, such compounds could have many potential unwanted side effects. For example, the antiproliferative effect of Hsp90 inhibitors shows clearly that tumor cells are dependent on elevated chaperone levels. Consistent with these studies, elimination of HSF1 in tumor cells induced by oncogenes protected mice and increased their survival [76]. Therefore, it is likely to be important to develop compounds that up-regulate the HSR selectively in affected cells and tissues and not nonspecifically.

DOWN-REGULATION OF AMYLOIDOGENIC PROTEIN EXPRESSION: GENE SILENCING BY RNA INTERFERENCE AS A POTENTIAL THERAPEUTIC APPROACH FOR PROTEIN MISFOLDING DISEASES

RNA interference (RNAi) has rapidly become a powerful tool to regulate gene expression. Based on a cellular mechanism of gene regulation found in all eukaryotes, this technique is used to silence specific genes in the genome. The development of RNAi as potential therapy has also generated considerable interest in treating protein misfolding diseases. Numerous protein misfolding diseases, such as FALS and HD, are caused by a specific mutant gene product that triggers disease through a pathogenic mechanism. In principle, RNAi could be used to target expression of the gene that encodes the misfolded protein. Therapeutic RNAi has shown promising results in rodent models of AD, PD, HD, and FALS.

PRINCIPLES OF RNAi

In 1998, Andrew Fire and Craig Mello described a novel RNA-based mechanism that silences genes in nematodes in a sequence-specific manner [77]. This remarkable discovery describes a highly conserved pathway of gene regulation. Only a small amount of double-stranded RNA (dsRNA) complementary to the mRNA of a particular gene is needed to suppress the expression of that gene by eliminating its messenger RNA. This dsRNA-mediated silencing is enzymatically regulated by two ribonuclease III enzymes called Drosha and Dicer, which process precursor dsRNA into smaller intermediates known as small interfering RNAs (siRNAs). siRNAs are composed of guide strands (e.g., antisense that is complementary to mRNA) and passenger strands. The antisense is incorporated into an RNA-induced silencing complex (RISC). RISC contains the endonuclease Argonaute2 (AGO2), which cleaves bound mRNA in a manner that is complementary to the antisense strand. The cleaved RNA is destroyed rapidly, leaving the RISC antisense complex available for a new cycle. The end result of this process is suppression of target gene expression [78].

There has been much interest in using RNAi to treat human diseases. In one approach, exogenous siRNA molecules complementary to the mRNAs of interest are introduced into cells or organs directly to suppress the expression of the corresponding target gene. Another approach is to introduce a plasmid or virus into the cell, leading to expression of intracellular short hairpin RNAs (shRNA). To understand this application, one must first understand microRNA (miRNA) biosynthesis. The human genome encodes hundreds of small dsRNA molecules, termed miRNAs, that regulate the expression of a large contingent of the protein-coding genes. miRNAs are mostly transcribed by RNA polymerase II, further processed in the nucleus and cytosol in multiple steps involving Dicer and Drosha, and finally loaded into RISC in the cytosol. In contrast to exogenous siRNA, endogenous miRNAs are not exactly complementary to their targets. Suppression of gene expression by miRNA is mediated by blocking the translation of target mRNA rather than by direct cleavage of mRNA. The application of viral or plasmid templates generates pre-miRNA or shRNA. Both are exogenous miRNAs and follow the endogenous miRNA pathway, which results in the blocking of mRNA translation.

In the following sections we summarize specific examples of toxic misfolded proteins in AD, HD, PD, and FALS that are being tested actively as targets to evaluate RNAi in preclinical models of disease. A more detailed description of RNAi as therapeutic strategy is available in excellent reviews [78,79].

Down-Regulation of Aβ and Tau by RNAi in AD

There are several ways to modulate levels of the amyloidogenic proteins Aβ or tau in AD. Aβ is generated by cleavage of the transmembrane protein amyloid precursor protein (APP) by β-site APP cleavage enzyme (BACE1) and γ-secretase, resulting in extracellular Aβ plaque formation. BACE1 is the better target for RNAi therapy because it is more specific to APP than γ-secretase, which is involved in many important cellular processes. Furthermore, the knockout of BACE1 inhibited Aβ plaque formation when crossed to an APP mouse [80]. These data showed that decreasing BACE1 diminished Aβ production and may be an effective strategy to treat AD. Indeed, RNAi-mediated suppression of BACE1 expression in primary neurons derived from a transgenic APP mouse model diminished Aβ production [81]. Moreover, lentiviral transduction of shRNA against BACE1 in the hippocampus of an APP mouse model suppressed BACE1 expression and inhibited formation of Aβ plaques [82] (Table 2). Interestingly, the injection of the viral vectors was

TABLE 2 Preclinical RNAi-Based Trials in Rodent Models of AD, PD, HD, and FALS

Disease	Animal Model	Therapeutic Molecule (Target)	Delivery Vehicle	Ref.
AD	APP double transgenic	shRNA (endogenous BACE1)	Lentivirus	82
	Lentivirus (APP)	shRNA (transgene)	HSV	88
PD	Lentivirus (α-synuclein)	shRNA (transgene)	Lentivirus	95
HD	HD-N171-82Q	shRNA (transgene)	rAAV1	99
	R6/1	shRNA (transgene)	rAAV5	100
	R6/2	siRNA (transgene)	Liposome (CSF infusion)	102
	HD190QG	shRNA (transgene)	rAAV5	101
FALS	SOD1(G93A)	shRNA (transgene)	Lentivirus	109
	SOD1(G93A)	shRNA (transgene)	Lentivirus	108
	SOD1(G93A)	shRNA (transgene)	Transgenic cross	106
	SOD1(G93A)	shRNA (transgene)	Transgenic cross	107

Source: Modified from [79].

performed after the onset of the disease and still led to a significant amelioration of neuronal loss, in contrast to control mice [82].

Another potential target for RNAi therapy is APP itself. Like BACE 1, APP is not an essential gene, and allele-specific silencing of mutant APP was demonstrated successfully in cell-culture models [83,84]. In addition, viral infection of vectors targeting APP in a cell-culture model and in a mouse model of AD down-regulated APP and inhibited Aβ accumulation [85].

Tau is another key protein in the pathogenesis of sporadic AD and inherited forms of frontotemporal dementia that is currently being targeted by RNAi approaches. Remarkably, reduction of endogenous tau by genetic approaches ameliorated Aβ-induced deficits in an AD mouse model [86]. Specifically, tau reduction prevented behavioral deficits in transgenic mice expressing human APP and protected both transgenic and nontransgenic mice against excitotoxicity. Although RNAi was not used to decrease tau expression, these results serve as a proof of concept for this approach. Toward this goal, a recent study showed that treatment with allele-specific RNA duplexes silenced mutant tau in a cell-culture model [83]. In summary, down-regulating multiple targets such as APP, BACE1, and tau by RNAi is being tested actively in preclinical mouse models of disease and may ultimately have a significant impact for treating AD and related disorders [78].

Down-Regulation of α-Synuclein by RNAi in PD

Protein aggregates containing α-synuclein, known as Lewy bodies, are a pathological hallmark in PD brains and several other diseases, known collectively as the synucleinopathies. Mutations in the gene encoding α-synuclein are also linked to dominantly inherited forms of PD [87]. In addition, gene duplication and triplication of the wild-type locus for α-synuclein causes PD [88]. These results suggest that the overexpression of α-synuclein is harmful and that suppressing its expression by RNAi might be beneficial. Interestingly, α-synuclein is not essential in mice [89,90], and α-synuclein knockout mice are resistant to dopaminergic cell death induced by toxins [90]. Recent studies identified a 21-nucleotide sequence in the coding region of human α-synuclein that constitutes an effective target for robust silencing by RNAi and demonstrated allele-specific silencing of the A53T mutant of human α-synuclein [91]. Furthermore, viral vector–based RNAi suppressed endogenous human α-synuclein expression in the human dopaminergic cell line SH-SY5Y and in rat brain [91,92]. Down-regulation of α-synuclein in dopaminergic neuroblastoma cells decreases dopamine transport, leading to 50% reduction of surface density of its transporter [92]. These results suggest that the RNAi-based reduction of α-synuclein affects cellular dopamine homeostasis.

Another group generated an adeno-associated viral (AAV) vector of an α-synuclein ribozyme (rAAV-SynRz) and injected it into the substantia nigra of rat models of PD. Ribozyme-mediated suppression of α-synuclein expression led to increased survival of dopaminergic neurons compared to uninjected controls [93]. Thus, down-regulation of α-synuclein expression could potentially be a suitable target for gene therapy in PD.

Down-Regulation of Huntingtin by RNAi in HD

HD is a dominantly inherited neurological disorder caused by a mutation (CAG repeats) in the gene that encodes the protein huntingtin. There are no effective treatments for HD. Several preclinical trials using RNAi against huntingtin in different animal models of HD have yielded encouraging results (Table 2). Davidson and colleagues demonstrated that AAV vector-delivered shRNA directed against huntingtin reduced mRNA and protein expression in cell culture and in mouse brains and concomitantly reduced the size and number of intraneuronal inclusion bodies. Moreover, gene silencing of huntingtin improved motor and pathological phenotypes in this transgenic mouse model (Table 2) [94]. The findings suggest that RNAi has the potential to improve HD-associated abnormalities dramatically. Similar results were demonstrated in R6/1 mice, which express a human mutated htt exon 1 with approximately 116 CAG repeats and have a more severe and progressive neurological phenotype. In this study, the injection of recombinant adeno-associated viral serotype 5 (rAAV5) into the striatum suppressed levels of mutant huntingtin and reduced the size and number of neuronal intranuclear inclusions [95]. In contrast to the first study, this model showed only a mild benefit on behavioral phenotypes. Postsymptomatic effects of RNAi treatment in the HD190QG transgenic mouse model were verified by Machida and co-workers [96]. The HD190QG transgenic mouse harbors mutant truncated N-terminal htt containing 190 CAG repeats fused with EGFP and shows progressive motor abnormality, neuropathology, and a shortened life span (Table 2). This study showed that rAAV-mediated delivery of RNAi into striatum after onset of the disease ameliorated neuropathological abnormalities. Furthermore, neuronal aggregates in the striatum were reduced after RNAi transduction. These results suggest that postsymptomatic therapy using RNAi against huntingtin leads to direct inhibition of mutant gene expression [96].

In a nonviral RNAi delivery system, siRNAs were used against the huntingtin gene in the R6/2 transgenic mouse model of HD [97]. This HD mouse model expresses the first exon of the human huntingtin gene with 150 CAG repeats under the control of its endogenous promotor, is more severe than R6/1, and has a progressive neurological phenotype. The intraventricular injection of siRNAs at an early postnatal period inhibited transgenic expression of huntingtin exon 1 in neurons and decreased the number and size of inclusion bodies in the striatum. Importantly, treatment with siRNA improved behavioral and pathological abnormalities and increased life span significantly in this mouse model of HD [97].

Overall, treatment using RNAi techniques suppressed mutant huntingtin expression and improved neurologic and pathologic deficits in mice brains. However, additional studies, such as those currently being performed in HD knock-in mice that may display a pattern of neurodegeneration that is more similar to that in HD patients, will be necessary before a clinical trial is initiated in humans. It is still unknown if reduction of both normal and mutant huntingtin alleles will be tolerated in humans. Therefore, to achieve allele specificity by RNAi in HD models, further experiments are being conducted, such as those that are examining polymorphisms in exons of mutant huntingtin or replacement of wt huntingtin.

Down-Regulation of Mutant SOD1 by RNAi in FALS

ALS is a neurodegenerative disease that causes motor neuron degeneration, skeletal muscle atrophy, and paralysis [98]. Research on this disease has focused primarily on FALS, caused by dominant mutations in the gene encoding superoxide dismutase1 (SOD1). Although SOD1-mediated FALS is responsible for only a small percentage of total ALS cases [79], the success of several preclinical mouse studies revealed that RNAi should be developed for this subset of disease.

The first studies describing allele-specific silencing of mutant SOD1 were performed in cell-culture models of FALS with siRNA [99]. Xia and colleagues developed a novel replacement RNAi strategy. After silencing of both wild-type and mutant SOD1 alleles, they replaced wtSOD1 with a single vector [100]. The same group crossed mice that express mutant SOD1 with mice expressing wild-type SOD1 and showed a specific shRNA-based silencing of the mutant but not of the wild-type SOD1 [101].

Saito and co-workers generated a mouse model with a modified siRNA and crossed it with a SOD1 mouse model [102]. In this study, siRNA against SOD1 prevented the development of the disease by suppressing mutant SOD1 in neurons [102]. Several RNAi constructs suppress mutant SOD1 in mouse models of FALS [103,104]. This SOD1 silencing delayed the onset of symptoms and extended life span [104]. These results are a clear proof of principle that siRNA-mediated gene silencing of SOD1 substantially extends survival in animal models of FALS.

RNAi THERAPY: RISKS AND CHALLENGES

The results of preclinical tests of RNAi in protein misfolding disease mouse models are promising. Nevertheless, many risks and challenges remain. One of the greatest hurdles to therapeutic use of RNAi is delivery into the human brain. In almost all preclinical studies in rodents, RNAi was expressed after viral transduction. Ongoing studies will determine if similar approaches may be tested in humans or whether alternative approaches are required, such as local delivery of siRNAs with osmotic minipumps and brain cannulas. Another major obstacle of specific shRNA delivery relates to limitations in siRNA stability and potency, which can be increased by chemically modifying the otherwise charged oligonucleotide so that it can pass the lipid biolayers. One strategy is to conjugate the siRNA with lipids, such as cholesterol, or to incorporate the siRNAs into liposomes. Another approach has been to modify the RNA phosphate backbone to minimize its charge. Recently, a clinical trial of a plasmid-expressed RNAi for hepatitis B was initiated using a proprietary cationic-lipid delivery system (http://www.nucleonicsinc.com/).

Another risk is that the off-target effects of RNAi are still unknown. Will the cell continue to perform its normal functions after foreign RNA is introduced, or will the excess of RNA affect cellular pathways? Recent studies have shown that virus-mediated RNAi delivery into hepatocytes affects endogenous miRNA and increases toxicity [105].

CONCLUSIONS

Over the past five years, several therapeutic strategies have been developed that target pathogenic amyloidogenic proteins. Strategies related to the aggregation pathway include the clearance of aggregated proteins, inhibition of protein aggregation with small molecules, and interference with posttranslational modification [73]. Combinations of drugs aimed at these targets are likely to be effective against protein misfolding diseases [73]. Two novel strategies that could be effective for treating these devastating diseases are up-regulation of molecular chaperones by modulating the HSR and down-regulation of specific pathogenic proteins by RNAi. Greater understanding of the mechanism of the

HSR will provide additional targets to develop more selective and specific drugs to induce chaperone expression in affected cells. The established potency and selectivity of RNAi combined with the development of strategies to increase the efficiency, and delivery of RNAi molecules to the brain may lead us to be cautiously optimistic that RNAi as therapy will be translated from the bench to the bedside for treating these brain diseases.

REFERENCES

1. Muchowski, P.J., Wacker, J.L. (2005). Modulation of neurodegeneration by molecular chaperones. *Nat Rev Neurosci, 6*, 11–22.
2. Ellis, R.J., Hartl, F.U. (1999). Principles of protein folding in the cellular environment. *Curr Opin Struct Biol, 9*, 102–110.
3. Agashe, V.R., Hartl, F.-U. (2000). Roles of molecular chaperones in cytoplasmic protein folding. *Semin Cell Dev Biol, 11*, 15–25.
4. Young, J.C., Agashe, V.R., Siegers, K., Hartl, F.U. (2004). Pathways of chaperone-mediated protein folding in the cytosol. *Nat Rev Mol Cell Biol, 5*, 781–791.
5. Feldman, D.E., Frydman, J. (2000). Protein folding in vivo: the importance of molecular chaperones. *Curr Opin Struct Biol, 10*, 26–33.
6. Lindquist, S. (1986). The heat-shock response. *Annu Rev Biochem, 55*, 1151–1191.
7. Westerheide, S.D., Morimoto, R.I. (2005). Heat shock response modulators as therapeutic tools for diseases of protein conformation. *J Biol Chem, 280*, 33097–33100.
8. Smith, D.F., Whitesell, L., Katsanis, E. (1998). Molecular chaperones: biology and prospects for pharmacological intervention. *Pharmacol Rev, 50*, 493–514.
9. Morimoto, R.I., Santoro, M.G. (1998). Stress-inducible responses and heat shock proteins: new pharmacologic targets for cytoprotection. *Nat Biotechnol, 16*, 833–838.
10. Schlesinger, M. (1986). Heat shock proteins: the search for functions. *J Cell Biol, 103*, 321–325.
11. Magrane, J., Smith, R.C., Walsh, K., Querfurth, H.W. (2004). Heat shock protein 70 participates in the neuroprotective response to intracellularly expressed β-amyloid in neurons. *J Neurosci, 24*, 1700–1706.
12. Cohen, E., Bieschke, J., Perciavalle, R.M., Kelly, J.W., Dillin, A. (2006). Opposing activities protect against age-onset proteotoxicity. *Science, 313*, 1604–1610.
13. Kim, S., Nollen, E.A.A., Kitagawa, K., Bindokas, V.P., Morimoto, R.I. (2002). Polyglutamine protein aggregates are dynamic. *Nat Cell Biol, 4*, 826–831.
14. Rujano, M.A., Kampinga, H.H., Salomons, F.A. (2007). Modulation of polyglutamine inclusion formation by the Hsp70 chaperone machine. *Exp Cell Res, 313*, 3568–3578.
15. Miller, V.M., Nelson, R.F., Gouvion, C.M., Williams, A., Rodriguez-Lebron, E., Harper, S.Q., Davidson, B.L., Rebagliati, M.R., Paulson, H.L. (2005). CHIP suppresses polyglutamine aggregation and toxicity in vitro and in vivo. *J Neurosci, 25*, 9152–9161.

16. Jana, N.R., Dikshit, P., Goswami, A., Kotliarova, S., Murata, S., Tanaka, K., Nukina, N. (2005). Co-chaperone CHIP associates with expanded polyglutamine protein and promotes their degradation by proteasomes. *J Biol Chem, 280*, 11635–11640.

17. Al-Ramahi, I., Lam, Y.C., Chen, H.-K., de Gouyon, B., Zhang, M., Perez, A.M., Branco, J., de Haro, M., Patterson, C., Zoghbi, H.Y., Botas, J. (2006). CHIP protects from the neurotoxicity of expanded and wild-type ataxin-1 and promotes their ubiquitination and degradation. *J Biol Chem, 281*, 26714–26724.

18. Wacker, J.L., Zareie, M.H., Fong, H., Sarikaya, M., Muchowski, P.J. (2004). Hsp70 and Hsp40 attenuate formation of spherical and annular polyglutamine oligomers by partitioning monomer. *Nat Struct Mol Biol, 11*, 1215–1222.

19. Warrick, J.M., Paulson, H.L., Gray-Board, G.L., Bui, Q.T., Fischbeck, K.H., Pittman, R.N., Bonini, N.M. (1998). Expanded polyglutamine protein forms nuclear inclusions and causes neural degeneration in *Drosophila. Cell, 93*, 939–949.

20. Warrick, J.M., Chan, H.Y.E., Gray-Board, G.L., Chai, Y., Paulson, H.L., Bonini, N.M. (1999). Suppression of polyglutamine-mediated neurodegeneration in *Drosophila* by the molecular chaperone HSP70. *Nat Genet*, *23*, 425–428.

21. Chaudhuri, T.K., Paul, S. (2006). Protein misfolding diseases and chaperone-based therapeutic approaches. *FEBS J, 273*, 1331–1349.

22. Muchowski, P.J. (2002). Protein misfolding, amyloid formation, and neurodegeneration: a critical role for molecular chaperones? *Neuron, 35*, 9–12.

23. Meriin, A.B., Sherman, M.Y. (2005). Role of molecular chaperones in neurodegenerative disorders. *Int J Hyperth, 21*, 403–419.

24. Holmberg, C.I., Tran, S.E.F., Eriksson, J.E., Sistonen, L. (2002). Multisite phosphorylation provides sophisticated regulation of transcription factors. *Trends Biochem Sci, 27*, 619–627.

25. Hietakangas, V., Ahlskog, J.K., Jakobsson, A.M., Hellesuo, M., Sahlberg, N.M., Holmberg, C.I., Mikhailov, A., Palvimo, J.J., Pirkkala, L., Sistonen, L. (2003). Phosphorylation of serine 303 Is a prerequisite for the stress-inducible SUMO modification of heat shock factor 1. *Mol Cell Biol, 23*, 2953–2968.

26. Kim, S.-A., Yoon, J.-H., Lee, S.-H., Ahn, S.-G. (2005). Polo-like kinase 1 phosphorylates heat shock transcription factor 1 and mediates its nuclear translocation during heat stress. *J Biol Chem, 280*, 12653–12657.

27. Ostling, P., Bjork, J.K., Roos-Mattjus, P., Mezger, V., Sistonen, L. (2007). Heat shock factor 2 (HSF2) contributes to inducible expression of hsp genes through interplay with HSF1. *J Biol Chem, 282*, 7077–7086.

28. Pirkkala, L., Nykanen, P., Sisotonen, L. (2001). Roles of the heat shock transcription factors in regulation of the heat shock response and beyond. *FASEB J, 15*, 1118–1131.

29. Vane, J.R., Botting, R.M. (2003). The mechanism of action of aspirin. *Thromb Res, 110*, 255–258.

30. Tsujii, M., Kawano, S., Tsuji, S., Sawaoka, H., Hori, M., DuBois, R.N. (1998). Cyclooxygenase regulates angiogenesis induced by colon cancer cells. *Cell, 93*, 705–716.

31. Lee, B., Chen, J., Angelidis, C., Jurivich, D., Morimoto, R. (1995). Pharmacological modulation of heat shock factor 1 by antiinflammatory drugs results in

protection against stress-induced cellular damage. *Proc Natl Acad Sci USA, 92*, 7207–7211.

32. Mortaz, E., Redegeld, F.A., Bloksma, N., Dunsmore, K., Denenberg, A., Wong, H.R., Nijkamp, F.P., Engels, F. (2006). Induction of HSP70 is dispensable for anti-inflammatory action of heat shock or NSAIDs in mast cells. *Exp* Hemat, 34, 414–423.
33. Jurivich, D., Sistonen, L., Kroes, R., Morimoto, R. (1992). Effect of sodium salicylate on the human heat shock response. *Science, 255*, 1243–1245.
34. Soncin, F., Calderwood, S.K. (1996). Reciprocal effects of pro-inflammatory stimuli and anti-inflammatory drugs on the activity of heat shock factor-1 in human monocytes. *Biochem Biophys Res Commun, 229*, 479–484.
35. Housby, J.N., Cahill, C.M., Chu, B., Prevelige, R., Bickford, K., Stevenson, M.A., Calderwood, S.K. (1999). Nonsteroidal anti-inflammatory drugs inhibit the expression of cytokines and induce Hsp70 in human monocytes. *Cytokine, 11*, 347–358.
36. Salminen, W.F., Jr., Voellmy, R., Roberts, S.M. (1998). Effect of *N*-acetylcysteine on heat shock protein induction by acetaminophen in mouse liver. *J Pharmacol Exp Ther, 286*, 519–524.
37. Lee, Y.J., Dewey, W.C. (1987). Effect of cycloheximide or puromycin on induction of thermotolerance by sodium arsenite in Chinese hamster ovary cells: involvement of heat shock proteins. *J Cell Physiol, 132*, 41–48.
38. Holmberg, C.I., Illman, S.A, Kallio, M., Mikhailov, A., Sistonen, L. (2000). Formation of nuclear HSF1 granules varies depending on stress stimuli. *Cell Stress Chaperones, 5*, 219–228.
39. Rossi, A., Elia, G., Santoro, M.G. (1998). Activation of the heat shock factor 1 by serine protease inhibitors. an effect associated with nuclear factor-kappa B inhibition. *J Biol Chem, 273*, 16446–16452.
40. Vigh, L., Literati, P.N., Horvath, I., Torok, Z., Balogh, G., Glatz, A., Kovacs, E., Boros, I., Ferdinandy, P., Parkas, B., et al. (1997). Bimoclomol: a nontoxic, hydroxylamine derivative with stress protein-inducing activity and cytoprotective effects. *Nat Med, 3*, 1150–1154.
41. Hargatai, J., Lewis, H., Boros, I., Racz, T., Fiser, A., Kurucz, I., Benjamin, I., Vigh, L., Penzes, Z., Csermely, P., Latchman, D.S. (2003). Bimoclomol, a heat shock protein co-inducer, acts by the prolonged activation of heat shock factor-1. *Biochem Biophys Res Commun, 307*, 689–695.
42. Kieran, D., Kalmar, B., Dick, J.R.T., Riddoch-Contreras, J., Burnstock, G., Greensmith, L. (2004). Treatment with arimoclomol, a coinducer of heat shock proteins, delays disease progression in ALS mice. *Nat Med, 10*, 402–405.
43. Vigh, L., Nakamoto, H., Landry, J., Gomez-Munoz, A., Harwood, J., L., Horvath, I. (2007). Membrane regulation of the stress response from prokaryotic models to mammalian cells. *Ann NY Acad Sci, 1113*, 40–51.
44. Torok, Z., Tsvetkova, N.M., Balogh, G., Horvath, I., Nagy, E., Penzes, Z., Hargatai, J., Bensaude, O., Csermely, P., Crowe, J.H., et al. (2003). Heat shock protein coinducers with no effect on protein denaturation specifically modulate the membrane lipid phase. *Proc Natl Acad Sci U S A, 100*, 3131–3136.
45. Vigh, L., Horvath, I., Maresca, B., Harwood, J.L. (2007). Can the stress protein response be controlled by "membrane-lipid therapy"? *Trends Biochem Sci, 32*, 357–363.

46. Kalmar, B., Burnstock, G., Vrbova, G., Greensmith, L. (2002). The effect of neonatal nerve injury on the expression of heat shock proteins in developing rat motoneurones. *J Neurotrauma, 19*, 667–679.
47. Kalmar, B., Kieran, D., Greensmith, L. (2005). Molecular chaperones as therapeutic targets in amyotrophic lateral sclerosis. *Biochem Soc Trans, 33*, 551–552.
48. Traynor, B.J., Bruijn, L., Conwit, R., Beal, F., O'Neill, G., Fagan, S.C., Cudkowicz, M.E. (2006). Neuroprotective agents for clinical trials in ALS: a systematic assessment. *Neurology, 67*, 20–27.
49. Bedlack, R.S., Traynor, B.J., Cudkowicz, M.E. (2007). Emerging disease-modifying therapies for the treatment of motor neuron disease/amyotropic lateral sclerosis. *Expert Opin Emerg Drugs, 12*, 229–252.
50. Aksenov, M.Y., Aksenova, M.V., Butterfield, D.A., Geddes, J.W., Markesbery, W.R. (2001). Protein oxidation in the brain in Alzheimer's disease. *Neuroscience, 103*, 373–383.
51. Butterfield, D.A. (2003). Amyloid beta-peptide 1–42-associated free radical–induced oxidative stress and neurodegeneration in Alzheimer's disease brain: mechanisms and consequences. *Curr Med Chem, 10*, 2651–2659.
52. Yatin, S., Varadarajan, S., Link, C.D., Butterfield, D.A. (1999). In vitro and in vivo oxidative stress associated with Alzheimer's amyloid β-peptide (1-42). *Neurobiol Aging, 20*, 325–330.
53. Abdul, H.M., Calabrese, V., Calvani, M., Butterfield, D.A. (2006). Acetyl-L-carnitine-induced up-regulation of heat shock proteins protects cortical neurons against amyloid-beta peptide 1–42-mediated oxidative stress and neurotoxicity: Implications for Alzheimer's disease. *J Neurosci Res, 84*, 398–408.
54. Westerheide, S.D., Bosman, J.D., Mbadugha, B.N.A., Kawahara, T.L.A., Matsumoto, G., Kim, S., Gu, W., Devlin, J.P., Silverman, R.B., Morimoto, R.I. (2004). Celastrols as inducers of the heat shock response and cytoprotection. *J Biol Chem, 279*, 56053–56060.
55. Sethi, G., Ahn, K.S., Pandey, M.K., Aggarwal, B.B. (2007). Celastrol, a novel triterpene, potentiates TNF-induced apoptosis and suppresses invasion of tumor cells by inhibiting NF-κB-regulated gene products and TAK1-mediated NF-κB activation. *Blood, 109*, 2727–2735.
56. Corson, T.W., Crews, C.M. (2007). Molecular understanding and modern application of traditional medicines: triumphs and trials. *Cell, 130*, 769–774.
57. Hieronymus, L., Lamb, J., Ross, K.N., Peng, X.P., Clement, C., Rodina, A., Nieto, M., Du, J., Stegmaier, K., Raj, S.M., et al. (2006). Gene expression signature-based chemical genomic prediction identifies a novel class of HSP90 pathway modulators. *Cancer Cell, 10*, 321–330.
58. Yan, D., Saito, K., Ohmi, Y., Fujie, N. Ohtsuka, K. (2004). Paeoniflorin, a novel heat shock protein–inducing compound. *Cell Stress Chaperones, 9*, 378–389.
59. Aggarwal, B.B., Sundaram, C., Malani, N. Ichikawa, H. (2007). Curcumin: the Indian solid gold. *Adv Exp Med Biol, 595*, 1–75.
60. Khar, A., Mubarak, A., Pardhasaradhi, V.V, Varalakshmi, C., Anjum, R., Kumari, A.L. (2001). Induction of stress response renders human tumor cell lines resistant to curcumin-mediated apoptosis: role of reactive oxygen intermediates. *Cell Stress Chaperones, 6*, 368–376.

61. Yang, F., Lim, G.P., Begum, A.N., Ubeda, O.J., Simmons, M.R., Ambegaokar, S. S., Chen, P.P., Kayed, R., Glabe, C.G., Frautschy, S.A., Cole, G.M. (2005). Curcumin inhibits formation of amyloid β oligomers and fibrils, binds plaques, and reduces amyloid in vivo. *J Biol Chem, 280*, 5892–5901.

62. Hsu, C.H., Cheng, A.L. (2007). Clinial studies with curcumin. *Adv Exp Med Biol, 595*, 471–480.

63. Zou, J., Guo, Y., Guettouche, T., Smith, D.F., Voellmy, R. (1998). Repression of heat shock transcription factor HSF1 activation by HSP90 (HSP90 complex) that forms a stress-sensitive complex with HSF1. *Cell, 94*, 471–480.

64. Miyata, Y. (2005). Hsp90 Inhibitor geldanamycin and its derivatives as novel cancer chemotherapeutic agents. *Curr Pharm Des, 11*, 1131–1138.

65. McCollum, A.K., TenEyck, C.J., Sauer, B.M., Toft, D.O., Erlichman, C. (2006). Up-regulation of heat shock protein 27 induces resistance to 17-allylamino demethoxygeldanamycin through a glutathione-mediated mechanism. *Cancer Res, 66*, 10967–10975.

66. Sittler, A., Lurz, R., Lueder, G., Priller, J., Hayer-Hartl, M.K., Hartl, F.U., Lehrach, H., Wanker, E. E. (2001). Geldanamycin activates a heat shock response and inhibits huntingtin aggregation in a cell culture model of Huntington's disease. *Hum Mol Genet, 10*, 1307–1315.

67. Hay, D.G., Sathasivam, K., Tobaben, S., Stahl, B., Marber, M., Mestril, R., Mahal, A., Smith, D.L., Woodman, B., Bates, G.P. (2004). Progressive decrease in chaperone protein levels in a mouse model of Huntington's disease and induction of stress proteins as a therapeutic approach. *Hum Mol Genet, 13*, 1389–1405.

68. Auluck, P.K., Meulener, M.C., Bonini, N.M. (2005). Mechanisms of suppression of α-synuclein neurotoxicity by geldanamycin in *Drosophila*. *J Biol Chem, 280*, 2873–2878.

69. McLean, P.J., Klucken, J., Shin, Y., Hyman, B.T. (2004). Geldanamycin induces Hsp70 and prevents α-synuclein aggregation and toxicity in vitro. *Biochem Biophys Res Commun, 321*, 665–669.

70. La Spada, A., Peterson, K., Meadows, S., McClain, M., Jeng, G., Chmelar, R., Haugen, H., Chen, K., Singer, M., Moore, D., et al. (1998). Androgen receptor YAC transgenic mice carrying CAG 45 alleles show trinucleotide repeat instability. *Hum Mol Genet, 7*, 959–967.

71. Waza, M., Adachi, H., Katsuno, M., Minamiyama, M., Sang, C., Tanaka, F., Inukai, A., Doyu, M., Sobue, G. (2005). 17-AAG, an Hsp90 inhibitor, ameliorates polyglutamine-mediated motor neuron degeneration. *Nat Med, 11*, 1088–1095.

72. Ansar, S., Burlison, J.A., Hadden, M.K., Yu, X.M., Desino, K.E., Bean, J., Neckers, L., Audus, K.L., Michaelis, M.L., Blagg, B.S.J. (2007). A non-toxic Hsp90 inhibitor protects neurons from Aβ-induced toxicity. *Bioorg Med Chem Lett, 17*, 1984–1990.

73. Rochet, J. (2007). Novel therapeutic strategies for the treatment of protein misfolding diseases. *Expert Rev Mol Med, 9*, 1–34.

74. Dickey, C.A., Dunmore, J., Lu, B., Wang, J.-W., Lee, W.C., Kamal, A., Burrows, F., Eckman, C., Hutton, M., Petrucelli, L. (2006). HSP induction mediates selective clearance of tau phosphorylated at proline-directed Ser/Thr sites but not KXGS (MARK) sites. *FASEB J, 20*, 753–755.

75. Dickey, C.A., Kamal, A., Lundgren, K., Klosak, N., Bailey, R.M., Dunmore, J., Ash, P., Shoraka, S., Zlatkovic, J., Eckman, C.B., et al. (2007). The high-affinity HSP90-CHIP complex recognizes and selectively degrades phosphorylated tau client proteins. *J Clin Invest, 117*, 648–658.

76. Dai, C., Whitesell, L., Rogers, A.B., Lindquist, S. (2007). Heat shock factor 1 is a powerful multifaceted modifier of carcinogenesis. *Cell, 130*, 1005–1018.

77. Fire, A., Xu, S., Montgomery, M.K., Kostas, S.A., Driver, S.E., Mello, C.C. (1998). Potent and specific genetic interference by double-stranded RNA in *Caenorhabditis elegans*. *Nature, 391*, 806–811.

78. Gonzalez-Alegre, P. (2007). Therapeutic RNA interference for neurodegenerative diseases: from promise to progress. *Pharmacol Ther, 114*, 34–55.

79. Gonzalez-Alegre, P., Paulson, H. L. (2007). Technology insight: therapeutic RNA interference—how far from the neurology clinic? *Nat Clin Pract Neurol, 3*, 394–404.

80. McConlogue, L., Buttini, M., Anderson, J.P., Brigham, E.F., Chen, K.S., Freedman, S.B., Games, D., Johnson-Wood, K., Lee, M., Zeller, M., et al. (2007). Partial reduction of BACE1 has dramatic effects on Alzheimer plaque and synaptic pathology in APP transgenic mice. *J Biol Chem, 282*, 26326–26334.

81. Kao, S.-C., Krichevsky, A.M., Kosik, K.S., Tsai, L.-H. (2004). BACE1 Suppression by RNA interference in primary cortical neurons. *J Biol Chem, 279*, 1942–1949.

82. Singer, O., Marr, R.A., Rockenstein, E., Crews, L., Coufal, N.G., Gage, F.H., Verma, I.M., Masliah, E. (2005). Targeting BACE1 with siRNAs ameliorates Alzheimer disease neuropathology in a transgenic model. *Nat Neurosci, 8*, 1343–1349.

83. Miller, V.M., Gouvion, C.M., Davidson, B.L., Paulson, H.L. (2004). Targeting Alzheimer's disease genes with RNA interference: an efficient strategy for silencing mutant alleles. *Nucleic Acids Res, 32*, 661–668.

84. Feng, X., Zhao, P., He, Y., Zuo, Z. (2006). Allele-specific silencing of Alzheimer's disease genes: the amyloid precursor protein genes with Swedish or London mutations. *Gene, 371*, 68–74.

85. Hong, C.-S., Goins, W.F., Goss, J.R., Burton, E.A., Glorioso, J.C. (2006). Herpes simplex virus RNAi and neprilysin gene transfer vectors reduce accumulation of Alzheimer's disease–related amyloid-β peptide in vivo. *Gene Ther, 13*, 1068–1079.

86. Roberson, E.D., Scearce-Levie, K., Palop, J.J., Yan, F., Cheng, I.H., Wu, T., Gerstein, H., Yu, G.-Q., Mucke, L. (2007). Reducing endogenous tau ameliorates amyloid β-induced deficits in an Alzheimer's disease mouse model. *Science, 316*, 750–754.

87. Polymeropoulos, M.H., Lavedan, C., Leroy, E., Ide, S.E., Dehejia, A., Dutra, A., Pike, B., Root, H., Rubenstein, J., Boyer, R., et al. (1997). Mutation in the α-synuclein gene identified in families with Parkinson's disease. *Science, 276*, 2045–2047.

88. Chartier-Harlin, M.C., Kacjergus, J., Roumier, C., Mouroux, V., Douay, X., Lincoln, S., Levecque, C., Larvor, L., Andrieux, J., Hulihan, M., et al. (2004). Alpha-synuclein locus duplication as a cause of familial Parkinson's disease. *Lancet, 364*, 1167–1169.

89. Abeliovich, A., Schmitz, Y., Farinas, I., Choi-Lundberg, D., Ho, W.-H., Castillo, P.E., Shinsky, N., Verdugo, J.M.G., Armanini, M., Ryan, A., et al. (2000). Mice lacking

α-synuclein display functional deficits in the nigrostriatal dopamine system. *Neuron, 25*, 239–252.

90. Dauer, W., Kholodilov, N., Vila, M., Trillat, A.-C., Goodchild, R., Larsen, K.E., Staal, R., Tieu, K., Schmitz, Y., Yuan, C.A., et al. (2002). Resistance of alpha-synuclein null mice to the parkinsonian neurotoxin MPTP. *Proc Natl Acad Sci U S A, 99*, 14524–14529.
91. Sapru, M.K., Yates, J.W., Hogan, S., Jiang, L., Halter, J., Bohn, M.C. (2006). Silencing of human α-synuclein in vitro and in rat brain using lentiviral-mediated RNAi. *Exp Neurol, 198*, 382–390.
92. Fountaine, T., Wade-Martins, R. (2007). RNA interference-mediated knockdown of alpha-synuclein protects human dopaminergic neuroblastoma cells from MPP^+ toxicity and reduces dopamine transport. *J Neurosci Res, 85*, 351–363.
93. Hayashita-Kinoh, H., Yamada, M., Yokota, T., Mizuno, Y., Mochizuki, H. (2006). Down-regulation of α-synuclein expression can rescue dopaminergic cells from cell death in the substantia nigra of Parkinson's disease rat model. *Biochem Biophys Res Commun, 341*, 1088–1095.
94. Harper, S.Q., Staber, P.D., He, X., Eliason, S.L., Martins, I.H., Mao, Q., Yang, L., Kotin, R.M., Paulson, H.L., Davidson, B.L. (2005). RNA interference improves motor and neuropathological abnormalities in a Huntington's disease mouse model. *Proc Natl Acad Sci U S A, 102*, 5820–5825.
95. Rodriguez-Lebron, E., Denovan-Wright, E.M., Nash, K., Lewin, A.S., Mandel, R.J. (2005). Intrastriatal rAAV-mediated delivery of anti-huntingtin shRNAs induces partial reversal of disease progression in R6/1 Huntington's disease transgenic mice. *Mol Ther, 12*, 618–633.
96. Machida, Y., Okada, T., Kurosawa, M., Oyama, F., Ozawa, K., Nukina, N. (2006). rAAV-mediated shRNA ameliorated neuropathology in Huntington disease model mouse. *Biochem Biophys Res Commun, 343*, 190–197.
97. Wang, Y.-L., Liu, W., Wada, E., Murata, M., Wada, K., Kanazawa, I. (2005). Clinico-pathological rescue of a model mouse of Huntington's disease by siRNA. *Neurosci Res, 53*, 241–249.
98. Rowland, L.P., Shneider, N.A. (2001). Amyotrophic lateral sclerosis. *N Engl J Med, 344*, 1688–1700.
99. Ding, H., Schwarz, D.S., Keene, A., Affar, E.B., Fenton, L., Xia, X., Shi, Y., Zamore, P.D., Xu, Z. (2003). Selective silencing by RNAi of a dominant allele that causes amyotrophic lateral sclerosis. *Aging Cell, 2*, 209–217.
100. Xia, X.G., Zhou, H., Zhou, S., Yu, Y., Wu, R., Xu, Z. (2005). An RNAi strategy for treatment of amyotrophic lateral sclerosis caused by mutant Cu,Zn superoxide dismutase. *J Neurochem, 92*, 362–367.
101. Xia, X., Zhou, H., Huang, Y., Xu, Z. (2006). Allele-specific RNAi selectively silences mutant SOD1 and achieves significant therapeutic benefit in vivo. *Neurobiol Dis, 23*, 578–586.
102. Saito, Y., Yokota, T., Mitani, T., Ito, K., Anzai, M., Miyagishi, M., Taira, K., Mizusawa, H. (2005). Transgenic small interfering RNA halts amyotrophic lateral sclerosis in a mouse model. *J Biol Chem, 280*, 42826–42830.
103. Raoul, C., Abbas-Terki, T., Bensadoun, J.-C., Guillot, S., Haase, G., Szulc, J., Henderson, C.E., Aebischer, P. (2005). Lentiviral-mediated silencing of SOD1

through RNA interference retards disease onset and progression in a mouse model of ALS. *Nat Med, 11*, 423–428.

104. Ralph, G.S., Radcliffe, P.A., Day, D.M., Carthy, J.M., Leroux, M.A., Lee, D.C.P., Wong, L.-F., Bilsland, L.G., Greensmith, L., Kingsman, S.M., et al. (2005). Silencing mutant SOD1 using RNAi protects against neurodegeneration and extends survival in an ALS model. *Nat Med, 11*, 429–433.

105. Grimm, D., Streetz, K.L., Jopling, C.L., Storm, T.A., Pandey, K., Davis, C.R., Marion, P., Salazar, F., Kay, M.A. (2006). Fatality in mice due to oversaturation of cellular microRNA/short hairpin RNA pathways. *Nature, 441*, 537–541.

45

UNDERSTANDING AND AMELIORATING THE TTR AMYLOIDOSES

STEVEN M. JOHNSON, R. LUKE WISEMAN, NATÀLIA REIXACH, JOHAN F. PAULSSON, SUNGWOOK CHOI, EVAN T. POWERS, JOEL N. BUXBAUM, AND JEFFERY W. KELLY

Departments of Chemistry and Molecular and Experimental Medicine and Skaggs Institute for Chemical Biology, The Scripps Research Institute, La Jolla, California

OVERVIEW

The long-term goal of our transthyretin (TTR) amyloid disease research is to understand the etiology of these disorders well enough to envision and apply new strategies for both therapy and prophylaxis. Although genetic and pathological evidence links extracellular TTR aggregation to the cause of these maladies, the precise structure of the toxic species, the mechanism by which proteotoxicity is conferred, and whether it is induced from outside the cell or inside the cell or both are, at best, understood only partially. In the first half of this chapter, we summarize our current knowledge of the mechanism of the TTR amyloidoses and consider that knowledge in the context of the normal functions of TTR. In the second half, we demonstrate that a reduction in the concentration of the amyloidogenic TTR monomers is sufficient to ameliorate amyloidosis. We discuss a chemical approach to prevent tetramer dissociation by utilizing ligand binding to raise the kinetic barrier of dissociation sufficiently to preclude monomer release and subsequent amyloid fibril formation. The hypothesis that small-molecule-mediated native-state kinetic stabilization of

Protein Misfolding Diseases: Current and Emerging Principles and Therapies,
Edited by Marina Ramirez-Alvarado, Jeffery W. Kelly, and Christopher M. Dobson
Copyright © 2010 John Wiley & Sons, Inc.

TTR will alleviate the TTR amyloidoses has recently been demonstrated by a phase II/III placebo-controlled clinical trial carried out by FoldRx Pharmaceuticals (www.foldrx.com).

INTRODUCTION TO AMYLOID DISEASES

The amyloid diseases are thought to be caused by the extracellular misfolding and/or aggregation of one of more than 30 distinct human peptides or proteins [1–12]. While numerous distinct quaternary structural misassembly intermediates and dead-end products are formed during these aggregation reactions, cross-β-sheet assemblies or amyloid fibrils appear to be the terminal assembly structure after which these diseases are named (Fig. 1) [13–17]. Although some still believe that tissue displacement by amyloid fibrils is the main cause of these maladies, the current consensus, based on the most recent data, favors the notion that smaller toxic oligomers formed as transient intermediates during aggregation (Fig. 1) lead to cellular dysfunction and death by a variety of mechanisms, including apoptosis [11,13,14,17–20]. Amyloid displacement of tissue appears to be secondary but may contribute to disease pathogenesis in some organs. The toxic oligomer hypothesis is difficult to prove, since unstructured spherical aggregates appear to progress to protofibrils and fibrils

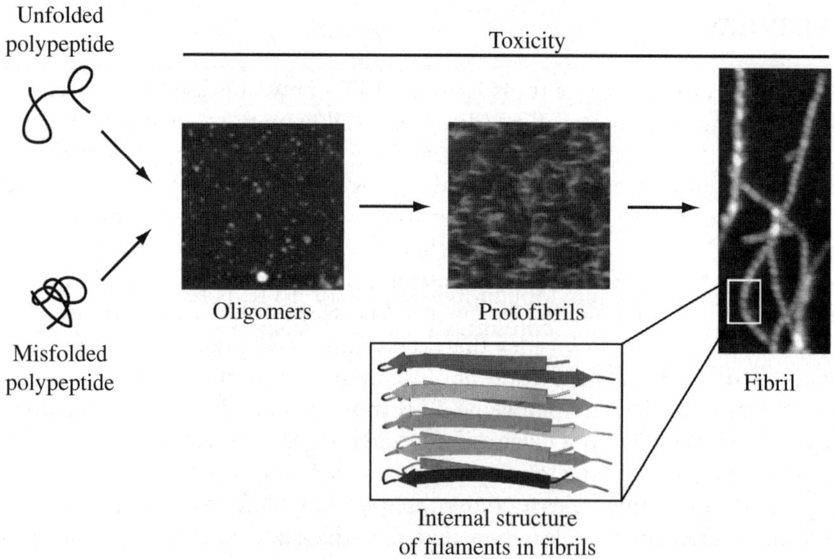

FIG. 1 Amyloid fibril formation arises from the concentration-dependent self-assembly of natively unfolded polypeptides (top) or partially denatured proteins (bottom) and ultimately proceeds through multiple intermediates to form the final cross-β-sheet amyloid fibril. (*See insert for color representation of figure.*)

in a highly dynamic, and not necessarily vectorial, process [13,15–17,21–27]. Further, in some instances, not only amyloid fibrils but amorphous aggregates, oligomers, and monomers are associated with tissue dysfunction [13,20,28,29]. It is also becoming clear from studies in cell culture that cells strongly influence the process of amyloidogenesis by either removing or rapidly remodeling aggregates, but at present the extent of cellular involvement in human amyloidogenesis is not clear [28]. The general observation of diverse assembly intermediates that self-assemble continuously into amyloid and other structures illustrates how difficult it can be to test the toxicity of a given intermediate in a cell- or tissue-based experiment requiring hours to days, especially when considering cellular influences on these processes [30].

Although there is much debate about the nature of the proteotoxic species and the mechanism of cytotoxicity, there is compelling genetic, biochemical, and pathological evidence supporting the amyloid hypothesis, the notion that the process of amyloidogenesis is the cause of the transthyretin (TTR) amyloid diseases and the amyloidoses in general [1,13,29,31–37]. With respect to sporadic Alzheimer disease, there is less acceptance of the hypothesis that the multistep process of protein aggregation, including cross-β-sheet fibril formation, causes the resulting neurodegeneration. The reason for the skepticism regarding the amyloid hypothesis in Alzheimer disease, and some of the other amyloidoses, is the lack of a mechanistic connection between the process of amyloidogenesis and degenerative processes. While amyloid fibrils are easily detected outside the cell in postmortem tissue, there is evidence that fibrillogenesis inside a living cell could also contribute to the pathology observed in some of these disorders [38–41]. The cellular origin of a particular amyloid precursor may be the determinant of whether aggregation is primarily intra- or extracellular, although our current inability to detect aggregates and amyloid within cells sensitively could bias our appreciation for the generality of intracellular amyloidogenesis [42–45].

Amyloidogenesis differs significantly from the sickle cell hemoglobin aggregation pathway familiar to many scientists, wherein surface mutations in the context of a natively folded hemoglobin protein facilitates its misassembly [46]. In contrast, partial unfolding of an initially folded and secreted extracellular protein, like the immunoglobulin light chains or TTR, is required for amyloid fibril formation. Partial unfolding exposes stretches of largely uncharged hydrophobic sequences that efficiently misassemble into cross-β-sheet amyloid structures [5,21–23,47–55].

In the cases where extracellular amyloid fibril formation arises from an intrinsically disordered or natively unfolded peptide, such as amyloid β (Aβ) linked to Alzheimer disease, the efficiency of amyloidogenesis is strongly influenced by concentration and whether a template or seed is present that recruits unfolded monomers to the fibril [24,56–60]. Although it is often said that intrinsically disordered peptides adopt a random coil conformation, most natively unfolded polypeptides/proteins adopt an ensemble of partially collapsed conformations to minimize the exposure of their side chains to the aqueous solvent. Hence, everything else being constant, amyloidogenesis proceeds faster under mildly denaturing conditions, even for natively unfolded proteins. Environmental changes can also trigger aggregation: for example, when the concentration of a membrane-derived aldehyde, resulting from inflammation/oxidative stress, is sufficient to modify a natively unfolded protein. This can significantly lower the concentration required for amyloidogenesis to begin [24,25,61–64]. Therefore, there are two types of amyloid proteins: those that need to partially unfold from a native state to misassemble and those that can misassemble from their intrinsically disordered conformational ensemble when the concentration reaches a critical level. Transthyretin, a plasma protein that is implicated in several amyloidoses, is an example of the former.

INTRODUCTION TO TRANSTHYRETIN

Transthyretin is a 55-kDa homotetrameric protein, composed of 127-amino acid subunits that adopt a β-sandwich fold (Fig. 2) [65–69]. TTR is present in both blood and cerebrospinal fluid (CSF). It is a well-established negative

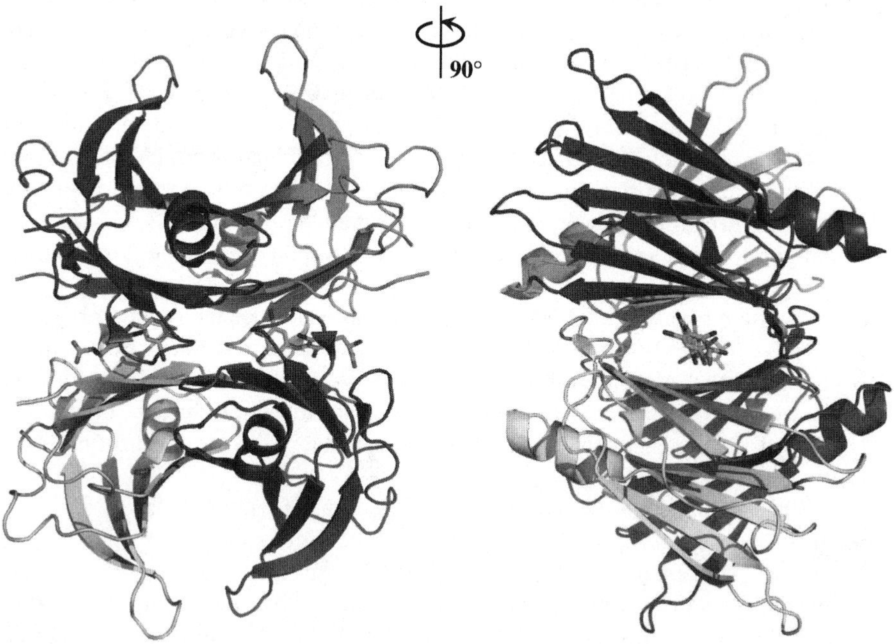

FIG. 2 Structure of the TTR tetramer, with each monomer depicted in a different color, with thyroxine bound along the crystallographic two fold axis in each of two symmetry related thyroxine-binding sites. (From [162], with permission. Copyright © 2008 American Chemical Society.) (*See insert for color representation of figure.*)

acute-phase protein; that is, serum TTR concentration is reduced in acute and chronic inflammation or during malnutrition as a result of decreased transcription in response to the inflammatory cytokines interleukin 6, interleukin 1, and TNFα, and mediated by hepatocyte nuclear factor 3α [70].

Although TTR is the primary carrier of thyroxine (T_4; Fig. 2) in the CSF, thyroid-binding globulin and albumin are the major carriers of T_4 in the blood. Thus, the vast majority (> 99.5%) of the two T_4-binding sites within each TTR tetramer in blood are unoccupied and available for small-molecule binding [71]. This can be exploited to ameliorate the TTR amyloidoses (see below). Transthyretin also carries approximately 0.5 equivalents of *holo*-retinol binding protein (RBP) per TTR tetramer, utilizing four partially overlapping binding sites, only two of which can be occupied at any one time [72–75]. There is no evidence that occupancy of the RBP sites influences T_4-binding affinity or that binding to the T_4 sites alters RBP binding affinity [75].

While the transport functions of TTR described above are well established, we suspect that TTR has another, yet relatively uncharacterized function in the brain. Cerebral TTR expression has been shown to rise dramatically during the course of experimental Alzheimer disease (AD) in mice and in response to the administration of druglike small molecules or mixtures of compounds such as *Gingko* extracts, or dietary fatty acids [76–78]. For many years, it was thought that the only site of cerebral TTR synthesis was the choroid plexus and the similarly derived leptomeningeal ependymal cells [79–81]. Recent experimental evidence suggests that neurons in some brain regions and peripheral nerves can also synthesize the protein, although at much lower levels than those produced by the choroid plexus epithelium [82,83]. The fact that overexpression of a wild-type human TTR transgene suppresses both the neuropathologic and behavioral changes seen in transgenic models of human Alzheimer disease and that the onset of these features is accelerated in mice in which the endogenous TTR gene has been silenced suggest a hitherto unsuspected central nervous system (CNS) function for TTR [77,78,84]. In AD mice overexpressing TTR, the hippocampal and cortical neurons stained with an antibody to TTR, as did the neurons in human AD and control brains. In addition, there was co-localization of Aβ and TTR in the residual extracellular Aβ plaques and in the vessels showing Congophilic angiopathy. It is not yet clear whether the intraneuronal TTR is the result of intraneuronal synthesis or endocytosis of protein synthesized in the choroid plexus. However, the notion of the choroid as the sole site of TTR synthesis in the brain can no longer be accepted as dogma.

Equally interesting is the defect in spatial learning exhibited by the TTR knockout mice, even in the absence of a human AD-associated Aβ transgene [85]. $TTR^{-/-}$ mice have also been described as having a defect in neuronal repair in response to injury and an abnormal ratio of proliferating to apoptotic cells in the subventricular zone of the adult mouse brain [86,87]. The latter effect is presumed to be related to its thyroxine transport function, perhaps due to its absence from the choroid plexus. The observations outlined in the last two paragraphs suggest that TTR is more than just a carrier of physiologically important small molecules in the nervous system.

MECHANISM OF TTR AMYLOIDOGENESIS

The process of transthyretin amyloidogenesis or amyloid fibril formation appears to cause a discrete set of human amyloid diseases, as outlined below and in Table 1. Although it remains unresolved precisely how TTR forms amyloid in patients, biophysical studies on wild-type (WT) TTR and several disease-associated variants reveal that tetramer dissociation is rate limiting for amyloidogenesis [35,36,48,49,52,71,88–95]. The folded monomer that results from the process of dissociation must subsequently undergo partial denaturation to misassemble into aggregate structures, including amyloid fibrils, assuming that the concentration is high enough (Fig. 3) [5,21–23,35,36,42, 48–52,93,95–101]. TTR forms numerous aggregate morphologies, including amorphous aggregates, spherical structures, porelike structures, protofilaments, ribbons, amyloid fibrils, and likely many other morphologies (Fig. 1) [15,17,22,28,92]. It is now apparent from studies on other amyloidogenic proteins that fibril quaternary structure can originate from the self-assembly of smaller multimeric intermediates having less well defined structures [13–15,17,22,29,49,102–107].

TTR amyloidogenesis in vitro occurs very inefficiently under physiological conditions but is accelerated under acidic or other partially denaturing conditions, where the thermodynamically linked equilibria between tetramer, natively folded monomer, and partially denatured monomer are shifted toward the latter, facilitating amyloidogenesis (Fig. 3) [22,23,48,49,52,93,98]. While these conditions accelerate amyloidogenesis and allow the process to be studied on a convenient laboratory time scale, the rate-limiting steps and mechanisms can change when amyloidogenesis takes place in vivo [56].

TTR AMYLOIDOGENESIS OCCURS BY A DOWNHILL POLYMERIZATION

Nucleation, the formation of a high-energy oligomeric intermediate, is often required before misassembly or aggregation reactions become self-sustaining or spontaneous (Fig. 4) [56–58]. Nucleated aggregation reactions are the norm for peptide amyloidogenesis, the clinically most important example being Aβ amyloid fibril formation associated with Alzheimer disease [58]. It is notable that nucleation does not appear to be required for TTR amyloidogenesis under any conditions thus far evaluated [93]. The initial bimolecular collisions between misfolded TTR monomers lead to lower-energy dimeric aggregates that further aggregate to yield even lower-energy structures (Fig. 4). Thus, the process of TTR aggregation appears to be under thermodynamic control from

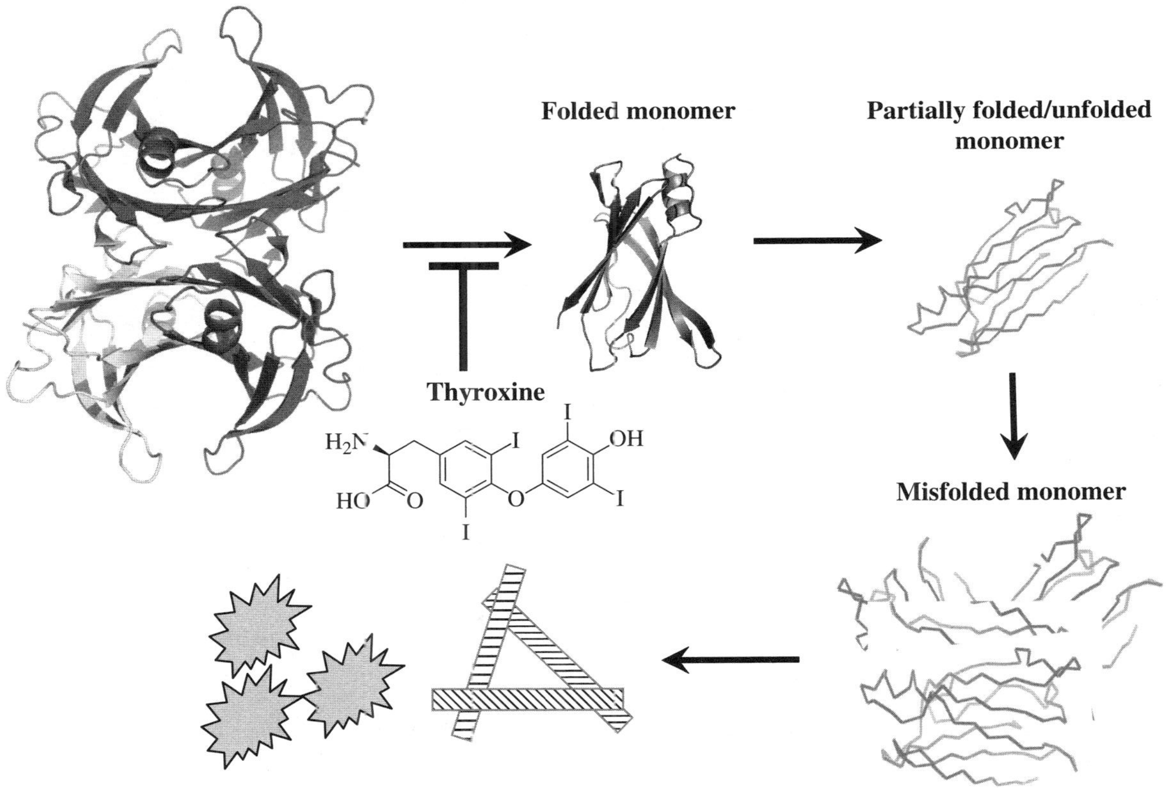

FIG. 3 TTR amyloidogenesis cascade. The TTR tetramer dissociates into four folded monomers which must undergo partial denaturation in order, subsequently, to misassemble into a spectrum of aggregate structures, including protofibrils, cross-β-sheet amyloid fibrils, and amorphous aggregates.

TABLE 1 Human Amyloid Diseases

Disease	Clinical Classification	Amyloid Constituent	Age of Onset (yr)	Population Affected
Senile systemic amyloidosis (SSA)	Cardiomyopathy	WT TTR	>60	10–25% of the general population over 80 years of age
Familial amyloid cardiomyopathy (FAC)	Cardiomyopathy	V122I TTR	>65	3–4% African-American population carriers; clinical penetrance probably high
		T60A TTR L111M TTR	>60	Geographic clusters
Familial amyloid polyneuropathy (FAP) V30M	Peripheral neuropathy	V30M TTR	30–80	Portugal; France; Japan high penetrance; Sweden low penetrance, late onset
Familial amyloid polyneuropathy (FAP) non V30M	Peripheral neuropathy, cardiomyopathy	One of >100 mutant TTRs	15–70	Worldwide
Central nervous system selective amyloidoses (CNSA)	CNS amyloidosis	Highly destabilized TTR mutants, e.g., A25T and D18G TTR	<50	Rare

the outset, and we call this mechanism a downhill polymerization [93]. Since nucleated polymerizations can convert to downhill aggregation reactions under certain circumstances (e.g., at high concentration), these mechanistic possibilities represent extremes of a continuum (Fig. 4) [56].

Although a high-resolution structure of TTR amyloid fibrils has proven elusive thus far, synchrotron diffraction studies of oriented transthyretin amyloid fibrils at a resolution of 2 Å have suggested a model of TTR amyloid featuring a cross-β-sheet structure (Fig. 1) [108]. The nuclear magnetic resonance- and mutagenesis-based high-resolution cross-β-sheet structural models of Aβ amyloid fibrils and the crystal structure-based cross-β-sheet structures of the hexa- and heptapeptide amyloids (derived from larger

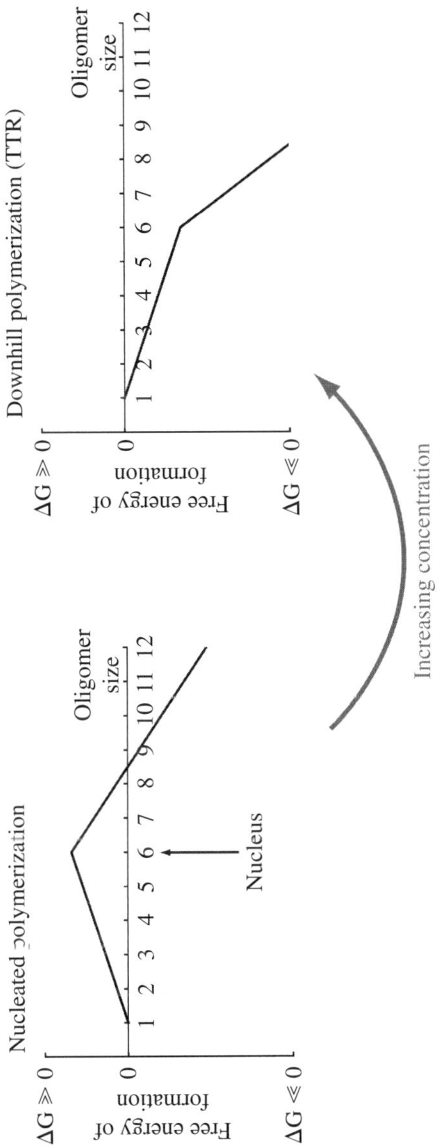

FIG. 4 Amyloidogenesis can occur by way of a nucleated polymerization or through a downhill polymerization mechanism. TTR always appears to aggregate through a downhill mechanism, whereas Aβ fibrillization proceeds through a high-energy nucleus at relatively low concentration ($< \mu M$) and can become a downhill polymerization at high concentration.

amyloidogenic proteins) support the hypothesis that TTR amyloid will probably have a cross-β-sheet structure, although heterogeneous amyloid structures appear likely. We know much less about the structure of the TTR amorphous aggregates, protofibrils, and the like.

CURRENT UNDERSTANDING OF THE ETIOLOGY OF THE TTR AMYLOID DISEASES

Substantial evidence links rate-limiting TTR tetramer dissociation, partial monomer misfolding, and misassembly or amyloidogenesis (Fig. 3) to the tissue dysfunction characterizing the TTR amyloidoses, including senile systemic amyloidosis (SSA), familial amyloid cardiomyopathy (FAC), familial amyloid polyneuropathy (FAP), and central nervous system–selective amyloidoses (CNSA) (Table 1) [5,21–23,28,35,36,42,48–52,93,95–101,109–111]. SSA is a late-onset sporadic disease associated with WT TTR deposition in the heart, affecting 10 to 25% of the population over 80 years of age and ultimately leading to congestive heart failure or sudden death related to arrhythmias [110–112]. TTR does not appear to be synthesized in the heart; therefore, it is likely that it is circulating plasma TTR, which is secreted by the liver, that undergoes amyloidogenesis in the heart. FAC is associated with the apparently preferential amyloid deposition of some TTR mutants in the heart, leading to congestive heart failure. One such mutant (TTR V122I) is found in 3 to 4% of African Americans. Although the precise clinical penetrance of V122I-asociated FAC is unknown, it appears to be high, possibly resulting in 100,000 to 150,000 symptomatic patients above the age of 65 in the United States alone [109,113,114].

Amyloid formation by one of the over 100 less common TTR mutations can lead to FAC or FAP [9,12,109,115–121]. FAP usually presents as a peripheral neuropathy, sometimes with autonomic involvement [122]. The most important FAP-associated mutation, V30M [115] is present in 5000 to 10,000 symptomatic patients worldwide. The penetrance of this mutation is above 80% in Portugal and Japan, but is much lower (perhaps 10%) in Sweden [123,124]. The reason for the wide variance in clinical penetrance in patients carrying the same amino acid mutation is not fully understood. At least two studies have suggested variation in the genetic backgrounds of different carriers could partially explain differential penetrance: variations in either their nuclear or mitochondrial genomes [125,126]. The number of patients associated with and the penetrance of the other 100+ rare FAP mutations is not well documented. The CNSA-associated mutants target the choroid plexus, and the similarly derived leptomeningeal cells lining the outside of the brain, probably because these highly destabilized TTR variants are substantially degraded in the liver, preventing their secretion into the blood and thus largely, but not completely, preventing systemic amyloidogenesis and amyloidosis [42]. The choroid plexus is able to secrete these highly destabilized TTR tetramers into the CSF because the high concentration of T_4 in this tissue can bind and stabilize the TTR variants. This enables their amyloidogenesis close to the site of secretion into the extracellular space [42]. The familial diseases (FAP, FAC, and CNSA) can appear as early as the second decade of life (e.g., the L55P TTR mutation associated with very early onset FAP) [15,127]. If left unchecked, the TTR amyloidoses generally lead to selective organ dysfunction and, ultimately, death within a decade.

Since TTR does not cross the blood–brain barrier, the liver appears to be responsible for almost all the circulating TTR (1.8 to 7.2 μM tetramer) that aggregates in tissues except the brain and the eye. The choroid plexus secretes TTR into the CSF, which bathes the brain (0.04 to 0.4 μM tetramer) [80,128–130]. Retinal epithelial cells secrete TTR in the eye [81]. Other tissues have been reported to secrete relatively small amounts of TTR, but it is not clear how much they contribute to circulating TTR levels and whether the concentrations of locally produced TTR contribute to local amyloidogenesis and the tissue selectivity of disease [83,131]. It is possible that TTR synthesis and secretion can be induced in certain tissues and that capacity could change with aging or upon activation of stress pathways.

The molecular basis for the susceptibility of certain tissues and why aging is a key risk factor in these maladies is only beginning to be understood [42,106,132]. The most intriguing observation in this regard is the failure to observe deposition in the liver, the major site of TTR synthesis, in patients with TTR deposition elsewhere. This phenomenon is also seen in mice transgenic for multiple copies of the human TTR gene, where serum TTR levels may be 5 to 10 times higher than in normal human serum. In these TTR transgenics, deposits are seen in the skin, gut, heart, and kidneys but not in the liver [27]. There are several potential explanations for the lack of hepatic deposits. It is possible that just as T_4 stabilizes mutant TTR synthesized in the choroid plexus and allows its secretion [42,133], binding of retinol-binding protein may stabilize the TTR tetramer in the liver sufficiently that it is secreted by the hepatocyte [75]. While only 25 to 50% of serum TTR can be shown to be complexed with retinol-binding protein, rendering TTR nonamyloidogenic, the decrease in TTR concentration that has the potential to be amyloidogenic could explain the lack of amyloid in and around the liver [75,93].

Another possible contributing factor is the highly developed protein homeostasis network within hepatocytes [132]. In the course of evolution, the liver has become a protein-synthesizing and protein-secreting factory for the entire organism; hence, its protein synthesis machinery must be able to handle large protein loads, a fraction of which must be in an unfolded or misfolded state at any moment in time. This hypothesis is supported by the observation that there appear to be few protein-misfolding diseases that result in liver damage secondary to intrahepatic aggregation, the exception being one of the serpinopathies, wherein the Z-variant of α1-anti-trypsin is retained in the endoplasmic reticulum of liver cells [134,135]. It is also possible that toxic oligomeric TTR aggregates form intracellularly, kill the hepatocytes, which are then replaced by active hepatic regeneration, and the aggregates are taken up and degraded by hepatic Kupffer cells, leaving a liver that is morphologically close to normal. However, there are no experimental or observational data supporting the last possibility even in transgenic animals producing large amounts of TTR.

What has become evident from studies in transgenic animals is that cardiac TTR deposition seems to occur in older animals; hence, the ability to maintain protein hoemostasis in the heart appears to become inadequate upon aging. This is inferred from reduced transcription of chaperone genes and those encoding molecules involved in the unfolded protein response (Buxbaum et al., unpublished)[132]. It is not clear why this occurs, but data from experiments in *Caenorhabditis elegans* and mice hemizygous for deletions of the insulin growth factor receptor indicate that some of the signaling pathways that control aging also influence the efficiency of the protein homeostasis network within the secretory pathway in an age-dependent fashion, and could have a significant effect on TTR secretion aptitude [106,132].

INFLUENCE OF THE CELLULAR PROTEOSTASIS NETWORK ON TTR AMYLOIDOGENESIS

As mentioned above, we have recently shown that different tissues have different TTR secretory aptitudes. Cultured choroid plexus epithelial cells appear to be more permissive secretors of highly destabilized, misfolding-prone TTR variants than cultured murine hepatocytes [42]. In vivo, this effect might be enhanced by high levels of T_4 in the choroid plexus, which when bound, stabilizes the tetrameric structure of TTR, allowing more properly folded TTR tetramer to engage the secretory machinery for export [42]. It is likely that the protein homeostasis (proteostasis) network within the cell, the capacity of which is cell-type dependent, determines whether destabilized TTR is folded and secreted or misfolded and targeted for degradation [132]. The proteostasis network is comprised of macromolecular chaperones, folding enzymes, trafficking machinery, and the components of the endoplasmic reticulum–associated degradation pathway, among others. Its capacity is controlled by numerous stress-responsive signaling pathways, including the unfolded protein response, principally influencing the proteostasis capacity of the endoplasmic reticulum, and the heat-shock response, mostly affecting the proteostasis capacity of the cytosol [132]. Thus, in an organismal context, proteostatic differences between hepatocytes and choroid cells are anticipated to influence the sites of TTR deposition by determining the amounts of TTR in the plasma and the CSF, respectively [42]. The proportion of TTR that is susceptible to misfolding and aggregation in a given tissue is further influenced by local factors, such as the extracellular matrix composition that dramatically influences the efficiency of amyloidogenesis [42,136].

DISEASE-ASSOCIATED TTR MUTANTS ARE THERMODYNAMICALLY LESS STABLE THAN WILD-TYPE TTR

In nearly every case, TTR tetramers composed of disease-associated subunits adopt a tetrameric structure indistinguishable from that exhibited by WT TTR [67–69,98]. Only the most destabilized variant associated with leptomeningeal

TTR deposition (D18G TTR) is incapable of tetramer formation in the absence of ligand binding, which can stabilize the tetramer dramatically [101]. Notably, all of the disease-associated TTR mutations characterized to date destabilize either the native tetrameric structure and/or the structure of the TTR monomer [42,92,96,137]. Since TTR quaternary and tertiary structural transitions are generally strongly thermodynamically linked, either mechanism of instability leads to a higher population of monomers that can take on an amyloidogenic misassembly-competent conformation [42,50,51,92,96]. In most mutants, thermodynamic instability seems to be the main risk factor for development of the TTR amyloidoses. V122I TTR seems to be one instance in which kinetic instability seems to strongly influence amyloidogenesis; it is not clear if this is related to its deposition relatively late in life [42,91].

The rate of TTR tetramer dissociation, generally rate limiting for amyloidogenesis, is also influenced by disease-associated mutations (as in the case of V122I TTR discussed above) and influences amyloidogenesis and the propensity for disease development [42]. Such kinetic influence does not appear to be as strong a determinant as thermodynamic stability because some tetramers comprised of disease-associated subunits dissociate faster in comparison to WT TTR, whereas others dissociate more slowly [35,36,42,91,96]. Generally speaking, consideration of both the thermodynamic and kinetic attributes of the disease-associated mutants is required to explain their relative amyloidogenicities in vivo [42].

NATURAL SUPPRESSION OF A TTR-ASSOCIATED AMYLOID DISEASE: STRONG SUPPORT FOR THE AMYLOID HYPOTHESIS

Our initial insight into a potential therapeutic approach for the TTR amyloidoses was bolstered by a clinical observation of T. Coelho and co-workers, who reported a compound heterozygous family in Portugal that expressed both the highly penetrant V30M FAP-associated and a nonpathogenic T119M TTR variant [37,138]. Members of the kindred were asymptomatic or developed a very mild form of FAP. The sera of the V30M/T119M compound heterozygotes that are resistant to disease contained heterotetramers composed of three different V30M and T119M subunit stoichiometries, as well as the V30M and T119M homotetramers at one-sixteenth the concentration of the heterotetramers. The inclusion of T119M subunits into tetramers otherwise composed of disease-associated subunits raises the kinetic barrier of tetramer dissociation substantially, protecting these individuals from amyloidogenesis and disease, by a process referred to as interallelic trans-suppression (Fig. 5) [35,36].

Biophysical studies reveal that T119M subunit incorporation into tetramers otherwise composed of disease-associated subunits proportionately reduces the rate of acid-mediated amyloidogenesis [35]. Thermodynamic comparisons of the WT and T119M homotetramers reveal very similar stabilities, even though the Met119 side chains increase the surface area of the contacts between the weaker dimer–dimer interface. The tetramer dissociation kinetics of the T119M homotetramer, in contrast, are markedly slower than those exhibited by the WT TTR homotetramer, demonstrating that the T119M TTR homotetramer has a much higher dissociation barrier than that of the WT TTR homotetramer (Fig. 5) [35]. Hence, T119M subunit inclusion into the tetramer destabilizes the dissociative transition state and effectively precludes dissociation and amyloidogenesis. Although this is difficult to explain in terms of a structural rationalization because the dissociative transition state is so distant in terms of energy, and therefore structure, relative to any of the currently available TTR crystal structures, the kinetic basis of T119M interallelic trans-suppression is supported by data from the analysis of other engineered mutations that perturb this quaternary structural interface, which also impose kinetic stability on the TTR tetramer [94,95]. Thus, although the V30M/WT heterozygotes and the V30M/T119M compound heterozygotes have equal concentrations of V30M, the latter are protected against disease because the V30M subunits cannot be liberated from the nonamyloidogenic V30M/T119M heterotetramer, providing strong support for the amyloid hypothesis, the concept that it is tetramer dissociation, misfolding, and misassembly of the V30M TTR subunits that results in pathology [35,36].

In vivo kinetic stabilization has also been confirmed experimentally in transgenic mouse models. For more than a decade, despite the efforts of numerous laboratories, it was difficult to generate transgenic models of TTR amyloid disease that exhibited an amyloidogenic phenotype [139]. When the human TTR gene was inserted into mice under the control of its own promoter, only two models resulted in tissue deposition [140,141]. These two models each had more than 30 copies of the human gene integrated into the mouse genome; models with fewer copies resulted in no tissue deposits. However, when low-copy-number inserts were crossed into a mouse strain in which the endogenous murine TTR gene had been disrupted, the frequency of tissue deposits increased tenfold. Examination of the sera of animals with the human transgene and an intact murine TTR gene revealed the presence of mixed tetramers [142]. Subsequent biophysical studies showed that incorporation of murine TTR subunits into a mixed human–murine TTR tetramer rendered the resulting tetramer kinetically stable, so that under physiological conditions there was essentially no monomer dissociation and thus no misfolded monomer to misassemble into amyloid [90]. The incorporation of murine TTR subunits duplicated the ability of T119M subunit incorporation to protect V30M subunit carriers from disease by imposing kinetic stability on the tetramers [35].

A SURGICAL FORM OF GENE THERAPY IS CURRENTLY UTILIZED TO TREAT FAP

Currently, FAP is treated by gene therapy mediated by liver transplantation, wherein the FAP variant TTR/WT TTR liver is replaced by a WT/WT TTR secreting liver. Surgery decreases disease-associated variant TTR plasma levels

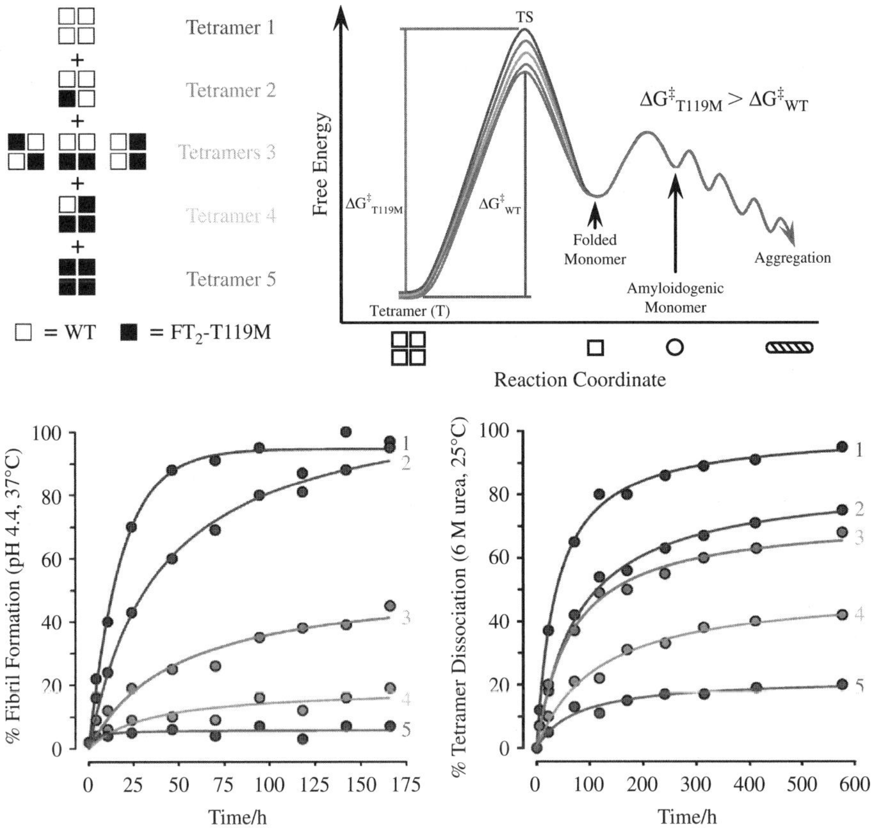

FIG. 5 Interallelic trans-suppression ameliorates TTR amyloid disease by making the TTR dissociation barrier increase proportional to the number of T119M TTR suppressor subunits in the tetramer otherwise composed of subunits that can engage in amyloidogenesis. (From [71], with permission. Copyright © 2005 American Chemical Society.) (*See insert for color representation of figure.*)

to less than 5% of pre-transplant levels, halting polyneuropathy progression in patients carrying the V30M TTR mutation [143–146]. Surprisingly, cardiac amyloidosis often progresses in V30M FAP patients after transplantation, due to the continued deposition of WT TTR in cardiac tissue [145–148]. From current evidence it seems unlikely that the heart produces TTR. Therefore, post-transplant cardiac amyloidosis appears to be caused by the dissociation and amyloidogenesis of wild-type TTR produced by the liver [131] (Buxbaum, unpublished results), suggesting an aging-associated process that renders WT TTR amyloidogenic in the heart. Aging-associated WT TTR amyloidogenicity could be related to aging-associated posttranslational changes in TTR or the molecular composition of the cardiac tissue (e.g., the extracellular matrix) or result from an aging-associated change in cardiac cell proteostasis capacity, including how efficiently the heart clears misfolded and/or aggregated TTR [132].

It has also recently been reported that TTR amyloidosis progresses post-transplantation in the eyes and CNS. Here the explanation is more apparent, as these tissues make their own TTR and thus the concentration of the amyloidogenic TTR variant is not altered by liver transplantation [149]. An additional observation made as a consequence of the follow-up of liver transplant patients has both clinical and biological significance. Since the livers from FAP patients appear to be functionally normal, they have been used as donor livers for "domino transplants" into patients with liver failure related to other, nonamyloid diseases [150–152]. The assumption underlying this strategy was that the domino recipients would not live long enough to experience deposition of the mutant TTR being synthesized by the donor liver since V30M TTR amyloidosis typically takes more than 30 years to develop. Recent reports have shown that approximately 5 to 10% of these recipients develop TTR amyloid deposits in tissues within five to eight years of the transplant [153–157]. The implication of these observations is that at the time of removal from some FAP patients, the livers either have an intrinsic defect in proteostatic capacity or a related problem that enables them to secrete TTR that is more amyloidogenic. Further studies of livers from FAP patients should reveal more about the liver's role in TTR amyloidogenesis, as well as provide insight into the risks of domino liver transplants.

Despite the aforementioned imperfections, liver transplantation as a therapeutic strategy for FAP has demonstrated that TTR amyloidosis can be treated by lowering the concentration of the more amyloidogenic TTR sequence in heterozygous FAP patients [158], consistent with biophysical data that TTR amyloidogenesis is dependent on TTR concentration [93]. It is becoming clear that liver transplantation is much less effective for non-V30M FAP, for reasons that remain unclear [146]. Since the liver also catabolizes TTR, it is essential to look into the role of reduced catabolism of the non-V30M variants (relative to wild-type TTR), a subject virtually unexplored as a possible explanation. It is also possible that mutant TTR synthesis and secretion from tissues other than the liver, choroid plexus, and retinal epithelium could contribute to non-V30M FAP. Given the limited supply of livers, the risk of death from transplantation complications and the limited applicability of liver transplantation, it follows that an alternative chemotherapeutic strategy would be highly desirable for intervention in SSA, FAP, FAC, and CNSA, either as primary treatment or in conjunction with liver transplantation [71].

SELECTIVE SMALL-MOLECULE BINDING TO TETRAMERIC TTR IMPOSES KINETIC STABILIZATION ON THE TETRAMER: A CHEMOTHERAPEUTIC STRATEGY FOR THE TTR AMYLOIDOSES

The TTR tetramer exhibits two different dimer–dimer interfaces (Fig. 6) [65–67,69]. The energetically weaker AB/CD interface creates the two T_4 binding cavities bisected by the crystallographic twofold axis or the *Z*-axis

(Fig. 6A). Small-molecule binding to the T_4 binding sites is expected to stabilize this interface. Fragmentation of the stronger AC/BD dimer–dimer interface bisected by the *X*-axis requires intermolecular β-sheet fracture (Fig. 6), which generally does not occur before fracture of the AB/CD interface [95,159,160]. Each T_4 binding site is characterized by an inner binding cavity and a larger outer binding cavity (Fig. 6B) [67,69,160–170]. The outer binding cavity, the interface between the outer and inner binding sites, and the inner binding cavity feature pairs of symmetric hydrophobic concave subsites referred to as the halogen-binding pockets, wherein the iodine atoms of T_4 reside (Fig. 6B) [171]. Small molecules comprised of two appropriately substituted aromatic rings and a hydrophobic linker are known to bind to the TTR tetramer with high affinity [71,96,133,161–169,172–182].

Alcohol or acid substituents on the aromatic ring occupying the inner binding cavity of TTR are capable of accepting and donating hydrogen bonds by interacting with the Ser117 and Thr119 side chains that can be revealed by dihedral angle changes within these side chains (Fig. 6B) [67,160–170]. Side-chain ammonium and carboxylate functional groups positioned at the periphery of the outer binding cavity can make electrostatic interactions with aromatic substituents bearing complementary charges (Fig. 6B) [67,163,167,169,170]. Thyroxine and most other ligands bind to TTR with negative cooperativity, which appears to result from conformational changes within the tetramer upon ligand binding to the first site—imparting lower-affinity binding to the second site within the TTR tetramer [71,183,184]. A few small molecules display noncooperative or positively cooperative binding; however, the structural and energetic basis for the cooperativity is unclear at this time and is being actively investigated [168,175].

Nearly 1000 aromatic small-molecule ligands exhibiting structural complementarity to the T_4-binding sites within TTR have been synthesized or procured, and representative examples are shown in Fig. 7 [71,96,133, 161–169,172–182]. Some were discovered through structure-guided screening [173,174], whereas the majority resulted from structure-based design utilizing numerous high-resolution TTR•(small molecule)$_2$ crystal structures to envision higher-binding-affinity ligands [67,160–165,167–170]. The aromatic rings of the TTR kinetic stabilizers are either linked directly (e.g., biphenyls) [168,169] or tethered through a variety of linkers, the best being short hydrophobic linkers such as an E-olefin or a $-CH_2CH_2-$ substructure (Fig. 7) [161,162]. A systematic study optimizing the aryl-X ring substituents and substitution pattern, the aryl-Z ring substituents and substitution pattern, as well as the linker substructure joining the rings revealed that there are numerous combinatorial solutions for occupying TTR's thyroxine-binding sites with high-affinity ligands [161,162]. As a consequence of these systematic studies and >100 TTR•(small molecule)$_2$ crystal structures, it is now possible to make accurate predictions regarding how well a small-molecule ligand will bind, its binding orientation, and its degree of TTR binding selectivity in plasma [67,160–170].

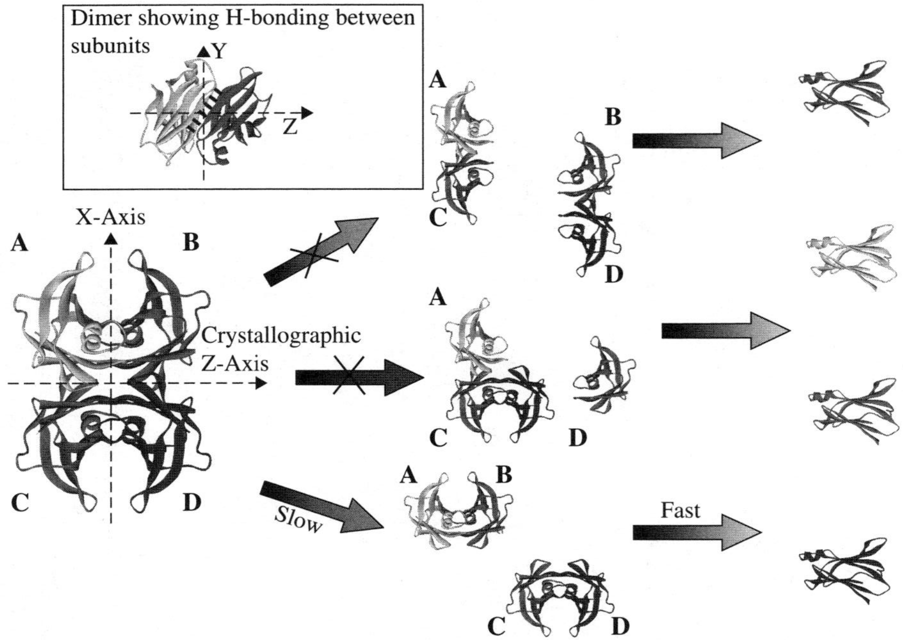

FIG. 6 (A) The TTR tetramer could dissociate through numerous mechanisms, many of which are shown. The operational mechanism is shown at the bottom, wherein the dimers dissociate from the tetramer about the interface incorporating the crystallographic *Z*-axis.

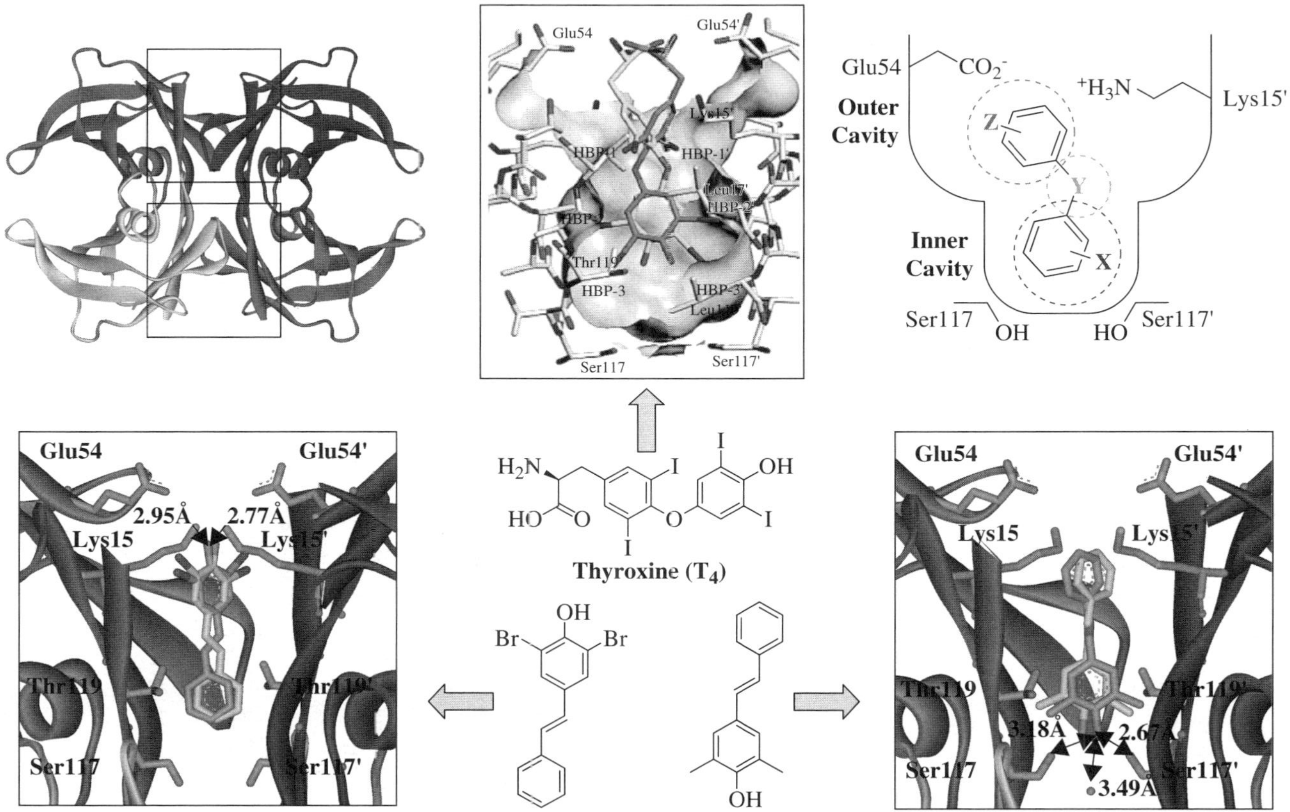

FIG. 6 (B) Dissociation through this mechanism is prevented through kinetic stabilization of the tetramer by binding of at least one ligand to one of the two thyroxine-binding sites. (From [162], with permission. Copyright © 2008 American Chemical Society.) (*See insert for color representation of figure.*)

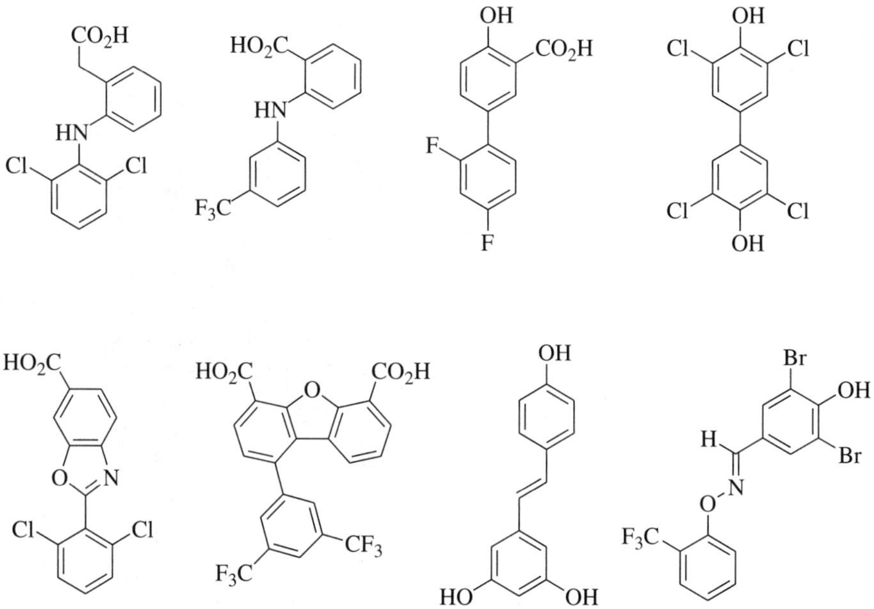

FIG. 7 Representative TTR kinetic stabilzers.

Since T119M TTR subunit inclusion into tetramers otherwise composed of disease-associated TTR subunits ameliorates FAP by kinetic stabilization of the native tetrameric state [35,36], we have focused on small molecules that can bind preferentially to the native state of TTR over the dissociative transition state, thereby similarly imposing kinetic stability on the tetramer (Fig. 8) [71]. Small molecule-mediated kinetic stabilization of the TTR tetramer is a conservative therapeutic strategy because it prevents the initiation of amyloidogenesis by raising the kinetic barrier to dissociation. Tissue culture studies suggest that the critical TTR conformation for cytotoxicity is that of a misfolded monomer or dimer generated after tetramer dissociation [28]; however, the basis for toxicity in a human may be different and remains unclear. Because tetramer dissociation is the critical first step toward misfolding and aggregation, the tetramer kinetic stabilization strategy should be effective regardless of the nature of the proteotoxic structure. It is important to recognize that because misassembly is concentration dependent, it will probably not be necessary to stabilize every tetramer kinetically to prevent amyloidogenesis [93]. Kinetically stabilizing half the tetramers and thus lowering the amyloidogenic TTR concentration by a factor of 2 should be sufficient to prevent TTR aggregation based on biophysical data [93].

Kinetic experiments in vitro demonstrate that kinetic stabilizer (KS) binding to the first (TTR•KS) and second (TTR•$(KS)_2$) T_4-binding sites within the TTR tetramer proportionally increase the activation barrier ($\Delta G^{\ddagger}$) associated with tetramer dissociation [i.e., $\Delta G^{\ddagger}$TTR•$(KS)_2 > \Delta G^{\ddagger}$TTR•KS$ > \Delta G^{\ddagger}$TTR]

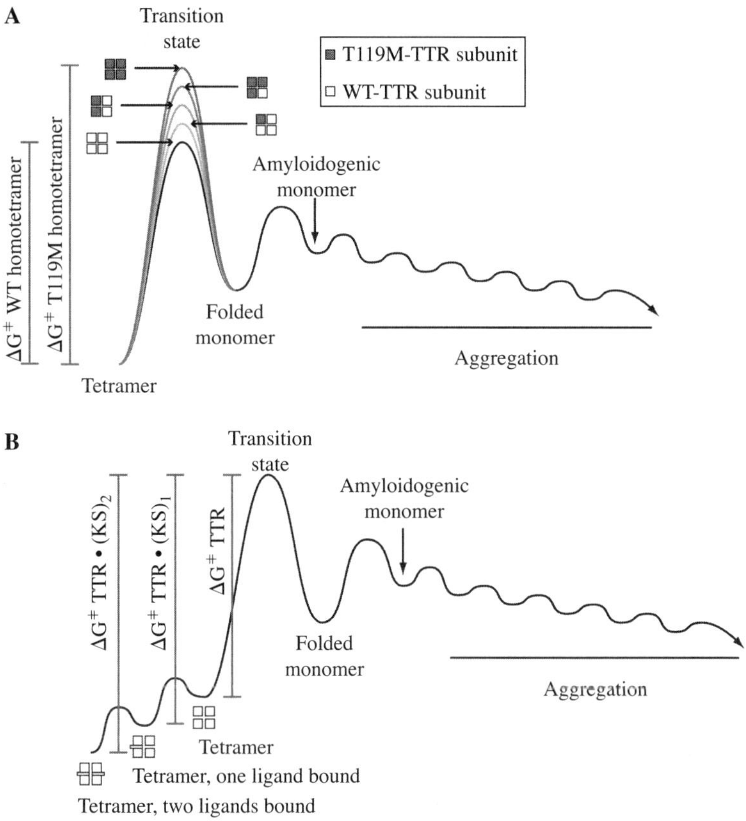

FIG. 8 Kinetic stabilization of TTR by way of interallelic trans-suppression through T119M subunit incorporation into the tetramer (A) or small-molecule binding (B) equivalently increase the tetramer dissociation barrier preventing amyloidogenesis and pathology. (From [71], with permission. Copyright © 2005 American Chemical Society.)

(Fig. 8). Specifically, the rate of tetramer dissociation decreases as the stoichiometry of the kinetic stabilizer increases [35,71,160,169]. Since TTR tetramer dissociation is rate limiting for amyloid fibril formation, small molecule–mediated kinetic stabilization of the TTR tetramer can also be demonstrated by evaluating the rate of TTR amyloidogenesis as a function of kinetic stabilizer concentration, with the caveat already noted that partial kinetic stabilization of a TTR tetramer population can completely block fibrillogenesis because stabilizing half of the TTR tetramers drops the concentration of the misfolding-competent TTR monomers sufficiently that amyloidogenesis cannot occur. The rate of acid-mediated TTR amyloidogenesis decreases dramatically as the stoichiometry of bound kinetic stabilizer increases (Fig. 8) [71,96,133,161–169,172–182].

That kinetic stabilizer binding preserves the native tetrameric structure of TTR (3.6 μM tetramer) under acidic denaturing conditions (pH 4.4, 37°C) can

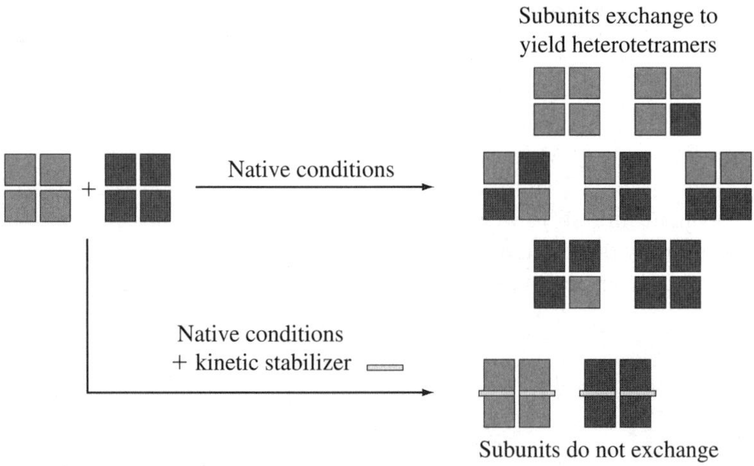

FIG. 9 Subunit exchange method to demonstrate that TTR kinetic stabilizers prevent TTR subunit exchange under physiological conditions.

also be demonstrated by equilibrium analytical ultracentrifugation methods. Under these conditions, TTR efficiently dissociates and self-assembles into high-molecular-mass aggregates within 72 hours in the absence of a kinetic stabilizer. TTR samples preincubated with one of the best kinetic stabilizers (7.2 to 10.8 μM) and then evaluated after several days of incubation at pH 4.4 maintain their tetrameric structure [22]. Kinetic stabilization can also be demonstrated under physiological conditions by monitoring the exchange of WT TTR and tagged WT TTR subunits from an initial mixture of homotetramers [35,36,89,97]. Rate-limiting TTR tetramer dissociation followed by reassembly of the resulting monomers ultimately yields TTR tetramers comprised of a statistical distribution of tagged and untagged WT TTR subunits (Fig. 9, top). Kinetic stabilizer–binding studies reveal that subunit exchange only arises from unliganded TTR; binding to one T_4 site in the TTR tetramer is sufficient to block tetramer dissociation required for subunit exchange. Since both the acid- and chaotrope-mediated denaturation methods have the potential to alter the kinetic stabilizer binding constants, the subunit exchange method seems to be best suited to rank order the efficacy of kinetic stabilizers under physiological conditions [97].

Covalent attachment of one small-molecule kinetic stabilizer to a TTR tetramer clearly establishes that occupancy of only one T_4-binding site is sufficient to impart kinetic stabilization on the entire TTR tetramer [160]. Covalent tethering of the A and C TTR subunits (Fig. 6A) using a polypeptide linker (the TTR quaternary structure is also made up of normal WT TTR B and D subunits) also creates a kinetically stabilized three-polypeptide-chain quaternary structure that is superimposable on the native tetrameric TTR structure, with the exception of the polypeptide covalently linking the A and C subunits [95,159]. These results suggest that a kinetic stabilizer exhibiting strong negatively cooperative binding to TTR would be ideal for therapeutic intervention in SSA, FAP, FAC, and CNSA.

TTR • (kinetic stabilizer)$_2$ x-ray crystal structures suggested that it would be possible to design a single bivalent kinetic stabilizer that occupies both of the TTR T_4-binding sites simultaneously (Fig. 10) [164]. Symmetric and asymmetric bivalent inhibitors were synthesized, the latter employing substructures comprising individual kinetic stabilizers whose binding orientation preferences were known from crystallographic studies [164]. Although these kinetic stabilizers do not bind to TTR that is already tetrameric, they are incorporated into the tetramer after subunit folding and before subunit assembly, probably templating tetramer formation, consistent with the structures revealed by x-ray crystallographic analysis (Fig. 10). Consistent with this, the TTR bivalent kinetic stabilizers are incorporated into the TTR tetramer during assembly within mammalian cells that secrete TTR [164]. Most significantly, the TTR•bivalent kinetic stabilizer complex does not dissociate or form amyloid on any reasonable biological time scale, demonstrating its very high kinetic stability. Many but not all of the compounds that are effective in the in vitro assays also inhibit TTR cytotoxicity in tissue culture [28].

TTR KINETIC STABILIZERS MUST BIND SELECTIVELY TO TTR OVER THE REMAINDER OF THE PROTEOME TO BE EFFECTIVE

For effective therapeutic intervention, transthyretin kinetic stabilizers must bind selectively to TTR relative to the remainder of the human proteome in tissues and biological fluids to impose kinetic stabilization on TTR and prevent TTR amyloidogenesis. A low-throughput method has been developed to measure the average binding stoichiometry of kinetic stabilizers to TTR in human plasma [185]. Preincubation of a candidate kinetic stabilizer (10 μM) with human plasma (TTR tetramer concentration 3 to 7 μM), followed by the capture of TTR, TTR • KS, and TTR • (KS)$_2$ by a resin-conjugated TTR antibody enables chromatographic analysis of the average kinetic stabilizer–binding stoichiometry. This assay is now routinely done in selecting clinical candidates, with the kinetic stabilizers exhibiting the highest average binding stoichiometry at the lowest plasma concentration being the most desirable.

Since transthyretin is the main carrier of thyroxine in the brain and the backup carrier in blood, some compounds that bind with high affinity to TTR can also bind with high affinity to the thyroid hormone receptor, which is an undesirable attribute for a chronic therapy. All potent TTR kinetic stabilizers are now also assessed for their ability to bind to the thyroid hormone receptor [161,162]. Ultimately, we envision prophylactic dosing of FAP- and FAC-associated mutant carriers or those with a family history of SSA (WT TTR amyloidogenesis) with a kinetic stabilizer to prevent the TTR amyloidogenicity associated with disease progression.

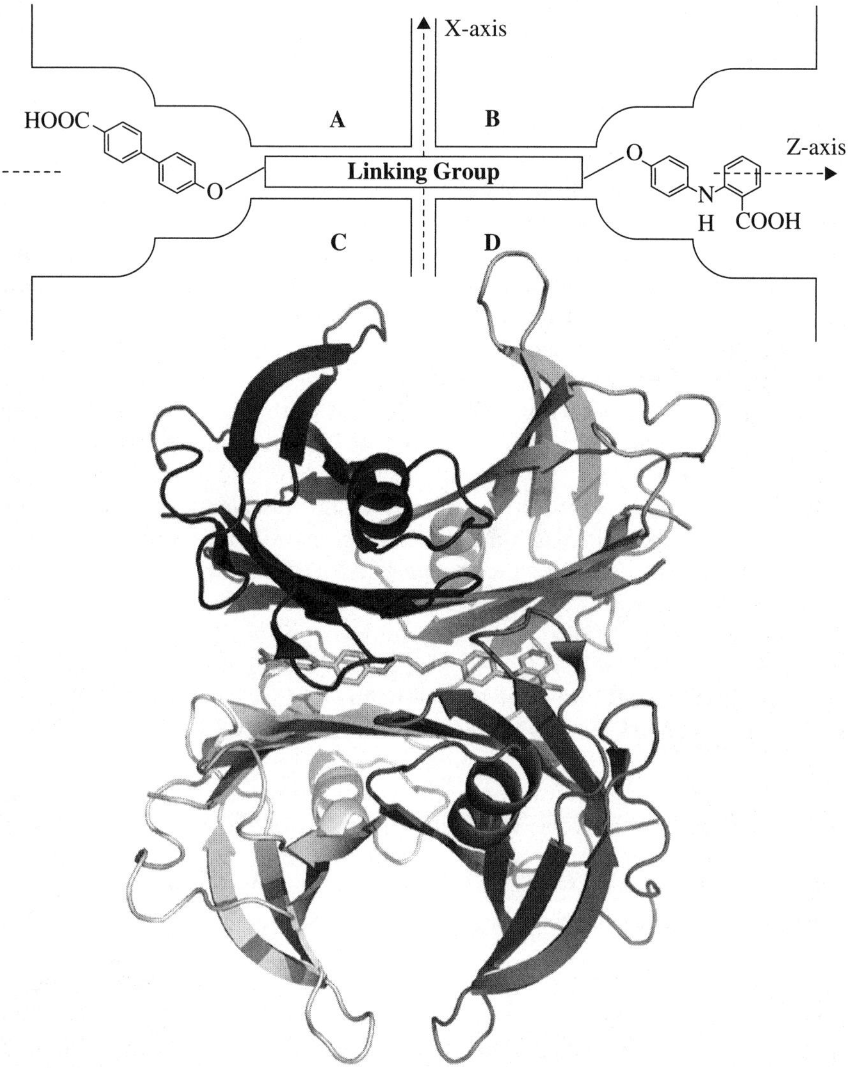

FIG. 10 Kinetic stabilization of the TTR tetramer can be accomplished by the design and synthesis of ligands that template TTR tetramer formation within the endoplasmic reticulum by simultaneously occupying both thyroxine-binding sites. (From [162], with permission. Copyright © 2008 American Chemical Society.)

The NSAID diflunisal exhibits good binding affinity to TTR in vitro, modest selectivity in binding to TTR over all other plasma proteins, and excellent oral bioavailability, all of which are independent of its intrinsic NSAID activity. Diflunisal kinetically stabilized TTR in human plasma after oral dosing at 250 mg twice a day [181,182]. When administered parenterally to human TTR transgenic mice lacking murine TTR, it stabilized the circulating human tetramer to urea denaturation in a manner similar to the effect of incorporation of murine monomer subunits into murine–human heterotetramers, confirming the results obtained in the human oral administration study [142]. These characteristics, along with its established safety profile (albeit in a different population), have enabled a placebo-controlled multicenter FAP clinical trial (directed by Professors Martha Skinner and John Berk at Boston University) with National Institutes of Health and U.S. Food and Drug Administration support that is enrolling patients. Whether the known NSAID effects of the compound compromise its therapeutic potential should become evident in this trial. The possibility of such side effects, particularly in patients with cardiac or renal compromise secondary to TTR deposition, have led to the investigation of other stabilizers without NSAID activity.

One of the kinetic stabilizers discovered by the Kelly Laboratory has been licensed to FoldRx Pharmaceuticals. FoldRx has successfully completed a placebo-controlled phase II/III clinical trial to evaluate the efficacy of this kinetic stabilizer (tafamidis, 20 mg once a day) for V30M TTR-associated FAP. Results from this pivotal clinical trial demonstrate that tafamidis halts disease progression in V30M FAP patients, reduces the burden of disease compared to placebo, and appears to be safe and well tolerated. This kinetic stabilizer, if approved by the European Union and the U.S. Food and Drug Administration, would be the first drug that addresses the underlying cause of a human amyloid disease and would be a first-in-class drug for the amelioration of an amyloid disease. Moreover, this is the first pharmacologic evidence that we are aware of that strongly supports the amyloid hypothesis. Motivated by this success, we are developing the next generation of therapeutic strategies against the TTR amyloidoses, especially those focused on enhancing the biology that protects the young human population from FAC, FAP, SSA, and CNSA [132].

Acknowledgments

We are grateful for the long-standing support of the National Institutes of Health (NIH) (DK046335 JWK)(AG030027 JB), the Lita Annenberg Hazen Foundation, the Skaggs Institute for Chemical Biology, and the W. M. Keck Foundation, which together funded the preclinical studies, and for FoldRx, the U.S. Food and Drug Administration (FD-R-002532-01), and the NIH (NS051306), which are funding the clinical trials. The progress outlined within would not have been possible without the hard work, determination, and creativity of numerous co-workers cited within, including those who coauthored this chapter. We thank Dr. Colleen Fearns for carefully editing this chapter.

REFERENCES

1. Selkoe, D.J. (2003). Folding proteins in fatal ways. *Nature, 426*, 900–904.

2. Westermark, P., Andersson, A., Westermark, G.T. (2005). Is aggregated IAPP a cause of beta-cell failure in transplanted human pancreatic islets? *Curr Diabetes Rep, 5*, 184–188.
3. Cohen, F.E., Kelly, J.W. (2003). Therapeutic approaches to protein-misfolding diseases. *Nature, 426*, 905–909.
4. Dobson, C.M. (2003). Protein folding and misfolding. *Nature, 426*, 884–890.
5. Kelly, J.W. (1996). Alternative conformations of amyloidogenic proteins govern their behavior. *Curr Op Struct Biol, 6*, 11–17.
6. Westermark, P., Benson, M.D., Buxbaum, J.N., Cohen, A.S., Frangione, B., Ikeda, S.I., Masters, C.L., Merlini, G., Saraiva, M.J., Sipe, J.D. (2007). A primer of amyloid nomenclature. *Amyloid, 14*, 179–183.
7. Sipe, J.D. (1992). Amyloidosis. *Ann Rev Biochem, 61*, 947–975.
8. Sipe, J.D. (1994). Amyloidosis. *Crit Rev Clin Lab Sci, 31*, 325–354.
9. Buxbaum, J.N., Tagoe, C.E. (2000). The genetics of the amyloidoses. *Annu Rev Med, 51*, 543–569.
10. Jahn, T.R., Radford, S.E. (2008). Folding versus aggregation: polypeptide conformations on competing pathways. *Arch Biochem Biophys, 469*, 100–117.
11. Pepys, M.B. (2001). Pathogenesis, diagnosis and treatment of systemic amyloidosis. *Philos Trans R Soc Lond Ser B, 356*, 203–211.
12. Benson, M.D. (1989). Familial amyloidotic polyneuropathy. *TIBS, 12*, 88–92.
13. Walsh, D.M., Selkoe, D.J. (2007). Abeta oligomers: a decade of discovery. *J Neurochem, 101*, 1172–1184.
14. Lambert, M.P., Barlow, A.K., Chromy, B.A., Edwards, C., Freed, R., Liosatos, M., Morgan, T.E., Rozovsky, I., Trommer, B., Viola, K.L., et al. (1998). Diffusible, nonfibrillar ligands derived from Abeta1–42 are potent central nervous system neurotoxins. *Proc Natl Acad Sci U S A, 95*, 6448–6453.
15. Lashuel, H.A., Wurth, C., Woo, L., Kelly, J.W. (1999). The most pathogenic transthyretin variant, L55P, forms amyloid fibrils under acidic conditions and protofilaments under physiological conditions. *Biochemistry, 38*, 13560–13573.
16. Lashuel, H.A., LaBrenz, S.R., Woo, L., Serpell, L., Kelly, J.W. (2000). Protofilaments, filaments, ribbons, and fibrils from peptidomimetic self-assembly: implications for amyloid fibril formation and materials science. *J Am Chem Soc, 122*, 5262–5277.
17. Lashuel, H.A., Hartley, D., Petre, B.M., Walz, T., Lansbury, P.T., Jr. (2002). Neurodegenerative disease: amyloid pores from pathogenic mutations. *Nature, 418*, 291.
18. Pepys, M.B. (1988). Amyloidosis. In *Immunological Diseases (*Samter, M., Ed.), Little, Brown, Boston.
19. Tan, S.Y., Pepys, M.B. (1994). Amyloidosis. *Histopathology, 25*, 403–414.
20. Sousa, M.M., Cardoso, I., Fernandes, R., Guimaraes, A., Saraiva, M.J. (2001). Deposition of transthyretin in early stages of familial amyloidotic polyneuropathy: evidence for toxicity of nonfibrillar aggregates. *Am J Pathol, 159*, 1993–2000.
21. Kelly, J.W., Colon, W., Lai, Z., Lashuel, H.A., McCulloch, J., McCutchen, S.L., Miroy, G.J., Peterson, S.A. (1997). Transthyretin quaternary and tertiary structural changes facilitate misassembly into amyloid. *Adv Protein Chem, 50*, 161–181.

22. Lashuel, H.A., Lai, Z., Kelly, J.W. (1998). Characterization of the transthyretin acid denaturation pathways by analytical ultracentrifugation: implications for wild-type, V30M, and L55P amyloid fibril formation. *Biochemistry, 37*, 17851–17864.
23. Liu, K., Cho, H.S., Lashuel, H.A., Kelly, J.W., Wemmer, D.E. (2000). A glimpse of a possible amyloidogenic intermediate of transthyretin. *Nat Struct Biol, 7*, 754–757.
24. Bieschke, J., Zhang, Q., Powers, E.T., Lerner, R.A., Kelly, J.W. (2005). Oxidative metabolites accelerate Alzheimer's amyloidogenesis by a two-step mechanism, eliminating the requirement for nucleation. *Biochemistry, 44*, 4977–4983.
25. Zhang, Q., Powers, E.T., Nieva, J., Huff, M.E., Dendle, M.A., Bieschke, J., Glabe, C.G., Eschenmoser, A., Wentworth, P., Jr., Lerner, R.A., Kelly, J.W. (2004). Metabolite-initiated protein misfolding may trigger Alzheimer's disease. *Proc Natl Acad Sci U S A, 101*, 4752–4757.
26. Kelly, J.W. (1998). The alternative conformations of amyloidogenic proteins and their multi-step assembly pathways. *Curr Opin Struct Biol, 8*, 101–106.
27. Teng, M.-H., Yin, J.-Y., Vidal, R., Ghiso, J., Kumar, A., Rabenou, R., Shah, A., Jacobson, D.R., Tagoe, C., Gallo, G., Buxbaum, J. (2001). Amyloid and nonfibrillar deposits in mice transgenic for wild-type human transthyretin: a possible model for senile systemic amyloidosis. *Lab Invest, 81*, 385–396.
28. Reixach, N., Deechongkit, S., Jiang, X., Kelly, J.W., Buxbaum, J.N. (2004). Tissue damage in the amyloidoses: transthyretin monomers and nonnative oligomers are the major cytotoxic species in tissue culture. *Proc Natl Acad Sci U S A, 101*, 2817–2822.
29. Shankar, G.M., Li, S., Mehta, T.H., Garcia-Munoz, A., Shepardson, N.E., Smith, I., Brett, F.M., Farrell, M.A., Rowan, M.J., Lemere, C.A., et al. (2008). Amyloid-beta protein dimers isolated directly from Alzheimer's brains impair synaptic plasticity and memory. *Nat Med, 14*, 837–842.
30. Silveira, J.R., Raymond, G.J., Hughson, A.G., Race, R.E., Sim, V.L., Hayes, S.F., Caughey, B. (2005). The most infectious prion protein particles. *Nature, 437*, 257–261.
31. Hardy, J.A., Higgins, G. (1992). Alzheimer's disease: the amyloid cascade hypothesis. *Science, 256*, 184–185.
32. Tanzi, R.E., Bertram, L. (2005). Twenty years of the Alzheimer's disease amyloid hypothesis: a genetic perspective. *Cell, 120*, 545–555.
33. Bertram, L., Tanzi, R.E. (2008). Thirty years of Alzheimer's disease genetics: the implications of systematic meta-analyses. *Nat Rev Neurosci, 9*, 768–778.
34. Selkoe, D.J. (2006). The ups and downs of Abeta. *Nat Med, 12*, 758–759.
35. Hammarstrom, P., Wiseman, R.L., Powers, E.T., Kelly, J.W. (2003). Prevention of transthyretin amyloid disease by changing protein misfolding energetics. *Science, 299*, 713–716.
36. Hammarstrom, P., Schneider, F., Kelly, J.W. (2001). Trans-suppression of misfolding in an amyloid disease. *Science, 293*, 2459–2462.
37. Coelho, T., Carvalho, M., Saraiva, M.J., Alves, I., Almeida, M.R., Costa, P.P. (1993). A strikingly benign evolution of FAP in an individual found to be a compound heterozygote for two TTR mutations: TTR Met 30 and TTR Met 119. *J Rheumatol, 20*, 179.

38. Golde, T.E., Janus, C. (2005). Homing in on intracellular Abeta? *Neuron, 45*, 639–642.
39. Gouras, G.K., Tsai, J., Naslund, J., Vincent, B., Edgar, M., Checler, F., Greenfield, J.P., Haroutunian, V., Buxbaum, J.D., Xu, H., et al. (2000). Intraneuronal Abeta42 accumulation in human brain. *Am J Pathol, 156*, 15–20.
40. LaFerla, F.M., Troncoso, J.C., Strickland, D.K., Kawas, C.H., Jay, G. (1997). Neuronal cell death in Alzheimer's disease correlates with apoE uptake and intracellular Abeta stabilization. *J Clin Invest, 100*, 310–320.
41. Wirths, O., Multhaup, G., Czech, C., Blanchard, V., Moussaoui, S., Tremp, G., Pradier, L., Beyreuther, K., Bayer, T.A. (2001). Intraneuronal Abeta accumulation precedes plaque formation in beta-amyloid precursor protein and presenilin-1 double-transgenic mice. *Neurosci Lett, 306*, 116–120.
42. Sekijima, Y., Wiseman, R.L., Matteson, J., Hammarstrom, P., Miller, S.R., Sawkar, A.R., Balch, W.E., Kelly, J.W. (2005). The biological and chemical basis for tissue-selective amyloid disease. *Cell, 121*, 73–85.
43. Puchtler, H., Sweat, F., Levine, M. (1962). On the binding of Congo Red by amyloid. *J Histochem Cytochem, 10*, 355.
44. LeVine, H., III. (1997). Stopped-flow kinetics reveal multiple phases of thioflavin T binding to Alzheimer beta(1–40) amyloid fibrils. *Arch Biochem Biophys, 342*, 306–316.
45. Nilsson, K.P.R., Aslund, A., Berg, I., Nystrom, S., Konradsson, P., Herland, A., Inganas, O., Stabo-Eeg, F., Lindgren, M., Westermark Gunilla, T., et al. (2007). Imaging distinct conformational states of amyloid-beta fibrils in Alzheimer's disease using novel luminescent probes. *ACS Chem Biol, 2*, 553–560.
46. Eaton, W.A., Hofrichter, J. (1990). Sickle cell hemoglobin polymerization. *Adv Protein Chem, 40*, 63–279.
47. Colon, W., Kelly, J.W. (1991). Transthyretin acid induced denaturation is required for amyloid fibril formation in vitro. In *Applications of Enzyme Biotechnology* (Kelly, J.W., Balwin, T.O., Eds.), Plenum Press, New York.
48. Colon, W., Kelly, J.W. (1992). Partial denaturation of transthyretin is sufficient for amyloid fibril formation in vitro. *Biochemistry, 31*, 8654–8660.
49. Lai, Z., Colon, W., Kelly, J.W. (1996). The acid-mediated denaturation pathway of transthyretin yields a conformational intermediate which can self-assemble into amyloid. *Biochemistry, 35*, 6470–6482.
50. McCutchen, S., Colon, W., Kelly, J.W. (1993). Transthyretin mutation Leu-55-Pro significantly alters tetramer stability and increases amyloidogenicity. *Biochemistry, 32*, 12119–12127.
51. McCutchen, S.L., Lai, Z., Miroy, G., Kelly, J.W., Colon, W. (1995). Comparison of lethal and non-lethal transthyretin variants and their relationship to amyloid disease. *Biochemistry, 34*, 13527–13536.
52. Jiang, X., Smith, C.S., Petrassi, H.M., Hammarstrom, P., White, J.T., Sacchettini, J.C., Kelly, J.W. (2001). An engineered transthyretin monomer that is nonamyloidogenic, unless it is partially denatured. *Biochemistry, 40*, 11442–11452.
53. Hurle, M.R., Helms, L.R., Li, L., Chan, W., Wetzel, R. (1994). A role for destabilizing amino acid replacements in light chain amyloidosis. *Proc Natl Acad Sci* USA, *91*, 5446–5450.

54. Wetzel, R. (1996). For protein misassembly, its the "I" decade. *Cell, 86*, 699–702.
55. Booth, D.R., Sunde, M., Bellotti, V., Robinson, C.V., Hutchinson, W.L., Fraser, P.E., Hawkins, P.N., Dobson, C.M., Radford, S.E., Blake, C.C.F., Pepys, M.B. (1997). Instability, unfolding and aggregation of human lysosozyme variants underlying amyloid fibrillogenesis. *Nature, 385*, 787–793.
56. Powers, E.T., Powers, D.L. (2008). Mechanisms of protein fibril formation: nucleated polymerization with competing off-pathway aggregation. *Biophys J, 94*, 379–391.
57. Ferrone, F. (1999). Analysis of protein aggregation kinetics. *Methods Enzymol, 309*, 256–274.
58. Jarrett, J., Lansbury, P.T. (1993). Seeding "one-dimensional crystallization" of amyloid: a pathogenic mechanism in Alzheimer's disease and scrapie? *Cell, 73*, 1055–1058.
59. Xue, W.-F., Homans, S.W., Radford, S.E. (2008). Systematic analysis of nucleation-dependent polymerization reveals new insights into the mechanism of amyloid self-assembly. *Proc Natl Acad Sci U S A, 105*, 8926–8931.
60. Lundmark, K., Westermark, G.T., Nystrom, S., Murphy, C.L., Solomon, A., Westermark, P. (2002). Transmissibility of systemic amyloidosis by a prion-like mechanism. *Proc Natl Acad Sci U S A, 99*, 6979–6984.
61. Bieschke, J., Siegel Sarah, J., Fu, Y., Kelly, J.W. (2008). Alzheimer's Abeta peptides containing an isostructural backbone mutation afford distinct aggregate morphologies but analogous cytotoxicity: evidence for a common low-abundance toxic structure(s)? *Biochemistry, 47*, 50–59.
62. Bieschke, J., Zhang, Q., Bosco, D.A., Lerner, R.A., Powers, E.T., Wentworth, P., Jr., Kelly, J.W. (2006). Small molecule oxidation products trigger disease-associated protein misfolding. *Acc Chem Res, 39*, 611–619.
63. Siegel, S.J., Bieschke, J., Powers, E.T., Kelly, J.W. (2007). The oxidative stress metabolite 4-hydroxynonenal promotes Alzheimer protofibril formation. *Biochemistry, 46*, 1503–1510.
64. Stewart, C.R., Wilson, L.M., Zhang, Q., Pham, C.L.L., Waddington, L.J., Staples, M.K., Stapleton, D., Kelly, J.W., Howlett, G.J. (2007). Oxidized cholesterol metabolites found in human atherosclerotic lesions promote apolipoprotein C-II amyloid fibril formation. *Biochemistry, 46*, 5552–5561.
65. Blake, C.C., Geisow, M.J., Oatley, S.J., Rerat, B., Rerat, C. (1978). Structure of prealbumin: secondary, tertiary and quaternary interactions determined by Fourier refinement at 1.8 Å. *J Mol Biol, 121*, 339–356.
66. Blake, C.C., Geisow, M.J., Swan, I.D., Rerat, C., Rerat, B. (1974). Structure of human plasma prealbumin at 2–5 Å resolution: a preliminary report on the polypeptide chain conformation, quaternary structure and thyroxine binding. *J Mol Biol, 88*, 1–12.
67. Klabunde, T., Petrassi, H.M., Oza, V.B., Raman, P., Kelly, J.W., Sacchettini, J.C. (2000). Rational design of potent human transthyretin amyloid disease inhibitors. *Nat Struct Biol, 7*, 312–321.
68. Hamilton, J.A., Benson, M.D. (2001). Transthyretin: a review from a structural perspective. *Cell Mol Life Sci, 58*, 1491–1521.

69. Hornberg, A., Eneqvist, T., Olofsson, A., Lundgren, E., Sauer-Eriksson, A.E. (2000). A comparative analysis of 23 structures of the amyloidogenic protein transthyretin. *J Mol Biol, 302*, 649–669.
70. Correa, C.R., Burini, R.C. (2000). Plasma acute-phase reactive proteins. *J Bras Patol, 36*, 26–34.
71. Johnson, S.M., Wiseman, R.L., Sekijima, Y., Green, N.S., Adamski-Werner, S.L., Kelly, J.W. (2005). Native state kinetic stabilization as a strategy to ameliorate protein misfolding diseases: a focus on the transthyretin amyloidoses. *Acc Chem Res,. 38*, 911–921.
72. Monaco, H.L., Rizzi, M., Coda, A. (1995). Structure of a complex of two plasma proteins: transthyretin and retinol-binding protein. *Science, 268*, 1039–1041.
73. Zanotti, G., Berni, R. (2004). Plasma retinol-binding protein: structure and interactions with retinol, retinoids, and transthyretin. *Vitam Horm, 69*, 271–295.
74. Kopelman, M., Mokady, S., Cogan, U. (1976). Comparative studies of human and chicken retinol-binding proteins and prealbumins. *Biochim Biophys Acta, 439*, 442–448.
75. White, J.T., Kelly, J.W. (2001). Support for the multigenic hypothesis of amyloidosis: the binding stoichiometry of retinol-binding protein, vitamin A, and thyroid hormone influences transthyretin amyloidogenicity in vitro. *Proc Natl Acad Sci U S A, 98*, 13019–13024.
76. Watanabe, C.M., Wolffram, S., Ader, P., Rimbach, G., Packer, L., Maguire, J.J., Schultz, P.G., Gohil, K. (2001). The in vivo neuromodulatory effects of the herbal medicine *Ginkgo biloba. Proc Natl Acad Sci U S A, 98*, 6577–6580.
77. Stein, T.D., Anders, N.J., DeCarli, C., Chan, S.L., Mattson, M.P., Johnson, J.A. (2004). Neutralization of transthyretin reverses the neuroprotective effects of secreted amyloid precursor protein (APP) in APPSw mice resulting in tau phosphorylation and loss of hippocampal neurons: support for the amyloid hypothesis. *J Neurosci, 24*, 7707–7717.
78. Stein, T.D., Johnson, J.A. (2002). Lack of neurodegeneration in transgenic mice overexpressing mutant amyloid precursor protein is associated with increased levels of transthyretin and the activation of cell survival pathways. *J Neurosci, 22*, 7380–7388.
79. Southwell, B.R., Tu, G.F., Duan, W., Achen, M., Harms, P.J., Aldred, A.R., Richardson, S.J., Thomas, T., Pettersson, T.M., Schreiber, G. (1992). Cerebral expression of transthyretin: evolution, ontogeny and function. *Acta Med Austriaca, 19* Suppl 1, 28–31.
80. Herbert, J., Wilcox, J.N., Pham, K.-T.C., Fremeau, R.T., Zeviani, M., Dwork, A., Soprano, D.R., Makover, A., Goodman, D.S., Zimmerman, E.A., et al. (1986). Transthyretin: a choroid plexus–specific transport protein in the human brain. *Neurology, 36*, 900–911.
81. Martone, R.L., Herbert, J., Dwork, A., Schon, E.A. (1988). Transthyretin is synthesized in the mammalian eye. *Biochem Biophys Res Commun, 151*, 905–912.
82. Hovatta, I., Zapala, M.A., Broide, R.S., Schadt, E.E., Libiger, O., Schork, N.J., Lockhart, D.J., Barlow, C. (2007). DNA variation and brain region–specific expression profiles exhibit different relationships between inbred mouse strains: implications for eQTL mapping studies. *Genome Biol, 8*, R25.

83. Murakami, T., Ohsawa, Y., Sunada, Y. (2008). The transthyretin gene is expressed in human and rodent dorsal root ganglia. *Neurosci Lett, 436*, 335–339.

84. Buxbaum, J.N., Ye, Z., Reixach, N., Friske, L., Levy, C., Das, P., Golde, T., Masliah, E., Roberts, A.R., Bartfai, T. (2008). Transthyretin protects Alzheimer's mice from the behavioral and biochemical effects of Abeta toxicity. *Proc Natl Acad Sci U S A, 105*, 2681–2686.

85. Sousa, J.C., Grandela, C., Fernandez-Ruiz, J., de Miguel, R., de Sousa, L., Magalhaes, A.I., Saraiva, M.J., Sousa, N., Palha, J.A. (2004). Transthyretin is involved in depression-like behaviour and exploratory activity. *J Neurochem, 88*, 1052–1058.

86. Fleming, C.E., Saraiva, M.J., Sousa, M.M. (2007). Transthyretin enhances nerve regeneration. *J Neurochem, 103*, 831–839.

87. Richardson, S.J., Lemkine, G.F., Alfama, G., Hassani, Z., Demeneix, B.A. (2007). Cell division and apoptosis in the adult neural stem cell niche are differentially affected in transthyretin null mice. *Neurosci Lett, 421*, 234–238.

88. Lai, Z., McCulloch, J., Kelly, J.W. (1997). GdnHCl-induced denaturation and refolding of transthyretin exhibits a marked hysteresis: equilibria with high kinetic barriers. *Biochemistry, 36*, 10230–10239.

89. Schneider, F., Hammarstrom, P., Kelly, J.W. (2001). Transthyretin slowly exchanges subunits under physiological conditions: a convenient chromatographic method to study subunit exchange in oligomeric proteins. *Protein Sci, 10*, 1606–1613.

90. Reixach, N., Foss, T.R., Santelli, E., Pascual, J., Kelly, J.W., Buxbaum, J.N. (2008). Human-murine transthyretin heterotetramers are kinetically stable and non-amyloidogenic. a lesson in the generation of transgenic models of diseases involving oligomeric proteins. *J Biol Chem, 283*, 2098–2107.

91. Jiang, X., Buxbaum, J.N., Kelly, J.W. (2001). The V122I cardiomyopathy variant of transthyretin increases the velocity of rate-limiting tetramer dissociation, resulting in accelerated amyloidosis. *Proc Natl Acad Sci U S A, 98*, 14943–14948.

92. Hurshman Babbes, A.R., Powers, E.T., Kelly, J.W. (2008). Quantification of the thermodynamically linked quaternary and tertiary structural stabilities of transthyretin and its disease-associated variants: the relationship between stability and amyloidosis. *Biochemistry, 47*, 6969–6984.

93. Hurshman, A.R., White, J.T., Powers, E.T., Kelly, J.W. (2004). Transthyretin aggregation under partially denaturing conditions is a downhill polymerization. *Biochemistry, 43*, 7365–7381.

94. Hammarstrom, P., Jiang, X., Deechongkit, S., Kelly, J.W. (2001). Anion shielding of electrostatic repulsions in transthyretin modulates stability and amyloidosis: insight into the chaotrope unfolding dichotomy. *Biochemistry, 40*, 11453–11459.

95. Foss, T.R., Kelker, M.S., Wiseman, R.L., Wilson, I.A., Kelly, J.W. (2005). Kinetic stabilization of the native state by protein engineering: implications for inhibition of transthyretin amyloidogenesis. *J Mol Biol, 347*, 841–854.

96. Hammarstrom, P., Jiang, X., Hurshman, A.R., Powers, E.T., Kelly, J.W. (2002). Sequence-dependent denaturation energetics: a major determinant in amyloid disease diversity. *Proc Natl Acad Sci U S A, 99* (*Suppl 4*), 16427–16432.

97. Wiseman, R.L., Green, N.S., Kelly, J.W. (2005). Kinetic stabilization of an oligomeric protein under physiological conditions demonstrated by a lack of subunit exchange: implications for transthyretin amyloidosis. *Biochemistry, 44*, 9265–9274.

98. Palaninathan, S.K., Mohamedmohaideen, N.N., Snee, W.C., Kelly, J.W., Sacchettini, J.C. (2008). Structural insight into pH-induced conformational changes within the native human transthyretin tetramer. *J Mol Biol, 382*, 1157–1167.

99. Liu, K., Cho, H.S., Hoyt, D.W., Nguyen, T.N., Olds, P., Kelly, J.W., Wemmer, D.E. (2000). Deuterium-proton exchange on the native wild-type transthyretin tetramer identifies the stable core of the individual subunits and indicates mobility at the subunit interface. *J Mol Biol, 303*, 555–565.

100. Liu, K., Kelly, J.W., Wemmer, D.E. (2002). Native state hydrogen exchange study of suppressor and pathogenic variants of transthyretin. *J Mol Biol, 320*, 821–832.

101. Hammarstrom, P., Sekijima, Y., White, J.T., Wiseman, R.L., Lim, A., Costello, C.E., Altland, K., Garzuly, F., Budka, H., Kelly, J.W. (2003). D18G transthyretin is monomeric, aggregation prone, and not detectable in plasma and cerebrospinal fluid: a prescription for central nervous system amyloidosis? *Biochemistry, 42*, 6656–6663.

102. Snyder, S.W., Ladror, U.S., Wade, W.S., Wang, G.T., Barrett, L.W., Matayoshi, E.D., Huffaker, H.J., Krafft, G.A., Holzman, T.F. (1994). Amyloid-β aggregation: selective inhibition of aggregation in mixtures of amyloid with different chain lengths. *Biophys J, 67*, 1216–1228.

103. Harper, J.D., Wong, S.S., Lieber, C.M., Lasbury, P.T. (1997). Observation of metastable A-beta amyloid protofibrils by atomic force microscopy. *Chem Biol, 4*, 119–125.

104. Wang, H.W., Pasternak, J.F., Kuo, H., Ristic, H., Lambert, M.P., Chromy, B., Viola, K.L., Klein, W.L., Stine, W.B., Krafft, G.A., Trommer, B.L. (2002). Soluble oligomers of beta amyloid (1–42) inhibit long-term potentiation but not long-term depression in rat dentate gyrus. *Brain Res, 924*, 133–140.

105. Gong, Y., Chang, L., Viola, K.L., Lacor, P.N., Lambert, M.P., Finch, C.E., Krafft, G.A., Klein, W.L. (2003). Alzheimer's disease-affected brain: presence of oligomeric A beta ligands (ADDLs) suggests a molecular basis for reversible memory loss. *Proc Natl Acad Sci U S A, 100*, 10417–10422.

106. Cohen, E., Bieschke, J., Perciavalle, R.M., Kelly, J.W., Dillin, A. (2006). Opposing activities protect against age-onset proteotoxicity. *Science, 313*, 1604–1610.

107. Serio, T.R., Cashikar, A.G., Kowal, A.S., Sawicki, G.J., Moslehi, J.J., Serpell, L., Arnsdorf, M.F., Lindquist, S.L. (2000). Nucleated conformational conversion and the replication of conformational information by a prion determinant. *Science, 289*, 1317–1321.

108. Blake, C., Serpell, L. (1996). Synchrotron X-ray studies suggest that the core of the transthyretin amyloid fibril is a continuous beta-sheet helix. *Structure, 4*, 989–998.

109. Jacobson, D.R., Pastore, R.D., Yaghoubian, R., Kane, I., Gallo, G., Buck, F.S., Buxbaum, J.N. (1997). Variant-sequence transthyretin (isoleucine 122) in late-onset cardiac amyloidosis in black Americans. *N Engl J Med, 336*, 466–473.

110. Westermark, P., Sletten, K., Johansson, B., Cornwell, G.G. (1990). Fibril in senile systemic amyloidosis is derived from normal transthyretin. *Proc Natl Acad Sci U S A, 87*, 2843–2845.

111. Cornwell, G.C., Sletten, K., Johansson, B., Westermark, P. (1988). Evidence that the amyloid fibril protein in senile systemic amyloidosis is derived from normal prealbumin. *Biochem Biophys Res Commun, 154*, 648–653.

INSTITUT MEDIATHEQUE PASTEUR

112. Tanskanen, M., Peuralinna, T., Polvikoski, T., Notkola, I.-L., Sulkava, R., Hardy, J., Singleton, A., Kiuru-Enari, S., Paetau, A., Tienari, P.J., Myllykangas, L. (2008). Senile systemic amyloidosis affects 25% of the very aged and associates with genetic variation in alpha2-macroglobulin and tau: a population-based autopsy study. *Ann Med, 40*, 232–239.

113. Afolabi, I., Asl, K.H., Nakamura, M., Jacobs, P., Hendrie, H., Benson, M.D. (2000). Transthyretin isoleucine-122 mutation in African and American blacks. *Amyloid, 7*, 121–125.

114. Yamashita, T., Asl, K.H., Yazaki, M., Benson, M.D. (2005). A prospective evaluation of the transthyretin Ile122 allele frequency in an African-American population. *Amyloid, 12*, 127–130.

115. Andrade, C. (1952). A peculiar form of peripheral neuropathy: familiar atypical generalized amyloidosis with special involvement of the peripheral nerves. *Brain, 75*, 408–427.

116. Saraiva, M.J., Sousa, M.M., Cardoso, I., Fernandes, R. (2004). Familial amyloidotic polyneuropathy: protein aggregation in the peripheral nervous system. *J Mol Neurosci, 23*, 35–40.

117. Suhr, O.B., Hastrup Svendsen, I., Andersson, R., Danielsson, A., Holmgren, G., Ranlov, P.J. (2003). Hereditary transthyretin amyloidosis from a Scandinavian perspective. *J Intern Med*, 254, 225–235.

118. Suhr, O.B., Herlenius, G., Friman, S., Ericzon, B.G. (2000). Liver transplantation for hereditary transthyretin amyloidosis. *Liver Transpl, 6*, 263–276.

119. Coelho, T. (1996). Familial amyloid polyneuropathy: new developments in genetics and treatment. *Curr Opin Neurol, 9*, 355–359.

120. Ikeda, S., Nakazato, M., Ando, Y., Sobue, G. (2002). Familial transthyretin-type amyloid polyneuropathy in Japan: clinical and genetic heterogeneity. *Neurology, 58*, 1001–1007.

121. Koike, H., Misu, K., Sugiura, M., Iijima, M., Mori, K., Yamamoto, M., Hattori, N., Mukai, E., Ando, Y., Ikeda, S., Sobue, G. (2004). Pathology of early- vs late-onset TTR Met30 familial amyloid polyneuropathy. *Neurology, 63*, 129–138.

122. Ando, Y., Suhr, O.B. (1998). Autonomic dysfunction in familial amyloidotic polyneuropathy (FAP). *Amyloid, 5*, 288–300.

123. Holmgren, G., Ando, Y., Wikstrom, L., Rydh, A., Suhr, O. (1997). Discordant symptoms in monozygotic twins with familial amyloidotic polyneuropathy (FAP) (TTR Met 30). *Amyloid, 4*, 178–180.

124. Sousa, A., Coelho, T., Barros, J., Sequeiros, J. (1995). Genetic epidemiology of familial amyloidotic polyneuropathy (FAP)-type I in Povoa do Varzim and Vila do Conde (north of Portugal). *Am J Med Genet, 60*, 512–521.

125. Soares, M.L., Coelho, T., Sousa, A., Batalov, S., Conceicao, I., Sales-Luis, M.L., Ritchie, M.D., Williams, S.M., Nievergelt, C.M., Schork, N.J., et al. (2005). Susceptibility and modifier genes in Portuguese transthyretin V30M amyloid polyneuropathy: complexity in a single-gene disease. *Hum Mol Genet, 14*, 543–553.

126. Helgadottir, A., Manolescu, A., Thorleifsson, G., Gretarsdottir, S., Jonsdottir, H., Thorsteinsdottir, U., Samani, N.J., Gudmundsson, G., Grant, S.F., Thorgeirsson, G., et al. (2004). The gene encoding 5-lipoxygenase activating protein confers risk of myocardial infarction and stroke. *Nat Genet, 36*, 233–239.

127. Jacobson, D.R., Buxbaum, J.N. (1991). Genetic aspects of amyloidosis. *Adv Hum Genet, 20*, 69–123.
128. Jahoor, F., Reeds, P.J. (1999). The measurement of the rate of synthesis of liver-derived plasma proteins. *Methods Invest Amino Acid Protein Metab*, 69–81.
129. Aldred, A.R., Brack, C.M., Schreiber, G. (1995). The cerebral expression of plasma protein genes in different species. *Comp Biochem Physiol B, 111*, 1–15.
130. Weisner, B., Roethig, H.J. (1983). The concentration of prealbumin in cerebrospinal fluid (CSF), indicator of CSF circulation disorders. *Eur Neurol, 22*, 96–105.
131. Santos, C.R., Anjos, L., Power, D.M. (2002). Transthyretin in fish: state of the art. *Clin Chem Lab Med, 40*, 1244–1249.
132. Balch, W.E., Morimoto, R.I., Dillin, A., Kelly, J.W. (2008). Adapting proteostasis for disease intervention. *Science, 319*, 916–919.
133. Miroy, G.J., Lai, Z., Lashuel, H., Peterson, S.A., Strang, C., Kelly, J.W. (1996). Inhibiting transthyretin amyloid fibril formation via protein stabilization. *Proc Natl Acad Sci U S A, 93*, 15051–15056.
134. Perlmutter, D.H. (2003). Alpha 1-antitrypsin deficiency: liver disease associated with retention of a mutant secretory glycoprotein in the endoplasmic reticulum. *Methods Mol Biol, 232*, 39–56.
135. Parfrey, H., Mahadeva, R., Lomas, D.A. (2003). Alpha 1-antitrypsin deficiency, liver disease and emphysema. *Int J Biochem Cell Biol, 35*, 1009–1014.
136. Suk, J.Y., Zhang, F., Balch, W.E., Linhardt, R.J., Kelly, J.W. (2006). Heparin accelerates gelsolin amyloidogenesis. *Biochemistry, 45*, 2234–2242.
137. Sekijima, Y., Hammarstrom, P., Matsumura, M., Shimizu, Y., Iwata, M., Tokuda, T., Ikeda, S., Kelly, J.W. (2003). Energetic characteristics of the new transthyretin variant A25T may explain its atypical central nervous system pathology. *Lab Invest, 83*, 409–417.
138. Coelho, T., Chorao, R., Sausa, A., Alves, I., Torres, M.F., Saraiva, M.J. (1996). Compound heterozygotes of transthyretin Met30 and transthyretin Met119 are protected from the devastating effects of familial amyloid polyneuropathy. *Neuromusc Disord, 6*, 27.
139. Buxbaum, J., Tagoe, C., Gallo, G., Reixach, N., French, D. (2003). The pathogenesis of transthyretin tissue deposition: lessons from transgenic mice. *Amyloid, 10*, 2–6.
140. Teng, M.H., Yin, J.Y., Vidal, R., Ghiso, J., Kumar, A., Rabenou, R., Shah, A., Jacobson, D.R., Tagoe, C., Gallo, G., Buxbaum, J. (2001). Amyloid and nonfibrillar deposits in mice transgenic for wild-type human transthyretin: a possible model for senile systemic amyloidosis. *Lab Invest, 81*, 385–396.
141. Kohno, K., Palha, J.A., Miyakawa, K., Saraiva, M.J., Ito, S., Mabuchi, T., Blaner, W.S., Iijima, H., Tsukahara, S., Episkopou, V., et al. (1997). Analysis of amyloid deposition in a transgenic mouse model of homozygous familial amyloidotic polyneuropathy. *Am J Pathol, 150*, 1497–1508.
142. Tagoe, C.E., Reixach, N., Friske, L., Mustra, D., French, D., Gallo, G., Buxbaum, J.N. (2007). In vivo stabilization of mutant human transthyretin in transgenic mice. *Amyloid, 14*, 227–236.
143. Holmgren, G., Ericzon, B.G., Groth, C.G., Steen, L., Suhr, O., Andersen, O., Wallin, B. G., Seymour, A., Richardson, S., Hawkins, P.N. (1993). Clinical

improvement and amyloid regression after liver transplantation in hereditary transthyretin amyloidosis. *Lancet, 341*, 1113–1116.

144. Suhr, O.B., Holmgren, G., Lundgren, E. (2004). Gene therapy: lessons learned from liver transplantation for transthyretin-amyloidosis. *Liver Transpl, 10*, 1551–1553.
145. Herlenius, G., Wilczek, H.E., Larsson, M., Ericzon, B.G. (2004). Ten years of international experience with liver transplantation for familial amyloidotic polyneuropathy: results from the Familial Amyloidotic Polyneuropathy World Transplant Registry. *Transplantation, 77*, 64–71.
146. Stangou, A.J., Hawkins, P.N. (2004). Liver transplantation in transthyretin-related familial amyloid polyneuropathy. *Curr Opin Neurol, 17*, 615–620.
147. Olofsson, B.O., Backman, C., Karp, K., Suhr, O.B. (2002). Progression of cardiomyopathy after liver transplantation in patients with familial amyloidotic polyneuropathy, Portuguese type. *Transplantation, 73*, 745–751.
148. Hornsten, R., Wiklund, U., Olofsson, B.O., Jensen, S.M., Suhr, O.B. (2004). Liver transplantation does not prevent the development of life-threatening arrhythmia in familial amyloidotic polyneuropathy, Portuguese-type (ATTR Val30Met) patients. *Transplantation, 78*, 112–116.
149. Ando, Y., Terazaki, H., Nakamura, M., Ando, E., Haraoka, K., Yamashita, T., Ueda, M., Okabe, H., Sasaki, Y., Tanihara, H., et al. (2004). A different amyloid formation mechanism: de novo oculoleptomeningeal amyloid deposits after liver transplantation. *Transplantation, 77*, 345–349.
150. Ando, Y., Ericzon, B.G., Suhr, O.B., Tashima, K., Ando, M. (1997). Reuse of a Japanese familial amyloidotic polyneuropathy patient's liver for a cancer patient: the domino liver transplantation procedure. *Intern Med, 36*, 847.
151. Azoulay, D., Samuel, D., Castaing, D., Adam, R., Adams, D., Said, G., Bismuth, H. (1999). Domino liver transplants for metabolic disorders: experience with familial amyloidotic polyneuropathy. *J Am Coll Surg, 189*, 584–593.
152. Ericzon, B.G., Larsson, M., Wilczek, H.E. (2008). Domino liver transplantation: risks and benefits. *Transplant Proc, 40*, 1130–1131.
153. Stangou, A.J., Heaton, N.D., Hawkins, P.N. (2005). Transmission of systemic transthyretin amyloidosis by means of domino liver transplantation. *N Engl J Med, 352*, 2356.
154. Bittencourt Paulo, L., Couto Claudia, A., Leitao Regina Maria, C., Siqueira Sheila, A., Farias Alberto, Q., Massarollo Paulo Celso, B., Mies, S. (2002). No evidence of de novo amyloidosis in recipients of domino liver transplantation: 12 to 40 (mean 24) month follow-up. *Amyloid, 9*, 194–196.
155. Obayashi, K., Yamashita, T., Ueda, M., Nakamura, M., Asonuma, K., Inomata, Y., Uchino, M., Tanaka, K., Ando, Y. (2008). Amyloid neuropathy in a Japanese domino liver-transplanted recipient. *Int Symp Amyloidosis, 11th*, 172–174.
156. Schmidt, H.H., Nashan, B., Propsting, M.J., Nakazato, M., Flemming, P., Kubicka, S., Boker, K., Pichlmayr, R., Manns, M.P. (1999). Familial amyloidotic polyneuropathy: domino liver transplantation. *J Hepatol, 30*, 293–298.
157. Takei, Y.-I., Gono, T., Yazaki, M., Ikeda, S.-I., Ikegami, T., Hashikura, Y., Miyagawa, S.-I., Hoshii, Y. (2007). Transthyretin-derived amyloid deposition on the gastric mucosa in domino recipients of familial amyloid polyneuropathy liver. *Liver Transpl, 13*, 215–218.

158. Anderson, O., Wallin, G. (1992). Stationary course during sixteen months after liver transplantation in familial amyloidotic polyneuropathy. In *Second International Symposium on Familial Amyloidotic Polyneuropathy and Other Transthyretin Related Disorders*, Skelleftea, Sweden.

159. Foss, T.R., Wiseman, R.L., Kelly, J.W. (2005). The pathway by which the tetrameric protein transthyretin dissociates. *Biochemistry, 44*, 15525–15533.

160. Wiseman, R.L., Johnson, S.M., Kelker, M.S., Foss, T., Wilson, I.A., Kelly, J.W. (2005). Kinetic stabilization of an oligomeric protein by a single ligand binding event. *J Am Chem Soc, 127*, 5540–5551.

161. Johnson, S.M., Connelly, S., Wilson, I.A., Kelly, J.W. (2008). Toward optimization of the linker substructure common to transthyretin amyloidogenesis inhibitors using biochemical and structural studies. *J Med Chem, 51*, 6348–6358.

162. Johnson, S.M., Connelly, S., Wilson, I.A., Kelly, J.W. (2008). Biochemical and structural evaluation of highly selective 2-arylbenzoxazole-based transthyretin amyloidogenesis inhibitors. *J Med Chem, 51*, 260–270.

163. Adamski-Werner, S.L., Palaninathan, S.K., Sacchettini, J.C., Kelly, J.W. (2004). Diflunisal analogues stabilize the native state of transthyretin: potent inhibition of amyloidogenesis. *J Med Chem, 47*, 355–374.

164. Green, N.S., Palaninathan, S.K., Sacchettini, J.C., Kelly, J.W. (2003). Synthesis and characterization of potent bivalent amyloidosis inhibitors that bind prior to transthyretin tetramerization. *J Am Chem Soc, 125*, 13404–13414.

165. Johnson, S.M., Petrassi, H.M., Palaninathan, S.K., Mohamedmohaideen, N.N., Purkey, H.E., Nichols, C., Chiang, K.P., Walkup, T., Sacchettini, J.C., Sharpless, K.B., Kelly, J.W. (2005). Bisaryloxime ethers as potent inhibitors of transthyretin amyloid fibril formation. *J Med Chem, 48*, 1576–1587.

166. Peterson, S.A., Klabunde, T., Lashuel, H., Purkey, H., Sacchettini, J.C., Kelly, J.W. (1998). Inhibiting transthyretin conformational changes that lead to amyloid fibril formation. *Proc Natl Acad Sci U S A, 95*, 12956–12960.

167. Petrassi, H.M., Klabunde, T., Sacchettini, J., Kelly, J.W. (2000). Structure-based design of *N*-phenyl phenoxazine transthyretin amyloid fibril inhibitors. *J Am Chem Soc, 122*, 2178–2192.

168. Purkey, H.E., Palaninathan, S.K., Kent, K.C., Smith, C., Safe, S.H., Sacchettini, J.C., Kelly, J.W. (2004). Hydroxylated polychlorinated biphenyls selectively bind transthyretin in blood and inhibit amyloidogenesis: rationalizing rodent PCB toxicity. *Chem Biol, 11*, 1719–1728.

169. Razavi, H., Palaninathan, S.K., Powers, E.T., Wiseman, R.L., Purkey, H.E., Mohamedmohaideen, N.N., Deechongkit, S., Chiang, K.P., Dendle, M.T., Sacchettini, J.C., Kelly, J.W. (2003). Benzoxazoles as transthyretin amyloid fibril inhibitors: synthesis, evaluation, and mechanism of action. *Angew Chem Int Ed Engl, 42*, 2758–2761.

170. Sacchettini, J.C., Kelly, J.W. (2002). Therapeutic strategies for human amyloid diseases. *Nat Rev Drug Discov, 1*, 267–275.

171. Wojtczak, A., Cody, V., Luft, J.R., Pangborn, W. (1996). Structures of human transthyretin complexed with thyroxine at 2.0 Å resolution and 3′,5′-dinitro-*N*-acetyl-L-thyronine at 2.2 Å resolution. *Acta Crystallogr D, 52*, 758–765.

172. Petrassi, H.M., Johnson, S.M., Purkey, H.E., Chiang, K.P., Walkup, T., Jiang, X., Powers, E.T., Kelly, J.W. (2005). Potent and selective structure-based dibenzofuran inhibitors of transthyretin amyloidogenesis: kinetic stabilization of the native state. *J Am Chem Soc, 127*, 6662–6671.

173. Baures, P.W., Oza, V.B., Peterson, S.A., Kelly, J.W. (1999). Synthesis and evaluation of inhibitors of transthyretin amyloid formation based on the non-steroidal antiinflammatory drug flufenamic acid. *Bioorg Med Chem, 7*, 1339–1347.

174. Baures, P.W., Peterson, S.A., Kelly, J.W. (1998). Discovering transthyretin amyloid fibril inhibitors by limited screening. *Bioorg Med Chem, 6*, 1389–1401.

175. McCammon, M.G., Scott, D.J., Keetch, C.A., Greene, L.H., Purkey, H.E., Petrassi, H.M., Kelly, J.W., Robinson, C.V. (2002). Screening transthyretin amyloid fibril inhibitors: characterization of novel multiprotein, multiligand complexes by mass spectrometry. *Structure, 10*, 851–863.

176. Miller, S.R., Sekijima, Y., Kelly, J.W. (2004). Native state stabilization by NSAIDs inhibits transthyretin amyloidogenesis from the most common familial disease variants. *Lab Invest, 84*, 545–552.

177. Oza, V.B., Petrassi, H.M., Purkey, H.E., Kelly, J.W. (1999). Synthesis and evaluation of biaryl amines and benzylamines as inhibitors of TTR amyloid fibril formation. *Book of Abstracts*, 217th ACS National Meeting, Anaheim, CAl, Mar. 21–25, ORGN-447.

178. Oza, V.B., Petrassi, H.M., Purkey, H.E., Kelly, J.W. (1999). Synthesis and evaluation of anthranilic acid–based transthyretin amyloid fibril inhibitors. *Bioorg Med Chem Lett, 9*, 1–6.

179. Oza, V.B., Smith, C., Raman, P., Koepf, E.K., Lashuel, H.A., Petrassi, H.M., Chiang, K.P., Powers, E.T., Sachettinni, J., Kelly, J.W. (2002). Synthesis, structure, and activity of diclofenac analogues as transthyretin amyloid fibril formation inhibitors. *J Med Chem, 45*, 321–332.

180. Razavi, H., Powers, E.T., Purkey, H.E., Adamski-Werner, S.L., Chiang, K.P., Dendle, M.T., Kelly, J.W. (2005). Design, synthesis, and evaluation of oxazole transthyretin amyloidogenesis inhibitors. *Bioorg Med Chem Lett, 15*, 1075–1078.

181. Sekijima, Y., Dendle Maria, A., Kelly, J.W. (2006). Orally administered diflunisal stabilizes transthyretin against dissociation required for amyloidogenesis. *Amyloid, 13*, 236–249.

182. Tojo, K., Sekijima, Y., Kelly, J.W., Ikeda, S.-I. (2006). Diflunisal stabilizes familial amyloid polyneuropathy-associated transthyretin variant tetramers in serum against dissociation required for amyloidogenesis. *Neurosci Res, 56*, 441–449.

183. Ferguson, R.N., Edelhoch, H., Saroff, H.A., Robbins, J. (1975). Negative cooperativity in the binding of thyroxine to human serum prealbumin. *Biochemistry, 14*, 282–289.

184. Neumann, P., Cody, V., Wojtczak, A. (2001). Structural basis of negative cooperativity in transthyretin. *Acta Biochim Pol, 48*, 867–875.

185. Purkey, H., Dorrell, M.I., Kelly, J.W. (2001). Evaluating the binding selectivity of transthyretin amyloid fibril inhibitors in blood plasma. *Proc Natl Acad Sci USA, 98*, 5566–5571.

INDEX

17-AAG, 949, 952
A-Bri, 334
Absorption, 98, 847, 856
Accessory molecules
  heparan sulfate in in vivo amyloidogenesis, 559–566
  metals in Alzheimer disease, 545–552
  oxidatively stress lipids in amyloid formation and toxicity, 585–594
  oxidative stress in protein misfolding and/or amyloid formation, 615–625
  serum amyloid P component, 571–580
Acetylation, 890
Acetyl-CoA:cholesterol acetyl transferase (ACAT) inhibitor, 739
Acetyl-L-carnitine, 949–951
Acetyl salicylic acid, 266
Acetyltransferase, 203
AchE inihibitors, 712, 722
ACh receptor agonists, 720
Acinar cell, pancreatic, 28
Acquired immunity, 117
Acridines, 269, 276–277, 284
Acrolein, 618
Actin, 61, 75, 86, 503
Actin-modulating proteins, 332
Activating transcription factor 4 (ATFR), 26
Active aggregation, 636, 638–639
Activity-dependent neuroprotective protein (ADNP), 742
AC-1202, 741
Acute leukemia, 781
Acute lymphoblastic leukemia (ALL), 454
Acylphosphatase, 859
ADAM17, 719
ADAM10, 719
Adaptor protein, 460
ADAS-COG scores, 285, 744
ADCS-ADI, 744
Adduct formation, 498
Adeno-associated viral (AAV) vector, 743, 956
Adenosine arabinoside, 271, 280
Adenosine diphosphate (ADP), 33, 35, 54, 57, 62, 35, 54
  functions of, 54, 57, 62
Adenosine triphosphate (ATP)
  ADP and, 33, 35, 54
  amyloid cytotoxicity, 101
  ATP/ADP
    exchange, 35, 54
    ratio, 33
  binding, 50, 52–54, 58
  -binding cassette (ABC) gene family, 427, 435–436
  functions of, 57–59, 62, 431–437, 721, 951
Adenoviral systems, 947
Adhesion molecules, 561
ADP-A1Fx, 61
Adrenergic receptor activation, 722
Adsorption, 94, 370, 703, 921
Advanced glycation end products (AGEs), 132, 139

Protein Misfolding Diseases: Current and Emerging Principles and Therapies,
Edited by Marina Ramirez-Alvarado, Jeffery W. Kelly, and Christopher M. Dobson
Copyright © 2010 John Wiley & Sons, Inc.

Adverse drug reactions, 453
Affinity binding, 547, 551
AFL, 695
Age-associated sporadic neurodegenerative diseases, 195
Age at onset
  of neurodegeneration, influences on, 639–640
  significance of, 632, 634
age-1, 184, 186, 633
Aggregate formation, kinetics of, 73
Aggregation
  Aβ, 939
  amyloid diseases, 73, 94
  amyloid formation, 133–139
  biochemical studies, 133, 140
  biophysical studies, 133, 139–140
  β-sheet propensity, 8, 94
  determinants of, from largely or partially unfolded states, 7–9
  diseases, 306
  downhill, 84, 974
  from equilibrium partially folded states, 4–5
  globular proteins, 11
  hazards of, 48
  hydrophobicity and, 8
  implications of, 461, 623–624
  inhibitors
    peptide-based, 724
    small-molecule, 723–724
  intrinsically disordered proteins, 7, 9,
  in vitro propensity of lysozyme, 876
  low net charge, 8
  modulation of, 96
  mutant SOD1, 384–387, 389–390
  mutation-induced, 9–10
  polyglutamine protein dynamics, 178–180
  polyQ, 316–317
  predictive methods, 9
  principles of, 74–82
  probability of, 12
  protein conformation disorders (PCDs), 125
  of proteins, 10–11, 181–182, 184, 186, 935, 938
  ribosome-binding chaperones, 52
  spontaneous folding, 59
Aggregopathies, 120, 125
Aggresome formation, 460, 463
Aging population, xxvii
Aging process
  aggregation-mediated proteotoxicity, 631–640
  counter-proteotoxicity mechanisms, 639
  health effects of, 631
  implications of, 10, 12–13, 62, 105, 405, 550, 824, 946
  manipulation of, 640
  oxidative stress and, 616
Akt
  counter polyQ toxicity effects, 636–637
  NRG-Akt pathway, 246
  phosphorylation of, 244, 730
  signaling, 892–893
Akt-1 phosphorylation, 196
Alanine, 156, 361, 521
Aldehyde oxidase (AO), 455
Aldehydes
  implications of, 549, 970
  lipid-derived
    amyloidogenesis, 619–625
    biological origins, 617–619
    chemical origins, 617–619
Aldose reductase, 490
Alkyl groups, 618
Allogeneic bone marrow transplantation, 779
Allosteric interdomain signaling, 54
Allozymes, TPMT, 459–460, 464
α_1-Antitrypsin
  characterized, 403–404, 977
  deficiency, *see* Alpha-1-antitrypsin deficiency
  replacement therapy, 414
α_1-Antitrypsin deficiency
  liver disease, 404–411
  lung disease, 404, 407
  overview of, 403–404
  pathobiology
    cellular, 407–413
    structural, 405–407
  prevalence of, 404
  therapeutic strategies, 125, 413–415
α-Crystallins, 491–493, 497–498, 502
α-Secretase, 241, 718–722
α-Synuclein
  aggregation, 631
  amyloid formation, 138–139
  characterized, 7, 10, 12, 34, 101, 122–123, 176–178, 182, 196–197, 545
  down-regulation by RNAi in Parkinson's disease, 955–956
  neural, degradation of, 828–833
  neurological syndromes, 821
  protein misfolding, 820–822
  protofibrils, 703
  sequence motifs, 822
Alzheimer disease (AD)
  age at onset, 238, 250
  aging and, 631
  amyloid
    beta, role in, 546–550
    cytotoxicity, 102

plaque, 176, 617, 917
precursor protein (APP), 237–241, 717–723, 738
amyloidoses, 844
anatomical changes in brain, 713
androgenic hormonal pathways, 739–740
animal models, 217–226
autophagy, 119, 123
autosomal dominant forms of, 234
biochemistry of, 234, 236–237, 239, 250
cholesterol pathway, 738–379
dementia, 221
diagnosis of, 216, 234, 250
down-regulation of Aβ and tau by RNAi, 954–955
early stages of, 226
environmental influences, 740–741
experimental manipulations, 245–249
features of, clinical and laboratory, xxii–xxiii, xxviii, 33–35, 49, 62, 93–94, 101, 175, 181, 192–193, 195, 233–235, 489, 526, 712–713, 933
fibrillogenic environment, 856
gender differences, 739–740
glucose metabolism, 741
high-cholesterol hypothesis, 96
incidence of, 235, 939
inhibition strategies, 906
lifestyle, impact of, 740–741
lipid aldehydes, 619
metals, role in, 545–552
mild to moderate, 741–742, 744
mouse models of, 716–718
neurofibrillar tangles, 134
neuroinflammation in, 736–738
neuropathology, 233–234, 236–237, 250, 735–744
onset
late-, 238
significance of, 712, 735
oxidative stress, 615, 617, 620
pathogenesis of, 226, 234–235, 238, 546, 934
posttranslational modifications, 134
progression of, 215, 335, 735
receptor-mediated pathways, 741–743
regenerative tissue implantation in, 743–744
risk factors for, 237–239
secretases, 240–241, 244–245
small-molecule drug screens, 183
sporadic, 237–238, 622, 631, 740, 969
taupathies, 99
tau-specific therapeutic targets, 730–735
therapeutic strategies
future challenges of, 744–745
gene therapy, 743–744
overview of, 245–249, 550–552, 711–745, 908
toxic misfolded proteins, 954
transthyretin (TTR) and, 971
trophic factor, 741–743
Alzheimer Disease and Frontotemporal Dementia Mutation Database, 238
Alzhemed, 936
Amantadine, 280
Amiloidogenicity, 849
Aminasine, 277
Amino acids, 457–458
amyloid formation, 137, 139
biosynthesis, 26
functions of, 549, 562
metabolism, 25
protein misfolding diseases, 364, 384, 407, 410, 427–428, 432, 457–458, 473, 478
protein misfolding principles, 8–9, 35, 48, 51, 57, 60, 113, 117, 152, 154–156, 163–164, 216
therapeutic treatments, 802, 843–844, 853, 909, 919–920, 938
Amino groups, 619–621, 623–624
Aminoguanidine, 623
Amioodarone, 271, 279
Amnestic form of MCIs (aMCIs), 215, 226, 235–236
AMPA receptors, 99, 741–742
Ampakine CX-717, 742
Amphotericin B, 269, 276, 286
Amylin, 517
Amyloid(s)
aggregates
cytotoxicity, 101
generic toxicity, 97–99
aggregation, 6–9
beta peptides. *See* Aβ
cascade, 220, 712
conformation, 158–159, 161, 163–165, 851
diseases
amyloid aggregates display generic toxicity, 97–99
cell functional state, 103–104
cellular context, 102–104
characterized, xxvii, 83, 93, 104–105
incidence of, 105
intracellular environment, 102–103
overview of, 968–970
shared biochemical modifications in cells exposed to toxic aggregates, 99, 101–102
triggers of amyloid precursor production, aggregation, and toxicity, 94–97

Amyloid(s) (*continued*)
extracellular, 339
fibrils, 3, *see* Fibril formation
formation, 131–134, 138–139, 150, 333, 521–522, 524–525, 530–531, 615–640, 945
kinetics of, 138–139
posttranslational modification and, 131–134
prohormones, 134
heart disease, 808
in hemodialyis-related amyloidosis, 350–353, 363–368
hypothesis, 93
inhibition
antibodies and immunotherapy, 906–907
using apolipoprotein E, 907
chemistry and biology of, 905–906
macromolecular, 906
mechanisms of action, 909–911
purpose of, 905–906
small-molecule, 907–909
peptide generation, 96
plaque, 176, 215–222, 233, 242, 617, 922, 924
pores, 385
precursor protein (APP), 636
in Alzheimer disease, 239–243, 245–247, 250, 550–551, 712, 716–718, 722, 739
characterized, 10, 96, 103, 123, 214, 216–217, 223, 225–226, 717–723, 738, 954–955
cholesterol levels and, 739
decreasing expression, 722–723
mutant, 237–239, 241, 248
removal of, 247
secretases, 239–240, 717
translation, 723
protofilament, 522–524
receptors, 97
secretases, 244–245
small-molecule inhibitors of, 937–938
Amyloid-beta (Aβ)
aggregation, 238, 712, 910
Alzheimer disease and, 546–550, 718–722, 740–741
amyloid formation, 138
amyloidosis, 242, 247, 250
-APP interactions, 97
binding
antibodies, 726–727
proteins, 936
decrease in, 740
degradation, 238, 246, 638
down-regulation of, 954–955
-enhancing factor (AEF), 925
familial forms of, 548
fibrillization, 623–625
fibrillogenesis, 741
fibrils, *see* Fibril growth, Aβ
immunotherapy, 248–249
implications of, 699–700
lipid aldehyde interactions, 620–622
metabolism, 238, 741
neuropathology, 742
neurotoxicity, 242
peptides (Aβ)
Aβ40, 721, 911
Aβ42, 101, 237–239, 241, 713, 716, 937, 911, 947
Aβ*56, 222–226, 241
aggregation, 909, 911
amyloid inhibitors, 907–908
antibodies against, 906
characterized, 7, 10, 12, 96, 104, 703–706
cleaving, 249
down-regulation, 954–955
immunotherapy, 248–249
sources of, removal of, 247
plaque, 236, 737, 936, 954
protein, 213, 217–226
posttranslational modifications, 134, 137, 140
protein aggregation, 623
proteotoxicity, 638
protofibrils, 634
real-time observation, by total internal reflection fluorescence microscopy, 700–702
spherulitic structure formation, 703–706
synthetic, 546
-TNFR1 interactions, 97
toxic protein aggregation, 634–635
Aβ-derived diffusible ligands (ADDLs), 222, 236, 241
Amyloid plaque, 617
Amyloidogenesis
atheronals, 618, 622–623, 625
characteristics of, 619
efficiency of, 978
in vivo, role of heparan sulfate, 559–566
native multimerization in, 335
process, 245, 561, 857, 925, 969, 975
TTR, 972–975, 978
Amyloidogenicity, 882
Amyloidoma, 919–922
Amyloidophilic compounds, 269, 274
Amyloidoses/amyloidosis (AL), *see specific types of amyloidosis*
Aβ models, 241–243
biomarkers identification, 689–696

cardiac, 681–682, 782, 800, 807, 981
classification, 677–679
defined, 843
diagnosis of, 776, 801
dialysis-related, 4, 10
etiology, 844
features of, xxviii, 235, 775, 860
hemodialysis-related, 347–370
immunotherapy, 918–919
light-chain, *see* Light-chain amyloidosis
localized, 775–776, 786–787
medullary, 856
sporadic, 615
survival of, 696
systemic, *see* Systemic amyloidosis
therapies
corticosteroid-based, 782–783
low-dose chemotherapy, 780–782
novel, 784
standard chemotherapy, 780
treatment decisions, prognostic factors for, 784, 786
unsuccesful, 783–784
Amyloid precursor-like proteins (APLP1/APLP2), 239–240, 720
Amylotopes, 875, 881
Amyotrophic lateral sclerosis (ALS), xxviii, 6, 9, 34, 175, 192, 197, 743, 821, 933, 950
AN1792, 744
Androgen receptors, 888, 952
Anemia, 845
Angiogenesis, 561
Angiotensin receptor blockers, 722
Anhidrosis, 804
Anilinonaphthalene sulfonic acid (ANS), 274
Animal models, *see specific types of animals*
advantages of, 191 192
α-syn-induced pathology, 832
Alzheimer disease, 103–104
biology of amyloid-β protein misfolding in Alzheimer disease
Aβ*56, memory impairment, 222–226
assaying effects on memory of Aβ, 215–217
differentiating roles of Aβ and amyloid plaques in memory loss, 217–222
overview, 213–215
cognitive deficits, 742
molecular chaperones, 947
polyglutamine diseases, 890
Anionic surfaces, 94
Annealing, 60, 414–415
Antiaggregation effects, 633
Anti-Aβ therapies, 735
Antiamyloid therapeutics, 235, 716
Antibiotics, polyene, 269, 276
Antibodies (Abs)
Aβ, 242, 249, 726–727
ADDL-specific, 222
against Aβ, 937
amyloid formation, 150
antiamyloid, 356
antigen-induced, 264
antioligomer, 357
heparan sulfates in amyloidogenesis, 565
immunotherapy and, 906–907, 918–919, 925, 927
lysozyme amyloidosis, 878, 881–882
memory deficits and, 220–221
monoclonal, 224, 243
polyclonal, 692
prion disease therapies, 263–265
SOD1, 384
synuclease activity, 830
toxic protein aggregation, 635
transgenic, 265
Antibody-catalyzed water oxidation pathway (ACWOP), 617
Antibody light-chain (LC) amyloidosis, 619
Anticancer agents, 734–735, 951–952
Anticoagulants, 859
Antifibrillization, 908
Antigens, in immunotherapy, 918–919
Antihypertensive therapy, 722
Anti-inflammatories
characteristics of, 722, 948
nonsteroidal, *see* Nonsteroidal anti-inflammatories
pathways, 740
therapy, 736
Antineurodegeneration treatment development, 640
Antioxidant response element (ARE), 26
Antioxidants, 37, 101–102, 183, 272, 282, 490, 741
Anti-oxidative stress, 25, 31, 33
Antisense oligonucleotides, 743, 828
Antisense RNA, 271, 280
Anti-tau therapy, 716
Antitrypsin (AT) gene, 411
Antivirals, 271, 280
Antraquinone, 735
A1-antichymotrypsin, 936
Aph-1, 237, 241, 245
Apolipoproteins
ApoAI mutants, 332, 337
ApoAII/mutants, 334, 337
ApoB, 738
ApoE

Apolipoproteins (*continued*)
alleles, 238, 526, 536
antibodies against, 906
binding domain, 724
cholesterol levels and, 738–739
implications of, 12, 936
systemic amyloidoses, 325
ApoE2 allele, 741, 907
ApoE3 allele, 741, 907
ApoE4, 234, 237–239, 250, 739, 907
ApoJ, 936
functions of, 338
Apomorphine, 183
Apoptosis, 25–26, 35, 99, 101–102, 104, 123, 616, 802
APP intracellular domain (AICD), 239–240, 717
Aptamers, 356–357, 370
Arachidonic acid, 617–618, 948
Arachnoiditis, 803
ARA70, 894
Arc48, 223
Archaea, 50–51, 53,56, 61
Arctic mutations, 223, 225
Arg3, 850, 852
Arginines, 151, 333
Argonaute2 (AGO2), 954
Aricept, 742
Arimoclomol, 949–950
Aromaticity, 909
Aromatics
aromatic-aromatic interactions, 521–522
functions of, 273, 359, 363, 521–522, 851, 985
Arthropathy, 844
Arundic acid, 737
Aryl group, 985, 909
ASH neurons, 184
Asn
deamidation, 135, 137
ladders, 524
Asn-Gly dipeptide, 135–136
Asn143, 499
Asparagines, 146–147, 160, 164, 473, 488, 499
Aspartate, 151
Asp-Pro bonds, 132
Assisted folding, 60
Astrocytosis, 737
Astroglial cells, 950
Ataxia
ataxin-1, 196, 201, 312, 893
ataxin, 2, 312
ataxia-3, 194–195, 197–199, 200–204, 312–313, 315
ataxin 7, 312
cerebral, 334
functions of, 890, 892
spinocerebellar, 947
ATF6
functions of, 25, 28–29
-mediated transcription, 28–29
ATG7, 461
Atheronals
amyloidogenesis, 617, 619, 622
atheronal-A/atheronal-B, 618, 623–625
Atherosclerosis, 36–37, 617, 619–620
Atorvastatin, 738
ATPase, 53–54, 57, 102, 114, 153
Atrial fibrillation, 807
Atrophins, 199, 312, 892
Atrophy, 821
Attenuation, 24–25, 735, 743, 825, 951
Autoantibodies, anti-tau, 735
Autoimmune diseases/disorders, 264, 454, 928
Autonomic nervous system, 263
Autophagic vacuole (AV), 115–116, 118, 120, 123–124
Autophagosomes, 116, 408–409
Autophagy
activation of, 125
α_1-antitrypsin deficiency, 408–410
blockage of, 125
in cancer, 118
defined, 115, 461
-dependent degradation, 461–463
disease and, 117–118
hepatic, 408–409
implications of, 21, 830, 833, 890
-lysosomal degradation system, 888–889
modulation of, 203
physiological functions of, 117
protein conformation disorders (PCDs), 118–125
protein misfolding, 119–121
reduction, using RNAi, 203
system failure, consequences of, 121–122
toxic protein aggregation, 636
types of, 115–117
Autophagy-related proteins (Atgs)
Atg5, 203, 408
Atg6, 408
Atg7, 203
characteristics of, 115–116
Autosomal dominant frontotemporal dementia with parkinsonism, 243
Autosomal dominant hereditary amyloidosis, 6
Avasimibe, 739
Axons
degeneration of, 236, 802

myelination, 802
transport, 237, 243, 735
Azathiopurine, 454–455
AZD340, 712
Azetidine, 949

Bacteria, 30, 50–51, 53, 99, 919
Bacterial proteins, 338
Bag-1 protein, 54
BAM-10 antibodies, 220–221
BAP31 activation, 409
Base pairs, 455
Biopsy, liver, 458
Basement membranes, vascular, 706
Batten disease, 125
Bax protein, 37
BBF2H7, 29
B cells, 33, 247, 273, 328
Bcl-2, 116, 118, 272
BDNF (brain-derived neurotropic factor) levels, 742
Beclin-1, 116, 118
Behavioral studies, 243–244
Beta-adrenergic receptor blockers, 808
β-amyloid cleaving enzyme (BACE)
Alzheimer disease therapies, 738–739
BACE1, 96, 237–241, 244, 246, 720, 736, 738, 954–955
BACE2, 240
inhibitors, 720–721
β-catenin pathway, 245
β-clam protein, 82–83
β-crystallins, 491, 493, 497
β-CTF, 738
β-Elimination, 132
β-galactosidase, 56, 497
β-glucocerebrosidase (GCase), 469–472, 474, 477–479
Betaine, 336
β_2-microglobulin (β_2m)
amyloidogenesis, 846
circulating, reduction of, 847–848
dialysis-related amyloidosis (DRA), 844–856, 859–860
in hemodialysis-related amyloidosis (HDRA)
aggregation, regions involved in, 359–363
amyloid deposits, constituents of, 350–353
amyloid precursor state, identification of, 363–368
fibril formation mechanisms, 353–356
fibril structure, 356–359
structure and function in vivo, 347–350
implications of, 4–6, 10, 12, 105, 330, 347, 783
interactors, identification of
molecular overview, 848–851
rational mutant designs, 851–856
variants and their effects, 361–363
wild-type, 364
β-propiolactone, 280
β-sandwich subdomain, 53–54
β-secretases, 239–240, 246, 718–722, 739
β-sheets
α_1-antitrypsin structure, 405–406
architecture, 358
breaker peptides, 270, 278–279, 724, 908, 938–939
characterized, 4, 8, 34, 94–95, 100, 145, 156, 158–161, 163, 264
conformations, 307, 414
fractures, 985
secondary structures, 919
β-synuclein, 822
bFGF, 827
Bimoclomal, 949
Binding motifs, 195
Bioavailability, 737
Biochemical studies
α_1-antitrypsin deficiency, 413
amyloids
cytotoxicity, 98–99, 101–102
diseases, 93, 99, 101–103, 845
familial amyotrophic lateral sclerosis (fALS), 385
posttranslational modifications, 133, 140
protein misfolding diseases, 936
Biocompatibility, 370, 845
Biogenesis, 53
Bioimaging studies, 711
Bioinformatic analysis, 52
Biological membranes, 11, 94–95
Biological systems, lipid peroxidation, 616–617
Biomarker studies
Alzheimer disease, 234–235, 250
cardiac, 776–777, 799
identification of biomarkers, 689–696
Biomolecules, 616
Biopharmaceutical stabilization, 133
Biophysical studies
amyloid diseases, 845
amyloid formation, 133, 139–140
amyloid inhibition, 911
systemic amyloidoses, 326
TTR amyloidosis, 979–980, 987
Biopolymers, 11
Biopsy, 458, 780, 786, 801–802
BiP
binding, 24–25, 29
chaperone, 410–411

BiP (*continued*)
  endoplasmic reticulum stress, 27–28
  functions of, 29, 35
Biquinoline, 270, 277
Birefringence, 327, 355, 619, 918, 926
Bisdesmethoxycurcumin, 738
Bleeding, intracranial, 796. *See also* Hemorrhage
Blood, *see* Red blood cells (RBCs)
  clotting, 859
  coagulation, 561
  flow, 235, 249
  glucose, 33
  transfusion, 260
Blood-brain barrier, 261, 264, 267, 270, 274, 278–279, 281, 480, 551, 735, 737, 832, 950–951, 976
B-lymphocyte-induced maturation protein-1 (BLIMP-1) molecule, 927
B lymphocytes, 27, 328
Bone marrow
  functions of, 330, 691–692, 776
  suppression, 457
Bone scan, 799
Bortezomib, 778, 784–785
Bovine papillomavirus 1, 264
Bovine spongiform encephalapathy (BSE), 105, 259, 262–263, 267, 276
Bradycardia, 784
Brain
  age-related changes, 214–215
  Aβ, 727
  amyloid imaging, 235
  anatomy
    amygdala, 236
    basal forebrain neurons, 235–236
    brainstem, 263, 285
    cerebellum, 721, 888
    cerebrum, 917
    cortex, 243, 305, 744, 831
    corticolimbo regions, 216
    entorhinal cortex, 235–236, 736
    forebrain, 225, 241, 817
    frontal cortex, 819, 826
    frontal lobe, 713
    hippocampus, 96, 216, 220, 235–236, 243, 246, 265, 744, 894, 954
    midbrain, 199, 818
    neocortex, 236
    parietal lobe, 235
    subcortical regins, 831
    substantia nigra, 826, 956
    temporal lobe, 235–236, 713
  hemorrhage, 248
  injury, traumatic, 821
  intracranial bleeding, 796
  lesions, 713, 737
  structure and Alzheimer disease, 96, 103
  synucleinopathy disorders, 818–820
  temporal lobe, 713
  thyroxine in, 989
Brain-derived neurotrophic factors (BDNF), 892
Brain naturietic factor (BNP), 776, 799
Branching, 86
BrdU labeling, 412
Breeding resistance, 283
Bri, 334
Bronchodilators, 413
Bryostatin1, 720
Bulk aqueous phase, 94
Bulk proteins, 52
bZIP, 28

CACNA1A, 312
Caenorhabdatis elegans
  α-synuclein metabolism, 823
  Alzheimer disease therapies, 741
  characteristics of, 12, 123, 123
  neurodegenerative diseases, 121
  polyQ expansion, 316
  posttranslational modifications, 632, 637
  protein aggregation/toxicity model
    aging effects, 184–186
    current neurodegenerative disease models, 177
    genetic screens for disease-related phenotype modifiers, 180–183, 186
    neurodegenerative diseases of protein folding, 176–178, 184, 186
    polyglutamine protein aggregation dynamics, 178–180
    protein homeostasis, 184–186
    small-molecule drug screens, 183–184
    stress effects, 184–186
CAG repeat diseases, 178, 193, 305, 312, 315, 887, 947, 956
Calcitonin family, 7, 520
Calcitonin gene-related peptide (CGRP), 517, 529
Calcium
  binding, 10
  Ca^{2+}, 22, 31, 99–100, 102
  channel
    blockers, 808
    glutamate-gated, 101
  characteristics of, 580, 803
  homeostasis deregulation, 101

ionophore, 412
Calnexin (CNX), 23–24
Calpain, 892
Calreticulin (CRT), 23–24, 27, 786
Camelid antibody, 878, 881
CaMKII kinase, 732
cAMP response element (CRE), 28
Cancer, *see specific types of cancers*
biology, 118
carcinogenesis, 409
cell lines, 951
etiologies, 118, 197
malignancy, 689, 691, 786, 805
therapies, 951. *See also* Chemotherapy
Candida albicans, 165
Cannabidiol, 272, 281–282, 742
Cannabinoid system, 742
Carbohydrate levels, 491
Carbon, C-C bonds, 618
Carbonylation, 616
Carcinogens, 269
Cardiac system
abnormal conduction, 805
cardiac amyloid, 796
cardiac amyloidosis, 681–682, 782, 800, 807, 981
cardiac muscle, 118
cardiomyocytes, 338–339
cardiomyopathy, 332, 781–782, 801, 974
heart failure
congestive, 783, 784, 796, 807
progressive, 807
Cardiovascular disease, 36, 739
Carnosine, 623
Carpal tunnel
amyloidosis, 787
syndrome, 369, 796, 807, 844
Caseins, 339
Caspases
α_1-antitrypsin deficiency, 409, 411
caspase-3 activation, 413
inhibition of, 735, 891–892
Catalase, 616
Cataract, as protein-aggregation disease, *see* Eye lens; Eye lens crystallin
classification of cataracts, 488
etiology of, xxviii, 489–492
formation, mechanistic models for
crystal cataracts, 499
nonnative disulfide bonds, association through, 500
phase separation, 499–500
formation process, 137
juvenile-onset, 491–492
mature-onset, 488–490
overview of, 487–488
properties of, 488–489
CathD, *see* Cathepsin D (*CTSD*) expression
deficiency, 830–831
up-regulation, 832–833
Catechol *O*-methyltransferase (COMT), 458
Cathepsin D (*CTSD*) expression
brain interactions, 831–832
synucleinase activity, 829–831
Cathepsins, 114
Caveolin-1 pathways, 284, 337
CCAAT/enhancer binding protein β (C/EBPβ), 827
CCT (chaperonin containing TCP-1), 61
CDK inhibitors
CDK1, 733
CDK2a, 411
CDK5, 730, 732–733, 743
cdk5-inhibitory peptide (CIP), 733
cDNA, 473, 829, 832
TPMT, 454
Celastrol, 949, 951
Cell(s), generally
cycling, 104
damage, xxviii
death, xxvii, 98–99, 117–118, 121. *See also* Apoptosis
differentiation, 104, 561, 721
membrane, cytotoxicity, 98–99
migration, 561
proliferation, 561
stress, 176
viability, 22, 117
Cell-cell adhesion, 561
Cell-cycle
hypothesis, 103–104
proteins, 243
Cellular environment, role of, 316–317. *See also* Environmental conditions
Cell-extracellular matrix interactions, 561
Cellular folding pathways, 47
Cellular response pathways, 409, 411
Cellular retinoic acid-binding protein I (CRABP I), 314, 316
Cellular signaling pathways, 716
Central hydrophobic cluster (CHC), 625
Central nervous system (CNS)
Aβ amyloidosis, 241
Alzheimer disease, 240, 244, 745
amyloidosis, 974
features of, 119, 796, 937, 971
lost functions of, 260
misfolding diseases of, 233

Central nervous system (CNS) (*continued*)
prion disease effects, 262, 286
systemic amyloidoses (CNSA), 326–327, 974, 976, 982, 989, 991
TTR amyloidosis, 982
Cereact, 737
Cerebral amyloid angiopathies (CAA), 334
Cerebrolysin, 742
Cerebrospinal fluid (CSF), 234–235, 718, 970–971
Ceredase, 477
Cerezyme, 477–479
Chain maker's cataract, 490
Channel hypothesis, 98–99
Chaperone(s), *see* Co-chaperones; Molecular chaperones
activity, 12, 22
aggregation/toxicity, 181
α_1-antitrypsin deficiency and, 408, 410, 414
Alzheimer disease and, 123, 734
Aβ, 725–276
apoptosis (CHOP), 25
autophagy and, 119
chemical, 414
copper, 547
crystallin, 638
cystic fibrosis, 430
deficiency, 49
downstream, 50, 52
ER, 29
in lens crystallin, 497–498
medical significance of, 62
mutations and, 104
network, protein flux through, 49–51
neurodegenerative diseases and, 176
pathological, 12
pathways, 49–51
pharmacological, 833
polyglutamine protein expression, 889
prion formation, 154
protein
degradation, 114
folding, xxi, 23
quality control systems, 113
saturation, 501
suppression of neuronal degeneration, 197
upstream, 50
zinc, 547
Chaperone-mediated autophagy (CMA), 115–123, 125, 638, 829–830
Chaperonins
cytosolic (CTT), 181
DNA, 50–56, 58–59
functions of, 56–57
GroEL, 50–51, 57–62
GroEL-Gro-ES reaction cycle, 56–59
GroES, 50–51, 57–61
Group I, 56–57
Group II, 56, 61–62
heat-shock, 498
substrates and folding mechanism, 59–60
Charge density, 94–95
Charge-state ions, 366–367
Chelation, 722
Chelators, functions of, 277–278, 548–552
Chemical genetics screens, 183–184
Chemical synthesis, 310–311, 907
Chemoattractants, 407
Chemokines, 37, 561, 736
Chemotherapy
agents, xxviii
cytotoxic, 779
high-dose, 776–779, 782
low-dose, 779–782
standard options, 779–780
Chimeras, 165–166, 927
Chinese hamster ovarian (CHO) cell model, 279, 477
CHIP (C-terminus of Hsp70 interacting protein), 55, 734–735, 947
Chloroplasts, 50, 54
Chlorpromazine, 270, 276–277
Cholesterol
-depleting agents, 271, 279
fibrillization process, 625
implications of, 36–37, 351
low-density lipoprotein, 338, 738
membrane, *see* Membrane cholesterol
metabolism, 736
oxidation, 617
synthesis, 103
Cholinergic deficit, 235–236
Cholinergic neurons, 196, 744
Cholinesterase inhibitors, 235
CHOP, 35–36
Choroid plexus, 329, 796, 971, 976–978
Chromatin remodeling, 890
Chromatographic analysis
high-performance liquid, 83, 135, 623
hydrophobic interaction, 477
immunoaffinity/size-exclusion, 225
implications of, 870
size-exclusion, 152, 307, 385
Chromatographic studies, high-performance liquid (HPLC), 135, 623
Chromosome 21, 382
Chronic liver failure, 404

Chronic wasting disease (CWD), 259, 261–262, 267
Chrysoidine, 270, 278
Ciliary neurotrophic factor (CNTF), 892
Circular dichroism (CD), 307, 311–312, 474, 623, 873–874
Cirrhosis, 406, 413
cis-proline, 364
9-*cis*-Retinoic acid receptor (RXR), 827
Citric acid cycle, 101
CK-1 kinase, 732
Class II proteins, 59
Class III phosphatidylinositol 3-kinase (PI3K), 116
Class III proteins, 59
Cleavage
  amyloid formation, 132, 134, 137
  proteolytic, 134
Clioquinol (CQ), 270, 278, 548, 551
CLK-2, 732
Clonal bone marrow
  plasma cells, 691
  plasmacytosis, 776
Clonal cell markers, 689
Clone plasma cells, 776
Clusterin, 338, 878–879
Co-chaperones, 48, 201, 335, 430
Cocktail therapy, 745
Coenzyme Q_{10}, 894
Coevolution, 104
Cognition
  cognitive deficits, 719
  cognitive function, 244, 246
  cognitive reserve hypothesis, 740
Co-immunoprecipitation, 50, 54, 59, 61, 411
Colchicine, 780–781, 787
Cold cataracts, 499
Collagen, 10, 351–352, 354, 560, 849, 851, 856–585, 860
Collar transcription factor, 26
Co-localization, 203, 407, 946–497, 971
Computed tomography (CT), 404
Computer software programs
  aggregate formation, 73
  Image Pro-Plus, 926
Concentration dependence, 85
Condensation, 139
Conduction
  abnormalities, cardiac,805
  delay, 799–800
  system, functions of, 807
Confinement theory, 60
Conformation
  conformational diseases, 33, 415
  conformational regulation, 50
  conformational remodeling, 56
  conformational states, amyloid diseases, 94, 846
  conformational stress, 48
  conformational switching, 48
  conformational transitions, 414–415
Congenital lysosomal storage disorders, 831
Congestive heart failure, 783, 784, 796, 807
Congophilic angiopathy, 248
Congo Red stain, 269, 273–274, 327, 489, 497, 619, 691, 905, 934, 937
Conjugation cascades, 115
Copaxone, 268, 273
Copper
  Alzheimer disease and, 546–548, 550
  chaperone for SOD1 (CCS)
    maturation of SOD1, 387
    mechanism of action model, 387–388
    structural properties of, 382–383
  characteristics of, 270, 277–278
  Cu^{2+}, 12, 355, 365, 368–369, 547, 551, 849
Copper-zinc superoxide dismutase (SOD1)
  genetics and models of, 382–384, 545
  immature pathogenic mutants and toxicity, 388 389
  maturation, 387
  mutant
    aggregation in fALS, 384–387, 389–390
    down-regulation by RNAi, 957–958
  stereo view of, 383
  structural properties of, 381–382, 389–390
  therapeutics, 389, 955
  wild-type, 382
Co-precipitation, 196
Cortical neurons, 101
Corticosteroid therapy, 413
COS-1 cells, 457, 459–460, 462
Co-translational processes
  compaction, 52
  folding, 429–433
  refolding, 61
Covalent conjugates, 114
CP49, 503
Creatine levels, 781, 894
CREB-binding protein (CBP), 203, 891
CREBH, endoplasmic reticulum stress, 28–29
Creutzfeldt-Jakob disease (CJD), 9, 105, 259–261, 271, 276, 281, 284–286, 705, 821
Cross β-sheet structures, 137, 619, 973–974
Cross β-structures, amyloid formation, 137
Cross-linking analyses, 11, 51–52, 132, 137–139, 158, 160, 265, 411, 739

Cross-linking reactions, amyloid formation, 132, 137–139
Cross-reactivity, 164
Cross-seeding, 152–153
Crosstalk, 115, 117, 122
Cryoglobulinemia, 689
Crystallin proteins, in eye lens, *see* Eye lens crystallin
Crystallographic studies, 192–193, 562, 909, 989
C-terminal/C-terminus
  protein misfolding diseases, 239, 241, 245, 249, 311, 350, 358–359, 364, 430, 439, 495–497, 518, 522, 526, 530–531
  protein misfolding principles, 7, 27, 50, 52, 54, 134–135, 146–147, 152, 159–161, 193
  therapeutic treatments, 620, 730, 853, 869, 919, 952
CTS-21166, 721
Culotta, Valeria, 388
Curcumin, 269, 274, 724, 735, 738, 741, 909, 949, 951
Cutaneous amyloidosis, 776
Cyclic AMP (cAMP), 427
Cyclic tetrapyrroles, 269, 275–276
Cyclo-oxygenases
  COX-2, 738
  functions of, 266, 948
  inhibition, 721
Cyclopentenone prostaglandins, 949
Cyclophosphamide, 784
Cyclosporine A, 413
CYP19, 459
Cys110, 491
Cystamine, 893
Cystathionine-β synthase (CBS), 36
Cystatin C, 692–693
Cysteine, 620
Cysteine-protease inhibitors, 271, 281
Cysteines, 30, 36, 119, 150–151, 157–158, 160, 162, 280, 491, 523
Cystic fibrosis (CF)
  conductance regulator, *see* Cystic fibrosis transmembrane conductance regulator (CFTR)
  etiology, xxviii, 62
  folding biology of, 426
Cystic fibrosis transmembrane conductance regulator (CFTR)
  CFTR-NBD1 structures, 436
  co-translational folding of, 429–433
  endoplasmic reticulum (ER), translocation into, 428–429
  folding correction strategies, 439–440
  genetic and clinical manifestations of, 426–427
  mutant, recognition and degradation of, 433–435
  stability and trafficking at cell surface, 429, 438–439
  structure of, 435–437
  topogenesis, 428
  trafficking to the cell surface, 437–438
  wild-type, 438
Cytidyl-guanyl oligodeoxynucleotides (CpG-ODNs), 264–266
Cytochrome(s)
  c, 101
  P450, 411, 459
Cytogenetics, 781
Cytokines
  implications of, 37, 561, 736, 828, 971
  inflammatory, 32, 352, 369–370, 407, 971
  pro-inflammatory, 32, 310–315, 327
Cytoplasm-to-vacuole trafficking (CVT) pathway, 462
Cytoscopic biopsies, 786
Cytoskeleton
  disturbances, 237
  features of, 234
  proteins, 243
Cytosolic proteins, 116, 119
Cytosol levels, 9, 24–25, 30, 47, 50–51, 53–54, 56, 102, 113, 717, 954
Cytotoxic T-cells (CTLs), 927
Cytotoxicity, 36, 62, 93, 97–99, 101–102, 260, 617, 779, 892, 945, 950, 986

daf-16, 184, 186, 633–635
DAF-16, 636, 638–639
daf-2, 184, 186, 633–634
DAF-2, 632
Danon disease, 118
Dapsone, 266
D-arginine, 283
Daunomycin, 735
DCIC, 949
dcr-1 mutation, 198–200
Deamidation, 488, 499
  amyloid formation, 132, 135–137
Decay process, 82
Declarative memories, 215
D18G TTR, 978–979
Degenerative diseases, 93
Degradation, *see specific types of degradation*
  accelerated, 459–460
  autophagy, 461–463
  impact of, 30, 113–114, 828–829

proteosome-mediated, 463
vacuolar, 461
Degrees of freedom, significance of, 136
Dehydration, 139
7-Dehydrocholesterol reductase, 271, 279
24-Dehydrocholesterol reductase inhibitors, 271, 279–280
Deletion-insertion mutation, 332
Deletions, 27–28, 52, 56, 152–153, 234, 244, 333, 455, 461, 632, 826, 978
Delta (Δ), 74, 85, 87
ΔF508 CFTR, 433–435, 438–439
Dementia, 215, 221, 243, 334, 619, 713, 821, 832
Demyelination, 720, 802–803
Denaturation, 151, 312, 489, 498, 809, 872, 920–921, 972, 988
SDS, 619
Dendritic cells, 264
De novo folding, 48, 50, 53
Dentatrubral-pallidoluysian atrophy (DRPLA), 177, 199, 887, 891
Deoxyoxorubicin, 783
Deoxyribonucleic acid. *See* DNA
Dephosphorylation, 734
Depolymerization, 74, 82, 358
Derlin 1, 430
Desferrioxamine, 550
Desipramine, 270, 284
Desmosterol reductase, 96, 103
Destabilized proteins, 946
Detoxification, age of onset of neurodegeneration, 639–640
Deuteroporphyrin n IX 2,4-bis(ethylene glycol) Iron(III) (DPG_2-FE^{3+}), 269, 275
Developmental timing, 197
Dex, 780
Dexamethasone, 779, 782–785
Diabetes, 118, 490, 801. *See also* Type 1 diabetes; Type 2 diabetes
Diagnostic sensitivity, 694–696
Dialysis-related amyloidosis (DRA)
β_2m
circulating, reduction of, 847–848, 859
interactors, identification of, 848–856, 860
classification of, 845
defined, 844
fibrillogenic environment modifications, 856–859
pathological properties of, 330, 846
prevalence of, 844
Diastolic dysfunction, 802
Dicer knockout, 198, 200
Diffusion, monomeric, 86
Diflunisal, 806, 990
Digestion, proteolytic, 358
Dimers
dimerization, 26, 29, 79, 152, 431, 548
functions of, 221–222, 335, 382–383, 388, 549, 986
Dimethyl sulfoxide (DMS), 270, 277, 787
Dipeptides, 908
Direct hydrolysis, 135
Direct misfolding, 620
Direct oxidation, 616
Disaccharides, 563–564
Disaggregation, 309
pathway, 636, 638–639
Disease proteins, specificity of, 196
Disulfide
binding, 385
bonds, 22, 30, 36, 358, 382, 388, 472, 488, 491, 500, 519, 870–872
formation, 132
loop, 382
Dityrosine, cross-links, 132
Diuretic therapy, 808
Divalproex, 733
Dj-1, 195
17-DMAG, 951
DNA
chaperonins
DnaJ, 50–55, 58
DnaK, 50–56, 58–59
double-stranded (dsDNA), 580
fragmentation, 891
immunization, 937
motifs, 826
sequencing, 333
synthesis, 243
DNAzymes, 828
Dock-and-lock scenario, 79–81
Domain swapping, 500–501
Donepezil, 712
Dopamine levels, 122, 956
Dopaminergic neurons, 178, 199, 205, 826, 956
Doppler analysis echocardiography, 802
DOSPA, 269, 275
Dot blot analysis, 620
Double-jeopardy model, 79–81
Double nucleation, 524
Double-stranded DNA (dsDNA), 632
Double-stranded RNA (dsRNA), 282–283, 953
Downhill polymerization, 78–79, 81–83, 85, 972–975
Down-regulation, 186, 888, 946, 953–958
toxic protein aggregation, 637
Down's syndrome, 237–238, 712–713

Doxycycline (Dox), 241–243
dp/dt, 75, 77, 80
DP109, 550
d-Penicillamine, 270, 278, 550
Driftscope plots, 367
Dronabinol, 742
Drosophila
  genetic screens, 205
  human degenerative disease model
    future research directions, 205
    modifier screens, 196–204
    overview, 191–192, 205
    protein-misfolding diseases, 192–196
    therapeutic compounds, effects of, 203
  melanogaster, 35, 104, 823, 830, 947
  protein-folding disease model, 176, 205
  tau aggregation, 732
  tauopathy model, 734
  toxic proteins in, 205
Drug(s), *see* Pharmacogenomics
  design, 415
  discovery, 745, 907, 911, 934
  labeling, 454, 457
  screens, small-molecule, 183–184
  trials, Alzheimer therapies, 712
DsbA/DsbB proteins, 30
D67H, 6
Duchenne dystrophy, 743
Duplications, 237–238, 334, 493, 716
Dynein, 462–464
Dysuria, 787

EBRT, 786
Echocardiography, 776, 805
ECOG performance status, 776–777
Edema, 803
Egb761, 741
eIF2 complex, 25–26, 32–33
Electrocardiograms, 807–808
Electron(s)
  acceptors, 30
  paramagnetic resonance, 359
  stacking, 359, 361
Electrophoretic analysis
  gel, 489
  immunofixation (IFE), 689–690, 692, 694–696
  protein electrophoresis (PEL), 690, 692, 695
Electrostatic(s), generally
  field, 94–95
  impact of, 851
  interactions, 985
  loop, 382
  mismatch, 435
  repulsion, 849
Elliot, Jeffery, 388
Embryogenesis, 561
Emotional function, 244
Emphysema, 403–404, 407, 413
Encapsulation, 57–61
Encephalitis, 249, 283
Endocytosis, 326
Endoplasmic reticulum (ER)
  α_1-antitrypsin deficiency, 404, 408–411
  amyloid aggregates, 98
  amyloid diseases, 102
  CFTR from, 428–429, 432
  characteristics of, 21–22, 50, 278, 977–978, 990
  ER-associated degradation (ERAD), 21–23, 25, 27, 34, 328, 433, 438
  future research directions, 37
  Gaucher disease, 480
  interluminal calcium, 22
  mutant secretory proteins, 125
  oxidative protein folding, 30–31
  oxidoreductases, 30
  protein folding and quality control in, 22–23
  protein trafficking from, 24
  stress
    oxidative stress and, 31–37
    reactions to, 24–28, 803
    response element (ERSE), 27
  translocation into CFTR, 428–429
  UPR signaling, 23–24, 29–30
Endoscopy, 801
End-stage renal disease, 330, 801
Energy landscape, 5–6, 60
Entropy, 11–12, 75–76
Environmental conditions, significance of, 234, 740–741, 946
Enzyme(s)
  Aβ-degrading, 724–725
  autophagy and, 119
  digestive, 114
  folding, 978
  functions of, generally, xxviii, 22, 37
  HDAC, 891
  intracellular, 280–281
  lysosomal, 412, 471
  myeloperoxidase, 616
  redox, 30
  replacement therapy, 477, 832
  synucleinase activity, 829–831
  synucleinopathy treatment, 832–833
  ubiquitin-activating, 888–889
Enzyme-linked immunosorbant (ELISA) screening, 920

Epidemiological studies, 937
Epifluorescence micrography, 179
(-)-Epigallocatechin-3-gallate (EGCG), 719, 722
Epigenetic protein modifications, 195–196
Episodic-like memory, 242
Epitopes, 356–358, 881, 918, 920, 925, 928
Eprodisate, 908
ErbB-4, 722
ERGIC53, 24
ERGL, 24
ERK
  ERK1, 732
  ERK2, 411, 732
  signaling, 893
Ero1p protein, 30–31
ERp57, 24
Erythroid cells, 826
Escherichia coli, 30, 50–52, 54, 59, 181, 278, 316, 852, 923
Estradiol, 740
Estrogen receptor β (Erβ), 740
Estrogen replacement therapy, 740
Eubacteria, 53
Eukarya, 53
Eukaryotes, xx, xxvii, 11, 21, 30, 50–51, 54–56, 176, 186, 460
Eukaryotic cells, 460
Eukaryotic model systems, protein-folding diseases, 176
EuLISA, 920
European Group for Blood and Marrow Transplantation (EBMT) registry, 779
Evolution, 49–50, 104
EXAFS, 387
Excitotoxicity, 235, 550, 742
Excluded-volume effect, 12
Exploratory Investigational New Drug (E-IND), 922–923
Export, 639
Expression profiling, 26
Extracellular matrix, 9, 134, 561, 981
  toxic protein aggregation, 636
Ex vivo studies
  fibril formation, 355
  oxidative stress, 625
  synucleinopathies, 828
Eye
  cataracts, *see* Cataracts
  iris, 490
  lens crystallin, *see* Eye lens crystallin
  retina, 490, 802, 805, 971, 977
  vitreous amyloid accumulation, 800, 805
Eye lens
  cells, cytoskeleton proteins of, 502–503
  characteristics of, 487–488
  crystallin, 137
    chaperone activities, 497–498
    covalent modifications of, 498–499
    folding, stability, and unfolding of, 495–496
    in vitro aggregation pathways, 496–497
    mutations, impact of, 491
    protein analysis, 137, 488
    structure and function of, 487–488, 492–494
  membrane proteins of cells, 503

Fabry disease, 469–470
Familial AD (FAD)
  early-onset, 237, 241, 712–713
  mouse studies, 234, 243–244, 716
  risk factors for, 237–241
Familial amyloid, generally
  cardiomyopathy (FAC), 974, 976, 982, 989, 991
  diseases, 93
  neuropathies, 844
  polyneuropathy (FAP)
    clinical studies, 991
    early-onset, 976
    future research directions, 991
    gene therapies, 980–982
    non V30M, 974
    overview of, 6, 9, 974, 976, 982, 989
    therapeutic strategies, 982
    TTR amyloidosis, 982
Familial amyloidosis, 867. *See also* Lysozyme amyloidosis
Familial amyloidosis of Finnish type (FAF), 332
Familial Amyloidotic Polyneuropathy World Transplant Registry (FAPWTR), 804–805
Familial amyotrophic lateral sclerosis (fALS)
  defined, 957
  down-regulation of SOD1 by RNAi, 957–958
  SOD1 mutations, 384–387, 389
  therapeutics, 389, 953
Familial British dementia (FBD), 334
Familial CJD (fCJD), 260, 262
Familial diseases, mutation-linked, 631
Familial neurodegenerations, 639
Familial neurohypophyseal diabetes insipidus, 118
Family history, significance of, 801, 989
Fas ligand, 37
Fast electron transfer, 490

Fast folding, 56, 60
Fatty acids, 32, 548, 617, 618–619, 740–741, 802, 820
Fe^{3+}, 547
Fenton reduction, 616
Ferulic acid, 909
Fes1p protein, 54
Fetal alcohol syndrome, 742
Fibril(s)
  Aβ growth, 699–704, 706, 974
  Alzheimer disease, 549
  amyloid diseases, 103
  in amyloidosis (AL), 691–692
  formation of, *see* Fibril formation
  growth, generally, 849
  hemodialysis-related amyloidosis (HRA)
    formation mechanisms, 353–356
    structure, 356–359
Fibril formation
  antigens and, 919
  atheronals, 622–623
  components of, 4–5, 12, 34, 805
  lysozyme amyloidosis, 876–879
  mechanisms, 353–356
  nonfibrillar aggregates, 86–88
  process
    amyloid, 843–844
    nucleated growth mechanism, 844
Fibrillary tangles, 93
Fibrillization, 99, 623–625, 908, 911, 975
Fibrillogenesis, 326, 351, 355, 497, 579–580, 741, 808, 848, 851–853, 855–858, 911, 919, 925, 935, 987
Fibrillogenic pathway, 846
Fibrillogenic tau, 735
Fibrinogen, 333, 409
Fibroblast
  Gaucher disease, 472–479
  growth factors, 561
Filamin, 337
Filensin, 503
Filipin, 269, 276
Fire, Andrew, 953
Firefly luciferase, 56
First-order kinetics, 88
FLAG-exon1-GFP fusion, 311
Flavin adenine dinucleotide (FAD), 31
Flavonoids, 741
Fluorescence recovery after photobleaching (FRAP) analysis, 178–179
Fluorescence resonance energy transfer (FRET)
  analysis, xxi, 178, 180, 193
  measurements, 52
Fluorescence studies
  amyloid formation, 151, 157–158, 162–163
  hemodialysis-related amyloidosis, 355, 357
  immunotherapy, 918
  lysozyme amyloidosis, 873, 880–881
  protein aggregation/toxicity, 178, 184
Fluorescence-polarization-based competitive binding assay, 261
Fluorophores, 162
Flupirtine maleate, 272, 281, 285
Flurizan,721
Folded states, 4, 13, 95
Folding, generally
  barriers to, 6
  co-translational, 429–433
  code, 47
  defined, xix-x
  process, 30
FoldRx, 991
Follicular dendritic cells (FDCs), 262–263, 265
Forced vital capacity (FEV1/FVC), 404, 413
fos, 411
Fourier transfer-infrared (FTIR) analysis, 619, 621
Fractionation, serial, 820
Fractures, 844, 985
Fragile X syndrome, 192
Fragmentation, 86
Fragmentation reactions, 139
Framingham Heart Study, 738
Free energy
  barrier diagram, 75–76
  implications of, 11–12, 75–76, 313, 844
Free fatty acids, 32
Free light chains (FLCs)
  functions fo, 329, 692–693
  immunoglobulin, 786
  monoclonal, 694–696
Free radicals, 101, 616–617
Frontotemporal dementia with parkinsonism linked with chromosome 17 (FTDP-17)
  characteristics of, 713
  tau mutation, 717, 732
Fucoidan, 267–268, 273
Fulguration, 786
Functional groups, 985
Fungal prions
  biochemical commonalities of, 147, 149, 166
  characteristics of, 145–147
  proof of prion hypothesis, 149
  protein modulators of, 152–154
Fungi, 53
Furin, 10, 332
Fusion proteins, 429
FVB/N mice, 216

GABA, 182
Gadolinium, 799
Gain-of-function
  diseases, 305–306
  process, 327, 906
Gain-of-toxic function, 403, 407–408
Galantamine, 712
γ-Crystallins, 491, 493–496, 498, 502
Gammagard, 923
γ-Glutamylcystein synthetase, 26
γ-Secretase, 10, 237–241, 245–247, 716–718, 739
Gamunex, 923
Gangliosides, 96
Gas transfer factor, low, 404
Gastrointestinal tract, 801–802
GATA (GATA-1/GATA-2/GATA-3), transcription factors, 826–828
Gaucher disease
  classification of, 469
  etiology, 469–470, 479, 821
  glucocerebrosidase, 470–472, 479
  G202R mutations, 476–477, 479
  incidence of, 469
  L444P mutation, 476–477, 479
  N370S mutation, 474–476, 479
  protein folding and, 472–474
  therapeutic strategies, 477–480
  variants of, 471
Geldanamycin, 205, 460, 734, 890, 949, 951–952
Gel filtration studies, 387
Gelsolin, 7, 10, 332
Gene expression, controlling therapies
  down-regulation of amyloidogenic protein expression, 953
  heat-shock response (HSR)
    pharmacological induction of, 948–952
    prospects/potential of, 953
    regulation of, 948, 958–959
  molecular chaperones, 946–948
  overview of, 947–948, 958–959
  RNAi
    principles, 953–958
    therapy, 958
Gene expression profile, 411
Gene silencing, 181, 953
Gene targeting studies, 239
Gene therapy, 265, 283, 414, 743–744, 934, 980–982
Genetic code, 47
Genetic diseases, 946
Genetic myopathies, 118
Genetic screening, 180–183, 186
Genetic studies, amyloid diseases, 93
Genome sequence, 147, 192
Geranylgeranylacetone, 890
Gerstmann-Straussler-Scheinker disease (GSS), 260, 285
G551D mutations, 427
Gingko Evaluation of Memory (GEM), 741
Gingko biloba, 183, 741
Glassblower's cataract, 490
Glaucoma, 800
GlcI/GlcII, 24
Glial cells, 102, 237
Glial cytoplasmic inclusions, 820
Glial-derived neurotrophic factor (GDNF), 892
Gliosis, 832
Gln deamidation, 135, 137
Gln3, 146
Globular proteins, xxii, 4, 7, 9, 11, 13, 136, 150, 157, 843, 845
Glucocerebrosidase gene (*GBA*), 470–472, 479, 832–833
Glucocorticoids, 281
Glucose utilization, 235
Glucuronic acid (GlcA) sugars, 562–564
Glutamate
  characteristics ofm 407, 720
  excitotoxicity, 235
  receptors, 101
Glutamatergic systems, 235
Glutaminergic synpase transmission, 742
Glutamines, 123, 146–147, 151, 164, 193, 488, 499, 633
Glutathione levels, 36, 101, 147, 272, 281
Glutathione peroxidase, 616
Glutathione S-transferase (GST), 26, 160
Glycans, 431
Glycation, 132, 139, 501–502, 737
Glycine, 146–147
Gly mutations, 853
Glycogen, 114
Glycolysis, 33, 502
Glycoprotein(s)
  antibodies against, 906
  characteristics fo, 24, 240, 431, 434
  mimetics, 908–909
Glycosaminoglycans (GAGs), xxviii, 12, 267, 340, 351, 370, 526, 528, 560, 564, 849, 858–860, 908
Glycosphingolipid metabolic pathway, 470
Glycosylation, 134, 139, 431, 470–471
Glycosylphosphatidyl-inositol (GPI), 259–260, 263
Glyoxal, 618
gmr-GAL4, 202
Golgi complex, 21, 27
G protein-coupled receptor (GPCR), 439, 722

Green fluorescent protein (GFP) stain, 160, 184, 200, 409, 461
GRP170, 32
GRP94, 22, 27, 36
GrpE, 54, 58
GSK-3β, 733, 743, 893
GSMs, 721
GTS-21, 712
G202R mutations, 476–479
Guanidine hydrochloride, 162
Guanidinium group, 849

Hac1p, 27
Hairpin RNA, 283
Half-life, significance of, 829
Halogen-binding pockets, 985
Hamster studies, 260, 268, 279, 477, 716
HDL-C, 738
HDM-SCT, 786
Head injury, 742
Heat denaturation, 809
Heat-shock, generally
  activation, 951
  elements (HSEs), 948
  factor 1 (HSF-1), 632–633, 636, 638, 947, 949, 951–953
  -organizing protein (HOP), 460
  proteins (HSPs), functions of, 48–50, 153, 197, 205, 833, 889–890, 951. *See also specific HSPs*
  response (HSR), 102, 176, 946–948, 958–959
  transcription factor (*hsf-1*), 98, 181, 184, 186
Hsp70, 460, 633
Hsp90, 460
HeLa cells, 199–200
Hemagglutination assays, 692
Hemagglutinin (HA), 462
Hematopoietic stem cell transplantation, 778
Hematoxyline, 819
Hematuria, 787
Heme group, 312
Heme metabolism genes, 826
Heme oxygenase I, 26, 101, 339
Hemine, 275–276
Hemodiafiltration, 847
Hemodialysis, *see* Hemodialysis-related amyloidosis (HDRA)
  complications of, 845
  drawbacks to, 845
  indications of, 801
  vanishing complications, 845
Hemodialysis-related amyloidosis (HDRA)
  β_2m effects, 347–363
  clinical manifestations of, 352
  CU^{2+}, 365, 368–369
  diagnosis, 352
  pathogenesis, 368
  therapeutic strategies, 369–370
Hemofiltration, 847
Hemoglobinopathy, 891
Hemoperfusion, 369
Hemorrhage, 868
Heparan sulfate
  implications of, 268, 340, 559–566
  proteoglycan (HSPG) perlecan, 526–527, 530
  proteoglycans, 325, 936
Heparanase, 859
Heparin, 273, 338, 561–562, 858–859
Hepatic genome analysis, 409
Hepatic genomic analysis, 411
Hepatitis B, 958
Hepatocellular carcinoma, 404, 412
Hepatocytes, 28, 329, 338, 405–406, 409, 412–413, 977
Hepatomegaly, 807
Herbal medicine, 951
Heritable mutations, 62, 824
HERP, 36
HET-s protein
  amyloids, 159, 161
  characteristics of, 147, 149
  nucleation and polymerization, 152
  prion formation, 153–154
Heterodimers, 387
Heterooligomeric complexes, 61
Heteropolyanion-23, 268
Heteropolymers, 406
HγC-crystallin, 492, 489–490, 493–498
HγS-crystallin, 493–494
High-affinity multivalent interactions, 57
High-density lipoprotein (HDL) levels, 332, 334
High-dose chemotherapy with peripheral blood stem cell support (HSCT), 776–778, 782, 786
High-fat diet, 32–33
High-flux polyacrylonitrile membranes, 847
High-molecular mass
  aggregates, 631, 635–636
  oligomers, 491
  polymers, 408, 489
  significance of, xxi
High-resolution computed tomography (HRCT), 404, 413
High-throughput screening (HTS), 183, 317, 735, 937, 951
His31, 848–852
Histamine *N*-methyltransferase (HNMT), 458
Histamines, 458, 712

Histidines, 362, 436, 549, 620–621, 6849
Histine deacetylase 6 (HDAC6), 460–464
Histochemistry, 560
Histone acetylase and deacetylase (HDAC) enzymes, 891
  HDAC6, 124, 203
  inhibitors of, 203, 890–891
HIV, 429
HLA heavy chain, 348
HMG-CoA reductase inhibitors, 276
Holoenzymes, SOD1, 382
Holo-retinol binding protein, 971
Homeostasis
  Alzheimer disease, 546, 550, 734
  protein misfolding diseases, 330
  protein misfolding principles 10, 12, 113, 118, 175–176, 181, 184–186, 200–201
  therapeutic treatments, 803, 889, 956, 977–978
Homocysteine (Hcy), 35–36, 740
Homodimerization, 29
Homodimers, 152, 493
Homologs, 29, 53, 199, 414, 918
Homo-oligomeric complexes, 61
Homopolymers, 310
Hormones
  androgenic, 739–740
  estrogen, 720, 740
  pancreatic, 134
  progesterone, 271, 279
  prohormones, 526
Host disease protein, neuronal degeneration, 193–195
Hsc70 chaperone, 829
hsf-1, 634–635
Hsp 42, 434
hsp-1 chaperone, 181
Hsp10 protein, 56
Hsp26, 435
Hsp27, 101, 435, 890, 952
Hsp40, 50–51, 53, 197, 947, 952
Hsp60 protein, 56
Hsp70 system
  characterized, 50–53, 734, 947–948, 952
  folding mechanisms, 54–56
  protein aggregation, 181
  protein misfolding diseases, 197, 201–202, 205, 430, 432, 435, 460, 734
  reaction cycle, 53–54
  structure, 53–54
  substrates, 54–56
  therapeutic treatments, 947–948, 952
Hsp90
  chaperone, 439
  characteristics of, 50–51, 433, 435, 460, 734, 948
  inhibitors, 949–952
Hsp104, 152–154, 637–638
Hsp110 proteins, 54
HspB8/Bag3 complex, 124
HspBP1 protein, 54
5-HT$_6$ antagonist, 712
hTRA2-β1, 732
Human cystatin C (hCC), 332
Human genome, 954
Human glycogen synthase kinase 3 (GSK-3), 195, 205, 730, 732, 743
Human interleukin-6 (hIL-6), 925. *See also* Interleukins
Human prion protein (hPrP), 6
Huntingtin (htt), 461, 463, 631, 634, 636
  Alzheimer disease, 545, 732
  autophagy, 123–124
  protein aggregation, 177, 182–183, 310, 312, 315–316, 461, 463, 631, 634, 636
  therapeutic treatments, 888, 890–893, 947, 956–957
Huntington disease (HD)
  aggregated simple polyQ sequences, 308–310
  aging and, 631
  altered aggregation of polyQ with flanking sequences, 310–315
  autophagy, 119, 123–124
  characterized, xxviii, 34, 49, 62, 177–178, 183–184, 193, 305–306, 551, 933
  cellular environment, 316–317
  classification of, 887
  inhibitors of, 940
  mouse models, 892–893
  neurodegeneration, 891, 894, 957
  pathogenesis of, 891
  prevalence of, 305
  protein degradation, 890
  protein expression, 888
  systemic amyloidoses, 327
  therapeutic strategies, 317–318
  toxicity mechanisms related to protein misfolding, 317
  toxic misfolded monomer model, 307–308
  toxic misfolded proteins, 954
  transglutaminase inhibition, 893
  treatment strategies, 893–894
Hyaline material, 517
Hydrocephalus, 803
Hydrogen
  bonding, 154, 909
  bonds, 223, 524, 616

Hydrogen (*continued*)
hydrogen/deuterium (H/D) exchange, 158, 161, 163
exchange (HX), 356, 358–360, 366
peroxide, 101, 548, 616
8-Hydroguanosine, 548
Hydrolysis, 54, 59, 62, 135–136, 431, 435–436
Hydroperoxides, 618
Hydrophobic interactions, 847, 909
Hydrophobicity, 8, 52, 54, 94–95, 522, 622, 625
4-Hydroxy-2-nonenal, oxidative stress, 132, 139
Hydroxylamine derivatives, 949–950
24S-Hydroxycholesterol, 96
3-Hydroxykynurinine, 490
Hydroxyl radicals, 548
Hydroxynonenal (HNE), 132, 139, 548, 616–622
5-Hydroxytryptamine$_{1A}$, 712
Hyperaggregation, 634
Hyperbilirubinemia, 404
Hypercholesteremia, 96, 617, 620
Hyperfibrinogenemia, 409
Hyperhomocysteinemia, 35–37
Hyperphosphorylation, 134
amyloid formation, 134
protein aggregation, 182
protein misfolding diseases, 195, 237
therapeutic treatments, 713, 716, 720, 730, 733–735
Hypertension, 404
Hypoalbuminema, 801
Hypohalous acids, 617
Hypomyelination, 246
Hypoxantine guanine phosphoriboxyltransferase (HPRT), 455
Hypoxia, 31

Iatrogenic amyloidosis, 845
Iatrogenic causes of CJD (iCJD), 260, 285
Ibuprofen, 266
Idebenone, 894
Idiopathic Bence Jones proteinuria, 695
Idoxuridine, 286
I56T, 6
IgG antibodies, 265
IgGκ, 689
Imaging of misfolded proteins
brain pathology, 659–662
early detection, 648–649
histology, 650
in vivo, 654–659
labeling of AD pathology in vivo, 651–654
overview of, 647–648, 662
reasons for, 648
Imatinib mesylate, 271
Immune-based therapies
antibodies, 263, 265
immunization, active and passive, 264–265
immunostimulation, 265–266
immunosuppression, 266
neuroinvasion, 263
peptide, interfering, 265
peripheral replication, 262–263
Immunization
active, 924–926, 928, 937
Aβ immunotherapy, 248–249
indications of, 917
passive, 919–923, 928, 937
prion disease treatment, 264
protein misfolding diseases, 936–937
Immunochemical studies, amyloid formation, 134
Immunofixation, 695
Immunogens, 925, 937
Immunoglobulins
characteristics of, 328, 844
immunoglobulin A (IgA), 692
immunoglobulin G (IgG), 338, 692, 923–924, 927
immunoglobulin M (IgM), 265, 692
light-chain variables, 333, 696
Immunohistochemical investigations, 527
Immunohistochemistry, 560, 819, 936
Immunoimaging, 922
Immunonephrelometry, 692
Immunostimulation, 265–266
Immunosuppression, 266, 890
Immunotherapy
active, 917
Aβ, 727–729
applications, generally, 248–249, 552, 719, 735, 906–907, 917
goals of, 917
IMP, 203
Implantation, regenerative tissue, 743–744
In bulk degradation, 116
Indiana mutations, 223, 225, 716
Indium(III) meso-tetra (4-sulfonatophenyl)porphine chloride (In-TSP) 269, 275
Indolinones, 732
Indomethacin, 266, 949
Infection(s), 260–261, 263–264, 266, 330, 561, 784
Infectious disease, 917
Infectious prion, 146
Inflammation, etiology, 617
Inflammatory bowel disease, 454

Inflammatory responses, 29, 617
Influenza virus, 429
Inheritance, 166
Injury/regeneration signals, 412
Innate immunity, 117
iNOS, 738, 742
Inositols, 908–909, 911
Insertions, 334, 455
In silico modeling, 261
Insulin
  amyloidosis, 105
  degrading enzyme (IDE), 249, 638, 719
  functions of, 27, 32, 118, 410, 518, 526, 528, 706
  receptor substrate 2 (IRS-2), 632, 636–637
  resistance, 32–33
  signaling, 409
Insulin growth factor receptor, 978
Insulin/IGF-1 signaling (IIS) pathway
  counter-proteotoxicity activities, 636
  defined, 632
  toxic protein aggregation, 633–637
Insulin-like growth factor 1 (IGF-1), 410, 412, 636, 638, 737, 892–893
Insulin-like signaling (ILS) pathway, 184–185
Interallelic trans-suppression, 979–981
Interferons (IFNs)
  characteristics of, 265–266, 286, 783
  INF α-2b, 783
  IFN-γ, 736
Interleukins
  IL-1, 971
  IL1α/IL1β, 736, 742
  IL-6, 736, 925, 971
  IL10, 736
Internalization, 338
Intestinal colitis, 27
Intraparenchymal therapy, 832
Intravenous immune globulin (IGIV), 923–924
Intrinsically disordered proteins (IDPs), 7, 9
Intuition, 74
Investigational New Drug (IND), 922–923, 928
In vitro studies
  Alzheimer disease, 237
  amyloid formation, 524–525
  fibrillogenesis, 849, 858
  polypeptide chains, 843
  polyQ sequences, 308
  posttranslational modifications, 132–133, 135–139
  prion proteins, 145, 155, 162–164
  protein misfolding, 4–7, 145
  refolding, 48
  synucleinopathies, 828
  systemic amyloidoses, 326
  toxic protein aggregation, 635
  TTR amyloidosis, 972, 990
  Z α_1-antitrypsin deficiency, 415
In vivo studies
  α-synuclein metabolism, 824
  α_1-antitrypsin deficiency, 406–415
  Alzheimer disease, 237
  amyloid cytotoxicity, 99
  amyloidogenesis, 846, 849
  dialysis-related amyloidosis, 848
  gene transcription modulators, 826
  hemodialysis-related amyloidosis, 347–350
  lipid aldehydes, 617
  misfolding inhibitors, 939–940
  oxidative stress, 625
  polypeptide chains, 843
  polyQ diseases, 306
  posttranslational modifications, 132–133, 135–139
  prion proteins, 145–146, 155, 166
  protein misfolding, 9–12
  synucleinopathies, 828–829
  systemic amyloidoses, 326
  TTR amyloidogenesis, 972
Ionic strength, significance of, 9, 135
Ion mobility spectrometry (IMS), 366
Ionotropic glutaminergic receptor agonist, 741
Ion permeabilization, 101
Iperparathyroidism, 845
Iron responsive elements (IRE)
  functions of, 722
  IRE1, 25–27, 29, 36, 328
IRS-1, 32
Ischemia, 118
Ischemic heart disease, 35, 332
Ischemic injury, 801, 803
Islet amyloid polypeptide (IAPP)
  analog role in treatment of type 1 diabetes, 531–532
  amyloid formation, 521–522, 530–531
  aromatic interactions and amyloid formation by, 521–522
  characteristics of, 7, 86, 517–518, 923
  helical intermediates, 530–531
  in vitro amyloid formation, kinetics of, 524–525
  islet cell transplantation, 526–528
  maturation, 743
  membrane interactions, 528–530
  normal physiological role of, 518–519
  posttranslational modifications, 134–135, 140
  primary sequence derivatization, 520–521

Islet amyloid polypeptide (IAPP) (*continued*)
structural models of amyloid protofilament, 522–524
synthesis and processing of, 519–520
in type 2 diabetes, 526–528
Isofagomine, 479
Isomerization, 7, 30, 368–369
Isoprinosine, 280

JNK pathway, 32, 35, 732
Joint tissue, 846
Josephin domain, 313, 315
J-proteins, 430
Jun-N terminal kinase signaling, 803–804

k, 80
Kar2p/BiP, 26–27
Keap1, 26
Kennedy disease, 177
Kenyon cells, 196
Ketasyn, 741
Ketones, 741
Kidneys, *see* Renal system
dialysis-related amyloidosis (DRA), 845
end-stage renal disease (ESRD), 330, 801
failure, 10, 349, 807, 844
functions of, 561, 807
light-chain degradation, 330–331
systemic amyloidoses, 336
Kinases
functions of, 181, 184, 237, 730
inhibitors, 745
stabilization, 980, 982, 991
Kinetic(s)
aggregation assays, 635
analyses, protein aggregation, 623–624
in vitro amyloid formation, 524–525
models for protein misfolding and association, 73–89
pharmacokinetics, 951
significance of, 11–12, 86
stabilization, of TTR, 985–991
tetramer dissociation, 979–980, 984
KLVFF, 908
Kupffer cells, 977
Kuru, 105

Laboratory markers, 820
Lactacystin, 949
Lag phase, 81, 85, 149–151, 356, 405, 525
Laminin, 936
Laminin receptor precursor protein (LRP/LR)-PrP interactions, 271, 280, 283
Laminins, 560
LAMP-2, 118
Laryngeal amyloid, 786
Laser(s)
infrared, 490
light scattering, 851
Late-onset Alzheimer disease (LOAD), 711
Late-onset degeneration, 193, 195
Late-onset neurodegenerative diseases, 193
Lattices, 619, 855
LC3, 120
LC3-II, 116
Learning deficits, 742
Learning process, components of, 240
Lenalidomide, 778, 784–785
Lens transparency, 498. *See also* Eye lens
Lentivirus, 955
Leptomeningeal amyloidosis, 803
Lesions
hippocampal, 216
neuropathological, 713, 736
Leucine, 156–157, 333, 473
Leukocytes, myeloperoxidase enzymes, 616
Leupeptin treatment, 479
Leuprorelin, 894
Leu-Val-Phe-Phe-Ala (LVFFA), 910
Lewy bodies, 12, 34, 93, 122, 192, 617, 818–821, 832, 955
Lewy body dementia, 619
L444P mutation, 476–477, 479
LGE_2, 618
Lichen amyloids, 787
Life span, 105
regulation of, 632
Ligand binding, 415, 889
Light-chain (LC), generally
amyloidosis
diagnosis, 775
immunotherapeutic strategies, 918–928
mutational diversity, 334–335
systemic amyloidoses, 327
deposition disease (LCDD), 336–337, 689, 695
multiple myeloma (LCMM), 695
proteins 918–920, 926
synthesis, 780
Light scattering, 83, 86, 150
Limbic system, 243
Linear polymerization, 83–88
Linoleic acid, 617–618
Lipid(s)
Alzheimer disease, 585–586, 593–594
binding, 822
composition, 103
fibril formation, thermodynamics of, 589–590

functions of, 114, 284
membranes
amyloid diseases, 94
synthetic, 97
metabolism, 411, 561, 907
oxidation, 548, 618–619, 625
oxidative damage, 132
oxidative stress and, 586–587
peroxidation, 101, 548, 616, 741
rafts, 96, 279, 739
synthesis, 25
vicious cycles involving, 593–594
Lipidation, 134
Lipoic acid, 101
Lipolysis, 859
Lipoprotein receptor-related protein (LRP), 249, 720
Lipoproteins, 562
Liposomes, 958
Lisuride, 183
Lithium, 124, 184, 205, 733, 893–894
Live-cell imaging, 178, 352
Liver
α_1-antitrypsin deficiency, 404–405, 408–413
chronic liver failure, 404
circulating TTR in, 976–977, 980
gene therapies, 980
hepatocytes, *see* Hepatocytes
mutant cells in, 403
transplantation, 413, 869–870, 982
treatment of ATTR, 804–805
Liver X receptor (LXR) ligands, 827
Livestock disease management, 261
Lixelle, 369, 847–848
Log-log plots, 87
Long-chain fatty acids (LCFAs), 548
Long, straight (LS) fibrils, 356–359, 370
Long-term potentiation (LTP), 634, 720, 742
Loop-sheet insertion, 500–501
Loss-of-function diseases, 3, 62
Lovastatin, 271, 279, 738
Low-density lipoprotein (LDL) cholesterol, 338, 738
Low-density lipoprotein receptor protein (LRP), 722
Low-density lipoproteins, 619
Low-flux cuprophane membranes, 847
Low-molecular-mass Hsp90 inhibitors, 949, 952
LSGGQ, 435
Lung disease, 404, 413
Lung transplantation, 413
Lung volume reduction surgery (LVRS), 413
LY404187, 741
LY450, 139, 721
Lymph node amyloidosis, 776
Lymphocytes, 37, 917
Lymphoma cells, 972
Lymphoproliferative disease, 691
Lymphotoxin (LT) $\alpha+\beta$, 263
Lymphotoxin-β immunoglobulin, 264
Lys58, 350, 368
Lysines, 151, 407, 549, 619
posttranslational modifications, 139
Lysosomal storage diseases (LSDs)
characterized, xxviii, 469–470, 821, 832–833
Gaucher disease, 469–480
Lysosomes, *see* Lysosomal storage diseases (LSDs); Lysozymes
α-synuclein degradation, 828–829
degradation, 479
functions of, 12, 274, 339, 829–830
lysosomal system, 113–115, 120–121
proteolysis, 439
Lysosome-associated membrane protein type 2A (LAMP-2A), 117, 122
Lysozyme amyloidosis
characteristics of, 9, 867–868
clinical, 868–869
effect of mutations
on folding of, 871–872
on in vitro aggregation propensity, 876
on stability and global cooperativity of, 873, 875–876, 881
on structure of, 871
fibril formation, molecular mechanisms of
from biochemical and biophysical characterization of proteins, 871–876
examination of ex vivo fibrils, 870
in vitro, inhibition of, 879–882
overview of, 876–879
therapies, 868–869
Lysozymes
amyloidosis, *see* Lysozyme amyloidosis
functions of, 337, 845
wild-type (WT), 869, 871–875, 877, 881

Machado-Joseph syndrome, 34
Macroautophagy, 116–118, 120–125, 829
Macroglobulinemia, 691
Macromolecules
amyloid diseases, 94
functions of, 7, 94, 114, 906
macromolecular chaperones, 978
macromolecular crowding, 10–13, 48, 102
Macrophages, 36–37, 264, 330, 340, 352, 470, 617, 737–738, 918
Macular amyloids, 787

Magnetic resonance imaging (MRI), applications of, 235, 682, 711, 736, 776
Maillard reaction, 498
Major histocompatibility complex (MHC)
  class I molecules (MHC-I), 844, 846–847, 850, 854–855, 927
  class II molecules (MHC-II), 117
  polypeptides, 330
Malabsorption, 796
Malarial circumsporozoite proteins, 562
MALDI-TOF mass spectrometry, 620
Malignant diseases, 689, 691, 786, 805
Mammalian disease models, 186
Mammalian prions, 146, 164–165. *See also* Prions
Mannose, 23–24
Mannose-6-phosphate, 471
MAP kinase (MAPK), 412, 732–733
Marinesco-SjÖgren syndrome, 34–35
Mass spectrometry analysis
  applications, 7, 223, 338, 562, 566, 851, 870
  MALDI-TOF, 620
  posttranslational modifications, 135
Mayo Clinic, 777–779
M cells, 262
MDR1, 432
Mean probe score (MPS), 219
Medical aspects of disease
  diagnosis techniques
    biomarker identification, 689–696
    of systemic amyloid diseases, 673–682
    total internal reflection fluorescence microscopy, 699–707
  imaging technologies, 647–662
  pathogenesis, familial and senile amyloidosis, 795–809
  therapeutic treatments, *see* Therapeutic treatments
Mefloquine, 270, 277
Mello, Craig, 953
Melphalan, 777–778, 780–784
Memantine, 734, 742, 892
Membrane, generally
  cholesterol, 96
  permeabilization, 99–100
  proteins, *see* Membrane proteins
Membrane proteins
  amyloid aggregates, 98
  characteristics of, 11, 94, 503
  cystic fibrosis, 428
Mementine, 235, 712
Memory
  deficit
    age-related, 216
    implications of, 243, 245
    spatial, 216–218, 223–224
  episodic, 242
  loss
    Aβ*56 and, 222–226
    amyloid plaques, 217–222
    influential factors, 213–217
    pre-dementia, 214
  process, components of, 240
  spatial, 219, 223–224, 242
MEM3454, 712
Meningitis, 248
Meningoencephalitis, 248, 264, 937
Mepartricin, 269, 276
6-Mercaptopurine (6-MP), 454–455, 457
Messenger RNA (mRNA)
  α-synuclein metabolism, 823
  biogenesis of CFTR, 432
  functions of, 197, 283, 560, 954–955
  splicing, 26–28
  TPMT metabolism, 455, 459
  transcription, 826
  translation, 24–25, 828
Metabolic disease, 32–33
Metabolism
  Alzheimer disease and, 736, 741
  amyloid inhibition, 907
  Drosophila, 197
  gene expression, 950
  in vivo amyloidogenesis and, 561
  implications of, 411
  Parkinson disease and, 823, 826
  protein aggregation, 181, 316
  RNA, 181, 316, 823
  thiopurine, 454–455
  TPMT, 455, 459
Metal(s), *see* Copper; Zinc
  binding, 369
  -catalyzed oxidation reactioons, 138
  chelation/chelators, 722, 909
  -complexing drugs, 551–552
  ions, 10, 12, 101
Metalloproteases, 28, 249, 551
Metalloprotease-2 (MMP)-2, 551
Metallothionein, 925
Metazoan cells, 24
Methanosarcina, 56
Methionine, 36, 138
Methisazone, 280
Methisoprinol, 280
Methylene blue stain, 909
Met99, 350
Met119, 979

Mevinolin, 271, 279
MG132, 462, 949
Micelles, 86
Michael addition, 132, 139, 620
Microaggregates, 460, 464
Microarray analysis, 164–165
Microautophagy, 116, 829
Microglia, 101, 242, 249, 272, 389, 736–738
Microhemorrhages, 937
MicroRNAs (miRNAs)
  bantam (*ban*), 197–199, 203–204
  functions of, 197, 954
  global loss of, 199
  neuronal degeneration, 197, 199–200, 204
Microscopic studies
  atomic force (AFM), 150, 225, 355, 620, 622, 851, 857–858
  cryoelectron microscopy analyses, 870
  differential interference contrast (DIC), 74
  electron (EM), 934
  epi-fluorescence, 700
  fluorescence, 462
  high-resolution electron (HREM), 560
  light microscopy, 306, 706
  scanning electron (SEM), 200
  total internal reflection fluorescence (TIRFM), 700, 704–706
  transmission electron (TEM), 149–151, 877
Microtubule(s)
  -associated protein (MAP), 237, 730
  autophagic degradation, 462–463
  dynamics, 502
  functions of, 98
  TPMT pharmacogenomics, 463–464
  transport alterations, 99
Mild cognitive impairment (MCI), 234–235, 250, 712, 744
Minocycline, 891
Misfolded proteins
  accumulation of, 49
  amyloid diseases, 103
  characteristics of, 101, 118–119
  endoplasmic reticulum and, 21
  neurogenerative diseases, 34
  production of, 62
  TPMT, 460–461, 463–464
Misfolding, *see* Misfolded proteins; Protein misfolding
  defined, xix-xx
  hazards of, 48
  impact of, 93, 145
  in vitro studies, 4–7
  in vivo studies, 9–12
  neurodegenerative diseases and, 175
  overview of, 3–4
  protein aggregation determinants, 7–9
  toxicity mechanisms, 317
Missense mutations, 241, 384
Mithramycin, 184
Mitochondria
  amyloid cytotoxicity, 98, 101
  functions of, 54, 121, 950
  phosphorylation, 616
  polyglutamine proteins, 889, 894
Mitochondrial aconitase, 60
Mitochondrial cell death, 37
Mitochondrial genomes, 976
Mitochondrial injury, 413
Mitochondrial membrane proteins, 50
Mitogen-activated protein kinases (MAPKs), 281, 803, 827
Mitotic cell cycle, 104
M* molecules, 405
Mnichinan mutation, 405
Molecular chaperones
  defined, 48, 946
  functions of, xx-xxi, 316, 433, 460, 936, 945–948
  role in protein folding, 47–62
Molecular crowding, 88, 820, 824
Molecular dynamics simulations, 9, 369
Monoclonal-free light chains, 695
Molecular genetics, 62
Molecular mechanics, 562, 566
Molecular modeling, 562, 566
Molecular oxygen, 30–31, 616
Monoamine oxidase (MAO) type B inhibitors, 742
Monoclonal antibodies (mAB), 224, 243, 407, 919–922, 927
Monoclonal gammopathy, 689–692, 695
Monoclonal gammopathy of undetermined significance (MGUS), 691–692
Monoclonal protein, 691
Monomers
  Aβ, 215, 242
  amyloid
    conformation, 159
    diseases, 94, 969
    inhibition, 910–911
  disease proteins, 203
  fibrillogenesis, 857
  folded, 94
  functions of, 74–78
  islet amyloid polypeptide, 525
  misfolded, 98, 103, 307–308
  peptide, 100
  prion species barriers, 164

Monomers (*continued*)
systemic amyloidoses, 338
toxic misfolded, 307–308
TTR amyloidosis, 972, 986
unfolded, 94
Morris Water Maze Task, 216–217, 222, 242
Motility defects, 181, 184
Motor neuron degeneration, 197
Mouse models/mouse studies
α_1-antitrypsin deficiency, 407, 409, 411–412, 954
Alzheimer disease, 234, 550, 734, 738–744
amyloid aggregate cytotoxicity, 101
amyloidosis
amyloidoma, 919–922
characteristics of, 859
amyotrophic lateral sclerosis, 197
atherosclerosis, 37
autophagy, 462
biology of amyloid-β protein misfolding in Alzheimer disease, 213–215
brain cells, 240
cholesterol studies, 96
down-regulation of
Htt by RNAi, 956–957
SOD1 expression, 957–958
endoplasmic reticulum stress, 27–28
familial amyotrophic lateral sclerosis (fALS), 384–385
host disease protein, 196
HSR
modulators, 949, 951
regulators, 950
Huntington disease, 203
immature pathogenic SOD1, 388–389
life span regulation, 632
metabolic disease, 32
neurodegenerative disease, 34–35
neuronal integrity, 199
polyglutamine protein expression, 888, 891–892
prion disease therapy, 260, 271
protein
degradation, 890
misfolding diseases (PMDs), 934
SNCA gene transcription, 825–826
SOD1 mutations, 388–389
synucleinase actvity, 831–832
systemic amyloidoses, 329
tauopathies, 243
toxic protein aggregation, 634, 637
transthyretin (TTR)
amyloidoses, 980, 991
characteristics of, 971
Movement disorders, 305
MP, 780
MPC-7869, 721
MPP^+, 183
MS-8209, 269, 276
MT1-matrix metalloprotease, 332
mTOR, 116, 118, 124, 203, 410
Mucopolysaccharide, 560
Multimodal therapy, 745
Multiple myeloma, 689–692, 695, 779,781, 952
Muscarinic agonists, 719–720
Muscle pathologies, 118
Mutagenesis, 9, 151, 158, 160, 163, 359, 974
Mutations, *see specific types of mutations*
aggregation and, 8–10
α-synuclein, 955–956
amyloidosis, 797–799
autosomal dominant, 491–492
class II, 434
disease-linked, 625, 631–632
fibril formation, 354
Gaucher disease, 474–477, 479
impact of, 60, 62, 104, 152, 155, 157, 162–163, 166, 234
lysozyme amyloidosis, 871–876
neurodegenerative disease and, 640
pathogenicity, 10
prion diseases, 6
SOD1, 384–387
myc, 411
Myelin/myelin sheath, 803, 832
Myelination, 244
Myelodysplasia, 781–782
Myeloma, 778
Myeloperoxidase, 616
Myocardium, 802

NAC (nascent chain-associated complex), 51–53
N-acetylglucosamine (GlcNac), 562–564
N-acetylglucsaminyltransferase III (GlcNAc-TIII), 738
NAD(P)H:quinone oxidoreductase, 26
NAP/AL108, 742
Naphthalene sulfonates, 909, 911
Nascent chains, 52, 61
National Cancer Institute, 923
National Institutes of Health (NIH), 522–523, 740–741
N-butyl-1-deoxynojirimycin (NN-DNJ), 478–479
Nct, 237, 245
Necrosis, 37, 99, 102, 616. *See also* Apoptosis

Nectrin receptors, 722
Nematode studies, 953–954
Neoepitopes, 919, 928
Nephelometry, 692
Nephrectomy, 787
Nephropathy, 332
Nephrotic syndrome, 780
Neprilysin, 638, 719, 740
Nerve growth factor (NGF), 742–743, 827
Net elongation, 74, 83
Neural cells, 824
Neuregulin (NRG), 244, 246, 740
Neurite/neuritic
  growth, 742
  Lewy, 820
  plaques, 236, 241, 721
  swellings, 236
Neuroaxonal dystrophy, 821
Neurobiology, 250, 546
Neuroblastomas, 96, 104, 739, 939
Neurochem, 908
Neurodegeneration, 10, 97, 184, 637, 639–640, 887
  familial, 639
  risk factors, 637
  sporadic, 639
Neurodegenerative amyloidosis, systemic amyloidosis compared with, 326–327
Neurodegenerative disease, *see specific neurodegenerative diseases*
  controlling gene expression, 952
  protein aggregation, 461
  protein misfolding principles, 33–35, 49, 104–105, 122, 145–146, 175
  proteotoxicity, 631–632
  role of metals, 545
  thiopurine *S*-methyltransferase pharmacogenomics, 461
Neurodegenerative disorders, 62, 118
Neurofibrillary tangles (NFTs)
  Alzheimer disease, 213, 215, 234, 236, 241, 243, 712–713,717, 730, 734
  posttranslational modifications, 134
  protein aggregation, 181
  protein misfolding diseases, 195, 234, 236
Neurogenesis, 736
Neuroinflammation, 731, 736–738, 742
Neuroinvasion, 262
Neurological deficits, 804, 821
Neurological examination, 235
Neuronal cells
  death, 35, 184
  functions of, 62
  loss of, 545
Neuronal ceroid lipofuscinosis (NCL), 125, 831–832
Neuronal cholinergic signaling, 183
Neuronal injury, 742
Neuronal loss, 215, 712, 716, 718, 955
Neuronal membranes, 94
Neuropathology, 250
Neuropathy, 720, 796, 799, 801–803, 806
Neuroprevention, 818
Neuroprotection
  neuroprotective agents, 235
  neuroprotective pathways, 740
  significance of, 272, 281–282, 946
Neuropsychological Test Battery (NTB), 551
Neurotoxicity, 35, 699, 739, 803, 820, 824, 947
Neurotoxins, 183
Neurotransmission, 547
Neurotransmitters, 182, 712
Neurotrophins, polyglutamate diseases, 892–893
Neurturin, 892
Neutrophils, 403, 407, 409
NFκB activation, 409, 411
NGX267, 720
Nicastrin (Nct), 240–241
Niemann-Pick disease, 821
Nitration, amyloid formation, 132, 138
Nitric oxide (NO), 35, 616–617
Nitrogen, 99, 146–147, 617
Nitrogen species, biologically relevant, 617
Nitrosylation, 101
N-methyl-D-aspartate (NMDA)
  Alzheimer disease, 550, 712
  receptors, 99, 272, 734, 742, 892
N-methyl groups, 939
Nodular amyloidosis, 787
Nomifensine,183
Nonamyloid diseases, 982
Nonsteroidal anti-inflammatories (NSAIDs), 247, 266, 369, 721, 741, 948, 950, 990–991
Nonsynonymous single-nucleotide polymorphisms (nsSNPs), xxviii, 454, 456–458, 460–461, 464–465
Notch proteins
  functions of, 247, 721–722
  Notch1 intracellular domain (NICD), 240, 245
Novobiocin analo, 949
NproIAPP, 527–528
NRF1/NRF2, 26
N-terminal brain naturietic factor (NT-BNP), 776

N-terminal/N-terminus, 123, 134, 137, 139, 146–147, 149, 151–152, 155, 159–161, 177, 195, 241, 249, 311, 315, 332–333, 358, 364, 430, 495–497, 526–527, 621, 848–849, 851–853, 856, 869, 920, 956
N370S mutation, 473, 474–476, 478–479
Nuclear magnetic resonance (NMR) analyses, 154–157, 159, 161, 163, 307, 347, 359, 364–365, 522–524, 529, 562, 566, 822, 848–849, 851–852, 855, 873, 881, 974
Nucleated conformational conversion process, 149, 855
Nucleation
  aggregation, 94–95, 102
  amyloid, 150
  barriers, 79
  complex, 115
  double, 86
  fibril, 86
  heterogenous, 86
  implications of, 83, 308–310, 314, 489, 524, 619, 975
  monomeric, 84
  prion species barriers, 166
  process, 75–78
  secondary reactions, 86
Nucleic acid
  oxidation, 616
  sequences, 828
Nucleotide(s)
  -binding domains (NBDs), 427, 436–437
  exchange factor (NEF), 54–55
  sequences, 455
  significance of, 436–437
Nucleus, monomeric, 78–79, 82–83
Nutritional deficiency, 550
Nutritional status, 804

OASIS (old astrocyte specifically induced substance), 29
Occupational accidents, 490
Off-pathway aggregates, 86–88
Olfactory cortex, 243
Oligodendrocyte cells, 827
Oligomannose carbonhydrate chains, 473
Oligomerization, 95, 100, 338, 369, 389
Oligomers
  Aβ, 215, 242
  Alzheimer disease therapies, 719
  amyloid
    diseases, 94, 968–969
    formation, 154–155
    inhibition, 906, 910–911
  amyloidogenesis, 851
  assemblies,48, 99
  disease proteins, 203
  fibril formation, 367–368
  fibrillogenesis, 857
  formation of, 62
  functions of, 75, 934
  IAPP, 530
  islet amyloid polypeptide, 530
  juuvenile cataracts, 491
  lysozyme amyloidosis, 878
  morphology of, 844
  peptide, 100
  polyglutamine proteins, 889
  prefibrillar, 820
  prion
    disease, 260
    generation and propagation, 150–151
    species barriers, 164
  protein misfolding diseases, 938
  spherical, 309
  transthyretin (TTR)
    amyloidosis, 972
    protein, 802–803
Oligonucleotides, 268, 273, 827–828, 833
Oligopeptide repeats, 146
OligoPro sequences, 311
Oligosaccharides, 438, 476
Oligosaccharyltransferase (OST), 23
Omega-3 fatty acids, 740–741
Oncogenes, 118
On-pathway aggregates, 86–88
Onset of disease, 9, 11. *See also* Age at onset
Open reading frame (ORF), 455, 458
Opsonization, 922, 928
Organ damage, mechanisms of, 338–340
Organelles, 9, 114, 123
Organic compounds, 909
Organ transplantation. *See* Transplantation
ORP150, 32, 35
Orthologs, biological counter-proteotoxicity activities, 637–638
Orthostatic hypotension, 796, 804
Osmolytes, 414
Osteoporosis, 859
Ovariectomy, 470
Overexpression, 153–154, 823
Overshoot, 86
Oxidation, 488, 498
  amyloid formation, 137–139
  nonenzymatic modifications, 135
  posttranslational modifications, 132, 135
Oxidative damage, 490, 500
  amyloid formation, 138–139
Oxidative degradation, 922

Oxidative injury, 548
Oxidative markers, 548, 550
Oxidative protein, 30
Oxidative stress
  Aβ proteins and, 587–589
  Alzheimer disease, 736, 741
  amyloid
    cytotoxicity, 101
    formation, 132, 139, 616, 619
  amyloidosis, 803, 950, 970
  defined, 616
  impact of, 616, 619
  influential factors, 616
  lipids and, 586–587
  lipoprotein E and, 591–592
  mouse models of, 592–593
  protein misfolding
    principles, 10–11, 13, 31, 33–35, 121, 195
    role in, 615–640
  proteomics in Alzheimer disease, 590–591
  systemic amyloidosis, 339
  therapeutic treatments, 282
Oxidoreductases, 430
9-Oxononeneyl cholesterol, 618
Oxygen
  ground-state, 616
  therapy, 413
Oxysterols, 548

Paclitaxel, 735
Paeniflorin, 949
PAGE, 307
Pain impairment, 796
Paired helical filaments (PHFs), 233, 236–237, 713, 730–731, 735
Pancreas, 27, 134, 527. *See also* Insulin
Paralysis, toxic protein aggregation, 634
PAR-1, 195
Parasthesis, 796
Parasympathetic nervous system, 263
Parenchymal amyloid, 238, 803
PARK7 gene, 195
Parkin gene, 34, 829
Parkinsonism, 195, 821
Parkinson disease (PD)
  age at onset, 824
  aging and, 631
  α-synclein and, 134
  amyloidoses, 844
  autophagy, 119, 122–123
  characteristics of, xxii, 12, 49, 62, 93, 101, 175, 193, 489, 551, 615, 617, 742–743, 817–818, 933, 937
  clinical drug trials, 737
  disease proteins, 196–197
  down-regulation of α-synuclein by RNAi, 955–956
  Lewy body-positive dementia, 832
  miRNA roles, 199
  pathogenesis, 822–824
  risk factors, 829
  sporadic, 631, 828
  tau protein and, 176–177, 183, 192
  toxic misfolded proteins, 954
Partially folded conformers, 500
Partial unfolding, 843–844
Partitioning, 60
Pathobiology, alpha-1-antitrypsin deficiency, 405–413
Patient history, significance of, 235
PBT2, 552
PC12 cells, 827
PDI (protein disulfide isomerase), 30, 34
Pedigrees, 494
Pen-2, 237, 241, 245
Pentosan polysulfate (PPS), 267–268, 275, 284–285
Pepechiae, 868
Peptide(s), *see specific types of peptides*
  Aβ, 239 241, 243, 245
  amyloids, 96, 151, 154–155
  amyloid inhibition, 907–908
  aptamers, 270, 278–279
  Asp-Pro bonds, 132
  backbone, amyloid diseases and, 94
  binding, 29
  bonds, 405
  chemistry, 940
  cleavage-generated, 234
  inhibitors, 907–910
  interpeptide binding, 622
  membrane cholesterol and, 96
  microarray analysis, 164–165
  mutations, 625
  neurotoxic, 250
  polyQ, 309–310
  -protein aggregates, 94
  synthetic, 706
  systematic mutagenesis, 9
Peptidyl groups, 852
Peptidyl-prolyl isomerase, 51
Peripheral amyloidoses, immunotherapy, 923–925, 928
Peripheral nerves, 802–803, 805–806
Peripheral nervous system (PNS), 244
Peripheral neuropathy, 97, 985–989
Peripheral sink hypothesis, 937
Peripheral vascular disease, 35

PERK, 25–26, 29, 32
Perlecan, 325
Permeabilization, 101
Peroxynitrite, 617
Perturbation expansion, 81–82
Peyer's patches, 262
PFD (prefoldin), 51
p53
  characteristics of, 118, 411
  inhibitors, 271, 281
pge-1 gene, 182
P glycoprotein, 737
pH, significance of, 4, 9, 135, 353–354, 363–364, 366, 497, 499, 857, 946
Phakinin, 503
Pharmaceuticals, chemical stability of, 133
Pharmacogenetics, 453
Pharmacogenomics
  clinical importance of, 453, 741
  defined, 453
  drug response, 453–454, 464
  TPMT, 454–464
Pharmacokinetics, 951
Pharmacological chaperones, 833
Pharmacological induction, of HSR, 948, 950
Phenathiazines, 269, 276–277
Phenserine, 722, 743–744
Phenylalanine, 156, 522
Phenylbutyrate, 891
Phenylethanolamine *N*-methyltransferase (PNMT), 458
Phe-Phe dipeptide, 908
Phosphatases, 181, 237, 730, 734
Phosphatidylinositols,908
Phosphatidylinositol 3-kinase/Akt signaling pathway, 893
Phosphoglycerate kinase, 497
Phospholipase
  A_2, 281, 949
  inhibitors, 271, 281
Phospholipid, generally
  anionic, 103
  bilayers, 94–95, 103
  functions of, 820
  surfaces, 94–95
Phosphonates, 799
Phosphonoacetic acid, 280
Phosphorus dendrimers, 269, 274
Phosphorylation, 636
  CathD, 830
  CFTR, 431–432
  host disease protein, 195–196
  implications of, 134, 184, 201, 616, 636, 730–731, 733, 824
  JNK, 32
  PERK-mediated, 24–26
  SNCA transcriptional regulation, 827
Photochemical degradation, 490
Photodamage, 490
Photoreceptor neurons, 198
Phthalocyanin tetrasulfonate (PcTS), 269, 275
Physical examination, 235
Pichia methanolica (PM), 164
Pick disease, 821
Pin1, 734
Pioglitazone, 738
PKA kinase, 732–733
PKC, 733
Plaque
  amyloid, *see* Amyloid, plaques
  extracellular, 192
  formation, 719
Plasma cells
  -directed immunotherapy, 926–928
  functions of, 329–330
  leukemia, 689
  proliferative disorders, 690–691, 693, 695
Plasma deficiency, 405–406
Plasmacytoma, 689, 691
Plasmids, 954
Platelet-activating factor (PAF), 281
Pleiotropic signaling pathways, 736
Podospora anserina, 147, 152
Point mutations, 155, 194–195
Poly(ADP-ribose) transferase/polymerase-1 (PARP-1), 826
Polyamidoamide, 269, 274
Polyamino acids, 910
Polyanions, 267–268, 273
Polycations, 269, 274–275
Polyethyleneimine (PEI), 703
Poly(Gln), 82–83
Polyglutamine (polyQ)
  aggregated simple sequences, 308–310
  aggregation, 313
  biological counter-proteotoxicity aggregation, 638
  defined, 12
  disease, *see* Polyglutamine disease
  disorders, 175, 177, 952
  expansion, 306, 313–315
  extension
    cytotoxicity, 101
    proteins, 62
  flanking sequences, 312
  functions of, 124, 928
  peptides, 309–310

protein aggregation, 181–183
toxicity, 316
toxic protein aggregation, 631, 633–634, 636
tract, 177–178, 182
Polyglutamine disease
characteristics of, 177–178, 193, 199, 203, 305–308, 310, 317, 887, 947
disease-specific treatment
Huntington disease, 893–894
SBMA, 894
SCA 1, 894
proteins, 196
treatment strategies
caspase inhibition, 891–892
histone deacetylase inhibition, 890–891
neurotrophic factors, 892–893
reducing protein expression, 888
targeting protein for degradation, 888–890
transglutaminase inhibition, 893
Polyglutamine repeats, 463
Polymer formation
components of, 406
primary, 75–82, 85
secondary, 84
Polymerization
α-1-antitripsin deficiency, 406, 408, 414
amyloidosis, 351, 622, 906, 975
protein aggregation, 309, 500
protein misfolding principles, 34, 78, 104, 149, 157
therapeutic treatments, 259, 274
Polymorphisms, 147, 197, 463, 736, 823, 957. *See also* Single-nucleotide polymorphisms (SNPs)
Polymorphonuclear
leukocytes (PMN), 918
neutrophils (PMNs), 618
Polyneuropathy, 923, 980
Polyols, 909
Polypeptides
amyloid diseases, 102
amyloidogenesis, 615
backbone, 48
binding to, 52, 54, 57
-binding proteins, 22
cataracts and, 500
chains
fibril formation, 359
functions of, 30, 523, 843 , 845
misfolding, 104
nascent, 61, 175–176
natural environment of, 9
physicochemical properties, 8
synthesis, 25
translation, 48
unfolded, 4, 6–7, 968
chaperones and, 50
endoplasmic reticulum conditions, 22, 24
extension of, 334
islet amyloid, 517–523
monomeric, 637
nascent chains, 61
posttranslational modifications, 134
toxic, 640
Polyphenols, 722
Polypropyleneimine, 269, 274
PolyQ. *See* Polyglutamine (polyQ)
PolyQ-YFP fusion proteins, 177, 179–182
PolyQ40 protein aggregation, 178
Polysaccharides, 562–564, 908
Polyubiquitin
binding motifs, 195
chain, 947
Polyunsaturated fatty acids (PUFAs), 618–619
Polyvinylsulfonate (PVS), 703
Positron emission tomography (PET), 235, 711, 736, 923
Posttranslational modifications
in amyloid formation
aberrant enzymatically catalyzed, 131, 134
components of, 131–134
nonenzymatic, 132, 135–139
potential effects of, 140
spontaneous, 132–134, 136, 139
deleterious, 133
enzymatically catalyzed, 131, 133–134, 140
implications of, 122, 195, 214, 387, 519, 824, 829, 948, 981
Posttranslation uptake, 50–51
PPARγ agonists, 737–738
PP1, 734, 908
PP2A, 734
PPIase activity, 52
Praecis Pharmaceuticals, 908
Precursor protein of prion diseases (PrP^c), 7, 260–264, 271–275, 278–283, 286. *See also* Prion
Prednisone therapy, 780–781, 783
Prefibrillar amyloid aggregates, 94, 98–99, 102
Prefoldin, 61
Premalignant syndromes, 689, 691
P-selectin glycoprotein ligand-1, 720
Presenilins (PS1/PS2), 237–238, 240–241, 243–246, 250, 716
Pressure, significance of, 4
Pre-synaptic proteins 831

Prion
  amyloid structure, 154–161, 166
  biology, 167
  conformation, 145–146, 149, 156, 161, 166
  diseases
    aging and, 631
    characteristics of, xxii, 6, 101, 103, 105, 192, 262, 821, 933, 937, 940
  disease therapy
    chemical-based and prophylaxis, 266–282
    combination therapy, 284
    human treatments, 284–286
    immune-based therapies, 262–266
    in vivo/in vitro tests, 260–262
    overview, 259–260, 286
    targeting PrP^c, 282–283
  polymerization, 154
  proof of prion hypothesis, 149
  proteins (PrP)
    amyloid structure, 154–161, 166
    antibodies, 265
    biochemical analysis, 166
    conformational isomerization, 580
    conversion reaction, 266–267
    functions of, 145–146, 326, 329
    fungal prions, 146–149
    generation and propagation, 148, 150–154, 166
    prion strains, 161–164, 167
    species barriers, 164–167
  PRP^{res}, 259–261, 263–265, 268–274, 276–284, 286
  PrP^{Sc}, 277
  self-perpetuating, 145–164
  species barriers, 146, 164–167
  strains, 146, 161–164, 167
  therapy
    diagnosis, 261
    therapeutic techniques, *see* Prion, disease therapy
Pro-amyloidogenic secretases, 240
proANF, 134
Procalcitonin, 134
Procedural learning, 216
Prodrugs, 454
Profibrils, 619
Progesterone, 271, 279
Programmed cell death, 117, 147, 197, 199, 891–892
Progressive supranuclear palsy, 733
Prohormones, 134, 526
ProIAPP processing, 134, 518, 526–527
Pro-inflammatory genes, 737–738
Prokaryotes, xx, 30, 53, 55, 176
Proliferative pathways, 740
Proline
  amyloid formation, 136
  functions of, 52, 359–361, 364, 520, 920, 939
  scanning mutagenesis, 158
Promoters, 243, 925
Proof of prion hypothess, 149
Prostaglandins, 948
Protease(s)
  domain mutation, 195
  functions of, 25, 28, 160, 181, 249, 638, 832
  inhibition/inhibitors, 318, 403
Proteasomes
  functions of, 12, 34, 101, 122, 408, 342
  inhibition/inhibitors, 461, 949, 951
  polyglutamine disease, 889
Protein(s)
  adaptors, 123
  aggregate toxicity, 103
  aggregation, *see* Aggregation
  aging, 133–134, 137
  biosynthesis, 181
  concentration, 10–11
  conformation disorders (PCDs)
    characteristics of, 118–119, 689
    late-stage, 124
    severe, 125
    toxic, 124
  databases, 9
  degradation, 50, 123, 181, 203, 824, 888–890
  deposition diseases, 5–6, 12
  disulfide isomerase (PDI), 22, 35
  fibrillization, 94
  folding
    cytotoxicity, 98
    efficient reactions, 22
    oxidative, 30–31
    quality control, 22–23, 102, 124
    significance of, 21, 181
    studies, 844
  homeostasis, 181, 186
  -inhibitor complexes, 905, 909
  internalization, 337–338
  kinases
    functions of, 26, 32
    protein kinase A (PKA), 427
    protein kinase C, 730
    stress-activated, 412
  misfolding, *see* Protein misfolding
  mutant, 175
  -peptide aggregation, 98
  -protein interaction, 938
  quality control system, 946
  rearrangement, 844, 848–849

synthesis
chaperone system, 50
de novo, 824–825
implications of, xxi, 124, 146, 824–825
inhibitors, 949–950
trafficking, 181, 243
Proteinases
inhibitors, 407
proteinase K, 820
Protein misfolding
defined, 3
diseases, *see* Protein misfolding diseases
misfolding disorders (PMDs)
aggregation and, 935
antibodies and, 936–937
characteristics of, 933–935
inhibitors of, 936–940
peptide inhibitors, 938–940
vaccinations against, 936–937
-misfolding modifiers, globality of, 201–203
principles of
amyloid-β, in Alzheimer disease, 213–226
autophagy, 113–125
biology of protein aggregation and toxicity, 175–187
Drosophila models of protein misfolding diseases, 191–205
endoplasmic reticulum stress, 21–37
kinetic models, 73–89
molecular chaperones, 47–62
oxidative stress, 21, 31–37
posttranslational modifications, 131–140
reasons for misfolding, 3–13
self-perpetuating prions, 145–167
toxicity in amyloid diseases, 93–105
Protein misfolding diseases
alpha-1-antitrypsin deficiency, 403–415
Alzheimer disease, 233–250
amyloidoses
hemodialysis-related, 347–370
systemic, 325–340
cystic fibrosis, 425–440
cataracts, 487–503
Drosophila models of, 191–205
familial amytrophic lateral sclerosis, 381–390
Gaucher disease, 469–480
Huntington disease, 305–318
islet amyloid polypeptide, 517–532
medical aspects of, *see* Medical aspects of disease
prion disease therapy, 259–286
thiopurine *S*-methyltransferase pharmacogenomics, 453–464
Proteinurea, 695, 780, 783, 796, 801
Proteoglycans, 134, 351–352, 527, 936
Proteolysis, 10–11, 25, 28–29, 356, 360, 434, 439, 498, 616, 735, 848, 870, 876, 892, 922
biological counter-proteotoxicity aggregation, 638–639
Proteolytic degradation, 48, 919
Proteolytic systems, 113–115, 119
Proteomes, disease initiation, 640
Proteomic studies, 59, 640
Proteosomes
α-synuclein degradation, 828–829
functions of, 21, 24, 113–114, 203
polyglutamine proteins, 889–890
Proteostasis
network, 978, 981
regulators, 439–440
Proteotoxicity
aggregation-mediated
age at onset of neurodegeneration, influences on, 639–640
biological counter-proteotoxicity activities, 637–639
insulin/IGF-1 signaling (IIS) pathway and, 633–637
life span regulation and aging, 632
overview of, 631–632
implications of, xxviii, 115
Proteotoxic stress, 183
Pro32, 849–853
Protofibrils, 86, 98, 309, 496, 524, 634, 636, 820, 844, 934, 968, 973
p62, 120
Psychiatric examination, 235
p-Tau, 952
PTEN, 118
p38, 828
Pulmonary amyloid, 786
Pulmonary disease, 403
Pulse chain studies, TPMT mechanisms, 459
Pulse-chase experiments, 459, 473, 476
Purkinje cells, 34–35, 196–197, 199–200, 893
Puromycin, 949
Purpura, 868
Putative sporadic disease, 237
Pyrazolone derivatives, 272, 282
Pyridine dicarbonitriles, 270, 278
Pyroglutamic acid, 132

QBP1, 313
Quality control systems, 22–23, 53, 113–114, 118, 124
Quantitative free light-chain assays
amyloidosis, 692

Quantitative free light-chain assays (*continued*)
monitoring disease activity, 695–696
overview of, 692–693
Quartz, amyloid-β fibril growth, 703
Quasi-domain swapping structure, 501
Quinacrine, 269–270, 276–278, 284–285
Quinine, 269–270, 276–277

Rabbit reticulocyte lysate (RRL) system, 459–460
Rabies virus glycoprotein (RVG), 283
Racemization, 132, 137
Radiation, infrared, 490
Radicicol, 949, 952
Radionucleotide imaging, 679–680
Raft markers, 739
Ramachandran profile, 853
Rapamycin, 116, 124, 203, 830, 890
Rapid Access to Intervention Development (RAID) program, 923
Rasagiline, 742
Rat studies
α_1-antitrypsin deficiency, 414
down-regulation of α-synuclein expression, 956
islet amyloid polypeptide, 529
memory deficits, 223–224, 243
prion disease therapies, 266
synucleinopathies, 832
toxic protein aggregation, 637
Reabsorption, 848
Reactive amyloidosis, 10
Reactive center loop, 414
Reactive loop peptides, 405, 414–415
Reactive nitrogen species (RNS), 99, 616
Reactive oxygen species (ROS), 11, 26, 30–31, 33–34, 36, 99, 101–102, 616–618
biologically relevant, 617
defined, 616
generation of, 616, 618
Rearrangements, 139, 872
Receptor for advanced glycation end (RAGE) products, 97, 737, 803
Recognition elements, 164–166
Recombinant adeno-associated viral serotype 5 (rAAV5), 956
Recombinant proteins, 146, 497, 832
Recombinant single-chain antibodies (scFvs), 271, 280
Recombinant technology, 326
Red blood cells (RBCs), TPMT activity, 454, 457
Redox activity, 30, 548–549
Refolding, 59, 61–62, 497
Reglucosylation, 23
Regulatory complex, 115
Renal system
renal amyloidosis, 334, 337
renal failure, 10, 349, 807, 844
renal hypertrophy, 118
renal mesangial cells, 338
renal transplants, 869
Repeat expansion diseases, 193
Rep-1 polymorphism, 823
Reporter gene expression, 183
Respiratory disease, 404
Respiratory tract, 786
Resveratrol, 183, 633
Retina
retinal epithelial cells, 977
retinopathy, 490
Retinoic acid, 414
Retinol binding protein (RBP), 971, 977
Retrotranslocation, 24
R-flurbiprofen, 721
RGS16 gene, 409, 412
Rhodanese, 59–60
Rhodospirillum rubrum, xxi
Ribonucleic acid. *See* RNA
Ribosome-associated complex (RAC), 53
Ribosomes, 50–53
Ribozymes, 806, 828
Rifampicin, 298
Risk factors, aging and aggregation-mediated proteotoxicity, 631–640
Rivastigmine, 712
RNA, *see specific types of RNA*
aptamers, 268, 370
binding proteins, 201
biosynthesis, 181
functions of, 11
-induced silencing complex (RISC), 954
metabolism, 181, 316
polymerase, 283
polymerase II, 954
RNase, 26–27
RNA interference (RNAi)
delivery strategies, 958
down-regulation
of α-synuclein expression in PD, 955–956
of Aβ and tau expression in AD, 954–955
of huntingtin expression in HD, 955–957
of mutant SOD1 in FALS, 957–958
gene silencing, 246, 828, 934, 953
principles, 953–954
screens, 180–182, 186
therapy risks and challenges, 958–959
toxic protein aggregation, 633, 635

as translation inhibitor, 828
Rnq1 protein
characteristics of, 146–147, 149
nucleation and polymerization, 152
in prion formation, 152–154
Rodent studies, *see* Hamster studies; Mouse studies/mouse models; Rat studies
Alzheimer disease therapies, 716–717, 744–745
IAPP, 520
polyglutamine diseases, 892
prion disease therapies, 264
transcriptional regulation, 827–828
Rodlike (RL) fibrils, 356
R114C mutation, 491
Rosiglitazone, 738
R3D1 mutation, 198–199
Rubisco fold, 59–60
Russell bodies, 93

Saccharomyces
cerevisiae, 26–27, 30, 50, 53, 145, 147–148, 164–165, 181, 461
pombe, 147
SAHA, 203
SAPK, 732
Saposin C, 472, 475
Scaffolds, 284
Schiff base formation, 139, 620–621, 623–624
Sciatic nerves, 803
Scission reaction, 618
SCNA gene
functions of, 818
mRNA translation, 828
transcription
contributing factors, overview of, 827–828
GATA transcription factors, 826–827
Rep-1 allele, 825–826, 828
SCNB gene, 822
Scrapie, 260–261, 264, 283, 329
Scriptaid, 460
SDS/PAGE analysis, 473, 476, 820
Secondary amyloidosis (AA), immunotherapeutic strategies, 918–925
Secondary nucleation, 84–85
Secretase(s)
functions of, 96
inhibitors, 552
types of, 240–241, 244–245
Secretory cells, 328
Seeding, 77, 83, 85
Seladin-1, 96, 103
Selective estrogen receptor modulators, 740
Selegiline, 183
Self-assembly, 3, 11–12, 437
Self-association, 364–365
Self-perpetuating prions, *see* Prion proteins
Self-stacking, 156, 159
Semiautomated cell-free conversion, 261
Senile cardiac amyloidosis, 807
Senile plaque formation, 699, 706
Senile systemic amyloidosis (SSA), 806–809, 974, 976, 982, 989, 991
Sensory neuropathy, 804
Sequence analysis, 201
Serine
functions of, 473, 730
proteases, 28
Ser117, 985
Ser129, 824, 830, 832
Ser307, 32
Ser776, 196
Seco-sterols, 617, 625
Segregation analysis, RBCs, 454
Serotonin, 719
Serpins, 403, 414, 501
Serum amyloid A (SAA), 10, 330, 337, 918–919, 925
Serum amyloid P (SAP)
amyloid fiber structure, 572–576
calcium-free, 580
characteristics of, 12, 28, 350–351, 370, 526, 918, 936
component scan, 680–681
defined, 325
overview, 571–572
structure of, 576–580, 869
Short hairpin RNAs (shRNAs), 828, 888, 954–955, 958
Short peptides, 521
SH-SY5Y neuroblastoma cells, 96, 104
SIB, 744
Sicca syndrome, 868
Sickle cell anemia, xxii, 489
Sickle hemoglobin (HbS), 86, 88
Side-chain interactions, 495
Signaling pathways, amyloidogenesis, 616
Signaling pathways, types of, 184–185, 616, 716, 736, 893, 948, 978
Signal recognition particle (SRP), 428
Signal transduction, 889
Signature sequence, 436
Siiyama mutation, 405
Silanes, 703
Simulations, 9, 369
Simvastatin, 270–271, 276–277, 279, 284, 738
Single nucleotide polymorphisms (SNPs), 455, 823, 828

Single-photon-emission computerized tomography (SPECT), 235, 921, 925
Single-point mutations, 503, 869
Singlet dioxygen, 617–618
Sirtuin (*sir-2*), 633
Site-1 protease (S1P), 25, 28–29
Site-2 protease (S2P), 25, 28–29
Skin cancer, 245
Small heat shock proteins (sHsps), 434–435, 492, 946, 948
Small-interfering RNA (siRNA)
  down-regulation of Htt expression, 957–958
  functions of, 316, 828, 888, 953
  gene therapy, 282–283, 743
  prion disease treatment, 262, 271, 282
  production, 199–200
Small molecule(s)
  lysozyme amyloidosis, 879
  therapy, 833
Smoking
  cessation, impact of, 413
  health effects of, 407
Smoldering myeloma, 691
sod1 gene, 197, 382. *See also* Copper-zinc superoxide dismutase (SOD1); Superoxide dismutase (SOD)
Sodium butyrate, 203
Sodium dodecyl sulfate (SDS), 619. *See also* SDS/PAGE
Sodium salycylate, 949
Soft tissue amyloid, 783
Soluble low-density lipoprotein receptor protein (sLRP), 249
Solution nonideality, 88
Sorbitol, 336
Sortilin 1 (SORL1), 238
Spatial learning, 216, 971–972
Spatial memory, 219, 223–224, 242
Species barriers. *See* Prion species barriers
Spectrometric analyses
  circular dichroism (CD), 312, 474, 873–874
  electrospray ionization mass spectrometry (ESI-MS), 365–368, 873–874
  electrospray mass spectrometric analysis, 848–849
  fluorescence correlation, 261, 307
  Fourier transform infrared (FTIR), 160, 358
  mass spectrometry, 7, 223, 338, 562, 566, 851, 870
Spectroscopic analysis, 405
Spermidine, 269, 274–275
Spermine, 269, 274–275
Spermiogenesis, 28
S-phase cells, 104
Spinal cord, 263, 385, 950
Spinobulbar muscular atrophy (SBMA)
  characteristics of, 177, 203, 952
  classification of, 887
  mouse models, 892, 894
  neurodegeneration, 891
  protein degradation, 890
  protein expression, 888
  transglutaminase inhibition, 893
  treatment strategies, 894
Spinocellular ataxia
Spinocerebellar ataxias (SCA)
  functions of, 177, 194–195
  SCA1
    characteristics of, 125, 192–197
    classification of, 887
    mouse models, 205, 893–894
    protein degradation, 890
    protein expression, 888
    transglutaminase inhibition, 893
    treatment strategies, 894
  SCA3, 34, 193–199
Splicing, 25–28, 828
Spongiform encephalopathies, xxviii, 125, 145
Spontaneous folding, 59–60
Sporadic amyloid diseases, 93
Sporadic CJD (sCJD), 260, 285
Sporadic disease, 93, 195, 237–238, 260, 285, 828, 946, 969
Squalestatin, 271, 279
SRN-003-556, 733
Ssb chaperones, 53
Staminal cells, 103–104
STA-9090, 952
Statin therapy, 279, 738, 741
Statistical mechanics, 8
Stem cells
  functions of, 934
  therapy, 743–744
  transplantation, 778–779, 784
Steric zipper structure, 154–155
Sterol response element-binding proteins (SREBPs), 28, 36
Stochastic polymerization, 88–89
Stoichiometry, 979, 989
ST1571, 280
Stop-and-run mechanism, 703
Stop codons, 146, 334
Stopped-flow kinetic analysis, 849
Stress
  impact of, 197
  pathways, 977
  oxidative, *see* Oxidative stress
  proteins, 48

response, 205
influential factors, 632
signaling pathways, 948
tolerance factor, 153
Stress-denatured proteins, refolding of, 48
Stressors, degradation and, 113
Stroke, 35, 737, 742
Structure-activity relationships (SAR) studies, 910–911
Substrate proteins, 114, 116, 119
Substrate reduction therapy, 478
Succinimide, 135
Sugar cataract, 490–491
Sugars, 139
Sulfation, 340
Sulfolobus solfataricus (Sso AcP), 7
Sulfonated dyes, 269, 273–274
Sulfotransferases (SULT) gene superfamily, 458–459
Sulfoxide formation, 138
Superoxide anions, 617–618
Superoxide dismutase (SOD)
activity, 282, 616
tauopathies, 99
type I (SOD-1), 6, 957–958
Sup45 protein, 148
Supportive therapy, 413
Sup35 protein
amyloids, 154–158, 161–162, 638
characteristics of, 146, 149
nucleation, 150–151
polymerization, 151
prion formation, 153
species barriers, 164
Suramin, 269, 273–274
Surface plasmon resonance, 261
Swedish mutations, 216, 223, 225, 716 717
Sympathetic nervous system, 263
Synapses
synaptic function, 234, 236, 239, 244, 549
synaptic loss, 215, 220, 712, 716
Syndecan-4 (Syn4), 564–565
Synucleinopathies (SNCAs), *see* α-Synuclein
characteristics of, 818, 820–824, 828, 831–832
gene transcription modulators, 825–828
mRNA translation modulators, 828
variable, 821
System amyloid diseases, diagnosis of
amyloid deposits, radionuclide imaging, 679–680
biopsy, of amyloid, 674–676
Congo Red stain, 673–674
localized distinguished with, 676–677
Systemic amyloidoses
amyloid deposition, causes of, 329–335
characteristics of, xxii, 10, 96–98, 103, 325–326, 809, 937
diagnosis, 673–682
late-onset, 334
mathematical model, 328–329
neurodegenerative amyloidosis compared with, 326–328
organ damage mechanisms, 338–340
prognostic categories of, 776–777
protein and secretion sources, 328–329
systemic deposition, tissue targeting, 335–338
tissue targeting, 335–338
treatment of, 777–786
truncations, 333

t^2, 78, 81–82, 85
TACE (TNFα converting enzyme), 241
Talin, 337
Talsaclidine, 720
Tanespimycin, *see* 17-DMAG
TANGO, 8
Tarenflurbil, 721
Target of rapamycin (TOR), 890
Tau
aggregation, 732
Alzheimer disease therapies, 717–718
chaperones, 734–735
characteristics of, 7, 10, 35
degeneration, 201
disease proteins and, 196
down-regulation of, 945–955
hyperphosphorylation, 182, 195, 237, 713, 716, 720, 730, 733–735
kinases, 745
metabolism, 741
phosphorylation, 327, 732–734, 952
production, 731–732
protein (p-tau), 123, 181, 213, 225–226, 234, 236, 713, 954–955
hyperphosphorylation, 134
proteotoxicity, 176
silencing of, 743
toxicity, 181
Tauopathies, 177, 243–244, 250, 734, 949
Taxol, 735
Tay-Sachs disease, 469–470
TBP, 312
T cell(s)
activator, 737
development, 247
lymphoma, 891
meningitis, 248

T cell(s) (*continued*)
  prion disease therapies, 264
  significance of, 347
Temperature, significance of, 4, 9, 48–49, 52, 135, 353, 857, 946
Teratogens, 269
Ternary complex, 25
Terpenoids, 741
TETA, 548
Tetanus, 264
Tetracyclic antibiotics, 270, 277
Tetra (4-*N,N,N*-trimethylanilinium)porphine (Fe-TAP), 269, 275
Tetra(4-sulfonatophenyl)porphine (Fe-TSP), 269, 275, 284
TfT fluorescence, 83
TGFβ, 352, 411, 736
Thalidomide, 784–785
Therapeutic treatments
  for Alzheimer disease, 711–745
  for amyloidosis
    dialysis-related, 843–860
    familial, 867–882
    light-chain, 775–787, 917–928
    secondary, 917–928
    TTR, 967–991
  amyloid inhibition, 905–911
  antifibrillization therapies, 933–940
  antimisfolding therapies, 933–940
  controlling gene expression, 945–959
  immunotherapy, 917–928
  Parkinson disease and related disorders, 817–833
  polyglutamine disease, 887–894
Thermodynamics, significance of, 11, 308, 329, 333, 336, 622, 849, 852, 873, 879–880, 910, 978–979
Thiamphenicol, 280
Thioflavin (ThT) fluorescence, 355, 357, 489, 531, 620, 700, 920
6-Thioguanine nucleotides (6-TGNs), 454–455, 457
Thiolactone, 36
Thiolation, 36
Thiols, 30
Thiopurine, *see* Thiopurine *S*-methyltransferase (TPMT)
  metabolism, 454–455
  methyltransferase, xxviii
Thiopurine *S*-methyltransferase (TPMT)
  alleles, 456–458
  clinical impact of, 454, 457, 463
  degradation, 462–463
  pharmacogenomics
    autophagy, 461–463
    clinical consequences, 456–457
    degradation mechanisms, 459–460
    discovery and clinical importance, 454–457
    molecular mechanisms, 457–459, 463–464
  significance of, 453
Thioredoxin
  A, 278
  reductase, 616
Threonine, 730
Thrombocytopenia, 859
Thrombosis, 36
Thr119, 985
Thy1 promoters, 243
Thyroid hormones, 732, 989
Thyroxine
  binding, 984
  functions of, 806, 989–990
  transport, 971
TIM barrel fold, 59
Time dependence, 85
Tisp40, 28
Tissue cultures, 335–338
Tissue degeneration, xxvii
Tissue plasminogen activator (tPA), 97
TLCK, 949
T-maze, 218
TMPP-Fe^{3+}, 269
TNF-receptor-1, 263
Tocopherol, 101
Toll-like receptors, 264, 738
T119M mutation, 805–806, 980–981, 987
Toxicity, 784, 889. *See also* Cytotoxicity; Excitoxicity; Proteotoxicity
Toxic oligomer hypothesis, 968
TPCK, 949
TPMT gene, 454–456
TPMT*3A protein, 456, 458, 460–461, 464. *See also* Thiopurine *S*-methyltransferase (TPMT), alleles
TPR (tetratricopeptide repeat), 55
Tpr2 protein, 201
Tracheobronchial amyloidosis, 775–776
Tramiprosate, 908
Trans-autophosphorylation, 26, 29
Transcription
  α-synuclein, 824–825
  chaperone genes, 977–978
  deregulation, 123
  factors, 123, 182, 632, 639
  implications of, 26
  SNCA gene, modulators of, 825–828
Transcriptional activation, 27

Transcriptional processes, 245
Transduction, 889, 957
Transfection, 200, 328, 733
Trans-folding mechanisms, 60
Transgenic antibodies, 265
Transgenic mice, *see* Mouse studies/mouse models
  Aβ amyloidosis, 241–243
  Alzheimer disease, 550, 717–718, 955
  assaying effects on memory of Aβ, 215–217
  behavioral studies, 242
  immunotherapy, 248–249
  SBMA models, 894
  tauopathies, 243–244
Transglutaminases, 137, 893
trans-glycine, 365
Translation, 25, 824–825, 954
Translocation, 23, 116, 464
Transmembrane (TM) proteins, 240–241, 247, 427–429
Transmissible amyloid diseases, 93
Transmissible spongiform encephalopathies (TSEs), 259–261, 267
Transmission barrier, 164–165
Trans-peptide bond, 852–853
Transplantation
  bone marrow, 779
  islet cell, 526–528
  liver, 413, 869–870, 982
  lung, 413
  nonmyeloablative allogeneic, 781
  orthotopic heart, 808–809
  orthotopic liver (OLT), 804–805
  stem cell, 744, 778–779, 784
Transport
  axonal, 237, 243
  convective, 847
  polyglutamine proteins, 889
  vesicular, 182, 461
Transporter proteins, 727
Transthyretin (TTR), *see* Transthyretin-associated amyloidosis (ATTR)
  amyloidosis
    cascade, 973
    cellular proteostasis network, 978
    characterized, 923, 967–968
    chemotherapeutic strategy for, 982, 985–989
    by downhill polymerization, 972, 974–975
    etiology of, 976–978
    mechanism of, 967, 972
    natural suppression of, 979–980
    systemic, 328–329
  characteristics of, 83, 97, 327, 735, 797–799, 845, 868, 936, 970–972, 978
  conformation, 986
  destabilized, 976
  disease-associated mutants compared with wild-type, 978–979
  kinetic stabilizers binding to, 989–991
  monomeric, 6
  mutations
    implications of, 335
    V30M TTR, 805–806, 976–977, 979–981, 991
    V122I TTR, 979
    protein, 796, 802
  synthesis, 971, 977, 982
  tetramer, 970–971, 976, 979, 983–985
  transgenics, 977
Transthyretin-associated amyloidosis (ATTR)
  amyloidomas, 923
  background of, 796
  clinical characteristics, 795, 799–801
  mutations, 796
  ocular manifestations, 800–801
  organ damage, 338–340
  pathogenesis, 802–804
  treatment of, 804–806
Trehalose, 124
TriC (TCP-1 ring complex), 51, 61–62, 638–639
TRiC/CCT, 50
Trigger factor (TF), 50–52
Triglycerides, 36
Trimers, 223
Tripterine, 951
Trisomy 21, 237–238
Trophic support, 736
Troponin I, 776–777, 799
Troponin T levels, 776, 784
Trp fluorescence, 490
TRP2, 201, 203
Trp60, 850–851, 853–856
Truncated proteins, 195, 333, 354, 362, 410, 488
Tryptophan fluorescence, 362, 497, 851
TTP488, 737
Tubulin, 61, 75
Tumor(s)
  cells, 953
  necrosis factor (TNF)
    -alpha (TNF-α), 241, 263, 736–737, 971
    receptors, 263
  skin, 245
  suppressors, 61, 118, 245, 247
Tunicamycin, 412
Two-photon imaging, 242

Type 1 diabetes, 531–532
Type 2 diabetes, xxii, xxviii, 32–33, 93, 518, 526–528, 532, 737, 844, 940
Type II trinucleotide expansion diseases, 192
Tyrosine (Tyr)
  functions of, 146, 156, 848
  kinase inhibitors, 271, 280–281
  mutations, 851–852

Ubiquilin 1 (UBQLN1), 238
Ubiquitin
  C-terminal esterase L1 (UCH-L1), 34
  functions of, 459–461, 463, 832
  hydrolase, 312
  interaction motifs (UIMs), 194
  pathways, 201
  proteasome pathway (UPP), 829
  -proteasome system (UPS), 50, 105, 113, 115, 119, 122–123, 125, 203, 459, 461, 735, 888–890
  response proteins, 316
Ubiquitination, 432, 439
Ubiquitinylation, 434
Ubp64E, 201
UchL-1 gene, 829
UDP-glucoronosyl transferase, 26
UDP-glucose:glycoprotein glucosyltransferase (UGT1), 24
Ultracentrifugation analyses, 382
Ultrasonication, 354
Ultraviolet (UV)
  irradiation, 498
  light, 11, 489–490
unc-30 gene, 182
Unfolded protein response (UPR)
  activation, 29, 31, 35–37
  α_1-antitrypsin deficiency, 410
  characteristics of, 327–328, 433
  defined, 936
  sensors, 29
  signaling, 22–24, 26, 32, 34, 37
  systemic amyloidoses, 327
  transcription, 27, 29
U.S. Food and Drug Administration (FDA), 124, 453, 457, 722, 740, 743, 828, 923–924
Untranslated region (UTR), 458
Up-down bonds, 76–77
Up-regulation
  aggregation, 637
  Alzheimer disease, 551, 734, 737, 739, 742
  amyloidogenic, 952–953
  amyloidoses, 317
  Parkinson disease, 832–833
  protein misfolding diseases, 272, 317, 409, 412, 890
  protein misfolding principles, 62, 114, 121, 124, 198, 202
  toxic protein aggregation, 637
Urea
  cycle, 891
  denaturation of, 336
Ure2 protein
  amyloids, 159–161
  characteristics of, 146–147, 149
  nucleation and polymerization, 151–152
  prion formation, 152–154
Urinary tract amyloidosis,775–776

Vaccination studies, 745, 907, 936–937. *See also* Immunization
VAD (vincristine, doxorubicin, and dexamethasone), 780, 782
Valine, 147
Valsartan, 722
van der Waals interactions, 155
Vanishing complications, 845
Variable number of tandem repeats (VNTRs), 455, 458
Variant CJD (vCJD), 260, 262, 267, 285
Vascular amyloid, 238
Vascular damage, 97
Vascular endothelial growth factor, 892
Vasopressin, 118
VBMCP (vincristine, carmustine, melphalan, cyclophosphamide, and prednisone), 780
Vesicular trafficking, 123, 461
VHL, 61
Vidarabine, 280, 286
VIPL, 24
Viral capsid proteins, 77, 561–562
Viral infections, 561
Viral Japanese encephalitis, 283
Viral vectors, 954–955
Virazole, 280
Viruses, 832, 954
Vision, *see* Eye
  loss of, 800
  visual acuity, 800
Vitamins
  B_6, 740
  B_{12}, 36, 740
  C, 741
  E, 740–741
Vitrectomy, 800
Vitreous amyloid accumulation, 800

Waldenström macroglobulinemia, 689
Walker A/B sequences, 436
Wallerian degeneration, 803
Western blot analysis, 202, 635, 692
Wildlife disease management, 261
Wild-type chimeras,166
Wild-type lysozyme, 869, 871–875, 877, 881
Wild-type proteins, 6, 349, 492, 497, 796, 851–857, 871, 873, 875, 881
Wild-type TTR (WT TTR) 809, 972, 976, 979–981, 988–989
Wolcott-Rallison syndrome (WRS), 32–33
Women's Health Initiative, 740
Wormlike (WL) fibrils, 355–360, 368, 370
Worm models, 947
Worm studies
  biological counter-proteotoxicity activities, 638
  stress-resistance, 632
  toxic protein aggregation, 633–634, 636–637
Wound repair and healing, 561, 890

Xanthine oxidase (XO), 455, 457
X-box-binding protein (XBP-1), 25–28, 32, 328
Xenogenetics, 921
Xenopus oocyte expression system, 415
X-linked myopathy, 118
X-ray crystallographic analyses, 29, 347, 871, 881, 989
X-ray diffraction studies, 154, 157, 160, 357, 562, 566, 619, 870, 877, 918
Xylose-galactose-galactose-glucuronate (Xyl-Gal-Gal-GlcA), 566

Yeast, 53, 388, 408, 459, 461, 638. *See also Saccharomyces cerevisiae*
Yellow fluorescent protein (YFP), 633

Z α_1-antitrypsin gene mutation, 404–408
Zavesca, 478
Zinc
  Alzheimer disease and, 547, 549–550
  loop, 382
  Zn^{2+} ions, 12
  ZnT3-mediated, 547